Prostate Cancer

47-030227

To our families
Jolie, Ariel, Suzanne, Ursi, Barbara and Nick

Prostate Cancer

Science and Clinical Practice

Jack H. Mydlo
Department of Urology,
Temple University Hospital, Philadelphia, USA

Ciril J. Godec
Department of Urology,
Long Island College Hospital, New York, USA

An imprint of Elsevier

Amsterdam • Boston • London • New York • Oxford • Paris
San Diego • San Francisco • Singapore • Sydney • Tokyo

This book is printed on acid-free paper

Academic Press
An Imprint of Elsevier
84 Theobald's Road, London WC1X 8RR, UK
http://www.academicpress.com

Academic Press
An Imprint of Elsevier
525 B Street, Suite 1900, San Diego, California 92101-4495, USA
http://www.academicpress.com

ISBN 0–12–286981–8

Library of Congress Catalog Number: 2002116455

A catalogue record for this book is available from the British Library

Typeset by Integra Software Services Pvt. Ltd, Pondicherry, India
www.integra-india.com
Printed and bound in China

03 04 05 06 07 08 RD 9 8 7 6 5 4 3 2

Contents

Foreword

This book is a comprehensive approach to the study of causes, detection, and the various treatment modalities of prostate cancer. It starts in the laboratory, takes us through the operating room, and finishes in clinics and doctors' offices all over the world.

The editors, Drs Jack Mydlo and Ciril Godec, have collected contributions from global leaders of their fields in an attempt to cover every aspect of this disease. As a survivor of prostate cancer, I know the devastation of first hearing about the diagnosis, and then the confusion of dealing with all the options, side-effects, and long-term outcomes.

This book is written in a way that clinicians, researchers, and in parts lay people, will understand and appreciate. Not only does it describe the standard radical surgery and radiation therapies, but also discusses the latest in laparoscopic surgery and cryoablation therapy. It is also unique in that it compares the differences in the theories of prostate cancer etiologies and therapies from various parts of the globe. Since this disease affects husbands, boyfriends, fathers, sons and uncles, it is of the utmost importance that not only all men over the age of 40 understand this disease, but women as well.

Prostate Cancer: Science and Clinical Practice should be required reading for people who want to understand and help themselves and the male loved ones in their lives.

Senator Bob Dole

Preface

Prostate cancer has become one of the greatest health care challenges of the industrialized world; this is based upon numbers of patients and families affected and the demands this places on ever shrinking health care budgets. This text represents an up-to-date look at the issues and answers regarding present knowledge and innovative ideas for the future.

The editors have assembled a stellar faculty to provide the latest information of an ever-changing disease. This includes tumor biology, biopsy techniques, heredity, dietary and environmental factors, prevention, nerve sparing surgery, sural nerve grafting, laparoscopic surgery, radiation and hormonal therapies, sexual aspects, personal testimonials and future treatments of the disease on the horizon.

Although this book is aimed at the clinician and scientist, we believe that even lay persons may appreciate certain sections of the text in order to help them understand and perhaps select their own treatment modality.

Jack H. Mydlo
Ciril J. Godec

Contributors

Clément-Claude Abbou
Service d'Urologie CHU Henri Mondor
51 Avenue Du M1. De Lattre de Tassigny
94010 Créteil-cedex
France

Neil A. Abrahams
Department of Anatomic Pathology
9500 Euclid Avenue
Cleveland Clinic Foundation
Cleveland
OH 44195
USA

Bulent Akduman
Department of Urology
University of Colorado Health Science Center
Campus Box F710
4200 E. Ninth Avenue
Denver
CO 80262
USA

Thomas Anderson, Jr.
Office of Community Relations
University Services Building
Suite 302
Temple University
1601 N. Broad Street
Philadelphia
PA 19122
USA

Gerald L. Andriole
Division of Urologic Surgery
Washington University School of Medicine
4960 Children's Place, Box 8242
St Louis
MO 63110-1002
USA

Ronald E. Anglade
Department of Urology
Boston University School of Medicine
720 Harrison Avenue
Suite 606
Boston
MA 02118
USA

Seetharaman Ashok
Department of Urology
Rhode Island Hospital
Brown Medical School
2 Dudley Street, Suite 174
Providence
RI 02905
USA

R. Joseph Babaian
Department of Urology
The University of Texas M.D. Anderson Cancer Center
1515 Holcombe Blvd. Box 446
Houston
TX 77030
USA
rbabaian@mail.mdanderson.org

Richard K. Babayan
Department of Urology
Boston University School of Medicine
720 Harrison Avenue
Suite 606
Boston
MA 02118
USA

David G. Bostwick
Bostwick Laboratories
2807 N. Parham Road
Richmond
VA 23294
USA
bostwick@bostwicklaboratories.com

Simon R.J. Bott
Institute of Urology and Nephrology
University College London
Royal Free and University College
Medical School
3rd Floor, Charles Bell House
67 Riding House Street
London W1W 7EY
UK

Steven C. Campbell
Department of Urology and The Cardinal Bernadin
Cancer Center
Loyola University Medical School
2160 S. 1st Avenue
Building 54, Room 237
Maywood
IL 60153
USA
scampb6@lumc.edu

Eduardo I. Canto
Scott Department of Urology
Baylor College of Medicine
Texas Medical Center
6560 Fannin, Suite 2100
Houston
TX 77030
USA

E. David Crawford
Department of Urology
University of Colorado Health Science Center
Campus Box F710
4200 E. Ninth Avenue
Denver
CO 80262
USA

Paul Crispen
Department of Urology
Temple University Hospital
Suite 350, Parkinson Pavilion
3401 N. Broad Street
Philadelphia
PA 19140
USA

Philipp Dahm
Duke University Medical Center
Division of Urology
Department of Surgery, Box 2977
Durham
NC 27710
USA

John W. Davis
Department of Urology and The Virginia Prostate
Center of the Eastern Virginia Medical School
and Sentara Cancer Institute
Yoo West Brambleton Avenue, Suite 100
Norfolk
VA 23501
USA

Frans M.J. Debruyne
Department of Urology – 426
University Medical Center Nijmegen
PO Box 9101
6500 HB Nijmegen
The Netherlands

Steven J. DiBiase
Department of Radiation Oncology
University of Maryland School of Medicine
22 S. Greene Street
Baltimore
MD 21201
USA

Bob Djavan
Department of Urology
University of Vienna
Wahringer Gurtel 18-20
1090 Vienna
Austria

Ehab A. El-Gabry
Department of Urology
Kimmel Cancer Center
Thomas Jefferson University
1025 Walnut Street
Philadelphia
PA 19107
USA

Lars Ellison
Johns Hopkins Hospital
600 North Wolfe Street
Carnegie 298
Baltimore
MD 21287
USA

Paul F. Engstrom
Department of Population Sciences
Fox Chase Cancer Center
Temple University School of Medicine
7701 Burholme Avenue
Philadelphia
PA 19111
USA

Abelardo Errejon
Department of Urology
University of Colorado Health Science Center
Campus Box F710
4200 E. Ninth Avenue
Denver
CO 80262
USA

John M. Fitzpatrick
Department of Surgery
Mater Misericordiae Hospital
Conway Institute
University College Dublin
47 Eccles Street
Dublin 7
Ireland

Neil Fleshner
Division of Urology
Princess Margaret Hospital
610 University Avenue, Room 3-130
Toronto
Ontario M5G 2M9
Canada

Eduard J. Gamito
Department of Urology
University of Colorado Health Science Center
Campus Box F710
4200 E. Ninth Avenue
Denver
CO 80262
USA

Ellen Gaynor
Department of Medicine
Loyola University Medical School and
The Cardinal Bernadin Cancer Center
2160 S. 1st Avenue
Building 54, Room 237
Maywood
IL 60153
USA

Glen Gejerman
Department of Radiation Oncology
Hackensack University Medical Center
30 Prospect Avenue
Hackensack
NJ 07601
USA

Inderbir S. Gill
Section of Laparoscopic and Minimally Invasive Surgery
Department of Urology, A-100
Cleveland Clinic Foundation
9500 Euclid Avenue
Cleveland
OH 44195
USA

Phillip C. Ginsberg
Department of Surgery
Division of Urology
Albert Einstein Medical Center
5401 Old York Road, Suite 500
Philadelphia
PA 19141
USA

Ciril J. Godec
Department of Urology
Long Island College Hospital
339 Hicks Street
Brooklyn
New York
NY 11201
USA

Kazuo Gohji
Department of Urology
Osaka Medical College
2-7 Daigakumatchi
Takatsuki
569-8686
Japan

Leonard G. Gomella
Department of Urology
Kimmel Cancer Center
Thomas Jefferson University
1025 Walnut Street
Philadelphia
PA 19107
USA

Richard Greenberg
Department of Urologic Oncology
Fox Chase Cancer Center
7701 Burholme Avenue
Philadelphia
PA 19111
USA

Richard C. Harkaway
Department of Surgery
Division of Urology
Albert Einstein Medical Center
5401 Old York Road, Suite 500
Philadelphia
PA 19141
USA

Beth A. Hellerstedt
University of Michigan, 7303 CCGC,
Box 0946
1500 East Medical Center Drive
Ann Arbor
MI 48109
USA

Eric M. Horwitz
Department of Radiation Oncology
Fox Chase Cancer Center
Temple University School of Medicine
7701 Burholme Avenue
Philadelphia
PA 19111
USA

András Hoznek
Service d'Urologie CHU Henri Mondor
51 Avenue Du M1. De Lattre de Tassigny
94010 Créteil-cedex
France

William B. Isaacs
Brady Urological Institute
Johns Hopkins Hospital
600 North Wolfe Street
Baltimore
MD 21287
USA

Jonathan I. Izawa
Department of Surgery and Oncology
Division of Urology
The University of Western Ontario
London Health Sciences Center
South Street, Victoria Campus
London
Ontario NGA 4G5
Canada

Stephen C. Jacobs
Department of Surgery
University of Maryland School of Medicine
22 S. Greene Street
Baltimore
MD 21201
USA
sjacobs@smail.umaryland.edu

Matthew Karlovsky
Department of Urology
Temple University Hospital
Suite 350, Parkinson Pavilion
3401 N. Broad Street
Philadelphia
PA 19140
USA

Michael W. Kattan
Department of Urology
Memorial Sloan–Kettering Cancer Center
1275 York Avenue
New York
NY 10021
USA
kattanm@mskcc.org

Aaron E. Katz
Department of Urology
College of Physicians and Surgeons
Columbia University
Atchley Pavilion 11th Floor
161 Fort Washington Avenue
New York
NY 10032
USA

Roger S. Kirby
Department of Urology
St George's Hospital
Blackshaw Road
London SW17 OQT
UK

Sohei Kitazawa
Department of Molecular Pathology
Kobe University Graduate School of Medicine
7-5-1 Kusonoki-cho
Chuo-ku, Kobe
650-0017
Japan

Laurence Klotz
Division of Urology
Sunnybrook and Women's College Health Sciences Centre
University of Toronto
2075 Bayview Avenue # MG 408
Toronto
Ontario M4N 3M5
Canada

Vladimir Kolenko
Department of Surgical/Urologic Oncology
Fox Chase Cancer Center
Temple University School of Medicine
7701 Burholme Avenue
Philadelphia
PA 19111
USA

Andre Konski
Department of Radiation Oncology
Fox Chase Cancer Center
Temple University School of Medicine
7701 Burholme Avenue
Philadelphia
PA 19111
USA

Richard E. Link
Scott Department of Urology
Baylor College of Medicine
Texas Medical Center
6560 Fannin, Suite 2100
Houston
TX 77030
USA

Stephan Madersbacher
Department of Urology
University of Vienna
Wahringer Gurtel 18-20
1090
Austria

S. Bruce Malkowicz
Division of Urology
Hospital of the University of Pennsylvania
3400 Spruce Street
Philadelphia
PA 19104
USA
Malkowic@mail.med.upenn.edu

Michael Marberger
Department of Urology
University of Vienna
Wahringer Gurtel 18-20
1090 Vienna
Austria

Fray F. Marshall
Department of Urology
Emory University School of Medicine
Building A, Rm 3225
Atlanta
GA 30322
USA

T. Casey McCullough
Department of Surgery
Division of Urology
Albert Einstein Medical Center
5401 Old York Road, Suite 500
Philadelphia
PA 19141
USA

Kevin McEleny
Department of Surgery
Mater Misericordiae Hospital
Conway Institute
University College Dublin
47 Eccles Street
Dublin 7
Ireland

Liza McLornan
Department of Surgery
Mater Misericordiae Hospital
Conway Institute
University College Dublin
47 Eccles Street
Dublin 7
Ireland

Anoop M. Meraney
Department of Urology
Cleveland Clinic Foundation
9500 Euclid Avenue
Cleveland
OH 44195
USA

Ronald A. Morton
Scott Department of Urology
Baylor College of Medicine
Texas Medical Center
6535 Fannin, MS FB403
Houston
TX 77030
USA

Judd W. Moul
Uniformed Services University of the Health Sciences
Center for Prostate Disease Research
1530 E. Jefferson Street
Rockville
MD 20852
USA

Mark A. Moyad
University of Michigan Medical Center
Department of Surgery (Section of Urology)
1500 East Medical Center Drive
Ann Arbor
MI 48109-0330
USA

Jack H. Mydlo
Department of Urology
Temple University Hospital
Suite 350, Parkinson Pavilion
3401 N. Broad Street
Philadelphia
PA 19140
USA

Charles E. Myers
Department of Urology
University of Virginia Health Science Center
Box 422
Charlottesville
VA 22908
USA

Vivek Narain
Department of Urology
Wayne State University
Harper Professional Building
4160 John R. Suite 1017
Detroit
MI 48201-2020
USA

Don W.W. Newling
VU Medical Center
Department of Urology
PO Box 7057
1007 MB Amsterdam
The Netherlands

Brian Nicholson
Department of Urology
University of Virginia Health Science Center
PO Box 422
Charlottesville
VA 22908
USA

Carl Olsson
College of Physicians and Surgeons
Columbia University
161 Fort Washington Avenue
New York
NY 10032-3702
USA

David F. Paulson
Division of Urology
Department of Surgery
Duke University Medical Center
PO Box 2977
Durham
NC 27710,
USA

Kenneth J. Pienta
University of Michigan, 7303 CCGC
Box 0946
1500 East Medical Center Drive
Ann Arbor
MI 48109
USA

Alan Pollack
Department of Radiation Oncology
Fox Chase Cancer Center
Temple University School of Medicine
7701 Burholme Avenue
Philadelphia
PA 19111
USA

Isaac J. Powell
Department of Urology
Wayne State University
Harper Professional Building
4160 John R. Suite 1017
Detroit
MI 48201-2020
USA

Timothy L. Ratliff
Department of Urology
University of Iowa
University of Iowa Hospital/Clinic
200 Hawkins Drive, 3243 RCP
Iowa City
IA 52242-1089
USA

Mesut Remzi
Department of Urology
University of Vienna
Wahringer Gurtel 18-20
1090 Vienna
Austria

Martin I. Resnick
Department of Urology
University Hospitals of Cleveland
Case Western Reserve University
School of Medicine
10900 Euclid Avenue
Cleveland
OH 44106
USA

Vincent S. Ricchiuti
NEO Urology Associates, Inc.
602 Parmalee Avenue, Suite #300
Youngstown
OH 44510-1653
USA
vricchiuti@yahoo.com

Eric S. Rovner
Division of Urology
1 Rhoads
Hospital of the University of Pennsylvania
3400 Spruce Street
Philadelphia
PA 19104
USA

Daniel B. Rukstalis
Division of Surgery
MCP Hahnemann University
3300 Henry Avenue
Philadelphia
PA 19129
USA

Ihor C. Sawczuk
Department of Urology
11th Floor, Columbia Presbyterian
Medical Center
161 Ft. Washington Avenue
NY 10032-3702
USA

Peter T. Scardino
Department of Urology
Memorial Sloan–Kettering Cancer Center
1275 York Avenue
NY 10021
USA

Paul F. Schellhammer
Department of Urology and The Virginia Prostate
Center of the Eastern Virginia Medical School
and Sentara Cancer Institute
Yoo West Brambleton Avenue, Suite 100
Norfolk
VA 23501
USA

Claude C. Schulman
Department of Urology
Erasme Hospital
University Clinics of Brussels
808 route de Lennik
B-1070 Brussels
Belgium

Ahmad Shabsigh
Department of Urology
11th Floor, Columbia Presbyterian Medical Center
161 Ft. Washington Avenue
NY 10032-3702
USA

Neil Sherman
Department of Urology
University of Medicine and Dentistry of New Jersey
NJ
USA

D. Robert Siemens
Department of Urology
Queen's University
76 Stuart Street
Kingston
Ontario K7L 2V7
Canada

Kevin M. Slawin
Scott Department of Urology
6560 Fannin, Suite 2100
Houston
TX 77030
USA

Barry Stein
Department of Urology
Rhode Island Hospital
Brown Medical School
2 Dudley Street, Suite 174
Providence
RI 02905
USA

Mitchell S. Steiner
Department of Urology
University of Tennessee Health Science Center
1211 Union Avenue, Suite 340
Memphis
TN 38104
USA

Chandru P. Sundaram
Division of Urologic Surgery
Washington University School of Medicine
4960 Children's Place, Box 8242
St Louis
MO 63110-1002
USA
sundaramc@msnotes.wustl.edu

Dan Theodorescu
Department of Urology
University of Virginia Health Science Center
Box 422
Charlottesville
VA 22908
USA

Edouard J. Trabulsi
Departments of Urology
Memorial Sloan-Kettering Cancer Center
1275 York Avenue
New York
NY 10021
USA

Aubrey R. Turner
Center for Human Genomics
Medical Center Blvd
Wake Forest University School of Medicine
Winston-Salem
NC 27157-1076
USA

Robert G. Uzzo
Department of Surgical/Urologic Oncology
Fox Chase Cancer Center
Temple University School of Medicine
7701 Burholme Avenue
Philadelphia
PA 19111
USA

Richard K. Valicenti
Director of Clinical Affairs
Thomas Jefferson University
111 S. 11th Street
Philadelphia
PA 19107
USA

Michael R. Van Balken
Department of Urology – 426
University Medical Center Nijmegen
PO Box 9101
6500 HB Nijmegen
The Netherlands

Deborah Watkins-Bruner
Department of Population Science
Fox Chase Cancer Center
Temple University School of Medicine
7701 Burholme Avenue
Philadelphia
PA 19111
USA

R. William G. Watson
Department of Surgery
Mater Misericordiae Hospital
Conway Institute
University College Dublin
47 Eccles Street
Dublin 7
Ireland
bwatson@mater.ie

Alan J. Wein
Division of Urology
Department of Surgery
University of Pennsylvania School of Medicine
Hospital University of Pennsylvania
Philadelphia
PA 19104
USA

Magali Williamson
Institute of Urology and Nephrology
University College London
Royal Free and University College
Medical School
3rd Floor, Charles Bell House
67 Riding House Street
London W1W 7EY
UK

David P. Wood
Department of Urology
University of Michigan
1500 E. Medical Center
Ann Arbor
MI 48109
USA

Jianfeng Xu
Center for Human Genomics
Medical Center Blvd
Wake Forest University School
of Medicine
Winston-Salem
NC 27157-1076
USA

Paul Matthew Yonover
Department of Urology
Loyola University Medical Center
2160 S. 1st Avenue
Building 54, Room 237
Maywood
IL 60153
USA

A.R. Zlotta
Department of Urology
Erasme Hospital
University Clinics of Brussels
808 route de Lennik
B-1070 Brussels
Belgium

Part I

Etiology, Pathology and Tumor Biology

Population Screening for Prostate Cancer and Early Detection in High-risk African American Men*

Judd W. Moul

Urology Service, Department of Surgery, Walter Reed Army Medical Center, Washington, DC 20307-5001, and Center for Prostate Disease Research, Department of Surgery, Uniformed Services University of the Health Sciences, Bethesda, MD 20814-4799, USA

Prostate cancer is the most common malignancy in American men, accounting for more than 29% of all diagnosed cancers and approximately 13% of all cancer deaths.[1] Nearly one of every six men will be diagnosed with the disease at some time in their lives.[1] In 2002 alone, an estimated 189 000 US men were diagnosed with prostate cancer, and more than 30 000 will die of the disease.[1] Despite the fact that population-based screening for prostate cancer has yet to be definitively proven or disproved to affect the disease-specific mortality, this summary explores changes in the 'PSA Era' (defined as the time in the USA when the prostate specific antigen (PSA) screening test came into widespread clinical use) and the prospects for targeted screening of certain groups of individuals who may be at particular risk of developing prostate cancer, namely African American men.

PROS AND CONS OF POPULATION-BASED SCREENING IN GENERAL

Early detection for prostate cancer has been practiced for many years in the form of the digital rectal examination (DRE). However, over the last 15 years with the advent of prostate specific antigen, or PSA, the topic of prostate cancer screening has become a hotly contested issue.[2,3] Early detection may take the form of the population-based screening or case finding. In the early 1990s, the American Cancer Society (ACS), the American Urological Association (AUA) and other organizations advocated screening for the early detection of prostate cancer.[4,5] These groups recommend a DRE and PSA test for men starting at age 50. Notably, however, other organizations had argued against screening, including the US Preventative Services Task Force, the American Academy of Family Physicians and the American College of Physicians.[6,7]

Several factors favor the use of screening for the early detection of prostate cancer. First, because patients do not experience symptoms during the early stages, they are unlikely to seek care until the disease has progressed. Second, improvements in detection methods have increased the prospects for identifying the disease in its early stages, when the cancer is still confined to the organ, and is more easily treatable and often curable. Third, early detection might mean the difference between life and death, as no cure has been found for the advanced disease.

To be of value, screening must lead to treatment that has a favorable impact on prognosis. Catalona et al. were one of the first groups to examine this issue by comparing disease stages in patients with prostate cancer who had or had not undergone or had PSA screening.[8] The screened group had a lower percentage of cases with advanced disease and no greater percentage with latent disease. The investigators concluded that screening reduces the incidence of advanced disease and implied that the death rate will ultimately decrease.[8,9] Since these reports, there have been a multitude of studies documenting the changes in the epidemiology of prostate cancer in the 'PSA-Era'.

*The opinion and assertions contained herein are the private views of the author and are not to be considered as reflecting the views of the US Army or the Department of Defense.

PSA-ERA CHANGES SUPPORTING GENERAL SCREENING

When an effective screening tool is introduced into a population, the following events should occur. There should be a transient increase in incidence owing to labeling of prevalent cancers. The cancers should be diagnosed in patients of a younger age. The cancer should be diagnosed at lower stages and there should be an apparent increase in disease-specific survival.[10,11] As of 2002, PSA testing has arguably fulfilled these criteria, and many clinicians practice screening.

Regarding the change in prostate cancer incidence, the SEER (Surveillance, Epidemiology and End Results Program of the National Cancer Institute) data as well as data from the Mayo Clinic and other sites document the change.[12–16] Specifically, starting in 1989 and peaking in 1992, there was a marked rise, with a subsequent decrease of approximately 30%. These trends exemplify the Cull effect, that is, with the introduction of any screening tool into a population, there will be a rise in incidence, with a subsequent fall. This fall is because of the depletion of the prevalence pool of previously undiagnosed cases. The steady-state incidence of prostate cancer in the PSA-Era will eventually be a reflection of both patients aging, and entering into the pool, and those being removed from the pool as a result of dying, or being diagnosed with prostate cancer.[12] It is unclear whether the incidence will decrease to prescreening levels, or remain higher than previous years.[16,17] Current statistics from the SEER data indicate that the steady-state incidence in the PSA-Era has yet to be reached.[1] Others have suggested that the fall in incidence is influenced by less aggressive screening by primary care physicians in response to data in the scientific and lay media that screening for prostate cancer does not effect morbidity and mortality.[18,19] In the Department of Defense Center for Prostate Disease Research (CPDR) program, the incidence of new cases continues to be above prescreening levels and appears to have leveled off.[20]

The mean and median age of men at diagnosis has decreased significantly in the PSA Era. The introduction of age-specific reference ranges for PSA screening, first by Osterling et al., and subsequently by race and age by Morgan et al., have shown that prostate cancer, detection and subsequent treatment is an age-dependent process.[21,22] The younger the patient is at the time of diagnosis, presumably, the earlier they are in the course of their disease. If curative-intent treatment is rendered at an earlier age, the more likely cure would occur. Chodak et al. have shown that men under 61 years of age had a statistically significant improved disease-specific survival.[23] Smith et al. from our group have also shown that younger age at diagnosis is an independent predictor of better prognosis.[24] Age greater than 65 years has been demonstrated to be an independent predictor of distant metastasis.[25] Carter et al. have recently suggested that detection of prostate cancer in younger men is likely to lead to a decrease in prostate cancer mortality.[26] In summary, there is strong evidence that, if the diagnosis of prostate cancer is made at a younger age, disease-specific mortality can be significantly reduced, and it is now demonstrated that, in the PSA-Era, the age at diagnosis had decreased significantly.

The decline in the age at the time of diagnosis from 72 to 69.4 years of age from 1990 to 1994 has been shown by the SEER data.[18] From the SEER data and other literature, a lead-time of between 3 and 5 years is derived.[13,15,18,27] In other words, with the aid of PSA, we are detecting cancers 3–5 years earlier than prior to screening. Thus the increase in survival (time of diagnosis to time of death) in the PSA-Era must be at least 3–5 years longer that the previous expected survival time to show any benefit in disease-specific survival with PSA screening.

There has also been a significant stage migration in the PSA-Era. Most strikingly, the percentage of patients presenting with metastatic disease decreased from 14.1% and 19.8% in 1988 and 1989, respectively, to 3.3% in 1998 from our CPDR database.[20] Before a survival benefit, a decrease in metastatic disease should be evident, which is clearly being shown on a national level by several studies.[12–14,28] These findings are more impressive in light of the fact that no curative treatment exists for patients with metastatic disease. Similarly, we have also shown a statistically significant decrease in the incidence of clinical T3 and T4 disease. Again, demonstrating a stage shift from disease that can merely be managed, to disease that is amenable to cure. These T3 and T4 tumors, as well as T2 tumors, are most likely being diagnosed several years prior to when they would previously become evident. The advent of PSA testing is reclassifying these tumors as T1c, which is associated with decreased recurrence rates, and increased disease-specific survival, when compared to other clinical stages.[29–31]

The level of PSA at the time of diagnosis is a general surrogate of tumor volume or cancer burden. A decline in PSA levels over time is further evidence of stage migration to support screening. In our CPDR studies, the median PSA at the time of diagnosis decreased from 11.8 ng/dl in 1990 to 6.3 ng/dl in 1998.[20] This is consistent with the findings of Vijayakumar et al. who have recently demonstrated a statistically significant drop in the PSA level of African American patients presenting with prostate cancer.[32] Partin et al. have shown that the lower the pretreatment PSA prior to prostatectomy, the lower the chance of extracapsular extension, seminal vesical involvement and lymph-node metastasis.[33] Moul et al. and many other investigators have shown that the PSA level is a significant contributor in determining the risk of recurrence after treatment with radical prostatectomy.[34] Similarly, in patients treated with external beam radiation, the lower the level of pretreatment PSA, the greater the disease-free survival.[29,35] Hopefully, this continued decrease in PSA at the time of diagnosis portends more successful treatment outcomes in the future.

Most of the tumors diagnosed in the PSA-Era are not indolent tumors by traditional grade and Gleason criteria. Multiple studies including SEER have shown that moderately differentiated or Gleason sum 5–7 tumors predominate in the PSA-Era.[12,13,36,37] One reason for the decrease in well-differentiated tumors is the decrease in the number of transurethreal resection of the prostate (TURP) performed. It is established that tumors arising in the transitional zone (area resected with a TURP) have a significantly higher incidence of tumors with lower Gleason sum than tumors discovered in the peripheral zones by needle biopsy.[38,39] Smith and Catalona have shown that 97% of the tumors detected through PSA screening are medically important (defined as palpable, multifocal or diffuse, and moderately or poorly differentiated).[31] Other recent studies confirm that moderately differentiated or mid-Gleason grade tumors are clinically significant, with higher recurrence and disease-specific mortality that lower grade tumors.[23,40–42]

Regarding survival, Gilliland et al. showed improved survival during a period of PSA screening in New Mexico using the SEER data.[43] The SEER 5-year relative cancer survival rates for patients diagnosed in the following years groups: 1974–1977, 1980–1982 and 1989–1995 were 67%, 73% and 92%, respectively; each of these changes were statistically significant ($P < 0.05$). Etzioni et al. have determined that the improved survival is not due simply to PSA testing, even if one considers a very short lead time of 3 years.[44] It is probably related to screening and more aggressive treatment. Nevertheless, our own data from CPDR also shows significant improvements in 5-year prostate cancer specific survival rates as the PSA Era has progressed.[45] Specifically, for patients diagnosed between 1988 and 1991 at the Walter Reed Army Medical Center, the cancer specific survival was 81.7% compared to 92.5% for men diagnosed between 1992 and 1994 and 98.3% for men diagnosed between 1995 and 1998. This was based on a cohort of 2042 men and the cause of death was determined prospectively using the National Death Index, direct deaths certificate review and choice point commercial service (**Fig. 1.1**).

Aside from these population and database trends, screening studies by Labrie et al.[46] and Bartsch et al.[47] suggest a benefit to population-based screening. Most notably, Bartsch et al. recently reported early results of the Tyrolean prostate cancer screening study in Austria.[47] By aggressively offering screening to all men in the Tyrolean state of Austria between 1993 and 1997, and testing two-thirds of eligible men between the ages of 45 and 75, they report a 42% decline in the disease-specific mortality for prostate cancer compared to the rest of the county where PSA testing was not commonly practiced.

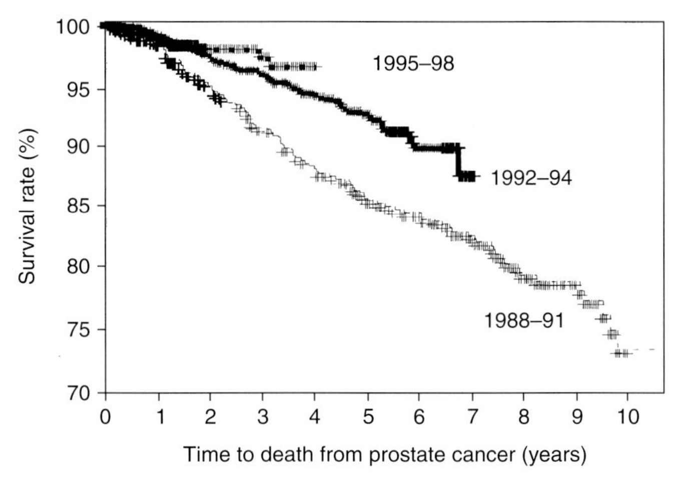

Fig. 1.1. Survival from prostate cancer by time period of diagnosis.

ARGUMENTS AGAINST POPULATION SCREENING

Opponents of screening point to the potential for side effects from treatment, the possibility that some men will be treated unnecessarily, the economic burden on the health care system and the lack of definitive scientific evidence that the screening will reduce overall disease-specific mortality.[19] Indeed, many men with prostate cancer do not die of the disease, whereas many other patients die of the disease despite our best efforts. No foolproof markers are currently available to differentiate these two groups. Therefore, a key objection to screening is that such efforts may uncover many cancers that, if left undetected, would never have caused morbidity or mortality; conversely, some cancers will cause death despite being detected by screening. Some observers have further recommended that screening should be avoided in men older than 70 or 75 years of age so as to reduce the detection of such 'incidental cancers' that are unlikely to affect life expectancy.

Aside from the 'true' screening, or population-based screening controversy, case finding for prostate cancer using PSA and DRE is widely practiced. Case finding is the process of evaluating men with symptoms where the testing is for differential diagnosis of urinary tract or other disease. In this setting, PSA and DRE are standard tools that physicians must have at their disposal.[48] The process of targeted screening for prostate cancer in a higher risk group of men, such as Blacks and those with a family history, could arguably, in my opinion, be considered similar to case finding.

Recognizing that population screening for prostate cancer using PSA and DRE remains disputed to reduce the morbidity and mortality of the disease, authorities currently recommend that physicians and health care organizations provide the pros and cons as outlined above and let the well-informed patient decide. Many would argue that it would be wrong to mandate screening; however, it would be just as wrong not to offer the option of early detection tests for prostate cancer, especially for high-risk targeted groups, especially in light of aforementioned positive trends in the PSA-Era. Recently, the ACS and AUA have updated their prostate cancer screening to this effect.[49,50] Most notably, the recommendations call for testing if the patient is undecided.[49]

SHOULD AFRICAN AMERICAN MEN UNDERGO SCREENING FOR PROSTATE CANCER?

For African American men specifically, do we recommend population screening based on current knowledge of greater incidence, greater stages at diagnosis and worse outcome or do we wait for more definitive screening efficacy studies? I believe we have reasonable basis to recommend routine testing now using screening guidelines *optimized* for Black men.

RACIAL DIFFERENCES

It has been widely recognized that there are observed differences in prostate cancers in White and Black men[51–53] and our own studies in the equal access US military health care system are no exception.[54] In 1995, we were the first to show that Black men with newly diagnosed prostate cancer had higher PSA values than White men, even after adjustment for age at diagnosis and grade and clinical stages of cancer.[55]

This study also documented that, even within the traditional clinical stage categories at diagnosis, Black men had significantly greater cancer volumes. Considering that the Black men were, on average, about 3 years younger than the White men, this tumor volume disparity was striking. In a more recent update with 226 cases, the higher cancer volume stage-for-stage in Black men persisted.[56]

What is accounting for the greater cancer burden in these Black men, who seemingly have equal access to screening as the White men? It could be that the Black men are not availing themselves to testing, have greater delay in diagnosis and larger tumors. We know that, in general, there has been less awareness of prostate cancer and less screening in the African American community.[57] Data from Prostate Cancer Awareness Week consistently show that only about 5% of participants are Black despite the fact that 12% of the population as a whole is Black.[58]

On the other hand, is there a biologic factor, or factors, such as testosterone, that is making these tumors grow bigger more quickly? Studies in younger men have shown that Blacks had testosterone levels 10–15% higher than age-matched Whites.[59] Furthermore, a more recent study has found genetic variation in the androgen receptor (AR) gene that may make the receptor more active in African American men.[60] Specifically, the CAG polymorphic repeat in exon 1 of AR is shorter in Black men and theoretically could potentiate the effect of testosterone in prostatic cells. Alternatively, is it a combination of behavior and biology that is responsible for the observed differences? Despite not knowing the exact cause, what can we do now?

RACE-ADJUSTED GUIDELINES

Working with the US military health care systems patients, our group has shown that Black men with newly diagnosed prostate cancer have higher serum PSA values than do Whites, even after correction for stage grade and tumor volume.[55] In view of these findings, our group has also studied the ability of PSA to detect prostate cancer in both Caucasian and African American men and developed age-adjusted PSA reference ranges for maximal cancer detection.[22] In this study, between January 1991 and May 1995, serum PSA concentration was determined for 3475 men without clinical evidence of prostate cancer (1802 Caucasians, 1673 African Americans) and 1783 men with this disease (1372 Caucasians, 411 African Americans). PSA concentration was analyzed as a function of age and race to determine operating characteristics of PSA for the diagnosis of prostate cancer. Serum PSA concentration correlated directly with age for both Black and White men ($r=0.40$, $P=0.0001$ for Blacks and $r=0.34$, $P=0.0001$ for Whites). African American men had significantly higher PSA concentrations than Caucasian men ($P=0.0001$). When sensitivity was plotted against 1-specificity, the area under the receiver operator characteristic (ROC) curve was 0.91 for Black men and 0.94 for White men, indicating that the PSA test was an excellent early detection tool. For comparison, the Papanicolaou smear for cervical cancer, which is an accepted clinical screening test, has an ROC value of 0.70. When we calculated age-specific reference ranges using identical methodology to Oesterling and colleagues in their 1993 study of primarily Caucasian patients from Olmstead County, Minnesota,[21] we found very similar values for White men but higher values for Black men. These ranges were 0–2.4 ng/ml for black men aged 40–49, 0–6.5 ng/ml for men aged 50–69, 0–11.3 ng/ml for men aged 60–69, and 0–12.5 ng/ml for men aged 70–79. We then tested these new ranges in our group of Black men with prostate cancer to determine how these ranges would have performed, if they had been used to detect their cancers. Unfortunately, these markedly higher ranges would have missed 41% of the cancer (only 59% sensitivity).

The reason these traditionally derived ranges performed so poorly is because they are simply the 95th percentile of values in the Black controls. Because there is more variability of PSA results in Blacks without the evidence of cancer, there is more skewness, which pushes the 95th percentile farther to the right (higher). This higher range, however, is not clinically useful. We, therefore, developed age-adjusted reference ranges for Black men with prostate cancer, selecting PSA upper limits of normal by decade in the men with prostate cancer by using the 5th percentile of PSA values. Only the lowest 5% of pre-diagnosis PSA values in the Black men with cancer are 'normal' and the remainder (95%) are above the normal (95% sensitivity).

We refer to these ranges as the Walter Reed/Center for Prostate Disease Research age-specific reference ranges for maximal cancer detection (**Table 1.1**). They maximize sensitivity (cancer detection) without undue loss of specificity [false-positive/unnecessary transrectal ultrasound (TRUS)/biopsy].

Reference ranges have been controversial especially for older men because they raise the 'normal' above 4.0 ng/ml. Advocates of screening have been concerned that 'important' cancers will be missed in these 'older' men who may be perfectly healthy and physiologically younger. By the same logic, the race-specific reference ranges that we have proposed[22] have been criticized in that many feel that it is inappropriate to raise the 'normal' above 4.0 ng/ml in a high-risk group regardless of age. Specifically, Littrup feels that our values are perhaps too complex and would favor only two PSA 'normals' >2.0 ng/ml for 'high-risk men' and >4.0 ng/ml for

Table 1.1. Comparison of Walter Reed/Center for Prostate Disease Research PSA reference ranges for maximal prostate cancer detection vs. traditional age-adjusted and original normal PSA values.

Age (years)	WR/CPDR ASRRs for maximum detection[22] (ng/ml) African American	Caucasian	Traditional PSA 'normal' for all men (ng/ml)	Traditional ASRRs based on Causcasian men Mayo Clinic[21] (ng/ml)
40–49	0–2.0	0–2.5	0–4.0	0–2.5
50–59	0–4.0	0–3.5	0–4.0	0–3.5
60–69	0–4.5	0–3.5	0–4.0	0–4.5
70–79	0–5.5	0–3.5	0–4.0	0–6.5

WR/CPDR, Walter Reed/Center for Prostatic Disease Research; ASRR, age-specific PSA reference ranges.

the 'general' population.[61] Furthermore, Powell et al. have similarly criticized our screening guidelines regarding the raising of the PSA screening threshold above that for Caucasian men.[62] They reason that, for any given PSA level, the prognosis for African American men is likely worse than Caucasian men and by raising the screening threshold, we will likely lower prognosis in this high-risk group. The key point is that our traditional age-specific PSA reference ranges (ASRRs) and reference ranges for maximal cancer detection provide compelling data that a PSA of 4.0 ng/ml is too high for many patients.

A PSA OF 4.0 NG/ML IS TOO HIGH ESPECIALLY IN YOUNG MEN

Despite the continuing controversy about the 'exact' proper PSA by age and race, the most important concept in my opinion is recognizing that a PSA of 4.0 ng/ml is too high a screening cut-point for younger men, particularly African American men between 40 and 49 years of age. Bullock et al. screened 214 Black men between 40 and 49 years of age and found a prevalence of prostate cancer of 0.9% (2 of 214) when a PSA of 4.0 ng/ml was used.[63] Interestingly, this prevalence increased to 5.6% (2 of 36) when the Black men also had a family history of prostate cancer. Conversely, Catalona et al. from the same university, found that the cancer detection rate was 38% in a small group of 16 Black men who had biopsy for PSA values between 2.6 and 4.0 ng/ml.[64] In a follow-up series, this same group found an even higher prevalence of 42% for African American men with a PSA between 2.6 and 4.0 ng/ml.[65] Until further data are available, I believe that African American men with a PSA >2.0 ng/ml between 40 and 49 years of age should have further evaluation. In older Black men, particularly those who are healthy and in their 50s and 60s, a clinician may use our age-adjusted sensitivity-based ranges[22] or may opt to be even more aggressive and use the 2.0 ng/ml advocated by Littrup[61,66] or the 2.5 ng/ml recommended by Catalona and associates.[64,65]

LOWERED PSA REFERENCE RANGES FOR CURABLE CANCER

The concept of developing PSA reference ranges not just for cancer but *curable* cancer is emerging. Reissigl et al. from Austria were the first group, to my knowledge, to define PSA cut-points by decade of age for curable prostate cancer.[67] These PSA values ranged from 1.25 ng/ml for men in their 40s to 3.25 ng/ml for men in their 70s. Curability definitions are somewhat arbitrary unless one waits the required 10 years or more to know exactly which patients were, in fact, cured. Instead, pathologic stage and grade of radical prostatectomy patients is a reasonable surrogate of curability. To this end, Carter et al. in a large study from Johns Hopkins, defined 'curability' in their series as organ confined with any grade of cancer or specimen confined (negative margins, seminal vesicles, and lymph nodes) with a Gleason sum less than or equal to 6.[68] I used this definition to show that only 45% of early PSA-Era Black men who underwent a radical prostatectomy at our hospital were 'curable'.[69] This compared to 74% for the predominantly White patients reported by Carter et al.[68] Furthermore, by lowering the pretreatment PSA value, the curability for both Black men and White men is strikingly better. Specifically, at Johns Hopkins 94% of men with a PSA value ≤4.0 ng/ml were curable and 83% of Black men from Walter Reed with this low PSA were curable.[25]

With this in mind, I have reported PSA reference ranges for curable prostate cancer in African American men who were considered cured and who had a pretreatment PSA value.[70] Age-adjusted 5th, 10th and 25th percentile of pretreatment PSA was calculated to define cut-points for screening PSA values to optimize curable prostate cancer in African American men (**Table 1.2**).

In other words, based on this preliminary study, one would biopsy Black men with a PSA value greater than approximately 1.0 ng/ml to have a 95% probability of diagnosing curable prostate cancer. Recognizing that lowering the PSA 'normal' is always a tradeoff of sensitivity versus specificity (unnecessary biopsies), the 90th or 75th percentile may be

Table 1.2. Age-adjusted PSA upper limits of normal to diagnose curable prostate cancer in African American men.

	40–59	60–79
5th percentile (95% sens.)	0.9	1.0
10th percentile (90% sens.)	1.7	2.7
25th percentile (75% sens.)	3.2	5.1

a better compromise. Recognize that three of four of these values are less than the traditional normal of 4.0 ng/ml. Conversely, Carter et al. have most recently further examined the concept of PSA values and curable cancer.[71] Unlike our concept of lowering the PSA, they feel that age is a more important determinant of curability and that the PSA threshold should remain at 4.0 ng/ml in younger men. They contend that simply by screening younger men, we will pick up a high proportion of curable disease.

WHAT IS THE 'NORMAL' PSA IN YOUNG MEN?

Our Department of Defense-funded CPDR has recently conducted studies showing that a PSA of 4.0 ng/ml is significantly higher than normal for young men. In 750 Black and 750 White military members between the age of 15 and 45 who had serum banked in the Army and Navy Serum Repository (ANSR), the mean PSA values were 0.52 and 0.47 ng/ml, respectively.[72] The 95th percentile ranged from 1.16 to 1.38 ng/ml for the Blacks stratified by decade of age (1.38 ng/ml in the 40–49 group); for the Whites, the corresponding values were 0.71–1.13 ng/ml. Based on these data, even using a PSA value of 2.0 ng/ml as a cut-point for both Black and White men is considerably higher than these 95th percentile values.

Our second study was a prospective screening study of healthy officers enrolled in the US Army War College at Carlisle Barracks, PA, USA and is a collaboration between CPDR and the Armed Forces Physical Fitness Institute. Starting in 1997, approximately 225 matriculating officers entering the school have been offered PSA testing as part of a comprehensive executive health screen. Three years of data are available for this interim analysis on 602 men between 40 and 49 years of age.[73] Only 10 of 602 (1.7%) had PSA $\geqslant$2.5 ng/ml and only 3 (0.5%) had PSA $\geqslant$4.0 ng/ml. One of 602 (0.17%) was diagnosed with prostate cancer (PSA of 15.5 ng/ml) and had pT3 disease at radical prostatectomy. This study was conducted in very healthy, primarily Caucasian, men and our conclusions to date suggest that screening below 50 in normal-risk individuals may not be needed. However, using a PSA screening normal cut-point of $\geqslant$2.5 ng/ml, less than 2% of men required further evaluation. Furthermore, baseline PSA data were obtained to compare with future screens, which likely will be of clinical value. A similar study is being initiated in African American military members to determine if the yield of screening would be higher.

Aside from PSA-reference ranges, Presti et al. have proposed lowering the cut-off for PSA density to 0.1 to optimize prostate cancer detection in African American men.[74] Even though it has recently been proposed that less than annual PSA screening may be appropriate for younger Caucasian men who start with a PSA <2.0 ng/nl,[75] PSA velocity data for black males are not fully established and, therefore, yearly testing remains prudent for this high-risk group.

In summary, despite the fact that we do not yet know the exact cause or causes for the observed differences in prostate cancer in Black men, we believe that screening is justified now to lessen the disparity. Starting testing at age 40 is reasonable in an attempt to catch tumors earlier when they are smaller. Using age- and race-adjusted PSA, particularly in the men between 40 and 49 in which the cut-off is 2.0 ng/ml, should also enhance the earlier detection at a curable stage.

Continued public awareness in the African American community is also needed – even a perfect screening test will not be effective if it is not recommended and no one gets it! Finally, these recommendations need confirmation by prospective clinical trials, and we should encourage the development of such trials and participation by our Black patients.

LATEST GOOD NEWS FOR AFRICAN AMERICAN MEN

Through our CPDR program, we have been keeping close tabs on racial differences in prostate cancer in our equal access US Military health care system. In the fall of 2001, we published exciting data showing remarkable improvements in pathologic stage in African American men undergoing radical prostatectomy.[76] Specifically, in 195 African American and 587 Caucasian men undergoing surgery between 1998 and 1999 at one tertiary center, the rate of extraprostatic extension (pT3 disease) declined from 100% to 34.8% ($P=0.007$) in the Blacks and from 56.9% to 43.2% in the Whites ($P=0.269$). Similarly, for positive surgical margins, the rate went from 100% to 26.1% in Blacks ($P<0.001$) and from 41.2% to 27.0% for whites ($P=0.021$). We believe this improvement represents equal availability of screening and better PSA testing practices for men of both races by military physicians. As noted previously, the literature supports that this improved pathologic stage end-point will eventually translate into improved disease-specific survival for these traditionally high-risk men.

CONCLUSIONS

Although randomized clinical trails have yet definitively to prove or disprove the efficacy of prostate cancer population-based screening, emerging data in the PSA-Era arguably support PSA testing in the early diagnosis of prostate cancer. Specifically, with public awareness of the disease and widespread PSA testing, smaller cancers are being detected in younger men and 5-year cancer-specific survivals are on the rise. Even though this lead-time effect may not translate into a long-term improvement, these changes are a necessary prerequisite to effective screening and are very promising. For high-risk African American men, a strategy consisting of an annual PSA blood test and DRE for men ⩾40 years old appears prudent. Use of age- and race-specific references ranges for PSA based on sensitivity, or maximal cancer detection, is my favored approach in this high-risk group. Specifically, for African American men between 40 and 49 years, those with a PSA value greater than 2.0 ng/ml should consider further evaluation. For older Black men (⩾50) entering screening, consideration should be given to further evaluation of men with a PSA greater than 4.0 ng/ml recognizing that any threshold chosen will be a balance between sensitivity and specificity.

ACKNOWLEDGMENTS

The author is supported by a grant from the Center for Prostate Disease Research, a program of the Henry M. Jackson Foundation for the Advancement of Military Medicine (Rockville, MD) funded by the US Army Medical Research and Material Command.

REFERENCES

1. Greenlee RT, Hill-Harmon MB, Murray T, Thun M. Cancer statistics 2001. *CA Cancer J. Clin.* 2001; 51:15–37.
2. Walsh PC. Using prostate-specific antigen to diagnose prostate cancer: sailing in uncharted waters. *Ann. Intern. Med.* 1993; 119:948–9.
3. Woolf SH. Screening prostate cancer with prostate-specific antigen. An examination of the evidence. *N. Engl J Med.* 1995; 333:1401–1405.
4. Mettlin C, Jones G, Avetett H et al. Defining and updating the American Cancer society guidelines for the cancer-related check-up: prostate and endometrial cancers. *Cancer. J. Clin.* 1993; 43:42–6.
5. American Urological Association: Early detection of prostate and cancer use of transrectal ultrasound. In AUA (ed.) *American Urological Association 1992 Policy Statement Book*, pp. 4–20. Baltimore, MD: American Urological Association, 1992.
6. US Preventive Service Task Force. Screening for prostate cancer. In *Guide to Clinical Preventive Services*, 2nd ed., p. 119. Baltimore, MD: Williams and Wilkins, 1996.
7. American College of Physicians. Screening for prostate cancer. *Ann. Intern. Med.* 1997; 126:480–4.
8. Catalona WJ, Smith DS Ratliff TL, Basler JW. Detection of organ-confined prostate cancer is increased through prostate-specific antigen-based screening. *JAMA* 1993; 270:948–54.
9. Catalona WJ. Screening for prostate cancer (Letter). *N. Engl. J. Med.* 1996; 334:666–7.
10. Morrison AS. The effects of early treatment, lead time, and length time bias the mortality experienced by cases detected by screening. *Int. J. Epidemiol.* 1982; 11:261.
11. Jacobson SJ, Katusic SK, Bergstralh EJ et al. Incidence of prostate cancer diagnosis in the eras before and after serum prostate specific-antigen. *JAMA* 1995; 274:1455.
12. Stephenson RA. Population-based prostate cancer trends in the PSA era: data from the Surveillance, Epidemiology, and End Results (SEER) Program. *Monogr. Urol.* 1998; 19:3–19.
13. Farkas A, Schneider D, Perotti M et al. National trends in the epidemiology of prostate cancer, 1973 to 1994: evidence for the effectiveness of prostate-specific antigen screening. *Urology* 1998; 52:444.
14. Schwartz KL, Serverson RK, Gurney JG et al. Trends in the stage specific incidence of prostate carcinoma in the Detroit metropolitan area. *Cancer* 1996; 78:1260.
15. Threlfall TJ, English DR, Rouse IL. Prostate cancer in Western Australia: trends in incidence and mortality from 1985 to 1996. *Med. J. Austral.* 1998; 169:21.
16. Roberts RO, Bergstralh EJ, Katusic SK et al. Decline in prostate cancer mortality from 1980 to 1997, and an update on incidence trends in Olmsted County, Minnesota. *J. Urol.* 1999; 161:529.
17. Feinstein AR, Sosin DM, Wells CK. The Will Rogers phenomenon. Stage migration and new diagnostic techniques as a source of misleading statistics for survival in cancer. *N. Engl. J. Med.* 1985; 312:1604.
18. Stephenson RA, Stanford JL. Population based prostate cancer trends in the United States: patterns of change in the era of prostate specific-antigen. *World J. Urol.* 1997; 15:331.
19. Lefevre ML. Prostate cancer screening: more harm than good? *Am. Fam. Physician* 1998; 58:432.
20. Sun L, Gancarczyk K, Paquette EL et al. Introduction to Department of Defense Center for Prostate Disease Research Multicenter National Prostate Cancer Database and analysis in the PSA-Era. *Urol. Oncol.* 2001; 6:203–9.
21. Oesterling JE, Jacobson SJ, Chute CG et al. Serum prostate-specific antigen in a community-based population of healthy men: established of age-specific reference ranges. *JAMA* 1993; 270:860.
22. Morgan TO, Jacobson SJ, McCarthy WF et al. Age-specific reference ranges for prostate-specific antigen in black men. *N. Engl. J. Med.* 1996; 335:304.
23. Chodak GW, Thisted RA, Gerber GS et al. Results of conservative management of clinically localized prostate cancer. *N. Engl. J. Med.* 1994; 330:242.
24. Herold DM, Hanlon AL, Movsas B et al. Age-related prostate cancer metastases. *Urology* 1998; 51:985.
25. Carter HB, Epstein JI, Partin AW. Influence of age and prostate-specific antigen on the change of curable prostate disease. *Urology* 1999; 53:126.
26. Litwiller SE, Djavan B, Klopukh BV et al. Radical retropubic prostatectomy for localized carcinoma of the prostate in a large metropolitan hospital: changing trends over a 10 year period (1984–1994). *Urology* 1995; 45:813.
27. Newcomer LM, Stanford JL, Blumenstein BA et al. Temporal trends in rates of prostate cancer: declining incidence of advanced stage disease, 1974 to 1994. *J. Urol.* 1997; 158:1427.
28. Perez CA, Hanks GE, Leibel, SA et al. Localized carcinoma of the prostate (stages T1b, T1c, T2, and T3). Review of Management with external beam radiation therapy. *Cancer* 1993; 72:3156.
29. Catalona WJ, Smith DS. 5-year recurrence rates after anantomic radical reptropubic prostectectomy for prostate cancer. *J. Urol.* 1994; 153:1837.
30. Smith DS, Catalona WJ. The nature of prostate cancer detected through prostate specific antigen based screening. *J. Urol.* 1994; 152:1732.
31. Smith DS, Catalona WJ. The nature of prostate cancer detected through prostate specific antigen based screening. *J. Urol.* 1994; 152:1732.
32. Vijayakumar S, Vaida F, Weichselbaum R et al. Race and the Will Rogers phenomenon in prostate cancer. *Cancer J. Sci. Am.* 1998; 4:27–34.
33. Partin AW, Kattan MW, Subong, EN et al. Combination of prostate-specific antigen, clinical stage, and Gleason score to predict pathologic stage of localized prostate cancer: a multi-institutional update. *JAMA* 1997; 277:1445.
34. Moul JW, Connelly RR, Lubeck DP et al. Predicting risk of prostate specific antigen recurrence after radical prostatectomy with the Center for Prostate Disease Research and Cancer of the Prostate Strategic Urologic Research Endeavor databases. *J. Urol.* 2001; 166:1322–27.

35. Preston DM, Bauer JJ, Connelly RR et al. Prostate-specific antigen to predict outcome of external beam radiation for prostate cancer: Walter Reed Army Medical Center experience, 1988–1995. *Urology* 1999; 53:131–8.
36. Schwartz KL, Grignon DJ, Sakr WA et al. Prostate cancer histologic trends in the Metropolitan Detroit area, 1982 to 1996. *Urology* 1999; 53:769.
37. Hankey BF, Feuer EJ, Clegg LX et al. Cancer surveillance series: interpreting trends in prostate cancer – Part I: evidence of the effects of screening in recent prostate cancer incidence, mortality, and survival rates. *J. Natl. Cancer Inst.* 1999; 91:1017.
38. McNeal JE, Redwine EA, Freiha FS et al. Zonal distribution of prostatic adenoncarcinoma. Correlation with histologic pattern and direction of spread. *Am. J. Surg. Pathol.* 1988; 12:897.
39. Lee F, Siders DB, Torp-Pedersen ST et al. Prostate cancer: transrectal ultrasound and pathology comparison. A preliminary study of outer gland (peripheral and central zones) and inner gland (transition zone) cancer. *Cancer* 1991; 67:1132.
40. Chodak GW. The role of watchful waiting in the management of localized prostate cancer. *J. Urol.* 1994; 152:1766.
41. Johansson JE. Expectant management of early stage prostatic cancer: Swedish experience. *J. Urol.* 1994; 152:1753.
42. Albertson PC, Hanley JA, Gleason DF et al. Competing risk analysis of men aged 55 to 74 years at diagnosis managed conservatively for clinically localized prostate cancer. *JAMA* 1998; 280:975.
43. Gilliland FD, Hunt WC, Key CR. Improving survival for patients with prostate cancer diagnosed in the prostate-specific antigen era. *Urology* 1996; 48:67.
44. Etzioni R, Legler JM, Feuer EJ et al. Cancer surveillance series: interpreting trends in prostate cancer – Part III: quantifying the link between population prostate-specific antigen testing and recent declines in prostate cancer mortality. *J. Natl. Cancer Inst.* 1999; 91:1033.
45. Paquette EL, Connelly RR, Sun L et al. African-American men undergoing radical retropubic prostatectomy improvements in pathologic staging for during PSA-ERA: implications for screening a high risk group for prostate cancer. *J. Urol.* 2001; 165(Suppl. 5):65(Abstract #266).
46. Labrie F, Candas B, Dupont A et al. Screening decreases prostate cancer death: first analysis of the 1988 Quebec Prospective Randomized Controlled Trail. *Prostate* 1999; 38:83.
47. Bartsch G, Horninger W, Klocker H et al. Prostate cancer mortality after introduction of prostate-specific antigen mass screening in the Federal State of Tyrol, Austria. *Urology* 2001; 58:417–24.
48. Barry MJ, Roberts RG. Indications for PSA testing. *JAMA* 1997; 277:955.
49. Smith RA, Von Eschenbach AC, Wender R et al. American Cancer Society Guidelines for Early Detection of Cancer: Update of Early Detection Guidelines for Prostate, Colorectal and Endometrial Cancers. *CA Cancer J. Clinicians* 2001; 51:38–75.
50. American Urologic Association's prostate-specific antigen (PSA) best practice policy. *Oncology* 2000; 14:267–86.
51. Boring CC, Squires TS, Heath CW. Cancer statistics for African Americans: 1992. *CA Cancer J. Clin.* 1993; 43:7–17.
52. Morton RA. Racial differences in adenocarcinoma of the prostate in North American men. *Urology* 1994; 44:637–45.
53. Pienta KT, Demers R, Hoff M et al. Effect of age and race on survival of men with prostate cancer in the metropolitan Detroit tri-county area, 1937 to 1987. *Urology* 1995; 45:93–102.
54. Moul JW. Increased risk of prostate cancer in African American men. *Mol. Urol.* 1997; 1:119.
55. Moul JW, Serterhenn IA, Connelly RR et al. Prostate-specific antigen values at the same time of prostate cancer diagnosis are higher in African American men. *JAMA* 1995; 274:1277–81.
56. Moul JW, Connelly RR, Mooneyhanm RM et al. Radical differences in tumor volume and prostate specific antigen among radical prostatectomy patients. *J. Urol.* 1999; 162:394–7.
57. Powell IJ. Prostate cancer and African-American men. *Oncology* 1997; 11:599.
58. Crawford ED. Prostate cancer awareness week – September 22 to 28, 1997, *CA Cancer J.* 1997; 47:288.
59. Ross R, Bernstein L, Judd H et al. Serum testosterone levels in healthy young black and white men. *J. Natl Cancer Inst.* 1986; 76:45.
60. Giovannucci E, Stampfer MJ, Krithivas K et al. The CAG repeat with the androgen receptor gene and its relationship to prostate cancer. *Proc. Natl Acad. Sci. USA* 1997; 94:3320.
61. Littrup PJ. Editorial: Prostate cancer in African American men. *The Prostate* 1997; 31:129–41.
62. Powell IJ, Bannerjee M, Novallo M et al. Should the age specific prostate specific antigen cutoff for prostate biopsy be higher for black than for white men older than 50 years? *J. Urol.* 2000; 163:146–8.
63. Bullock AD, Harmon T, Smith DS et al. Prostate screening in younger men. *J. Urol.* 1997; 157(Suppl):66 (abstract #252).
64. Catalona WJ, Smith DS, Ornstein DK. Prostate cancer detection in men with serum PSA concentrations of 2.6 to 4.0 ng/ml and benign prostate examination. *JAMA* 1997; 277:1452–55.
65. Smith DS, Catalona WJ, Bullock AD. Lower total PSA cutoffs for cancer screening in African American men. *J. Urol.* 1997; 157(Suppl.):160 (abstract #618).
66. Littrup PJ, Sparschu RA. Transrectal ultrasound and prostate cancer risks: the 'tailored' prostate biopsy. *Cancer* 1995; 75(Suppl):1805–13.
67. Reissigl A, Horninger W, Ennemoser O et al. Measurement of the ratio F/T PSA enhances localized prostate cancer (PCA) detection in men with low total PSA levels (1.25 to 3.25 ng/ml) and negative rectal examination. *J. Urol.* 1997; 157(Suppl. 4):365 (abstract # 1430).
68. Carter HB, Epstein JI, Chan DW et al. Recommended prostate-specific antigen testing intervals for the detection of curable prostate cancer. *JAMA* 1997; 277:1456–60.
69. Moul JW. PSA thresholds for prostate cancer detection. *JAMA* 1997; 278:699.
70. Moul JW. Curability of prostate cancer in African American Implications for neoadjuvant/adjuvant hormonal therapy. *Mol. Urol.* 1998; 2:209–13.
71. Carter HB, Epstein JI, Partin AW. Influence of age and prostate-specific antigen on the chance of curable prostate cancer among men with nonpalpable disease. *Urology* 1999; 53:126–30.
72. Preston DM, Levin LI, Jacobson DJ et al. Prostate-specific antigen levels in young white and black men 20 to 45 years old. *Urology* 2000; 56:812–16.
73. Moul JW, Connelly RR, Barko WF et al. Should healthy men between the age of 40–49 be screened for prostate cancer: a Department of Defense (DoD), Center for Prostate Disease Research (CPDR), and Army Physical Fitness Research Institute (APFRI) Prospective Study at the US Army War College (USAWC). *J. Urol.* 2000; 163(Suppl.):90(Abstract # 393).
74. Presti JC Jr, Hovey R, Bhargava V et al. Prospective evaluation of prostate specific antigen and prostate specific antigen density in the detection of carcinoma of the prostate: ethnic variations. *J. Urol.* 1997; 157:907–11.
75. Carter HB, Epstein JI, Chan DW et al. Recommended prostate-specific antigen testing interval for the detection of curable prostate cancer. *JAMA* 1997; 14:277(18):1456–60.
76. Paquette EL, Connelly RR, Sesterhenn IA et al. Improvements in pathologic staging for African-American men undergoing radical retropubic prostatectomy during the prostate specific antigen era. *Cancer* 2001; 92:2673–9.

Molecular Mechanism of Prostate Cancer Invasion and Metastasis

Kazuo Gohji
Department of Urology, Osaka Medical College, Takatsuki, Japan

Sohei Kitazawa
Department of Molecular Pathology, Kobe University Graduate School of Medicine, Kobe, Japan

INTRODUCTION

Prostate cancer is the most common malignant tumor among men in the USA: more than 40 000 die of the disease annually.[1] It is the most prevalent tumor in men and, despite increasing efforts at early detection, 10–20% of the cases present bone metastases at diagnosis. Most deaths from the disease are still caused by widespread metastases that are resistant to conventional treatment in spite of improved surgical techniques and local and systemic therapies. Bone is one of the common metastatic sites of malignant neoplasms, including those from carcinoma of the breast, prostate, thyroid, kidney and lung. Normal bone is continuously being remodeled with new bone formation by osteoblasts and bone degradation by osteoclasts; at bone metastatic sites, however, fine balance between the two processes is disturbed. When bone destruction dominates, net loss of bone mass occurs and the lesion is described as osteolytic; when an excessive amount of new bone formation takes place with less bone destruction, the lesion is described as osteoblastic or osteosclerotic. Human prostate cancer is one of the rare cancers that consistently produces osteoblastic metastasis to bone in approximately 95% of cases. Several growth factors including transforming growth factor (TGFβ)[2], basic fibroblast growth factor (bFGF)[3] and bone morphogenetic protein (BMP)[4], which stimulate osteoblast growth and bone matrix formation, are known in benign and malignant prostate cells.

The molecular basis for tumor progression, invasion and metastasis is still unclear, although recent reports have emphasized the importance of secreted proteases, cellular molecules,[5] and the presence of mitogenic and angiogenic growth factors at the site of tumors, including at that of prostate cancer.[5–8]

In this chapter, we describe the general mechanism of cancer invasion and metastasis, and discuss the relationship between cancer cells, including prostatic carcinoma, and host stromal cells in cancer growth, invasion and metastasis. We then focus on the mechanism of and consider several growth factors related to osteoblastic metastasis from prostate cancer.

PROCESS OF CANCER METASTASIS

The process of cancer metastasis consists of a series of sequential interrelated steps, with the outcome depending on both the intrinsic properties of the tumor cells and the host.[6,8–11] Bone is one of the major metastatic sites for prostate, lung and breast cancer and, in principle, the steps or events in the pathogenesis of metastasis are similar in all tumors (**Fig. 2.1**).

Recently, new therapeutic modalities have improved the survival of cancer patients; the outcome of the treatment of cancer at the primary site and of lung and liver metastasis has improved. As a result, although bone metastasis alone is not fatal, it poses clinically crucial problems caused by pain or the disturbance of the quality of life. Human bone consists of three major parts: the cortical bone, trabecular bone and bone marrow. The cortical bone contains a small number of osteocytes and hard, richly mineralized tissue. It plays an important role in the structure of the bony frame; therefore, its destruction by cancer cells induces pathological fractures, severe pain and

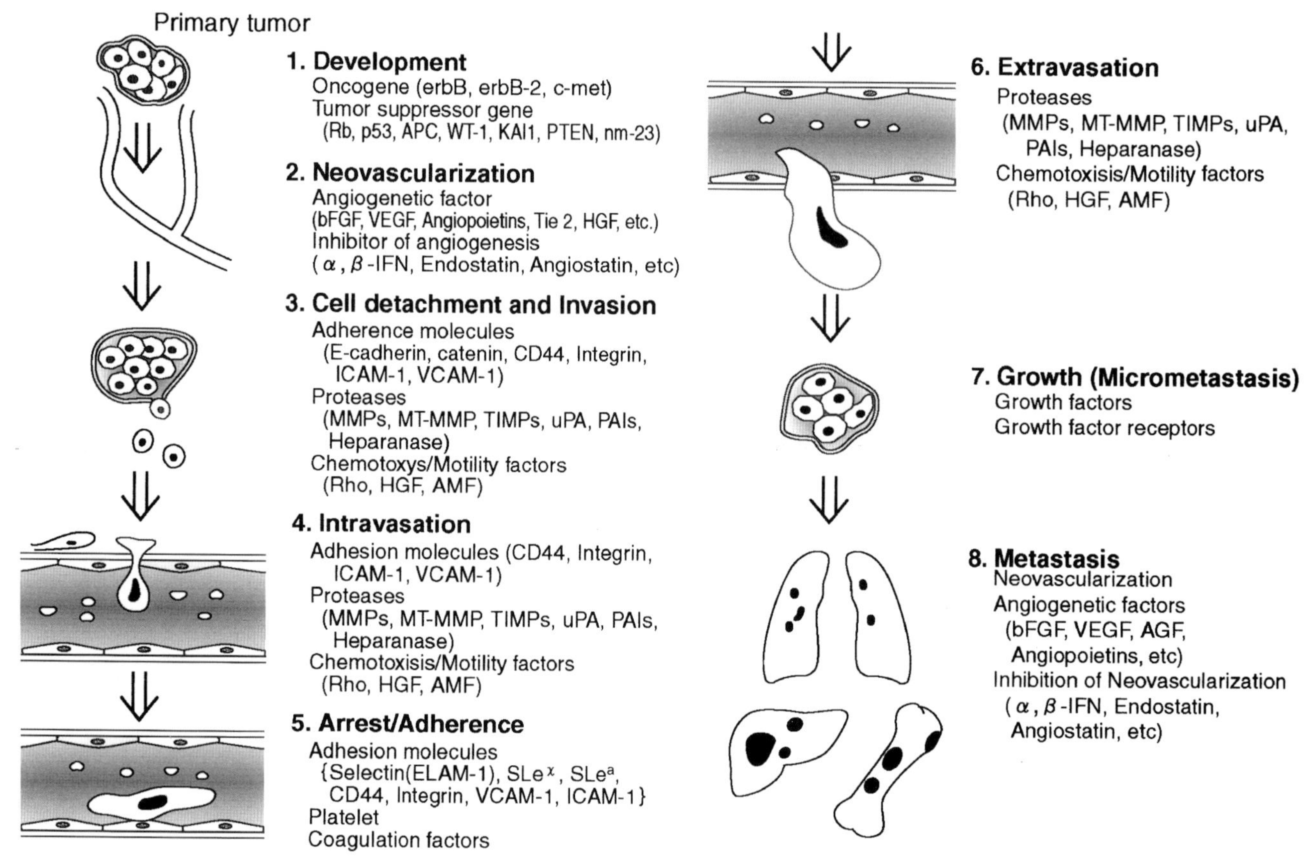

Fig. 2.1. General mechanism of hematogenous metastasis.

movement disturbances. On the other hand, several cytokines and growth factors produced and secreted by osteoblast are also stored in cortical bone tissue: insulin-like growth factor II (IGF-II) is the most abundant of cytokines, followed by TGFβ, IGF-I, platelet-derived epidermal growth factor (PDEGF) and bFGF. The incidence of pure osteoblastic metastasis is approximately 95% of prostate cancer skeletal metastases; the remaining 5% is of the mixed type (osteoblastic and osteolytic). Osteolysis is also an important step in the establishment of osteoblastic metastasis from prostate cancer. Two possible mechanisms of osteolysis prior to the osteoblastic metastasis are a cell-mediated mechanism and a non-cell-mediated one.[12] The direct release of proteases (serine proteases, cystine proteases or metalloproteases) by tumor cells generally resulting in extracellular matrix degradation and host-cell lysis, and facilitating tumor invasion is the so-called non-cell-mediated mechanism. The capacity to degrade the mineralized matrix, however, requires the involvement of osteoclasts. On the other hand, several cytokines and growth factors – TGFβ, parathyroid hormone-related protein (PTHrP), interleukin-6 (IL-6), tumor necrosis factor (TNFα), PDEGF, bFGF and prostaglandin secreted by tumor cells and host immune cells under cell-to-cell interaction,[13] stimulate osteoclastic activity, osteoclast growth and differentiation (cell-mediated osteolysis).[13] Cancer cells that also produce some stimulatory factors, including the above cytokines and growth factors, stimulate the expression of RANKL on the cell surface of osteoblasts. Thus, increased RANKL reacts with the receptor (RANK) on the cell surface of osteoclast-precusor cells and induces cell differentiation to osteoclasts. Trabecular bone tissue, which is much softer than cortical bone tissue, is thought to be an important site for bone remodeling. Osteoclasts absorb bone tissue and release cytokines stored in cortical bone tissue; the cytokines and growth factors then stimulate new bone formation by osteoblasts, which in turn stimulate cancer cell growth.

Bone is a suitable microenvironment for cancer cell growth because large amounts of several cytokines are secreted by bone tissue. Bone marrow, the site of blood cell production, has two types of stem cells: blood stem cells that have the potential of multidifferentiation to all blood cells and osteoclasts, and stromal cells that differentiate to osteoblasts, muscular cells and adipose cells. Cancer cells pass through the sinus vein and enter the bone marrow, invade the trabecular bone and then grow in cortical bone in close conjunction with two types of stem cells and immunocytes. The interactions of cancer cells, bone tissues and stem cells are shown in **Fig. 2.2**. Thus, in the development of osteoblastic metastasis from prostate cancer, osteoblasts might play a fundamental role as a site of attachment, as a mediator of skeletal invasiveness by facilitating osteoblastic osteolysis, and as a target for growth-promoting agents to induce bone-forming lesions.

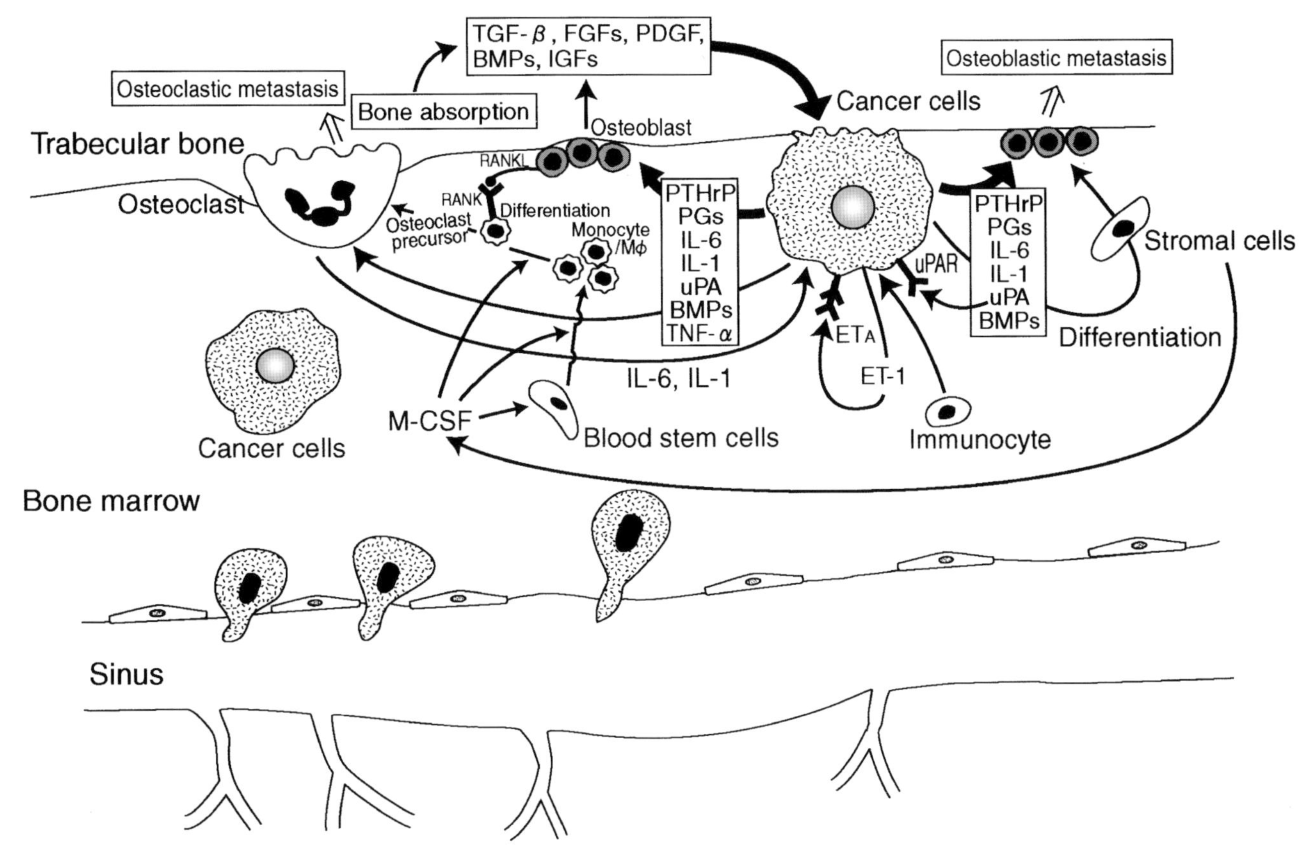

Fig. 2.2. Interaction between cancer cells and bone microenvironment.

INFLUENCE OF MICROENVIRONMENT IN PROSTATE CANCER INVASION AND METASTASIS

Cancer metastasis is established by not only cancer cells but also host tissue cells (seed and soil theory).[14] For example, kidney cancer metastasizes mostly to the lung, breast and prostate cancer mostly to the bone, and colon cancer with high incidence to the liver (**Fig. 2.3**). The influence of organ environment on the growth, invasion and metastasis of

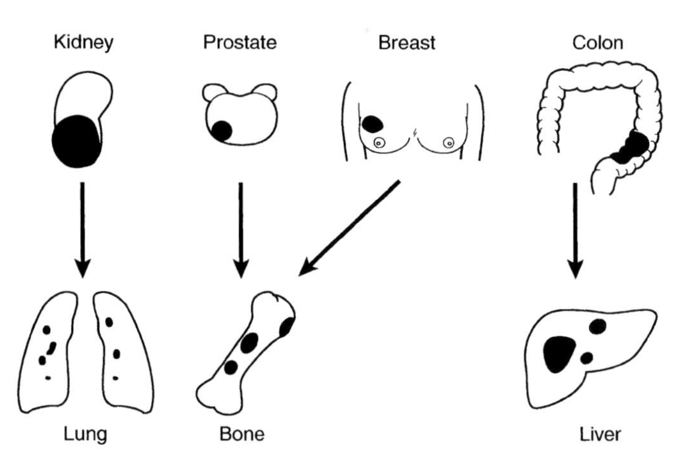

Fig. 2.3. 'Seed and soil' hypothesis (Paget, 1889).[14]

human renal cell carcinoma cells,[8,10] bladder carcinoma cells[15] and prostate carcinoma cells[16,17] has been examined in animal models by injecting these cells at orthotopic sites. Renal and bladder cancer cells injected subcutaneously do not produce visceral or lymph node metastasis but, when injected at orthotopic sites, they produce lymph node or visceral metastasis, or both.[8,15] Both prostate and bone (not lung or kidney) stromal cells are capable of LNCaP tumor growth *in vivo* by co-inoculating LNCaP cells with these organ-specific fibroblasts,[17] on the speculation that LNCaP cell growth is regulated by growth factors such as bFGF and hepatocyte growth factor (HGF) secreted from organ-specific stromal cells.[17,18] We have also demonstrated that PC3M prostate cancer cells established from liver metastasis of PC3 cells inoculated at orthotopic sites produce large tumors with lymph node metastasis and seminal vesicle invasion[16] (**Fig. 2.4**). PC3M cells inoculated subcutis, however, produce only small tumors with no metastasis. Metastatic variant cells established by *in vivo* selection also showed high metastatic potential *in vivo*; high metastatic clones of human renal cancer, bladder cancer and prostate cancer show a higher incidence of invasion and metastasis than parental cells do.[15–16,19] The highly metastatic renal cell carcinoma produces larger amounts of extracellular matrix degradative enzymes (MMP-2 or uPA) and angiogenetic factors (bFGF) than parental cells do. Moreover, the high metastatic variant of human bladder cancer cells produces larger amounts of epidermal growth factor (EGF)-receptor

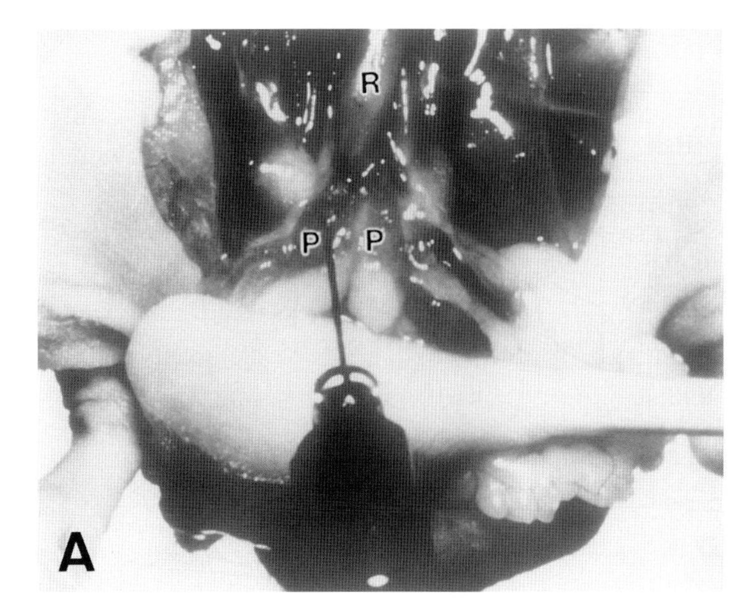

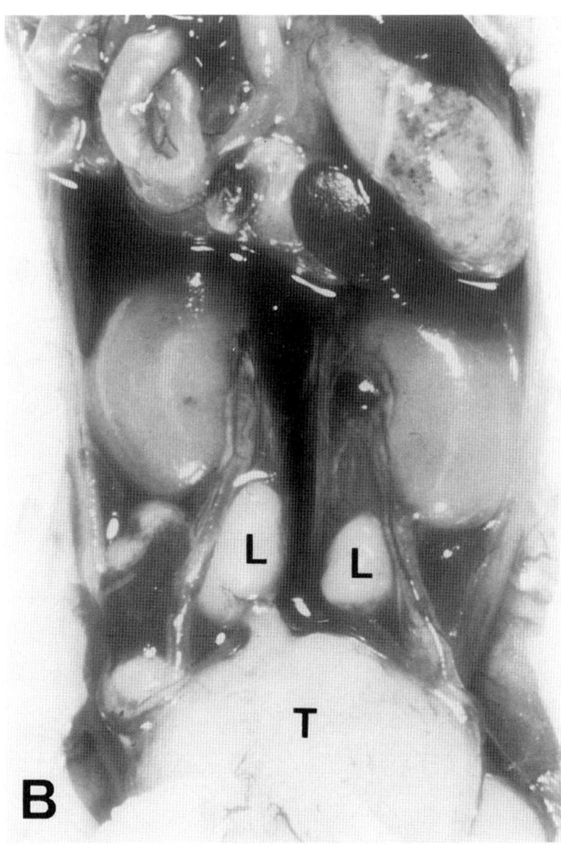

Fig. 2.4. Orthotopic animal model of prostate cancer. (A) Inoculation of PC3M cells into the prostate of nude mice. P, prostate; R, rectum. (B) PC3M prostate cancer cells growing at orthotopic site (T) and lymph node metastasis (L).

than parental cells do.[15] These findings suggest that the crucial factors for growth, invasion and metastasis are different for different cell types, and the capability for the production of these crucial factors could be induced by the micronenvironment at the tumor growing site, indicating that organ environment influences the growth and progression of human cancer. Factors related to organ-specific metastasis are shown in **Table 2.1**.

Table 2.1. Factors related to organ-specific metastasis.

Factors determined by tumor cell
1. Attachment to vascular endothelial cells
2. Response to motility factors
3. Degradation of vascular basement membrane and extracellular matrix
4. Escape from host immune system
5. Response to growth factors
Factors determined by host metastatic organ
1. Vascular endothelial cells
2. Extracellular matrix, stromal cells
3. Motility factors
4. Growth factors

LYMPHATIC METASTASIS

Tumor cells are spread by both the lymphatic route and the bloodstream. The lymphatic and vascular systems have numerous connections, and disseminating tumor cells may pass from one system to the other. The mechanism of lymphatic metastasis is very similar to hematogenous spread, which is described below.

General mechanism of hematogenous metastasis

The general mechanism of hematogenous metastasis is shown in **Fig. 2.1**.

Gene alteration for tumor cell growth and metastasis

Several oncogenes and tumor suppressor genes have a role in cancer development.[20] KAI1, initially isolated as an important metastasis suppressor gene of prostate cancer,[21] is located on human chromosome 11p11.2–13 and encodes a protein of 267 amino acids with a molecular mass of 29 610 Da.[21] It belongs to a structurally distinct family of membrane glycoproteins that includes ME491/CD63, MRP-1, TAPA-1, CD37 and CD53, most of which have been identified as leukocyte surface proteins.[21] These proteins have four hydrophobic (and presumably transmembrane) domains and one large extracellular *N*-glycosylated domain. They function in cell–cell and cell–extracellular matrix interactions, thereby potentially influencing the ability of cancer cells to invade tissue and to metastasize.[21] When human prostate cancer samples are inoculated into nude mice, KAI1 suppress the ability of prostate cancer to metastasize.[21] Moreover, KAI1 expression decreases in human prostate cancer cells derived from metastatic prostate cancer compared with that in normal prostate tissue,[21] indicating that KAI1 is crucial in the progression of prostate cancer.

Neovascularization

Tumor cells growing at the primary site need a vascular structure, which is induced when tumor mass exceeds 1–2 mm in diameter.[22] Several angiogenetic factors, such as bFGF[22], vascular endothelial growth factor (VEGF)[23], PDEGF[24], IL-8[25] and HGF,[26] are a prerequisite to the establishment of a capillary network from the surrounding host tissue. Angiogenin is also an important angiogenetic factor,[27] and we have demonstrated that its expression in high-stage or high-grade bladder cancer is significantly higher than that in low-stage and low-grade cancer.[28] Moreover, a second family of growth factors, like VEGF, specific to the vascular endothelium has been identified, with its members termed the angiopoietins and its specific receptors Tie 1 and 2.[29] All angiopoietins bind to Tie 2, but whether they ultilize the closely related receptor Tie 1 is still unclear. The structure of angiopoietins differs from that of known angiogenetic factors or other ligands for receptor tyrosine kinases. Angiopoietin-1 binds to and induces the tyrosine phosphorylation of Tie 2; it does not directly stimulate endothelial cell growth, but is required for development of the vascular structure. Since the role of angiopoietins and the receptors in angiogenesis, invasion and metastasis of prostate cancer are unknown, further investigation into their role is necessary.

Invasion

Tumor cells released from the primary foci must overcome the influence of some attachment factors such as E-cadherin; they then also infiltrate the surrounding tissue and vascular basement membrane by degrading extracellular matrices or vascular basement membranes, which are composed of type IV collagen, proteogrycans, fibronectin and laminin. In this step, extracellular matrix degradative enzymes [matrix metalloproteinase (MMP), urokinase-type plasminogen activator (uPA)] are requisite.[5,6] MMPs are secreted from both cancer and stromal cells in non-active pro-forms, and MMP-2 is activated by membrane-type MMP (MT-MMP) and MMP-9 is activated by other proteases increasing of uPA (**Fig. 2.5**).

At present, 19 subtypes of MMP are known (**Table 2.2**). The expression of extracellular matrix enzymes is associated with cancer invasion, metastasis and poor patient survival.[30,31] The expression of MMP-2 and MMP-9 in high-grade and high-stage bladder cancer tissue is significantly higher than that in low-grade and low-stage cancer.[32] The expression of MMP-2 in high-grade and high-stage colorectal cancer is higher than that in the low-grade and low-stage one.[31] On the other hand, three types of tissue inhibitors of metalloproteinase (TIMPs), endogenous inhibitors of MMPs, are known: TIMP-1 predominantly inhibits the activity and activation of MMP-9; and TIMP-2 predominantly inhibits the activity of MMP-2.[33] The balance between MMP and TIMP plays a crucial role in cancer invasion and metastasis.[34] Cancer cell metastasis is inhibited by transfection of TIMP-2 cDNA.[35] Moreover, TIMPs not only inhibit the invasion and metastasis of tumor cells, but also decrease tumor growth *in vivo*.[35] Overexpression of TIMP-2 by retroviral-mediated gene transfer inhibits tumor cell growth and invasion.[34] We have shown that the ratio of MMP-2 to TIMP-2 determines the metastatic

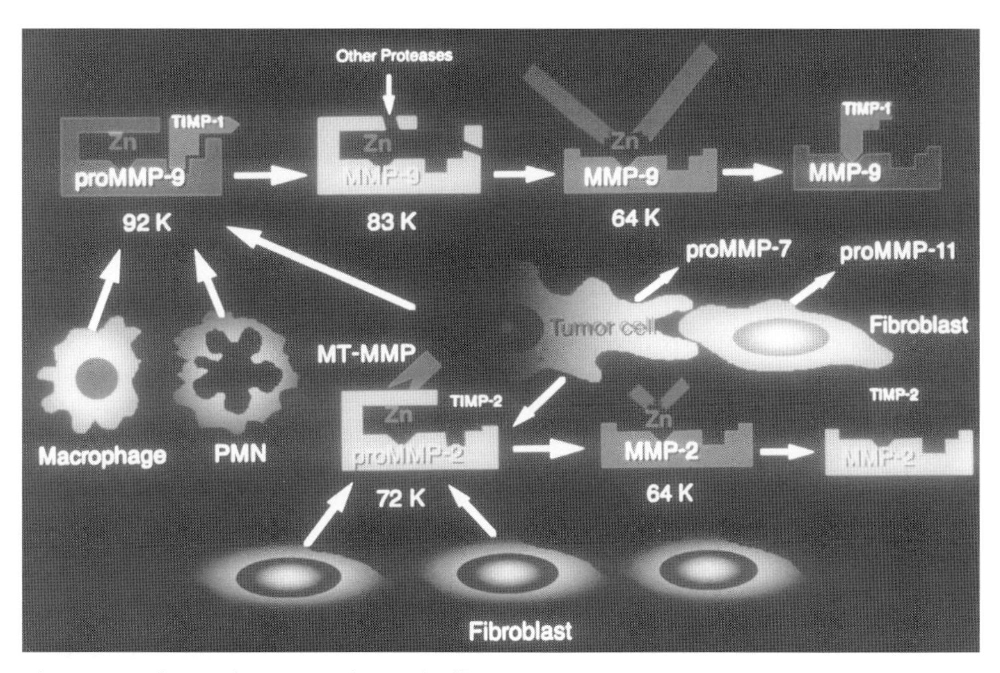

Fig. 2.5. Production and activation of MMPs by tumor and normal cells.

Table 2.2. Types of matrix metalloproteinase (MMPs).

MMP family	Enzyme	Alternate name
Collagenases	MMP-1	Fibroblast collagenase-1
	MMP-8	Neutrophil collagenase-2
	MMP-13	Collagenase-3
	MMP-18	Collagenase-4
Stromelysins	MMP-3	Stromelysin-1
	MMP-10	Stromelysin-2
	MMP-11	Stromelysin-3
	MMP-7	Matrilysin
Gelatinases	MMP-2	Gelatinase A (72 kDa)
	MMP-9	Gelatinase B (92 kDa)
Membrane Type	MMP-14	MT1-MMP
	MMP-15	MT2-MMP
	MMP-16	MT3-MMP
	MMP-17	MT4-MMP
Others	MMP-12	Metalloelastase
	MMP-19	
	MMP-20	Enamelysin
	MMP-21	*Xenopus laevis*
	MMP-ABT	Abbott, database

potential of renal cancer.[36] Compared with parental cells, mouse renal cell carcinoma (Renca cell) cells transfected with MMP-2 cDNA demonstrate enhanced metastatic potential, but those transfected with low MMP-2 cDNA and high TIMP-2 cDNA demonstrate decreased potential. Moreover, the serum ratio of MMP-2 to TIMP-2 levels is capable of predicting the recurrence of bladder cancer; the recurrence-free survival of patients with high MMP-2/TIMP-2 ratios who undergo radical cystectomy is significantly lower than that of patients with low ratios.[37] In prostate cancer, the expression of MMP-2 in high-stage and high-grade cancer is significantly higher than in organ-confined cancer.[38] Also, the expression of MMP-7mRNA and MMP-7mRNA/TIMP-1mRNA ratio in high-stage, high-grade or metastatic prostate cancer is significantly higher than that in low-stage, low-grade or non-metastatic cancer, indicating that MMP-7 may play an important role in the invasion and metastasis of prostate cancer, and that the balance between MMP-7 and TIMP-1 expression may relate to the invasive ability of prostate cancer.[39]

uPA is also an important extracellular matrix degradative enzyme in several types of malignant diseases.[40,41] The expression of uPA in muscular invasive bladder cancer is higher than that in superficial cancer.[42] The expression of uPA and the uPA receptor is associated with the progression and is a significant prognostic factor in renal cancer. The roles of uPA and the uPA receptor in prostate cancer invasion and metastasis are mentioned in the latter part of this paper.

Heparanase, an endo-beta-D-glucuronidase, degrades heparan sulfate (HS) and heparan sulfate proteoglycans (HSPGs) which are also major components of the extracellular matrix or the vascular basement membrane. Moreover, HSPGs are implicated in a number of cellular processes, including cell adhesion (**Fig. 2.6**).[43] Heparanase cleaves HS/HSPGs into characteristic, large-molecular-weight fragments and modulates biological features of HS/HSPGs-binding proteins.[44,45] Among the HS/HSPGs-binding proteins, several growth factors and cytokines, such as bFGF, VEGF, PDEGF, IL-8, HGF and TGF-β, are potent mitogens and chemotactic factors for endothelial cells that play crucial roles in angiogenesis and metastasis.[44,45] Moreover, the products of HS or HSPGs degraded by heparanase supress the activation of T cells. Thus, metastatic tumor cells are found to degrade HS, and a good correlation has been found between metastatic potential and heparanase activity.[43,46,47] We have recently determined that heparanase expression in high-grade and high-stage bladder cancer is significantly higher than that in low-grade and low-stage cancer at both protein and mRNA levels.[48,49] Interestingly, heparanase expression is much higher than that of MMP-2 and MMP-9.[48] Moreover, the neovascuralization in bladder cancer tissue correlates well with heparanase expression,[48] which is associated with patient survival; the microvessel count in 200× field in bladder cancer with positive heparanase expression (32.3 ± 18.2, range 3–49) is significantly higher than that in bladder cancer with negative expression (5.5 ± 6.1, range 1–16) ($P = 0.0008$), and both cancer-specific and overall survival of patients with positive heparanase expression is significantly lower than those of patients with negative expression ($P = 0.0001$ and $P = 0.0008$, respectively).[48] These results suggest that heparanase also plays important roles in invasion, angiogenesis and metastasis of bladder cancer, and thus this molecule could be a new molecule to inhibit invasion, angiogenesis and metastasis of bladder cancer. Moreover, prostate cancer cells, PC3M and LNCaP C4-2, have been shown to produce heparanase.[50,51] Further systematic investigations are needed, however, to determine the role of heparanase in prostate cancer growth, invasion and metastasis.

CD44 is an important molecule, which plays crucial roles in cancer invasion and metastasis.[52] CD44 variant form 6 (CD44v6) has been cloned and, when its cDNA is transfected into low-metastatic pancreatic carcinoma cells, metastatic potential is stimulated and the monoclonal antibody for CD44v6 inhibits the metastasis, indicating the crucial role of CD44v6 in cancer metastasis. Several investigators have demonstrated the highest incidence of a CD44 isoform containing variant forms of v8, v9 and v10 exon (CD44v8–10) in colon cancer, non-small-cell lung cancer and bladder cancer.[53] We have shown that the expression of CD44v8–10 in metastatic non-seminomatous testicular cancer is significantly higher than that in non-metastatic seminoma.[54] On the other hand, PC3 prostate cancer cells transfected with CD44v8–10 decrease tumor growth and invasion by increasing the connection with cancer cells and hyarulonic acid.[55] These findings indicate that CD44 alternative splicing alters the metastatic potential of cancer cells in relation to cancer cells, including prostate cancer and the extracellular matrix.

HGF, initially isolated as a growth factor of hepatocyte, shows several important biological effects, such as stimulation of cell motility and as an angiogenetic factor *in vivo*. HGF is

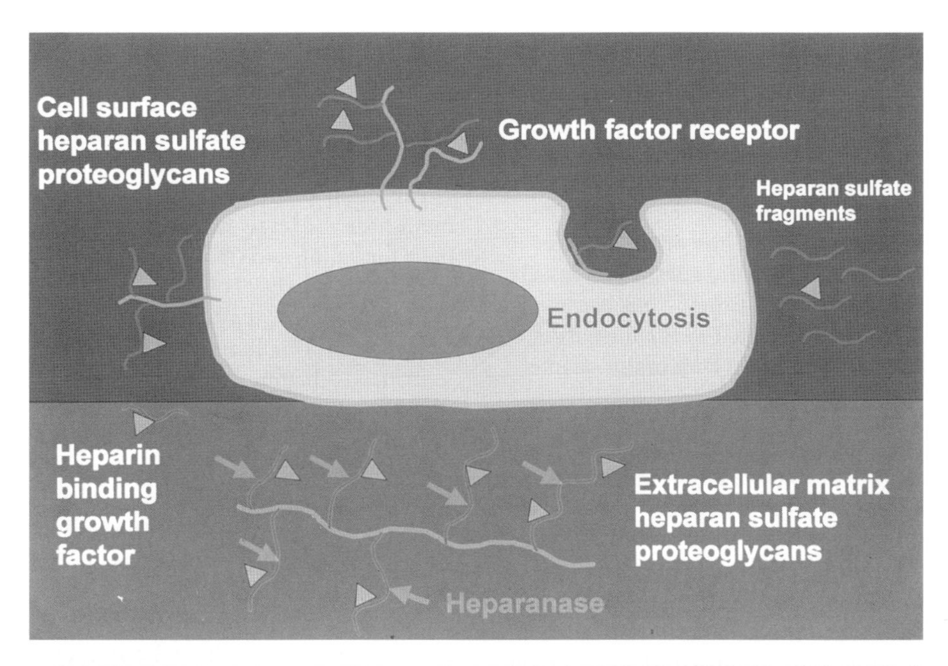

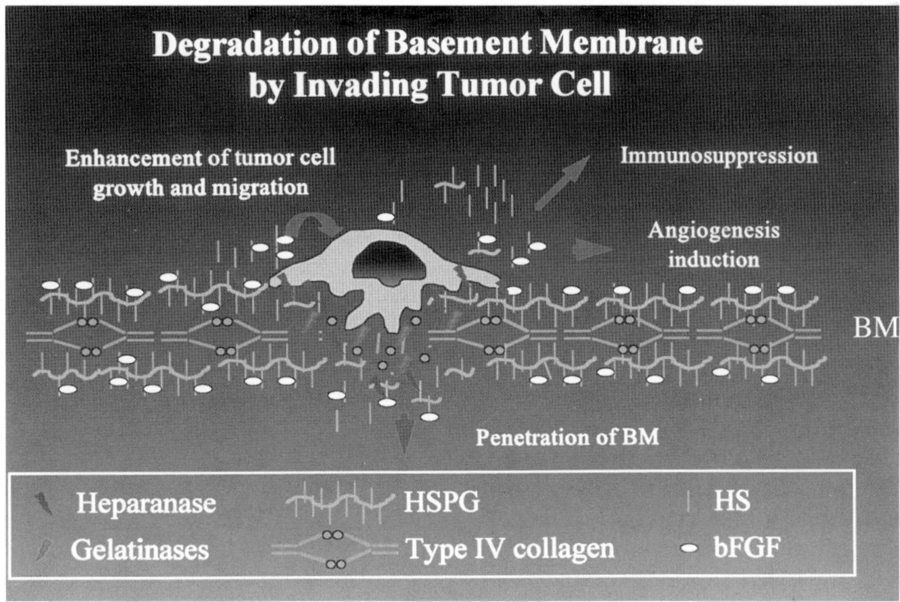

Fig. 2.6. Biological function of HS/HSPG, and role of heparanase in cancer invasion and metastasis. BM, basement membrane.

expressed in human prostatic stromal myofibroblasts, not in cancer cells; however, c-met transcripts, which produce an HGF receptor, have been identified by Northern blot in DU-145 and PC3 androgen-insensitive human prostate cancer cells.[56] c-met protein has also been detected by immunostaining in 45% of radical prostatectomy specimens,[56] and its expression in metastatic foci is significantly higher than that at the primary site (75% vs. 39%, $P < 0.005$).[56] Exogenous recombinant HGF stimulates cell growth with a high expression of c-met; moreover, the highest c-met expression is found in androgen-insensitive subclones of Dunning R-3327 rat prostatic cancer where the subclones show the highest metastatic potential, indicating that c-met is up-regulated by androgen deprivation, and that c-met expression is higher in androgen-insensitive metastatic carcinoma cells.[56] c-met expression could thus play a crucial role in prostate cancer progression.[56]

Survival

Tumor cells survive circulation, although the vast majority of circulating tumor cells are rapidly destroyed.

Arrest

Tumor cells arrest the capillary bed of distant organs by adhering either to capillary endothelial cells or to the subendothelial basement membrane. Establishment of this step is influenced by several factors such as CD44 and selectin.

Extravasation
This step is similar to Invasion.

Growth at metastatic foci
At the metastatic site, small tumor foci (micrometastasis) proliferate, grow and complete the metastasic process (metastasis) by developing a vascular network and growth factors of metastatic organs. The cells can then invade blood vessels, enter the circulation and produce additional metastases. In prostate cancer, the growth factors related to the formation of osteoblastic bone metastasis are mentioned later in this chapter.

MECHANISM OF THE DEVELOPMENT OF OSTEOBLASTIC METASTASES

Several neoplasms may produce osteoblastic metastases. Prostate cancer cells commonly metastasize to the most heavily vascularized parts of the skeleton, particularly the red bone marrow of the axial skeleton and the proximal ends of the long bones, the ribs and the vertebral columns. The interaction of prostate carcinoma with the skeleton is influenced by the microenvironment and a major determinant of the site of skeletal metastasis is blood flow. The most common metastatic site of prostate cancer is the vertebral column, probably because prostate cancer cells are transported to the spine via Batson's plexus, which is a low-pressure, high-volume system of vertebral veins that runs up the spine with extensive intercommunication with other major venous systems, such as the pulmonary, caval and portal systems.

In experimental animal models with skeletal metastasis, prostate cancer cells are injected via the intracardiac route. Bone metastases are not produced by injection via venous or subcutaneous routes because passage through the lung after intravenous or subcutaneous inoculation alters the surface adhesive properties of the tumor and impairs its capacity to interact with bone elements.[57] Thus, several animal models with skeletal metastasis from prostate cancer have been established by inoculation via the interarterial route.[58–60] Spontaneous development of osteoblastic metastasis has been demonstrated in transgenic mice where the transgene is a regulatory element for the rat probasin gene linked to the SV40 T antigen.[61] The factors related to osteoblastic metastasis from prostate cancer are mentioned below.

Parathyroid hormone-related protein (PTHrP)

PTHrP has been purified from human lung, breast and renal cell[62] carcinoma as a hypercalcemic factor simultaneously by several investigators over the past 10 years. The protein has 70% homology of the first 13-amino acids of the N-terminal protein of PTH,[63] binds to PTH receptors[64] and shows biologic activity similar to that of PTH.[65] High levels of PTHrP in tumor tissues have been demonstrated in squamous, breast and renal cell carcinoma patients with hypercalcemia.[66] PTHrP binds to PTH receptors, increases osteoclastic bone resorption and renal tubular reabsorption of calcium, and induces hypercalcemia. Its production is stimulated by EGF,[67] IGF-I[68] and IGF-II,[68] TGFα[69] and TGFβ,[69,70] which are released in the microenvironment of tumor-growing sites, including that of the bone.

Although prostate cancer is characterized by osteoblastic metastasis, several investigators have suggested that both bone formation and resorption are accelerated at bone metastatic sites.[71,72] Usually, bone scintigrams used for detection of metastatic bone disease are incapable of disclosing bone degradation. Biochemical markers of bone metabolism, such as urinary calcium, urinary hydroxyproline and serum alkaline phosphatase, studied as indicators of bone response, may be influenced by the rate of bone resorption and are a non-specific reflection of bone function. Recently developed osteolytic markers, such as pyridinoline and deoxypyridinoline in urine, reflect an increased rate of bone resorption and are sensitive and specific indexes of the rate of cartilage and bone breakdown.[73] In breast cancer patients with osteolytic bone metastasis, the serum levels of these markers are significantly higher than those in patients without bone metastasis. Also in patients with prostate cancer, they increase in urine in association with skeletal metastasis.[74,75] These studies show that the levels in prostate cancer patients with bone metastasis are significantly higher (89.3%) than in patients with organ-confined prostate cancer (5.9%). However, the level of urinary pyridinoline in patients with benign prostatic hyperplasia (BPH) is 0%. The increase in these bone resorption markers associated with bone metastasis is determined by nuclear bone scintigraphy. Moreover, the levels of urinary pyridinoline and deoxypyridinoline correlate with a positive response to treatment and with clinical progression of disease before the detection of new bone lesions by bone scintigrams.[74] These findings indicate that osteolysis plays an important role in the development of bone metastasis, even when new bone (osteoblastic metastasis) is being formed.

The levels of PTHrP in the serum and tissue of prostate cancer has been determined by several investigators.[76] In a study of organ-confined prostate cancer, all 33 cases resected by retropubic radical prostatectomy have shown some degree of immunoreactivity in the cytoplasm of the tumor cells,[76] with the intensity of the staining correlating directly with increasing tumor grade.[76] Although the study has not examined the expression in PTHrP of prostate cancer cells with bone metastasis, it suggests that PTHrP produced by prostate cancer cells rarely flows into the circulation, and that calcitonin, frequently detected in neuroendocrine cells of the prostate gland, may regulate the action of PTHrP because of the discrepancy between the high expression rate of PTHrP in prostate cancer cells and the low incidence of hypercalcemia ($<2\%$) in prostate cancer patients.[76] Moreover, PTHrP stimulates the DNA synthesis of prostate cancer cells in an autocrine manner; thus, PTHrP may be a potential autocrine growth factor in prostate cancer cells.[76] Further examination of the expression of PTHrP in prostate cancer cells in bone is needed to determine the role of PTHrP in prostate cancer metastasizing to bone.

Bone morphogenetic proteins (BMPs)

BMPs, first identified as factors included in bone and cartilage[77] and belonging to the TGF-β superfamily[78] (except BMP-1), play important roles in embryogenesis, organogenesis and morphogenesis during fetus development. They induce ectopic bone formation *in vivo* and *in vitro*,[79] are actively involved in bone formation and have the capacity to induce differentiation of mesenchymal cells into cartilage and then into bone.[77] BMPs are expressed at sites of bone morphogenesis and limb bud formation, aid in fracture repair and repair of bone defects. Osteoblasts in normal skeletal tissue express BMP as they differentiate to mineralized bone and it has been suggested that BMPs may be involved in directing cells along the osteoblast lineage. Of three BMPs (BMP-2, -4 and -6), only BMP-6 stimulates a number of rat osteoblasts *in vivo*, suggesting that BMP-6 might have a crucial role in osteoblastic reaction found in bone metastasis.[80] BMPs are produced by several types of tumors, such as those of breast, prostate, esophagus and pancreatic adenocarcinoma. Examination of BMP-6 mRNA and protein expression in malignant and benign prostates has revealed a very high positive expression rate of both mRNA and protein in specimens of prostate cancer patients with metastasis (95%), compared with those with localized cancer (18%); however, the expression is absent in benign samples.[81] Moreover, the expression of BMP-6 mRNA in matched prostatic primary and secondary bony lesions and in isolated skeletal metastases from prostatic adenocarcinomas, as well as in other common human malignancies, by *in situ* hybridization, is strongly demonstrated in prostatic adenocarcinomas, both in the primary tumor and in bone metastasis.[82] The expression in bone metastasis from other malignancies is less frequent, however, suggesting that BMP-6 may hold potential as an attractive marker and a possible mediator of skeletal metastases in prostate cancer.[82] On the other hand, normal human prostate expresses BMP-4 mRNA predominantly, whereas human prostate cancer cell line PC3 expresses BMP-3 mRNA in larger amount than the normal prostate and the LNCap prostate cancer cell lines do. These studies indicate that the expression of a subtype of BMPs may also be important in normal prostate tissue physiology. It may play a role in the organization of the glandular structure of the prostate. The expression of BMPs is regulated by sonic (Shh) and Indian hedgehog (Ihh), which are important factors regulating skeletal tissue development.

Alterations of the DNA methylation status have been observed in many human cancers.[83] Current interest in the role that methylation plays has been focused on abnormal methylation events that activate by demethylation or silence by hypermethylation the genes that are important for development, progression and metastasis of tumors.[84] We have cloned and characterized the human BMP-6 gene promotor and found that it lacks a TATA-box but contains a CpG island with three Spl recognition sites (**Fig. 2.7A**).[85] Moreover, studies on the relationship between the methylation status of the BMP-6 gene promotor and its steady-state expression in prostate cancer cell lines have shown that the methylation status of the CpG loci around the Spl site of the BMP-6 promotor is related to its steady-state expression and to an alternative splicing of mRNA in prostate cancer cell lines.[86] The expression of MMP-6 mRNA was not determined in BPH, and that was weak in organ-confined prostate cancer. However, the expression in non-organ-confined disease was much stronger than organ-confined disease, especially in metastatic cancer (**Fig. 2.7B**). Moreover, no expression of Shh/Ihh mRNA was shown in BPH and organ-confined prostate cancer (**Fig. 2.7B**). The methylated cytosine was commonly shown in the 11th, 16th, 25th, 34th and 35th CpG loci in prostate cancer. Moreover, BMP-6 expression is high at both primary and secondary sites in advanced prostate cancer, and the demethylation of the CpG loci around the Spl binding site has been shown in cases with high BMP-6 expression (**Fig. 2.7B**).[86] For example, three methylated and one hemimethylated site was shown in the primary site of metastatic prostate cancer; however, there was no methylated site in the metastatic sites of lung and bone (**Fig. 2.7C**). All this suggests that, during cancer progression, including metastasis, besides inactivation of tumor suppressor genes by hypermethylation, activation of BMP-6 by selective demethylation, not by expression of Shh/Ihh, is a common epigenetic event giving a variable character to the invading and metastasizing prostate cancer cells.

Urokinase-type plasminogen activator

Urokinase-type plasminogen activator (uPA) is a one of the serine proteases which also play important roles in cancer invasion, metastasis and angiogenesis.[40,41,87] uPA is produced and secreted by various types of malignant tumors and binds with high affinity to its specific cell surface receptor (uPAR).[88,89] uPA exerts its effect by converting the proenzyme plasminogen into the widely acting serine protease plasmin, which cleaves the extracellular components, including fibronectin, laminin, and collagen and other extracellular matrices, either directly or through the activation of other zymogens, such as matrix metalloproteinases.[87] The proteolytic action is shown by the binding uPA and uPAR; thus the expression of not only uPA but also that of uPAR is important in cancer invasion and metastasis.[90] The level of uPA in prostate cancer correlates with an aggressive and invasive phenotype.[60,91] Moreover, the displacement of uPAR-bound uPA with an active-site mutant uPA blocks spontaneous metastasis of human prostate cancer cells in nude mice. The introduction of the uPA gene into prostate cancer cells enhances the skeletal metastasis of the cells.[60] We have demonstrated that the mean serum levels of uPA and uPAR in patients with prostate cancer are significantly higher than those in healthy controls or in BPH (**Fig. 2.8**); also that serum uPA and uPAR levels in prostate cancer with skeletal metastasis or with pathologically non-organ-confined disease are significantly higher than those in the cancer without these accompanying pathologies (**Fig. 2.8**).[91] The survival of prostate cancer patients with elevated serum levels of uPA or uPAR, or both, is significantly lower than that of patients with normal serum levels of both (**Fig. 2.9**).[91] Tese data suggest that the elevation of one or both could be a new predictor of prostate cancer progression. A unique mitogen for cells of the osteoblast phenotype has been isolated from the conditioned culture medium of the human prostate-cancer cell line PC-3, the so-called amino-terminal

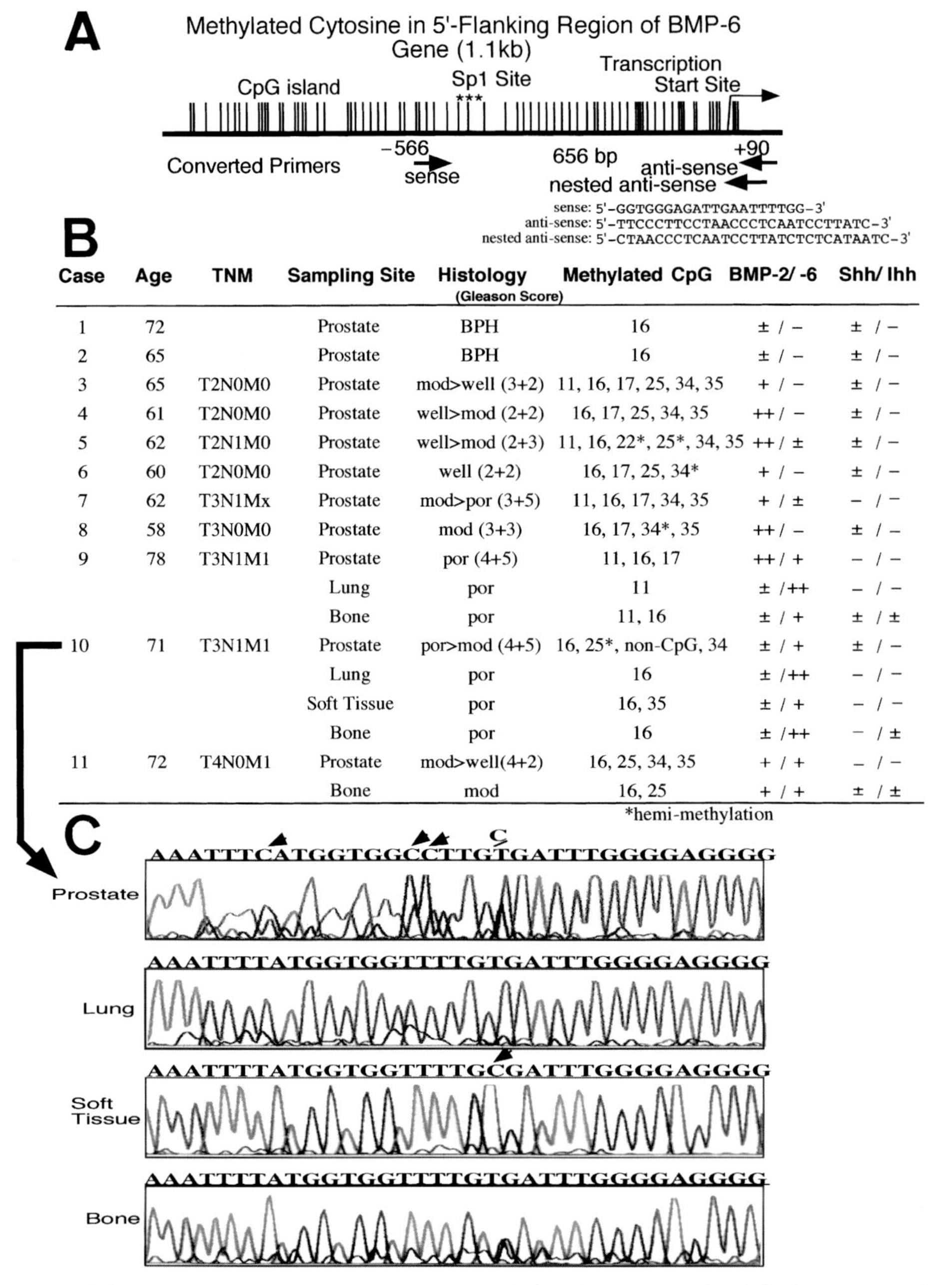

Case	Age	TNM	Sampling Site	Histology (Gleason Score)	Methylated CpG	BMP-2/ -6	Shh/ Ihh
1	72		Prostate	BPH	16	± / –	± / –
2	65		Prostate	BPH	16	± / –	± / –
3	65	T2N0M0	Prostate	mod>well (3+2)	11, 16, 17, 25, 34, 35	+ / –	± / –
4	61	T2N0M0	Prostate	well>mod (2+2)	16, 17, 25, 34, 35	++ / –	± / –
5	62	T2N1M0	Prostate	well>mod (2+3)	11, 16, 22*, 25*, 34, 35	++ / ±	± / –
6	60	T2N0M0	Prostate	well (2+2)	16, 17, 25, 34*	+ / –	± / –
7	62	T3N1Mx	Prostate	mod>por (3+5)	11, 16, 17, 34, 35	+ / ±	– / –
8	58	T3N0M0	Prostate	mod (3+3)	16, 17, 34*, 35	++ / –	± / –
9	78	T3N1M1	Prostate	por (4+5)	11, 16, 17	++ / +	– / –
			Lung	por	11	± / ++	– / –
			Bone	por	11, 16	± / +	± / ±
10	71	T3N1M1	Prostate	por>mod (4+5)	16, 25*, non-CpG, 34	± / +	± / –
			Lung	por	16	± / ++	– / –
			Soft Tissue	por	16, 35	± / +	– / –
			Bone	por	16	± / ++	– / ±
11	72	T4N0M1	Prostate	mod>well(4+2)	16, 25, 34, 35	+ / +	– / –
			Bone	mod	16, 25	+ / +	± / ±

*hemi-methylation

Fig. 2.7. Epigenetic regulation of the BMP-6 gene in prostate cancer. (A) Methylated cytosine in the 5′-flanking region of BMP-6. (B) Summary of prostate cancer patients. BPH, benign prostatic hyperplasia; mod, moderately differentiated adenocarcinoma; well, well-differentiated carcinoma; por, poorly differentiated carcinoma; Shh, sonic headghog; Ihh, Indian hedgehog. (C) Sequence analysis of genomic DNA from metastatic sites from prostate cancer. Arrow, methylated cytosines; C/, hemi-methylated cytosine.

fragment (ATF) of uPA (**Fig. 2.10A**).[12,92] uPA is synthesized as a single chain precusor termed pro-uPA converted by serine protease to a two-chain entity, and high-molecular-weight uPA (HMW-uPA), proteolytic activity; also HMW-uPA has been cleaved by metalloproteinase and divided into ATF, which contains growth factor domain (GFD) and a short residual chain terminal low-molecular-weight uPA (LMW-uPA) that retains proteolytic activity (**Fig. 2.10A**).[12,92]

HMW-uPA reacts with uPAR on the surface of prostate cancer cells, and demonstrates proteolytic activity and cancer cell invasion and metastasis. On the other hand, ATF contains, near its amino terminus, a GFD with a structure similar to that of EGF. The uPAR cloned in osteoblastic cells binds the ATF containing GFD, which is essential for osteoblastic activation (**Fig. 2.10B**). Indeed, rat prostate cancer cell (Mat-LyLu) variants with overexpressed uPA produce multiple metastatic lesions in both soft tissue and bone; however, prostate cancer cells with low expression of uPA do not metastasize in soft tissue and produce only little skeletal metastasis.[60] Moreover, histological examination shows that skeletal metastasis

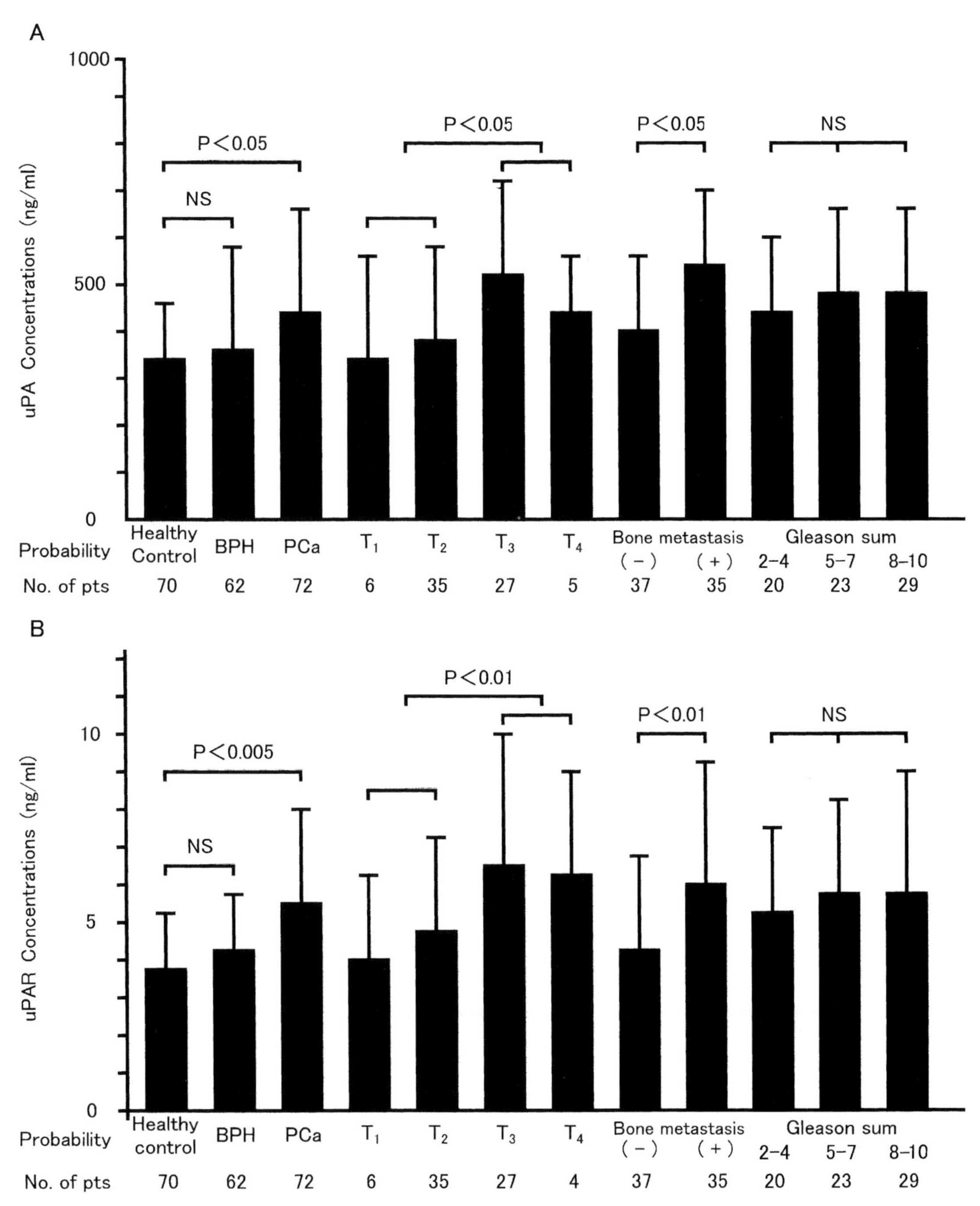

Fig. 2.8. Serum levels of uPA and uPA receptor in healthy controls, benign prostatic hyperplasia (BPH) and prostate cancer. (A) uPA. (B) uPA receptor.

produced by prostate cancer cells with overexpressed uPA is observed in new bone formation only; however, that with low-expressed uPA is observed in both osteolytic and osteoblastic lesions.[60] These results indicate that uPA plays crucial roles in prostate cancer skeletal metastasis with characteristic osteoblastic formation, and that ATP might stimulate osteoblast differentiation and growth. These phenomena do not, however, rule out the role of other mitogen factors, such as TGFβ and IGF-I on osteoblasts.

Endothelin and its receptor

Endothelin (ET) is a potent vasoconstrictor initially isolated from endothelial cells,[93] and includes three subtypes, ET-1–3.[94] ET-1, an important factor in the pathophysiology of prostate cancer and produced by the prostatic epithelium, is also a mitogenic factor to a variety of cell types including osteoblasts.[95] Human prostate cancer cells produce ET-1 mRNA and secrete immunoreactive ET, and the plasma concentration of ET in patients with metastatic prostate cancer is significantly higher than in

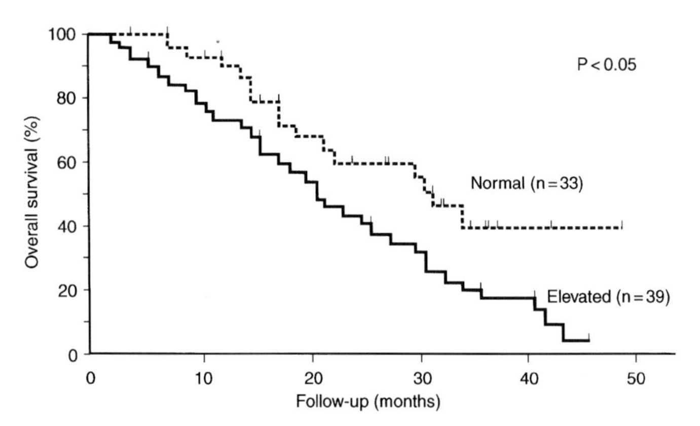

Fig. 2.9. Overall survival of prostate cancer patients according to serum uPA and uPA receptor status.

patients without metastasis.[95] Also exogenous ET-1 increases prostate cancer cell proliferation and alkaline phosphatase activity in new bone formation, indicating that ET-1 might be a mediator of the osteoblastic response of bone to metastatic prostate cancer.[95] On the other hand, of the different high-affinity ET receptors, ET receptor A (ET_A) and B (ET_B), ET_A has a high and equivalent affinity for ET-1 and ET-2 but little cross-reactivity with ET-3, whereas ET_B is non-selective with a similar affinity for the three subtypes.[96,97] The ET_A-selective receptor antagonist A-127722 inhibits ET-1 stimulated growth but the ET_B-selective antagonist BQ-788 does not.[98] The expression of ET_A is significantly higher than that of ET_B in prostate cancer tissue (71% vs. 24%, $P < 0.0001$) (**Figs 2.11A,B and 2.12**).[99] Also the expression of ET_A in non-organ confined and non-metastatic prostate cancer is significantly higher than that in organ-confined cancer (87% vs. 29%, $P = 0.0003$), the positive rate for ET_A of patients with lymph node metastasis is significantly higher than patients without lymph node metastasis (80% vs. 57%, $P = 0.0306$), and all patients with bone metastasis show positive expression of ET_A (**Fig. 2.12**).[99] Moreover, the positive staining rate for ET_A in patients with high Gleason sum is significantly higher than in those with low sum ($P < 0.00001$) (**Figs 2.11A,C and 2.12A**). The positive staining rate for ET_B is comparatively low; however, there is no significant difference among cancer stages and Gleason sum (**Fig. 2.12B**). Interestingly, the positive staining for anti-ET_A of cancer cells penetrating to the prostate capsule is stronger than for those growing at the primary site (**Fig. 2.11D**). These findings indicate that ET_A expression may have an important role in prostate cancer progression and metastasis. Recently, a randomized double-blind study of placebo and ET_A antagonist (atrasentan) for hormone refractory prostate cancer has shown that atrasentan significantly delays time to clinical progression and PSA progression; with placebo, 2.5 mg and 10 mg atrasentan, the median time to clinical progression is 129, 184 ($P = 0.035$), 196 ($P = 0.02$) days, respectively.[100] Moreover, with the same doses, the median time to PSA progression is 71, 134 ($P = 0.013$) and 155 ($P = 0.002$) days, respectively, and the corresponding 6-month progression-free rates are 35%, 53% and 54% ($P = 0.022$ and 0.018 compared to placebo, respectively).[100] Although a much longer follow-up is necessary, these results suggest that the ET_A antagonist, atrasentan, would be effective for hormone refractory prostate cancer.

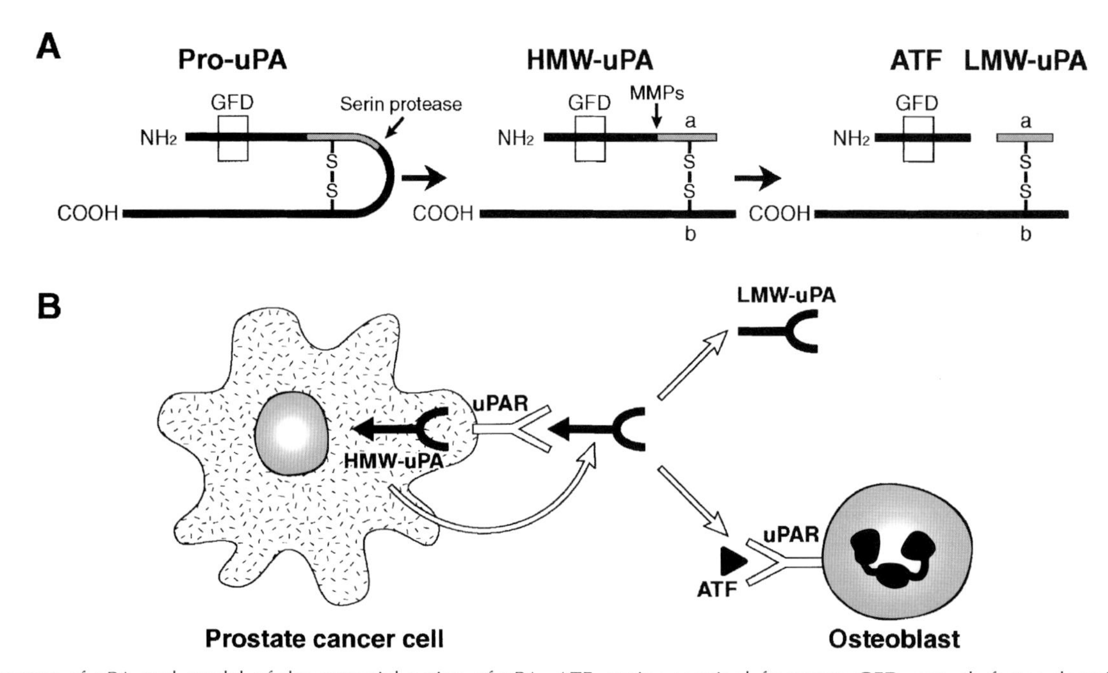

Fig. 2.10. Structure of uPA and model of the potential action of uPA. ATF, amino-terminal fragment; GFD, growth factor domain; HMW-uPA, high-molecular-weight uPA; LMW-uPA, low-molecular-weight uPA; uPA, urokinase-type plasminogen activator; uPAR, uPA receptor. *Cancer*, Vol. 80, 1997, pp. 1581–7.

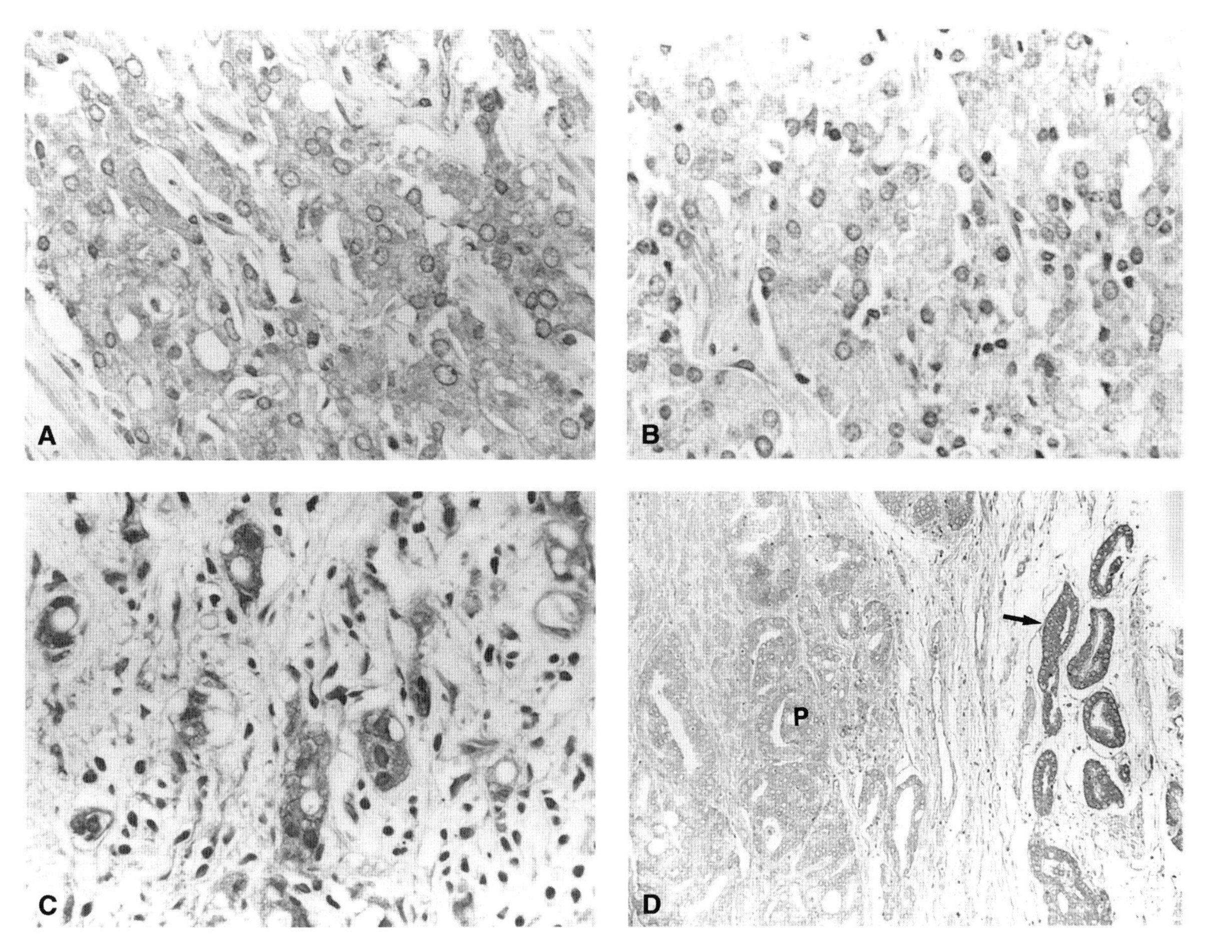

Fig. 2.11. Immunohistochemical staining of endothelin receptors in prostate cancer. (A) Cytoplasm of Gleason sum 4 neoplastic cells stained positive for antiendothelin receptor A antibody. Reduced from ×400. (B) Neoplastic cells stained negative for antiendothelin receptor B antibody. (C) Stage D2, Gleason sum 8 prostate cancer stained intensely positive for endothelin receptor A. (D) Endothelin receptor A in moderately differentiated, Gleason sum 4 prostate cancer cells penetrating the prostate capsule (arrow). Cancer cells penetrating the capsule stained more strongly than those at the primary foci (P). Reduced from ×100.

BISPHOSPHONATES IN PROSTATE CANCER

Bisphosphonates are agents that selectively inhibit osteoclastic bone resorption in ways that currently are only partially understood. A direct cytotoxic effect on osteoclasts has been shown for clodronate, whereas aminobisphosphonates impede the attachment of osteoclasts to the bone surface and promote their apoptosis.[101,102] Also, bisphosphonates induce the apoptosis of osteoblasts and might inhibit the attachment of tumor cells to bone matrix. These agents have been shown to reduce bone resorption due to malignancy, and to be useful for patients with hypercalcemia and bony pain due to bone metastasis.[103]

A synchronous increase in bone resorption in osteoblastic metastasis has recently come to light. More specific urine markers of bone resorption increase in patients with osteosclerotic prostate cancer metastasis.[73–75] Moreover, histopathological investigations have demonstrated that the number of active osteoblasts and indices of bone resorption increase significantly at sites of skeletal bone metastasis from prostate cancer, indicating that bone resorption plays a crucial role in the appearance and growth of osteosclerotic lesions and that its inhibition at the level of tumor infiltration might inhibit the progression of metastasis.

The clinical usefulness of bisphosphonates has been examined in bone metastasis from prostate cancer, typically osteoblastic lesions,[104,105] and with a relatively small number of patients, pamidronate has been shown to inhibit osteoblastic metastasis from prostate cancer.[105] Bisphosphonate has a favorable effect on bone pain in patients with osteoblastic bone metastasis from prostate cancer possibly as a result of inhibition of bone absorption, which precedes the apposition of excess woven bone and the appearance of osteosclerotic skeletal lesions. Also, bisphosphonates might inhibit cancer cell growth by decreasing growth factors released from bone tissue induced by osteoclasts. Further investigation is needed, however, to determine whether bisphosphonates inhibit growth of cancer cells in bone and osteoblastic metastasis.

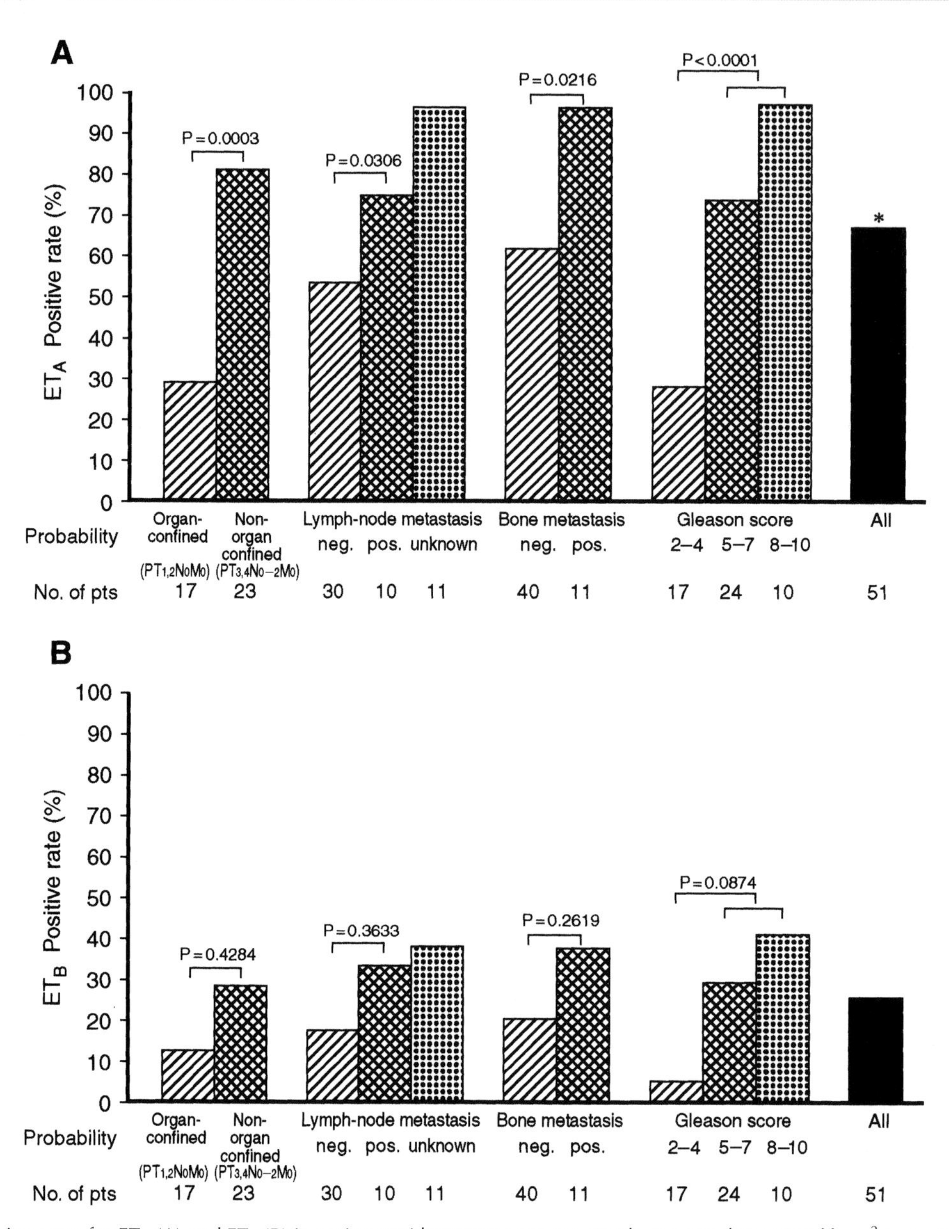

Fig. 2.12. Positive staining rates for ET_A (A) and ET_B (B) in patients with prostate cancer. *P* values were determined by χ^2 tests. *ET_A vs. ET_B, $P < 0.0001$.

CONCLUSIONS

Cancer metastasis is established by a multistep mechanism, which is not understood in terms of single molecules. Prostate cancer shows unique metastatic behavior, chiefly osteoblastic bone metastasis, prior to other visceral metastases. Bone metastasis itself is not a very serious condition. However, it significantly decreases the quality of life of patients by pathological fracture, bony pain and gait disturbance, owing to compression of the spinal cord. Unfortunately, the mechanism of osteoblastic metastasis from prostate cancer is still unclear because of the paucity of useful animal models. It is essential that animal models with spontaneous osteoblastic bone metastasis from prostate cancer are developed. Understanding the mechanism of osteoblastic bone metastasis is important for developing useful methods for the treatment of prostate cancer patients. Recently, some new therapies for cancer metastasis have been developed and clinically evaluated. Indeed, some anti-MMP [Marimastat (BB-2516), BAY12-9566 and AG-3340], which inhibit MMP-2, MMP-9 and MMP-7, and antiangiogenic drugs (endostain, interferon α, suramin) and platelet-derived growth factor receptor (PDGFR)-tyrosine kinase inhibitors are used for patients with metastasis from some malignant tumors, such as colorectal, breast and esophagial cancer. In prostate cancer, suramin in combination with hydrocortisone has also been used for the treatment of more than 400 patients with advanced prostate cancer where a rate of more than 50% reduction in PSA levels in patients so treated (32%) has been significantly higher than that in patients treated with hydrocortisone and placebo (16%). However, the clinical results have not been considered satisfactory. Therefore, more investigation is needed for establishing new therapies for prostate cancer metastasis.

REFERENCES

1. Wingo PA, Tong T, Boldens S. Cancer statistics, 1995. *CA Cancer J. Clin.* 1995; 45:8–30.
2. Steiner MS, Barrack ER. Transforming growth factor-β1 overproduction in prostate cancer: effects on growth in vivo and in vitro. *Mol. Endocrinol.* 1992; 61:15–25.
3. Mansson PE, Adams P, Kan M, McKeehan WL. HBGF1 gene expression in normal rat prostate and two transplantable rat prostate tumors. *Cancer Res.* 1989; 46:2485–94.
4. Harris SE, Harris M, Mahy M et al. Expression of bone morphogenetic proteins by normal rat and human prostate and prostate cancer cells. *The Prostate* 1994; 24:204–11.
5. Liotta LA, Stetler-Stevenson WC. Metalloproteinases and cancer invasion. *Semin. Cancer Biol.* 1990; 1:99–106.
6. Nakajima M, Chop AM. Tumor invasion and extracellular matrix degradative enzymes: regulation and activity by organ factors. *Semin. Cancer Biol.* 1991; 2:115–27.
7. Folkman J. The role of angiogenesis in tumor growths. *Semin. Cancer Biol.* 1992; 3:65–71.
8. Gohji K, Nakajima M, Boyd D et al. Organ-site dependence for the production of urokinase-plasminogen activator and metastasis by human renal cell carcinoma cells. *Am. J. Pathol.* 1997; 151:1655–61.
9. Nakajima M, Morikawa K, Fabra A et al. Influence of organ environment on extracellular matrix degradative activity and metastasis of human colon carcinoma cells. *J. Natl Cancer Inst.* 1990; 82:1890–98.
10. Gohji K, Nakajima M, Fabra A et al. Regulation of gelatinase production in metastatic human renal cell carcinoma by organ-specific fibroblasts. *Jpn J. Cancer Res.* 1994; 85:152–60.
11. Fujimaki T, Fan D, Staroselsky AH et al. Critical factors regulating site-specific brain metastasis of murine melanoma. *Int. J. Oncol.* 1993; 3:789–99.
12. Goltzman D. Mechanisms of the development of osteoblastic metastases. *Cancer* 1997; 80:1581–7.
13. Mundy GR, Yoneda, T. Mechanisms of bone metastases. In: Orr FW, Singh GN (eds) *Bone Metastasis Mechanisms and Pathophysiology*, pp. 1–16. Austin, TX: R. G. Landes Co., 1996.
14. Paget S. The distribution of secondary growths in cancer of the breast. *Lancet* 1889; 1:571–3.
15. Dinney CPN, Fishbeck R, Singh RK et al. Isolation and characterization of metastatic variants from human transitional cell carcinoma passaged by orthotopic implantation in athymic nude mice. *J. Urol.* 1995; 154:1532–8.
16. Stephenson RA, Dinney CPN, Gohji K et al. Metastatic model for human prostate cancer using orthotopic implantation in nude mice. *J. Natl Cancer Inst.* 1992; 84:951–7.
17. Gleave M, Hsieh J-T, Gao O et al. Acceleration of human prostate cancer growth in vivo by factors produced by prostate and bone fibroblasts. *Cancer Res.* 1991; 51:3753–61.
18. Chung LWK. Fibroblasts are critical determinants in prostatic cancer growth and dissemination. *Cancer Metastasis Rev.* 1991; 10:263–74.
19. Gohji K, Nakajima M, Dinney CPN et al. The importance of orthotopic implantation to the isolation and biological characterization of a metastatic human clear cell renal carcinoma in nude mice. *Int. J. Oncol.* 1993; 2:23–32.
20. Quinn DI, Henshall SM, Head DR et al. Prognostic significance of p53 nuclear accumulation in localized prostate cancer treated with radical prostatectomy. *Cancer Res.* 2000; 60:1585–94.
21. Dong JT, Lamb PW, Rinker-Schaeffer CW et al. KAI1, a metastasis suppressor gene for prostate cancer on human chromosome 11p11.2. *Science (Washington, DC)* 1995; 268:884–6.
22. Folkman J. How is blood vessel growth regulated in normal and neoplastic tissue? G. H. A. Clowes Memorial Award Lecture. *Cancer Res.* 1986; 46:467–73.
23. Leung DW, Cachianes G, Kuang WJ et al. Vascular endothelial growth factor is secreted angiogenetic mitogen. *Science (Washington DC)* 1989; 246:1306–9.
24. Kubota Y, Miura T, Moriyama M et al. Thymidine phosphorylase activity in human bladder cancer: difference between superficial and invasive cancer. *Clin. Cancer Res.* 1997; 3:973–6.
25. Koch AE, Polverini PJ, Kunkel SL et al. Interleukin-8 as a macrophage-derived mediator of angiogenesis. *Science* 1992; 258:1798–801.
26. Grant DS, Kleinman HK, Goldberg ID et al. Scatter factor induces blood vessel formation in vivo. *Proc. Natl Acad. Sci. USA* 1993; 90:1937–41.
27. Fett JW, Strydom DJ, Lobb RR et al. Isolation and characterization of angiogenin, an angiogenic protein from human carcinoma cells. *Biochemistry* 1985; 24:5480–6.
28. Miyake H, Hara I, Yamanaka K et al. Increased angiogenin expression in the tumor tissue and serum of urothelial carcinoma patients is related to disease progression and recurrence. *Cancer* 1999; 86:316–24.
29. Davis S, Aldrich TH, Jones PF et al. Isolation of angiopoietin-1, a ligand for the TIE2 receptor, by selection-trap expression cloning. *Cell* 1996; 87:1161–9.
30. Liotta LA, Steeg PS, Stetler-Stevenson WG. Cancer metastasis and angiogenesis: an imbalance of positive and negative regulation. *Cell* 1991; 64:327–36.
31. Liabakk NB, Talbot I, Smith RA et al. Matrix metalloproteinase 2 (MMP-2) and matrix metalloproteinase-9 (MMP-9) type IV collagenases in colorectal cancer. *Cancer Res.* 1996; 56:190–6.
32. Davies B, Waxman J, Wasan H et al. Levels of matrix metalloproteinases in bladder cancer correlate with tumor grade and invasion. *Cancer Res.* 1993; 53:5365–9.
33. DeCleark YA, Imren S. Protease inhibitors: role and potential therapeutic use in human cancer. *Eur. J. Cancer* 1994; 30A:2170-80.
34. Imren S, Kohn DB, Shimada H et al. Overexpression of tissue inhibitor of metalloproteinases-2 by retroviral-mediated gene transfer in vivo inhibits tumor growth and invasion. *Cancer Res.* 1996; 56:2891–5.
35. DeClearck YA, Peretz N, Shimada H et al. Inhibition of invasion and metastasis in cells transfected with an inhibitor of metalloproteinases. *Cancer Res.* 1992; 52:701–8.
36. Miyake H, Hara I, Gohji K et al. Relative expression of matrix metalloproteinase-2 and tissue inhibitor of metalloproteinase-2 in mouse renal cell carcinoma cells regulates their metastatic potential. *Clin. Cancer Res.* 1999; 5:2824–9.
37. Gohji K, Fujimoto N, Fujii A et al. Prognostic significance of circulating matrix metalloproteinase-2 to tissue inhibitor of metalloproteinases-2 ratio in recurrence of urothelial cancer after complete resection. *Cancer Res.* 1996; 56:3196–8.
38. Boag AH, Young ID. Increased expression of the 72-kd type IV collagenase in prostatic adenocarcinoma. Demonstration by immunohistochemistry and in situ hybridization. *Am. J. Pathol.* 1994; 144:585–91.
39. Hashimoto K, Kihara Y, Matuo Y et al. Expression of matrix metalloproteinase-7 and tissue inhibitor of metalloproteinase-1 in human prostate. *J. Urol.* 1998; 160:1872–6.
40. Cantero D, Friess H, Deflorin J et al. Enhanced expression of urokinase plasminogen activator and its receptor in pancreatic carcinoma. *Br. J. Cancer* 1997; 75:388–95.
41. Hofmann R, Lehmer A, Buresch M et al. Clinical relevance of urokinase plasminogen activator, its receptor, and its inhibitor in patients with renal cell carcinoma. *Cancer* 1996; 78:487–92.

42. Hasui Y, Marutsuka K, Suzumiya J et al. The content of urokinase-type plasminogen activator antigen as a prognostic factor in urinary bladder cancer. *Int. J. Cancer* 1992; 50:871–3.
43. Vlodavshy I, Friedmann Y, Elkin M et al. Mammalian heparanase: gene cloning, expression and function in tumor progression and metastasis. *Nature Med.* 1999; 7:793–802.
44. Vlodavsky I, Korner G, Ishai-Michaeli R et al. Extracellular matrix-resident growth factors and enzymes: possible involvement in tumor metastasis and angiogenesis. *Cancer Metastasis Rev.* 1990; 9:203–26.
45. Hileman RE, Fromm JR, Weiler JM et al. Glycosaminoglycan protein interacts: definition of consensus sites in glycosaminoglycan binding proteins. *BioEssays* 1998; 20:156–67.
46. Friedmann Y, Vlodavsky I, Aingorn H et al. Expression of heparanase in normal, dysplastic, and neoplastic human colonic mucosa and stroma: evidence for its role in colonic tumorigenesis. *Am. J. Pathol.* 2000; 157:1167–75.
47. Nakajima M, Irimura T, Nicolson GL. Heparanase and tumor metastasis. *J. Cell Biochem.* 1988; 36:157–67.
48. Gohji K, Hirano H, Okamoto M et al. Expression of three extracellular matrix degradative enzymes in bladder cancer. *Int. J. Cancer* 2001; 95:295–301.
49. Gohji K, Okamoto M, Kitazawa S et al. Heparanase protein and gene expression in bladder cancer. *J. Urol.* 2001; 166:1286–90.
50. Kosir MA, Wang W, Zukowski KL et al. Degradation of basement membrane by prostate tumor heparanase. *J. Surg. Res.* 1999; 81:42–7.
51. Kosir MA, Quinn CC, Zukowski KL et al. Human prostate carcinoma cells produce extracellular heparanase. *J. Surg. Res.* 1997; 67:98–105.
52. Gunthert U, Hofmann M, Rudy W et al. A new variant of glycoprotein CD44 confers metastatic potential to rat carcinoma cells. *Cell* 1991; 65:13–24.
53. Miyake H, Okamoto I, Hara I et al. Highly specific and sensitive detection of malignancy in urine samples from patients with urothelial cancer by CD44v8-10/CD44v10 competitive RT-PCR. *Int. J. Cancer* 1998; 79:560–4.
54. Miyake H, Hara I, Yamanaka K et al. Expression patterns of CD44 adhesion molecule in testicular germ cell tumors and normal testes. *Am. J. Pathol.* 1998; 152:1157–60.
55. Miyake H, Hara I, Okamoto I et al. Interaction between CD44 and hyaluronic acid regulates human prostate cancer development. *J. Urol.* 1998; 160:1562–6.
56. Humphrey PA, Zhu X, Zamegar R et al. Hepatocyte growth factor and its receptor (c-MET) in prostatic carcinoma. *Am. J. Pathol.* 1995; 147:386–96.
57. Kamenov B, Longenecker M. Further evidence of the existence of 'homing' receptor on murine leukemia cells which mediate adherence to normal bone marrow stromal cells. *Leukemia Res.* 1985; 9:1529–37.
58. Arguello F, Baggs RB, Frantz CN. A murine model of experimental metastasis to bone and bone marrow. *Cancer Res.* 1988; 48:6876–81.
59. Haq M, Goltzman D, Trembalay G et al. Rat prostate adenocarcinoma cells disseminate to bone and adhere preferentially to bone marrow-derived endothelial cells. *Cancer Res.* 1992; 52:4613–19.
60. Achbarou AA, Kaiser S, Tremblay G et al. Urokinase overproduction results in increased skeletal metastasis by prostate cancer cells in vivo. *Cancer Res.* 1994; 54:2372–7.
61. Gingrich JR, Barrios RJ, Morton RA et al. Metastatic prostate cancer in a transgenic mouse. *Cancer Res.* 1996; 56:4096–102.
62. Stewler GJ, Stern P, Jacobs JW et al. Parathyroid hormonelike protein from human renal carcinoma cell: structural and functional homology with parathyroid hormone. *J. Clin. Invest.* 1987; 80:1803–7.
63. Suva LJ, Winslow GA, Wettenhall REH et al. A parathyroid hormone-related protein implicated in malignant hypercalcemia: cloning and expression. *Science* 1987; 237:893–6.
64. Abou-Samra A, Juppner H, Force T et al. Expression cloning of a common receptor for parathyroid hormone and parathyroid hormone-related peptide from rat osteoblast-like cells: a single receptor stimulates intracellular accumulation of both cAMP and inositol triphosphates and increases intracellular free calcium. *Proc. Natl. Acad. Sci. USA* 1992; 89:2732–6.
65. Horiuchi N, Caulfeld MP, Fisher JE et al. Similarity of synthetic peptide from human tumor to parathyroid hormone in vivo and in vitro. *Science* 1987; 238:1566–8.
66. Danks JA, Ebeling PR, Hayman J et al. Parathyroid hormon-related protein: immunohistochemical localization in cancer and in normal skin. *J. Bone Miner. Res.* 1989; 4:273–8.
67. Ferrari SL, Rizzoli R, Bonjour J-P. Effects of epidermal growth factor on parathyroid hormone-related protein production by mammary epithelial cells. *J. Bone Miner. Res.* 1994; 9:639–44.
68. Sebag M, Henderson J, Goltzman D et al. Regulation of parathyroid hormone-related peptide production in normal human mammary epithelial cells in vivo. *Am. J. Physiol.* 1992; 9:634–44.
69. Burton PBJ, Knight DE. Parathyroid hormone-related peptide can regulate growth of human lung cancer cells. *FEBS Lett.* 1992; 305:220–32.
70. Merryman JI, Dewille JW, Werkmeister JR et al. Effects of transforming growth factor-β on parathyroid hormone-related protein production and ribonucleic acid expression by a squamous carcinoma cell line in vitro. *Endocrinology* 1994; 134:2424–30.
71. Urwin GH, Percival RC, Harris S et al. Generalized increase in bone resorption in carcinoma of the prostate. *Eur. J. Urol.* 1985; 57:721–3.
72. Clarke NW, McClure J, George NJR. Morphometric evidence for bone resorption and replacement in prostate cancer. *Br. J. Urol.* 1991; 68:74–80.
73. Uebelhut D, Gineyts E, Chapuy MC et al. Urinary excretion of pyridinium crosslinks: a new marker of bone resorption in metabolic disease. *Bone Miner.* 1990; 8:87–96.
74. Takeuchi S, Arai K, Saitoh H et al. Urinary pyridinoline and deoxypyridinoline as potential marker of bone metastasis in patients with prostate cancer. *J. Urol.* 1996; 156:1691–5.
75. Ikeda I, Miura T, Kondo I. Pyridinium cross-links as urinary markers of bone metastases in patients with prostate cancer. *Br. J. Urol.* 1996; 77:102–6.
76. Iwamura M, di Saint'Agnese A, Wu G et al. Immunohistochemical localization of parathyroid hormone-related protein in human prostate cancer. *Cancer Res.* 1993; 53:1724–6.
77. Urist MR, DeLange RJ, Finerman GA. Bone cell differentiation and growth factors. *Science* 1983; 220:680–6.
78. Wozney JM, Rosen V, Celeste AJ et al. Novel regulators of bone formation: molecular clones and activities. *Science* 1988; 242:1528–34.
79. Urist MR. Bone formation by autoinduction. *Science (Washington, DC)* 1965; 150:893–9.
80. Hughes FJ, Collyer J, Stanfield M et al. The effects of bone morphogenetic protein-2, protein-4 and protein-6 on differentiation of rat osteoblast cells in vitro. *Endocrinology* 1995; 136:2671–7.
81. Hamdy FC, Autzen P, Robinson MC et al. Immunolocalization and messenger RNA expression of bone morphogenetic protein-6 in human benign and malignant prostatic tissue. *Cancer Res.* 1997; 57:4427–31.
82. Autzen P, Robson CN, Bjartell A et al. Bone morphogenetic protein 6 in skeletal metastases from prostate cancer and other common human malignancies. *Br. J. Cancer* 1998; 78:1219–23.
83. Jones PA, Laird PW. Cancer epigenetics comes of age. *Nature Genet.* 1999; 21:163–7.
84. Leung SY, Yuen ST, Chung LP et al. hMLH1 promoter methylation and lack of hMLH1 expression in sporadic gastric carcinomas with high-frequency microsatellite instability. *Cancer Res.* 1999; 59:159–64.

85. Tamada H, Kitazawa R, Gohji K et al. Molecular cloning and analysis of 5′-flanking region of the bone morphogenetic protein-6 (BMP-6). *Biochem. Biophys. Acta* 1998; 1395:247–51.
86. Tamada H, Kitazawa R, Gohji K et al. Epigenetic regulation of human bone morphogenetic protein 6 gene expression in prostate cancer. *J. Bone Miner. Res.* 2001; 16:487–96.
87. Danø K, Anderson PA, Grongahl-Hansen J et al. Plasminogen activators, tissue degradation, and cancer. *Adv. Cancer Res.* 1985; 44:139–266.
88. Gaylis FD, Keer HN, Wilson MJ et al. Plasminogen activators in human prostate cancer cell lines and tumors. Correlation with aggressive phenotype. *J. Urol.* 1988; 142:193–8.
89. Kerr HN, Gaylis FD, Kozlowski JM et al. Heterogeneity in plasminogen activator (PA) levels in human prostate cancer cell lines: increased PA activity correlates with biologically aggressive behavior. *Prostate* 1991; 18:201–14.
90. Needham GK, Sherbet GV, Farndon JR et al. Binding of urokinase and its receptor in human colon and breast cancer metastasis. *Br. J. Cancer* 1987; 55:13–16.
91. Miyake H, Hara I, Yamanaka K et al. Elevation of serum levels of urokinase type plasminogen activator and its receptor is associated with disease progression and prognosis in patients with prostate cancer. *Prostate* 1999; 39:123–9.
92. Rabbani SA, Desjardins J, Bell AW et al. An amino-terminal fragment of urokinase isolated from a prostate cancer cell line (PC3) in mitogenic for osteoblast-like cells. *Biochem. Biophys. Res. Commun.* 1990; 173:1058–64.
93. Yanagisawa M, Kurihara H, Kimura S et al. A novel potent vasoconstrictor peptide produced by vascular endothelial cells. *Nature* 1988; 332:411–15.
94. Yanagisawa M, Inoue A, Ishikawa T et al. Primary structure, synthesis, and biological activity of rat endothelin, an endothelium-derived vasoconstrictor peptide. *Proc. Natl Acad. Sci. USA* 1988; 85:6964–7.
95. Nelson JB, Hedicom SP, George DJ et al. Identification of endothelin-1 in the pathophysiology of metastatic adenocarcinoma of the prostate. *Nature Med.* 1995; 1:944–9.
96. Nakajo S, Sugiura M, Snajdar RM et al. Solubilization and identification of human placental endothelin receptor. *Biochem. Biophys. Res. Commun.* 1989; 164:205–11.
97. Watanabe H, Miyazaki H, Kondoh K et al. Two distinct types of endothelin receptors are present in chick cardiac membranes. *Biochem. Biophys. Res. Commun.* 1989; 161:1252–9.
98. Nelson JB, Chan-Tack K, Hedican SP et al. Endothelin-1 production and decreased endothelin B receptor expression in advanced prostate cancer. *Cancer Res.* 1996; 56:663–8.
99. Gohji K, Kitazawa S, Tamada H et al. Expression of endothelin receptor A associated with prostate cancer progression. *J. Urol.* 2001; 165:1033–6.
100. Nelson JB, Carducci MA, Zonnenberg B et al. The endothelin-A receptor antagonist atrasentan improves time to clinical progression in hormone refractory prostate cancer patients: a randomized, double-blind, multi-national study. *J. Urol.* 2001; 165(Suppl.):168.
101. Hughes DE, Wright KR, Uy HL Bisphosphonates promote apoptosis in murine osteoclasts in vitro and in vivo. *J. Bone Miner. Res.* 1995; 10:1478–87.
102. Rogers MJ, Watts DJ, Russell RGG et al. Inhibitory effects of bisphosphonates on growth of amoebae of the cellular slime mold *dictyostelium discoideum*. *J. Bone Miner. Res.* 1994; 9:1029–39.
103. Paterson AHG, Powels TJ, Kanis JA et al. Double-blind controlled trial of clodronate in patients with bone metastases from breast cancer. *J. Clin. Oncol.* 1993; 11:59–65.
104. Adami S. Bisphosphonates in prostate cancer. *Cancer* 1997; 80:1674–9.
105. Clarke NW, Holbrook IB, McClure J et al. Osteoblastic inhibition by pamidronate in metastatic prostate cancer: a preliminary study. *Br. J. Cancer* 1991; 63:420–3.

Does Prostate Cancer Represent More than One Cancer?

Chapter 3

Richard Greenberg

Department of Urologic Oncology, Fox Chase Cancer Center, Philadelphia, PA, USA

INTRODUCTION

The increase in the incidence of prostate cancer in the USA in the last 15 years has approached epidemic proportions. It has currently reached its prominent stature as the most frequently diagnosed solid tumor and second leading cause of cancer-related deaths in men. This is despite the fact that deaths from prostate cancer have actually decreased by approximately 20% since the beginning of the last decade.[1] This fact, presumably as a consequence of prostatic specific antigen (PSA) screening, is a topic to be discussed in great detail elsewhere in this text. Prostate cancer also appears to be exceptional in that it exists in two apparent forms, a purely histologic or latent preclinical variety, and a clinically evident array, capable of demonstrating significant pathologic potential. The latent form is identified in 30% of men over the age of 50 years and in as much as 70% of men over 80 years old. The pathologic entity affects approximately 15% of all American males in their lifetime.[2] The question to be considered in this chapter is 'do these facts implicate more than one proper prostate cancer'? In order to attempt to answer this fascinating question, I will review the historical perspective, the clinical presentation and the known natural history as well as specific pathological 'oddities' associated with prostate cancer. I will also examine some of the many genetic and molecular biological issues under recent scrutiny, including briefly the topic of ethnicity. In actuality, one disease is quite sufficient in its tests of physicians and the societies in which we practice. No other specific disease seems to be quite as adept at challenging those who have chosen both the fields of medicine and politics.

HISTORICAL PERSPECTIVE

Clinical prostate cancer is clearly a disease of variable and generally unpredictable behavior, especially for the individual patient. Any clinician who has seen more than even a handful of patients with carcinoma of the prostate will attest to this phenomenon. It is also quite clear that the appropriate treatment for an individual patient remains even more controversial now than at the beginning of the last century. In 1926, Hugh Hampton Young and David M. Davis, in the first urology text published in the USA, detailed several notions about prostate cancer that were novel for their era and remarkably still insightful.[3] Some of these conceptions construe that prostate cancer is symptomless at first and there is a long lag period from the time of initial symptom complex until clinical presentation for treatment. Cancer should be suspected in all men over the age of 50 years, when presenting with obstructive voiding symptoms and, additionally, obstructive symptoms are the hallmark for prostate cancer in most men. Persistent pain usually implied metastatic disease, yet in many men with metastatic prostate cancer, they actually remained symptom free for extended periods of time, measured in years.

At the turn of the 20th century, the treatment options for the management of prostate cancer were surprisingly not dissimilar to those we now still utilize 75 years later, although, obviously, many technical improvements have been made. They included radical prostatectomy and interstitial radiation therapy. In the 1920s, radium was used as the radiation source. Local surgery to manage obstructive symptomatology was common and external beam radiation therapy was being discussed as an alternative, but interestingly, in his commentary, Young states, as of 1926, he had yet to attempt this 'new' technology himself. This 'deep x-ray' treatment was being touted for use in palliative management of advanced prostate cancer metastases to bone, primarily for pain control. Suggestion that multimodal therapy may improve the outcome for locally advanced disease was also argued. Probably the most intuitive of Young's commentary was his realization that the effectiveness of localized treatments was likely to benefit only a small minority of prostate cancer patients without improvements in 'screening techniques' for early prostate cancer.

In addition to the clinical experience documented by Young and Davis, in their text, Young reviews the available pathologic considerations of prostate cancer, at the commencement of that new century. He credits Albarran and Halle with the first pathologic description of a malignant process found among 14 out of 100 patients with benign prostatic hyperplasia, in the 1850s, it was his belief that, rather than many histologically different types of prostate cancer existing, all prostate cancers were indeed adenocarcinomatous in nature. Although not entirely accurate, this notion was relatively revolutionary in his time. Moreover, Young was the first to postulate that these theoretical differences seen microscopically, in various cases of prostate cancer, were purely manifestations of different growth patterns within the tumor rather than distinctly different histologic tumors and that various combinations of the patterns may be present throughout an individual adenocarcinoma of the prostate.

We now know that Young's revolutionary vision has been validated. Clearly, although not all prostate cancers are glandular in origin, arising from the acinar structures of the prostate gland, 98% are adenocarcinomas. The remaining other malignancies arise from both acinar and ductal elements. Although characteristic of the vast minority of clinical cases, they are still quite clinically important because of their distinctly different biological behavior when contrasted against the more customary prostatic adenocarcinomas and may represent a therapeutic dilemma for the clinician, if not properly recognized. Squamous cell carcinoma from prostatic ductal origin was the first of these rare tumors to be identified in 1939. Since that time, other tumors of ductal origin comprise transitional cell carcinoma, and endometrial carcinoma. The uncommon tumors of acinar origin are neuroendocrine or carcinoid tumor, adenoid cystic carcinoma and, lastly, small cell undifferentiated carcinoma. The microscopic basis for the diagnosis of adenocarcinoma of the prostate is primarily the distinctive features of glandular formation and pattern, and only secondarily based upon the cytologic features. Variability of all features within different areas of the same tumor is so common as to be regarded as the rule. Only rarely, small tumors found incidentally may demonstrate little histologic and cytologic variability.[4]

CLINICAL PRESENTATION

Prostate carcinoma with all its significant prevalence is also quite heterogeneous in its clinical forms. Prostate cancer rarely causes symptoms early in its course, since the malignancy arises in most patients posteriorly, in the peripheral portion of the gland, away from the prostatic urethra. Actually, once symptoms are noticed, it usually implies locally advanced or even metastatic disease. As the prostate carcinoma involves the urethra or bladder neck, obstructive voiding symptoms general develop. These symptoms include hesitancy, slowing of the urinary stream and intermittent flow. Irritative voiding symptoms, including frequency, nocturia, urgency and urge-related incontinence, also may occur. These symptoms are not significantly different than those symptoms associated with progressive benign prostatic hyperplasia, although they usually occur with a more rapid onset in the prostate cancer patients, with the duration of symptoms measured in months as opposed to years. As the tumor continues its local progression, patients may notice either hematospermia or decreased ejaculatory volume secondary to ejaculatory duct obstruction. Erectile dysfunction as a consequence of extension beyond the capsule of the prostate by the malignant process and involvement of the neurovascular bundle(s) with local tumor progression may occur. In later stages of locally advanced prostate cancer, even the corpora cavernosa may be directly involved.

Signs of metastatic prostate cancer are most often related to bony pain as a result of tumor involvement within the axial or appendicular skeleton. Anemia may result if sufficient bone marrow stores are replaced by the carcinoma. Genital and lower extremity edemas are related to the invasion and subsequent obstruction of the local lymphatics as well as compression of the iliac veins by lymph node enlargement. With advanced disease, any organ or system may be involved. In rare cases, isolated metastatic lesions to various organs including the central nervous system may be the first clinical manifestation of prostate cancer. Yet, the percentage of patients with symptomatic prostate cancer diagnosed because of these symptoms has declined proportionally to the vast majority of patients in the USA now being diagnosed based upon PSA screening.[5]

NATURAL HISTORY

From a purely clinical perspective it becomes relatively easy to answer the questioned posed by the title of this chapter. Surely prostate cancer must be more than a single disease entity. How could a disease that is occult at the time of death from natural causes in an 85-year-old man be the same disease that rapidly kills his 50-year-old son? Yet, I believe there are data to support the hypothesis that this is indeed the same disease at opposite ends of the clinical and biological spectrum. Obviously, the complex interactions of both genetics and environment must play a role in explaining the widely divergent clinical course noted in the above example, even though the exact mechanism has yet to be evoked. What other disease varies so greatly as it travels from an occult biologically inactive process through a localized and then metastatic state, and then further from androgen dependence to independence frequently over a period of decades?

Prostate cancer is a disease noted for its extremely high prevalence relative to its clinical incidence. Interestingly, using data from the pre-PSA era, the Surveillance, Epidemiology and End Results (SEER) program of the National Cancer Institute derived estimates for the average duration of asymptomatic disease to be between 11 and 12 years for Caucasians, and a year shorter for African American men. Their conclusion was that, comparing the lifetime risks of preclinical and clinical disease, 75% of prostate cancers will never become clinically apparent.[6]

Yet, once clinically apparent, prostate cancer may indeed be significant, not only for the individual patient but also the society responsible for that patient's care. In Denmark, the traditional therapeutic approach to the management of prostate cancer regardless of stage has been limited to palliative intent. This is despite a significant prostate cancer incidence in Danish men. Because of the socialistic nature of medical care in Denmark, this philosophy has provided an opportunity to investigate further the still controversial natural history of prostate cancer. I believe that this study is quite important because it demonstrates a modern cohort far different from that which is customary in other Western countries and conspicuously different from the cohort of patients at first diagnosis, especially here in the USA. The majority of the new cases of prostate cancer here is primarily dependent upon PSA screening and, at least clinically, usually represents early-stage prostate cancer.

In contrast to the Etzioni study described above, 45% of the cases were diagnosed incidentally regardless of the lack of formal screening programs. Only approximately 30% of these patients had clinically organ-confined disease. Both stage and grade of tumor were statistically significant in multivariate analyses with respect to prostate cancer specific death. Overall, more than 60% of these patients suffered and died primarily from prostate cancer. The disease specific survival rates at 1, 5 and 10 years, were 80%, 38% and 17% respectively. Approximately one-third of these patients, however, neither suffered nor died of their malignancy.[7] Clearly, one would further expect that, as this general unscreened population also ages, these statistics would become even more negatively skewed compared with the cohort of men actively screened and treated with intent to cure.

Using more modern, post-PSA data, Merrill concludes that lifetime and age-conditional risk measures are reflective of both changes in the disease incidence rates and age distribution over time. It is, therefore, possible for the overall cancer burden to increase due to the aging population, even if cancer rates remain stable or actually decrease. Of note, despite African American males having a higher incidence of invasive cancer, the lifetime risk for Caucasian males to develop cancer is higher, because of the increased life expectancy among Caucasians.[1]

Certainly, quality of life issues should always be taken into account when counseling a patient with newly diagnosed prostate cancer. It is interesting that, as a clinician, this is not easy to 'sell' to the patient, or frequently his family. This is not unique to my personal practice and yet, in trying to define the role of surveillance for the patient and his family, one is confronted with a vast amount of information, not necessarily actual scientific data, much of which is contradictory. The most widely utilized current prognostic factors for survival of patients with prostate cancer include clinical and pathologic (where available) stage, histological grade, PSA level, patient age and comorbidity.[8,9] DNA ploidy is not uniformly available, nor is it always helpful when available secondary to limitations in methodology and clinical understaging that are pervasive in any discussion of prostate cancer. Permutations in PSA values, which are under investigation and beginning to demonstrate clinical usefulness, are PSA velocity and doubling time.[10] Also percentage-free PSA and more recently complexed PSA values have been used to predict timing for therapy.

Are these more modern laboratory values sufficient to determine which patients will benefit from intervention or, indeed, merely a marker of those patients with an aggressive biologic phenotype destined to have progressive disease and succumb to their malignant disease? A study reported from Buffalo on clinically palpable prostate cancers in the pre-PSA era and patients initially refusing treatment resulted in a high rate of local progression requiring intervention within 36 months, in all patients with clinical T_3 tumors, all moderately well-differentiated T_2 tumors and even 5 of the 13 patients with well-differentiated T_2 tumors. Systemic disease developed in 50% of the cohort. The conclusion of the authors was that observation should be limited to early stage, non-palpable tumors detected by PSA alone.[11]

A more recent study from Miami looking primarily at elderly patients with prostate cancer in the post-PSA era, where deferred treatment was chosen but careful monitoring of these patients in order to assess the clinical behavior of clinically localized prostate cancer, was undertaken. These patients were agreeable to treatment at the time of documented progression, however. Progression in this group was defined as local stage progression, biochemical progression or the development of clinical metastases. Treatments offered were either definitive radiation therapy or hormonal manipulation, although all of the 52% that progressed chose hormonal therapy. The mean time to progression was 35 months. Using multivariate analysis, the conclusion of the study was that treatment of patients aged over 75 years should not be deferred if they are of good performance status and the Gleason score is ≥ 6 and the PSA level is ≥ 10 ng/ml, as they were most likely to progress.[12] A similar result has been reported from Sweden where a cohort was followed conservatively. They deduced that patients with localized tumors have a favorable progress, even without initial treatment, as long as they are followed regularly and subsequently receive appropriate therapy. In their report, as in most others, pathological grade had a significant influence of survival.[13]

PATHOLOGICAL CONSIDERATIONS

Prostate cancer is a remarkably prevalent tumor, possibly representing histologically the most common malignant tumor in the world. Grading systems have been used by pathologists to help urologists since the 1920s. The intent of these protocols is to convey valuable pathologic information of potentially prognostic value to clinicians. Broders' classification was the first to attempt to stratify prostate adenocarcinoma.[14] More recently, proposed grading systems have included those presented by Mostofi,[15] Gleason[16] and Brawn et al.,[17] with the Gleason grading protocol eventually becoming the most accepted and utilized. In fact, a recent article in *JAMA*[18] and two recently reported studies in the *Journal of Urology*[19,20] support the continued prognostic importance of the Gleason score in both univariate and multivariate analyses.

Significant research in the field of prostate cancer pathology over the last several decades has increased our overall knowledge of the varied nature of this malignancy. Several important impressions have been focused on the problem of unpredictability, as the dictum for this ever increasing public health issue. First and possible foremost, prostate cancer is multifocal, with an average of at least two anatomically distinct foci of adenocarcinoma. These various foci may have completely different histological characteristics and, therefore, Gleason scores. This predisposes to the possibility of sampling errors and subsequent clinical as well as obvious pathological understaging. Another example of this sampling error problem and understaging is evaluation of the DNA ploidy status. Both cancers that are localized as well as those that are extracapsular can demonstrate either diploid or aneuploid DNA. Although in several studies, ploidy status is an important independent variable in the expected natural history and subsequent management of patients with prostate cancer, these sampling errors plague its usefulness in the general population.[21]

In addition, by definition, all prostate cancers can invade and metastasize, regardless of their histological appearance and grade. In addition, it seems that capsular penetration is unnecessary for this to occur. This reality impacts on our concept of observation in low-volume, early-stage prostate cancer and is responsible for the failure to cure these patients in a minority of cases, despite apparently adequate local therapy. Lymphovascular invasion has been documented in prostate cancers of less than 0.5 cm^3 in volume. Indeed, as many as 20% of patients with pathologically organ-confined disease have been shown to have circulating tumor cells in their marrow by reverse transcriptase polymerase chain reaction for the PSA messenger RNA. Prostatic intraepithelial neoplasia (PIN) is a special pathologic entity that will be addressed much more completely in subsequent chapters and may have little place in the scope of this present chapter. However, I feel compelled to at least touch upon the general issue at the conclusion of the section on pathology as we prepare to enter the realm of the genetics of prostate cancer. Initially thought to be a precursor of all prostate cancer, it now appears that PIN may indeed suffer from even a greater degree genetic alteration than malignant prostate cancer. The key to why some PIN remains in a non-invasive state while some PIN does progress to carcinoma of the prostate remains hidden to date.[22] However, PIN, once considered a way to distinguish different forms of prostate cancer as well as biological potential and possible prognosis, appears to be another incompletely understood non-prognostic histological variable in the complex world of prostate adenocarcinoma pathology.

Neuroendocrine differentiation of prostatic tissue has been reported in both benign and malignant conditions. This is distinctly different from the very rare clinicopathological entities of both small cell prostatic carcinoma and carcinoid-like tumors. It can be measured by immunohistochemical staining for chromogranin A. Immunoreactivity demonstrated a differential density with regard to histology, although the neuroendocrine cells have similar morphology and distribution in normal prostate, benign prostatic hyperplasia and prostate cancer. It does appear that the density of chromogranin A immunoreactive cells is higher in neoplastic tissues than benign prostate conditions. Also, the highest concentration of these cells was noted in poorly differentiated prostatic adenocarcinomas with Gleason scores of 8–10. Their density increased further following hormonal manipulation and the level of chromogranin A may indeed be a useful marker in assessing the development of hormone-independent prostate cancer, assuming the presence of appropriate subsequent therapy.[23,24] There does not appear to be any useful prognostic significance to this neuroendocrine differentiation in localized early-stage prostate cancer.[25] Attempts to utilize other neuropeptides besides chromogranin A, such as serotonin and neuron-specific endolase have not been as successful, with only chromogranin A being related to Gleason score, clinical stage, PSA level and patient survival based upon immunoreactivity and statistical analyses.[26]

NORMAL PSA (?)

In a study from the Prostate Cancer Risk Assessment Program of the Fox Chase Cancer Center in Philadelphia, men considered at increased risk for prostate cancer based upon race or family predisposition were screened in an attempt to improve their chances for prostate cancer cure. This was under the qualification that early diagnosis should affect improved curability. Utilizing a lower limit of normal at 27% free component of the total PSA, in all cohorts of African American males, and males with at least one first-degree relative with prostate cancer, who after the screening process were noted to have 'normal' PSA levels between 2.0 and 4.0 ng/ml and a non-suspicious digital rectal examination (DRE), biopsies were obtained using a standard transrectal ultrasound-guided technique (TRUSBx). Notably, 6 of 12 patients who were counseled and chose to undertake TRUSBx were found to have prostate cancer, with Gleason scores 6/7 in five of the six men. Only one of these men had a biopsy interpreted as a Gleason score 5. The median age of these patients was 52 years. Median PSA results were total PSA 2.9 ng/ml and 16% free component. Four of the six men underwent radical prostatectomy and pathology demonstrated significant rather than incidental prostate cancer, since all of the prostates were stage T_2B, N_0, M_0.[27] This modification of PSA determination has greatly improved the cancer detection rate from that originally reported by Catalona, where the prevalence of prostate cancer was less than 8% in men with a normal PSA.[28]

These results support the previously published report of Catalona[29] and more recently that of Schroder et al.[30] reporting on the outcome of the European Randomized Study of Screening for Prostate Cancer. In the European study, approximately 50% of the patients with benign feeling prostate examinations and PSA levels less than 4.0 ng/ml had organ-confined prostate cancer that recognized histologically aggressive characteristics, specifically Gleason scores 7 or greater. The issues of screening and the ability to distinguish clinically relevant prostate cancer from incidental, clinically harmless prostate cancer continues.

GENETIC CONSIDERATIONS

Clearly, the pathogenesis of prostate cancer reflects complex interactions among both environmental and genetic factors. There are few areas of research in prostate cancer that have increased as quickly as the area of susceptibility gene research, nor have the results been as promising, especially in this new era of the human genome project. A recent Scandinavian study of twins has attempted to elucidate the relative contributions of these two factors. This study definitely increases our prior estimation of the genetic factor in the development of prostate cancer, as 42% of the risk of prostate cancer was attributed to the genetic factors. Interestingly, this significant genetic predisposition was greater than both colorectal carcinoma (35%) and breast cancer (27%), within this same cohort.[31] This value is much higher than previously thought, as recent as the last decade. Monozygotic twins have a four fold increase in likelihood of developing prostate cancer when compared with dizygotic twins.[32]

Ethnicity and race undoubtedly are also part of this genetic equation. It has been long known that, with regard to both incidence and mortality data, that clinical prostate cancer is rare in Oriental men and relatively more common in Scandinavian men.[33] The issue of African American men is somewhat more controversial. In the early 1990s Baquet[34] and then Demers[35] reported several similar findings. African American men living in the USA have a higher incidence of prostate cancer than Caucasian males, regardless of educational level and socioeconomic class. In addition, African American men routinely present with higher stage disease and, more importantly, survival rates even when corrected for stage were consistently lower in the African American cohort. Yet, in contrast, several more recent reviews seem to contradict some of these earlier analyses. In a multivariate analysis, preoperative Gleason score and PSA remained independent variables for nodal status at the time of radical prostatectomy for clinically localized disease. In this study, there was no demonstrable impact of race on nodal status.[36] In those patients who have undergone radical prostatectomy or who were treated for cure with definitive radiation therapy, there was no statistical difference in the biochemical failure rate based on race analysis.[37] In those African American men who subsequently develop postprostatectomy biochemical evidence of recurrence, the rate of tumor growth as measured by PSA velocity kinetics does not support the hypothesis that prostate cancer in these African American males is more aggressive due to an enhanced growth rate.[38] Lastly, at an equal access center in Los Angeles, no differences were found in patient age or clinical stage at the time of initial diagnosis, comparing Caucasians and African Americans. The African Americans did present with higher serum PSA values and minimally higher Gleason scores.[39]

Further understanding of the genetic factors in the development and progression of adenocarcinoma of the prostate recently has been advanced through the use of microsatellite arrays. The genetic alterations demonstrated are characterized by loss of heterozygosity (LOH) in a widely heterogeneous chromosomal pattern, not unlike the variable growth patterns seen in clinical prostate cancer, ranging from indolent, incidentally noted tumors to rapidly progressive malignancies with a propensity for early metastasis. The most common genetic modifications identified to date, in prostate cancer, include LOH of chromosome arms 1p, 7q, 8p, 13q, 16q and 18q.[40] By continuing to explore these molecular fingerprinting studies, we may soon be able to distinguish the clinical phenotypes of prostate cancer, in effect distinguishing those clinical important tumors from the relatively harmless prostate cancers that do not require aggressive therapy.

CONCLUSION

At the recent, second annual meeting of The Society of Urologic Oncology, December 1–2, 2001, Rodrigo Chuaqui, MD, reported on the molecular pathology of prostate cancer. Exploiting the newer techniques of molecular biology, looking at prostate cancer compared with normal prostatic tissue, there are 136 different genes identified to date. When the National Cancer Institute evaluated ‘high grade’ versus ‘moderate grade’ prostate cancers, 21 different genes were noted. Some of these genes alterations result in up-regulation, while others are involved in down-regulation. There are oncogenes that function as facilitators and still suppressor genes, all within the complex process of ‘normal’ genetic expression. Although there can be little doubt that prostate cancer does indeed represent more than one phenotypic disease, I believe that prostate adenocarcinoma exists as a spectrum of a single disease. It clearly embodies one of the most common clinical problems faced throughout the industrialized world despite its unique and unpredictable presentation and clinical course. Because of this wide-ranging disease process, the cure for prostate cancer is unlikely to be any single agent or technological advancement, but will no doubt be identified through a multidisciplinary integrated approach spearheaded by a partnership of both clinicians and basic scientists working in concert. It will undoubtedly require an understanding of the interaction of genetics and molecular biology, the normal aging process, and our ever-changing environment.

REFERENCES

1. Merrill RM, Weed DL. Measuring the public health burden of cancer in the United States through lifetime and age-conditional risk estimates. *Ann. Epidemiol.* 2001; 11:547–53.
2. Carter HB, Coffey DS. The prostate: an increasing medical problem. *Prostate* 1990; 16:39–48.
3. Young HH, Davis DM. Neoplasms of the prostate and seminal vesicles. In: Young HH, Davis DM (eds) *Young's Practice of Urology*, pp. 613–71. Philadelphia: W.B. Saunders, 1926.
4. Petersen RO. Neoplasms of the prostate and seminal vesicales. In: *Urologic Pathology*, pp. 611–12. Philadelphia: J.B. Lippincott Company, 1992.

5. Carter HB, Partin AW. Diagnosis and staging of prostate cancer. In: Walsh PC, Retik AB, Vaughan ED et al. (eds) *Campbell's Urology*, pp. 2519–37. Philadelphia: W.B. Saunders, 1998.
6. Etzioni R, Cha R, Feuer EJ et al. Asymptomatic incidence and duration of prostate cancer. *Am. J. Epidemiol.* 1998; 148:775–85.
7. Borre M, Nerstrom B, Overgaard J. The natural history of prostate cancer based upon a Danish population treated with no intent to cure. *Cancer* 1997; 80:827–33.
8. Montie JE. Current prognostic factors for prostate carcinoma. *Cancer* 1996; 78:341–4.
9. Palmer JS, Chodak GW. Defining the role of surveillance in the management of localized prostate cancer. *Urol. Clinics North Am.* 1996; 23:551–6.
10. Nam RK, Klotz LH, Jewett MA et al. Prostate specific antigen velocity as a measure of the natural history of prostate cancer: defining a 'rapid riser' subset. *Br. J. Urol.* 1998; 81:100–4.
11. Allison RR, Schulsinger A, Vongtama V et al. If you 'watch and wait', prostate cancer may progress dramatically. *Int. J. Radiat. Oncol. Biol. Phys.* 1997; 39:1019–23.
12. Neulander EZ, Duncan RC, Tiguert R et al. Deferred treatment of localized prostate cancer in the elderly: the impact of the age and stage at the time of diagnosis on the treatment decision. *Br. J. Urol. Int.* 2000; 85:699–704.
13. Sandblom G, Dufmats M, Varenhorst E. Long-term survival in a Swedish population-based cohort of men with prostate cancer. *Urology* 2000; 56:442–47.
14. Broders AC. Carcinoma: grading and practical application. *Arch. Pathol.* 1926; 2:376–81.
15. Mostofi FK. Grading of prostate cancer. *Cancer Chemother.* 1975; 59:111–16.
16. Gleason DF. Histologic grading and clinical staging of prostatic carcinoma. In: Tannenbaum M (ed.) *Urologic Pathology: The Prostate*, pp. 171–88. Philadelphia: Lea and Febiger. 1977.
17. Brawn PN, Ayala AG, Von Eshenbach AC et al. Histologic grading study of prostate adenocarcinoma: the development of a new system and comparison with other methods. *Cancer* 1982; 49:525–31.
18. Albetson PC, Hanley JA, Gleason DF et al. Competing risk analysis of men aged 55 to 74 years at diagnosis managed conservatively for clinically localized disease. *JAMA* 1998; 280:975–80.
19. Lau WK, Blute ML, Bostwick DG et al. Prognostic factors for survival of patients with pathological Gleason score 7 prostate cancer: differences in outcome between primary Gleason grades 3 and 4. *J. Urol.* 2001; 166:1692–7.
20. Grossfeld GD, Chang JJ, Broering JM et al. Understaging and undergrading in a contempory series of patients undergoing radical prostatectomy: results from the cancer of the prostate strategic urologic research endeavor database. *J. Urol.* 2001; 165:851–6.
21. So MJ, Cheville JC, Katzman JA. Factors that influence the measurement of prostate cancer DNA ploidy and proliferation in paraffin embedded tissue evaluated by flow cytometry. *Mod. Pathol.* 2001; 14:906–12.
22. Crawford DE, Miller GJ, Labrie F et al. Prostate cancer pathology, screening and epidemiology. *Rev. Urol.* 2001; (Suppl.) 3:2–10.
23. Guate JL, Escaf S, Mendenez CL et al. Neuroendocrine cells in benign prostatic hyperplasia and prostate cancer: effect of hormone treatment. *Urol. Int.* 1997; 59:149–53.
24. Cussenot O, Villette JM, Cochand-Priollet B et al. Evaluation and clinical value of neuroendocrine differentiation in human prostatic tumors. *Prostate* Suppl. 1998; 8:43–51.
25. Abrahamsson PA, Cockett AT, di Sant'Agnese PA. Prognostic significance of neuroendocrine differentiation in clinically localized prostate cancer. *Prostate* Suppl. 1998; 8:37–42.
26. Yu D, Hsiel D, Chen H et al. The expression of neuropeptides in hyperplastic and malignant prostate tissue and its possible clinical implications. *J. Urol.* 2001; 166:871–5.
27. Pinover WH, Brunner D, James J et al. Screening with free PSA benefits men at high risk of developing prostate cancer. *Int. J. Radiat. Oncol.* 1999; 45(Suppl.):354–55.
28. Colberg JW, Smith DD, Catalona WJ. Prevalence and pathological extent of prostate cancer in men with prostate specific antigen levels of 2.9 to 4.0 ng/ml. *J. Urol.* 1993; 149:507–9.
29. Catalona WJ, Smith DS, Ornstein DK. Prostate cancer detection in men with serum PSA concentrations of 2.6 to 4.0 ng/ml and benign prostate examination: enhancement of specificity with free PSA measurements. *JAMA* 1997; 277:1452–7.
30. Schoder FH, van der Cruijsen-Koeter I, de Koning HJ et al. Prostate cancer detection at low prostate specific antigen. *J. Urol.* 2000; 163:806–12.
31. Lichtenstein P, Holm NV, Verkosalo PK et al. Environmental and heritable factors in the causation of cancer-analyses of cohort twins from Sweden, Denmark and Finland. *N. Engl. J. Med.* 2000; 343:78–85.
32. Steinberg GD, Carter BS, Beaty TH et al. Family history and the risk of prostate cancer. *Prostate* 1990; 17:337–47.
33. Pienta KJ, Esper PS. Risk factors for prostate cancer. *Ann. Intern. Med.* 1993; 118:793–803.
34. Baquet CR, Horn JW, Gibbs T et al. Socioeconomic factors and cancer incidence among blacks and whites. *J. Nat. Cancer Inst.* 1991; 83:551–7.
35. Demers RY, Swanson GM, Weiss LK et al. Increasing incidence of cancer of the prostate. *Arch. Intern. Med.* 1994; 154:1211–16.
36. Barroso U, Oskanian P, Tefilli MV et al. Population-based study of lymph node positivity in clinically localized prostate cancer: a study comparing African-Americans and whites. *Urology* 1999; 53:187–91.
37. Sohayda CJ, Kupelian PA, Altsman KA et al. Race as an independent predictor of outcome after treatment for localized prostate cancer. *J. Urol.* 1999; 162:1331–6.
38. Bissonette EA, Fulmer BR, Petroni GR et al. Prostate specific antigen kinetics at tumor recurrence after radical prostatectomy do not suggest a worse prognosis in black men. *J. Urol.* 2001; 166:1328–31.
39. Freedland SJ, Sutter ME, Naitoh J et al. Clinical characteristics in black and white men with prostate cancer in an equal access medical center. *Urology* 2000; 55:387–90.
40. Bochner BH. Commentary: Genetic alterations in prostate cancer. *J. Urol.* 1999; 162:1543.

High-grade Prostatic Intraepithelial Neoplasia

Mitchell S. Steiner
Department of Urology, University of Tennessee, Memphis, TN, USA

David G. Bostwick*
University of Virginia, Charlottesville, VA and Bostwick Laboratories, Richmond, VA, USA

INTRODUCTION

High-grade prostatic intraepithelial neoplasia (PIN) is accepted as a premalignant lesion that has the potential to progress to prostate adenocarcinoma.[1–16] Prostatic intraepithelial neoplasia is the abnormal proliferation within the prostatic ducts of premalignant foci of cellular dysplasia and carcinoma *in situ* without stromal invasion (**Fig. 4.1**).[17–19] The diagnostic term 'prostatic intraepithelial neoplasia' has been endorsed at multiple multidisciplinary and pathology consensus meetings,[5,13,15,20–25] and the interobserver agreement between pathologists has been determined to be 'good to excellent'[26,27] for high-grade PIN.

Prostatic intraepithelial neoplasia is graded from 1 to 3. Grade 1 is defined as low-grade PIN, whereas grades 2 and 3 are currently considered together as high-grade PIN; currently, conventional use of the term 'PIN' without qualification refers to only high-grade PIN. High-grade PIN is a standard diagnosis that must be included as part of the reported pathologic evaluation of prostate biopsies, transurethral resected prostate chips and radical prostatectomy specimens.[15,22,24]

In the USA, approximately 1 300 000 prostate biopsies are performed annually to detect 198 500 new cases of prostate cancer.[28] The incidence of high-grade PIN averages 9% (range 4–16%) of prostate biopsies representing 115 000 new cases of high-grade PIN without cancer diagnosed each year.[29] High-grade PIN is the earliest accepted stage in carcinogenesis, possessing most of the phenotypic, biochemical and genetic changes of cancer without invasion of the basement membrane of the acini (**Fig. 4.2**).[30]

A role for high-grade PIN in the development of prostate adenocarcinoma is based on the fact that the prevalence of both high-grade PIN and prostate cancer increases with patient age, and that high-grade PIN precedes the onset of prostate cancer by less than one decade (**Table 4.1**).[31–34] The severity and frequency of high-grade PIN in prostates with histological cancer is greatly increased (73% of 731 specimens) when compared to prostates without cancer (32% of 876 specimens).[35–38] When high-grade PIN is found on needle biopsy, there is a 50% risk of finding carcinoma on subsequent biopsies over 3 years.[24] There is also convincing evidence to suggest that high-grade PIN may represent a precursor to a more aggressive form of prostate cancer phenotype than to those that are more likely to remain indolent.[9,39,40]

PATHOLOGY OF HIGH-GRADE PIN AND OTHER POSSIBLE PREMALIGNANT LESIONS OF THE PROSTATE

Strong support for high-grade PIN as the main premalignant lesion of prostate cancer is based on several lines of evidence derived from prostate cancer animal models, epidemiological, morphologic, genetic and molecular studies.

Animal models of prostate cancer

Several different animal models of prostate cancer have demonstrated that high-grade PIN is in the direct causal pathway to prostate cancer.[41] The transgenic mouse model of prostate cancer (TRAMP) has been shown to mimic human prostate cancer.[42,43] In the TRAMP model, the Probasin promoter-SV40 large T antigen (PB-Tag) transgene is expressed specifically in the epithelial cells of the murine prostate under the control of the probasin promoter. The probasin promoter is

*To whom correspondence should be addressed.

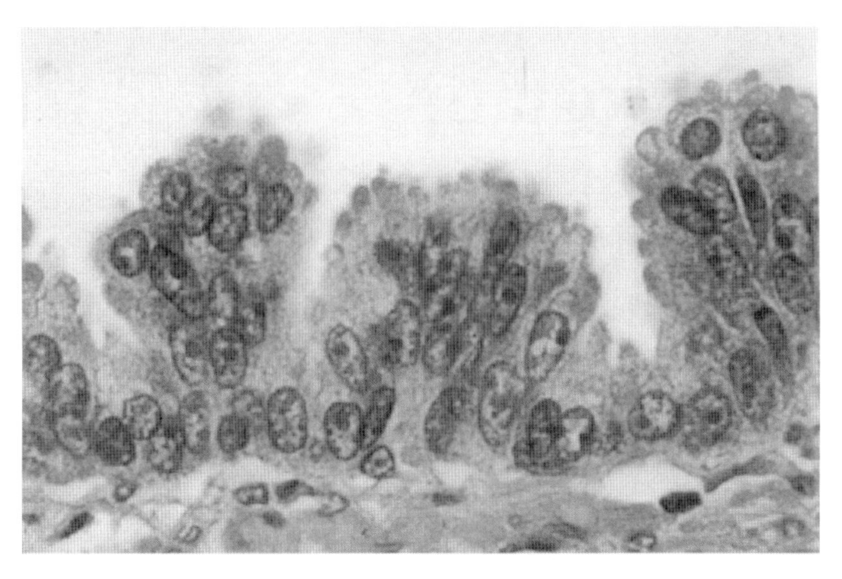

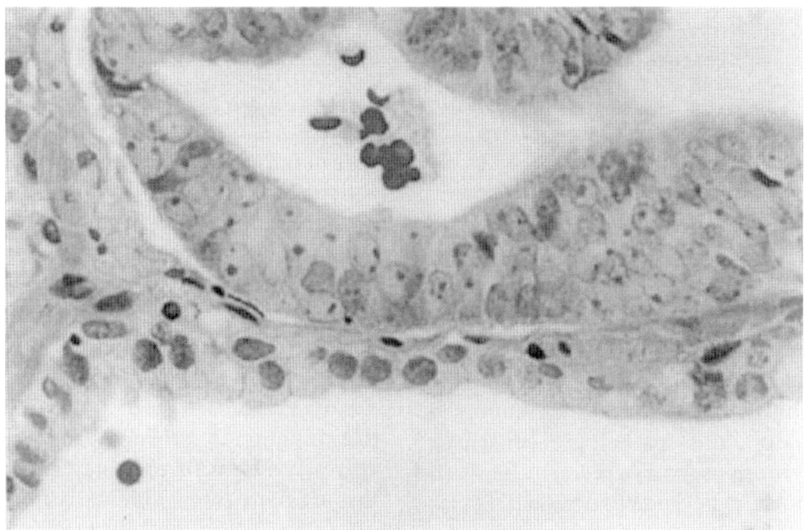

Fig. 4.1.

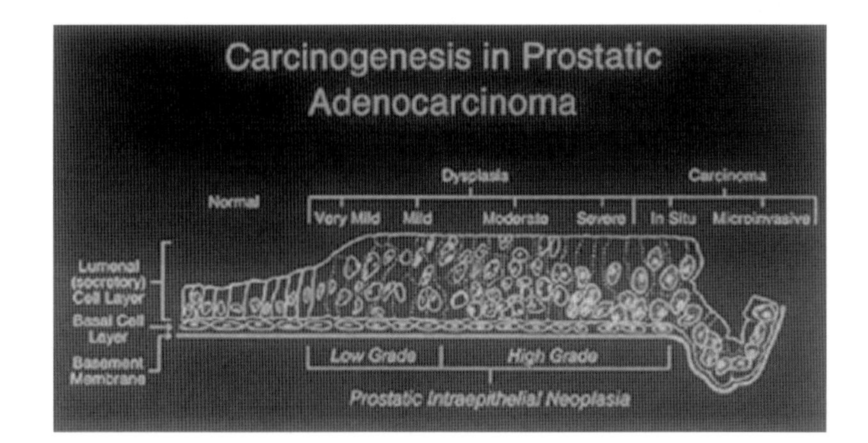
Carcinogenesis in Prostatic Adenocarcinoma
Dysplasia
Carcinoma
Normal
Very Mild
Mild
Moderate
Severe
In Situ
Microinvasive
Lumenal (secretory) Cell Layer
Basal Cell Layer
Basement Membrane
Low Grade
High Grade
Prostatic Intraepithelial Neoplasia

Fig. 4.2.

Table 4.1. Estimated frequency of men harboring high-grade PIN in the USA.

Age (years)	High-grade % PIN	US population* (thousands)	Number of PINs
40–49	15.2	20 550	3 123 600
50–59	24.0	14 187	3 404 880
60–69	47.3	9 312	4 404 576
70–79	58.4	6 926	4 044 784
80–89	70.0	2 664	1 864 800
	Total	**53 639**	**16 842 640**

*1990 US census.

androgen dependent. As a result, this model has several advantages over currently existing models:

1. Mice develop progressive forms of prostatic epithelial hyperplasia and high-grade PIN as early as 10 weeks and invasive prostate adenocarcinoma around 18 weeks of age.[42]
2. The pattern of metastatic spread of prostate cancer mimics that of human prostate cancer with common sites of metastases being lymph node, lung, kidney, adrenal gland and bone.
3. The development as well as the progression of prostate cancer can be followed within a relatively short period of 10–30 weeks.
4. Spontaneous prostate tumors arise with 100% frequency.
5. Animals may be screened for the presence of the prostate cancer transgene prior to the onset of clinical prostate cancer.

Another animal model is the transgenic mouse model that contains a probasin promoter that controls the ECO:R1 gene. This gene product has been implicated in the induction of genomic instability.[44] Prostates from these animals were followed prospectively from 4 to 24 months of age and showed the progressive presence of mild to severe hyperplasia, low-grade PIN, high-grade PIN and then well-differentiated adenocarcinoma of the prostate.[44] Stanbrough et al. (2001) demonstrated that transgenic mice that have prostatic overexpression of androgen receptor (AR) protein develop focal areas of high-grade PIN.[45]

The mechanism of prostate carcinogenesis appears to be involve estrogenic signaling. Wang et al. (2001) treated wild-type mice with testosterone propionate and estradiol for 4 months.[46] These mice developed prostatic hyperplasia, high-grade PIN and invasive prostate cancer. When α-ERKO mice, mice that have the ERα genetically knocked out, are treated the same way, they develop prostatic hyperplasia, but not high-grade PIN or invasive prostate cancer.[46] Similarly, a prospective, placebo-controlled study of TRAMP mice treated with an antiestrogen, Acapodene (toremifene), was performed to pharmacologically antagonize ERα. These Acapodene-treated TRAMP mice had a reduction in high-grade PIN, significant decrease in prostate cancer incidence and an increase in animal survival. Thus, estrogenic signally through ERα may play a key role[29] in prostate carcinogenesis and that high-grade PIN was observed to be in the direct causal pathway to prostate cancer.

The dog is the only non-human species in which spontaneous prostate cancer occurs and, like humans, the rate of canine prostate cancer increases with aging.[47–51] High-grade PIN has been also observed in the prostates of these animals.[48–51] Canine high-grade PIN shows cytological features identical to the human counterpart, including cell crowding, loss of polarity, and nuclear and nucleolar enlargement. Like prostatic adenocarcinoma, high-grade PIN also increases with aging.[48] High-grade PIN appears to represent an early event in prostate carcinogenesis that occurs with high frequency within the prostates of pet dogs sharing the same environment as humans. In this model, high-grade PIN was determined to be an intermediate step between benign epithelium and invasive carcinoma. Thus, like the transgenic mouse models, the canine model supports high-grade PIN as part of a continuum in the progression of prostate cancer.

Epidemiologic evidence

A role for high-grade PIN in the development of prostate adenocarcinoma is based on the fact that the prevalence of both high-grade PIN and prostate cancer increases with patient age and that high-grade PIN precedes the onset of prostate cancer by less than one decade.[20,34] The severity and frequency of high-grade PIN in prostates with histological cancer is greatly increased (73% of 731 specimens) when compared to prostates without cancer (32% of 876 specimens).[4,18,52–54] When high-grade PIN is found on needle biopsy, there is a 30–50% risk of finding carcinoma on subsequent biopsies over 3–5 years, and most patients with high-grade PIN develop carcinoma within 10 years.[4] One retrospective study revealed that the amount of high-grade PIN for 42 patients followed over 3 years was 12.5% in 1997, 33% in 1998 and 47% in 1999, demonstrating that the volume of high-grade PIN progressively increases over time.

In general, the rate of high-grade PIN appears to be similar for all men across all ages regardless of race and geographical location.[4] However, if specific age groups are compared between races, there are significant differences in the frequency of high-grade PIN. African American men have the highest incidence of prostate cancer that is about 50% more than Caucasians.[20,31,33,55] Consistent with high-grade PIN being the precursor lesion of prostate cancer, African American men have a greater prevalence of high-grade PIN when compared to Caucasians in the 50–60 year age group, the decade preceding the manifestation of most clinically detected prostate cancers.[20,31,33,56] In contrast, Asians have the lowest clinically detected rate of prostate cancer. Several studies from Japan have shown that Japanese men living in Osaka, Japan, have a significantly lower incidence of high-grade PIN compared to men residing in the USA.[57,58] Interestingly, those Japanese men diagnosed with high-grade PIN also had an increased likelihood of developing prostate cancer, suggesting that high-grade PIN is a precursor of clinical prostate cancer in

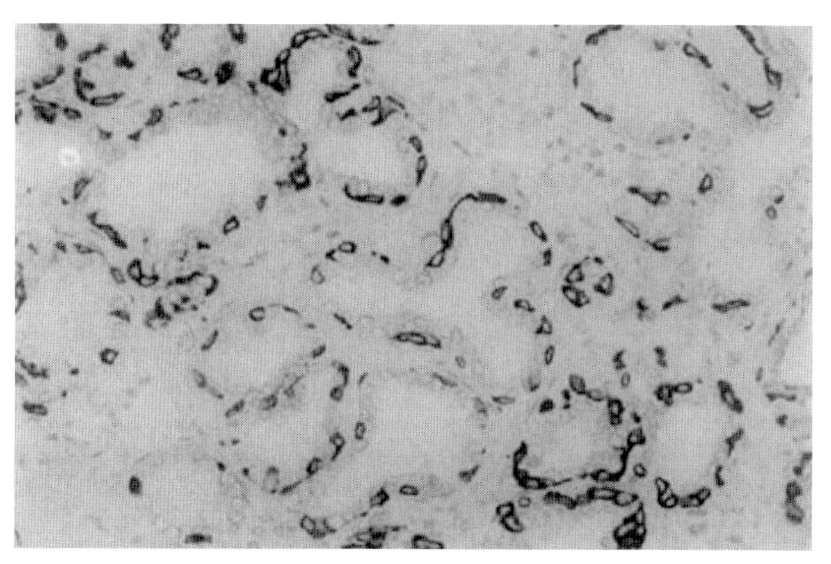

Fig. 4.3. Flat pattern of high-grade PIN showing some fragmentation of the basal cell layer (antibodies directed against high-molecular-weight cytokeratin).

Asian men too.[40] Thus, the differences in the frequency of high-grade PIN in the 50–60 age group across races essentially mirror the rates of clinical prostate cancer observed in the 60–70 age group.[31,57] Finally, there is also convincing evidence to suggest that high-grade PIN may represent a precursor to a more aggressive prostate cancer phenotype than to those that are more likely to remain indolent.[31,39,40,57]

Morphologic data

High-grade PIN and prostate cancer are morphometrically and phenotypically similar. High-grade PIN occurs primarily in the peripheral zone and is seen in areas that are in continuity with prostate cancer.[5,37,38,56,59,60] High-grade PIN and prostate cancer are multifocal and heterogeneous.[37,61,62] Increasing rates of aneuploidy and angiogenesis as the grade of PIN progresses are further evidence that high-grade PIN is a precancer.[30,60,63–66] Prostate cancer and high-grade PIN also have similar proliferative and apoptotic indices.[2,30,57,67–72] Histologically, the cytologic abnormalities observed in high-grade PIN is virtually indistinguishable from invasive prostate cancer except that, in high-grade PIN, the basal cell layer is still at least partially intact (**Fig. 4.3**).[18] As high-grade PIN progresses, the likelihood of basal cell layer disruption increases, very much like that which is observed for carcinoma *in situ* (CIS) of the urinary bladder. CIS of the urinary bladder, like PIN, may become invasive and is treated aggressively. The standard of care for management of CIS of the bladder is intravesical instillation of chemotherapy or BCG, and, in some cases, radical cystectomy.

Genetic and molecular changes

High-grade PIN and prostate cancer share similar genetic alterations (**Fig. 4.4**).[30,73–76] For example, the frequent 8p12–21 allelic loss commonly found in prostate cancer was also found in microdissected PIN.[73] Other examples of genetic changes found in carcinoma that already exist in PIN include loss of heterozygosity (LOH) at 8p22, 12pter-p12, 10q11.2,[20,73] and gain of chromosomes 7, 8, 10 and 12.[77] Alterations in oncogene bcl2 expression and RER+ phenotype are similar for PIN and prostate cancer.[78,79]

In summary, these clinical and molecular studies taken together provide strong evidence that high-grade PIN is the

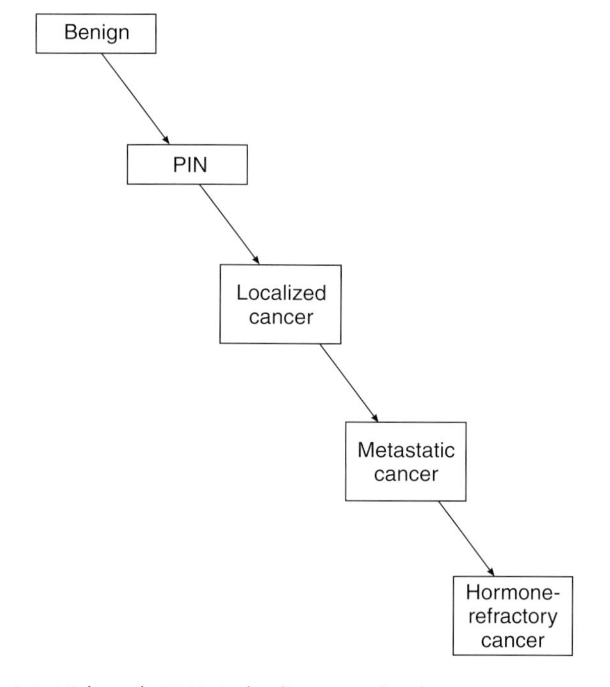

Fig. 4.4. High-grade PIN is in the direct causal pathway to prostate cancer.

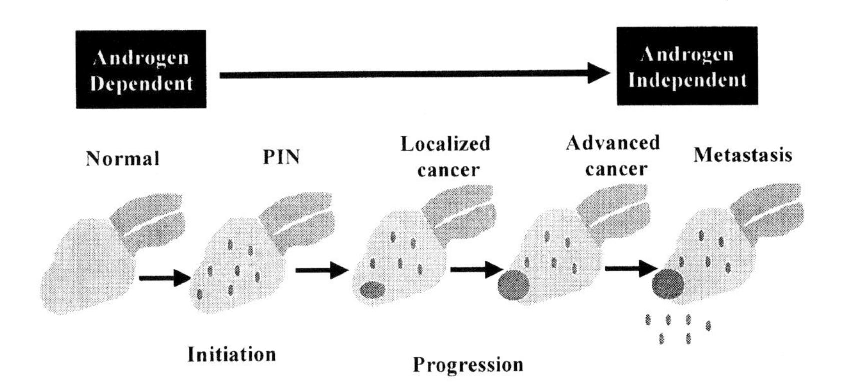

Fig. 4.5. High-grade PIN.

main precursor of prostate cancer. The development of high-grade PIN represents an important step in prostate cancer: the normal prostate epithelium develops high-grade PIN (carcinogenesis), invasive clinically localized prostate cancer, regionally advanced prostate cancer, and then metastatic prostate cancer (**Fig. 4.5**). The presence of high-grade PIN alerts both the clinician and the patient that progression to clinically significant prostate cancer is likely.

PIN IS A DISEASE

Premalignant lesions that affect other organs have been identified and currently are treated when diagnosed such that the premalignant lesions itself is a disease. Treatment of these precancerous lesions would appear to be of clinical benefit notwithstanding the potential for cancer prevention. These clinical benefits would reduce morbidity, enhance quality of life, delay surgery or radiation, and increase the interval for surveillance requiring invasive procedures.[25] Examples of treated premalignant lesions include:

- Cervix – cervical intraepithelial neoplasia.
- Breast – ductal carcinoma *in situ*.
- Colon – adenomatous polyps.
- Bladder – carcinoma *in situ*.
- Skin – actinic keratosis.
- Esophagus – Barrett's esophagus.
- Oral mucosa – dysplastic oral leukoplakia.

Like high-grade PIN, these all represent types of intraepithelial neoplastic lesions.[25]

High-grade PIN is also a premalignant lesion that merits treatment. The clinical implications of high-grade PIN are very different from that for an elevated serum prostate specific antigen (PSA). Serum PSA is prostate-specific, not prostate cancer-specific. Other conditions besides prostate cancer more commonly raise serum PSA concentration, including benign prostatic hyperplasia (BPH), ejaculation, prostatitis and prostatic infarct. More than 70–80% of men with an elevated PSA over 4 ng/ml will ultimately be found to have no evidence of prostate cancer. Interestingly, high-grade PIN does not contribute to serum PSA or serum free PSA, which is not surprising, since high-grade PIN, unlike prostate cancer, has not yet invaded the vasculature of the prostate to leak secreted PSA into the bloodstream.[80–84] Thus, high-grade PIN appears to precede even prostate cancer-related serum PSA elevations.

High-grade PIN is in the direct causal pathway to prostate cancer and its presence specifically portends an increased risk of prostate cancer. Consequently, men diagnosed with high-grade PIN have dramatic changes in their quality of life, since there is great patient and physician anxiety about the concern that prostate cancer may be imminent. Moreover, the patient must be subjected to more frequent biopsies and physician visits, since currently the only way to diagnose high-grade PIN is by prostate biopsy. Although there is no medical consensus for the exact standard of care of high-grade PIN, urologists recognize that high-grade PIN is a dangerous lesion and it should be aggressively managed. Coexistent cancer (prostate cancer found within 2 years of diagnosis of high-grade PIN) is present in approximately 22–64% of cases; differences in predictive value reflect variance in patient populations and indications for biopsy. Although some urologists have advocated saturation biopsies of the prostate following the diagnosis of high-grade PIN, the more common recommendation is repeat prostate biopsies every 3–6 months for 2 years, then annually.[4–6,8,39,85–93] Since high-grade PIN does not contribute to PSA, the use of serial serum PSA may not help to detect prostate cancer soon enough; thus, serial prostate biopsies are currently the only sure way to follow patients who have high-grade PIN. Interestingly, PIN coexisting with cancer was not predictive of PSA (biochemical) failure at 32 months in patients undergoing radical prostatectomy and androgen deprivation therapy.[94]

CAN PIN BE TREATED?

Currently, there is no treatment available for patients who have high-grade PIN. Prophylactic radical prostatectomy or radiation is not an acceptable treatment for patients who have high-grade PIN only.[95] The development and identification of acceptable agents to treat high-grade PIN would fill a greatly needed therapeutic void.[23] There is evidence that all modes of androgen deprivation that induce acinar atrophy and apoptosis cause regression of high-grade PIN.[68,94–100] Neoadjuvant hormone deprivation with monthly leuprolide and flutamide 250 mg PO TID for 3 months resulted in a 50% reduction in high-grade PIN.[22] Longer therapy with 6 months of neoadjuvant androgen deprivation therapy prior to radical prostatectomy in the European Randomized Study of Screening for Prostate Cancer (ERSPC) study reduced high-grade PIN even more.[24] Flutamide, but not finasteride, decreased the prevalence and extent of high-grade PIN and induced epithelial atrophy.[101] There is also evidence that cessation of flutamide resulted in the return of high-grade PIN.[15] Similarly, radiation reduced high-grade PIN.[101]

The mechanism of action for androgen deprivation is epithelial hyperplasia, cytoplasmic clearing and prominent acinar atrophy, with a decreased ratio of acini to stroma that indicates the hormone dependence of high-grade PIN. Treated dysplastic epithelial cells characteristic of high-grade PIN separate from the basal cell layer and exfoliate into the acinar lumen to be cleared with normal prostatic secretions (**Fig. 4.6**).[101] Radiation appears to have the same mechanism to resolve high-grade PIN. The long-term efficacy of radiation treatment may depend on eradication of cancer as well as these high-grade PIN precancerous lesions, which otherwise would lead to metachronous invasive prostate cancer.[101] This mechanism of treatment is similar to that of other precancerous lesions, like adenomatous polyps of the colon, which, when removed, allow normal epithelium to be restored.

Fig. 4.6. The effects of androgen deprivation on high-grade PIN. The percentage of radical prostatectomies with PIN is greater in untreated prostrates (bar on left) than in those receiving up to 3 months of therapy. (bar on the right). Data from Vailancourt et al. (1996).[98] Androgen deprivation induces apoptosis of hormone-dependent high-grade PIN that separates and exfoliates into the prostatic ducts to be carried away by the prostatic secretions. The prostatic acinus is then hypothetically replaced by the normal prostate epithelium.

Chronic therapy, however, would most likely be required to prevent new high-grade PIN lesions from invading and becoming clinical prostate cancer. Although more toxicity is likely to be tolerated for the treatment agents targeted to regress or inhibit high-grade PIN,[22] as compared to treating healthy patients to reduce prostate cancer incidence, androgen deprivation therapy has too many adverse effects in men to be clinically useful. Newer agents with a better safety and lower side-effect profile are greatly needed since patients may be taking the agent at least until they attain 70 years of age.[95] Acapodene is currently in a Phase IIb multicenter, randomized, prospective placebo-controlled human clinical trial to determine if it can treat high-grade PIN and reduce prostate cancer incidence.[29]

SUMMARY

High-grade PIN is a disease that, if left untreated, would be clinically important to the patient. This is not surprising, as other intraepithelial neoplasias affecting other organ systems are routinely treated with clear clinical benefit. As high-grade PIN is hormone sensitive, antiandrogen therapy causes regression of high-grade PIN within months. The process involves separation of PIN from the prostatic acini, exfoliation into the prostatic ducts and removal of these cells through the prostate fluids secreted into the urethra. Treatment of high-grade PIN would appear to be a clinical benefit notwithstanding the potential for cancer prevention. These clinical benefits would reduce morbidity, enhance quality of life, delay surgery or radiation, and increase the interval for surveillance requiring invasive procedures.[25] Availability of agents less toxic than antiandrogens, like Acapodene, may become the mainstay treatment for high-grade PIN and prostate cancer prevention.

REFERENCES

1. Alcaraz A, Barranco MA, Corral JM et al. High-grade prostate intraepithelial neoplasia shares cytogenetic alterations with invasive prostate cancer. *Prostate* 2001; 47:29–35.
2. Sakr WA, Partin AW. Histological markers of risk and the role of high-grade prostatic intraepithelial neoplasia. *Urology* 2001; 57(4 Suppl. 1):115–20.
3. Foster CS, Bostwick DG, Bonkhoff H et al. Cellular and molecular pathology of prostate cancer precursors. *Scand. J. Urol. Nephrol. Suppl.* 2000; 205:19–43.
4. Bostwick DG, Norlen BJ, Denis L. Prostatic intraepithelial neoplasia: the preinvasive stage of prostate cancer. Overview of the prostate committee report. *Scand. J. Urol. Nephrol. Suppl.* 2000; 205:1–2.
5. Bostwick DG, Montironi R, Sesterhenn IA. Diagnosis of prostatic intraepithelial neoplasia: Prostate Working Group/consensus report. *Scand. J. Urol. Nephrol. Suppl.* 2000; 205:3–10.

6. Zlotta AR, Schulman CC. Clinical evolution of prostatic intraepithelial neoplasia. *Eur. Urol.* 1999; 35:498–503.
7. Bostwick DG. Prostatic intraepithelial neoplasia is a risk factor for cancer. *Semin. Urol. Oncol.* 1999; 17:187–98.
8. Algaba F. Evolution of isolated high-grade prostate intraepithelial neoplasia in a Mediterranean patient population. *Eur. Urol.* 1999; 35:496–7.
9. Sakr WA. High-grade prostatic intraepithelial neoplasia: additional links to a potentially more aggressive prostate cancer? *J. Natl Cancer. Inst.* 1998; 90:486–7.
10. Bostwick DG, Shan A, Qian J et al. Independent origin of multiple foci of prostatic intraepithelial neoplasia: comparison with matched foci of prostate carcinoma. *Cancer* 1998; 83:1995–2002.
11. Montironi R, Mazzucchelli R, Pomante R. Preneoplastic lesions of the prostate. *Adv. Clin. Path.* 1997; 1:35–47.
12. Haggman MJ, Macoska JA, Wojno KJ et al. The relationship between prostatic intraepithelial neoplasia and prostate cancer: critical issues. *J. Urol.* 1997; 158:12–22.
13. Montironi R, Bostwick DG, Bonkhoff H et al. Origins of prostate cancer. *Cancer* 1996; 78:362–5.
14. Shin HJ, Ro JY. Prostatic intraepithelial neoplasia: a potential precursor lesion of prostatic adenocarcinoma. *Yonsei Med. J.* 1995; 36:215–31.
15. Prostatic Intraepithelial Neoplasia and the Origins of Prostatic Carcinoma. Proceedings of the First International Consultation Meeting. Ancona, Italy, September 11–12, 1994. *Pathol. Res. Pract.* 1995; 191:828–959.
16. Bostwick DG. Premalignant lesions of the prostate. *Semin. Diagn. Pathol.* 1988; 5:240–53.
17. McNeal JE, Yemoto CE. Spread of adenocarcinoma within prostatic ducts and acini. Morphologic and clinical correlations. *Am. J. Surg. Pathol.* 1996; 20:802–14.
18. McNeal JE, Bostwick DG. Intraductal dysplasia: a premalignant lesion of the prostate. *Hum. Pathol.* 1986; 17:64–71.
19. Bostwick DG, Brawer MK. Prostatic intra-epithelial neoplasia and early invasion in prostate cancer. *Cancer* 1987; 59:788–94.
20. Sakr WA, Grignon DJ, Haas GP et al. Epidemiology of high grade prostatic intraepithelial neoplasia. *Pathol. Res. Pract.* 1995; 191:838–41.
21. Algaba F, Epstein JI, Aldape HC et al. Assessment of prostate carcinoma in core needle biopsy – definition of minimal criteria for the diagnosis of cancer in biopsy material. *Cancer* 1996; 78:376–81.
22. Guidelines on the Management of Prostate Cancer. A document for local expert groups in the United Kingdom preparing prostate management policy documents. The Royal College of Radiologists' Clinical Oncology Information Network. British Association of Urological Surgeons. *B. J. Urol. Int.* 1999; 84:987–1014.
23. Precursors of prostatic adenocarcinoma: recent findings and new concepts. *Eur. Urol.* 1996; 30:131–279.
24. International consultation on prostatic intraepithelial neoplasia and pathologic staging of prostatic carcinoma. Rochester, Minnesota, November 3–4, 1995. *Cancer* 1996; 78:320–81.
25. Project NPCD. Prostatic intraepithelial neoplasia: significance and correlation with prostate-specific antigen and transrectal ultrasound. Proceedings of a workshop of the National Prostate Cancer Detection Project. March 13, 1989, Bethesda, Maryland. *Urology* 1989; 34(6 Suppl.):2–69.
26. Epstein JI, Grignon DJ, Humphrey PA et al. Interobserver reproducibility in the diagnosis of prostatic intraepithelial neoplasia. *Am. J. Surg. Pathol.* 1995; 19:873–86.
27. Allam CK, Bostwick DG, Hayes JA et al. Interobserver variability in the diagnosis of high-grade prostatic intraepithelial neoplasia and adenocarcinoma. *Mod. Pathol.* 1996; 9:742–51.
28. Greenlee R, Hill-Harmon M, Murray T et al. Cancer statistics, 2001. *CA Cancer J. Clin.* 2001; 51:15–36.
29. Steiner MS, Raghow S, Neubauer BL. Selective estrogen receptor modulators for the chemoprevention of prostate cancer. *Urology* 2001; 57(4 Suppl. 1):68–72.
30. Bostwick DG, Pacelli A, Lopez-Beltran A. Molecular biology of prostatic intraepithelial neoplasia. *Prostate* 1996; 29:117–34.
31. Sakr WA, Billis A, Ekman P et al. Epidemiology of high-grade prostatic intraepithelial neoplasia. *Scand. J. Urol. Nephrol. Suppl.* 2000; 205:11–18.
32. Sakr WA, Grignon DJ, Crissman JD et al. High grade prostatic intraepithelial neoplasia (HGPIN) and prostatic adenocarcinoma between the ages of 20–69: an autopsy study of 249 cases. *In Vivo* 1994; 8:439–43.
33. Sakr WA, Grignon DJ, Haas GP et al. Age and racial distribution of prostatic intraepithelial neoplasia. *Eur. Urol.* 1996; 30:138–44.
34. Sakr WA, Haas GP, Cassin BF et al. The frequency of carcinoma and intraepithelial neoplasia of the prostate in young male patients. *J. Urol.* 1993; 150:379–85.
35. Helpap B, Bonkhoff H, Cockett A et al. Relationship between atypical adenomatous hyperplasia (AAH), prostatic intraepithelial neoplasia (PIN) and prostatic adenocarcinoma. *Pathologica* 1997; 89:288–300.
36. Bostwick DG, Aquilina JW. Prostatic intraepithelial neoplasia (PIN) and other prostatic lesions as risk factors and surrogate endpoints for cancer chemoprevention trials. *J. Cell. Biochem. Suppl.* 1996; 25:156–64.
37. Qian J, Wollan P, Bostwick DG. The extent and multicentricity of high-grade prostatic intraepithelial neoplasia in clinically localized prostatic adenocarcinoma. *Hum. Pathol.* 1997; 28:143–8.
38. Qian J, Bostwick DG. The extent and zonal location of prostatic intraepithelial neoplasia and atypical adenomatous hyperplasia: relationship with carcinoma in radical prostatectomy specimens. *Pathol. Res. Pract.* 1995; 191:860–7.
39. O'Dowd GJ, Miller MC, Orozco R et al. Analysis of repeated biopsy results within 1 year after a noncancer diagnosis. *Urology*. 2000; 55:553–9.
40. Fujita M, Shin M, Yasunaga Y et al. Incidence of prostatic intraepithelial neoplasia in Osaka, Japan. *Int. J. Cancer* 1997; 73:808–11.
41. Bostwick DG, Ramnani D, Qian J. Prostatic intraepithelial neoplasia: animal models 2000. *Prostate* 2000; 43:286–94.
42. Gingrich JR, Barrios RJ, Kattan MW et al. Androgen-independent prostate cancer progression in the TRAMP model. *Cancer Res.* 1997; 57:4687–91.
43. Greenberg NM, DeMayo F, Finegold MJ et al. Prostate cancer in a transgenic mouse. *Proc. Natl Acad. Sci. USA* 1995; 92:3439–43.
44. Voelkel-Johnson C, Voeks DJ, Greenberg NM et al. Genomic instability-based transgenic models of prostate cancer. *Carcinogenesis* 2000; 21:1623–7.
45. Stanbrough M, Leav I, Kwan PW et al. Prostatic intraepithelial neoplasia in mice expressing an androgen receptor transgene in prostate epithelium. *Proc. Natl Acad. Sci. USA* 2001; 98:10823–8.
46. Wang Y, Sudilovsky D, Zhang B et al. A human prostatic epithelial model of hormonal carcinogenesis. *Cancer Res.* 2001; 61:6064–72.
47. Aquilina JW, McKinney L, Pacelli A et al. High grade prostatic intraepithelial neoplasia in military working dogs with and without prostate cancer. *Prostate* 1998; 36:189–93.
48. Waters DJ. High-grade prostatic intraepithelial neoplasia in dogs. *Eur. Urol.* 1999; 35:456–8.
49. Waters DJ, Bostwick DG. The canine prostate is a spontaneous model of intraepithelial neoplasia and prostate cancer progression. *Anticancer Res.* 1997; 17:1467–70.
50. Waters DJ, Bostwick DG. Prostatic intraepithelial neoplasia occurs spontaneously in the canine prostate. *J. Urol.* 1997; 157:713–16.

51. Waters DJ, Hayden DW, Bell FW et al. Prostatic intraepithelial neoplasia in dogs with spontaneous prostate cancer. *Prostate* 1997; 30:92–7.
52. Helpap B, Riede C. Nucleolar and AgNOR-analysis of prostatic intraepithelial neoplasia (PIN), atypical adenomatous hyperplasia (AAH) and prostatic carcinoma. *Pathol. Res. Pract.* 1995; 191:381–90.
53. Helpap BG, Bostwick DG, Montironi R. The significance of atypical adenomatous hyperplasia and prostatic intraepithelial neoplasia for the development of prostate carcinoma. An update. *Virchows Arch.* 1995; 426:425–34.
54. Helpap B. Cell kinetic studies on prostatic intraepithelial neoplasia (PIN) and atypical adenomatous hyperplasia (AAH) of the prostate. *Pathol. Res. Pract.* 1995; 191:904–7.
55. Fowler JE Jr, Bigler SA, Lynch C et al. Prospective study of correlations between biopsy-detected high grade prostatic intraepithelial neoplasia, serum prostate specific antigen concentration, and race. *Cancer* 2001; 91:1291–6.
56. Sakr WA, Grignon DJ, Haas GP, Pathology of premalignant lesions and carcinoma of the prostate in African-American men. *Semin. Urol. Oncol.* 1998; 16:214–20.
57. Sakr WA. Prostatic intraepithelial neoplasia: a marker for high-risk groups and a potential target for chemoprevention. *Eur. Urol.* 1999; 35:474–8.
58. Watanabe M, Fukutome K, Kato H et al. Progression-linked overexpression of c-Met in prostatic intraepithelial neoplasia and latent as well as clinical prostate cancers. *Cancer Lett.* 1999; 141:173–8.
59. Montironi R, Diamanti L, Pomante R et al. Subtle changes in benign tissue adjacent to prostate neoplasia detected with a Bayesian belief network. *J. Pathol.* 1997; 182:442–9.
60. Montironi R, Scarpelli M, Sisti S et al. Quantitative analysis of prostatic intraepithelial neoplasia on tissue sections. *Anal. Quant. Cytol. Histol.* 1990; 12:366–72.
61. Qian J, Jenkins RB, Bostwick DG. Detection of chromosomal anomalies and c-myc gene amplification in the cribriform pattern of prostatic intraepithelial neoplasia and carcinoma by fluorescence in situ hybridization. *Mod. Pathol.* 1997; 10:1113–19.
62. Qian J, Jenkins RB, Bostwick DG. Genetic and chromosomal alterations in prostatic intraepithelial neoplasia and carcinoma detected by fluorescence in situ hybridization. *Eur. Urol.* 1999; 35:479–83.
63. Montironi R, Diamanti L, Thompson D et al. Analysis of the capillary architecture in the precursors of prostate cancer: recent findings and new concepts. *Eur. Urol.* 1996; 30:191–200.
64. Montironi R, Mazzucchelli R, Algaba F, Lopez-Beltran A. Morphological identification of the patterns of prostatic intraepithelial neoplasia and their importance. *J. Clin. Pathol.* 2000; 53:655–65.
65. Montironi R, Scarpelli M, Galluzzi CM et al. Aneuploidy and nuclear features of prostatic intraepithelial neoplasia (PIN). *J. Cell Biochem. Suppl.* 1992; 16H:47–53.
66. Qian J, Jenkins RB, Bostwick DG. Determination of gene and chromosome dosage in prostatic intraepithelial neoplasia and carcinoma. *Anal. Quant. Cytol. Histol.* 1998; 20:373–80.
67. Qian J, Jenkins RB, Bostwick DG. Potential markers of aggressiveness in prostatic intraepithelial neoplasia detected by fluorescence in situ hybridization. *Eur. Urol.* 1996; 30: 177–84.
68. Montironi R, Pomante R, Diamanti L et al. Apoptosis in prostatic adenocarcinoma following complete androgen ablation. *Urol. Int.* 1998; 60(Suppl. 1):25–9; discussion 30.
69. Montironi R, Magi-Galluzzi C, Fabris G. Apoptotic bodies in prostatic intraepithelial neoplasia and prostatic adenocarcinoma following total androgen ablation. *Pathol. Res. Pract.* 1995; 191:873–80.
70. Montironi R, Magi-Galluzzi CM, Marina S et al. Quantitative characterization of the frequency and location of cell proliferation and death in prostate pathology. *J. Cell Biochem. Suppl.* 1994; 19:238–45.
71. Montironi R, Magi Galluzzi C, Scarpelli M et al. Occurrence of cell death (apoptosis) in prostatic intra-epithelial neoplasia. *Virchows Arch. A Pathol. Anat. Histopathol.* 1993; 423:351–7.
72. Montironi R, Filho AL, Santinelli A et al. Nuclear changes in the normal-looking columnar epithelium adjacent to and distant from prostatic intraepithelial neoplasia and prostate cancer. Morphometric analysis in whole-mount sections. *Virchows Arch.* 2000; 437:625–34.
73. Emmert-Buck MR, Vocke CD, Pozzatti RO et al. Allelic loss on chromosome 8p12–21 in microdissected prostatic intraepithelial neoplasia. *Cancer Res.* 1995; 55:2959–62.
74. Kim NW, Hruszkewycz AM. Telomerase activity modulation in the prevention of prostate cancer. *Urology* 2001; 57(4 Suppl. 1): 148–53.
75. Willman JH, Holden JA. Immunohistochemical staining for DNA topoisomerase II-alpha in benign, premalignant, and malignant lesions of the prostate. *Prostate* 2000; 42:280–6.
76. Ge K, Minhas F, Duhadaway J et al. Loss of heterozygosity and tumor suppressor activity of Bin1 in prostate carcinoma. *Int. J. Cancer* 2000; 86:155–61.
77. Qian J, Bostwick DG, Takahashi S et al. Chromosomal anomalies in prostatic intraepithelial neoplasia and carcinoma detected by fluorescence in situ hybridization. *Cancer Res.* 1995; 55:5408–14.
78. Miet SM, Neyra M, Jaques R et al. RER(+) phenotype in prostate intra-epithelial neoplasia associated with human prostate-carcinoma development. *Int. J. Cancer* 1999; 82:635–9.
79. Baltaci S, Orhan D, Ozer G et al. Bcl-2 proto-oncogene expression in low- and high-grade prostatic intraepithelial neoplasia. *B. J. Urol. Int.* 2000; 85:155–9.
80. Alexander EE, Qian J, Wollan PC et al. Prostatic intraepithelial neoplasia does not appear to raise serum prostate-specific antigen concentration. *Urology* 1996; 47:693–8.
81. Ramos CG, Carvahal GF, Mager DE et al. The effect of high grade prostatic intraepithelial neoplasia on serum total and percentage of free prostate specific antigen levels. *J. Urol.* 1999; 162:1587–90.
82. Morote J, Encabo G, Lopez M et al. Influence of high-grade prostatic intra-epithelial neoplasia on total and percentage free serum prostatic specific antigen. *B. J. Urol. Int.* 1999; 84:657–60.
83. Morote J, Lopez M, de Torres IM et al. [Influence of prostatic intraepithelial neoplasia on blood levels of PSA and free PSA percentage]. *Actas Urol. Esp.* 1999; 23:342–9.
84. Morote J, Raventos CX, Encabo G et al. Effect of high-grade prostatic intraepithelial neoplasia on total and percent free serum prostatic-specific antigen. *Eur. Urol.* 2000; 37:456–9.
85. Davidson D, Bostwick DG, Qian J et al. Prostatic intraepithelial neoplasia is a risk factor for adenocarcinoma: predictive accuracy in needle biopsies. *J. Urol.* 1995; 154(4):1295–9.
86. Zlotta AR, Raviv G, Schulman CC. Clinical prognostic criteria for later diagnosis of prostate carcinoma in patients with initial isolated prostatic intraepithelial neoplasia. *Eur. Urol.* 1996; 30:249–55.
87. Shepherd D, Keetch DW, Humphrey PA et al. Repeat biopsy strategy in men with isolated prostatic intraepithelial neoplasia on prostate needle biopsy. *J. Urol.* 1996; 156:460–2; discussion 462–3.
88. Kronz JD, Allan CH, Shaikh AA et al. Predicting cancer following a diagnosis of high-grade prostatic intraepithelial neoplasia on needle biopsy: data on men with more than one follow-up biopsy. *Am. J. Surg. Pathol.* 2001; 25:1079–85.
89. Stewart CS, Leibovich BC, Weaver AL, Lieber MM. Prostate cancer diagnosis using a saturation needle biopsy technique after previous negative sextant biopsies. *J. Urol.* 2001; 166:86–91; discussion 91–82.

90. Perachino M, di Ciolo L, Barbetti V et al. Results of rebiopsy for suspected prostate cancer in symptomatic men with elevated PSA levels. *Eur. Urol.* 1997; 32:155–9.
91. Newling D. PIN I–III: when should we interfere? *Eur. Urol.* 1999; 35:504–7.
92. Feneley MR. Does screening for prostate cancer identify clinically important disease? *Ann. R. Coll. Surg. Engl.* 1999; 81:207–14.
93. Feneley MR, Busch C. Precursor lesions for prostate cancer. *J. R. Soc. Med.* 1997; 90:533–9.
94. Balaji KC, Rabbani F, Tsai H et al. Effect of neoadjuvant hormonal therapy on prostatic intraepithelial neoplasia and its prognostic significance. *J. Urol.* 1999; 162:753–7.
95. Abbas F, Hochberg D, Civantos F et al. Incidental prostatic adenocarcinoma in patients undergoing radical cystoprostatectomy for bladder cancer. *Eur. Urol.* 1996; 30:322–6.
96. Bostwick DG, Qian J. Effect of androgen deprivation therapy on prostatic intraepithelial neoplasia. *Urology* 2001; 58:91–3.
97. Ferguson J, Zincke H, Ellison E et al. Decrease of prostatic intraepithelial neoplasia following androgen deprivation therapy in patients with stage T3 carcinoma treated by radical prostatectomy. *Urology* 1994; 44:91–5.
98. Vailancourt L, Ttu B, Fradet Y et al. Effect of neoadjuvant endocrine therapy (combined androgen blockade) on normal prostate and prostatic carcinoma. A randomized study. *Am. J. Surg. Pathol.* 1996; 20:86–93.
99. Montironi R, Magi-Galluzzi C, Muzzonigro G et al. Effects of combination endocrine treatment on normal prostate, prostatic intraepithelial neoplasia, and prostatic adenocarcinoma. *J. Clin. Pathol.* 1994; 47:906–13.
100. Montironi R, Pomante R, Diamanti L et al. Evaluation of prostatic intraepithelial neoplasia after treatment with a 5-alpha-reductase inhibitor (finasteride). A methodologic approach. *Anal. Quant. Cytol. Histol.* 1996; 18:461–70.
101. Alers JC, Krijtenburg PJ, Vissers KJ et al. Interphase cytogenetics of prostatic adenocarcinoma and precursor lesions: analysis of 25 radical prostatectomies and 17 adjacent prostatic intraepithelial neoplasias. *Genes Chromosomes Cancer* 1995; 12:241–50.

Update on the Regulation of Apoptosis in Prostate Cancer

R. William G. Watson, Liza McLornan, Kevin McEleny and John M. Fitzpatrick

Department of Surgery, Mater Misericordiae Hospital, Conway Institute, University College Dublin, Ireland

INTRODUCTION

The death of cells is not always as a result of damage but may occur spontaneously to allow for further development and growth. Homeostatic control of cell numbers occurs as a result of the dynamic balance between cell proliferation and cell death and is essential to maintain a steady volume.[1] Cancer cells have lost their ability to control this balance, resulting in an accumulation of cells.[2,3]

Cells die by one of two processes, necrosis or programmed cell death (apoptosis).[4,5] Necrotic death occurs as a result of severe injury to the cell, in a sudden and uncontrolled process. It is characterized by the swelling and rupture of the cell, owing to uncontrolled regulation of fluids and ions (sodium and calcium). Programmed cell death or apoptosis, as the name suggests, is a controlled process resulting in the death of a cell and removal by surrounding phagocytes, mainly the macrophage.[6] *Caenorhabditis elegans*, a 1-mm nematode, demonstrated changes in cell numbers. This worm loses exactly 131 cells during its development. All these process are programmed and result in the process of apoptosis.

DEFINITIONS AND CHARACTERISTICS OF APOPTOSIS

Cells undergoing apoptosis all demonstrate more or less the same characteristics; however, individual cell types differ in the extent by which they express these changes. A number of morphologically identifiable stages have been reported. These include nuclear changes, exuberant cell surface and breaking up of the nucleus to form multiple fragments and compacted chromatin. These changes are accompanied by flipping of phosphatidylserine from the inner plasma membrane to the cell surface, which can be detected by the phosphatidylserine-binding protein, Annexin V. Finally, the cell-surface protuberances separate to produce membrane-enclosed apoptotic bodies of varying size in which the closely packed cytoplasmic organelles remain well preserved. The nuclear collapse that is the hallmark of apoptosis has as its biochemical correlate the fragmentation of DNA by endonucleases, producing fragments in the range 300–50 kbp. The DNA cleavage continues with internucleosomal double-stranded cutting to produce the familiar ladder on agarose gel electrophoresis. The cleave of these proteins is centrally dependent on proteolytic enzymes called the caspases, which are activated by death receptors or factors released from the mitochondria. **Fig. 5.1** outlines the apoptotic pathway and sites of inhibition.

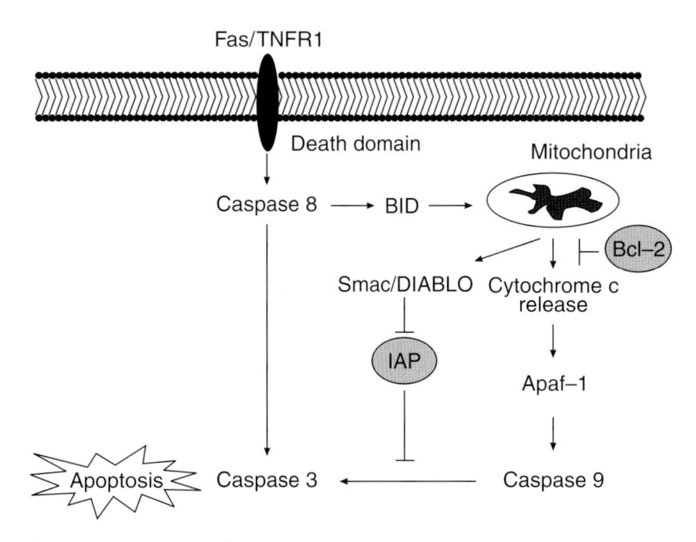

Fig 5.1. Apoptotic pathway.

REGULATION OF APOPTOSIS

Cancer is associated with uncontrolled proliferation and loss of apoptotic potential leading to the development of tumor masses. Two theories exist in relation to the survival and death of cells and, subsequently, to the activation of a central pathway in the induction of apoptosis. It is believed that cells die by default by the process of apoptosis. Cells constantly produce survival factors to inhibit this mechanism and maintain viability. Alternatively, it has been suggested that the apoptotic pathway is triggered by environmental stimuli that are produced when the cell is required to die. Prostate epithelial cells are fundamentally dependent on the presence of androgen for growth and differentiate, eventually dying by the process of apoptosis.[7] Removal of androgen results in their death via the process of apoptosis.[8] This demonstrates that androgens represents a critical survival factor in prostate epithelial cells. However, prostate cancer cells, specifically androgen-independent tumors, have developed mechanisms to overcome the loss of androgen.

This chapter will discuss some of the more recent proteins implicated in the regulation of prostate cancer cell death, specifically the antiapoptotic proteins (inhibitors of apoptosis proteins), growth factors that activate the intracellular signaling AKT survival pathways and the pro-apoptotic proteins (caspases).

PATHWAYS PREVENTING APOPTOSIS

Inhibitors of apoptosis proteins

The inhibitors of apoptosis proteins (IAPs) are a recently identified group of antiapoptotic proteins. So far, five members have been identified in human cells: NAIP, cIAP-1, cIAP-2, XIAP and survivin.

The first human IAP discovered was NAIP,[9] which was absent in most cases of severe spinal muscle atrophy, a rare autosomal recessive disorder of infancy, characterized by progressive apoptosis of spinal muscle motor neurons.[10,11] Subsequent work confirmed that NAIP was cytoprotective and homologous sequence tagging led to the discovery of X-linked IAP (XIAP/hILP).[12] Human IAP-1 (hIAP-1/cIAP-2) and human IAP-2 (hIAP-2/cIAP-1) were subsequently identified.

IAP structure

The IAP structure consists of between one and three imperfect amino-acid repeats, each 60–80 amino acids in length, which fold into a hydrophobic core with hydrophilic surface residues.[13,14] The core contains a zinc-binding site that is capable of supporting protein–protein interactions, as well as possible sites of phosphorylation. XIAP, cIAP-1 and cIAP-2 contain zinc RING (really interesting new gene) finger at the C-terminus. This motif is not unique and present in many proteins, where it is required for interactions with other proteins. Survivin lacks the Zn RING finger but instead possesses a particular three-dimensional 'coiled coil' protein arrangement. Removal of this blocks the antiapoptotic effect by dissociating survivin from the mitotic spindle, where it is required at the G2/M phase of the cell cycle.[15]

cIAP-1 and cIAP-2 also possess a caspase recruitment domain (CARD). This motif, often found in apoptotic signaling proteins, is thought to have a role in substrate recognition. While all the IAPs except NAIP have been shown to inhibit caspases, XIAP has the strongest binding affinity for the caspases and may well be the physiological inhibitor.[16]

Mechanism of action

IAPs have been shown to protect cells from a wide range of apoptotic triggers including Fas and tumor necrosis factor α (TNFα) ligation, Bax-mediated mitochondrial disruption, caspases activation, cytochrome *c* release, some chemotherapeutic agents, viral infection and radiation.[17–22] All IAP (except NAIP whose antiapoptotic mechanism remains unclear) have been shown to be able to bind to and inhibit activated caspases 3 and 7, furthermore, cIAP-1, cIAP-2 and XIAP have also been shown to inhibit the activation of caspase 9.[23,24]

Other antiapoptotic mechanisms have also been described. In response to TNFα, nuclear factor κB (NFκB) is activated, and this leads to increased levels of cIAP-1 and 2, and also TRAF1 and TRAF2[25] (TNF receptor associated factor). cIAP-1 and 2 can bind TRAF 1 and 2, and these molecules can associate at the cytosolic death domain, and thereby prevent caspase 8 activation in response to the death signal. Other interactions reported include cIAP-2 with RIP kinase[26] and XIAP with the BMP receptor.[27] (The latter interaction is mediated via the RING finger.)

Regulation

Expression of some IAPs is known to be regulated by NFκB.[28,29] This transcription factor that is key to so many cellular processes is believed to have an important role in inflammatory and neoplastic diseases. Inhibitors of NFκB activation, such as pyrrolidinedithiocarbamate (PDTC) and genestein, have been shown to cause an increase in prostate cell apoptosis in response to agents such as TNFα and DNA-damaging agents.[30,31] It is tempting to think that this could in part be due to the blockade of IAP activation. However, NFκB exerts its antiapoptotic effects via a number of different molecules including the Bcl-2 homologs A1/Bfl-1[32] and Bcl-X_L.[33]

With regards to what actually inactivates IAP, the evidence is limited; however, it has been shown that, in response to an apoptotic trigger, cIAP-1 and XIAP can undergo ubiquitination and degradation in a process requiring an intact RING domain.[34] Also there is some published evidence, at least with regard to XIAP, that the IAPs may themselves be targets for activated caspases.[35]

IAP and cancer

Survivin is expressed only in fetal tissue under normal conditions,[36] although mRNA can also be detected in adult thymus and, to a lesser degree, in placenta. Survivin has also been shown to be strongly expressed in endothelial cells in granulation tissue. This suggests that survivin may have an important role in angiogenesis, another key event in carcinogenesis.

Survivin expression has been demonstrated using immunohistochemistry to be present in the majority of carcinomas tested including breast,[37] gastric[38] and colon.[39] It has also been found in high-grade (but not in low-grade) lymphomas. Interestingly, transfection of HeLa cells with both survivin and EPR-1 abrogates the antiapoptotic effect of survivin transfection alone. A number of studies have looked at survivin expression using immunohistochemistry. In neuroblastoma, expression has been shown to correlate with a more aggressive grade and unfavorable histology. In gastric cancer, survivin expression was also associated with Bcl-2 expression and a reduced apoptotic index, as well as expression of mutated p53. Survivin expression in colorectal cancer contributed to a worse 5-year survival. cIAP-1, cIAP-2 and XIAP have been identified in 12 different malignant glioma cell lines.[40]

With regards to prostate cancer, survivin expression has been demonstrated in the PC3 and DU145 cell lines. Expression of survivin protein and mRNA has also been discussed in five formalin-fixed, paraffin-embedded sections using immunohistochemistry and *in situ* hybridization, but no published data are available. In addition, immunohistochemistry has also been used to demonstrate the presence of cIAP-1 and cIAP-2 in prostate cancer.[41] Data have been presented on 23 patients who had undergone neoadjuvant androgen ablation therapy prior to radical prostatectomy. The percentage of cells demonstrating positive expression of cIAP-1 increased from 4.2% prior to androgen ablation to 65.2% after androgen ablation. Also the expression of cIAP-2 increased from 4.2% to 73.9% after treatment. The observation that IAP expression actually increases with androgen blockade, implies a potentially important role for IAPs in the development of apoptotic resistance in androgen-independent prostate cancer. Recent studies have demonstrated that IAP expression is associated with apoptotic resistance in a number of prostate cancer cell lines.[42]

GROWTH FACTORS AND PROSTATE CANCER

The insulin-like growth factor (IGF) system has been shown to play an important role in proliferation and differentiation of tissues. IGF promotes cancer cell growth in an autocrine/paracrine manner via the IGF-1 receptor. It has also been implicated in the process of transformation to human epithelial prostate carcinoma. Recent studies suggest that there is a strong association between serum IGF-1 levels and prostate carcinoma, and that this growth factor may even be an important predictor of risk for prostate cancer.[43,44] These growth factors activate their corresponding receptors leading to phosphorylation of second-messenger signaling systems and initiation of their individual downstream effects.

Insulin-like growth factor

IGF-I has been demonstrated to induce apoptotic resistance in various cell types following serum and interleukin-3 withdrawal, etoposide, adriamycin and anoikis-induced apoptosis.[45] Its molecular effects on prostate cancer cells have yet to be determined. However, hormone manipulation induces the expression of insulin-like growth factor binding protein (IGFBP)-2, 3, 4 and 5, which can sequester away free IGF-I. IGF-I is known to act downstream of the c-*myc* oncogene and so block entry into the apoptotic pathway.[46] c-*myc* expression is raised after castration and, therefore, IGF-I may abrogate castration-induced apoptosis.[46] In other cell types, IGF-I stimulates the expression of Bcl-2, Bcl-X and Mcl-1.[47] Overexpression or up-regulation of Bcl-2 occurs 2–3 months after androgen ablation. Current work in our laboratory has demonstrated that IGF-I does induce expression of Bcl-X and Mcl-1 in PC-3 and LNCaP prostatic cell lines, thereby inducing a resistance to both chemical and radiation induced apoptosis. Our work supports the theory that IGF-I may be involved in the progression or development of prostate cancer, in particular androgen independence. How IGF-1 signals for this response may have more important implications in the regulation of prostate cancer resistance to apoptosis.

Ligation of the IGF-1 receptor activates an intracellular tyrosine kinase domain, resulting in phosphorylation of several substrates. Two important intracellular signal transduction pathways are known to be activated, namely the Ras mitogen-activated protein kinase (MAP kinase) pathway and the phosphatidylinositol 3-kinase (PI3K)/Akt pathway.[45] Activation of Ras induces an increase in MAP kinase activity, resulting in protein expression. Activation of PI3K stimulates Akt kinase activity, resulting in the phosphorylation of proteins including Bad, a pro-apoptotic protein. PI3K is also known to phosphorylate focal adhesion kinase (FAK) involved in cell adhesion signaling.[45]

AKT SURVIVAL SIGNALS

Phosphorylation

Phosphorylation is a covalent modification frequently used by eukaryotic organisms to handle responses to extracellular signals, e.g. growth factors or hormones. The process of phosphorylation can result in either the activation or deactivation of proteins. Alternatively, dephosphorylation can activate or deactivate protein function. Phosphorylation is a reversible process and involves addition of a phosphate molecule to the side chains of serine, threonine or tyrosine residues. The introduction and removal of the phosphate group is catalyzed by separate enzymes; phosphorylation by a protein kinase and dephosphorylation by a protein phosphatase, which are themselves generally under metabolic regulation.[48] On reaching tissue targets, extracellular signals initiate a chain of events that led to the activation of a protein kinase. The kinase then catalyzes the phosphorylation of one or more specific proteins depending on the tissue. For a protein to be regulated by phosphorylation, the maximal activities of the protein kinases and phosphatases acting on a particular site must be clearly in balance; otherwise the protein would be either fully phosphorylated or completely dephosphorylated. This precise balance of kinase and phosphatase activity plays a major role in receptor-mediated signaling pathways and cell cycle control.

The importance of protein kinases in regulating cellular activities is underscored by the large number of protein kinase genes that are present in the eukaryotic genome. Phosphorylation of the protein may involve an allosteric mechanism that affects subunit dimerization or that alters the catalytic machinery of the substrate cleft through conformational changes.

Serine/threonine protein kinases

Serine/threonine protein kinases are an important family of proteins that modulate the phosphorylation of many key effectors of the apoptotic process.[46] Akt, also known as protein kinase B, which is a member of this family, has been implicated in cell survival. Akt was first discovered in 1991 and is activated downstream of phosphatidylinositol 3′-kinase (PI3-K), a key signal transduction molecule, which is growth factor and is receptor mediated.

Regulation of Akt

PI3-K

Phosphatidylinositol 3′-kinases are a ubiquitously expressed lipid kinase family. PI3-K consists of a regulatory subunit (p85) that binds to an activated growth factor/cytokine receptor and undergoes phosphorylation, which results in the activation of its catalytic subunit (p110). Once activated, PI3-K catalyzes the addition of a PO4 molecule to the 3′-position of the inositol ring of phosphoinositides to generate second-messenger lipids. PI3-K can be divided into three classes according to the molecules they preferentially utilize as substrate. The lipid products of PI3-K include the singly phosphorylated form PtdIns-3-P, the doubly phosphorylated forms PtdIns-3,4-P_2 and PtdIns-3,5-P_2, and the triply phosphorylated form PtdIns-3,4,5-P_3. PtdIns-3,4-P_2 facilitates the recruitment of the protein kinase Akt to the plasma membrane and subsequent activation.

The PI3-Kinase/Akt pathway is a general mediator of growth-factor-induced survival and has been shown to suppress the apoptotic death of a number of cell types, induced by a variety of stimuli, including growth-factor withdrawal, cell-cycle discordance, loss of cell adhesion and DNA damage.

Akt is a polypeptide of approximately 55–60 kDa containing an N-terminal pleckstrin homology (PH) domain and a serine/threonine kinase catalytic domain.[49] The PH domain exhibits an affinity for 3′-phosphorylated lipids. Activation of Akt requires an intact PH domain but this is not sufficient for function. For catalytic activation the enzyme must undergo a conformational shift that occurs upon phosphorylation of two residues, a threonine Thr308 in the catalytic loop and a serine close to the carboxyl terminus Ser473. Phosphorylation of both sites is required for full activation. Phosphorylation of these two residues is dependent on PI3K activity. PI3K enzymes are normally present in the cytosol and can be activated directly by recruitment to an activated receptor tyrosine kinase receptor or indirectly through activated ras. After the binding of lipid, Akt is translocated from the cytoplasm to the inner surface of the plasma membrane, which brings the kinase into close proximity with its activators. The kinases that phosphorylate and activate Akt, the 3-phospho-inositol-dependent protein kinases (PDK1 and PDK2) are themselves regulated by phospholipids.[50] Thus the lipid products generated by PI3K enzymes control the activity of Akt by regulating its location and activation.

Recent papers have shown that Akt is also regulated by the tumor suppressor gene PTEN.[51] Loss of function mutations in this tumor suppressor gene have been shown to dysregulate signal transduction pathways leading from growth factors and their cognate receptor tyrosine kinases to PI3K.

PTEN

PTEN/MMAC1 is a tumor suppressor gene located on chromosome 10q23 that encodes a protein and phospholipid phosphatase, which acts to inhibit Akt and promote apoptosis.[52] PTEN has been implicated in multiple human malignancies including breast,[53] glioblastoma[54] and prostate cancer.[55] Loss of PTEN function can occur through homozygous gene deletion, point mutation or loss of expression. The human PTEN gene encodes a 403-amino-acid polypeptide with a high degree of homology to protein phosphatases. PTEN is capable of dephosphorylating both phosphotyrosine and phosphoserine/threonine-containing residues. This gene functions through dephosphorylation of the PI3-K substrate Ptdins-3,4,5-P_3, thereby preventing activation of Akt.[51] Loss of PTEN function leads to constitutive activation of Akt, which in turn leads to resistance to apoptosis and cell survival.

Targets of Akt and the regulation of apoptosis

With the knowledge that activated Akt is sufficient to confer a resistance to apoptosis in a number of different tumor types, there has been intensive research into how Akt mediates its antiapoptotic effects. To date, several downstream targets of Akt have been identified (**Fig. 5.2**). One of the first to be identified is the Bcl-2 related protein, Bad.

Bad

Akt phosphorylates the pro-apoptotic Bcl-2 family member Bad at Ser136. Bad is thought to induce cell death by heterodimerizing with the antiapoptotic protein Bcl-X_L and the concomitant generation of Bax heterodimers. Phosphorylation of Bad facilitates its binding to the 14-3-3 cytosolic proteins. As a result, Bad is dissociated from Bcl-X_L, allowing this protein to exert its antiapoptotic effects at the level of the mitochondrial membrane.[47]

Caspase 9

Activation of Akt causes phosphorylation of pro-caspase 9 at Ser196, preventing its conversion to active caspase 9.[56] Cytochrome *c* release from the mitochondria serves as a critical step in the activation of downstream caspases. Caspase 9 acts as an 'initiator' caspase that cleaves and activates 'effector' caspases, such as caspase 3. Caspase 3 and 7, ultimately appear to function downstream in the death pathway and are associated with the cleavage of many critical cellular substances. Therefore, phosphorylation and inactivation of caspase 9 by Akt inhibits a critical step in the apoptotic pathway leading to a inhibition of apoptosis and cell survival.

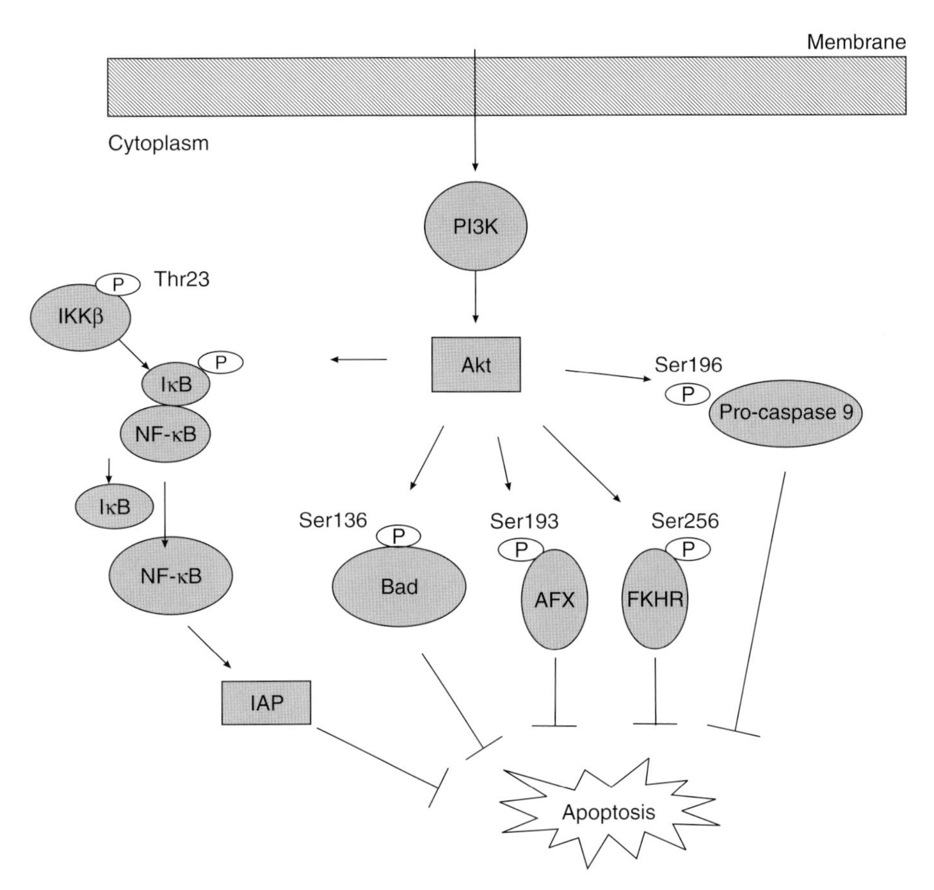

Fig. 5.2. AKT antiapoptotic pathways.

Forkhead
Another antiapoptotic target of Akt identified to date is the Forkhead, or winged helix, family of transcription factors. These are characterized by the presence of a conserved 100-amino-acid DNA binding or forkhead domain. Studies have demonstrated the consistent involvement of Forkhead in chromosomal translocation found in human cancer, suggesting a critical role in the regulation of cellular proliferation and/or differentiation.[57] Among the genes induced by these factors are various pro-death molecules, including Fas ligand. PKB/Akt directly phosphorylates FKHR1 at Thr32 and Ser253 promoting its association with 14-3-3 protein.[58] This phosphorylation negatively regulates FKHR1 by promoting export from the nucleus and transcriptional inactivation. This leads to a decrease in the expression of Fas on the surface of the cell.

NFκB
Another target molecule regulated by Akt is NFκB. NFκB controls the expression of numerous genes involved in inflammation, tumor development and immune responses.[59] NFκB, in unstimulated cells, is retained in the cytoplasm through an interaction with the inhibitory protein IκB. Akt phosphorylates and activates IKK-β, resulting in IκB degradation, which causes translocation of NFκB to the nucleus and transcription activation of target genes, such as the IAPs.[60] IAP promote cell survival by inhibiting caspase 3, 7 and 9.

Akt and prostate cancer
A number of recent papers have investigated the role of PI3-K/Akt in prostate cancer. Her-2/neu has been implicated in the activation of the androgen receptor (AR) and in inducing androgen-resistant prostate cancer cell growth. Her-2/neu is a transmembrane receptor tyrosinse kinase with homology to members of the epidermal growth factor (EGF) family. This tyrosine kinase is known to activate the PI3K/Akt pathway constitutively without extracellular stimulation. A recent paper by Wen et al. demonstrates that Her-2/neu promotes survival and growth in androgen-deprived LNCaP prostate cancer cells via the Akt pathway.[61]

Other studies have also implicated interleukin (IL)-6 as an essential regulator in prostate cancer growth. IL-6 can induce either growth-inhibitory or growth-stimulatory effects, depending on the target cell. IL-6 appears to function as a paracrine growth inhibitor in LNCaP cells, but as an autocrine growth stimulator in PC3 and DU145 cells.[62] IL-6 has been identified as an activator of PI3-K in the human prostate cancer cell lines LNCaP and PC-3, and it is proposed that the anti-apoptotic effect of endogenous IL-6 may be one of the means by which prostate cancer escapes androgen dependence.

Previous studies have investigated the apoptotic activity of the cyclooxygenase-2 (COX-2) inhibitor celecoxib in prostate carcinoma cells. COX-2 is constitutively expressed in androgen-responsive LNCaP and androgen-non-responsive PC-3 cells.

Exposure of these cells to celecoxib induces significant characteristic features of apoptosis, including morphological changes, DNA laddering and caspase-3 activation, whereas piroxicam, a COX-1-specific inhibitor, displays no appreciable effect on either cancer cell line even after prolonged exposure. The underlying mechanism shown is that celecoxib treatment blocks the phosphorylation of Akt. This correlation is supported by studies showing that overexpression of constitutively active Akt protects PC-3 cells from celecoxib-induced apoptosis.

Whilst the PI3K/Akt pathway has been shown to be a dominant growth factor-activated pathway in LNCaP human prostate carcinoma cells, its role in androgen-resistant cells is still under investigation.[63] LNCaP cancer cells express high constitutive levels of Akt, even on serum deprivation, while PC-3 and DU145 express very low levels.[64] This expression of Akt in the different cell lines correlates with defective expression of the tumor suppressor gene PTEN.

PTEN has recently been demonstrated to act as a negative regulator of the PI3-K/Akt pathway in prostate cancer cells. Wu et al. have shown that higher levels of Akt activation are observed in human prostate cancer cell lines (LNCaP/PC3/LAPC-9), and xenografts lacking PTEN/MMAC1 expression when compared with PTEN/MMAC1-positive prostate tumors (DU145/LAPC-4) or normal prostate cells.[65] These higher levels of PTEN correlate with increased angiogenesis and progression to hormone-refractory metastatic disease.[66] Loss of PTEN expression in paraffin-embedded primary prostate cancer specimens has also been demonstrated to correlate with a high Gleason score and advanced stage.[67] Recently Nakamura et al. have shown that members of the Forkhead transcription family, FKHR and FKHRL1, are critical effectors of PTEN-mediated tumor suppression. FKHR and FKHRL1 are aberrantly localized to the cytoplasm and cannot activate transcription in PTEN-deficient cells.[68] Restoration of PTEN function restores these transcription factors to the nucleus where they can exert their function.

PTEN-mutant prostate cancer cells, LNCaP and PC3, show increased expression of integrin-linked kinase (ILK). ILK has been identified as responsible for phosphorylation of the AKT Ser473 position and its subsequent activation.[69] Recently diminished expression of the cell-cycle regulator, p27kip1, has been shown to be associated with increased Akt activity. Akt is thought to inhibit AFX/Forkhead-mediated transcription of p27kip1 directly. Decreased p27kip1 expression has been repeatedly associated with prostate cancer progression.[70]

Nesterov et al. have demonstrated that LNCaP cells are highly resistant to TRAIL (tumor-necrosis factor-related apoptosis-inducing ligand).[71] These cells express highest constitutively activated Akt, which appears to block TRAIL-induced apoptosis at the level of BID cleavage. By inhibiting the PI3-K/Akt pathway with LY294002 (a specific PI3K inhibitor), it was possible to sensitize these cells to the effects of TRAIL and induce cell death, thus overcoming the antiapoptotic effects of PI3-K/Akt and inducing cell death.

The role of PI3-K/Akt signaling pathways in prostate cancer is presently an area of intensive scientific research. As knowledge of their involvement in cancer growth and survival increases, these pathways appear to be an increasingly attractive target for drug development.

PATHWAYS LEADING TO APOPTOSIS

Caspases

The main cellular regulators of the cell death pathway are the caspases, which are a family of cysteine-dependent death proteases.[72,73] They contain a variable pro-domain and a large (~17–20 kDa) and small (~10 kDa) functional enzyme subunit separated, by an interlinker region flanked by Asp residues.[74] As with most zymogen forms, generation to the active form requires limited proteolysis. As caspases are capable of cleaving at Asp residues, certain caspases once activated can sequentially cleave either their own or other zymogen forms in a cascade manner. Inspection of the internal Asp target sites demonstrate that caspases preferentially activate others in a hierarchical manner.[75] The importance of caspase pro-domains in the regulation of caspase activity has recently become evident by the recognition of a family of adapter inhibitory apoptotic proteins capable of binding the caspases and inhibiting their proteolytic activity.[76]

Caspases implicated in cell death can be loosely classified into two groups, initiators and executioners. Depending on the size of the pro-domains, caspases can be further classified into two functional subgroups. Caspases-2, 8 and 10 have long pro-domains, whereas caspases-3, 6, 7 and 9 have short domains. The size of the pro-domain has functional importance. Caspases with the longer pro-domains are implicated in targeting and regulating activation, while caspases with shorter domains appear to function more downstream in the apoptotic pathway cleaving critical substrates. Cytochrome *c* release from the mitochondria serves as a critical step in the activation of the downstream caspases, such as caspases-9.[77] Activation of caspase-9 in turn leads to the activation of caspase-3. Caspase-3 and caspase-7 appear to function ultimately downstream in the death pathway, and are associated with the cleavage of many critical cellular substrates.

Caspase substrates: their significance in the apoptotic phenotype

Caspases cleave both structural proteins involved in cell architecture and functional proteins and those involved in cell-cycle regulation and DNA repair.[78] The inactivation of DNA repair enzymes, such as poly (ADP-ribose) polymerase (PARP) and activation of a caspase-activated DNase (CAD), serve as crucial events in the commitment of the cell to undergo apoptosis.[79] Likewise, caspase-3-mediated cleavage of p21/Waf1 has been shown to convert cancer cells from growth arrest to apoptosis, leading to the acceleration of chemotherapy-induced apoptotic process in these cancer cells.[80] In many instances caspase cleavage generates a constitutive kinase activity responsible for transduction of the apoptotic signal.[81] Identification of these kinases, in particular their deregulation, allows for novel targets for gene therapy in prostate cancer.[82] Certain antiapoptotic proteins of the Bcl-2 family, when cleaved by the caspases, serve to amplify the apoptotic signal.[83] Caspase-8 cleavage of the antiapoptotic protein Bid into an active carboxyl fragment has been shown to induce the release of cytochrome *c* from the mitochondria.[84]

Caspases: their relevance in prostate cell apoptosis

The role the caspases play in the normal process of prostatic glandular self-renewal and in the etiology of prostate cancer is becoming a major area of prostate cancer research. An understanding as to how deregulation of the caspases contributes to cancer development, is likely to help in therapeutic strategies aimed at preventing progression to the androgen-insensitive state.

Immunohistochemistry of the prostate has demonstrated the expression of caspase-3 and caspase-1, both of which correlate to cell type and physiology.[85] Staining for caspase-3 shows low basal cell expression compared to glandular secretory cells. High expression of caspase-3 in the luminal secretory cells correlates to a high cell turnover, explaining how a steady-state equilibrium process of glandular self-renewal is maintained. Recent studies have also demonstrated decreased protein expression of caspase 3[86,87] and caspase 1[87] with increased Gleason grade.

Caspases have also been shown to be involved during prostate regression following castration. Prostate tissue sections obtained from castrated mice and rats stain specifically for processed caspases.[88] Likewise inhibition of the caspases by the over-expression of CrmA, a viral inhibitor of caspase activity, prevents androgen-withdrawal-induced apoptosis *in vitro*.[89] The continued growth of androgen-insensitive tumors after androgen withdrawal may, therefore, result from altered apoptotic signaling, mediated by caspase inhibition. Likewise, the potential for prostate cancer to metastasize may involve caspase inhibitory mechanisms. Using a non-metastatic variant of the LNCaP prostate cancer cell line (LNCaP-Pro5), it was found that these cells were more sensitive to apoptotic induction than their metastatic counterparts.[90] Such studies imply, at least *in vitro*, that apoptotic resistance (caspase inhibition) increases with the metastatic potential of human prostate cancer cell lines.

These studies would implicate a role for the therapeutic manipulation of the caspases in prostate cancer, either through their increased activation or a decrease in their inhibition via the expression of the IAP or activation of the AKT pathway.

CONCLUSION

Understanding the cellular mechanisms that regulate the cell-death pathway has important implications in the detection and treatment of prostate cancer as well as other related diseases. Upstream regulators of the apoptotic pathway, such as Bcl-2, have proved useful in our understanding of apoptotic resistance and antisense treatment strategies have been developed, which may prove useful in the future. As we explore further into the apoptotic pathway, we begin to learn that this is a completed pathway with many regulatory mechanisms. This chapter has reviewed some of the most recent proteins of interest in the regulation of prostate cancer resistance to treatment and will represent future targets for manipulation in the future.

REFERENCES

1. Kerr JFR, Wyllie AH, Currie AR. Apoptosis: a basic biological phenomenon with wide-ranging implications in tissue kinetics. *Br. J. Cancer* 1972; 26:239–57.
2. Carson DA, Ribeiro JM. Apoptosis and disease. *Lancet* 1993; 341:1251–4.
3. Kerr JFR, Winterford CM, Harmon BV. Apoptosis: its significance in cancer and cancer therapy. *Cancer* 1994; 73:2013–26.
4. Gerschenson LE, Rotello RJ. Apoptosis: a different type of cell death. *FASEB J.* 1992; 6:2450–5.
5. Cohen JJ. Apoptosis. *Immunol. Today* 1993; 14:126–30.
6. Savill JS, Wyllie AH, Henson JE et al. Macrophage phagocytosis of aging neutrophils in inflammation. Programmed cell death in the neutrophil leads to its recognition by macrophages. *J. Clin. Invest.* 1989; 83:865–75.
7. Denmeade SR, Lin XS, Isaacs JT. Role of programmed (Apoptotic) cell death during the progression and therapy for prostate cancer. *Prostate* 1996; 28:251–65.
8. Berges RR, Vukanovic J, Epstein JI. Implication of the cell kinetic changes during the progression of human prostatic cancer. *Clin. Cancer Res.* 1995; 1:473–80.
9. Roy N, Mahadervan MS, Melean M et al. The gene for neuronal apoptosis inhibitory protein is partially deleted in individuals with spinal muscular atrophy. *Cell* 1995; 80:167–78.
10. Biros I, Forrest S. Spinal muscular atrophy: untying the knot? *J. Med. Genet.* 1999; 31:1–8.
11. Iannaccone ST. Spinal muscular atrophy. *Semin. Neurol.* 1998; 18:19–26.
12. Liston P, Roy N, Tamai K et al. Suppression of apoptosis in mammalian cells by NAIP and a related family of IAP genes. *Nature* 1996; 379:349–53.
13. Miller LK. An exegesis of IAPs: salvation and surprises from BIR motifs. *Trends Cell Biol.* 1999; 9:323–8.
14. Takahashi R, Deveraux Q, Tamm I et al. A single BIR domain of XIAP sufficient for inhibiting caspases. *J. Biol. Chem.* 1998; 273:7787–90.
15. Li F, Ambrosini G, Chu EY et al. Control of apoptosis and mitotic spindle check point by survivin. *Nature* 1998; 396:580–4.
16. Huang Q, Deveraux QL, Maeda S, Salvesen GS et al. Evolutionary conservation of apoptosis mechanisms: lepidopteran and baculoviral inhibitor of apoptosis proteins are inhibitors of mammalian caspase-9. *Proc. Natl Acad. Sci. USA* 2000; 97:1427–32.
17. Tamm I, Wang Y, Sausville E et al. IAP-Family protein survivin inhibits caspase activity and apoptosis induced by Fas (CD95), Bax, Caspases, and anticancer drugs. *Cancer Res.* 1998; 58:5315–20.
18. Suzuki A, Tsutomi Y, Akahane K et al. Resistance to Fas-mediated apoptosis: activation of caspase 3 is regulated by cell cycle regulator $p21^{WAF1}$ and IAP gene family ILP. *Oncogene* 1998; 17:931–9.
19. Lee R, Collins T. Nuclear factor-kappa B and cell survival: IAP call for support. *Circ. Res.* 2001; 88:262–4.
20. Chu ZL, McKinsey TA, Liu L et al. Suppression of tumour necrosis factor-induced cell death by inhibitor of apoptosis c-IAP-2 is under NF-kappa B control. *Proc. Natl Acad. Sci.* 1997; 94:10057–62.
21. Stehlik C, de Martin R, Kumabashiri I et al. Nuclear factor (NF)-κB-regulated X-chromosome-linked iap gene expression protects endothelial cells from tumour necrosis factor α-induced apoptosis. *J. Exp. Med.* 1998; 188:211–16.
22. Manji GA, Hozak RR, LaCount DJ et al. Baculovirus inhibitor of apoptosis functions at or upstream of the apoptotic suppressor P35 to prevent programmed cell death. *J. Virol.* 1997; 71:4509–16.

23. Roy N, Deveraux QL, Takahashi R et al. The c-IAP-1 and c-IAP-2 proteins are direct inhibitors of specific caspases. *EMBO J.* 1997; 16:6914–25.
24. Deveraux Q, Roy N, Stennicke HR et al. IAPs block apoptotic events induced by caspase-8 and cytochrome *c* by direct inhibition of distinct caspases. *EMBO J.* 1998; 17:2215–23.
25. Wang C-Y, Mayo MW, Kornduk RG et al. NF-κB antiapoptosis: induction of TRAF1 and TRAF2 and c-IAP-1 and c-IAP-2 to suppress caspase 8 activation. *Science* 1998; 281:1680–3.
26. McCarthy JV, Dixit VM. RIP2 is a novel NF-KappaB-activating and cell death-inducing kinase. *J. Biol. Chem.* 1998; 273:16968–75.
27. Yamaguchi K, Nagai S, Ninomiya-Tsuji J et al. XIAP, a cellular member of the inhibitor of apoptosis protein family, links the receptors to TAB1-TAK1 in the BMP signalling pathway. *EMBO J.* 1999; 18:179–87.
28. Erl W, Hansson GK, de Martin R et al. Nuclear factor-kappa B regulates induction of apoptosis and inhibitor of apoptosis protein-1 expression in vascular smooth muscle cells. *Circ. Res.* 1999; 84:668-77.
29. Stehlik C, de Martin R, Binder BR et al. Cytokine induced expression of porcine inhibitor of apoptosis protein (iap) family member is regulated by NF-kappa. *Biochem. Biophys. Res. Commun.* 1998; 243:827–32.
30. Sumitomo M, Tachibana M, Nakashima J et al. An essential role for nuclear factor kappa B in preventing TNF-α-induced cell death in prostate cancer cells. *J. Urol.* 1999; 161:674–9.
31. Davis JN, Kucuk O, Sarkar FH. Genistein inhibits NF-κB activation in prostate cancer cells. *Nutr. Cancer* 1999; 35:167–74.
32. Wang C-Y, Guttridge C, Mayo MW et al. NF-kappaB induces expression of the Bcl-2 homologue A1/Bfl-1 to preferentially suppress chemotherapy-induced apoptosis. *Mol. Cell Biol.* 1999; 19:5923–9.
33. Chen C, Edelstein LS, Gilinas C. The Rel/NF-kappa B family directly activates expression of the apoptosis inhibitor Bcl-X_L. *Mol. Cell Biol.* 2000; 8:2687–95.
34. Yang Y, Fang S, Jensen JP et al. Ubiquitin protein ligase activity of IAPs and their degradation in proteasomes in response to apoptotic stimuli. *Science* 2000; 288:874–7.
35. Johnson DE, Gastman BR, Wieckowshi E et al. Inhibitor of apoptosis protein hILP undergoes caspase-mediated cleavage during T lymphocyte apoptosis. *Cancer Res.* 2000; 60:1818–23.
36. Ambrosini G, Adida C, Altieri D. A novel anti-apoptosis gene, survivin, expressed in cancer and lymphoma. *Nature Med.* 1997; 3:917–21.
37. Tanaka K, Iwamoto S, Gon G et al. Expression of survivin and its relationship to loss of apoptosis in breast carcinomas. *Clin. Cancer Res.* 2000; 6:127–34.
38. Lu C-D, Altieri DC, Tanigawa N. Expression of a novel anti-apoptosis gene, survivin, correlated with tumour cell apoptosis and p53 accumulation in gastric carcinomas. *Cancer Res.* 1998; 58:1808–12.
39. Kawasaki H, Alterieri DC, Lu CD et al. Inhibition of apoptosis by survivin predicts shorter survival rates in colorectal cancer. *Cancer Res.* 1998; 58:5071–4.
40. Wagenknecht B, Glaser T, Naumann U et al. Expression and biological activity of X-linked inhibitor of apoptosis (XIAP) in human malignant glioma. *Cell Death Diff.* 1999; 6:370–6.
41. Hiromitsu M. Expression of the IAP (inhibitor of apoptosis protein) gene family in prostate cancer. *Eur. Urol.* 2000; 37(Suppl. 2):104.
42. McEleny K, Watson RWG, O'Neill A et al. The inhibitors of apoptosis proteins in prostate cancer cell lines. *Prostate* 2002; 51:131.
43. Cohen P. Serum insulin-like growth factor-I levels and prostate cancer risk – interpreting the evidence. *J. Natl Cancer Inst.* 1998; 90:876–9.
44. Koutsilieris M, Tzanela M, Dimopoulos T. Novel concept of antisurvival factor (ASF) therapy produces an objective clinical response in four patients with hormone-refractory prostate cancer: case report. *Prostate* 1999; 38:313–16.
45. Coffer PJ, Jin J, Woodgett JR. Protein kinase B (c-Akt): a multifunctional mediator of phosphatidylinositol 3-kinase activation. *Biochem. J.* 1998; 335:1–13.
46. Cross TG, Scheel-Toellner D, Henriquez NV et al. Serine/threonine protein kinases and apoptosis. *Exp. Cell Res.* 2001; 256:34–41.
47. Datta SR, Dudek H, Tao X. Akt phosphorylation of BAD couples survival signals to the cell intrinsic death machinery. *Cell* 1997; 91:231–41.
48. Ostman A, Bohmer FD. Regulation of receptor tyrosine kinase signaling by protein tyrosine phosphatases. *Trends Cell Biol.* 2001; 11:258–66.
49. Kandel ES, Hay N. The regulation and activities of the multifunctional serine/threonine kinase Akt/PKB. *Exp. Cell Res.* 1999; 253:210–29.
50. Alessi DR, James SR, Downes CP et al. Characterization of a 3-phosphoinositide-dependent protein kinase which phosphorylates and activates protein kinase B alpha. *Curr. Biol.* 1997; 7:261–9.
51. Li J, Simpson L, Takahashi M et al. The PTEN/MMAC1 tumor suppressor induces cell death that is rescued by the AKT/protein kinase B oncogene. *Cancer Res.* 1998; 58:5667–72.
52. Simpson L, Parsons R. PTEN: life as a tumor suppressor. *Exp. Cell Res.* 2001; 264:29–41.
53. Ghosh AK, Grigorieva I, Steele R et al. PTEN transcriptionally modulates c-myc gene expression in human breast carcinoma cells and is involved in cell growth regulation. *Gene* 1999; 235:85–91.
54. Wick W, Furnari FB, Naumann U et al. PTEN gene transfer in human malignant glioma: sensitization to irradiation and CD95L-induced apoptosis. *Oncogene* 1999; 18:3936–43.
55. Davies MA, Koul D, Dhesi H et al. Regulation of Akt/PKB activity, cellular growth, and apoptosis in prostate carcinoma cells by MMAC/PTEN. *Cancer Res.* 1999; 59:2551–6.
56. Cardone MH, Roy N, Stennicke HR et al. Regulation of cell death protease caspase-9 by phosphorylation. *Science* 1998; 282:1318–21.
57. Tang ED, Nunez G, Barr FG et al. Negative regulation of the forkhead transcription factor FKHR by Akt. *J. Biol. Chem.* 1999; 274:16741–6.
58. Brunet A, Bonni A, Zigmond MJ et al. Akt promotes cell survival by phosphorylating and inhibiting a Forkhead transcription factor. *Cell* 1999; 96:857–68.
59. Yamamoto Y, Gaynor RB. Therapeutic potential of inhibition of the NF-kappaB pathway in the treatment of inflammation and cancer. *J. Clin. Invest.* 2001; 107:135–42.
60. Romashkova JA, Makarov SS. NF-kappaB is a target of AKT in anti-apoptotic PDGF signalling. *Nature* 1999; 401:86–90.
61. Wen Y, Hu MC, Makino K et al. HER-2/neu promotes androgen-independent survival and growth of prostate cancer cells through the Akt pathway. *Cancer Res.* 2000; 60:6841–5.
62. Chung TD, Yu JJ, Spiotto MT et al. Characterization of the role of IL-6 in the progression of prostate cancer. *Prostate* 1999; 38:199–207.
63. Lee C, Sintich SM, Mathews EP et al. Transforming growth factor-beta in benign and malignant prostate. *Prostate* 1999; 39:285–90.
64. Page C, Huang M, Jin X et al. Elevated phosphorylation of AKT and Stat3 in prostate, breast, and cervical cancer cells. *Int. J. Oncol.* 2000; 17:23–8.
65. Wu X, Senechal K, Neshat MS et al. The PTEN/MMAC1 tumor suppressor phosphatase functions as a negative regulator of the phosphoinositide 3-kinase/Akt pathway. *Proc. Natl Acad. Sci. USA* 1998; 95:15587–91.
66. Giri D, Ittmann M. Inactivation of the PTEN tumor suppressor gene is associated with increased angiogenesis in clinically localized prostate carcinoma. *Hum. Pathol.* 1999; 30:419–25.

67. McMenamin ME, Soung P, Perera S et al. Loss of PTEN expression in paraffin-embedded primary prostate cancer correlates with high Gleason score and advanced stage. *Cancer Res.* 1999; 59:4291–6.
68. Nakamura N, Ramaswamy S, Vazquez F et al. Forkhead transcription factors are critical effectors of cell death and cell cycle arrest downstream of PTEN. *Mol. Cell Biol.* 2000; 20:8969–82.
69. Persad S, Attwell S, Gray V et al. Inhibition of integrin-linked kinase (ILK) suppresses activation of protein kinase B/Akt and induces cell cycle arrest and apoptosis of PTEN-mutant prostate cancer cells. *Proc. Natl Acad. Sci. USA* 2000; 97:3207–12.
70. Graff JR, Konicek BW, McNulty AM et al. Increased AKT activity contributes to prostate cancer progression by dramatically accelerating prostate tumor growth and diminishing p27Kip1 expression. *J. Biol. Chem.* 2000; 275:24500–5.
71. Nesterov A, Lu X, Johnson M et al. Elevated akt activity protects the prostate cancer cell line lncap from trail-induced apoptosis. *J. Biol. Chem.* 2001; 276:10767–74.
72. Kidd VJ. Proteolytic activities that mediate apoptosis. *Annu. Rev. Physiol.* 1998; 60:533–73.
73. Martin SJ, Green DR. Protease activation during apoptosis: death by a thousand cuts. *Cell* 1995; 82:349–52.
74. Stennicke HR, Salvesen GS. Properties of the caspases. *Biochim. Biophys. Acta* 1998; 1387:17–31.
75. Salvesen GS. Caspases: opening the boxes and interpreting the arrows. *Cell Death Diff.* 2002; 9:3–5.
76. Clem RJ, Duckett CS. The iap genes: unique arbitrators of cell death. *Trends Cell Biol.* 1997; 7:337–45.
77. Zhivotovsky B, Hanson KP, Orrenius S. Back to the future: the role of cytochrome c in cell death. *Cell Death Diff.* 1998; 5:459–60.
78. Stroh C, Schulze-Osthoff K. Death by a thousand cuts: an ever increasing list of caspase substrates. *Cell Death Diff.* 1998; 5:997–1000.
79. Kaufman S, Desnoyers S, Ottaviano Y et al. Specific proteolytic cleavage of poly(ADP-ribose) polymerase: an early marker of chemotherapy-induced apoptosis. *Cancer Res.* 1993; 53:3976–82.
80. Zhang C, Fujita N, Tsuruo T. Caspase-mediated cleavage of p21Waf1/cip1 converts cancer cells from growth arrest to undergoing apoptosis. *Oncogene* 1999; 18:1131–7.
81. Lee N, MacDonald H, Reinhard C et al. Activation of hPAK65 by caspase cleavage induces some of the morphological and biochemical changes of apoptosis. *Proc. Natl Acad. Sci. USA* 1997; 94:13642–7.
82. Cornford P, Evans J, Dodson A et al. Protein kinase C isoenzyme patterns characteristically modulated in early prostate cancer. *Am. J. Pathol.* 1999; 154:137–44.
83. Cheng E, Kirsch D, Clem R et al. Conversion of Bcl-2 to a Bax-like death effector by caspases. *Science* 1997; 278:1966–8.
84. Li H, Zou H, Slaughter C et al. Cleavage of BID by caspase 8 mediates the mitochondrial damage in the Fas pathway of apoptosis. *Cell* 1998; 94:491–8.
85. Krajewska M, Wang HG, Krajewski S et al. Immunohistochemical analysis of in vivo patterns of expression of CPP32 (Caspase-3), a cell death protease. *Cancer Res.* 1997; 57:1605–13.
86. O'Neill A, Coffey RNT, Hegarty NJ et al. Caspase 3 expression in benign prostatic hyperplasia and prostate carcinoma. *Prostate* 2001; 47:183–8.
87. Winter RN, Kramer A, Borkowski A et al. Loss of caspase-1 and caspase-3 protein expression in human prostate cancer. *Cancer Res.* 2001; 61:1227–32.
88. Marti A, Jaggi R, Vallan C et al. Physiological apoptosis in hormone-dependent tissues: involvement of caspases. *Cell Death Diff.* 1999; 6:1190–200.
89. Srikanth S, Kraft AS. Inhibition of caspases by cytokine response modifier A blocks androgen ablation-mediated prostate cancer cell death in vivo. *Cancer Res.* 1998; 58:834–9.
90. McConkey DJ, Greene G, Pettaway CA. Apoptosis resistance increases with metastatic potential in cells of the human LNCaP prostate carcinoma line. *Cancer Res.* 1996; 56:5594–9.

Prostate Cancer with Other Primary Malignancies

Jack H. Mydlo and Matthew Karlovsky

Department of Urology, Temple University School of Medicine, Philadelphia, PA, USA

The American Cancer Society has recently stated that one out of five Americans will develop cancer in his or her lifetime. In addition, for those patients who develop a tumor, there is a one out of three chance of developing a second tumor in their lifetime. This suggests that there may be an underlying problem with the person's immune surveillance system, or something inherently wrong with their genetic self-regulation and/or tumor suppressor genes.[1,2]

Within the sample of patients who have more than one cancer in their lifetime, there is a subset of patients who present for the diagnosis and treatment of one tumor, and a second primary tumor is detected during the work-up of the first primary tumor. Multiple primary malignancies are defined as those tumors that present themselves as distinct primary entities, and are themselves not metastases of the other lesion.[3]

This finding of two simultaneous primary tumors is a particularly unusual and sometimes difficult dilemma. Treatment will usually be initially determined by the more aggressive lesion, age and medical condition of the patient.[4]

Among the secondary tumors that occur with a higher incidence in cancer patients are melanoma and lung cancer, and lymphoma and renal cell carcinoma.[5,6] Greenberg et al. reported an increased incidence of bladder cancer in those patients with prostate cancer.[7] Liskow et al. also reported an increase in bladder cancer in those patients with prostate cancer. They also found an increase in lymphoma in these patients as well.[8] However, Kawakami et al. found no increase in multiple primary malignancies in prostate cancer patients,[9] which has also been shown by Isaacs et al.[10] Furthermore, in a large Swiss study by Levi et al. examining 4503 cases, they actually found a reduced incidence of neoplasms in men who were diagnosed with prostate cancer. In addition, they did not find an association between cigarette smoking and prostate cancer.[11]

Brenner et al. reported a small but statistically significant risk of developing secondary tumors, most commonly bladder cancer, after radiotherapy for prostate cancer. This was especially noted in the earlier days of radiation therapy, prior to today's more focused radiotherapy ports.[12]

In a retrospective chart review of 2339 patients with prostate cancer, Moyer et al. reported 222 on patients (9.5%) who developed a second primary malignancy. Sixty-nine of these 222 patients (31%) presented with a synchronous lesion, while the other 153 patients presented with metachronous lesions.[13] They found that, of the 153 patients with metachronous lesions, 86 (56%) were diagnosed with prostate cancer first before their other tumor, whereas 67 (44%) were diagnosed with prostate cancer after the diagnosis of a prior primary tumor. When stratified among race, the most common second malignancies among Whites with localized prostate cancer were bladder and colon cancer. However, when stratified for stage, bladder cancer was much more common in Whites with localized prostate cancer, whereas colon cancer was more common in Whites with metastatic prostate cancer. In Black patients with localized prostate cancer, they found a greater incidence of colon and bladder cancers; for those with metastatic disease, they found a greater incidence of lung cancer. These findings were statistically significant with regard to observed versus expected cases of multiple malignancies, both in the general population, and when stratified by race.[11] Also, second primary malignancies were more likely to be detected in White patients.

Overall, White patients developed more cases of colon and bladder carcinoma while Black patients developed more cases of lung carcinoma. With respect to race and prostate cancer stage in their study, Whites with localized and metastatic disease were more likely to develop bladder and colon carcinoma, respectively, while Blacks were more likely to develop bladder and colon carcinoma when localized and lung carcinoma when metastatic, suggesting different risks depending upon patient race and stage of prostate cancer.

Moyer et al. concluded that Black patients were more likely to present with metastatic disease compared to Whites. Also, when prostate cancer was diagnosed first, a statistically significant increase was seen in the number of colon, bladder and kidney cancers. However, when race was considered, they discovered a statistically significant increase in the number of colon and bladder cancer cases in Whites, and lung and esophageal cases in Blacks. Their data suggested that patient race and initial stage of prostate cancer may influence the likelihood and site of a second primary malignancy.[13]

Various other studies have found the incidence of multiple primary tumors in prostate cancer patients to range from 11.5% in a series by Liskow et al.[8] to 20% in a series by Lynch et al.[14] and even to 27% in autopsy series by Hadju and Hadju.[15] The exact etiology of this incidence is uncertain. It may be related to the overall increased awareness of prostate cancer in the general population, and the medical community, especially since the advent of prostate specific antigen (PSA) screening. The earlier diagnosis of prostate cancer has led to its detection prior to any potential development of a second malignancy.

Considering race and the detection of a second primary malignancy, the data of Moyer et al. also revealed that 147 out of 1240 (11.9%) of White patients had a second primary detected, while only 69 out of 1035 (6.7%) of Blacks patients had detection of a second primary.[13] This contradicts several previous studies. The study by Hadju and Hadju noted a higher incidence of second primary malignancies in Blacks over Whites.[15] Studies by Liskow[8] and Paulish[16] found a similar incidence of second primaries in both Blacks (approx. 11%) and Whites (approx. 11%). The etiology behind this difference between races is not known. One possible cause could have been that certain factors (dietary, environmental, genetic) were present in the White population and not in the Black population.[17] Another possible reason could have been that the White patients underwent better and longer follow-up compared to the Black patients. Also, smoking may be more common among certain Black populations than their White cohorts.

Another possible reason that certain studies have demonstrated more aggressive tumors in Blacks than Whites, or more metastatic lesions, may be at the hormonal and molecular level. This has been reported by Mydlo et al., who demonstrated a greater expression of mutant p53 expression, higher levels of serum testosterone and a higher incidence of positive pathologic margins in Black patients versus White patients with prostate cancer.[18,19] Adding to the notion of increased aggressive tumor behavior may have been the relative decreased prostate cancer awareness earlier on in the Black community. The percentage of advanced prostate cancer might have been decreased, if these patients underwent earlier routine screening.

As mentioned previously, one of the main reasons for the increased detection of prostate cancer, either as a primary tumor, or as a secondary tumor, has been the use of increased PSA testing. However, along with the increased utilization of sonography, computed tomography (CT) scanning and magnetic resonance of patients, it may be possible to detect secondary or multiple tumors in cancer patients, either at initial presentation or at follow-up.[12] In addition, because the specialist may focus more closely on the involved organ system, it is reasonable to expect the urologist to detect prostate cancer and synchronous bladder or renal cancers with an increased frequency, or the gynecologist to detect cervical and ovarian cancers with increased frequency. The detection of other primary malignancies outside the organ system of specialty concern may also be 'unmasked' during preoperative evaluation.

Mydlo et al. described patients with urological cancers and malignancies of non-urologic origin.[4] Over a 6-year period, 515 patients were evaluated for a urological malignancy. Twenty-three (4.3%) were found to have multiple tumors. Three of these patients had concomitant prostate and bladder cancer. One developed bladder cancer after external-beam radiotherapy for his prostate cancer.

It is important to note that those patients with multiple primary malignancies had been categorized differently from those patients who developed second primaries as a result of chemotherapy and/or radiotherapy of the primary tumor.[20,21] This would include those patients who develop bladder cancer years after they had treatment for prostate cancer by external beam radiation therapy. The other 19 patients (3.5%) were found to also have a non-urological primary malignancy.

Thirteen patients were found to have a second primary malignancy during the work-up of their urological tumor. All the prostate cancer patients were diagnosed after scanning PSA's were elevated. All the renal tumor patients had their tumors detected from a hematuria work-up, except for one. This one patient had her renal tumor detected incidentally during her colon cancer work-up (**Fig. 6.1a,b**). One bladder cancer patient had her colon tumor diagnosed from a preoperative CT scan. The other urological cancer patients had gynecological tumors detected by physical examination, CT examination and cervical cytology preparations. Four prostate cancer patients had heme-positive stools; this led to a diagnosis of colon cancer. The last patient had renal cancer as well (patient 19, **Table 6.1**).

In the study by Mydlo et al., one patient was referred who was previously diagnosed with two primary malignancies (gastric cancer was detected first and then his prostate cancer was detected from a screening PSA). Another patient with prostate cancer had a lung mass detected on his preoperative chest X-ray, which revealed primary adenocarcinoma of the lung. After undergoing a successful wedge resection, the patient underwent a radical retropubic prostatectomy without incident. Another prostate cancer patient was treated for his prostate tumor and had a sigmoid tumor resection simultaneously. The latter was detected during a digital rectal examination for his prostate, which was positive for occult blood (**Table 6.1**).

While some surgeons may be reluctant to combine such complicated surgeries together, which would overlie suture lines, these patients were found to have no more complications than the general population.[4]

Many patients with presumed localized prostate cancer in the USA with PSA values below 10 do not undergo nuclear bone scans or abdominal CT scans, especially if surgery is planned. This is because numerous studies have demonstrated the low incidence of positive bone scans for PSA levels below this value. However, for those patients undergoing either brachytherapy or external beam radiation therapy, these imaging modalities may be used for volume assessment of the gland and may, therefore, detect additional pathology.[20]

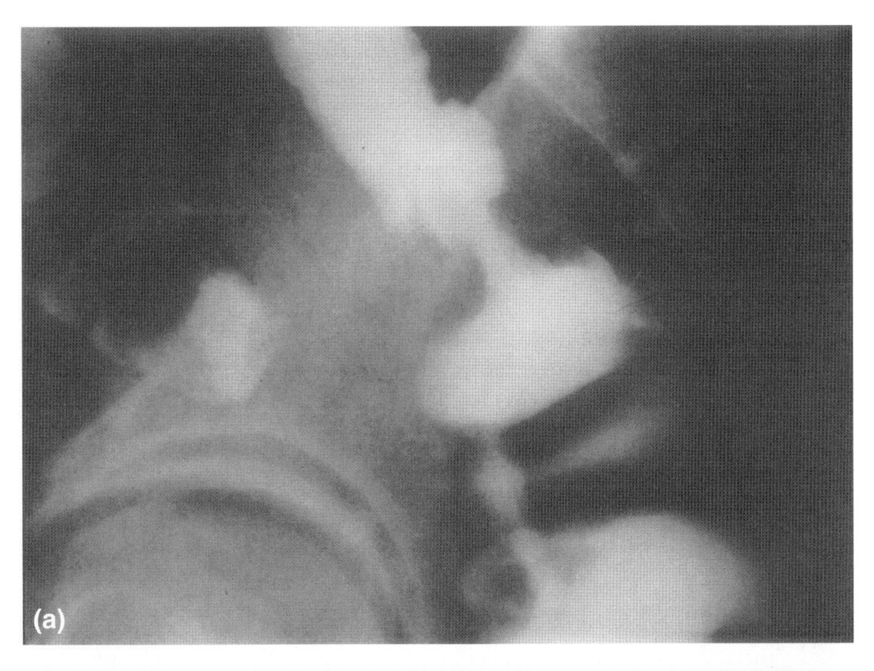

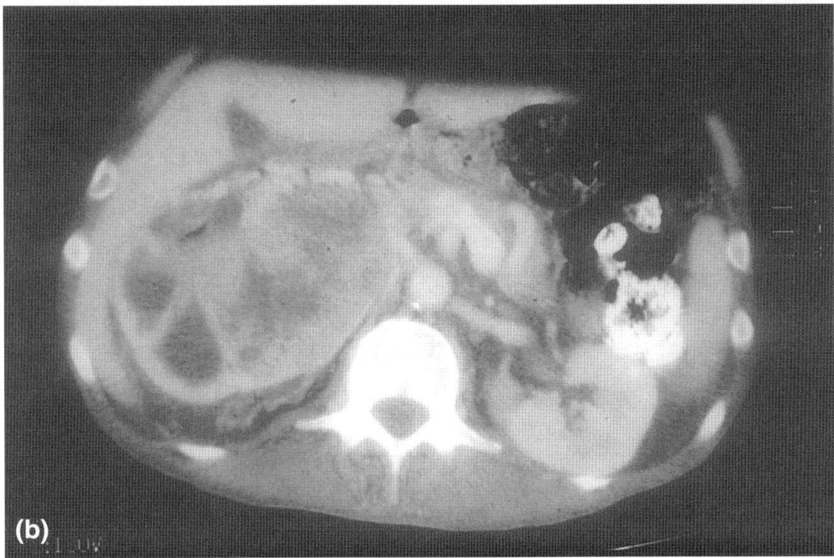

Fig. 6.1. (a) A 63-year-old male (patient # 19) underwent a prostate examination for an elevated PSA. His stool was positive for occult blood and a barium enema revealed an 'apple core' lesion, consistent with colon cancer. Further work-up with a CT scan revealed a large right renal tumor. (b) His colon and renal tumors were resected first in one operation, and then the patient underwent subsequent brachytherapy for his prostate cancer.

However, Miller et al. reported a low incidence of coexistent disease in pretreatment CT scanning for radiation therapy in patients with prostate carcinoma.[20] Only two patients out of 77 patients (3%) had any major abnormalities that required intervention, and these were not tumors. Nakata et al. analyzed the clinical features of multiple primary cancers, which included prostate cancer. They found an incidence of 15.2% of multiple primary malignancies in their review. The organ most commonly involved was the stomach, followed by the bladder, colon and lungs.[21]

Other reports described the incidental pathological findings of prostate cancer along with bladder cancer after radical cystoprostatectomies. Moutzouris et al. reported a 27% incidence of finding prostate cancer in cystectomy specimens. These secondary tumors occurred most likely at the apex.[22]

Cannon et al. reported on the increased occurrence of primary cancers in association with multiple myeloma and Kaposi's sarcoma.[23] With the advent of effective pharmacotherapy for erectile dysfunction in the elderly, there is speculation that there may be a corresponding increase in the incidence of sexually transmitted diseases among this age group, including the human immunodeficiency virus, or HIV.[24] This may have many ramifications in the future in the treatment of patients with prostate cancer.

Hepatitis C has been considered a high-risk factor for secondary primary malignancies besides hepatocellular carcinoma.

Table 6.1. Nineteen patients with urologic cancer and another primary malignancy. Reprinted from *Urology*, Vol. 58, Mydlo and Gerstein, Urologic cancer patients with another primary malignancy, pp. 243–7. Copyright (2001), with permission from Elsevier Science.

Patient	Age/Sex	Urologic tumor	Treatment	Stage	Other tumor	Treament
1	68 M	Prostate	Prostatectomy	pT2aNo	Lung	Wedge
2	65 M	Prostate	Prostatectomy	pT2bNo*	Lymphoma	Chemotherapy
3	67 F	Renal	Nephrectomy	pT1No	Colon	Resection
4	70 M	Renal	Nephrectomy	pT2No	Colon/thyroid	Resection
5	63 F	Renal	Nephrectomy	pT2No	Uterine	Hysterectomy
6	72 M	Prostate	Brachytherapy		Colon	Resection†
7	57 M	Prostate	Prostatectomy	pT2bNo	Esophageal*	Resection
8	63 F	Bladder	Cystectomy	pT3aNo	Pancreas*	(Advanced)
9	59 F	Bladder	Cystectomy	pT2bNo	Colon	Resection
10	69 F	Renal	Nephrectomy	pT2No	Colon	Resection
11	71 F	Bladder	BCG		Cervical	Resection
12	61 M	Prostate	Prostatectomy	pT2bNo	Gastric	Resection†
13	63 M	Prostate	Brachytherapy		Colon	Resection†
14	67 M	Prostate	Prostatectomy	pT2bNo	Sigmoid	Resection
15	52 F	Renal	Nephrectomy	pT1No	Breast	Lumpectomy
16	61 M	Prostate	Prostatectomy	pT2bNo	Lymphoma	Chemotherapy
17	71 M	Prostate	Prostatectomy	pT2bNo	Colon	Colectomy
18	54 F	Bladder	Cystectomy	pT2No	Melanoma	Resection
19	70 M	Prostate	Brachytherapy	pT2No	Colon/renal	Resection

*Second tumors appeared within 1 year of primary urological cancer surgery.
†Gastric resection performed before prostatectomy, and colectomy performed before brachytherapy.

Bruno et al. reported on the development of renal, breast and prostate carcinoma in patients with a history of hepatitis C who developed hepatocellular carcinoma.[25]

There are several studies that looked at the incidence of multiple primary malignancies in patients with renal cancer. Although this may appear to be beyond the scope of this particular chapter, it may be of interest to the urological readership and, therefore, will be addressed somewhat for completeness.

Tihan and Filippa reported an increased coexistence of renal cell carcinoma and malignant lymphoma.[26] In their report, all the nephrectomies performed for suspicion of renal tumors were pathologically confirmed to be renal cell carcinoma, and then the patients were subsequently treated for their lymphoma. Although there was no genetic or other predisposing factor for the simultaneous evolution of both these malignancies, since the renal cell carcinoma pathologic stage was higher than was generally thought preoperatively, they suggest that the concommitant lymphoma may have had an overall suppressive immunologic effect.

Onishi et al. reported that, of 804 patients that were being treated for renal cell carcinoma, 38 of these patients had another primary tumor (25 male and 13 female). Gastric cancer proved to be the most frequent (14) followed by cancer of the lung (3), prostate (3), bladder (3), uterus (3), rectum (2) and thyroid (2). There were no common genetic loci or environmental factors.[27]

Rabbani et al. examined the incidence of multiple primary malignancies in patients with renal cell carcinoma (RCC). Out of 551 renal tumor patients, they found at least one second primary in 148 (26.9%), two second primaries in 34 (6.2%), three separate primaries in six (1.1%) and four primaries in one patient (0.2%). The most common other tumors were breast, prostate, colorectal, bladder and non-Hodgkin's lymphoma.[28] Men with RCC had an increased risk of bladder cancer.

The study by Sugiyama et al. found an incidence of multiple cancers in their urological cancer patients of 6.6%.[29] In another study by Wegner over a 19-year period, 4353 patients treated for urologic cancers from the University of Berlin Hospital had a secondary tumor incidence of 3.3%.[30] Nogueras et al. found an incidence of 6.1% of secondary tumors from their urological cancer patients[31] and Mikata et al. had an incidence of 6.4%.[32]

In the report by Mydlo et al., several urologic oncologists from major medical centers in the USA were questioned as to how to handle multiple lesions.[4] They summarized their results, which were fairly consistent. Their range of urologic cancer patients with other primary non-urologic tumors was from 5% to 15%, with the most common associations being kidney with colon, lymphoma and skin. For primary and secondary urological tumors the most common finding was prostate and bladder cancer, followed by prostate and kidney cancer. Most agreed that renal tumors were the most common genitourinary tumor detected during the work-up for a non-urologic primary tumor, usually due to the use of CT scans. This was followed by prostate cancer, usually due to the increased use of screening PSA testing. Colon cancer was the most common non-urologic tumor detected during the work-up for a urologic tumor, also due to the use of CT scanning, followed by lymphoma and skin cancer. Most of the urologists added that the secondary tumor is usually asymptomatic, that there is usually no strong family history and agreed that operating on one tumor does not change the natural history of the second.[4]

As far as treatment recommendations are concerned, most of the urologists responded that they prefer to do the tumor resections simultaneously unless one tumor is more aggressive than the other; then they prefer to stage surgery by doing the more aggressive tumor first. If patient cure is evident, they will treat the second neoplasm. However, in younger patients, they prefer to do both, since waiting may make the second tumor grow to incurability.

They all agreed that treatment should be performed simultaneously, especially if the lesions are relatively small, require a single incision and the patient's medical condition allows for longer anesthesia exposure. If these prerequisites are not met, the respondents agreed that treatment should be directed at the more aggressive lesion first, which may improve the patient condition and/or survival, and thus if a second operation is warranted.[4]

Of interest in this review of patients with multiple primary malignancies is that no patient was found to have an underlying immunological disorder that could have been a factor for these lesions. They also reported that there was minimal blood loss during combined surgeries and there was no significant postoperative morbidity, taking into account that operative time was longer with simultaneous surgeries by 1–2 hours.[4]

In examining the possible etiology and/or risk factors for multiple primary malignancies, several theories exist. One report demonstrated an increased incidence of multiple tumors due to environmental factors, such as the association of bladder cancer and lung cancer due to arsenic in drinking water, which leads to faulty DNA repair mechanisms.[33] While tobacco may also account for the increased incidence of bladder and lung cancers, other cancers do not necessarily have a simplistic causative agent other than a predisposition for tumor formation.[34]

Dietary fat intake and obesity have also been associated with independent increased rates of prostate, breast and colon cancer.[35,36] Adipose tissue is composed of triglycerides and cholesterol, the latter of which forms the building blocks for estrogen and testosterone. This can enhance proliferation of such hormone-sensitive tumors, such as prostate and breast cancer. Adipose tissue is also replete with numerous growth factors that can stimulate tumor progression. On a more immunological level, lipid-laden macrophages do not function as well as normal ones and, therefore, a break in the immune surveillance system may predispose to tumor formation and or progression.[36,37]

Thus, combining the effects of aging with obesity should create a fertile ground for neoplastic development. However, many of the studies examining multiple primary malignant tumors did not take body mass index (BMI) into account as factors or comorbidities. Furthermore, tumor development is multifactorial and it is difficult to stratify the numerous confounders.

On a psychoneurobiological level, mental depression and obesity have a high association. It has been demonstrated that mentally depressed animals and humans may have a suppressed immune system function.[38,39] Obesity and mental depression also have been associated together and, therefore, may act as a synergistic mechanism to further weaken the immune surveillance system, as described above.[39]

Another question in this field is related to the possibility that the second non-urological malignancy may be nothing more than a coincidental phenomenon. Although one may speculate that these cancers may be related to some genetic, environmental or other factor, one should still ask whether a similarly age-matched population might show a similar incidence of non-urological cancers if subjected to similar intense scrutiny. Furthermore, different societies have more prevalent tumors than others. For example, the high incidence of gastric cancer but low incidence of prostate cancer in Japan may not be a valid comparison for US patients.[40] Varying links have been detected between genetic and environmental factors and prostate cancer in different populations. In the study by Ekman et al., they reported varying lengths of GAG repeats on the androgen receptors between Japanese and Swedish prostate cancer patients, as well as stronger vitamin D and A receptor staining in the former.[34]

It is difficult to ascertain whether having a urological tumor is a risk factor for having another type of tumor, since this would require analysis of large population studies and not merely tumor registries. However, novel genes have been detected in different organ tumors, for example, UROC28 in prostate, breast and bladder tumors,[41] and databases of human expressed sequence tags (dbEST) have revealed a common gene on chromosome 17 for both prostate and colon tumors.[42] Further research will determine if more associations will arise and if urologic cancer specifically is a risk factor for other tumors.

As mentioned previously, there are numerous confounding factors that could allow for the appearance of several tumors during one's lifetime, especially during the later years of life: accumulation of free radicals, which would lead to mistakes in DNA replication, impaired immune surveillance system, environmental, smoking and dietary effects, etc. Klippel et al. examined a sample of 55 patients with solitary and multiple neoplasias that underwent immunostaining of B and T lymphocytes. They found a difference of diminished immunocompetence among both cancer groups compared to a non-malignant control group of patients.[43] Interestingly, though, they found no significant differences between the multiple malignancy group and a group of patients with solitary tumors.[43] However, if we couple this finding with the increased chance of DNA mistakes during replication because of free radicals, one can speculate about other possible mechanisms for the increased incidence of malignancies one can develop as one ages.[1–3] Considering that 20% of Americans will develop cancer in their lifetime, the range of 3–6% in the review of the literature of multiple primary malignancies (MPM) might not be unexpected and thus could represent coincidental detection (**Table 6.2**).

In conclusion, there may be a small cohort of urologic cancer patients with other primary malignancies because of the

Table 6.2. Studies of urological cancer patients with another primary tumor. Reprinted from *Urology*, Vol. 58, Mydlo and Gerstein, Urologic cancer patients with another primary malignancy, pp. 243–7. Copyright (2001), with permission from Elsevier Science.

Investigator	Year	Length of study	Secondary/primary tumor	Incidence
1. Mikata et al.	1983	5 years	11/172	6.4%
2. Rabbani et al.	1998	8 years	148/551	26.9%*
3. Sugiyama et al.	1984	8 years	26/397	6.5%
4. Tihan et al.	1996	10 years	15/1262	1.2%
5. Onishi et al.	1998	10 years	38/804	4.7%
6. Wegner	1992	19 years	144/4353	3.3%
7. Nogueras et al.	1992	9 years	23/377	6.1%
8. Salminen et al.	1994	36 years	652/10014	6.5%
9. Mydlo et al.	2001	6 years	18/515	3.5%
10. Moyer et al.	2001	20 years	222/2339	9.5%

*Includes antecedent, synchronous and subsequent tumors.

increased aging population, increased world-wide incidence of obesity, increased exposure to numerous environmental causative agents, smoking and certain genetic predispositions, which may be detected due to advanced technology in imaging.

The importance of not having ‘tunnel’ vision cannot be overemphasized. In many reports of multiple tumors, the second lesion was detected during the primary preoperative work-up.

Some reports which questioned urologists as to the treatment of these patients suggested that they preferred to treat both tumors in one operation, especially if they are small and/or in close proximity, require one excision, or require a reasonable amount of additional anesthesia. However, if the first tumor is more aggressive or advanced than the other tumor, they will excise this first and observe the outcome before performing surgery for the other lesion. In those cases where survival is sufficiently poor, it is very unlikely that the incidentally found tumor will be clinically significant. However, in younger patients, both tumors may be excised simultaneously because additional waiting may make the second tumor grow to incurability.

Vigilance for these multiple malignancies should rest initially with the primary care doctor. However, the subspecialist should also be on the alert, since it is usually the preoperative work-up that will uncover these additional tumors. If the status of a patient deteriorates after surgery for a primary tumor, the physician should consider a secondary primary tumor within the differential.

REFERENCES

1. Trichopoulos D, Li FP, Hunter DJ. What causes cancer? *Sci. Am.* 1996; 275:80–7.
2. Chandra RK. Graying of the immune system. *JAMA* 1997; 277:1398–9.
3. Warren S, Gates O. Multiple primary tumors. *Am. J. Cancer* 1932; 16:1358–414.
4. Mydlo JH, Gerstein M. Urologic cancer patients with another primary malignancy. *Urology* 2001; 58:243–7.
5. Moertel CG, Dockerty MB, Baggenstoss AH. Multiple primary malignant neoplasms. *Cancer* 1977; 14:221–48.
6. Teppo L, Pukkala E, Saxen E. Multiple cancer – an epidemiologic excercise in Finland. *J. Natl Cancer Inst.* 1985; 75:207–17.
7. Greenberg RS, Rustin ED, Clark WS. Risk of genitourinary malignancies after cancer of the prostate. *Cancer* 1988; 61:396–401.
8. Liskow AS, Neugut AI, Benson M et al. Multiple primary neoplasms in association with prostate cancer in black and white patients. *Cancer* 1987; 59:380–4.
9. Kawakami S, Fukui I, Yonese J et al. Multiple primary malignant neoplasms associated with prostate cancer in 312 consecutive cases. *Urol. Int.* 1997; 59:243–7.
10. Isaacs SD, Kiemeney LA, Baffoe-Bonnie A et al. Risk of cancer in relatives of prostate cancer probands. *J. Natl Cancer Inst.* 1995; 87:991–6.
11. Levi F, Randimbison L, Te VC et al. Second primary tumors after prostate carcinoma. *Cancer* 1999; 86:1567–70.
12. Brenner DJ, Curtis RE, Hall EJ et al. Second malignancies in prostate carcinoma patients after radiotherapy compared with surgery. *Cancer* 2000; 88:398–406.
13. Moyer C, Ginsberg PC, Gerboc J et al. Multiple primary malignancies associated with prostate cancer. (In press).
14. Lynch HT, Larsen AL, Magnusin CU et al. Prostatic carcinoma and multiple primary malignancies. Cancer 1966; 19:1891–7.
15. Hadju SI, Hadju EO. Multiple primary malignant tumors. *J. Am. Geriatr. Soc.* 1968; 16:16–26.
16. Paulish K, Schottenfeld D, Severson R et al. Risk of multiple primary cancers in prostate cancer patients in the Detroit metropolitan area: a retrospective cohort study. Prostate 1997; 33:75–86.
17. Kleinerman RA, Lieberman JV, Li FP. Second cancer following cancer of the male genital system in Connecticut, 1935–82. *Natl. Cancer Inst. Monogr.* 1985; 139–47.
18. Mydlo JH, Kral JG, Volpe MA et al. An analysis of microvessel density, androgen receptors, p53 and HER-2/neu expression and Gleason score in prostate cancer: preliminary results and therapeutic implications. *Eur. Urol.* 1998; 34:426–32.
19. Mydlo JH, Tieng NL, Volpe MA et al. A pilot study analyzing PSA, serum testosterone, lipid profile, body mass index, and race in a small sample of patients with and without carcinoma of the prostate. *Pros. Can. Pros. Dis.* 2001; 4:1–5.

20. Miller JS, Puckett ML, Johnstone PA. Frequency of coexistent disease at CT in patients with prostate carcinoma selected for definitive radiation therapy: is limited treatment-planning CT adequate? *Radiology* 2000; 215:41–4.
21. Nakata S, Takahashi H, Takezawa Y et al. Clinical features of multiple primary cancers including prostate cancer. *Hinyokika Kiyo – Acta Urol. Jap.* 2000; 46:385–91.
22. Moutzouris G, Barbatis C, Plastiras D et al. Incidence and histological findings of unsuspected prostatic adenocarcinoma in radical cystoprostatectomy for transitional cell carcinoma of the bladder. *Scand. J. Urol. Nephrol.* 1999; 33:27–30.
23. Cannon MJ, Flanders WD, Pellet PE. Occurrence of primary cancers in association with multiple myeloma and Kaposi's sarcoma in the United States, 1973–1995. *Int. J. Cancer* 2000; 85:453–6.
24. Karlovsky M, Lebed B, Mydlo JH. Incidence of HIV and sexually transmitted diseases in the elderly population using pharmacotherapy for erectile dysfunction. (In press).
25. Bruno G, Andreozzi P, Graf U et al. Hepatitis C virus: a high risk factor for a second primary malignancy besides hepatocellular carcinoma. Fact or fiction? *Clin. Therapeut.* 1999; 150:413–18.
26. Tihan T, Filippa DA. Coexistence of renal cell carcinoma and malignant lymphoma. A causal relationship or coincidental occurrence? *Cancer* 1996; 77:2325–31.
27. Onishi T, Ohishi Y, Suzuki H et al. Study of double cancers with renal cell cancer. *Jap. J. Urol.* 1998; 89:808–15.
28. Rabbani F, Grimaldi G, Russo P. Multiple primary malignancies in renal cell carcinoma. *J. Urol.* 1998; 160:1255–9.
29. Sugiyama T, Park YC, Iguchi M et al. Double cancer in urology. *Acta Urol. Jap.* 1984; 30:1427–30.
30. Wegner HE. Multiple primary cancers in urologic patients. Audit of 19-year experience in Berlin and review of the literature. *Urology* 1992; 39:231–4.
31. Nogueras Gimeno MA, Espuela Orgaz R, Abad Menor F et al. Incidence and characteristics of multiple neoplasms in urologic patients. *Actas Urol. Espanol.* 1992; 16:316–19.
32. Mikata N, Kinoshita K. Primary multiple cancers related to urological malignancies. *Jap. J. Cancer Clinics* 1983; 29:A-12,183.
33. Smith AH, Goycolea M, Haque R et al. Marked increase in bladder and lung cancer mortality in a region of Northern Chile due to arsenic in drinking water. *Am. J. Epidemiol.* 1998; 147:660–9.
34. Ekman P, Gronberg H, Matsuyama H et al. Links between genetic and environmental factors and prostate cancer risk. *Prostate* 1999; 39:262–8.
35. Deslypere JP, Verdonck L, Vermeulen A. Fat tissue: a steroid reservoir and site of steroid metabolism. *J. Clin. Endocrinol. Metab.* 1985; 61:564–70.
36. Mydlo JH, Kral JG, Macchia RJ. Differences in prostate and adipose tissue basic fibroblast growth factor (FGF-2): analysis of preliminary results. *Urology* 1997; 50:472–8.
37. Erickson KL, Hubbard NE. A possible mechanism by which dietary fat can alter tumorigenesis: lipid modulation of macrophage function. *Adv. Exp. Med. Bio* 1994; 364:67–81.
38. Irwin M, Daniels M, Smith TL et al. Impaired natural killer cell activity during bereavement. *Brain Behavior Immun.* 1987; 1:98–104.
39. Ben-Eliyahu S, Page GG, Yirmiya R et al. Evidence that stress and surgical interventions promote tumor development by suppressing natural killer cell activity. *Int. J. Cancer* 1999; 80:880–8.
40. Tomoda H, Taketomi A, Baba H et al. Multiple primary colorectal and gastric carcinoma in Japan. *Oncol. Rep.* 1998; 5:147–9.
41. An G, Ng AY, Meka CS et al. Cloning and characterization of UROC28, a novel gene overexpressed in prostate, breast and bladder cancers. *Cancer Res.* 2000; 60:7014–20.
42. Liu XF, Olsson P, Wolfgang CD et al. PRAC: A novel small nuclear protein that is specifically expressed in human prostate and colon. *Prostate* 2001; 47:125–31.
43. Klippel KF, Hutschenreiter G, Jacobi G et al. Double urologic tumors: reduced immunocompetence? *Onkologie* 1979; 2:12–17.

Angiogenesis, Growth Factors, Microvessel Density, p53 and p21 in Prostate Cancer

Jack H. Mydlo and Paul Crispen

Department of Urology, Temple University School of Medicine, Philadelphia, PA, USA

GROWTH FACTORS

Growth factors are regulatory proteins that control the response of a cell to growth, differentiation and apoptosis. Since these peptides are responsible for a major role in the growth and development of the organism, their aberrant expression, or mistakes in their receptors, may increase the chances of unihibited growth and/or neoplasia.[1]

Growth factors may stimulate cells in an autocrine (self-stimulating), paracrine (stimulation of neighboring cells, but not itself), or endocrine or hormonal manner (stimulation of distant cells). The growth-factor-receptor activation sets in motion a cascade of events leading to the turning off or turning on of numerous genes.[1]

Although there are many different types of growth factor, we will limit our discussion to those involved with prostate growth. Epidermal growth factor (EGF) was first isolated from the mouse submaxillary gland. The group of epidermal growth factor-like proteins includes transforming growth factor α (TGFα) and heparin-binding epidermal growth factor.[2]

EGF and TGFα are involved in cellular differention and angiogenesis. Angiogenesis is the development of a vascular network to support a tumor once it reaches the size where simple diffusion of nutrients is no longer possible.[3] This concept will be discussed in detail subsequently. The family of fibroblast growth factors (FGFs) includes acidic and basic FGF, int-2, hst/KS3, FGF-5 and keratinocyte growth factor.[4] All members of the FGF family can attach to common receptors. Although basic FGF is a mitogen for mesodermal- and ectodermal-derived cells, it is a potent angiogenic factor.

Insulin-like growth factor I and II have a similar homology to insulin, and they may also signal via the insulin receptor. They also induce osteoblasts to form bone. Insulin-like growth factor is produced primarily by stromal cells and uses a paracrine stimulatory pathway to bind to insulin-like growth factor I receptors on prostatic epithelial cells.[5]

Several reports have demonstrated the possibility of using interleukin (IL)-1 and/or IL-2 as tumor markers for prostate cancer; however, more research in this arena needs to be done before this can be considered valid.[6]

The family of transforming growth factor β consists of five isoforms. TGFβ may be a stimulator or inhibitor of cells depending on the cell type. TGFβ is also a potent angiogenic factor, may regulate cell adhesion and cell differentiation, and suppresses the immune system. TGFβ also appears to inhibit normal prostate epithelial cells, and can induce cell death.[7,8]

The normal prostate has been found to contain EGF, FGF, interleukin growth factor (ILGF) and TGFβ. Although there is much interaction between androgen and growth factor expression in the prostate, it has been demonstrated that αFGF and βFGF were able to support prostate epithelial cell growth in culture without the addition of androgen.[9,10] However, the adult prostate has low levels of αFGF, but high levels of βFGF.[10]

Since the presence of androgen induces the production of EGF, FGF and insulin-like growth factor (INGF) from both prostate stroma and epithelium, castration may cause a decrease in the expression of these factors. However, there is an increase in EGF and INGF receptor expression. In addition, castration induces an increase of TGFβ, which can induce cellular apoptosis.[11]

In prostate cancer, in the androgen-sensitive LNCaP cell line, and the androgen-insensitive PC-3 and DU145 cell lines, TGFα is secreted. Furthermore, upon the addition of androgen, there is increased production of TGFα and the EGF receptor. Thus, there is an autocrine stimulation of the prostate cancer cell via this mechanism. Furthermore, it has been demonstrated that an EGF receptor antibody can inhibit this androgen-induced cell growth, both in sensitive and insensitive cell types.[4,10,11]

Although much research has shown the importance of βFGF in rat prostate cancer, its role in human prostate cancer is less well defined. The androgen-insensitive lines of DU145 and androgen-sensitive LNCaP are responsive to βFGF stimulation, but the androgen-insensitive PC-3 cell line is not.[12]

It has also been demonstrated that PC3 and DU145 cells secrete insulin-like growth factor II and overexpress insulin-like growth factor-I receptor, which would act as an autocrine stimulation loop. Furthermore, because this stimulatory mechanism works in the absence of androgens, it may account for prostate cancer cell growth after castration.[13]

TGFβ1 is overexpressed in rat prostates and, furthermore, is not susceptible to hormonal manipulation. The human prostate cancer cell lines LNCaP, PC3 and DU145, were not inhibited by TGFβ1. However, because TGFβ1 is a potent angiogenic factor, it may still feed the growing tumor by supporting a vascular network. The inhibition of the immune system by TGFβ1 may act further in a synergistic manner to help promote tumor growth. TGFβ1 may also play a role in the osteoblastic lesions seen in advanced prostate cancer.[14–16]

ANGIOGENESIS AND MICROVESSEL DENSITY

'Angiogenesis' is the term used to describe the development of neovascularity to support the growth of tumors after they have reached a certain size, where simple diffusion of nutrients is not possible. This vascularity helps the tumor to grow, invade and metastasize.[3,17] Most of this concept was described by Judah Folkman in the 1960s and 1970s, and he still continues to make strides in this field. It is theorized that a single millimeter of tumor capillary length can support about 10 000 tumor cells. This extrapolates to about 400 tumor cells per endothelial cell of growth can be supported. Hypervascular tumors have been shown to be more aggressive than their hypovascular counterpart, such as renal cell carcinoma, compared to papillary renal adenocarcinoma of the kidney.[18–20]

Recent investigations have shown that assessment of vascularity of breast carcinomas via microvessel density (MVD) has prognostic value in clinical outcome. Several reports demonstrated the use of MVD, prostate specific antigen (PSA) and Gleason score to obtain an optimized microvessel density (OMVD) value to explain tumor biology and predict extracapsular extension.[21–23] They suggested that a patient with a higher OMVD value on prostate needle biopsy may require adjuvant therapy, and may not necessarily be treated optimally with radical surgery alone. Although several companies tried to capitalize on this potential prognostic indicator, many other factors exist, especially at the molecular level, which have independent predictive value.[21–23]

Others have reported a significant correlation between MVD in histologic sections of adenocarcinoma of the prostate and the incidence of metastases.[24] However, the amount of tumor angiogenesis quantified in colorectal cancer has provided only limited predictive value of determining lymphatic involvement compared to prostate cancer.[25]

While MVD can be quantified in histological specimens of prostate cancer, one can demonstrate the effects of hormonal ablation in terms of a direct effect on cell death, or apoptosis, and a slower death due to the ablation of nutrient supplying blood vessels (**Fig. 7.1a–c**). However, this same concept of

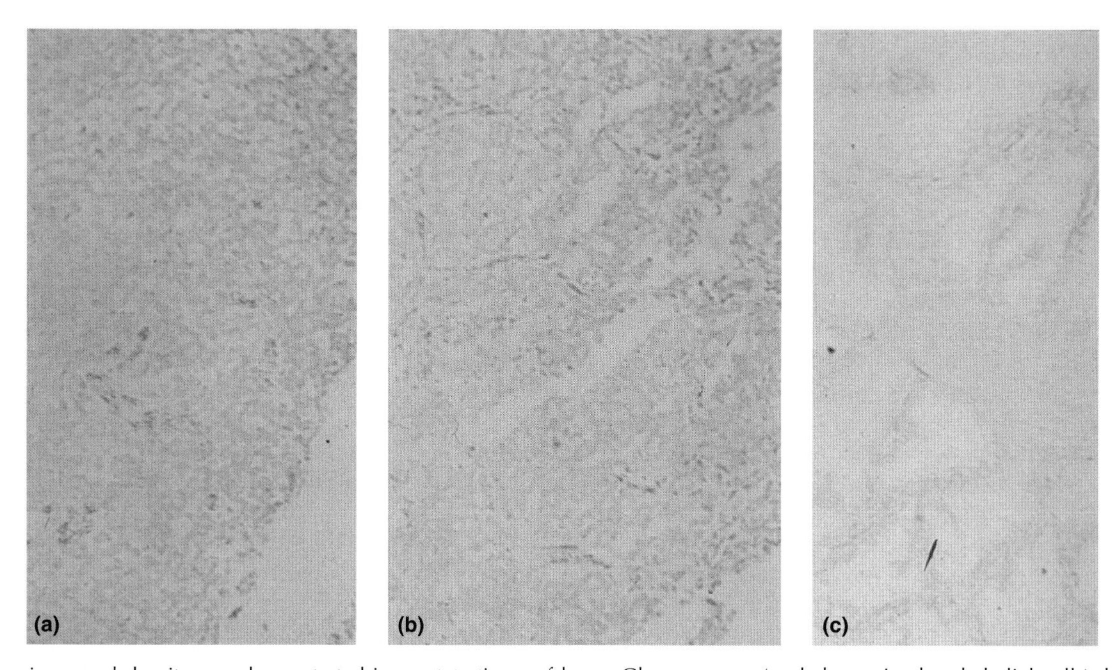

Fig. 7.1. Less microvessel density was demonstrated in prostate tissue of lower Gleason score (a, darker stained endothelial cells) than in prostate cancer tissue of higher Gleason score (b). Last panel (c) demonstrates ablation of vascularity after radiation and/or hormonal therapy, one of the mechanisms involved in cell death. Reproduced from Mydlo et al. *Eur. Urol.* 1998; 34:426–32, with permission from Karger, Basel. (*See Color plate 1.*)

vascular ablation suggested possible therapeutic interventions using angiogenic inhibitors. While initial results were encouraging, side effects, as well as the continued use of the inhibitory agent, needs much to be desired. However, ongoing trials using these antiangiogenic trials with other agents are being explored.[26,27]

TREATMENT STRATEGIES

Since stimulation of the signal transduction pathway of the EGF receptor (EGFR) enhances cellular proliferation, prevents apoptosis, and promotes tumor-cell motility, adhesion and invasion, strategies for treatment of prostate cancer include blockage of the receptor and/or inhibition or the cascade of events.

Androgens exert prostate growth and development by binding to intracellular androgen receptors, which regulate the transcription of androgen-sensitive genes. Without androgens, these receptors are inactive and are bound to heat-shock proteins.[9] In cell culture, prostate cells require growth factors to grow. Androgens alone cannot induce proliferation. However, it is androgens that stimulate stromal cells to secrete growth factors, which in turn stimulate epithelial cells.[2]

Many modalities that are used today to treat prostate cancer have been learned from breast cancer therapies.[17] Signaling from steroids/hormones and growth factors are very closely linked. Furthermore, overexpression of growth factors and/or their receptors are found in many breast cancers.[17] However, in prostate cancer, while the evidence for overexpression of the growth factor receptor is conflicting, there is evidence for autocrine activation of the growth factor receptor and its subsequent cascade of events.[10]

Some of the novel therapies for prostate cancer, which are directed against the EGFR family of receptors and the subsequent cascade of events leading to transduction of signals to the nucleus are outlined in **Table 7.1**.

The p53 tumor suppressor gene is the most widely mutated gene in human cancer. Mutations of p53 arise at certain times during the progression of tumors, for example, in the emergence of carcinoma *in situ* from benign lesions of the colon, testis and prostate. Mutations of p53 sometimes coincide with more aggressive neoplasms, which have been described to be resistant to chemotherapy, radiotherapy and hormonal therapy.[28–30] Wild-type p53 suppresses angiogenesis, whereas mutant p53 is associated with enhanced angiogenesis.[31] Expression of the p53 suppressor protein has been shown to have prognostic ability in prostate cancer; however, Shurbaji et al. only found this in low to intermediate grade cancer,[28] while Kallakury et al. found prognostic potential in high-grade prostate adenocarcinomas, with particular overexpression prior to metastasis.[32]

Table 7.1. Antibodies to receptors and/or their mechanisms.

1. ISIS 3521 (Isis Pharmaceuticals, Carlsbad, CA)
2. Thalidomide (Thalomid, Celgene Corporation, Warren, NJ)
3. IMC-C225 (ImClone Systems, Inc., New York, NY)
4. ABX-EGF, R155777, CEP-701 (Cephalon, Inc. West Chester, PA)
5. Trastuzumab (Herceptin) MDX-210 (Medarex, Inc., Princeton, NJ) – a monoclonal antibody that binds to the HER2/neu receptor
6. CEP-701 – a selective tyrosine kinase inhibitor of neutrophin-specific trk receptors

Another mechanism that can affect the biological behavior of certain tumors and may also be considered a tumor marker is p21. P21 (wafl/cip 1) is a protein, which is a cyclin-dependent kinase inhibitor able to arrest the cell cycle at the G1 phase by inhibiting DNA replication. The expression of p21 and its prognostic value in prostate cancer has also been explored. Aaltomaa et al. used immunohistochemistry to analyze the expression of p21 in 213 prostate cancer cases, and followed patient survival. They found the expression of p21 was significantly associated with a high Gleason score, DNA aneuploidy, and expression of Ki-67 and bcl-2. Furthermore, they found that the expression of p21 protein was a significant independent predictor of prostate cancer patient survival.[33] Baretton et al. found that p21 overexpression before and after total androgen deprivation was significant associated with prostate cancer recurrence and overall patient survival.[34] They reported that 5-year distant metastatic-free survival and cancer-specific survival were 71% and 82%, respectively, for those patients with low expression of p21, compared with 94% and 100%, respectively, for those patients with high expression of p21.[34]

Hypoxia limits tumor growth but selects for high metastatic potential. Salnikow and colleagues reported that loss of sensitivity to p21 inhibition is a part of hypoxic phenotype associated with aggressive cancer behavior.[35] Furthermore, there may be ethnic differences among p21 expression and survival. Sarkar et al. examined the expression of p21 in Black and White prostate cancer patients. They reported that p21 expression in 34 White patients served as a predictor in biochemical recurrence, but it had no predictive value in 28 Black patients.[36]

Lacombe et al. suggested that p21 overexpression is an independent predictor of PSA failure after radical surgery and suggested that this marker may help clinicians identify patients who may require adjuvant treatment strategies following surgery.[37]

HER-2/neu is an oncogene with some reports demonstrating predictive value for the progression of prostate cancer. According to one report, expression of this oncogene was found in about one-third of all localized prostate cancers, was significantly correlated with high-grade tumor and aneuploid status, but was not an independent predictor of metastases.[38]

Androgen receptor (AR) concentration has also been demonstrated to be prognostic in a subgroup of patients with higher Gleason scores.[39,40] In one study a subgroup of patients whose tumor specimens lacked AR receptors in the Gleason

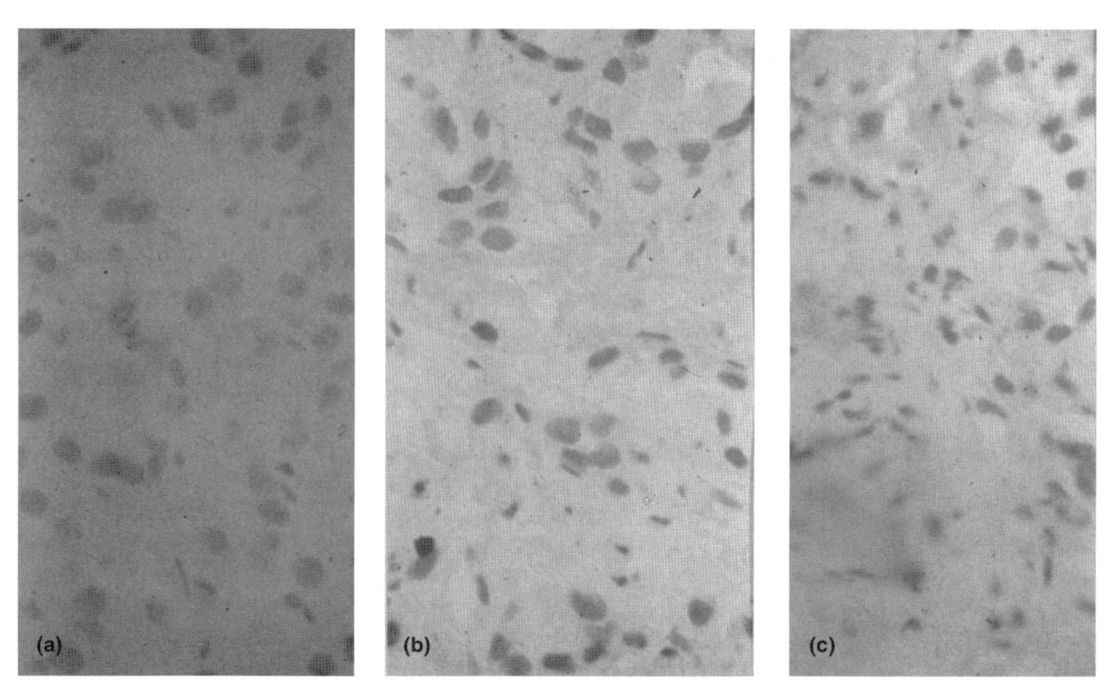

Fig. 7.2. Normal prostate tissue does not demonstrate expression of mutant p53 (a) compared to Gleason 6 prostate cancer tissue (b, brown avitin-biotin stained cells). Last panel (c) demonstrates the persistence of expression of mutant p53 even after radiation therapy, suggesting that these tumors are resistant to such therapy. Reproduced from Mydlo et al. *Eur. Urol.* 1998; 34:426–32, with permission from Karger, Basel. (*See Color plate 2.*)

score range between 7 and 9 had a significantly poorer prognosis. Furthermore, a positive AR status correlated with endocrine sensitivity and improved clinical outcome.[40] The fact that more aggressive tumors are associated with a lack of AR suggests that these tumors may be less sensitive to hormonal manipulation.

DIAGNOSTIC AND/OR THERAPEUTIC OPTIONS

Presently, our evaluation for treatment options for patients with prostate cancer consists of the age and general health of the patient, the Gleason score (GS) and the PSA value. Microvessel density, expression of p53, p21, HER-2/neu or androgen receptor status is usually not done on the prostate biopsy specimens, except in some academic protocols, and these parameters are not usually considered in the equation for patient therapy.

In one report, Mydlo et al. examined the findings of MVD, p53 and HER-2/neu, and androgen receptor expression in samples of prostate cancer tissue. They found an almost four-fold increase of MVD in cancerous prostate specimens (40.3 + 5.9) compared to normal surrounding prostate controls (11.7 + 1.3; $P < 0.001$). These findings showed that GS correlated with MVD ($r = 0.40$, $P = 0.06$; **Fig. 7.1**) and with mutant p53 expression ($r = 0.57$, $P < 0.05$), but they did not find a correlation between MVD and p53. Furthermore, they found persistent mutant p53 expression in prostate cancer biopsy samples after radiation therapy, suggestive of radioresistance[41] (**Fig. 7.2**). In addition, they found greater expression of mutant p53 in prostate cancer tissue from Black patients compared to White patients (**Fig. 7.3**). This may account for more aggressive tumor biology, stage for stage and stratified for GS, in Blacks versus Whites.[41] Although they also found a lower AR concentration in prostate cancer tissue compared to benign prostate tissue, they did not find any definite correlation with GSs. Similarly, there were no correlations with HER-2/neu.[41]

Their results, which correlate well with other published reports, demonstrate that there is a correlation between greater MVD and GSs, and a greater prevalence of mutant p53 expression in patients' specimens with higher GSs and in more advanced stages. This may further explain prostate tumor biology. It suggests that the lesion probably grows slowly initially because its neovascularity is being suppressed by wild-type p53. When the tumor progresses to the next phase,

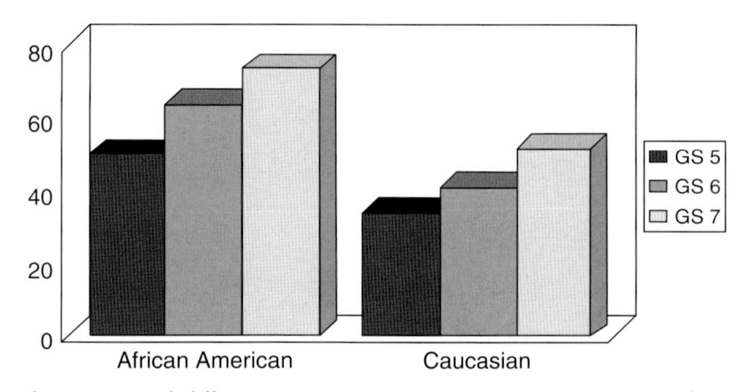

Fig. 7.3. Racial differences in p53 expression in prostate cancer. There was a greater expression of mutant p53 in prostate cancer specimens, stratifed for Gleason score, in Black patients compared to White patients.

Table 7.2. Several antiangiogenesis substances for treatment of prostate cancer.

Endogenous substances
Interferons α, β and γ and interferon-inducible proteins
Interleukins 1, 4 and 12
Platelet factor 4
Thrombospondins 1 and 2
Pigmented epithelial-derived factor
Encrypted molecules (angiostatin, endostatin)
Anticancer treatments
Antiestrogens (tamoxifen)
Retinoids
Chemotherapeutic agents: methotrexate, bleomycin and paclitaxel
Radiotherapy, hyperthermia
Antibiotics
TNP-470
Linomide
Suramin (antiparasitic)
Factors that affect matrix interactions
Protease inhibitors
Inhibitors of collagen synthesis
Vascular endothelial cell growth Factor (VEGF)-targeted approaches
VEGF receptor-signaling antagonists
Monoclonal antibodies (Mabs) against VEGF
Antisense oligonucleotides constructs against VEGF
Soluble VEGF receptors
Miscellaneous
Aspirin and cyclooxygenase-2 inhibitors
Genestein and other soy products
Captopril
Thalidomide
Vitamin D_3 analogs
Squalamine (derived from shark cartilage)

the expression of mutant p53 enhances more angiogenesis, which promotes more tumor growth, more potential for invasion and metastasis, and other changes of the cell to resist apoptosis from chemotherapy, radiotherapy or hormone therapy. Clinical trials need to be structured to compare these markers to outcome after different treatment options.[42–45]

It appears that increased MVD and increased p53 expression are associated with the more aggressive tumors. Since these factors promote more cell division, increase the potential for metastasis and may alter the cells response to certain therapies, it may suggest adjuvant therapy after primary tumor treatment.

In summary, it is still unclear which of these parameters, individually or in combination, are interrelated and which may further add to our understanding of prostate tumor biology. Such information could be used to design clinical trials to determine which parameters are most useful for predicting outcome after treament by surgery, radiotherapy, chemotherapy and/or hormonal therapy. However, it is unlikely that a single inhibitory approach to treating prostate cancer is in the future. It will most likely be a multimodality approach, including hormones, chemotherapy, radiotherapy as well as growth factor inhibitors, that will be the future of treatment (**Tables 7.1** and **7.2**).

REFERENCES

1. Aaronson SA. Growth factors and cancer. *Science* 1991; 254:1146–53.
2. Traish AM, Wotiz HH. Prostatic epidermal growth factor receptors and their regulation by androgens. *Endocrinology* 1987; 121:1461–7.
3. Folkman J. The vascularization of tumors. *Sci. Am.* 1976; 234:58–68.
4. Burgess WH, Maciag T. The heparin-binding (fibroblast) growth factor family of proteins. *Ann. Rev. Biochem.* 1989; 58:575–606.
5. Iwamura M, Sluss PM, Casamento JB et al. Insulin-like growth Factor I: action and receptor characterization in human prostate cancer cell lines. *Prostate* 1993; 22:243–52.
6. Waldmann TA. The interleukin-2 receptor on malignant cells: a target for diagnosis and therapy. *Cell Immunol.* 1986; 99:53–9.
7. Derynck R, Lindquist PB, Bringman TS et al. Expression of the transforming growth factor-alpha gene in tumor cells and normal tissue. *Cancer Cells* 1989; 7:297–303.
8. Steiner MS, Barrack ER. Transforming growth factor beta 1 overproduction in prostate cancer: effects on growth in vivo and in vitro. *Mol. Endocrinol.* 1992; 6:15–21.
9. Liu XH, Wiley HS, Meikle AW. Androgens regulate proliferation of human prostate cancer cells in culture by increasing transforming growth factor alpha (TGFa) and epidermal growth factor (EGF)/TGF-alpha receptor. *J. Clin. Endocrinol. Metab.* 1993; 77:1472–8.
10. Steiner MS. Review of peptide growth factors in the prostate: a review. *Urology* 1993; 42:99–110.
11. Steiner MS. Review of peptide growth factors in benign prostatic hyperplasia and urological malignancy. *J. Urol.* 1995; 153:1085–96.
12. Nakamoto T, Chang CS, Li AK et al. Basic fibroblast growth factor in human prostate cancer cells. *Cancer Res.* 1992; 52:571–7.
13. Kaplan PJ, Mohan S, Cohen P et al. The insulin-like growth factor axis and prostate cancer: lessons from the transgenic adenocarcinoma of mouse prostate (TRAMP) model. *Cancer Res.* 1999; 59:2203–9.
14. Ritchie CK, Andrews LR, Thomas KG et al. The effects of growth factors associated with osteoblasts on prostate carcinoma proliferation and chemotaxis. Implications for the development of metastatic disease. *Endocrinology* 1997; 158:1145–50.
15. Wikstrom P, Bergh A, Damber JE. Transforming growth factor beta 1 and prostate cancer. *Scand. J. Urol. Nephrol.* 2000; 34:85–94.
16. Duivenvoorden WC, Hirte HW, Singh G. Transforming growth factor beta 1 acts as an inducer of matrix metalloproteinase expression and activity in human bone metastasizing cancer cells. *Clin. Exp. Metastasis* 1999; 17:27–34.
17. Fox SB, Leek RD, Smith K et al. Tumor angiogenesis in node negative breast carcinoma: relationship with epidermal growth factor receptor, estrogen receptor, and survival. *Breast Cancer Res. Treatment* 1994; 29:109–16.
18. Mydlo JH, Bard R. An analysis of papillary adenocarcinoma of the kidney. *Urology* 1987; 30:529–34.
19. Mydlo JH, Heston WDW, Fair WR. Characterization of a heparin-binding growth factor from adenocarcinoma of the kidney. *J. Urol.* 1988; 140:1575–9.

20. Mydlo JH, Fair WR. *Growth Factors and Angiogenesis in Urology*. AUA Update Series, Vol. 7, Lesson 39. Houston, TX: American Urological Association, 1988.
21. Bostwick DG, Wheeler TM, Blute M et al. Optimized microvessel density analysis improves prediction of cancer stage from prostate needle biopsies. *Urology* 1996; 48:47–57.
22. Brawer MK, Deering RE, Brown M et al. Predictors of pathologic stage in prostate carcinoma: the role of neovascularity. *Cancer* 1994; 73:678–87.
23. Weidner N, Carroll P, Flax J et al. Tumor angiogenesis correlates with metastasis in invasive prostate carcinoma. *Am. J. Pathol.* 1993; 143:401–11.
24. Hughes JH, Cohen MB, Robinson RA. p53 immunoreactivity in primary and metastatic prostatic adenocarcinoma. *Mod. Path.* 1995; 8:462–6.
25. Vermeulen PB, Roland L, Mertens V et al. Correlation of intratumoral microvessel density and p53 protein overexpression in human colorectal adenocarcinoma. *Microvasc. Res.* 1996; 51:164–74.
26. Campbell SC. Advances in angiogenesis research: relevance to urological oncology. *J. Urol.* 1998; 160:134–5.
27. Logothetis CJ, Wu KK, Finn LD et al. Phase I trial of the angiogenesis inhibitor TNP-470 for progressive androgen-independent prostate cancer. *Clin. Cancer Res.* 2001; 7:1198–203.
28. Shurbaji MS, Kalfleisch JH, Thurmond TS. Immunohistochemical detection of p53 protein as a prognostic indicator of prostatic cancer. *Human Pathol.* 1995; 26:106–9.
29. Albers P, Orazi A, Ulbright TM et al. Prognostic significance of immunohistochemical proliferation markers (Ki-67/MIB-1 and proliferation-associated nuclear antigen), p53 protein accumulation and neovascularization in clinical stage A nonseminomatous testicular germ cell tumors. *Mod. Pathol.* 1995; 8:492–7.
30. Ruley HE. p53 and response to chemotherapy and radiotherapy. In: DeVita VT, Hellman S, Rosenberg SA (eds) *Important Advances in Oncology 1996*, pp. 37–56. Philadelphia: Lippincott-Raven Publishers, 1996.
31. Mukhupadhy D, Tsiokas L, Sukhatme VP. Wild type p53 and v-src exert opposing influences on human vascular endothelial growth factor gene expression. *Cancer Res.* 1995; 55:6161–5.
32. Kallakury BV, Figge J, Ross JS et al. Association of p53 immunoreactivity with high gleason tumor grade in prostate cancer. *Human Pathol.* 1994; 25:92–7.
33. Aaltomaa S, Lipponen P, Eskelinen M et al. Prognostic value and expression of p21(wafl/cipl) protein in prostate cancer. *Prostate* 1999; 39:8–15.
34. Baretton GB, Klenk U, Diebold J et al. Proliferation and apoptosis-associated factors in advanced prostatic carcinomas before and after androgen deprivation therapy: prognostic significance of p21/WAF1/C1P1 expression. *Br. J. Cancer* 1999; 80:546–55.
35. Salnikow K, Costa M, Figg WD et al. Hyperinducibility of hypoxia-responsive genes without p53/p21-dependent checkpoint in aggressive prostate cancer. *Cancer Res.* 2000; 60:5630–4.
36. Sarkar FH, Li Y, Sakr WA et al. Relationship of p21 (WAF1) expression with disease free survival and biochemical recurrence in prostate adenocarcinoma (Pca). *Prostate* 1999; 40:256–60.
37. Lacombe L, Maillette A, Meyer F et al. Expression of p21 predicts PSA failure in locally advanced prostate cancer treated by prostatectomy. *Int. J. Cancer* 2001; 95:135–9.
38. Kuhn EJ, Kurnot RA, Sesterhenn IA et al. Expression of the c-erbB-2 (HER-2/neu) oncoprotein in human prostatic carcinoma. *J. Urol.* 1993; 150:1427–000.
39. Pertschuk LP, Macchia RJ, Feldman JG et al. Immunocytochemical assay for androgen receptors in prostate cancer: a prospective study of 63 cases with long term followup. *Ann. Surg. Oncol.* 1994; 1:495–503.
40. Pertschuk LP, Schaffer H, Feldman JG et al. Immunostaining for prostate cancer androgen receptor in parrafin identifies subset of men with a poor prognosis. *Lab. Invest.* 1995; 73:302–9.
41. Mydlo JH, Kral JG, Volpe M et al. An analysis of microvessel density, androgen receptors, p53 and HER-2/neu expression and Gleason score in prostate cancer: preliminary results and therapeutic implications. *Eur. Urol.* 1998; 34:426–32.
42. Schumacher G, Bruckheimer EM, Beham AW et al. Molecular determinations of cell death induction following adenovirus-mediated gene transfer of wild-type p53 in prostate cancer cells. *J. Cancer* 2001; 91:159–66.
43. Eastham JA, Grafton W, Martin CM et al. Suppression of primary tumor growth and the progression to metastasis with p53 adenovirus in human prostate cancer. *J. Urol.* 2000; 164:814–19.
44. Osman I, Drobnjak M, Fazzari M et al. Inactivation of the p53 pathway in prostate cancer: impact on tumor progression. *Clin. Cancer Res.* 1999; 5:2082–8.
45. Montironi R, Diamanti L, Thompson D et al. Analysis of the capillary architecture in the precursors of prostate cancer: recent findings and new concepts. *Eur. Urol.* 1996; 30:191–200.

Staging of Prostate Cancer, PSA Issues Leading up to Prostate Biopsy and Biopsy Technique

Vincent S. Ricchiuti* and Martin I. Resnick

Department of Urology, University Hospitals of Cleveland, Case Western Reserve University School of Medicine, Cleveland, OH, USA

INTRODUCTION

Prostate cancer is the most common cancer (excluding skin cancer) of American men. It is estimated by the American Cancer Society that approximately 198 000 new cases of prostate cancer will be diagnosed in the USA, and 31 500 men will have died from this disease in 2001.[1] Ninety-five per cent of patients with prostate cancer are diagnosed between 45 and 89 years of age, with a median age at diagnosis of 72 years.[2] After the age of 50, an almost exponential rise in the incidence and mortality rates confronts a significant number of men. Many issues challenge the urologist and those patients diagnosed with prostate cancer. The most important are determining viable treatment options and estimating the potential for achieving a cure from the disease. Once the diagnosis is established, the decision to treat (or even not to treat) is typically based upon generated evidence of organ-confined or metastatic disease.

The focus of this chapter will be to outline certain issues facing the clinical urologist in order to assist in making treatment decisions. Specifically, it will address the role of staging in both clinical and pathologic prostate adenocarcinoma. Further, the 'PSA era' has altered the contemporary management of this disease, and this chapter will evaluate the utility of prostate-specific antigen (PSA) assessment. Finally, it will discuss variations on prostatic biopsy techniques and methods.

The decisions each urologist makes regarding the management of prostate cancer are extremely complex. The recommendations made to each patient can have significant impact on patient's lives in terms of both morbidity and mortality. Proper consideration of patient quality of life issues and potential treatment outcomes is mandatory. Successful cancer prevention strategies can be developed with proper understanding of the disease etiology and contributing factors associated with its pathogenesis. It should also help to define those patients who are likely to benefit from treatment, and those who portend a poorer prognosis.

STAGING

It is essential to the management of patients with prostate cancer that accurate clinical staging is achieved. The characterization of features associated with prostate cancer is improving progressively and newer, more innovative techniques, are currently being developed. The most important and challenging task for the clinical urologist is to distinguish patients with pathologically organ-confined disease from those with non-organ-confined disease. Therefore, the goal of staging patients with prostate cancer is to determine the extent of disease as precisely as possible in order to evaluate prognosis and subsequently guide management recommendations.

Several staging systems have been introduced over the years for prostate cancer. The staging system most commonly used by urologists in the USA was described by Whitmore in 1956.[3] This was later revised and modified by Jewett in 1975.[4] Patients are grouped into four categories (letters A–D) according to clinical and pathologic information in the Whitmore–Jewett system.

*Present address: NEO Urology Associates, Inc., Youngstown, OH, USA.

The American Joint Commission (AJC) and International Union Against Cancer (UICC) have also devised tumor, node and metastases (TNM) systems to classify prostate cancer. Many differences, however, existed between these latter two systems. In 1987, these two staging systems were combined, and the AJC/UICC TNM classification for prostate cancer resulted (**Table 8.1**).[5] Yet, even today, many urologists believe that the TNM classification is flawed and unacceptable. Owing to controversy in the available staging systems, the National Cancer Institute developed a committee in 1986 whose task was to create a prognostic, reproducible and therapeutically relevant prostate cancer staging system. The distinguished committee members reported their results in 1988. Named the OSCC staging system, it essentially incorporated features from both the Whitmore–Jewett and TNM systems.[6]

Patients in whom the diagnosis has been established can be staged in two different ways, depending on the modality of treatment they choose. All patients are first clinically staged. This is through a combination of information such as serum PSA, digital rectal examination (DRE), tumor grade and imaging results. Local extent of disease is typically determined by DRE. However, DRE considerably understages local tumor extent with clinically organ-confined prostate cancer in approximately 30% of patients.[7] Once determined, the clinical T stage is established.

Pathologic stage, on the other hand, involves histologic evaluation of pelvic lymph nodes and the prostate following surgical removal. It is a much more accurate assessment of the extent of cancer and is extremely useful for helping predict prognosis. Although pathologic stage is more useful than

Table 8.1. 1987 AJC/UICC TNM prostate cancer staging classification.

Stage	Substage	Description
Primary tumor (T)		
TX		Primary tumor cannot be assessed
T0		No evidence of primary tumor
T1		Clinically inapparent tumor not palpable or visible by imaging
	T1a	Tumor incidental histologic finding in 5% or less of tissue resected
	T1b	Tumor incidental histologic finding in more than 5% of tissue resected
	T1c	Tumor identified by needle biopsy (e.g. because of elevated PSA)
T2		Palpable tumor confined within prostate
	T2a	Tumor involves one lobe
	T2b	Tumor involves both lobes
T3		Tumor extends through the prostatic capsule
	T3a	Extracapsular extension (unilateral or bilateral)
	T3b	Tumor invades seminal vesicle(s)
T4		Tumor is fixed or invades adjacent structures other than seminal vesicles: bladder neck, external sphincter, rectum, levator muscles and/or pelvic wall
Primary tumor, pathologic (pT)		
pT2		Organ confined
	pT2a	Unilateral
	pT2b	Bilateral
pT3		Extraprostatic extension
	pT3a	Extraprostatic extension
	pT3b	Seminal vesicle invasion
pT4		Invasion of bladder, rectum
Regional lymph nodes (N)		
NX		Regional lymph nodes cannot be assessed
N0		No regional lymph node metastasis
N1		Metastasis in regional lymph node or nodes
Distant metastasis (M)		
MX		Distant metastasis cannot be assessed
M0		No distant metastasis
M1		Distant metastasis
	M1a	No regional lymph nodes
	M1b	Bone(s)
	M1c	Other site(s)

clinical stage in the prediction of prognosis, it cannot be determined with presurgical information. Essentially, pathologic stage differentiates between organ-confined or non-organ-confined disease.

The most common modalities currently used by clinicians as prognostic markers for prostate cancer are DRE, histologic tumor grade (Gleason score), serum markers (PSA and prostatic acid phosphatase) and other radiologic studies. Even though most of these clinical staging modalities are widely used, none are sensitive or specific enough when used alone as prognostic markers for prostate cancer. Before the advent of PSA, the DRE was the principal method of prostate cancer detection and staging. This, however, is limited by inter-observer error and clinical experience.[8] The most frequently used system for grading the tumor is the Gleason score. This identifies and scores the first and second most common glandular pattern of the tumor, respectively. Narayan et al. correlated the biopsy Gleason score with final pathologic stage and found that, when the biopsy score was ≤6, 70% of patients had confined lesions. Conversely, only 34% had tumors confined to the prostate when the score was ≥7.[9]

The serum PSA level depends upon the size of the tumor, its differentiation and stage. It will be discussed in more detail later in this chapter but, for an individual, it is not a good predictor of clinical or pathologic stage. In general, men with PSA levels <4 ng/ml have organ-confined disease, whereas almost 50% of patients with PSA levels greater than 10 ng/ml have extracapsular extension.[10] Prostatic acid phosphatase (PAP), especially the enzymatic method, is one of the first oncologic tumor markers and has proven to be very useful in clinical practice as evidence of bony metastatic disease.

Currently no imaging studies are available that can reliably demonstrate evidence of early extraprostatic spread. The reported staging accuracy of transrectal ultrasound of the prostate (TRUS) has reported to range from 46% to 66%, and can be affected by many variables. The utility of MRI as a staging tool has also been evaluated. Although no advantage in local staging with body coil magnetic resonance imaging (MRI) has been noted, endorectal coil MRI has been shown to enhance the periprostatic soft tissue and prostate resolution when compared with conventional MRI.[11]

Radioimmune localization with Prostascint scan remains suspect as a standard staging modality. It has demonstrated clinical usefullness in imaging nodal metastases in patients with newly diagnosed prostate cancer. The radiolabeled antibody is directed against a glycoprotein expressed by prostatic epithelium. Its primary use is as an imaging modality when others have failed to detect metastatic disease in those patients at high risk, or in those with an increasing PSA level following radical prostatectomy or radiation therapy.[8]

Although computed tomography (CT) cannot identify intrinsic disease of the prostate, it has been utilized for the detection of nodal metastases. The detection of lymph node metastases is based on size criteria, with a nodal size of ≥1.0 cm often used as the upper limit of normal. The diagnostic sensitivity and specificity of this modality are 25–78% and 90%, respectively.[11] Pelvic lymphadenectomy remains the gold standard for the detection of nodal metastases, however, and can be performed at the time of radical prostatectomy with low associated morbidity.

Because prostate cancer metastasizes most commonly to bone and because early bone metastases are frequently asymptomatic, a radionuclide bone scan is very useful and accurate as an imaging modality to assess the axial skeleton for metastases. Further, plain films are poor at detecting bone metastases because a density change of ≥50% must occur before they can be seen radiographically. Bone scans may not be necessary to obtain for staging purposes in all newly diagnosed patients, however. With the availability of serum PSA, most urologists would agree that a bone scan is not necessary in patients whose PSA is ≤20 ng/ml. Many investigators have determined that the incidence of skeletal metastases is low when using this PSA value as a cut-off.[11]

SERUM PSA ISSUES

Over the past 15 years, serum PSA has developed into a significant marker for the diagnosis and management of prostate cancer. It was first discovered in human serum by Wang et al.[12] in 1979, and later isolated from prostate tissue by Papsidero et al.[13] in 1980. PSA is the first tumor marker to be approved by the Food and Drug Administration as an aid for diagnosing prostate cancer in population screening programs.[14] One of the major advantages of PSA as a tumor marker is its tissue specificity. However, the use of PSA as a prognostic marker has been rather limited because it is produced in large amounts by cells of benign prostatic hyperplasia (BPH) and prostate adenocarcinoma.

Produced almost exclusively by prostatic ductal and acinar epithelial cells, as well as male periurethral glands, it is primarily responsible for liquefaction of the seminal coagulum. PSA is a serine protease of the kallikrein family with a molecular weight of 35 kDa, and is secreted in high concentrations into seminal fluid. It is composed of a single polypeptide chain of 240 amino acids. Most PSA exists within serum bound to protease inhibitors such as α_1-antichymotrypsin (ACT) and α_2-macroglobulin. The rest exists unbound or free. Pathologic processes within the prostate, such as inflammation, hyperplasia and neoplasia, lead to progressive disruption of physiologic barriers. This, in turn, increases passage of PSA into the circulation.[15] This allows for its rather simple quantification in serum samples by modern immunoassay. The serum half-life of PSA is approximately 2–3 days.

The use of PSA testing in symptomatic men for the detection of prostate cancer is controversial, owing to the lack of evidence from randomized trials demonstrating the efficacy of early detection and treatment. Despite this controversy, PSA testing is currently widely used in the USA for early prostate cancer detection, and yearly PSA testing of men at greatest risk is recommended.[16] None the less, there are a variety of clinical uses of PSA as a test for screening the general population for prostate cancer. It has been thoroughly demonstrated that PSA offers an increased detection rate of prostate cancer compared to DRE alone, and further enhances the predictive value of DRE as well. It has also been described to improve the detection of clinically significant, organ-confined cancers.

In an effort to optimize the diagnostic value of serum PSA, several useful concepts and clinical algorithms have been developed. Some of these newer instruments include calculation of PSA density, PSA velocity, development of age-specific reference ranges and the quantification of per cent free PSA. Each of these studies were devised to make clinical decision-making simpler for the urologist when attempting to distinguish between patients with BPH and those with early, potentially curable prostate cancer.

The use of PSA density in the assessment of prostate cancer as a screening study was first utilized in 1992 by Benson et al.[17] PSA density is calculated by dividing the serum PSA value (ng/ml) by total prostate volume (cc). Prostate volume is typically determined at the time of TRUS. Since PSA is created by prostate epithelial cells, it is likely that variation in the amount of epithelium may be more important than the size of the prostate when determining serum PSA levels. The accuracy of this study, however, is limited by high variability in prostate gland shape, and standard volume equations may incorrectly measure prostate volume. It has been suggested that a PSA density of ⩾0.15 is significant in the context of a serum PSA level between 4.1 and 10 ng/ml. Using these guidelines would reduce the number of unnecessary TRUS prostate biopsies by about 60%. This would potentially miss 10% of prostate cancers and should, therefore, not be used as a standard screening instrument.

The advent of PSA velocity is credited to Carter et al. who first described its use in 1992.[18] It is defined as the rate of change of serum PSA value over a minimum time length of 18–24 months. Three repeat PSA measurements over this time period optimize the accuracy of this test. An increase of more than 0.75 ng/ml per year in PSA level was determined to be a specific marker for prostate cancer. Less than 5% of men without prostate cancer will demonstrate a PSA velocity >0.75 ng/ml. The broadest area of clinical usefulness of this measurement is in men followed long term with repeatedly normal results on DRE and serum PSA levels routinely <4 ng/ml.

Age-specific reference ranges have been developed as an additional modality in an attempt to improve the sensitivity and specificity of serum PSA in prostate cancer screening. It is well established that serum PSA increases with age, which is attributed to concurrent increase in prostatic size.[19] Oesterling et al. estimated ranges of normal PSA values for specific age groups in a prospective study of 471 men who ranged in age from 40 to 79 years old. None had evidence of prostate cancer by PSA, DRE or TRUS. Serum PSA was convincingly found to correlate directly with age over the entire age range and age-specific reference ranges were determined using the 95th percentile to define the appropriate upper limit of normal for age (**Table 8.2**).[20] Use of age-specific PSA ranges may lead to the increased detection of prostate cancer in younger men and therefore those patients more likely to benefit from early intervention. However, it is criticized by others because of the decreased detection that will occur in older men.

Table 8.2. Age-adjusted reference ranges for prostate specific antigen (PSA).

Age range (years)	PSA reference range (ng/ml)
40–49	0.0–2.5
50–59	0.0–3.5
60–69	0.0–4.5
70–79	0.0–6.5

Shortly after the use of serum PSA became routine in clinical practice, Christensson et al. found that the concentrations of total and free PSA may vary with the extent of the disease state. They were able to determine that a larger quantity of serum PSA existed complexed to ACT and a much lower percentage of that total PSA was free PSA.[21] The clinical utility of free PSA is best saved for those patients with an indeterminate serum PSA range between 4 and 10 ng/ml. Free PSA is able to provide independent predictive information for prostate cancer and does so better than any other clinical indices. When using a free/total PSA cutoff of 0.18 or less, the detection of prostate cancer is significantly improved when compared with the use of serum PSA alone.[22] Yet, much of its ability to detect cancer relies on prostate size. In men with enlarged prostates with or without prostate cancer, the percentage of free PSA greatly overlaps. Therefore, the percentage of free PSA that is most accurate in detecting prostate cancer has yet to be determined. Nevertheless, the calculation of free and complexed PSA proportions appears to be a promising study in the detection of early, organ-confined prostate cancer in patients with associated benign prostatic hyperplasia.

BIOPSY TECHNIQUE

Prostate biopsy can be performed either transperineally or transrectally, and there are advocates of both approaches. Transrectal digitally directed needle biopsy of the prostate gland was first performed in 1937.[23] It was not until the early 1980s that this procedure was performed using ultrasound guidance as it is commonly done today. Since that time, many technological advances have been developed that have revolutionized the ability to examine this gland. These include the development of hand-held high-resolution probes with multi-axial planar imaging capabilities and spring-loaded biopsy needle guns. This has proven to be a relatively rapid and well-tolerated procedure with low morbidity, which can be performed in the office setting.

The exact role of TRUS in cancer diagnosis has not been clearly established. The most important contribution is that it has drastically simplified the process of obtaining prostate biopsies. The ultimate objective is to detect clinically significant cancers and obtain a histologic diagnosis. This is where TRUS has proven to be most useful. Based upon the yearly incidence of new prostate cancer cases detected in the USA, it was estimated that more than half a million TRUS biopsy procedures were performed in the year 2000,[24] thus demonstrating its utility and feasibility. It allows for excellent visualization of both the prostate and seminal vesicles, and can provide invaluable information pertaining to the pathologic process.

Transrectal ultrasound has received increasing attention recently because of its potential for the early detection of prostate cancer.[25] Indications for TRUS in combination with needle biopsy generally include either an elevation in the serum PSA, or the finding of an abnormality on digital rectal examination (DRE). When DRE and PSA are used in combination, detection rates are higher than with either modality alone. Yet, TRUS alone is not recommended as a first-line screening test, owing to its relatively low predictive value for early detection of prostate cancer and the high cost of the examination.

Patient preparation is extremely important in order to achieve an adequate and successful evaluation. Several steps should be taken in order to optimize accuracy of the study and to minimize complications. First and foremost, informed consent should be properly obtained by the physician. A thorough DRE should also be performed prior to the procedure to establish baseline information on the prostate size and to evaluate for presence of prostatic nodules. Patients should be questioned about their history of oral anticoagulant (i.e. warfarin), non-steroidal anti-inflammatory agents or aspirin usage. These medications should be discontinued several days prior to the procedure to limit the potential for bleeding.

There exists significant variability in patient preparation between physicians. The urologic literature is saturated with various pre-prostate biopsy preparation regimens. This discordance among urologists is chiefly controversial in the area of antibiotic used, timing and duration, as well as if a cleansing enema is necessary. Shandera et al. surveyed 568 randomly selected urologists with regard to their routine pre-biopsy patient preparation. Interestingly, 98.6% prescribed some form of antibiotic prophylaxis. Oral quinolones were chosen by 91.6%. However, the duration of prophylactic treatment varied from one dose to 17 days. A rectal cleansing enema was included in the preparation by 81%. TRUS was used by 97% of the respondents, but only 91.5% used TRUS to guide the prostate biopsy.[26] For a procedure that is one of the most commonly performed by urologists, it is intriguing that a standardized protocol does not exist.

Yet, what constitutes the most effective regimen? In an attempt to answer this question, Shandera et al., in a non-randomized and prospective study, evaluated 150 consecutive patients who underwent TRUS prostate biopsy using a simple pre-biopsy regimen. Each patient received a minimal preparation consisting of one self-administered rectal enema, and one oral flouroquinolone tablet 1 hour prior to biopsy. This proved effective with an infectious complication rate of 0.75%.[27] In addition, the total patient cost of the preparation was $10.06. In this era of modern medicine, this demonstrates that cost containment while maintaining quality can be achieved. Therefore, 1-day antibiotic treatment can probably be proposed for prostate needle biopsies without any major risks and at a low cost to the patient.

The procedure is routinely performed with the patient in the left lateral decubitous position, with the knees flexed toward the chest. This position allows for maximum relaxation for the patient, and for easy insertion of the ultrasound probe. The dorsal lithotomy position can also be utilized, especially if cystoscopic examination is also planned.

Historically, no anesthetic is required, and the discomfort associated with the procedure is often minimal. However, Issa et al. demonstrated that 52% of patients report significant discomfort during biopsy without anesthesia. This was determined after more careful interrogation of patients. Discomfort was indicated by a pain score of 5 or greater on a scale of 1 to 10. The authors proposed that intrarectal instillation of 2% lidocaine gel is a simple, safe and effective approach toward limiting the discomfort during TRUS prostate biopsy.[24] Others have advocated performing TRUS-guided lidocaine nerve blockade of the inferior hypogastric plexus as they course inferolateral to the prostate. This, however, is technically more difficult and has demonstrated variable success in achieving anesthesia during the procedure. The theoretical risk of intravascular lidocaine injection also exists, which can lead to the onset of seizures.

Standard ultrasound equipment in the urologist's office should include a 7.0–7.5 MHz probe, which allows multiplanar imaging of the prostate gland. Prior to rectal insertion of the probe, it should be fitted with a well-lubricated jelly-filled condom. Next, a thorough evaluation of the prostate should commence from the apex to its base. Note should also be made of prostate volume, the presence of hypoechoic lesions and any areas of discernable asymmetry. Gray scale ultrasound equipment makes it possible for visualization of the internal echo pattern of the prostate. Early reports demonstrated that malignant tissue contains few sonographically detectable interfaces and, therefore, appears hypoechoic. Not all tumors are able to be identified with ultrasonography.[25] For patients who are not candidates for transrectal biopsy, owing to prior rectal surgery, D'Amico et al. described a new technique using real-time MRI-guided biopsy. However, its clinical usefulness has yet to be determined.[28]

After imaging is complete, transrectal biopsy is then performed. An 18-guage Biopty gun is inserted through a puncture guide attached to the ultrasound probe. This can be guided by manipulating the rectal probe, and switching between longitudinal and transverse images. The gun is then activated under real-time sonographic imaging. The biopsy course is easily visualized by a strong reflection from the metallic needle.

There exists significant debate regarding the appropriate number of cores, which should be obtained at biopsy. Hodge et al. defined the current standard of sextant biopsies in 1989.[29] A criticism of systemic biopsies is the possibility of detecting what may amount to small, insignificant cancer foci. The false-negative rate of standard sextant biopsy has been reported to range from 15% to 31%[30] and, in order to reduce the number of false-negative studies, Stamey recommended shifting the sextant biopsy regimen more laterally to increase

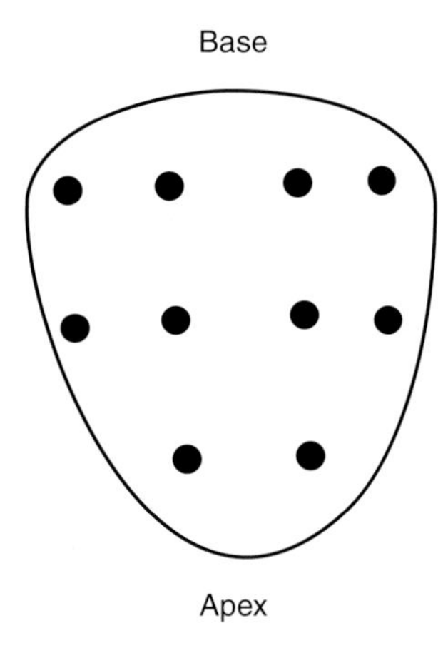

Fig. 8.1. An example of a ten-core prostate biopsy scheme.

the sampling of the peripheral zone, since most cancers arise in this area.[31] Recently, it has convincingly been demonstrated that adding 2–4 laterally placed peripheral zone biopsies to the traditional sextant technique improves the cancer detection rate by 42–44%.[32,33] It appears that the most important of the added biopsies are those directed more laterally at the base and mid-gland (**Fig. 8.1**). Presti et al. have suggested that this new systemic biopsy technique be considered as the new standard of practice.[32] When biopsy specimens are obtained from multiple areas, it is recommended that they be submitted in separate containers to the pathologist.

This issue becomes even more challenging in the presence of negative core biopsies in patients with high clinical suspicion of prostate cancer. Especially when palpable abnormalities exist, most urologists would mandate repeat biopsy under TRUS guidance. Babian et al. recently proposed an 11-core technique for men undergoing repeat biopsy. They incorporated five biopsies from three alternate sites in addition to the traditional sextant approach. This technique improved cancer detection when compared to conventional sextant biopsies by 37%.[34] In men undergoing repeat biopsy after a negative result, this appears to be a reasonable and well-tolerated approach. This issue will be covered in more detail in the next chapter.

With increasing numbers of core biopsies, the incidence of procedure morbidity increases as well. Reported complications rates in the literature range from 2.1% to 7.2%.[35] Most complications following TRUS biopsy are minor and limited to rectal bleeding, urinary tract infection or urinary retention. However, prostatitis, sepsis, hematuria and hematospermia have all been reported. There has been only one documented case of tumor implantation following transrectal biopsies.[36] Most complications are easily treated but prevention remains the best practice policy.

CONCLUSION

Prostate cancer is an extremely perplexing disease. One reason as to why this disease is so confusing to both doctors and patients is that it is sometimes difficult to distinguish men who will benefit from treatment from those in whom the side effects of therapy will outweigh any benefits. This requires a definition of what constitutes clinically significant prostate cancer. Unfortunately, there is no universal agreement on this issue at the present time.

Prostate specific antigen testing, in combination with annual digital rectal examination, is essential for early prostate cancer detection. Many clinical studies are available to urologists to establish a diagnosis and to assist in recommendations of therapy. The goal of cancer staging is to assess the extent of disease as precisely as possible. The use of serum PSA and many modern imaging modalities are currently available to gather information that ultimately help to guide management options. TRUS-guided biopsy is the mainstay for achieving a histologic diagnosis of prostate cancer. An optimal protocol for the detection of prostate cancer with transrectal ultrasound-guided biopsy has yet to be determined. Once determined, patients can be counseled appropriately and informed decision-making regarding therapy can be initiated.

REFERENCES

1. American Cancer Society, Inc. *Prostate Cancer: Treatment Guidelines For Patients, Version II*. Atlanta, GA: American Cancer Society, Inc., January 2001.
2. Pienta KJ. Etiology, epidemiology, and prevention of carcinoma of the prostate. In: Walsh PC, Retik AB, Vaughn, WD Jr et al. (eds) *Campbell's Urology*, 7th edn, pp. 2489–96. Philadelphia: W.B. Saunders Company, 1998.
3. Whitmore WF Jr. Hormone therapy in prostate cancer. *Am. J. Med.* 1956; 21:697.
4. Jewett HJ. The present status of radical prostatectomy for stages A and B prostatic cancer. *Urol. Clinics North Am.* 1975; 2:105.
5. Hermanek P, Sobin LH. Urological tumours. TNM Classification of Malignant Tumours, 4th edn, pp. 121–6. New York: Springer-Verlag, 1987.
6. Whitmore WF Jr, Catalona WJ, Grayhack JT et al. Organ systems program staging classification for prostate cancer. In: Coffey DS, Resnick MI, Dorr FA et al. (eds) A Multidisciplinary Analysis of Controversies in the Management of Prostate Cancer, pp. 295–7. New York: Plenum Press, 1988.
7. Scardino PT. Early detection of prostate cancer *Urol. Clinics North Am.* 1989; 16:635–55.
8. Wilkinson BA, Hamdy FC. State-of-the-art staging in prostate cancer. *Br. J. Urol.* 2001; 87:423–30.
9. Narayan P, Gajendran V, Taylor S et al. The role of transcretal ultrasound-guided biopsy-based staging, preoperative serum prostate-specific antigen, and biopsy Gleason score in prediction of final pathological diagnosis in prostate cancer. *Urology* 1995; 46:205–12.
10. Presti JC Jr. Prostate cancer: assessment of risk using digital rectal examination, tumor grade, prostate-specific antigen and systemic biopsy. *Radiol. Clinics North Am.* 2000; 38:49–58.
11. Perrotti M, Allan P, Rabbani F et al. Review of staging modalities in clinically localized prostate cancer. *Urology* 1999; 54:208–14.
12. Wang MC, Valenzuela LA, Murphy GP et al. Purification of a human prostate specific antigen. *Invest. Urol.* 1979; 17:159–63.
13. Papsidero LD, Wang MC, Valenzuela LA et al. A prostate antigen in sera of prostatic cancer patients. *Cancer Res.* 1980; 40:2428–32.

14. Diamandis EP. Prostate-specific antigen: a cancer fighter and a valuable messenger? *Clin. Chem.* 2000; 46:896–900.
15. Tchetgen MB, Oesterling JE. The role of prostate-specific antigen in the evaluation of benign prostatic hyperplasia. *Urol. Clinics North Am.* 1995; 22:333–44.
16. Carter HB, Epstein JI, Chan DW et al. Recommended prostate-specific antigen testing intervals for the detection of curable prostate cancer. *JAMA* 1997; 277:1455–60.
17. Benson MC, Wang IS, Olsson CA. The use of prostate specific antigen density to enhance the predictive value of intermediate levels of serum prostate specific antigen. *J. Urol.* 1992; 147:815–16.
18. Carter HB, Pearson JD, Metter EJ et al. Longitudinal evaluation of prostate-specific antigen levels in men with and without prostate disease. *JAMA* 1992; 267:2215–20.
19. Collins GN, Lee RJ, McKelvie GB et al. Relationship between prostate-specific antigen prostate volume and age in the benign prostate. *Bri. J. Urol.* 1993; 71:445.
20. Oesterling JE, Jacobsen SJ, Chute CG et al. Serum prostate-specific antigen in a community-based population of healthy men: establishment of age-specific reference ranges. *JAMA* 1993; 270:7.
21. Christensson A, Bjork T, Nilsson O et al. Serum prostate-specific antigen complexed to alpha$_1$-antichymotripsin as an Indicator of Prostate Cancer. *J. Urol.* 1993; 150:100–5.
22. Catalona WJ, Smith DS, Wolfert RL et al. Evaluation of percentage of free serum prostate-specific antigen to improve specificity of prostate cancer screening. *JAMA* 1995; 274:1214–20.
23. Renfer LG, Vaccaro JA, Kiesling VJ et al. Digitally-directed transrectal biopsy using biopty gun versus transrectal needle aspiration: comparison of diagnostic yield and comfort. *Urology* 1991; 38:108–12.
24. Issa MM, Bux S, Chun T, Petros JA et al. A randomized prospective trial of intrarectal lidocaine for pain control during transrectal prostate biopsy: the Emory University experience. *J. Urol.* 2000; 164:397–9.
25. Shinohara K, Wheeler TM, Scardino PT. The appearance of prostate cancer on transrectal ultrasonography: correlation of imaging and pathologic examinations. *J. Urol.* 1989; 142:76–82.
26. Shandera KC, Thibault GP, Deshon GE Jr. Efficacy of one dose fluoroquinolone before prostate biopsy. *Urology* 1998; 52:641–3.
27. Shandera KC, Thibault GP, Deshon GE Jr. Variability in patient preparation for prostate biopsy among American urologists. *Urology* 52:644–6.
28. D'Amico AV, Tempany CM, Cormack R et al. Transperineal magnetic image guided prostate biopsy. *J. Urol.* 2000; 164:385–7.
29. Hodge KK, McNeal JE, Terris MK et al. Random systemic versus directed ultrasound guided transrectal core biopsies of the prostate. *J. Urol.* 1989; 142:71–5.
30. Borboroglu PG, Comer SW, Riffenburgh RH et al. Extensive repeat transrectal ultrasound guided prostate biopsy in patients with previous benign sextant biopsies. *J. Urol.* 2000; 163:158–62.
31. Stamey TA. Making the most out of six systemic sextant biopsies. *Urology* 1995; 45:2.
32. Presti JC, Chang JJ, Bhargava V et al. The optimal systemic prostate biopsy scheme should include 8 rather than 6 biopsies: results of a prospective clinical trial. *J. Urol.* 2000; 163:163–7.
33. Chang JJ, Shinohara K, Bhargava V et al. Prospective evaluation of lateral biopsies of the peripheral zone for prostate cancer detection. *J. Urol.* 1998; 160:2111–14.
34. Babian RJ, Toi A, Kamoi K et al. A comparative analysis of sextant and an extended 11-core multisite directed biopsy strategy. *J. Urol.* 2000; 163:152–7.
35. Desmond PM, Clark J, Thompson IM et al. Morbidity with contemporary prostate biopsy. *J. Urol.* 1993; 150:1425–6.
36. Blight EM. Seeding of prostate adenocarcinoma following transrectal needle biopsy. *Urology* 1992; 34:297–8.

Prostate Biopsy: Who, How and When?

Bob Djavan, Mesut Remzi and Michael Marberger
Department of Urology, University of Vienna, Vienna, Austria

INTRODUCTION

The widespread use of digital rectal examination (DRE) coupled with measurement of prostate specific antigen (PSA) for prostate cancer screening has led to a dramatic increase in the number of men undergoing prostate biopsy guided by transrectal ultrasound (TRUS). Systematic parasagittal sextant biopsies have been widely adopted as the standard protocol for prostate biopsy. Significant numbers of men undergo repeat biopsy because of clinical indications (abnormal DRE, elevated PSA) or atypical findings on the initial biopsy or because it is well known that prostate cancer is often multifocal.[1] Frequently, urologists are faced with the dilemma of treating a patient with a high index of suspicion of prostate cancer but an initial set of negative biopsies.

In recent studies, cancer detection rate on repeat prostate biopsy was found to be between 10% and 20%. If cancer is still present in 1 out of 10 to 1 out of 5 of patients with a negative initial biopsy, one has to re-evaluate the biopsy policy. On the other hand, one needs to consider when to stop the repeat prostate biopsy. The answer to this dilemma lies in a prospective evaluation of cancers detected on repeat biopsy. If cancers detected on repeat biopsy have 'significant' features or features similar to cancers detected on initial biopsy, then repeat biopsy should be advocated and balanced against biopsy related costs and patient morbidity.

A number of diagnostic dilemmas arise regarding repeat biopsies:

- Should we perform lesion-directed or random biopsies?
- How many biopsy cores should be obtained for optimal diagnostic yield, to reduce the incidence of false-negative biopsies?
- What areas of the prostate should be biopsied to give the best diagnostic results?
- If the initial biopsy fails to detect cancer, who should undergo repeat biopsy, and how often should we repeat it?
- What role do PSA and PSA parameters, such as PSA density (PSAD), PSA density of the transition zone (PSA-TZ), PSA velocity (PSAV) and % free PSA, play in findings upon the outcome of repeat biopsy?
- Are cancers detected on multiple repeat biopsies clinically 'significant' and thus do they deserve therapy?

THE IMPACT OF PROSTATE VOLUME

The impact of prostate volume in the decision of prostate biopsy technique and of when to perform a repeat prostate biopsy is still a matter of debate. Given that sextant prostatic systematic biopsies sample only about 90 mm of prostate tissue (6 × 15 mm cores), increased prostatic volume may significantly reduce the chance of detecting cancer.[2–4]

Owing to the wide variation in gland sizes and shapes, it seems logical that in smaller glands the prostate biopsy leads to a more extensive sampling and less extensive or suboptimal sampling of larger prostates with accompanying significant differences in biopsy yields.[5,6] Several groups have proposed new biopsy strategies, often increasing the number of biopsies and the sectors to be sampled or performing biopsies more laterally, but controversial data have also been reported about the real advantage of these new techniques.[7–10] Concerned by the fact that the original sextant method may not include adequate sampling of the prostate, Norberg et al. found in a prospective study including 512 patients that the standard method left 15% of cancers undetected as compared with the results of a more extensive procedure using 8–10 biopsies.[11]

Recently, the influence of the total and transition zone (TZ) prostate volumes on prostate cancer detection was prospectively analyzed in 1018 men using two successive sets of sextant biopsies plus two TZ biopsies.[12] As compared to patients diagnosed with prostate cancer after the first set of biopsies, patients diagnosed after the second set had larger total prostate and TZ volumes ($43.1 \pm 13.0\,cm^3$ vs. $32.5 \pm 10.6\,cm^3$, $P < 0.0001$ and $20.5 \pm 8.3\,cm^3$ vs. $12.8 \pm 6.0\,cm^3$, $P < 0.0001$). Receiver operator characteristic (ROC) curves showed that total and TZ volumes of $45\,cm^3$ and $22.5\,cm^3$, respectively, provided the best combination of sensitivity and specificity for discriminating between patients diagnosed with prostate cancer after the first from those diagnosed after a second set. In patients with total prostate volume above 45 cm^3 and TZ above 22.5 cm^3, a single set of sextant biopsies was not sufficient to rule out prostate cancer and a repeat biopsy was to be considered in case of a negative first biopsy.

Evaluating the variation of cancer detection in relation to prostate size, through random systematic sextant biopsies, Uzzo et al. found 23% of the patients had cancer and a large prostate $\geqslant$50 cm^3 compared to 38% in patients with smaller prostates ($P < 0.01$).[13] This group concluded that significant sampling errors may occur in men with large glands, suggesting the need for repeat biopsies.

One of the central difficulties elucidating the relationship of prostate size and biopsy yield is the fact that prostates without cancer are virtually unavailable for pathologic analysis. Chen and colleagues approached this problem by developing a novel computer simulation that allowed comparison of biopsy results for given prostate and cancer volumes. The authors first evaluated 180 whole-mount radical prostatectomy specimens. The prostates were weighed, step-sectioned and digitized for computer modelling. Tumor volumes were calculated and compared with prostate volumes. Overall 607 tumours in 180 prostates were quantified. Computer-simulated biopsy runs were performed on the digitized prostates. Sextant biopsies in glands weighing $\leqslant$50 g and glands >50 g were positive in 67% and 48% of cases, respectively. Small-volume cancers were more prevalent in larger prostates. The authors concluded that biopsy rates in large glands were lower than in small glands because biopsy of larger glands is often driven by elevations of PSA that may be produced by benign prostate tissue. They argued that, if sampling error were the primary reason for the consistent finding of lower biopsy yields in larger prostates, larger volume cancers would preferentially be found in large prostates. Chen and associates recommended against obtaining extra biopsies solely because of larger prostate size, arguing that this would be likely to detect a disproportionate number of small-volume cancers. The question of optimal biopsy sites and number of cores (independent of prostate volume) is still unanswered. Chen et al., following a stochastic computer-simulation model, developed a 10-case biopsy scheme incorporating midline peripheral zone (PZ), inferior portion of the anterior horn of the PZ, significantly improving cancer detection to 96% and thus recommending the sampling of these zones in the repeat biopsy strategy after prior negative biopsies.[14] Eskew et al. have recommended the use of a five-region sampling technique to improve the cancer detection rate, since the 11-core technique increased the percentage of prostate cancer detected from 26% to 40% compared to the usual sextant biopsies.[8]

The importance of the total volume on prostate cancer yielded by sextant biopsies is generally accepted. The importance of the transition zone was less well investigated.[15] Many authors have already stressed the importance of the TZ and especially of the TZ density for predicting prostate cancer in men with serum PSA levels below 10 ng/ml.[16,17]

Recently, another group has used two consecutive sets of transrectal ultrasound-guided sextant biopsies for improving prostate cancer detection.[18] These two sets were performed during the same session. Prostate cancer was detected in 43%, 27% and 24% of men with prostate volumes <30 cm^3, 30–50 cm^3, and >50 cm^3, respectively. Analyzing the second set of biopsies, performed in the same session, the probability of detecting a prostate cancer was approximately two-fold greater in men with large prostates compared to men with smaller and intermediate-sized prostates. Letran et al. and Djavan et al. suggested that the total and PZ volumes significantly affected the biopsy yield only when both were above the 75th percentile.[16,19]

Interestingly, the debate over the need of more lateral biopsies is open. When tumor foci from the 40 cases in which sextant biopsies did not reliably detect tumor were mapped, Chen et al. found that the foci were distributed in areas not biopsied by the sextant method, that is the TZ, midline PZ and inferior portion of the anterior horn of the PZ. A 10-core biopsy scheme incorporating these areas as well as the posterolateral prostate reliably detected cancer in 141 of 147 patients (96%) with total tumor volumes greater than 0.5 cm^3.[14]

A systematic five-region biopsy technique has been proposed where, in addition to random sextant biopsies, two sets of lateral PZ and three midline biopsies were obtained.[8] A total of 12–15 cores were taken when the prostate volume was less than 50 cm^3 and up to 18 cores if the prostate was larger than 50 cm^3. An improvement in diagnostic yield of 35% over the systematic sextant biopsy method alone was noted, with 83% of the additional tumors identified having a Gleason score of six or more.[8] Levine and coworkers noted that cancer detection rates could be increased by 30% by performing two consecutive sets of sextant biopsies at a single office visit.[18]

Creating an accurate model for prostate biopsy requires the ability to determine prostate gland volume and the minimum number of cores required to detect cancers with a high degree of certainty. Clinically, patient's age is the major determinant of life-threatening tumor volume at diagnosis. Subsequently, Vashi et al. and Djavan et al. proposed a nomogram concerning the number of cores for biopsy, required to ensure a 90% certainty of cancer detection as a function of prostate gland size and life-threatening volume.[20,21] The Vienna nomogram was based on patient's age and gland volume (**Table 9.1**). The authors evaluated the minimum number of cores needed to detect cancer accurately. This model was based on the findings of the European Prostate Cancer Detection Study (EPCDS) and a three-dimensional model of virtual biopsies taken from prostatectomy specimens.

- Increasing (>6) biopsy cores are needed.
- Biopsies should be directed more laterally (peripheral zone).
- Repeat biopsies should be performed when the initial biopsy shows no prostate cancer in total prostate volumes greater than 45 cm^3 and/or greater transition zone volumes than 25 cm^3.
- There is a significant sampling error in greater prostate glands (>50 cm^3), therefore, a repeat biopsy is needed, when initial biopsy yield showed no prostate cancer.

Table 9.1. Vienna nomogram: the number of cores needed per biopsy to ensure 90% certainty of cancer detection as a function of prostate gland size and age.

Size of prostate gland (cm^3)	Age (years)			
	<50	50–60	60–70	>70
20–29	8	8	8	6
30–39	12	10	8	6
40–49	14	12	10	8
50–59	16	14	12	10
60–69	—	16	14	12
>70	—	18	16	14

- Small volume cancers are more frequent in larger prostates.
- Performing two sets of sextant biopsy, patients with greater prostates (>50 cm^3) have twofold higher probability of detecting prostate cancer.

THE IMPACT OF PSA DERIVATES

As we know from PSA-based screening studies, approximately 9% of all asymptomatic men will have elevated serum PSA values but only about a third will have cancer detected on initial biopsy.[22,23] The question is whether the 66% men with an initially negative prostate biopsy have an elevated serum PSA value because of benign prostatic hyperplasia, prostatitis or undetected cancer in their peripheral or TZ of the prostate. Keetch et al. have previously shown that cancer detection rates at biopsies two or three were 19% and 8%, respectively. However, their patients underwent serial biopsies of the PZ only.[24]

In a recent study, the ability of % free PSA, PSAD and PSA-TZ to increase the sensitivity and specificity of PSA screening was evaluated prospectively.[25] Of the 1051 men with a total PSA level of 4–10 ng/ml, the initial biopsy was positive for prostate cancer in 22% (231 out of 1051) of the subjects. All 820 subjects diagnosed with benign prostatic hyperplasia (BPH) after the initial biopsy underwent a repeat prostatic biopsy 6 weeks later. Prostate cancer was detected in 10% (83 out of 820) of these subjects. Compared to the 231 subjects in whom prostate cancer was detected from the initial biopsy, both total prostate volume and TZ volume were significantly higher ($P < 0.001$) in the 83 subjects with prostate cancer detected in the repeat biopsy sample. The majority of cancers (84%) were detected in PZ, compared to 16% detected in the TZ. Total PSA, PSAD and PSA-TZ were all significantly higher in subjects diagnosed with prostate cancer in initial and repeat biopsy ($P < 0.01$).

Based on the results of a previous study,[16] cut-off values improving specificity with a sufficiently high level of cancer detection (sensitivity) were selected. At a cut-off of 30% and 0.26 ng/ml per cm^3, respectively, % free PSA and PSA-TZ were the most accurate predictors of a positive repeat biopsy result. Although a similar number of unnecessary repeat biopsies would have been eliminated by either % free PSA or PSA-TZ (approximately 50%), prostate cancer detection was much lower for PSA-TZ (78%) compared to % free PSA (90%).

In both the overall group (initial biopsy results plus repeat biopsy results) and the repeat biopsy group, % free PSA and PSA-TZ were the best predictors of prostate cancer ($r=0.2150$, $P < 0.001$). The respective areas under curve (AUCs) for % free PSA and PSA-TZ were 74.2% and 82.7% in the overall group, and 74.5% and 69.1% in the repeat biopsy group ($P < 0.001$). Although Zlotta et al. previously reported that measurement of the TZ volume is accurate and reproducible,[26] all TRUS measurements are operator dependent and, therefore, subject to possible variability. Furthermore, the measurement of the TZ in small or large prostates sometimes might be difficult. It was reported previously that PSA-TZ was not a significant predictor of biopsy results when the total prostate volume was <30 cm^3.[16]

In a multicenter study of 773 men with a total PSA level between 4 and 10 ng/ml (379 with prostate cancer and 394 with BPH), a % free PSA cut-off of 25% had a sensitivity of 95% and a specificity of 20%.[27] Another multicenter study of 317 men with a total PSA level between 4.0 to 10.0 ng/ml found that a % free PSA cut-off of 26% detected 95% of subjects with prostate cancer and eliminated 29% of negative biopsies.[28] In another study of 308 volunteers with an elevated total PSA level (2.5–10.0 ng/ml), a % free PSA cut-off of >20% would have eliminated 45.5% of negative biopsies. When % free PSA was combined with a PSA-TZ density cut-off of >0.22 ng/ml per cm^3, 54.2% of negative biopsies could have been avoided.[29]

Although some previous studies focused on repeat-biopsy results, most have been retrospective trials conducted in relatively small patient populations.[24,30–39] In one study of 193 men with a negative initial biopsy, 51 (26%) were found to have prostate cancer on repeat biopsy. Total PSA and volume-referenced PSA had the highest sensitivity. Another retrospective study in 51 men with a total serum PSA level of 2–15 ng/ml demonstrated a lower median % free PSA value in patients with a positive repeat biopsy compared with those with a negative biopsy (15% vs. 19%, respectively; $P=0.05$).[30] A % free PSA cut-off of 22% yielded a sensitivity of 95% and a specificity of 44% for predicting repeat biopsy results. In a prospective study of 67 men with persistent total PSA elevations and normal DRE, a low % free PSA (<10%) was a powerful predictor of prostate cancer, even after two negative biopsies. The AUC of the ROC curve was 0.93 for % free PSA, compared with 0.69, 0.66 and 0.51 for free PSAD, PSAD and total PSA, respectively.[32] In another study that used % free PSA to predict repeat biopsy results, prostate cancer was detected in 20 (20%) of 99 men with a total PSA level between 4.1 and 10.0 ng/ml, and an initial biopsy that was negative for prostate cancer.[34] Percentage free PSA cut-offs of 28% and 30% had a sensitivity of 90% and 95%, respectively, and a specificity of 13% and 12%, respectively. For PSAD cut-offs of 0.10 and 0.08 ng/ml/cm^3, the respective sensitivities were 90% and 95%, and the respective specificities were 31% and 12%.

In three of the previous studies, the positive repeat biopsy rate was 17% after an interval of 14.7 months between the first and second biopsies,[30] 29% after an interval of 19.1 months[31] and 30% after 12.8 month.[33]

- PSA, PSA-TZ and PSAD are significantly higher in patients detected for prostate cancer in second prostate biopsy.
- Percentage free PSA (cut-off 30%) is better than PSA-TZ (cut-off 0.26 ng/ml per cm^3) for prostate cancer detection in repeat biopsy.
- Using % free PSA, PSAD, PSA-TZ leads to higher specificity at 95% sensitivity for the first prostate biopsy, and increases sensitivity and specificity in repeat prostate biopsy.

PATHOLOGICAL FEATURES OF PROSTATE BIOPSY: WHEN TO STOP?

Little was reported on the differences in pathological stage, grade and cancer behavior of cancers detected on initial and repeat prostate biopsy. Although optimal predictors of cancer detection on repeat biopsy are crucial, one can spare or delay a repeat biopsy if the cancers detected are 'insignificant'. Certainly the dilemma of prostate cancer is that only a small proportion of men with untreated cancer will die from it, especially if these are small in volume, well differentiated and detected on repeat biopsy.

In a recent study, Djavan et al. showed that of cancers detected on initial ($n = 231$) and repeat biopsy ($n = 83$), 148 out of 231 (64%) and 56 out of 83 (67.5%) had clinically localized disease, respectively, and were offered radical prostatectomy or radiation therapy.[25] Watchful waiting was not offered as a primary option. Ten out of 148 (6.7%) and 3 out of 56 (5.3%), respectively, opted for radiation therapy, and thus 138 out of 148 (93.2%) and 53 out of 56 (94.6%), respectively, underwent radical retropubic prostatectomy. All specimen underwent histopathological evaluation by a single pathologist. Overall, 58.0% and 60.9% had organ-confined disease in both groups, respectively. No differences were noted with respect to organ confinement ($P = 0.15$), extracapsular extension ($P = 0.22$) and seminal vesical invasion ($P = 0.28$). Positive margins were noted in 23% and 18%, respectively ($P = 0.23$). No differences were noted between both cancer groups (initial vs. repeat) in the biopsy Gleason score (6.0 vs. 5.7; $P = 0.252$) as well as in Gleason score of the surgical specimen (5.3 vs. 4.9; $P = 0.358$). The same accounted for the % Gleason grade 4/5 (31.1% vs. 29.8%; $P = 0.10$). In contrast, cancers detected on initial biopsy expressed a higher rate of multifocality ($P = 0.009$), whereas overall cancer volume was identical ($P = 0.271$) in both groups.[25]

Recently, Stamey et al. challenged the 'traditional' predictors of cancer progression, such as stage, capsular penetration and surgical margins.[40] In a retrospective analysis of 379 men treated by radical prostatectomy only, eight morphologic variables were analyzed and associated with cancer progression, defined by an increasing PSA level (≥0.07 ng/ml). They identified % Gleason score 4/5, cancer volume, positive lymph node findings and intraprostatic vascular invasions as independent predictors of cancer progression.[40]

In contrast, cancers detected on initial biopsy expressed a higher rate of multifocality ($P = 0.0009$), whereas overall cancer volume was identical ($P = 0.271$) in both groups. Based on these findings, Djavan et al. concluded that cancers detected on repeat biopsy exhibit similar characteristics as cancers detected initially.[25] Thus, repeat biopsies do detect significant cancers and a repeat biopsy policy should be advocated in case of a negative initial biopsy. This conclusion, however, is limited to cases in which initial and repeat biopsies are performed in a similar fashion, as was done in the current study. If the biopsy technique is modified, cancers detected may differ and the conclusion may differ as well.

In a recent study, Djavan et al. presented the results of a prospective study of the pathological features find in first, second, third and fourth prostate biopsy. Of those with benign prostatic tissue on the first, second and third biopsy, 820 out of 829, 737 out of 756 and 94 out of 101 agreed to undergo repeat biopsy. Cancer detection rates on first, second, third and fourth biopsy were 22% (231 out of 1051), 10% (83 out of 820), 5% (36 out of 737) and 4% (4 out of 94), respectively. Overall, of patients with clinically localized disease (67% of cancers detected), 86% underwent radical prostatectomy and 14% opted for watchful waiting or radiation therapy. Of cancers detected on initial ($n = 231$), repeat ($n = 83$), third ($n = 36$) and fourth biopsy ($n = 4$), 148 out of 231 (64%), 56 out of 83 (67.5%), 33 out of 36 (91.6%) and 4/4 (100%) had a clinically localized disease, respectively, and were offered radical prostatectomy or radiation therapy. Watchful waiting was not offered as a primary option. Ten out of 148 (6.7%), 3 out of 56 (5.3%), 1 out of 33 (3%) and 0 out of 4 (0%), respectively, opted for radiation therapy, and thus 138 out of 148 (93.3%), 53 out of 56 (94.7%), 32 out of 33 (97%) and 4 out of 4 (100%), respectively, underwent radical retropubic prostatectomy. All specimens underwent histopathological evaluation by a single pathologist at each institution. Overall, 58.0%, 60.9%, 86.3% and 100% had organ-confined disease on first, repeat, third and fourth biopsy, respectively. No differences were noted with respect to organ confinement (OC) ($P = 0.15$), extracapsular extension (ECE) ($P = 0.22$) and seminal vesical invasion (SV) ($P = 0.28$) between first and repeat biopsy, whereas the same parameters were significantly different (higher values for organ confinement and lower for all other parameters) for cancers on third versus first biopsy ($P = 0.001$, $P = 0.02$, $P = 0.01$, respectively) as well as cancers on fourth versus first biopsy ($P = 0.001$, $P = 0.01$, $P = 0.001$, respectively). Positive margins (M+) were noted in 23%, 18% ($P = 0.23$), 8% ($P = 0.03$) and 0%, respectively. No differences were noted between cancers detected on initial versus repeat biopsy in the biopsy Gleason score (6.0 vs. 5.7; $P = 0.252$) as well as in Gleason score of the surgical specimen (5.3 vs. 4.9; $P = 0.358$). The same accounted for the % Gleason grade 4/5 (31.1% vs. 29.8%; $P = 0.10$). Cancers detected on initial biopsy expressed a higher

Table 9.2. Cancer characteristics and grading of prostate cancers detected on first, second, third and fourth repeat biopsy.

	First biopsy	Second biopsy	*P*	Third biopsy	*P*	Fourth biopsy	*P*
Gleason score Bx	6.0±0.7	5.7±0.5	0.25	4.6±0.4	0.02	4.4±0.7	0.01
Gleason score RPE	5.3±0.5	4.9±0.8	0.36	4.2±0.3	0.001	4.0±0.4	0.001
% grade 4/5	31.1%	29.8%	0.1	8.2%	0.02	0%	—
Multifocality	2.6±0.4	1.8±0.2	0.009	1.6±0.4	0.009	1.9±0.1	0.008
Cancer volume (cm^3)	4.2±0.7	4.9±0.8	0.27	0.83±0.5	0.001	0.79±0.4	0.001

RPE, radical prostatectomy.

rate of multifocality ($P=0.009$), whereas overall cancer volume was identical ($P=0.271$) on first and second biopsy. In contrast, cancers detected on third and fourth biopsy had a significantly lower biopsy Gleason score (4.6, $P=0.02$ and 4.4, $P=0.01$), Gleason score of the specimen (4.2, $P=0.001$ and 4.0, $P=0.001$), grade 4/5 cancer (8.2%, $P=0.02$ and 0%), rate of multifocality ($P=0.009$ and $P=0.008$), cancer volume (0.83 cm^3, $P=0.001$ and 0.79 cm^3, $P=0.001$) and stage ($P=0.001$ and $P=0.001$), respectively, when compared to cancers detected on first biopsy (**Table 9.2**).

- Cancers detected after second prostate biopsy have a significantly lower Gleason score, Gleason 4/5 grade, multifocality, cancer volume and pathological stage. Therefore, these prostate cancers can be called 'insignificant'. However, based on these findings, a systematic prostate biopsy after second repeat biopsy is not needed.

BIOPSY TECHNIQUES

After introducing the widely practiced technique of random systematic sextant prostatic biopsies by the Stanford group in 1989,[41] the optimal biopsy technique is a matter of debate. In a study of 273 men with suspected prostate cancer, while sextant biopsies alone detected 82% of cancers and lateral PZ biopsies alone detected 70% of cancers, when both techniques were combined, the cancer detection rate increased to 96%. A cancer detection rate of 59% was achieved, if only hypoechoic lesions were targeted for biopsy. The authors concluded that combining both systematic sextant and lateral PZ biopsies would detect the majority of prostate cancers and eliminate the need for lesion-directed biopsies.[10]

Perhaps the most sensible approach to this issue is that proposed by Presti and coworkers, who performed lateral PZ biopsies of the base (two) and midgland (two) in addition to routine sextant biopsies in 483 consecutive men referred for biopsy with an abnormal DRE and/or an elevated serum PSA.[42] A total of 202 cancers (42%) were detected, the majority (96%) by the combination of lateral PZ and sextant biopsies. Of the eight 'missed' cancers, five were detected by lesion-directed biopsies. Furthermore, three cancers were found in the transition zone of prostates >50 cm^3. If midlobar base biopsies were omitted from the protocol, the resultant eight-biopsy PZ regimen detected 95% of tumors. The systematic sextant biopsy method, practiced for a decade now, is inadequate and more extensive biopsy protocols obtaining a minimum of eight tissue cores, particularly from the lateral PZs, should be performed.

The detailed maps of consecutive radical prostatectomies have shown that prostate cancers expand mostly in the transverse direction across the posterior surface of the capsule, followed by the cephalocaudal direction.[2] Directing the biopsies more lateral to the midparasagittal plane might enable sampling of the large group of cancers located more laterally in the PZ.[2]

The posterior and posterolateral border of compressed fibromuscular tissues caused by the expansion of transition zone hyperplastic nodules, observed on TRUS, serves as an excellent marker for placement of PZ biopsies where more than 70% of cancers originate.[2] Based on a three-dimensional (3D) computer-assisted prostate biopsy simulator and mounted step-sectional radical prostatectomy specimens, Bauer et al. found that all the biopsy protocols that use laterally placed biopsies based on the five-region anatomic model are superior to the routinely used sextant prostate biopsy technique.[9]

Although the majority of tumors arise in the PZ, up to 24% of prostate cancers originate in the TZ.[43] Other studies have reported rates significant lower than this. Of 847 men who underwent TRUS-guided systematic sextant and TZ biopsies, only eight (2.9%) patients (4.1% of only state T1c considered) had solitary TZ tumors.[44] Terris noted that patients undergoing routine TZ biopsies, in addition to sextant biopsies, had positive biopsies involving the TZ alone in 1.8% of cases.[45] In a similar study, isolated TZ cancer was found in only 2.6% of men biopsied.[46] The Johns Hopkins group performed repeat sextant and TZ biopsies in 193 radical prostatectomy specimens. They found that the TZ biopsy by itself was positive in only 2.1% of cases, concluding that routine TZ biopsies are not justified in light of such low detection rates.[47] The low diagnostic yield of systematic TZ biopsy at the time of initial biopsy argues strongly against their routine use for detection of early-stage prostate cancer. However, performing TZ biopsies as part of a protocol of repeat systematic biopsies in selected men with prior negative biopsies may sometimes provide important information, which is additional to that obtained by repeating sextant biopsies alone.

To evaluate cancer location on initial and repeat biopsy in the EPCD, all cancers were entered in a 3D spatial model and areas of highest cancer density measured by means of a computer-based 3D image reconstruction.[9] Whereas cancers detected on initial biopsy are distributed homogeneously over the entire prostate, cancers on repeat biopsy are found in a more apicodorsal location. A comparison of tumor density between both groups shows significant differences in apical tumor density ($P=0.001$), in dorsal tumor density ($P=0.02$) and especially in apicodorsal tumor density ($P<0.001$).

This may explain the lower cancer detection rate on repeat biopsies (biopsies are rarely directed apicodorsally) as well as the fact that these cancers are commonly missed on initial biopsy. Thus, upon repeat biopsy, the biopsy technique should be modified and needles should be directed to a more apicodorsal location.[9]

- A minimum of eight prostate cores (particularly from the lateral PZ zone) is needed.
- The impact of systematic TZ zone biopsies is still unclear.
- Repeat prostate biopsies should be directed in a more apicodorsal orientation.

PIN/ATYPIA AND PROSTATE BIOPSY

High-grade prostatic intraepithelial neoplasia (PIN) is most likely a precursor of prostate cancer and is frequently associated with it, whereas the direct link between low-grade PIN and cancer is not established. The clinical evolution of isolated high-grade PIN has been the object of much concern because of the possibility of undiagnosed prostate cancer or the evolution of this premalignant lesion in invasive carcinoma.

High-grade PIN has been identified in approximately 4–14% of prostate needle biopsies.[48–50] Evaluation pathology trends show that, of 62 537 initial prostate needle-core biopsies submitted by office-based urologists, processed at a single pathology laboratory, isolated high-grade PIN was diagnosed in 4.1% of the biopsies, a number which probably reflects the real incidence of this entity in general practice as opposed to reference centres.[47] According to Zlotta et al. and Raviv et al.,[51–53] analyzing 93 patients with diagnosis of isolated PIN without concurrent carcinoma, the cancer detection rate on repeat biopsy or operated specimen increased with PIN grades. In each PSA subgroup (0–4, 4.1–10, >10 ng/ml) the subsequent cancer detection rate was higher for high-grade PIN than for low-grade PIN. Nearly 50% of patients with isolated high-grade PIN were found to have prostate cancer on repeat biopsy, whereas only 13% of patients with low-grade PIN did so.

Weinstein and Epstein noted that serum PSA levels were elevated in 90% of patients with high-grade PIN and cancer compared to 50% of those with PIN without associated cancer at the time of initial diagnosis, suggesting that serum PSA may be useful in distinguishing patients with PIN who will have cancer detected on repeat biopsy.[54] In the study by Raviv et al.,[51,52] the group of patients that later developed cancer had significantly higher PSA compared to patients without cancer on repeat biopsy. In high-grade PIN, the incidence of later cancer was 33% and 62%, respectively, when PSA was below 4 ng/ml or above 10 ng/ml.[52] Therefore, all subgroups of patients with high-grade PIN, but especially those with elevated PSA, should undergo repeat biopsy. Low-grade PIN was associated with subsequent cancer in 42.8% of cases when PSA was above 10 ng/ml, in 10.7% when PSA was between 4 and 10 ng/ml and in none of the cases when PSA was ≤4 ng/ml.[52] Low grade should cause no further action unless other factors, such as an elevated PSA, increase the suspicion of prostate cancer and should prompt repeat biopsies. The high detection rate of later cancer when PSA is over 10 ng/ml in low-grade PIN is similar to Cooner's report on the detection of prostate cancer without reference to PIN.[53] When PSA is above 10 ng/ml and whatever PIN grade, it seems clear that the high detection rate is more a reflection of the undetected presence of cancer than of the presence of PIN.

Most experts agree that high-grade PIN is clearly a preneoplastic lesion because it fulfils all requirements for such a lesion. Thus, high-grade PIN, especially when associated with high PSA or abnormal DRE or TRUS, should be taken as a signal of high to extremely high probability of prostate cancer and repeat biopsies should be performed.[53]

- High-grade PIN in first prostate biopsy requires repeat biopsy.

MORBIDITY OF PROSTATE BIOPSY

Generally, TRUS-guided biopsies are considered safe and are commonly performed in an outpatient setting. Djavan et al. found only two major complications in 1871 biopsies performed.[25] In contrast, minor complications were frequent with 69.7% of all patients experiencing at least one complication. Although these complications generally do not need intervention (neither conservative nor surgical), patients need to be informed adequately.

When evaluating 81 patients undergoing TRUS-guided biopsy, Irani et al. reported a mean visual analog pain scale (VAS score) of 3.[55] Nineteen per cent of patients refused a repeat biopsy without some sort of anesthesia, and the authors questioned the safety and morbidity of repeat biopsies.[55] Except for moderate to severe vasovagal episodes (2.8% vs. 1.4%, $P=0.03$), no differences were noted in pain apprehension (VAS score 2.4 vs. 2.6, $P=0.09$), and patient discomfort (moderate to severe in 8% vs. 11%, $P=0.29$) during first and repeat biopsy, respectively.[25] In contrast, in series reported by Collins et al.[56] and Clemens et al.,[57] 22% and 30% of patients, respectively, considered the procedure significantly painful. Djavan et al. found an age-dependent pattern in pain apprehension with the younger patient (age <60 years), reporting a higher discomfort rate.[25] Seventy-eight per cent of patients under 60 reported significant discomfort as compared to 33%, 11% and 8% for those in the age ranges 60–69, 70–80 and >80 years, respectively. Thus, younger patients need to be counselled adequately, and local or topical anaesthesia used. Rodriguez et al. confirmed also a higher discomfort rate in the younger age group as well.[58]

A review of the literature shows that the majority of the studies evaluated infectious and bleeding complications only at first biopsy. Several studies have demonstrated a significant decrease in infectious complications when prophylactic antibiotics were used.[25,59–63] Thompson et al. reported asymptomatic bacteraemia, commonly with *Bacteroides* species and *Enterococcus*.[64] Symptomatic infections were most commonly caused by *Escherichia coli* and *Enterococcus*, suggesting the use of fluoroquinolones and metronidazole for antibiotic prophylaxis.[60,64–66] Sieber et al. in a retrospective review of 4439 patients reported an infection rate of 0.1%.[67] Infectious complications are rare and, when needed, oral therapy is sufficient. Thus, cost issues need to be taken into consideration when selecting the type of antibiotic prophylaxis to be used. Urinary tract infections with fever were seen in 2.1% and 1.9% ($P=0.02$) of patients at first and repeat biopsy, respectively.[25] In one case (0.1%), a patient had to be admitted for prolepsis and was managed with intravenous antibiotics and discharged on day 6. In the immediate postbiopsy period, rectal bleeding (2.1% vs. 2.4%, $P=0.13$), mild hematuria (62 vs. 57%, $P=0.06$) and severe hematuria (0.7 vs. 0.5%, $P=0.09$) were reported for the first and repeat biopsy, respectively. Delayed hemorrhagic morbidity included hematospermia (9.8% vs. 10.2%, $P=0.1$), recurrent mild hematuria (15.9 vs. 16.6%, $P=0.06$), respectively. Again no difference was noted between first and repeat biopsy. The rate of persistent hematuria was lower than previously reported (30–50%).[59] This may be due to the fact that we routinely performed sextant biopsies and two additional TZ biopsies, whereas others had various numbers and sites of biopsy cores. Hematospermia was obviously reported for those patients who had ejaculation prior to the biopsy procedure ($n=671$). The lower rate in our population (as compared to 59% reported in the literature) was similar to that reported by Rodriguez et al.[59]

Transrectal ultrasound-guided biopsy of the prostate is generally well tolerated with minor pain and morbidity. Repeat biopsies can be performed 6 weeks later with no significant difference in pain apprehension, infectious or hemorrhagic complications.

- The morbidity of prostate biopsy in general is very low.
- Prophylactic antibiotic decreases morbidity.
- A systematic repeat biopsy has no more morbidity.
- Younger patients less than 60 years need to be counselled with respect to a higher pain and discomfort level during repeat biopsy, and local or topical anaesthesia offered, if desired.

SUMMARY

It seems that traditional sextant biopsies might become obsolete in future designed trials, since a body of evidence supports the findings that this technique is far from optimal in patients with large prostates. Whether the future lies in performing biopsies more laterally, increasing the number of biopsies to 10, 12, 15 or even 18, or simply repeating the sextant biopsy scheme still needs to be verified. Current data suggest that cancers detected on repeat biopsy have a similar stage and grade distribution and total PSA values compared to cancers found on initial biopsy. Moreover, specific biological determinants such as % Gleason grade 4/5, Gleason score and cancer volume were identical in both groups, suggesting similar biological properties and at least identical characteristics, thus, opting in favor of a repeat biopsy policy. Location and density measurements of cancers detected suggest that cancers missed on initial biopsy and subsequently detected on repeat biopsy are located in a more apicodorsal location. Repeat biopsies should thus be directed to this rather spared area in order to improve cancer detection rates.

However, cancers detected on third and fourth systematic biopsy show characteristics of ‘insignificant’ cancers. Therefore, a systematic repeat biopsy after the second biopsy in a PSA range of 2.5–10 ng/ml is not indicated and PSA-based watchful waiting should be advocaded. Percentage free PSA and PSA-TZ are the best markers for repeat prostate biopsy, increasing sensitivity and specificity. In prostate cancer patients detected on repeat biopsy, PSAD and PSA-TZ levels were higher. In all patients with high-grade PIN a repeat biopsy should be performed.

Repeat biopsies performed as early as 6 weeks after the initial biopsy were generally well tolerated with minor pain and morbidity. Vasovagal episodes were more frequent after the first biopsy. Antibiotic prophylaxis seems to be warranted and limited to 4 days at least. Younger patients less than 60 years need to be counselled with respect to a higher pain and discomfort level during repeat biopsy, and local or topical anesthesia offered, if desired.

REFERENCES

1. Djavan B, Susani M, Bursa B et al. Predictability and significance of multifocal prostate cancer in the radical prostatectomy specimen. *Tech. Urol.* 1999; 5:139–42.
2. Stamey TA. Making the most out of six systematic sextant biopsies. *Urology* 1995; 45:2–12.
3. Billebaud T, Villers A, Astier L et al. Advantage of systematic random ultrasound-guided biopsies, measurement of serum specific antigen levels and determination of prostate volume in the early diagnosis of prostate cancer. *Eur. Urol.* 1992; 21:6–14.
4. Vashi AR, Wojno KJ, Gillespie B et al. A model for the number of cores per prostate biopsy based on patient age and prostate gland volume. *J. Urol.* 1998; 159:920–4.
5. Karakiewicz PI, Aprikian AG, Meshref AW et al. Computer-assisted comparative analysis of four-sector and six section biopsies of the prostate. *Urology* 1996; 48:747–50.
6. Karakiewicz PI, Hanley JA, Bazinet M. Three-dimensional computer-assisted analysis of sector biopsy of the prostate. *Urology* 1998; 52:208–12.
7. Ravery V, Billeband T, Toublanc M et al. Diagnostic value of the systematic trus-guided prostate biopsies. *Eur. Urol.* 1999; 35:298–303.
8. Eskew AL, Bare RL, Mc Cullongh DL. Systematic 5-region prostate biopsy is superior to sextant method for diagnosing carcinoma of the prostate. *J. Urol.* 1997; 157:199–202.

9. Bauer JJ, Zeng J, Weir J et al. Three-dimensional computer-simulated prostate models: lateral prostate biopsies increase the detection rate of prostate cancer. *Urology* 1999; 53:961–7.
10. Chang JJ, Shinohara K, Bhargava V et al. Prospective evaluation of lateral biopsies of the peripheral zone for prostate cancer detection. *J. Urol.* 1998; 160:2111–14.
11. Norberg M, Egevad L, Holmberg L et al. The sextant protocol for ultrasound-guided core biopsies of the prostate underestimates the presence of cancer. *Urology* 1997; 50:562–6.
12. Djavan B, Zlotta AR, Ekane S et al. Is onset of sextant biopsies enough to rule out prostate cancer? Influence of transition and total prostate volumes on prostate cancer yield. *Eur. Urol.* 2000; 38:218–24.
13. Uzzo RG, Wei JT, Waldbaum RS et al. The influence of prostatic size on cancer detection. *Urology* 1995; 46:831–6.
14. Chen ME, Troncoso P, Johnston DA et al. Optimization of prostate biopsy strategy using computer based analysis. *J. Urol.* 1997; 158:2168–75.
15. Djavan B, Zlotta AR, Remzi M et al. Total and transition zone prostate volume and age: How do they affect the utility of PSA based diagnostic parameters for early prostate cancer detection? *Urology* 1999; 54:846–52.
16. Djavan B, Zlotta AR, Byttebier G et al. Prostate specific antigen density of the TZ for early detection of prostate cancer. *J. Urol.* 1998; 160:411–18.
17. Zlotta AR, Djavan B, Marberger M et al. Prostate specific antigen density of the transition zone: a new effective parameter for prostate cancer prediction. *J. Urol.* 1997; 157:1315–21.
18. Levine MA, Ittman M, Melamed J et al. Two consecutive sets of transrectal ultrasound guided sextant biopsies of the prostate for the detection of prostate cancer. *J. Urol.* 1998; 159:471–6.
19. Letran JL, Meyer GE, Loberiza FR et al. The effect of prostate volume on the yield of needle biopsy. *J. Urol.* 1998; 160:1718–21.
20. Vashi AR, Wojno KJ, Gillespie B et al. A model for the number of cores per prostate biopsy based on patient age and prostate gland volume. *J. Urol.* 1998; 159:920–4.
21. Djavan B, Rovery V, Zlotta A et al. Prospective evaluation of prostate cancer on biopsies 1, 2, 3 and 4: when should we stop. *J. Urol.* 2001; 166:1673–83.
22. Catalona WJ, Smith DS, Ratliff TD et al. Measurement of prostate specific antigen in serum as a screening test for prostate cancer. *N. Engl. J. Med.* 1991; 324:1156–61.
23. Brawer MK, Chetner MP, Beatie J et al. Screening for prostatic carcinoma with prostate specific antigen. *J. Urol.* 1992; 147:841–5.
24. Keetch DW, Catalona WJ. Prostatic TZ biopsies in men with previous negative biopsies and persistently elevated serum prostate specific antigen values. *J. Urol.* 1995; 154:1795–7.
25. Djavan B, Zlotta AR, Remzi M et al. Optimal predictors of prostate cancer in repeat prostate biopsy: a prospective study in 1,051 men. *J. Urol.* 2000; 163:1144–8.
26. Zlotta AR, Djavan B, Roumeguere T et al. Transition zone volume on transrectal ultrasonography is more accurate and reproducible than the total prostate volume. *Br. J. Urol.* 1997; Suppl. 80:A926.
27. Catalona WJ, Partin AW, Slawin KM et al. Use of the percentage of free prostate-specific antigen to enhance differentiation of prostate cancer from benign prostatic disease. *JAMA* 1998; 279:1542–7.
28. Chan DW, Sokoll LJ, Partin AW et al. The use of % free PSA to predict prostate cancer probabilities: an eleven center prospective study using an automated immunoassay system in a population with nonsuspicious DRE. *J. Urol.* 1999; 161 (Suppl. 4):A353.
29. Horninger W, Reissigl A, Klocker H et al. Improvement of specificity in PSA-based screening by using PSA transition zone density and percent free PSA in addition to total PSA levels. *Prostate* 1998; 37:133–9.
30. Ukimura O, Durrani O, Babaian J. Role of PSA and its indices in determining the need for repeat prostate biopsies. *Urology* 1997; 50:66–72.
31. Letran JL, Blasé AB, Loberiza FR et al. Repeat ultrasound guided prostate needle biopsy: use of free to total PSA ratio in predicting prostatic carcinoma. *J. Urol.* 1998; 60:426–9.
32. Morgan TO, McLeod DG, Leifer ES et al. Prospective use of free prostate-specific antigen to avoid repeat prostate biopsies in men with elevated total prostate-specific antigen. *Urology* 1996; 48:76–80.
33. Fleshner NE, O'Sullivan M, Fair WR. Prevalence and predictors of a positive repeat transrectal ultrasound guided needle biopsy of the prostate. *J. Urol.* 1997; 158:505–9.
34. Catalona WJ, Beiser JA, Smith DS. Serum free prostate specific antigen and prostate specific antigen density measurements for predicting cancer in men with prior negative prostatic biopsies. *J. Urol.* 1997; 158:2162–7.
35. Keetch DW, McMurtry JM et al. Prostate specific antigen density versus prostate specific antigen slope as predictors of prostate cancer in men with initially negative prostatic biopsies. *J. Urol.* 1996; 156:428–31.
36. Rietbergen JBW, Boeken Kruger AE, Hoedemaeker RF et al. Repeat screening for prostate cancer after 1-year followup in 984 biopsied men: clinical and pathological features of detected cancer. *J. Urol.* 1998; 160:2121–5.
37. Durkan GC, Greene DR. Elevated serum prostate specific antigen levels in conjunction with an initial prostatic biopsy negative for carcinoma: who should undergo a repeat biopsy? *Br. J. Urol. Int.* 1999; 83:34–8.
38. Noguchi M, Yahara J, Koga H et al. Necessity of repeat biopsies in men for suspected prostate cancer. *Int. J. Urol.* 1999; 6:7–12.
39. Keetch DW, Catalona WJ, Smith DS. Serial prostatic biopsies in men with persistently elevated serum prostate specific antigen values. *J. Urol.* 1994; 151:1571–4.
40. Stamey TA, McNeal JE, Yemoto CM et al. Biological determinants of cancer progression in men with prostate cancer. *JAMA* 1999; 281:1395–400.
41. Hodge KK, McNeal JE, Terris MK et al. Random systematic versus directed ultrasound guided transrectal core biopsies of the prostate. *J. Urol.* 1989; 142:71–4.
42. Presti JR, Chang JJ, Bhargava V et al. The optimal systematic prostate biopsy scheme should include 8 rather than 6 biopsies: results of a prospective clinical trial. *J. Urol.* 2000; 163:163–7.
43. McNeal JE, Redwine EA, Freiha FS et al. Zonal distribution of prostatic adenocarcinoma: correlation with histologic pattern and direction of spread. *Am. J. Surg. Path.* 1988; 12:897–906.
44. Bazinet M, Karakiewicz PI, Aprikian AG et al. Value of systematic TZ biopsies in the early detection of prostate cancer. *J. Urol.* 1996; 155:605–6.
45. Terris MK, Pham TQ, Issa MM et al. Routine TZ and seminal vesicle biopsies in all patients undergoing transrectal ultrasound giuded prostate biopsies are not indicated. *J. Urol.* 1997; 157:204–6.
46. Morote J, Lopes M, Encabo G, de Torres I. Value of routine TZ biopsies in patients undergoing ultrasound-guided sextant biopsies for the first time. *Eur. Urol.* 1999; 35:294–7.
47. Epstein JI, Walsh PC, Sauvageot J et al. Use of repeat sextant and transistion zone biopsies for assessing extent of prostate cancer. *J. Urol.* 1997; 158:1886–90.
48. Orozco R, O'Dowd GJ, Kunnel B et al. Observations on pathology trends in 62,537 protstate biopsies obtained from urology private practices in the United States. *Urology* 1998; 51:186–95.
49. Bostwick DG, Qian J, Frankel K. The incidence of high grade prostatic intraepithelial neoplasia in needle biopsies. *J. Urol.* 1995; 154:1791–4.
50. Green J, Feneley MR, Young M et al. The prevalence of prostatic intraepithelial neoplasia (PIN) in biopsies from hospital practice and pilot screening: clinical implications. *J. Urol.* 1996; 155(Suppl.):A1260.

51. Raviv G, Janssen TH, Zlotta AR et al. Prostatic intraepithelial neoplasia: influence of clinical and pathological data on the detection of invasive prostate cancer, in patients initially diagnosed on previous needle biopsy. *J. Urol.* 1996; 156:1050–5.
52. Raviv G, Zlotta AR, Janssen TH et al. Does prostate-specific antigen and prostate-specific antigen density enhance the detection of prostate cancer in patients initially diagnosed to have prostatic intraepithelial neoplasia? *Cancer* 1996; 77:2103–8.
53. Zlotta AR, Raviv G, Schulman CC. Clinical prognostic criteria for later diagnosis of prostate carcinoma in patients with initial isolated prostatic intraepithelial neoplasia. *Eur. Urol.* 1996; 30:249–55.
54. Weinstein MH, Epstein JI. Significance of high grade prostatic intraepithelial neoplasia on needle biopsy. *Human Pathol.* 1993; 24:624–9.
55. Irani J, Fournier F, Bon D et al. Patient tolerance of transrectal ultrasound-guided biopsy of the prostate. *Br. J. Urol.* 1997; 79:608–10.
56. Collins GN, Lloyd SN, Hehir M et al. Multiple transrectal ultrasound-guided prostatic biopsies – true morbidity and patient acceptance. *Br. J. Urol.* 1993; 71:460–3.
57. Clements R, Aideyan OU, Griffiths GJ et al. Side effects and patient acceptability of transrectal biopsy of the prostate. *Clin. Radiol.* 1993; 47:125–6.
58. Rodriguez LV, Terris MK. Risks and complications of transrectal ultrsound guided prostate needle biopsy: a prospective study and review of the literature. *J. Urol.* 1998; 160:2115–20.
59. Cooner WH, Mosley BR, Rutherford CL et al. Prostate cancer detection in a clinical urological practice by ultrasonography, digital rectal examination and prostate specific antigen. *J. Urol.* 1990; 143:1146–52.
60. Crawford ED, Haynes AL Jr, Story MW et al. Prevention of urinary tract infection and sepsis following transrectal prostatic biopsy. *J. Urol.* 1982; 127:449–51.
61. Davison P, Malament M. Urinary contamination as a result of transrectal biopsy of the prostate. *J. Urol.* 1971; 105: 545–6.
62. Fawcett DP, Eykyn S, Bultidue MI. Urinary tract infection following trans-rectal biopsy of the prostate. *Br. J. Urol.* 1975; 47:679–81.
63. Ashby EC, Rees M, Dowding CH. Prophylaxis against systemic infection after transrectal biopsy for suspected prostatic carcinoma. *Br. Med. J.* 1978; 2:1263–4.
64. Thompson PM, Talbot Rw, Packham DA et al. Transrectal biopsy of the prostate and bacteremia. *Br. J. Surg.* 1980; 67:127–8.
65. Thompson PM, Prior JP, Williams JP et al. The problem of infection after prostatic biopsy: the case for the transperineal approach. *Br. J. Urol.* 1982; 54:736.
66. Gustafsson O, Norming U, Nyman CR et al. Complications following combined transrectal aspiration and core biopsy of the prostate. *Scand. J. Urol. Nephrol.* 1990; 24:249–51.
67. Sieber PR, Rommel FM, Agusta VE et al. Antibiotic prophylaxis in ultrasound guided transrectal prostate biopsy. *J. Urol.* 1997; 157:2199–200.

Part II

Genetic Susceptibility and Hereditary Predisposition, Screening and Counseling

Prostate Cancer Prevention: Strategies and Realities

Robert G. Uzzo*, Deborah Watkins-Bruner, Eric M. Horwitz, Andre Konski, Alan Pollack, Paul F. Engstrom and Vladimir Kolenko

Department of Surgical/Urologic Oncology (RGU, VK), Department of Population Science (DWB, PFE), Department of Radiation Oncology (EMH, AK, AP), Fox Chase Cancer Center, Temple University School of Medicine, Philadelphia, PA, USA

INTRODUCTION

The management of solid malignancies has evolved over the last half century. Historically, cancer treatments were administered only when patients presented with acute events or developed overt manifestations of their disease. This reactive approach has largely been supplanted by a proactive one, whereby the goal of clinical oncology is increasingly early detection and prevention of long-term negative outcomes. The point of first intervention has been pushed forward by the development of sophisticated imaging techniques, flexible fiber optics and increasingly sensitive biomarkers, allowing most malignancies to be detected much earlier in their natural history. Additionally, important advances in therapeutics, including infection control, blood banking, effective anesthesia and pain control, invasive monitoring and acute postsurgical care permit complex patient management.

Prostate cancer represents a relevant case in point. Prior to the widespread use of prostate specific antigen (PSA), nearly three-quarters of prostate cancer patients presented with symptomatic locally advanced disease or metastases.[1] In contrast, the most common presentation today is non-palpable asymptomatic disease detected due to elevation in the serum PSA level.[2] Clinical evidence suggests this tendency toward early diagnosis and intervention will increase cancer-specific survival rates.[3] Given the potential for improved outcomes, oncologic interventions often begin with the diagnosis of asymptomatic microscopic disease, incipient disease or increasingly no disease at all.

The practice of 'proactive medicine' in clinical oncology can be separated into cancer screening and prevention strategies. Screening involves detection of established disease early in its course, whereas prevention relates to patients with no clinically identifiable disease. Preventative efforts have been classified as primary, secondary or tertiary, depending on the targeted population. Primary cancer prevention efforts involve interventions in otherwise healthy individuals so as to prevent disease occurrence. This means intervening prior to the initiating cellular or molecular events. Secondary prevention strategies involve intervention after the initiating cellular events have already occurred in an attempt to prevent the promotional phase of carcinogenesis. Since these patient may be difficult to identify, on a practical level, secondary cancer prevention targets patients 'at risk' for a given disease because of known inheritable characteristics such as germline mutations. Tertiary prevention strategies involve intervention in an attempt to alter the course of already established disease. These efforts target patients with 'premalignant' disease, such as *in situ* carcinoma, or in the case of prostate cancer, prostatic intraepithelial neoplasia (PIN).[4]

Most prevention strategies focus on the identification of risk factors, public education and reinforcement of avoidance

*To whom correspondence should be addressed.

behaviors. Non-cancer examples include the association between dietary fat and obesity with heart disease, alcohol abuse with liver disease and promiscuity with sexually transmitted diseases. The concept of prevention is relatively new in clinical oncology and has been patterned after these and other successful public health initiatives. Oncologic examples include smoking cessation to reduce the incidence of lung carcinoma, limiting sun exposure for patients at risk for melanoma, removing asbestos to limit mesothelioma risk and decreasing contact with other known industrial carcinogens.

While early cancer prevention programs have stressed avoidance, more recent cancer prevention strategies include direct intervention with substances believed to inhibit one or more steps involved in malignant transformation. The administration of agents in this manner has been termed chemoprevention. To date, there have been relatively few well-controlled cancer chemoprevention studies. Examples include the Breast Cancer Prevention Trial (BCPT) randomizing over 13 000 women at risk for developing breast cancer to receive either tamoxifen or placebo,[5] the α-tocopherol–β-carotene (ATBC) lung cancer prevention trial evaluating over 29 000 male smokers,[6] the Prostate Cancer Prevention Trial (PCPT)[7,8] and the SELECT[9] trial following over 18 000 and 32 000 men, respectively, in an attempt to decrease the risk of developing prostate cancer with 5α-reductase inhibitors (PCPT) or selenium and/or vitamin E (SELECT). Trials of this magnitude require immense resources and intense collaboration to avoid confounding variables, which can limit data interpretation. The National Institutes of Health and the National Cancer Institute now recognize the importance of chemoprevention trials and are actively involved in their development and oversight.

Successful preventative agents must be sufficiently effective, minimally invasive, offer a low side-effect profile and be administered early enough in the course of the disease to affect outcome. Implicit in the design of preventative strategies is an understanding of the natural history of the disease they are intended to prevent. Malignancies most likely to benefit from preventative measures must have early events associated with a slow progression to malignant transformation. It is in these types of tumors that the interaction between environmental influences and genetic ones are likely to be greatest and, therefore, potentially benefited by avoidance or interactive preventative techniques. This makes carcinoma of the prostate an ideal neoplasm for preventative efforts.

PROSTATE EVOLUTION, GROWTH AND DIFFERENTIATION

Prostate cancer is the most common non-cutaneous malignancy in American men.[10] It is also primarily a disease of men beyond their reproductive years, thereby severely limiting the evolutionary selective process against cancer.[11] Observational data suggest that environment plays an important role in determining the risk of prostate cancer. First, there is tremendous species specificity of prostate cancer, with a significant incidence of spontaneous prostate cancer observed only in humans and dogs.[11] None of man's closest primate relatives are susceptible to carcinoma of the prostate. Second, there is tissue specificity for cancer among men's reproductive organs. Despite sharing similar genetics and microenvironments, the human vas deferens and seminal vesicle rarely undergo malignant transformation. In addition, ethnic and geographical variations exist in the incidence of prostate cancer. Compared to the USA, the risk of prostate cancer is as much as tenfold less in Asian countries, a tendency that is lost upon immigration and adoption of a Western diet.[11,12] Lastly, there is a high incidence of early pathological changes in the prostate epithelium of both American and Asian men, suggesting that differences in molecular promotion and not initiation explain the observed variations in incidence and mortality.[11,13] Taken together, these data support the influence of the environment and epigenetic events on the clinical progression of prostate cancer.

The prostate originates from solid epithelial outgrowths that emerge from the endodermal urogenital sinus below the developing bladder at approximately 10 weeks gestation. The prostatic buds grow into the adjacent mesenchyme, lengthen to form ducts, arborize and canalize.[14] By 13 weeks approximately 70 primary ducts exist, which exhibit secretory cytodifferentiation.[15] Further embryonic growth involves ductal proliferation and the establishment of prostate zonal anatomy. Considerable cellular heterogeneity exists across the prostatic ductal epithelium (**Fig. 10.1**). Signals derived from the developing prostatic stroma are believed to control the rate and fate of proliferating prostate epithelial cells.[16]

The human prostate is known to undergo several growth phases: fetal growth, followed by limited regression after birth and growth cessation, then ultimately prostatic regrowth accompanying the androgen surge of puberty until adulthood at which time the process stops[17] (**Fig. 10.2**). In man and dog, prostatic growth is reiterated during aging as part of benign or neoplastic proliferative processes. This has led McNeal to theorize a 'reawakening' of embryonic growth within the prostate.[18] Studies of prostatic cellular proliferation demonstrate that prostate epithelial growth occurs early in a man's life, peaking in men 30–40 years old when prostate doubling times are 4–5 years. However, clinical benign prostatic hyperplasia (BPH) and prostate cancer are much more common during the seventh and eighth decades of life, when prostate epithelial growth is minimal and doubling times are significantly longer.[17,19] These data suggest that for chemopreventative strategies to be maximally effective, they must be applied to men much younger than those typically at risk for developing prostate cancer.

EPIDEMIOLOGY RISK FACTORS AND INITIATING EVENTS

Pathological studies reveal that the development of prostate cancer is a near inevitable consequence of aging in men. Its prevalence is substantial in autopsies of men in their 50s and 60s (29%) and increases progressively with each succeeding

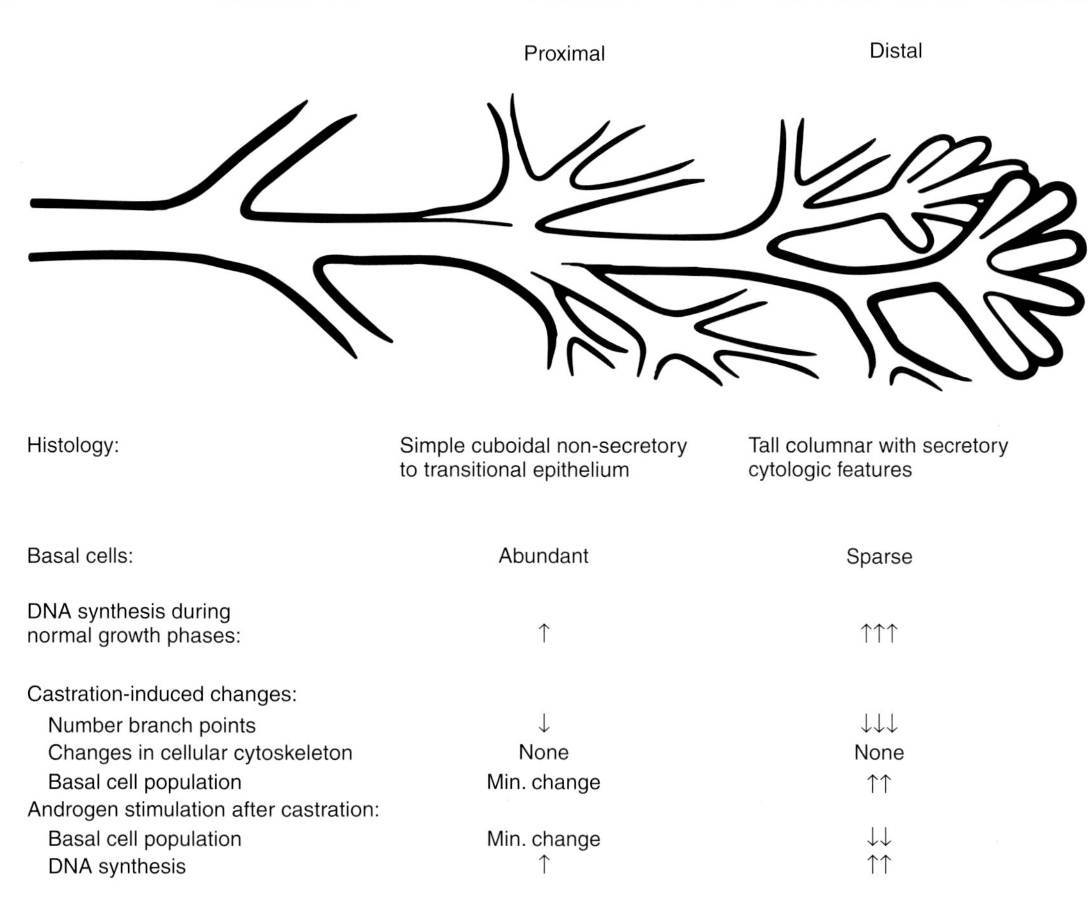

Fig. 10.1. Considerable heterogeneity exists across the prostatic ductal epithelium that may provide valuable insight into clinical relevant mechanisms, such as cellular initiation, disease progression, and response or resistance to therapies. Adapted from Uzzo et al.[14]

decade.[20] In autopsied males, 40–50% of men in their 70s and nearly 70% of men in their 80s have histologic evidence of the disease.[21] More recent data suggest that 6% of Asian men in their 30s who died from unrelated causes had histologic changes

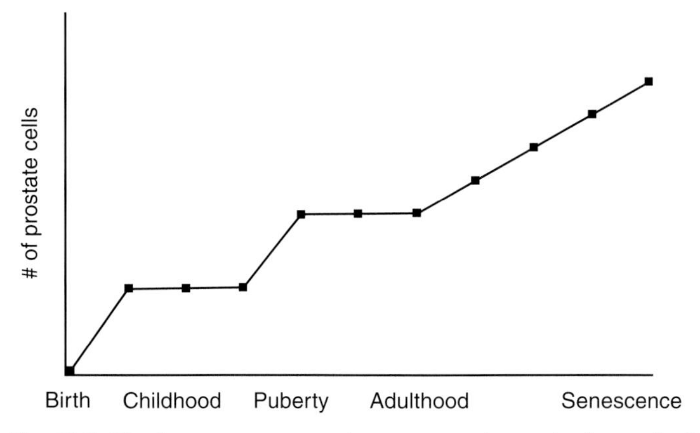

Fig. 10.2. The human prostate undergoes several growth phases: fetal growth, followed by limited regression after birth and growth cessation, then ultimately prostatic regrowth accompanying the androgen surge of puberty until adulthood, at which time the process stops.[17] Studies of prostatic epithelial doubling times demonstrate higher levels of cellular growth several decades before the appearance of clinical disease, suggesting that prostate cancer prevention efforts should be started early in patients at risk.

consistent with prostate cancer.[22] Race and geographic area do not seem to influence this autopsy prevalence.[23] In contrast to histologic evidence of latent prostate cancer, the age-adjusted incidence and death rates from prostate cancer vary dramatically between countries and ethnic groups with the highest rates in African Americans (149/100 000 person years), intermediate rates in USA Whites (107/100 000 person years) and the lowest rates in Japanese (39/100 000 person years) and Chinese (28/100 000 person years).[24] It has been recognized for some time that there are at least two distinct populations of men with prostate cancer: those that will die with the disease and those that will die as a consequence of it. Understanding the genetic and epigenetic factors distinguishing these two groups is imperative and has immediate clinical implications.

A number of modifiable and non-modifiable risk factors have been associated with the development of prostate cancer. The most well established of these are non-modifiable including age, race and family history.[25] However, possible modifiable risk factors are beginning to emerge including intake of dietary fat, calcium, lycopene and soy, cigarette smoking, androgen levels and receptor expression.[25] Many other potential risk factors have been reported but not validated. Each may offer a potential point of dietary or pharmacological intervention.

Despite these associations, the etiology of prostate cancer remains elusive. A number of somatic genetic lesions have been identified in the DNA from prostate cancer cells using sophisticated molecular techniques, including comparative genome hybridization, microsatellite repeat polymorphism allelotyping and fluoresence *in situ* hybridization (FISH).[26–29] These studies demonstrate that most cases of prostate cancer exhibit somatic genome abnormalities; however, there is striking heterogeneity in the patterns displayed.[26] These differences reflect the variable clinical behavior of most prostate cancers. The basis of genetic heterogeneity is poorly understood but may represent an accumulation of multiple environmental assaults on the DNA. New and compelling data suggest hypermethylation of the GSTP1 'CpG island' sequence in prostate epithelial cells may lead to increased susceptibility to genomic damage by inflammatory processes and dietary mutagens offering potential targets for chemopreventative agents.[26]

Unfortunately, the study of these early processes has been limited by the lack of good preclinical animal models of prostate cancer and surrogate end-points to measure progression.[30] The natural history of prostate cancer extends over decades and existing parameters, including PSA, lack sufficient specificity for early progression. Additional clinical measures of local or advanced disease are notoriously difficult to ascertain by history, physical examination or conventional radiographic studies. Furthermore, prostate cancer does not typically grow as a focal mass lesion within the prostate, but rather is infiltrative, making isolation of prostate cancer epithelia difficult and subsequent purification of its DNA challenging. Proposed intermediate biomarkers to identify progression include measures of cellular proliferation (PCNA, Ki-67), angiogenesis [vascular endothelial growth factor (VEGF), microvessel density], differentiation (loss of high-molecular-weight keratin), apoptosis (TUNEL assay, *bcl-2/bax*), genetic damage (cytogenetic changes such as on 8p, 9p or 16q) and regulatory molecular defects (c-*erb*B-2, TGF, p53, androgen receptor).[30] Currently, none of these parameters exhibit sufficient specificity, making the efficacy of chemopreventative therapies difficult to ascertain. Ongoing Phase II trials may help establish appropriate intermediate measures as surrogate cancer end-points.[30,31]

MOLECULAR TARGETS AND CANDIDATE PROSTATE CANCER PREVENTION AGENTS

Discoveries and technologies associated with the National Institute of Health Human Genome Project[32] and the National Cancer Institute (NCI) Cancer Genome Anatomy Project (C-GAP)[33] promise innovative molecular targets for prostate cancer prevention. When evaluating the role of potential new agents, several critical questions must be considered.[34] For example, does the agent inhibit growth or cause regression in existing models of prostate cancer? What is the agent's toxicity profile? Can it be given effectively in an oral dose? Is cost a factor? Have phase I trials in humans been conducted? Is the mechanism of action known and can its effects be ? Is there a suitable target population for testing? The answer to these and other pressing questions regarding chemoprevention agents are critically important and can only be answered through systematic study.

Existing putative prostate cancer prevention agents can be classified into those intended to inhibit ongoing cellular mechanisms, such as sex steroid signaling, differentiation or proliferation, pro-apoptosis and angiogenesis, and those intended to reverse or prevent progressive DNA damage, such as gene therapy, growth factor therapy or antioxidant therapy[35] (**Table 10.1**). Observational data suggest that the hormonal milieu during early prostate development is an important determinant of subsequent cancer formation. Castration early in life nearly eliminates the risk of BPH and prostatic cancer in subsequent years. Additionally, because of the known hormone dependency that normal prostate epithelia and prostate cancer cells exhibit, sex steroid hormone signaling has been aggressively targeted for chemotherapeutic and chemopreventative strategies. Recent studies evaluating the role of 5α-reductase inhibition and androgen deprivation for patients with premalignant changes (PIN) are yet to be completed.[36] Many 'natural' remedies for prostate cancer are believed to contain phytoestrogens that target sex steroid pathways.[37] Several promising agents thought to affect cell differentiation and proliferation are also being explored including retinoids and vitamin D analogs.[38] The goal of other novel strategies including inhibition of growth signaling and antiapoptotic pathways is to target transformed cells without adversely affecting normal cellular turnover. It remains to be seen if this is clinically feasible.

Perhaps a more attainable strategy is to prevent initial or progressive DNA damage. Epidemiologic studies provide encouraging preliminary data suggesting antioxidant nutrients, such as selenium, vitamin E and lycopene, are associated with decreased prostate cancer risk.[39–42] Oxidative stress has been implicated in the early events initiating prostate cancer and there is strong evidence that endogenous or exogenous reactive oxygen species (ROS) damage lipids, proteins and nucleic acids. Oxidative modifications are thought to be directly mutagenic and may alter gene function including p53, c-*fos* and c-*jun*.[43,44] The effects of two antioxidants are actively being studied as part of the Selenium and Vitamin E Cancer Prevention Trial (SELECT), which aims to enrol over 32 000 men throughout the USA and Canada over 7–12 years.[9] Further prostate cancer prevention approaches, including the use of gene therapy and/or growth factors, await more convincing proof of the principle in the adjuvant setting.

TARGET POPULATIONS

Target populations appropriate for the study of prostate cancer prevention should be stratified based on their risk of developing the disease given current epidemiologic evidence.[45] Populations can be divided into low-, intermediate- and high-risk groups. Those at highest risk include patients with known premalignant changes in their prostate, such as PIN. This subgroup provides patients with the most immediate likelihood of progression to cancer; however, it represents

Table 10.1. Potential prostate cancer prevention agents (adapted from Lieberman et al.[35]).

Inhibiting ongoing cellular events	
Sex steroid signaling	5α-reductase inhibitors Antiandrogens (agonists)
Proliferation/differentiation	Vitamin D Retinoids Ornithine decarboxylase inhibitors
Angiogenesis	VEGF, PDGF, FGF receptor inhibition Farnesyl transferase inhibitors
Pro-apoptosis mechanisms	Cyclooxygenase inhibitors Other anti-inflammatory agents Apoptotic signal transduction (NFκB)
Reverse existing or progressing DNA damage	
Antioxidants	Selenium Vitamin E Carotenoids Polyphenols
Growth factors	Endothelin Metalloproteinases PSA proteases
Gene therapy	Vaccines Cytotoxicity or suicide genes Immunostimulatory genes (GM-CSF)

PDGF, Platelet-derived growth factor; FGF, Fibroblast growth factor.

a rather small percentage of men and there is considerable controversy among pathologists regarding the spectrum of abnormal early prostate epithelial histology.[46] Those at intermediate risk include African Americans, patients with a strong family history and those who can be identified with hereditary prostate cancer gene mutations such as HPC-1 linkage.[47] Low-risk populations include the general public; however, recruitment requires large study groups and long follow-up intervals.[45] Those at low, intermediate and high risk are candidates for primary, secondary and tertiary preventative strategies, respectively.

THE PROSTATE RISK ASSESSMENT PROGRAM

Given the need to identify appropriate target populations for prevention and screening studies, many institutions have begun to establish registries for patients at risk. Here we review the findings from one of the first institutional programs designed to collect such information. The Fox Chase Cancer Center (FCCC) Prostate Cancer Risk Registry (PCRR) and Prostate Cancer Risk Assessment Program (PRAP) were founded in 1996. The FCCC is an NCI-designated Comprehensive Cancer Center committed to the support of cancer prevention and early detection. The PRAP has been developed building on the successful history of the FCCC Family Risk Assessment Programs for breast and ovarian cancers. PRAP is a multidisciplinary effort that includes clinicians across multiple disciplines, basic and behavioral scientists, genetic counselors, health educators, biostatisticians and others. This team has developed a program for prostate cancer risk analysis, risk education, early detection, and testing of risk reduction and cancer prevention strategies.

Objectives and aims

The primary objectives of the PCRR and PRAP are to establish a registry and screening clinic for families at high risk for prostate cancer, to further our understanding of the complex mechanisms of prostate carcinogenesis, and to provide prevention, early detection and psychosocial interventions to unaffected individuals at high risk. Secondary objectives include the study of prostate cancer genetics through the creation of a high-risk specimen bank, the development and testing of tools for intervention and prevention strategies, and the evaluation of behavioral interventions to assist in educating, processing and coping with risk-related information.

PRAP eligibility

PRAP eligibility criteria for at-risk populations include African Americans, men with at least one first-degree relative or two or more second-degree relatives with prostate cancer on the same

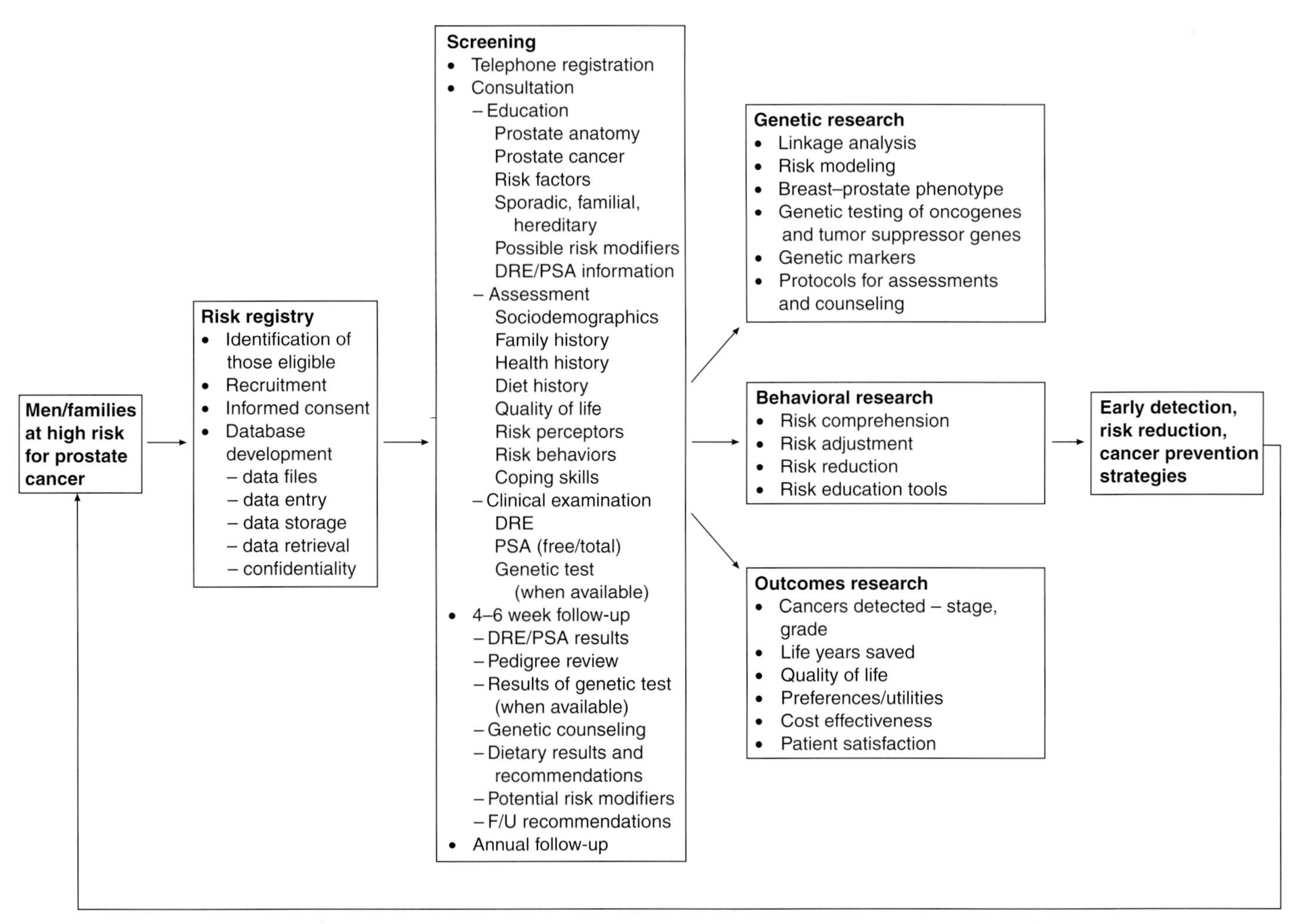

Fig. 10.3. The Fox Chase Cancer Center model for a unified multidisciplinary Prostate Risk Assessment Program (PRAP).[49]

side of the family, or men who have tested positive for the BRCA1 gene mutation. PRAP eligibility is limited to men aged 35–69 years of age. The lower bound age was based on reports of early age of onset for at-risk men and the upper bound age was based on the PRAP goal of targeting men at-risk prior to the mean age of onset of prostate cancer for the general population.[48] The PCRR includes affected family members of PRAP participants.

The logistics of the PRAP and PCRR, including recruitment, registration, the screening appointment, feedback and follow-up are outlined in **Fig. 10.3** and are described in detail elsewhere.[49] The program includes an extensive assessment of family history, pertinent medical history, dietary habits, preventive health care behaviors, risk perceptions, psychological status and quality of life. An educational session and interview with a genetic counselor is mandatory prior to screening and informed consent. Most importantly, the health educator and genetic counselor work to answer questions and keep concerns in perspective. Following education, the participant has blood drawn for PSA (total and % free) and research purposes after which the physician takes a targeted health history and performs the digital rectal examination (DRE). An appointment for the feedback of results is made for approximately 4 weeks postscreening. In the interim, a pedigree is generated and reviewed, the PSA results are obtained, and a computer printout of nutritional intake is generated and analyzed. Men with abnormal results, particularly an elevated total or low percentage % PSA, are contacted by the physician for consultation and appropriate referrals for diagnostic follow-up are made.

Participants with a normal PSA or DRE return to PRAP for their feedback visit where they meet with the genetic counselor to review their pedigree and receive the physician's follow-up instructions, which generally consist of annual or, when appropriate, semiannual screening. Other familial or hereditary illnesses discovered on pedigree review are discussed along with appropriate follow-up and referral.

Implications of the findings for the participant and family members are discussed in addition to dietary recommendations and potential preventive health behaviors, including prevention clinical trials when appropriate.

Preliminary results

PRAP/PCRR accrual

Since its inception in October 1996 through August 2001, PRAP accrual has accrued nearly 400 men at increased risk for prostate cancer (mean age 48 years). Since African American men have the highest risk of prostate cancer, PRAP specifically targets recruitment efforts to this population. Our ongoing minority recruitment efforts have yielded a 57% accrual of African American men. This is over 40% greater than the 13.5% African American representation in the FCCC primary service area. This is of twofold significance, since African American men have the greatest risk of developing prostate cancer and are often difficult to recruit into screening programs and clinical trials.[50]

Preliminary PRAP screening results

A total of 71 out of 323 (22%) at-risk men who were screened have had abnormal findings. Of these, 9 out of 71 (13%) had abnormal DREs, 25 out of 71 (35%) had PSAs >4 ng/ml (range 4.5–10.1 ng/ml), and another 37 out of 71 (52%) had PSAs between 2 and 4 ng/ml with % free PSA at 27% or lower. Compliance with biopsy recommendations has been approximately 70%.

To date, 19 pathologically confirmed prostate cancers have been detected. All were clinically significant tumors with Gleason scores of 6 or higher. Twelve of the 19 men (63%) with pathologically confirmed cancers had both normal DREs and total PSAs, and were only detected based on their low % free PSA (<27%). Of those diagnosed with cancer, ten men (53%) were African American and nine (47%) were Caucasian.

From these initial findings of screening men at high-risk for prostate cancer participating in the FCCC PRAP, we have documented a 6% cancer detection rate of clinically significant prostate cancers in screened men age 39–69 years. This is twice the expected rate compared with our benchmark, the American Cancer Society–National Prostate Cancer Detection Project (NPCDP). Published reports from the NPCDP indicated that we could expect a 3% cancer detection rate had the screening included a general population of men age 55–70 years, in whom the disease is highly prevalent due to age alone.[51] Our findings indicate that we have been successful at early detection in those men truly at high risk for prostate cancer.

Prostate cancer knowledge, risk perceptions and preferences

In addition to screening and early detection, PRAP provides the opportunity to assess prostate cancer knowledge, screening and treatment preferences, and risk perceptions, as well as test risk communication strategies. While men with a family history of prostate cancer are generally aware that they are at an increased risk for prostate cancer, they are often misinformed about the magnitude of their risk. There is often confusion about the interaction of environmental and lifestyle risk factors and hereditary risk,[52] and about the appropriateness of genetic risk assessment. PRAP offers the opportunity to examine men's risk perceptions and prostate cancer knowledge, preferences for screening and treatment as well as adjustment to risk and screening behaviors.

Most of the educational and psychosocial interventions developed in oncology have targeted patients already diagnosed with cancer and focus on coping with the illness and its treatment. In contrast, little attention has been given to the informational and psychological needs of participants in cancer risk counseling programs. There is currently no consensus on the optimal way of delivering prostate cancer risk information or on the true impact of counseling programs on participants' risk comprehension, psychological adaptation or adoption of recommended health practices. The anticipated widespread availability of genetic testing for prostate cancer susceptibility genes in the near future mandates that we learn the most appropriate ways to evaluate these outcomes and to define the essential elements of psychosocial support needed to maximize positive outcomes to cancer risk counseling. The development of the PRAP and the PCRR provide the opportunity to develop and evaluate educational and psychological strategies to optimize prostate cancer risk counseling in both the cancer center and the community.

In a series of studies to assess prostate cancer risk perceptions in order to enhance risk communications, we have begun to identify cognitive–affective profiles driving risk perceptions and risk behaviors.[53] For example, although first-degree relatives (FDRs) of patients with prostate cancer have reported a statistically significant greater personal risk for prostate cancer than non-FDRs, 45% of the FDRs still indicated that they did not feel at elevated risk for developing the disease. FDRs agreed more strongly than non-FDRs that prostate cancer was inherited and FDRs believed that there was less that people could do to control the development of prostate cancer than non-FDRs. FDRs reported less negative expectations about the physical impact of prostate cancer than non-FDRs. Additionally, FDRs reported higher anxiety and marginally higher levels of intrusive ideation regarding prostate cancer. These findings suggest that FDR status may be a predictor of an individual's cognitive–affective pattern of perceived risk and play an important role in encoding risk information.

Studies of preference and utility

Our work on cognitive–affective profiles indicate that there are two possible routes to suboptimal preferences for screening and treatment, either unrealistically low- or high-risk perceptions. To understand better what preferences men at risk have for prostate cancer treatment and to assess if their preferences differ from men with established prostate cancer, we recently completed a study which aimed to: (1) assess preferences for the treatment of early-stage prostate cancer; (2) evaluate sociodemographic and clinical factors as predictors of subject preferences; and (3) calculate utility values (preference weights between 0 and 1 that can be used in cost–utility analyses). Fifty men, 25 at-risk PRAP participants and 25 PCRR participants previously treated with radiotherapy for prostate cancer, were interviewed. Using the Time Trade-Off (TTO) technique, preferences and utilities for disease states and treatments were collected. Preferences for therapy differed widely among men (but not between groups) in this study. The men, regardless of group (at-risk vs. cancer), showed an increased preference and slightly higher utility for surgery versus radiotherapy or hormonal therapy, as compared to observation. Higher education and being married were the only factors identified that predicted a preference for any prostate cancer treatment that prolonged survival, despite quality of life impairments, over observation. Calculated utility values elicited from these participants were closer in value to previous samples of men with cancer than healthy men or physicians. Studies in progress are seeking to correlate preferences with risk perceptions.

Assessing genetic predisposition

The second major objective of PRAP is to study markers of genetic predisposition to prostate cancer. Multiple genetic markers are under investigation for their role in prostate cancer initiation and or progression including ELAC2,[54,55] CYP3A4,[56,57] and SRD5A2.[58,59] In addition, a growing body of research has begun to evaluate the association among levels of the insulin-like growth factor (IGF) binding proteins and susceptibility to prostate cancer.[60] In fact, researchers speculate that some variations in endogenous factors, such as sex steroids or IGF-1 circulating levels, may partly explain differences in risk observed between different populations.[61] However, an at-risk population has yet to be examined, and PRAP is facilitating this effort. We recently examined the overall plasma levels of IGF-1 and IGF binding protein-3 (IGFBP-3) in 105 PRAP participants (63 African American, 42 Caucasian). We investigated the relationships between demographics and IGF levels, particularly with regard to race. Our results demonstrate that IGFBP-3 plasma levels are lower in African American men than in Caucasian men. Since IGFBP-3 can control IGF-1 bioavailability, the lowered IGFBP-3 could explain in part the increased risk of prostate cancer in African American men.[62]

Tools for interventions

The third and final major objective of PRAP is to develop tools for intervention and primary prevention of prostate cancer in men at high risk for the disease.

Protocols for genetic assessment

Physicians can have a positive impact on their patients by encouraging them to reduce high-risk behaviors and to adopt healthy lifestyles.[63] However, there is a gap between the knowledge of, and general approval for, preventive health guidelines by physicians, and their actual performance of these practices.[64] Several barriers to the incorporation of cancer control activities into the primary care setting exist, including lack of physician training, lack of confidence in their ability to deliver preventive services and time constraints.[65] As a result, methods to evaluate and reduce risk are often underutilized in the primary care setting. In order to select individuals and families for closer genetic investigation and counseling, PRAP has also been involved with developing protocols to identify, notify and recruit individuals at risk for prostate cancer. An important task is to develop informed consent documents appropriate for the various stages of genetic assessment and eventual testing in language amenable to the lay person. Identification of genetic cancer risk protocols are being determined by establishing genetic risk criteria, including the number of affected relatives, the age at onset of the disease, the presence of other related cancers in the family and the apparent degree of penetrance. A case review panel consisting of a medical oncologist, molecular geneticist and genetic counselor periodically review selected family histories to determine the appropriateness of further genetic investigation. Protocols have been developed for use in FCCC Network community hospitals to contact eligible individuals, to explain the purpose of genetic risk assessment and to arrange an initial counseling visit. PRAP educational tools assist primary care practitioners in counseling patients regarding prostate screening, risk assessment, risk reduction, coping skills enhancement and anxiety reduction.[53] PRAP training of community physicians and nurses focuses on integrating these counseling protocols into the mainstream of medical care without jeopardizing the care and confidentiality of high-risk family members.

Developing models for prostate cancer risk

Necessary tools for risk assessment include multivariate models that help identify individual risk of prostate cancer. Initial work has focused on developing clinically useful models for calculating prostate cancer risk. Risk factors for prostate cancer include age, race, serum PSA (free/total); serum IGF-1/IGF binding protein (IGFBP-3); family history of prostate cancer; carriers of prostate cancer susceptibility genes, such as ELAC2, CYP3A4 and SRD5A2; and histology such as atypia and high-grade prostatic intraepithelial neoplasia.[66]

A multidisciplinary team of PRAP researchers has begun a systematic study of factors that may contribute to a multivariate model of prostate cancer risk. Pilot work began with translating relative risk (RR) associated with family history into clinically useful risk categories to assist with education, counseling and decision-making. Risk categories were determined by an extensive review of the literature in collaboration with a genetic counselor. Using RR for specific risk factors, four categories were defined from low to very high. The categories were then used to assess prostate cancer outcomes for 264 men (mean age 47.7 years, 52% African American, 48% Caucasian) participating in PRAP. By nature of study eligibility, only three (1%) men were in the low-risk category. According to study criteria, 54 (20%) were in the moderate-risk, 129 (49%) in the high-risk, and 62 (24%) in the very high-risk categories; 16 (6%) were unknown. Prostate cancer was diagnosed in 0 of the low and unknown, 7% of the moderate, 5% of the high and 10% of the very high-risk groups. The categories proposed in this pilot study showed a trend toward clinically meaningful categorization of prostate cancer risk. A meta-analysis risk associated with family history of the disease is under way to refine the categories.

Table 10.2. Highlight summary of prostate cancer prevention strategies.

1. Prostate cancer imposes an immense burden on an aging population.
2. The molecular and genetic events underlying prostate cancer begin decades before clinically significant disease is detected.
3. The interaction of genetics and environmental influences in the development of prostate cancer is significant given the prolonged natural history of the disease.
4. Most risk factors associated with significant prostate cancers are non-modifiable including age, race and genetics; however, there is a growing awareness of modifiable risk factors including diet and environmental concerns requiring systematic study.
5. Most previous disease prevention strategies have focused primarily on avoidance behaviors. We have entered a pro-active era in cancer prevention with the use of chemopreventative agents intended to disrupt one or more steps involved in carcinogenesis.
6. Preventative efforts must address factors responsible for cellular initiation as well as progression of transformed prostatic epithelia.
7. The molecular targets for candidate prostate cancer prevention agents are now becoming known and will continue to grow as results from the Human Genome Project and Cancer Anatomy Projects become available. Current targets include sex steroid signaling differentiation and proliferation, angiogenesis and pro-apoptotic pathways.
8. Large-scale population studies funded and sponsored through the National Institutes of Health and NCI are mandatory to perform chemoprevention trials such as SELECT and PCPT adequately.
9. Prostate cancer risk assessment programs such as PRAP and PCRR provide integrated multidisciplinary approaches to screening, counseling and researching preventative efforts.

CONCLUSION

Prostate cancer imposes an enormous burden upon us. Pathological evidence suggests that neoplastic changes of the prostate epithelium begin early in a man's adult life but do not become clinically evident or relevant until decades later. The natural history of this enigmatic disease is heterogeneous. The genetic and epigenetic events responsible for differences in prostate cancer initiation and promotion remain poorly understood. Conventional oncologic wisdom suggests early treatment will improve cancer control and longevity. In the case of prostate cancer, mounting epidemiological evidence supports this contention.[3] Given the potential for improved outcomes, cancer interventions now begin with the diagnosis of asymptomatic microscopic disease, incipient disease or increasingly no disease at all, underscoring the importance of cancer prevention and screening efforts.

For prostate cancer prevention efforts to be successful, they must be studied systematically across multiple medical, social and economic disciplines. Chemopreventative strategies must start early in the disease process and parallel advances in our molecular understanding of prostate epithelial transformation. Surrogate oncologic end-points must be established and target populations defined. Follow-up must be rigorous and extend over decades. Established and emerging programs, such as PCRR and PRAP, must be integrated into NCI-sponsored studies, such as SELECT and PCPT. The development of high-risk specimen banks should allow collaboration and cross-fertilization among investigators in different disciplines and at different institutions. Finally, chemoprevention and risk assessment programs must serve as a catalyst for the development of educational training tools directed toward community-based health care professionals and their patients.

REFERENCES

1. Noldus J, Graefen M, Haese A et al. Stage migration in clinically localized prostate cancer. *Eur. Urol.* 2000; 38:74.
2. Carter HB, Sauvageot J, Walsh PC et al. Prospective evaluation of men with stage T1C adenocarcinoma of the prostate. *J. Urol.* 1997; 157:2206.
3. Hankey BF, Feuer EJ, Clegg LX et al. Cancer surveillance series: interpreting trends in prostate cancer – part I: Evidence of the effects of screening in recent prostate cancer incidence, mortality, and survival rates. *J. Natl Cancer Inst.* 1999; 1:1017.
4. Byar DP. The importance and nature of cancer prevention trials. *Semin. Oncol.* 1990; 17:413.
5. Wickerham DL, Tan-Chiu E. Breast cancer chemoprevention: current status and future directions. *Semin. Oncol.* 2001; 28:253.
6. Khuri FR, Lippman SM. Lung cancer chemoprevention. *Semin. Surg. Oncol.* 2000; 18:100.
7. Coltman CA Jr, Thompson IM Jr, Feigl P. Prostate Cancer Prevention Trial (PCPT) update. *Eur. Urol.* 1999; 35:544.
8. Thompson IM Jr, Kouril M, Klein EA et al. The Prostate Cancer Prevention Trial: current status and lessons learned. *Urology* 2001; 57:230.
9. Klein EA, Thompson IM, Lippman SM et al. Select: the next prostate cancer prevention trial. *J. Urol.* 2001; 166:1311.
10. Greenlee RT, Murray T, Bolden S et al. Cancer statistics, 2000. *CA Cancer J. Clin.* 2000; 50:7.
11. Coffey DS. Similarities of prostate and breast cancer: evolution, diet, and estrogens. *Urology* 2001; 57:31.

12. Angwafo FF. Migration and prostate cancer: an international perspective. *J. Natl Med. Assoc.* 1998; 90:S720.
13. Benoit RM, Naslund MJ. Detection of latent prostate cancer from routine screening: comparison with breast cancer screening. *Urology* 1995; 46:533.
14. Uzzo RG, Herzlinger D, Vaughan ED Jr. Prostate development: hormonal and cellular considerations. *American Urologic Association Update Series* 1996; 15:1.
15. Shapiro E. Embryonic development of the prostate. Insights into the etiology and treatment of benign prostatic hypertrophy. *Urol. Clin. North Am.* 1990; 17:487.
16. Cunha GR. Role of mesenchymal–epithelial interactions in normal and abnormal development of the mammary gland and prostate. *Cancer* 1994; 74:1030.
17. Coffey DS, Berry SJ, Ewing LL. *An Overview of Current Concepts in the Study of Benign Prostatic Hyperplasia*, pp. 1–14. Washington, DC: United States General Printing Office, 1987.
18. McNeal JE. Origin and evolution of benign prostatic enlargement. *Invest. Urol.* 1978; 15:340.
19. Cunha GR, Alarid ET, Turner T et al. Normal and abnormal development of the male urogenital tract. Role of androgens, mesenchymal–epithelial interactions, and growth factors. *J. Androl.* 1992; 13:465.
20. Franks LM. Proceedings: Etiology, epidemiology, and pathology of prostatic cancer. *Cancer* 1973; 32:1092.
21. Kozlowski JM, Grayhack JT. *Carcinoma of the Prostate*, 3rd edn, pp. 1575–1713. Philadelphia: Mosby, 1996.
22. Shiraishi T, Watanabe M, Matsuura H et al. The frequency of latent prostatic carcinoma in young males: the Japanese experience. *In Vivo* 1994; 8:445.
23. Catalona WJ. *Prostate Cancer.* New York: Grune and Stratton, Inc., 1984.
24. Pienta KJ. Etiology, epidemiology, and prevention of carcinoma of the prostate. In: Walsh PC, Retik AB, Vaughan ED et al. (eds) *Campbell's Urology*, 7th edn, Vol. 3, pp. 2489–96. Philadelphia: W.B. Saunders, 1998.
25. Plantz EA, Kantoff PW, Giovannucci E. Epidemiology of risk factors for prostate cancer. In: Klein EA (ed.) *Management of Prostate Cancer*, pp. 19–45. Totowa: Humana Press, 2000.
26. Nelson WG, De Marzo AM, DeWeese TL. The molecular pathogenesis of prostate cancer: implications for prostate cancer prevention. *Urology* 2001; 57:39.
27. Cunningham JM, Shan A, Wick MJ et al. Allelic imbalance and microsatellite instability in prostatic adenocarcinoma. *Cancer Res.* 1996; 56:4475.
28. Visakorpi T, Kallioniemi AH, Syvanen AC et al. Genetic changes in primary and recurrent prostate cancer by comparative genomic hybridization. *Cancer Res.* 1995; 55:342.
29. Macoska JA, Trybus TM, Sakr WA et al. Fluorescence in situ hybridization analysis of 8p allelic loss and chromosome 8 instability in human prostate cancer. *Cancer Res.* 1994; 54:3824.
30. Kelloff GJ, Lieberman R, Steele VE et al. Agents, biomarkers, and cohorts for chemopreventive agent development in prostate cancer. *Urology* 2001; 57:46.
31. Kelloff GJ, Hawk ET, Crowell JA et al. Strategies for identification and clinical evaluation of promising chemopreventive agents. *Oncology* 1996; 10:1471.
32. Collins FS. The human genome project and the future of medicine. *Ann. N.Y. Acad. Sci.* 1999; 882:42.
33. Kuska B. Cancer genome anatomy project set for takeoff. *J. Natl Cancer Inst.* 1996; 88:1801.
34. Nelson WG, Wilding G. Prostate cancer prevention agent development: criteria and pipeline for candidate chemoprevention agents. *Urology* 2001; 57:56.
35. Lieberman R, Nelson WG, Sakr WA et al. Executive summary of the National Cancer Institute Workshop: highlights and recommendations. *Urology* 2001; 57:4.
36. Bostwick DG, Qian J. Effect of androgen deprivation therapy on prostatic intraepithelial neoplasia. *Urology* 2001; 58:91.
37. Wiseman H. The therapeutic potential of phytoestrogens. *Expert Opin. Invest. Drugs* 2000; 9:1829.
38. Ekman P, Gronberg H, Matsuyama H et al. Links between genetic and environmental factors and prostate cancer risk. *Prostate* 1999; 39:262.
39. Clark LC, Combs GFJ, Turnbull BW et al. Effects of selenium supplementation for cancer prevention in patients with carcinoma of the skin. A randomized controlled trial. Nutritional Prevention of Cancer Study Group. *JAMA* 1996; 276:1957.
40. Moyad MA, Brumfield SK, Pienta KJ. Vitamin E, alpha- and gamma-tocopherol, and prostate cancer. *Semin. Urol. Oncol.* 1999; 17:85.
41. Gann PH, Ma J, Giovannucci E et al. Lower prostate cancer risk in men with elevated plasma lycopene levels: results of a prospective analysis. *Cancer Res.* 1999; 59:1225.
42. Heinonen OP, Albanes D, Virtamo J et al. Prostate cancer and supplementation with alpha-tocopherol and beta-carotene: incidence and mortality in a controlled trial. *J. Natl Cancer Inst.* 1998; 90:440.
43. Wei H. Activation of oncogenes and or inactivation of anti-oncogenes by reactive oxygen species. *Med. Hypotheses* 1992; 39:267.
44. Hainaut P, Miller J. Redox modulation of p53 conformation and sequence-specific DNA binding. *Cancer Res.* 1993; 53:4469.
45. Klein EA, Meyskens FL Jr. Potential target populations and clinical models for testing chemopreventive agents. *Urology* 2001; 57:171.
46. Algaba F, Epstein JI, Fabus G et al. Working standards in prostatic intraepithelial neoplasia and atypical adenomatous hyperplasia. *Pathol. Res. Pract.* 1995; 191:836.
47. Cussenot O, Valeri A, Berthon P et al. Hereditary prostate cancer and other genetic predispositions to prostate cancer. *Urol. Int.* 1998; 60:30.
48. Cupp MR, Oesterling JE. Prostate-specific antigen, digital rectal examination, and transrectal ultrasonography: their roles in diagnosing early prostate cancer. *Mayo Clin. Proc.* 1993; 68:297.
49. Bruner DW, Baffoe-Bonnie A, Miller S et al. Prostate cancer risk assessment program. A model for the early detection of prostate cancer. *Oncology (Huntingt)* 1999; 13:325.
50. Catalona WJ, Richie JP, Ahmann FR et al. Comparison of digital rectal examination and serum prostate specific antigen in the early detection of prostate cancer: results of a multicenter clinical trial of 6,630 men. *J. Urol.* 1994; 151:1283.
51. Mettlin C, Jones G, Averette H et al. Defining and updating the American Cancer Society guidelines for the cancer-related checkup: prostate and endometrial cancers. *CA Cancer J. Clin.* 1993; 43:42.
52. Smith GE, DeHaven MJ, Grundig JP et al. African-American males and prostate cancer: assessing knowledge levels in the community. *J. Natl Med. Assoc.* 1997; 89:387.
53. Grumet SC, James J, Bruner DW et al. Risk perceptions and attitudes regarding genetic testing in men at risk for prostate cancer. *J. Gen. Counseling* 1999; 8:378.
54. Rokman A, Ikonen T, Mononen N et al. ELAC2/HPC2 involvement in hereditary and sporadic prostate cancer. *Cancer Res.* 2001; 61:6038.
55. Wang L, McDonnell SK, Elkins DA et al. Role of HPC2/ELAC2 in hereditary prostate cancer. *Cancer Res.* 2001; 61:6494.
56. Rebbeck TR, Jaffe JM, Walker AH et al. Modification of clinical presentation of prostate tumors by a novel genetic variant in CYP3A4. *J. Natl Cancer Inst.* 1998; 90:1225.
57. Walker AH, Jaffe JM, Gunasegaram S et al. Characterization of an allelic variant in the nifedipine-specific element of CYP3A4: ethnic distribution and implications for prostate cancer risk. Mutations in brief no. 191. Online. *Human Mutat.* 1998; 12:289.
58. Jaffe JM, Malkowicz SB, Walker AH et al. Association of SRD5A2 genotype and pathological characteristics of prostate tumors. *Cancer Res.* 2000; 60:1626.

59. Nam RK, Toi A, Vesprini D et al. V89L polymorphism of type-2, 5-alpha reductase enzyme gene predicts prostate cancer presence and progression. *Urology* 2001; 57:199.
60. Chan JM, Stampfer MJ, Giovannucci E et al. Plasma insulin-like growth factor-I and prostate cancer risk: a prospective study. *Science* 1998; 279:563.
61. Cussenot O, Valeri A. Heterogeneity in genetic susceptibility to prostate cancer. *Eur. J. Intern. Med.* 2001; 12:11.
62. Tricoli JV, Winter DL, Hanlon AL et al. Racial differences in insulin-like growth factor binding protein-3 in men at increased risk of prostate cancer. *Urology* 1999; 54:178.
63. Schwartz JS, Lewis CE, Clancy C et al. Internists' practices in health promotion and disease prevention. *Ann. Intern. Med.* 1991; 114:46.
64. Valente CM, Sobal J, Muncie HL et al. Health promotion: physicians' beliefs, attitudes, and practices. *Am. J. Prev. Med.* 1986; 2:82.
65. Beck JR, Kattan MW, Miles BJ. A critique of the decision analysis for clinically localized prostate cancer. *J. Urol.* 1994; 152:1894.
66. Lieberman R. Androgen deprivation therapy for prostate cancer chemoprevention: current status and future directions for agent development. *Urology* 2001; 58:83.

Genetic Changes and Their Prognostic Significance in Prostate Cancer

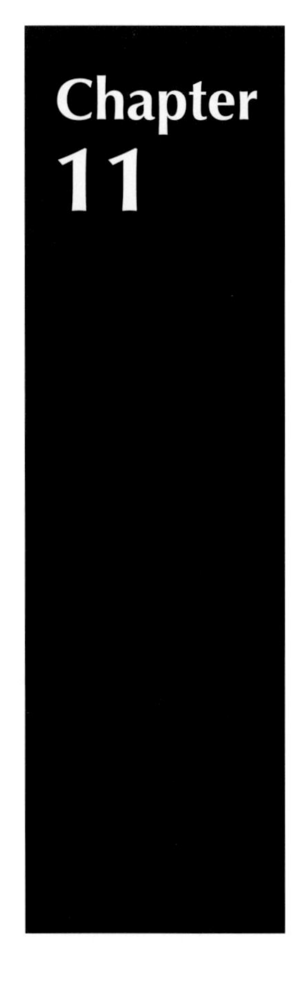

Simon R.J. Bott
Institute of Urology and Nephrology, University College London, Royal Free and University College Medical School, London, UK

Magali Williamson
Institute of Urology and Nephrology, University College London, Royal Free and University College Medical School, London, UK

Roger S. Kirby
Department of Urology, St George's Hospital, Tooting, London, UK

INTRODUCTION

Prostate cancer, like all other cancers, develops as a consequence of genetic changes. While several putative genes have been isolated for the development of breast, ovarian (*BRCA1, BRCA2*) and colon cancer (*hMLH1, hMSH2*), the etiology and pathogenesis of prostate cancer remain poorly understood. This is because prostate cancer is such a heterogeneous disease. A total of 50% of men aged 80 on autopsy studies have foci of prostate cancer, and have had full and healthy lives oblivious to what their prostate is harboring.[1] On the other hand, 13% of cases are diagnosed in men in their pre-retirement years and, of these, 20–30% will have metastases at presentation, with a median life expectancy of 18 months.

The diversity of prostate cancer as well as the stepwise transition from benign cells through prostatic intraepithelial neoplasia (PIN), invasive carcinoma, the development of metastases to hormone refractory disease make this a fascinating disease to research, if at times a difficult disease to treat. A great deal of work has been published on clinical prognostic factors. However, with advances in molecular genetics, it will be the analysis of the genetic code that will in the future provide both prognostic information and targets for therapeutic intervention. Currently available prognostic factors have recently been ranked by a multidisciplinary group of clinicians, pathologists and statisticians, and subsequently endorsed by The World Health Organization's second international consultation on prostate cancer.[2] Category I are factors of prognostic importance and are currently used on a daily basis in patient management. These include preoperative prostate specific antigen (PSA), TNM stage, Gleason grade and surgical margin status. Category II factors are those that have been extensively studied but whose importance remains to be validated statistically, e.g. tumor volume, histological type and DNA ploidy. Category III includes all factors not sufficiently well studied to demonstrate their prognostic value: oncogenes, tumor suppressor genes and apoptosis genes, as well as perineural invasion, neuroendocrine differentiation, microvessel density, nuclear roundness and chromatin texture.[2] This chapter will address the genetic factors as prognostic markers.

THE PATHOGENESIS OF CANCER

The behavior of every cell in the body is dependent on its DNA. Changes in this DNA affect the expression and function of critical genes. Altered genes result in altered mRNA expression and subsequently altered proteins. Many changes may occur; however, 5–10 specific alterations are required to convert a benign into a fully malignant prostate cell.[3] In a small number of prostate cancers, some of these alterations may be inherited. In the majority, however, these are acquired somatic mutations.[4,5] There are two classes of gene associated with cancer: oncogenes and tumor suppressor genes. The problem for researchers is to identify which changes in these genes are important in cancer progression, which can be used to predict prognosis and which can act as targets for therapeutic intervention.

Oncogenes

A proto-oncogene is a normal regulatory gene. If its activity is increased as a consequence of genetic alteration, it becomes an oncogene. Oncogenes are described as dominant as only one allele needs to be changed for a biological effect to occur. Modification of the proto-oncogene to form an oncogene leads to a gain in function. This may be due to a qualitative or quantitative change in protein product. Alteration in the regulatory sequence brings about quantitative change in a normal product. Qualitative change, on the other hand, generates abnormal protein product. These occur as a result of mutation in the coding region or by rearrangement (translocation) of two genes, resulting in the production of a fusion protein made up of parts of each participating gene.

Oncogenes implicated in prostate cancer include the *RAS* oncogenes, *MYC, ERBB2, PSCA, EIF3S3* and the *AR* gene. The *RAS* oncogene is activated by point mutation, the *MYC* gene is activated by amplification, and *ERBB2* is either amplified or modified through transcriptional or post-transcriptional deregulation. Amplification occurs relatively late in the development of prostate cancer; it is found in metastatic (*MYC, PSCA, EIF3S3*) and hormone refractory disease (*AR*). Altered regulation may be seen in the early as well as the later stages of cancer development.

Tumor suppressor genes

A tumor suppressor gene (TSG) encodes for an inhibitory protein whose function is lost in cancer. TSGs are inactivated by deletion of one or both genes, or by a point mutation and deletion. Three types of TSG have been described. First, recessive TSGs, which are inactivated only if both gene copies of a diploid cell are lost, e.g. *RB*. Second, the dominant negative TSG, where a mutation in one allele generates an abnormal protein that inactivates the normal product from the unaffected allele. *p53* may be inactivated in this manner. Finally, haplo-insufficiency is where inactivation of one allele copy gives rise to a change in the phenotype, e.g. *PTEN*.

The majority of TSGs code for proteins that complex with other effector proteins and inhibit their actions. Where an individual has a hereditary mutation of one TSG he has an increased susceptibility to develop cancer and for developing cancer at a younger age.

Abnormal methylation, including hypermethylation, demethylation and redistribution of methyl groups, have all been reported in prostate cancer.[6] The significance of demethylation in prostate cancer is unknown;[7] however, hypermethylation is of importance. The promotor and exons at the 5′-end of some recognized TSG contain a high density of CG (cytosine–guanine) repeats, called CpG islands. Hypermethylation of these CpG islands is believed to result in transcriptional inactivation of the associated gene. Hypermethylation is not only confined to malignant tissues. The *HIC1* gene (hypermethylation in cancer gene) has been found in non-malignant as well as prostate cancer tissues.[8]

HEREDITARY PROSTATE CANCER

In 1895, Dr Alfred Warthin's seamstress told him she was convinced she would die from cancer as so many of her family had. Indeed, she died at a young age from endometrial cancer. This prompted Warthin to publish a description of her family, which included relatives dying prematurely from gastric, endometrial and colorectal cancers. Progress was made in understanding familial cancers in the 1960s by Lynch,[9] Li[10] and Knudson[11] who systematically examined at-risk families. Knudson went on to predict the 'two-hit' hypothesis based on his study of a rare eye tumor in children – retinoblastoma (RB). He postulated the presence of regulatory TSGs that are normally involved in the cell cycle, apoptosis and proliferation. Humans normally have two copies of these TSG, one from each parent. Where inherited mutations occur, one copy is affected. The remaining TSG maintains sufficient control of its regulatory function; however, should a mutation occur in the other copy of the TSG in one cell, a malignant potential is acquired. This hypothesis was confirmed 20 years later when germline mutations in the *RB* gene were detected in hereditary retinoblastoma.[12]

The problem with studying familial prostate cancer is that, as prostate cancer is so common, there is a high incidence of sporadic cases amongst the familial cluster. Furthermore, these families are screened for prostate cancer and these PSA-detected tumors may behave differently from the rest of the population. Nevertheless, a man with a first-degree relative with prostate cancer has a relative risk of 2.0 (1.2–3.3, 95% confidence interval), with a second-degree relative risk of 1.7 (1.0–2.9) and with both a first- and second-degree relative an 8.8 (2.8–28.1) relative risk of developing prostate cancer.[13]

Further evidence to support a hereditary predisposition is that the incidence of prostate cancer is higher in men who have relatives with breast cancer.[14,15] Conversely, the risk of breast cancer is doubled in families with a history of prostate cancer.[16,17] Hereditary prostate cancer is usually defined by the pedigree as no associated genes have yet been identified. This definition, therefore, includes nuclear families with three cases of prostate cancer, the presence of prostate cancer in each of three generations in the maternal or paternal lineage and a cluster of two relatives diagnosed with prostate cancer before the age of 55 years.[18] Using this definition, 3–5% of patients with prostate cancer have hereditary disease.[18] This definition, however, excludes families with a hereditary susceptibility to prostate cancer (with mutations in autosomal dominant susceptibility genes), so the true proportion is likely to be 5–10%.[18]

Candidate genes in hereditary prostate cancer

Prostate cancer susceptibility loci have been reported at 1q24–25, 1p36, 1q42, 20q13, Xq27–28,[19] 16q23[20] and 17p[21] (**Table 11.1**). Several authors have reported a linkage (the inheritance of a recognized genetic marker with the disease) on the long arm

Table 11.1. Chromosomal loci reported to contain hereditary prostate cancer genes.

Locus	Putative gene	Number of families	Two-point lod score*	Multipoint lod score*
1q24–25	HPC1	91	3.7	5.4
1q42	PCAP	47	2.7	3.1
1p36	CAPB	12	3.7	2.2
17p	ELAC2	33	4.5	
20q13	HPC20	162	2.7	3.0
Xq27–28	HPCX	360	4.6	3.9

*The lod score is a statistical estimate of whether two (two-point) or several (multipoint) loci are likely to lie near each other on a chromosome and are, therefore, likely to be inherited together.

of chromosome 1.[22–24] The exact site of the susceptible genes on this chromosome region remains unclear; 1q24–25, named *HPC1*,[22] and 1q42, named *PCAP*,[25] have been implicated. Two further studies confirmed linkage at the *HPC1* locus on 1q24–25,[23,24] three other studies did not.[25–27]

Other candidate genes include the breast cancer susceptibility genes *BRCA1* (17q21) and *BRCA2* (13q12.3). These confer a relative risk of prostate cancer of 3.0 and 2.6–7.0, respectively. The UK/Canadian/Texan Consortium found up to 30% of familial clusters may be linked to *BRAC1/2*, although the confidence intervals were wide and included zero. Indeed the UK Familial Prostate Cancer Study was unable to find any mutations in the *BRAC1* gene but did find two germline mutations in *BRCA2*. They postulated the *BRCA2* mutations may be a coincidental finding or may play a role in modifying the expression of another cancer gene.

Schaid et al.,[28] using segregation analysis, found prostate cancer was 1.5 times more common in brothers than in fathers of men with prostate cancer. This may indicate that prostate cancer is inherited in some families in a recessive or an X-linked way. A study of 360 families in North America, Sweden and Finland strongly implicated a region on the long arm of the X-chromosome at Xq27–28, which the authors named *HPCX*.[29] This gene accounted for 15–16% of cases in North America and 41% of Finnish cases of hereditary prostate cancer.

Tavitigian et al., working on the Utah Population database, mapped an inherited predisposition to prostate cancer to the *ELAC2* gene on chromosome 17p in 33 families.[21] When the number of families studied increased to 127, the linkage was no longer significant. The authors conclude this was due to the heterogeneity of the families. The prevalence of the mutation is low; however, the gene is thought to code for a protein involved in interstrand cross-link repair mechanisms, which is consistent with a possible role in prostate cancer.

Confirmatory studies showing weak or no linkage to these regions and a putative susceptibility gene has yet to be identified. This has led investigators to conclude that hereditary prostate cancer is a heterogeneous disease. Furthermore, there may be no major susceptibility genes for prostate cancer as there is for say breast or colon cancer.[13,19]

Low penetrance polymorphisms

As discussed, mutations in high penetrance susceptibility genes result in a significantly increased risk of prostate cancer and are called hereditary cancers. These are relatively uncommon. On the other hand, polymorphisms (the simultaneous occurrence in a population of genomes showing allelic variations) in low penetrance genes increase the risk of developing the disease only modestly but occur with greater frequency in a population. Low penetrance polymorphisms may, therefore, have a greater impact on the frequency of prostate cancer in the population as a whole.

The polymorphic CAG (cytosine–adenosine–guanine) repeat in the androgen receptor gene has been studied extensively. This codes for a polyglutamine tract of varying lengths (dependent on the number of CAG repeats in the gene). Workers have shown an inverse relationship between the CAG repeat length and prostate cancer risk.[30–32] In normal healthy males there are 13–30 CAG repeats; however, African American men have a higher prevalence of <22 CAG repeats in the *AR* gene.[33] Furthermore, Chinese men have a higher prevalence of long CAG repeats.[34] The prevalence of prostate cancer is higher in African Americans and lower in Chinese men, which may be partly explained by the number of CAG repeats.[3] Giovannucci et al. showed that, if an individual had $\leqslant 18$ CAG repeats, they had an increased relative risk of 1.52 for developing prostate cancer compared with men with $\geqslant 26$ CAG repeats.[35]

Polymorphisms in the *SRD5A2*, the gene that codes for the enzyme 5α-reductase type II, have been reported.[3] 5α-reductase is responsible for the conversion of testosterone to its active metabolite dihydrotestosterone in the prostate. Polymorphisms in the *SRD5A2* gene result in increased catalytic activity of this enzyme and are associated with an increased risk of advanced prostate cancer.[36]

The vitamin D receptor (*VDR*) gene is also a polymorphic steroid hormone receptor gene implicated in prostate cancer. Vitamin D is antiproliferative to prostate cancer cell lines[37,38] and low serum vitamin D may be associated with an increased risk of prostate cancer.[39,40] Whether the *VDR* polymorphisms confer an increased susceptibility to prostate cancer remains contested.[41–46] However, a number of vitamin D analogs are currently undergoing clinical trials for prostate cancer treatment and prevention.[47]

Other polymorphisms reported to influence the risk of developing prostate cancer include the cytochrome P450 family genes (*CYP3A4* and *CYP17*) and the *ELAC2* gene. These may be low, or alternatively possibly high penetrance genes.[48–50] Larger studies are required to corroborate these findings.

SOMATIC CHANGES IN PROSTATE CANCER

While some tumors may contain inherited mutations, all tumors contain acquired or somatic alterations. Advances in the techniques used to examine these changes resulted in the identification of chromosomal regions deleted or amplified that may play a role in tumorigenesis. Chromosomal changes have been identified in specific stages of disease progression with the average number of alterations per case significantly higher in distant metastases than primary tumors.[51] As the technology improves, specific gene sequences and their protein products are being analyzed, which will improve our understanding of the mechanism of disease. It is hoped that our improved understanding will enable biomarkers to be developed, which will enable more accurate prediction of patient outcome as well as act as targets for drug treatments.

One of the most widely used techniques for examining chromosomal alterations is comparative genomic hybridization (CGH). This allows the detection of DNA sequence copy number changes throughout the genome and can, therefore, identify regions where deleted TSG or amplified oncogenes may be harbored. Researchers in Finland have used this technique and shown that locally recurrent hormone refractory prostate cancers contain almost four times as many alterations than the untreated primary tumors.[52] Their findings are summarized in **Table 11.2** and suggest the early development of prostate cancer is as a result of inactivation of TSG, whereas later progression, including the development of hormone refractory disease, is associated with oncogenic activation (**Fig. 11.1**).

CGH and LOH (loss of heterozygosity) studies have demonstrated the most common chromosomal aberrations in prostate cancer are deletions in chromosome regions 3p, 6q, 7q, 8p, 9p, 10q, 13q, 16q, 17q and 18q and gains in 7p, 7q, 8q and Xq.[3,51,54–57] The important alterations and putative critical gene changes to date will be discussed, in particular where they are believed to alter the pathogenesis of prostate cancer.

Early genetic changes

Hypermethylation of the *GSTP1* gene is seen in 70% of cases of high-grade PIN (HGPIN) and over 90% of prostate cancers, but is a rare event in benign prostate tissue.[58,59] The *GSTP1* gene on 11q13, encodes for glutathione S-transferase, which conjugates electrophilic and hydrophobic environmental carcinogens with glutathione, thus protecting the cell.[60] Hypermethylation of the CpG islands of the promoter region prevents the transcription of the gene, thus removing the cell's in-built protection mechanism against potential environmental carcinogens, including dietary factors.[47] As *GSTP1* inactivation occurs in the vast majority of cases of HGPIN and prostate cancer early in the disease course, it may be used, in the future, as a molecular marker.[62] Indeed, studies have already reported measurement of *GSTP1* methylation in cells in the urine of men with prostate cancer.[61,62] Cairns et al. detected *GSTP1* methylation in the urine of 27% of patients with *GSTP1* methylation in their primary tumor.[61] Goessl et al. improved the sensitivity of this technique to 73% by looking at *GSTP1* methylation in the urine sediment after prostatic massage. They found *GSTP1* methylation in 1 (2%) of 45 patients with BPH, 2 (29%) of 7 patients with HGPIN, 15 (68%) of 22 patients with localized prostate cancer and 14 (78%) of 18 with locally advanced or disseminated disease.[62] Furthermore, augmentation of GST activity, using pharmaceutical GST inducers, may have a role in prostate cancer prevention.[63]

Two of the most common deletions found using CGH are on 8p and 13q.[64] Other laboratory techniques including

Table 11.2. Chromosomal changes associated with the progression of prostate cancer during hormonal therapy. Reproduced from Nupponen and Visakorpi (1999).[56]

	Tumors showing alteration (%)		Minimal region
Locus	Primary cancer	Recurrent cancer	of alteration
Losses			
1p	7	54	1p36-pter
10q	10	46	10cen-q21, 10q26
19p and 19q	7	43	19pter-q13.1
20q	0	22	20cen-q22
Gains			
8q	6	73	8q21, 8q23-qter
18q	0	30	18q12
Xq	0	35	Xcen-q13

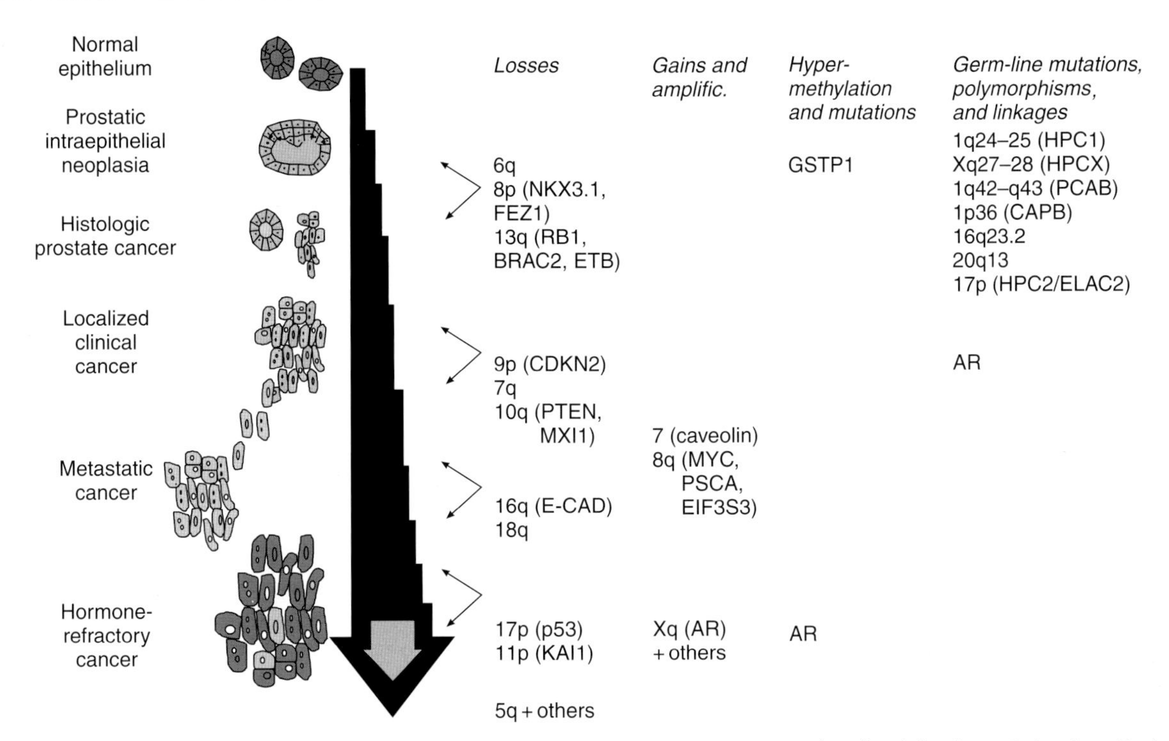

Fig. 11.1. The genetic changes underlying development and progression of prostate cancer. (Reproduced with kind permission from Tapio Visakorpi).

fluorescence *in situ* hybridization (FISH) and LOH have found that losses in these regions are frequently found in HGPIN.[65,66] It would appear that inactivation of an as yet unidentified TSG in 8p and 13q is an early event in prostate cancer pathogenesis. Several regions of loss on 8p have been identified (8p12–21, 8p22) using deletion mapping, which implies several important TSG may exist.[51] *NKX3.1*[67] at 8p12–21, and *FEZ1*[68] have been implicated. *FEZ1* codes for a leucine zipper protein and alterations in this gene have been reported in esophageal and breast as well as prostate cancer.[68]

TSGs on the long arm of chromosome 13 include the retinoblastoma gene *RB1* (13q14), *DBM* (13q14) and the breast cancer susceptibility gene *BRCA2* (13q12–13). However, mutations in these genes are uncommon in prostate cancer.[54,69]

The *CDKN2* (p16) gene located on 9p21 is also inactivated by hypermethylation of CpG islands, as well as by deletion and point mutation.[70,71] The gene codes for a cyclin-dependent kinase inhibitory protein that controls passage through the G1 phase of the cell cycle. Inactivation of *CDKN2* gene may facilitate progression through the cell cycle. Jarrard et al. reported 3 (13%) of 24 primary and 1 (8%) of 12 metastatic tumor samples had hypermethylation in the promoter region.[70] Deletions near the *CDKN2* gene were detected in 12 (20%) of 60 primary tumors and in 13 (46%) of metastatic lesions. Gu et al. looked at *CDKN2* in early prostate cancers and could find no evidence of methylation, although deletions close to the gene were identified.[71] These findings suggest *CDKN2* may be inactivated in early prostate cancer by deletion and hypermethylation of the promoter and is important later in the disease course.

Up to 50% of prostate cancers have been found to harbor deletions on 10q.[64,73] Putative TSGs include *PTEN* at 10q23[74] and *MXI1* on 10q25.[75,76] *PTEN* induces $p27^{kip1}$ expression, which negatively regulates the cell cycle. Mice who are deficient in one *Pten* allele and both *Cdkn1b* alleles (which code for $p27^{kip1}$) all develop prostate cancer within 3 months of birth.[77] Germline mutations are found in *PTEN* in cancer syndromes, such as Cowden's disease (breast and thyroid cancer)[78] and Bannayan–Zonana syndrome (multiple lipomas and hemangiomas).[79] *PTEN* mutations have also been found in 5–27% of early and 30–58% of metastatic prostate cancers.[51,74,80–84] *MXI1* is a negative regulator of the oncogene *MYC* and may, therefore, also have a role as a tumor suppression gene.[76]

Metastatic prostate cancer

In order for invasive cancer to metastasize, the malignant cells need to acquire the potential to detach themselves, survive in the blood or lymphatic system, implant and then divide in a distant tissue. This requires several critical genetic changes, and these changes may act as molecular markers and targets for therapeutic intervention.

CGH and LOH studies have shown losses of 16q in late-stage disease.[56,57,73] Furthermore, loss of 16q23 and q24 is associated with metastatic potential.[85,86] A candidate gene, *CDH1* coding for the E-cadherin protein, is located on 16q22.1. Decreased expression of this gene is associated with high-grade prostate cancer;[56,80–83,85–88] in addition, decreased expression is associated with metastatic potential in the primary tumor.[87] Although no mutations have been described, hypermethylation of the CpG island has been reported in cancer cell lines. E-cadherin is calcium dependent and is responsible for cell to cell recognition and adhesion.[89] α- and β-catenins form part of the same adhesion mechanism and decreased expression of α-catenin has been demonstrated in prostate cancers, which have normal E-cadherin expression.[88] About 5% of prostate cancers contain mutations in β-catenin.[90,91] β-Catenin has two known roles: first, in cell adhesion with α-catenin and, second, as part of the *Wnt* signalling pathway.

Mutations in the TSG *p53* have been described in the majority of cancers, including breast, colon and lung.[92] Located at 17p13.1, *p53* codes for a nuclear phosphoprotein that has a negative effect on cell growth. In the event of DNA damage, the cell cycle is arrested or apoptosis is induced. If *p53* is inactivated and loses its function, damaged DNA may be transcribed unchecked.[9]

The frequency of *p53* mutations in prostate cancer ranges from 1% to 42%,[93–98] but a consistent finding is its association with high-grade and high-stage disease.[97–99] This finding precludes *p53* from being an effective target for curative therapeutic intervention.[94,100] *p53* may, however, play a role in radiotherapy resistance along with *BCL*, *MYC* and *RAS*. These genes influence cellular commitment to apoptosis. Therapeutic agents, including radiotherapy, that induce apoptosis, may be less effective in patients who have defective apoptotic pathways as a result of mutations in these genes.[101] Likewise, the *p53* gene mutation predicts adversely overall survival for patients treated with antiandrogen therapy in locally advanced cases of prostate cancer after radiotherapy.[102] The presence of mutations in the *p53* gene may, therefore, be used before radiotherapy and antiandrogen therapy to predict the likelihood of treatment failure.[101,103]

Mutations in the *ERBB2* gene have been widely reported in a number of tumors, including breast and ovarian.[104] *ERBB2* or *HER-2/neu*, codes for a transmembrane tyrosine kinase growth factor, which is involved in interleukin-6 (IL-6) signaling through the MAP kinase pathway. IL-6 induces phosphorylation in the ErbB2 and 3 receptors in prostate cancer cell lines. Phosphorylation inactivates the receptors, resulting in abrogation of the MAP kinase pathway.[105]

The recent development of a monoclonal antibody to the epidermal growth factor receptor, anti-ERBB2 (Herceptin, Genetech Inc, South San Francisco, CA) for the use in advanced breast cancer[106] has led researchers to refocus on the role of this oncogene in prostate cancer. Trials on the use of Herceptin in prostate cancer are ongoing[107] but it is clear that high-level *ERBB2* amplification does not occur in this disease.[5,108] However, Herceptin does inhibit growth in prostate cancer cell lines, including LNCaP[109] and *ERBB2* enhances androgen receptor signaling in the presence of low androgen levels.[110–111]

The *KAI1* gene, found on 11p11.2, has been shown to suppress metastatic potential in highly metastatic Dunning rat models.[112] In addition, expression of this gene is reduced in metastatic prostate cancer cell lines,[113] but no mutation in this gene has yet been found. The *KAI1* gene is in close proximity to *CD44* (11p13), this latter gene codes for a membrane glycoprotein, which is also present on lymphocytes and participates in cell to cell interactions.[114] Support for the role of *CD44* in the metastatic progression of prostate cancer was reported by Dong et al.[114] They transfected *CD44* into highly metastatic rat prostate cells and demonstrated suppression of metastatic potential, without suppression *in vivo* of growth rate or tumorigenicity.[114] Moreover, down-regulation of *CD44* at a protein and mRNA level correlates with metastatic potential in the Dunning rat model.[114]

Finally, loses of 18q have been reported to be associated with metastatic prostate cancer.[115] Two TSGs implicated in cancer progression are located in this region, *DCC* (deleted in colon cancer) and *DPC4* (deleted in pancreatic cancer); however, no abnormalities have been found in these genes in prostate cancer.[115,116]

Hormone refractory prostate cancer

During embryological development and puberty, androgens play a key role in prostate development. Testosterone is converted to its more potent metabolite dihydrotestosterone by the enzyme 5α-reductase. There are no reported cases of prostate cancer in men who have a congenital deficiency of 5α-reductase. This enzyme can be blocked pharmacologically by finasteride (Proscar®) and this drug is currently undergoing a phase III trial in over 18 000 men to evaluate a possible role in prostate cancer chemoprevention.[117] Dihydrotestosterone and testosterone bind to the androgen receptor (AR), which initiates phosphorylation and dimerization of the receptor. The AR can then bind to specific DNA sequences, called the androgen responsive elements, located in the promoters of androgen-responsive genes. The AR complex, in collaboration with cofactor proteins, is able to up- or down-regulate the transcriptional activity of the androgen response genes. The removal of androgens by surgical castration or therapeutic blockade initiates apoptosis in prostate cancer cells, resulting in a clinical response in up to 95% of men.[118] However, some prostate cancer cells maintain the ability to grow even in low or absent androgen levels[119] and, consequently, the median response time to androgen withdrawal is 18 months.[120]

The androgen receptor is amplified in 30% of hormone refractory prostate cancers but is rarely amplified in untreated cases (**Fig. 11.2**).[3,121] This suggests AR amplification is selected following androgen withdrawal. Amplification is associated with overexpression of the *AR* gene in hormone refractory disease.[108,122,123] Linja et al. in Finland have also shown that some hormone refractory tumors highly express AR even in the absence of *AR* gene amplification.[121] Overexpression allows even very low levels of androgen to promote androgen-dependent growth and patients with AR amplification may respond better to maximum androgen blockade.[122,124]

However, Kinoshita et al. recently reported 10–15% of hormone refractory metastatic lesions express only low AR levels as a result of methylation in the promoter CpG islands.[125] The authors' explanation for this disparity was that the Linja et al. group looked at locally recurrent disease rather than metastases and, as only 13 specimens were examined by the Finns, they may not have detected such a small subgroup with low AR levels.

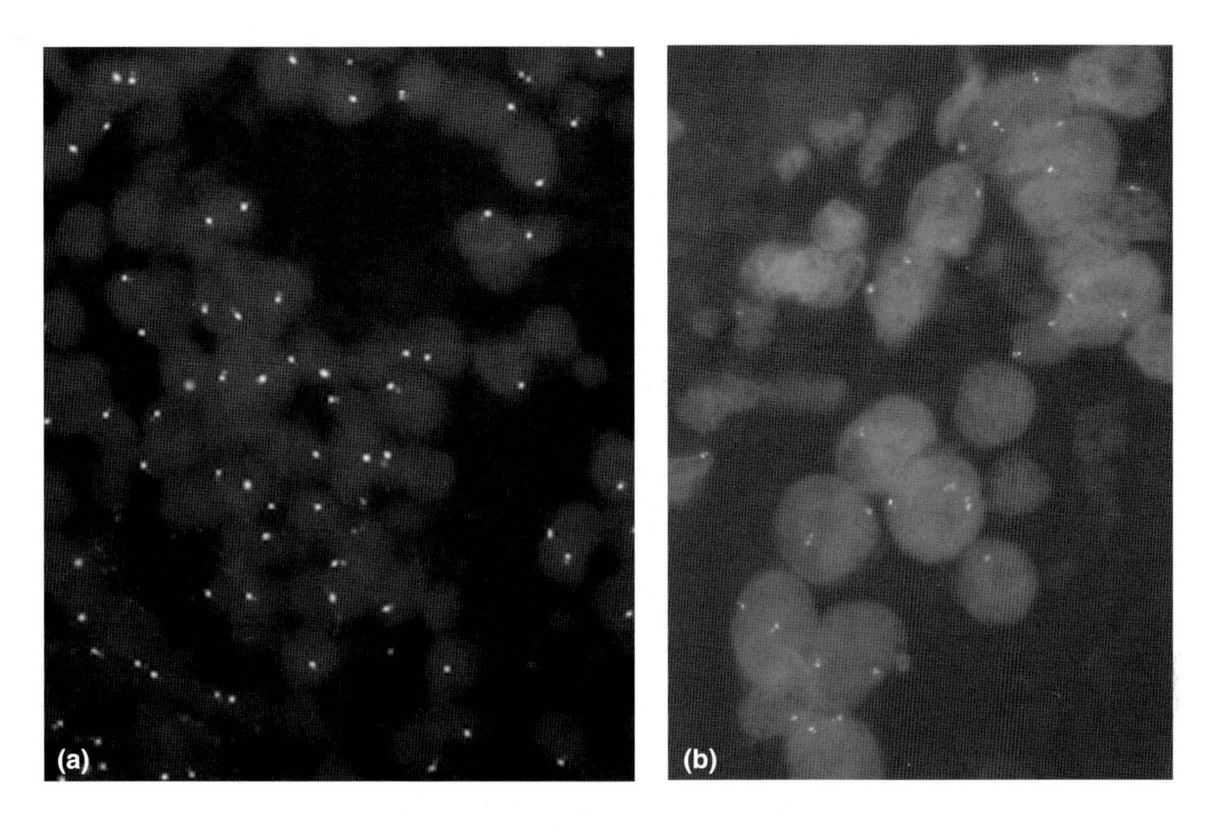

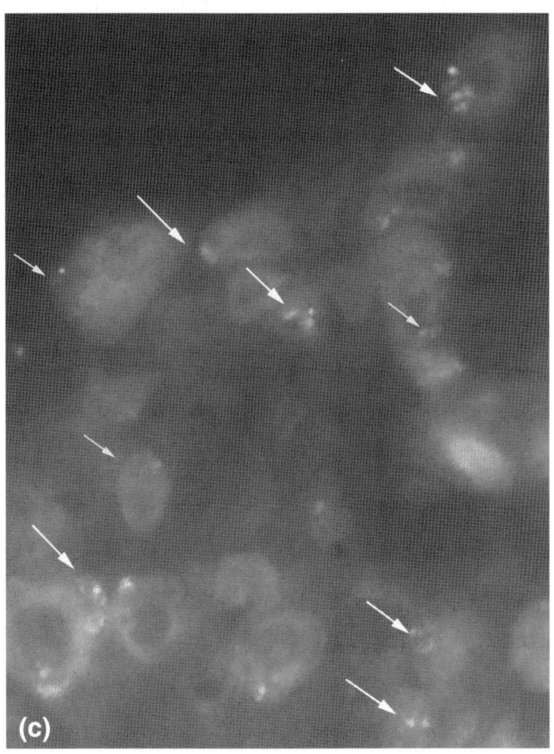

Fig. 11.2. Section of primary prostate cancer showing nuclei stained with DAPI. AR gene and X chromosome copy numbers were analyzed by FISH using red (AR) and green (X chromosome) probes. Most nuclei contain one green and one red signal, indicating that each nucleus contains one X chromosome, on which there is one AR gene. (b) FISH section showing most cells in a primary tumor specimen containing two red (AR gene) and green (X chromosome) signals, indicating disomy. (c) FISH section from an HRPC bone marrow sample, showing multiple red AR gene signals (white arrows) associated with a single X chromosome green signal. This indicates AR gene amplification. Note the glandular structure in the lower right corner of the picture and other areas of AR gene amplification not in the current plane of focus.

AR gene amplification may enable low levels of androgen (such as adrenal androgen levels after surgical or medical castration) to stimulate androgen-dependent functions. There is also evidence that growth factors including insulin-like growth factor 1 (IGF-1), keratinocyte growth factor (KGF) and epidermal growth factor (EGF) may be able to activate the AR through a ligand-independent pathway.[126] For example, the cAMP, protein kinase A and protein kinase C pathway may be able to activate the androgen receptor.[127] This path may be a target for therapeutic intervention in patients with hormone refractory disease.

As well as amplification of the AR, several hundred mutations have been reported (www.mcgill.ca), although not all these reported mutations are associated with prostate cancer. These mutations may alter the AR function and ligand specificity.[128] Mutations are uncommon in untreated prostate cancer and, in cases treated with surgical castration, they are found in patients treated with the testosterone antagonist flutamide.[119,129] Approximately one-third of patients treated with flutamide have specific mutations that result in AR activation.[130] This mutation, T887A, found only in patients treated with flutamide, changes the AR response from an antagonist to an agonist. This may explain why a minority of patients paradoxically show a temporary clinical improvement when flutamide therapy is withdrawn.

As stated, the AR complex exerts its effect on the target gene in conjunction with various cofactor proteins. Several of these proteins have been cloned, including SNURF, ARIP3, ARA54,55,70,160 and ANPK.[131–135] Their role in cancer progression *in vivo* remains unknown at present.

Several studies using CGH have shown that gain in 8q is a characteristic of hormone refractory disease.[64,136,137] Almost 90% of samples of hormone refractory prostate cancer and distant metastases compared with 5% of untreated primary tumors contained gains of the whole of 8q. Alers et al. found the number of individuals with 8q gains increases with advancing tumor stage.[51] Furthermore, gain in 8q, as well as loss of 6q, is significantly less frequently encountered in regional lymph node metastases than in distant metastases.[51] This implies differential genetic changes for hematogenous versus lymphatic spread.[51] Two sites within 8q have been identified as containing critical oncogenes, 8q21 and 8q 23–q24,[136,137] suggesting several target genes. The well-recognized oncogene *MYC* is located at 8q24.1 and amplification of this oncogene has been found in 8% of primary and 11–30% of hormone refractory or advanced prostate cancers.[108,137,138] Furthermore, amplification of this gene is associated with poorer prognosis in patients with locally advanced disease.[138] *MYC* codes for a nuclear phosphoprotein whose function is in the promotion of DNA replication, regulation of the cell cycle and control over cell differentiation.[139]

Other sites of amplification on 8q include 8q23 and 8q24.2, which harbors the putative oncogenes *EIF3S3* and *PSCA*.[122,140] Using FISH, Nupponen et al. found *EIF3S3* in 30% of cases of hormone refractory prostate cancers. Moreover, *EIFS3S* was frequently, but not exclusively, coamplified with *MYC*.[122] *PSCA* is often coamplified with *MYC*, and overexpression of *PSCA* has been reported in high-grade and metastatic prostate cancer.[140]

TELOMERASE ACTIVITY

Telomeres are the non-coding repeat base sequences that are found at the end of all eukaryotic chromosomes. With each cell division the telomeric sequence is shortened as DNA polymerase starts at an RNA primer that is subsequently removed. The telomere thus protects the coding regions until it has been shortened to a critical length. Then it may act as a signal to initiate cell death or apoptosis. Telomerase is an enzyme that adds telomeric sequences to newly formed DNA, therefore, compensating for loss of sequences as a result of replication. This prevents the telomere reaching its critical shortened length and allows unlimited cell division.[9]

Telomerase activity has not been found in normal prostate tissue, it has, however, been found in 172 (89%) of 194 cases of prostate cancer.[141–144] Increased levels of telomerase activity correspond more frequently to poor tumor differentiation.[145] Furthermore, telomerase activity was found in HGPIN in 14 (74%) of 19 samples[146] and 5 (12%) of 42 samples of normal prostate tissue adjacent to prostate cancers.[141,144] These latter findings imply that 'normal' adjacent tissue may have genotype changes before malignant phenotype changes can be identified. Attempts have been made to use telomerase activity in prostate fossa biopsies at radical prostatectomy to predict local recurrence. Increased telomerase activity was found in the prostatic fossa samples of 5 (10%) of 48 men with pT2 and 7 (15%) of 47 men with >pT2.[146] Recurrence rates in these trials are awaited, although telomerase activity may enable more accurate molecular staging of prostate cancer, which will allow earlier therapeutic intervention.

THE FUTURE

The pathogenesis of prostate cancer remains poorly understood as a result of the heterogeneous nature of the disease and the limitations of current scientific techniques. The scientific search for the molecular pathogenesis of prostate cancer has been greatly enhanced by the advent of DNA microarray technology and the mapping of the human DNA sequence – the genome. Microarrays consist of thousands of spots of DNA on a 'chip', each coding for part of a known or unidentified gene. Radiolabeled or fluorescently labeled DNA or RNA of interest is then applied to the chip and binds (hybridizes) where there is a complementary sequence. The site on the chip and intensity of the radioactive or fluorescent signal is proportional to the amount of a particular DNA/RNA.

Rather than studying individual genes, this technique allows examination of complete sets of genes and RNA expression. Examining the RNA, using gene expression profiling, gives an indication of which genes are being expressed at a particular time.[147] This can be used to compare which genes are expressed, for example, in metastases compared with primary tumors.[148] These techniques are already in use at least in research, where clinically similar diseases can be subclassified into distinct conditions, for example, acute myeloid leukemia and acute lymphocytic leukemia,[149] malignant melanoma[150] and breast tumors.[151]

Advances in the molecular biology laboratory will have a fundamental effect on our understanding, diagnosis and treatment of many clinical conditions, including prostate cancer.

REFERENCES

1. Scardino PT, Weaver R, Hudson MA. Early detection of prostate cancer. *Human Pathol.* 1992; 23:211–22.
2. Bostwick DG, Grignon DJ, Hammond ME et al. Prognostic factors in prostate cancer. College of American Pathologists Consensus Statement 1999. *Arch. Pathol. Lab. Med.* 2000; 124: 995–1000.
3. Elo JP, Visakorpi T. Molecular genetics of prostate cancer. *Ann. Med.* 2001; 33:130–41.
4. Fearon ER, Vogelstein B. A genetic model for colorectal tumorigenesis. *Cell* 1990; 61:759–67.
5. Kallioniemi OP, Visakorpi T. Genetic basis and clonal evolution of human prostate cancer. *Adv. Cancer Res.* 1996; 68:225–55.
6. Ruijter E, van de KC, Miller G et al. Molecular genetics and epidemiology of prostate carcinoma. *Endocrinol. Rev.* 1999; 20:22–45.
7. Bedford MT, van Helden PD. Hypomethylation of DNA in pathological conditions of the human prostate. *Cancer Res.* 1987; 47:5274–6.
8. Sakr WA, Grignon DJ, Haas GP et al. Age and racial distribution of prostatic intraepithelial neoplasia. *Eur. Urol.* 1996; 30:138–44.
9. Lynch HT, Shaw MW, Magnuson CW et al. Hereditary factors in cancer. Study of two large midwestern kindreds. *Arch. Intern. Med.* 1966; 117:206–12.
10. Li FP, Fraumeni JF Jr. Soft-tissue sarcomas, breast cancer, and other neoplasms. A familial syndrome? *Ann. Intern. Med.* 1969; 71:747–52.
11. Knudson AG Jr. Mutation and cancer: statistical study of retinoblastoma. *Proc. Natl Acad. Sci. USA* 1971; 68:820–3.
12. Hogg A, Bia B, Onadim Z et al. Molecular mechanisms of oncogenic mutations in tumors from patients with bilateral and unilateral retinoblastoma. *Proc. Natl Acad. Sci. USA* 1993; 90:7351–5.
13. Steinberg GD, Carter BS, Beaty TH et al. Family history and the risk of prostate cancer. *Prostate* 1990; 17:337–47.
14. Thiessen EU. Concerning a familial association between breast cancer and both prostatic and uterine malignancies. *Cancer* 1974; 34:1102–7.
15. Tulinius H, Egilsson V, Olafsdottir GH et al. Risk of prostate, ovarian, and endometrial cancer among relatives of women with breast cancer. *Br. Med. J.* 1992; 305:855–7.
16. Anderson DE, Badzioch MD. Breast cancer risks in relatives of male breast cancer patients. *J. Natl Cancer Inst.* 1992; 84:1114–17.
17. Sellers TA, Potter JD, Rich SS et al. Familial clustering of breast and prostate cancers and risk of postmenopausal breast cancer. *J. Natl Cancer Inst.* 1994; 86:1860–5.
18. Carter BS, Bova GS, Beaty TH et al. Hereditary prostate cancer: epidemiologic and clinical features. *J. Urol.* 1993; 150:797–802.
19. Ostrander EA, Stanford JL. Genetics of prostate cancer: too many loci, too few genes. *Am. J. Human Genet.* 2000; 67:1367–75.
20. Suarez BK, Lin J, Burmester JK et al. A genome screen of multiplex sibships with prostate cancer. *Am. J. Human Genet.* 2000; 66:933–44.
21. Tavtigian SV, Simard J, Teng DH et al. A candidate prostate cancer susceptibility gene at chromosome 17p. *Nature Genet.* 2001; 27:172–80.
22. Smith JR, Freije D, Carpten JD et al. Major susceptibility locus for prostate cancer on chromosome 1 suggested by a genome-wide search. *Science* 1996; 274:1371–4.
23. Cooney KA, McCarthy JD, Lange E et al. Prostate cancer susceptibility locus on chromosome 1q: a confirmatory study. *J. Natl Cancer Inst.* 1997; 89:955–9.
24. Hsieh CL, Oakley-Girvan I, Gallagher RP et al. Re: prostate cancer susceptibility locus on chromosome 1q: a confirmatory study. *J. Natl Cancer Inst.* 1997; 89:1893–4.
25. Berthon P, Valeri A, Cohen-Akenine A et al. Predisposing gene for early-onset prostate cancer, localized on chromosome 1q42.2-43. *Am. J. Human Genet.* 1998; 62:1416–24.
26. Eeles RA, Durocher F, Edwards S et al. Linkage analysis of chromosome 1q markers in 136 prostate cancer families. The Cancer Research Campaign/British Prostate Group U.K. Familial Prostate Cancer Study Collaborators. *Am. J. Human Genet.* 1998; 62:653–8.
27. McIndoe RA, Stanford JL, Gibbs M et al. Linkage analysis of 49 high-risk families does not support a common familial prostate cancer-susceptibility gene at 1q24-25. *Am. J. Human Genet.* 1997; 61:347–53.
28. Schaid DJ, Rowland C. Use of parents, sibs, and unrelated controls for detection of associations between genetic markers and disease. *Am. J. Human Genet.* 1998; 63:1492–506.
29. Xu J, Meyers D, Freije D et al. Evidence for a prostate cancer susceptibility locus on the X chromosome. *Nature Genet.* 1998; 20:175–9.
30. Chamberlain NL, Driver ED, Miesfeld RL. The length and location of CAG trinucleotide repeats in the androgen receptor N-terminal domain affect transactivation function. *Nucleic Acids Res.* 1994; 22:3181–6.
31. Kazemi-Esfarjani P, Trifiro MA, Pinsky L. Evidence for a repressive function of the long polyglutamine tract in the human androgen receptor: possible pathogenetic relevance for the (CAG)n-expanded neuronopathies. *Human Mol. Genet.* 1995; 4:523–7.
32. Stanford JL, Just JJ, Gibbs M et al. Polymorphic repeats in the androgen receptor gene: molecular markers of prostate cancer risk. *Cancer Res.* 1997; 57:1194–8.
33. Irvine RA, Yu MC, Ross RK et al. The CAG and GGC microsatellites of the androgen receptor gene are in linkage disequilibrium in men with prostate cancer. *Cancer Res.* 1995; 55:1937–40.
34. Hsing AW, Gao YT, Wu G et al. Polymorphic CAG and GGN repeat lengths in the androgen receptor gene and prostate cancer risk: a population-based case-control study in China. *Cancer Res.* 2000; 60:5111–16.
35. Giovannucci E, Stampfer MJ, Krithivas K et al. The CAG repeat within the androgen receptor gene and its relationship to prostate cancer. *Proc. Natl Acad. Sci. USA* 1997; 94:3320–3.
36. Jaffe JM, Malkowicz SB, Walker AH et al. Association of SRD5A2 genotype and pathological characteristics of prostate tumors. *Cancer Res.* 2000; 60:1626–30.
37. Skowronski RJ, Peehl DM, Feldman D. Vitamin D and prostate cancer: 1,25 dihydroxyvitamin D3 receptors and actions in human prostate cancer cell lines. *Endocrinology* 1993; 132:1952–60.
38. Peehl DM, Skowronski RJ, Leung GK et al. Antiproliferative effects of 1,25-dihydroxyvitamin D3 on primary cultures of human prostatic cells. *Cancer Res.* 1994; 54:805–10.
39. Corder EH, Guess HA, Hulka BS et al. Vitamin D and prostate cancer: a prediagnostic study with stored sera. *Cancer Epidemiol. Biomarkers Prev.* 1993; 2:467–72.

40. Ahonen MH, Tenkanen L, Teppo L et al. Prostate cancer risk and prediagnostic serum 25-hydroxyvitamin D levels (Finland). *Cancer Causes Control* 2000; 11:847–52.
41. Taylor JA, Hirvonen A, Watson M et al. Association of prostate cancer with vitamin D receptor gene polymorphism. *Cancer Res*. 1996; 56:4108–10.
42. Habuchi T, Suzuki T, Sasaki R et al. Association of vitamin D receptor gene polymorphism with prostate cancer and benign prostatic hyperplasia in a Japanese population. *Cancer Res*. 2000; 60:305–8.
43. Blazer DG III, Umbach DM, Bostick RM et al. Vitamin D receptor polymorphisms and prostate cancer. *Mol. Carcinog*. 2000; 27:18–23.
44. Ingles SA, Coetzee GA, Ross RK et al. Association of prostate cancer with vitamin D receptor haplotypes in African-Americans. *Cancer Res*. 1998; 58:1620–3.
45. Watanabe M, Fukutome K, Murata M et al. Significance of vitamin D receptor gene polymorphism for prostate cancer risk in Japanese. *Anticancer Res*. 1999; 19:4511–14.
46. Furuya Y, Akakura K, Masai M et al. Vitamin D receptor gene polymorphism in Japanese patients with prostate cancer. *Endocrinol J*. 1999; 46:467–70.
47. Nelson WG, Wilding G. Prostate cancer prevention agent development: criteria and pipeline for candidate chemoprevention agents. *Urology* 2001; 57(4 Suppl 1):56–63.
48. Rebbeck TR, Jaffe JM, Walker AH et al. Modification of clinical presentation of prostate tumors by a novel genetic variant in CYP3A4. *J. Natl Cancer Inst*. 1998; 90:1225–9.
49. Habuchi T, Liqing Z, Suzuki T et al. Increased risk of prostate cancer and benign prostatic hyperplasia associated with a CYP17 gene polymorphism with a gene dosage effect. *Cancer Res*. 2000; 60:5710–13.
50. Rebbeck TR, Walker AH, Zeigler-Johnson C et al. Association of HPC2/ELAC2 genotypes and prostate cancer. *Am. J. Human Genet*. 2000; 67:1014–19.
51. Alers JC, Rochat J, Krijtenburg PJ et al. Identification of genetic markers for prostatic cancer progression. *Lab. Invest*. 2000; 80:931–42.
52. Nupponen N, Visakorpi T. Molecular biology of progression of prostate cancer. *Eur. Urol*. 1999; 35:351–4.
53. Cooney KA, Wetzel JC, Consolino CM et al. Identification and characterization of proximal 6q deletions in prostate cancer. *Cancer Res*. 1996; 56:4150–3.
54. Cooney KA, Wetzel JC, Merajver SD et al. Distinct regions of allelic loss on 13q in prostate cancer. *Cancer Res*. 1996; 56:1142–5.
55. Gao X, Zacharek A, Grignon DJ et al. Localization of potential tumor suppressor loci to a <2 Mb region on chromosome 17q in human prostate cancer. *Oncogene* 1995; 11:1241–7.
56. Latil A, Cussenot O, Fournier G et al. Loss of heterozygosity at chromosome 16q in prostate adenocarcinoma: identification of three independent regions. *Cancer Res*. 1997; 57:1058–62.
57. Li C, Berx G, Larsson C et al. Distinct deleted regions on chromosome segment 16q23–24 associated with metastases in prostate cancer. *Genes Chromosomes Cancer* 1999; 24:175–82.
58. Lee WH, Isaacs WB, Bova GS et al. CG island methylation changes near the GSTP1 gene in prostatic carcinoma cells detected using the polymerase chain reaction: a new prostate cancer biomarker. *Cancer Epidemiol. Biomarkers Prev*. 1997; 6:443–50.
59. Brooks JD, Weinstein M, Lin X et al. CG island methylation changes near the GSTP1 gene in prostatic intraepithelial neoplasia. *Cancer Epidemiol. Biomarkers Prev*. 1998; 7:531–6.
60. Mannervik B, Alin P, Guthenberg C et al. Identification of three classes of cytosolic glutathione transferase common to several mammalian species: correlation between structural data and enzymatic properties. *Proc. Natl Acad. Sci. USA* 1985; 82:7202–6.
61. Cairns P, Esteller M, Herman JG et al. Molecular detection of prostate cancer in urine by GSTP1 hypermethylation. *Clin. Cancer Res*. 2001; 7:2727–30.
62. Goessl C, Muller M, Heicappell R et al. DNA-based detection of prostate cancer in urine after prostatic massage. *Urology* 2001; 58:335–8.
63. Wilkinson J, Clapper ML. Detoxication enzymes and chemoprevention. *Proc. Soc. Exp. Biol. Med*. 1997; 216:192–200.
64. Visakorpi T, Kallioniemi AH, Syvanen AC et al. Genetic changes in primary and recurrent prostate cancer by comparative genomic hybridization. *Cancer Res*. 1995; 55:342–7.
65. Emmert-Buck MR, Vocke CD, Pozzatti RO et al. Allelic loss on chromosome 8p12–21 in microdissected prostatic intraepithelial neoplasia. *Cancer Res*. 1995; 55:2959–62.
66. Saric T, Brkanac Z, Troyer DA et al. Genetic pattern of prostate cancer progression. *Int. J. Cancer* 1999; 81:219–24.
67. Brothman AR, Maxwell TM, Cui J et al. Chromosomal clues to the development of prostate tumors. *Prostate* 1999; 38:303–12.
68. Ishii H, Baffa R, Numata SI et al. The FEZ1 gene at chromosome 8p22 encodes a leucine-zipper protein, and its expression is altered in multiple human tumors. *Proc. Natl Acad. Sci. USA* 1999; 96:3928–33.
69. Li C, Larsson C, Futreal A et al. Identification of two distinct deleted regions on chromosome 13 in prostate cancer. *Oncogene* 1998; 16:481–7.
70. Jarrard DF, Bova GS, Ewing CM et al. Deletional, mutational, and methylation analyses of CDKN2 (p16/MTS1) in primary and metastatic prostate cancer. *Genes Chromosomes Cancer* 1997; 19:90–6.
71. Gu K, Mes-Masson AM, Gauthier J et al. Analysis of the p16 tumor suppressor gene in early-stage prostate cancer. *Mol. Carcinog*. 1998; 21:164–70.
72. Serrano M, Hannon GJ, Beach D. A new regulatory motif in cell-cycle control causing specific inhibition of cyclin D/CDK4. *Nature* 1993; 366:704–7.
73. Cher ML, Lewis PE, Banerjee M et al. A similar pattern of chromosomal alterations in prostate cancers from African–Americans and Caucasian Americans. *Clin. Cancer Res*. 1998; 4:1273–8.
74. Cairns P, Okami K, Halachmi S et al. Frequent inactivation of PTEN/MMAC1 in primary prostate cancer. *Cancer Res*. 1997; 57:4997–5000.
75. Di Cristofano A, Pandolfi PP. The multiple roles of PTEN in tumor suppression. *Cell* 2000; 100:387–90.
76. Schreiber-Agus N, Meng Y, Hoang T et al. Role of Mxi1 in ageing organ systems and the regulation of normal and neoplastic growth. *Nature* 1998; 393:483–7.
77. Di Cristofano A, De Acetis M, Koff A et al. Pten and p27KIP1 cooperate in prostate cancer tumor suppression in the mouse. *Nature Genet*. 2001; 27:222–4.
78. Liaw D, Marsh DJ, Li J et al. Germline mutations of the PTEN gene in Cowden disease, an inherited breast and thyroid cancer syndrome. *Nature Genet*. 1997; 16:64–7.
79. Marsh DJ, Dahia PL, Zheng Z et al. Germline mutations in PTEN are present in Bannayan–Zonana syndrome. *Nature Genet*. 1997; 16:333–4.
80. Dong JT, Sipe TW, Hyytinen ER et al. PTEN/MMAC1 is infrequently mutated in pT2 and pT3 carcinomas of the prostate. *Oncogene* 1998; 17:1979–82.
81. Feilotter HE, Nagai MA, Boag AH et al. Analysis of PTEN and the 10q23 region in primary prostate carcinomas. *Oncogene* 1998; 16:1743–8.
82. Pesche S, Latil A, Muzeau F et al. PTEN/MMAC1/TEP1 involvement in primary prostate cancers. *Oncogene* 1998; 16:2879–83.
83. Suzuki H, Freije D, Nusskern DR et al. Interfocal heterogeneity of PTEN/MMAC1 gene alterations in multiple metastatic prostate cancer tissues. *Cancer Res*. 1998; 58:204–9.
84. Vlietstra RJ, van Alewijk DC, Hermans KG et al. Frequent inactivation of PTEN in prostate cancer cell lines and xenografts. *Cancer Res*. 1998; 58:2720–3.

85. Elo JP, Harkonen P, Kyllonen AP et al. Three independently deleted regions at chromosome arm 16q in human prostate cancer: allelic loss at 16q24.1–q24.2 is associated with aggressive behaviour of the disease, recurrent growth, poor differentiation of the tumour and poor prognosis for the patient. *Br. J. Cancer* 1999; 79:156–60.
86. Suzuki H, Komiya A, Emi M et al. Three distinct commonly deleted regions of chromosome arm 16q in human primary and metastatic prostate cancers. *Genes Chromosomes Cancer* 1996; 17:225–33.
87. Umbas R, Isaacs WB, Bringuier PP et al. Decreased E-cadherin expression is associated with poor prognosis in patients with prostate cancer. *Cancer Res.* 1994; 54:3929–33.
88. Richmond PJ, Karayiannakis AJ, Nagafuchi A et al. Aberrant E-cadherin and alpha-catenin expression in prostate cancer: correlation with patient survival. *Cancer Res.* 1997; 57:3189–93.
89. Takeichi M. Cadherin cell adhesion receptors as a morphogenetic regulator. *Science* 1991; 251:1451–5.
90. Voeller HJ, Truica CI, Gelmann EP. Beta-catenin mutations in human prostate cancer. *Cancer Res.* 1998; 58:2520–3.
91. Chesire DR, Ewing CM, Sauvageot J et al. Detection and analysis of beta-catenin mutations in prostate cancer. *Prostate* 2000; 45:323–34.
92. Hollstein M, Sidransky D, Vogelstein B et al. p53 mutations in human cancers. *Science* 1991; 253:49–53.
93. Castagnaro M, Yandell DW, Dockhorn-Dworniczak B et al. [Androgen receptor gene mutations and p53 gene analysis in advanced prostate cancer]. *Verh. Dtsch. Ges. Pathol.* 1993; 77:119–23.
94. Brooks JD, Bova GS, Ewing CM et al. An uncertain role for p53 gene alterations in human prostate cancers. *Cancer Res.* 1996; 56:3814–22.
95. Schlechte HH, Schnorr D, Loning T et al. Mutation of the tumor suppressor gene p53 in human prostate and bladder cancers – investigation by temperature gradient gel electrophoresis (TGGE). *J. Urol.* 1997; 157:1049–53.
96. Voeller HJ, Sugars LY, Pretlow T et al. p53 oncogene mutations in human prostate cancer specimens. *J. Urol.* 1994; 151:492–5.
97. Navone NM, Troncoso P, Pisters LL et al. p53 protein accumulation and gene mutation in the progression of human prostate carcinoma. *J. Natl Cancer Inst.* 1993; 85:1657–69.
98. Shi XB, Bodner SM, deVere White RW et al. Identification of p53 mutations in archival prostate tumors. Sensitivity of an optimized single-strand conformational polymorphism (SSCP) assay. *Diagn. Mol. Pathol.* 1996; 5:271–8.
99. Aprikian AG, Sarkis AS, Fair WR et al. Immunohistochemical determination of p53 protein nuclear accumulation in prostatic adenocarcinoma. *J. Urol.* 1994; 151:1276–80.
100. Stattin P. Prognostic factors in prostate cancer. *Scand. J. Urol. Nephrol. Suppl.* 1997; 185:1–46.
101. Prendergast NJ, Atkins MR, Schatte EC et al. p53 immunohistochemical and genetic alterations are associated at high incidence with post-irradiated locally persistent prostate carcinoma. *J. Urol.* 1996; 155:1685–92.
102. Grignon DJ, Caplan R, Sarkar FH et al. p53 status and prognosis of locally advanced prostatic adenocarcinoma: a study based on RTOG 8610. *J. Natl Cancer Inst.* 1997; 89:158–65.
103. Huang A, Gandour-Edwards R, Rosenthal SA et al. p53 and bcl-2 immunohistochemical alterations in prostate cancer treated with radiation therapy. *Urology* 1998; 51:346–51.
104. Thompson TC. Growth factors and oncogenes in prostate cancer. *Cancer Cells* 1990; 2:345–54.
105. Qiu Y, Ravi L, Kung HJ. Requirement of ErbB2 for signalling by interleukin-6 in prostate carcinoma cells. *Nature* 1998; 393:83–5.
106. Pegram MD, Slamon DJ. Combination therapy with trastuzumab (Herceptin) and cisplatin for chemoresistant metastatic breast cancer: evidence for receptor-enhanced chemosensitivity. *Semin. Oncol.* 1999; 26(4 Suppl. 12):89–95.
107. Scher HI. HER2 in prostate cancer – a viable target or innocent bystander? *J. Natl Cancer Inst.* 2000; 92:1866–8.
108. Bubendorf L, Kononen J, Koivisto P et al. Survey of gene amplifications during prostate cancer progression by high-throughout fluorescence in situ hybridization on tissue microarrays. *Cancer Res.* 1999; 59:803–6.
109. Agus DB, Scher HI, Higgins B et al. Response of prostate cancer to anti-Her-2/neu antibody in androgen-dependent and -independent human xenograft models. *Cancer Res.* 1999; 59:4761–4.
110. Craft N, Shostak Y, Carey M et al. A mechanism for hormone-independent prostate cancer through modulation of androgen receptor signaling by the HER-2/neu tyrosine kinase. *Nature Med.* 1999; 5:280–5.
111. Yeh S, Lin HK, Kang HY et al. From HER2/Neu signal cascade to androgen receptor and its coactivators: a novel pathway by induction of androgen target genes through MAP kinase in prostate cancer cells. *Proc. Natl Acad. Sci. USA* 1999; 96:5458–63.
112. Behrens J, Mareel MM, Van Roy FM et al. Dissecting tumor cell invasion: epithelial cells acquire invasive properties after the loss of uvomorulin-mediated cell-cell adhesion. *J. Cell Biol.* 1989; 108:2435–47.
113. Gao AC, Lou W, Dong JT et al. CD44 is a metastasis suppressor gene for prostatic cancer located on human chromosome 11p13. *Cancer Res.* 1997; 57:846–9.
114. Dong JT, Lamb PW, Rinker-Schaeffer CW et al. KAI1, a metastasis suppressor gene for prostate cancer on human chromosome 11p11.2. *Science* 1995; 268:884–6.
115. Ueda T, Komiya A, Emi M et al. Allelic losses on 18q21 are associated with progression and metastasis in human prostate cancer. *Genes Chromosomes Cancer* 1997; 20:140–7.
116. Gao X, Porter AT, Honn KV. Involvement of the multiple tumor suppressor genes and 12-lipoxygenase in human prostate cancer. Therapeutic implications. *Adv. Exp. Med. Biol.* 1997; 407:41–53.
117. Feigl P, Blumenstein B, Thompson I et al. Design of the Prostate Cancer Prevention Trial (PCPT). *Control Clin. Trials* 1995; 16:150–63.
118. Palmberg C, Koivisto P, Visakorpi T et al. PSA decline is an independent prognostic marker in hormonally treated prostate cancer. *Eur. Urol.* 1999; 36:191–6.
119. Taplin ME, Bubley GJ, Shuster TD et al. Mutation of the androgen-receptor gene in metastatic androgen-independent prostate cancer. *N. Engl. J. Med.* 1995; 332:1393–8.
120. Crawford ED. Challenges in the management of prostate cancer. *Br. J. Urol.* 1992; 70(Suppl. 1):33–8.
121. Linja MJ, Savinainen KJ, Saramaki OR et al. Amplification and overexpression of androgen receptor gene in hormone-refractory prostate cancer. *Cancer Res.* 2001; 61:3550–5.
122. Visakorpi T, Hyytinen E, Koivisto P et al. In vivo amplification of the androgen receptor gene and progression of human prostate cancer. *Nature Genet.* 1995; 9:401–6.
123. Koivisto P, Kononen J, Palmberg C et al. Androgen receptor gene amplification: a possible molecular mechanism for androgen deprivation therapy failure in prostate cancer. *Cancer Res.* 1997; 57:314–19.
124. Palmberg C, Koivisto P, Kakkola L et al. Androgen receptor gene amplification at primary progression predicts response to combined androgen blockade as second line therapy for advanced prostate cancer. *J. Urol.* 2000; 164:1992–5.
125. Kinoshita H, Shi Y, Sandefur C et al. Methylation of the androgen receptor minimal promoter silences transcription in human prostate cancer. *Cancer Res.* 2000; 60:3623–30.
126. Culig Z, Hobisch A, Cronauer MV et al. Androgen receptor activation in prostatic tumor cell lines by insulin-like growth

factor-I, keratinocyte growth factor, and epidermal growth factor. *Cancer Res.* 1994; 54:5474–8.
127. Ikonen T, Palvimo JJ, Kallio PJ et al. Stimulation of androgen-regulated transactivation by modulators of protein phos phorylation. *Endocrinology* 1994; 135:1359–66.
128. Zhao XY, Malloy PJ, Krishnan AV et al. Glucocorticoids can promote androgen-independent growth of prostate cancer cells through a mutated androgen receptor. *Nature Med.* 2000; 6:703–6.
129. Wallen MJ, Linja M, Kaartinen K et al. Androgen receptor gene mutations in hormone-refractory prostate cancer. *J. Pathol.* 1999; 189:559–63.
130. Taplin ME, Bubley GJ, Ko YJ et al. Selection for androgen receptor mutations in prostate cancers treated with androgen antagonist. *Cancer Res.* 1999; 59:2511–15.
131. Moilanen AM, Poukka H, Karvonen U et al. Identification of a novel RING finger protein as a coregulator in steroid receptor-mediated gene transcription. *Mol. Cell Biol.* 1998; 18:5128–39.
132. Kang HY, Yeh S, Fujimoto N et al. Cloning and characterization of human prostate coactivator ARA54, a novel protein that associates with the androgen receptor. *J. Biol. Chem.* 1999; 274:8570–6.
133. Fujimoto N, Yeh S, Kang HY et al. Cloning and characterization of androgen receptor coactivator, ARA55, in human prostate. *J. Biol. Chem.* 1999; 274:8316–21.
134. Yeh S, Chang C. Cloning and characterization of a specific coactivator, ARA70, for the androgen receptor in human prostate cells. *Proc. Natl Acad. Sci. USA* 1996; 93:5517–21.
135. Hsiao PW, Chang C. Isolation and characterization of ARA160 as the first androgen receptor N-terminal-associated coactivator in human prostate cells. *J. Biol. Chem.* 1999; 274:22373–9.
136. Cher ML, Bova GS, Moore DH et al. Genetic alterations in untreated metastases and androgen-independent prostate cancer detected by comparative genomic hybridization and allelotyping. *Cancer Res.* 1996; 56:3091–102.
137. Nupponen NN, Kakkola L, Koivisto P et al. Genetic alterations in hormone-refractory recurrent prostate carcinomas. *Am. J. Pathol.* 1998; 153:141–8.
138. Sato K, Qian J, Slezak JM et al. Clinical significance of alterations of chromosome 8 in high-grade, advanced, nonmetastatic prostate carcinoma. *J. Natl Cancer Inst.* 1999; 91:1574–80.
139. Kato GJ, Wechsler DS, Dang CV. DNA binding by the Myc oncoproteins. *Cancer Treatment Res.* 1992; 63:313–25.
140. Reiter RE, Sato I, Thomas G et al. Coamplification of prostate stem cell antigen (PSCA) and MYC in locally advanced prostate cancer. *Genes Chromosomes Cancer* 2000; 27:95–103.
141. Sommerfeld HJ, Meeker AK, Piatyszek MA et al. Telomerase activity: a prevalent marker of malignant human prostate tissue. *Cancer Res.* 1996; 56:218–22.
142. Kim NW, Piatyszek MA, Prowse KR et al. Specific association of human telomerase activity with immortal cells and cancer. *Science* 1994; 266:2011–15.
143. Scates DK, Muir GH, Venitt S et al. Detection of telomerase activity in human prostate: a diagnostic marker for prostatic cancer? *Br. J. Urol.* 1997; 80:263–8.
144. Zhang W, Kapusta LR, Slingerland JM et al. Telomerase activity in prostate cancer, prostatic intraepithelial neoplasia, and benign prostatic epithelium. *Cancer Res.* 1998; 58:619–21.
145. Lin Y, Uemura H, Fujinami K et al. Telomerase activity in primary prostate cancer. *J. Urol.* 1997; 157:1161–5.
146. Straub B, Muller M, Krause H et al. Molecular staging of surgical margins after radical prostatectomy by detection of telomerase activity. *Prostate* 2001; 49:140–4.
147. Williamson M, Naaby-Hansen S, Masters JR. 21st century molecular biology in urology. *BJU Int.* 2001; 88:451–7.
148. Clark EA, Golub TR, Lander ES et al. Genomic analysis of metastasis reveals an essential role for RhoC. *Nature* 2000; 406:532–5.
149. Golub TR, Slonim DK, Tamayo P et al. Molecular classification of cancer: class discovery and class prediction by gene expression monitoring. *Science* 1999; 286:531–7.
150. Bittner M, Meltzer P, Chen Y et al. Molecular classification of cutaneous malignant melanoma by gene expression profiling. *Nature* 2000; 406:536–40.
151. Perou CM, Sorlie T, Eisen MB et al. Molecular portraits of human breast tumours. *Nature* 2000; 406:747–52.

Epidemiology

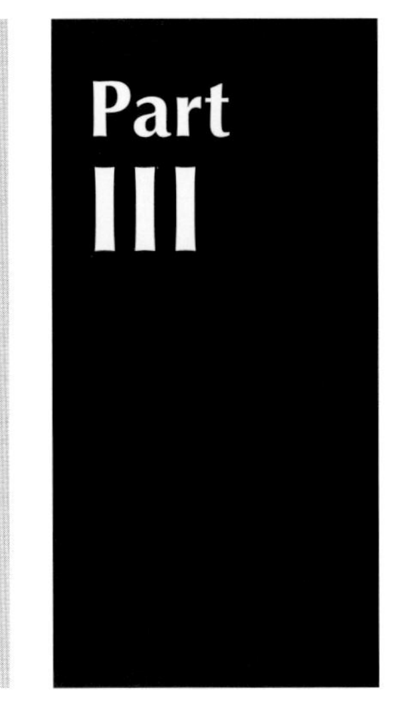

Cancer of the Prostate: Incidence in the USA

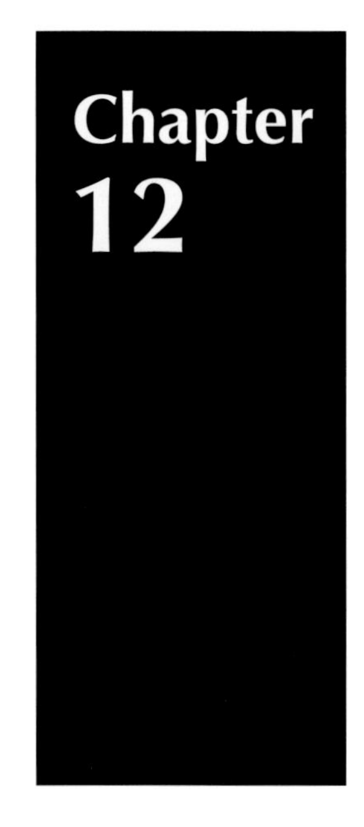

Ihor C. Sawczuk

Department of Urology, Hackensack University Medical Center/ College of Physicians and Surgeons, Columbia University

Ahmad Shabsigh

Department of Urology, College of Physicians and Surgeons, Columbia University

Prostate cancer is a major public health problem, which over a lifetime will affect an estimated one in five American men. The risk of a 50-year-old man with a life expectancy of 25 years of having a microscopic cancer is 42%, of having a clinically significant cancer is 9.5% and of dying of prostate cancer 2.9%.[1] In fact, only a small percentage of prostate cancers become clinically evident: more men die with prostate cancer than of it.[2] Despite its distinction as the most frequently diagnosed non-cutaneous cancer and the second leading cause of cancer deaths in men, little is known about the causes of this malignancy.[3] Prostate cancer is the fourth most commonly diagnosed cancer in men world-wide. It is the second most common malignancy in the European Union, with 200 000 new cancers and 40 000 deaths each year. In England and Wales alone, over 15 000 new cases and 8000 deaths are registered every year. In the USA, it is the cause of 41 000 deaths. As a consequence of PSA screening, the number of new cases increased significantly to about 317 000 in 1992 (**Table 12.1**). Subsequently, it declined to 180 000 new cases in 2000.[4] Years of epidemiological studies showed a highly variable incidence throughout the world. For example, the incidence of prostate cancer in China is nearly 200 times less than that for African Americans. Although this may be due to differences in screening methodology between countries, it does not explain the discrepancy in mortality rates.[4]

Because prostate cancer is seen in an older population, it is expected that the number of diagnosed prostate cancers will increase as life expectancy increases.

Multiple studies have demonstrated that immigration from areas of low risk to areas of high risk increases the risk of prostate cancer in migrants compared to their counties of origin. This might be related to environmental and dietary factors.[5] Astonishingly, the incidence of prostate cancer has been on the rise in the last two decades in Far East Asian states such as Japan and China, most likely due to the Westernization of dietary and environmental norm.[5]

The fact that African Americans have a higher chance of developing prostate cancer suggests that genetic factors play an important role in prostate cancer etiology.[6] This also is apparent in men of African heritage in Jamaica and Southern American counties.[7–8] In addition, a family history in first-degree relatives may be associated with a specific inherited gene abnormality, which may account for up to 9% of cases.[9] For example, a 40-year-old man with three affected close relatives will have a 30–40% lifetime risk of developing prostate cancer.[10,11]

Table 12.1. Estimated cancer deaths of five leading sites, USA.

	Estimated cancer deaths
Lung	31%
Prostate	11%
Colon and rectum	10%
Pancreas	5%
Non-Hodgkin's Lymphoma	5%

INCIDENCE

Since the introduction of the prostate specific antigen (PSA) test to monitor progression and recurrence of prostate cancer in 1986, there have been dramatic changes in the epidemiology of this disease. The absolute number of diagnosed prostate cancer cases increased rapidly to a peak in 1992, then declined subsequent years. PSA also changed the patterns of prostate cancer stage and grade at the time of diagnosis.[3]

In a recent study of the World Health Organization mortality database, researchers were able to discern lower median age

Table 12.2. Incidence and mortality among different ethnic groups in the USA[3] (per 100 000 and age-adjusted; SEER cancer statistics review 1973–1995).

	White	Black	Asian/Pacific Islands	American Indians	Hispanic
Incidence	145.8	225.0	80.4	45.8	101.6
Mortality	23.2	54.1	10.4	14.2	16.2

Table 12.3. Trends in 5-year relative cancer survival among White and Black Americans (SEER cancer statistics review 1973–1995).

	1974–1976	1980–1982	1989–1996
White	68%	75%	94%
Black	59%	65%	87%

of diagnosis in the USA and Canada as compared to other countries. However, the median age at death is very similar in all countries. A decline in the mortality rate has been recorded for men older than 50 years in the USA, the UK, Canada, France and Austria. Rates are beginning to stabilize or are still on the rise in other countries.[12–14]

In the USA, between 1986 and 1992, the overall incidence among White males increased from 86 per 100 000 to 179 per 100 000. On the other hand, the incidence rate among African Americans increased from 124 per 100 000 to 250 per 100 000. In other words, the incidence of prostate cancer increased by 108% for Whites and 102% for African Americans (**Fig. 12.1**). This increase peaked in 1992 for whites and a year later in African American.

The data from the population-based survey of the Surveillance, Epidemiology and End Results (SEER) of the National Cancer Institute do not cover the period after 1995. Therefore, more modern statistics are currently not available.[3]

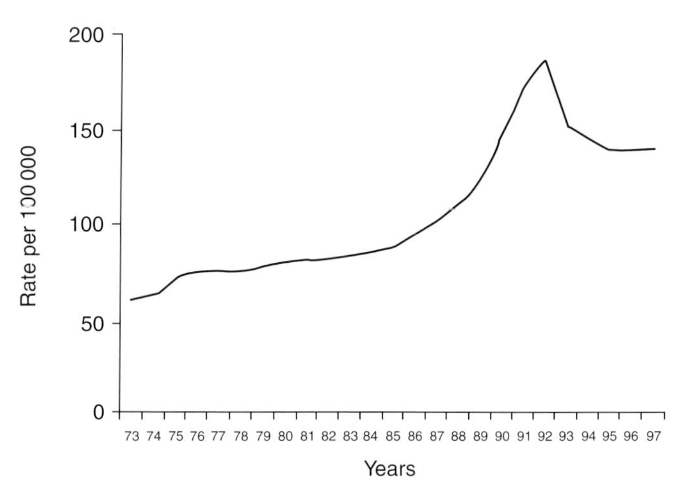

Fig. 12.1. Age-adjusted prostate cancer incidence rate, USA.

In the USA, it is estimated that 198 100 new cases of prostate cancer will be diagnosed in the year 2001. The probability of developing prostate cancer increases with age. Less than 1 in 10 000 of men between birth and 39 will develop this disease, and 2.06 in 10 000 or 1 in 49 of men between the ages of 40 and 49 and 13.42 in 10 000 (1 in 7) of men older than 60. A total of 31 500 men are estimated to have died from prostate cancer in 2001.[3]

Recent studies have showed comparable mortality and recurrence rates for African Americans and Whites adjusting for socioeconomic factors, stage and grade.[15,16]

FAMILY HISTORY AND THE RISK OF PROSTATE CANCER

In a recent study by Kalish and associates using data from the Massachusetts Male Aging Study (MMAS), covering 1149 men with an average of 8.7 person-years of follow-up, the age-adjusted relative risk of prostate cancer incidence associated with prostate cancer family history was 3.78 [95% confidence interval (CI) 1.96–7.28]. This association was independent of environmental factors, such as smoking, alcohol use, body mass index, physical activity, education, sexually transmitted diseases, diet and hormone levels.[17] In addition, several case-control studies identified a familial pattern of prostate cancer. For example, men with one first-degree relative with prostate cancer have a 2.1–2.8-fold greater risk of being diagnosed with prostate cancer compared to men in the same age without a family history. As mentioned before, approximately 9–10% of all diagnosed cases are attributed to familial genetics.[10]

In the last two decades, researchers targeted different genes, seeking to explain the familial predisposition of prostate cancer. At present, the vast majority of men with clinically significant tumors have no known risk factors (with the exception of gender and age). No shared genetic factor has been identified.

Variation of the enzyme 5α-reductase, which converts testosterone to the more potent metabolite of dihydrotestosterone, may stabilize the mRNA of the enzyme, resulting in prolonged androgenic stimulation of the prostatic tissue. This stimulation, although not proven, may increase the risk of prostate cancer.

Polymorphisms in exon 1 of the androgen receptor gene have been implicated in prostate cancer risk. Ingles et al. found a positive correlation with a lower number of CAG repeats and a younger age of diagnosis and higher grade and

Table 12.4. Changes of pathological characteristics of radical prostatectomy between 1990 and 2000[38].

		1990 (%)	1992 (%)	1994 (%)	1996 (%)	1998 (%)	2000 (%)
pTumor	Well	36.1	18.9	9.1	8.8	12.5	8.35
	Moderate	44.3	58.1	57.0	67.1	62.5	69.4
pT stage	≤T2	59.7	56.8	54.9	61.8	63.2	68.0
	≥T3	40.3	43.2	44.1	38.2	36.8	32.0
pCapsule	Positive	35.7	36.8	41.6	36.0	33.8	25.3
pMargin	Positive	33.8	35.1	37.7	27.9	26.3	22.7

more aggressive type of cancer. African Americans have the highest prevalence of short CAG repeats, followed by Whites, while Asians have the lowest. Whether this is only a coincident or true relationship, is left for further research.[18]

Mutations in two genes that are involved in breast cancer, BRCA1 and BRCA2, correlate with an increased risk of prostate cancer. Mutation of BRCA1 or an adjacent gene on chromosome 17 has a stronger relationship with the incidence of prostate cancer than mutation of BRCA2, which is located on chromosome 13.[19,20] Study by Easton showed an increased risk of more aggressive prostate cancer in 16 affected families.[20] Spencer and associates showed that men with no family history appeared to have higher grade tumors. In his study, patients with a positive family history of prostate cancer had well-differentiated tumors.

AGE AT DIAGNOSIS

Between 1991 and 1995, 75–80% of prostate cancers were diagnosed in men over 65 years old. By race, the age-adjusted incidence rate of White men under age 65 was 45.5 per 100 000 and 81.5 per 100 000 for African American men. However, by age 65, the incidence rate for White males in the USA was 1180 per 100 000 and for African Americans 1698; the median age of diagnosis of whites was 71, and for African Americans 69 years.[3]

In a recent study by the Defense Center for Prostate Disease Research, the percentage of men diagnosed with prostate cancer over 65 years of age decreased from 53% in 1990 to 27.8% in 1996 and remained stable thereafter. The number of patients with prostate cancer less than 60 years of age increased from 18.6% in 1991 to 40.7% in 2000.[21]

STAGE AND GRADE AT DIAGNOSIS

Between 1973 and 1995, the proportion of local and regional cancers at diagnosis rose, and the number of metastatic and advanced prostate cancers declined. This was confirmed in a study by Labrie published in 1999.[22] Now, it is estimated that more than 67% of men are diagnosed with localized and early tumors. Still, one-third of patients diagnosed with apparently local disease and treated with radical retropubic prostatectomy relapse biochemically within 5 years. In other words, these men can be considered understaged at the time of diagnosis.

Less than 10% of men diagnosed in the USA have insignificant tumors, which do not cause death or serious medical problems. The chance of a patient dying from prostate cancer at age 70–80 is much lower than dying from comorbid factors.[23]

The data from the Defense Center for Prostate Disease Research showed an increased number of well-differentiated cancer detected from 1991 to 2000, with stage T1 being 16.3% and T2 82.2% in 1991 to 49.8% T1 vs. 48.9 T2 in 1995, and 55.6 T1 vs. 43.8% T2 in 2000. This migration of stage and grade was reported in several studies over the past few years.

A higher proportion of poorly differentiated cancers was found in older men than in younger men. In 1997, Freeman et al. reported a higher proportion of higher grade cancers in African Americans compared to Whites in the same age group, with the possible implication that prostate cancer is more aggressive in African Americans than in Whites.[24]

SURVIVAL AND MORTALITY

Prostate cancer screening with PSA, digital rectal examination and ultrasound-guided biopsy is changing the natural history of prostate cancer, the age of diagnosis, the stage and grade at diagnosis, the socioeconomic factors and, most importantly, the survival patterns and the mortality for this disease.[25,26] Because of lead time and length biases of screening, an increase in survival time after diagnosis does not necessarily lead to a decrease in prostate cancer specific mortality. For example, lead-time bias due to early diagnosis causes artificial increase of survival time. On the other hand, length-time bias occurs when we detect less aggressive tumors during screening. The false impression that our therapies are more effective will result, while in fact a large number of insignificant prostate cancers are diagnosed, which are unlikely to cause death. Also, detecting clinically important tumors through screening in an advanced age will not change the fact that the patient will most likely die from his comorbidities.

Table 12.5. Survival rates between 1989 and 1994.

	Local disease (%)	Regional disease (%)	Distant disease (%)
White Americans	99.9	99.8	33.1
African Americans	92	81.2	29.7

Between 1989 and 1994 the 5-year survival rates among White Americans was 95.1%, compared to 81.2% of African Americans (Table 12.5). In 2000, the 5-year survival rate for prostate cancer in white Americans was 100% for localized disease and 93% for all stages. For African Americans, it was 95% and 84%, respectively.[3]

There is a controversy as to whether prostate cancer has more aggressive biological characteristics among African Americans. The wide perception is that prostate cancer among African Americans is less responsive to aggressive intervention.[27] An interesting study by Underwood and colleagues on 1074 men with localized prostate cancer who underwent radical prostatectomy revealed similar responsiveness of African and White Americans after stratification for age, stage and grade.[28]

In general, prostate cancer specific survival rates have been improving over the last 25 years, most likely due to the stage/grade migration, public awareness and considerable improvement of different therapeutic modalities, specifically surgical management. There are only few large published studies of patients treated with radical prostatectomy that have had prospective, continuous follow-up for 15 years. Even in these series, relatively few patients were followed for 10 years or more. Moreover, all series that presented a 15-year interval began to lose relevance as the natural history of prostate cancer detected changed since 1986.[29]

About 70% of men with clinically localized disease are cured with radical prostatectomy. In a recent analysis of 1778 patients who had pathologically confirmed organ-confined disease (these percentages have increased in recent years with PSA screening), 4% had extraprostatic tumor extension with negative (tumor-free) surgical margins, 21% had extraprostatic tumor with positive (cancerous) margins, 9% had seminal vesicle invasion and 2% had lymph node metastasis. With follow-up ranging between 1 and 15 years, 19% of these men have evidence of cancer recurrence. Most recurrences occurred within 2–3 years of surgery. The 7-year all-cause survival rate was 90% and the cancer-specific survival rate was 97% (i.e. only 3% of patients died of prostate cancer within 7 years of surgery). This cancer-specific survival rate is consistent with the results reported from other contemporary radical prostatectomy series.[7,30,31] It is also higher than the cancer-specific survival rates reported with conservative management.[32,33]

The prostate cancer mortality among White males in the USA was 20.3 per 100 000 in 1973. It peaked in 1991 at 24.7 per 100 000 and decreased to 22.9% per 100 000 in 1995 and 21.7 per 100 000 in 1998. The mortality rate for African American was 39.5, 56.2, 53.5 and 52.1, respectively.[3]

SOCIOECONOMIC FACTORS

Early studies of prostate cancer prevalence and poverty indicated that men of lower socioeconomic status are diagnosed with more advanced disease.[34–36] However, there was no solid evidence until recently that associated poverty and risk of prostate cancer among African America men. Robbins et al. found that differences associated with socioeconomic status do not explain why Black men die from prostate cancer at a higher rate when compared with White men with this condition.[37]

REFERENCES

1. Whitmore WF Jr. Localised prostatic cancer: management and detection issues. *Lancet* 1994; 343:1263–7.
2. Bubendorf L, Schopfer A, Wagner U et al. Metastatic patterns of prostate cancer: an autopsy study of 1,589 patients. *Human Pathol.* 2000; 31:578–83.
3. Ries LAG, Kosary CL, Hankey BF et al. (eds) *SEER Cancer Statistics Review 1973–1995*. Bethesda, MD: National Cancer Institute, 1998.
4. Greenlee RT, Hill-Harmon MB, Murray T et al. Cancer statistics, 2001. *Ca: Can. J. Clin.* 2001; 51:15–36.
5. Tomatis L, Aitio A, Day NE et al. Cancer: causes, occurrence and control. *IARC Sci. Publ.* 1990; 100:1–344.
6. Farkas A, Marcella S, Rhoads GG. Ethnic and racial differences in prostate cancer incidence and mortality. *Ethn. Dis.* 2000; 10:69–75.
7. Glover FE Jr, Coffey DS, Douglas LL et al. Familial study of prostate cancer in Jamaica. *Urology* 1998; 52:441–3.
8. Glover FE Jr, Coffey DS, Douglas LL et al. The epidemiology of prostate cancer in Jamaica. *J. Urol.* 1998; 159:1984–6; discussion 1986–7.
9. Carter BS, Bova GS, Beaty TH et al. Hereditary prostate cancer: epidemiologic and clinical features. *J. Urol.* 1993; 50:797–802.
10. Steinberg GD, Carter BS, Beaty TH et al. Family history and the risk of prostate cancer. *Prostate* 1990; 17:337–47.
11. Arason A, Barkardottir RB, Egilsson V. Linkage analysis of chromosome 17q markers and breast–ovarian cancer in Icelandic families, and possible relationship to prostatic cancer. *Am. J. Human Genet.* 1993; 52:711–17.
12. Oliver SE, May MT, Gunnell D. International trends in prostate-cancer mortality in the 'PSA ERA'. *Int. J. Cancer* 2001; 92:893–8.
13. Vercelli M, Quaglia A, Marani E et al. Prostate cancer incidence and mortality trends among elderly and adult Europeans. *Crit. Rev. Oncol. Hematol.* 2000; 35:133–44.
14. Oliver SE, Gunnell D, Donovan JL. Comparison of trends in prostate-cancer mortality in England and Wales and the USA. *Lancet* 2000; 356:1278.
15. Robbins AS, Whittemore AS, Thom DH. Differences in socioeconomic status and survival among white and black men with prostate cancer. *Am. J. Epidemiol.* 2000; 151:409–16.
16. Conlisk EA, Lengerich EJ, Demark-Wahnefried W et al. Prostate cancer: demographic and behavioral correlates of stage at diagnosis among blacks and whites in North Carolina. *Urology* 1999; 53:1194–9.
17. Kalish LA, McDougal WS, McKinlay JB. Family history and the risk of prostate cancer. *Urology* 2000; 56:803–6.

18. Ingles SA, Ross RK, Yu MC et al. Association of prostate cancer risk with genetic polymorphisms in vitamin D receptor and androgen receptor. *J. Natl Cancer Inst*. 1997; 89:166–70.
19. Sigurdsson S, Thorlacius S, Tomasson J et al. BRCA2 mutation in Icelandic prostate cancer patients. *J. Mol. Med*. 1997; 75: 758–61.
20. Easton DF, Steele L, Fields P et al. Cancer risks in two large breast cancer families linked to BRCA2 on chromosome 13q12–13. *Am. J. Human Genet*. 1997; 61:120–8.
21. The Department of Defense Center for Prostate Disease. *Oncology (Huntingt.)* 1999; 13:1336, 1339–40.
22. Labrie F, Candas B, Dupont A et al. Screening decreases prostate cancer death: first analysis of the 1988 Quebec prospective randomized controlled trial. *Prostate* 1999; 38:83–91.
23. Helgesen F, Holmberg L, Johansson JE et al. Trends in prostate cancer survival in Sweden, 1960 through 1988: evidence of increasing diagnosis of nonlethal tumors. *J. Natl Cancer Inst*. 1996; 88:1216–21.
24. Freeman VL, Leszczak J, Cooper RS. Race and the histologic grade of prostate cancer. *Prostate* 1997; 30:79–84.
25. Prorok PC, Potosky AL, Gohagan JK et al. Prostate cancer screening: current issues. *Cancer Treatment Res*. 1996; 86:93–112.
26. Legler JM, Feuer EJ, Potosky AL et al. The role of prostate-specific antigen (PSA) testing patterns in the recent prostate cancer incidence decline in the United States. *Cancer Causes Control* 1998; 9:519–27.
27. Merrill RM, Lyon JL. Explaining the difference in prostate cancer mortality rates between white and black men in the United States. *Urology* 2000; 55:730–5.
28. Underwood W, Rubin MA, Montie JE et al. Prostate cancer among African American are similarly responsive to radical prostatectomy as are Caucasian prostate cancers of matched stage and grade. *J. Urol*. 2000; 165:64.
29. Akazaki K, Stemmerman GN. Comparative study of latent carcinoma of the prostate among Japanese in Japan and Hawaii. *J. Natl Cancer Inst*. 1973; 50:1137–44.
30. Bouchardy C, Mirra AP, Khlat M et al. Ethnicity and cancer risk in Sao Paulo, Brazil. *Cancer Epidemiol. Biomarkers Prev*. 1991; 1:21–7.
31. Catalona WJ, Smith DS, Ratliff TL et al. Measurement of prostate-specific antigen in serum as a screening test for prostate cancer. *N. Engl J. Med*. 1991; 324:1156–61.
32. Brawley OW. Prostate carcinoma incidence and patient mortality: the effects of screening and early detection. *Cancer* 1997; 80:1857–63.
33. Wingo PA, Ries LA, Rosenberg HM, Miller DS, Edwards BK. Cancer incidence and mortality, 1973–1995: a report card for the U.S. *Cancer* 1998; 82:1197–207.
34. Hakky SI, Chisholm GD, Skeet RG. Social class and carcinoma of the prostate. *Br. J. Urol*. 1979; 51:393–6.
35. Buell P, Dunn JE, Breslow L. The-occupational-social class risk of cancer mortality in men. *J. Chronic Di*. 1960; 12:600–621.
36. Ernster VL, Winkelstein W Jr, Selvin S et al. Race, socioeconomic status, and prostatic cancer. *Cancer Treat Rep*. 1977; 61:187–91.
37. Robbins AS, Whittemore AS, Thom DH. Differences in socioeconomic status and survival among white and black men with prostate cancer. *Am. J. Epidemiol*. 2000; 151:409–16.
38. Moul JW, Sun L, Kane CJ et al. Epidemiologic features of 2699 radical prostatectomy cases between 1990 and 2000: a review of the Department of Defense Center for Prostate Cancer Research Multicenter prostate cancer database. *J. Urol*. 2000; 165:62.

Race, Ethnicity, Religion, Marital Status and Prostate Cancer in the USA

Ronald E. Anglade and Richard K. Babayan
Department of Urology, Boston University School of Medicine, Boston, MA, USA

INTRODUCTION

Prostate cancer is the most commonly diagnosed cancer of men in the USA. In 2001, the American Cancer Society estimates there will be 198 100 new cases of prostate cancer accounting for 31% of cancer diagnosed in American men. In addition, prostate cancer deaths (31 500) will account for 11% of cancer deaths in the same group.[1] According to data from the National Cancer Institute, over a lifetime, the disease will affect one in five American men.[2] There are many factors that influence the incidence, severity and time to diagnosis of this disease. Race and ethnicity have been the subject of numerous investigations and, although the differences between races are durable and reproducible, their root causes have been difficult to elucidate.

During the years 1989–1992, US men experienced a rise in prostate cancer incidence rates of approximately 20% or more. Since 1992, these rates have been on a steady decline. These trends are seen across the board regardless of race. The rise in the early 1990s was felt to have been caused by the initiation of intensive screening which became possible with the approval and widespread use of prostate specific antigen (PSA) testing (**Fig. 13.1**).

To date, only age, race and family history have been established as significant risk factors for the development of prostate cancer. Ethnicity, religion and marital status have also been postulated to influence prostate cancer development and diagnosis directly as well as effect its treatment and prognosis. In this chapter, these factors will be addressed and their influence on the epidemiology of prostate cancer examined. Furthermore, our discussion of race and ethnicity will include the prevailing theories to explain the observed differences as well as current clinical observations.

RACE AND ETHNICITY

Prostate cancer in the USA is the most often diagnosed non-cutaneous cancer of males and the second leading cause of cancer death.[3] The number of prostate cancer cases varies widely among different racial and ethnic groups. It has been well established that African Americans have a greater incidence of prostate cancer than any other racial or ethnic group. Furthermore, the mortality is much greater in this group. The Surveillance, Epidemiology, and End Results (SEER) Program of the National Cancer Institute is the most authoritative source of information on cancer incidence and survival in the USA. SEER collects and publishes data on cancer incidence and survival from 14 cancer registries across the USA. These data cover approximately 14% of the US population, and have been refined to represent an accurate subset of America's diverse racial and ethnic make-up.

In the USA, cancer incidence is generally high (almost 135 cases per 100 000) among White men, lower (around 89 cases per 100 000) amongst Hispanic and Japanese men, and lowest of all in other groups whose ancestors came from Asia.[4] Second-generation Asian Americans have incidence rates that are lower than those of African and Caucasian Americans, but remain higher than their counterparts who had migrated to the USA or continued to reside in their homelands.[5] The lowest incidences of prostate cancer are seen in the Asian populations. Most recent epidemiological data demonstrate an over 65-fold greater risk of prostate cancer in African Americans over that of Mainland Chinese. Besides having a greater incidence of disease, African Americans are usually diagnosed at an age 2 years younger than their Caucasian counterparts.[6] Even when controlled for socioeconomic factors and access to health care, the findings of increased incidence and mortality persist.[7]

The most recent data estimate that there will be 25 300 new cases of prostate cancer diagnosed in African American males in 2001 accounting for 37% of all cancer diagnosed in this racial group. In addition, there are 6100 expected deaths, which makes prostate cancer the second leading cause of cancer death in African Americans after lung cancer. African American men have the highest incidence of prostate cancer (225.0 per 100 000) world-wide. They are at least 50% more likely to develop this disease than any other US racial or ethnic

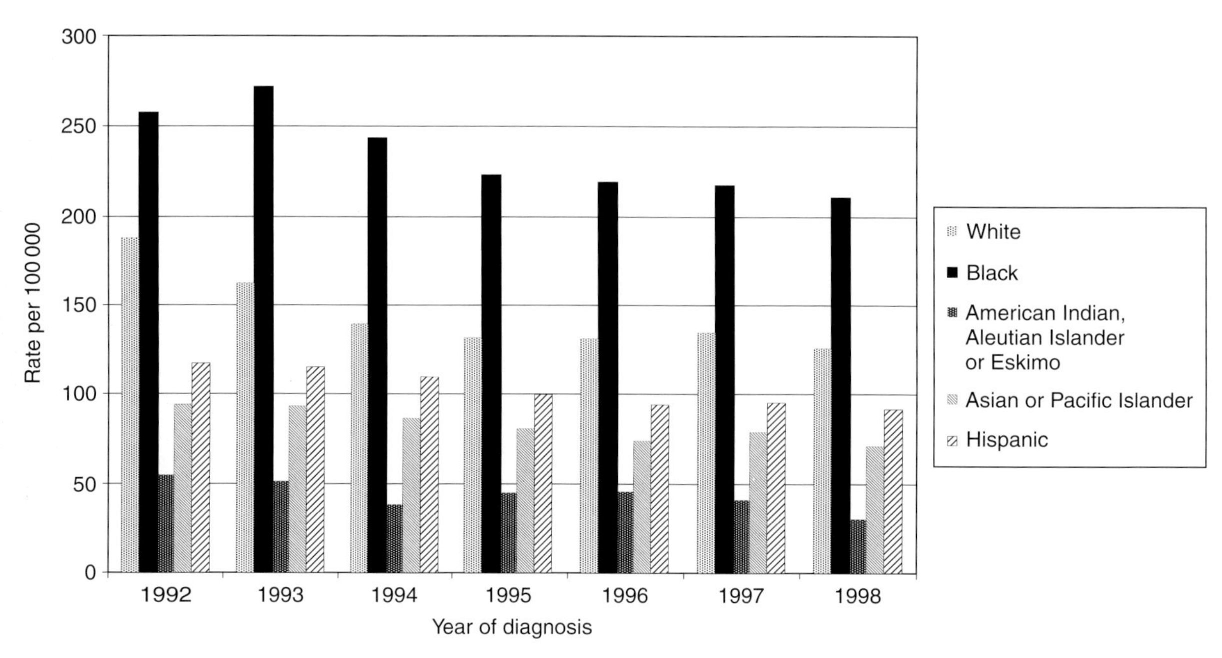

Fig. 13.1. Prostate cancer age adjusted incidence rates (SEER 1992–98).

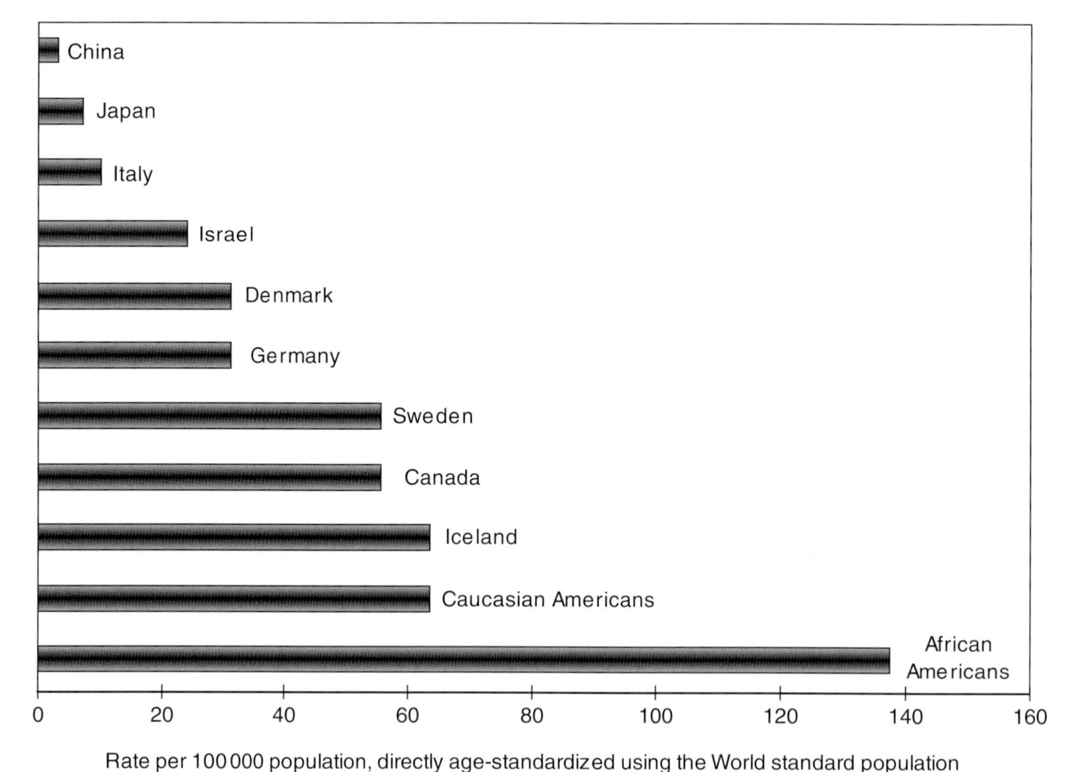

Fig. 13.2. International incidence rates of prostate cancer 1988–92.

group. African American men have the highest mortality rate for prostate cancer at 54.1 per 100 000 (almost double that of any other racial/ethnic group)[8] (**Fig. 13.2**).

The gaps between rates of African Americans and other racial and ethnic groups are wider for mortality than for incidence. These statistics are sobering and clearly illustrate the disproportionate morbidity that prostate cancer has for men of African descent. Genetic causes which have always seemed to be the most likely factor do not appear sufficient to explain these striking differences and they will be discussed later. It appears that these discrepancies must be due to genetics as well as environmental factors. The majority of prostate cancer research has been centered on defining the importance of both of these elements (**Table 13.1**).

African American males as previously stated are generally younger at their initial presentation for prostate cancer. When

Table 13.1. Prostate Cancer Statistics 2001. (Based on data from The American Cancer Society Surveillance Research.)

	New prostate cancer cases (% of all new cancer diagnoses)	Deaths from prostate cancer (% of deaths from all cancers)
US males	198 100 (31%)	31 500 (11%)
Hispanic males	6 700 (26%)	1 200 (11%)
African American males	25 300 (37%)	6 100 (19%)

Fig. 13.3. Prostate cancer US mortality rates (2000).

they are diagnosed, they generally have higher prostate specific antigen (PSA) values, owing to larger tumor volumes.[9] In addition, they are more likely to be diagnosed with advanced or metastatic disease at initial presentation.[10] Taking these observations into account, investigators have tried to modify PSA ranges to capture disease in African Americans at earlier stages. Morgan and colleagues (1996) suggested adjusting the age-specific ranges to better identify African American males with prostate cancer. They advocated further investigation for Black men in their 40s with PSA values greater than 2.0 ng/dl. This reference range improved the sensitivity of the test to greater than 95% in their cohort of men. At the same time, their cut-off for white males remained greater than 4.0 ng/dl.[11] The current American Urological Association (AUA) guidelines recommend routine screening, PSA test and Digital Rectal Examination (DRE), for men above the age of 50 years. This recommendation changes for African Americans and men with a family history of disease. The screening begins at 40 years for men in these categories.

As previously stated, initial observations suggested that the disparate impact of prostate cancer might be related to socio-economic factors that adversely affect African Americans. Reduced access to health care, poorer diet, limited insurance coverage and lower educational level were all felt to play a role in the progression and virulence of the disease. Dayal and colleagues included over 2500 African American and Caucasian men in their study, and found that socioeconomic status and the substantially higher percentage of African Americans at the lower end of this scale led to poorer survival. This was found to be a stronger prognostic factor independent of race itself.[12] Platz et al. studied approximately 46 000 men of different races but of similar socioeconomic circumstances, and revealed a persistently elevated incidence of prostate cancer among African Americans, not explained by dietary and lifestyle factors.[13] Although these social inequalities may contribute to the excess burden of prostate cancer in African Americans, they are not the most important determinant. When traditional socioeconomic, clinical and pathologic factors were controlled for by Hoffman and colleagues, they appeared to account for the increased relative risk for presenting with advanced disease in Hispanic males, but not in African Americans[14] (**Fig. 13.3** and **Table 13.2**).

Table 13.2. 5 year Survival Statistics.

	US males	African American males
Overall survival	93%	87%
Organ confined	100%	98%
Distant disease	30%	33%

Recent studies have attempted to demonstrate that the increased incidence and mortality seen in African Americans is shared by others of African descent. According to two published studies, the incidence rates of prostate cancer among Jamaicans living in Jamaica and Sub-Saharan Africans approach or exceed that of African Americans.[15] Glover et al. reported that the age-adjusted incidence rate in Kingston, Jamaica, was actually higher than that of African Americans during the same time period.[16] Osegbe reported in 1997 that the previously reported clinical prostate cancer rate in Nigeria was grossly underestimated. Although the study size was small, the data suggested that, once Nigeria began to employ screening and testing protocols similar to the USA, the incidence and mortality rates would be similar.[17] These findings would suggest that the likelihood of their being a genetic basis for the elevated rates of prostate cancer seen in those of African descent is valid irrespective of societal and environmental factors, although they may still play a role. Differences can also be seen in Caucasians from northern European countries when compared to their US counterparts. The factors that account for these differences have not been clearly identified, but environmental factors, such as diet, exposure to toxins and screening practices for the disease, have been implicated.

In discussing race and prostate cancer among Americans, it should be noted that, other than for Whites and Blacks, racial data are sparse and limited to small numbers of cases and studies. Since the introduction and widespread use of the PSA test, approved by the Food and Drug Administration in 1986, we have witnessed significant changes in the way prostate cancer is screened for.

One study by Gilliland and Key (1998) examined the impact of prostate cancer on American Indians of New Mexico. They found that carcinoma of the prostate was not only the most frequently diagnosed cancer of this group but was also the leading cause of cancer death. Their data showed that Native Americans had incidence and mortality rates that were greater than their non-Hispanic White counterparts, but were still lower than those of African Americans. There were geographic as well at tribal differences within the Native American communities, with the highest incidence rates seen in an Alaskan tribe when compared to the same tribe located in New Mexico.[18]

Another interesting finding regarding risk factors for prostate cancer is that a positive family history of disease is far more important than one's race. With a positive family history some studies have indicated one's risk for prostate cancer increases as much as twofold. A recent study by Catalona and colleagues (2002) revealed that African American men with a family history of prostate cancer are at a 75–80% increased risk for the disease. In addition, in high risk men (Black or family history) in their forties, approximately 50% had medically important tumors.[19] These data indicate that both race and history are significant prognostic factors, but when taken together, they can be extremely predictive of disease.

Understanding the implications of race and ethnicity are paramount to understanding the behavior of prostate cancer. If the factors that cause those of African descent to suffer from this disease while those of Asian lineage fare relatively quite well can be determined, this information may be useful in designing future treatment and prevention strategies.

GENETIC THEORIES

Besides having the highest rates in the USA, African American men suffer the highest incidence rate of prostate cancer in the world. It has been postulated that these findings may be due to a genetic predisposition of African Americans to the development of this disease. Thus far, studies have been unable clearly to identify a gene or genes responsible for the development of prostate cancer. It appears that, whatever genetic predisposition there exists for prostate cancer, there also exists a concomitant environmental factor responsible for the development of neoplasia.

Studies looking for possible genetic causes have demonstrated linkage to the site of the HPC-1 (human prostate cancer) gene, mapped to 1q 24–25. This locus is associated with an increased risk of developing the hereditary form of prostate cancer. This finding was first noted by Issacs and colleagues in 1996.[19] Linkage analysis revealed that 34% of 66 North American pedigrees where three or more men were affected by prostate cancer were linked to the HPC gene. These studies suggest that derangements in this gene may be associated with up to one-third of hereditary prostate cancer cases, which amounts to only about 3% of total prostate cancer cases.[20]

Subsequently, additional linkage studies have been performed.[21] Studies by Thibodeau et al.,[22] Eeles et al.[23] and McIndoe et al.[24] failed to demonstrate the same linkage to the HPC-1 gene locus in families with hereditary prostate cancer seen in Issacs' work. Further studies on linkage to HPC-1 demonstrated that positive linkage was seen in African American families with the age of onset greater than 65 years of age. It appears in these studies that African Americans disproportionately contribute to the finding of positive linkage. The linkage studies to HPC-1 thus far undertaken have shown marginal evidence for linkage to this site. None the less, the strongest evidence comes from the African American community.

Another avenue of interest in the study of prostate cancer has been the effect of circulating androgens. Testosterone and its most active metabolite, dihydrotestosterone, are the principal steroid hormones involved in the growth and development of the prostate gland. Serum levels of these hormones have been shown to influence the development of neoplasia and hypertrophy. Testosterone levels of young adult Black and White men have been previously compared, and the African American males are found to have statistically higher levels of the circulating hormone.[25] Animal and human studies have demonstrated the responsiveness of prostate cancer to androgens.[26] To evaluate the root cause of these discrepancies, the genetics involved in the regulation of these hormones has been studied.

The androgen receptor genome has recently become a focus of study. As a steroid receptor, it consists of three structural and functional domains, the DNA binding domain, the hormone binding domain and the *trans*-activation domain. Genetic studies involving this receptor have revealed that abnormalities in the length of CAG (glutamine) repeats in the 5′-transactivation domain affects the function of this receptor. Apparently, a shorter number of repeats causes the receptor to become more sensitive to circulating androgens, and has been associated with high grade and advanced stage cancers.[27] A study by Sator and colleagues has shown that African Americans have significantly shorter lengths of CAG repeats.[28] One study has shown that patients with shorter CAG repeat lengths are diagnosed with prostate cancer at an earlier age. Also, Caucasian men with prostate cancer have shorter CAG alleles than their cancer-free counterparts. Another group has shown that individuals with short CAG repeats are twofold more likely to develop prostate cancer.[29,30] Although these findings are not the sole explanation for the increased susceptibility of African American to prostate cancer, they may begin to explain this finding.

Another candidate gene that may influence the development and progression of prostate cancer is SRD5A2 gene on chromosome 2. This gene regulates expression of the Type II 5α-reductase enzyme. Type II 5α-reductase is responsible for catalyzing testosterone to its active metabolite dihydrotestosterone. Studies of this gene have identified different distributions with characteristic racial expression. Although not thoroughly characterized, the SRD5A2 gene may play a role in the racial predisposition of African Americans and their development of this disease.[31] When circulating testosterone levels were measured in African Americans versus Caucasian and Japanese Americans, no significant difference could be identified. Surprisingly, when 5α-reductase metabolites were measured in addition, African Americans and Caucasians were found to have significantly higher levels than their Japanese counterparts.[32]

Elevated serum levels of vitamin D have been long thought to have a protective effect against the development of prostate cancer. There have been studies examining the genes responsible for the vitamin D receptor. There are multiple polymorphisms of the vitamin D receptor and many studies have looked for reproducible defects that may be associated with the higher likelihood of cancer development. At this juncture, none of the studies (Taylor et al., Ingles et al., Ma et al. and others) have been able to demonstrate reproducible associations between polymorphisms of this gene and increased risk for development of disease in regard to race. More research, including studies with larger sample sizes, may produce more durable results.[33–35]

DIET

African Americans consume more fat and less fiber in their diets than their Caucasian or Asian counterparts.[36] When saturated fat was examined as an independent risk factor for disease, higher intakes were found to be a more significant risk for Asian Americans than for Blacks and Whites. This finding was seen as a partial explanation for the increased incidence of disease in second-generation Asian Americans. It has also been shown that African Americans consume less lycopene, selenium and soy products. These substances are thought to provide a protective effect against the development of prostate cancer.

A recently published study looked at obesity, measured by body mass index (BMI), and its relationship to race and prostate cancer. Of the patients studied, African Americans were found to have higher grade tumors than their Caucasian counterparts at the time of diagnosis. They were also found to have the highest average BMI. In this study, BMI was found to be an independent predictor of Gleason grade.[37]

MARITAL STATUS

Owing to the inherent difficulty in organizing a study that can evaluate the effect of marriage on prostate cancer as an independent variable, few studies have been completed. Krongard et al.[38] examined SEER data from 1973 to 1990 and observed that married patients enjoyed a significantly longer median survival when compared to non-married patients regarding their prostate cancer diagnosis. Their results showed that, independent of age, race, stage and treatment, married individuals had the longest median survival time, while separated and widowed patients had the shortest. This effect was seen even in those with advanced and distant disease. A lower risk of mortality was also found. From these data and other studies, it appears that being married, as an independent factor, results in earlier treatment and better outcomes.

A study from the *Journal of the National Cancer Institute* failed to demonstrate an association between widowhood and increased risk for development of prostate cancer.[39] Single Black men aged 55–74 had significantly increased risk when compared to their married counterparts. A study of approximately 30 000 Norwegian men noted a statistically significant lower incidence of prostate cancer in those who were never married compared to those who were married. In addition, this study, in agreement with others in the literature, demonstrated that married individuals had improved survival statistics at 5 years postdiagnosis.[40]

Theories that have been offered to explain the beneficial effects of marriage have suggested that there exists a psychological effect on the biology of prostate disease and that stress may be an independent modulator of prostate cancer growth.[41] Also, there may exist a lead-time bias (diagnosis at earlier stages) in married individuals. This is due to another variable: the social support that is generally found between married individuals.

Another aspect of marriage that has been examined as it relates to prostate cancer is sexual activity. Although the data are inconclusive, some hypothesized that the increased sexual activity that most married individuals enjoy, imparted a protective effect against the development of prostate cancer.[42] Mandel and Schuman (1987) conducted a case-control study of approximately 700 men and concluded that those with prostate cancer had higher rates of venereal disease in their partners, more sexual encounters with prostitutes and an overall lower frequency of intercourse than their controls.[43]

Apparently married individuals were more likely to be diagnosed with their disease and, subsequently, fare better than their single counterparts. This social benefit is seen in other cancers as well. Breast cancer patients with the same perceived social support have prolonged survival when compared to their unmarried counterparts.[44]

Clearly, marital status plays a role in the expected outcome of cancer patients. Anecdotally, married patients appear to enjoy better outcomes and this has been shown in the literature. What it is about marriage that imparts this advantage is not clear, and will require additional study so that future treatment strategies and recommendations can include these findings.

RELIGION

There have been no studies that have revealed religion to be an independent factor related to the incidence of prostate cancer. What we have noted in the past is that those who practice a specific religion may benefit from the dietary and social restrictions placed upon them by their religious beliefs. Another effect that has been realized is the social and spiritual support patients receive from their religious groups. This effect has been repeatedly noted but is extremely difficulty to quantify.

In examining religious groups in Los Angeles County, Roman Catholics were found have lower risk to developing cancer of the prostate than their Protestant and Adventist counterparts.[45]

Another study that addressed religion as it relates to the incidence of prostate cancer was performed in New York State. In addition to examining Roman Catholicism, this study investigated the importance of the celibate lifestyle of Catholic priests. Previous studies had suggested that sexual variables were implicated in the development of prostate cancer. By examining this group of men, they were able to assess religious as well as sexual variables. Their study revealed that priests died from prostate cancer at a statistically significant lower rate than the general population.[48] One may ask, is this due to the religion these men practiced or is it due to their celibate lifestyle? These questions are extremely difficult to answer and merit further investigation.

CONCLUSION

The epidemiology of prostate cancer has been increasingly the subject of study since the introduction of serum PSA as a screening and diagnostic tool. The widespread initiative for early diagnosis and treatment has revealed differences in the patterns that are seen with screening and treatment. The fact that diagnosis, treatment and responses to treatment differ when correlated with race, religion and marital status were originally just observations. Subsequently, studies were performed and epidemiologic data gathered to substantiate these observations. Now we must begin to explain these findings and determine the causes behind the discrepancies that are observed.

Investigations that examine aspects of the biology of prostate cancer that extend beyond stage and grade of the tumor itself are ongoing. Current research is now mainly targeted at genetic and microbiologic levels, looking for areas of increased susceptibility. Diets and cultural customs, when investigated, may provide a clue to environmental factors.

While searching for the more scientific and most direct causes for the observed trends, we have not ignored the influences that marital status and religion may have on one's well-being and attitude towards the diagnosis of prostate cancer and its implications. In the future, our therapies and investigations will be able to clearly identify the genetic risk factors for this disease and, with further study, characterize the environmental and social factors that influence the development and severity of prostate cancers. With this knowledge, we may direct novel therapies to treat and eventually cure prostate cancer.[47]

REFERENCES

1. *Cancer Facts and Figures 2001*, p. 11. Atlanta; GA: American Cancer Society, 2001.

2, 3. Stanford JL, Stephenson RA, Coyle LM et al. *Prostate Cancer Trends 1973–1995*. SEER Program. NIH Pub. No. 99-4543. Bethesda, MD: National Cancer Institute, 1999.

4. Smith JR, Carpten J, Kallioniemi O et al. Major susceptibility locus for prostate cancer on chromosome 1 revealed by a genome-wide search. *Science* 1996; 274:1371–4.

5. Shimizu H, Ross RK, Bernstein I et al. Cancers of the prostate and breast among Japanese and white immigrants in Los Angeles County. *Br. J. Cancer* 1991; 63:963–6.

6. Stanford JL, Stephenson RA, Coyle LM et al. *Prostate Cancer Trends 1973–1995*. SEER Program. NIH Pub. No. 99-4543. Bethesda, MD: National Cancer Institute, 1999.

7. Browley OW, Knopf K, Thompson I. The epidemiology of prostate cancer, Part II: The risk factors. *Semin. Urol. Oncol.* 1998; 16:193–201.

8. *Cancer Facts and Figures for African Americans 2000–2001*. American Cancer Society, 2001.

9. Moul JW et al. Prostate specific antigen values at the time of prostate cancer diagnosis in African American men. *JAMA* 1995; 274:1277–81.

10. Powell IJ. Prostate cancer and African-American men. *Oncology* 1997; 11:599.

11. Moul JW, Sesterhenn IA, Connelly RR et al. Prostate specific antigen values at the time of prostate cancer diagnosis in African–American men. *JAMA* 1995; 274:1277–81.

12. Morgan TO, Jacobsen SJ et al. Age specific reference ranges for serum prostate specific antigen in black men. *N. Engl. J. Med.* 1996; 335:304–10.
13. Dayal HH, Polisar L, Dahlberg S. Race, socioeconomic status, and other prognostic factors for survival from prostate cancer. *J. Natl Cancer Inst.* 1985; 74:1001.
14. Platz EA, Rimm EB, Willett, WC et al. Racial variation in prostate cancer incidence and in hormonal system markers among male health professionals. *J. Natl Cancer Inst.* 2000; 92:2009–17.
15. Hoffman RM et al. Racial and ethnic differences in advanced-stage prostate cancer: the prostate cancer outcomes study. *J. Natl Cancer Inst.* 2001; 93:388–95.
16. Osegbe DN. Prostate cancer in Nigerians: facts and nonfacts. *J. Urol.* 1997; 157:1340–3.
17. Glover FE et al. The epidiemiology of prostate cancer in Jamaica. *J. Urol.* 1998; 159:1984–86.
18. Osegbe DN. Prostate cancer in Nigerians: facts and nonfacts. *J. Urol.* 1997; 157:1340–3.
19. Catalona WJ, Antenor JA, Roehl KA, Moul JW. Screening for prostate cancer in high risk populations. *J. Urol.* 2002; 168:1980–3; discussion 1983–4.
20. Gilliland FD, Key CR. Prostate cancer in American Indians, New Mexico, 1969 to 1994. *J. Urol.* 1998; 893–8.
21. Isaacs SD, Kiemeny LM et al. Risk of cancer in relatives of prostate cancer probands. *J. Natl Cancer Inst.* 1995; 87:991–6.
22. Miller BA, Kolonel LN, Bernstein L et al. (eds) *Racial/Ethnic Patterns of Cancer in the United States 1988–1992*. NIH Pub. No. 96-4104. Bethesda, MD: National Cancer Institute, 1996.
23. Narod S. Genetic epidemiology of prostate cancer. *Biochim. Biophys. Acta* 1998; 1423: F1–F13.
24. Eeles RA, Durocher F, Edwards S et al. Linkage analysis of chromosome 1q markers in 136 prostate cancer families. *Am. J. Human Genet.* 1998; 62:653–8.
25. McIndoe RA, Stanford JL, Gibbs M et al. Analysis of 49 high risk families does not support a common familial prostate cancer susceptibility gene at 1q 24–25. *Am. J. Human Genet.* 1997; 61:347–53.
26. Nupponen NN, Carpten JD. Prostate cancer susceptibility genes: many studies, many results, no answers. *Cancer and Metastasis Reviews* 2001; 20:155–64.
27. Ross RK, Bernstein L, Judd H et al. Serum testosterone levels in young black and white men. *J. Natl Cancer Inst.* 1986; 76:45–8.
28. Noble RL. The development of prostatic adenocarcinoma in Nb rats following prolong sex hormone administration. *Cancer Res.* 1977; 37:1929–33.
29. Giovannuci E, Stampfer MJ, Krithivas K et al. The CAG repeat within the androgen receptor gene and its relationship to prostate cancer. *Proc. Natl Acad. Sci. USA* 1997; 94:3320–3.
30. Sator O, Zheng Q, Eastham JA. Androgen receptor CAG repeat length varies in a race-specific fashion in men without prostate cancer. *Urology* 1999; 378–80.
31. Hardy DO, Scher HI, Bogenreider T et al. Androgen receptor CAG repeat lengths in prostate cancer: correlation with age of onset. *J. Endocrinol. Metab.* 1996; 81:4400–5.
32. Ingles SA, Ross RK, Yu MC et al. Association with prostate cancer risk with genetic polymorphism in Vitamin D receptor and androgen receptor. *J. Natl Cancer Inst.* 1997; 89:166–70.
33. Reichardt JKV et al. Genetic variability of the human SRD5A2 gene: implications for prostate cancer risk. *Cancer Res.* 1995; 55:3973–5.
34. Ross RK et al. 5-alpha-Reductase activity and risk of prostate cancer among Japanese and US white and Black males. *Lancet* 1992; 339:887–9.
35. Taylor J. Association of prostate cancer with vitamin D receptor gene polymorphism, *Cancer Res.* 1996; 56:4108–10.
36. Ingles SA. Association of prostate cancer with vitamin D receptor haplotypes in African-Americans, *Cancer Res.* 1998; 58: 1620–3.
37. Ma et al. Vitamin D receptor polymorphisms, circulating vitamin D metabolites, and risk of prostate cancer in United States physicians. *Cancer Epidemiol. Biomarkers Prev.* 1998; 7:385–90.
38. Baquet CR, Horm JW, Gibbs T et al. Socioeconomic factors and cancer incidence among blacks and whites. [see comments]. *J. Natl Cancer Inst.* 1991; 83:551–7.
39. Amling CL et al. Relationship between obesity and race in predicting adverse pathologic variables in patients undergoing radical prostatectomy. *Urology* 2001; 58:723–8.
40. Krogard A, Lai H et al. Marriage and mortality in prostate cancer. *J. Urol.* 1996; 156:1696–700.
41. Newell GR, Pollack ES, Spitz MR et al. Incidence of prostate cancer and marital status. *J. Natl Cancer Inst.* 1987; 79:259–62.
42. Harvei S, Kravdal O. The importance of marital and socioeconomic status in incidence and survival of prostate cancer. *Prev. Med.* 1997; 26:623–32.
43. Miller HC. Stress prostatitis. *Urology* 1988; 32:507.
44. Brody S. Marriage and mortality in prostate cancer (letter). *J. Urol.* 1997; 158:552.
45. Mandel JS, Schuman LM. Sexual factors and prostate cancer: results from a case-control study. *J. Gerontol.* 1987; 3:259–64.
46. Litwin MS, Melmed GY, Nakazon T. Life after radical prostatectomy: a longitudinal study. *J. Urol.* 2001; 166:587–92.
47. Reynolds P, Boyd PT et al. The relationship between social ties and survival among black and white breast cancer patients. NCI Black/White Cancer Survival Study Group. *Cancer Epidemiol. Biomarkers Prev.* 1994; 3:253.
48. Mack TM. *Religion and Cancer in Los Angeles County*. NCI Monograph No. 69 235-45. Bethesda, MD: National Cancer Institute, 1985.
49. Michalek AM. Prostate cancer mortality among Catholic priests. *J. Surg. Oncol.* 1981; 17:129–33.
50. Kaplan RS. Overview of clinical problems in prostate cancer. State of the Science Lecture *Prostate Cancer: Molecular Targets*, Nov. 1999.

Prostate Cancer: Detection and Biopsy Strategies

Jonathan I. Izawa
Department of Surgery and Oncology, Division of Urology, The University of Western Ontario, London, Ontario, Canada

R. Joseph Babaian
Department of Urology, The University of Texas M. D. Anderson Cancer Center, Houston, TX, USA

DETECTION

Epidemiology

Prostate cancer is a significant health care problem and the second most common cause of cancer death among American men. It is estimated that in the USA in the year 2001 prostate cancer will be diagnosed in approximately 198 100 men and that 31 500 men will die of this disease.[1] With current screening methods, a man's lifetime risk of being diagnosed with prostate cancer is approximately 16% with approximately a 3.4% risk of dying from the disease. Therefore, many men will die with prostate cancer but not of prostate cancer. Therein lies the dilemma of determining which patients need to be screened, diagnosed and treated for this malignancy.

Digital rectal examination

An abnormal digital rectal examination (DRE) is the second most common finding that initiates further investigation for this malignancy.[2–5] Abnormal DRE findings that might indicate prostate cancer include a hard mass or nodule, induration or asymmetry. However, patients with these findings are usually asymptomatic. Although the DRE has a good positive predictive value,[6,7] few cancers detected solely on the basis of an abnormal DRE are organ confined.[8] One problem with DRE is the low reproducibility of interpreted results owing to interexaminer variability.[9] The sensitivity of DRE is low: approximately 17% of men who have a normal DRE are found to have prostate cancer following a prostatic biopsy for other urologic reasons other than an elevated prostate specific antigen (PSA) concentration. Because of its low sensitivity, low specificity and high associated understaging error rate, DRE should not be used as the sole screening method.

Transrectal ultrasound

Transrectal ultrasonography (TRUS) is another component of the diagnostic triad used in the detection of prostate cancer. The most common associated cancer finding is a hypoechoic lesion located in the peripheral zone.[10] TRUS has a higher sensitivity than does DRE; however, it is less sensitive than DRE and PSA combined as demonstrated in a study[3] in which TRUS missed 40% of prostate cancers detected by either an abnormal DRE and/or PSA.[3] In addition, the combination of DRE and PSA has a positive predictive value that is superior to that of the combination of TRUS and DRE and equivalent to that of TRUS, DRE and PSA. TRUS cannot detect lesions less than 0.5 cm in diameter.[3,4,11–13] In addition, prostate cancer can be isoechoic.[13,14] Consequently, TRUS has both a low specificity and positive predictive value.[15] Cost is also unnecessarily increased when TRUS is used as a primary detection test with DRE and PSA, since this triad has not been shown to increase the detection rate significantly compared with using DRE and PSA alone.[16] TRUS, therefore, should be used as a secondary screening test rather than an initial screening tool.

Prostate-specific antigen

The most common finding leading to further investigation for possible prostate cancer is an elevated PSA level.[2–5,17–19] PSA, a glycoprotein, is an endogenous serine protease and a member of the human kallikrein family. PSA is secreted primarily by the prostatic epithelial cells and epithelial lining of the periurethral glands. The half-life has been determined to be between 2.2 and 3.2 days. Mechanisms responsible for the detectability of PSA in the serum include its production from the prostatic epithelial cells, changes in the prostatic glandular architecture interfering with secretion into the ducts, changes in basement membrane permeability and alterations in metabolism. Specific factors that can affect PSA measurement in the serum, other than the presence of prostate cancer, include benign prostatic hyperplasia, prostatitis, urinary retention, finasteride, cystoscopy, prostatic needle biopsy, transurethral resection of the prostate, ejaculation and vigorous prostatic massage.[20] The management in these situations varies depending on the etiology, but one recommendation is to repeat the PSA determination in a few days to 6 weeks to obtain an accurate measurement of the baseline value.[20] A routine DRE can elevate the serum concentration of PSA in individual patients; however, the elevations were not considered to be clinically significant when the median values were evaluated.[21]

Therefore, there are many factors other than prostate cancer that can increase serum levels of PSA. PSA has also been shown to correlate with age as well as gland volume.[22] Formulas have been developed to estimate the serum PSA levels using age and prostate volume as variables.[22] Based upon a cohort of men without clinical evidence of prostate cancer, these formulas will allow one to estimate the serum PSA level changes per decade of age and 10 g increments in gland volume. Even in the presence of prostate cancer, non-cancerous prostatic tissue can significantly affect the serum level of PSA, which decreases the usefulness of PSA in terms of estimating tumor volume or pathologic stage of disease for individual patients, especially in patients with glands larger than 30 cm^3.[23] Furthermore, PSA values can have significant variability with serial measurements, and this can result from assay and biologic variability.[24–28] The exact metabolism of PSA is not well characterized, and the mechanisms involved with this process may explain the observed biologic variability that is not explained by the presence of prostate cancer.

PSA, however, has gained popularity and is commonly utilized in screening because it is an objective measurement, less expensive than TRUS and because it can detect more prostate cancers than DRE or TRUS.[17,18,29,30] In addition, cancers detected by PSA are more likely to be organ confined compared with those detected by DRE alone.[3,5,30–32] Currently, 70–80% of prostate cancers are organ confined versus 20–30% in the pre-PSA era.[31,32] Furthermore, it has been observed that PSA has improved the detection of organ-confined disease over DRE by 31–78% and that PSA offers approximately a 5.5-year detection lead time over DRE.[6,8,30] PSA, however should not be used alone in screening, as this has been shown to result in missing 18–28% of prostate cancer that would have been detected if PSA using the 4 ng/ml cut-off had been used in conjunction with a DRE.[30,33]

There is controversy as to what may be the optimal PSA cut-off to direct further investigations for the detection of prostate cancer. Approximately 20% of men diagnosed with prostate cancer have a serum PSA level of less than 4 ng/ml.[34] It has been shown[35] that 22% of men with normal DRE and PSA values between 2.6 and 4.0 ng/ml had cancer on biopsy, with 81% of the cancers being organ confined. Similarly, another study[36] found that 24.5% of men with a PSA value between 2.5 and 4.0 ng/ml had prostate cancer; 70% of the cancers were deemed clinically significant. Although decreasing the cut-off point may result in the detection of more prostate cancers with a higher proportion of organ-confined disease, the true biological significance of these tumors is not known. It has been suggested that the optimal screening interval is every 2 years for men with an initial PSA value of less than 2 ng/ml and a normal DRE, whereas for men with a PSA greater than 2 ng/ml and a normal DRE, annual screening seems appropriate.[37,38] This screening strategy has been supported by the findings that there is a 1% risk of detecting prostate cancer within 4 years, if the PSA is less than 2.5 ng/ml, and a 27–36% risk of the serum PSA value progressing to greater than 4 ng/ml within 4 years, if the current PSA level is between 2.1 and 4.0 ng/ml.[38] New screening strategies in Europe have been implemented whereby a biopsy is performed if the serum PSA concentration is $\geqslant$3.0 ng/ml, and no further screening tests are performed if the PSA is less than 3.0 ng/ml.[39] For this screening strategy, the rate of prostate cancer detection (4.7%) was no different from that for the previous screening regimen (4.8%) in which men underwent PSA, DRE, TRUS and subsequent biopsy, if the PSA value was greater than 4.0 ng/ml or if they had abnormal results for DRE or TRUS.[39] The 4 ng/ml threshold has been challenged in several studies in terms of its lower sensitivity in detecting clinically important prostate cancers in younger men and its reduced specificity in detecting clinically important prostate cancers in older men.[34–36,40] There have been significant improvements (25–90%) in the detection of organ-confined prostate cancer over the past 30 years.[41,42]

Enhancements

A PSA density (PSAD) value of 0.15 was found to improve the detection rate for prostate cancer when the PSA value was between 4.0 and 10 ng/ml;[43] however, this has not been a consistent finding, as others have found no advantage over the use of the PSA assay alone.[44] In retrospective studies, the use of PSAD of the transition zone has been observed to improve upon the specificity of PSAD in PSA ranges both from 2.5 to 4.0 ng/ml and from 4.0 to 10.0 ng/ml.[45–48] The volume of benign prostatic tissue as well as the volume of prostate cancer affects the serum PSA value and the zone of cancer origin affects the pathologic stage of disease. Therefore, the serum PSA alone is not useful in accurately predicting either tumor volume or pathologic stage for individual patients.[49]

Age-specific reference ranges have also been described; the suggested ranges are as follows: age 40–49 years, PSA range 0–2.5 ng/ml; age 50–59 years, PSA range 0–3.5 ng/ml; age 60–69 years, PSA range 0–4.5 ng/ml; and age 70–79 years, PSA range 0–6.5 ng/ml. Although there may be some utility applying age-referenced PSA for younger men, there is no evidence overall that age-specific reference ranges will significantly benefit patients in terms of detecting the clinically important prostate cancers, while minimizing the detection of the insignificant prostate cancers. It has been demonstrated that, using these reference ranges, one can miss more organ-confined prostate cancers in older men.[50] Although the use of PSA age reference ranges may eliminate 15% of biopsies in

65–69-year-old men, this would be associated with missing 8% of prostate cancers. This incidence of cancers missed increases to 47% for men 70 years old or older.[50] Others have shown that age-referenced PSA may save 13% of biopsies but will result in 14% of prostate cancers being missed without any improvement in the positive predictive value.[51] There are also data supporting different age-specific reference ranges for PSA in African American men.[51,52] Therefore, different reference ranges are likely to be required for different ethnic groups depending on their risk of prostate cancer.

The Baltimore Longitudinal Study for Aging found that 72% of men with prostate cancer had a PSA velocity (requires a minimum of three PSA values over a 2-year period to calculate) ≥0.75 ng/ml.[53] When studied prospectively on another dataset, PSA velocity had a sensitivity of 63% and a specificity of 62% when the PSA was >4 ng/ml. Sensitivity was better at 79% when the PSA was below 4 ng/ml.[54] However, the biologic variability in PSA may affect the utility of PSA velocity.

PSA isoforms

The free-to-total PSA ratio has been utilized in an attempt to improve biopsy selection criteria of men who are being screened for prostate cancer. The lower the free-to-total PSA ratio, the greater the likelihood of prostate cancer. Different cut-points have been studied to determine the optimal cut-point. The free-to-total PSA ratio may be most helpful for men with a total PSA value in the 4.0–10 ng/ml range. A cut-point of 25% or less free PSA (fPSA) provides a 95% sensitivity and 20% specificity.[55] This cut-point may spare 20% of men with benign causes for their PSA elevation from undergoing biopsy. In a prospective multicenter clinical trial,[56] prostate cancers associated with a % fPSA greater than 25% were infrequent and more prevalent in older men generally of low tumor grade and volume. When the percentage of fPSA was less than 25%, the risk of CaP ranged from 8% to 56%. The diagnostic value of % fPSA for men with PSA levels less than 3.0 ng/ml was also studied in a retrospective study; in all men with prostate cancer identified (9 of 209 patients), the % fPSA was less than 18%.[57] The % fPSA may be used in two ways: as a single cut-point to perform biopsies for all patients below a cut-off of 25%, or as an individual patient risk assessment on which to base biopsy decisions for individual patients.

Complexed PSA has been studied and may have utility as a marker to improve the sensitivity and specificity in the selection of men who should undergo prostatic biopsy. Over 65% of immunoreactive PSA occurs complexed to α1-antichymotrypsin (ACT), whereas 5–30% of PSA is bound to α2-macroglobulin (A2M). Only 2–5% is bound to α1-protease inhibitor. A greater proportion of PSA is bound to ACT in men with prostate cancer.[58] The specificity for complexed PSA was 26.7% compared with 21.8% for total PSA and 15.6% for % free PSA, at 95% sensitivity.[57] In a retrospective study, Brawer et al. found that one patient with prostate cancer would have been missed and 35 of 225 men would have been spared an unnecessary biopsy if complexed PSA levels were utilized.[58] A subsequent study[59] of men with total serum PSA values between 4.0 and 10.0 ng/ml found that, if a complexed PSA cut-off of 3.75 ng/ml was used to recommend a prostatic biopsy, specificity could be improved by 21%. In the same study, complexed PSA outperformed % fPSA. Advocates of complexed PSA feel its clinical utility may be greatest for patients considering re-biopsy of the prostate when their total PSA value is between 4.0 and 6.0 ng/ml. In a retrospective study, Stamey et al.[60] found no statistically significant difference between total and complexed PSA used for patients with and without prostate cancer. Therefore, further studies of complexed PSA on various datasets appear necessary.[61]

Newer isoforms of PSA, benign prostatic hyperplasia-associated PSA (BPSA) and precursor PSA (pPSA) have recently been identified, and assays to measure these forms are currently being developed. Both of these are isoforms of free PSA; BPSA appears to be associated with benign prostatic hyperplasia, and pPSA may be a more specific marker for prostate cancer.

Other markers

PSA and hK2 have 80% amino acid sequence homology. Both have their production stimulated by androgens. Poorly differentiated prostate cancer expresses higher levels of hK2 than does well-differentiated prostate cancer. Interestingly, hK2 has been observed to be elevated in prostatic intraepithelial neoplasia relative to normal prostates. Partin et al.[62] observed that a higher hK2 level associated with lower fPSA increased the likelihood for the presence of prostate cancer. They observed a 13–62% increase in prostate cancer detection in men with a % fPSA of less than 25% and an hK2/fPSA ratio of over 0.18.[62] Higher serum concentrations of hK2 have been associated with extraprostatic extension relative to organ-confined disease in radical prostatectomy specimens.[63] In a series of 68 men who underwent radical prostatectomy, preoperative serum testing indicated that the sensitivity and specificity for detecting organ-confined disease were 37% and 100%, respectively, using hK2 measurements compared with a sensitivity and specificity of 14% and 100%, respectively, for total PSA.[63] Serum hK2 compared favorably to total and fPSA to improve the preoperative risk assessment of organ-confined versus non-organ-confined prostate cancer. Its clinical utility may be greatest for men with lower fPSA levels and in differentiating organ-confined status; however, further studies are necessary before widespread use of this assay can be recommended.

Genetic markers offer promise in the early detection of prostate cancer. Several genes have been associated with hereditary prostate cancer, including HPC1, HPCX, BRCA1 and BRCA2.[64] Genetic instability or loss of heterozygosity has been identified at a number of chromosomal sites, including 5q, 6q, 7q, 10q, 13q, 16q, 17q, 18q and Xq. In addition, a number of clonal abnormalities have been identified in chromosomes 1, 4, 12, 20 and Y.[65] Sites of chromosome abnormalities associated with disease progression include 7q, 8p, 10q, 16q, X and Y, and genes associated with metastasis have been observed at 16q, as having amplification of the androgen receptor gene on X.[65–68] For men identified as having these genetic alterations before prostate cancer is detectable clinically by standard strategies, genetic analysis may prompt close surveillance and subsequent early detection. In addition, genetic markers may offer better predictive value and an improved cost–benefit ratio: routine frequent screening may be indicated only for those patients with identifiable genetic susceptibility, with less frequent screening for others.

Insulin-like growth factor (IGF) has also been shown to be detectable in prostate cancer, and in combination with PSA may be an adjunctive strategy in early detection programs.[69] However, the studies on IGF are preliminary and further investigation is necessary. Prostate-specific membrane antigen (PSMA) is a membrane protein expressed in all types of prostatic tissue.[70,71] Strategies employing PSMA in the detection of prostate cancer include radiographic examinations linking a radionuclide to a monoclonal antibody against PSMA (e.g. ProstaScint scan) and serum measurements that use reverse transcriptase-polymerase chain reaction (RT-PCR). RT-PCR is a powerful assay that allows detection of small amounts of a molecule like PSMA in the circulation. PSMA is consistently expressed in prostate cancer regardless of hormonal status and is a promising marker for diagnosis, staging, prognosis and possibly therapy. Studies have found that PSMA may have specificity problems.[72] Therefore, it has not demonstrated additional utility for clinical staging for clinically localized prostate cancer or staging for treatment failures, and it has not gained widespread use clinically in the detection of prostate cancer.[72]

Neural networks are an exciting method by which many input variables are studied and the interaction between each can be learned by using computers to detect complex non-linear relationships between the independent and dependent variables.[73] Many models are currently being studied but, to date, none has proven reliable enough to gain confidence for widespread clinical use.

PSA-based screening has allowed the detection of prostate cancer at an earlier stage.[74] The positive predictive values of various tests and combination of tests for prostate cancer detection as well as sensitivity and specificity of PSA are listed in **Table 14.1** and **Table 14.2**, respectively. Several prospective trials are ongoing to determine whether prostate cancer screening significantly affects disease specific mortality. Thus far, there are no prospective studies that have mature data to answer this question.

Table 14.1. The American Cancer Society National Prostate Cancer Detection Project reported positive predictive value of various screening tests for prostate cancer.

Test	Number with prostate cancer	Positive predictive value
DRE	9/116	7.8%
TRUS	24/399	6.0%
DRE+PSA	13/41	31.7%
DRE+TRUS	16/123	13.0%
PSA+TRUS	50/157	31.8%
DRE+TRUS+PSA	34/66	51.5%

Table 14.2. The sensitivity and specificity of PSA.

Reference	Sensitivity	Specificity
Babaian et al.[51]	80%	54%
Catalona et al.[50]	81.8%	48%
Labrie et al.*[109]	80.7%	89.6%
Mettlin et al.[110]	69.2%	89.5%

*Included transrectal ultrasound with prostate specific antigen.

Conclusions

Patients need to be informed of the risks and benefits of screening for prostate cancer prior to any formal testing. They need to be informed that, thus far, there are no definitive data that demonstrate a significant decrease in disease mortality as a consequence of the early detection and treatment of prostate cancer; however, retrospective data support a significant stage migration and an improved survival advantage.

BIOPSY STRATEGIES

Role of biopsy

Prostate biopsy, which provides pathologic confirmation of the presence or absence of prostate cancer, is the cornerstone test in the detection of this malignancy. Advances in the field of prostate biopsy in terms of technology and techniques have not only affected the early detection of prostate cancer but may also provide information regarding pathologic stage.[75–77]

Patient preparation

Informed consent is obtained from the patient before TRUS-guided biopsy. A DRE should be performed to rule out any anal and rectal pathology that may preclude insertion of the ultrasound probe. One study found that bacteremia and bacteriuria occurred in 44% and 16% of men, respectively, following TRUS-guided biopsy of the prostate. In one double-blind, randomized controlled trial,[78] prophylactic antibiotic administration significantly reduced the complications of fever and urinary tract infection compared with placebo. More than two doses of a fluoroquinolone antibiotic can also significantly decrease the infection rate,[79] which can be as low as 1.7% when antibiotics have been administered.[80] Therefore, it appears prudent to administer antibiotic prophylaxis for this procedure, with treatment begun before biopsy and continuing for 24–48 hours following biopsy. Hematuria, the most common complication, usually is minor and resolves spontaneously. To be prudent, platelet-inhibiting medications should be discontinued 7 days prior to the procedure. Although mechanical bowel preparation does not seem to be necessary to decrease significant clinical complications, we routinely administer a Fleet enema, while others[81] suggest it may not be necessary.

It has been demonstrated using a prospective randomized double-blind study that TRUS-guided prostatic nerve blockade can significantly reduce the discomfort experienced with TRUS-guided prostate biopsies.[82] The local anesthetic used was 5 ml of 1% lidocaine injected with a 7-inch 22-gauge spinal needle under TRUS guidance into the region of the prostatic vascular pedicle at the base of the prostate, just lateral to the junction between the prostate and seminal vesicle. This local anesthetic treatment may become increasingly important to patients who undergo the extended biopsy schemes. Another more recent report, however, did not confirm these findings.

Sextant and digital directed

Digitally directed biopsies have been shown to be inferior to TRUS-guided biopsies.[83] Systematic sextant biopsy of the prostate under TRUS guidance was first shown to outperform directed biopsies in 1989 and has significantly affected our ability to detect prostate cancer.[84] The sextant biopsy strategy involves obtaining biopsies in the midsagittal plane from the base, midgland and apex of both sides of the prostate from the peripheral zone. If a hypoechoic lesion is visualized, it is targeted and the biopsy needle is directed through the widest diameter of the lesion as determined on the TRUS.

Extended strategies

In 1995, it was recommended that utilizing the sextant biopsy method but directing the biopsies in a more lateral direction would enhance the cancer biopsy yield.[85,86] Subsequently, it was observed that the sextant biopsy strategy missed 15% of prostate cancers compared with a more extensive biopsy strategy involving additional biopsies of both hypoechoic lesions as well as irregular echogenic lesions with an indistinct border between the peripheral and transition zones.[87] In this same study, an additional 8–10 systematic biopsies were taken, depending on the gland size. An immediate set of repeated sextant biopsies in a single office visit can also increase the number of prostate cancers detected by 30%.[88] Other investigators have observed prostate cancer detection rates in the range of 23–26% in men undergoing repeated biopsies using the sextant strategy.[89–91] Furthermore, a five-region prostate biopsy strategy that included far lateral and mid-region biopsies also detected more prostate cancers than did the standard sextant technique.[92] This extended biopsy strategy did not result in any significant increase in morbidity.

Studies targeting the transition zone, where approximately 20% of prostate cancers originate, have been performed. The addition of transition zone biopsies to an initial biopsy strategy increased detection rates by only 1.8–4.3%.[93–95] Routine biopsies of the transition zone for prostates larger than 50 cm^3 increased detection by 13%.[96] The reported incidences of positive results for transition zone biopsy in men undergoing repeated biopsy range from 10% to 13%.[97,98]

Further data have found that, in addition to the number of biopsies, the exact location of the biopsies in the prostate affects the cancer detection rate, regardless of gland size. Chen et al.[99] observed that an 11-core multisite-directed biopsy technique performed the best when a number of biopsy schemes were compared using computer simulation.[99] This 11-core biopsy scheme included the conventional sextant biopsies, two anterior horn biopsies (one each from the left and right sides of the prostate), two transition zone biopsies just lateral and anterior to the urethra (one each from the left and right sides of the prostate) and one midline biopsy (**Fig. 14.1**). This 11-core biopsy scheme improved cancer detection rates and correlated better with the tumor volume found in the radical prostatectomy specimen. In addition, a comparative analysis of this 11-core multisite-directed biopsy strategy versus the sextant method found that prostate cancer detection was significantly enhanced and that the anterior horn area in the peripheral zone was the most frequently positive site.[100] Presti et al.[101] concluded that a six-sample systematic biopsy strategy of the peripheral zone is inadequate and that a minimum of eight biopsies (including the apex, midlobar midgland, lateral midgland and lateral base) should be routinely performed to optimize the detection rate. Improved detection has also been described using a saturation biopsy technique whereby 14–45 biopsies were performed in men with previous negative biopsy results.[102] These investigators demonstrated that the number of previous negative biopsies was not predictive of subsequent cancer detection and prostate cancer was detected in 34% (77 out of 224) of patients. Over 85% of these prostate cancers were considered clinically significant.

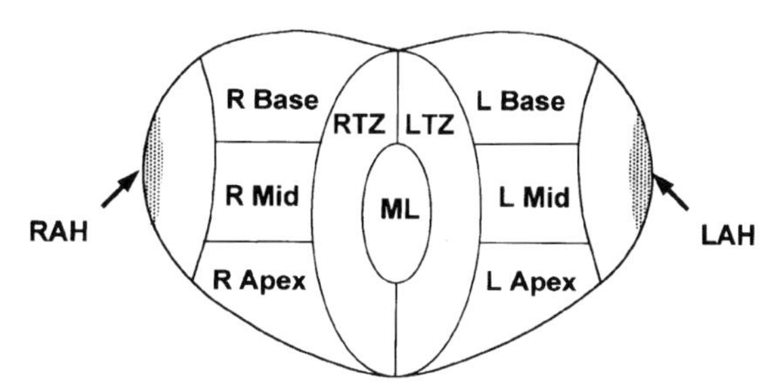

Fig. 14.1. The 11-core biopsy scheme includes the conventional sextant biopsies, two anterior horn biopsies (one each from the left and right sides of the prostate), two transition zone biopsies just lateral and anterior to the urethra (one each from the left and right sides of the prostate), and one midline biopsy. Reprinted with permission from Babaian et al. (2000).[100]

Factors that may affect the number of biopsy cores

The question of whether the number of biopsies should be varied based upon the volume of the prostate remains unanswered. Even more specifically, the question of whether to perform more biopsies of the transition zone based upon volume estimates of this region of the prostate also remains unanswered. Cancer detection rates of 38% were achieved using sextant biopsies when the prostate volume was less than 50 cm^3.[103] An inverse relationship between prostate volume and yield of sextant biopsy has also been observed by others.[104] Using Bayer's conditional probability theorem, Stricker et al.[105] determined that to achieve a detection rate sensitivity of 95%, a cancer would have to occupy 40% of the gland volume to be detected by six biopsies. This level of sensitivity could be attained for prostate cancers occupying only 15% of the prostate, if 18 biopsies were performed.[105] In a retrospective study, Chen et al.[106] observed a significant ($P = 0.03$) association between tumor volume and prostate size ($0.5\ cm^3$ or less prostate cancers were twice as frequent in prostate glands greater than $50\ cm^3$). In this study, 39% of all prostate cancers with a volume of $0.5\ cm^3$ or less occurred in prostates that were larger than $50\ cm^3$, and only 18% of all prostate cancers larger than $0.5\ cm^3$ occurred in prostates larger than $50\ cm^3$. They concluded that biopsies in men with prostates larger than $50\ cm^3$ were driven by an elevated serum PSA that was secondary to benign prostatic hyperplasia and that the lower cancer detection rate in larger volume prostates was a consequence of the higher proportion of small volume cancer in these glands.[106] This study implied that large prostates are more likely to be biopsied because of an elevated PSA concentration resulting from benign prostate tissue and not from a significant prostate cancer and that increasing the number of biopsies to compensate for incresed prostate volume runs the risk of increasing the detection of clinically insignificant tumors. Whether the improved detection rates provided by alternate biopsy strategies will result in better patient outcome awaits the results of prospective trials.

In summary, there are no definitive data that ultimately support efforts to screen for prostate cancer in order to improve survival. However, there are detection modalities that appear to optimize the detection of organ-confined prostate cancer. DRE and PSA appear to be adequate and the most cost-efficient screening tests. A number of PSA indices may improve our specificity in detecting prostate cancer, but currently only % free PSA has been accepted for widespread clinical use for this purpose. There is a potential role for complexed PSA. There are multiple biopsy strategies, and the results of comparative studies indicate that detection rates can be significantly improved utilizing directed biopsies and utilizing more than a six-sample systematic biopsy technique. The ultimate biopsy strategy has not been determined but an extended schema without transition zone biopsies is strongly recommended for an initial biopsy. Biopsy strategies will likely be further refined and individualized based upon whether a man is undergoing an initial or repeat biopsy and upon the total PSA, % of PSA isoforms, DRE findings, TRUS findings, age and ethnicity.[100,107] The reader is urged to read the screening algorithm recommended by the National Cancer Center Network.[108] This algorithm represents a consensus of the participating institutions and is updated annually.

REFERENCES

1. Greenlee RT, Hill-Harmon MB, Murray T et al. Cancer statistics, 2001. *CA Cancer J. Clin.* 2001; 51:16–36.
2. Cooner WH, Mosley BR, Rutherford CL Jr et al. Prostate cancer detection in a clinical urological practice by ultrasonography, digital rectal examination and prostate specific antigen. *J. Urol.* 1990; 143:1146–54.
3. Catalona WJ, Richie JP, Ahmann FR et al. Comparison of digital rectal examination and serum prostate specific antigen in the early diagnosis of prostate cancer: results of a multicenter clinical trial of 6630 men. *J. Urol.* 1994; 151:1283–90.
4. Ellis WJ, Chetner MP, Preston SD et al. Diagnosis of prostatic carcinoma: the yield of serum prostate specific antigen, digital rectal examination and transrectal ultrasonography. *J. Urol.* 1994; 52:1520–5.
5. Stone NN, DeAntoni EP, Crawford ED. Screening for prostate cancer by digital rectal examination and prostate-specific antigen: results of prostate cancer awareness week. *Urology* 1994; 44:18–25.
6. Jewett HJ. Significance of a palpable prostatic nodule. *JAMA* 1956; 160:838.
7. Sika JV, Lindquist HD. Relationship of needle biopsy diagnosis of prostate cancer to clinical signs of prostate cancer: an evaluation of 300 cases. *J. Urol.* 1963; 84:737.
8. Mueller EJ, Crain TW, Thompson IM et al. An evaluation of serial digital rectal examinations in screening for patients for prostate carcinoma. *J. Urol.* 1988; 140:1445–7.
9. Smith DS, Catalona WJ. Interexaminer variability of digital rectal examination in detecting prostate cancer. *Urology* 1995; 45:70.
10. Lee FR, Gray JM, McLeary RD et al. Prostatic evaluation by transrectal sonography: criteria for early detection of carcinoma. *Radiology* 1986; 158:91–5.
11. Fornage BD, Babaian RJ, Troncosa P. Diagnosis of early prostate cancer. Correlation of in vitro echography and histopathological mapping. *Ann. Radiol.* 1989; 32:415–19.
12. Carter HB, Hamper UM, Sheth S et al. Evaluation of transrectal ultrasound in the early detection of prostate cancer. *J. Urol.* 1989; 142:1008–10.
13. Shinohara K, Scardino PT, Carter SSC et al. Pathological basis of the sonographic appearance of the normal and malignant prostate. *Urol. Clin. North Am.* 1989; 16:675–91.
14. Gilliland F, Becker TM, Smith A et al. Trends in prostate cancer incidence and mortality in New Mexico are consistent with an increase in effective screening. *Cancer Epidemiol. Biol. Prev.* 1994; 3:105–11.
15. Mettlin C, Murphy GP, Babaian RJ et al. The results of a five year early prostate cancer detection intervention. *Cancer* 1996; 77:150–9.
16. Babaian RJ, Dinney CPN, Ramirez EI et al. Diagnostic testing for prostate cancer detection: less is best. *Urology* 1993; 41:421–5.
17. Catalona WJ, Smith DS, Ratliff TL et al. Measurement of prostate-specific antigen in serum as a screening test for prostate cancer. *N. Engl J. Med.* 1991; 324:1156–61.
18. Brawer MK, Chetner MP, Beatie J et al. Screening for prostatic carcinoma with prostate specific antigen. *J. Urol.* 1992; 147:841–5.
19. Gann PH, Hennekens CH, Stampfer MJ. A prospective evaluation of plasma prostate-specific antigen for detection of prostate cancer. *JAMA* 1995; 273:289–94.
20. Carter HB, Partin AW. Diagnosis and staging of prostate cancer. In: Walsh PC, Retik AB, Vaughn ED et al. (eds) *Campbell's Urology* 7th edn, pp. 2519–37. Philadelphia: W.B. Saunders Co., 1998.
21. Chybowski FM, Bergstrahl EJ, Oesterling JE. The effect of digital rectal examination on the serum prostate specific antigen concentration: results of a randomized study. *J. Urol.* 1992; 148:83–6.
22. Babaian RJ, Miyashita H, Evans RB et al. The distribution of prostate specific antigen in men without clinical or pathologic evidence of prostate cancer: relationship to gland volume and age. *J. Urol.* 1992; 147:837–40.
23. Kojima M, Troncoso P, Babaian RJ. Influence of noncancerous prostatic tissue volume on prostate-specific antigen. *Urology* 1998; 51:293–9.
24. Prestigiacomo AF, Stamey TA. Physiological variation of serum prostate specific antigen in the 4.0 to 10.0 ng./ml range in male volunteers. *J. Urol.* 1996; 155:1977–80.

25. Roehrborn CG, Pickens GJ, Carmody T III. Variability of repeated serum prostate-specific antigen (PSA) measurements within less than 90 days in a well-defined patient population. *Urology* 1996; 47:59–66.
26. Riehmann M, Rhodes PR, Cook TD et al. Analysis of variation in prostate-specific antigen values. *Urology* 1993; 42:390–7.
27. Ornstein DK, Smith DS, Rao GS et al. Biological variation of total, free and percent free serum prostate specific antigen levels in screening volunteers. *J. Urol.* 1997; 157:2179–82.
28. Nixon RG, Wener MH, Smith KM et al. Biological variation of prostate specific antigen levels in serum: an evaluation of day-to-day physiological fluctuations in a well-defined cohort of 24 patients. *J. Urol.* 1977; 157:2183–90.
29. Babaian RJ, Camps JL, Feangos DN et al. Monoclonal antibody PSA in untreated prostate cancer: relationship to clinical stage and grade. *Cancer* 1991; 67:2200–6.
30. Crawford DE, Deantoni EP, Etizoni R et al. Serum prostate-specific antigen for early detection of prostate cancer in a national community based program. *Urology* 1993; 46:863–9.
31. Mettlin C, Murphy GP, Lee F et al. Characteristics of prostate cancers detected in a multimodality cancer detection program. *Cancer* 1993; 72:1701–8.
32. Catalona WJ, Smith DS, Ratliffe TL et al. Detection of organ confined prostate cancer is increased through prostate specific antigen based screening. *Cancer* 1993; 270:948–54.
33. Babaian RJ, Camps JL. The role of prostate-specific antigen as part of the diagnostic triad and as a guide when to perform a biopsy. *Cancer* 1991; 68:2060.
34. Andriole GL, Catalona WJ. Using PSA to screen for prostate cancer: the Washington University Experience. *Urol. Clin. North Am.* 1993; 20:647–51.
35. Catalona WJ, Smith DS, Ornstein DK. Prostate cancer detection in men with serum PSA concentrations of 2.6 to 4.0 ng/mL and benign prostate examination: enhancement of specificity with free PSA measurements. *JAMA* 1997; 277:1452–5.
36. Babaian RJ, Johnston DA, Naccarato W et al. The incidence of prostate cancer in a screening population with a serum prostate specific antigen between 2.5 and 4.0 ng/ml: relation to biopsy strategy. *J. Urol.* 2001; 165:757–60.
37. Smith DS, Catalona WJ, Herschman JD et al. Longitudinal screening for prostate cancer with prostate-specific antigen. *JAMA* 1996; 276:1309.
38. Carter HB, Epstein JI, Chan DW et al. Recommended prostate-specific antigen intervals for the detection of curable prostate cancer. *JAMA* 1997; 277:1456–60.
39. Schroder FH, Kranse R, Reitbergen J et al. The European Randomized Study of Screening for Prostate Cancer (ERSPC): an update. *Eur. Urol.* 1999; 35:539–43.
40. Oesterling JE, Jacobsen SJ, Chute CG. Serum prostate-specific antigen in a community-based population of healthy men: establishment of age-specific reference ranges. *JAMA* 1993; 270:860–4.
41. Murphy GP, Natarajan N, Pontes JE. The National Survey of prostate cancer in the United States by the American College of Surgeons. *J. Urol.* 1982; 127:928–34.
42. Mettlin C, Murphy GP, Lee F et al. Characteristics of prostate cancer detected in the American Cancer Society-National Prostate Cancer Detection Project. *J. Urol.* 1994; 152:1737–40.
43. Benson MC, McMahon DJ, Cooner WH et al. An algorithm for prostate cancer detection in a patient using prostate-specific antigen and prostate-specific antigen density. *World J. Urol.* 1993; 11:206.
44. Brawer MK, Arambur EA, Chen GL et al. The inability of prostate specific antigen index to enhance the predictive value of prostate specific antigen in the diagnosis of prostatic carcinoma. *J. Urol.* 1993; 150:369.
45. Taneja SS, Tran K, Lepor H. Volume-specific cutoffs are necessary for reproducible application of prostate-specific antigen density of the transition zone in prostate cancer detection. *Urology* 2001; 58:222–6.
46. Kikuchi E, Nakashima J, Ishibashi E et al. Prostate specific antigen adjusted for transition zone volume: the most powerful method for detecting prostate cancer. *Cancer* 2000; 89:842–9.
47. Djavan B, Zlotta A, Kratzik C et al. PSA, PSA density, PSA density of transition zone, free/total PSA ratio, and PSA velocity for early detection of prostate cancer in men with serum PSA 2.5 to 4.0 ng/ml. *Urology* 1999; 54:517–522.
48. Djavan B, Zlotta AR, Byttebier G et al. Prostate specific antigen density of the transition zone for early detection of prostate cancer. *J. Urol.* 1998; 160:411–18.
49. Kojima M, Troncoso P, Babaian RJ. Influence of noncancerous prostatic tissue volume on prostate-specific antigen. *Urology* 1998; 51:293–9.
50. Catalona WJ, Hudson MA, Scardino PT et al. Selection of optimal prostate specific antigen cutoffs for early detection of prostate cancer: receiver operating characteristic curves. *J. Urol.* 1994; 152:2037–42.
51. Babaian RJ, Kojima M, Ramirez EI. Comparative analysis of prostate specific antigen and its indexes in the detection of prostate cancer. *J. Urol.* 1996; 156:432–7.
52. Morgan TO, Jacobean SJ, McCarthy WF et al. Age-specific reference ranges for serum prostate-specific antigen in black men. *N. Engl. J. Med.* 1996; 335:304–10.
53. Carter HB, Pearson ID, Metter EJ et al. Longitudinal evaluation of prostate specific antigen levels in men with and without prostate cancer. *JAMA* 1992; 267:2215–20.
54. Smith DS, Catalona WJ. Rate of change of serum prostate specific antigen levels as a method of prostate cancer detection. *J. Urol.* 1994; 152:1163.
55. Catalona WJ, Smith DS, Wolfert RL et al. Evaluation of percentage of free serum PSA to improve specificity of prostate cancer screening. *JAMA* 1995; 274:1214–20.
56. Catalona WJ, Partin AW, Slawin KM et al. Use of the percentage of free prostate-specific antigen to enhance differentiation of prostate cancer from benign prostatic disease. A prospective multicenter clinical trial. *JAMA* 1998; 279:1542–7.
57. Tornblom M, Norming U, Adolfsson J et al. Diagnostic value of percent free prostate-specific antigen: Retrospective analysis of a population-based screening study with emphasis on men with PSA levels less than 3.0 ng/mL. *Urology* 1999; 53:945–50.
58. Brawer MK, Meyer GE, Letran JL et al. Measurement of complexed PSA improves specificity for early detection of prostate cancer. *Urology* 1998; 52:372–8.
59. Brawer M, Cheli CD, Neaman IE et al. Complexed prostate specific antigen provides significant enhancement of specificity compared with total prostate specific antigen for detecting prostate cancer. *J. Urol.* 2000; 163:1476–80.
60. Stamey TA, Yemoto CE, McNeal JE et al. Examination of the three molecular forms of serum prostate specific antigen for distinguishing negative from positive biopsy: relationship to transition zone volume. *J. Urol.* 2000; 163:119–26.
61. Okihara K, Fritsche HA, Ayala, A et al. Can complexed prostate specific antigen and prostate volume enhance prostate cancer detection in men with total prostate specific antigen between 2.5 and 4.0 ng/mL. *J. Urol.* 2001; 165:1930–6.
62. Partin AW, Catalona WJ, Finlay JA et al. Use of human glandular kallikrein 2 for the detection of prostate cancer: preliminary analysis. *Urology* 1999; 54:839–45.
63. Haese A, Becker C, Noldus J et al. Human glandular kallikrein 2: a potent serum marker for predicting the organ confined versus the non-organ confined growth of prostate cancer. *J. Urol.* 2000; 163:1491–7.
64. Bratt O. Hereditary prostate cancer. *Br. J. Int.* 2000; 85:588–98.
65. Brothman AR, Maxwell TM, Cui J et al. Chromosomal clues to the development of prostate tumors. *Prostate* 1999; 38:303–12.
66. Qian J, Jenkins RB, Bostwick DG et al. Genetic and chromosomal alterations in prostatic intraepithelial neoplasia and carcinoma detected by fluorescence in situ hybridization. *Eur. Urol.* 1999; 35:479–83.

67. Verma RS, Manikal M, Conte RA et al. Chromosomal basis of adenocarcinoma of the prostate. *Cancer Invest.* 1999; 17:441–7.
68. Yin Z, Spitz MR, Babaian RJ et al. Limiting the location of a putative human prostate cancer tumor suppressor gene at chromosome 13q14.3. *Oncogene* 1999; 18:7576–83.
69. Djavan B, Bursa B, Seitz C et al. Insulin-like growth factor 1 (IGF-1), IGF-1 density, and IGF-1/PSA ratio for prostate cancer detection. *Urology* 1999; 54:603–6.
70. Chang SS, Gaudin PB, Reuter VE et al. Prostate-specific membrane antigen: present and future applications. *Urology* 2000; 55:622–9.
71. Seiden MV, Kantoff PW, Krithivas K et al. Detection of circulating tumor cells in men with localized prostate cancer. *J. Clin. Oncol.* 1994; 12:2634–9.
72. Millon R, Jacqmin D, Muller D et al. Detection of prostate-specific antigen or prostate-specific membrane antigen-positive circulating cells in prostate cancer patients: clinical implications. *Eur. Urol.* 1999; 36:278–85.
73. Wei JT, Zang Z, Barnhill SD et al. Understanding artificial neural networks and exploring their applications for the practicing urologist. *Urology* 1998; 52:161–72.
74. Smith DS, Catalona WJ. The nature of prostate cancer detection through prostate specific antigen based screening. *J. Urol.* 1994; 152:1732–6.
75. Noguchi M, Stamey TA, McNeal JE et al. Relationship between systematic biopsies and histopathological features of 222 radical prostatectomy specimens: lack of prediction of tumor significance for men with non-palpable disease. *J. Urol.* 2001; 166:104–9.
76. Sebo TJ, Bock BJ, Cheville JC et al. The percent of cores positive for cancer in prostate needle biopsy specimens is strongly predictive of tumor stage and volume at radical prostatectomy. *J. Urol.* 2000; 163:174–8.
77. Partin AW, Kattan MW, Subong EN et al. Combination of prostate-specific antigen, clinical stage, and Gleason score to predict pathological stage of localized prostate cancer. A multi-institutional update. *JAMA* 1997; 277:1445–51.
78. Crawford ED, Haynes AL Jr, Story MW et al. Prevention of urinary tract infection and sepsis following transrectal prostatic biopsy. *J. Urol.* 1982; 127:449–51.
79. Aus G, Hermansson GG, Hugosson J et al. Transrectal ultrasound examination of the prostate: complications and acceptance by patients. *Br. J. Urol.* 1993; 71:457–9.
80. Rodriguez LV, Terris MK. Risks and complications of transrectal ultrasound guided prostate needle biopsy: a prospective study and review of the literature. *J. Urol.* 1998; 160:2115–20.
81. Carey JM, Korman HJ. Transrectal ultrasound guided biopsy of the prostate. Do enemas decrease clinically significant complications? *J. Urol.* 2001; 166:82–5.
82. Nash PA, Bruce JE, Indudhara R et al. Transrectal ultrasound guided prostatic nerve blockade eases systematic needle biopsy of the prostate. *J. Urol.* 1996; 155:607–9.
83. Clements R, Griffiths GJ, Peeling WB et al. How accurate is the index finger? A comparison of digital and ultrasound examination of the prostatic nodule. *Clin. Radiol.* 1988; 39:87–9.
84. Hodge KK, McNeal JE, Terris MK et al. Random systematic versus directed ultrasound guided transrectal core biopsies of the prostate. *J. Urol.* 1989; 142:71–5.
85. Stamey TA: Making the most out of six sytematic sextent biopsies. *Urology* 1995; 45:2–12.
86. Terris MK, Wallen EM, Stamey TA. Comparison of mid lobe versus lateral systematic sextant biopsies in the detection of prostate cancer. *Urol. Int.* 1997; 59:239–42.
87. Norberg M, Egevad L, Holmberg L et al. The sextant protocol for ultrasound-guided core biopsies of the prostate underestimates the presence of cancer. *Urology* 1997; 50:562–6.
88. Levine MA, Ittman M, Melamed J et al. Two consecutive sets of transrectal ultrasound guided sextant biopsies of the prostate for the detection of prostate cancer. *J. Urol.* 1998; 159:471–6.
89. Roehrborn CG, Pickens GS, Sanders JS. Diagnostic yield of repeated transrectal ultrasound-guided biopsies stratified by specific histopathologic diagnoses and prostate specific antigen levels. *Urology* 1996; 47:347–52.
90. Ukimura O, Durrani O, Babaian RJ. Role of PSA and its indices in determining the need for repeat prostate biopsies. *Urology* 1997; 50:66–72.
91. Keetch DW, Catalona WJ, Smith DS. Serial prostatic biopsies in men with persistently elevated serum prostate specific antigen values. *J. Urol.* 1994; 151:1571–4.
92. Eskew LA, Bare RL, McCullough DL. Systematic 5 region prostate biopsy is superior to sextant method for diagnosing carcinoma of the prostate. *J. Urol.* 1997; 157:199–203.
93. Fleshner NE, Fair WR. Indications for transition zone biopsy in the detection of prostatic carcinoma. *J. Urol.* 1997; 157:566–8.
94. Karakiewicz PI, Bazinet M, Aprikian AG et al. Value of systematic transition zone biopsies in the early detection of prostate cancer. *J. Urol.* 1996; 155:605–6.
95. Terris MK, Pham TQ, Issa MM et al. Routine transition zone and seminal vesicle biopsies in all patients undergoing transrectal ultrasound guided prostate biopsies are not indicated. *J. Urol.* 1997; 157:204–6.
96. Chang JJ, Shinohara K, Hovey RM et al. Prospective evaluation of systemic sextant transition zone biopsies in large prostates for cancer detection. *Urology* 1998; 52:89–93.
97. Lui PD, Terris MK, McNeal JE et al. Indications for ultrasound-guided transition zone biopsies in the detection of prostate cancer. *J. Urol.* 1996; 153:1000–3.
98. Keetch DW, Catalona WJ. Prostatic transition zone biopsies in men with previous negative biopsies and persistently elevated serum prostate specific antigen values. *J. Urol.* 1995; 154:1795–7.
99. Chen ME, Troncoso P, Tang K et al. Comparison of prostate biopsy schemes by computer simulation. *Urology* 1999; 53:951–60.
100. Babaian RJ, Toi A, Kamoi K et al. A comparative analysis of sextant and an extended 11-core multisite directed biopsy strategy. *J. Urol.* 2000; 163:152–7.
101. Presti JC, Chang JJ, Bhargava V et al. The optimal systematic prostate biopsy sheme should include 8 rather than 6 biopsies: results of a prospective clinical trial. *J. Urol.* 2000; 163:163–7.
102. Stewart CS, Leibovich BC, Weaver AL et al. Prostate cancer diagnosis using a saturation needle biopsy technique after previous negative sextant biopsies. *J. Urol.* 2001; 166:86–92.
103. Uzzo RG, Wei JT, Waldbaum RS et al. The influence of prostate size on cancer detection. *Urology* 1995; 46:831–6.
104. Karakiewicz PI, Basinet M, Aprikian AG et al. Outcome of sextant biopsy according to gland volume. *Urology* 1997; 49:55–9.
105. Stricker HJ, Ruddock LJ, Wan J et al. Detection of non-palpable prostate cancer. A mathematical and laboratory model. *Br. J. Urol.* 1992; 71:43–51.
106. Chen ME, Troncosa P, Johnston D et al. Prostate cancer detection: relationship to prostate size. *Urology* 1999; 53:764–8.
107. Petteway CA, Troncoso P, Steelhammer L et al. Prostate specific antigen and tumor volume in Africian American and Caucasian men: a comparative study based on radical prostatectomy specimens. *J. Urol.* 1996; 155:443A.
108. Potter S, Partin AW. NCCN 1999 Guidelines for Early Detection of Prostate Cancer. *Oncology* 1999; 10:99–115.
109. Labrie F, Dupont A, Suburu R et al. Serum prostate specific antigen as pre-screening test for prostate cancer. *J. Urol.* 1992; 147:846–52.
110. Mettlin C, Murphy GP, Babaian RJ et al. Observations of the early detection of prostate cancer from the American Cancer Society National Prostate Cancer Detection Project. *Cancer* 1997; 80:1814–17.

Prostate Cancer Treatment Outcomes between African Americans and Caucasians

Isaac J. Powell
Department of Urology, Wayne State University, Detroit, MI, USA

INTRODUCTION

It is estimated that approximately 198 000 men will have been diagnosed with prostate cancer in 2001. It is estimated that approximately 31 700 men will have died from prostate cancer in 2001. Incidence and mortality differ between races and the highest incidence of prostate cancer occurs among African Americans. Prostate cancer incidence and mortality is more prevalent among Caucasians in comparison to Hispanics, American Indians, Asians and Pacific Islanders.[1] The clinical incidence of prostate cancer reported by the American Cancer Society in 1997 was 66% greater among African American men than Caucasian men. Recently reported prostate cancer age-adjusted mortality rates between 1992 and 1997 were approximately two times worse for African American men than for Caucasians.[2] The age-specific mortality rate for prostate cancer is reported to be three times greater among African American compared to Caucasian men between the ages of 40 and 60 years. This disparity decreases with advancing age.[3] The focus of this chapter will compare African Americans and Caucasians and treatment outcomes.

The overall 5-year relative survival rate for prostate cancer is worse among African Americans compared to Caucasians for local disease, regional and distant disease based on the Surveillance, Epidemiology and End Results Program (SEER) data. The 5-year relative survival rate by race and stage at diagnosis from 1989 to 1996 is 94% for Caucasians and 87% for African Americans for all stages.[3] A recently reported study has examined prostate cancer survival by age in an equal access health system. This study was retrospective and observed overall survival; they demonstrated a worse survival among African Americans compared to Caucasians men under age 65 but a trend toward a better survival above 65 years.[4] At present, the cause for this phenomenon is purely speculative; however, it does suggest a possible age base difference in the natural history of prostate cancer among African American men compared to Caucasian men. This difference in survival in addition to the disproportionate mortality among young African American men suggest that there may be a more aggressive biology of prostate cancer than among young Caucasian men.

The disease-specific survival of prostate cancer is impacted on by several factors, in addition to treatment. Some of those factors include health-seeking behaviors, dietary factors and social economic factors, as well as the biology of the disease. Polymorphic associated genes that include androgen receptor CAG repeat polymorphism and single nucleotide polymorphic (snp) genes that impact on 5α-reductase activity may be responsible for aggressive prostate cancer among African American men versus Caucasian men.[5,6] It has been reported that men who consume a high-fat diet have a worse outcome from prostate cancer. African American men have been found to have a higher content of fat in their diet.[7,8] It has also been reported that African Americans are diagnosed with more advanced prostate cancer than Caucasians. Poor health-seeking behavior may be partially responsible for the late stage in diagnosis. It is known that African Americans are less likely to be screened for prostate cancer than Caucasians.[9–13] Recent data have demonstrated no association between socioeconomic factors and prostate cancer outcome among African Americans but a strong association among Whites. Whites of lower socioeconomic status demonstrated a worse outcome than those of higher socioeconomic status.[14] All of the above factors may have a confounding impact on treatment outcomes and these factors should be kept in mind when reading this chapter.

SURGICAL TREATMENT OUTCOMES

Several reports have examined outcomes after radical prostatectomy among African American and Caucasian men. There has been variation in the reports of findings, but close examination of the studies that include large sample sizes of African Americans and Caucasians reveals that the findings are similar. The objective of one study was to determine the outcome between African American and Caucasian men with prostate cancer stage for stage for a large cohort of men. They examined 848 consecutive patients who underwent radical prostatectomy between 1991 and 1995. The mean follow-up was 34 months (range 1.5–75 months). They included men with a Gleason score of (4 + 3) with those men with a Gleason score of 8–10 for racial/ethnic comparison. The results demonstrate that African American men and White men diagnosed with organ-confined prostate cancer had similar prostate specific antigen (PSA) levels, Gleason grade and biochemical recurrence. However, African American men diagnosed with pathologically non-organ-confined disease demonstrated higher PSA levels and a higher incidence of recurrence than did White men with non-organ confined disease[15] (**Fig. 15.1**). There was a trend towards African American men having a greater proportion of high-grade lesions than White men when prostate cancer was not organ confined.

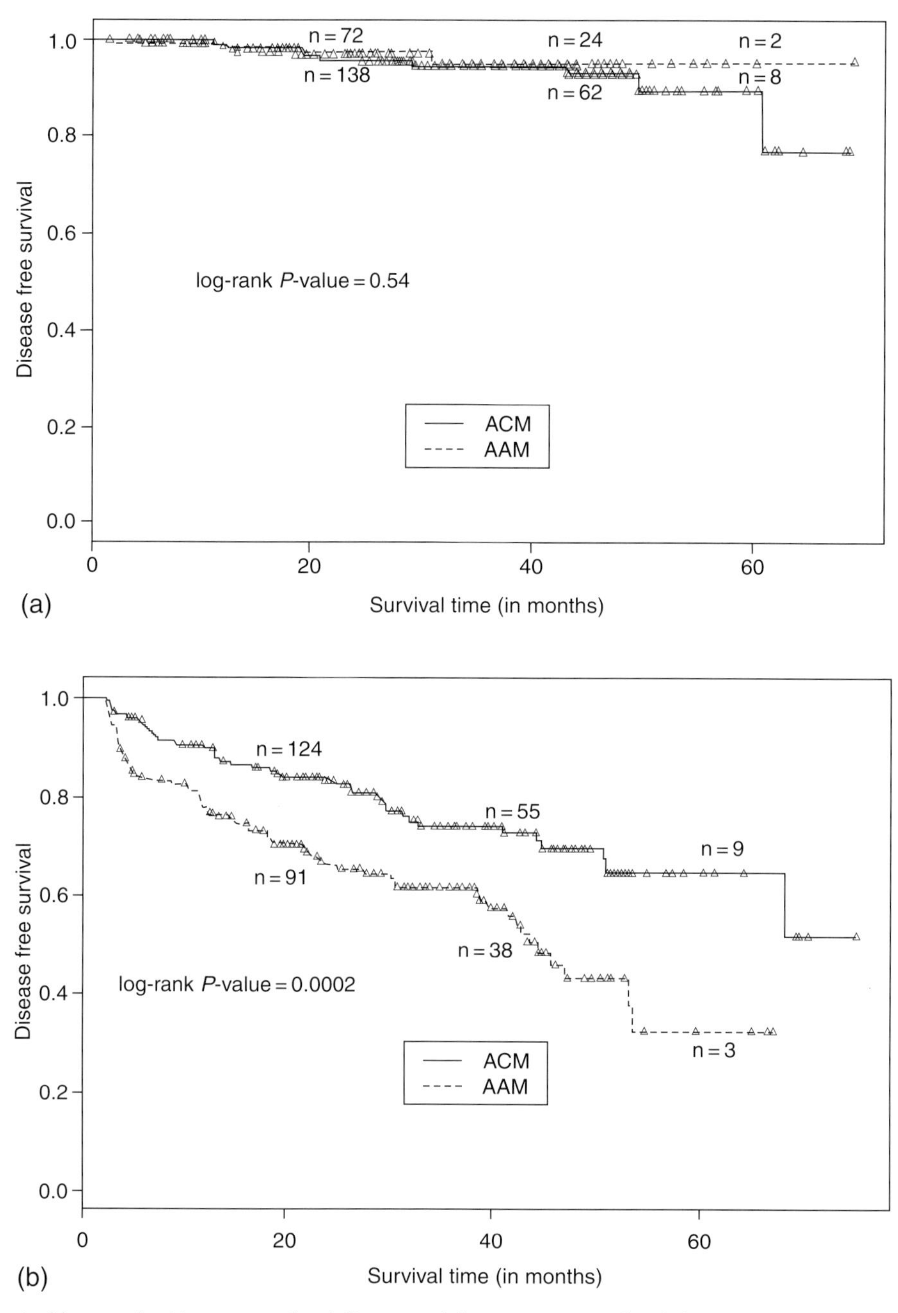

Fig. 15.1. Disease-free survival by race for (a) organ-confined disease and (b) non-organ-confined disease. Reprinted from *Urology*, 55, Powell et al., Prostate cancer biochemical recurrence stage for stage is more frequent among African American than White men with locally advanced but organ-confined disease, pp. 246–51. Copyright 2000, with permission from Elsevier Science.

Excluded from this study were men who had received preoperative and immediate postoperative hormone therapy or who underwent salvage prostatectomy, and men with lymph node metastasis. The clinical T stage of all men who underwent radical prostatectomy was from T1A to T2C. The distribution of men in the entire cohort diagnosed with organ-confined disease was 100 out of 263 (38%) African American men and 178 out of 348 (51%) White men ($P=0.001$). The results from the multivariable analysis among men with organ-confined prostate cancer demonstrate that race was not significant in predicting disease-free survival after adjusting for Gleason grade and PSA. However, among men diagnosed with pathologically non-organ-confined prostate cancer, race was a predictor of outcome even after adjusting for Gleason grade and PSA.[15] Multivariable analysis of a expanded cohort at this institution, where the majority of African Americans were diagnosed with non-organ-confined prostate cancer, demonstrate that race was a significant predictor of disease-free survival.[16]

At another institution, biostatistical models to predict the risk of recurrence after radical prostatectomy of clinically localized prostate cancer were developed to identify high-risk patients shortly after surgery. Their models suggest that race, preoperative prostate specific antigen, postoperative Gleason score and pathological stage were important independent prognosticators of recurrence after radical prostatectomy. They suggest race should be considered in future models that attempt to predict the likelihood of recurrence after surgery. Also, in this study, the cohort was diagnosed with clinically localized disease before radical prostatectomy. However, a larger percentage (57.1%) of these radical prostatectomy specimens revealed extracapsular disease. The majority of patients (76.9%) have Gleason sums of between 5 and 7. The above study was done in an equal access health care setting.[17,18]

However, another study performed in an equal access medical center demonstrated that race is not an independent predictor of biochemical recurrence after radical prostatectomy. They reported on 273 patients (125 Black, 148 White) who underwent radical prostatectomy in a Veteran Affairs Medical Center between 1991 and 1999. A multivariable analysis was used to determine the clinical and pathological variables that were significant in predicting biochemical recurrence after radical prostatectomy and to determine whether race was an independent predictor of biochemical failure. No significant difference was found among preoperative factors or pathological features of radical prostatectomy specimens between Black and White men. In addition, no difference was found between Black and White men, in the PSA recurrence rates after radical prostatectomy using Kaplan–Meier survival curves ($P=0.651$). Multivariable analysis revealed that serum PSA, biopsy Gleason score, younger age, surgical Gleason score, lymph node involvement were all independent predictors of biochemical recurrence. Race was not a significant predictor of biochemical failure in this multivariables analysis ($P=0.199$). The mean follow-up was 29 months for both Black and White patients. The overall biochemical recurrence rates were similar between Black and White men (22% and 24%; $P=0.657$). It is important to note that in this study 80% of Black and 81% of White men had pathologically organ-confined prostate cancer. Also more than 50% of Black and White men were found to have pathological Gleason score of less than 7. The characteristics of this population selection of men illustrated an earlier diagnosis of prostate cancer than the two previous studies.[19]

In the first study presented in this chapter comparing surgical outcome, the patients were stratified between pathologically organ-confined verses pathologically non-organ-confined prostate cancer. Among those patients diagnosed with pathologically organ-confined cancer African Americans versus Caucasians, there was no difference in disease-free survival; when these data were examined in multi variable analysis, race was not a predictor of biochemical recurrence. The second study also presented data showing there was no difference in disease-free survival among African Americans compared to Caucasians when the cancer was pathologically confined to the prostate gland. Therefore, even though the conclusions reached in the third study versus the first two studies would appear to be different, when examined closely, the outcomes are similar.

These studies suggest that, for all men, the more advanced the stage of prostate cancer when diagnosed, the more likely African Americans will have a worse outcome. Conversely, the earlier prostate cancer is diagnosed, the more likely the disparity in outcome will be eliminated. This is a strong argument for early detection, especially for African American men.

A more recent study examined stage shift in prostate cancer to see if it would correlate with improved disease-free survival among African American compared to Caucasian men, who have undergone radical prostatectomy. In this study, 1042 consecutive patients underwent radical prostatecomy and were divided by year of surgery. Group A underwent surgery from 1990 through 1995 and Group B 1996 thorough 1999. There were 585 men in Group A and 457 in Group B. Disease-free survival was stratified among the two races/ethnicities.

Improvements in clinical stage, preoperative PSA and biopsy Gleason score were observed for Group B ($P=0.0001$). Pathological organ-confined disease increased for Group B compared to Group A in both races, 58% (89 out of 153) from 37% (66 out of 178) in African American men, and 62% (189 out of 304) from 48% (194 out of 407) in Caucasian men ($P=0.003$ and 0.001, respectively). The calculated cancer recurrence-free median probability for Group A at 42 months was 81% and 68% for Caucasians and African American men (log rank$=$0.001). These differences have become insignificant for Group B patients with median probability at 42 months, of 90% and 88% from Caucasian men and African American men (log rank$=$0.39), a net increase of disease-free survival of 20% in African American men. The specimen Gleason score, PSA and pathological stage were independent predictors of survival for both groups. In contrast, race was an independent predictor only in Group A.[20] These early data suggest that the survival gap between African American men and Caucasian men is narrowing and becoming statistically insignificant.

THE IMPACT OF AGE ON TREATMENT OUTCOME

In a study conducted in an equal access veteran affairs system, we discovered the phenomenon of survival cross-over based on age among African Americans compared to Caucasians. Cases for this study included only those histologically confirmed, newly diagnosed prostate cancers treated at this Veterans Administration Hospital between 1973 and 1992. Trained abstractors from the SEER registry determined the stage at diagnoses, according to SEER criteria. The distribution of race and annual income of all male patients seen at this institution was similar. Over the entire 20-year period (1972–1992), there were a total of 358 prostate cancers in White patients and 383 in Black patients. The ages of Black and White patients were comparable. The proportion of White and Black men presenting with localized disease is similar (57% and 54%, respectively). The age distributions for Black and White prostate cancer patients were similar within each stage and across the stages. The average age of Black men with localized disease was 67 years (range 47–89 years) and 69 years for those with distant disease (range 49–96 years). White men with localized disease were an average of 67 years old (range 43–90 years) and 68 years old for remote disease (range 52–95 years).

Survival analysis was performed using the 340 patients (160 White, 180 Black) for which at least 5 years of follow-up was available. Other criteria for inclusion in survival analysis included known stage, known age and survival greater than 1 month. When all patients were compared, survival plots of Black and White men were similar. Stratifying for stage, the plots for Black and White men were also similar for each stage. However, stratifying for age (<65 years and >65 years), we found that, in patients less than 65 years, White men experience significantly better survival (log rank, $P=0.001$) whereas, in men 65 years and older, Blacks tended to experience better survival ($P=0.164$). When the two age groups were further stratified by stage (local, regional and distant), there was no significant difference found between Black and White men. However, in patients younger than 65 years, Whites tended to have better survival than Blacks for each stage, whereas in patients 65 years and older, there was a trend for Black men to have a better survival for each stage.[4]

In another study that examined this issue outside the VA Hospital, they reported poor survival in Black men under the age of 70 years also, but only those with local and regional disease. Among the 70 year and older group, Black men generally had better survival within the regional disease category. Black and White men of any age with distant prostate cancer had similar survival rates. Racial differences in survival among younger men might be explained in theory by early malignant transformation to clinically significant prostate cancer among Black men.[21]

One autopsy study revealed a higher volume of prostate cancer in Blacks compared with Whites among men 40–50 years who have died from other causes.[22] Another autopsy study reveals a statistically significant greater percentage of high-grade prostate intraepithelial neoplasias among Blacks compared with Whites between 40 and 50 years of age.[23] These two autopsy studies may indicate that the disease process of prostate cancer is more aggressive, with greater volumes in young Black men compared to White men of a similar age. Aggressiveness of the disease early in the natural history may then account for the trend of poor survival in younger Black men, as well as the advanced stage of presentation.

How does one explain the finding that Black men in the oldest age group have a better survival rate than White men? Black men diagnosed with prostate cancer at 70 years and older may represent a population with slowly growing prostate cancer who remain alive compared with those of rapidly growing prostate cancer who were selected out by death at an earlier age. White men may experience malignant transformation to clinically significant prostate cancer at a later age, resulting in worse survival at a later age than Black men.

Examination of consecutive radical prostatectomy specimens from African Americans and Caucasians was performed from January 1991 to June 1996 among African Americans and Caucasians at Wayne State University, Detroit, Michigan. The authors examined biochemical recurrence of prostate cancer in this cohort of men. In a multivariable analysis, a contingency table was performed, using a χ^2-test to assess the correlation between stage and race after stratification of patients by age group. Biochemical recurrence was analyzed using the Kaplan–Meier method and the log rank test. A total of 759 patients (303 African Americans and 426 Caucasian) were examined for PSA, Gleason grade and stage. Biochemical recurrence outcome was analyzed based on follow-up data from 655 patients who underwent radical prostatectomy.

Ethnic/racial groups were examined by pathological stage (organ-confined vs. non-organ-confined) and age by decade. African American men aged 50–59 years had a 56% chance of extraprostatic disease compared with 41% for Caucasian men ($P=0.03$). Between the ages of 60 and 69, African American men continued to demonstrate a worse pathological stage than Caucasian men (63% vs. 51%; $P=0.013$). A difference was not found among African American men versus Caucasian men aged 70–79 years. A recurrence after radical prostatectomy among men treated was examined. Among men aged 50–59 years, progression was significantly higher among African Americans compared with Caucasian men: 17 of 63 patients (27%) versus 15 of 102 patients (15%), respectively ($P=0.03$). At age 60–69 years, progression was significantly higher among African American men (31%) than among Caucasian men (16%), $P=0.0001$. Differences in disease-free survival were not detected among African American men compared with Caucasian men between 70 and 79 years[24] (**Fig. 15.2**).

TREATMENT OUTCOMES FROM RADIATION THERAPY

We examined prostate cancer outcome based on age and ethnicity in the previous section and we reported that young African American men have a worse survival than Caucasian

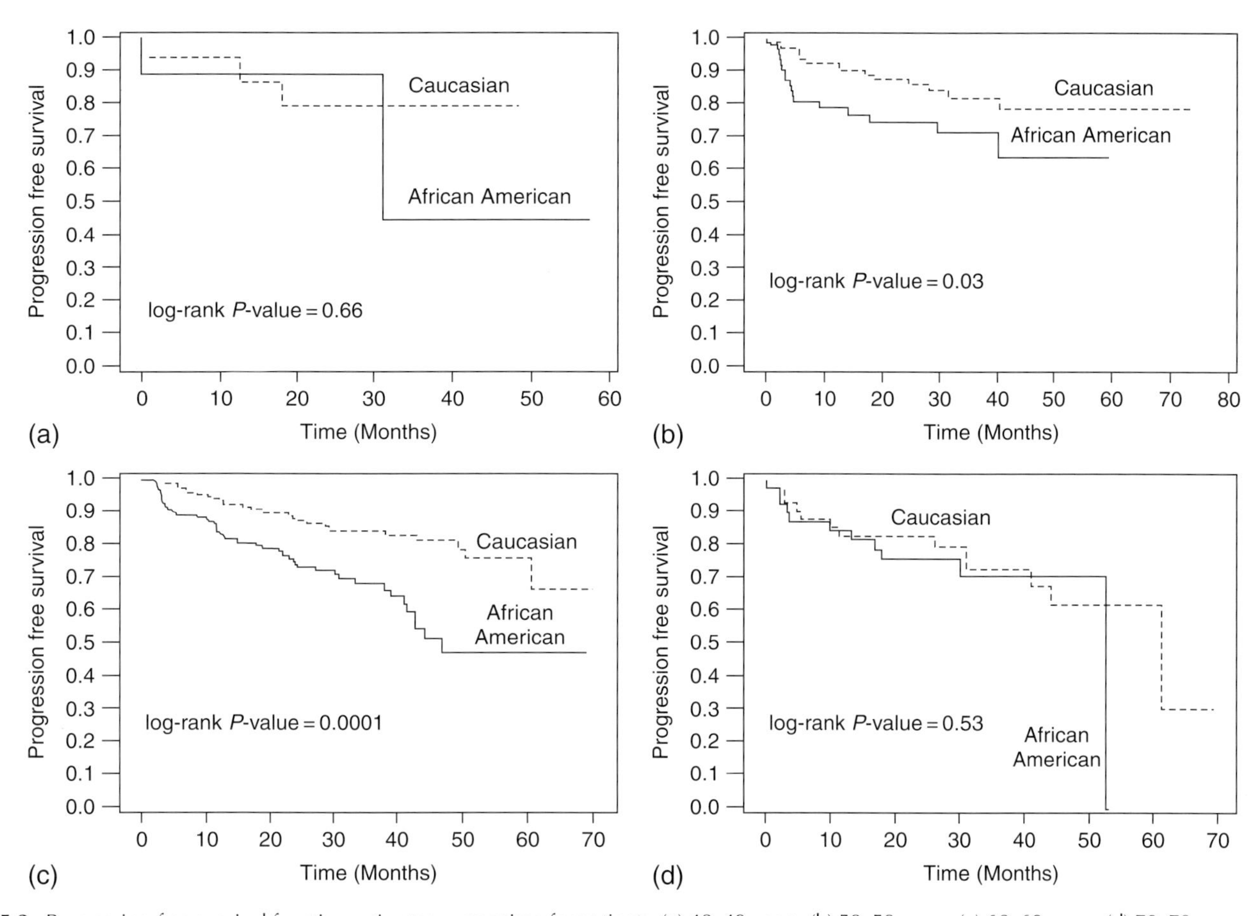

Fig. 15.2. Progression-free survival function estimates versus time for patients: (a) 40–49 years; (b) 50–59 years; (c) 60–69 years; (d) 70–79 years. *Cancer* Vol. 85, No. 2, 1999, 472–7. Copyright © 1999 American Cancer Society. Reprinted by permission of Wiley-Liss, Inc. a subsidiary of John Wiley & Sons. Inc.

men but among older men there is no difference. Recent publications have reported different racial survival outcomes among men treated by radical prostatectomy and radiation therapy. Recently, Hart and others reported a study to examine the effect of race on the outcome of patients treated curatively with external beam radiation for carcinoma of the prostate gland.[25] The study was performed between January 1980 and December 1993 and included 1529 men diagnosed with prostate cancer. Similar percentages of Caucasian men and African American men were diagnosed with clinically localized disease (stages T1 and T2) and advance staged disease (stage T3). There was no difference in crude survival by race ($P=0.13$). At 5 years, crude survivals by race were 75% for Caucasian men and 73% for African American men. At 10 years, the crude survivals were 50% and 40%, respectively, for Caucasians men and African American men. Disease-specific survival rates were equivalent for African American men and Caucasians men ($P=0.66$). The 5-year disease-specific survival was 83% for Caucasians men and 85% for African American men. At 10 years, the disease-specific survival was 65% for Caucasians and 69% for African American men. There was no difference in disease-specific survival by race when stage for stage comparisons were made. Among those patients referred for curative radiation therapy, African American men and Caucasian men were diagnosed with similar age, stage and grade of distribution.[25]

This study demonstrated there was no difference in disease-specific survival between Caucasian men and African American men treated curatively with radiation for prostate cancer. However, the mean ages of men treated for radiation therapy were 68 and 69 for Caucasian and African American men, respectively. At the same institution the mean ages of men treated for radical prostatectomy were 61 and 63 for Caucasian and African American men, respectively. The ages among men undergoing radiation therapy versus those undergoing radical prostatectomy at the same institution are statistically significantly different.

As stated earlier in this chapter, there has been no demonstration of a difference in outcome or survival in men over the age of 70 among African American men compared to Caucasian men. Therefore, the results of radiation therapy based on this evidence is similar to men treated with radical prostatectomy for this subrange of prostate cancer. In another study conducted to assess the impact of race on survival of patients treated with radiation therapy for localized prostate cancer by the Radiation Therapy Oncology Group (RTOG), it was concluded that race is not an independent predictor of disease-specific long-term survival among African American men compared to Caucasian men.[26] Between the years 1975 and 1992, 2048 men were treated for clinically localized prostate cancer on one of four consecutive prospective phase III randomized trials.

After excluding non-Black and non-White patients, 2012 patients were left for this analysis. The mean pretreatment PSA was 68 ng/ml and 35.2 ng/ml among Black and White patients, respectively. Cox proportional hazard models were used to identify the impact of previously defined risk groups on overall survival and disease-specific survival. Multivariable analysis for the significance of race was carried out using stratified Cox models. The median follow-up times for patients treated on early and late studies exceeded 11 and 6 years, respectively. In a univariable analysis, Black race was associated with a lower overall survival and disease-free survival ($P=0.04$, RR = 1.24 and 0.016, RR = 1.41, respectively). After adjusting for risk group and treatment type, race was no longer associated with outcome.[26] They concluded that there was no evidence that race has independent prognostic significance for patients treated for prostate cancer on RTOG prospective randomized trials. However, the mean age of men treated for radiation therapy in this study is 72 years for both groups.

AGE, RACE AND RADIATION THERAPY

Survival rates of 117 African American and Caucasian patients treated by primary radiation therapy for carcinoma of the prostate gland were analyzed according to age and race by Austin et al.[27] In addition, stage, grade and delayed time seeking medical attention were also analyzed. The mean ages were 68.9 years for Caucasian patients and 66.4 years for African American patients. Patients were stratified by <60 and >60 years of age. In the study, African Americans tended to have a poorer 5-year survival rate than their Caucasian counterparts (35% and 48%, respectively). When survival for each race was individually analyzed by age, they noted that younger African American patients tended to do worse than older African American patients (36% vs. 41%, 5-year survival, NS). The younger Caucasians patients, however, tended to do better than their older Caucasians patients (62% vs. 45% 5-year survival, NS). When examining age and grade distribution for African American and Caucasians patients ≤60 and >60 years, 64% of young Blacks had a high-grade prostate cancer compared to 11% of young Caucasians, and the differences were statistically significant. Among older African Americans and Caucasians, 42% of African Americans had high-grade prostate cancer compared to 50% of Caucasians and this difference was not significant. They demonstrated a worse survival of young African Americans, with a median survival of 3.9 years as compared with a median survival of 6.0 years among young Caucasians. There was no difference in median survival between older African Americans and older Caucasians. The limiting factor, however, in this study was a small sample size. Again the total number of patients was 117, of which 56 were Caucasian and 61 African American.[27] None the less, this study, in addition to the surgical study presented, does suggest that age should be considered when comparing African American and Caucasian patient prostate cancer outcomes.

SURVIVAL IN MEN WITH METASTATIC PROSTATE CANCER

Results of Southwest Oncology Group (SWOG) study (8894), a randomized, prospective phase III trial comparing orchiectomy and anti-antigen with orchiectomy and placebo in men with metastatic prostate cancer provided an opportunity to analyze the difference in survival among men with metastatic disease by ethnic background in a rigidly controlled treatment regimen. The large sample size and the lack of uncontrolled variables made this trial a good setting for evaluating whether differences in survival between African American men and Caucasian men can be explained by differences in prognostic variables or whether there may be a difference in the biological activity of metastatic prostate cancer among African American men. Using data from 288 African American men and 975 White men in the trial, they conducted a proportional hazard regression analysis to determine if ethnicity was an independent predictor of survival. The study was double blinded and the primary end-point of the study was death from any cause, with the secondary end-point being progression-free survival.

No differences were noted in treatment assignment by ethnicity: 49.3% African American men and 50.2% of White men were randomly assigned to receive flutamide. The results of the study show that, compared to White men in the study, African American men in the study were generally younger, had higher PSA levels, more extensive disease, higher Gleason scores, more frequent bone pain and a worse performance status. African American men in the study also had poorer survival than White men in the study, perhaps because of their later stage of diagnosis. Of the 198 African American and 719 White men for whom complete information about covariants was available at study entry, the median survival is 26 months and 35 months, respectively (a log rank test, $P=0.001$).

To determine whether the poorer survival among African American men relative to White men was a reflection of the poor prognostic factors, they controlled for the effect of these factors in a multivariable proportional hazard regression model. After adjustment for the potential confounding variables, the African American patients had a hazard ratio for death relative to White patients of 1.23 [95% confidence intervals (CI) = 1.042–1.47], and this increase risk was statistically significant ($P=0.018$). The likelihood ratio test for the addition of the ethnicity to the model with other prognostic variables showed that the effect of ethnicity was statistically significant. The conclusion of this study was that African American men with metastatic prostate cancer have a statistically significant worse prognosis than White men that cannot be explained by the prognostic variable explored in this study. Further, they state that their data suggest that the existence of ethnic differences in the biology of the disease may be the cause for these outcomes and should be further investigated.[28]

SUMMARY

We have analyzed prostate cancer outcomes after various treatments among African Americans compared to Caucasians. We have identified two factors that determined differences and similarities of outcomes between African Americans and Caucasians. One is the stage at diagnosis and treatment and the other is age. The earlier the stage at diagnosis and treatment, the greater will be the similarity in outcome. However, as prostate cancer advances in stage at diagnosis and treatment, the greater the differences in survival outcomes between African Americans and Caucasians become. This suggest that the natural history of prostate cancer may be more progressive among African American compared to Caucasians. This is a concept that has been promoted but the evidence to support it is early in its development, such as hormonal regulation and epigenetic factors, i.e. diet.

The second factor is age at diagnosis and treatment. It has been reported that older Caucasians have more aggressive prostate cancer at diagnosis than younger Caucasians and that age is a predictor of poor outcome among this cohort. This has been consistent with our findings. However, among African Americans this finding is not apparent. In fact, young African Americans (<65 years of age) present with equally if not more aggressive prostate cancer as older African Americans. Of course, this suggests that malignant transformation to aggressive cancer occurs at an earlier age among African Americans than Caucasians. These findings and concepts imply that the biological factors responsible for the phenotypic presentation of the prostate may be ethnically/racially associated and different or more enhanced at an earlier age. Considerable biological research will be necessary to explain these outcomes. Even though there may be biological factors responsible for what may be a rapidly growing prostate cancer among African Americans, early detection can eliminate outcome differences between African Americans and Caucasians.

REFERENCES

1. Greenlee RT, Hill-Harmon MB, Murray T et al. Cancer statistics, 2001. *CA Cancer J. Clin.* 2001; 51:15–26.
2. Von Eschenbach A, Ho R, Murphy G et al. American Cancer Society Guidelines for the Early Detection of Prostate Cancer Update 1997. *CA Cancer J. Clin.* 1997; 47:261–4.
3. Ries LAG, Kosary CL, Hankey BF et al. *SEER Cancer Statistics Review 1973–1997*. Bethesda, MD: National Cancer Institute, 2000.
4. Powell IJ, Schwartz K, Hussain M. Removal of the financial barrier to health care: does it impact on prostate cancer at presentation and survival? A comparative study, between black and white men in a Veteran Affairs System. *Urology* 1995; 46:825–30.
5. Coetzee GA, Ross RK. Prostate cancer and the androgen receptor. *J. Natl Cancer Inst.* 1994; 86:872–3.
6. Makridakis NM, Ross RK, Pike MC et al. Association of mis-sense substitution in SRD5A2 gene with prostate cancer in African American and Hispanic men in Los Angeles USA. *Lancet* 1999; 354:975–8.
7. Whittemore AS, Kolonel LN, Wu AH et al. Prostate cancer in relation to diet physical activity, and body size in Blacks, Whites, and Asians in the United States and Canada. *J. Natl Cancer Inst.* 1995; 87:652–61.
8. Wang Y, Corr JG, Thaler HT et al. Decreased growth of established human prostate LNCAP tumors in male mice fed a low-fat diet. *J. Natl Cancer Soc. Inst.* 1995; 87:1456–62.
9. Natarajan N, Murphy GP, Metllin C. Prostate cancer in Blacks. *J. Surg. Oncol.* 1989; 40:232–6.
10. Brawn PN, Johnson EH, Kohl DL et al. Stage of presentation and survival of white and black patients with prostate carcinoma. *Cancer* 1993; 71:2569–3.
11. Hill-Hermon MB, Greenlee R, Thun MJ. *Cancer facts and figures for African Americans 2000–2001*. Atlanta, GA: American Cancer Society, 2001.
12. Powell IJ, Gelfand DE, Parzuchowski J et al. A successful recruitment process of African American men for early detection of prostate cancer. *Cancer* Suppl. 1995; 75:1880–4.
13. Catalona W, Richie J, Ahmann F et al. Comparison of digital rectal examiniation and serum prostate specific antigen in early detection of prostate cancer of a multi center clinical trial of 6,630 men. *J. Urol.* 1994; 151:1283–90.
14. Powell IJ, Nelson J, Severson R. The Impact of social economic factors on PCa Outcome for Whites and African Americans. *J. Urol.* 2001; 165:Abstract #262.
15. Powell IJ, Banerjee M, Novallo M et al. Prostate cancer biochemical recurrence stage for stage is more frequent among African American than White men with locally advanced but not organ-confined disease. *Urol.* 2000; 55:246–51.
16. Powell IJ, Dey J, Dudley J et al. Disease-free survival difference between African Americans and whites after radical prostatectomy for local prostate cancer: a multivariate analysis. *Urology* 2002; 59:907–12.
17. Bauer JJ, Connelly RR, Seterhenn IA et al. Biostatistical modeling using traditional preoperative and pathological prognostic variables in the selection of men at high risk for disease recurrence after radical prostectectomy for prostate cancer. *J. Urol.* 1998; 159:929–33.
18. Moul JW, Douglas TH, McCarthy WF et al. Black race is an adverse prognostic factor for prostate cancer recurrence following radical prostatectomy in an equal access health care setting. *J. Urol.* 1996; 155:1667–73.
19. Freeland SJ, Jalkut M, Dorey F et al. Race is not an independent predictor of biochemical recurrence after radical prostatectomy in an equal access medical center. *Urology* 2000; 56:87–91.
20. Bianco FJ, Wood DP, Pontes JE et al. Prostate stage shift has eliminated the gap in survival between African American and Caucasian men after radical prostectomy. *J. Urol.* 2001; 165:Abstract # 955.
21. Pienta KJ, Demers R, Hoff M et al. Effect of age and race on the survival of men with prostate cancer in the Metropolitan Detroit tricounty area, 1973 to 1987. *Urology* 1995; 45:93–102.
22. Whittemore AS, Keller JB, Betensky R. Low grade, latent prostate cancer, volume: predictor of clinical cancer incidence? *J. Natl Cancer Inst.* 1991; 83:1231–5.
23. Sakr WA, Grignon DJ, Haas GP et al. Epidemiology of high grade prostate intraepithelial neoplasia. *Pathol. Res. Pract.* 1995; 191:838.
24. Powell IJ, Banerjee M, Sakr W et al. Should African American men be tested for prostate cancer at an earlier age than white men? *Cancer* 1999; 85:472–7.
25. Hart KB, Porter AT, Shamsa F et al. The influence of race on the efficacy of culture radiation therapy for carcinoma of the prostate. *Semin. Urol. Oncol.* 1998; 16:227–31.
26. Roach M, Lu J, Pilepich MV et al. Race is not an independent predictor of disease specific long term survival among men treated for prostate cancer on radiation therapy oncology (RTOG) trials. *J. Urol.* 2000; 163:Abstract #241.
27. Austin JP, Aziz H, Potters L et al. Diminished survival of young blacks with adenocarcinoma of the prostate. *Am. J. Clin. Oncol.* 1990; 13:465–9.
28. Thompson IM, Tangen C, Tolcher A et al. Association of African American ethnic background with survival in men with metastatic prostate cancer. *J. Natl Cancer Inst.* 2001; 93:219–25.

Hereditary Prostate Cancer

Aubrey R. Turner
Center for Human Genomics, Wake Forest University School of Medicine, Winston-Salem, NC, USA

William B. Isaacs
Brady Urological Institute, Johns Hopkins Hospital, Baltimore, MD, USA

Jianfeng Xu
Center for Human Genomics, Wake Forest University School of Medicine, Winston-Salem, NC, USA

BACKGROUND: GENETICS AND INHERITANCE

It could be said that all cancer has a genetic component, yet only a fraction of cancers are hereditary. To define hereditary cancer, we can rely on a basic understanding of traits that are inherited in a family due to DNA changes or mutations. However, study of the genetics of cancer includes, in addition to inherited mutations, somatic mutations that may be acquired in any cell of our body over a lifetime. While hereditary mutations are present in every cell of the body, somatic mutations are present only in cells that arise from a specific somatic cell. Many inherited mutations do not unilaterally cause cancer, but typically increase the risk of developing cancer due to subsequent somatic events. Finally, all inherited mutations must begin with some type of somatic event that leads to the formation of gametes destined to pass increased cancer risk and/or increased cancer severity to progeny. Regardless of the source, if specific mutations are present in specific cell types, a cancer process can begin.

This chapter specifically focuses on the hereditary components of prostate cancer. We will document the inherited contribution to prostate cancer and review the ongoing search for hereditary prostate cancer genes.

INHERITED VERSUS ENVIRONMENTAL FACTORS: TWIN STUDIES

As a simple analogy to a card game, we could say inheritance deals us a deck, but environment plays the game. Environment can be described by diet, lifestyle and features of geographic location. There have been a number of studies that examine the role of inherited versus environmental factors in the development of prostate cancer.

Epidemiology studies have shown that the risk of prostate cancer among Asian men increases within two generations after immigration to North America, presumably due to adoption of a 'Western' diet and lifestyle. However, the definitive quantification of genetic versus environmental factors in prostate cancer has been provided by studies of cancer in twins. Monozygotic (MZ) twins have 100% of their genes in common. Dizygotic (DZ) twins share an average of 50% common genes. As the contribution of heritable factors increases, the observed concordance of MZ twins increases relative to DZ twins. In fact, every reported study of cancer in MZ and DZ twins reveals a higher concordance rate for prostate cancer among MZ twins, than for DZ twins. When 1649 MZ and 2983 DZ unselected Swedish male twin pairs were studied, 19% versus 4% of the twin pairs were concordant for prostate cancer, respectively.[1] Another large twin study followed over 31 000 US veteran twins, and found 27% concordance among MZ twins pairs, compared to 7.1% in DZ twins pairs.[2] Most recently, a study among nearly 45 000 twins in the Danish, Swedish and Finnish cancer registries found that 21% of MZ twin pairs, versus 6% of DZ pairs, were concordant for prostate cancer.[3] Twin studies have solidly established that prostate cancer has a stronger hereditary component than any other human malignancy, including breast cancer and colon cancer, with an estimated 42% of prostate cancer due to inherited factors (**Fig. 16.1**).

In some cases, an inherited predisposition may be so strong that even the most preventative of lifestyles cannot stall the

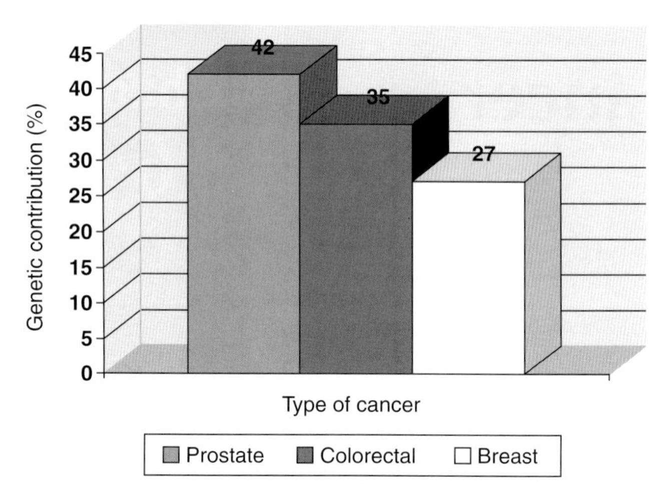

Fig. 16.1. Hereditary contribution to common cancers.

development of prostate cancer. In other cases, environmental factors may completely override an inherited resistance to prostate cancer. For most people, the balance of genetics and environment probably falls within some middle ground, such that heritable and environmental factors combine to yield an individual prostate cancer risk.

RECOGNIZING AND QUANTIFYING THE RISK: FAMILY STUDIES

Family studies have utilized the variable balance of genetics and environment to quantify the risk of prostate cancer, thus providing the basis for family history risk assessment. Although we can now use family history information to provide general risk estimates to family members of men with prostate cancer, our current understanding stems from clinical observations made nearly half a century ago. Family studies provided the original impetus to conduct detailed studies of the inheritance of prostate cancer.

In 1956, Morganti et al. found that patients with prostate cancer reported having affected relatives more frequently than did controls who did not have prostate cancer.[4] Likewise, in 1960 Woolf compared 228 prostate cancer cases to age-matched controls, and found a threefold increased risk of prostate cancer death in brothers and fathers of the cases, compared to the general population.[5] Although these first studies were subject to biases, including recall bias, and were certainly less than definitive, they provided a crucial first attempt to document clinical observations that suggested an inherited component of prostate cancer.

Family studies began to provide more thorough quantification of these familial risks in the 1980s. Foreshadowing the more definitive studies with MZ and DZ twins, a study of the Utah Mormon population in 1982 by Cannon et al. reported that prostate cancer shows stronger familial clustering than both breast cancer and colon cancer.[6] Using the same Utah Mormon population, Meikle and Stanish observed a fourfold increased risk of prostate cancer among the brothers of prostate cancer probands, compared to brothers-in-law and population controls.[7]

Studies of families with multiple cases of prostate cancer have provided direct evidence of an inherited etiology. Consistent with the original observations of familial clustering by Morganti et al. and Woolf, a retrospective case-control study by Steinberg et al. in 1990 showed that almost twice as many prostate cancer probands reported a father or brother affected with prostate cancer, compared to controls (15% versus 8%, respectively).[8] A key finding of their study was the quantification of the risk associated with an increasing number of affected family members. Steinberg reported that the presence of one, two or three affected family members, increases the risk of prostate cancer to first-degree (immediate blood relation) male relatives by 2-, 5- and 11-fold, respectively. Individuals who reported an affected second-degree relative were at a marginally increased risk, if any, of prostate cancer, although the presence of one first degree and one second-degree affected increased the risk by nearly ninefold. These risk estimates have withstood the critique of time, as they represent the average of risk estimates provided by a number of additional retrospective case-control studies and prospective cohort studies among a variety of populations. However, these risk estimates were based on prostate cancer incidence prior to widespread use of the prostate specific antigen (PSA) screening test, when the estimated incidence was 9% in White US males and 10% in Black US males. Although the PSA era may limit the absolute accuracy of these estimates, the trends remain intact. Accordingly, a 1996 population based case-control study found that the risk of prostate cancer is increased by twofold in men with one affected family member, and fourfold with two or more affected relatives.[9] Even more recent confirmation of increased familial risk was provided by a 1997 study of Canadian men in Montreal, Toronto and Vancouver, in which 15% of cases, versus 5% of controls, reported a positive family history.

The effect of family history in conjunction with age of onset has also been examined. An epidemiological evaluation by Carter et al. revealed that the proband age of onset is inversely related to hereditary prostate cancer risk, and a diagnosis at less than 50 years effectively doubles the risk of hereditary prostate cancer among other family members.[11] Specifically, among families with at least two affected first-degree relatives, Carter reported a risk ratio of 3.8 when the proband onset was 70 years, increasing to 5.2 at an onset of 60 and peaking at 7.1 for an onset of 50. The family study by Lesko et al. also demonstrated that the risk of prostate cancer increases with earlier age of onset in affected relatives. Although family members were found to have a 4.1-fold increased risk of prostate cancer when a relative had onset prior to age 65 years, this risk dropped to 0.76 when the age of onset was greater than 74 years.[9]

Whittemore et al. confirmed that early age of onset increases the risk of prostate cancer in families with at least one affected first-degree relative, and also defined the combined effects of family history, age of onset and race.[12] Epidemiologists have clearly documented the fact that, compared to all other populations, Black Americans have the highest risk of prostate cancer in the world. Furthermore, we know the risk of prostate cancer among Black American men is double the risk observed in White American men. In the general population, a family history of early onset prostate cancer accounts for an increase of approximately 30% above the nearly doubled risk observed in families with prostate cancer. In White families, onset at ages 80 and 60 confers risk increases of 1.9 and 2.5 fold, respectively. In Black families, these increases are 3.0 and 4.3 fold, respectively.[12] These findings suggest that race and age of onset contribute independently to the risk of hereditary prostate cancer.

Family studies have also provided clues regarding the inheritance of prostate cancer. Studies of men with affected fathers or affected brothers consistently show that men with an affected brother are at greater risk to develop prostate cancer than men with an affected father. In a study of Blacks, Whites, Japanese and Hispanics, the age-adjusted risk of prostate cancer was doubled for men with an affected brother, compared to men with an affected father, regardless of race.[13] Another study, with a similar experimental design, also observed this trend, with a risk increase of 2.5-fold for men with an affected father, versus 5.3 with an affected brother.[14]

Just as the original family studies of prostate cancer by Morganti and by Woolf were subject to recall bias, modern studies are likewise vulnerable to this source of misleading data. In addition, improved screening sensitivity with PSA has enhanced the problem of detection bias, and the use of age of onset as a high risk screening tool may be severely limited as a result. Recall bias and detection bias may potentiate unequal comparisons between the awareness of members of hereditary prostate cancer families versus controls. Indeed, family studies can quantify the risk of prostate cancer given a variety of familial characteristics; however, these designs do not capture the individual toll on members of a family with hereditary prostate cancer. Under such conditions, one might always predict some degree of extant bias.

Despite the potential biases, after decades of work, there is a remarkable degree of consistency and agreement among the results provided by the overwhelming majority of prostate cancer family studies. The credibility of retrospective family studies has been enhanced by several reports of prostate cancer cohort family studies, as these prospective studies minimize the biases inherent to retrospective family studies. In summary, cohort studies by Goldgar, Gronberg, Schuurman, and Cerhan have confirmed many of the findings of the retrospective family studies, including the observed trends and quantification of increased prostate cancer risk.[15–18] Specifically, cohort studies have shown an increased risk due to an affected brother compared to an affected father, an increased risk for family members of early onset cases, and an increased risk with increasing number of affected first-degree family members.

In retrospect, the findings from family studies, particularly the initial findings, might have been explained by an unknown balance of genetic and shared environmental factors, and then discounted due to biased methods. However, our current knowledge of twin studies provides a solid basis from which the family studies can be interpreted, because the role of heritable factors in prostate cancer is now clearly established. Therefore, we can safely state that the observed increase in risk due to family history is a direct result of an increase in the proportion of heritable factors. Further definition of the role of inheritance in prostate cancer has been provided by complex segregation analyses.

UNDERSTANDING THE MODES OF INHERITANCE: COMPLEX SEGREGATION ANALYSIS

With the role of genetics firmly established by twin studies and the increased risk quantified by family studies, several complex segregation analyses were undertaken to identify the most likely mode of inheritance for prostate cancer genes. These analyses examine family structures, and use these to measure the segregation of a disease or trait (in this case prostate cancer) as it is passed from parent to child. Based on the observed segregation, multiple inheritance models are evaluated ultimately to identify the most likely mode(s) of inheritance for the trait or disease in question. For hereditary prostate cancer, a total of five reported segregation studies have proposed various inheritance modes.

Most segregation analyses support the trends observed in the initial study by Carter et al., who evaluated Mendelian inheritance, including autosomal dominant, autosomal recessive and X-linked, as well as an age-related sporadic model and a familial environment model.[19] These analyses suggested that the inheritance of prostate cancer among families with multiple affected first-degree relatives is best explained by a rare (population frequency = 0.36%) high-risk (88% lifetime penetrance) allele that is inherited in an autosomal dominant fashion, accounting for 43% of early onset (= 55 years) and 9% of all prostate cancer cases.[19] Confirmatory segregation analyses have proposed increased allele frequencies and decreased penetrance, compared to the initial report.[20–22] These differences may be due to differences in the populations studied. The segregation analyses of prostate cancer in Australia by Cui et al. found that an autosomal dominant model is the best fit to early onset families, while an X-linked model provides the best fit to late onset families.[22] Importantly, this study found that two-locus models provided a better explanation of the segregation analysis, compared to one-locus models, implicating more than one major gene.

Complex segregation analyses provided the impetus to begin a search for hereditary prostate cancer genes that fit the profile of a major susceptibility gene. Following the model for gene discovery employed in mapping a variety of human disease genes, linkage approaches were the obvious next step.

IDENTIFICATION OF LOCI: LINKAGE STUDIES

Linkage analyses identify and define specific chromosomal regions associated with inheritance of a disease of interest, such as prostate cancer. This involves tracing the inheritance of multiple genetic landmarks (genetic markers) within families, to identify landmarks associated with the disease and to measure the extent of this linkage. Linkage analyses provide results in the form of an individual log of odds (LOD) score for each molecular landmark studied, or a multipoint LOD score for a group of nearby markers. LOD scores are logarithmic, such that LOD of 1.0, 2.0 and 3.0, represent a likelihood of false linkage equal to 1:10, 1:100 and 1:1000, respectively. In traditional gene searches, a LOD score greater than 3.0 is interpreted as a significant linkage finding. In studies of disease with complex genetic etiology, such as prostate cancer, lower LOD scores are typically considered informative, and values greater than 1.0 might generally warrant further study. When linkage analysis identifies a region of the genome that is likely to harbor a causative gene but a specific gene has yet to be identified in the region, the region is generally referred to as a locus (loci for plural). Although linkage studies can suggest loci that are most likely to contain genes of interest, they cannot independently refute the existence of a causative gene within a locus. In simplest terms, linkage analysis helps to focus efforts to identify genes.

Genomewide linkage screens

Genomewide linkage screens are simply linkage studies that include markers distributed throughout the genome. Because the human genome contains about 3300 centimorgan (cM), ~350 evenly spaced markers are included in a typical genomewide screen to obtain a coverage rate of 1 marker per 10 cM. Genomewide screens are a first step in gene discovery when no clear candidate gene is available or when multiple candidate genes are available.

The first genomewide screen for inherited prostate cancer susceptibility genes provided a likely chromosomal location for the typified inherited prostate cancer gene that had been suggested by the segregation analysis of Carter.[19] Such a gene would be rare, follow autosomal dominant inheritance and confer high risk. This first study included 66 families ascertained at Johns Hopkins Hospital prior to widespread implementation of PSA screening.[23] Each family met an operational definition of hereditary prostate cancer, known as the Hopkins Criterion, that included families with no evidence of bilineal inheritance, and either (1) three or more first-degree diagnosed relatives, or (2) two or more brothers with diagnoses ≤55 years. The average age at diagnosis in these families was 65 years, falling more than 5 years below the national average in the USA. A total of 341 dinucleotide repeat markers were genotyped and analyzed in these families, covering the genome with ~10 cM resolution. The overall analyses assumed dominant inheritance, a 15% phenocopy rate and that unaffected men over the age of 75 are not likely to be gene carriers. Within this analytical framework, two-point parametric linkage analysis identified seven regions with a LOD score >1, including 1q24–25, 1q33–42, 4q26–27, 5p12–13, 7p21, 13q31–33 and Xq27–28. The highest LOD score was 2.75 at marker D1S218, which maps to the long arm of chromosome 1 (1q24–25). When the linkage peak at 1q24–25 was evaluated in detail with additional genetic markers in a combined collection of 91 US and Swedish families, a LOD score of 5.43 was detected by accounting for heterogeneity under the assumption that 34% of families were linked to this region.[23] The locus at chromosomal segment 1q24–25 has come to be known as HPC1. The impressive LOD scores reported by this study fueled optimism that a major hereditary prostate cancer gene would be identified in short order, and thus a large share of the efforts to detect such a gene have been focused on chromosome 1q24–25. However, follow-up linkage studies have provided mixed results.

The second genomewide screen included 504 brothers from 230 multiplex sibships that were collected by Washington University School of Medicine in St Louis. This study employed non-parametric linkage analyses, free of specific inheritance models, to avoid the potential for misleading results due to genetic model mispecification. Their approach identified five regions with 'nominal' evidence for linkage, including 2q, 12p, 15q, 16q and 16p.[24] Among these regions, the strongest evidence of linkage was a multipoint LOD of 3.15 on chromosome 16q23. Interestingly, the 16q23 region includes a putative tumor suppressor gene that may be involved in breast, ovarian and prostate cancer. Although this study reported linkage to 1q, they found no evidence of linkage at the HPC1 locus previously reported on 1q24–25. Instead, this study reported linkage to two regions that flank 1q24–25, and only when they stratified based on the Hopkins Criterion. Linkage was detected at 1q41–43 among sibships that met the Hopkins Criterion for HPC, and at 1q21–22 among familial prostate cancer sibships that did not meet the HPC Criterion.

Goddard et al. published a genomewide study that expanded directly upon the study reported by Suarez et al.[24,25] This study included a modest increase in the total number of study participants, to include 564 men from 254 families with both prostate cancer and available Gleason scores. However, this report was distinguished by a study design that included four covariates in the analyses: (1) the sum of the sibling-pair Gleason scores; (2) age at onset (family mean age at diagnosis); (3) presence or absence of male-to-male transmission in the nuclear family; and (4) the number of affected first-degree relatives in the nuclear family. They did not detect strong linkage to any region, except when they incorporated clinical and family history variables in the analyses. A chromosome 14 linkage of 2.74 LOD was identified based on age at onset, and this covariate also provided significant linkage to chromosomes 4, 6, 7 and 20. They detected a large and significant linkage peak of 3.25 LOD near HPC1 after they stratified by high Gleason score. Additional linkage regions identified in the Gleason score strata were reported on chromosomes 2, 5, 8, 16 and X. The use of male-to-male transmission as a covariate increased insignificant linkage at chromosome 21 to a LOD of 3.12, and also implicated chromosomes 1–5. The analyses yielded a LOD score of 2.84 at the 1q41–43 region of chromosome 1 when both low Gleason score and male-to-male transmission were included, supporting the linkage originally reported by Berthon (1998) as detailed under the heading of 'PCAP' later in this chapter.[25]

Shortly after the publication by Suarez,[24] a group at Fred Hutchinson Cancer Center published the third genomewide screen for prostate cancer.[27] Among 94 families, they found two-point LOD scores ≥1.5 for several chromosomal regions. When an autosomal dominant model of inheritance was evaluated, loci on 10q, 12q and 14q were implicated. Chromosomal regions on 1q, 8q, 10q and 16p were suggested as the most likely regions to contain an HPC gene when a recessive model of inheritance was assessed. A late onset locus was suggested by a multipoint LOD score of 3.02 on chromosome 11 when families were stratified by age of diagnosis. This study did not confirm linkage to the locus at 1q24–25. However, a linkage result of 1.99 LOD was found at 1q41–43 in their total set of samples, thus moderately supporting the linkage originally reported by Berthon et al.,[26] as detailed under the heading of 'PCaP' later in this chapter.

The most recently reported genomewide linkage analysis included 98 families from the USA and Canada.[28] Each family had at least three verified diagnoses of prostate cancer among first- and second-degree relatives. Positive linkage signals of 'nominal' statistical significance were found in the 5p–q and 12p regions. The strongest finding of this study was the LOD score of 2.87 at 19p. The 1q24–25 locus was not confirmed by their results, nor were any of the other proposed HPC loci from previous genomewide linkage screens.

To summarize the genomewide linkage screens, among the findings with at least marginal significance, the regions at 5p and 12p were the only specific linkages reported by more than one genomewide linkage study, except when detailed covariate analyses were performed. It is worth noting that, to date, each of these linkage studies have provided a different overall result. Specifically, the strongest LOD score findings from each study were not replicated by any of the other linkage studies. However, many of the regions implicated by genomewide linkage studies have been observed commonly in loss of heterozygosity (LOH) studies, including 7p and 13q by Smith,[26] 16q by Suarez,[24] 10q and 8q by Gibbs[27] and, finally, 5p–q by Hsieh,[28] thus providing additional support for the existence of important prostate cancer genes within these linkage regions.

Linkage confirmation and fine mapping

Each of the most significant findings from genomewide linkage studies has been followed by locus-specific studies that utilize dense DNA markers. The use of more markers within a limited interval allows for confirmation of initial genomewide linkage results, as well as additional refinement from a broad linkage region that might contain thousands of genes to a narrow region manageable for the study of each included gene.

Interestingly, a review of the literature reveals that every chromosome in humans, except for the Y chromosome, has been implicated in prostate cancer, at least marginally, by linkage analyses. The regions that have received the most attention in detailed linkage and gene-mapping efforts include HPC1 at 1q24–25, PCaP at 1q42–43, CaPB at 1p36, HPCX at Xq27–28, HPC20 at 20q13, ELAC2/HPC2 on 17p, and also linkage at 8p22–23.

HPC1

The initial linkage result in the region of 1q24–25 stands as the highest LOD score among all published studies of hereditary prostate cancer. The locus was quickly termed HPC1. Following the original study, analysis of HPC1 linkage by other research groups has been variable. While several groups have published results ranging from nominal to strong evidence for linkage to this region, there have been no studies that quantitatively reproduce the very strong initial findings. In addition, based on linkage analyses limited to the HPC1 region, there have been no results that clearly narrow down this large region, and some studies have even suggested the HPC region extends beyond the bounds originally reported by Smith et al.[23] Many genes lie within this interval and detailed study of each of these genes is a daunting task.

In fact, four groups have reported no significant evidence for linkage of HPC1 in their study populations after detailed studies of the region. Among 49 high-risk prostate cancer families, McIndoe et al. found no evidence for linkage in this region when their data were evaluated with either a non-parametric analysis or a parametric LOD score approach.[29] Stratified analysis of these families by age at diagnosis was performed, and the 18 families with early age at diagnosis (<65 years) provided no evidence for linkage at HPC1.[29] When this initial pool of 49 families was later extended to include a total of 150 HPC families, linkage to HPC1 was strongly rejected.[30] A second group reported results of a genomewide screen and also specific results from the 1q24–25 region in 47 French and German families.[26] For three markers in the 1q24–25 region, this group found negative two-point LOD scores assuming a dominant inheritance model, thus rejecting the HPC1 linkage. A report by Eeles et al. also targeted the 1q24–25 region, and included 136 prostate cancer families ascertained in the UK, Quebec and Texas.[31] Their samples were enriched for hereditary prostate cancer, as 76 of the total 136 families had three or more affected individuals. Although the total sample provided negative non-parametric linkage scores in the HPC1 region, a subset of 35 families with four or more affected members positive non-parametric linkage scores.[31] In study of 230 multiplex sibships by Suarez et al.,[24] they reported no evidence for linkage at the HPC1 locus, although positive linkage results in adjacent regions were observed when the analyses were stratified by family history, defined as an affected family member in a previous generation. Suarez et al. also reported negative findings for HPC1 in 45 new multiplex sibships and four very large HPC families.[32] When Goddard et al. reanalyzed this total collection of 189 families, they reported a significant linkage at 1q24–25 (LOD score=3.25, P=0.0001) after including Gleason score as a covariate.[25] This most recent result suggests a role of HPC1 in modulating the aggressiveness of prostate cancer in inherited cases.

Despite multiple reports of no linkage or negative linkage, multiple studies have independently substantiated linkage to HPC1. Several studies have provided interesting insights and at least weak confirmation of HPC1, including those by Cooney et al.,[33] Berry et al.,[34] Hsieh et al.[28,35] Cooney et al.

reported a linkage study of HPC1 in 59 prostate cancer families with two or more affected individuals, and enriched for affected siblings.[33] The peak non-parametric linkage score in the 1q24–25 region was 1.58 ($P=0.057$) in the analysis of all 59 families, but was increased to 1.72 ($P=0.045$) in the subset of 20 families that met the more stringent criteria for hereditary prostate cancer families as defined in the original linkage report of Smith et al.[23] Hsieh et al. reported further evidence to support HPC1.[35] In 92 unrelated families having three or more affected individuals, the non-parametric linkage score was 1.71 ($P=0.046$). The evidence for linkage was stronger in the 46 families with mean age at diagnosis <67. The non-parametric linkage score was 2.04 ($P=0.023$). Finally, among the studies that moderately support HPC1, Berry et al. reported linkage results of 144 HPC families collected at the Mayo Clinic.[34] They did not find evidence for linkage at the HPC1 region in the total sample, but a peak non-parametric linkage score of 1.99 ($P=0.03$) at HPC1 was provided by a subset of 102 families with male-to-male disease transmission.

In a recent chromosome-wide linkage study to evaluate different prostate cancer susceptibility loci on chromosome 1 in 159 hereditary prostate cancer (HPC) families, a report by Xu et al. reported evidence for linkage in a broad region from 1p13 to 1q32.[36] The evidence for linkage was stronger in families with five or more affected men (allele sharing LOD=2.22, $P=0.001$), and in families with mean age of onset >65 years (allele sharing LOD=1.45, $P=0.01$). In another genomewide scan for prostate cancer susceptibility loci, Goddard et al. reported a LOD score of 3.25 ($P=0.0001$) at 1p13, when Gleason score was included as a covariate.[25]

Key reports that have provided stronger support for HPC1 and a major HPC susceptibility locus include those by Neuhausen et al.,[37] Goode et al.,[38] Xu and ICPCG,[39] and Xu et al.[36] Neuhausen found positive evidence for linkage in 41 large HPC families ascertained in Utah.[37] Although the peak two-point LOD was 1.73 ($P=0.005$) when HPC1 linkage was evaluated among the total collection of families, the major finding of this study was a two-point LOD of 2.82 ($P=0.0003$) in early onset families.

To clarify the variable results from HPC1 replication studies, members of the International Consortium for Prostate Cancer Genetics (ICPCG) reported results from a combined analysis of six markers in the 1q24–25 region among 772 HPC families. These families were collected in North America, Australia, Finland, Norway, Sweden and the UK.[39] Overall, there was some evidence for linkage, with a peak parametric multipoint LOD score assuming heterogeneity (hlod) of 1.40 ($P=0.01$). The evidence for linkage was stronger in families with male-to-male disease transmission, as the peak hlod was 2.56 ($P=0.0006$) in the subset of 491 families with male-to-male disease transmission, compared with a hlod of 0 in the remaining 281 families. Within the male-to-male disease transmission families, the contribution of HPC1 to prostate cancer risk increased with early mean age of diagnosis (<65) and with a greater number of affected family members (≥ 5). The highest risk was observed for the 48 families that met the criteria of all three strata (peak hlod=2.25, $P=0.001$, $a=29\%$). The results from parametric and non-parametric analyses were consistent, as the peak non-parametric linkage score of 1.14 was observed in the total 772 HPC families. The most convincing evidence for linkage at this region was observed when the 491 families with male-to-male disease transmission were evaluated free of a specific inheritance model, as the peak non-parametric linkage reached 2.3 ($P=0.01$). These results support the existence of a prostate cancer susceptibility gene, or genes, within the interval within or near chromosomal segment 1q24–25.

PCaP

Berthon et al. reported results from a combined genomewide screen and fine mapping study in a collection of 47 French and German families.[26] Their linkage analysis found a prostate cancer susceptibility locus at 1q42–43, titled PCaP, with a maximum two-point LOD score of 2.7. Multipoint parametric analysis yielded a LOD score assuming heterogeneity of 2.2. Non-parametric analysis yielded a non-parametric linkage score of 3.1. They estimated 50% of the 47 families were linked to the locus. The greatest evidence for linkage was a LOD of 3.31 among the families with early-onset prostate cancer (<60 years). In 2001, Cancel-Tassin et al. revisited the original sample, plus 17 newly recruited families, for a total of 64 families, and reported a non-parametric LOD score of 2.8 at the PCaP region. Their article suggested that PCaP is the most important HPC locus among individuals from Southwest Europe.[40] There has been no independent confirmation of the strong original PCaP findings among Southwest European populations.

Two replication studies in samples collected in North America did not confirm the locus overall.[41,42] Four studies have provided weak confirmation of PCaP in specific HPC strata.[25,32,34,36,41] Moderate confirmation of linkage to this region has been provided by a two-point LOD score of 1.99 reported by Gibbs et al.[27] Although Xu et al. found some evidence to support the existence of the locus at 1q42–43 among 159 HPC families when they performed a linkage study of 50 evenly distributed chromosome 1 markers that included PCaP,[36] conditional analysis revealed that evidence for linkage at 1q24–25 and 1q42–43 were related. This result suggested the linkage to this region was provided by families most closely linked to the HPC1 locus, and that the evidence of linkage seen at PCaP was primarily a statistical anomaly in this study. The mixed reports of linkage to PCaP based on various strata, and the reports of shifting HPC1 interval boundaries, may both reflect an unappreciated complexity of the role of the q arm of chromosome 1 in HPC.

CaPB

A 1995 study by Isaacs et al. sought to assess whether hereditary prostate cancer might be associated with other types of cancer.[43] Interestingly, cancers such as breast cancer and colon cancer were not associated, and tumors of the central nervous system (CNS) were the only cancers found to be in excess among a study of other cancers in multiplex prostate cancer families. The impact of this surprising result was difficult to interpret at the time. However, Gibbs et al. provided results that appeared to explain the previous association findings.[44] Based on initial results of a genomewide screen in 70 HPC families and candidate-region mapping in a total set of 141 families, they reported linkage to 1p36. Further evaluation of family subsets revealed an association between families linked to 1p36 and the presence of brain cancer in the family. An overall two-point LOD score of 3.22 was provided by analysis of 12 families with a history of prostate cancer and at least one blood relative with primary brain cancer. In the younger age group (mean age at diagnosis <66 years), a maximum two-point LOD of 3.65 was observed. However, the evidence for linkage was weaker in the multipoint analyses, as the peak multipoint LOD score fell to 0.81, assuming heterogeneity. This study was able to reject linkage to 1p36 in all strata without a history of brain cancer. The group at the Johns Hopkins University and Wake Forest University investigated this linkage, and provided results that are consistent with a prostate cancer locus at 1p36.[36] An additional study by Berry et al. did not replicate the finding.[34] Badzioch et al. found evidence of linkage to CaPB in families with early onset prostate cancer, although no association with other cancer was seen.[45]

HPCX

In a combined study population of 360 prostate cancer families collected at four different sites in North America, Finland and Sweden, linkage to Xq27–28 was observed, which was termed HPCX.[46] The peak two-point LOD score was 4.6, and the peak multipoint LOD score was 3.85. Significant evidence for locus heterogeneity was observed, suggesting the contribution of multiple genes within this chromosomal region to the observed linkage results. The proportion of families linked to *HPCX* was estimated to be 16% in the combined study population and was similar in each separate family collection. This study was important not only because it was the first to identify linkage to the X chromosome, but also because this was the first study to perform stratified analyses based on the presence or absence of male to male inheritance. In fact, the bulk of the linkage evidence at HPCX was provided by 129 families with no evidence of male-to-male inheritance. The linkage of a prostate cancer gene to the X-chromosome is consistent with the results of several population-based studies suggesting an X-linked mode of inheritance of prostate cancer and the increased risk of prostate cancer among men with an affected brother versus an affected father.[5,13,14,17,18,47] Two subsequent studies have provided weak confirmatory evidence of linkage to HPCX[48,49] and one study did not confirm the linkage.[50]

HPC20

Berry et al. identified a HPC locus at 20q13 in a collection of 162 North American families with at least three members affected with prostate cancer.[51] The highest two-point LOD score was 2.69 and the maximum multipoint non-parametric linkage score was 3.02 ($P = 0.002$). Strong and significant linkage was reported among families with any evidence of male-to-male disease transmission, <5 family members affected with prostate cancer or average age of diagnosis greater than 66 years. The subset of 19 families with all three of these characteristics had a non-parametric linkage score of 3.69 ($P = 0.0001$).

While one study has rejected linkage to HPC20 altogether, two other studies have since provided weak confirmative evidence, although not statistically significant.[52–54] A report by Cancel-Tassin et al. rejected linkage to the HPC20 region amongst all strata of a Southwest European collection of HPC families, including covariates from the original report.[52] When Bock et al. performed a replication study, 16 Black families provided a LOD of 0.86 and a non-parametric LOD of 1.99; however, these were the strongest evidence for linkage and even this result did not achieve statistical significance.[53] The report by Zheng et al. is the only study that has confirmed HPC20 linkage based on the strata of the original report.[54] Specifically, they found non-parametric LODs of 1.94, 1.74 and 1.01, based on respective later age at diagnosis, fewer than five affected family members, or absence of male-to-male inheritance.

ELAC2/HPC2

The year 2001 brought the first report of a prostate cancer susceptibility gene identified among HPC families using a linkage approach. Based on a genomewide scan that included a limited number of large pedigrees from Utah, linkage to chromosome arm 17p was detected when a dominant mode of inheritance was assumed (Tavtigian et al. 2001). Specifically, a 2-point LOD score of 4.5 was reported at 17p11. However, when the analyses were extended to include a substantial number of families, the evidence for linkage faded and was not significant. Positional cloning within this linkage region led to the discovery of the ELAC2 gene as a candidate for further study.[55]

The initial excitement surrounding ELAC2 has subsided following the absolute lack of confirmatory evidence reported by other linkage studies. Xu et al. reported a linkage study of the ELAC2 region.[56] When markers at 17p11.2 were studied among 159 HPC families, both the non-parametric and multipoint analyses provided no evidence for linkage, with overall LOD scores of 0 across the region. Furthermore, covariates that included age at onset, number of affected members, male-to-male disease transmission, or race, did not provide evidence for linkage. A report by Suarez et al. was not able to confirm linkage to 17p11.2 among any family strata.[57] In fact, published LOD scores were negative throughout most of the linkage region.[57] Both the study by Xu et al.[56] and by Suarez et al.[57] suggest ELAC2 may play a limited role in prostate cancer, particularly in HPC families.

Linkage at 8p22–23

Among 66 HPC pedigrees originally studied by Smith et al., there were statistically significant positive linkage scores at 8p22–23.[23] Findings included two-point parametric LOD of 0.7 and a multipoint LOD of 0.81. However, these findings were clearly not the focus of researchers in the field because much stronger LOD scores were reported at loci such as HPC1.

More recently, a prostate cancer susceptibility linkage at 8p22–23 was highlighted by an analysis of 159 HPC families that included the 66 families originally studied by Smith.[58] In the complete set of 159 families, a significant LOD of 1.84 was reported, with an estimated 14% of the HPC families linked to this region. A LOD score of 2.64 was found among HPC families with late onset disease (>65). Interestingly, the group of 11 Ashkenazi Jewish pedigrees that were included in this study contributed disproportionately to the linkage evidence, while 14 African American pedigrees did not demonstrate 8p22–23 linkage.

The prostate cancer linkage at the 8p22–23 region was also independently supported by a genomewide screen performed in 94 HPC families ascertained in the Seattle-based Prostate Cancer Genetic Research Study (PROGRESS).[27] The likelihood of a prostate cancer susceptibility gene in this region is further strengthened by the accumulated evidence that 8p is the site of the most frequent loss of heterozygosity in prostate cancer tumors.[59–66] Considered as a whole, the LOH studies and linkage studies at 8p suggest genes in this region may be of general importance in prostate tumor progression, and also specifically in hereditary prostate cancer risk.

Challenges to linkage studies of prostate cancer

Paralleling studies of other complex diseases, segregation analyses have suggested a heterogeneous genetic etiology of inherited prostate cancer that is characteristic of complex diseases. The suggestion of multiple modes of inheritance implicates the presence of multiple genes in hereditary prostate cancer susceptibility. In linkage studies, the presence of genetic heterogeneity dramatically decreases the power to detect the effect of any single major gene when the LOD scores of multiple families (presumably with a spectrum of inherited mutations) are tabulated.

As if genetic variability were not enough of a barrier to the detection of prostate cancer susceptibility genes, linkage analysis of prostate cancer has also proven difficult because the natural history and clinical characteristics of this disease beget many pedigrees that are only marginally informative for linkage analysis. Commonly samples from affected men in multiple generations of a family are not available because, relative to the proband, individuals in the parental generations are more likely to be deceased, and individuals in the offspring generations are usually too young to manifest the phenotype. This creates a narrow range of available affected individuals within a given family and may be further limited by the clinical misclassification intrinsic to prostate cancer within the context of a family history. Men from high-risk families who carry a high-risk susceptibility gene, and who truly have a high lifetime risk for prostate cancer, may avoid identification due to death from an unrelated disease or accident at a relatively young age. All of this may be further confounded by the presence of phenocopies (non-gene carriers with disease) within a family. Given the high prevalence of prostate cancer and the fact that twin studies have established a significant contribution of environmental factors to prostate cancer, the presence of phenocopies within hereditary prostate cancer families may pose a significant problem for linkage analyses. The late age of onset, incomplete penetrance, age-dependent penetrance and high prevalence of the disease may collectively result in misclassified inheritance patterns and loss of power in linkage studies. The complexity of prostate cancer genetics was suspected when the first genomewide screen for inherited prostate cancer susceptibility genes were initiated, although this complexity may not have been fully appreciated by many researchers in this field until relatively recently. Indeed, as of the writing of this capter, there appears to be no definitive prostate cancer gene of any clinical utility.

EVALUATION OF HPC CANDIDATE GENES: ASSOCIATION STUDIES

The regions identified as the potential locations of prostate cancer susceptibility genes by linkage studies tend to be broad, which causes great difficulty for detailed mapping and cloning the relevant genes. To supplement fine mapping efforts in linkage studies, association studies can be undertaken as an alternative approach.

There are essentially two different association study designs, consisting of population-based and family-based approaches. Population-based association studies are generally classic case-control studies that compare the frequency of a specific genetic variant (allele) in unrelated cases and normal controls. A significant difference in the allele frequency between cases and controls can be expected if the allele sequence itself is causal, or if the allele is in linkage disequilibrium (LD) with a disease-causing mutation. Because LD can only be observed when two markers are closely linked, a significant finding of LD can pinpoint the location of the disease gene. In contrast, family-based association studies calculate the observed transmission frequency of a specific genetic variant to affected sons versus the expected frequency of 50%. Because family-based association studies do not use population controls, this approach effectively eliminates the potential for population stratification.

Association studies are commonly used for testing the relevance of candidate genes as disease susceptibility genes and for testing LD of anonymous marker alleles in the regions implicated through linkage studies. This approach is becoming increasingly popular as more candidate genes are identified and denser maps of single nucleotide polymorphisms (SNPs) become available. Overall, association studies represent a complementary approach to linkage studies because association studies evaluate the relationship of a specific genetic variant with disease status in an individual, whereas linkage studies survey and evaluate the inheritance of chromosomal regions.

ELAC2/HPC2

The initial report of linkage directly to the ELAC2 gene among several families with hereditary prostate cancer generated a great deal of excitement in the research community, particularly because it appeared to describe the first clinically relevant genetic alterations involved in prostate cancer risk.[55,67]

Mutation screening led to the identification of two common missense variants, Ser217Leu and Ala541Thr, that were reported to be associated with prostate cancer via a case-control approach.[55] When the frequencies of variant genotypes were compared amongst nearly 400 cases and 150 population controls, they reported that homozygous variant 217Leu increased prostate cancer risk by 2.4 times, the heterozygous variant 541Thr increased risk by 3.1 times, and the combination of both variants increased the risk by 2.9 times.

Rebbeck et al. provided independent confirmation of the association study segment of the original report by Tavtigian et al.[55,68] Amongst 266 controls and 266 matched controls in a population-based study, they reported that the 217Leu/541Thr variants increased the risk of prostate cancer by 2.37 times. These variants were estimated to cause 5% of prostate cancer in the general population.

The initial association results, however, were not confirmed in a report by Xu et al.[56] This family-based and population-based association study compared the frequency of the two common ELAC2 missense variants among 159 probands from HPC families, 249 sporadic prostate cancer cases, and 222 unaffected male control subjects, and found no evidence to support a significant role of these variants in HPC or sporadic prostate cancer. Specifically, family-based tests did not reveal excess transmission of the Leu217 and/or Thr541 alleles to affected offspring, and population-based tests failed to reveal any statistically significant difference in the allele frequencies of the two polymorphisms between patients with prostate cancer and control subjects.

An additional association study reported by Suarez et al. was also not able to confirm the initial association findings.[57] This study included 257 prostate cancer cases randomly sampled from HPC families and 355 unrelated controls. Although they reported no significant difference in the frequency of the Ser217Leu alleles between cases and controls, the frequency of the Thr allele of the Ala541Thr variant was significantly greater among HPC cases. If the 541Thr variant substantially increased prostate cancer risk, the authors reasoned that increased clustering of this variant would be observed among HPC families; yet this phenomenon was not observed. Overall, Suarez et al. found little evidence to support a major role of the ELAC2 common missense variants in HPC.[57]

Wang et al. reported findings similar to Suarez et al.[57,69] Among 446 patients with a family history of prostate cancer and 502 population-based controls, they found no association between prostate cancer and either the Ser217Leu or Ala541Thr variant. This report concludes that ELAC2 is not implicated in a significant proportion of HPC cases and the previously described variants do not increase the risk of HPC.[69]

To investigate the use of ELAC2 variants as an adjunct to clinical PSA screening, Vesprini et al. compared the Ser217Leu and Ala541Thr genotypes of 944 men who underwent a prostate biopsy, versus a control population of 922 healthy, unselected women from the same population.[70] There were no significant differences in the prevalence of the Ala541Thr allele among men diagnosed with prostate cancer (6.3%), men with other prostate conditions (6.8%) and healthy women in the control group (6.3%). The authors report that the ELAC2 variants are not likely to increase the sensitivity of PSA screening among men with sporadic prostate cancer in the general population.

Although we might learn more about ELAC2 in the future, specifically regarding any influence on clinical characteristics of prostate cancer and/or any increased risk for other cancers, the role of ELAC2/HPC2 in prostate cancer susceptibility appears to be very limited. Alterations in this gene are unlikely to be a valuable indicator of prostate cancer risk in a clinical setting.

Androgen synthesis and metabolism pathways

As androgen ablation therapy has long been one of the most commonly employed treatments for prostate cancer, there has been a great deal of attention focused on the androgen pathway in regard to prostate cancer. However, the majority of androgen pathway studies have been focused on sporadic prostate cancer. Of the limited number of HPC focused studies, there has been variability in the results, perhaps due to the populations studied, statistical methods employed and the genetic variants included in the analyses. To date, at least four genes involved in the androgen pathway have received substantial attention regarding HPC risk, including the androgen receptor, CYP17, HSD3B1 and HSD3B2.

Androgen receptor

The androgen receptor (AR) gene has garnered particular interest as an HPC candidate gene. Although the AR gene is located outside of the HPCX linkage region defined by Xu et al. on the X chromosome,[46] this gene has been implicated in hereditary prostate cancer by two genomewide linkage studies that found significant LOD scores at the AR gene. This gene contains polymorphisms that result in variable AR activity. Specifically, there is a widely studied polymorphic triplet repeat in exon 1 of the AR gene that codes for polyglutamine (CAG) repeats of varying lengths ranging from 11 to 31 residues.[71–74]

A number of studies have documented a correlation between shorter AR glutamine repeats and increased prostate cancer risk.[75–77] Functional studies have shown that the length of the glutamine repeat is inversely related to the ability of the androgen receptor to stimulate androgen-specific transcriptional activity.[78–80] This is of particular interest because the African American population is observed to have the shortest average glutamine repeat length, and also the highest incidence and mortality rates reported for prostate cancer. In contrast, the Asian population has the lowest risk for prostate cancer and also the longest average glutamine repeat lengths.[71,72]

Exon 1 of the AR gene contains an additional polymorphic trinucleotide repeat that codes for a polyglycine (GGC) tract that is typically 10–22 residues in length. The glycine repeats have not been studied as extensively as the glutamine repeats; however, studies that include both of these repeats have detected stronger associations with increased prostate cancer risk. Stanford et al. reported that men with short repeat lengths for both polymorphisms (CAG <22 and GGC ≤16) had a twofold increase in risk compared to men with two long repeats. Platz et al. and Irvine et al. have also described increased prostate cancer risk associated with the GGC repeat.[72,81]

Although several studies have suggested that shorter CAG and/or GGC repeat lengths may increase the risk of developing more aggressive prostate cancer,[75,82] there have been an equal number of studies that found no evidence of an association. Even the supportive association studies have, for the most part, detected only marginally significant evidence for an association. However, it is important to recognize that very few association studies of the CAG repeats, and none of the studies of GGC repeats, have included HPC probands. This is problematic if inherited variation in AR repeat length is more important than sporadic changes.

In an attempt to address the inconsistent findings reported thus far, a study by Chang et al. employed both linkage and association approaches to examine CAG and GGC repeats among HPC families, sporadic prostate cancer cases and controls.[83] This study found that the CAG repeats contributed very little to the evidence for linkage and was not significantly associated with prostate cancer risk. Significant evidence for linkage to AR was found among the HPC families with early onset disease and among families with short GGC repeats. In addition, significantly increased frequencies of the ≤16 GGC repeat alleles in 159 independent hereditary cases (71%) and 245 sporadic cases (68%), compared with 211 controls (59%), suggested that GGC repeats are associated with prostate cancer ($P = 0.02$). Likewise, a significant association was detected between the short GGC repeat (≥16 repeats) and prostate cancer risk, with increases by factors of 1.69 and 1.51 among men with HPC and sporadic prostate cancer, respectively. This study was remarkable because it suggested that the conflicting association reports from CAG studies might reflect proximity to the GGC repeats. The consistent results from both linkage and association studies strongly implicate the GGC repeats in the AR as a prostate cancer susceptibility gene. Further study will be necessary to confirm these recent findings, and ultimately to establish the overall role of these AR polymorphisms in determining or modifying prostate cancer risk.

CYP17

The cytochrome P450c17α (CYP17) gene is one of several genes in the androgen metabolism pathway that have been evaluated as a candidate gene for prostate cancer. It catalyzes several key reactions for both sex steroid and cortisol biosynthesis. Picado-Leonard and Miller mapped the CYP17 gene to chromosomal region 10q24.3,[84] and this is near an HPC linkage interval at 10q25 that was defined by the genomewide screen of Gibbs et al.[27] Because mutations in the CYP17 gene can result in disrupted testosterone synthesis, many population-based association studies have been conducted to investigate the possible effects of this polymorphism on the risk of hormone-related cancers, including prostate cancer. However, studies of CYP17 have thus far provided very mixed results. A single T to C transition in CYP17 has been subject to the most detailed study.

While Lunn et al. and Gsur et al. both reported that the C allele increases the risk of prostate cancer,[85,86] studies by Wadelius et al. and Habuchi et al. found that the T allele is the prostate cancer risk allele.[87,88] To clarify the debate, the unique study design of Chang et al. employed multiple analytical approaches within a single report that examined HPC probands, sporadic prostate cancer cases and unaffected controls.[89] This paper found no evidence of an association between the C/T variant and prostate cancer risk, but was the first to report evidence of HPC linkage to the CYP17 gene. By combined use of a genetic linkage study, family-based association study and population-based association study, the authors suggested that the variable findings from previous reports of standard association studies probably reflect the linkage evidence, suggesting another mutation(s) in CYP17 are likely to be more important with regard to prostate cancer.

HSD3B1 and HSD3B2 genes

The enzyme 3β-hydroxysteroid dehydrogenase (HSD3B) is another critical component of the androgen metabolism pathway that has been evaluated with regard to prostate cancer and HPC. The *HSD3B* gene family has two genes that map to chromosome 1p13 (17–19), titled *HSD3B1* and *HSD3B2*. The study of chromosome 1 linkage by Xu et al. expanded the original HPC1 region to include 1p13 and, therefore, significantly increased the likelihood that *HSD3B* genes play an important role in HPC susceptibility.[36]

There are very few studies on the relationship between *HSD3B2* sequence variants and prostate cancer. A complex $(TG)_n(TA)_n(CA)_n$ repeat has been described and studied in intron 3 of *HSD3B2*.[90,91] However, there are no published reports that describe evaluation of the associations between this repeat and other sequence variants in *HSD3B1* and HPC risk. To evaluate the possible role of *HSD3B* genes in prostate cancer, Chang et al. screened a panel of DNA samples, collected from 96 men with or without prostate cancer, for sequence variants throughout the *HSD3B1* and *HSD3B2* genes by direct sequencing.[92] Eleven SNPs were identified, four of which, including a missense change (*B1-N367T*), were informative. These four SNPs were further genotyped in a total of 159 HPC probands, 245 sporadic prostate cancer cases and 222 unaffected controls. Although a weak association between prostate cancer risk and a missense SNP (*B1-N367T*) was found, stronger evidence for association was found when the joint effect of the two genes was considered. Men with the variant genotypes at either *B1-N367T* or *B2-c7519g* had a significantly higher risk to develop prostate cancer (adjusted risk increase = 1.6), and especially HPC (adjusted risk increase = 2.17). Most importantly, the subset of HPC probands whose families provided evidence for linkage at 1p13 predominantly contributed to the observed evidence for association. Similar to the other recent gene findings relating to HPC, these findings from the HSD3B gene studies may potentially explain a reasonable portion of HPC risk if confirmed in additional data sets.

Challenges to association studies of prostate cancer

Although association studies have been performed successfully in many simple Mendelian diseases, this approach encounters some difficulty in complex diseases. As discussed above in relation to linkage analysis, one of the major problems is heterogeneity. Association studies are not only susceptible to etiological (genetic or environmental), inheritance (dominant, recessive or X-linked) and locus heterogeneity; they are also very sensitive to allelic (same gene but different mutations) and founder (same mutation exists in different genetic background) heterogeneity. Despite these difficulties, a large number of association studies have been reported.

To date, the bulk of reported association studies have focused on a straightforward design that compares genotype frequency among cases versus controls. However, it is unclear whether the traditional association approaches are likely to reveal hereditary prostate cancer genes. Rather, this approach may be more likely to reveal genes involved in prostate tumor progression. With regard to hereditary prostate cancer and the assessment of risk among unaffected men who may simply be presymptomatic, the most promising association studies focus on genes located within linkage regions, and include probands (the first individual ascertained) from HPC families, along with non-HPC cases and controls in the analyses.

CURRENT CLINICAL APPLICATION . . . AND THE POTENTIAL FOR SO MUCH MORE

As the search for HPC genes has transitioned from linkage analyses to gene studies, thousands of genes and genetic variants have been studied. There are many published reports that describe modest increases in prostate cancer risk associated with specific variants. However, the identification of one or more major prostate cancer genes remains elusive. This has led to considerable speculation, including development of the idea that multiple gene combinations influence prostate cancer risk. For example, one can envision a system in which 4–10 genes each carry 'minimal' risk and then produce a synergistic risk elevation in specific combinations, and perhaps these risks are additionally inflated by specific environmental triggers, such as sunlight exposure or caloric intake. While there is most certainly some degree of genetic heterogeneity, the fact remains that segregation analyses have provided strong evidence for one or two major effect genes. It is important to recognize that, in the presence of biological redundancy, the major gene model and the multiple gene model are not mutually exclusive, and both may ultimately prove important in prostate cancer etiology. Thus the search for genes of significant clinical value continues.

Most recently, advances in molecular biology, laboratory automation and analytical capabilities have accelerated the search for hereditary prostate cancer genes. The completion of the human genome draft sequence now fuels high-throughput laboratories with information to streamline genetic research. Amidst the backdrop of 'genomic' science, it is important to review the goals and potential rewards of such an effort, and to differentiate promising science clearly from clinical applications. We must ask 'How does an understanding of the genetics of prostate cancer help people and families with prostate cancer and/or hereditary prostate cancer?' – 'Now, in the near future, or in the distant future?'.

The only well-established clinical product of the efforts to understand HPC is the use of risk assessment screening. Within risk assessment, the factors most consistently shown to increase risk are early age of onset, race/ethnicity and number of affected family members. A query of these factors allows identification of individuals at highest risk of prostate cancer, and the subsequent offer of more aggressive PSA and DRE screening. Although most patients or physicians can recognize inheritance patterns of traits and diseases within a family, intentional dialogue and well-founded information are prerequisite to the effective use of family history risk assessment. While both the American Cancer Society (ACS) and the American Urological Association (AUA) recommend beginning prostate screening earlier in men with a family history, we do not currently know the degree to which family risk assessment is consistently, accurately and appropriately implemented in clinical settings, and particularly in primary care practices. Furthermore, attempts to elucidate prostate cancer family history may be particularly limited among practitioners who adhere to the guidelines of both the American College of Physicians and/or the United States Preventative Services Task Force, which recommend against the use of PSA as a screening method. Just as the implementation of screening, and the funding for research, lag behind the health care burden of prostate cancer relative to other cancers, it is reasonable to think that use of family history screening for prostate cancer may also lag behind the documented utility of such an approach.

When major susceptibility genes, and/or specific combinations of minor effect genes, are in fact identified and characterized, it will be possible to offer gene testing to high-risk individuals among HPC families, in addition to risk assessment and genetic counseling. This will also allow for an individualized measure of prostate cancer risk, following a model of genetic assessment similar to that currently used in hereditary breast cancer care. Such a measure will allow individuals at highest risk to pursue more frequent screening and more aggressive preventative measures, while also helping to ease the concerns of individuals who are at low risk and do not carry an HPC gene. This means the ultimate in patient specific preventative medicine for prostate cancer is on the near horizon, while the more distant horizon may hold dramatic changes in the way we treat prostate cancer by including molecular medicine and pharmacogenetics.

Viewed within the big picture of society, the success of this search for HPC genes is very important because it is expected that the aging US population will present an increased burden on the health care system in coming years, and Medicare is expected to be bankrupt by the year 2025. Thus, identification of individuals who are truly at the highest risk of HPC offers the opportunity to allocate limited medical resources most appropriately, and offers the promise of earlier detection at lower overall fiscal cost. Such an approach most certainly offers the hope of lower overall human cost and, therefore, better lives for many. So, although the search for HPC genes has yet to significantly impact the clinical care of men and families with prostate cancer, there is a great deal to hope for.

REFERENCES

1. Gronberg H, Damber L, Damber JE. Studies of genetic factors in prostate cancer in a twin population. *J. Urol.* 1994; 152:1484–7.
2. Page WF, Braun MM, Partin AW et al. Heredity and prostate cancer: a study of World War II veteran twins. *Prostate* 1997; 33:240–5.
3. Lichtenstein P, Holm NV, Verkasalo PK et al. Environmental and heritable factors in the causation of cancer – analyses of cohorts of twins from Sweden, Denmark, and Finland. *N. Engl. J. Med.* 2000; 343:78–85.
4. Morganti G, Gianferrari L, Cresseri A et al. Recherches clinico-stastistiques et genetiques sur les neoplasies de la prostate. *Acta Genet.* 1956; 6:304–5.
5. Woolf CM. An investigation of familial aspects of carcinoma of the prostate. *Cancer* 1960; 13:739–44.
6. Cannon L, Bishop DT, Skolnick M et al. Genetic epidemiology of prostate cancer in the Utah Mormon genealogy. *Cancer Surveys* 1982; 1:47–69.
7. Meikle AW, Stanish WM. Familial prostatic cancer risk and low testosterone. *J. Clin. Endocrinol. Metab.* 1982; 54:1104–08.
8. Steinberg GD, Carter BS, Beaty TH et al. Family history and the risk of prostate cancer. *Prostate* 1990; 17:337–47.
9. Lesko SM, Rosenberg L, Shapiro S. Family history and prostate cancer risk. *Am. J. Epidemiol.* 1996; 144:1041–7.
10. Ghadirian P, Howe GR, Hislop TG et al. Family history of prostate cancer: a multi-center case-control study in Canada. *Int. J. Cancer* 1997; 70:679–81.
11. Carter BS, Bova GS, Beaty TH et al. Hereditary prostate cancer: epidemiologic and clinical features. *J. Urol.* 1993; 150:797–802.
12. Whittemore AS, Wu AH, Kolonel LN et al. Family history and prostate cancer risk in black, white, and Asian men in the United States and Canada. *Am. J. Epidemiol.* 1995; 141:732–40.
13. Monroe KR, Yu MC, Kolonel LN et al. Evidence of an X-linked or recessive genetic component to prostate cancer risk [see comments]. *Nature Med.* 1995; 1:827–9.
14. Hayes RB, Liff JM, Pottern LM et al. Prostate cancer risk in U.S. blacks and whites with a family history of cancer. *Int. J. Cancer* 1995; 60:361–4.
15. Goldgar DE, Easton DF, Cannon-Albright LA et al. Systematic population-based assessment of cancer risk in first-degree relatives of cancer probands. *J. Natl Cancer Inst.* 1994; 2; 86:1600–8.
16. Gronberg H, Damber L, Damber JE. Familial prostate cancer in Sweden. A nationwide register cohort study. *Cancer* 1996; 77:138–43.
17. Schuurman AG, Zeegers MP, Goldbohm RA et al. A case-cohort study on prostate cancer risk in relation to family history of prostate cancer. *Epidemiology* 1999; 10:192–5.
18. Cerhan JR, Parker AS, Putnam SD et al. Family history and prostate cancer risk in a population-based cohort of Iowa men. *Cancer Epidemiol. Biomarkers Prev.* 1999; 8:53–60.
19. Carter BS, Beaty TH, Steinberg GD et al. Mendelian inheritance of familial prostate cancer. *Proc. Natl Acad. Sci. USA* 1992; 89:3367–71.
20. Schaid DJ, McDonnell SK, Blute ML et al. Evidence for autosomal dominant inheritance of prostate cancer. *Am. J. Human Genet.* 1998; 62:1425–38.
21. Gronberg H, Damber L, Damber JE et al. Segregation analysis of prostate cancer in Sweden: support for dominant inheritance. *Am. J. Epidemiol.* 1997; 146:552–7.
22. Cui J, Staples MP, Hopper JL et al. Segregation analyses of 1,476 popultaion-based Australian families affected by prostate cancer. *Am. J. Human Genet.* 2001; 68:1207–18.
23. Smith JR, Freije D, Carpten JD et al. Major susceptibility locus for prostate cancer on chromosome 1 suggested by a genome-wide search [see comments]. *Science* 1996; 274:1371–4.
24. Suarez BK, Lin J, Burmester JK et al. A genome screen of multiplex sibships with prostate cancer. *Am. J. Human Genet.* 2000; 66:933–44.
25. Goddard KA, Witte JS, Suarez BK et al. Model-free linkage analysis with covariates confirms linkage of prostate cancer to chromosomes 1 and 4. *Am. J. Human Genet.* 2001; 68:1197–206.
26. Berthon P, Valeri A, Cohen-Akenine A et al. Predisposing gene for early-onset prostate cancer, localized on chromosome 1q42.2–43. *Am. J. Human Genet.* 1998; 62:1416–24.
27. Gibbs M, Stanford JL, Jarvik GP et al. A genomic scan of families with prostate cancer identifies multiple regions of interest. *Am. J. Human Genet.* 2000; 67:100–9.
28. Hsieh Cl, Oakley-Girvan I, Balise RR et al. A genome screen of families with multiple cases of prostate cancer: evidence of genetic heterogeneity. *Am. J. Human Genet.* 2001; 69:148–58.
29. McIndoe RA, Stanford JL, Gibbs M et al. Linkage analysis of 49 high-risk families does not support a common familial prostate cancer-susceptibility gene at 1q24–25. *Am. J. Human Genet.* 1997; 61:347–53.
30. Goode EL, Stanford JL, Chakrabarti L et al. Linkage analysis of 150 high-risk prostate cancer families at 1q24–25. *Genet. Epidemiol.* 2000; 18:251–75.
31. Eeles RA, Durocher F, Edwards S et al. Linkage analysis of chromosome 1q markers in 136 prostate cancer families. The Cancer Research Campaign/British Prostate Group U.K. Familial Prostate Cancer Study Collaborators. *Am. J. Human Genet.* 1998; 62:653–8.
32. Suarez BK, Lin J, Witte JS et al. Replication linkage study for prostate cancer susceptibility genes. *Prostate* 2000; 45:106–14.
33. Cooney KA, McCarthy JD, Lange E et al. Prostate cancer susceptibility locus on chromosome 1q: a confirmatory study [see comments]. *J. Natl Cancer Inst.* 1997; 89:955–9.
34. Berry R, Schaid DJ, Smith JR et al. Linkage analyses at the chromosome 1 loci 1q24–25 (HPC1), 1q42.2–43 (PCAP), and 1p36 (CAPB) in families with hereditary. *Am. J. Human Genet.* 2000; 66:539–46.
35. Hsieh CL, Oakley-Girvan I, Gallagher RP et al. Re: prostate cancer susceptibility locus on chromosome 1q: a confirmatory study [letter; comment]. *J. Natl Cancer Inst.* 1997; 89:1893–4.
36. Xu J, Zheng SL, Chang B et al. Linkage of prostate cancer susceptibility loci to chromosome 1. *Human Genet.* 2001; 108:335–45.
37. Neuhausen SL, Farnham JM, Kort E et al. Prostate cancer susceptibility locus HPC1 in Utah high-risk pedigrees. *Human Mol. Genet.* 1999; 8:2437–42.
38. Goode EL, Stanford JL, Peters MA et al. Clinical characteristics of prostate cancer in an analysis of linkage to four putative susceptibility loci. *Clin. Cancer Res.* 2001; 7:2739–49.
39. Xu J, and International Consortium for Prostate Cancer Genetics (ICPCG). Combined analysis of hereditary prostate cancer linkage to 1q24–25: results from 772 hereditary prostate cancer families from the international consortium for prostate cancer genetics. *Am. J. Human Genet.* 2000; 66:945–57.
40. Cancel-Tassin G, Latil A, Valeri A et al. PCAP is the major known prostate cancer predisposing locus in families from south and west Europe. *Eur. J. Human Genet.* 2001; 9:135–42.
41. Gibbs M, Chakrabarti L, Stanford JL et al. Analysis of chromosome 1q42.2–43 in 152 families with high risk of prostate cancer. *Am. J. Human Genet.* 1999; 64:1087–95.

42. Whittemore AS, Lin IG, Oakley-Girvan I et al. No evidence of linkage for chromosome 1q42.2–43 in prostate cancer. *Am. J. Human Genet.* 1999; 65:254–6.
43. Isaacs SD, Kiemeney LA, Baffoe-Bonnie A et al. Risk of cancer in relatives of prostate cancer probands. *J. Natl Cancer Inst.* 1995; 87:991–6.
44. Gibbs M, Stanford JL, McIndoe RA et al. Evidence for a rare prostate cancer-susceptibility locus at chromosome 1p36. *Am. J. Human Genet.* 1999; 64:776–87. .
45. Badzioch M, Eeles R, Leblanc G et al. Suggestive evidence for a site specific prostate cancer gene on chromosome 1p36. The CRC/BPG UK Familial Prostate Cancer Study Coordinators and Collaborators. The EU Biomed Collaborators. *J. Med. Genet.* 2000; 37:947–9.
46. Xu J, Meyers D, Freije D et al. Evidence for a prostate cancer susceptibility locus on the X chromosome. *Nature Genet.* 1998; 20:175–9.
47. Narod SA, Dupont A, Cusan L et al. The impact of family history on early detection of prostate cancer [letter]. *Nature Med.* 1995; 1:99–101.
48. Lange EM, Chen H, Brierley K et al. Linkage analysis of 153 prostate cancer families over a 30-cM region containing the putative susceptibility locus HPCX. *Clin. Cancer Res.* 1999; 5:4013–20.
49. Peters MA, Jarvik GP, Janer M et al. Genetic linkage analysis of prostate cancer families to Xq27–28. *Human Hered.* 2001; 51:107–13.
50. Bergthorsson JT, Johannesdottir G, Arason A et al. Analysis of HPC1, HPCX, and PCaP in Icelandic hereditary prostate cancer. *Human Genet.* 2000; 107:372–5.
51. Berry R, Schroeder JJ, French AJ et al. Evidence for a prostate cancer-susceptibility locus on chromosome 20. *Am. J. Human Genet.* 2000; 67:82–91.
52. Cancel-Tassin G, Latil A, Valeri A et al. No evidence of linkage to HPC20 on chromosome 20q13 in hereditary prostate cancer. *Int. J. Cancer* 2001; 93:455–6.
53. Bock CH, Cunningham JM, McDonnell SK et al. Analysis of the prostate cancer-susceptibility locus HPC20 in 172 families affected by prostate cancer. *Am. J. Human Genet.* 2001; 68:795–801.
54. Zheng SL, Xu J, Chang B et al. Evaluation of linkage of HPC20 in 159 hereditary prostate cancer pedigrees. *Human Genet.* 2001; 108:430–5.
55. Tavtigian SV, Simard J, Teng DH et al. A candidate prostate cancer susceptibility gene at chromosome 17p. *Nature Genet.* 2001; 27:172–80.
56. Xu J, Zheng SL, Carpten JD et al. Evaluation of linkage and association of HPC2/ELAC2 in familial and unrelated prostate cancer patients. *Am. J. Human Genet.* 2001; 68:901–11.
57. Suarez BK, Gerhard DS, Lin J et al. Polymorphisms in the prostate cancer susceptibility gene HPC2/ELAC2 in multiplex families and healthy controls. *Cancer Res.* 2001; 61:4982–4.
58. Xu J, Zheng SL, Chang B et al. Linkage and association studies of prostate cancer susceptibility gene on 8p22–23. *Am. J. Human Genet.* 2001; 69:341–50.
59. Macoska JA, Trybus TM, Benson PD et al. Evidence for three tumor suppressor gene loci on chromosome 8p in human prostate cancer. *Cancer Res.* 1995; 15:5390–5.
60. Bova GS, Carter BS, Bussemakers MJ et al. Homozygous deletion and frequent allelic loss of chromosome 8p22 loci in human prostate cancer. *Cancer Res.* 1993; 53:3869–73.
61. Bova GS, Isaacs WB. Review of allelic loss and gain in prostate cancer. *World J. Urol.* 1996; 14:338–46.
62. MacGrogan D, Levy A, Bova GS et al. Structure and methylation-associated silencing of a gene within a homozygously deleted region of human chromosome band 8p22. *Genomics* 1996; 35:55–65.
63. Vocke CD, Pozzatti RO, Bostwick DG et al. Analysis of 99 microdissected prostate carcinomas reveals a high frequency of allelic loss on chromosome 8p12–21. *Cancer Res.* 1996; 56:2411–16.
64. Deubler DA, Williams BJ, Zhu XL et al. Allelic loss detected on chromosomes 8, 10, and 17 by fluorescence in situ hybridization using single-copy P1 probes on isolated nuclei from paraffin-embedded prostate tumors. *Am. J. Pathol.* 1997; 150:841–50.
65. Prasad MA, Trybus TM, Wojno KJ et al. Homozygous and frequent deletion of proximal 8p sequences in human prostate cancers: identification of a potential tumor suppressor gene site. *Genes Chromosomes Cancer* 1998; 23:255–62.
66. Oba K, Matsuyama H, Yoshihiro S et al. Two putative tumor suppressor genes on chromosome arm 8p may play different roles in prostate cancer. *Cancer Genet. Cytogenet.* 2001; 124:20–6.
67. Ostrander EA, Stanford JL. Genetics of prostate cancer: too many loci, too few genes. *Am. J. Human Genet.* 2000; 67:1367–75.
68. Rebbeck TR, Walker AH, Zeigler-Johnson C et al. Association of HPC2/ELAC2 genotypes and prostate cancer. *Am. J. Human Genet.* 2000; 67:1014–19.
69. Wang L, McDonnell SK, Elkins DA et al. Role of HPC2/ELAC2 in hereditary prostate cancer. *Cancer Res.* 2001; 61:6494–9.
70. Vesprini D, Nam RK, Trachtenberg J et al. HPC2 variants and screen-detected prostate cancer. *Am. J. Human Genet.* 2001; 68:912–17.
71. Edwards A, Hammond HA, Jin L et al. Genetic variation at five trimeric and tetrameric tandem repeat loci in four human population groups. *Genomics* 1992; 12:241–53.
72. Irvine RA, Yu MC, Ross RK et al. The CAG and GGC microsatellites of the androgen receptor gene are in linkage disequilibrium in men with prostate cancer. *Cancer Res.* 1995; 55:1937–40.
73. Macke JP, Hu N, Hu S et al. Sequence variation in the androgen receptor gene is not a common determinant of male sexual orientation. *Am. J. Human Genet.* 1993; 53:844–52.
74. Sleddens HF, Oostra BA, Brinkmann AO et al. Trinucleotide (GGN) repeat polymorphism in the human androgen receptor (AR) gene. *Human Mol. Genet.* 1993; 2:493.
75. Giovannucci E, Stampfer MJ, Krithivas K et al. The CAG repeat within the androgen receptor gene and its relationship to prostate cancer [published erratum appears in *Proc. Natl Acad. Sci. USA* 1997; 94:8272]. *Proc. Natl Acad. Sci. USA* 1997; 94:3320–3.
76. Kantoff P, Giovannucci E, Brown M. The androgen receptor CAG repeat polymorphism and its relationship to prostate cancer. *Biochim. Biophys. Acta* 1998; 1378:C1–C5.
77. Stanford JL, Just JJ, Gibbs M et al. Polymorphic repeats in the androgen receptor gene: molecular markers of prostate cancer risk. *Cancer Res.* 1997; 57:1194–8.
78. Chamberlain NL, Driver ED, Miesfeld RL. The length and location of CAG trinucleotide repeats in the androgen receptor N-terminal domain affect transactivation function. *Nucleic Acids Res.* 1994; 22:3181–86.
79. Kazemi-Esfarjani P, Trifiro MA, Pinsky L. Evidence for a repressive function of the long polyglutamine tract in the human androgen receptor: possible pathogenetic relevance for the (CAG)n-expanded neuronopathies. *Human Mol. Genet.* 1995; 4:523–7.
80. Sobue G, Doyu M, Morishima T et al. Aberrant androgen action and increased size of tandem CAG repeat in androgen receptor gene in X-linked recessive bulbospinal neuronopathy. *J. Neurol. Sci.* 1994; 121:167–71.
81. Platz EA, Giovannucci E, Dahl DM et al. The androgen receptor gene GGN microsatellite and prostate cancer risk. *Cancer Epidemiol. Biomarkers Prev.* 1998; 7:379–84.
82. Hakimi JM, Rondinelli RH, Schoenberg MP et al. Androgen-receptor gene structure and function in prostate cancer. *World J. Urol.* 1996; 14:329–37.
83. Chang B, Zheng SL, Hawkins GA et al. Polymorphic GGC repeats in the androgen receptor gene are associated with

hereditary and sporadic prostate cancer risk. *Human Genet.* 2002; 110:122–9.
84. Picado-Leonard J, Miller WL. Cloning and sequence of the human gene for P450c17 (steroid 17 alpha-hydroxylase/17,20 lyase): similarity with the gene for P450c21. *DNA* 1987; 6:439–48.
85. Lunn RM, Bell DA, Mohler JL et al. Prostate cancer risk and polymorphism in 17 hydroxylase (CYP17) and steroid reductase (SRD5A2). *Carcinogenesis* 1999; 20:1727–31.
86. Gsur A, Bernhofer G, Hinteregger S et al. A polymorphism in the CYP17 gene is associated with prostate cancer risk. *Int. J. Cancer* 2000; 87:434–7.
87. Wadelius M, Andersson AO, Johansson JE et al. Prostate cancer associated with CYP17 genotype. *Pharmacogenetics* 1999; 9:635–9.
88. Habuchi T, Liqing Z, Suzuki T et al. Increased risk of prostate cancer and benign prostatic hyperplasia associated with a CYP17 gene polymorphism with a gene dosage effect. *Cancer Res.* 2000; 60:5710–13.
89. Chang B, Zheng SL, Isaacs SD et al. Linkage and association of CYP17 gene in hereditary and sporadic prostate cancer. *Int. J. Cancer* 2001; 95:354–9.
90. Verreault H, Dufort I, Simard J et al. Dinucleotide repeat polymorphisms in the HSD3B2 gene. *Human Mol. Genet.* 1994; 3:384.
91. Devgan SA, Henderson BE, Yu MC et al. Genetic variation of 3 beta-hydroxysteroid dehydrogenase type II in three racial/ethnic groups: implications for prostate cancer risk. *Prostate* 1997; 33:9–12.
92. Chang B, Zheng SL, Hawkins GA et al. Joint effect of *HSD3B1* and *HSD3B2* genes is associated with hereditary and sporadic prostate cancer susceptibility. *Cancer Res.* 2002; 62:1784–9.

Breast and Prostate Cancer: a Comparison of Two Common Endocrinologic Malignancies

Paul Matthew Yonover
Department of Urology, Loyola University Medical Center, Maywood, IL, USA

Ellen Gaynor
Department of Medicine, Loyola University Medical School and The Cardinal Bernadin Cancer Center, Maywood, IL, USA

Steven C. Campbell*
Department of Urology, Loyola University Medical School and The Cardinal Bernadin Cancer Center, Maywood, IL, USA

INTRODUCTION

A comparison between prostate cancer in men and breast cancer in women reveals intriguing similarities. Together, they represent a major burden on the health care system and have captured the attention of clinicians and basic scientist alike. New insights into their overlapping behavior and biology will allow for novel treatments and more effective management strategies. Prevention and early detection can play a crucial role in both of these cancers, which can be greatly enhanced through a better understanding of the disease processes. Progress in clinical and scientific research in breast cancer has arguably preceded that in prostate cancer, reflecting traditional funding priorities that have only recently begun to change. Urologists and their basic scientist colleagues can thus look to the breast cancer field for clues about promising lines of investigation that may translate to prostate cancer, thereby facilitating progress against this all too often lethal malignancy.

An analysis of the similarities and differences between breast and prostate cancers provides a fascinating picture of the natural history, biological basis, and clinical features of these two common endocrinologic cancers (**Table 17.1**).

EPIDEMIOLOGY

There are striking similarities between the epidemiology of prostate and breast cancer. With an estimated 193 000 and 198 000 newly diagnosed cases of breast and prostate cancer, respectively, in 2001 in the USA alone, these neoplasms represent the most commonly diagnosed cancer in each sex.[1] A nearly equivalent rate of cancer-related deaths exists for each cancer in the USA, with 40 000 deaths estimated for breast cancer and 31 000 deaths for prostate cancer in 2001. For men 60 years or older, prostate cancer is the second most common cause of cancer-related mortality, while breast cancer is the leading cause of cancer-related mortality for women aged 40–59. Overall, the mortality rates from both of these cancers have been decreasing – prostate cancer mortality rates decreased an average of 4.4% annually from 1994 through 1997, and breast cancer mortality rates decreased an average of 2.2% annually between 1990 and 1997.[1]

When race and ethnicity are considered, interesting differences emerge. In breast cancer, incidence is highest in Caucasians and lowest in American Indians, but death rates are highest in African Americans. In prostate cancer, incidence and mortality has traditionally been highest for African

*To whom correspondence should be addressed.

Table 17.1. Breast and prostate cancer: similarities and differences.

	Breast cancer		**Prostate cancer**
Epidemiology	193 000 cases per year Most common cancer in women More common in Caucasians 40 000 cancer-related deaths per year Mortality rate declining 2.2% per year		198 000 cases per year Most common cancer in men More common in African Americans 31 000 cancer-related deaths per year Mortality rate declining 4.4% per year
Risk factors	Prolonged exposure to estrogen	Advancing age Family history Nutritional: dietary fat	Prolonged exposure to androgens
Screening	Survival benefit proven		Survival advantage suggested but not proven
Prevention	Retinoids may benefit premenopausal patients Tamoxifen	Dietary: low-fat diet	Finasteride, vitamin E, selenium, lycopenes, soy products all under study
Genetics		5–10% familial, autosomal dominant Crossover suggested by many studies Families with breast cancer with increased incidence prostate cancer, and vice versa	
	BRCA1 and 2 increase risks of breast and ovarian cancer		BRCA1 and 2 or closely linked genes increase risk of prostate cancer
Adjuvant therapies	Proven role for adjuvant therapies Tamoxifen for all with ER+ status Chemotherapy for most patients Radiation therapy for patients at high risk of local recurrence		Adjuvant therapies remain controversial Anti-androgen monotherapy under study Adjuvant chemotherapy under study Adjuvant vs. salvage radiation therapy – an ongoing controversy
Metastatic profile		Bone metastasis common	
	Osteoclastic pattern (typically)		Osteoblastic pattern
		Beneficial role for bisphosphanates suggested for both cancers	
Molecular features		Vitamin D_3 pathways contribute to tumor biology Apoptotic pathways important Role for Bcl-2 and NFκB	

Americans. Mortality rates are lowest for Asians but increase after relocation into Western cultures for both cancers, which suggests that nutritional or other environmental factors may play an important role for both breast and prostate cancer.[1–3]

RISK FACTORS

For breast cancer, the major risk factor is prolonged and unopposed exposure to estrogen. Well-established risk factors thus include early onset of menarche (before age 14), nulliparity, completion of first live birth after age 30, and delayed menopause.[2] The incidence of breast cancer increases with age, doubling with every decade until menopause. Smoking, either directly or through second-hand exposure, has also been implicated as a possible risk factor.[2,4,5] Women with a family history of breast cancer and those who carry germline mutations of BRCA1 or BRCA2 are also at high risk for developing breast cancer. For instance, women with one first-degree relative with a history of breast cancer have a 1.5–2.0 relative risk of developing the disease, but the risk is even further increased (relative risk approximately 20.0) if the first-degree relative had bilateral premenopausal onset of the cancer. Relative risk has been estimated at 4.0–6.0 if two first-degree relatives have had breast cancer. In addition, women carrying germline mutations for BRCA1 or BRCA2 have a 56–85% lifetime risk of developing breast cancer, much higher than the general population.[6–8]

For prostate cancer, the most important risk factor is also prolonged exposure to steroid hormones, i.e. androgens. Early castration or genetic defects blocking androgen synthesis essentially eliminate the risk of prostate cancer, while more subtle abnormalities, such as qualitative differences in the levels of circulating androgens or qualitative differences in androgen receptor function or related signal transduction pathways due to various genetic polymorphisms may also alter the risk of prostate cancer in certain subpopulations.[9] For example, genetic variations of 5α-reducatase may create more productive enzymatic conversion of testosterone to dihydrotestosterone (DHT), which has been suggested as one mechanism of increased prostate cancer risk.[10] Studies have looked at the potential role of vitamin D receptor (VDR) polymorphisms in prostate cancer risk, particularly in African American men.[11] It has also been noted that Black men of college age have 13% higher serum free testosterone levels than White men of the same age.[12] In addition, advancing age, race (African American) and positive family history also confer an increased risk for the development of prostate cancer.[13] A single first-degree relative with prostate cancer increases the relative risk by a factor of 2.1–2.8, and having a first-degree and a second-degree relative with prostate cancer may increase the relative risk by as much as 4–6-fold when compared to the general population.[14] The corresponding relative risks may be even higher in African Americans – in this population a history of two or more first-degree relatives with prostate cancer may increase the relative risk up to ten fold.[15]

The role of diet in cancer risk has been under intense scrutiny, especially given the strong association between high dietary fat intake and prostate cancer risk. Although the exact mechanism is unclear, high levels of fat intake may increase the bioavailability of steroid sex hormones, perhaps by interfering with their ability to bind to serum globulins.[16] Conversely, vegetarian diets have been shown to lead to lower levels of circulating androgens and estrogens. Data also demonstrate a strong correlation between dietary fat and significantly different incidence rates of prostate cancer in Western vs. Eastern cultures.[9] Several highly suggestive studies in animal and *in vitro* model systems also support an important role for dietary fat in the promotion and progression of prostate cancer. Dietary nutrients may also have a function in the development of prostate cancer; however, this is difficult to elucidate and is currently under study.

While body mass and various occupational exposures are currently being studied as putative risk factors, other risk factors, such as socioeconomic status, smoking and history of prostatitis or sexually transmitted diseases, probably play little if any role in prostate cancer development.[9,13] Recent evidence has also failed to support vasectomy as a significant risk factor for prostate cancer, and any association between the two may simply be related to health care-seeking behavior.

SCREENING AND PREVENTION

Survival rates for breast cancer and prostate cancer are significantly improved when the disease is discovered in an early, localized stage, and this has prompted screening efforts for both of these cancers. For breast cancer, recent data suggest that screening should begin early, at age 40, with a yearly mammogram and physical examination.[17] In addition, women are encouraged to begin monthly self-examination at age 20 and a clinical breast examination is recommended every 3 years between the ages of 20 and 39. Women with a strong family history of breast cancer, particularly early onset or bilateral disease, should begin formal screening with mammography and physical examination earlier in life, although the exact age for this has been controversial. Policy for screening for prostate cancer remains a topic of debate, but most urologists have followed the recommendations of the American Urologic Association, screening with digital rectal examination (DRE) and serum PSA beginning at age 50, unless there is a family history of prostate cancer, African American race or prior vasectomy, in which screening would begin at age 40.

The data in favor of screening for breast cancer are certainly more direct and mature than that for prostate cancer. As mentioned earlier, screening mammography has been shown to improve survival for women over age 50, and more recent studies suggest that women between 40 and 49 years of age will also benefit.[17] It is interesting to note that the studies of screening mammography for breast cancer started back in the 1960s, almost 30 years before any truly analogous studies were initiated for prostate cancer. A survival advantage for prostate screening has never been demonstrated in a prospective, randomized manner, but the declining mortality rates for prostate cancer observed over the last few years are thought to be due to the widespread use of PSA testing beginning in the early 1990s. Several other lines of evidence argue in favor of screening for prostate cancer, but as with many important issues in the field, irrefutable evidence has not yet been obtained.

Clearly, prevention of disease is more desirable than simply early detection, and both of these cancers have been targeted for cancer prevention efforts. For breast cancer, retinoids have been investigated with experimental evidence suggesting that they may prevent breast cancer. When almost 3000 women with breast cancer were randomized to fenretinoid vs. control, the incidence of contralateral breast cancer was not significantly altered. However, this study suggested a possible benefit for premenopausal patients [adjusted hazard ratio 0.66, 95% confidence interval (CI) 0.14–1.07].[2] Tamoxifen has also been studied extensively to assess its ability to prevent breast cancer, initially prompted by data showing a reduced incidence of contralateral breast cancer in patients enrolled in adjuvant trials of this agent. Several prevention trials with tamoxifen have been reported, unfortunately yielding conflicting results. The National Surgical Adjuvant Breast and Bowel Project P-1 (NSABP) study randomized 3338 women with a moderate risk for developing breast cancer and demonstrated a 47% reduction in the incidence of invasive breast cancer in women receiving tamoxifen when compared to controls.[18] However, two other large prevention studies failed to confirm these findings, and one must keep in mind that chronic use of tamoxifen is not without morbidity, as it increases the risk of endometrial cancer by 2.5-fold and can predispose patients to deep venous thrombosis, pulmonary embolism and cerebral vascular accident.[8,19,20]

For prostate cancer, current studies are evaluating the role of micronutrients and antioxidants, such as vitamin E, selenium, lycopenes and soy products.[9] In addition, the prostate cancer prevention trial testing finasteride versus placebo is now maturing with data expected in the next few years.

GENETICS

Overall, hereditary cases are thought to account for only about 5–10% of prostate cancer cases, with the overwhelming majority believed to be sporadic in occurrence.[9,21,22] Interestingly, familial breast cancer is also thought to represent a minority of cases, again approximately 5–10%, mirroring prostate cancer.[2] When prostate and breast cancer occur in a familial pattern, they are both characterized by: (1) early onset of disease (age <50 years for breast cancer and <55 years for prostate cancer); (2) multiple affected family members; and (3) autosomal dominant inheritance with high penetrance. In the case of familial breast cancer, the disease is also frequently bilateral.

The parallels do not end with their patterns of inheritance; perhaps just as important from a clinical standpoint is the potential crossover between the two cancers. Studies of families with breast cancer reveal an increased risk of prostate cancer in male relatives, a relationship that was first recognized by Thiessen in 1974.[23] In the Utah Mormon database the relative risk of developing prostate cancer if there was a family history of breast cancer was 2.21, and Tulinius independently estimated this relative risk at 1.50.[24] Similarly, many studies indicate that a family history of prostate cancer can significantly increase the risk of breast cancer in female relatives. The exact degree by which this risk is increased, however, is still in dispute. Up to 11% of women with breast cancer have a first-degree relative with prostate cancer, and Anderson and colleagues report a relative risk of 4.7 (95% CI 3.1–6.9; no P value given) of breast cancer risk in families with a history of prostate cancer.[25,26] A recent case control cohort study also suggested an increased risk of breast cancer in families with a history of prostate cancer, although statistical significance was not reached.[27] In this study, the risk of breast, ovarian and prostate cancer was evaluated in families of men with prostate cancer. When prostate and breast cancers only were considered, there was a significantly increased risk of these tumors in relatives of men with prostate cancer compared to controls, with an odds ratio of 2.35 (95% CI 1.43–3.85, $P < 0.001$). For female relatives of patients with prostate cancer, the odds ratio of developing breast cancer was 1.51, but the sample size was relatively small ($n = 209$), and statistical significance was not achieved (95% CI 0.87–2.6). The data thus can be considered suggestive but not conclusive.[27] Most such studies in the literature are marred by methodological defects, and some have focused on sporadic cancer rather than the familial cases that are more likely to yield informative data. Nevertheless, several authors have suggested that a family history of breast cancer should be taken into consideration when screening for prostate cancer and vice versa, but the data in support of this appear to be preliminary and suggestive at best, and further studies will be required to resolve these important issues. Current recommendations for screening for prostate cancer from the American Urologic Association do *not* take into account a family history of breast cancer, and the analogous situation is true for breast cancer screening policies.

This, however, may change in the future, as the molecular genetics of breast cancer and prostate cancer come into better focus. A complete review of the genetics of prostate and breast cancer are beyond the scope of this chapter; however, there are several interesting areas of overlap that warrant mentioning. For example, frequent chromosomal deletions at 7q31 and 16q22 are common to both sporadic prostate and breast tumors, suggesting that these regions may be involved in tumorigenesis, and ongoing studies are attempting to identify and characterize potential tumor suppressor genes in these and other regions.[28,29]

Perhaps a more important link may exist between these cancers and ovarian cancer via the BRCA1 and BRCA2 genes.[7,30,31] Mutation or inactivation of either of these genes predisposes women to breast and ovarian cancers, and most current evidence also points to an increased risk of prostate cancer in male carriers. BRCA1, located on chromosome 17q, encodes for a protein of 1863 amino acids, while BRCA2, located on chromosome 13q, is approximately twice the size of BRCA1.[8,30] BRCA1 is normally located in the nucleus, where it is thought to regulate gene transcription. One recent study suggests that BRCA1 functions to repress estrogen receptor-α transcriptional activity by binding to and neutralizing the activation domain of the receptor.[32] Other data suggest that BRCA1 may regulate cellular proliferation and/or the response to DNA damage, and it is likely that BRCA1 may have a multitude of functions that could impact on the tumor biology of breast cancer cells.[33] BRCA2 is homologous to BRCA1 and there appears to be significant overlap in the expression and function of the two genes. Both genes confer an increased risk of breast cancer of approximately 80% by the age of 80 years, although penetrance for BRCA1 starts to increase slightly earlier than for BRCA2.[6–8] Both genes also confer an increased risk of ovarian cancer, with a lifetime risk of 60% for BRCA1 carriers and 27% for BRCA2 carriers. Overall, about 5–10% of breast cancer is thought to be familial, and mutations of BRCA1 and BRCA2 are thought to contribute to approximately 50% of these cases.

An increased incidence of prostate cancer has also been reported in families with BRCA mutations, with several studies supporting this association. Ford and colleagues estimate a 3.3-fold increased relative risk of prostate cancer in male carriers of mutations of the BRCA1 gene, and other studies suggest a 3–7-fold increased relative risk for male carriers of mutations of the BRCA2 gene.[6] In a population study of Ashkenazi Jews that more commonly carry mutations of these genes, two specific mutations of BRCA1 were defined that conferred an approximately four fold increased lifetime risk for prostate cancer, with early onset being the norm.[7] Other studies have shown that families that carry mutations of

BRCA1 and BRAC2 are more likely to have a first-degree relative with prostate cancer when compared to controls.[34] Park and colleagues have recently shown that BRCA1 may modulate androgen receptor signaling, although the functional implications of this have not yet been defined.[33]

In another fascinating study showing a potential link between the endocrinologic cancers, 143 families with three or more cases of ovarian cancer were analyzed. Male relatives were found to have a 4.5-fold increased risk for prostate cancer, and female relatives a 2.5-fold higher risk for breast cancer and a five fold increased risk for uterine cancer.[35] The role of BRCA1 or BRCA2 was not defined, but most likely contributed in at least some of these cases.

More recent studies have called into question the role of BRCA mutations in the etiology of prostate cancer. For instance, Vazina and colleagues found that the incidence of BRCA1 or BRCA2 mutations was relatively low in patients with sporadic prostate cancer and not significantly different from the general population.[36] In addition, they also found that mutational inactivation of BRCA1 was rare in patients with familial prostate cancer. Similarly, Nastiuk and colleagues reported that mutations of the BRCA genes were rare in Ashkenazi Jewish prostate cancer patients, many with a family history of prostate cancer.[37] One possible explanation of these somewhat divergent findings is to postulate that there are other genes closely linked to the BRCA genes on chromosomes 13 and 17 that alter the risk for development of prostate cancer. The increased incidence of prostate cancer observed in many families with BRCA abnormalities may be due to coinheritance of these closely linked genes.[38] Further studies will be required to clarify these issues.

Until this has been resolved, a prudent policy would be to offer families with early onset and clustering of endocrinologic cancers of the prostate, breast and/or ovary screening for each of these cancers in addition to molecular genetic analysis. Of course, this should be pursued within the context of careful counseling regarding the currently available database, allowing for informed decision-making. While such screening efforts may eventually save lives, many families may elect against a proactive approach due to concerns about the related costs and anxiety that is often associated with such surveillance protocols. In addition, there are also very real concerns about discrimination by health and life insurance providers, should a well-defined cancer diathesis be documented.

ADJUVANT THERAPIES

Over the past several years, the role of adjuvant therapies have become well defined for breast cancer patients, contributing to the declining mortality rates for this cancer. Generally accepted prognostic factors after surgery for breast cancer include patient age, tumor size, lymph node status, grade, mitotic rate and hormonal receptor status. As summarized by a recent NIH Consensus Statement, adjuvant hormonal therapy should be offered to all women whose tumors stain positive for expression of the estrogen receptor.[39] A 5-year treatment course with tamoxifen is considered standard for adjuvant therapy for this group of women, although ovarian ablation can be considered for selected premenopausal patients. Adjuvant multiagent chemotherapy is also recommended for the majority of women with localized breast cancer independent of nodal, menopausal or hormone receptor status, since it has also been shown to improve survival in this setting. Recent data suggest that incorporation of the anthracyclines enhances outcomes with adjuvant chemotherapy for this malignancy; an analogous role for the taxanes has not been determined. Finally, there is also fairly strong evidence that women with a high risk of locoregional recurrence after mastectomy may benefit from adjuvant radiotherapy. This high-risk group includes women with four or more positive lymph nodes or a locally advanced primary cancer. The role of adjuvant radiation therapy for women with 1–3 positive lymph nodes has not yet been defined. Overall, the role of adjuvant therapies for women with breast cancer is now well defined, with a beneficial effect now demonstrated in several subpopulations that in total represent a substantial majority of women with the disease.[39]

Adjuvant treatments for prostate cancer are much more controversial, reflecting a lack of randomized prospective data, and again indicative of the contrast between the two malignancies, with prostate cancer unfortunately lagging behind in many respects. Adjuvant radiotherapy for men with high-risk features, such as positive margins, remains highly controversial, with many centers opting for observation rather than active treatment. A substantial proportion of such patients may remain disease-free even without adjuvant treatment, and our ability to salvage a significant number of men who exhibit a rising prostate specific antigen (PSA) level after surgery through the timely administration of salvage radiation therapy have encouraged this more conservative approach.[40] Still, there are several studies suggesting an advantage to radiation therapy in an adjuvant setting for high-risk patients and this more proactive approach should be considered in younger patients with good functional status after surgery. Similarly, a role for early administration of chemotherapy for patients at high risk for systemic recurrence (patients with a high Gleason score, seminal vesical involvement, etc.) remains uncertain, and is now being tested in several randomized prospective studies.

Finally, the role of adjuvant hormonal blockade also remains controversial, although the recent Eastern Cooperative Oncology Group (ECOG) study from Messing and colleagues did show a survival benefit for early androgen deprivation therapy for men with positive lymph nodes after radical prostatectomy.[41] One interesting trend has been the recent use of anti-androgen monotherapy as an adjuvant treatment in patients with a wide variety of disease stage and treatment status, analogous to the use of tamoxifen for breast cancer patients. Early results suggest a delay in clinical progression, but current studies are not yet mature enough to evaluate the effect on survival.[42,43] As with tamoxifen, anti-androgen monotherapy is not without potential morbidity, with breast tenderness and/or enlargement reported in over

50% of the patients enrolled in the bicalutamide study (150 mg per day).

METASTATIC PROFILE: OSSEOUS METASTASES

It has long been recognized that both prostate and breast cancer belong to a well-defined group of tumors that readily metastasize to bone. One autopsy series reported that 64% and 66% of patients with metastatic breast and prostate cancer, respectively, had metastases to the bone.[44] There are important differences, however, most notably being the pattern of bone metastasis seen in each of these cancers, i.e. typically osteoclastic in breast cancer and osteoblastic in prostate cancer. In an effort to further characterize this phenomenon, researchers have identified several osteoblast-stimulating factors produced by prostate cancer cells, such as bone morphogenic proteins, osteoblast mitogenic factor, prostatic osteoblastic factor and bone phosphatase-elevating factor. Other studies have suggested that insulin-like growth factor binding proteins sequestered in the bone matrix, which normally inhibit the osteoblastic activity of insulin-like growth factors, may undergo PSA-dependent proteolysis, thereby resulting in the osteoblastic phenotype often observed in metastatic prostate cancer.[45] Other evidence suggests that osteoclastic reactions similar to those seen in breast cancer may also be involved in prostate cancer, at least initially. Furthermore, just as clinical studies have demonstrated elevated parathyroid hormone-related protein levels in metastatic breast cancer, similar proteins have also been detected in patients with prostate cancer.[44,46,47] It is interesting to note that recent data suggest that patients with osseous metastases from both of these cancers may benefit from treatment with bisphosphanates, although these agents are thought to act primarily by inhibiting osteoclastic processes.[48]

MOLECULAR FEATURES

Breast and prostate cancer also share in common other molecular features that may contribute to their tumor biology. Recent studies have focused on the role of another steroid hormone in prostate and breast cancer, namely 1,25-dihydroxyvitamin D_3 ($1\alpha,25(OH)_2D_3$).[49] This vitamin D receptor ligand has the ability to inhibit cellular proliferation and induce differentiation in various cell lines, such as MCF-7 breast cancer cells and LNCaP prostate cancer cells. However, hormone-independent MDA-MA-436 breast cancer cells and DU-145 prostate cancer cells are relatively resistant to the antineoplastic effects of $1\alpha,25(OH)_2D_3$, despite expression of a functionally active vitamin D receptor. Campbell and colleagues have shown that the antiproliferative effects of $1\alpha,25(OH)_2D_3$ are mediated, at least in part, by induction of BRCA1 gene expression in both breast and prostate cancer cells. In their study, sensitivity to the antiproliferative effects of $1\alpha,25(OH)_2D_3$ was strongly correlated with the ability to modulate BRCA1 protein expression.

The apoptosis pathway has also been an active area of research in both breast and prostate cancer. For example, in breast cancer upregulation of the proto-oncogene Bcl-2, which blocks apoptosis, is commonly found in estrogen receptor positive tumors and may be controlled by estrogen, thereby promoting cancer cell survival.[50] Experimentally, breast cancer cell lines pretreated with estradiol are protected from apoptosis. Likewise, many advanced or hormone refractory prostate cancers also overexpress Bcl-2.[51] Modulation of Bcl-2 with antisense oligonucleotides *in vitro* can enhance apoptotic responses and may improve results with cytotoxic treatments for these and other cancers.[52–54]

Resistance to apoptosis has also been associated with activation of NFκB in both breast and prostate cancer. NFκB is a transcription factor that regulates cell proliferation, differentiation and apoptosis. NFκB has been shown to repress transcription of the androgen receptor and to inhibit its function, and an inverse relationship has also been reported between estrogen receptor levels and the function of NFκB. Constitutively activated NFκB has been reported in both hormone-independent PC-3 and DU-145 human prostate cancer cell lines and in several estrogen receptor negative human breast cancer cells lines.[55,56] Activated NFκB may represent a mechanism for enhancing cell survival in these hormone-independent cells, preventing them from completing the apoptotic pathway. Modulation of NFκB with inhibitors sensitizes cancer cell lines to cytotoxic drugs and radiation therapy, and may improve response rates for patients with advanced breast or prostate cancer.

Other potential links between breast and prostate cancer include the expression of unique genes that are hormonally regulated including KLK-L1, a member of the human kallikrein gene family, and fibroblast growth factor-8 (FGF-8).[57,58] KLK-L1, also known as prostase, is a serine protease that maps to chromosome 19q in the same region as PSA and related genes. This gene is expressed by both breast and prostate cancer cell lines, in both cases in an androgen- and progestin-dependent manner. Its role in the tumor biology of these cancers remains undefined, although some have speculated that the kallikreins including PSA may modulate apoptotic pathways. FGF-8, a growth factor that can induce malignant transformation, is also expressed in both breast and prostate cancer in a hormone-dependent manner.

CONCLUSIONS

A striking number of similarities exist between the endocrinologic tumors of the breast and prostate, including clinical, epidemiological and molecular factors. The genetics of these two malignancies also share some interesting features, with the BRCA family of genes, or related and closely linked genes, potentially serving as a tangible link that may explain certain clinical associations between the cancers. Urologists and basic scientists with an interest in prostate cancer should be cognizant of the great advances that have been made in breast cancer research, as they may serve as a useful template to foster progress in the prostate cancer field.

REFERENCES

1. Greenlee RT, Hill-Harmon MB, Murray T et al. Cancer statistics 2001. *CA Cancer J. Clin.* 2001; 51:15–35.
2. McPherson K, Steel CM, Dixon JM. Breast cancer – epidemiology, risk factors, and genetics. *Br. Med. J.* 2000; 321:624–8.
3. Ziegler RG, Hoover RN, Pike MC et al. Migration patterns of breast cancer risk in Asian-American Women. *J. Natl Cancer Inst.* 1993; 85:1819–27.
4. Johnson K, Hu J, Mao Y. Passive and active smoking and breast cancer risk in Canada, 1994–97. *Cancer Causes Control* 2000; 11:211–21.
5. Working Party on Breast-Cancer Etiology for the Nordic Cancer Union. Breast cancer etiology. *Int. J. Cancer* (Suppl.) 1990; 5:22–39.
6. Ford D, Easton DF, Peto J. Estimates of the gene frequency of BRCA1 and its contribution to breast and ovarian cancer incidence. *Am. J. Human Genet.* 1995; 57:1457–62.
7. Struewing J, Hartage P, Wacholder S et al. The risk of cancer associated with specific mutations of BRCA1 and BRCA2 among Ashkenazi Jews. *N. Engl. J. Med.* 1997; 336:1401–8.
8. Eeles RA, Powles TJ. Chemoprevention options for BRCA1 and BRCA2 mutation carriers. *Clin. Oncol.* 2000; 18(21 Suppl.): 93S–99S.
9. Brawley OW, Knopf K, Thompson I. The epidemiology of prostate cancer part II: the risk factors. *Semin. Urol. Oncol.* 1998; 16:193–201.
10. Kantoff PW, Febbo PG, Giovannucci E et al. A polymorphism of the 5 alpha reductase gene and its association with prostate cancer: a case-controlled analysis. *Cancer Epidemiol. Biomarkers Prev.* 1997; 6:189–92.
11. Ingels SA, Coetzee GA, Ross RK et al. Association of prostate cancer with vitamin D receptor haplotypes in African-Americans. *Cancer Res.* 1998; 58:1620–3.
12. Ross RK, Bernstein L, Judd H et al. Serum testosterone levels in healthy black and white men. *J. Natl Cancer Inst.* 1986; 76:45–48.
13. Gallager RP, Fleshner N. Clinical basics: prostate cancer: individual risk factors. *Can. Med. Assoc. J.* 1998; 159:807–13.
14. Steinberg GS, Carter BS, Beaty TH et al. Family history and the risk of prostate cancer. *Prostate* 1990; 17:337–40.
15. Whittemore AS, Wu A, Kolonel LN et al. Family history and prostate cancer risk in black, white, and Asian men in the United States and Canada. *Am. J. Epidemiol.* 1995; 141:732–40.
16. Street C, Howell RJS, Perry L. Inhibition of binding of gonadal steroids to serum binding proteins by non-esterified fatty acids: the influence of chain length and degree of unsaturation. *Acta Endocrinol.* 1989; 120:175–9.
17. Paley P. Screening for the major malignancies affecting women: current guidelines. *Am. J. Obstet. Gynecol.* 2001; 84:1021–30.
18. Fisher B, Constantino JP, Wickerman DL et al. Tamoxifen for the prevention of breast cancer: report of the National Surgical Adjuvant Breast and Bowl Project P-1 study. *J. Natl Cancer Inst.* 1998; 90:1371.
19. Powles TJ, Eeles R, Ashley S et al. Interim analysis of the incident breast cancer in the Royal Marsden Hospital tamoxifen randomised chemoprevention trial. *Lancet* 1998; 362:98.
20. Veronesi U, Maisonneuve P, Costa A et al. Prevention of breast cancer with tamoxifen: preliminary findings from the Italian randomised trial among hysterectomised women. *Lancet* 1998; 362:93.
21. Carter BS, Beaty TH, Steinberg GD et al. Mendelian inheritance of familial prostate cancer. *Proc. Natl Acad. Sci. USA* 1992; 89:3367–71.
22. Bratt O. Hereditary prostate cancer. *Br. J. Urol.* 2000; 85:588–98.
23. Thiessen EU. Concerning a familial association between breast cancer and both prostatic and uterine malignancies. *Cancer* 1974; 34:1102–7.
24. Tulinius H, Egilsson V, Olafsdóttir BGH et al. Risk of prostate, ovarian, and endometrial cancer among relatives of women with breast cancer. *Br. Med. J.* 1992; 305:855–7.
25. Anderson DE, Badzioch MD. Familial breast cancer risks. *Cancer* 1993; 72:114–19.
26. Sellers TA, Potter JD, Rich SS et al. Familial clustering of breast and prostate cancers and risk of postmenopausal breast cancer. *J. Natl Cancer Inst.* 1994; 86:1860–5.
27. McCahy PJ, Harris CA, Neal DE. Breast and prostate cancer in the relatives of men with prostate cancer. *Br. J. Urol.* 1996; 78:552–6.
28. Huang H, Qian C, Jenkins RB et al. FISH mapping of YAC clones at human chromosomal band 7q31.2: identification of YACS spanning FRA7G within the common region of LOH in breast and prostate cancer. *Genes Chromosomes Cancer* 1998; 21:152–9.
29. Filippova GN, Lindblom A, Meincke LJ et al. A widely expressed transcription factor with multiple DNA sequence specificity, CTCF, is localized at chromosome segment 16q22.1 within one of the smallest regions of overlap for common deletions in breast and prostate cancers. *Genes Chromosomes Cancer* 1998; 22:26–36.
30. Rosen E, Fan S, Goldberg I. BRCA1 and prostate cancer. *Cancer Invest.* 2001; 19:396–412.
31. Gayther S, de Foy K, Harrington P et al. The frequency of germline mutations in the breast cancer predisposition genes BRCA1 and BRCA2 in familial prostate cancer. *Cancer Res.* 2000; 60:4513–18.
32. Fan S, Ma Y, Wang, C et al. Role of direct interaction in BRCCA1 inhibition of estrogen receptor activity. *Oncogene* 2001; 20:77–87.
33. Park H, Irvine R, Buchanan G et al. Breast cancer susceptibility gene 1 (BRCA1) is a coactivator of the androgen receptor. *Cancer Res.* 2000; 60:5946–9.
34. Warner E, Foulkes W, Goodwin P et al. Prevalence and penetrance of BRCA1 and BRCA2 gene mutations in unselected Ashkenazi Jewish women with breast cancer. *J. Natl Cancer Inst.* 1999; 91:1241–7.
35. Jishi MF, Itnyre JH, Oakley-Girvan IA et al. Risks of cancer in the Gilda Radner Familial Ovarian Cancer Registry. *Cancer* 1995; 76:1416–21.
36. Vazina A, Baniel J, Yaacobi Y et al. The rate of the founder Jewish mutations in BRCA1 and BRCA2 in prostate cancer patients in Israel. *Br. J. Cancer* 2000; 83:463–6.
37. Nastiuk K, Mansukhani N, Terry M et al. Common mutations in BRCA1 and BRCA2 do not contribute to early prostate cancer in Jewish men. *Prostate* 1999; 40:172–7.
38. Williams BJ, Jones E, Zhu X et al. Evidence for a tumor suppressor gene distal to BRCA1in prostate cancer. *J. Urol.* 1996; 155:720–5.
39. National Institutes of Health (NIH). Adjuvant therapy for breast cancer. *NIH Concensus Statement* 2000; 17:1–35.
40. Forman JD, Meetze K, Pontes E et al. Therapeutic irradiation for patients with an elevated post-prostatectomy prostate specific antigen level. *J. Urol.* 1997; 154:1436–9.
41. Messing EM, Manola J, Sarosdy M et al. Immediate hormonal therapy compared with observation after radical prostatectomy and pelvic lymphadenopathy in men with node-positive prostate cancer. *N. Engl. J. Med.* 1999; 341:1781–8.
42. Wirth M, Tyrrell C, Wallace M et al. Bicalutamide (Casodex) 150 mg as immediate therapy in patients with localized or locally advanced prostate cancer significantly reduces the risk of disease progression. *Urology* 2001; 58:146–51.
43. Kolvenbag G, Iversen P, Newling D. Antiandrogen monotherapy: a new form of treatment for patients with prostate cancer. *Urology* 2001; 58(Suppl. 2A):16–22.
44. Yoneda T. Cellular and molecular mechanisms of breast and prostate cancer metastasis to bone. *Eur. J. Cancer* 1998; 34:240–5.
45. Smith GL, Doherty AP, Mitchell H et al. Inverse relation between prostate-specific antigen and insulin-like growth factor-binding

protein 3 in bone metastases and serum of patients with prostate cancer. *Lancet* 1999; 354:2053–4.
46. Iwamura M, di Sainte Agnese Pa, Wu G et al. Immunohistochemical localization of parathyroid hormone-related protein in human prostate cancer. *Cancer Res*. 1992; 53:1724–6.
47. Guise TA, Yin JJ, Taylor SD et al. Evidence for a casual role of parathyroid hormone-related protein in the pathogenesis of human breast cancer-meditated osetolysis. *J. Clin. Invest*. 1996; 98:1544–9.
48. Smith MR, McGovern FJ, Zietman A et al. Pamidronate to prevent bone loss during androgen-deprivation therapy for prostate cancer. *N. Engl. J. Med*. 2001; 345:948–55.
49. Campbell MJ, Gombart AF, Kwok SH et al. The anti-proliferative effects of $1\alpha,25(OH)_2D_3$ on breast and prostate cancer cells are associated with induction of BRCA1 gene expression. *Oncogene* 2000; 19:5091–7.
50. Nass SJ, Davidson NE. Advances in breast cancer therapy. *Hematol. Oncol. Clinics North Am*. 1999; 13:311.
51. McDonnell TJ. Expression of bcl-2 oncoprotein and p53 protein accumulation in bone marrow metastases of androgen independent prostate cancer. *J. Urol*. 1997; 15:569–74.
52. Webb A, Cunningham D, Cotter F et al. Bcl-2 antisense therapy in patients with non-Hodgkin lymphoma. *Lancet* 1997; 349:1137–41.
53. Teixeira C, Reed JC, Pratt MAC. Estrogen promotes chemotherapeutic drug resistance by a mechanism involving bcl-2 proto-oncogene expression in human breast cancer cells. *Cancer Res*. 1995; 55:3209.
54. Piché A, Grim J, Rancourt C et al. Modulation of Bcl-2 protein levels by an intracellular anti-bcl-2 single-chain antibody increases drug-induced cytotoxicity in the breast cancer cell line MCF-7. *Cancer Res*. 1998; 58:2134–40.
55. Nakshatri H, Poornima BK, Martin DA et al. Constitutive activation of NK-kB during progression of breast cancer to hormone-independent growth. *Mol. Cell Biol*. 1997; 17:3629–9.
56. Palayoor ST, Youmell MY, Calderwood SK et al. Constitutive activation of IκB kinase and NFκB in prostate cancer cells is inhibited by ibuprofen. *Oncogene* 1999; 18:7389–94.
57. Yousef GM, Obienza CV, Luo LY et al. Prostase/KLK-L1 is a new member of the human kallikrein gene family, is expressed in prostate and breast tissues, and is hormonally regulated. *Cancer Res*. 1999; 59:4252–6.
58. Tanaka A, Furuya A, Yamasaki M et al. High frequency of fibroblast growth factor (FGF) 8 expression in clinical prostate cancers and breast tissues, immunohistochemically demonstrated by a newly established neutralizing monoclonal antibody against FGF8. *Cancer Res*. 1998; 58:2053–6.

Artificial Neural Networks for Predictive Modeling in Prostate Cancer

Eduard J. Gamito, E. David Crawford and Abelardo Errejon

Department of Urology, University of Colorado Health Science Center, Denver, Co, USA

INTRODUCTION

In medical decision-making, rarely does one factor, marker or test provide the sufficient predictive value to be a definitive predictor of an outcome of interest. For example, serum prostate specific antigen (PSA) level has been shown to be useful in identifying patients at risk for prostate cancer. However, this test is limited by appreciable false-positive and false-negative rates and, therefore, is not a definitive test for carcinoma of the prostate.[1–3] As a result, patients with elevated PSA levels may be subjected to unnecessary procedures and concomitant anxiety.[4] In order to improve the predictive value of a clinical parameter, it is often useful to analyze it in combination with other clinical variables. A number of multifactorial analysis methods are available to investigators and have been used to develop predictive models. The majority of predictive models developed for prostate cancer to date have been developed using traditional statistical techniques, such as multivariate logistic regression. Perhaps the best known example of such a model is represented in a set of tables produced by Partin and coworkers.[5] The Partin tables provide a prediction of pathologic stage based on clinical stage, PSA level and biopsy Gleason sum for men with clinically localized prostate cancer. Utilizing tools like the Partin tables, physicians may offer their patients a prognosis based on their predicted pathological stage. Other models have been developed to predict outcomes of varying treatments directly, also using traditional statistical techniques.[6–9]

In the past 10 years, a relatively new methodology – collectively called artificial neural networks – has emerged that holds promise for improving predictive models and multifactorial analysis. Artificial neural networks (ANNs) are a software construct loosely based on concepts in human synaptic physiology that fall under the broad rubric of artificial intelligence. Although they are greatly simplified computer-based versions of their biological counterpart, the brain, ANNs have demonstrated the ability to learn from experience and solve problems in various applications including pattern recognition, signal processing and predictive modeling in fields as varied as engineering, meteorology, finance and medicine. This chapter will describe the theory behind ANNs, how they function and how they have been used to develop predictive models for prostate cancer outcomes.

ARTIFICIAL NEURAL NETWORK THEORY AND FUNCTION

Research into the workings of the brain suggests that the brain learns and stores information by altering the pathways of electrical activity within its neuronal structures. Changes in the relative connection strengths between neurons and signal conduction thresholds within neurons provide a mechanism by which information can be processed and stored. With this idea in mind, computer scientists have developed software that simulates selected aspects of the form and function of the brain including the connectivity between neurons, the changeable connection strengths between neurons and the activation thresholds (fire, not fire) of individual neurons.

Many types of ANNs have been developed since their inception in the 1950s and continue to arise from ongoing artificial intelligence research. ANNs can be classified based on their architecture, information flow and learning algorithms. The most commonly used ANN class is the multilayer perceptron (MLP), which is schematically represented in **Fig. 18.1**. In MLPs and other types of ANNs, simple processing units

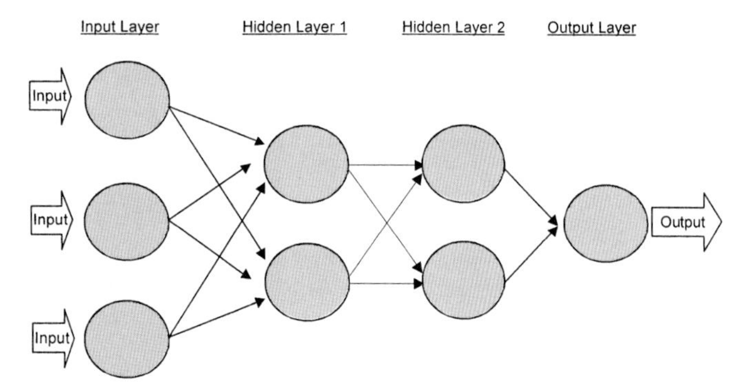

Fig. 18.1. Schematic representation of a generic multilayer perceptron.

called nodes (which simulate neurons) are linked via weighted interconnections. The interconnection weights function as multipliers that simulate the connection strengths between neurons. The nodes (represented by shaded circles in **Fig. 18.1**) are commonly arranged in three or more layers. An input layer accepts the values of the predictor variables presented to the network (e.g. clinical variables) while one or more output nodes represent(s) the predicted output(s), e.g. prediction of treatment outcome. One or more hidden layers of nodes link(s) the input and the output layers.

The processing of information within an ANN occurs at the hidden and output nodes, while the information learned by the ANN is stored in the weights. **Figure 18.2** is a schematic representation of a generic hidden node. As shown in this figure, values from the variables entering the input layer (X_1, X_2 and X_3) are multiplied by the weights (w_1, w_2 and w_3) associated with each interconnection. The weighted values from the inputs are summed along with a weighted bias value. This summed value is then subjected to a transfer function, usually a logistic function, to produce a scaled value (y). This value will, in turn, serve as an input to the next node layer.

An ANN model is typically developed in three phases, a design phase, a training phase and a validation phase. The design phase involves selection of an ANN architecture, which can vary in the number of nodes and the arrangement and number of node layers. Selection of the optimal architecture for

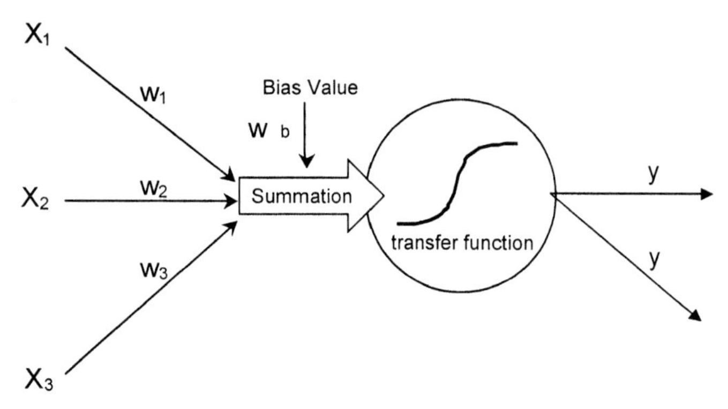

Fig. 18.2. Schematic representation of a generic hidden node.

a given problem can be somewhat subjective and is sometimes achieved through trial and error. More sophisticated methods are available, such as genetic algorithms, which use concepts in genetics and evolutionary theory to find the optimal ANN architecture.

An ANN will produce an output in a way analogous to a mathematical function even before training. The ANN inputs are analogous to independent variables and the output(s) analogous to dependent variable(s). At the beginning of a typical training regimen, the interconnection weights are small, randomly set values. Thus, initially, the output values produced by an ANN will be somewhat arbitrary and the overall error will be high. In a commonly employed method for training ANNs, called supervised learning, the actual output of the ANN is compared with the expected output. A training algorithm adjusts the weights based on the calculated error between the actual and expected outputs. Training algorithms often use a gradient descent method of adjusting the weights and minimizing error. A common algorithm is *error back-propagation*.[10,11] In the example of a feed-forward, error back-propagation ANN, information passes from the input nodes to the output node(s) (left to right in **Fig. 18.1**). As its name implies, the back-propagation algorithm calculates the error between the actual and expected output, and adjusts the weights back along the node layers (from right to left in **Fig. 18.1**). Thus, for example, the weights for the first hidden layer in **Fig. 18.1** are adjusted based on an error calculation from the second hidden layer.

During training, cases with known output values are presented to the ANN sequentially and repeatedly. Each complete cycle in which all the training cases are presented to the ANN is known as a training epoch or iteration. The training algorithm will adjust the weights incrementally and, over time, a matrix of weights emerges that produces outputs for the training set that approach the lowest global error. It is important to emphasize that the training algorithm produces a set of weights that 'fit' the *training* data, which does not necessarily mean that the ANN model will generalize to new data. It is possible for an ANN to 'over-fit' or 'memorize' the training set. This phenomenon, also known as overtraining, can limit the ability of an ANN model to generalize and, therefore, limit its usefulness.

A number of methods are available to help prevent overfitting, including a technique known as *early stopping*. In a variation of this method, a test set is randomly drawn from the general data set under study. The test set is used periodically to test the ANN model during training, but is not allowed to affect the weights. As the ANN begins to train, the overall error – measured by the mean squared error or root mean squared error – for the training set and the test set will begin to drop. At some point during training, the error for the test set will stop dropping and will begin to rise, while the error for the training set continues to drop. This indicates that the ANN is continuing to improve its fit for the training data, but has begun to lose its ability to generalize to the test data. **Figure 18.3** illustrates how the error for the training and testing sets can be used to select an early stopping point.

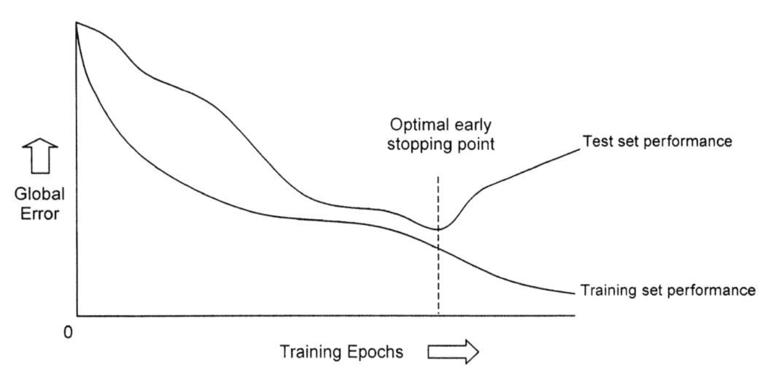

Fig. 18.3. A generic training curve showing the drop in error for both the training and testing sets as the number of training epochs increase. The dashed line indicates where the error in the testing set begins to increase and where training should be stopped to avoid overfitting.

MEASURES OF ANN PERFORMANCE

In order to provide meaningful and objective information about an ANN's performance, results on a validation set – and not the training set – should be reported. The validation data set(s) should not be used during training and, therefore, should not have had an influence on the ANN's weights. An ANN's performance can be quantified using measures such as sensitivity, specificity, positive predictive value (PPV) and negative predictive value (NPV). Receiver operating characteristic (ROC) curve analysis can provide an objective measure of a model's sensitivity and specificity over a range of output cutoffs.[12,13] Overall performance can be expressed as the area under the receiver operating characteristic curve (AUROC).[12] Also, the mean squared error (MSE) and root mean square error (RMSE) may be used. In the case of reporting sensitivity and specificity (also PPV and NPV), it is important that these values be 'matched', that is, both sensitivity and specificity for a given ANN output cut-off should be reported together. Reporting the maximum sensitivity and specificity over a range of cut-offs may be misleading.

POTENTIAL ADVANTAGES AND DISADVANTAGES OF ANNs

There are advantages as well as disadvantages to the use of ANNs in predictive modeling. A powerful advantage of ANNs over other techniques is the ability of ANNs to automatically resolve non-linear complex relationships between variables without the need for prior assumptions about the nature of these relationships. ANNs have demonstrated the ability to approximate arbitrarily complex functions or systems, regardless of the presence of multiple variables or the complexity of their interactions.[14–18] Another advantage to ANNs is that, because of the automated nature of most available software, they require little statistical training to operate them.

Despite these advantages, ANNs should be used with caution when developing predictive models that may ultimately impact prostate cancer management. A prime disadvantage to ANNs is that they are prone to overfitting. Also known as overtraining, this phenomenon can lead an investigator to misinterpret an ANN's good performance on a training data set. To avoid this pitfall, validation of the ANN model with a data set not used during training is essential. Techniques such as early stopping (described briefly above), cross-validation and bootstrapping can be used to lessen the likelihood that an ANN has been overtrained. Another disadvantage is the perception that ANNs produce models that are 'black boxes'. This is due, in part, to the difficulty encountered in trying to interpret the ANN weights. Although, the weights are roughly analogous to coefficients in logistic regression, they are much more difficult to decifer. For example, quantifying the importance of an independent variable, x, in relation to a dependent variable (output), y, is problematic with an ANN using currently available techniques. Extraction of rules based on ANN weights is also difficult. Some attempts have been made to develop special ANN designs for rule extraction, but are not widely used to date.[19–21]

APPLICATIONS OF ANNs IN PROSTATE CANCER PREDICTIVE MODELING

ANNs have been used in a variety of clinical applications including diagnosis, staging, image analysis, pathological specimen classification and treatment outcome prediction.[22–29] The majority of reports of the application of ANNs in prostate cancer are the diagnosis, prognosis and treatment of the disease. A brief review of selected applications of ANNs in prostate cancer follows.

ANNs IN THE DIAGNOSIS OF PROSTATE CANCER

Snow and coworkers are among the first to report the application of ANNs to predictive modeling in prostate cancer. In 1994, Snow investigated the use of an ANN to predict positive biopsy in 1787 men from a screening population with PSA levels greater than 4 ng/ml.[23] The inputs included age, PSA level, digital rectal examination (DRE) findings and transrectal ultrasound (TRUS) findings. They reported a sensitivity of 84% and a specificity of 88% for predicting biopsy result at an ANN output cut-off of 0.35. The model's AUROC was not reported. The authors concluded that ANNs may be useful in reducing the number of unnecessary biopsies and detection of clinically insignificant tumors, and thus reducing costs and morbidity. However, the ability of the model by Snow and coworkers to generalize to new patient data may suffer because, according to the report, the test set used to select the best weight matrix was also used to validate the model. In a subsequent study,[30] Snow and coworkers used data from another screening population to develop an ANN model for predicting positive biopsy in 1500 men who had an abnormal PSA or DRE examination. They reported a sensitivity of 72% and a specificity of 78%. Stamey and coworkers reported the results of a study[31] using a commercially available ANN known as ProstAsure (Horus Therapeutics, Inc., Savananah, GA, USA). This model was developed using the following variables from 416 men: age, PSA level, prostatic acid phosphatase and creatine kinase isoenzymes (CK-MM, CK-MB, and CK-BB). The model's sensitivity was reported to be 81% and the specificity was reported to be 92%. In a separate report, Babaian reported a statistically significant advantage for the ProstAsure® Index compared to free PSA in detecting prostate cancer.[24]

In 2000, Finne and colleagues compared the performance of an ANN model to a logistic regression model for predicting biopsy outcome.[32] Data from 656 men with PSA levels above 4.0 ng/ml were analyzed using both techniques. The variables used were PSA level, % free PSA, prostate volume, DRE findings and family history of prostate cancer. Finne reported that the ANN had a higher accuracy than the logistic regression model and had a sensitivity and specificity of 89% and 46%, respectively.

ANNs IN THE STAGING OF PROSTATE CANCER

Pathologic stage is an important predictor of prostate cancer outcome and several ANN models have been developed to predict pathologic stage from clinical variables. In 1998, Tewari and Narayan reported a pilot study using an ANN to develop a staging tool for clinically localized prostate cancer.[33] Data from 1200 men from four institutions were used to develop the model to predict three possible outcomes: positive surgical margins, seminal vesicle involvement and lymph node involvement. The input variables used included race, age, DRE findings, tumor size by ultrasound, serum PSA level, biopsy Gleason sum, perineural infiltration and biopsy staging findings, such as the number of positive cores. The resulting ANN model had a sensitivity of 81%, specificity of 75% and an AUROC of 0.79 for predicting positive margins. For predicting seminal vesicle involvement, the model produced a sensitivity of 100%, specificity of 72.1% and an AUROC of 0.80. The ANN's performance on predicting lymph node spread was 73% and 83% for sensitivity and specificity, respectively. The AUROC for predicting lymph node spread was 0.77. The authors concluded that the ANN model was accurate enough to miss less than 10% of patients with positive margins, less than 2% with positive lymph nodes, and none with seminal vesicle involvement. Further, they concluded that implementation of such a model could preclude unnecessary staging tests at a significant cost saving.

In 2000, Crawford and coworkers developed an ANN staging model to predict lymph node spread in men with clinically localized prostate cancer using data from 6454 patients from two institutions.[34] The inputs used were clinical stage, biopsy Gleason sum and PSA level. Two validation sets from separate institutions were used. The model's sensitivities on each of the validation sets were 64% and 44%, specificities were 82% and 81%, and the AUROCs were 0.81 and 0.77. The authors concluded that their results suggest a role for ANNs in the accurate staging of prostate cancer. Further work by this group includes a model to predict the risk of non-organ-confined prostate cancer in men who are clinically localized.[35] This work resulted in a model with a sensitivity of 72% and a specificity of 67%. The AUROC was 0.76. Another ANN model by this group was able to predict capsular penetration (CP) in 83% of those patients with CP and had a false-negative rate of 16%.[36] Several working ANN models published by Crawford and coworkers have been made available on the internet for use by patients and physicians (www.annsincap.org).

ANNs FOR PREDICTING PROSTATE CANCER RECURRENCE

Determination of whether or not cancer will recur or progress is an important goal in prostate cancer management. Snow and coworkers reported a model for predicting recurrence[23] using data from 240 men undergoing radical prostatectomy. The inputs to the ANN were age, PSA level, clinical stage, tumor grade, potency and race. The outcome of interest was cancer recurrence, characterized by biochemical (PSA) failure, local recurrence in the prostate bed or distant metastasis. The reported sensitivity was 67% and the specificity was 100% at an output cut-off of 0.5. Although four bootstrap validation and training sets were used, the validation sets were small (5% of the data base). In addition, the validation sets were used to pick the best weight matrices and, thus, the ability of this model to generalize to new patient data is questionable. Douglas and coworkers reported a sensitivity of 100% and specificity of 96% for predicting recurrence in radical prostatectomy patients using 40 clinical and pathological variables.[37] Mattfeldt and associates reported a feasibility study to predict cancer progression after radical prostatectomy.[38] Input variables from 40 patients included the histopathological variables: Gleason sum, World Health Organization (WHO) grade and maximum diameter of the tumor transects. Morphometric variables included volume and surface area of the epithelial tumor component and the surface area of the lumina of the neoplastic glands per unit tissue volume. An ANN using the three histopathological variables correctly predicted progression in 85% of validation cases. An ANN model using four morphometric variables correctly predicted progression in 93% of the cases.

In a study comparing ANN methods to logistic regression and Cox regression, Potter and colleagues developed models to predict progression in a selected group of patients ($n = 214$) at intermediate risk of progression after radical prostatectomy.[39] The input variables used included age, Gleason sum, extra prostatic extension, surgical margin status, quantitative nuclear grade and DNA ploidy. The ANN models outperformed the regression models with a maximum average AUROC of 0.74 ± 0.04 for the ANN method versus maximum average AUROC of $0.68 + 0.06$ for the logistic regression methods. Paired sensitivities and specificities were also significantly higher in the ANN models. The authors concluded, in part, that ANNs hold promise for predicting progression in select groups of patients undergoing radical prostatectomy and that the ANN models were superior to the logistic and Cox regression modeling for predicting progression in this study. In another study applying ANN modeling to a selected set of patients, Han and associates compared an ANN to logistic regression for predicting biochemical treatment failure in 564 men with Gleason sums of seven in a retrospective study.[40] In addition, Han and coworkers were interested in determining whether there was a difference in treatment outcome for patients undergoing radical prostatectomy with Gleason grades 3+4 versus those with grades 4+3. Cox proportional hazards and Kaplan–Meier analysis indicated that Gleason 7 status was an important predictor of biochemical failure with Gleason 4+3 patients fairing worse. The ANN performed better than logistic regression in this study for predicting biochemical recurrence at 3 and 5 years. The sensitivity and specificity at 3 years for the logistic regression model were 16% and 90%, respectively. The AUROC for the regression model was 0.68. The sensitivity and specificity for the ANN model were 37% and 90%, respectively, with an AUROC of 0.81.

ANNs FOR PREDICTING PROSTATE CANCER SURVIVAL

Predicting survival in prostate cancer patients is a difficult task for any modeling method. Traditional survival modeling, such as Kaplan–Meier analysis and Cox proportional hazards methods, are established but rely heavily on assumptions, especially with regard to censored patients. Although the ANN methodology is also limited by censored data, the ability of ANNs to analyze large numbers of variables that would be impractical for traditional methods may provide ANNs with a theoretical advantage. Research is ongoing to develop ANN methods to cope with censored data.[41–43]

A number of investigators have attempted using ANNs in non-prostate cancer survival analysis with promising results.[44–46] The use of ANNs in prostate cancer survival is relatively rare but promising results have been achieved in the field. A case in point is the work of Tewari and coworkers, who used predictive modeling techniques including a genetic adaptive ANN to predict 10-year overall survival in patients with localized prostate cancer.[47] The variables used in the study included age, PSA, race, biopsy grade, clinical stage, comorbidity (Charlson index) and socioeconomic status. The outcomes of interest were overall survival and cancer specific survival. The resulting model achieved a sensitivity of 80%, a specificity of 79% and an AUROC of 0.85 for predicting 10-year survival. The end result of this work was a set of comprehensive tables providing predictions of survival based on race, presence of comorbidities, age, cancer grade and PSA level. This study illustrates the potential for the use of ANNs in prostate cancer survival prediction. It is anticipated that the tables produced by Tewari et al. will garner a significant amount of attention from the prostate cancer research community as well as prostate cancer patients.

CONCLUSIONS

Artificial neural networks represent a methodology that can complement traditional statistical techniques for the development of predictive models in prostate cancer. A number of reports suggest that ANNs can perform on a par with various forms of traditional statistical methods, and in some cases, ANNs have been found to perform better than traditional techniques. Like all methods, ANNs have inherent advantages and disadvantages. An important advantage to ANNs is that they are capable of automatically resolving relationships between variables without the need for prior assumptions about the nature of these relationships. Prime disadvantages include the tendency of ANNs to overfit the data used to train them and their lack of transparency relative to other statistical methods. The last 10 years have seen an increasing number of applications of ANNs in medicine. The majority of studies done with ANNs in prostate cancer have been of an exploratory nature and many have yielded promising results. The ultimate role of ANNs in predictive modeling for prostate cancer has not yet been fully determined. A critical approach that combines ANNs in a complementary fashion with other techniques will most likely be the long-term role for ANNs.

REFERENCES

1. Gann PH, Hennekens CH, Stampfer MJ. A prospective evaluation of plasma prostate-specific antigen for detection of prostate cancer. *JAMA* 1995; 273:289–94.
2. Ruckle HC, Klee GG, Oesterling JE. Prostate-specific antigen. *Mayo Clinic Proc*. 1994; 69:59–68.
3. Smith DS, Catalona WJ. The nature of prostate cancer detected through prostate specific antigen based screening. *J. Urol.* 1994; 152:1732–6.
4. Zisman A, Liebovich D, Kleinman J et al. The impact of prostate biopsy on patient well-being: a prospective study of pain, anxiety and erectile function. *J. Urol*. 2001; 165:445–54.
5. Partin AW, Kattan MW, Subing ENP et al. Combination of prostate-specific antigen, clinical stage and Gleason sum to predict pathological stage of localized prostate cancer: a multi-institutional update. *JAMA* 1997; 277:1445–51.
6. Ross PL, Scardino PT, Kattan MW. A catalog of prostate cancer nomograms. *J. Urol*. 2001; 165:1562–8.
7. Kattan MW, Eastham JA, Stapleton AMF et al. A preoperative nomogram for disease recurrence following radical prostatectomy for prostate cancer. *J. Natl Cancer Inst*. 1998; 90:766–71.
8. Kattan MW, Wheeler TM, Scardino PT. A postoperative nomogram for disease recurrence following radical prostatectomy for prostate cancer. *J. Clin. Oncol*. 1999; 17:1499–507.
9. Kattan MW, Zelefsky MJ, Kupelian PA et al. Pretreatment nomogram for predicting the outcome of three-dimensional conformal radiotherapy in prostate cancer. *J. Clin. Oncol*. 2000; 18:3252–9.
10. Werbos PJ. Beyond regression: new tools for prediction and analysis in the behavioral sciences. Ph.D. thesis, Harvard University, 1974.
11. Rumelhart DE, McClelland JL. *Parallel Distributed Processing*, Vols 1 and 2. Cambridge, MA: MIT Press, 1986.
12. Hanley JA, McNeil BJ. The meaning and use of the area under a receiver operating characteristic (ROC) curve. *Radiology* 1982; 143:29–36.

13. Hanley JA, McNeil NJ. A method of comparing the areas under the receiver operating characteristic curves derived from the same cases. *Radiology* 1983; 148:839–43.
14. Chen T, Chen H. Universal approximation to nonlinear operators by neural networks with arbitrary activation functions and its application to dynamical systems. *Neural Networks* 1995; 6:911–17.
15. Hornik K, Stinchcomb M, White H. Multilayer feedforward networks are universal approximators. *Neural Networks* 1989; 2:359–66.
16. Park J, Sandberg IW. Universal approximation using radial-basis function networks. *Neural Computation* 1993; 3:246–57.
17. Dayhoff JE, DeLeo JM. Artificial neural networks: opening the black box, *Cancer* 2001; 91(S8):1615–35.
18. Dayhoff J. *Neural Networks Architectures: An Introduction*. Boston: International Thompson Computers Press, 1996.
19. McMillan C, Mozer MC, Smolensky P. Rule induction through integrated symbolic and subsymbolic processing. In: Moody JE, Hanson SJ, Lippman RP (eds) *Advances in Neural Information Processing Systems 4*, p. 969. San Mateo, CA: Kaufmann 1991.
20. Alexander JA, Mozer MC. Template-based procedures for neural network interpretation, *Neural Networks* 1999; 12:479–98.
21. Towell G, Shavlik JW. Interpretation of artificial neural networks: mapping knowledge-based neural networks into rules. In: Moody JE, Hanson SJ, Lippman RP (eds) *Advances in Neural Information Processing Systems 4*, p. 977. San Mateo, CA: Kaufmann, 1991.
22. Lacson RC, Ohno-Machado L. Major complications after angioplasty in patients with chronic renal failure: a comparison of predictive models. In: *Proceedings/AMIA Annual Symposium*, pp. 457–61. Bethesda, MD: American Medical Informatics Association. 2000.
23. Snow PB, Smith DS, Catalona WJ. Artificial neural networks in the diagnosis and prognosis of prostate cancer: a pilot study. *J. Urol.* 1994; 152:1923–6.
24. Babaian RJ, Fritsche HA, Zhang KH et al. Evaluation of a prostasure index in the detection of prostate cancer: a preliminary report. *Urology* 1998; 51:132–6.
25. Ronco AL, Fernandez R. Improving ultrasonographic diagnosis of prostate cancer with neural networks. *Ultrasound Med. Biol.* 1999; 25:729–33.
26. Loch T, Leuschner I, Genberg C et al. Artificial neural network analysis (ANNA) of prostatic transrectal ultrasound. *Prostate* 1999; 39:198–204.
27. Feleppa EJ, Fair WR, Liu T et al. Three-dimensional ultrasound analyses of the prostate. *Mol. Urol.* 2000; 4:133–9.
28. Boon M, Kok CP. Neural network processing can provide means to catch errors that slip through human screening of Pap smears. *Diagn. Cytopathol.* 1993; 9:411–16.
29. Baxt WG. Application of artificial neural networks to clinical medicine. *Lancet* 1995; 346:1135–8.
30. Snow P, Crawford ED, DeAntoni EP et al. Prostate cancer diagnosis from artificial neural networks using the Prostate Cancer Awareness Week (PCAW) database. *J. Urol.* 1997; 157 (Suppl.):365.
31. Stamey TA, Barnhill SD, Zhang Z et al. Effectiveness of ProstAsure in detecting prostate cancer and benign prostatic hyperplasia in men age 50 and older. *J. Urol.* 1996; 155(Suppl.):436A.
32. Finne P, Finne R, Auvinen A et al. Predicting the outcome of prostate biopsy in screen-positive men by multilayer perceptron network. *Urology* 2000; 56:418–22.
33. Tewari A, Narayan P. Novel staging tool for localized prostate cancer: a pilot study using genetic adaptive neural networks. *J. Urol.* 1998; 160:430–6.
34. Batuello JT, Gamito EJ, Crawford ED et al. Artificial neural network model for the assessment of lymph node spread in patients with clinically localized prostate cancer. *Urology* 2001; 57:481–5.
35. Crawford ED, Gamito EJ, O'Donnell C et al. Artificial neural network model to predict risk of non-organ-confined disease and risk of lymph node spread in men with clinically localized prostate cancer. *J. Urol.* 2001; 165(Suppl.):233.
36. Gamito EJ, Stone NN, Batuello JT et al. Use of artificial neural networks in the clinical staging of prostate cancer: implications for prostate brachytherapy. *Tech. Urol.* 2000; 6:60–3.
37. Douglas TH, Connelly RR, McLeod G et al. Neural network analysis of pre-operative and post-operative variables to predict pathologic stage and recurrence following radical prostatectomy. *J. Urol.* 1996; 155:487A.
38. Mattfeldt T, Kestler HA, Hautmann R et al. Prediction of prostatic cancer progression after radical prostatectomy using artificial neural networks: a feasibility study. *Br. J. Urol. Int.* 1999; 84:316–23.
39. Potter SR, Miller MC, Mangold LA et al. Genetically engineered neural networks for predicting prostate cancer progression after radical prostatectomy. *Urology* 1999; 54:791–5.
40. Han M, Snow PB, Epstein JI et al. A neural network predicts progression for men with gleason score 3+4 versus 4+3 tumors after radical prostatectomy. *Urology* 2000; 56:994–9.
41. Zupan B, Demsar J, Kattan MW et al. Machine learning for survival analysis: a case study on recurrence of prostate cancer. *Artif. Intell. Med.* 2000; 20:59–75.
42. Biganzoli E, Boracchi P, Mariani L et al. Feed forward neural networks for the analysis of censored survival data: a partial logistic regression approach. *Stats. Med.* 1998; 17:1169–86.
43. Anand SS, Hamilton PW, Hughes JG et al. On prognostic models, artificial intelligence and censored observations. *Meth. Information Med.* 2001; 40:18–24.
44. Bryce TJ, Dewhirst MW, Floyd CE Jr et al. Artificial neural network model of survival in patients treated with irradiation with and without concurrent chemotherapy for advanced carcinoma of the head and neck. *Int. J. Radiat. Oncol. Biol. Phys.* 1998; 41:339–45.
45. Harbeck N, Kates R, Ulm K et al. Neural network analysis of follow-up data in primary breast cancer. *Int. J. Biol. Markers* 2000; 15:116–22.
46. Lundin M, Lundin J, Burke HB et al. Artificial neural networks applied to survival prediction in breast cancer. *Oncology* 1999; 57:281–6.
47. Tewari A, Peabody J, Stricker H et al. Genetic adaptive network to predict long term survival in patients with clinically localized prostate cancer. *J. Urol.* 2001; 165(Suppl.):389.

Prevention of Prostate Cancer

Obesity, Aging and Immunity in Prostate Cancer*

Jack H. Mydlo

Department of Urology, Temple University School of Medicine, Philadelphia, PA, USA

INTRODUCTION

Obesity, the excess accumulation of adipose tissue, has reached epidemic proportions in the USA and the prevalence is increasing. Several large studies have demonstrated associations between obesity and high-fat diets and malignant and non-malignant diseases. Approximately 35% of all cancers may be caused by dietary factors and, therefore, may be preventable.

In Dr Schulman's chapter on diet and prostate cancer, much evidence has been demonstrated to associate diet and prostate cancer (see Chapter 20). In this chapter, we plan to analyze the association of obesity alone, separate from diet, and prostate cancer. We think it is important though that we show the major effects on obesity on other health-related issues, including cancers of other organ systems, before we focus on urological cancers, and specifically prostate cancer.

Diet and obesity are inseparably intertwined with socio-economic status, genetic susceptibility and environmental factors, all of which individually or in consort, can affect tumor biology. Adipose tissue may have a central role in carcinogenesis. It is a store for lipids that may alter immune competence by interfering with macrophage function and may serve as a source for mutagenic peroxidized lipids as well as gonadal steroid precursors, which have been implicated in several types of tumors. Furthermore, vasoactive peptides and growth factors have been identified in abundance in adipose tissue.

Just as increased energy content and fat content of the diet may influence adipose tissue accumulation and carcinogenesis, energy restriction and reduced fat intake leading to weight loss can reduce the risk and virulence of numerous cancers. In addition, vitamins and other naturally occurring substances are associated with reduced risk and may have protective effects. Thus, following dietary guidelines like those formulated by the US government should, over the next 20 years, lead to reductions in the incidence of new cancers and prolongation of survival of patients with established cancers.

OBESITY AND ITS COMORBIDITIES

In the USA, the majority of the top ten leading causes of death are associated with obesity. Cardiovascular diseases, cancer, stroke, respiratory failure and diabetes are all linked to obesity, making it, in aggregate, the leading preventable cause of death in the USA, remarkably exceeding smoking according to recent statistics. Obese patients have numerous other associated diseases or comorbidities, such as hypertension, thrombo-embolism, low back pain, osteoarthritis and depression, all of which contribute substantially to overall health care expenditures.[1] Even mild degrees of obesity are associated with amenorrhea or irregular menstrual periods,[2] sleep apnea and toxemia of pregnancy.

In the area of urology, there are several obesity-related diseases (**Table 19.1**). Stress urinary incontinence (SUI) in women, caused by increased intra-abdominal pressure, is associated with truncal or upper-body obesity.[3] Varicoceles are prevalent in obese men and may be a cause of infertility.[4] Obese men also have a higher incidence of impotence, which

Table 19.1. Obesity-related urologic diseases.

Benign prostatic hypertrophy
Erectile dysfunction
Stress urinary incontinence
Varicoceles

*Sections of this chapter were published previously in Mydlo JH, Kanter JL, Kral JG and Macchia RJ. Obesity, diet and other factors in urological malignancies: a review. *Br. J. Urol.* 1999; 83:225–34 and are reproduced with permission from Blackwell Science Ltd.

is multifactorial, including non-insulin-dependent diabetes mellitus (NIDDM), gonadal steroid imbalance and psychological effects of low self-esteem and poor self-image.[5] There is one report of increased incidence of benign prostatic hyperplasia (BPH) among overweight men.[6]

Obesity is, by definition, an excess accumulation of adipose tissue. Although obesity represents one arm of a distribution curve with no sharp cut-off point, 20–40% overweight is defined as 'mild' obesity, while moderate obesity is 41–100% overweight and severe obesity is defined as a weight greater than 100% above actuarial weight for height standards.[7] According to the National Center for Health Statistics, the body mass index (BMI) is a more appropriate method to define obesity than weight alone. BMI is calculated as weight in kilograms divided by height in meters, squared (kg/m^2). For adults, a BMI >30 indicates obesity, 26–29 is overweight and <26 is considered normal.

Between 1976 and 1980 and 1988 and 1994, the prevalence of obesity (BMI >30) increased markedly in the USA in agreement with trends elsewhere in the world.[8] Although the measurement of skinfold thickness to assess fatness has been widely used in epidemiological studies, clinicians have generally not found it helpful and it is, therefore, rarely used in practice.[9]

BMI and weight are not the only predictors of risk: the distribution of fat is also important. Upper body (android) fat distribution in both men and women, as opposed to lower (gynecoid) distribution, predisposes to cardiovascular and cerebrovascular diseases, hypertension, diabetes mellitus and a host of other obesity-related conditions. Fat distribution is very easy to estimate by measuring the abdominal circumference midway between the costal margin and the iliac crest (= waist) and the largest circumference below the iliac crest (= hip) in a standing patient to calculate the waist:hip ratio (WHR). Levels above 0.90 in men and 0.85 in women indicate increased risk.[7–10]

Obesity is the most common and costly nutritional problem in the USA, affecting approximately 33% of adults. Health care costs attributable to obesity account for $68 billion per year, or 7% of the total health care budget. Americans spend about $30 billion per year on weight reduction programs. Unfortunately, 90–95% of this money is wasted because of weight regain. Although severe obesity is equally present in women and men, women represent more than 75% of people seeking and obtaining treatment for obesity.[11]

In the USA, obesity is commonly found in lower socioeconomic groups.[12] There are also significant ethnic differences when controlling for poverty and education, and within races there are sex differences. Black and Hispanic men, who make up a large part of the manual labor force, are leanest. On the other hand, Hispanic and Black women are the most obese, with prevalences of 50% in some age strata. In addition to the influence of education and financial status, it is clear that cultural and genetic factors are also important determinants of obesity.[12]

To ascribe defined macronutrient choices to socioeconomic groups might be overly simplistic. It is difficult to determine fat and dietary intake in populations because of a lack of precision in questionnaires and surveys: there is about a 30% margin of error between what people state they eat and what they really eat, and objective studies are extremely complicated and prohibitively expensive.[13]

In many societies, overweight people are actually embarrassed to be seen in public. They are shunned or even persecuted. This is hardly the case in the USA where some overweight persons flaunt their obesity. Indeed, obesity has been termed 'the American disease'. Such typically American inventions as television, assembly-line production and 'fast food' contribute to this epidemic. Added to this are the large selection of high-fat foods, which generally are inexpensive and accessible. Unfortunately, the American diet and culture are conquering other cultures with healthier dietary habits. Thus, 'le Big Mac' is found in Paris and fast food at baseball games in Tokyo.[14]

In summary then, obesity is a prevalent disease in the USA, accounting for a large part of the health care costs in associated diseases. Owing to the increase in the number of cancers and the recognition of obesity and diet as modifiable risk factors for cancer, it is timely to review the associations between malignancies, diet and obesity.

OBESITY AND CARCINOGENESIS

The American Cancer Society followed more than 750 000 individuals for 12 years and showed that obese men and women have an increased risk of certain types of cancer (**Table 19.2**). For obese women, there was an increase in cancer of the endometrium, cervix, ovary and breast. In obese men, there was an increased prevalence of prostate and colorectal cancer.[14]

The development of cancer is influenced by three main mechanisms: (1) exposure to exogenous factors damaging genes that regulate cell proliferation and migration; (2) selective enhancement of ambient tumor cells and their precursors; and (3) loss of natural inhibition of cell growth exemplified by dysregulation of programmed cell death (apoptosis).[15]

How many cancer cases would be expected to occur in a healthy population without exposure to dietary or environmental carcinogens? It is theorized that perhaps 25% of tumors arise through natural biological variability in the body's normal accumulation of 'mistakes'.[15] Aging plays a role by increasing the amount of free radicals, which subsequently may affect the cell's DNA causing mutations.[16] As we will discuss, diet and obesity may also affect these mechanisms.

Table 19.2. Obesity-related cancers.

Biliary
Breast
Cervical
Colorectal
Endometrial
Ovarian
Prostate
Renal

Other agents that seem to play a role in neoplasia are polypeptide growth factors found in most tissues in varying amounts.[17] They have been demonstrated to promote angiogenesis and mitogenesis in various cell types and thus to play a role in wound healing and organ formation. One of the earliest to be identified was fibroblast growth factor (FGF). It has an affinity for heparin-binding, and can signal through common receptors. Basic fibroblast growth factor (bFGF or FGF-2), a 146-amino-acid polypeptide, is a potent mitogen for mesodermal and ectodermal derived cells and is a potent angiogenic factor, stimulating endothelial cell proliferation.[17] Growth factors homologous to FGF-2 have been isolated and purified from numerous urological tissues and have been shown to be mitogenic for urological epithelial and stromal cells.[18,19] Thus, it is possible that growth factors support tumor growth and development in an autocrine manner.

Since the end-result of high-fat or high-caloric diets is an increase in the amount of adipose tissue or body fat, it is not possible to separate the effects of diet from those of obesity *per se* in epidemiological studies. To approach this problem, we investigated the premise that adipose tissue in itself might contribute to the carcinogenic effects of obesity via the production or binding of growth factors. Inspired by the work of Folkman et al.,[17] Mydlo et al. analyzed the angiogenic properties of adipose tissue, comparing it to tumor tissue. They demonstrated a greater recovery of basic fibroblast growth factor (FGF-2) as well as a greater angiogenic/mitogenic specific activity of this growth factor in adipose tissue compared to malignant or benign prostate and renal tissues[20,21] (**Figs 19.1** and **19.2**). They suggested that this may be one of several mechanisms for obesity-related tumor biology of these lesions. They further speculated that adipose tissue, and prostate and/or renal cancers may be associated through several mechanisms, so that reduction of obesity and dietary fat may be beneficial.

In an attempt to disentangle the effects of dietary fat and obesity, Weber et al. demonstrated an increased incidence of chemically induced colon cancer in genetically obese Zucker rats on a low-fat diet compared to genetically lean Zuckers on a high-fat diet.[22] Preliminary differences in FGF-2 levels imply involvement of the growth factor.[22]

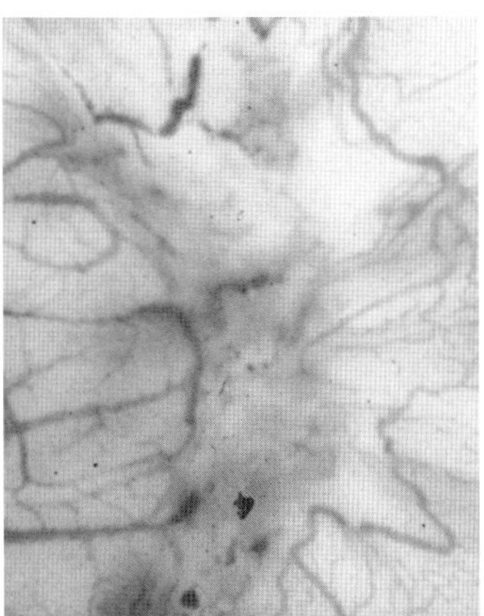
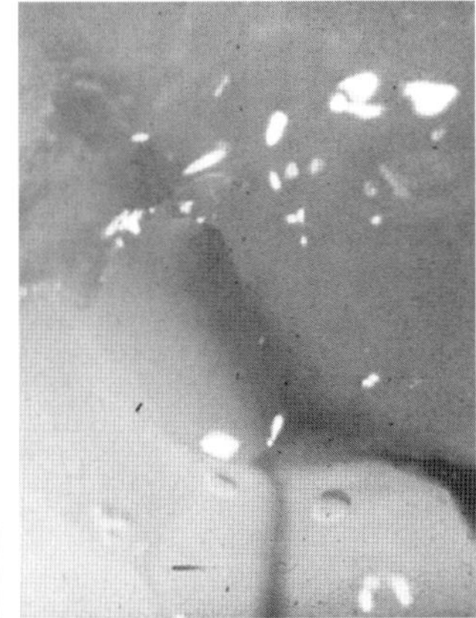

Fig. 19.1. Greater angiogenic response, as measured by neovascularity on the chorioallantoic membrane (CAM) assay, is demonstrated from purified adipose tissue FGF-2 (bFGF, left) than from benign or cancerous prostate tissue FGF-2 (right). Reprinted from *Journal of Urology*, 1988; 140:1575–9 with permission from Lippincott, Williams and Wilkins. (*See Color plate 3.*)

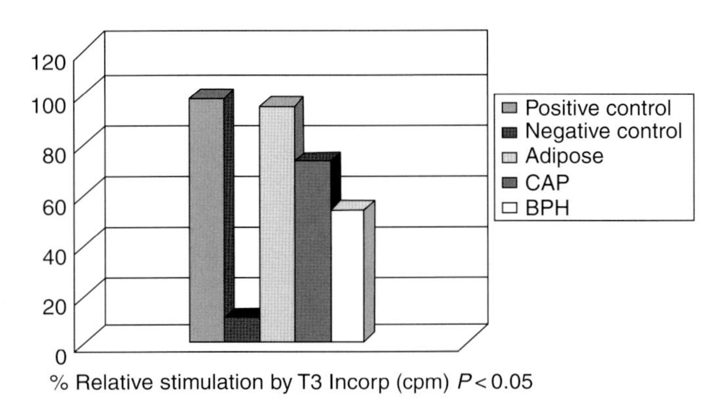

Fig. 19.2. Greater mitogenic response on human umbilical vein endothelial cells (HUVEC) is demonstrated from equal amounts of purified adipose-derived FGF-2 (bFGF) compared to cancerous or benign prostate tissue FGF-2. Reprinted with permission from Elsevier Science (in press).

Mydlo et al. reported on 53 patients who underwent radical retropubic prostatectomies and demonstrated an increase in stage and positive margin status in both Black and White patients with increasing BMI.[23] Histologically, the specimens from those patients with a higher BMI demonstrated a greater microvessel density, or MVD. This suggests that perhaps the adipose tissue is having a stimulatory effect on prostate cancer growth via angiogenic growth factors. Other reports demonstrated that a low-fat, exercise intervention lifestyle reduced the growth of prostate cancer cells, presumably by a decrease in adipose tissue and/or growth factors[23,24] (**Fig. 19.3**).

Recently associations between serum FGF-2 levels and cancer outcome have been shown, providing further support for a role of growth factors in carcinogenesis. One investigative team reported that the levels of some growth factors in primary tumors as well as in urine and/or serum are associated with cancer progression and may act as markers.[25] Insulin-like growth factor (IGF) has also been found to stimulate prostate cancer cells in the laboratory, while serum IGF has been found to be a clinical marker for prostate cancer.[26]

Adipose tissue is not only a source of angiogenic growth factors, but it is also a large repository of cholesterol and triglycerides. Metabolites of cholesterol, such as testosterosterone and androstenedione, may stimulate prostate epithelial and stromal cell growth via regulation of androgen receptors.[27] Although increased levels of adipose tissue FGF-2 in obesity may hypothetically be one mechanism involved in neoplasia, increased gonadal steroid levels from large adipose tissue stores in obese patients may be another equally important link between obesity and cancer.[27] One may influence initiation, while the other is more important for the progression of cancer (**Fig. 19.4**).

Another possible mechanism linking obesity and carcinogenesis is the effect of lipids on macrophages. Their function is impaired, which may also increase the incidence and/or progression of tumors via poor immune surveillance.[28]

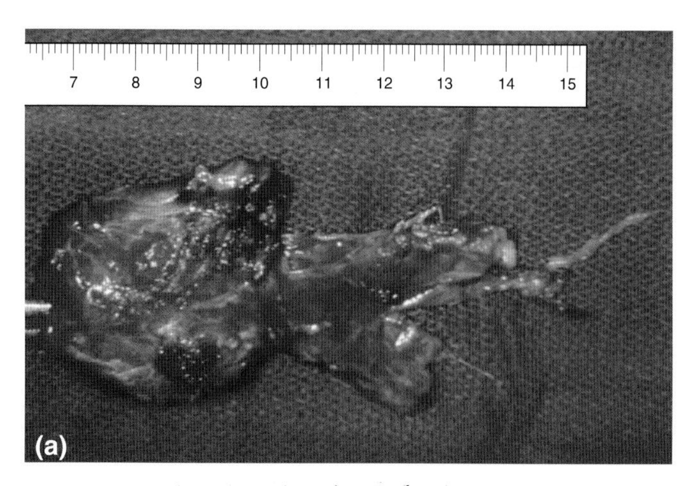

Fig. 19.3. (a) There were greater positive margins on the pathology specimens, as well as a greater microvessel density (MVD), as demonstrated by the darker stained endothelial cells, (b) in prostate cancer tissue from more obese patients than more lean patients. These were stratified for PSA and Gleason score. (*See Color plate 4.*)

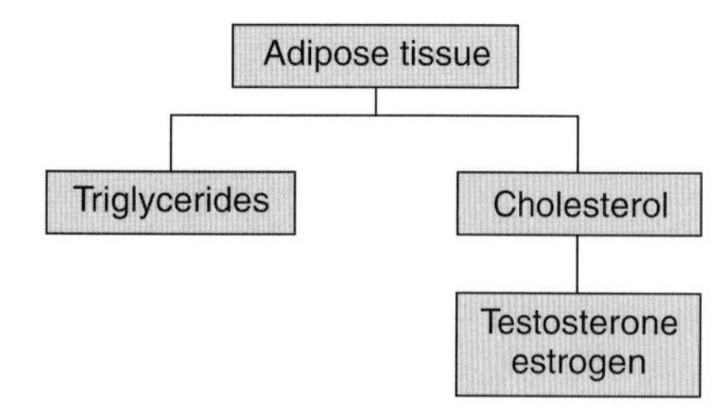

Fig. 19.4. This possible adipose tumor biology mechanism suggests that the breakdown of adipose tissue into cholesterol may yield additional hormones, which could stimulate neoplasia.

AGING AND TUMOR INITIATION AND PROGRESSION

Aging is associated with a general reduction in many immune responses. Immune dysfunction, as demonstrated by, or caused by, the presence of autoantibodies, is also increased in the elderly.[16] One theory of aging suggests that every mother cell that gives rise to two daughter cells has slight alterations in its DNA, which consequently affect the RNA and protein translations. These slightly altered proteins may then participate in the production of other proteins, which may be even further altered, thus giving rise to malignancies. This theory of aging has been suggested to explain the increased incidence of tumors in the aged.[16]

Another mechanism that may be responsible for the initiation and/or progression of cancer may be related to the increase of oxidation products in the immune system during aging. Mitochondria in cells from older patients are less efficient at metabolizing lipids, and are thus more likely to accumulate oxidative damaging agents. Examples of such damaging free radicals are superoxide, hydroxy radical, hydrogen peroxide and peroxynitrite [the result of nitric oxide (NO) interacting with superoxide].[29,30] One can speculate that the more circulating lipids, as in obesity, the greater the stress on these aging mitochondria.[29,30]

Thus, the interactions between adipose tissue, aging, malnutrition, impaired immune defense and simply prolonged exposure over time, which increases the risk of mutations and free radical formation, may all contribute to promote cancer.

TECHNICAL FACTORS INFLUENCING OUTCOME

Technical factors may influence tumor recurrence and progression. Several studies have shown the presence of thousands of circulating tumors cells in the peripheral vasculature after surgery for organ-confined disease, especially in channel TURPs and radical prostatectomies.[31,32] It is reasonable to expect that such 'shedding' of tumor cells may have a greater impact in patients with a compromised immune system, but this does not yet seem to have been studied. Other factors that may contribute to the 'spillage' of tumor cells might include tumor size, compromise in technique because of obesity and proximity to vessels. This may account for the differences seen in tumor recurrences among patients with similar grades and stages of tumors[31,32] and may be one explanation for poorer outcomes in obese patients.

DIETARY PROMOTERS OF CANCER

Diet is thought to be responsible for approximately 35% of all cancer deaths in the USA, with a range varing from as little as 10% to a surprising 70%.[33] Dietary causes of cancer are not only based on the content of the diet, but also on its deficiencies. The publication 'Diet, nutrition, and cancer' in 1982 from the National Research Council was the first US government publication that suggested that cancer risk might be reduced by diet restrictions and decreases in obesity.[34] However, in examining the relationship of dietary fat intake and cancer in humans, many of the studies have been based on inadequate dietary methodology and, therefore, there are many inconsistencies among the findings.[35–38]

The cancer-promoting property of dietary fat may be twofold: a general effect from excess calories and a specific carcinogenic effect of lipids, such as the essential fatty acid, linoleic acid. Linoleic acid, a polyunsaturated fat, comprises about 40% of the fatty acids of corn oil.[39] It enhances carcinogen-induced breast, pancreas and colon cancer in rodents.[40] One of the proposed mechanisms is that it affects macrophage function, which can in turn, decrease immunity.[28,38] There is, however, no evidence that linoleic acid has adverse effects in man.

DIETARY PROMOTERS IN UROLOGIC CANCER

Although the rate of initiation of prostate cancer is equal worldwide, the rates of progression differ greatly in different regions of the world.[37] In the USA, the prevalence of prostate cancer between the ages of 60 and 79 is 16%, but only 4% of the deaths of men over 55 years of age are attributable to prostate cancer.[37] Whites in the USA have an eightfold higher incidence of prostate cancer than Japanese in their native country, and African Americans have a 15 times higher incidence than Japanese men.[41] Interestingly, this difference decreases in second-generation Japanese Americans, who have grown up on western diets.[41]

Andersson et al. reported that total caloric intake was a risk factor for prostate cancer in a population-based case-control study in Sweden.[42] Another study examining 414 cases reported a 75% increase in prostate cancer risk in the uppermost quartile level of energy intake with no clear association between total fat and monounsaturated fat and prostate cancer, although there was an inverse relationship association with saturated fat intake.[43]

Giovannucci et al. reported that animal fat is the cause of the increased incidence of prostate cancer rather than other fats, such as vegetable or fish oils.[44,45] Gann et al., in a prospective clinical study, reported that dietary increase of linoleic acid was not associated with an increased risk of prostate cancer,[46] as was previously hypothesized because of findings in breast, pancreas and colon cancer mentioned above.[39,40]

Investigators have shown an increase in renal cell cancer (RCC) among men consuming high-fat diets.[47–49] Wolk et al. reported on the increased risk of RCC with increased energy intake, especially from increased consumption of fried meats.[50] Chow et al. reported that high protein consumption was associated with other chronic renal diseases that may predispose to RCC.[51] Some studies revealed an association between dietary fat intake and bladder cancer,[52] while others did not.[53] Increased consumption of chlorination byproducts also appears to increase the risk.[52–56] The role of alcohol in increasing the risk of prostate, renal and bladder cancer is still uncertain, because of the confounding variables of smoking and dietary fat.[57]

PROTECTIVE FACTORS

It has been demonstrated in numerous animal experiments that caloric restriction prolongs survival.[58,59] Studies have shown that caloric restriction is far more important in reducing the risk of mammary tumors in rodents than restricting the intake of fat or even linoleic acid.[58–60] In several cancers, it is still controversial whether dietary energy or fat is more important.[61–63]

Low-fat diets and dietary factors

Just as high-fat and high-caloric diets seem to be promoters of some cancers, low-fat diets are protective. This has been demonstrated for prostate cancer in the laboratory. Wang et al. reported that there was less growth of prostate cancer cells implanted in nude mice fed low-fat diets.[64] A recent clinical report described that dietary modification of fat intake could lower prostate specific antigen (PSA) levels without affecting overall testosterone levels, although it is not clear whether this influenced the prognosis.[65]

Interestingly, although linoleic acid has been shown to be a promoter of carcinogenesis, as mentioned previously, conjugated dienoic derivatives of linoleic acid (CLA) have been shown to be anticarcinogenic or ‘protectors’. As little as 0.5% of CLA in the diet significantly reduced carcinogen-induced mammary neoplasia in rats.[66] CLA is thought to be an antioxidant, protecting other membrane fatty acids from oxidation damage, an important promoter of carcinogenesis.[67]

Giovannucci et al. reported a reduced risk of prostate cancer associated with increased overall intake of fruits and vegetables,[44] and lycopene, an ingredient in tomato-based foods, was associated with a lower risk of prostate cancer in an epidemiologic study.[45] The presence of lycopene in the prostate at concentrations that are biologically active caused cell inhibition in laboratory studies.[65]

Isoflavonoid extracts from soybean foods, such as genistein, have been shown to inhibit prostate cancer cells *in vitro*.[68] This agrees with the observation of a lower incidence of prostate cancer in Asians, with their soy diets.[37] However, there have been no studies which show that soybean supplementation decreases the incidence of prostate cancer in humans. The active ingredients in garlic, alliinn and allicin have also been shown to decrease tumor cell growth *in vitro*, but have not shown any benefit in clinical studies.[69]

Increased consumption of dietary fiber is considered to have salutory effects, particularly with respect to preventing colon cancer and possibly breast cancer.[70] Interestingly, some foods behave differently depending on subtle differences in preparation. For example, if potatoes are eaten hot the starch is totally digested in the small bowel, but behaves as fiber if the potatoes are eaten cold.[70,71] However, there is no evidence for a protective effect of dietary fiber on urological malignancies.

Antioxidants

If oxidative damage in the immune system of aging/obese individuals plays an important role in prostate cancer, then antioxidants should inhibit the process. Selenium has been suggested to reduce prostate cancer progression. Clark et al., in examining the effects of selenium in skin cancer, reported an incidental 63% decrease in prostate cancer incidence using selenium supplements in the diet.[72] β-Carotene and vitamin E were also evaluated for their cancer-inhibiting potential, but only vitamin E was associated with a decrease in the incidence of prostate cancer.[73]

Vitamin E supplementation was shown to improve immunocompetence in the elderly, therefore, requirements for vitamin E may be greater in this group.[74] However, there is an inverse dose–response relationship. At high doses it antagonizes the effects of other fat-soluble vitamins, reduces liver storage of vitamin A, decreases bone mineralization and causes disorders of coagulation. Very high doses of vitamin E have been shown to decrease immunologic responses in the elderly.[74]

Daviglus et al. reported that overall survival of patients with prostate cancer was positively associated with the intake of β-carotene and vitamin C.[75] This, however, does not prove that vitamin C alone improves survival.

Vitamin D

Vitamin D has not been proven to be beneficial in patients with prostate cancer, but there are some interesting laboratory findings. First, vitamin D receptors are present on the cells of the prostate cancer cell line LNCaP, and physiologic doses of 1,25-dihydroxyvitamin D_3 ($1,25(OH)_2D_3$) inhibited these prostate cancer cells in culture.[76] Corder and colleagues found low serum levels of $1,25(OH)_2D_3$ in men above 57 years of age with palpable tumors but not in younger men with tumors diagnosed incidentally.[77] Northern USA has a greater incidence of prostate cancer than the South. Since sunlight is known to activate vitamin D, it has been speculated that the reduced sunlight found in the North could play a role via absence of vitamin D activation.[78]

Although fruits and vegetables were apparently protective against RCC, the levels of vitamin C and β-carotene did not appear to affect risk.[49–51] Bruemmer et al. reported that multivitamin supplements over a 10-year period decreased the incidence of bladder cancer.[53]

Confounders

An important question is whether isolated vitamin, mineral and antioxidant supplements are as effective as those naturally present in fruits and vegetables. There are increasing reports that the interaction of the natural vitamins and antioxidants make them more effective compared to the isolated compounds.[79] Thus, there has been an increased awareness of the benefits of eating fruits and vegetables to prevent or hinder cancer progression. However, there is also an increasing awareness of the relationships between the use of pesticides and the etiology of certain tumors, such as bladder cancer.[80,81] Thus, a nationwide effort was established to decrease the use of DDT and other toxic pesticides known to be carcinogenic. Unfortunately, approximately 50% of all the fruit and vegetable produced in the USA is imported from countries that do not have restrictions on pesticides.

Exercise

Prospective studies have shown that overweight patients who exercise have less morbidity and mortality than those who are sedentary.[82–84] Although there have been no well-controlled studies comparing cancer incidence in sedentary and active overweight persons, exercise seems to be 'protective' against breast cancer.[84] No studies have reported any protection of exercise against prostate cancer.[85]

Weight loss

Some reports have found no change in the risk of cancer in obese patients who have lost weight, while other studies found that the risk ratio for cancer mortality did not alter if patients lost weight, gained weight or showed no change in body weight.[84,86] Although weight loss provides many benefits, it may also have some undesirable consequences, such as gallstones and nutrient deficiencies.[87]

TREATMENT OF OBESITY

Although it is conceptually easy to treat obesity: exhort the patient to 'push away from the table' decreasing calorie intake and to adopt a healthy program of physical activity increasing energy expenditure, it is notoriously difficult to achieve sustained medically significant weight loss. Without extraordinary measures such as prison-like conditions or surgical methods, the failure rate is in excess of 90% after 1 year. Diets, exercise and behavior modification, all 'difficult methods', fail miserably over the long term.[88]

In the food industry, it is cheaper to use low-cost fillers such as fats and oils in food processing, than reducing fat content, which could affect profits. Furthermore, fat is palatable, which is exploited by restaurants, cafeterias and fast-food establishments. The image of 'healthy' food being unpleasant, bad tasting or no fun to eat is deeply rooted in the American psyche.[89,90] It is of interest that snack foods with reduced calories and/or saturated fat are usually eaten in greater quantities, to compensate for the reduction, in part because the consumer believes the food is healthy. This behavior thus cancels out any benefit of using these low-fat products. There are many reasons why weight gain is easy, and also why weight loss is difficult. Thus, people are forever looking for the 'easy way out'. Hypnotherapy, acupuncture, vibrators, creams, massage and even liposuction have not been demonstrated to achieve lasting weight loss in clinical trials.[87,88]

Since obesity is so difficult to treat, it is impossible to study the impact of weight loss on the initiation or progression of cancer in a clinical setting. In fact, most 'beneficial' practices, whether adherence to a low-fat diet, exercise, reduction of total energy intake or increased consumption of 'healthy' nutrients are so difficult to maintain for a sufficient number of individuals over a sufficiently long period of time that conclusive evidence (requiring an intervention study), is not available. Any data are circumstantial at best and are derived from observations of cohorts differing in their healthful practices within well-characterized populations.

These caveats notwithstanding, based on the circumstantial evidence presented earlier, a case can be made for treating obesity and for adopting reductions in dietary fat, increases in consumption of fruits and vegetables, and a physically active lifestyle with the goal of reducing the prevalence of cancer and improving survival of patients with cancer. Over the last 20–30 years, cardiovascular mortality, hypertension and hypercholesterolemia have decreased, in part from the success of public health intervention. This is encouraging because it demonstrates the feasibility of achieving populationwide changes in unhealthy behavior. However, massive investments, both public and private, are necessary to curb the trend of increasing prevalence of obesity in industrialized nations.

Dietary modification has been increasingly gathering momentum in the medical community. Medical school curricula now include courses on nutrition and diet. Many urological surgeons now routinely give their patients dietary guidelines after radical prostatectomy.[90] Patients are generally more satisfied when family members and physicians give them guidance, such as dietary counselling.[91–93] Thus, there are an increasing number of physicians and surgeons who encourage dietary changes in their patients after cancer surgery has been performed, giving the patients a sense of well-being.[92,93] Moreover, it has been shown that spouses respond positively to the other's dietary modification. Therefore, two people in a household may mutually benefit from proper nutrition, dietary modification and/or lifestyle changes.

In addition to public health efforts, it is also important to recognize the contribution of improved pharmacotherapy and the increased awareness of the medical profession for identifying and treating these diseases. Only recently have government drug-regulating agencies understood the need for chronic, indeed, lifelong medication for obesity in analogy with hypertension or diabetes. Not surprisingly, this change in policy coincides with the development of new drugs for treating obesity. Unfortunately for obese subjects, one of these drugs, d-fenfluramine, was inappropriately and 'off-label' combined with phentermine ('fen-phen'), giving rise to a rash of cases with valvular heart disease, resulting in the withdrawal of this otherwise promising and relatively innocuous drug.[94]

Two other drugs have recently been approved: sibutramine, a centrally acting transmitter blocker, and orlistat, an intraluminal lipase inhibitor preventing absorption of fat calories.[95] The effects of both these drugs are modest and they are only recommended for use within programs for lifestyle changes.

Progress has also been made toward identifying predictors of success in behavioral weight programs. Lavery and Loewy described eight characteristics of individuals successful in reducing their weights: (1) feeling in control of eating habits; (2) moderate excess weight at enrollment; (3) early weight loss response (first 8 weeks); (4) frequency of weight measurement; (5) increase in physical activity; (6) absence of emotional eating; (7) recent weight gain; and (8) occupation.[96] The optimal strategy would be to prevent rather than treat obesity. Unfortunately, there are only few studies on this topic and very little is known about preventing obesity. Childhood obesity is extremely difficult to manage. Success requires the participation of the whole family.[97]

The most successful methods for treating obesity to date are surgical. Candidates are individuals with a BMI in excess of 35 in the presence of comorbidity. Exclusion criteria include serious psychopathology, inability to cooperate with treatment plans in follow-up, and lack of understanding of the rationale and methods for treatment after failure of non-surgical approaches.[98] Urologists treating obesity-related benign conditions should consider referring patients for surgical management of obesity prior to operating.

PUBLIC HEALTH MESSAGE

The goals are to prevent and to improve the management of patients with urologic diseases including cancers. Steps to educate the public about the risks of a high-fat diet and obesity are crucial.[99–101] Examples of dietary modification recommendations are as follows.

1. Reduce fat intake to less than 30% of total calories (in accordance with AHA guidelines) with no more than 10% of total calories from saturated fats, 6–8% as polyunsaturated fats, and the remainder as monounsaturated fats. Particularly hydrogenated or trans-fats should be avoided.
2. Increase consumption of leafy vegetables and citrus fruits.
3. Adjust exercise and food intake for a healthy body weight.
4. Limit the use of dietary supplements, since the majority of 'normal' diets are complete. Many of the products sold in health food stores impart a false sense of security and many people 'overdose' in the belief that, if something is good, more is better. Taking a supplement but not reducing fat may negate any benefit from the former. Furthermore, the combination of vitamins in fruits and vegetables may account for activity differences due to interaction, compared to taking single, isolated supplements. Thus, fruits and vegetables may offer more benefits than isolated compounds.
5. Reduce the consumption of alcoholic beverages.

Directions for future investigations should include the following.

1. Identifying foods and nutrients that alter the risk of cancer and determining mechanisms of action.
2. Improving the database and the methodology for assessing human exposure to foods and dietary supplements that may alter the risk of cancer.
3. Identifying indicators of cancer risk.
4. Evaluating non-traditional and innovative interventions purported to reduce the risk of cancer.
5. Developing methods to educate the public about diet, obesity and cancer.

CONCLUSION

The incidence and progression of prostate cancer, as well as many other cancers, depend on many inter-related factors, such as obesity, diet, genetics, environment, age and the immune system. Numerous publications have demonstrated that a high dietary intake of fat increases prostate cancer risk, although the mechanisms are not clear. Although some reports may demonstrate an association between obesity and prostate cancer, it may be hard to establish, since, in general, obese men have an increased high-fat diet. Using genetically bred obese and lean animals may eventually give us the answer of the inter-relationships between diet, obesity and prostate cancer.

Obesity, recurrent urinary tract infections, increased intake of protein and fried foods as well as female sex appear to increase the risk of renal cancer. Environmental toxins seem to predominate as the major factors affecting the incidence of bladder cancers.

Thus, dietary modification and other public health measures directed at environmental carcinogens have the potential for reducing the incidence of urological malignancies. More studies are necessary to determine the therapeutic effects of weight loss and dietary modification on urological tumors, once they have developed.

REFERENCES

1. Wolf AM, Colditz GA. Current estimates of the economic cost of obesity in the United States. *Obesity Res.* 1998; 6:97–106.
2. Norman RJ, Clark AM. Obesity and reproductive disorders: a review. *Reprod. Fert. Dev.* 1998; 10:55–63.
3. Bump RC, Sugerman HJ, Fantl JA et al. Obesity and lower urinary tract function in women: effect of surgically induced weight loss. *Am. J. Obst. Gynecol.* 1992; 167:392–9.
4. Mulcahy JJ. Scrotal hypothermia and the infertile man. *J. Urol.* 1984; 132:469–70.
5. McKendry JB, Collins WE, Silverman M et al. Erectile impotence: a clinical challenge. *Can. Med. Assoc. J.* 1983; 128:653–63.
6. Soygur T, Kupeli B, Aydos K et al. Effect of obesity on prostatic hyperplasia: its relation to sex steroid levels. *Int. Urol. Nephrol.* 1996; 28:55–9.
7. Najjar MF, Rowland M. National Center for Health Statistics. Anthropometric reference data and prevalence of overweight, United States. 1976–80. *Vital Health Stat.* 1987; 11:238.
8. Flegal KM, Carroll MD, Kuczmarski RJ et al. Overweight and obesity in the United States: prevalence and trends 1960–1994. *Int. J. Obesity* 1998; 22:39–47.
9. Rosenbaum M, Leibel RL, Hirsch J. Obesity. *N. Engl. J. Med.* 1997; 337:396–407.
10. Demark-Wahnefried W, Conaway MR, Robertson CN et al. Anthropometric risk factors for prostate cancer. *Nutr. Cancer* 1997; 28:302–7.
11. Rodin J. Cultural and psychosocial determinants of weight concerns. *Ann. Intern. Med.* 1993; 119:643–5.
12. Sobal J, Stunkard AJ. Socioeconomic status and obesity: a review of the literature. *Psychol. Bull.* 1989; 105:260–75.
13. Brownell KD, Fairburn CG (eds). *Eating Disorders and Obesity: A Comprehensive Handbook.* New York: Guilford Press, 1995.
14. Van Itallie T. Health implications of overweight and obesity in the United States. *Ann. Intern. Med.* 1985; 103:983–8.
15. Trichopoulos D, Li FP, Hunter DJ. What causes cancer? *Sci. Am.* 1996; 275:80–7.
16. Chandra RK. Graying of the immune system. *JAMA* 1997; 277:1398–9.
17. Yancopoulos GD, Klagsbrun M, Folkman J. Vasculogenesis, angiogenesis, and growth factors. *Cell* 1998; 93:741–53.
18. Mydlo JH, Bulbul MA, Richon VM et al. Heparin-binding growth factor isolated from human prostatic extracts. *Prostate* 1988; 12:343–55.
19. Mydlo JH, Heston WDW, Fair WR. Characterization of a heparin-binding growth factor from adenocarcinoma of the kidney. *J. Urol.* 1988; 140:1575–9.
20. Mydlo JH, Kral JG, Macchia RJ. Differences in prostate and adipose tissue basic fibroblast growth factor: an analysis of preliminary results. *Urology* 1997; 50:472–8.
21. Mydlo JH, Kral JG, Macchia RJ. Preliminary results comparing the recovery of basic fibroblast growth factor (FGF-2) in adipose, benign and malignant renal tissue. *J. Urol.* 1998; 159:2159–63.
22. Weber RV, Stein DE, Scholes J, Kral JG. The role of obesity vs high fat diet in rat colon cancer. *Proc. Am. Assoc. Cancer Res.* 1998; 39:813.
23. Mydlo JH, Tieng N, Volpe MA et al. A pilot study analyzing PSA, serum testosterone, lipid profile, body mass index and race in a small sample of patients with and without carcinoma of the prostate. *Prostate Cancer Prostate Dis.* 2001 4:101–5.
24. Tymchuk CN, Barnard RJ, Heber D et al. Evidence of an inhibitory effect of diet and exercise on prostate cancer cell growth. *J. Urol.* 2001; 166:1185–9.
25. Nguyen M, Watanabe H, Budson AE et al. Elevated levels of an angiogenic peptide bFGF in the urine of patients with a wide spectrum of cancers. *J. Natl Cancer Inst.* 1994; 86:356–64.
26. Cohen P, Peehl DM, Stamey TA et al. Elevated levels of insulin-like growth factor binding protein-2 in the serum of prostate cancer patients. *J. Clin. Endocrinol. Metab.* 1993; 76:1031–6.
27. Deslypere JP, Verdonck L, Vermeulen A. Fat tissue: a steroid reservoir and site of steroid metabolism. *J. Clin. Endocrinol. Metab.* 1985; 61:564–70.
28. Erickson KL, Hubbard NE. A possible mechanism by which dietary fat can alter tumorigenesis: lipid modulation of macrophage function. *Adv. Exp. Med. Biol.* 1994; 364:67–81.
29. Miller AB, Berrino F, Hill M et al. Diet in the aetiology of cancer: a review. *Eur. J. Cancer* 1994; 30A:133–46.
30. Hietanen E, Bartsch H, Bereziat JC et al. Diet and oxidative stress in breast, colon and prostate cancer patients: a case control study. *Eur. J. Nutr.* 1994; 48:575–86.
31. Mansfield JT, Stephenson RA. Does transurethral resection of the prostate compromise the radical treatment of prostate cancer? *Semin. Urol. Oncol.* 1996; 14:174–7.
32. Oefelein MG, Kaul K, Herz B et al. Molecular detection of prostate epithelial cells from the surgical field and peripheral circulation during radical prostatectomy. *J. Urol.* 1996; 155:238–42.
33. Willet WC. Diet, nutrition, and avoidable cancer. *Environ. Health Perspect.* 1995; 103:165–70.
34. McDowell A, Engel A, Massey JT et al. Plan and operation of the Second National Health and Nutrition Examination Survey 1976–80. *Vital Health Stat.* 1982; 1:15.
35. Morton MS, Griffiths K, Blacklock N. The preventive role of diet in prostatic disease. *Br. J. Urol.* 1996; 77:481–93.
36. Chyou PH, Nomura AM, Stemmermann GN. A prospective study of diet, smoking, and lower urinary tract cancer. *Ann. Epidemiol.* 1993; 3:211–16.
37. Pienta KJ, Goodson JA, Esper PS. Epidemiology of prostate cancer: molecular and environmental clues. *Urology* 1996; 48:676–83.
38. Boyle P, Zaridze DG. Risk factors for prostate cancer and testicular cancer. *Eur. J. Cancer* 1993; 29A:1048–55.
39. Crevel RW, Saul JA. Linoleic acid and the immune response. *Eur. J. Clin. Nutr.* 1992; 46:847–55.
40. De Vries CE, van Noorden CJ. Effects of dietary fatty acid composition on tumor growth and metastasis. *Anticancer Res.* 1992; 12:1513–22.
41. Kodama M, Kodama T. The interference of geographical changes of cancer risk in tumor etiology in Japan. *Anticancer Res.* 1993; 13:1035–42.
42. Andersson SO, Wolk A, Bergstrom R et al. Energy, nutrient intake and prostate cancer risk: a population-based case control study in Sweden. *Int. J. Cancer* 1996; 68:716–22.

43. Rohan TE, Howe GR, Burch JD, Jain M. Dietary factors and risk of prostate cancer: a case-control study in Ontario, Canada. *Cancer Causes and Control*. 1995; 6:145–53.
44. Giovannucci E. How is individual risk for prostate cancer assessed? *Hematol. Oncol. Clin. North Am*. 1996; 10:537–48.
45. Giovannucci E, Ascherio A, Rimm EB et al. Intake of carotenoids and retinoids in relation to risk of prostate cancer. *J. Natl Cancer Inst*. 1995; 87:1767–76.
46. Gann PH, Hennekens CH, Sacks FM et al. Prospective study of plasma fatty acids and risk of prostate cancer. *J. Natl Cancer Inst*. 1994; 86:281–9.
47. Mellemgaard A, McLaughlin JK, Overvad K et al. Dietary risk factors for renal cell carcinoma in Denmark. *Eur. J. Cancer* 1996; 32A:673–82.
48. Kreiger N, Marrett LD, Dodds L et al. Risk factors for renal cell carcinoma: results of a population based case-control study. *Cancer Causes and Control* 1993; 4:101–10.
49. Lindblad P, Wolk A, Bergstrom R et al. The role of obesity and weight fluctuations in the etiology of renal cell cancer: a population-based case-control study. *Cancer Epidemiol. Biomark. Prev*. 1994; 3:631–9.
50. Wolk A, Gridley G, Niwa S et al. International renal cell cancer study: the role of diet. *Int. J. Cancer* 1996; 65:67–73.
51. Chow WH, Gridley G, McLaughlin JK et al. Protein intake and risk of renal cell cancer. *J. Natl Cancer Inst*. 1994; 86:1131–9.
52. Vena JE, Graham S, Freudenheim J et al. Diet in the epidemiology of bladder cancer in western New York. *Nutr. Cancer* 1992; 18:255–64.
53. Bruemmer B, White E, Vaughan TL et al. Nutrient intake in relation to bladder cancer among middle-aged men and women. *Am. J. Epidemiol*. 1996; 144:485–95.
54. Shirai T, Fradet Y, Huland H et al. The etiology of bladder cancer – are there any new clues or predictors of behavior? *Int. J. Urol*. 1995; 2:64–76.
55. Pirastu R, Iavarone I, Comba P. Bladder cancer: a selected review of the epidemiological literature. *Ann. Istit. Sup. San*. 1996; 32:3–20.
56. La Vecchia C, Negri E. Nutrition and bladder cancer. *Cancer Causes and Control*. 1996; 7:95–100.
57. De Stefani E, Fierro L, Barrios E et al. Tobacco, alcohol, diet and risk of prostate cancer. *Tumori* 1995; 81:315–20.
58. Rogers AE, Zeisel SH, Groopman J. Diet and carcinogenesis. *Carcinogenesis* 1993; 14:2205–17.
59. Greenwald P, Clifford C, Pilch S et al. New directions in dietary studies in cancer: The National Cancer Institute. *Adv. Exp. Med. Biol*. 1995; 369:229–39.
60. Tisdale MJ. Mechanism of lipid mobilization associated with cancer cachexia: interaction between the polyunsaturated fatty acid, eicosapentaenoic acid, and inhibitory guanine nucleotide-regulatory protein. *Prostagland. Leukotri. Essen. Fat Acids* 1993; 48:105–9.
61. La Vecchia C. Dietary fat and cancer in Italy. *Eur. J. Clin. Nutr*. 1993; 47(Suppl. 1):S35–8.
62. Nixon DW. Cancer prevention clinical trials. *In Vivo* 1994; 8:713–16.
63. Deslypere JP. Obesity and cancer. *Metab. Clin. Exp*. 1995; 44:24–7.
64. Wang Y, Corr JG, Thaler HT et al. Decreased growth of established human prostate LNCaP tumors in nude mice fed a low fat diet. *J. Natl Cancer Inst*. 1995; 87:1456–67.
65. Fair WR, Fleshner NE, Heston WDW. Cancer of the prostate: A nutritional disease? *Urology* 1997; 50:840–8.
66. Istfan NW, Wan JM, Bistrian BR. Nutrition and tumor promotion: in vivo methods for measurement of cellular proliferation and protein metabolism. *J. Parent. Ent. Nutr*. 1992; 16(Suppl.): 76S–82S.
67. Kolonel LN. Nutrition and prostate cancer. *Cancer Causes and Control* 1996; 7:83–7.
68. Wang Y, Heston WDW, Fair WR. Soy isoflavones decrease the high fat promoted growth of human prostate cancer: results of in vitro and animal studies. *Proc. Am. Urol. Assoc*. 1995; 153:269A.
69. Lee ES, Steiner M, Lin R. Thioallyl compounds: Potent inhibitors of cell proliferation. *Biochem. Biophys. Acta* 1994; 1221:73–7.
70. Schapira DV. Nutrition and cancer prevention. *Prim. Care Clin. Off. Prac*. 1992; 19:481–91.
71. Rose DP. Diet, hormones, and cancer. *Ann. Rev. Pub. Health* 1993; 14:1–17.
72. Clark LC, Combs DF Jr, Turnbull BW et al. Effects of selenium supplementation for cancer prevention in patients with carcinoma of the skin: a randomized controlled trial. *JAMA* 1996; 276:1957–64.
73. The Alpha-Tocoferol Beta-Carotene Cancer Prevention Study Group. The effect of vitamin E and beta carotene on the incidence of lung cancer and other cancers in male smokers. *N. Engl. J. Med*. 1994; 330:1029–35.
74. Eichholzer M, Stahelin HB, Gey KF et al. Prediction of male cancer mortality by plasma levels of interacting vitamins: 17-year follow-up of the prospective Basel study. *Int. J. Cancer* 1996; 66:145–50.
75. Daviglus ML, Dyer AR, Persky V et al. Dietary beta-carotene, vitamin C, and the risk of prostate cancer: results from the Western Electric Study. *Epidemiology* 1996; 7:472–7.
76. Schwartz GG, Hill CC, Oeler TA et al. 1,25-Dihydroxy-16-ene-23-ync-vitamin D3 and prostate cell proliferation in vivo. *Urology* 1995; 46:365–9.
77. Corder EH, Guess HA, Hulka BS et al. Vitamin D and prostate cancer: a prediagnostic study with stored sera. *Epidemiol. Biomark. Prev*. 1993; 2:467–72.
78. Hanchette CL, Schwartz GG. Geographic patterns of prostate cancer mortality: evidence for a protective effect of UV radiation. *Cancer* 1992; 70:2861–9.
79. Halpern GM, Trapp CL. Nutrition and immunity: where are we standing? *Alleg. Immunopath*. 1993; 21:122–6.
80. Sorensen G, Morris DM, Hunt MK et al. Work-site nutrition intervention and employees' dietary habits: the Treatwell program. *Am. J. Pub. Health* 1992; 82:877–80.
81. Viel JF, Challier B. Bladder cancer among French farmers: does exposure to pesticides in vineyards play a part? *Occup. Environ. Med*. 1995; 52:587–92.
82. DiPietro L, Mossberg HO, Stunkard AJ. A 40 year history of overweight children in Stockholm: lifetime overweight, morbidity and mortality. *Int. J. Obesity Relat. Metab. Dis*. 1994; 8:585–90.
83. Shephard RJ. Exercise and cancer: linkages with obesity? *Int. J. Obesity Metab. Disor*. 1995; 19(Suppl. 4):562–8.
84. Shephard RJ. Physical activity and reduction of health risks: how far are the benefits independent of fat loss? *J. Sports Med. Phys. Fit*. 1994; 34:91–8.
85. Whittemore AS, Kolonel LN, Wu AH et al. Prostate cancer in relation to diet, physical activity, and body size in blacks, whites, and Asians in the United States and Canada. *J. Natl Cancer Inst*. 1995; 87:652–62.
86. Moller H, Mellemgaard A, Lindvig K et al. Obesity and cancer risk: a Danish record-linkage study. *Eur. J. Cancer* 1994; 30A:344–50.
87. Gorsky RD, Pamuk E, Williamson DF et al. The 25 year health care costs of women who remain overweight after 40 years of age. *Am. J. Prevent. Med*. 1996; 12:388–94.
88. Spielman AB, Kanders B, Kienholz M et al. The cost of losing: an analysis of commercial weight loss programs in a metropolitan area. *J. Am. Coll. Nutr*. 1992; 11:36–43.
89. Heston WDW. Gene therapy or fruits and vegetables? A Biased perspective on Prostate Carcinogenesis and Hypothetical Approaches to Its Control. *Molec. Urol*. 1997; 1:11–18.
90. McGuire MS, Fair WR. Prostate cancer and diet: investigations, interventions, and future considerations. *Molec. Urol*. 1997; 1:3–9.
91. Kropiunigg U. Basics in psychoneuroimmunology. *Ann. Med*. 1993; 25:473–9.

92. Cohen S, Manuck SB. Stress, reactivity and disease. *Psychosomat. Med.* 1995; 57:423–6.
93. Stein M. Stress, depression and the immune system. *J. Clin. Psych.* 1989; 50:35–40.
94. Williamson DF. Pharmacology for obesity. *JAMA* 1999; 281:278–80.
95. Davidson MH, Hauptman J, DiGirolamo M et al. Weight control and risk factor reduction in obese subjects treated for 2 years with orlistat. *JAMA* 1999; 281:235–42.
96. Lavery MA, Loewy JW. Identifying predictive variables for long term weight change after participation in a weight loss program. *J. Am. Diet. Assoc.* 1993; 93:1017–24.
97. Epstein LH, Valoski A, Wing RR et al. Ten-year follow-up of behavorial, family based treatment for obese children. *JAMA* 1990; 264:2519–23.
98. Kral JG. Surgical treatment of obesity. *Med. Clin. North Am.* 1989; 73:251–64.
99. Greenwald P, Kramer B, Weed D. Expanding horizons in breast and prostate cancer prevention and early detection. The 1992 Samuel C. Harvey Lecture. *J. Cancer* 1993; 8:91.
100. Pienta KJ, Esper PS. Risk factors for prostate cancer. *Ann. Intern. Med.* 1993; 118:793–803.
101. Oliver MF. Should we treat hypercholesterolemia in patients over 65? *Heart* 1997; 77:491–2.

Prevention of Prostate Cancer: The Role of Diet

Chapter 20

Claude C. Schulman and Alexandre R. Zlotta
Department of Urology, Erasme Hospital, University Clinics of Brussels, Brussels, Belgium

INTRODUCTION

The epidemiology of prostate cancer gives some indications that its etiology is likely to be both environmental and genetic. Indeed, large international variations in rates of prostate cancer incidence but also mortality suggest that environmental factors have an influence on the development of the first neoplasm in man.[1,2]

International variations in rates of prostate cancer are considerable. These variations in incidence of prostate cancer range from 0.5 per 100 000 in Qidong (China) to 102.1 per 100 000 in the USA to 135.5 per 100 000 in Sweden.[3] The large variability of the tumor in different geographical regions suggests the possibility of nutritional influences regarding the stimulation and/or inhibition of clinical cancer, since there is a similar prevalence worldwide of the precursor lesion.

Pathological studies on prostatic tissue from various populations support the fact that non-infiltrating lesions have similar distributions, whereas small infiltrating lesions discovered incidentally at autopsy follow global distributions of invasive disease. The implication of this observation is that invasive cancer has a distinct etiologic basis in which it is believed diet plays a key role.[4] Support for the relationship between diet and cancer is found in studies of migrants moving from countries of birth, such as China and Japan through Hawaii to North America, where their incidence of prostate cancer increases from an initial low rate to almost equal that of the indigenous population within a few generations.[5,6]

There is also a growing body of evidence from a basic science point of view showing that several of these dietary compounds demonstrate antineoplastic activities. A significant number of publications have dealt with various nutritional factors, including fat, phytoestrogens, vitamins (especially vitamin E) and minerals, such as selenium and calcium. Although evidence is accumulating to suggest a link between diet, lifestyle, several compounds and prostate cancer, so far there are no conclusive results or study outcomes to achieve the scientific bases of our accepted standards of evidence.[7] In order to establish the role of dietary compounds on prostate cancer genesis or promotion, clinical epidemiological data and large-scale prospective trials are required to support the experimental data.

Prostate cancer lends itself to chemoprevention owing to a number of characteristics specific to this disease. These include a high prevalence, long latency time, hormone dependency, the availability of an ideal marker (prostate specific antigen) and last, but not least, the availability of a defined precursor lesion (prostatic intraepithelial neoplasia) among the pathways leading to clinical disease.

DIET

Energy intake and fat

The risk of prostate cancer was found to be about 70% greater in men in the upper quartile of energy intake than those in the lower quartile.[8] The results of a number of dietary intake surveys support the concept that a high-fat diet, especially of animal fat, may increase the risk of clinically significant prostate cancer. Animal dietary fat presumably is converted to androgens with resultant increased androgenic stimulation of the prostate.

Sources of polyunsaturated fats (e.g. linolenic acid) include corn oil and sunflower oil. These fats are thought to have a damaging effect on DNA and other cell components and to affect cell proliferation, immune defenses, tissue invasiveness and tumor metastatic spread. They have equally been shown to alter 5α-reductase activity.[9]

Polyunsaturated and vegetable fat have been investigated equally in cohort studies. These authors observed an increased risk of prostate cancer in men with higher intakes of α-linolenic acid, with adjusted relative risk of 3.4 ($P=0.002$ for trend). α-Linolenic acid is present essentially in red meat, butter and vegetable oils (soya bean oil, rapeseed oil).[4] Consistent evidence on the possible association of high monounsaturated and polyunsaturated fat intakes with prostate cancer is yet to be reported.[10]

Omega-3 fatty acids have been shown to inhibit prostate cancer cell lines. Omega-3 fatty acids obtained essentially from fatty fish and eicosanoid synthesis inhibitors are found to block cancer cell invasion by regulating tumor cell proteolytic enzyme activity *in vitro*.[11] Acids measured before diagnosis in the donors' serum was examined. Data showed no definitive conclusion can be drawn at this point.

VITAMINS AND MICRONUTRIENTS (TABLES 20.1 AND 20.2)

Carotenoids

These are a group of complex unsaturated hydrocarbons occurring as pigments in plants, for example, carrots. Some carotenoids are precursors of vitamin A, whereas others, such as lycopene, have a different structure and are not convertible to vitamin A. These compounds have been shown to have antioxidant potential, which is particularly marked with lycopene.[12,13]

A population-based case-control study carried out in Auckland, New Zealand, in 1996–1997 and recruiting 317 prostate cancer cases and 480 controls investigated associations between prostate cancer risk and dietary intake of the carotenoids β-carotene and lycopene and their major plant food sources, including carrots, green leafy vegetables and tomato-based foods. Dietary intake of β-carotene and its main vegetable sources was largely unassociated with prostate cancer risk, whereas intake of lycopene and tomato-based foods was weakly associated with a reduced risk.[14]

A prospective study was designed to examine the relationship between plasma concentrations of several major antioxidants and risk of prostate cancer, using plasma samples obtained in 1982 from healthy men enrolled in the Physicians' Health Study, a randomized, placebo-controlled trial of aspirin and β-carotene. Subjects included 578 men who developed prostate cancer within 13 years of follow-up and 1294 controls matched for age and smoking status. Lycopene was the only antioxidant found at significantly lower mean levels in cases than in matched controls ($P=0.04$ for all cases). There was no evidence for a trend among those assigned to β-carotene supplements. None of the associations for lycopene were confounded by age, smoking, body mass index, exercise, alcohol, multivitamin use or plasma total cholesterol level.[15]

The current literature regarding intake of tomatoes and tomato-based products and blood lycopene (a compound derived predominantly from tomatoes) level in relation to the risk of various cancers was recently reviewed.[16] Among 72 studies identified, 57 reported inverse associations between tomato intake or blood lycopene level and the risk of cancer at a defined anatomic site; 35 of these inverse associations were statistically significant. Lycopene may account for or contribute to these benefits, but this possibility is not yet proven and requires further study. Numerous other potentially beneficial compounds are present in tomatoes and, conceivably, complex interactions among multiple components may contribute to the anticancer properties of tomatoes.

Vitamin C

Vitamin C is a water-soluble antioxidant obtained from fruits and vegetables in food. Most recent cohort and case-control studies showed no significant association between vitamin C intake and risk of prostate cancer.[17,18] The lack of activity of vitamin C, which is known to be a powerful antioxidant, underlines the complexity of prostate cancer biology and prevention.

Table 20.1. Vitamins and prostate cancer. Adapted from Thompson and Coltman (2000).[41]

Dietary constituent	Source	Evidence	Mechanism
Carotenoids			
β-Carotene	Carrots	Insufficient	Unclear
α-Carotene	Carrots	No association	
Lycopene	Tomatoes	Inhibition	Unclear
Vitamin C	Fruits Vegetables	Insufficient	Antioxidant
Retinoids (retinol, retinoin, isotretinoin, fenretinide)	Vegetables Synthetic		Inhibit cell proliferation
Vitamin E (α-tocopherol)	Lettuce, water cress Cotton seed oil, hemp seed oil	Inhibition	Inhibition of tumor progression (prolongs latent phase)
Vitamin D [1,25 $(OH_2D_3]$	Vegetables Milk UV radiation	Inhibition	↓ cell proliferation ↑ cell differentiation

Table 20.2. Minerals/trace elements and prostate cancer. Adapted from Thompson and Coltman (2000).[41]

Element	Source	Evidence	Molecular target effect
Calcium	Milk Cheese	Promotion	Vitamin D synthesis inhibition
Zinc	Meat	Insufficient	
Selenium	Bread Cereals Fish Meat	Inhibition	Antioxidant Apoptosis inducer Catalase enhancer Cytochrome P450 modifier Immunostimulant
Isoflavonoids	Peas Beans Soy beans	Inhibition	Angiogenesis inhibition Antioxidant Apoptosis induction Oncogene expression inhibition EGFR inhibition Antioestrogens ODC synthesis inhibition ↓ cholesterol and LDH
Fenretinide (4HPR)	Carrots	Inhibition	Antiproliferative Apoptosis induction Angiogenesis inhibition Cellular differentiation IGF-1 inhibition Immunostimulation ODC synthesis inhibition Protein kinase inhibition

Retinoids (retinol or vitamin A)

Vitamin A (retinol) is found in foods of animal origin, such as animal liver, fish oil, with eggs and milk as low concentration sources. The precursor β-carotene is found in carrots and green vegetables (spinach, broccoli) and transformed into vitamin A in the gut.[19,20]

Cohort and case-control studies have shown differences in risk estimates between younger and older men based on dietary retinoid levels with association in some reports and protective role in others. The major setbacks to the use of these agents in clinical trials reside in their dose-related side effects including hepatotoxicity, central nervous system changes and mucocutaneous dryness.[21]

Vitamin E (α-tocopherol)

Vitamin E is one of the best-researched compounds in medicine. Vitamin E is actually a general name for different compounds, so supplements can contain several forms.[22] Vitamin E in the diet also differs from the form found over the counter. There has been a strong interest in this supplement in the prostate cancer arena, primarily because of a Finnish study that demonstrated a lower morbidity and mortality from this disease in men taking 50 mg of synthetic (α-tocopherol) vitamin E daily.[21,23] A trial of 29 133 subjects aged 58–69 years in Finland with α-tocopherol 50 mg and/or β-carotene 20 mg daily for 5–8 years on lung cancer unexpectedly showed a 32% decrease in incidence of prostate cancer among men receiving α-tocopherol.[21] It is noteworthy that β-carotene, when given alone, was shown to be associated with 23% and 15% higher incidence and mortality, respectively.

In 1986, 47780 US male health professionals, free from diagnosed cancer, completed a dietary and lifestyle questionnaire, and supplemental vitamin E and prostate cancer incidence were updated through 1996.[24] Supplemental vitamin E was not associated with prostate cancer risk generally, but a suggestive inverse association between supplemental vitamin E and the risk of metastatic or fatal prostate cancer among current smokers and recent quitters was consistent with the Finnish trial among smokers.

Caution should be exercised in interpreting data on the role of vitamin E because of a possible bias in the end-point assessment with respect to the effect of α-tocopherol on prostate cancer, and differences both in diagnostic procedures but also on the types of vitamin E analyzed.

Vitamin D

Studies have demonstrated an inverse correlation between ultraviolet radiation, the main source of vitamin D and prostate cancer mortality.[25] Recent studies have suggested that vitamin D is an important determinant of prostate cancer risk, and inherited polymorphisms in the 3′-untranslated region (3′UTR) of the vitamin D receptor (VDR) gene are associated with the risk and progression of prostate cancer.[26] Clinical trials in patients with advanced hormone refractory prostate cancer showed a rapid drop in levels of prostate specific antigen (PSA). In patients with minimal recurrence, tumor doubling time was increased by a mean of 45% during treatment with hydroxy-1,25-vitamin D_3 [1,25(OH_2)D_3] assessed by PSA values.[27,28]

Solid clinical and epidemiological data about the place of vitamin D in prostate cancer promotion and prevention are, however, lacking.

FRUIT AND VEGETABLE INTAKE

Fruit and vegetable intake have been hypothesized to be associated with a decreased risk of many cancers, but results for prostate cancer are sparse. A case-control study including over 1200 patients has investigated the association between fruit and vegetable intake and prostate cancer risk. Although no association was found between fruit intake and prostate cancer risk, high consumption of vegetables, particularly cruciferous vegetables (with four leaves), was associated with a reduced risk of prostate cancer.[29]

Calcium

It is hypothesized that dietary and supplemental calcium intakes or diets high in milk, the main source of calcium, are consistently associated with prostate cancer risk by lowering the serum levels of bioactive metabolites of vitamin D [1,25$(OH)_2D_3$], which plays a role in reducing the development and/or progression of prostate cancer. Low serum calcium levels stimulate the secretion of parathyroid hormone, which promotes the conversion of vitamin D into 1,25$(OH)_2D_3$. A clinical study on 47781 men found higher consumption of calcium to be related to risk of advanced prostate cancer [relative risk (RR)2.97] and metastatic cancer (RR: 4.57).[30]

Selenium

Selenium is a trace element found in the soil as selenide, is widely obtained from bread, cereals, fish, chicken and meat. Differences in bioavailability of selenium reflect large geographical dietary intake variations. Selenium enters the food chain through plants.[31,32] Commercial sources of selenium come from copper ores refinement. Selenium is a key component of a number of functional selenoproteins required for normal health, such as glutathione peroxidase enzymes, which are antioxidants that remove hydrogen peroxide and damaging lipid and phospholipid hydroperoxides generated *in vivo* by free radicals. A clinical double-blind study on 974 men in the USA showed that selenium reduced overall cancer incidence by 37% and that of prostate cancer by 50%.[33]

The association between risk of prostate cancer and prediagnostic level of selenium in toenails, a measure of long-term selenium intake, was investigated in 51529 male health professionals aged 40–75 years. The selenium level in toenails varied substantially among men. When matched case-control data were analyzed, higher selenium levels were associated with a reduced risk of advanced prostate cancer. After additionally controlling for family history of prostate cancer, body mass index, calcium intake, lycopene intake, saturated fat intake, vasectomy and geographical region, the odds ratio (OR) was 0.35 ($P = 0.03$).[34]

Given this significant reduction in prostate cancer incidence and mortality associated with selenium supplementation at an initial daily dose of 250 μg reduced to 80–90 μg daily as recommended doses, larger trials are on the way. High doses of selenium have been shown to be hepatotoxic and toxic to the nervous system in animals.[31]

Isoflavonoids, flavonoids and soy proteins

One of the major differences in diet between Asian and Western countries is the consumption of soy-derived products. Soy bean plant has been cultivated by the Chinese for at least 4500 years and there are more than 2500 known varieties.[35] Soy is a well-recognized source of phytoestrogens, also known as isoflavones. A limited amount of clinical evidence points to a beneficial role of soy in reducing hormonal levels and exhibiting weak estrogen and antiestrogen-like qualities.

The beneficial effects of soy diet have been attributed to isoflavonoids. These are compounds or plant pigments found in legumes (peas, beans) with soya bean as major source. Genistein and daidzein are the major isoflavones shown to inhibit the growth of prostate cancer cell lines. The lignan enterolactone and the soya-derived isoflavone, genistein, are inhibitors of several steroid-metabolizing enzymes, such as aromatase or 5α-reductase.

Although there are many experimental data regarding the antitumoral effects of several soy-derived products, large-scale epidemiological studies are sparse. A recently published work aimed to identify variables for prostate cancer mortality from data collected in 59 countries. Prostate cancer was inversely associated with estimated consumption of cereals, nuts and oil-seeds, fish and, above all, with soy products. The effect size per calory for soy products was at least four times as large as for any other dietary factor.[36]

COMMON CONCEPTS OF CANCER CHEMOPREVENTION

Chemoprevention consists in the administration of drugs or other agents aimed at preventing induction or inhibiting or delaying progression of cancer. Development of chemopreventive strategies require proper knowledge of the mechanisms of carcinogenesis in prostate disease and identification of agents that can interfere with these mechanisms.

Prostate cancer represents a prime target for chemoprevention studies because known risk factors, hormonal dependency and precursor lesions, such as prostatic intraepithelial neoplasm (PIN) are well documented.[37,38]Chemotherapeutic agents need be given basically for short periods of time or in precise cycles to patients under the care of a physician to allow early evaluation and management of their side effects[37] (**Tables 20.3** and **20.4**).

Chemopreventive agents are usually intended for relatively healthy subjects, many without detectable histological lesions, thus without evident clinical benefits at evaluation of intervention. Cancer incidence is not usually a feasible end-point for chemoprevention trials because long time periods for carcinogenesis and large cohorts are needed for assessable studies. The identification and characterization of intermediate biomarkers and their validation as surrogate end-points for cancer incidence in clinical chemoprevention trials are significant components in the development of chemoprevention agents and strategies.

Several biomarkers can be used to monitor carcinogenesis in the prostate, such as PSA, PIN, nuclear and nucleolar morphometry, DNA ploidy, etc. Other potentially useful biomarkers are associated with cellular proliferation kinetics (proliferating cell nuclear antigen and apoptosis): differentiation (blood group antigens, vinentins); genetic damage (loss of heterozygosity on chromosome 8); signal transduction (tumor growth factor – TGFα, TGFβ, insulin-like growth factor I, c-erb B-2 expression); and angiogenesis and biochemical changes (PSA changes).[37]

A significant proportion of the promising chemopreventive agents for the prostate used in chemoprevention trials are phytonutrients in food, the anticancer effects of which have been demonstrated in a variety of tumor types.

Table 20.3. Prostate cancer chemoprevention and chemoactive target population. Adapted from Rohan et al.[8]

Target population	Major advantage	Major disadvantage
Chemoprevention		
1. Healthy men	Applicable to general population	Requires a large number of subjects Requires a long study period Expensive May require biopsy at end of study to establish status
2. High-risk groups (e.g. strong family history)	Findings directly applicable to the high-risk group studied	Findings may not be applicable to general population
3. High-grade PIN	Greatly decreases required sample size, study time and expense	Possibility of coexisting malignancy may be decreased by requiring second biopsy before randomization
	Easily identied on subsequent biopsies	Findings may not be applicable to general population
Chemoactive		
1. Cancer on biopsy (treated during a 3–12-week period before radical prostatectomy)	Ability to evaluate whole mounted pathology specimen	Only able to evaluate short-term effects of the chemopreventive agent
		Would require subsequent biopsies Findings may not be applicable to general population
2. Cancer on biopsy treated by watchful waiting	Results would evaluate long-term effects of the chemopreventive agent on malignancy	Findings may confound the heterogeneity of prostate cancer

Table 20.4. Differences between trials for chemoprevention and therapy of cancer. Adapted from Thompson and Coltman.[40]

Characteristic	Chemoprevention	Therapy
Population	Healthy men	Men with disease
Study sample size	Very large (often tens of thousands)	Large (perhaps 500–2000)
Acceptable toxicity of intervention	None to minimal	Moderate to large
Cost of trial	Very expensive	Expensive
Duration of accrual and follow-up	Very long (often 10–20 years)	Intermediate length

Cohorts in clinical trials should be suitable for measuring the chemopreventive activity and efficacy of agents as well as assessing of the intermediate biomarker. Chemoprevention might also be considered in patients with rising PSA after radical surgery. Results from the first ever randomized double-blind placebo-controlled study of dietary supplements in tertiary prevention of prostate using a soy extract, tea extract, carotenoids, phytosterols, selenium and vitamin E were recently revealed. The slope of the normal PSA significantly improved in the nutritional supplement group with a delay in PSA rise of 8 weeks with a 6-week course of supplements.[39]

CONCLUSION

Preventing rather than treating already established neoplasms is an attractive concept in oncology. Primary chemoprevention of prostate cancer is a relatively new concept and could be a promising strategy for preventing and arresting the development of the first neoplasm in men.

Chemoprevention could be performed using new pharmaceutical drugs with minimal long-term toxicity but it has become more evident each year that diet plays a major role in prostate cancer promotion or inhibition. Indeed, large-scale studies tend to favor environmental rather than genetic factors as key determinants of the development of prostate cancer[40] and, among these environmental factors, nutrition has a leading role.

Definitive proof of the direct link between nutrition and prostate cancer is difficult to establish because of methodological problems, and because etiologies are obviously complex and multifactorial. Retrospective studies are much more numerous than prospective trials and can be fraught with unsuspected confounding factors. To establish the influence of dietary compounds firmly on prostate cancer biology and behavior, we urgently need well-performed adequately powered prospective trials. We also need experimental research laboratory work, which unequivocally supports the influence of a dietary product on prostate cancer biology.

We also have to address the patient population itself, since mentalities have to be changed. A growing subset of the population is aware of the potential link between dietary habits and prostate cancer, but it is not obvious that they are necessarily ready to change their lifestyle.[41] When all these issues are addressed, ultimately the fate of millions of individuals may be changed in the future.

REFERENCES

1. Parkin DM, Muir CS. Cancer incidence in five continents: comparability and quality of data. *IARC Sci. Publ.* 1992; 120:45–173.
2. Brawley OW, Thompson IM. Chemoprevention of prostate cancer. *Urology* 1994; 43:594–9.
3. Heber D, Fair WR, Ornish D. *Nutrition and Prostate Cancer: A Monograph from the CAPSURE Nutrition Project*, 2nd edn. 1998 CAPSURE.
4. *Food, Nutrition and Prevention of Cancer: A Global Perspective*, pp. 310–23. American Institute for Cancer Research, 1997. CAPSURE.
5. Muir CS, Nectoux J, Staszewski J. The epidemiology of prostatic cancer: geographical distribution and time trends. *Acta Oncol.* 1991; 30:133–40.
6. Morton MS, Turkes A, Denis L et al. Can dietary factors influence prostatic disease? *Br. J. Urol. Int.* 1999; 84:549–54.
7. Schulman CC, Ekane S, Zlotta AR. Nutrition and prostate cancer: evidence or suspicion? *Urology* 2001; 58:318–34.
8. Rohan TE, Howe GR, Burch JD et al. Dietary factors and risk of prostate cancer: a case-control study in Ontario, Canada. *Cancer Causes Control* 1995; 6:145–54.
9. Montironi R., Mazzucchelli R, Marshall JR et al. Prostate cancer prevention: review of target populations, pathological biomarkers, and chemopreventive agents. *J. Clin. Pathol.* 1999; 52:793–803.
10. Bairati I, Meyer F, Fradet Y et al. Dietary fat and advanced prostate cancer. *J. Urol.* 1998; 159:1271–5.
11. Pandali PK, Pilat MJ, Yamazaki K et al. The effects of omega-3 and omega-6 fatty acids on in vitro prostate cancer growth. *Anticancer Res.* 1996; 16:815–20.
12. Giovannucci E, Ascherio A, Rimm EB et al. Intake of carotenoids and retinols in relation to risk of prostate cancer. *J. Natl Cancer Inst.* 1995; 87:1767–76.
13. Krisnsky NI. Overview of lycopene, carotenoids and disease prevention. *Proc Soc Exp Biol Med* 1998; 218:95–7.
14. Norris AE, Jackson RT, Sharpe SJ et al. Prostate cancer and dietary carotenoids. *Am. J. Epidemiol.* 2000; 151:119–23.
15. Gann PH, Ma J, Giovannucci E et al. Lower prostate cancer risk in men within elevated plasma lycopene levels: results of a prospective analysis. *Cancer Res.* 1999; 59:1225–30.
16. Giovannucci E. Tomatoes, tomato-based products, lycopene, and cancer: review of the epidemiologic literature. *J. Natl Cancer Inst.* 1999; 91:317–31.
17. Eichholzer M, Stahelin HB, Gey KF et al. Prediction of male cancer mortality by plasma levels of interacting vitamins: 17 years of follow-up of the prospective based study. *Int. J. Cancer* 1996; 66:145–50.
18. Fair WR, Fleshner NE, Heston W. Cancer of the prostate: a nutritional disease? *Urology* 1997; 50:840–8.
19. Peto R, Doll R, Buckley JB et al. Can dietary beta-carotene materially reduce human cancer rates? *Nature* 1981; 290:201–8.
20. Machlin LJ, Bendich A. Free radical tissue damage: protective role of antioxidant nutrients. *FASEB J.* 1987; 1:441–5.
21. Heinonen OP, Albanes D, Virtano J et al. Prostate cancer and supplementation with α-tocopherol and β-carotene: incidence and mortality in a controlled trial. *J. Natl Cancer Inst.* 1998; 90:440–6.
22. Moyad MA. Soy, disease, prevention and prostate cancer. *Semin. Urol. Oncol.* 1999; 17:97–102.
23. Bonn D. Vitamin E may reduce prostate cancer incidence. *Lancet* 1998; 351:961.
24. Chan JM, Stampfer MJ, Ma J et al. Supplemental vitamin E intake and prostate cancer risk in a large cohort of men in the United States. *Cancer Epidemiol. Biomarkers Prev.* 1999; 8:893–9.
25. Hanchette CL, Schwarz GG. Geographic patterns of prostate cancer mortality: evidence for a preventive effect of ultraviolet radiation. *Cancer* 1992; 70:2861–9.
26. Habuchi T, Suzuki T, Sasaki R et al. Association of vitamin D gene polymorphism with prostate cancer and benign prostatic hyperplasia in Japanese population. *Cancer Res.* 2000; 60:305–8.
27. Peehl DM. Vitamin D and prostate cancer risk. *Eur. Urol.* 1999; 35:392–4.
28. Konety BR, Johnson CS, Trump DL et al. Vitamin D in the prevention and treatment of prostate cancer. *Semin. Urol. Oncol.* 1993; 17:77–84.

29. Cohen JH, Kristal AR, Stanford JL. Fruit and vegetable intakes and prostate cancer risk. *J. Natl Cancer Inst.* 2000; 92:61–8.
30. Giovannucci E, Rimm EB, Wolk A et al. Calcium and fructose intake in relation to risk of prostate cancer. *Cancer Res.* 1998; 58:442–7.
31. Nelson MA, Porterfield BW, Jacobs ET et al. Selenium and prostate cancer prevention. *Semin. Urol. Oncol.* 1999; 7:91–6.
32. Yip I, Heber D, Aronson W. Nutrition and prostate cancer. *Urol. Clin. North Am.* 1999; 26:403–11.
33. Clark LC, Dalkin B, Krongrad A et al. Decreased incidence of prostate cancer with selenium supplementation: results of a double-blind cancer prevention trial. *Br. J. Urol.* 1998; 81:730–4.
34. Yoshizawa K, Willett WC, Morris SJ et al. Study of prediagnostic selenium levelin toenails and the risk of advanced prostate cancer. *J. Natl Cancer Inst.* 1998; 90:1219–24.
35. Miller GJ. Prostate cancer among the Chinese: pathologic, epidemiologic and nutritional considerations. In: Resnick, Thompson IM (eds) *Advanced Therapy of Prostate Disease*, pp. 18–27. Hamilton, Ontario: BC Decker Inc., 2000.
36. Hebert JR, Hurley TG, Olendzki BC et al. Nutrition and socioeconomic factors in relation to prostate cancer mortality: a cross-national study. *J. Natl Cancer Inst.* 1998; 90:1637–47.
37. Kelloff GJ, Lieberman R, Steele VE et al. Chemoprevention of prostate cancer: concepts and strategies. *Eur. Urol.* 1996; 35:342–50.
38. Montironi R, Schulman CC. Precursors of prostatic cancer: progression, regression and chemoprevention. *Eur. Urol.* 1996; 30:133–7.
39. Schröder FH, Kranse R, Dijk MA et al. Tertiary prevention of prostate cancer by dietary intervention: results of a randomized double blind, placebo controlled, cross-over study. *Eur. Urol.* 2000; 37(Suppl. 2):A96.
40. Lichtenstein P, Holm NV, Verkasalo PK et al. Environmental and heritable factors in the causation of cancer – analyses of cohorts of twins from Sweden, Denmark, and Finland. *N. Engl. J. Med.* 2000; 343:135–6.
41. Demark-Wahnefried W, Peterson B, McBride C et al. Current health behaviors and readiness to pursue life-style changes among men and women diagnosed with early prostate and breast carcinomas. *Cancer* 2000; 88:674–84.

What are the Effects of Dietary Supplements (Selenium and Vitamin E) or Aspirin/NSAIDs on Prostate Cancer?

Mark A. Moyad

Department of Surgery, Section of Urology, University of Michigan Medical Center, Ann Arbor, MI, USA

INTRODUCTION

Selenium and vitamin E supplements have received considerably more attention than probably any other in the area of prostate cancer. The overall interest is reflected in the National Institutes of Health sponsoring one of the largest chemoprevention trials in prostate cancer history utilizing only these two agents and a placebo.[1] This clinical trial is called SELECT, which stands for the selenium and vitamin E chemoprevention trial. It will consist of four arms or intervention groups that include: (1) selenium at 200 µg/day; (2) vitamin E at 400 mg/day; (3) selenium (200 µg/day) and vitamin E (400 mg/day); and (4) placebo.

A total of 32 000 patients from approximately 300 different sites are expected to participate and the trial will last 7–12 years. Patients without a history of prostate cancer or high-grade prostatic intraepithelial neoplasia (PIN) who are aged 55 or older will be eligible for the study. African Americans who are 50 years of age or older will be eligible. This should be the best determination of where these two supplements stand in the area of prostate cancer prevention or possibly even progression. However, several questions should still be considered. For example, are researchers acting too quickly or do these supplements have enough data right now to be part of this 150–175 million dollar clinical trial? How did researchers become so interested in these two supplements exclusively over the hundreds of others available on the market today? Should either, or both of these supplements, be recommended by physicians who are currently dealing with patients who are inquiring about them? Is it possible to find so many men who are not truly taking any or both of these supplements currently? Are there other potential over the counter (OTC) supplements or products that appear promising for the prevention and/or progression of prostate cancer? For example, what role does aspirin or the non-steroidal anti-inflammatory drugs (NSAIDs) play in prostate carcinogenesis? It is difficult right now to answer most of these questions, but this chapter will attempt to answer a few of these and many others.

SELENIUM SUPPLEMENTS AND DIETARY SOURCES

Lower serum levels of the mineral selenium have been linked to numerous cancers over the past few decades.[2] In fact, it was this and other findings that were the impetus for the Nutritional Prevention of Cancer Study.[3,4] This was a double-blind trial of dietary selenium supplementation whose primary goal was to establish whether or not selenium supplements have any role in reducing the recurrence of skin cancer in individuals at high risk for this disease. The primary end-points for this trial were the incidence of basal and squamous cell carcinoma. The first patients in this trial were randomized in 1983. In 1990, after the start of this trial, other secondary end-points were added, such as mortality from all causes and cancer, and the incidence of lung, colon and prostate cancers. Patients either received 200 µg of selenium daily or a placebo. When the trial was completed, patients randomized to the selenium group had a statistically significant lower incidence (63%) of prostate cancer compared to the placebo group. In fact, the selenium group had a lower incidence of localized disease (RR = 0.42) and advanced disease (RR = 0.27). However, this particular study did reveal some other important findings. Examples are listed below.

- Selenium had no effect on the primary end-point (basal and squamous cell cancers). A higher number of both types of cancer were actually found in the selenium group (RR = 1.14 and RR = 1.10).
- Selenium had a significant effect on reducing several secondary end-points (prostate, lung, colon, total carcinoma incidence and mortality...). Several other end-points such as bladder, breast and leukemia/lymphoma cancers were non-significantly higher in the selenium group. However, clinicians should remember that, when a variety of end-points or subgroups are analyzed in studies, the potential of finding chance differences increases.[5]
- A pathology report was available to confirm only 76% of the cancer diagnoses.[3]
- The proportion of subjects reporting other previous cancers in the selenium group was 35% lower than the placebo group. However, the actual percentage of men with several other baseline risk factors (age, prostate specific antigen levels) for prostate cancer were approximately equal between the two compared groups. Family history of prostate cancer was not mentioned.
- Individuals in this study were recruited from selenium-deficient areas. The mean plasma selenium level at baseline (114 ng/ml) was in the lower range of normal levels reported in the USA. It is possible that numerous individuals had suboptimal glutathione peroxidase activity, which would allow more free radicals to initiate and promote cancer, primarily in subjects with low selenium levels.[6] However, this important hypothesis was not tested in this trial.
- Patients in the lowest (less than 106.4 ng/ml) and middle (106.4–121.2 ng/ml) tertile of baseline plasma selenium had a significant benefit or effect, while those in the highest tertile (greater than 121.2 ng/ml) did not.[4] The selenium supplement was protective only for individuals who began the study with lower serum selenium levels. Recent updates from this investigation, which was continued for 3 more years after its initial publication, and has added more cases during this time, demonstrated some interesting results.[7,8] For example, men in the higher baseline category of plasma selenium actually experienced an increase in prostate cancer risk with selenium suppplementation.[7] Additionally, the significant reduction in lung cancer risk initially reported (RR = 0.54, $P = 0.04$) failed to persist after the most recent follow-up (RR = 0.74, $P = 0.23$).[8]
- The 200 μg of selenium utilized in this trial came from a 0.5 g high-selenium brewers yeast tablet, which consists of a mixture of different forms of selenium.[4] Therefore, is it possible that a non-yeast-based selenium supplement would have had a similar effect?
- Only 25% of the participants in this trial were women, so researchers are not exactly sure what role, if any, this mineral plays in women's health. However, one large prospective trial of toenail selenium levels in women did not reveal any benefit of this mineral (primarily from the diet) for the prevention of cancer, including breast carcinoma.[9] Interestingly, higher compared to lower selenium levels were associated with a non-significant greater risk of cancer. Women who reported taking selenium supplements also experienced a non-significant increase in overall cancer risk. Therefore, selenium does not have adequate clinical data right now to espouse its use in women. This is important because it is not uncommon for the spouse of a prostate cancer patient to ask 'Should I be taking this supplement?'.

These overall results and the patient population in the selenium supplement study encourage the question 'Are patients with a deficiency of selenium the only ones who could potentially benefit from a supplement?'. This is difficult to say at this time, but the potential exists. The Nutritional Prevention of Cancer Study suggests normalizing selenium levels from a dietary or supplement source is what provides protection to this high-risk group from cancer. The proper choice of repletion should depend on the individual history, plasma selenium status and lifestyle changes of the individual through proper monitoring with a health care professional. The actual study population utilized in the selenium supplement trial had a rate of cancer incidence higher than the general US population, and selenium supplements appeared to reduce this risk to what is normally observed in the USA.[10] A number of past studies, which associate selenium intake with lower cancer rates, were conducted in Finland, which was known to have lower levels of this mineral (versus the USA) before it was added to fertilizers in 1985.[11–13] It has been noted that the highest quintile of plasma selenium in the past Finnish studies overlaps with the lowest quintile in the US study groups.[14] Selenium is not necessarily easy to obtain from food sources,[15] and the actual content of selenium in food varies greatly depending on which region of the country you live in.[16] **Table 21.1** provides a partial list of healthy food sources of selenium.

Table 21.1. A partial list of the approximate selenium content of some healthy foods (in micrograms per 3.5-oz or 100-g serving).

Healthy food	Selenium content
Wheatgerm	110
Brazil nuts	105
Bran	65
Whole-wheat bread	65
Oats	55
Brown rice	40
Barley	25
Garlic	25
Turnips	25
Orange juice	20

Note: other good sources of selenium include fish, meat, mushrooms and poultry. The RDA for selenium is 70 μg/day for men and 55 μg for women.

Theoretically speaking, those individuals who consume healthy food sources of selenium should be able to achieve adequate blood levels of selenium. An added option would be to include a multivitamin daily because most contain the RDA (70 μg/day for men and 55 μg/day for women) for selenium, or another option is initially to have a blood test to determine selenium levels. Japan, which is noted for its past low incidence of prostate cancer has a high average intake (about 130 μg/day per person) of this mineral from dietary sources.[17]

Other smaller past studies that have used serum measurements of selenium status and its correlation to prostate cancer have not revealed an adequate amount of data for espousing higher intakes of selenium. Three prospective studies included only a total of 75 prostate cancer cases.[18–20] The two smaller studies (from the USA) demonstrated a non-significant inverse association,[18,19] while the largest study conducted in Finland, which is noted for its low levels of selenium, showed no association.[20] A more recent study provided some evidence for the link between selenium and a lower risk of advanced prostate cancer.[21] Toenail clippings were collected from over 33 000 men in the Health Professional Follow-Up Study. Toenails have been considered a useful marker of the average intake of selenium over several months.[22] The researchers found that higher concentrations of selenium in the toenails (after a 7-year observation) were associated with a reduction of advanced prostate cancer in 181 patients.[21]

The most recent study was a nested case-control study in a cohort of over 9000 Japanese American men examined between 1971 and 1977.[23] The strength of this study was that it included 249 cases identified during 20 or more years of follow-up. The only strong association between increased selenium levels and a lower incidence of prostate cancer were found in current and past smokers. This finding was unexpected because the majority of large cohort studies of smokers have not demonstrated an association with prostate cancer.[24] Although two fairly recent prospective studies found a slight increase in prostate cancer risk among smokers,[25,26] and another prospective study found that recent smokers had an increased risk for metastatic or fatal prostate carcinoma,[27] lung cancer, which is strongly related to smoking, has had mixed results when looking at selenium status.[20,28,29] Some studies report a slight inverse association with selenium,[20,28] while others do not.[29] In reality, smoking and/or aging can significantly lower serum levels of numerous antioxidants, but whether or not this includes selenium requires further study.[30] Although smokers tend also to differ from non-smokers with respect to several lifestyle behaviors, for example, eating less healthful diets, exercising less, etc.,[31] the problem with the results of the most recent trials lies in the fact that the two previous studies that showed an association between higher levels of selenium and lower prostate cancer risk (selenium supplement trial for skin carcinoma and the Health Professional Follow-up Study) were not analyzed separately by smoking history.[3,21] Approximately 60% or more of the overall participants or cases and controls in the past three trials were current or past smokers. This leads to the conundrum that currently exists when it comes to selenium for prostate cancer. Does selenium reduce the risk of this disease, and if it does, is this reduction only found in current and recent smokers? This is the crucial question because, as described later in this chapter, vitamin E supplementation has been caught in this same dilemma.

VITAMIN E SUPPLEMENTS, DIETARY SOURCES AND SMOKERS

Vitamin E did not receive much attention as an anti-prostate cancer agent until the results of the Alpha-Tocopherol, Beta-Carotene Cancer Prevention Study (ATBC Study) were released in 1998.[32] Over 29 000 male smokers were involved in this 5–8-year study that was designed primarily to determine if either vitamin E and/or β-carotene supplements could prevent lung cancer. Again, as in the selenium supplement trial, prostate cancer was not the primary end-point. There was a 32% reduction in the incidence of prostate cancer among the subjects receiving vitamin E and a 41% decrease in mortality from prostate cancer. However, the results of this study have provided several other interesting observations as listed below.

- Almost all of the participants were current (60%) or past (39%) heavy smokers (median 20 cigarettes/day) with a long duration of smoking prior to study entry (median 36 years), so do these results have any bearing on non-smokers?
- The vitamin E supplement used in this study only contained 50 mg of synthetic vitamin E, not the higher levels (400 mg/day) that will be taken during the SELECT trial. In fact, it is difficult if not impossible to find a commercially available individual vitamin E supplement that contains only 50 mg of synthetic vitamin E.
- Vitamin E had no impact on the time interval between prostate cancer diagnosis and death. Many clinicians and patients believe that vitamin E has demonstrated some impact on prostate cancer progression because in the ATBC trial there was a lower number of prostate cancer deaths in the vitamin E group. However, these data were misconstrued because a lower number of cases being diagnosed will automatically be tantamount to a lower rate of death from prostate cancer. This does not necessarily mean that vitamin E will significantly or slightly impact the progression of already diagnosed prostate cancer.
- The men with the highest serum levels of vitamin E had significantly lower testosterone, androstenedione, sex hormone-binding globulin and estrone levels.[33] The precise difference in hormone concentration per milligram of α-tocopherol was approximately 1.8–2.6% for these four hormones. So, it could be possible that, if vitamin E is preventing prostate cancer in smokers, it is doing so by lowering androgen levels (partial androgen suppression).
- Men in the highest weight category had a 39% increase in the risk of prostate cancer.[34] This was a fascinating result because it was so dramatic and because it received so little

media attention, while the supplement results seemed to dominate headlines across the country.

- Men receiving β-carotene supplements had 23% increase in prostate cancer incidence and a 15% increase in mortality.[32] Does this mean that the secondary end-points are not a quality finding because with one supplement it demonstrated a benefit and with the other it demonstrated a negative finding. This could also mean that β-carotene supplements are indeed a potential hazard in smokers, as is the case from this trial and others. One of the largest and longest large-scale prospective studies of β-carotene (Physicians' Health Study) found that men with the lowest baseline plasma levels of β-carotene or men within the highest body mass index (BMI) quartile were the only subgroup that demonstrated a lower risk of prostate cancer after 12 years of β-carotene supplementation (RR = 0.68 and RR = 0.8).[35,36] Lower levels of baseline β-carotene were also associated with younger, heavier, smoking men with greater alcohol intakes. This would lend some further support to the concept of supplements providing a benefit only for those with initially lower levels of certain plasma antioxidants or men at a higher risk of prostate cancer because of their current lifestyle. A review of 10 past cohort studies that analyzed prediagnostic serum samples of selenium, vitamin E, β-carotene or retinol found a fairly consistent association of an increased cancer risk related to lower nutrient levels.[37] Again, those who are somewhat deficient or deficient in certain antioxidants may have the most to gain from ingesting greater dietary sources of these compounds and/or taking a supplement. On the other hand, men with the highest baseline β-carotene plasma levels in Physicians' Health Study who were assigned to the β-carotene supplement were found to have an increased risk of prostate cancer and a higher risk for aggressive disease (RR = 1.33). This would lend some further support to the concept of supplements providing a detrimental effect for those with normal or higher levels of plasma antioxidants from dietary sources.
- Men who combined vitamin E with aspirin had a greater risk of oral bleeding compared to those who only took aspirin.[38]
- Men taking vitamin E had a slight (non-significant) increase in hemorrhagic stroke. Is it possible that such small quantities of vitamin E supplements in smokers could demonstrate such a hazardous side effect?[38]
- Men in this trial had lower mean selenium levels compared to men in the selenium supplement trial. So, is it possible that the only smokers who could potentially benefit from taking vitamin E are those with overall low levels of other key antioxidants? A small prospective study from Switzerland with a 17-year follow-up found a significantly increased risk of prostate cancer deaths among smokers with low serum levels of vitamin E (RR = 8.3).[39] However, there were only 30 deaths from prostate cancer among the almost 3000 men followed. Another similar correlation was found in the prospective Physicians' Health Study, which had a 13-year follow-up period and 259 cases of aggressive prostate cancer diagnosed.[40] This study found that smokers with the highest levels of serum α-tocopherol experienced an almost 50% decrease in the risk of aggressive prostate cancer, while non-smokers did not demonstrate any association. More recently, the US Health Professional Follow-up Study examined prospectively the intake of vitamin E supplements and prostate cancer risk.[41] There were 1896 total cases of prostate cancer in this study, and 522 of these cases were extraprostatic, while 232 were metastatic or fatal. There was no overall association between vitamin E supplements and prostate cancer for non-smokers, although non-smoking men ingesting the largest amounts of vitamin E supplements (more than 100 IU/day) had an increased risk of advanced prostate cancer. However, among current and recent smokers, those men who consumed more than 100 IU of supplemental vitamin E daily had a 56% decrease risk of advanced or fatal prostate cancer compared with non-users. Therefore, four separate trials (ATBC, the Swiss study, the Physicians' Health Study and the US Health Professionals Follow-up Study) have all found a link between vitamin E and a reduction of prostate cancer progression and mortality in current or past/recent smokers.

Dietary vitamin E is generally found in various nuts, seeds and oils.[16] **Table 21.2** provides details of some dietary sources of vitamin E.

A large proportion of the vitamin E found in the diet is γ-tocopherol, not α-tocopherol, which is the type found in most bottles of supplemental vitamin E and is the type used in most past clinical trials.[42] Why is this important? Several studies of heart disease have found a lower risk of death from heart disease for individuals obtaining dietary vitamin E compared with supplemental vitamin E. For example, the Iowa Women's Health Study is a prospective cohort study of almost 35 000 postmenopausal women.[43] In this study, dietary vitamin E was inversely associated with the risk of death from heart disease. This correlation was especially evident in a subgroup of almost 22 000 women who did not take any vitamin E supplements. Women in the highest dietary intake of vitamin E had a 58% decrease in the risk of death from heart disease. Additionally, women who consumed higher amounts of dietary vitamin E were less likely to smoke. Other past

Table 21.2. Selected healthy dietary sources of vitamin E.

Dietary source	Total vitamin E (mg)
Wheatgerm oil (1 tablespoon)	35
Almonds and other nuts (½ cup)	10–15
Soybean oil (1 tablespoon)	10–15
Sunflower oil (1 tablespoon)	8–9
Milk non-fat (1 cup)	7.5
Peas (1 cup)	3.5
Olive oil (1 tablespoon)	2.0
Salmon (3 oz)	1–1.5

Note: the RDA for vitamin E is 15 IU or 10 mg/day

prospective studies have also found an association between higher intakes of dietary vitamin E and a lower risk of heart disease in men and women.[44,45]

We have recently reported that γ-tocopherol (from dietary sources) demonstrated greater inhibitory capacity on prostate cancer cell lines versus supplemental (α-tocopherol) vitamin E.[42] Interestingly, a recent look at men obtaining higher levels of dietary vitamin E (γ-tocopherol) reported a greater reduction of prostate cancer risk versus that of α-tocopherol and/or selenium.[46] In this nested case-control study of a cohort from Maryland, men in the highest fifth of plasma γ-tocopherol had an 80% decrease in the risk of prostate cancer compared to those in the lowest fifth. In addition, statistically significant reductions for men with high levels of selenium and α-tocopherol were also demonstrated, but only when γ-tocopherol levels were high. Higher levels of γ-tocopherol are especially found in soybeans, soybean oil and a variety of other sources.[16] These early reports are fascinating but they need to be confirmed in larger prospective studies.

Studies of vitamin E supplements to reduce morbidity and mortality from other causes seem pertinent to the issues discussed with prostate cancer. For example, although past observational studies may support the use of vitamin E supplements to reduce cardiovascular risk,[47,48] recent large-scale prospective studies suggest a different situation.[49–51] For example, the Heart Outcomes Prevention Evaluation (HOPE) trial demonstrated no apparent effect after 4.5 years on cardiovascular outcomes with a 400 IU daily supplement of vitamin E versus placebo in high-risk patients.[49,50] Although an angiotensin-converting-enzyme inhibitor (ramipril) versus placebo demonstrated a significant reduction in the rates of death, myocardial infarction and stroke,[51] a subsequent meta-analysis of all of the previous large prospective trials of vitamin E have failed to demonstrate a reduction in overall cardiovascular events. The most recent trial published (not included in the above meta-analysis) was the Primary Prevention Project, which failed to reduce the prevention of cardiovascular events with 300 IU daily of vitamin E over a mean follow-up of 3.6 years.[52] However, this trial was stopped early because low-dose aspirin (100 mg/day) was found to reduce similar events significantly. Aspirin had no effect on cancer incidence, but previous studies suggest that regular use, higher dosages, longer durations of treatment and preferably high-risk individuals are needed to observe large reductions.[53] The benefit with aspirin versus a dietary supplement is consistent with other trials such as the Physicians' Health Study.[54] Perhaps, low-dose aspirin and other NSAIDS may be the best weapon to utilize for cardiovascular and prostate or other cancer risk reduction?[55,56] Critics of the vitamin E trials might argue that vitamin E was not used in a large enough dosage and for a duration that was adequate, or that other cofactors are needed along with vitamin E to demonstrate a benefit.[57,58] This is a reasonable argument, but the early results have not been promising in the longest trials and they might be providing some insight into what could be expected with future trials of vitamin E.

SELENIUM AND VITAMIN E SUPPLEMENTS – ANALYZING THE SUM OF THE CLINICAL DATA

This review of past selenium and vitamin E dietary studies leads this author to numerous conclusions that can currently benefit clinicians discussing these supplements with their patients. These are listed below.

- Supplements of selenium and vitamin E need more clinical data to espouse their use for the prevention of prostate cancer. The SELECT trial will ultimately decide the potential role of these supplements in prostate cancer, but several concerns over the SELECT trial seem to exist (see later).
- There are some data to suggest that supplemental selenium at 200 μg/day has a role in prevention and in potentially slowing the progression of prostate cancer, especially in men with low baseline levels of serum selenium.
- There are also some clinical data to suggest that dietary selenium may also prevent and/or slow the progression of prostate cancer.
- There are some data to suggest that men with low baseline levels of selenium are the ones who could benefit most from selenium supplements or higher intakes of dietary selenium. Higher dietary intakes appear to be safe and may reduce the risk of prostate cancer, while selenium supplements in individuals with normal to high selenium levels may have an adverse effect. This has been observed in other trials of different dietary supplements, such as β-carotene.
- There are no strong clinical data to suggest whether or not increasing selenium intake from diet or supplements benefits non-smokers and/or smokers and/or recent smokers.
- There are no strong clinical data to suggest that supplemental vitamin E has a role in the prevention or progression of prostate cancer in non-smokers, and supplements of vitamin E may even have an adverse effect on non-smokers.
- There are some adequate clinical data to suggest a benefit for taking supplemental vitamin E for prostate cancer prevention and progression, for current or recent smokers.
- There is some evidence to suggest that, if supplemental vitamin E is effective in smokers, then it may be working through a partial hormonal suppression mechanism.
- There is some recent evidence to suggest that dietary vitamin E (γ-tocopherol) may reduce the risk of prostate cancer and high levels of supplemental vitamin E may reduce serum levels of γ-tocopherol. Prospective studies of vitamin E have not been impressive when dealing with the issue of cardiovascular risk reduction. However, low-dose aspirin continues to have favorable results.[52,54] Studies with low-dose aspirin should receive more research for reducing the risk of prostate cancer. It is of interest that several past cohort studies have demonstrated a significant reduction in colon cancer risk with aspirin intake, but attention to dosage, frequency, duration of treatment and established risk factors are needed.[58] Patient selection is critical because aspirin use long-term may be associated with serious side effects, such as gastrointestinal problems and/or hemorrhagic stroke.[59,60]

- Other significant clinical findings have been discovered from the past selenium and vitamin E trials. For example, extremely obese men may have a large increase in prostate cancer risk. In fact, the negative impact (approximately 40% increase in risk) of extreme obesity in the ATBC trial was greater than the positive impact (32% decrease in risk) of taking the vitamin E supplement. It seems as if every media outlet covered the favorable dietary supplement data from this trial, but few if any sources seemed to report the unfavorable effect of obesity.[32,34]

SIDE EFFECTS OF SELENIUM AND VITAMIN E

No discussion of selenium and vitamin E supplements would be complete without mentioning the potential side effects of these popular supplements. The literature reminds us of the potential harm that could exist if someone is selenium deficient. For example, Keshan's disease is a dilated cardiomyopathy that is found in individuals with little to no selenium intake (less than 19 μg/day) in certain areas of China and other countries.[61] However, this disease is a rare finding in most countries, and should not be used as evidence to espouse the use of selenium supplements. In fact, one might equate this problem with that of scurvy and vitamin C, which is difficult if not impossible to find these days in many countries.

Excess levels of selenium ingestion can be quite toxic. Selenium supplements have been associated with gastrointestinal problems, pathologic nailbed changes and the loss of fingernails, temporary hair loss and even fatigue.[62] However, at 200 μg/day these side effects, apart from some gastrointestinal upset were not observed in a past trial,[3] but for the patient who believes in the 'more is better' theory, this could represent a potential problem. Clinicians need to advise patients on the potential harm that could result from taking too much supplemental selenium. A CDC report from 1983 highlights the dangers of high selenium intake.[62] Individuals were taking a selenium supplement that actually contained hundreds of more micrograms and milligrams than was reported on the purchased bottle. This resulted in temporary and permanent nail and hair loss, and fatigue that was mentioned earlier.

Vitamin E, when combined with blood thinners of all types, may increase the risk of internal bleeding.[38] It is notable that smokers in the ATBC trial had an increased risk of hemorrhagic stroke at only 50 mg of vitamin E daily. Other clinical studies have demonstrated that obtaining over 1000 mg of supplemental vitamin E daily may be harmful.[63–65] For example, it can reduce the body's absorption of other antioxidants (like vitamin C) and may act as a pro-oxidant at these levels rather than an antioxidant, especially if you are not a smoker.[66] It seems that, if any of these supplements have an impact on health, then they do so within a certain window period. Too little does not do anything and too much has either a negative or no impact. This will be a real challenge for future supplement trials: deciding on the right supplement, at the right dosage, and which can give maximal efficacy and minimal side effects.

THE SELECT TRIAL AND ASPIRIN (NSAIDS)

Does it seem adequate at this time to spend such a large sum of money on a prospective 12-year study of vitamin E and/or selenium to reduce prostate cancer risk? Some laboratory and observational studies would suggest that this is a reasonable study. However, after closely evaluating the sum of the previous prospective data in prostate cancer and cardiovascular disease, this conclusion is not necessarily reasonable. There are several reasons for this concern, as listed below.

- There is the issue of past and current smokers who only seem to benefit at this point with vitamin E supplements.
- Other forms of vitamin E (γ-tocopherol, etc.) may also provide a benefit but they are not being tested. The supplement being utilized is α-tocopherol.
- The dosage of vitamin E in this trial is 400 IU. Why is this dosage being utilized when the largest randomized prospective study of a specific vitamin E dosage that demonstrated a potential benefit to reduce prostate cancer risk utilized 50 IU? When in medical research did preliminary studies demonstrate a benefit with a specific dosage (50 IU), and then the larger follow-up prospective study to validate these results utilized a dosage eight times (400 IU) that of what seemed to be effective? It would be difficult to believe that any conventional study would even be permitted or interested in changing the dosage to such an enormous degree. If a specific intervention looks promising in preliminary trials at 50 mg, for example, researchers would not validate these results by utilizing only a 400 mg dose in a subsequent validation trial.
- Some may argue that higher dosages of vitamin E is what has looked promising in other prospective studies of vitamin E and prostate cancer. However, these specific dosages were not tested but only observed to be effective after the data were analyzed, and smokers were the only group of men who benefited. Non-smokers actually had a higher risk of prostate cancer, including aggressive disease or vitamin E at the higher dosages had little to no impact on risk. Regardless, where is the clinical data to precisely support such a dosage? Other critics might argue that cardiovascular trials utilize higher dosages. However, the primary end-point of the SELECT trial is prostate cancer and not cardiovascular disease. In addition, if researchers examine the results of prospective trials of vitamin E and cardiovascular disease, it would seem unlikely that they support such a dosage or even such a trial at this time.
- Research from cardiovascular trials seem to support the notion that aspirin or other NSAIDS may be a better choice for testing the potential to reduce prostate cancer risk through an over the counter product.[67,68] In addition, aspirin or similar products have demonstrated a potential benefit in other studies to reduce colon cancer. Already, there is a strong precedence to utilize aspirin for cardiovascular disease and prostate cancer prevention in future studies. Additionally, the number 1 or 2 cause of death in men at risk or diagnosed

with prostate cancer is ischemic heart disease.[69] Thus, aspirin seems to be one of the most logical choices for any future long-term clinical trial for prostate cancer prevention or to reduce the progression of this disease. This is not necessarily the case with vitamin E. Other cancers and cardiovascular disease have not demonstrated such a substantial benefit in taking vitamin E to reduce risk.

- The issue of also utilizing a selenium supplement at this point is not free of controversy. The dosage of 200 μg does make sense from the past Nutritional Prevention of Cancer Study, but other findings raise some serious concerns. Only men with lower levels of plasma selenium benefited and men with higher levels had an increased risk. These individuals were also at a higher risk for cancer. A newer prospective study only suggests that smokers may benefit. Does this substantiate the use of this supplement for men in a large prospective trial? Other large-scale prospective trials only seem to support the notion that unless lower plasma levels of certain antioxidants exist it is difficult to favor specific individual supplements. For example, the Physicians' Health Study failed to provide any noticeable benefit after 12 years of study. However, possibly the men with the lowest baseline levels of this compound received a benefit. Otherwise, no benefit was found. It could be argued that more data exist for the selenium supplement over vitamin E, but overall the data seem weak to support either supplement for such a large-scale prevention trial, unless researchers utilize the sum of this information learned from past trials. This does not seem to be the goal of the SELECT trial. Rather, its purpose is to recruit thousands of individuals from approximately 300 sites who fit the minimal criteria for study entry. Timing seems more of an issue than a reliance on proper past methodology.
- Why is there such an immediate concern to begin such a trial? Dozens of prospective trials are in the process and have been completed. Every passing year seems to provide greater clinical insight into the possibility of which antioxidants or over the counter products may provide the best benefit. What about the possibility of patiently waiting for or conducting retrospective looks at prospective selenium and vitamin E trials to decide whether or not enough evidence exists to support the use of either supplement for a 12-year costly trial. The cost would be lower and the information gleaned would be valuable. Are there better ways to spend this money now and in the future?
- Another current reality and concern is that numerous participants in these newer trials will probably supplement or have recently supplemented with extra or unique sources of other antioxidants. Will this be regularly determined and recorded, and what impact will this have on the results? For example, recent results from the ongoing Colorectal Adenoma Prevention Study (CAPS), which is a trial of aspirin to determine if it has an effect on reducing new colorectal adenomas among individuals who already have had at least one colorectal carcinoma, demonstrates the concern.[70] Supplement use in this trial is common among the participants. One or more supplements were used at some point by 55% of the sample subjects. Among those taking supplements, 66% took more than one and more than 10% took five or more. These researchers have expressed the concern that investigators need to record more precisely supplement use in these studies and that it may also be necessary to increase the size of these trials if many of the subjects are taking supplements.
- Researchers working in the so-called field of 'alternative medicine' seem to be appropriately reminded of the fact that medicine is medicine, and research is research. It is not 'alternative' because either proper evidence exists or it does not, and following the normal course of the scientific method is the only sound way to determine future clinical recommendations. However, conventional medicine cannot become immune from this same scientific method when dealing with dietary supplements and potential objective criticism. The current SELECT trial seems to be somewhat guilty of similar accusations that have been delivered by certain individuals to alternative researchers. Large sums of money are being spent and enormous resources are being allocated in an area where a lack of adequate evidence seems to exist. The timing of such a trial does not seem appropriate when evaluating the sum of the data. More selective criteria and a proper attention to design or dosages should be employed, or probably more appropriately, the trial should be delayed until a better understanding of the proper scientific method in this situation can be determined. There is little doubt that something will be learned from this trial, but the question that needs an answer is 'Could substantially more information been gleaned if the trial used patients, dosages, and other interventions that properly fit the design from past successful studies of these supplements?.' Past and current academic mentors have always reminded us that this is what separates the proper from the improper or the 'conventional' from the 'alternative' scientific method. Regardless of the outcome, the current SELECT trial may elicit more controversy than clarity.

ASPIRIN/NSAIDS AND PROSTATE CANCER

Again, there are some investigations, albeit preliminary, that aspirin may have a potential role in preventing prostate cancer or the progression of this disease. Studies have suggested that men with ischemic heart disease may be at an increased risk for prostate cancer.[69,71] Thus, the potential for investigating this agent for both conditions concurrently seems worthwhile.[72,73] A variety of laboratory studies have noted an increased production of prostaglandins in human prostate cancer tissue,[74–82] and an overall inhibitory effect of NSAIDs on the cellular growth of both androgen-sensitive and androgen-insensitive human prostate cancer cell lines.[83–86] Collectively, these studies suggest that NSAIDs may have a favorable role in the early or latter stages of carcinogenesis.[83] Two cohort studies investigating the use of aspirin in relation with multiple cancer sites observed weak inverse relationships for prostate cancer risk (RR = 0.90–0.95).[87,88] A third cohort studied aspirin use and mortality outcomes in association to male genital system cancers (mostly prostate).[89] Cancer mortality risk decreased (RR=0.82) for men taking the most frequent dosage of aspirin. However, these past three studies had limited power to analyze prostate cancer risk specifically and contained several questions regarding other confounding variables.

Recently, the first epidemiologic study to analyze the association between specific prostate cancer risk and use of NSAIDs was published. This was a population-based case-control study from Auckland, New Zealand.[90] A total of 317 recently diagnosed prostate cancer cases including 192 advanced cases were age-matched to 480 controls. Approximately 66% of the total cases and 65% of the advanced cases were 67 years of age or older. Most of the cases were ex-smokers, and less than 10% of the men had a family history of this disease. An inverse relationship (not significant) was found for regular use of total NSAIDs (RR = 0.73) and total aspirin (RR = 0.71) and a reduction in risk for advanced prostate cancer. Interestingly, low-dose daily aspirin for cardiovascular disease prophylaxis was related with the lowest risk of advanced prostate cancer (RR = 0.69). A smaller inverse association was noted for all prostate cancers that included small-volume low-grade cancers (RR=0.84–0.88). The findings of this investigation and others suggested that NSAIDs may exert their largest effect on more aggressive or rapidly growing carcinomas and may play a smaller role in inhibiting small, slower growing and less aggressive tumors.[78,91] In addition, compounds produced from the COX pathway may be released in higher concentrations in more aggressive cancers. Several other recent retrospective and prospective studies continue to support the role of NSAIDs for preventing or inhibiting the progression of prostate cancer. For example, a case-control study of approximately 400 prostate cancer cases and a similar number of controls found that regular daily use of NSAIDs, ibuprofen or aspirin was associated with a significant 66% decrease in prostate cancer risk.[92] Men on prescription NSAIDs also experienced a significant 65% decrease in risk.

A cohort of men from Olmstead County, MN (n=1362), aged 50–79 years were analyzed to determine if NSAIDs had any effect on prostate cancer risk.[93] Approximately 4% (n=23) of the 569 NSAIDs users and 9% (n=68) of the non-NSAIDs users were diagnosed with prostate cancer, which was a significant difference between the groups. This was tantamount to a 55% reduction in the NSAIDs users versus the non-users. Although, the apparent benefit of NSAIDs increased with increasing age of the study participants. Men aged 50–59 years of age had a risk reduction of only 10%, whereas those 60–69 years old had a reduction of 60%, and men aged 70–79 years experienced an 80% reduction if they utilized NSAIDs regularly versus the controls.

A prospective cohort presented concurrently as the previous study also examined the relationship of NSAIDs with prostate cancer risk from the Baltimore Longitudinal Study of Aging.[94] The median follow-up was 9.4 years, and the median age after the final follow-up was 68.6 years. The total relative risk of prostate cancer was lower (RR=0.76) for men using NSAIDs other than aspirin, and also for men utilizing aspirin (RR=0.85). No inverse relationship was found for men using acetaminophen. The findings from this particular study suggested a non-significant decrease in risk that was related to the strength of the anti-inflammatory product. Final analyses revealed that each year of NSAID use was associated with a 6% overall reduction in prostate cancer risk.

More recently, a brief summary of results from the Health Professional Follow-up Study was presented.[95] This is a large cohort of over 35 000 men who were 46–81 years old and with no history of prostate cancer in 1992. Six years of follow-up resulted in 1265 total cases, 239 of which were advanced and 80 metastatic. No significant relationship was found for aspirin use and total or advanced prostate cancer. However, a significant inverse association was found for metastatic prostate cancer and aspirin use. Overall relative risks for men using aspirin 0, 1–4, 5–14, 15–21, 22+days/month were 1.0, 0.35, 0.69, 0.65 and 0.42. Additionally, the total number of aspirin tablets taken/day was also inversely related with the risk of metastatic prostate cancer. The overall relative risk for men taking 0, 0.5, 1, 2+aspirin tablets/day were 1.0, 0.45, 0.44 and 0.54. Thus, the combined effect of the preliminary past investigations suggest that aspirin/NSAIDs may lower the risk of prostate cancer and may have a more pronounced effect on more aggressive tumors, and they may have the greatest benefit in older men.

Thus far, only one study has not demonstrated any benefit for NSAIDs on some aspect of prostate cancer, but an effect on other cancers were noted.[96] This was a case-control study of over 600 analgesic abusers (aspirin, phenacetine, caffeine or codeine) autopsied between 1968 and 1983, and compared with a control group without overt evidence of analgesic abuse. Analgesic abusers had a 60% reduction in tumor development and, when urinary tract tumors were eliminated from this group, this risk decreased to 72%. No significant benefit was noted for analgesic abusers and prostate cancer risk, but this was not a study that exclusively analyzed this outcome, nor were NSAIDs exclusively studied in this investigation. Again, the overall preliminary evidence thus far suggests that aspirin/NSAIDs may have an effect on lowering the risk and slowing the progression of prostate cancer, but this is still controversial because only a small number of studies have been published.[97] More studies examining this potential relationship are needed. Interestingly, the prostate has been found to have the highest concentration of COX-2 mRNA among all other tissues in humans.[98] Therefore, it may be sensitive to the effects of the non-selective or selective COX-2 inhibitory agents. Several studies have found that COX-2, and possibly COX-1 expression is increased in prostate cancer tissue. For example, 31 specimens of prostate cancer and ten specimens of benign prostatic hyperplasia (BPH) were analyzed with murine anti-human COX-1 and COX-2 monoclonal antibodies.[99] COX-1 expression was higher in prostate cancer specimens (63% of samples), as was COX-2 expression (87% of samples). COX-2 expression was minimal in BPH samples, but it was higher in the epithelial cells of high-grade PIN. A similar investigation utilizing specimens from 28 prostate cancer, eight BPH, one PIN and eight men with normal prostate tissue found minimal expression of COX-1 in tumor cells compared to a higher expression of COX-2 in this same tissue.[100] Both COX enzyme isoforms were minimally expressed in the BPH and normal prostate tissues. A recent investigation of 82 prostate cancer and 30 BPH specimens noted a similar effect of a up-regulation of COX-2 in cancerous specimens.[101] There was also a larger and significant increase in the concentration of COX-2 in more aggressive or high-grade cancers compared to the less aggressive or low-grade type. Approximately four times greater expression of COX-2 was found in cancerous compared to BPH tissue. COX-1 expression was approximately equal in the BPH and cancerous specimens. Future investigations of COX-2 inhibitors and NSAIDs in patients with and without prostate carcinoma should provide potentially valuable therapeutic insight. Men with chronic prostate infections or prostatitis have experienced considerable relief of their symptoms with NSAIDs or non-selective COX inhibitors.[102] It is also of considerable interest that other preliminary laboratory investigations utilizing several different human cancer specimens, such as pulmonary, colon, mammary, etc., also found an increase in COX isoforms (especially COX-2).[103–109]

CONCLUSIONS

There seems to be a number of patients and clinicians who are taking or recommending large amounts of supplemental selenium and/or vitamin E for prostate cancer without knowing the precise results of the various clinical studies on these two supplements. Too much is not good, and too little is not good, but in the end the dosage recommended (if any) should come from past clinical studies and trials. There is little doubt that these supplements have a wealth of laboratory data to espouse their use but, beyond that, the data are weak at best. In fact, the only men that should probably be taking selenium and especially vitamin E supplements for prostate cancer prevention or to slow the progression of this disease are current or recent smokers and/or those at high-risk for this disease. Apart from these men, there are currently no strong data to espouse their use. In fact, there seems to be as much data, if not more, to espouse the use or greater intake of better dietary sources of this mineral and vitamin for overall health until more clinical data are published. It could easily turn out in several years that those individuals who focus more on dietary sources right now stand to benefit the most in terms of prostate cancer prevention or reducing progression. Other critical findings from these past studies need to be reiterated to the patient; for example, that men who maintain an adequate body weight may benefit more than taking any supplement could potentially offer.

Perhaps another entirely different approach should be taken with these supplements because they are not chemoprevention agents *per se*. The more logical method from past studies may involve measuring the plasma level of certain antioxidants (selenium, vitamin E, etc.) and, if in the lower range, dietary recommendations should be implemented. If these levels remain low after dietary change, then supplements could be suggested. This method is identical to the strategy utilized for cholesterol or blood pressure reduction to lower cardiovascular risk. If diet and other lifestyle changes are not effective, then a prescribed agent may be suggested. This method also reduces potential adverse effects because restoring normal antioxidant levels has not created such a situation. However, increasing levels beyond normal has the potential to create such an adverse situation for the patient. The concept that dietary supplements operate in a similar fashion to other effective chemoprevention agents (such as tamoxifen) seems somewhat false. These agents reduce the risk in high-risk patients through other unique mechanisms and not by restoring inadequate nutrient or antioxidant levels to normal. For example, by blocking the effect of estrogen or inhibiting the cyclooxygenase pathway and not through restoration of abnormal levels during various times. Also, the concept of reducing birth defects through folic acid supplements makes sense because of the increasing demands for such a compound during the time of pregnancy.

Other high-risk patients may cause researchers to redefine normal levels of these nutrients in these specific individuals only. For example, the possibility exists that higher supplement intakes of calcium in patients at high risk for colon adenomas would be beneficial, or higher intakes of calcium and vitamin D for elderly individuals at risk for osteoporosis seems appropriate. The medical history of the individual, along with up-to-date laboratory and clinical assessments are critical, and should provide the pathway to proper dietary supplement or nutritional recommendations. This seems hardly the case today, but with greater regulation, involvement, objective education and proper research this ideal scenario should be achievable. However, it must be treated as an absolute priority,[110] otherwise patients seem destined to depend on other less credible sources primarily because these same sources seem to treat the issue of dietary supplements as a priority. Finally, it does seem likely that many patients will continue to take these supplements regardless of the present data because they do not want to wait 10–15 years until this supplement question is hopefully resolved. Clinicians should be able to sympathize and understand this position because of the enormous impact that cancer continues to have on our society. Numerous patients have a strong cancer history and probably have a serious desire to reduce this risk in any way that is practical. Therefore, selenium at 200 μg/day and vitamin E at 50 mg/day or more (especially if a current or past smoker) may currently make sense for some specific patients. However, taking more of these supplements would translate into a complete disregard for the past studies on these supplements and prostate cancer.

In conclusion, it is important to remember and reiterate that the dietary sources of certain antioxidants should be considered initially and supplements secondarily. The dietary sources of these and other supplements seem to remain the greater protector of overall health rather than a specific disease entity. In other words, from a probability standpoint, it seems that patients who consume a moderately healthy diet, exercise and incorporate other positive lifestyle changes will always be in a potential win–win situation.[111,112] However, those who depend mostly on supplements find themselves in a possible win, a possible lose and a possible no-benefit scenario. Selenium and vitamin E have no doubt acquired some intriguing evidence over the past several years but these past data also seem to be have been seriously misconstrued.

RECENT UPDATES

During the time of this chapter's publication, several laboratory and clinical studies have provided some updates that seem relevant to the topic of antioxidant use and prostate cancer. Some of these findings include the following.

- The selenium serum concentration in 235 prostate cancer cases and 456 matched controls were analyzed from the past Carotene and Retinol Efficacy Trial (CARET) randomized trial.[113] Serum samples were collected a mean of 4.7 years before diagnosis. Significant predictors of low serum selenium concentration in the controls were current smoking status and East Coast US locations. Overall, no significant differences in mean serum selenium in prostate cancer cases versus controls were found (11.48 μg/dl vs. 11.43 μg/dl). No significant trend in the odds ratio was observed across quartiles of serum selenium for prostate cancer ($P=0.69$). In addition, in a subgroup of diagnosed prostate cancer patients ($n=174$), there were no relationships between serum selenium and Gleason score or clinical or pathological stage. Similarly, in this study, no associations between serum selenium and lung cancer risk were found. Interestingly, the authors of this study concluded by mentioning that 'Our results suggest that in current and former smokers, a selenium supplementation trial that modifies the serum concentration of selenium to the upper quartile of that seen in an unsupplemented population is unlikely to affect or reduce lung cancer or prostate cancer incidence'.
- A recent laboratory study found that dietary vitamin E or γ-tocopherol, but not supplementary vitamin E (α-tocopherol), was found to inhibit the enzyme cyclooxygenase and thus may contain some anti-inflammatory properties somewhat similar to aspirin and the other NSAIDs.[114,115]
- A 3-year, double-blind trial of 160 patients with heart disease, low high-density lipoprotein (HDL) cholesterol levels, and normal low-density lipoprotein (LDL) cholesterol levels were randomly assigned to receive one of four treatments: statin cholesterol-lowering drug plus niacin, antioxidants, the combination or placebo.[116] The clinical end-points of the study were arteriographic evidence of any change in coronary blockage and the occurrence of any first cardiovascular event that included death, myocardial infarction, stroke or revascularization. The antioxidant supplements utilized in this study were 800 IU vitamin E/day plus 100 μg selenium/day plus 1000 mg of vitamin C/day, and 25 mg/day of β-carotene. Interestingly, numerous findings were found that may be relevant to the clinician discussing supplements with any patient (prostate cancer or not) taking any of the statin cholesterol-lowering drugs with or without niacin. Mean levels of LDL and HDL cholesterol did not change in the antioxidant or placebo group. However, these levels did change dramatically and favorably in the statin drug plus niacin group. The overall protective rise in HDL2 with the statin drug plus niacin was actually reduced with the addition of the antioxidant supplements. The only group to experience an actual regression in stenosis was also the statin drug plus niacin group. Also, the frequency of the clinical end-point was found to be 24% with the placebo, 21% in the antioxidant group, 14% in the statin drug plus niacin plus antioxidant group, and only 3% with the statin drug plus niacin alone.

Therefore, in light of recent research, clinicians must be careful about recommending antioxidant supplements to prevent or to reduce the progression of prostate cancer until more research has accumulated. Lifestyle changes should be emphasized first and foremost. In addition, those patients on cholesterol-lowering drugs, especially with niacin, should probably not be encouraged to take certain dietary supplements, especially in higher dosages. However, aspirin continues to result in numerous benefits for those patients that qualify for this OTC based on their medical history.[117]

REFERENCES

1. Brawley OW, Parnes H. Prostate cancer prevention trials in the USA. *Eur. J. Cancer* 2000; 36:1312–15.
2. Willett WC, Stampfer MJ. Selenium and cancer. *Br. Med. J.* 1988; 297:573–4.
3. Clark LC, Combs GF Jr, Turnbull BW et al. For the Nutritional Prevention of Cancer Study Group. Effects of selenium supplementation for cancer prevention in patients with carcinoma of the skin: a randomized clinical trial. *JAMA* 1996; 76:1957–66.
4. Clark LC, Dalkin B, Krongrad A et al. Decreased incidence of prostate cancer with selenium supplementation: results of a double-blind cancer prevention trial. *Br. J. Urol.* 1998; 81:730–4.
5. Ingelfinger J, Mosteller F, Thibodeau L et al. *Biostatistics in Clinical Medicine*, 3rd edn. New York: McGraw-Hill Inc., 1994.
6. Fleet JC, Mayer J. Dietary selenium repletion may reduce cancer incidence in people at high risk who live in areas with low soil selenium. *Nutrition Rev.* 1997; 55:277–9.
7. Dalkin BW, Lillico AJ, Reid ME et al. Selenium and chemoprevention against prostate cancer: an update on the Clark results. *AACR Ann. Meeting* 2001; 42:460 (abstract 2478).
8. Reid ME, Clark LC, Combs GF et al. Lung cancer and selenium supplementation: an update of a clinical trial. *AACR Ann. Meeting* 2001; 42:827 (abstract 4443).
9. Garland M, Steven Morris J, Stampfer MJ et al. Prospective study of toenail selenium levels and cancer among women. *J. Natl Cancer. Inst.* 1995; 87:497–505.
10. Kuller LH. Selenium supplementation and cancer rates. *JAMA* 1977; 277:880.
11. Knekt P, Aromaa A, Maatela J et al. Serum selenium and subsequent risk of cancer among Finnish men and women. *J. Natl Cancer. Inst.* 1990; 82:864–8.
12. Salonen JT, Salonen R, Lappetelainen R et al. Risk of cancer in relation to serum concentrations of selenium and vitamins A and E: matched case-control analysis of prospective data. *Br. Med. J. (Clin. Res. Ed.)* 1985; 290:417–20.
13. Salonen JT, Alfthan G, Huttunen JK et al. Association between serum selenium and the risk of cancer. *Am. J. Epidemiol.* 1984; 120:342–9.

14. Willett WC, Stampfer MJ, Hunter D et al. The epidemiology of selenium and human cancer. In: Aitio A, Aro A (eds) *Trace Elements in Health and Disease*, pp. 141–55. Cambridge: Royal Society of Chemistry, 1991.
15. Geering HR, Car EE, Jones LPH, Allaway WH. Solubility and redox criteria for the possible forms of selenium in soils. *Soil Sci. Soc. Am. Proc.* 1988; 32:35–47.
16. Moyad MA. *The ABCs of Nutrition and Supplements for Prostate Cancer*. Chelsea, MI: Sleeping Bear Press, 2000.
17. Kumpulainen JT. Selenium in foods and diets of selected countries. *J. Trace Elem. Electrolytes Health. Dis.* 1993; 7:107–8.
18. Coates RJ, Weiss NS, Daling JR et al. Serum levels of selenium and retinol and subsequent risk of cancer. *Am. J. Epidemiol.* 1988; 128:515–23.
19. Willett WC, Polk BF, Morris JS et al. Prediagnostic serum selenium and risk of cancer. *Lancet* 1983; 2:130–4.
20. Knekt P, Aromaa A, Maatela J et al. Serum selenium and subsequent risk of cancer among Finnish men and women. *J. Natl Cancer. Inst.* 1990; 82:864–8.
21. Yoshizawa K, Willett WC, Morris SJ et al. Study of prediagnostic selenium level in toenails and the risk of advanced prostate cancer. *J. Natl Cancer Inst.* 1998; 90:1219–24.
22. Longnecker MP, Stram DO, Taylor PR et al. Use of selenium concentration in whole blood, serum, toenails, or urine as a surrogate measure of selenium intake. *Epidemiology* 1996; 7:384–90.
23. Nomura AMY, Lee J, Stemmermann GN, Combs GF. Serum selenium and subsequent risk of prostate cancer. *Cancer Epidemiol. Biomark. Prev.* 2000; 9:883–7.
24. Nomura AMY, Kolonel LN. Prostate cancer: a current perspective. *Am. J. Epidemiol.* 1991; 13:200–27.
25. Cerhan JR, Torner JC, Lynch CF et al. Association of smoking, body mass, and physical activity with risk of prostate cancer in the Iowa 65+ Rural Health Study (United States). *Cancer Causes Control* 1997; 8:229–38.
26. Hiatt RA, Armstrong MA, Klatsky AL et al. Alcohol consumption, smoking, and other risk factors and prostate cancer in a large health plan cohort in California (United States). *Cancer Causes Control* 1994; 5:66–72.
27. Giovannucci E, Rimm EB, Ascherio A et al. Smoking and risk of total and fatal prostate cancer in United States health professionals. *Cancer Epidemiol. Biomark. Prev.* 1999; 8:277–82.
28. Ringstad J, Jacobsen BK, Tretli S et al. Serum selenium concentration associated with risk of cancer. *J. Clin. Pathol.* 1988; 41:454–7.
29. Menkes MS, Comstock GW, Vuilleumier JP et al. Serum beta-carotene, vitamins A and E, selenium, and the risk of lung cancer. *N. Engl. J. Med.* 1986; 315:1250–4.
30. Handelman GJ, Packer L, Cross CE. Destruction of tocopherols, carotenoids, and retinol in human plasma by cigarette smoke. *Am. J. Clin. Nutr.* 1996; 63:559–65.
31. Castro FG, Newcomb MD, McCreary C et al. Cigarette smokers do more than just smoke cigarettes. *Health Psychol.* 1989; 8:107–29.
32. Heinonen OP, Albanes D, Virtamo J et al. Prostate cancer and supplementation with alpha-tocopherol and beta-carotene: incidence and mortality in a controlled trial. *J. Natl Cancer. Inst.* 1998; 90:440–6.
33. Hartman TJ, Dorgan JF, Virtamo J et al. Association between serum alpha-tocopherol and serum androgens and estrogens in older men. *Nutrition Cancer* 1999; 35:10–15.
34. Aziz NM, Hartman T, Barrett M et al. Weight and prostate cancer in the Alpha-Tocopherol Beta-Carotene Cancer Prevention (ATBC) Trial. *ASCO Ann. Meeting* 19:647a (abstract 2550).
35. Cook NR, Stampfer MJ, Ma J et al. Beta-carotene supplementation for patients with low baseline levels and decreased risks of total and prostate carcinoma. *Cancer* 1999; 86:1783–92.
36. Cook NR, Le IM, Manson JE et al. Effects of beta-carotene supplementation on cancer incidence by baseline characteristics in the Physicians' Health Study (United States). *Cancer Causes Control* 2000; 11:617–26.
37. Comstock GW, Bush TL, Helzlsouer K. Serum retinol, beta-carotene, vitamin E, and selenium as related to subsequent cancer of specific sites. *Am. J. Epidemiol.* 1992; 135:115–21.
38. Liede KE, Haukka JK, Saxen LM et al. Increased tendency towards gingival bleeding caused by joint effect of alpha-tocopherol supplementation and acetylsalicylic acid. *Ann. Med.* 1998; 30:542–6.
39. Eichholzer M, Stahelin HB, Gey KF et al. Prediction of male cancer mortality by plasma levels of interacting vitamins: 17-year follow-up of the prospective Basel study. *Int. J. Cancer* 1996; 66:145–50.
40. Gann PH, Ma J, Giovannucci E et al. Lower prostate cancer risk in men with elevated plasma lycopene levels: results of a prospective analysis. *Cancer Res.* 1999; 59:1225–30.
41. Chan JM, Stampfer MJ, Ma J et al. Supplemental vitamin E intake and prostate cancer risk in a large cohort of men in the United States. Cancer. *Epidemiol. Biomark. Prev.* 1999; 8:893–9.
42. Moyad MA, Brumfield SK, Pienta KJ. Vitamin E, alpha- and gamma-tocopherol, and prostate cancer. *Semin. Urol. Oncol.* 1999; 17:85–90.
43. Kushi LH, Folsom AR, Prineas RJ et al. Dietary antioxidant vitamins and death from coronary heart disease in postmenopausal women. *N. Engl. J. Med.* 1996; 334:1156–62.
44. Prineas RJ, Kushi LH, Folsom AR et al. Walnuts and serum lipids. *N. Engl. J. Med.* 1993; 329:359.
45. Fraser GE, Sabate J, Beeson WL, Strahan TM. A possible protective effect of nut consumption on risk of coronary heart disease: the Adventist Health Study. *Arch. Intern. Med.* 1992; 152:1416–24.
46. Helzlsouer KJ, Huang HY, Alberg AJ et al. Association between alpha-tocopherol, gamma-tocopherol, selenium, and subsequent prostate cancer. *J. Natl Cancer Inst.* 2000; 92:2018–23.
47. Stampfer MJ, Hennekens CH, Manson JE et al. Vitamin E consumption and the risk of coronary disease in women. *N. Engl. J. Med.* 1993; 328:1444–9.
48. Rimm EB, Stampfer MJ, Ascherio A et al. Vitamin E consumption and the risk of coronary heart disease in men. *N. Engl. J. Med.* 1993; 328:1450–6.
49. The Heart Outcomes Prevention Evaluation Study Investigators. Vitamin E supplements and cardiovascular events in high-risk patients. *N. Engl. J. Med.* 2000; 342:154–60.
50. Lonn E, Yusuf S, Dzavik V et al. Effects of ramipril and vitamin E on atherosclerosis: the study to evaluate carotid ultrasound changes in patients treated with ramipril and vitamin E (SECURE). *Circulation* 2001; 103:919–25.
51. Yusuf S, Sleight P, Pogue J et al. Effects of an angiotensin-converting-enzyme inhibitor, ramipril, on cardiovascular events in high-risk patients. The Heart Outcomes Prevention Evaluation Study Investigators. *N. Engl. J. Med.* 2000; 342:145–53.
52. Collaborative Group of the Primary Prevention Project (PPP). Low-dose aspirin and vitamin E in people at cardiovascular risk: a randomized trial in general practice. *Lancet* 2001; 357:89–95.
53. Janne PA, Mayer RJ. Chemoprevention of colorectal cancer. *N. Engl. J. Med.* 2000; 342:1960–8.
54. The Steering Committee of the Physicians' Health Study Research Group. Final report on the aspirin component of the ongoing Physicians' Health Study. *N. Engl. J. Med.* 1989; 321:129–35.
55. Norrish AE, Jackson RT, McRae CU. Non-steroidal anti-inflammatory drugs and prostate cancer progression. *Int. J. Cancer* 1998; 77:511–15.
56. Neugut AI, Rosenberg DJ, Ahsan H et al. Association between coronary heart disease and cancers of the breast, prostate, and colon. *Cancer Epidemiol. Biomark. Prev.* 1998; 7:869–73.
57. Steinberg D. Clinical trials of antioxidants in atherosclerosis: are we doing the right thing? *Lancet* 1995; 346:36–8.

58. Upston JM, Terentis AC, Stocker R. Tocopherol-mediated peroxidation of lipoproteins: implications for vitamin E as a potential antiatherogenic supplement. *FASEB J.* 1999; 13:977–94.
59. Boissel J-P. Individualizing aspirin therapy for the prevention of cardiovascular events. *JAMA* 1998; 280:1949–50.
60. He J, Whelton PK, Vu B, Klag MJ. Aspirin and risk of hemorrhagic stroke: a meta-analysis of randomized controlled trials. *JAMA* 1998; 280:1930–5.
61. Yang GA, Ge K, Chen J, Chen X. Selenium-related endemic diseases and the daily requirement of humans. *World Rev. Nutr. Dietet.* 1988; 55:98–152.
62. Leads from the MMWR. Selenium intoxication – New York. *JAMA* 1984; 251:1938.
63. Cohen HM. Fatigue caused by vitamin E. *N. Engl. J. Med.* 1973; 281:980.
64. Roberts HJ. Perspective on vitamin E as therapy. *JAMA* 1981; 246:129–31.
65. Klingman AM. Vitamin E toxicity. *Arch. Dermatol.* 1982; 118:289.
66. Brown KM, Morrice PC, Duthie GG. Erythrocyte vitamin E and plasma ascorbate concentrations in relation to erythrocyte peroxidation in smokers and nonsmokers: dose response to vitamin E supplementation. *Am. J. Clin. Nutr.* 1997; 65:496–502.
67. Moyad MA. An introduction to aspirin, NSAIDs, and COX-2 inhibitors for the primary prevention of cardiovascular events and cancer and their potential preventive role in bladder carcinogenesis: Part I. *Semin. Urol. Oncol.* 2001; 19:294–305.
68. Moyad MA. An introduction to aspirin, NSAIDs, and COX-2 inhibitors for the primary prevention of cardiovascular events and cancer and their potential preventive role in bladder carcinogenesis: Part II. *Semin. Urol. Oncol.* 2001; 19:306–16.
69. Newschaffer CJ, Otani K, McDonald KM, Penberthy LT. Causes of death in elderly prostate cancer patients and in a comparison nonprostate cancer cohort. *J. Natl Cancer Inst.* 2000; 92:613–21.
70. Sandler RS, Halabi S, Kaplan EB et al. Use of vitamins, minerals, and nutritional supplements by participants in a chemoprevention trial. *Cancer* 2001; 91:1040–5.
71. Neugut AI, Rosenberg DJ, Ahsan H et al. Association between coronary heart disease and cancers of the breast, prostate, and colon. *Cancer Epidemiol. Biomark. Prev.* 1998; 7:869–73.
72. Peterson HI. Tumor angiogenesis inhibition by prostaglandin synthetase inhibitors. *Anticancer Res.* 1986; 6:251.
73. Campbell SC. Advances in angiogenesis research: relevance to urological oncology. *J. Urol.* 1997; 158:1663–74.
74. Smith BI, Wills MR, Savory J. Prostaglandins and cancer. *Ann. Clin. Lab. Sci.* 1983; 13:359–65.
75. Shaw MW, Albin RJ, Ray P et al. Immunology of the Dunning R-3327 rat prostate adenocarcinoma sublines: plasma and tumor effusion prostaglandins. *Am. J. Reprod. Immunol. Microbiol.* 1985; 8:77–9.
76. Rubenstein M, Shaw MW, McKiel CF et al. Immunoregulatory markers in rats carrying Dunning R3327 H.G. or MAT-LyLu prostatic adenocarcinoma variants. *Cancer Res.* 1987; 47:178–82.
77. Dunzendorfer U, Zahradnik HP, Grster K. 13,14-Dihydro-15-keto-prostaglandin F_2alpha in patients with urogenital tumors. *Urol. Int.* 1980; 35:171–5.
78. Khan O, Hensby CN, Williams G. Prostacyclin in prostatic cancer: a better marker than bone scan or serum acid phosphatase? *Br. J. Urol.* 1982; 54:26–31.
79. Albin RJ, Shaw MW. Prostaglandin modulation of prostate tumor growth and metastases. *Anticancer Res.* 1986; 6:327–8.
80. Karmali RA. Eicosanoids in neoplasia. *Prev. Med.* 1987; 16:493–502.
81. Chaudry AA, Wahle KWJ, McClinton S et al. Arachidonic acid metabolism in benign and malignant prostatic tissue in vitro: effects of fatty acids and cyclooxygenase inhibitors. *Int. J. Cancer* 1994; 57:176–80.
82. Faas FH, Dang AQ, Pollard M et al. Increased phospholipid fatty acid remodeling in human and rat prostatic adenocarcinoma tissues. *J. Urol.* 1996; 156:243–8.
83. Rose DP, Connolly JM. Effects of fatty acids and eicosanoid synthesis on the growth of two prostate cancer cell lines. *Prostate* 1991; 18:243–54.
84. Rose DP, Connolly JM. Dietary fat, fatty acids and prostate cancer. *Lipids* 1992; 27:798–803.
85. Viljoen TC, Van Aswegen CH, Du Plessis DJ. Influence of acetylsalicylic acid and metabolites on DU-145 prostatic cancer cell proliferation. *Oncology* 1995; 52:465–9.
86. Tjandrawinata RR, Dahiya R, Hughes-Fulford M. Induction of cyclo-oxygenase-2 mRNA by prostaglandin E2 in human prostatic carcinoma cells. *Br. J. Cancer* 1997; 75:1111–18.
87. Paganini-Hill A, Chao A, Ross RK et al. Aspirin use and chronic diseases: a cohort study of the elderly. *Br. Med. J.* 1989; 299:1247–50.
88. Schreinemachers DM, Everson RB. Aspirin use and lung, colon and breast cancer incidence in a prospective study. *Epidemiology* 1994; 5:138–46.
89. Thun MJ, Namboodiri MM, Calle EE et al. Aspirin use and the risk of fatal cancer. *Cancer Res.* 1993; 53:1322–7.
90. Norrish AE, Jackson RT, McRae CU. Non-steroidal anti-inflammatory drugs and prostate cancer progression. *Int. J. Cancer* 1998; 77:511–15.
91. Badawi AF. The role of prostaglandin synthesis in prostate cancer. *Br. J. Urol. Int.* 2000; 85:451–62.
92. Nelson JE, Harris RE. Inverse association of prostate cancer and non-steroidal anti-inflammatory drugs (NSAIDs): results of a case-control study. *Oncol. Rep.* 2000; 7:169–70.
93. O Roberts R, Jacobsen DJ, Lieber MM et al. Prostate cancer and non-steroidal anti-inflammatory drugs: a protective association. *J. Urol.* 2001; 165:62 (abstract 256).
94. Pearson JD, Watson DJ, Corrada MM et al. A prospective cohort study of nonsteroidal anti-inflammatory drugs and risk of prostate cancer: Baltimore Longitudinal Study of Aging. *J. Urol.* 2001; 165:63 (abstract 259).
95. Leitzmann MF, Rimm EB, Stampfer MJ et al. Aspirin use in relation to risk of prostate cancer. *Am. J. Epidemiol.* 2001; 153: abstract 391.
96. Bucher C, Jordan P, Nickeleit V et al. Relative risk of malignant tumors in analgesic abusers: effects of long-term intake of aspirin. *Clin. Nephrol.* 1999; 51:67–72.
97. Swan DK, Ford B. Chemoprevention of cancer: review of literature. *Oncol. Nurs. Forum* 1997; 24:719–27.
98. O'Neill GP, Ford-Hutchinson AW. Expression of mRNA for cyclooxygenase-1 and cyclooxygenase-2 in human tissue. *FEBS Lett.* 1993; 330:156–60.
99. Kirschenbaum A, Klausner AP, Lee R et al. Expression of cyclooxygenase-1 and cyclooxygenase-2 in the human prostate. *Urology* 2000; 56:671–6.
100. Yoshimura R, Sano H, Masuda C et al. Expression of cyclooxygenase-2 in prostate carcinoma. *Cancer* 2000; 89:589–6.
101. Madaan S, Lalani EN, Chaudhary KS et al. Cyclo-oxygenase-2 up-regulation in human prostate cancer. *Br. J. Urol. Int.* 2000; 86:385–6 (abstract).
102. Magoha GA. Ten years experience with chronic prostatitis in Africans. *East Afr. Med. J.* 1996; 73:176–8.
103. Soslow RA, Dannenberg AJ, Rush D et al. COX-2 is expressed in human pulmonary, colonic, and mammary tumors. *Cancer* 2000; 89:2637–45.
104. Wolff H, Saukkonen K, Anttila S et al. Expression of cyclooxygenase-2 in human lung carcinoma. *Cancer Res.* 1998; 58:4997–5001.
105. Hida T, Yatabe Y, Achiwa H et al. Increased expression of cyclooxygenase 2 occurs frequently in human lung cancers, specifically adenocarcinomas. *Cancer Res.* 1998; 58:3761–4.
106. Chan G, Boyle JO, Yang EK et al. Cyclooxygenase-2 expression is up-regulated in squamous cell carcinoma of the head and neck. *Cancer Res.* 1999; 59:991–4.

107. Zimmermann KC, Sarbia M, Weber A-A et al. (1999) Cyclooxygenase-2 expression in human esophageal carcinoma. *Cancer Res.* 1999; 59:198–204.
108. Koga H, Sakisaka S, Ohishi M et al. Expression of cyclooxygenase-2 in human hepatocellular carcinoma: relevance to tumor dedifferentiation. *Hepatology* 1999; 29: 688–96.
109. Tucker ON, Dannenberg AJ, Yang EK et al. Cyclooxygenase-2 expression is up-regulated in human pancreatic cancer. *Cancer Res.* 1999; 59:987–90.
110. Blendon RJ, DesRoches CM, Benson JM. Americans' views on the use and regulation of dietary supplements. *Arch. Intern. Med.* 2001; 161:805–10.
111. Kant AK, Schatzkin A, Graubard BI et al. A prospective study of diet quality and mortality in women. *JAMA* 2000; 283:2109–15.
112. Stampfer MJ, Hu FB, Manson JE et al. Primary prevention of coronary heart disease in women through diet and lifestyle. *N. Engl. J. Med.* 2000; 343:16–22.
113. Goodman GE, Schaffer S, Bankson DD et al. Predictors of serum selenium in cigarette smokers and the lack of association with lung and prostate cancer risk. *Cancer Epidemiol. Biomark. Prev.* 2001; 10:1069–76.
114. Jiang Q, Elson-Schwab I, Courtemanche C, Ames BN. Gamma-tocopherol and its major metabolite, in contrast to alpha-tocopherol, inhibit cyclooxygenase activity in macrophages and epithelial cells. *Proc. Natl Acad. Sci. USA* 2000; 97: 11494–9.
115. Jiang Q, Christen S, Shigenaga MK et al. Gamma-tocopherol, the major form of vitamin E in the US diet, deserves more attention. *Am. J. Clin. Nutr.* 2001; 74:714–22.
116. Brown BG, Zhao X-Q, Chait A et al. Simvastatin and niacin, antioxidant vitamins, or the combination for the prevention of coronary disease. *N. Engl. J. Med.* 2001; 345:1583–92.
117. Gum PA, Thamilarasan M, Watanabe J et al. Aspirin use and all-cause mortality among patients being evaluated for known or suspected coronary artery disease: a propensity analysis. *JAMA* 2001; 286:1187–94.

Effects of Smoking, Alcohol, Exercise and Sun Exposure on Prostate Cancer

Chapter 22

Neil Fleshner

Division of Urology, University Health Network, Princess Margaret Hospital, Toronto, Ontario, Canada

BACKGROUND AND INTRODUCTION

Variations in the risk of disease in different geographical locations suggest that the environment plays a vital role in disease causation. As clinicians we are often faced with questions from our patients about various environmental risk factors and their effects on a particular disease. The events that trigger these questions usually relate to recent newspaper articles or television programs that the patient or his/her family has encountered. The purpose of this chapter is to review the evidence critically for associations between alcohol, smoking, sun exposure and exercise and the development of prostate cancer.

The primary goal of epidemiological investigation is to determine risk factors for disease causation or prevention. Once these risk factors are determined, primary prevention programs can be initiated in order to limit disease incidence. The tremendous success over the past two decades with respect to cardiovascular disease prevention is an excellent example of this. Other epidemiologically discovered risk factors cannot be modified (e.g. family history); however, their importance in disease causation is still relevant as they can lead to early detection and treatment of disease (so-called 'secondary prevention') for individuals at high risk. Non-preventable risk factors can also provide clues for basic science endeavors that aim to discover disease mechanisms (e.g. BRCA1[1]).

On a superficial level, it may seem trivial to determine these risk factors; in practice, it is often very difficult. Unlike laboratory science, epidemiology aims to study free living persons with diverse genetic, geographical and cultural traits. Owing to this lack of experimental control, minor errors in study design or analysis can have profound impact on study outcome, often leading to spurious associations or inconsistencies in the published literature. In addition, the labeling of a specific risk factor (also known as exposures or agents) as a definite causal factor for a specific disease rarely results from one study. It is invariably a slow accumulation of knowledge that arises from many studies (of various epidemiological designs) carried out in different study populations. For example, the causal association between smoking and lung cancer took decades to prove.[2]

ECOLOGIC STUDIES

Ecologic studies (also referred to as correlation studies) are epidemiological studies in which groups of people, rather than individuals, form the basis of the study. Three types of correlation studies exist: international, regional and time trend.

International studies correlate disease incidence (or mortality) rates and level of exposure in various countries. Regional studies compare international disease incidence (or mortality) and correlate them with exposure level.[3] One example of this type of study relevant to this chapter is the association between geographical latitude and prostate cancer risk.[4] Time trend studies examine changing levels of exposure within the same population over time and attempt to correlate this with changing disease incidence.

On the surface, ecologic studies appear compelling; they are also easy to perform usually with readily available data. They are, however, generally regarded as the *weakest of all* epidemiologic methodologies in terms of causal inference. The reasons for this include the following:

1. There is no information about exposure at the level of the individual.
2. Proxy measures (such as food production), rather than true measures (i.e. food consumption) are often used for exposure classification.
3. These studies cannot assess if other exposures rather than that of interest is responsible for the association. This is referred to as confounding.

Despite their limitations, these studies can signal the presence of a disease exposure association worthy of study. Therefore, they should only be viewed as hypotheses generating.

CASE-CONTROL STUDIES

Case-control studies are investigations that compare exposures in persons with a particular disease (case) and compare them to a similar group of persons who are not diseased (controls). The scientific basis of case-control studies makes them the most efficient of all epidemiological studies. In addition, they can be completed in a relatively short time. Associations from case-control studies are usually expressed as an odds ratio (OR; **Table 22.1**), which compares the odds of exposure in diseased individuals compared to the odds of exposure in the controls. The major limitation of case-control studies is the possibility of bias, in particular, recall bias, which refers to the inaccurate reporting of exposure in cases compared to controls. This can create an artificial association. It is very easy to imagine how, for example, a prostate cancer patient overestimates his exposures to environmental carcinogens (smoking, etc.) as he may be self-blaming compared to a healthy control.

In general, however, case-control studies are efficient and, if done correctly, can provide powerful insight into risk factors for disease causation. The well-accepted association between transplacental diethylstilbestrol exposure and clear cell vaginal cancer was derived from case-control studies.[5]

COHORT STUDIES

Cohort studies are the most robust of observational epidemiological studies and most closely simulate a controlled trial. In cohort studies, disease-free individuals are questioned about environmental exposures. They may also have tissue (e.g. blood, toenails, hair) procured for laboratory analysis. The individuals are followed forward in time. As time elapses and persons begin to develop disease, one compares the rates of disease in the exposed versus the unexposed persons. The measure of association is referred to as a relative risk (RR; **Table 22.1**).

The obvious advantage of cohort studies is that exposure is determined *prior* to the development of the disease, thus, it is free of recall bias. The major limitations of these studies are their cost and length of time to complete. Loss of participants over time is another limitation of these studies. The Physicians Health Study and Health Professionals Follow-Up Study and Framingham study are well-known examples of cohort studies.[6]

CAUSAL CRITERIA

It is important to realize that elevated risks of disease in persons with a particular exposure is merely an association. In epidemiology we are not interested in general associations but causal associations. Hill developed a set of criteria that can be used for causal inference in observational epidemiologic studies.[7]

Strength

Strength refers to the degree of association between an exposure and a disease. Strong associations (RR or OR >3) are more likely to be indicative of causation because weak associations may be alternatively explained by bias. None the less, the fact that an association is weak does not eliminate the potential for that exposure to be causal.

Consistency

Consistency refers to the degree with which different studies done by various methodologies, investigators and populations conclude that an exposure and disease are associated. It is

Table 22.1. Basic design and analysis of case-control and cohort studies.

	Exposure		
	Present	Absent	
Diseased	*a*	*b*	N1
Non-diseased	*c*	*d*	N2
	M1	M2	Total

Case-control study: estimate of effect is the *odds ratio* (OR), which represents the odds of exposure between cases (diseased) and control (non-diseased). It is calculated by the formula:

$$OR = \frac{a \times d}{b \times c}$$

Cohort study: estimate of effect is the *relative risk* (RR). This compares the disease rates in exposed persons versus the disease rates in unexposed persons. It is calculated by the formula:

$$RR = \frac{a/M1}{b/M2}$$

important to realize that not every observational epidemiological study may reveal a hypothesized association. Limitations due to methodology in study design, bias and chance may prevent 100% consistency.

Specificity

This criterion stipulates that a single exposure can only produce one disease. This criterion is not consistent with a modern view of disease. For example, cigarette exposure is causally related to many diseases. Many investigators now find this criterion unnecessary and misleading.[8]

Time order

This criterion is intuitive. It refers to the necessity that the exposure to the causal agent precedes the development of the disease. This is the only criterion that must be satisfied to infer causality.

Dose–response

This criterion suggests that those individuals exposed to more of a causal agent are more likely to be afflicted with the disease. Not all causal associations, however, follow a traditional dose–response relationship. Other models of dose–response, such as threshold, critical period (e.g. particular week of gestation) or parabolic responses, have been hypothesized.[9]

Biologic plausibility

This criterion mandates that a hypothesized association be scientifically plausible. The limitation of this criterion is that a discovered association may precede scientific knowledge.

Coherence

This refers to the fact that the hypothesized association should not conflict with the well-accepted facts about the natural history of the disease under study. Absence of coherence, however, should not be regarded as evidence against causality.

Analogy

Analogy suggests that, if similar mechanisms of causation exist in other diseases (e.g. diet and cancer), it can enhance the credibility of a newly discovered potential causal agent. The limitation of this criterion is that it constrains inventive and alternate ways of thinking about risk factors for disease.

Experimental evidence

This criterion suggests that a controlled study should exist to confirm the relationship. In practice this is rarely available. This criterion is usually applicable when the exposure is a preventative agent (e.g. vitamins and cancer) as it is not ethical to carry out studies of risk factors for disease.

USING THE HILL CRITERIA

It is important to understand that the aforementioned criteria are not meant to hinder the discovery of disease risk factors, but to provide general guidelines for investigators and public health policy makers. The only criterion that must be satisfied is the time-order association. For most risk factors, many of these criteria will never be satisfied.

In this review, many associations will be listed and described. Associations will be categorized as follows.

1. Sufficient evidence: this suggests that there is ample evidence of a causal association.
2. Limited evidence: this suggests that, although evidence exists for causality, alternative explanations are credible.
3. Inadequate evidence: this indicates that the data are either consistently negative, or that the studies are seriously biased, confounded or underpowered.

PROSTATE CANCER

Prostate cancer is the most common malignant tumor and the second leading cause of cancer deaths in men in the developed world.[10] The descriptive epidemiology of prostate cancer is unique among human neoplasms. In particular, the high, but geographically consistent, incidence of latent carcinoma coupled with the tremendous global variation in disease incidence and mortality suggests that environmental factors play a pivotal role.[11] In addition, 'risk factors' in the traditional sense of prostate cancer are not likely factors for disease initiation/promotion but risk factors for disease progression.

Little is known about risk factors for prostate cancer progression. Only family history, race and dietary fat are accepted risk factors.[12] In addition, racial differences may be primarily environmental as evidenced by migration studies.[13] Historically, case-control and cohort studies in prostate cancer have been plagued by contamination of the control group by individuals with significant prostate cancer. This tends to bias the study towards a null finding when evaluating potential risk factors.

SMOKING AND PROSTATE CANCER

Cigarette smoke contains hundreds of carcinogenic substances that have demonstrable effects on all phases of cancer development.[14] Tumor initiation, promotion and progression have been associated with compounds found in tobacco smoke in a variety of experimental tumor model systems.[15] The well-accepted associations between smoking and neoplasms of the lung, bladder, and head and neck do not necessarily rule out a potential causal association between smoking and prostate cancer. It must be remembered that weaker (but nevertheless causal) associations between a particular agent and a disease may only be apparent with robust investigations and, indeed, may be almost impossible to prove given the limitations of modern epidemiology. From a methodological viewpoint, it is becoming more and more difficult to assess associations between tobacco and cancers as a larger proportion of the population are ex-smokers. Thus, not only is cumulative exposure data potentially important but so are data about smoking cessation and its temporal association with diagnosis. For

Table 22.2. Potential associations between tobacco and prostate cancer incidence or progression: influence of the causation effect.

Mechanism	Incidence	Progression
Screening bias	Protective	None
Hormonal effects	Increase	Increase
Genotoxic	Increase	Increase
Oxidative stress	Increase	Increase
Metastatic/progression pathways or inhibits therapy	No association	Increase

example, if smoking were primarily associated with disease progression (as opposed to incidence), then smoking status at the time of diagnosis would be most crucial in sorting out this important question. There exist few data on the association between prostate cancer and other forms of tobacco exposure, such as cigars, chewable tobacco and second-hand smoke.

There are a host of hypothetical mechanisms by which cigarette smoke may adversely affect either risk or progression of prostate cancer. **Table 22.2** lists some of these mechanisms and the effect that would be realized in terms of data interpretation. It is also important to determine where and when the studies were performed and whether the study attempts to distinguish between incident cases and advanced (or fatal) cases of prostate cancer. For example, the prevalence of screening for prostate cancer has risen dramatically since 1994, particularly in the USA.[16] Thus more contemporary studies of US-based populations represent a different case mix than in older non-US-based studies, as they are likely to contain patients with a higher proportion of clinically insignificant disease.[17] One potential example and outcome of this screening bias is that, if non-smokers are more likely to possess health-seeking behavior (such as cancer screening), then it is plausible that smoking could be seen as protective of prostate cancer. In this setting, however, smoking should not be a risk factor for advanced or fatal disease, as cancer-related death is free of detection bias.

Several levels of evidence suggest that sex hormones play an important role in the etiology and progression of prostate cancer. There are well-documented physiological effects of cigarette smoking on hormonal physiology. English and colleagues reported that smoking influences levels of sex hormone binding globulins.[18] Serum dehydroepiandrostenedione, estrogen and testosterone levels were found to be significantly different between current and former smokers among 52 healthy Greek men.[19]

Genetic mutation is a hallmark of all cancers. Cigarette smoke contains many known mutagens, including benzopyrenes and 4-aminobiphenyl.[20] If cigarette smoke were important as a cause of genetic mutation associated with disease initiation or promotion, then a positive association would exist between tobacco use and prostate cancer incidence. If certain progression-associated genes were preferentially affected by tobacco exposure (e.g. p53),[21] then it is plausible that the association would exist only among fatal or advanced cases.

Oxidative pathways are gaining increased credence as an important mechanism for prostate cancer progression.[22] Oxidative stress refers to the generation of reactive oxygen species (ROS), which then interact with both genetic and epigenetic pathways to facilitate mutation, cell cycle progression and neovascularization.[22–24] Reactive oxygen species are either generated endogenously or via exogenous sources, such as dietary fat consumption or sedentary lifestyle. Smoking is also a known strong inducer of ROS.[22] Once generated, ROS induce cellular damage or are neutralized by antioxidants. Rao and colleagues have shown that levels of the potent antioxidant lycopene can be reduced by smoking.[25] Smokers are known to have lower levels of the antioxidant selenium[26] and vitamin E.[27] Many of these antioxidants are being studied as potential chemopreventive therapies for prostate cancer.[28] It is thus plausible that smoking, via interacting with antioxidants, is playing an important role in disease incidence and progression.

Molecular and cellular pathways, influenced by both genetic and epigenetic mechanisms, are important in determining the metastatic phenotype of a particular cancer. Tumor cell invasiveness, motility, angiogenesis induction, and paracrine properties may also be associated with tobacco exposure. For example, E cadherin, a molecule important in cell adhesion is known to be negatively influenced by smoking.[29] Once again, in this case, cigarette smoking may be associated with case fatality and not incidence.

Tables 22.3 and **22.4** examine the results of case-control and cohort studies that have examined associations between smoking and prostate cancer.[30–56] The list represents many well-recognized studies but is by no means complete. Of the 15 case-control studies only two suggest an association between smoking and risk of prostate cancer. In both of these studies, the number of cases was relatively small and in the Scandinavian study by Andersson et al., no dose–response was observed, arguing against a causal association. In the study by Honda and colleagues, a dose–response was observed, however, only 216 cases were examined and cases were limited to men under age 60. This raises the possibility that early onset cancer could be associated with smoking. None the less, the overwhelming totality of these data suggests that there is inadequate evidence for an association between tobacco exposure and prostate cancer incidence. One recent observation that deserves further future examination is the possible interaction between body mass index and smoking in the context of prostate cancer risk. Sharpe and colleagues reported over a two-fold increased risk of prostate cancer among obese men who smoke and may involve endocrine changes.

Among the cohort studies, only one of seven that tested for associations between smoking and incidental cases revealed an association, thereby confirming the results of the case-control studies. In the one positive study, diet was not controlled for raising the possibility of confounding. An interesting and quite opposite relationship exists for smoking and its association with prostate cancer progression. Of the existing cohort studies, eight have examined, as either exclusive outcomes or as subset analyses, patients with fatal prostate cancer. Among this

Table 22.3. Smoking and prostate cancer: case-control studies.

Author	Reference	Cases	Incident or advanced	Significant association	Comments
Giles	30	1498	Incident	No	Former and current smokers
Van der Gulden	31	345	Incident	No	
Hayes	32	981	Incident	No	Former and current smokers
Distefani	33	156	Incident	No	
Hsieh	34	320	Incident	No	
Honda	35	216	Incident	Yes	
Oishi	36	100	Incident	No	
Yu	37	1162	Incident	No	
Sharpe	38	399	Incident	Yes/no	High BMI only
Furuya	39	329	Incident	No	
Rohan	40	408	Incident	No	
Lumey	41	1097	Incident	No	
Andersson	42	256	Incident	Yes	No dose response
Slattery	43	362	Incident	No	
Gronberg	44	406	Incident	No	nested

BMI, body mass index.

Table 22.4. Smoking and prostate cancer: cohort studies.

Author	Reference	Cases	Incident or fatal	Significant association	Comments
LeMarchand	45	198	Incident	No	
Hiatt	46	238	Incident	Yes	No diet adjustment
Engeland	47	712	Incident	No	
Putnam	48	101	Incident	No	
Giovannucci	49	1369	Incident	No	
			Fatal	Yes	Recent use
Adami	50	2368	Incident	No	
		709	Fatal		
Lotufo	51	996	Incident	No	
			Fatal		
Eichholzer	52	15	Fatal	Yes	Lower vitamin E levels
Rodriguez	53	1748	Fatal	Yes	Former smokers – negative
Hsing	54	4607	Fatal	Yes	
Hsing	55	149	Fatal	Yes	
Coughlin	56	826	Fatal	Yes	Dose response

group, six consistently demonstrate an association with smoking. This observation is particularly interesting as current smoking seems to be most associated with case fatality, suggesting an effect on disease progression separate from a priori smoking-induced mutations could confer a more aggressive disease phenotype. In general, risk of death appears increased by 30–75%. These data are further supported by the studies of Daniell who demonstrated that smokers with metastatic prostate cancer had a worse 5-year survival rate than non-smokers.[57] Johansson and colleagues have also demonstrated that the Dunnning R3327 rat prostate cancer grows faster among animals exposed to cigarette smoke.[58] In summary, it would thus appear that there is sufficient evidence linking smoking and fatal prostate cancer.

ALCOHOL CONSUMPTION AND PROSTATE CANCER

Ethanol is a known carcinogen and its consumption is causally associated with many neoplasms including head and neck, esophageal and hepatic cancers.[3] Alcohol consumption has been shown to alter sex steroid metabolism and thus could theoretically play a role in prostate cancer causation or even protection.[59–62] Alcohol is also known to clear serum androgens.[63]

Table 22.5 details many of the epidemiological studies that have assessed associations between alcohol use and prostate cancer. As alluded to earlier, case-control studies are naturally biased towards an association between alcohol use and prostate cancer. Despite this, only two of nine reviewed studies suggest an association. In the study by Andersson, users of hard liquor were at 40% increased risk of prostate cancer development, but beer and wine consumers were not at increased risk.[42] In addition, no dose–response was noted. In contrast, DeStefani reported that beer consumers and not hard liquor users were at higher risk of prostate cancer.[33] This lack of specificity combined with data inconsistency makes it likely that this observation is spurious and not causal in nature.

Six cohort studies have examined alcohol use and prostate cancer. All but one study demonstrated no association. In the recent study by Putnam, consumers of >22 g of alcohol per week were at elevated prostate cancer risk. Users of over 96 g per week had more than a threefold increase. In the Putnam study, type of alcohol was not associated with risk and appropriate confounders were measured and adjusted for aside from dietary fat.

In summary, the lack of data consistency and specificity make it unlikely that a significant association exists with alcohol use and prostate cancer. There is, therefore, inadequate evidence for a causal association between alcohol use and prostate cancer.

EXERCISE AND PROSTATE CANCER

Physical activity is beneficial to health. Demonstrable effects in the realm of cardiovascular health and aging are well accepted.[64] In the context of prostate cancer risk, more needs to be known about the effects of exercise on endocrine physiology and other mechanisms that could modulate prostate cancer risk. There are some hypothetical physiological mechanisms by which exercise could influence prostate carcinogenesis. For example, acute bouts of exercise can lower androgen levels.[65] Exercise can also improve the physiologic response to oxidative stress[66] – a process associated with prostate cancer development.[22] Tymchuk and colleagues demonstrated that serum from men on an exercise program inhibited the growth of cultured human prostate cancer cells.[67]

Table 22.6 outlines the major published studies that have assessed the association of exercise and prostate cancer risk.[68–76] It is somewhat intuitive and important to realize that exercise is highly correlated with dietary consumption and smoking. Therefore, well-conducted studies will have statistically adjusted for these covariates. It is also important to emphasize that measurement of physical activity can be disparate. In some studies, the level of work-associated physical activity was used as surrogate measure and, in others, the number of times per week sweat-induced activity was reached was used. The ideal way to measure physical activity has yet to be developed and could lead to misclassification error. Misclassification error has the potential to bias study outcome quite significantly.

Table 22.5. Alcohol and prostate cancer.

Author	Reference	Cases	Case-control or cohort	Significant association	Comments
Liu	59	982	Cohort	No	
Tavani	60	599	Case-control	No	
Sharpe	38	400	Case-control	No	
Giovanucci	61	300	Cohort	No	
Hiatt	46	238	Cohort	No	
Le Marchand	45	198	Cohort	No	
Putnam	48	101	Cohort	Yes	Positive for trend
Hsing	55	149	Cohort	No	
Andersson	42	256	Case-control	Yes	No dose–response Hard liquor only
Yu	37	1162	Case-control	No	
Hsieh	34	320	Case-control	No	
Destefani	33	156	Case-control	Yes	Beer only
Van der Gulden	31	345	Case-control	No	
Gronberg	44	406	Case-control	No	
Slattery	43	362	Case-control	No	

Table 22.6. Exercise and prostate cancer risk.

Author	Reference	Case-control or cohort	Number of cases	Significant association	Comments
Sung	68	Case-control	90	Yes	Increased risk!
Bairati	69	Case-control	64	Yes	Protective Stage A cancers only
Albanes	70	Cohort	214	No	
Thune	71	Cohort	220	No	
Cerhan	72	Cohort	71	Yes	Increased risk
Hartman	73	Cohort	317	Yes	Protective No diet data
Giovannucci	74	Cohort	1362	Yes – met only	Protective Health professionals with advanced disease only
Liu	75	Cohort	982	No	Fully adjusted
Wannamethee	76	Cohort	969	Yes	Protective Adjusted

Among the reviewed literature, two of two case-control and four of seven cohort studies demonstrate some degree of association between physical activity and risk of prostate cancer. However, two of these studies suggest that physical activity increases the risk of prostate cancer, whereas the others suggest that it decreases the risk. In addition, in many of the studies that reveal an association, little statistical adjustment for diet has occurred. Given the inconsistent directions of the associations and the lack of controlling for confounding variables, one can only conclude that there is inadequate evidence to suggest a causal association between levels of physical activity and prostate cancer.

SUN EXPOSURE AND PROSTATE CANCER

Exposure to sunlight can have significant effects on human endocrinology. In particular, vitamin D is synthesized via exposure of the skin to ultraviolet light, which in turn, becomes subsequently hydroxylated in the liver and kidney to its active form, 1,25-dihydroxy vitamin D. In 1990, Schwartz and Hulka published their hypothesis about vitamin D and prostate cancer.[77] They rationalized that many of the known risk factors for prostate cancer could potentially be explained by vitamin D deficiency. **Table 22.7** lists some of the risk factors associated with increased or decreased risk of prostate cancer and how these mechanistically could relate to Vitamin D.

Age is a well-accepted risk factor for prostate cancer. Data have shown that synthesis of vitamin D diminishes with advancing age.[78] In addition, the elderly have less sun exposure, in part due to residence in homes for the aged. African American's are at highest risk for prostate cancer and tend towards a more aggressive disease phenotype.[79] Although, numerous genetic, epigenetic and environmental hypotheses may explain this observation, melanin skin pigmentation can inhibit vitamin D synthesis.[80] Asian's are relatively protected from prostate cancer development,[81] in this population oily fish consumption is relatively high. Oily fish is the only known dietary source of vitamin D[82] (prior to milk supplementation,

Table 22.7. Sun exposure and prostate cancer risk.

Risk factor	Mechanism
Age	Vitamin D synthesis decreases with age Elderly are less sun exposed
African descent (high risk)	Melanin in skin inhibits vitamin D synthesis
Northern latitudes	Less sunlight
Asian (low risk)	Fish oils – natural dietary source of vitamin D
High serum calcium	Inhibits vitamin D hydroxylation
Low fruit intake	Fructose raises vitamin D synthesis

Table 22.8. Studies of vitamin D, sunlight and prostate cancer.

Author	Reference	Type of study	Number of cases	Measure of exposure	Significant association	Comments
Giovannucci	84	Cohort	1792	Calcium intake	Yes	Health professionals study
Corder	88	Cohort	181	Vitamin D levels	Yes	
Chan	89	Cohort		Calcium intake	Yes	Physicians health study
Chan	90	Case-control	526	Calcium intake	Yes	
Tavani	91	Case-control	288	Calcium levels	No	Did not have power to assess a wide range of calcium intake
Nomura	92	Case-control	136	Calcium levels	No	?Underpowered
Hanchette	83	Ecologic	N/a	Ultraviolet exposure	Yes	

which is a relatively new phenomenon). A north–south gradient in prostate cancer risk also exists.[83] One potential reason for this observation relates to the increased sun exposure among residents of the southern areas in the northern hemisphere. Calcium intake is also associated with prostate cancer incidence. Data from the Health Professional cohort study demonstrated that high calcium intake, which suppresses the conversion of 25-hydroxy vitamin D to its active form 1,25-dihydroxy vitamin D, is associated with a two fold increased risk of prostate cancer.[84] The same study also demonstrated that fruit intake was protective of prostate cancer. The hypothesized association between fruit intake and prostate cancer lies in fructose's ability to reduce plasma phosphate and thus stimulate 1,25-dihydroxy vitamin D synthesis.[85] The data supporting a link between vitamin D and prostate cancer are not limited to epidemiological observations. A large body of basic laboratory work has now demonstrated the vitamin D can inhibit the growth and induce apoptosis of human prostate cancer cells.[86,87]

Table 22.8 lists some of the seminal studies that have assessed associations between either sunlight or biological measures of vitamin D (serum levels, polymorphisms, etc.) activity and prostate cancer.[83,84,88–92] On reviewing these data, the best methodological studies have consistently demonstrated an association between calcium intake, levels or dairy intake and prostate cancer risk. Smaller studies and those unable to control for confounders have been negative. Furthermore, the range of calcium intake in the large US-based cohort studies is more diverse and not easy to duplicate in smaller studies. Data on different VDR polymorphisms are largely negative. Although one must exert caution in interpreting these data, there does appear to be limited evidence for a causal association between dietary calcium and prostate cancer. These observations have now been reinforced by clinical data suggesting an antitumor effect on vitamin D and its analogs.[93,94] Numerous studies are underway to test this hypothesis further.

Another potential link between prostate cancer and sun exposure is melatonin. Melatonin is produced primarily at night time and is suppressed by exposure to light. Melatonin has demonstrated antiproliferative effects on human prostate cancer cell lines and can inhibit androgen-mediated actions at the nuclear level.[95] Further evidence of a significant effect lies in the fact that one study has suggested that blind individuals have a lower incidence of prostate cancer (31% protection).[96] If, in fact, this were true, then studies of sunlight and cancer would indicate a higher incidence of prostate cancer among individuals who lived in more Southern latitude in the northern hemisphere. These data thus conflict with the vitamin D hypothesis. Clearly more study is needed on this subject.

REFERENCES

1. Narod SA, Feunteun J, Lynch HT et al. Familial breast–ovarian cancer locus on chromosome 17q12–q23. *Lancet* 1991; 338:82–3.
2. Rothman KJ, Poole C. Science and policy making. *Am. J. Publ. Health* 1985; 75:340–1.
3. Breslow NE, Enstrom JE. Geographic correlations between cancer mortality rates and alcohol–tobacco consumption in the United States. *J. Natl Cancer Inst.* 1974; 3:631–9.
4. Schwartz GG. Geographic trends in prostate cancer mortality: an application of spatial smoothers and the need for adjustment *Ann. Epidemiol.* 1997; 7:430.
5. Herbst AL, Ulfelder H, Poskanzer DC. Adenocarcinoma of the vagina. Association of maternal stilbestrol therapy with tumor appearance in young women. *N. Engl. J. Med.* 1971; 284:878–81.
6. Kannel WB, Abbot RD. Incidence and prognosis of unrecognized myocardial infarction. An update on the Framingham study. *N. Engl. J. Med.* 1984; 311:1144–7.
7. Hill AB. The environment and disease: association or causation? *Proc. R. Soc. Med.* 1965; 58:295–300.
8. Rothman KJ. *Modern Epidemiology*, 1st edn. Boston, MA: Little, Brown and Company, 1986.
9. Willet W. *Nutritional Epidemiology*, p.14. New York: Oxford University Press, 1990.
10. Pirtskhalaishvili G, Hrebinko RL, Nelson JB. The treatment of prostate cancer: an overview of current options. *Cancer Pract.* 2001; 9:295–306.
11. Fair WR, Fleshner NE, Heston W. Cancer of the prostate; a nutritional disease? *Urology* 1997; 50:840–8.
12. Brawley OW, Knopf K, Thompson I. The epidemiology of prostate cancer part II: the risk factors. *Semin. Urol. Oncol.* 1998; 16:193–201.

13. Migration and prostate cancer: an international perspective. *J. Natl Med. Assoc*. 1998; 90:S720–3.
14. Hecht SS. Tobacco and cancer: approaches using carcinogen biomarkers and chemoprevention. *Ann. N.Y. Acad*. 1997; 9:91–111.
15. Rodgman A, Smith CJ, Perfetti TA. The composition of cigarette smoke: a retrospective, with emphasis on polycyclic components. *Human Exp. Toxicol*. 2000; 19:573–95.
16. Fossa SD, Eri L, Sovlund E et al. No randomized trial of prostate-cancer screening in Norway. *Lancet Oncol*. 2001; 2:746–9.
17. Hsing AW, Devessa SS. Trends and patterns of prostate cancer: what do they suggest? *Epidemiol. Rev*. 2001; 23:3–13.
18. English KM, Pugh PJ, Parry H et al. Effect of cigarette smoking on levels of bioavailable testosterone in healthy men. *Clin. Sci*. 2001; 100:661–5.
19. Hsieh CC, Signorello LB, Lipworth L et al. Predictors of sex hormone levels among the elderly: a study in Greece. *J. Clin. Epidemiol*. 1998; 51:837–41.
20. Shields PG. Epidemiology of tobacco carcinogenesis. *Curr. Oncol. Rep*. 2000; 2:257–62.
21. Larue H, Allard P. Simoneau M et al. P53 point mutations in initial superficial bladder cancer occur only in tumors from current or recent cigarette smokers. *Carcinogenesis* 2000; 21:101–6.
22. Fleshner NE, Klotz L. Diet, androgens, oxidative stress and prostate cancer susceptibility. *Cancer Metastasis Rev*. 1998–99; 17:325–30.
23. Venkateswarran V, Klotz L, Fleshner NE. Selenium modulation of cell proliferation and cell cycle biomarkers in human prostate carcinoma cell lines. *Cancer Res*. 2002; 62:2540–5.
24. Lu J. Apoptosis and angiogenesis in cancer prevention by selenium. *Adv. Exp. Med. Biol*. 2001; 492:131–45.
25. Agarwal S, Rao AV. Tomato lycopene and its role in human health and chronic disease. *Canadian Medical Association Journal* 2000; 163:739–44.
26. Kocyigit A, Erel O, Gur S. Effects of tobacco smoking on plasma selenium, zinc, copper and iron concentrations and related antioxidative enzyme activities. *Clin. Biochem*. 2001; 34:629–33.
27. Van der Berg H, van der Gaag M, Hendriks H. Influence of lifestyle on vitamin bioavailability. *Int. J. Vitamin Nutr. Res*. 2002; 72:53–9.
28. Klein EA, Thompson IM, Lippman SM et al. SELECT: the next prostate cancer prevention trial. *J. Urol*. 2001; 166:1311–15.
29. Field JK. Oncogenes and tumor-supressor genes in squamous cell carcinoma of the head and neck. *Eur. J. Cancer* 1992; 28:67–76.
30. Giles GG, Severi G, McCredie MR et al. Smoking and prostate cancer: findings from an Australian case-control study. *Ann. Oncol*. 2001; 12:761–5.
31. Van der Gulden JW, Verbeek Al, Kolk JJ. Smoking and drinking habits in relation to prostate cancer. *Br. J. Urol*. 1994; 73:382–9.
32. Hayes RB, Pottern LM, Swanson GM et al. Tobacco use and prostate cancer in blacks and whites in the Unites States. *Cancer Causes Control* 1994; 5:221–6.
33. Distefani E, Fierro L, Barrios E et al. Tobacco, alcohol, diet and risk of prostate cancer. *Tumori* 1995; 8:315–20.
34. Hsieh CC, Thanos A, Mitropoulos D et al. Risk factors for prostate cancer: a case control study in Greece. *Int. J. Cancer* 1999; 80:699–703.
35. Honda GD, Bernstein L, Ross RK. Vasectomy, cigarette smoking and age at first sexual intercourse as risk factors for prostate cancer in middle-aged men. *Br. J. Cancer* 1988; 57:326–31.
36. Oishi K, Okada K, Toshida O. Case-control study of prostatic cancer in Kyoto, Japan: demographic and some lifestyle risk factors. *Prostate* 1989; 14:117–22.
37. Yu H, Harris RE, Wynder EL. Case control study of prostate cancer and socioeconomic factors. *Prostate* 1988; 13:317–25.
38. Sharpe CR, Siemiatycki J, Parent ME. Activities and exposures during leisure and prostate cancer risk. *Cancer Epidemiol. Biomark. Prev*. 2001; 10:855–60.
39. Furuya Y, Akimoto S, Akakura K et al. Smoking and obesity in relation to the etiology and disease progression of prostate cancer in Japan. *Int. J. Urol*. 1998; 5:134–7.
40. Rohan TE, Hislop TG, Howe GR et al. Cigarette smoking and risk of prostate cancer: a population based case-control study in Ontario and British Columbia, Canada. *Eur. J. Cancer Prev*. 1997; 6:382–8.
41. Lumey LH, Pittman B, Zang EA et al. Cigarette smoking and prostate cancer: no relation with six measures of lifetime smoking habits in a large case-control study. *Prostate* 1997; 33:195–200.
42. Andersson SO, Baron J, Bergstrom R et al. Lifestyle factors and prostate cancer risk: a case-control study in Sweden. *Cancer Epidemiol. Biomark. Prev*. 1996; 5:509–13.
43. Slattery ML, West DW. Smoking, alcohol, coffee, caffeine, and threobromine: risk of prostate cancer in Utah. *Cancer Causes Control* 1993; 4:559–63.
44. Gronberg H, Damber L, Damber JE. Total food consumption and body mass index in relation to prostate cancer risk: a case-control study in Sweden with prospectively collected exposure data. *J. Urol*. 1996; 155:969–74.
45. LeMarchand L, Kolonel LN, Wilkens LR et al. Animal fat consumption and prostate cancer: a prospective study in Hawaii. *Epidemiology* 1994; 5:271–3.
46. Hiatt RA, Armstrong MA, Klatsky AL et al. Alcohol consumption, smoking and other risk factors and prostate cancer in a large health plan cohort in California. *Cancer Causes Control* 1994; 5:66–72.
47. Engeland A, Andersen A, Haldorsen T et al. Smoking habits and risk of cancers other than lung cancer: 28 years' follow up of 26 000 Norwegian men and women. *Cancer Causes Control* 1996; 7:497–506.
48. Putnam SD, Cerhan JR, Parker AS et al. Lifestyle and anthropometric risk factors for prostate cancer in a cohort of Iowa men. *Ann. Epidemiol*. 2000; 10:361–9.
49. Giovannucci E, Rimm EB, Ascherio A et al. Smoking and risk of total and fatal prostate cancer in Unites States health professionals. *Cancer Epidemiol. Biomark. Prev*. 1999; 8:277–82.
50. Adami HO, Bergstrom R, Engholm G et al. A prospective study of smoking and risk of prostate cancer. *Int. J. Cancer* 1996; 67:764–8.
51. Lotufo PA, Lee IM, Ajani UA et al. Cigarette smoking and risk of prostate cancer in the physician health study. *Int. J. Cancer* 2000; 87:141–4.
52. Eichholzer M, Stahelin HB, Ludin E et al. Smoking, plasma vitamins C, E, retinal and carotene, and fatal prostate cancer: seventeen year follow up of the prospective Basel study. *Prostate* 1999; 38:189–98.
53. Rodriguez C, Tatham LM, Thun MJ et al. Smoking and fatal prostate cancer in a large cohort of adult men. *Am. J. Epidemiol*. 1997; 145:466–75.
54. Hsing AW, McLaughlin JK, Schuman LM et al. Diet, tobacco use, and fatal prostate cancer: results from the Lutheran Brotherhood Cohort Study. *Cancer Res*. 1990; 50:6836–40.
55. Hsing AW, McLaughlin JK, Hrubec Z et al. Tobacco use and prostate cancer: 26 year follow-up of US veterans. *Am. J. Epidemiol*. 1991; 133:437–41.
56. Coughlin SS, Neaton JD, Sengupta A. Cigarette smoking as a predictor of death from prostate cancer in 384,874 men screened for the Multiple Risk Factor Intervention Trial. *Am. J. Epidemiol*. 1996; 143:1002–6.
57. Daniell HW. A worse prognosis for smokers with prostate cancer. *J. Urol*. 1995; 154:153–7.
58. Johansson S, Landstrom M, Bjermer L et al. Effects of tobacco smoke on tumor growth and radiation response of dunning R3327 prostate adenocarcinoma in rats. *Prostate* 2000; 42:253–9.
59. Liu S, Lee IM, Linson P et al. A prospective study of physical activity and risk of prostate cancer in US physicians. *Int. J. Epidemiol*. 2000; 29:29–35.

60. Tavani A, Negri E, Franceschi S et al. Alcohol consumption and risk of prostate cancer. *Nutr. Cancer* 1994; 21:24–31.
61. Giovannucci E, Ascherio A, Rimm EB et al. A prospective cohort study of vasectomy and prostate cancer in US men. *JAMA* 1993; 269:873–7.
62. Martinez FE, Laura IA, Martinez M et al. Morphology of the ventral lobe of the prostate and seminal vesicles in an ethanol-drinking strain of rats. *J. Submicrosc. Cytol. Pathol.* 2001; 22:99–106.
63. Sarkola T, Aldercreutz H, Heinonen S et al. Alcohol intake, androgen and glucocorticoid steroids in premenopausal women using oral contraceptives: an intervention study. *J. Steroid Biochem. Mol. Biol.* 2001; 78:157–65.
64. Houde SC, Melillo KD. Cardiovascular health and physical activity in older adults: an intergrative review of research methodology and results. *J. Adv. Nurs.* 2002; 38:219–34.
65. Jurimae J, Jurimae T, Purge P. Plasma testosterone and cortisol responses to prolonged sculling in male competitive rowers. *J. Sports Sci.* 2001; 19:893–8.
66. Ji LL. Exercise-induced modulation of antioxidant defense. *Ann. N.Y. Acad. Sci.* 2002; 959:82–92.
67. Tymchuk CN, Barnard RJ, Heber D et al. Evidence of an inhibitory effect of diet and exercise on prostate cancer cell growth. *J. Urol.* 2001; 166:1185–9.
68. Sung JF, Lin RS, Pu YS et al. Risk factors for prostate carcinoma in Taiwan: a case-control study in a Chinese population. *Cancer* 1999; 86:484–91.
69. Bairati I, Larouche R, Meyer F et al. Lifetime occupational physical activity and incidental prostate cancer. *Cancer Causes Control* 2000; 11:759–64.
70. Albanes D, Blair A, Taylor PR. Physical activity and risk of cancer in NHANES 1 population. *Am. J. Public Health* 1989; 79:744–50.
71. Thune I, Lund E. Physical activity and the risk of prostate and testicular cancer: a cohort study of 53,000 Norwegian men. *Cancer Causes Control* 1994; 5:549–64.
72. Cerhan JR, Torner JC, Lynch CF et al. Association of smoking, body mass, and physical activity with risk of prostate cancer in the Iowa 65+ rural health study. *Cancer Causes Control* 1997; 8:229–38.
73. Hartman TJ, Albanes D, Rautalahti M et al. Physical activity and prostate cancer in the Alpha-Tocopherol, Beta Carotene (ATBC) Cancer Prevention Study. *Cancer Causes Control* 1998; 9:11–18.
74. Giovannucci E, Leitzmann M, Spiegelman D et al. A prospective study of physical activity and prostate cancer in male health professionals. *Cancer Res.* 1998; 58:5117–22.
75. Liu S, Lee IM, Linson P et al. A prospective study of physical activity and risk of prostate cancer in US physicians. *Int. J. Epidemiol.* 2000; 29:29–35.
76. Wannamethee SG, Shaper AG, Walker M. Physical activity and risk of cancer in middle-aged men. *Br. J. Cancer* 2001; 85:1311–16.
77. Schwartz GG, Hulka BS. Is vitamin D deficiency a risk factor for prostate cancer? *Anticancer Res.* 1990; 10:1307–11.
78. Lee WP, Lin LW, Yeh SH et al. Correlations among serum calcium, vitamin D, and parathyroid hormones levels in the elderly in southern Taiwan. *J. Nurs. Res.* 2002; 10:65–72.
79. Piffath TA, Whiteman MK, Flaws JA et al. Ethnic differences in cancer mortality trends in the US, 1950–1992. *Ethn. Health* 2001; 6:105–19.
80. Jablonski NG, Chaplin G. The human evolution of skin coloration. *J. Human Evol.* 2000; 39:57–106.
81. Nutritional aspects of prostate cancer. *Can. J. Urol.* 2000; 7:927–35.
82. Savige GS. Candidate foods in the Asia-Pacific region for cardiovascular protection: fish, fruit and vegetables. *Asia Pac. J. Clin. Nutr.* 2001; 10:134–7.
83. Hanchette CL, Schwartz GG. Geographic patterns of prostate cancer mortality. Evidence for a protective effect of ultraviolet radiation. *Cancer* 1992; 70: 2861–9.
84. Giovannucci E, Rimm EB, Wolk A et al. Calcium and fructose intake in relation to risk of prostate cancer. *Cancer Res.* 1998; 58:442–7.
85. Kapur S. A medical hypothesis: phosphorous balance and prostate cancer. *Cancer Invest.* 2000; 18:664–9.
86. Blutt SE, Weigel NL. Vitamin D and prostate cancer. *Proc. Soc. Exp. Biol. Med.* 1999; 221:89–98.
87. Crescioli C, Maggi M, Luconi M et al. Vitamin D3 analogue inhibits keratinocyte growth factor signaling and induces apoptosis inhuman prostate cancer cells. *Prostate* 2002; 50:15–26.
88. Corder EH, Guess HA, Hulka BS et al. Vitamin D and prostate cancer; a prediagnostic study witrh stored sera. *Cancer Epidemiol. Biomark. Prev.* 1993; 2:467–72.
89. Chan JM, Pietinen P, Virtanen M et al. Diet and prostate cancer risk in a cohort of smokers with a specific focus on calcium and phosphorous. *Cancer Causes Control* 2000; 11:859–67.
90. Chan JM, Giovannucci E, Andersson SO et al. Dairy products, calcium, phosphorous, vitamin D and risk of prostate cancer (Sweden). *Cancer Causes Control* 1998; 9:559–66.
91. Tavani A, Gallus S, Franceschi S et al. Calcium, dairy products and risk of prostate cancer. *Prostate* 2001; 48:118–21.
92. Nomura AM, Stemmermann GN, Lee J et al. Serum vitamin D metabolite levels and subsequent devlopment of prostate cancer (Hawaii, United States). *Cancer Causes Control* 1998; 9:425–32.
93. Gross C, Stamey T, Hancock S et al. Treatment of early recurrent prostate cancer with 1,25t-dihydroxyvitamin D3 (calcitriol). *J. Urol.* 1998; 159:2039–40.
94. Van Veldhuizen PJ, Taylor SA, Williamson S et al. Treatment of vitamin D deficiency with metastatic prostate cancer may improve bone pain and muscle strength. *J. Urol.* 2000; 163:187–90.
95. Gilad E, Laufer M, Matzkin H et al. Melatonin receptors in PC3 human prostate tumor cells. *J. Pineal Res.* 1999; 26:211–20.
96. Feychting M, Osterlund B, Ahlbom A. Reduced cancer incidence among the blind. *Epidemiology* 1998; 9:490–4.

Prostate Cancer on the Internet: Review 2002

Neil A. Abrahams

Department of Anatomic Pathology, Cleveland Clinic Foundation, Cleveland, OH, USA

David G. Bostwick*

Bostwick Laboratories, Richmond, VA, and the Department of Pathology, University of Virginia, Charlottesville, VA, USA

INTRODUCTION

The use of the Internet as a portal for medical information has resulted in an explosion of interest, with more than 20 000 medically related websites in the year 2000. This wealth of information has had a substantial and expanding impact on the practice of medicine as related to the prevention, diagnosis and treatment of prostate cancer. Within the USA, use of the Internet has tripled over the last 4 years, from 26 million users in 1996 to over 78 million in 1999.[1] Remarkably, in a recent survey, the most common search word introduced into search engines on the Internet was 'cancer'.[2]

Given the unfettered access to the Internet that a website creator has, there is great concern regarding the accuracy and bias of information available. Efforts such as the Health on the Net Foundation (HON) standards are a good first step in providing reassurance of credibility to some, but concerns persist about the ease of obtaining HON approval and the appropriateness of self-proclaimed internet 'watchdog' sites (e.g. 'Who polices the policeman?'). Thus, contemporary reviews of specific searches may provide greater utility for clinicians and patients.

The urologist may be involved with prostate cancer-related websites in multiple ways. First, there is the ability to obtain current information on any relevant aspects of prostate cancer, such as screening, treatment options and current research. Second, the urologist or health care provider may use the Internet as a tool for patient education, referring the patient to sites that discuss treatment options or emotional support. Finally, empowered patients are obtaining their own on-line information from credible and not so credible sources and providing it to the urologist for verification or rebuttal. It is critical that the urologist be aware of many of these websites and be able to determine the accuracy of the information provided.

Many published articles have systematically reviewed websites in other fields of medicine, such as pathology,[3–5] radiology,[6,7] breast cancer,[8] and molecular medicine,[9] but there is a paucity of articles specifically evaluating websites that provide information on prostate cancer. In order to address this need, we undertook a structured search of prostate cancer on the Internet in January 2002.

We undertook a thorough review of top results for the search term 'prostate cancer' provided by widely used and well-known search engines such as Yahoo, Northern Light, AltaVista and others. Interestingly, many sites were out of date and had not been updated for at least 12 months, and we excluded these stagnant sites from further consideration. The sites under review were stratified according to sponsor's level of commercialization (not-for-profit and for-profit), and then graded subjectively according to ease of use, content accuracy, relevance for prostate cancer and updated content. We excluded all sites requiring fees for access to information.

Other benefits and resources of the Internet were beyond the scope of our review, such as e-mail, electronic patient charting and internet results reporting, and are described elsewhere.[3,10–15]

The rapid and often unheralded appearance of new websites on the Internet has been accompanied by the silent disappearance of some existing websites. Our contemporary review reflects the current Internet sites on prostate cancer at the time the article was submitted for publication in January 2002.

*To whom correspondence should be sent.

MATERIALS AND METHODS

Search for 'prostate cancer'

During the month of January 2002, we utilized 13 widely used search engines to identify all sites found by the key words 'prostate cancer' (**Table 23.1**). This was done to ensure that the search was broad enough to include websites that focused specifically on urogenital disorders as well as those that had relevant content concerning prostate cancer. Some of these search engines have their own on-site information regarding prostate cancer (usually modest and generic information), and we excluded this from consideration; instead the search key words were introduced directly into the search 'box' on each search engine. America On-Line (AOL) was excluded as not being considered a search engine. The number of sites are presented in **Table 23.1**. For those search engines that returned more than 100 site results, only the first 70 were evaluated. If the search engine returned less than 100 results, all were evaluated. While performing the search and accessing the results generated, we subjectively evaluated the advantages and

Table 23.1. Search engines listed in alphabetical order used to seek out relevant websites on 'prostate cancer'.*

Search engine	URL	Initial results	Advantages	Disadvantages
AltaVista	www.altavista.com	4 708 915	Most exhaustive results list	Bias towards sponsored sites, which are listed at the beginning of the results list
Ask Jeeves	www.askjeeves.com	Not stated	Categorizes results into six groups, which narrows search further	Non-focused sites in abundance
Direct Hit	www.directhit.com	61	Results are listed from the most popular to the least. Popularity defined as the number of people visiting a website	Since medically relevant sites were not accessed by a large proportion of the population these were considered 'not popular' and appeared at the end of the results list
Excite	www.excite.com	Not stated	Easy to use search engine	Majority of websites were not specifically focused to prostate cancer at all
Google	www.google.com	556 000	Medically relevant sites at top of results list	Intermixed websites of no interest with websites of interest
Hot Bot	www.hotbot.com	Not stated	Results displayed in large easy-to-read font	Many non-relevant sites were also listed
Infoseek	www.infoseek.com	Not stated	Results displayed in large font with brief description of website	Many non-relevant sites on first three pages of results
Lycos	www.lycos.com	348 000	Categorizes results in terms of treatment, diagnosis and prevention	Bias towards sponsored sites, which are listed at the beginning of the results list
Microsoft Network	http://home.microsoft.com	239	Overall comprehensive and concise results list	Medically relevant sites were at the end of the results list
Netscape Netcenter	http://home.netscape.com	1639	Concise and relevant results	None appreciated
Northern Light	www.northernlight.com	340 352	Non-profit websites listed at beginning of results	Majority of websites were not specifically focused to prostate cancer at all
Webcrawler	www.webcrawler.com	Not stated	Overall comprehensive and relevant results	None appreciated
Yahoo	www.yahoo.com	107	Quick focused results obtained	Medically relevant sites are at end of list

*Keywords for search were 'prostate cancer'; search performed in January 2002.

Table 23.2. Criteria used to score websites.

	Grade 1	2	3
Accuracy	Some statements are incorrect compared to standard medical literature*	Most statements are correct but not referenced	Statements are correct; material provided is referenced
Ease of navigation	Data are presented in a disorganized fashion, many pop-ups and commercial banners Too much information on one page, need to scroll through many screen pages	Systematic organization of the data presented Commercial banners excluded to one side No search index Navigational links on bottom of page have to scroll to find them	Short concise screen pages with easily readable and concise display of information Visible search index Navigational links on top or side of page
Relevance	Of little relevance to either clinicians or patients diagnosed with prostate cancer	Of some relevance to either clinicians or patients diagnosed with prostate cancer	Relevant to both clinicians or patients diagnosed with prostate cancer
Updates	Quarterly or less	Weekly	Daily
Completeness	Information provided is superficial and only discusses isolated issues	Information provided is of general depth, various issues are discussed but no links are provided to other websites of possible interest	Information provided is in-depth, covers a range of relevant issues and provides links to other Internet sites of possible interest

*Usually owing to outdated information.

disadvantages of each search engine (summarized in **Table 23.1**) based on ease of use.

Inclusion criteria

In order for a website to be included in our review, the following criteria had to be met: (1) free access with no fee required; (2) minimal log-in required (no more than a user name and password); and (3) site updated within the last 12 months. If these criteria were met, the website was then evaluated and scored (see later). We excluded those sites in which the authors had a direct or indirect interest, including cancerhome.com, cancerfacts.com and bostwicklaboratories.com.

The websites were categorized into three groups. The first, not-for-profit sites, included all websites in the .edu, .org, .gov, and .net domains. These usually were universities, non-profit organizations or foundations, government-sponsored research institutions and networking websites. The second group, for-profit sites, included all websites in the .com domain. Generally, these were private organizations that are involved in diagnosis, treatment or education of prostate cancer, and have direct or indirect commercial interests associated with the website. This does not indicate, however, that accessing the website required a fee. The third group, support groups, included websites that were either hosted by cancer support groups or individuals, and that were targeted more for patients than for physicians. Sites specifically designated as chat rooms were not evaluated.

Grading of websites

Each website was graded subjectively by the authors according to five distinct features: accuracy, ease of navigation, relevance, timeliness of updates and completeness. The criteria used were modified from previously used guidelines,[16,17] and are defined in **Table 23.2**. Each category was graded using a three-tiered numerical system (1–3) and the sum provided an overall score (range 5–15). Subsequently, the sites were rank-ordered from highest to lowest.

RESULTS

Search engines

A total of 13 search engines were used. The results generated varied slightly between search engines. At the top of the results list were personal websites of patients diagnosed with prostate cancer, support groups, cancer treatment centers, charity organizations and occasional law firms specializing in prostate cancer malpractice suits. Medically relevant websites on prostate cancer were usually at or near the end of the results list. This probably represents a bias in favor of groups that pay to have their websites listed and thus are placed at the top of the lists. This was most noticeable, to the authors, with Yahoo, Lycos and AltaVista.

Table 23.3. Not-for-profit websites with specific content related to prostate cancer* (ranked in descending order based on score).

Website URL	Sponsor	Accuracy	Ease	Relevance	Updates	Completeness	Market	Score	Rank
http://www.UroHealth.org	Media Health Online	3	3	3	2	3	U	14	1
http://www.uroreviews.org	British Urology Institute	3	3	3	2	3	U/P	14	1
http://cancernet.nci.nih.gov/Cancer_Types/Prostate_Cancer	National Cancer Institute	2	3	3	2	3	U/P	13	2
http://www.dfci.harvard.edu/	Dana Farber Cancer Institute	2	3	3	2	3	U/P	13	2
http://www.hon.ch	Health on the Net Foundation	3	2	3	2	3	P/U	13	2
http://www.prostatecalculator.org/	Institute for Clinical Research	2	3	2	3	2	U/P	12	3
http://www.urology.jhu.edu	Johns Hopkins Medical Institutions	2	1	3	3	3	P/U	12	3
www.nlm.nih.gov/medlineplus/prostatecancer	National Library of Medicine/National Institutes of Health	2	3	3	2	2	U/P	12	3
http://cancertrials.nci.nih.gov/	National Cancer Institute	3	2	3	1	2	U/P	11	4
http://www.prostate-cancer.org	Prostate Cancer Research Institute	2	2	2	1	3	P/U	10	5
http://www.capcure.org	Association for the Cure of Prostate Cancer	2	3	3	1	1	P	10	5
www.cdc.gov/cancer/prostate	Center for Disease Control	2	3	2	1	2	P	10	5
http://www.auanet.org	American Urological Association	2	2	3	1	2	U	10	5
http://www.clevelandclinic.org/health/	Cleveland Clinic Foundation	2	3	2	1	2	P/U	10	5
http://www.brachytherapy.org.uk	Princess Grace Hospital, London	2	3	2	1	2	U/P	10	5
http://www.prostatefoundation.org	Robert Benjamin Ablin Foundation for Cancer Research	2	1	2	1	3	U/P	9	6
http://www.cpdr.org	Center for Prostate Disease Research	2	2	2	1	2	P/U	9	6
http://pcrn.org	Prostate Cancer Resource Network	2	2	2	1	1	P	8	7
http://www.urology.med.umich.edu/	University of Michigan Department of Urology	2	1	2	1	2	P/U	8	7
http://www.uronet.org	Astra-Zeneca Pharmaceuticals	2	1	2	1	2	U	8	7

www.cancer.org	American Cancer Society	2	1	2	1	2	P	8	7
http://www.imperialcancer.co.uk	Imperial Cancer Research Institute	2	2	2	1	1	P/U	8	7
http://www.prostate-cancer.org.uk	Prostate Cancer Charity	2	2	2	1	1	P	8	7
http://www.mskcc.org/mskcc/html/403.cfm	Memorial Sloan Kettering Cancer Center	2	2	2	1	1	P/U	8	7
http://www.mdanderson.org/diseases/prostate/	MD Anderson Cancer Center	2	1	1	1	2	P/U	7	8
http://www.cancernews.com/male.htm	Cancer News on the Net	2	1	1	1	2	P	7	8
http://www.roswellpark.org/	Roswell Park Cancer Institute	2	2	1	1	1	P	7	8
http://www.hopkinsprostate.com/	Johns Hopkins Hospital	2	1	1	1	2	P	7	8
www.mayohealth.org	Mayo Clinic	2	1	2	1	1	P	7	8
http://www.afud.org/pca/pcaindex.html	American Foundation for Urologic Disease	2	1	1	1	1	P	6	9
http://www.prostatecancernet.org	Prostate Cancer Knowledge Center	2	1	1	1	1	P/U	6	9
http://www.cancer.med.umich.edu/prostcan/prostcan.html	University of Michigan Cancer Center	2	1	1	1	1	P/U	6	9
http://www.fccc.edu/clinicalresearch/prostateriskassessment/prostate.html	Fox Chase Cancer Center	1	1	1	1	1	P	5	10
http://www.ameripros.org/	American Prostate Society	1	1	1	1	1	P	5	10
http://prostate.tju.edu	Thomas Jefferson University	1	1	1	1	1	P	5	10

U denotes that the content of the website contains material that is specifically of interest to urologists; P denotes content that is designed for patients diagnosed with prostate cancer; U/P denotes content of interest to both patient and urologist, but especially focused for urologists; P/U indicates content of interest to patients and urologists, but especially focused for patients.

*See 'Materials and methods' for a description of how these sites were chosen.

Not-for-profit groups

Of the 60 sites that met our inclusion criteria, 35 were ranked and scored (**Table 23.3**); the remaining 25 websites scored a total of less than five points and are not presented. Very few websites had grossly inaccurate information posted: when this was found, it was more due to a lack of updating. Only three websites were updated daily and the majority were updated monthly or bimonthly. The top ten ranked websites are presented with a brief description of the contents in **Table 23.4.**

Table 23.4. Top ten not-for-profit websites about prostate cancer for urologists.

Website URL	Sponsor	Summary of website
http://www.UroHealth.org	Media Health Online	Comprehensive information on the latest research, treatment and diagnostic developments in PCa. Listing of all urology meetings. Requires registration but is free, some information can be accessed without registration.
http://www.uroreviews.org	British Urology Institute	Selects articles from various urology journals on specific topics of PCa and posts the abstracts. Has links to the journal websites for possible access to the entire article. Updated and a quick way to stay on top of latest developments.
http://cancernet.nci.nih.gov/ Cancer_Types/Prostate_Cancer	National Cancer Institutes	National Institutes of Health website with in-depth and comprehensive aspect of all aspects of PCa, including patient education materials clinical trials listings, and support groups information. Good source of credible information.
http://www.dfci.harvard.edu/	Dana Farber Cancer Institute	User-friendly website with listing of clinical trials, patient education material and treatment options at their institute.
http://www.hon.ch	Health on the Net Foundation	Non-for-profit foundation that rates internet resources that provide medical information. Provides direct links to 'outstanding' websites providing medical information. Useful studies as to use of Internet by patients and follows trends of Internet usage by patients and monitors what content is accessed.
http://www.prostatecalculator.org/	Institute for Clinical Research	Allows patients to calculate the probability of organ-confined disease, lymph node metastasis and possible treatment options based on age, PSA, biopsy Gleason score, percentage of cancer in the biopsy and clinical stage. Provides results in useful histograms and suggests options to explore based on the information provided.
http://urology.jhu.edu	Department of Urology, Johns Hopkins University	Johns Hopkins urology website; comprehensive information on all aspects of PCa including treatment options at their institution. Useful newsletter that you can subscribe to that discusses in general terms latest developments in the field.
www.nlm.nih.gov/medlineplus/ prostatecancer	National Library of Medicine/ National Institutes of Health	National Institutes of Health website with dictionary of medical terms and information on drugs and treatment options. Highly illustrated patient education information section.
http://cancertrials.nci.nih.gov/	National Institutes of Health	National Institutes of Health sponsored website with information on participation criteria of clinical trials and location of centers that participate in the trials. Additional information is available on prostate cancer and current screening initiatives.
http://www.prostate-cancer.org	Prostate Cancer Research Institute	Selection of patient education materials on PCa as well as article summaries on a wide variety of subjects concerning PCa. Recent seminars on PCa are available for the user to download in Powerpoint format.

PCa, prostate cancer.

Table 23.5. For-profit organizations' websites with specific content related to prostate cancer* (ranked in descending order based on score).

Website URL	Sponsor	Accuracy	Ease	Relevance	Updates	Completeness	Market	Score	Rank
http://www.psa-rising.com	PSA Rising Magazine	3	2	3	3	3	U/P	14	1
http://www.veritasmedicine.com	Veritas Medicine	2	3	3	2	3	U/P	13	2
http://www.mdchoice.com	MD Choice	3	2	3	1	3	U/P	12	3
http://www.prostatehealthdirectory.com /	Active Health Inc.	2	2	2	3	3	P	12	3
www.prostatedoctor.com	Prostate Doctor.Com	2	2	3	1	3	P	11	4
www.library.utoronto.ca/medicine/prostate	Medical Library of the University of Toronto	2	3	2	1	3	P/U	11	4
http://wmfurology.com	WMF Urology Associates	3	2	2	1	3	U/P	11	4
http://webmd.com	WebMD	2	3	2	2	2	U	11	4
http://medscape.com	Medscape	2	3	2	1	2	U	10	5
http://www.prostatehealth.com/	Merck Pharmaceutical	2	3	1	2	1	P	9	6
http://www.urologychannel.com	Health Communities	2	2	2	1	1	U/P	8	7
http://www.intellihealth.com	Intellihealth Inc. (AETNA)	2	1	2	1	2	P	8	7
http://www.centerwatch.com	Center Watch Inc.	2	2	1	1	2	U/P	8	7
http://www.prostatematters.com/	Uro Matters Inc.	1	2	1	1	1	P	6	8
www.cancerlinksusa.com/prostate	Cancer Links USA	1	1	1	1	2	P	6	8
http://www.drkoop.com	Dr Koop Life Care Company	2	1	1	1	1	P	6	8

U denotes that the content of the website contains material that is specifically of interest to urologists; P denotes content that is designed for patients diagnosed with prostate cancer; U/P denotes content of interest to both patient and urologist, but especially focused for urologists; P/U indicates content of interest to patients and urologists, but especially focused for patients.

*See 'Materials and methods' for a description of how these sites were chosen.

For-profit groups

Among more than 100 websites that met the inclusion criteria, 16 are ranked and scored in **Table 23.5** The remaining 82 websites scored a total of less than five points and are not presented. Inaccurate information that was posted usually resulted from the lack of timely updates. The top ten ranked websites are presented in **Table 23.6** with a brief description of the contents.

Support groups

Surprisingly few support group websites were included in the results generated by the search engines, and this may be

Table 23.6. Top ten for-profit websites about prostate cancer for urologists.

Website URL	Sponsor	Summary of website
http://www.psa-rising.com	PSA Rising Magazine	Latest news on PCa by category, treatment options, diagnosis, etc. Editorial reviews on certain topics and also has patient education articles. Material is referenced and is one of the few sites that updates daily.
http://www.veritasmedicine.com	Veritas Medicine	Free access to information regarding location and inclusion criteria of clinical trials for PCa. Section that describes in depth what the different components of a clinical trial are.
http://www.mdchoice.com	MD Choice	Access to folders that contain collections of peer-reviewed articles on select topics of PCa treatment and patient education. Fee charged for certain information.
http://www.prostatehealthdirectory.com/	Active Health Inc.	Comprehensive daily update of on all aspects of PCa with direct links to websites with information on latest clinical trials and research developments. Sign up for a very useful newsletter.
www.prostatedoctor.com	Prostate Doctor.Com	Provides information on treatment options of PCa with useful links to other websites of interest. Alternative non-traditional therapies are also briefly described.
www.library.utoronto.ca/ medicine/prostate	Medical Library of the University of Toronto	Incredibly well-illustrated website that provides general information on all aspects of PCa with streaming videos of prostate biopsy procedures, as well as photographs of the instruments used in the diagnosis of prostate cancer. Links for support groups predominantly in Canada but some US sites are listed.
http://wmfurology.com	WMF Urology Associates	Urology associates web page that provides in-depth coverage of a wide range of information concerning urological disorders. One of the few websites* that provide information about high-grade prostatic intraepithelial neoplasia (high grade PIN).
http://webmd.com	Medscape	Access to recent articles on focused topics of PCa. Requires registration and restricted to physicians only.
http://medscape.com	Web MD	Site divided into two sections, one for physicians and the other for consumers (patients). There are numerous subcategories on diagnosis, treatment and alternative therapies for prostate cancer. Listing of schedules of national urological associations meetings.
http://www.prostatehealth.com/	Merck Pharmaceuticals	Illustrative patient education section as well as referenced articles of interest to urologists and patients contemplating hormone-based therapy.

PCa, prostate cancer.
*Other websites of interest that contain information regarding high-grade prostatic intraepithelial neoplasia are not rated due to the authors' direct affiliation with these websites (e.g. www.bostwicklaboratories.com).

Table 23.7. Websites characterized as providing support for patients diagnosed with prostate cancer* (ranked in descending order based on score).

Website URL	Sponsor	Accuracy	Ease	Relevance	Updates	Completeness	Market	Score	Rank
http://www.prostate-online.com/	Virgil Simons personal website	2	2	3	1	3	P	11	1
http://www.healingwell.com/prostatecancer	Healingwell.Com	3	2	2	1	3	P	11	1
http://www.pcansupportscot.f9.co.uk	Scottish Association of Prostate Cancer Support Groups	2	2	2	1	3	P	10	2
http://www.cancerlineuk.net	Astra Zeneca Pharmaceuticals	2	3	2	1	2	P	10	2
http://www.orchid-cancer.org.uk	The Orchid Cancer Appeal	2	3	2	1	2	P	10	2
http://www.phoenix5.org	Phoenix 5 Organization	2	2	2	1	2	P	9	3
http://www.cancernews.com	Net Ventures Inc.	2	2	1	1	1	P	7	4
http://www.malecare.org/	Male Care Inc.	2	1	1	1	2	P	7	4
http://www.ustoo.com	US TOO International.Com	2	1	2	1	1	P	7	4
http://www.prostatecancersupport.co.uk	Prostate Cancer Support Association	1	2	1	1	2	P	7	4
http://www.prostatecancerguide.com/	Wallace Messer personal website	1	2	1	1	1	P	6	5
http://www.pcaw.com	Prostate Cancer Education Council	1	1	1	1	2	P	6	5

P denotes content that is designed for patients diagnosed with prostate cancer.
*See 'Materials and methods' for a description of how these sites were chosen.

Table 23.8. Top five support group websites for prostate cancer.

Website URL	Sponsor	Summary of content of website
http://www.prostate-online.com/	Virgil Simmons personal website	Website hosted by a survivor of PCa and provides information on treatment options reflecting upon personal experiences. Fairly comprehensive with links to national support groups.
http://www.healingwell.com/prostatecancer	Healingwell.Com	Good source of information on all aspects of PCa with patient education section and links to support groups.
http://www.pcansupportscot.f9.co.uk	Scottish Association of Prostate Cancer Support Groups	Scottish support group for PCa aimed at supporting a network for PCa patients in Scotland, but has an enormous amount of relevant information for PCa patients with links to websites in the USA.
http://www.cancerlineuk.net	Astra Zeneca Pharmaceuticals	Provides information on all types of cancer with patient education newsletter available to subscribers via e-mail.
http://www.orchid-cancer.org.uk	The Orchid Cancer Appeal	Charity organization dedicated to providing support to PCa patients and funding PCa research. Has downloadable PDF* format files on select topics of PCa.

PCa, prostate cancer.

directly related to the search criteria used. A far greater number of 'hits' would be generated if the search criteria were modified to 'prostate cancer support groups' (data not shown). Twelve websites having information considered to be of support to patients diagnosed with prostate cancer are ranked and scored in **Table 23.7**. Websites based in the UK tended to rank better than those in the USA. The top five ranked websites are presented in **Table 23.8**, with a brief description of the contents.

DISCUSSION

The Internet is a rich and valuable source of current information for patients and physicians. The volume of information available is staggering, but there is great heterogeneity in terms of accuracy, relevance and, most importantly, timeliness of updates. We confirmed our impression that caution is needed ('buyer beware') when using the Internet as a source of information on prostate cancer. Apart from these drawbacks, the Internet is today the fastest and most efficient means of diffusing medically relevant information.

The medical community has begun to address concerns of accuracy and quality of information on the Internet.[18–20] In 2000, the American Medical Association published a series of guidelines for medical and health information sites,[21] but these guidelines remain unenforceable and few websites adhere to them. A number of web-based tools have been developed to evaluate the quality of health-related information on the Internet; however, these have been found to be complicated and difficult to use.[22] We propose a simple three-tiered graded system based on five variables. These five variables, we feel, allow a quick analysis of the website in order to determine if the information is accurate, complete, relevant, updated and easy to access. This does not require input of data into tables or web-based analysis tools, and are modified criteria previously used to evaluate health-related websites.[16,17] These variables allow interactive participation of the user and should form the foundation for the evaluation of any medically relevant website.

- *Accuracy* is determined by comparing the information presented on the website with current published peer-reviewed medical literature. Preferably, all material presented should be referenced as to source.
- *Ease of navigation* – while this is quite subjective, there are some universal features that make a website more user friendly. Web pages should contain a single screen, navigational toolbar either at the top or the side of the page (so the user does not have to scroll down to the bottom of the page to access it), systematic organization of the data, and large text that is well spaced and easy to read. If commercial banners are displayed, they should not be of the 'pop-up' type and preferably limited to no more than three per page so as to avoid cluttering of the page. Not-for-profit websites scored higher as they had either no commercial banners or a small icon noting an institutional affiliation. (See the Dana Farber Cancer Institute web page (www.dfci.harvard.edu/)).
- *Relevance* – this is directly linked to the type of information desired by the browser. We focused on prostate cancer and its relevance to urologists.
- *Updates* – if a website has not been updated in 6 months, this is considered a stagnant site. If 12 months have passed without an update, the website should probably be passed over.
- *Completeness* – this is an indication of the depth as well as the breadth of the discussion provided (e.g. whether the

information only discusses one type of treatment option or if a number of options are discussed) and provision for more in-depth information either by URL links or referencing other sources of information on the topic.

Additionally, the Health on the Net Foundation (www.hon.ch) and other groups were created with the explicit purpose of monitoring medically related websites and endorsing sites that have accurate and reliable medical information. This endorsement stamp can be found on such websites as the National Institutes of Health and the American Cancer Society. This certification program by the Health on the Net Foundation remains, however, voluntary.

'Buyer beware' also applies to the search engines that are used to locate information. The authors noted varying degrees of commercial bias of the search engine to place certain sponsor sites at the top of the results list. This was noticeable especially with Yahoo (www.yahoo.com), Lycos (www.lycos.com) and AltaVista (www.altavista.com). Others have noted a similar bias.[17] Furthermore, search engines that return a list of results that is over 100 sites, while extremely comprehensive, defeats the object of a focused search: to find relevant information on the topic desired in a short amount of time. This cannot be done if the browser has to sift through over 100 pages analyzing the results rendered. The authors noted that highly useful sites could be obtained in the first 50–60 results of abbreviated lists provided by search engines such as Yahoo (www.yahoo.com).

Direct access

- Accessing the latest developments with regards to the diagnosis, screening and treatment of prostate cancer. With more than 100 urological specialty journals published worldwide, it is difficult for the urologist to keep up with the massive influx of medical information. A number of Internet sites such as the American Urological Association (www.auanet.org), PSA rising (www.psa-rising.com) and Uro Reviews (www.uroreviews.org) provide summaries of the most recent articles and abstracts grouped by topic. This provides an economical way to remain in step with the latest scientific developments without subscribing to many urological journals. The reader, however, must rely on the websites editorial choices.
- Current national and international meeting schedules as well as focused topic seminars are presented on a number of websites.
- Information on new clinical trials, their location and eligibility criteria are well presented on such sites as the National Institutes of Health Cancer Trials site (http://cancertrials.nci.nih.gov/) and the Dana Farber Cancer Institute (www.dfci.harvard.edu/).
- There are a number of sites that calculate the probability of organ-confined prostate cancer and/or survival based on Gleason score, preoperative prostate specific antigen (PSA) and clinical and pathologic tumor stage (e.g. www.cancerhome.com). These calculations are made using look-up tables and artificial neural networks, and provide graphical information on the percentage probability of organ-confined cancer, extraprostatic extension, seminal vesicle involvement and lymph node metastasis, e.g. CancerFacts.Com (www.cancerfacts.com) and CancerHome.Com (http://cancerhome.com). Other websites such as the Department of Defense Center for Prostate Disease Research (www.cpdr.org) and Prostate Calculator (www.prostatecalculator.org) provide a prostate cancer decision site that educates patients on prostate cancer research and also provides urologists with information concerning the probability of success if radical prostatectomy is chosen as the treatment option.

Indirect access

- Patients who are better educated about their disease make better treatment decisions and have superior outcomes.[23–25] Often, physician discussions about treatment options may provide too much information to absorb or digest in one sitting for the patient.[26–28] Many patients can be referred to excellent patient education sites that discuss a variety of topics on prostate cancer. For example, risks and benefits of various treatment options for prostate cancer is available from the National Institutes of Health (http://cancernet.nci.nih.gov/). For the patient that wants a more visual experience, the Medical Library of the University of Toronto (www.library.utoronto.ca/medicine/prostate) provides a prostate cancer education website with video clips of a prostate biopsy procedure, radical prostatectomy highlights, and detailed anatomy and grading information.
- Patients can be referred to treatment decision websites for answers to questions about treatment options, risks, benefits, probability of recurrence, and success of radical prostatectomy or brachytherapy (http://www.brachytherapy.org.uk/).
- Patients diagnosed with any type of cancer often inquire about support groups and the urologist needs to provide local or national support group information.[29] Numerous support groups post information on the Internet, including the Orchid Foundation Appeal (www.orchid-cancer.org.uk), which has an artistic website.

Internet technology continues to change the patient–physician relationship. The resources that are available on the Internet can be used as a powerful tool to fully embrace this relationship. There are still many obstacles to be overcome in terms of the credibility of the information provided, but the Internet is here to stay and, as physicians, we will be called upon to use and evaluate these resources effectively.

REFERENCES

1. 8 F. *USA Today* 2000; p. 1.
2. 20 M. *USA Today* 2000; p. 1.
3. Turbett G. *Adv. Anat. Pathol.* 1996.
4. Weinstein S. Broadsheet number 46: Internet for pathologists. *Pathology* 1998; 30:364–8.
5. Wheeler D. Pathology and the internet. *Adv. Anat. Pathol.* 2001; 8:111.

6. Smart JM, Burling D. Radiology and the internet: A systematic review of patient information resources. *Clin. Radiol.* 2001; 56:867–70.
7. Rolland Y, Bousquet C, Pouliquen B et al. Radiology on internet: Advice in consulting websites and evaluating their quality. *Eur. Radiol.* 2000; 10:859–66.
8. Hoffman-Goetz L, Clarke JN. Quality of breast cancer sites on the world wide web. *Can. J. Public Health* 2000; 91:281–4.
9. Killeen AA. Overview of current internet resources for molecular pathology. *J. Clin. Lab. Anal.* 1996; 10:375–9.
10. Smith MP, Tetzlaff JE, Sheplock GJ. An introduction to the world wide web. *Reg. Anesth. Pain Med.* 1999; 24:369–74.
11. Sjogren D. Urology sites on the internet and search tips for internet browsers. *Aorn J.* 1999; 69:270–1.
12. DiGiorgio CJ, Richert CA, Klatt E et al. E-mail, the internet, and information access technology in pathology. *Semin. Diagn Pathol.* 1994; 11:294–304.
13. Laniado ME, Feneley MR. Urologists on the internet: Practical applications are now a reality. *Br. J. Urol.* 1997; 80:12–15.
14. London JW, Gomella LG. Overview of the internet and prostate cancer resources. *Semin. Urol. Oncol.* 2000; 18:245–53.
15. Ahmed M. Information technology, the internet and urology. *Br. J. Urol.* 1997; 80:1.
16. Smith D. What makes a good web site? *Br. J. Urol.* 1997; 80:16–19.
17. Sacchetti P, Zvara P, Plante MK. The internet and patient education – resources and their reliability: Focus on a select urologic topic. *Urology* 1999; 53:1117–20.
18. Silberg WM, Lundberg GD, Musacchio RA. Assessing, controlling, and assuring the quality of medical information on the internet: caveant lector et viewor – let the reader and viewer beware. *JAMA* 1997; 277:1244–5.
19. Sikorski R, Peters R. Oncology asap. Where to find reliable cancer information on the internet. *JAMA* 1997; 277:1431–2.
20. Wyatt JC. Commentary: Measuring quality and impact of the world wide web. *Br. Med. J.* 1997; 314:1879–81.
21. Winker MA, Flanagin A, Chi-Lum B et al. Guidelines for medical and health information sites on the internet: Principles governing ama web sites. American medical association. *JAMA* 2000; 283:1600–6.
22. Jadad AR, Gagliardi A. Rating health information on the internet: Navigating to knowledge or to babel? *JAMA* 1998; 279:611–14.
23. Bergeron BP. Where to find practical patient education materials. Empowering your patients without spending a lot of time and money. *Postgrad. Med.* 1999; 106:35–8.
24. Lowes RL. Here come patients who've Studied? Medicine on-line. *Med. Econ.* 1997; 74:175–84.
25. Pautler SE, Tan JK, Dugas GR et al. Use of the internet for self-education by patients with prostate cancer. *Urology* 2001; 57:230–3.
26. Moul JW, Esther TA, Bauer JJ. Implementation of a web-based prostate cancer decision site. *Semin. Urol. Oncol.* 2000; 18:241–4.
27. Sandrick K. The internet as a patient/medical education tool. *Bull. Am. Coll. Surg.* 2000; 85:47–9, 68.
28. Kunkel EJ, Myers RE, Lartey PL et al. Communicating effectively with the patient and family about treatment options for prostate cancer. *Semin. Urol. Oncol.* 2000; 18:233–40.
29. Smith J. 'Internet patients' turn to support groups to guide medical decisions. *J. Natl Cancer Inst.* 1998; 90:1695–7.

Treatment

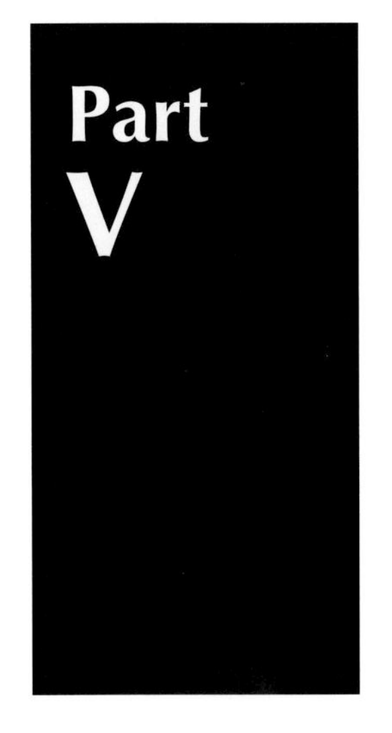

The Decision-making Process for Prostate Cancer

Edouard J. Trabulsi, Peter T. Scardino, and Michael W. Kattan*

Departments of Urology and Epidemiology and Biostatistics, Memorial Sloan–Kettering Cancer Center, New York, NY, USA

INTRODUCTION

Prostate cancer is the most common solid tumor in men, with an estimated 198 100 new cases diagnosed and an estimated 31 500 prostate cancer deaths in 2001 alone.[1] There are several treatment options for men diagnosed with clinically localized prostate cancer, including radical retropubic prostatectomy (RRP), external beam radiotherapy (XRT), brachytherapy, hormonal therapy and watchful waiting. Each of these treatment options has different short- and long-term risks and complications, all with similar cancer control rates, which makes the treatment decision particularly vexing for the individual patient. This patient dilemma has gained increased awareness amongst clinicians. There is an obvious need for decision-making tools that individual patients can apply to the specific parameters of their disease to reach an informed decision.

Outcomes research for prostate cancer is complicated by several factors, which make clinical trials difficult to prepare, accrue and complete. The slow-growing nature of prostate cancer necessitates prolonged follow-up,[2] with clinical trials lasting many years, in order to accrue enough data to compute Kaplan–Meier curves and median survival times. During the period of prolonged follow-up that such clinical trials require, many non-disease-related events occur owing to competing comorbidities in this generally older population, complicating statistical analysis and emphasizing the need for large sample sizes.[2] The operative technique for radical prostatectomy has been greatly modified and improved over the past 20–25 years, so that the outcomes of long-term studies, which have been published contemporaneously, represent patients treated years or decades ago and whom were unlikely to have had the same operative technique as offered today.

Similarly, external beam radiotherapy delivery and localization has improved markedly over the past 10 years, through the use of three-dimensional conformal radiotherapy and intensity-modulated radiotherapy. These newer techniques allow dose escalation with decreased adjacent tissue damage and lower toxicity, revolutionizing external beam radiotherapy compared with conventional therapy of the past. Additionally, the use of prostate specific antigen (PSA)-based screening, as well as the prognostic significance of preoperative PSA has only been realized during the past 10–15 years, making comparisons with older studies difficult or impossible. All of these aspects of the treatment of prostate cancer make the execution of well-designed, statistically rigorous clinical trials very cumbersome, expensive and difficult to complete. Thus, the majority of published studies on the treatment of prostate cancer consist of single institution non-randomized studies, which are prone to bias and error, making the decision-making process for localized prostate cancer difficult for both the patient and the clinician.

In order to make accurate and informed decisions about treatment choices for prostate cancer, the patient and clinician must follow an orderly decision-making process. The patient must know how the various treatment options compare, in terms of cancer control, survival, and acute and chronic complications. The patient must then factor his own particular clinical parameters into the overall equation, to gain an improved predictive model from which to base decisions. If, for instance, a young patient with clinically localized disease has significant medical comorbidities, such that the normally very low-risk surgical treatment of prostate cancer becomes more risky for him, then he may favor XRT or brachytherapy over RRP, despite the evidence that any of the three treatments are equally beneficial in large cohorts of clinically localized patients. Lastly, the patient must consider the impact that the risks and complications would have on him personally, within his own frame of reference. If the patient

*To whom correspondence should be addressed.

considers the cancer control for all treatments equal but he is very concerned about any degree of incontinence after treatment, then his decision-making process would reflect this particular concern when evaluating his options. Therefore, the most important starting point for decision analysis is to review the available literature for each treatment modality and elucidate the respective success rates in cancer control, including PSA recurrence, local and distant recurrence, and cancer-specific survival, and ascertain significant predictors of treatment success or failure. Beyond cancer control, acute and chronic morbidities for each modality must be assessed and predictive factors calculated for each complication when possible.

RADICAL PROSTATECTOMY

Description

Nerve sparing versus non-nerve sparing

Radical prostatectomy is the gold standard for surgical treatment of prostate cancer, having been revived and popularized by Walsh in the late 1970s and early 1980s.[3] Since Walsh's landmark description of the local anatomy of the prostate gland in relation to the adjacent vasculature and neurovascular bundles, the surgical complication rate has dramatically decreased.[4] With the introduction of the nerve-sparing radical retropubic approach, markedly reduced rates of postoperative incontinence and impotence have been achieved, making radical prostatectomy the most common treatment of localized prostate cancer offered to men younger than 70.[5,6]

Cancer control

Prostate cancer survival after radical prostatectomy is defined in a variety of ways, each of which deserves explanation. Common terms used include overall survival, cancer-specific survival, PSA non-progression and long-term progression-free probability. Because prostate cancer grows at such a slow rate with prolonged doubling times, use of overall survival as a clinical end-point in clinical trials is cumbersome and expensive. Thus several other end-points in prostate cancer research are used, including rate of PSA recurrence, rate of local or distant recurrence and cancer-specific survival.

PSA recurrence

Several studies have looked at the actuarial probability of PSA recurrence after radical prostatectomy over the past 30 years, with remarkable concordance. The risk of PSA recurrence after RRP is significantly impacted by preoperative predictive factors, the most important of which are the preoperative serum PSA, biopsy Gleason score and clinical stage.[7] With PSA recurrence variably defined as serum PSA >0.2 ng/dl to >0.6 ng/dl by different groups, as shown in **Table 24.1**, the 5- and 10-year PSA non-progression rate for clinically localized (T1–T2Nx) prostate cancer ranges from 77% to 80% and 54% to 75%, respectively.[7–11] Risk factors most strongly predictive of PSA failure include clinical stage, the pathologic Gleason grade and the preoperative serum PSA. Adverse features on the final pathologic specimen are also predictive of PSA failure; these include extracapsular extension of disease, seminal vesicle or regional nodal involvement, and positive surgical margins.[7] However, the strongest predictors of risk of progression on multivariate analysis of preoperative markers remains clinical stage, Gleason grade and preoperative serum PSA.[7]

Combining all of these predictive factors (PSA, biopsy Gleason score, clinical stage, etc.) into a cogent summary for the patient to make his decision on treatment is the goal of decision analysis research in prostate cancer. If there is a significant chance of extraprostatic disease, nodal involvement and, ultimately, PSA recurrence after RRP, the patient is better able to make an informed decision. He can tailor his treatment approach to his own personal concerns about the different treatment modalities, and their respective risks and possible complications. Thus, several nomograms have developed to provide this type of information to patients and assist in their decision-making process. The two most prominent are the Partin tables and the Kattan nomogram.

Partin et al., using the Hopkins database of RRP patients, examined preoperative serum PSA, clinical stage and biopsy Gleason score to predict the final pathologic stage after RRP. This series was first reported on over 700 patients from the

Table 24.1. PSA progression-free probability after radical retropubic prostatectomy. Adapted from Eastham and Scardino (2002).[105]

Group	Number of patients	Stage	Years	PSA non-progression (%)	
				5 years	10 years
Pound et al. (1997)[9]	1623*	T1–2NX	1982–95	80	68
Trapasso et al. (1994)[8]	425†	T1–2N0	1987–92	80	—
Zincke et al. (1994)[10]	3170†	T1–2NX	1966–91	77	54
Catalona et al. (1994)[11]	925‡	T1–2NX	1983–93	78	—
Hull et al. (2002)[7]	1000†	T1–2NX	1983–98	78	75

*Progression defined as serum PSA >0.2 ng/ml.
†Progression defined as serum PSA >0.4 ng/ml.
‡Progression defined as serum PSA >0.6 ng/ml.

pre-PSA era in 1993,[12] expanded in a multi-institutional study involving over 4000 patients in 1997[13] and updated on over 5000 patients from Hopkins in 2001.[14] Partin et al. found that the combination of PSA, Gleason score and clinical stage together were more accurate in predicting the final pathologic stage than any one of these three parameters alone. The resulting tabulation is a categorization, stratified for preoperative PSA, biopsy Gleason score and clinical stage, giving the chance of organ-confined disease, extracapsular extension, seminal vesicle involvement and nodal involvement after RRP. This allowed the patient and clinician the opportunity to derive the percentage chance of postoperative pathologic findings based on the individual biopsy data of the patient. This data set was validated on an external prostate database and had a classification accuracy of 0.724 – within 0.10.[13,15] However, one drawback of this analysis is that the end-point is final pathologic stage and not a clinical parameter, such as recurrence or survival, which limits its usefulness.

Using a similar approach, Kattan et al. evaluated the Baylor RRP database to find independent predictors of both pathologic stage and PSA recurrence after RRP based on preoperative information.[16] Using Cox proportional hazards regression analysis, the clinical information and outcomes for 983 men undergoing RRP were modeled and a nomogram was developed. Incorporating clinical data similar to Partin, including preoperative PSA, biopsy Gleason and clinical stage, a point system was devised to categorize individual patients, as shown in **Fig 24.1.** The number of points increase as the PSA, stage or Gleason score increases, and the total points determine the 60-month recurrence-free probability. This model was validated with a separate cohort, giving a receiver operating characteristic curve (ROC), which is the comparison of predicted probability with the actual outcome, of 0.79. This approach allows the clinician and the individual patient a more rigorous prediction of recurrence based on the specific clinical scenario.

Local recurrence

Local recurrence after radical prostatectomy is defined as recurrent prostate cancer present in the prostatic fossa or pelvis, and is diagnosed by radiographic documentation of pelvic disease on computed tomography (CT), magnetic resonance imaging (MRI) or nuclear imaging, or by a positive biopsy of the prostatic fossa. Rising PSA after surgical treatment alone is not diagnostic; distant disease, either in regional lymph nodes or distant metastasis must be ruled out. Determining true local recurrence from subclinical micrometastatic disease is difficult and the majority of patients with locally recurrent pelvic disease will have concomitant distant metastatic disease. Han et al. examined the Hopkins experience in men undergoing RRP, which included 2404 men operated on from 1982 to 1999, with a mean follow-up of 76.6 months (range 12–204 months).[17] A total of 412 of 2404 men experienced disease recurrence (17%) during follow-up. Of these, only 40 of 2404 men had isolated local recurrence (1.6%); the remainder had distant disease ± local recurrence or PSA recurrence only. Thus isolated local recurrence represented only 9.8% of all men with

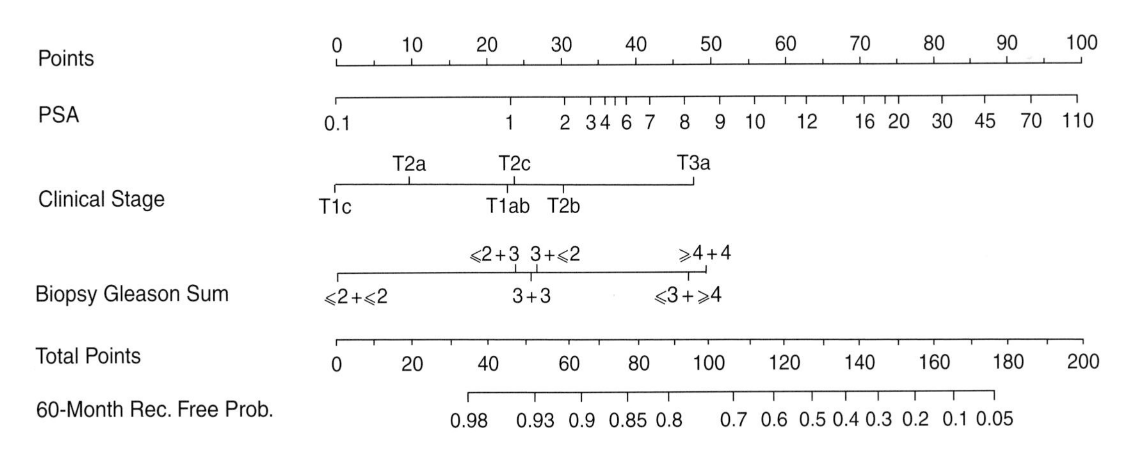

Instructions for Physician: Locate the patient's PSA on the PSA axis. Draw a line straight upwards to the Points axis to determine how many points towards recurrence the patient receives for his PSA. Repeat this process for the Clinical Stage and Biopsy Gleason Sum axes, each time drawing straight upward to the Points axis. Sum the points achieved for each predictor and locate this sum on the Total Points axis. Draw a line straight down to find the patient's probability of remaining recurrence free for 60 months assuming he does not die of another cause first.

Note: This nomogram is not applicable to a man who is not otherwise a candidate for radical prostatectomy. You can use this only on a man who has already selected radical prostatectomy as treatment for this prostate cancer.

Instruction to Patient: 'Mr. X, if we had 100 men exactly like you, we would expect between <predicted percentage from nomogram – 10%> and <predicted percentage + 10%> to remain free of their disease at 5 years following radical prostatectomy, and recurrence after 5 years is very rare.'

Fig. 24.1. Preoperative nomogram for prostate cancer recurrence. Reprinted with permission from Kattan et al. (1998).[16]

recurrence and only 1.6% of all men overall.[17] Men putatively at high risk for local recurrence were specifically examined, namely those with positive surgical margins and Gleason 7 or greater in the absence of seminal vesicle involvement or nodal disease. In this subcategory of patients, the Hopkins group reported an isolated local recurrence rate of only 6% of patients. The majority of patients in this cohort had both locally recurrent and distant metastases.[18] Isolated local recurrence after surgical treatment of prostate cancer represents an uncommon outcome and local cancer control is excellent after RRP.

Distant recurrence

The risk of distant recurrence, as with PSA recurrence, is dependent on the preoperative serum PSA, biopsy Gleason score and clinical stage. Distant recurrence after RRP is defined as the development of clinically evident distance metastases, with either soft tissue lesions or more commonly bony lesions. The rate of clinically evident distant recurrence is difficult to ascertain from published series, as many studies report only PSA progression or cancer-specific survival and do not explicitly distinguish those with clinically evident distant metastases from isolated PSA recurrence. One of the largest series to report on this topic, by Pound et al. from Johns Hopkins, reported on 1997 RRP, of which there were 315 men (15%) with PSA recurrence and 103 men (5.2%) with distant metastatic disease with a median follow-up of 5.3 years. In this series, the vast majority of men with PSA recurrence did not receive adjuvant radiotherapy or hormonal therapy until the development of documented distant metastasis. The median time from PSA recurrence to distant recurrence was 8 years and the median time from distance recurrence to death was 5 years. Factors predictive of PSA and distant recurrence and death from disease were Gleason score, time from RRP to PSA recurrence (< or >2 years) and PSA doubling time (< or >10 months).[19]

Survival

One of the only randomized studies to evaluate radical prostatectomy in the treatment of prostate cancer is the VA Cooperative Urological Research Group (VACURG) study, which after 23 years of follow-up found no difference in survival between radical prostatectomy and placebo.[20] Many have criticized this study for its small sample size (143 patients) and incomplete follow-up, as well its study period from the late 1960s and early 1970s, before the innovations in RRP popularized by Walsh and before the introduction of radionuclide bone scans to better categorize patients. Since this study, however, several contemporary single institution studies in the PSA era have demonstrated excellent results with RRP as monotherapy for localized prostate cancer. Hull et al. evaluated 1000 consecutive RRP performed at Baylor from 1983 to 1998, with a median follow-up of 53.2 months. The overall survival for this cohort at 5 and 10 years was 95.5%±1.7% and 86.6%±4.4%, respectively; the actuarial cancer-specific survival at 5 and 10 years in this cohort was 99.1%±0.78% and 97.6%±1.7%, respectively. Only 11 patients (1.1%) died of prostate cancer in this study, with 40 patients dying of other causes (4.0%).[7] Similarly, Han et al. reviewed the Hopkins experience with RRP from 1982 and 1999; a total of 2404 patients were evaluated, with a mean follow-up of 6.3 years and follow-up extending as long as 17 years.[17] The overall actuarial cancer-specific survival rates at 5, 10 and 15 years in this group were 99%, 96% and 90%, respectively. Thus, in large single-institution series encompassing the PSA era, excellent cancer control has been demonstrated for RRP in prostate cancer.

One concern with these types of analyses has been the impact of stage migration, changes in surgical technique, patient population and pathological interpretation over the past decade. Another study from the Johns Hopkins group evaluated 2370 patients who underwent RRP between 1982 and 1999, and were stratified by different years of treatment.[21] When examined by era, the pre-PSA era (1982–1988), PSA introduction (1989–1991) and the PSA era (1992–1998), there was significant improvement in PSA recurrence-free survival during the PSA introduction and PSA eras compared with the pre-PSA era. Concurrently, a stage migration was noted from more advanced to lower stage disease over these time periods.[21] In agreement with this, Scardino and Kattan have noted that year of surgery is an independent predictor of PSA-free progression (PFP) rates, with PFP improving as the year of surgery increased from 1991 to present (unpublished). Thus, in evaluating survival data for prostate cancer after RRP, the time period that RRP was performed must be considered.

Short-term complications

Anesthetic and operative complications

The morbidity and mortality rate for radical retropubic prostatectomy has declined since the delineation of the periprostatic anatomy and vasculature in the late 1970s and early 1980s.[4] The most common intraoperative complication is that of hemorrhage, with mean estimated blood loss (EBL) ranging from 600 ml to 1400 ml in various series.[22] In the Baylor series, using modified and improved operative techniques, the mean EBL declined from 1200 ml in the period of 1983–1988 to 800 ml during 1991–1992. During the same period, the rate of autologous blood donation increased from 13% to 74%, with the numbers autologous donors receiving blood bank blood decreasing from 46% to 4%. After this time period, the blood donation and transfusion policies were made more stringent after finding that 75% of patients had EBL less then 1000 ml and only 21% of autologous donated units were transfused. In an effort to contain costs and prevent unneeded blood donations, patients after 1991 were no longer recommended for autologous blood donation, with the rate of non-autologous transfusions in donors and non-donors remaining constant at 3% and 10%, respectively.[23]

Other major operative complications are fortunately far less common. The rate of operative mortality is extremely low, with a rate of 11 in 3834 patients (0.3%) pooled from five contemporary series worldwide,[22] including Washington University, Baylor, the Mayo Clinic, University of Ulm (Germany) and Toulouse (France).[24–28] Similarly, the rates of rectal injury and colostomy were very low in these pooled studies at 0.7% and 0%, respectively.[22] Non-urologic complications were also very rare. Myocardial infarction occurred in 22 patients

(0.6%) and pulmonary embolism and deep vein thrombosis occurred in 44 and 49 patients, respectively (1.1% and 1.3%).[22] In a multivariate analysis at Baylor of 472 patients undergoing RRP, risk factors for perioperative complications were identified.[25] The patient's American Society of Anesthesiologists (ASA) classification score and operative blood loss were both significantly associated with major complications; patient age, operative time and year of operation were not significantly associated. ASA classification is determined by medical comorbidity, and patients with more severe medical comorbidities, designated by ASA class 3 or greater, accounted for a threefold increase in major complications, prolonged hospital stay, increased use of intensive care admission and more frequent blood transfusions.[25] In a more recent review, Lepor et al. reported on 1000 consecutive RRP procedures by one surgeon from 1994 to 2000, and found an overall complication rate (intraoperative or postoperative) of 2% for all causes.[29] Thus, while RRP is a major surgical procedure, it is generally well tolerated in the majority of patients in centers of excellence.

Catheter duration

Traditionally, a urinary catheter is left in place after RRP for 2–3 weeks to allow the vesicourethral anastomosis to heal. With the average length of hospital stay after RRP continuing to decline with improvements in technique and health cost containment, this represents a significant burden for patients and their families at home. Patients usually tolerate the catheter well and learn how to handle the drainage bag, but will occasionally experience discomfort or bladder spasms from the catheter. Recent experience suggests that the urinary catheter can be removed earlier. Several centers have examined early removal of urinary catheters between 4 and 7 days and found no adverse events with earlier removal.[30–33] Lepor et al. described their algorithm for early catheter removal, in which a cystogram is performed on postoperative day 3 or 4, with the catheter removed if there is no extravasation noted; if there is leakage visualized, then the catheter is removed on day 14.[34]

One advantage of the laparoscopic radical prostatectomy is the magnified view of the operative field with improved visualization of the vesicourethral anastomosis. Nadu et al. reported on their laparoscopic technique, in which a running suture is used for the anastomosis, theoretically improving the watertight closure and allowing earlier catheter removal. Cystograms were performed on postoperative days 2–4 and no leak was seen in 85% of patients. One potential pitfall of this approach is that 10% of the patients in this series experienced urinary retention shortly after catheter removal, presumably due to edema at the anastomosis or postoperative pain. In all of these patients a catheter was replaced manually at the bedside without difficulty. The rates of urinary incontinence and bladder neck contracture were no different in this study.[35]

Long-term complications

Bladder neck contracture (BNC)

Stricture or contracture formation at the vesicourethral anastomosis after RRP has been reported in as many as 17.5% of patients[36] in published single institution series. In a study of the Medicare claims, 19.5% of patients identified to have had RRP by claim underwent one or more procedures for bladder neck obstruction or stricture after RRP.[37] This contrasts with other studies reporting lower rates of postoperative BNC after RRP.[38–40] Several studies have reported risk factors for the development of postoperative BNC, including previous prostate procedures or transurethral resections, excessive intraoperative blood loss, postoperative urinary extravasation and asymptomatic bacteriuria.[39,41,42] In contrast, other studies found no relationship between development of BNC and previous transurethral resection of the prostate (TURP), or pathologic features including cancer volume, positive surgical margins, lymph node or seminal vesicle involvement.[36,42] Patient comorbidities have also been implicated as risk factors for the development of BNC after RRP, including cigarette smoking, coronary artery disease, diabetes mellitus and hypertension, suggesting a microvascular component to the development of BNC. In all series, the majority of patients were treated with one or more transurethral dilations or incisions without requiring major reconstruction, which did not appear to impair continence. Thus, while the reported rate of BNC varies widely in the literature, the subsequent treatment and impact on patient quality of life appears to be minimal.

Incontinence

Urinary incontinence is one of the dreaded and most feared complications of RRP, and one that causes significant bother to patients.[43] The rate of urinary incontinence varies widely in published series. As shown in **Table 24.2**, the published rates vary depending on how incontinence is measured, either from the surgeon's interview with the patient, a written survey from the institution or a population-based survey. Thus, obtaining an accurate measure of postoperative urinary incontinence remains difficult.

Eastham et al. looked at risk factors for urinary incontinence after RRP.[44] This study examined the Baylor database from 1983 to 1994, overlapping a procedural change in RRP in 1990 in which meticulous care was taken to avoid traction on the urethra with minimal suture bites taken through the urethral stump, and to stomatize the bladder neck hiatus completely prior to performing the vesicourethral anastomosis. In a multivariate analysis, where incontinence is defined as leakage with moderate activity, the most important independent predictors of postoperative continence were younger age, improved modification of operative technique, preservation of both neurovascular bundles and the absence of postoperative BNC. With the modification of their operative technique, the time until regaining continence was significantly decreased from 5.6 months to 1.5 months, with a concomitant increase in overall continence from 82% to 95%.[44]

Impotence

Examination of postoperative impotence after RRP is difficult for several reasons. Objective measure of potency, both before and after treatment, is difficult to assess. The definition of potency also varies widely in reported series, without a consensus definition. Definitions include the degree of tumescence

Table 24.2. Incontinence after radical retropubic prostatectomy. Adapted from Eastham and Scardino (2000).[105]

Series	Number of Patients	Incontinence (%)	Definition of incontinence
Interview by treating physician at center of excellence			
Steiner et al. (1991)[106]	593	8	Leaks with moderate activity
Leandri et al. (1992)[28]	398	5	Leaks with moderate activity
Zincke et al. (1994)[10]	1728	5	Requires 3 or more pads/day
Catalona et al. (1999)[107]	1325	8	No pads required
Geary et al. (1995)[108]	458	20	Requires pads
Eastham et al. (1996)[44]	581	9	Leaks with moderate activity
Series from patient surveys			
Murphy et al. (1994)[109]	1796	19	Requires pads
Litwin et al. (1995)[43]	98	25	'Bother' score
Walsh et al. (2000)[110]	62	5	Requires pads
Population-based studies using patient surveys			
Fowler et al. (1993)[111]	738	31	Requires pads or clamps
Stanford et al. (2000)[112]	1291	8.4	Severe incontinence
		21.6	Requires pads

achieved, frequency of successful penetration, achieving erections with the use of oral aids such as sildenafil, and patient and partners' degree of satisfaction. The psychometric questionnaire used most commonly to assess erectile dysfunction is the International Index of Erectile Function (IIEF), a 15-question survey, which assesses the patient's opinion of the strength of erections, success of penetration, sexual satisfaction and sexual desire.[45] While this survey has been validated and shown to be sensitive and specific for erectile dysfunction, its use has not been consistent in the RRP literature.

Rabbani et al. prospectively examined patients undergoing RRP to determine the most important predictive factors of postoperative potency.[46] In this study of 314 men between 1993 and 1996, preoperative and postoperative potency were assessed in an interview with the surgeon and graded on a five-point scale. If patients had erections sufficient for intercourse (grade of 1, 2 or 3), they were considered potent. The operative technique was also evaluated, including the degree of neurovascular bundle preservation for each side, defined as definite neurovascular bundle resection, definite preservation, or degrees of suspected neurovascular bundle resection or preservation. With a median follow-up of 25.4 months, several independent factors were predictive of postoperative potency on multivariate analysis. These included age, preoperative potency and degree of neurovascular bundle preservation. Overall, 25% of men recovered satisfactory erections by 4.5 months and 50% by 30.1 months. Men younger then 60 had a 61% potency rate; 42% of men 60–65 and 31% of men older then 65 regained potency. When both neurovascular bundles were spared (Grade 1 preservation), 55% of men recovered potency versus 41% where one or both nerves had partial damage, or 21% where one bundle was definitely resected. Not surprisingly, preoperative erectile function was also a strong predictor of postoperative potency. Fifty-four per cent of patients with full erections (Grade 1) prior to surgery regained potency versus 37% with Grade 2 and 22% with Grade 3. Pathologic stage of disease was significant on univariate analysis but was not significant on multivariate analysis. Therefore, when evaluating potency after RRP, preoperative potency, the age of the patient and the surgeon's ability to preserve both neurovascular bundles, considering both technical ability and extent of disease, must be considered to make a prediction. Clearly there is a need for better predictive models or nomograms incorporating these factors.

EXTERNAL BEAM RADIOTHERAPY

Description

External beam radiotherapy has evolved and improved tremendously over the past 30 years. Treatment was initially given with conventional radiotherapy in the 1960s using cobalt 60 and linear accelerators. There has been tremendous improvement in both radiation delivery and localization since then. With the introduction of three-dimensional CT-guided beam localization and intensity-modulated radiotherapy (IMRT), the radiation oncologist is able to deliver higher radiation doses to the prostate, while minimizing damage to surrounding normal tissue, thereby improving cancer control and morbidity.[47,48]

Comparison of XRT with surgical management of localized prostate cancer is challenging on several levels. The accurate pathologic information afforded by surgical removal of the prostate is missing after XRT and thus pathologic stage-matched comparisons are impossible. Additionally, the improvements in radiation delivery and ability to escalate radiation doses in the contemporary era make analysis of older series with lengthy follow-up treated with conventional radiotherapy obsolete. Lastly, there have been no randomized clinical trials comparing radiotherapy with RRP, so that biases in patient selection are inevitable. As with RRP, the best data available are single or pooled multi-institutional prospective studies investigating XRT with or without hormonal therapy stratified for known risk factors, namely serum PSA, biopsy Gleason score and clinical stage.

The use of neoadjuvant hormonal therapy adds yet another level of complexity when evaluating XRT for prostate cancer. Strong evidence exists that the addition of neoadjuvant hormonal therapy to XRT improves survival for locally advanced or high-risk prostate cancer patients[49] and, therefore, is used widely. Neoadjuvant hormonal therapy's benefit for clinically localized, low- or intermediate-risk prostate cancer remains uncertain. Evidence for the use of neoadjuvant hormonal therapy in combination with RRP for high-risk or locally advanced patients is less compelling.[50] Use of neoadjuvant hormonal therapy confounds comparison of biochemical recurrence rates, however, and impairs accurate comparison between groups in the decision-making process.

Cancer control

PSA recurrence

PSA recurrence after external beam radiotherapy, as with RRP, is a commonly used surrogate end-point used to judge treatment efficacy. The definition of PSA recurrence, as defined by the American Society for Therapeutic Radiology and Oncology (ASTRO), of three consecutive rises in PSA above nadir PSA after XRT, is the accepted standard in the radiation oncology community. This definition sets the time of recurrence at the midpoint between the PSA nadir and first rise in PSA.[51] While controversial, this definition has been shown to correlate well with other commonly used end-points, including distant metastatic or local recurrence and cause-specific survival.[52]

Predictors of PSA recurrence after XRT are similar to preoperative prognostic factors used for RRP, including serum PSA, biopsy Gleason score and clinical stage.[53,54] Using these criteria, several models to predict PSA recurrence have been developed for patients receiving XRT.[53,55–61] These models stratify patients using these prognostic factors and typically combine them into one of three groups: low-, intermediate- and high-risk categories. These groups are given group-level predictions for different end-points, which may or may not be accurate. For example, a patient with a serum PSA of 4.1 is typically grouped with a patient having a PSA of 9.9 in the <10 PSA group, who are unlikely to have the same prognosis. This type of patient categorization leads to heterogeneous groupings and imperfect disease modeling. Kattan et al. have developed a continuous risk approach using the standard prognostic factors of serum PSA, biopsy Gleason score and clinical stage to create a nomogram to predict the PSA recurrence after XRT.[62] Another important factor included in this nomogram is the radiation dose delivered. Numerous

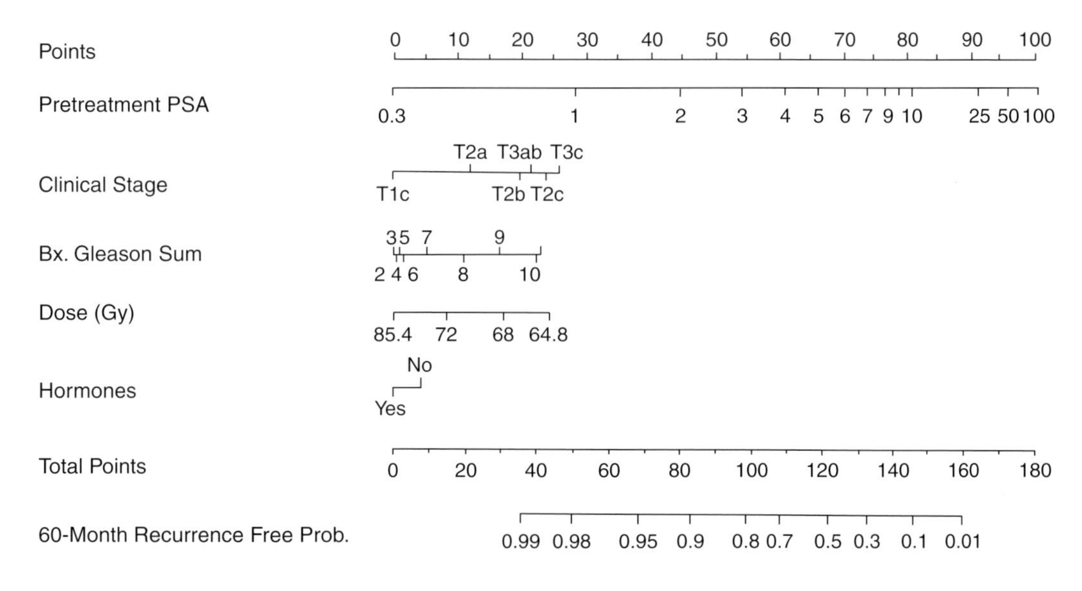

Instructions for Physician: Locate the patient's PSA on the PSA axis. Draw a line straight upwards to the Points axis to determine how many points towards recurrence the patient receives for his PSA. Repeat this process for the other axes, each time drawing straight upward to the Points axis. Sum the points achieved for each predictor and locate this sum on the Total Points axis. Draw a line straight down to find the patients probability of remaining recurrence free for 60 months assuming he does not die of another cause first.

Note: This nomogram is not applicable to a man who is not otherwise a candidate for radiation therapy. You can use this only on a man who has already selected radiation therapy as treatment for his prostate cancer.

Instruction to Patient: 'Mr. X, if we had 100 men exactly like you, we would expect between <predicted percentage from nomogram – 10%> and <predicted percentage + 10%> to remain free of their disease at 5 years following conformal radiation therapy, and recurrence after 5 years is very rare.'

Fig. 24.2. Three-dimensional conformal radiation therapy nomogram for PSA recurrence. Reproduced with permission from Kattan et al. (2000).[62]

studies have suggested a benefit to radiation dose-escalation to doses approaching 80 Gy, especially in the higher risk patients.[59,63–66] Using a continuous model of PSA and radiation dose delivered together with clinical stage and biopsy Gleason, the nomogram shown in **Fig. 24.2** gives the 5-year freedom from PSA recurrence.[62] When compared with other predictive models, this nomogram delivered significantly more accurate recurrence probabilities, and better predictive information for the clinician and patient to use in decision making.

Local recurrence

Local recurrence after XRT, as in after RRP, is usually diagnosed because of a rising PSA after treatment. Distinguishing local recurrence from distant metastatic disease can be challenging. The majority of published series do not make this distinction and do not use local recurrence as a clinical end-point. In a study of 938 men treated from 1987–1995 at MD Anderson, Zagars et al. reported an overall 6-year actuarial incidence of local recurrence of 27%.[53] These patients all had ultrasound-guided biopsies of the prostate to diagnose local recurrence and the majority were treated by conventional radiotherapy with total radiation doses below 70 Gy. Ninety-five patients (10%) received conformal XRT with doses ranging from 74 to 78 Gy (median 78 Gy). On multivariate analysis, clinical stage, biopsy Gleason grade and serum PSA correlated with local recurrence. Prognostic categories were also developed in this analysis, stratified for clinical stage (T1–2 vs. T3–4), as shown in **Table 24.3**; within each stage, patients were subcategorized by Gleason grade (2–6, 7, 8–10) and PSA (4–10, 10–20, ⩾20) to give risk groupings (Group I, II, III, IV, V and unfavorable) for PSA recurrence, local recurrence and distant recurrence. Local recurrence in the T1–2 group was 3%, 24%, 26% and 43% for categories I, II, III and unfavorable, respectively; and in the T3–4 group was 29%, 45% and 42% for categories IV, V and unfavorable, respectively.[53]

In a more contemporary series from Memorial Sloan–Kettering Cancer Center (MSKCC) utilizing IMRT dose escalation, a total of 1100 patients were treated from 1988 to 1998.[67] Sixty-seven per cent of patients received 75.6 Gy or higher total radiation dose, with some receiving as high as 81 Gy. Post-treatment sextant prostate biopsies were offered to patients at least 2½ years after treatment to document local recurrence, and was performed in 43% of 586 evaluable patients. Overall local recurrence was documented on biopsy in 27% of patients, with a dose–response noted. Only 10% of patients receiving 81 Gy had positive biopsies, compared with 23%, 34% and 54% receiving 75.6 Gy, 70.2 Gy and 64.8 Gy, respectively. Patients were also stratified into risk categories, based on serum PSA, clinical stage and biopsy Gleason score (PSA <10 or >10, clinical stage T1–2 or >T2, Gleason score ⩽6 or >6). Patients were considered to have a favorable prognosis when all three parameters were favorable; intermediate or unfavorable prognostic groups were categorized if one or two parameters, respectively, were unfavorable. Using this stratification, the local recurrence rates based on post-treatment biopsies were 13%, 23% and 37% for favorable, intermediate and unfavorable, respectively.[67]

Distant recurrence

As with local recurrence, distant recurrence, or the development of clinically apparent metastatic disease after XRT is not an end-point of treatment routinely reported or measured in studies evaluating XRT. MD Anderson has reported a large contemporary series that tabulated the risk of developing distant metastatic disease after XRT and found similar prognostic risk factors as that with PSA recurrence and local recurrence, namely serum PSA, clinical stage and biopsy Gleason grade.[53] The overall incidence of the development of metastatic disease was 6% in this series, ranging from 0% to 24% in the risk groups shown in **Table 24.3.**[53]

Survival

With the use of PSA recurrence as a common end-point in radiation series, and with a generally short follow-up in the PSA era, the rate of overall or cancer-specific survival has not been reported in many large contemporary series in the PSA era. Subsequently, predictive models or paradigms for survival after XRT have been sparse. One of the largest series in the

Table 24.3. MD Anderson prognostic categories for PSA recurrence, local recurrence and distant recurrence after external beam radiotherapy in prostate cancer. Adapted from Zagars et al. (1997).[53]

Gleason score	PSA ⩽ 4	4 < PSA ⩽ 10	10 < PSA ⩽ 20	PSA > 20
T1 and T2				
2–6	I	II	III	Unfav
7	II	II	III	Unfav
8–10	II	III	Unfav	Unfav
T3 and T4				
2–6	IV	IV	V	Unfav
7	IV	IV	V	Unfav
8–10	IV	IV	Unfav	Unfav

Unfav, unfavorable.

PSA era to incorporate survival data is the multi-institutional analysis compiled by Shipley et al.[58] This retrospective study of XRT for clinically localized prostate cancer totaled 1765 patients from six centers from 1988 to 1995 with a median follow-up of 4.1 years. Just over half (51%) of patients were treated using the three-dimentional (3D) conformal technique. The overall and cancer-specific survival rates for the entire cohort was 85% [95% confidence interval (CI) = 82.5–87.6%] and 95.1% (95% CI = 94.0–96.2%), respectively. Serum PSA and biopsy Gleason score were both independent predictors of PSA recurrence in this series, but were not analyzed for survival.[58]

The Radiation Therapy Oncology Group (RTOG) is currently conducting two prospective, randomized phase III clinical trials, which should give better predictive information for survival after XRT.[68,69] RTOG 9408, which has enrolled over 2000 patients, stratifies patients by serum PSA, biopsy Gleason score and clinical stage, and is a double-armed study examining XRT alone versus XRT with neoadjuvant hormonal therapy (NHT). RTOG 9910 will include either standard or extended course of NHT for intermediate risk patients and is designed to accrue 1600 patients. Both studies will use cancer-specific survival as the primary end-point and should allow more accurate predictive modelling for XRT in prostate cancer.

Complications

Urinary complications

Urinary difficulties after XRT include a spectrum of complications from mild irritative voiding symptoms to severe incontinence, bladder neck or urethral strictures and hematuria. Gauging urinary complications has included standard toxicity grades on a four-point scale (mild, moderate, severe, life-threatening) to patient quality of life data by questionnaire, which do not always correlate. The desire to lower toxicity and complications to adjacent tissues like the bladder and rectum has energized the development of more precise imaging and dose-delivery mechanisms, such as 3D conformal XRT and IMRT.

In a prospective randomized controlled study from MD Anderson examining conventional XRT and 3D conformal XRT, 101 patients completed a questionnaire to assess late bladder function at least 2 years after treatment.[70] The overall reported rate of incontinence was 29% for all patients, with 36% and 8% having urge and stress incontinence, respectively. The reported use of pads or diapers, however, was only 2% overall, highlighting the disparity in accurate measurement of this problem. Between the conventional and 3D XRT arms, there was no difference in reported incontinence rates or use of protective devices, but patients receiving 3D XRT were significantly less likely to report daily leakage of urine.[70] In the MSKCC series utilizing IMRT in 1100 patients, 73% of patients developed no late urinary toxicity.[67] In the remainder, 16%, 10% and 2% had Grade 1, Grade 2 and Grade 3 toxicity, respectively, with no Grade 4 toxicity reported. Seventeen patients developed Grade 3 urinary toxicity, including 15 with urethral strictures requiring dilation and two with hematuria. The 5-year actuarial risk of grade 2 urinary toxicity was 10% overall, with a significantly higher risk (13%) in patients receiving 75.6 Gy versus a lower risk (4%) receiving lower radiation doses. There was no difference in late urinary toxicity between patients receiving 3D conformal XRT or IMRT. This study confirmed an earlier study performed at MSKCC examining 3D conformal XRT, in which the 5-year actuarial risk of developing urinary toxicity was calculated at 10% and 3% for Grade 2 and Grade 3 urinary toxicity, respectively.[71] In this study, higher doses (75.6 Gy) and development of urinary symptoms acutely, including irritative voiding symptoms and dysuria, were significantly associated with the development of late urinary toxicity.

Impotence

The natural history of impotency after XRT differs from that of impotency resulting after RRP. Like with RRP, impotence after XRT is difficult to quantify accurately due to its subjective nature. Impotence occurring after XRT, however, tends to develop slowly and progress after treatment,[72] unlike surgery where impotence occurs acutely but can recover over time. Thus measurement of potency after surgical and radiation treatment of prostate cancer must have adequate follow-up, to encompass both late recovery of potency (after RRP) and late development of impotence (after XRT). Accurate measure of potency after XRT is further confounded by the natural age-related decline of potency occurring in the generally older population of prostate cancer patients, so that the long-term impotency rates after XRT may be overstated, and may not be due solely to the treatment given. Nevertheless, potency after treatment remains a large concern to the individual patient and may disproportionately influence their ultimate treatment choice, so accurate measures and predictive models are paramount.

When investigating potency after XRT, several factors must be noted. First, it is important to verify the pretreatment level of potency. The use of neoadjuvant or adjuvant hormonal therapy must also be considered, as its use with XRT is common and will obviously affect potency. Moreover, the duration of NHT may also impact potency, although a short course, in and of itself, is unlikely to cause any lasting impotence. Finally, the manner in which potency is measured may lead to considerable variation in reported potency rates after XRT. As with RRP, the post-treatment potency rate varies based on the metric used and the manner in which the patient is queried, whether it be in the office by the treating physician, by a research questionnaire, or by telephone survey.

In a meta-analysis of reported rates of erectile function after treatment for prostate cancer, Robinson et al. examined 40 articles published from 1981 to 1998, including 15 reports on XRT, 22 on RRP, two on cryotherapy, and one comparative study of XRT and RRP. Using logistic regression, a significantly better probability of maintaining erections was found for XRT (0.69) than RRP (0.42).[73] This study confirmed earlier studies favoring XRT over RRP,[74] but included both nerve-sparing and non-nerve-sparing RRP together in the analysis. Additionally, the definition of potency was not defined or standardized, and the follow-up times of the various studies varied widely. Importantly, of 600 reported studies examined, only 40 had adequate pretreatment erectile data to allow use in a meta-analysis.[73] Conversely, in a single institution study comparing XRT and RRP involving 802 men, no difference in post-treatment potency was seen after RRP or XRT.[75]

In a retrospective study of high-dose 3D conformal XRT with 124 patients, Chinn et al. found 33 of 47 patients (70%) that were fully potent and 4 of 13 patients (31%) marginally potent before treatment maintained erectile function sufficient for intercourse after treatment, with a medial follow-up of 21 months.[76] They also found that patients with a history of a major urologic surgical procedure prior to treatment were significantly less likely to regain potency after XRT; the most common procedure performed was transurethral resection of the prostate. Conversely, in a prospective trial at MSKCC, 743 patients treated with high-dose 3D conformal XRT with a much longer follow-up of 42 months were examined for treatment morbidities, and TURP was not found to be a risk factor for impotence.[71] In this study, 211 of 544 patients who were potent prior to treatment developed impotence after treatment (39%), yielding a 5-year actuarial risk of loss of potency of 60%. Risk factors for impotence were radiation dose delivered (≥75.6 Gy) and use of NHT; age, clinical stage and a history of TURP or diabetes mellitus were not independent predictors of post-treatment impotence.[71] Other studies have examined the effect of XRT on potency, but these studies do not always clearly delineate the radiation technique used or the dose given, magnifying the uncertainty in creating accurate predictors of XRT-related impotence.

In a small study looking at the radiation dose delivered to the bulb of the penis after 3D XRT, Fisch et al. found a dose–volume relationship between post-treatment potency and radiation dose delivered to the bulb of the penis.[77] All patients were potent before treatment, and the radiation dose covering 70% of the bulb of the penis was measured. Radiation doses totaling of <40 Gy were associated with significantly less impotence then doses >70 Gy. As the bulb of the penis is adjacent to the urogenital diaphragm and in the anatomic vicinity of the neurovascular bundles of the prostate, this study, while small, demonstrates the importance of accurate dosimetry and dose delivery to minimize adjacent toxicity, including impotence.[77]

Bowel toxicity

Another common complication after XRT for prostate cancer is bowel toxicity, owing to its close proximity to the prostate. As with urinary difficulties, bowel toxicity can vary from mild to severe, including proctitis, tenesmus, urgency, rectal bleeding, rectal stenosis, necrosis, and rectovesical or rectourethral fistula formation. In the randomized study examining conventional and 3D conformal XRT from MD Anderson, the majority of patients (78%) in both groups had minimal or no late rectal toxicity.[70] While 27% reported urgency of bowel movements (BMs), 90% had good control without accidents, and only 1% took antidiarrheal agents daily or weekly. The patients receiving 3D conformal XRT were significantly less symptomatic, including less frequent BMs and less urgent BMs. Hematochezia or rectal bleeding was uncommon is both groups, occurring daily in 4% and weekly in 7%.[70] In the MSKCC comparison of 3D conformal XRT and IMRT in 1100 patients, IMRT had significantly less rectal toxicity.[67] In this study, 3D conformal XRT had a dose-dependent increase in the occurrence of late rectal toxicity, with the 5-year actuarial incidence of Grade 2 rectal toxicity calculated at 14% for radiation doses ≥75.6 Gy compared with 5% for lower doses. IMRT significantly reduced the incidence of late Grade 2 rectal toxicity, with a 3-year actuarial incidence of 2% for patients receiving 81 Gy compared with 14% at the same dose level by 3D conformal XRT. The overall rate of late Grade 3 rectal toxicity was only 1%, with 12 patients requiring at least one transfusion or cauterization procedure for rectal bleeding.[67]

BRACHYTHERAPY

Description

Brachytherapy for prostate cancer was first implemented on a large scale at MSKCC using an open surgical technique. Through a retropubic approach, the prevesical space was opened and, after performing a staging pelvic lymphadenectomy, permanent interstitial ^{125}I implants were placed freehand into the prostate.[78] The dosimetry proved inaccurate and often inadequate using this technique, and the local recurrence proved higher then expected on long-term follow-up.[79] Enthusiasm for prostate brachytherapy waned after this experience, but was again popularized using a transperineal approach in the late 1980s. Using iodine or palladium implants, transperineal prostate brachytherapy under ultrasonic or fluoroscopic guidance has become a common treatment choice for clinically localized prostate cancer. Prostate brachytherapy offers several advantages over surgical treatment or XRT. Unlike RRP, which is performed under general anesthesia and carries the risks of a major operation, brachytherapy is generally performed under local or regional anesthesia in an outpatient setting with much shorter recuperation. In contrast with XRT, which requires daily treatment for several weeks, brachytherapy monotherapy is a 1-day treatment, which offers great appeal to patients. Prostate brachytherapy is currently being used as monotherapy for low-risk patients or in conjunction with adjuvant XRT for intermediate or high-risk patients, and its use is expected to grow as it is offered in more centers across the country.

Cancer control

PSA recurrence

The longest experience with modern techniques of prostate brachytherapy comes from Seattle, for which Ragde reported 12-year long-term follow-up on 229 patients.[80] The majority of patients (64%) were categorized as low risk and treated with brachytherapy as monotherapy; the remaining patients (36%) were considered to be higher risk and received adjuvant XRT. PSA recurrence was defined as PSA >0.5 with a median follow-up of 122 months. For the patients receiving brachytherapy monotherapy, the 10-year PSA progression-free rate was 66%, and for brachytherapy with adjuvant XRT was 79%.[80]

Other studies from MSKCC and Arkansas, with shorter follow-up, have reported 5-year PSA recurrence-free rates for brachytherapy monotherapy that are comparable, using the ASTRO definition of three consecutive PSA rises as a PSA recurrence. Zelefsky et al. reported on 248 patients receiving ^{125}I implant monotherapy, with a 5-year PSA recurrence-free rate of 71%.[81] From Arkansas, Storey et al. examined their experience with ^{125}I implant monotherapy in 206 patients, and found a 5-year PSA recurrence-free rate of 63%.[82] Storey found the 5-year PSA recurrence-free rate for patients with a pretreatment PSA ≤10 was significantly better then PSA >10, at 76% and 51%, respectively. Patients with a pretreatment PSA ≤4 had a PSA recurrence-free rate of 84%.[82] Biopsy Gleason score or clinical stage were not predictive of outcome, but all patients were preselected for Gleason scores ≤7 and clinically localized (T1–2) in this study. The MSKCC series included 60 patients (24%) with Gleason scores ≥7, and found similar prognostic indicators, with favorable-risk, intermediate-risk and high-risk patients having 5-year PSA-recurrence free rates of 88%, 77% and 38%, respectively. On multivariate analysis, pretreatment PSA >10 and Gleason scores >6 were significantly associated with PSA recurrence.[81]

As part of a comparative study involving RRP, XRT and brachytherapy, D'Amico reported on the 5-year outcome of 218 men receiving prostate brachytherapy in Boston.[57] In this study, the majority of men (70%) in the brachytherapy arm received neoadjuvant hormonal therapy of varying length, which limits its comparison with other published studies. The patients were categorized into low-, intermediate- and high-risk groups based on PSA, biopsy Gleason score and clinical stage, with low-risk patients defined as PSA ≤10 and Gleason ≤6 and clinical stage T1c or T2a; intermediate-risk defined as 10 < PSA ≤ 20 or Gleason 7 or stage T2b) and high risk defined as PSA >20 or Gleason ≥8 or stage T2c. Patients in the low-risk category had similar PSA recurrence rates for all treatments, i.e. RRP, XRT and brachytherapy with or without NHT. For high-risk patients, however, patients receiving brachytherapy with or without NHT had significantly worse PSA recurrence rates than RRP. The addition of NHT to brachytherapy for intermediate-risk patients gave comparable results to RRP and XRT, but brachytherapy alone for intermediate-risk patients also did significantly worse. From this study, D'Amico concluded that brachytherapy may be ineffective treatment for intermediate- and high-risk patients.

In an attempt to quantify the risk of PSA progression better using a continuous risk model, Kattan et al. created a pretreatment nomogram for freedom from PSA recurrence for permanent prostate brachytherapy.[83] This nomogram, shown in **Fig. 24.3**, is similar to the Kattan nomograms for PSA recurrence after RRP or XRT (**Figs 24.1** and **24.2**). A continuous risk estimate model is used based on the prognostic variables shown to be significant, including PSA, biopsy Gleason score, clinical stage and the use of adjuvant XRT, to give a pretreatment estimate of the chance of PSA recurrence at 5 years after treatment. As with the other proposed nomograms, this model avoids grouping presumably heterogeneous risk patients into rigid risk categories, to improve the predictive information for a given patient. The concordance index of this predictive model, when applied to external data sets for validation, was 0.61 for the Seattle Prostate Institute and 0.64 for the Arizona Oncology Services, and the calculated confidence intervals were +5% to −30%. While this model may be imperfect, it appears to model the available prostate brachytherapy data better then other published prognostic models.[83]

Local recurrence

As with the other treatment modalities described, calculation of the local recurrence rate after prostate brachytherapy can be difficult to distinguish from distant recurrence. In a study of 490 patients from Mayo Scottsdale, the reported 5-year local control rate was 98%, including five patients with local failure only and three patients with both local and distant recurrence.[84] Importantly, however, this series was a heterogeneous mixture of patients, including 36 receiving NHT and 72 receiving adjuvant XRT, of which the local failures in each group were not delineated. In the Seattle series reporting 10-year follow-up on 152 patients receiving implant monotherapy or implant with adjuvant XRT, 23 patients (15%) had a local recurrence documented on prostate biopsy.[85] Beyer et al. in their report of 489 patients treated with ^{125}I implant monotherapy, noted a 5-year actuarial local control rate of 83%.[86] Looking at predictors of local recurrence, Beyer found that clinical stage and Gleason grade were predictive of local recurrence, but did not evaluate PSA as a predictor.

Distant recurrence

As with the development of local recurrence, the development of bony metastases or distant recurrence after prostate brachytherapy is not routinely tabulated, and defining predictive parameters of distant recurrence is problematic. In his series with 10-year follow-up, Ragde reported the development of bony metastasis in 9 of 152 patients (6%), all of whom also had local recurrence as determined on prostate biopsy.[85] Grado et al. in his series from Mayo Scottsdale reported 33 distant failures in 490 patients (6%) treated with both implant monotherapy and combined implant and adjuvant XRT with or without NHT.[84] As with the local recurrence numbers in this series, the failures in the adjuvant and/or NHT patients were not stratified.

Survival

The relatively short follow-up and small numbers in the published series of prostate brachytherapy limit the calculation of overall and cancer-specific survival. In the longest series by Ragde, the overall 10-year survival rate of 215 evaluable patients was 60%, with only four patients dying of prostate cancer, giving a cancer-specific survival of 98% at 10 years.[80] Owing to these small numbers and significant intercurrent deaths from other diseases in 86 of the initial 215 patients (40%), no definitive predictive information can be gathered until larger long-term studies have matured.

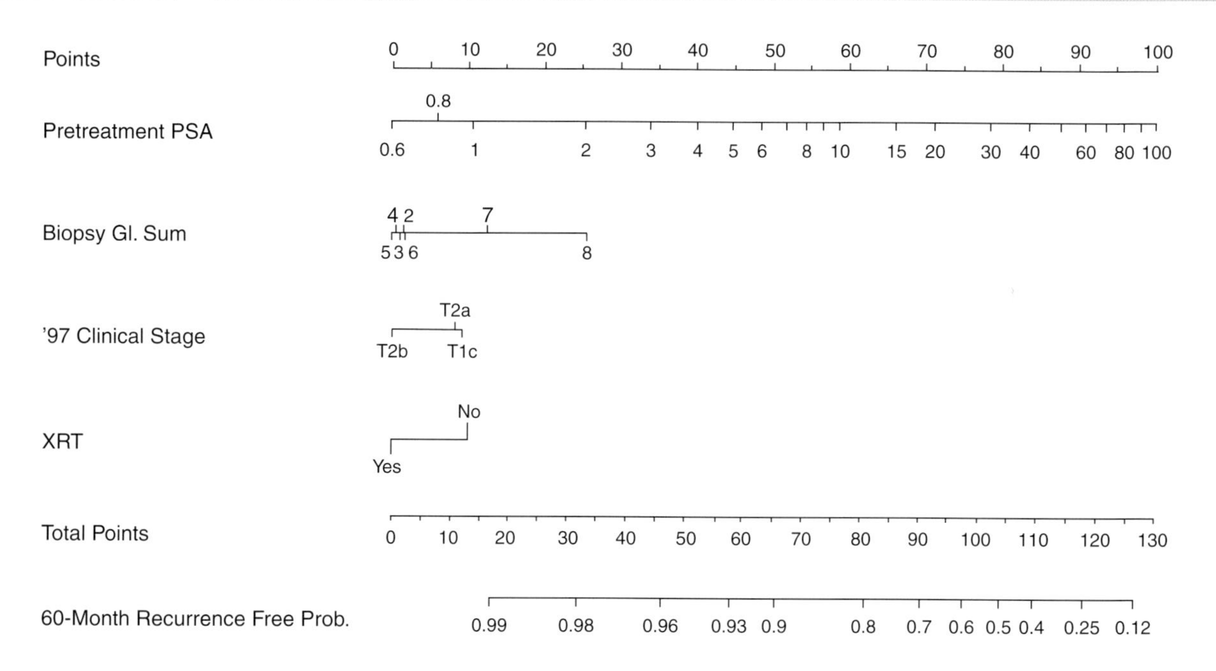

Instructions for Physician: Locate the patient's PSA on the Pretreatment PSA axis. Draw a line straight upwards to the Points axis to determine how many points towards recurrence the patient receives for his PSA. Repeat this process for the other axes, each time drawing straight upward to the Points axis. Sum the points achieved and locate this sum on the Total Points axis. Draw a line straight down to find the patient's probability of remaining recurrence free for 60 months assuming he does not die of another cause first.

Note: This nomogram is not applicable to a man who is not otherwise a candidate for permanent prostate brachytherapy. You can use this only on a man who has already selected permanent prostate brachytherapy as treatment for his prostate cancer. You must decide upon use of adjuvant XRT prior to consulting this nomogram.

Instruction to Patient: 'Mr. X, if we had 100 men exactly like you, we would expect between <predicted percentage from nomogram – 30%> and <predicted percentage + 5%> to remain free of their disease at 5 years following permanent prostate brachytherapy, and recurrence after 5 years is very rare.'

Fig. 24.3. Nomogram for freedom from recurrence after prostate brachytherapy. Reprinted, with permission from Kattan et al. (2001).[83]

Complications

Urinary toxicity

One of the major perceived benefits of prostate brachytherapy is lower long-term complications, because of the presumably lower radiation doses to adjacent normal tissue. Whether this perception is confirmed by published series is unclear, however, with at least one study showing no benefit to urinary, bowel or sexual domains of quality of life (QOL) surveys with brachytherapy compared with RRP or XRT.[87] Nearly all patients will develop some urinary symptoms acutely after implantation, including urinary urgency and frequency. In the MSKCC series, 55% of patients developed acute Grade 2 urinary toxicity, defined as irritative voiding symptoms requiring medication, and 3% developed acute Grade 3 urinary toxicity, defined as urinary retention requiring catheterization.[81]

The long-term urinary toxicity and complications require more attention. At MSKCC, late urinary toxicity, defined as symptoms persisting or developing *de novo* longer then 1 year after treatment, occurred in nearly half of patients. Forty per cent of patients developed late Grade 2 urinary toxicity with a 5-year actuarial rate of 41%.[81] Late Grade 3 urinary toxicity, defined as urethral obstruction or stricture requiring transurethral resection, occurred in 9% of patients, with a 5-year actuarial incidence of 10%; the median time to development of stricture was 18 months. One patient (0.4%) developed a late Grade 4 urinary toxicity requiring urinary diversion and colostomy in the MSKCC series. Beyer et al. reported slightly better late urinary toxicity results, with 4% of patients developing Grade 2 or 3 urinary toxicity.[86] Seven patients (1%) in this series were incontinent, with over half (four patients) incontinent after TURP.

TURP has been implicated as a risk factor for incontinence after brachytherapy. In an earlier study from Ragde and Blasko of implant monotherapy and implant plus adjuvant XRT, 551 patients were evaluated.[88] Severe late urinary complications requiring urinary diversion occurred in nine patients, and 28 patients developed mild stress incontinence. In this group of 37 patients, 36 had undergone prior TURP. Conversely, in a small study from MSKCC totalling 19 patients who had a history of prior TURP, only one patient developed incontinence after brachytherapy treatment, giving a 3-year actuarial freedom from incontinence of 94%.[89] With the awareness that the central, peri-urethral region of the prostate does not require as high a radiation dose, the incontinence rate decreased in the

Seattle series, and the prognostic effect of prior TURP on incontinence after prostate brachytherapy appears less significant.[88]

Impotence
In a large prospective study examining 482 men receiving prostate brachytherapy who were potent prior to treatment, the 5-year actuarial impotency rate was 47%.[90] When stratified for implant monotherapy, implant plus adjuvant XRT, implant plus NHT or implant plus adjuvant XRT and NHT, the 5-year actuarial impotency rates were 76%, 56%, 52% and 29%, respectively. Multivariate analysis identified pretreatment NHT use and patient age as independent predictors of post-treatment impotence in this study. In a subset analysis of 84 men who tried sildenafil for erectile dysfunction after treatment, patients who did not receive NHT had a significantly better response then those treated with NHT. These results are similar to a comparative study examining 3D conformal XRT and ^{125}I prostate brachytherapy in the favorable risk patient by Zelesky.[91] This study noted a 5-year likelihood of erectile dysfunction of 53% after implantation in men who were potent before treatment, and was slightly higher but not significantly different then the XRT group (43%). A multivariate analysis to identify predictors of post-treatment erectile dysfunction recognized only radiation dose delivered (>75.6 Gy for XRT or ⩾160 Gy for brachytherapy) as an independent predictor of impotence; age younger or older then 60 years, prior TURP, use of NHT and mode of treatment (XRT or brachytherapy) were not significant predictors of post-treatment impotence.[91] The potency rate for brachytherapy, unlike after RRP, appears to worsen over time, with Stock et al. noting a decrease in potency from 3 years to 6 years of 79% to 59% after brachytherapy.[92]

Bowel toxicity
Bowel toxicity after prostate brachytherapy includes similar symptoms to those after XRT, namely proctitis, tenesmus, urgency, rectal bleeding, rectal stenosis, necrosis, and rectovesical or rectourethral fistula formation. In Zelefsky's series at MSKCC, severe bowel complications were rare, with a 5-year actuarial rate of Grade 2 rectal bleeding of 9%, and one patient (0.4%) developing a Grade 4 rectourethral fistula requiring urinary diversion and colostomy.[81] A slightly higher Grade 4 rectal complication rate of 1% was noted in the series of Grado et al.[84] Although these complications involved only five of 490 patients, less serious bowel toxicities were not mentioned.[84] Overall, severe bowel toxicity resulting from implant monotherapy is rare and appears less common than in XRT.

SALVAGE THERAPY

Description
Amongst the considerations for deciding on a treatment modality, patients commonly ask about the feasibility of subsequent salvage therapy, if the primary treatment fails. While the literature for this question is not as developed as that concerning primary treatment, this consideration deserves inclusion into the decision analysis. If a particular treatment modality makes subsequent salvage therapy prohibitively morbid or dangerous, the patient may use this information to opt for different initial treatment choice. While these concerns may be overblown, patient concern over this issue warrants a closer examination of the available salvage literature.

RRP after radiotherapy
Salvage RRP after radiotherapy generally encompasses both XRT and brachytherapy in the literature, because of the similar concerns and potential complications with either radiation modality. One of the largest studies, from the Mayo Clinic, reported on 108 patients receiving salvage RRP over 30 years from 1966 to 1996.[93] The overall and cancer-specific survival at 10 years was 70% and 43%, respectively. Predictors for cancer-specific survival on multivariate analysis were preoperative PSA and DNA ploidy. Importantly, however, 19% of patients in this series had lymph node metastases at the time of surgery, highlighting the difficulty in distinguishing local recurrence from regional or distant spread.

One of the major concerns with salvage surgery has been the high reported rate of complications, including the risk of rectal injury and incontinence, as well as high rates of positive surgical margins.[94] In the Mayo Clinic series mentioned above, the rates of rectal injury, bladder neck contracture and urinary incontinence were 6%, 21% and 50%, respectively.[93] In fact, some have advocated radical cystoprostatectomy in this setting because of these concerns.[95] In a more recent review from MSKCC, rectal injury is now uncommon, but incontinence remains problematic.[96] One potential advance is that of preoperative endorectal MRI evaluation; urethral length measured on coronal MRI sections has been correlated with postoperative continence (unpublished). Patient selection for salvage RRP remains paramount; despite the reduction in morbidity and complications, the incidence of advanced disease on final pathologic examination of the prostate and pelvic lymph nodes remains high.[94] Patients must be counseled initially that effective surgical salvage of definitive radiotherapy failure is unlikely and carries significant risk.

XRT after RRP
As with salvage RRP after XRT, the patients most likely to benefit from salvage XRT after surgical treatment are those with local recurrence only, with the absence of regional or distant metastatic spread. This distinction is difficult to make, since the majority of patients typically have PSA elevation without clinically evident local or distant recurrence. Several factors have been identified to indicate local versus distant recurrence, specifically time to PSA recurrence (longer then 2 years postoperatively), and slow PSA doubling time (greater then 10 months).[19] In another study specifically examining response to salvage XRT after RRP, the most important predictors on multivariate analysis of a durable PSA response were PSA doubling time (< or >11.8 months) and pre-XRT PSA (< or >2.1).[97] Other factors postulated to be important, such

as palpable recurrence on digital rectal examination (DRE) or positive biopsy of prostatic fossa, were not predictive. For patients receiving salvage XRT in this study, the actuarial 3- and 5-year PSA relapse-free probabilities were only 43% and 24%, respectively. This again demonstrates that patients with isolated local recurrence after definitive local therapy, the group that should respond well to salvage local therapy, were the exception rather then the rule, with the majority of patients having distant disease at time of disease recurrence diagnosis.

XRT after brachytherapy

While XRT is often used in conjunction with brachytherapy as initial treatment for intermediate- and high-risk patients, there is no literature examining the role of salvage XRT after primary brachytherapy failure.

Brachytherapy after XRT

Salvage brachytherapy after XRT failure has been reported in two small studies[98,99] but remains controversial. The biologic effect of additional radiation on a presumably radioresistant tumor remains unknown. The reported 5-year PSA recurrence-free rates were 34% and 53%, respectively. A postimplant PSA nadir <0.5 was predictive of PSA recurrence-free survival.[98] The reported complications were not insignificant, however, with a 24% incontinence rate in one series,[99] and a TURP rate of 14% and rectal ulceration rate of 6% in the other series.[98] In an editorial comment, D'Amico suggested that use of salvage brachytherapy should be limited to patients with clinically low-risk disease prior to initial XRT (pre-XRT PSA of <10, biopsy Gleason score ≤6, and clinical stage T1c or T2a), with a post-XRT PSA failure occurring >1 year from XRT and a pre-implant PSA of <2.0.[100]

DECISION MAKING

Evaluating treatment options

Advantages and disadvantages

Examination of the treatment options for prostate cancer requires an analysis of the specific advantages and disadvantages of each modality. As shown in **Table 24.4**, each modality has specific issues which may make the modality more or less advantageous. RRP has an advantage in that it offers accurate pathologic prognostic information, which is lacking with both radiotherapy modalities, and offers the most data on long-term (>10 year) cancer control. Conversely, XRT and brachytherapy carry much less risk of peritreatment morbidity or mortality, with quicker recovery and return to work. The issues of incontinence and erectile dysfunction, as mentioned previously, differ between modalities, with acute occurrence and delayed recovery for RRP, and delayed occurrence in either radiotherapy modality.

Indications and contraindications

Apart from patient preference, there are several clinical parameters that may exclude a treatment modality from the decision-making process. Several commonly held indications and contraindications for RRP, XRT and brachytherapy are listed in **Table 24.5**. While precise data regarding each of these issues are lacking, most practitioners abide by these indications when offering treatment options for prostate cancer. Relative indications and contraindications are denoted, for which the data or commonly held opinions are not concrete. Thus, some urologists would not feel comfortable offering RRP to anyone refusing a potential blood transfusion, on the grounds that operative blood loss may potentially be high, which would place the patient at risk. Additionally, the age of the patient is used commonly as a deciding factor between treatment modalities, but these 'loose' age ranges may change as medical care improves with improved life expectancies in the future. Specific patient comorbidities are listed as well, including inflammatory bowel disease as a contraindication for radiotherapy, owing to concerns of exacerbating this condition. An irreversible bleeding diathesis is also considered a contraindication for radiotherapy, owing to concerns over the risk of chronic bowel or bladder bleeding after treatment.

Availability of services

The services available to a patient in his local area are another factor to be considered in the decision-making process. As mentioned previously, XRT dose and technique, specifically 3D conformal XRT and IMRT, are important parameters for good outcomes in prostate cancer, including the ability to deliver higher radiation doses accurately with less toxicity to surrounding normal tissue, but are not universally available nationwide. Similarly, recent data confirm that complications after RRP inversely correlate with hospital and surgeon volume of RRP.[101] Lastly, there are wide variations in outcome after prostate brachytherapy, owing to practitioner experience and variations in technique, impacting on the precise placement of seeds. Thus, a patient living in an area with a busy well-trained surgeon operating at a high-volume hospital, without access to the newest radiotherapy techniques, may be well served to opt for RRP. Conversely, in another part of the country possessing good radiotherapy facilities and the availability of IMRT or brachytherapy, with no busy surgical practices, a patient may get a better outcome opting for one of the radiotherapy modalities.

Determining Utilities

After reviewing the available literature regarding cancer control and treatment morbidity, the patient and clinician then have to derive order and understanding from all of the statistics, advantages and indications in order to make a decision. One way to order or categorize different scenarios is to assign value or utility to different states of health associated with prostate cancer. Markov models, as shown in **Fig. 24.4**, are used to simulate decision-making, by categorizing each exclusive health state that a patient may be in at a point in time.[102] Thus, for prostate cancer in **Fig. 24.4**, a patient moves from one health state to the next as his life progresses. Patients will not value each of these health states equally, allowing quality

Table 24.4. Advantages and disadvantages of prostate cancer treatment modalities. Adapted from Ohori and Scardino (in press).[113]

Advantages	Disadvantages
Radical prostatectomy	
Excellent long-term cancer control	Risk of perioperative morbidity and mortality
Accurate prognostic predictions based on final pathologic examination	Hospitalization and time lost from work
Pelvic lymph-node dissection through same incision	Risk of incomplete excision (positive surgical margins)
	Risk of long-term incontinence and erectile dysfunction
	Visible scar with possible delayed morbidity (ventral hernia)
External beam radiotherapy	
Ease of outpatient administration	Long course of therapy
Avoidance of anesthesia and surgery	Difficulty in precise targeting of prostate
No immediate incontinence or erectile dysfunction	Delayed risk of incontinence or erectile dysfunction
Maintenance of normal work schedule	Lack of universal availability of 3D conformal XRT and IMRT techniques
	Large gland or locally advanced cancer may require hormonal deprivation, increasing risk of erectile dysfunction
	Risk of long-term bowel and bladder toxicity
	Uncertainty of complete eradication of local tumor (PSA does not become undetectable)
	Uncertain long-term cancer control (>10 years)
	Difficulty interpreting PSA level and/or biopsy after treatment
	Salvage surgery risky if attempted for local recurrence
	No information about pathologic stage for accurate prognosis
Brachytherapy	
Convenient with single day treatment	Requires anesthesia
No immediate incontinence or erectile dysfunction	Variable results depending on quality of implant and experience of implanter
Low risk of immediate complications	'Cold spots' are common
Quick recovery and return to work	Applicable only to favorable-risk cancers
	Uncertain long-term cancer control (>10 years)
	Symptomatic urethritis is common
	Requires neoadjuvant hormonal deprivation for large glands
	No information about pathologic stage for accurate prognosis
	Recurrence hard to detect
	Salvage therapy is risky
	Risk of acute urinary retention requiring TURP with risk of incontinence

of life adjustments, or utilities, to be applied to each health state or circle in the Markov model. Patients who consider themselves as in perfect health would be given a utility of 1.0, while a utility of 0 would signify death. For instance, if the patient is free of disease (no evidence of disease; NED) after treatment without any complications, then he may assign a utility of 0.9 to his health state; conversely, a patient who develops bony metastases after treatment may assign a utility of 0.4 to his health state. For each modification of this scenario, health state utilities can be assessed.

Calculating cohort-based health state utilities can be performed to gauge patient preference for different modalities. Utilities can be measured in a variety of ways. Anonymous questionnaires, telephone surveys and waiting room computer surveys have all been used to survey patients' opinions of quality and quantity of life, and their importance to individual patients. When comparing different health states, utilities can be calculated using algorithms, such as standard gamble or time trade-off. These are research tools that quantify a given cohort of men's values of different health states, in an attempt to make objective comparisons of quality of life. The time trade-off technique is shown in **Fig. 24.5**. Patients are queried about their current health state, and asked whether they would trade that health state for a perfect state

Table 24.5. Indications and contraindications for treatment modalities in prostate cancer.

Indications (absolute/relative)	Contraindications (absolute/relative)
Radical prostatectomy	
Good overall health	Poor health with life expectancy <10 years
Age <70–75 years old	Advanced age
Life expectancy >10 years	Locally advanced (relative) or metastatic disease
Young patients (<50–55 years old) with life expectancy >15 years	Refusal to accept blood transfusion (e.g. Jehovah's Witness) (relative)
Severe voiding symptoms/BPH refractory to medical management (relative)	Prior perineal/pelvic surgery or injury to external sphincter
External beam radiotherapy	
Clinically localized disease	Pelvic metallic foreign bodies preventing accurate CT-based imaging
Any age (if can tolerate simulation and positioning)	Irreversible bleeding diathesis
	Prior radiation to pelvis (relative)
	Inflammatory bowel disease
	Metastatic disease
Brachytherapy	
Clinically localized, low-risk disease	Intermediate or high-risk disease
Any age	Known or suspected metastatic disease
	Significant pubic arch interference
	Irreversible bleeding diathesis
	Inflammatory bowel disease (relative)
	Severe voiding symptoms or prior TURP (relative)
	Significant BPH >60 g prostate size
	Lumbar disc disease or inability to tolerate exaggerated lithotomy position

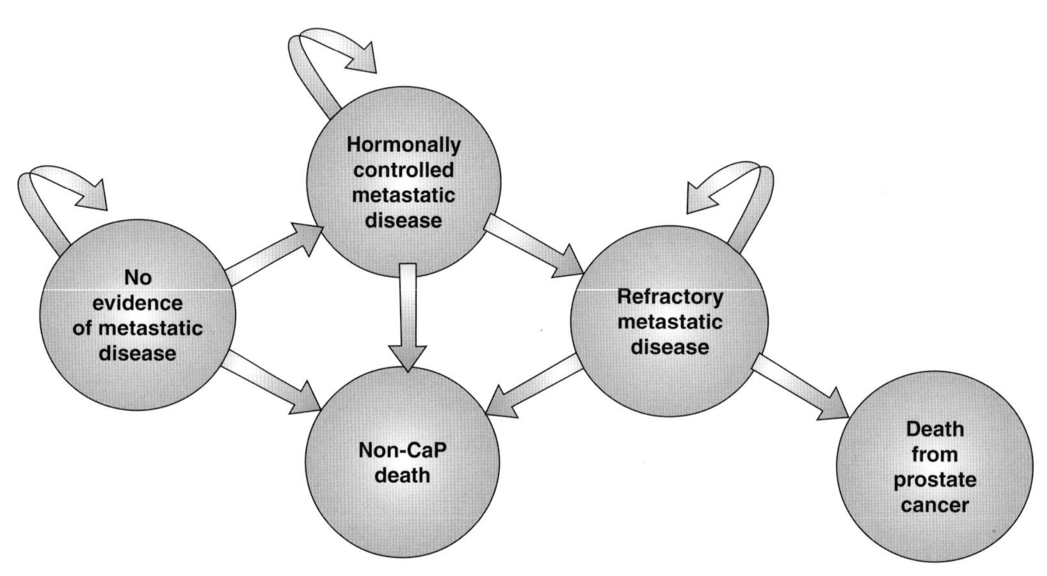

Fig. 24.4. Markov model for prostate cancer. Reprinted with permission from Kattan et al. (in press).[114]

of health for a defined, guaranteed time period, followed by immediate death, or whether they would be indifferent to a trade. As this guaranteed time period of perfect health is varied from shorter to longer, patients will at some point be indifferent. At the point of indifference, the value of the current health state can be quantified, as the number of perfect health years divided by the life expectancy. This utility, calculated for each health state, such as alive with metastatic disease,

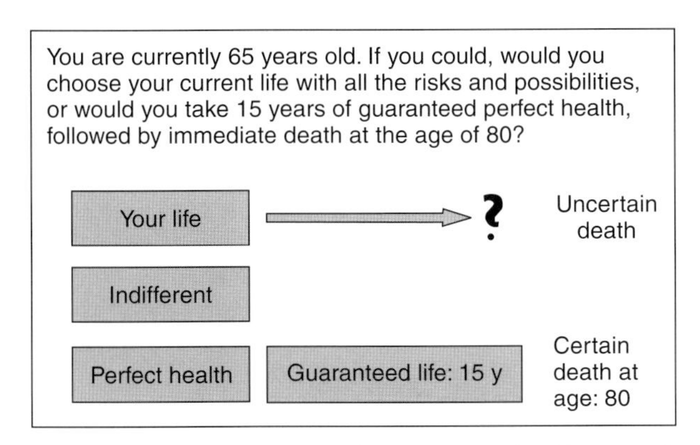

Fig. 24.5. Time trade-off technique for determining health state utilities. Reprinted with permission from Kattan et al. (in press).[114]

NED with incontinence, NED with impotence, etc., can give a numeric quality of life value to each health state. Another common metric used to determine health state utilities is the standard gamble, as shown in **Fig. 24.6**. This technique uses a magic pill, which will either cure the patient and give him perfect health or kill him. As the probability of perfect health versus instant death is varied, a quantitative assessment of the quality of the patient's current health state is determined and defined as the utility of that health state. These health state utility determination techniques allow accurate quantitative determinations of the quality of life associated with different health states.

After health state utilities are calculated for a randomly collected cohort of men, these can be extrapolated to the individual patient for use in his own decision-making process. They can also be used quantitatively to give objective scores for each treatment option and possible combination, to allow an orderly and logical decision. Using available data on the probability of resulting in a specific health state for each treatment modality, survival-adjusted quality of life can be calculated, by multiplying the health state utility by the probability of achieving that health state.[103] If the probability of a patient achieving NED status after RRP, based on his own individual prognostic variable parameters, is 0.7, and the utility of achieving NED after RRP is 0.9, then his quality-adjusted life expectancy (QALE) is $0.7 \times 0.9 = 0.63$. Conversely, if based on his prognostic parameters for XRT, his chance of achieving NED after XRT is 0.6, and the utility of achieving NED after XRT is 0.85, then his QALE for NED after XRT would be $0.6 \times 0.85 = 0.51$. With this type of analysis, a patient can use his own individual prognostic information to make a rational comparison between treatment modalities.

Group-level determination

Using group-level determinations derived from a representative cohort of patients to a specific patient can be problematic, however. Cowen et al.[104] looked at this very question in a study of 31 men in an attempt to validate the health state utilities reported by Kattan et al.[102] from 32 men with prostate cancer in an effort to maximize QALE. In this validation study, each of the 31 patients had two treatment recommendations: one from the previous group-level determination, and one based on the individual patient's own determination of utilities. In

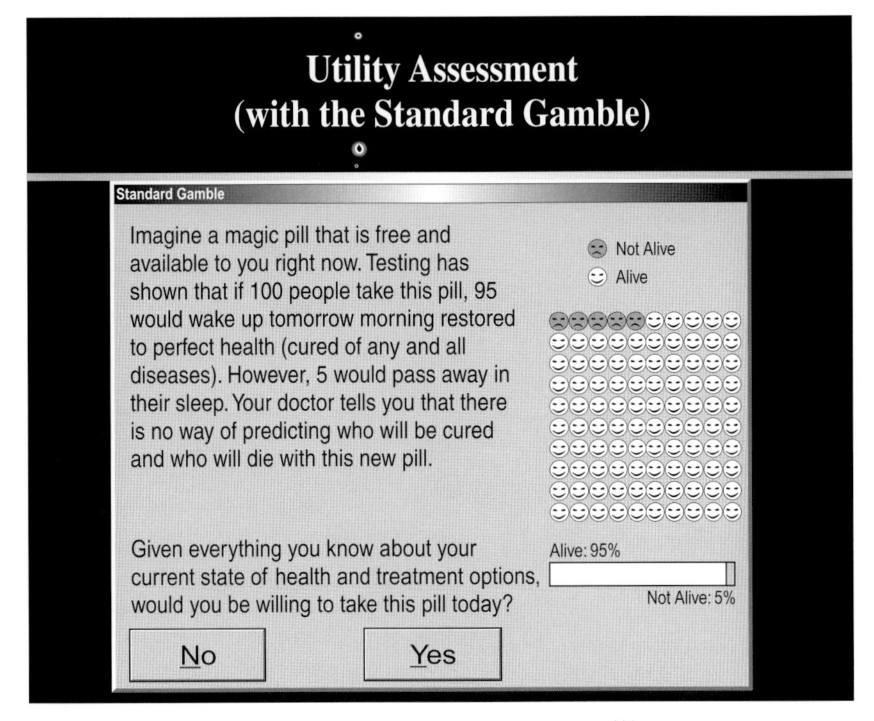

Fig. 24.6. Standard gamble technique. Reprinted with permission from Kattan et al. (in press).[114]

this analysis, the group-level utilities misrepresented between 25% and 48% of the individual patient preferences based on the clinical scenario, and the average QALE lost because of this error was as high as 1.7 quality-adjusted life-years. As with prognostic nomograms, this study highlights the difficulty in categorizing heterogeneous patients into groups without accounting for individual preference and variability.

CONCLUSION

The decision-making process for prostate cancer is not straightforward. There are multiple treatment modalities with similar cancer control, especially for low-risk patients. The risks and complications for each differ in the short and long term. The patient and clinician must reconcile the published literature with the individual clinical information and patient

Table 24.6. Summary of best available studies.

	Primary therapy		
Parameter	RRP	XRT	Brachytherapy
Disease control			
PSA recurrence	See Fig. 24.1[16]	See Fig. 24.2[62]	See Fig. 24.3[83]
Local recurrence	1.6%[17]	13–37%[67]	15%[85] 17%[86]
Distant recurrence	5.2%[19]	6% overall (0–24% in Table 24.3 categories)[53]	6%[84,85]
Survival (cancer-specific)	99.1% (5 years) 97.6% (10 years)[7]	95.1% (5 years)[58]	98% (10 years)[80]
Complications			
Peri-operative	0.4% mortality[25] 2% overall complications[29]	N/A	N/A
Urinary	See Table 24.2[105]	29% incontinence: 36% urge 8% stress[70] 10% Grade 2 toxicity[67]	Acute: 58% Grade 2–3 Tox Late (5 years): 41% Grade 2 Tox 10% Grade 3 Tox[81]
Impotence	50% overall 45% both nerves spared 79% one nerve spared[46]	60% (5 years)[71]	53% (5 years)[91]
Bowel	N/A	22% overall: 10% fecal incontinence 7% rectal bleeding[70] Grade ≥2 toxicity: 14% ≥ 75.6 Gy dose, 5% lower doses[67]	9% Grade 2 rectal bleeding (5 years)[81] 0.4–1% Grade 4[81, 84]
Salvage therapy			
RRP	N/A	43% 10-year cancer-specific survival 6% rectal injury 21% bladder neck contracture 50% urinary incontinence[93]	
XRT	24% PSA recurrence-free probability (5 years)[97]	N/A	None available
Brachytherapy	N/A	34–53% PSA recurrence-free probability (5 years)[98, 99] 24% incontinence[99] 6% Rectal ulceration[98]	N/A

preferences to arrive at an individualized treatment plan. Summarized in **Table 24.6** are representative published series concerning cancer control, survival, short- and long-term complications and salvage options for each treatment modality. Armed with such a body of knowledge, the patient and clinician can work together to tailor their treatment approach quickly and efficiently. Fortunately, the field of decision analysis and outcomes research is expanding rapidly, and will only improve the decision-making process in the years to come. Information dissemination should be greatly facilitated by the World Wide Web. In fact, each of the nomograms displayed in the figures in this chapter are currently available for downloading on to hand-held computers (www.nomograms.org). The greatest need, however, for accurate comparative analyses between treatment modalities, is for randomized clinical trials. Two important trials that should answer some of these questions are now accruing patients (trial information courtesy of the National Cancer Institute). The Prostate Intervention Versus Observation Trial (PIVOT) is a two-armed trial comparing RRP with expectant management. It was opened in 1994 and has accrued 700 patients to date, with an expected follow-up of 15 years. This study hopes to show, for the first time, a survival benefit with surgery over observation. A second trial, initiated by the American College of Surgeons Oncology Group (ACOSOG), is accruing patients for a randomized trial of RRP versus brachytherapy for Stage 1 and 2 prostate cancer. This study hopes to accrue nearly 1000 patients per treatment arm and follow them for at least 5 years. Once completed, these studies should greatly clarify the benefits of prostate cancer treatment and simplify the decision-making process.

REFERENCES

1. Greenlee RT, Hill-Harmon MB, Murray T et al. Cancer statistics, 2001. CA *Cancer J. Clin.* 2001; 51:15–36.
2. Lu J. Statistical aspects of evaluating treatment and prognostic factors for clinically localized prostate cancer. *Semin. Urol. Oncol.* 2000; 18:83–92.
3. Walsh PC, Donker PJ. Impotence following radical prostatectomy: insight into etiology and prevention. *J. Urol.* 1982; 128:492–4.
4. Walsh PC. Anatomic radical prostatectomy: evolution of the surgical technique. *J. Urol.* 1998; 160:2418–24.
5. Lu-Yao GL, Yao SL. Population-based study of long-term survival in patients with clinically localised prostate cancer. *Lancet* 1997; 349:906–10.
6. Yan Y, Carvalhal GF, Catalona WJ et al. Primary treatment choices for men with clinically localized prostate carcinoma detected by screening. *Cancer* 2000; 88:1122–30.
7. Hull GW, Rabbani F, Abbas F, Wheeler TM et al. Cancer control with radical prostatectomy alone in 1,000 consecutive patients. *J. Urol.* 2002; 167:528–34.
8. Trapasso JG, deKernion JB, Smith RB et al. The incidence and significance of detectable levels of serum prostate specific antigen after radical prostatectomy. *J. Urol.* 1994; 152:1821–5.
9. Pound CR, Partin AW, Epstein JI et al. Prostate-specific antigen after anatomic radical retropubic prostatectomy. Patterns of recurrence and cancer control. *Urol. Clinics North Am.* 1997; 24:395–406.
10. Zincke H, Oesterling JE, Blute ML et al. Long-term (15 years) results after radical prostatectomy for clinically localized (stage T2c or lower) prostate cancer [see comments]. *J. Urol.* 1994; 152:1850–7.
11. Catalona WJ, Smith DS. Five-year tumor recurrence rates after anatomical radical retropubic prostatectomy for prostate cancer [see comments]. *J. Urol.* 1994; 152:1837–42.
12. Partin AW, Yoo J, Carter HB et al. The use of prostate specific antigen, clinical stage and Gleason score to predict pathological stage in men with localized prostate cancer [see comments]. *J. Urol.* 1993; 150:110–14.
13. Partin AW, Kattan MW, Subong EN et al. Combination of prostate-specific antigen, clinical stage, and Gleason score to predict pathological stage of localized prostate cancer. A multi-institutional update [see comments] [published erratum appears in *JAMA* 1997; 278:118]. *JAMA* 1997; 277:1445–51.
14. Partin AW, Mangold LA, Lamm DM et al. Contemporary update of prostate cancer staging nomograms (Partin Tables) for the new millennium. *Urology* 2001; 58:843–8.
15. Blute ML, Bergstralh EJ, Partin AW et al. Validation of Partin tables for predicting pathological stage of clinically localized prostate cancer. *J. Urol.* 2000; 164:1591–5.
16. Kattan MW, Eastham JA, Stapleton AM et al. A preoperative nomogram for disease recurrence following radical prostatectomy for prostate cancer. *J. Natl Cancer Inst.* 1998; 90:766–71.
17. Han M, Partin AW, Pound CR et al. Long-term biochemical disease-free and cancer-specific survival following anatomic radical retropubic prostatectomy. The 15-year Johns Hopkins experience. *Urol. Clinics North Am.* 2001; 28:555–65.
18. Han M, Pound CR, Potter SR et al. Isolated local recurrence is rare after radical prostatectomy in men with Gleason 7 prostate cancer and positive surgical margins: therapeutic implications. Journal of Urology 2001; 165:864–6.
19. Pound CR, Partin AW, Eisenberger MA et al. Natural history of progression after PSA elevation following radical prostatectomy. *JAMA* 1999; 281:1591–7.
20. Iversen P, Madsen PO, Corle DK. Radical prostatectomy versus expectant treatment for early carcinoma of the prostate. Twenty-three year follow-up of a prospective randomized study. *Scand. J. Urol. Nephrol.* Suppl. 1995; 172:65–72.
21. Han M, Partin AW, Piantadosi S et al. Era specific biochemical recurrence-free survival following radical prostatectomy for clinically localized prostate cancer. *J. Urol.* 2001; 166:416–19.
22. Eastham JA, Scardino PT. Radical prostatectomy for clinical stage T1 and T2 prostate cancer. In: Vogelzang NJ, Scardino PT, Shipley WU et al. (eds) *Comprehensive Textbook of Genitourinary Oncology*. Philadelphia, PA: Williams & Wilkins, 2000:722–38.
23. Goad JR, Scardino PT. Modifications in the technique of radical retropubic prostatectomy to minimize blood loss. *Atlas Urol. Clin. North Am.* 1994; 2:65–80.
24. Andriole GL, Smith DS, Rao G et al. Early complications of contemporary anatomical radical retropubic prostatectomy [see comments]. *J. Urol.* 1994; 152:1858–60.
25. Dillioglugil O, Leibman BD, Leibman NS et al. Risk factors for complications and morbidity after radical retropubic prostatectomy. *J. Urol.* 1997; 157:1760–7.
26. Lerner SE, Blute ML, Lieber MM et al. Morbidity of contemporary radical retropubic prostatectomy for localized prostate cancer. *Oncology* 1995; 9:379–82.
27. Hautmann RE, Sauter TW, Wenderoth UK. Radical retropubic prostatectomy: morbidity and urinary continence in 418 consecutive cases. *Urology* 1994; 43:47–51.
28. Leandri P, Rossignol G, Gautier JR et al. Radical retropubic prostatectomy: morbidity and quality of life. Experience with 620 consecutive cases. *J. Urol.* 1992; 147:883–7.
29. Lepor H, Nieder AM, Ferrandino MN. Intraoperative and postoperative complications of radical retropubic prostatectomy in a consecutive series of 1,000 cases. *J. Urol.* 2001; 166:1729–33.

30. Lau KO, Cheng C. Feasibility of early catheter removal after radical retropubic prostatectomy. *Techniques Urol.* 2001; 7:38–40.
31. Coogan CL, Little JS, Bihrle R et al. Urethral catheter removal prior to hospital discharge following radical prostatectomy. *Urology* 1997; 49:400–3.
32. Little JS Jr, Bihrle R, Foster RS. Early urethral catheter removal following radical prostatectomy: a pilot study. *Urology* 1995; 46:429–31.
33. Souto CA, Teloken C, Souto JC et al. Experience with early catheter removal after radical retropubic prostatectomy. *J. Urol.* 2000; 163:865–6.
34. Lepor H. Radical retropubic prostatectomy. *Urol. Clinics North Am.* 2001; 28:509–19, viii.
35. Nadu A, Salomon L, Hoznek A et al. Early removal of the catheter after laparoscopic radical prostatectomy. *J. Urol.* 2001; 166:1662–4.
36. Geary ES, Dendinger TE, Freiha FS et al. Incontinence and vesical neck strictures following radical retropubic prostatectomy. *Urology* 1995; 45:1000–6.
37. Benoit RM, Naslund MJ, Cohen JK. Complications after radical retropubic prostatectomy in the medicare population. *Urology* 2000; 56:116–20.
38. Shelfo SW, Obek C, Soloway MS. Update on bladder neck preservation during radical retropubic prostatectomy: impact on pathologic outcome, anastomotic strictures, and continence. *Urology* 1998; 51:73–8.
39. Surya BV, Provet J, Johanson KE et al. Anastomotic strictures following radical prostatectomy: risk factors and management. *J. Urol.* 1990; 143:755–8.
40. Popken G, Sommerkamp H, Schultze-Seemann W et al. Anastomotic stricture after radical prostatectomy. Incidence, findings and treatment. *Eur. Urol.* 1998; 33:382–6.
41. Tomschi W, Suster G, Holtl W. Bladder neck strictures after radical retropubic prostatectomy: still an unsolved problem. *Br. J. Urol.* 1998; 81:823–6.
42. Borboroglu PG, Sands JP, Roberts JL et al. Risk factors for vesicourethral anastomotic stricture after radical prostatectomy. *Urology* 2000; 56:96–100.
43. Litwin MS, Hays RD, Fink A et al. Quality-of-life outcomes in men treated for localized prostate cancer. *JAMA* 1995; 273:129–35.
44. Eastham JA, Kattan MW, Rogers E et al. Risk factors for urinary incontinence after radical prostatectomy. *J. Urol.* 1996; 156:1707–13.
45. Rosen RC, Riley A, Wagner G et al. The international index of erectile function (IIEF): a multidimensional scale for assessment of erectile dysfunction. *Urology* 1997; 49:822–30.
46. Rabbani F, Stapleton AM, Kattan MW et al. Factors predicting recovery of erections after radical prostatectomy. *J. Urol.* 2000; 164:1929–34.
47. Purdy JA, Michalski JM. Does the evidence support the enthusiasm over 3D conformal radiation therapy and dose escalation in the treatment of prostate cancer? [letter; comment]. *Int. J. Radiat. Oncol. Biol. Phys.* 2001; 51:867–70.
48. Perez CA, Michalski JM, Purdy JA et al. New trends in prostatic cancer research. Three-dimensional conformal radiation therapy (3-D CRT), brachytherapy, and new therapeutic modalities. *Rays* 2000; 25:331–43.
49. Bolla M, Gonzalez D, Warde P et al. Improved survival in patients with locally advanced prostate cancer treated with radiotherapy and goserelin. *N. Engl. J. Med.* 1997; 337:295–300.
50. Scolieri MJ, Altman A, Resnick MI. Neoadjuvant hormonal ablative therapy before radical prostatectomy: a review. Is it indicated? [see comments]. *J. Urol.* 2000; 164:1465–72.
51. Anonymous. Consensus statement: guidelines for PSA following radiation therapy. American Society for Therapeutic Radiology and Oncology Consensus Panel. *Int. J. Radiat. Oncol. Biol. Phys.* 1997; 37:1035–41.
52. Horwitz EM, Vicini FA, Ziaja EL et al. The correlation between the ASTRO Consensus Panel definition of biochemical failure and clinical outcome for patients with prostate cancer treated with external beam irradiation. American Society of Therapeutic Radiology and Oncology. *Int. J. Radiat. Oncol. Biol. Phys.* 1998; 41:267–72.
53. Zagars GK, Pollack A, von Eschenbach AC. Prognostic factors for clinically localized prostate carcinoma: analysis of 938 patients irradiated in the prostate specific antigen era. *Cancer* 1997; 79:1370–80.
54. Kuban DA, el-Mahdi AM, Schellhammer PF. Prostate-specific antigen for pretreatment prediction and posttreatment evaluation of outcome after definitive irradiation for prostate cancer. *Int. J. Radiat. Oncol. Biol. Phys.* 1995; 32:307–16.
55. D'Amico AV, Whittington R, Malkowicz SB et al. Pretreatment nomogram for prostate-specific antigen recurrence after radical prostatectomy or external-beam radiation therapy for clinically localized prostate cancer. *J. Clin. Oncol.* 1999; 17:168–72.
56. Zagars GK, Geara FB, Pollack A et al. The T classification of clinically localized prostate cancer. An appraisal based on disease outcome after radiation therapy. *Cancer* 1994; 73:1904–12.
57. D'Amico AV, Whittington R, Malkowicz SB et al. Biochemical outcome after radical prostatectomy, external beam radiation therapy, or interstitial radiation therapy for clinically localized prostate cancer [see comments]. *JAMA* 1998; 280:969–74.
58. Shipley WU, Thames HD, Sandler HM et al. Radiation therapy for clinically localized prostate cancer: a multi-institutional pooled analysis. *JAMA* 1999; 281:1598–604.
59. Zelefsky MJ, Leibel SA, Gaudin PB et al. Dose escalation with three-dimensional conformal radiation therapy affects the outcome in prostate cancer. *Int. J. Radiat. Oncol. Biol. Phys.* 1998; 41:491–500.
60. Duchesne GM, Bloomfield D, Wall P. Identification of intermediate-risk prostate cancer patients treated with radiotherapy suitable for neoadjuvant hormone studies. *Radiother. Oncol.* 1996; 38:7–12.
61. Pisansky TM, Kahn MJ, Bostwick DG. An enhanced prognostic system for clinically localized carcinoma of the prostate. *Cancer* 1997; 79:2154–61.
62. Kattan MW, Zelefsky MJ, Kupelian PA et al. Pretreatment nomogram for predicting the outcome of three-dimensional conformal radiotherapy in prostate cancer. *J. Clin. Oncol.* 2000; 18:3352–9.
63. Bey P, Carrie C, Beckendorf V et al. Dose escalation with 3D-CRT in prostate cancer: French study of dose escalation with conformal 3D radiotherapy in prostate cancer-preliminary results. *Int. J. Radiat. Oncol. Biol. Phys.* 2000; 48:513–7.
64. Vicini FA, Abner A, Baglan KL et al. Defining a dose–response relationship with radiotherapy for prostate cancer: is more really better? *Int. J. Radiat. Oncol. Biol. Phys.* 2001; 51:1200–8.
65. Valicenti R, Lu J, Pilepich M et al. Survival advantage from higher-dose radiation therapy for clinically localized prostate cancer treated on the Radiation Therapy Oncology Group trials. *J. Clin. Oncol.* 2000; 18:2740–6.
66. Pollack A, Zagars GK, Smith LG et al. Preliminary results of a randomized radiotherapy dose-escalation study comparing 70 Gy with 78 Gy for prostate cancer. *J. Clin. Oncol.* 2000; 18:3904–11.
67. Zelefsky MJ, Fuks Z, Hunt M et al. High dose radiation delivered by intensity modulated conformal radiotherapy improves the outcome of localized prostate cancer. *J. Urol.* 2001; 166:876–81.
68. Sandler H, Shipley WU, Gomella L et al. Radiation Therapy Oncology Group. Research Plan 2002–2006. Genitourinary Cancer Committee. *Int. J. Radiat. Oncol. Biol. Phys.* 2001; 51:28–38.
69. Pisansky TM, Davis BJ. Predictive factors in localized prostate cancer: implications for radiotherapy and clinical trial design. *Semin. Urol. Oncol.* 2000; 18:93–107.
70. Nguyen LN, Pollack A, Zagars GK. Late effects after radiotherapy for prostate cancer in a randomized dose–response

study: results of a self-assessment questionnaire. *Urology* 1998; 51:991–7.

71. Zelefsky MJ, Cowen D, Fuks Z et al. Long term tolerance of high dose three-dimensional conformal radiotherapy in patients with localized prostate carcinoma. Cancer 1999; 85:2460–8.
72. Turner SL, Adams K, Bull CA et al. Sexual dysfunction after radical radiation therapy for prostate cancer: a prospective evaluation. *Urology* 1999; 54:124–9.
73. Robinson JW, Dufour MS, Fung TS. Erectile functioning of men treated for prostate carcinoma. *Cancer* 1997; 79:538–44.
74. Wasson JH, Cushman CC, Bruskewitz RC et al. A structured literature review of treatment for localized prostate cancer. Prostate Disease Patient Outcome Research Team. *Arch. Family Med.* 1993; 2:487–93.
75. Siegel T, Moul JW, Spevak M et al. The development of erectile dysfunction in men treated for prostate cancer. *J. Urol.* 2001; 165:430–5.
76. Chinn DM, Holland J, Crownover RL et al. Potency following high-dose three-dimensional conformal radiotherapy and the impact of prior major urologic surgical procedures in patients treated for prostate cancer. *Int. J. Radiat. Oncol. Biol. Phys.* 1995; 33:15–22.
77. Fisch BM, Pickett B, Weinberg V et al. Dose of radiation received by the bulb of the penis correlates with risk of impotence after three-dimensional conformal radiotherapy for prostate cancer. *Urology* 2001; 57:955–9.
78. Hilaris BS. Brachytherapy in cancer of the prostate: an historical perspective. *Semin. Surg. Oncol.* 1997; 13:399–405.
79. Zelefsky MJ, Whitmore WF, Jr. Long-term results of retropubic permanent 125-iodine implantation of the prostate for clinically localized prostatic cancer. *J. Urol.* 1997; 158:23–9; discussion 29–30.
80. Ragde H, Korb LJ, Elgamal AA et al. Nadir BS. Modern prostate brachytherapy. Prostate specific antigen results in 219 patients with up to 12 years of observed follow-up. *Cancer* 2000; 89:135–41.
81. Zelefsky MJ, Hollister T, Raben A et al. Five-year biochemical outcome and toxicity with transperineal CT-planned permanent I-125 prostate implantation for patients with localized prostate cancer. *Int. J. Radiat. Oncol. Biol. Phys.* 2000; 47:1261–6.
82. Storey MR, Landgren RC, Cottone JL et al. Transperineal 125-iodine implantation for treatment of clinically localized prostate cancer: 5-year tumor control and morbidity. *Int. J. Radiat. Oncol. Biol. Phys.* 1999; 43:565–70.
83. Kattan MW, Potters L, Blasko JC et al. Pretreatment nomogram for predicting freedom from recurrence after permanent prostate brachytherapy in prostate cancer. *Urology* 2001; 58:393–9.
84. Grado GL, Larson TR, Balch CS et al. Actuarial disease-free survival after prostate cancer brachytherapy using interactive techniques with biplane ultrasound and fluoroscopic guidance. *Int. J. Radiat. Oncol. Biol. Phys.* 1998; 42:289–98.
85. Ragde H, Elgamal AA, Snow PB et al. Ten-year disease free survival after transperineal sonography-guided iodine-125 brachytherapy with or without 45-gray external beam irradiation in the treatment of patients with clinically localized, low to high Gleason grade prostate carcinoma. *Cancer* 1998; 83:989–1001.
86. Beyer DC, Priestley JB Jr. Biochemical disease-free survival following ^{125}I prostate implantation. *Int. J. Radiat. Oncol. Biol. Phys.* 1997; 37:559–63.
87. Wei JT, Dunn RL, Sandler HM et al. Comprehensive comparison of health-related quality of life after contemporary therapies for localized prostate cancer. *J. Clin. Oncol.* 2002; 20:557–66.
88. Ragde H, Blasko JC, Grimm PD et al. Brachytherapy for clinically localized prostate cancer: results at 7- and 8-year follow-up. *Semin. Surg. Oncol.* 1997; 13:438–43.
89. Wallner K, Lee H, Wasserman S et al. Low risk of urinary incontinence following prostate brachytherapy in patients with a prior transurethral prostate resection. *Int. J. Radiat. Oncol. Biol. Phys.* 1997; 37:565–9.
90. Potters L, Torre T, Fearn PA et al. Potency after permanent prostate brachytherapy for localized prostate cancer. *Int. J. Radiat. Oncol. Biol. Phys.* 2001; 50:1235–42.
91. Zelefsky MJ, Wallner KE, Ling CC et al. Comparison of the 5-year outcome and morbidity of three-dimensional conformal radiotherapy versus transperineal permanent iodine-125 implantation for early-stage prostatic cancer. *J. Clin. Oncol.* 1999; 17:517–22.
92. Stock RG, Kao J, Stone NN. Penile erectile function after permanent radioactive seed implantation for treatment of prostate cancer. *J. Urol.* 2001; 165:436–9.
93. Amling CL, Lerner SE, Martin SK et al. Deoxyribonucleic acid ploidy and serum prostate specific antigen predict outcome following salvage prostatectomy for radiation refractory prostate cancer. *J. Urol.* 1999; 161:857–62; discussion 862–3.
94. Rogers E, Ohori M, Kassabian VS et al. Salvage radical prostatectomy: outcome measured by serum prostate specific antigen levels. *J. Urol.* 1995; 153:104–10.
95. Ahlering TE, Lieskovsky G, Skinner DG. Salvage surgery plus androgen deprivation for radioresistant prostatic adenocarcinoma. *J. Urol.* 1992; 147:900–2.
96. Wei DC, Eastham JA, Scardino PT. Modern salvage radical prostatectomy (SalvRP) after radiotherapy: safe and effective in selected patients. *J. Urol.* 2001; 165:351.
97. Leventis AK, Shariat SF, Kattan MW et al. Prediction of response to salvage radiation therapy in patients with prostate cancer recurrence after radical prostatectomy. *J. Clin. Oncol.* 2001; 19:1030–9.
98. Grado GL, Collins JM, Kriegshauser JS et al. Salvage brachytherapy for localized prostate cancer after radiotherapy failure. *Urology* 1999; 53:2–10.
99. Beyer DC. Permanent brachytherapy as salvage treatment for recurrent prostate cancer. *Urology* 1999; 54:880–3.
100. D'Amico AV. Analysis of the clinical utility of the use of salvage brachytherapy in patients who have a rising PSA after definitive external beam radiation therapy. *Urology* 1999; 54:201–3.
101. Begg CB, Riedel ER, Bach PB et al. Variations in morbidity after radical prostatectomy. *N. Engl. J. Med.* 2002; 346:1138–44.
102. Kattan MW, Cowen ME, Miles BJ. A decision analysis for treatment of clinically localized prostate cancer. *J. Gen. Intern. Med.* 1997; 12:299–305.
103. Kattan MW, Fearn PA, Miles BJ. Time trade-off utility modified to accommodate degenerative and life-threatening conditions. *Proc. AMIA Symp.* 2001; 304–308.
104. Cowen ME, Miles BJ, Cahill DF et al. The danger of applying group-level utilities in decision analyses of the treatment of localized prostate cancer in individual patients. *Med. Decision Making* 1998; 18:376–80.
105. Eastham JA, Scardino PT. Radical Prostatectomy. In: Walsh PC, Retik AB, Vaughan EDJ et al. (eds) *Campbell's Urology*. Philadelphia: W.B. Saunders, 2002.
106. Steiner MS, Morton RA, Walsh PC. Impact of anatomical radical prostatectomy on urinary continence. *J. Urol.* 1991; 145:512–14.
107. Catalona WJ, Carvalhal GF, Mager DE et al. Potency, continence and complication rates in 1,870 consecutive radical retropubic prostatectomies. *J. Urol.* 1999; 162:433–8.
108. Geary ES, Dendinger TE, Freiha FS et al. Nerve sparing radical prostatectomy: a different view [see comments]. *J. Urol.* 1995; 154:145–9.
109. Murphy GP, Mettlin C, Menck H et al. National patterns of prostate cancer treatment by radical prostatectomy: results of a survey by the American College of Surgeons Commission on Cancer. *J. Urol.* 1994; 152:1817–19.
110. Walsh PC, Marschke P, Ricker D et al. Patient-reported urinary continence and sexual function after anatomic radical prostatectomy. *Urology* 2000; 55:58–61.
111. Fowler FJ Jr, Barry MJ, Lu-Yao G Patient-reported complications and follow-up treatment after radical prostatectomy.

The National Medicare Experience: 1988–1990 (updated June 1993). *Urology* 1993; 42:622–9.

112. Stanford JL, Feng Z, Hamilton AS et al. Urinary and sexual function after radical prostatectomy for clinically localized prostate cancer: the Prostate Cancer Outcomes Study. *JAMA* 2000; 283:354–60.

113. Ohori M, Scardino PT. *Localized Prostate Cancer: A Monograph.* Current Problems in Surgery (in press).

114. Kattan MW, Miles BJ. The patient's choice for surgery for clinically localized prostate cancer. In: Vaughn ED Jr, Perlmutter AP (eds) *Atlas of Clinical Urology*. London: Churchill Livingstone (in press).

Expectant Management in 2002: Rationale and Indications

Chapter 25

Laurence Klotz

Division of Urology, Sunnybrook and Women's College Health Sciences Centre, Toronto, Ontario, Canada

INTRODUCTION

The optimal management of clinically localized prostate cancer remains unresolved. Management options are diverse, varying from a conservative approach (expectant management) to definitive treatment (radical prostatectomy or radiotherapy). Several studies have suggested that expectant management provides similar 10-year survival rates and quality-adjusted life years compared with radical prostatectomy or radiotherapy.[1–8] Expectant management alone, however, clearly deprives some patients with potentially curable life-threatening disease of the opportunity for curative therapy. Lu-Yao reported in a population-based study that particularly patients with a high Gleason score who had undergone radical prostatectomy or radiotherapy had improved 5-year overall and disease-specific survival compared with those managed by expectant management alone.[a]

The dilemma of management stems from the heterogeneity of the natural history of prostate cancer. Estimates from autopsy studies indicate that 30% of men over the age of 50 have prostate cancer. However, only 10% of men over 50 years old will have clinical progression of prostate cancer resulting in a diagnosis. Among those with clinically diagnosed prostate cancer, the likelihood of death from prostate cancer is one in three. While these statistics suggest a high incidence of 'latent' prostate cancer and a slow natural history of prostate cancer in many patients, they also indicate that the risk of dying from clinically diagnosed prostate cancer is substantial. This conundrum is the rationale for both conservative management and radical treatment.

The surveillance studies in the published literature are summarized in **Table 25.1**.[1–8, 10–16] A number of observations can be made from these studies. Mortality from other causes is common in all cohorts, likely reflecting the average age of patients at entry. Cause-specific survival varies substantially from 30% to 80% at 15 years. This reflects patient selection at study entry. All reflect natural history from the pre-prostate specific antigen (PSA) era. The stage migration phenomenon of the last decade had not occurred when these studies were carried out. Secondly, none offered salvage radical therapy for local progression. Watchful waiting in these series consisted of no active treatment until symptomatic metastases developed, at which point androgen ablation was offered. Additionally, these series are characterized by problems of selection bias to varying degrees. Confounding issues include the use of aspiration cytology for diagnosis, exclusion of higher risk patients, elderly cohorts and inclusion of T1a patients.

None the less, one striking similarity stands out: every series contains a substantial subset of long-term survivors, particularly in the group with favorable clinical parameters. This is a critical observation. In the absence of treatment, a substantial subset of patients with prostate cancer are not destined to die of the disease. The challenge, of course, is to identify that subset accurately.

RATIONALE FOR AN EXPECTANT APPROACH

Since the advent of PSA in 1989, substantial resources have been directed towards the early detection and treatment of prostate cancer. Mortality rates have fallen about 20% during that period. Whether this improvement in mortality is due to these efforts, or to other causes, is the subject of intense controversy. Other factors, including dietary and lifestyle modification, and a trend towards earlier initiation of androgen ablation for recurrent disease, may explain some or all of the fall in mortality. Indeed, Albertsen has demonstrated that the fall in mortality in Connecticut, where screening is uncommon, is equivalent to the reduction in Oregon, a highly screened population.[17] Thus it remains uncertain whether our efforts at early diagnosis and local treatment have resulted in a decline in prostate cancer mortality.

Table 25.1. Summary of surveillance studies in the published literature.

Reference	Stage	Year last patient accrued	*N*	%survival		
				5 years	10 years	15 years
Hanash (1972)[3]	A	1942	50	86	52	22
	B		129	19	4	1
Lerner (1991)[6]	T1b–T2	1982	279	88	61	
				95 CSS	80 CSS	
Adolfsson (1992)[8]	T1–2	1982	122	82	50	
				99 CSS	84 CSS	
Johansson (1997)[2]	T1–2	1984	223		41/86 CSS	21
						81 CSS
Albertsen (1998)[16]		1984				
Handley (1988)[7]	278	1985				
Waaler (1993)[10]	T2	1985	28	94 CSS		
Whitmore (1991)[11]	T2	1986	37	95	90	62
George (1988)[12]	Tx	1986	120	86	66	66
Aus (1995)[13]	T1–4	1991	301	80 CSS	50 CSS	30 CSS
Holmberg (2002)	T1–2	1999	348	87 (92 CSS)	–	–

The prevalence of prostate cancer far exceeds the incidence. In the PSA era, increasing efforts at screening, and the consequent rise in incidence, has resulted in about one of seven men being diagnosed. The mortality rate has fallen as well. Thus, the chance of dying of prostate cancer in patients who are diagnosed has decreased steadily, from about one in three to one in five. While this may reflect improved treatment, it may also reflect increased diagnosis of indolent disease.[18,19]

Prostate cancer is typically slow growing. Work by Sakr has indicated that the disease develops in the 30s in the typical patient, and takes 20 years to become clinically detectable.[20] Studies by Pound demonstrate that, in patients who fail radical prostatectomy and go on to die of prostate cancer, a median of 16 years elapses from surgery until death.[21] The watchful waiting studies also demonstrate that disease-related mortality in populations of prostate cancer patients only becomes substantial after 10 years. In addition, it is particularly clear that low-grade prostate cancer is associated with low progression rates and high survival rates in the intermediate term.

One indirect piece of evidence supporting the long window of curability can be derived from nomograms predicting the likelihood of biochemical recurrence from PSA, grade and stage. Using the Kattan nomograms of a patient with T1c prostate cancer and Gleason 6 prostate cancer, with PSA of 5, the 5-year biochemical disease free survival is 95%.[22] If one were to delay intervention until the PSA reaches 10, the 5-year DFS is still 90%; and with further delay until the PSA is 15, it is 85%. Thus, following such a patient during a period of PSA doubling or tripling is associated with a 5–10% reduction in the risk of progression.

Widespread use of PSA testing has also resulted in a profound stage migration. Most patients newly diagnosed with prostate cancer have clinically impalpable, stage T1c disease. Additionally, these patients typically have a PSA that is only mildly elevated (<10). These patients usually have slowly growing cancer with a long window of curability. This is also supported by the Albertson data (**Table 25.2**).[16]

A meta-analysis of six surveillance series comprising 828 patients reported by Chodak indicated that, at 10 years, disease-specific survival was 87% for well and moderately differentiated cancers, and metastasis-free survival was 81% and 58%, respectively.[19] These studies also incorporated an 'either-or' approach, and reflected a pre-stage migration population. Thus, many patients with favorably prognostic factors, diagnosed considerably earlier in their disease process than the average patient in this surveillance population, are likely to have an incredibly long natural history.

The PIVOT trial, comparing radical prostatectomy to watchful waiting, has been an ambitious effort to compare these two approaches in a randomized design.[18] This trial, against all expectations, is close to reaching its accrual target. The outcome will be of profound importance. However, one

Table 25.2. Prostate cancer mortality in a watchful waiting cohort according to grade. Patients with low-grade prostate cancer, by and large do not die of their disease. As grade increases, the risk of death also goes up, but for a Gleason score of 6, this remains at only 18–30%.

Gleason score	Prostate cancer mortality at 15 years
2–4	4–7%
5	6–11%
6	18–30%
7	42–70%
8–10	60–87%

limitation of the trial is that the observation arm does not have an option for intervention for the subset with evidence of rapid biochemical or local progression early on in the course of the disease.

The art of the management of localized prostate cancer is to differentiate patients with biologically aggressive disease, for whom curative therapy is strongly warranted, from those with indolent malignancy, for whom conservative management is equally efficacious. A blanket policy of observation for all results in undertreatment for some; similarly, a policy of treatment for all results in overtreatment for a subset.

Traditionally, watchful waiting has meant no treatment until progression to metastatic or locally advanced disease, followed by androgen ablation therapy. Today, in the PSA era, patients who are managed conservatively are typically still followed with periodic PSA tests. This raises the question: 'Can treatment of favorable prostate cancer be deferred indefinitely in many, while effective, albeit delayed therapy is offered to those who progress rapidly?'.

RESULTS OF A WATCHFUL WAITING WITH SELECTIVE INTERVENTION APPROACH

We have been conducting a clinical study to evaluate a novel approach in which the choice between a definitive therapy and conservative policy is determined by the rate of PSA increase or the development of early, rapid clinical and/or histologic progression. This strategy, which has never been previously described or evaluated, offers the powerful attraction of individualizing therapy according to the biological behavior of the cancer. This would mean that patients with slowly growing malignancy would be spared the side effects of radical treatment, while those with more rapidly progressive cancer would still benefit from curative therapy.

This prospective study consisted of 250 patients followed with watchful waiting with selective delayed intervention. Patients had PSA of <15, Gleason $\leqslant 7$ and T $\leqslant$2b. Patients were followed with watchful waiting until they met specific criteria defining rapid or clinically significant progression. These criteria were as follows.

1. PSA progression, defined by all of the following three conditions:
   (a) PSA doubling time <2 years, based on at least three separate measurements over a minimum of 6 months;
   (b) final PSA >8 ng/ml;
   (c) P value <0.05 from a regression analysis of ln (PSA) on time.
2. Clinical progression when one of the following conditions was met:
   (a) more than twice increase in the product of the maximum perpendicular diameters of the primary lesion as measured digitally;
   (b) local progression of prostate cancer requiring transurethral resection of the prostate (TURP);
   (c) development of ureteric obstruction;
   (d) radiological and/or clinical evidence of distant metastasis.
3. Histologic progression: Gleason score $\geqslant 8$ in the rebiopsy of prostate at 12–18 months.

Most of the patients in this series fulfilled the criteria for favorable disease (PSA <10, Gleason score $\leqslant 6$, T $\leqslant$2a). Eighty per cent of patients had a Gleason score of 6 or less, and the same proportion had a PSA <10. With a median follow-up of 42 months, 60 patients (30%) came off watchful observation while 140 have remained on surveillance. Of the patients coming off surveillance, 8% came off because of rapid biochemical progression, 8% for clinical progression and 8% due to patient preference. The remaining 6% came off surveillance for a variety of other reasons.

The distribution of PSA doubling times (PSA dt) is seen in **Fig 25.1**. The median PSA dt was 5.13 years. One third had a PSA dt >10 years, one half had a PSA dt >5 years, one third had a PSA dt between 2 and 5 years, and only one seventh had a PSA dt <2 years.

Patients were rebiopsied 1.5–2 years after being placed on the surveillance protocol. Grades remained stable in 92%; only 8% demonstrated significant (>2 Gleason score) rise. This is also consistent with the recent publication by Epstein and Walsh, demonstrating a 4% rate of grade progression over 2–3 years.

Nine patients (of 200) had a radical prostatectomy after they manifested a PSA doubling time of less than 2 years. All had Gleason scores of 5–6, PSA <10, pT1–2 at study entry. Final pathology was as follows: three out of nine were pT2, five were pT3a–c and one was N1. For a group of patients with favorable clinical characteristics, this is a high rate of locally advanced disease. This supports the view that a short PSA dt is associated with a more aggressive phenotype. A PSA dt <2 years, in patients with otherwise favorable clinical features, portents a high likelihood of advanced disease. Fortunately, this scenario is uncommon. This also suggests that,

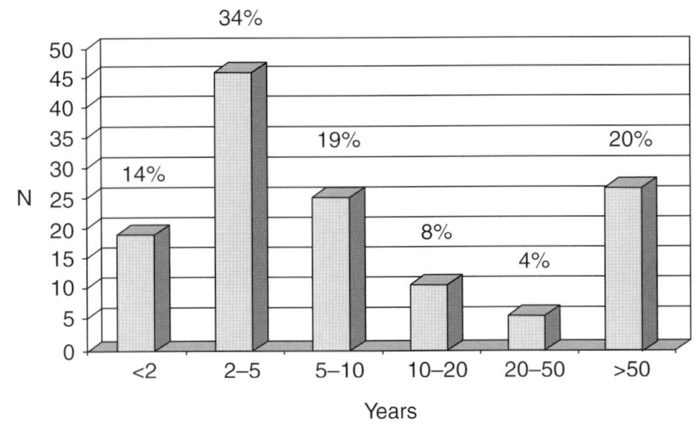

Fig. 25.1. Doubling times of PSA in patients on a watchful waiting protocol. The data are based on 134 patients with a median follow-up of 42 months (range 18–78 months). Median number of PSA measurements = 7 (range 3–15); median PSA doubling time was 5.13 years.

insofar as cure of the patients with early rapid biochemical progression is a goal, the optimal PSA dt threshold for intervention should be greater than 2 years. The optimal threshold is likely in the range of 3 years. In our series, that constituted 22% of patients.

Zeitman and Schellhammer recently published a retrospective review of 199 men with T1–2 prostate cancer and PSA <20 ng/ml, managed with watchful waiting.[23] Median follow-up was 3.4 years. Overall survival at 5 and 7 years was 77% and 63%, and disease-specific survival was 99% and 99%. At 5 and 7 years, the proportion of patients who were alive and untreated was 43% and 26%. A total of 63 patients were treated radically. The median PSA rise from diagnosis to treatment was 2.9 ng/ml in the treated cohort, compared to 0.9 in the untreated group.

This study raises the concern that watchful waiting may simply be a version of delayed therapy, unless patients die of comorbid illness in the interim. However, the indication for intervention in this series was a mild rise in PSA (<3 ng) over a prolonged period. This emphasizes that conservative management in the modern PSA era requires 'buy-in' by the patient and the doctor. This involves an understanding that PSA will likely progress slowly over time, but that slow progression is not a reason for intervention.

One assumption on which the selective intervention approach is based is that the PSA rate of rise remains relatively stable over time. This is not the case in some patients; rapid rises in PSA after long periods of stability (PSA acceleration) have been clearly documented. The critical unanswered questions with respect to PSA acceleration are: When does it occur in the natural history of prostate cancer; in particular, relative to the point at which the disease becomes metastatic or locally incurable? How common is a sudden rapid increase? Are patients who manifest PSA acceleration still curable?

Who are the best candidates for watchful waiting? It is helpful to segregate patients into three groups, depending on the expectations and rationale for a watchful waiting approach. The first and least controversial group consists of patients for whom no intervention is almost certainly the optimal approach due to diminished life expectancy. This includes patients with early, low-grade, small-volume prostate cancer whose life expectancy is less than 10 years (i.e. >age 74, or significant comorbidity). For most of these patients, the risk of prostate cancer death and morbidity is low. The benefits of treatment are very likely to be outweighed by the risks.

The second group consists of patients for whom watchful waiting offers the benefit of improved quality of life, with the understanding that a minor trade-off with respect to quantity of life may be involved. The approach of selective delayed intervention should they demonstrate rapid biochemical progression is appealing in this group. This includes younger, healthier patients with favorable-risk prostate cancer, and patients whose life expectancy is in the 5–10-year range with intermediate-risk prostate cancer (i.e. Gleason score of 7, PSA >10). In these patients, expectant management involves a trade-off but may be an attractive option.

The third group consists of intermediate-risk patients who are anxious to maintain quality of life with respect to erectile function, and willing to sacrifice some quantity of life for this; or high-risk patients who, recognizing that local therapy is unlikely to be curative, elect a conservative approach. For these patients as well, an approach of initial expectant management with the option of delayed definitive therapy in the case of rapid biochemical or local progression may be appealing. Clearly, in these patients, the likelihood of progression is high, but quality of life will be maintained until progression occurs.

CONCLUSIONS

The approach of watchful waiting with selective intervention for patients with rapid biochemical or clinical progression is feasible. Most patients, who understand the basis for the approach, will remain on observation long term. Doubling time varies widely, and was not predicted by grade, stage or baseline PSA. A total of 33% have a PSA doubling time (T_d) >10 years. Doubling time appears to be a useful tool to guide treatment intervention for patients managed initially with expectant management. A doubling time of less than 2 years appears to identify patients at high risk for local progression in spite of otherwise favorable prognostic factors. The appropriate threshold for initiation of definitive therapy is a doubling time of around 3 years; approximately 20% of patients will fall into this category. The remainder have a high chance of remaining free of recurrence and progression for years.

Watchful waiting is clearly appropriate for patients who are elderly, have significant comorbidity and have favorable clinical parameters. The use of comorbidity indices like the Index of Co-Existent Disease (ICED) facilitate the identification of patients whose life expectancy is diminished relative to the natural history of their prostate cancer. The likelihood of a prostate cancer death in these patients is low. Many patients, however, fall into a gray zone where the benefits of treatment are unclear. In these patients, a policy of close monitoring with selective intervention for the 15% who progress rapidly is appealing. This approach is currently the focus of several clinical trials.

REFERENCES

1. Johansson JE, Adami HO, Andersson SO et al. High 10 year survival rate in patients with early, untreated prostatic cancer. *JAMA* 1992; 267:2191–6.
2. Johansson JE, Holmberg L, Johansson S et al. Fifteen-year survival in prostate cancer. A prospective, population-based study in Sweden. *JAMA* 1997; 277:467–71.
3. Hanash KA, Utz DC, Cook EN et al. Carcinoma of the prostate: a 15 year followup. *J. Urol.* 1972; 107:450–3.
4. Cook GB, Watson FR. Twenty nodules of prostate cancer not treated by total prostatecotmy. *J. Urol.* 1968; 100:672–4.
5. Barnes R, Hirst A, Rosenquist R. Early carcinoma of the prostate: comparison of stages A and B. *J. Urol.* 1976; 115:404–5.
6. Lerner SP, Seale-Hawkins C, Carlton CE et al. The risk of dying of prostate cancer in patients with clinically localized disease. *J. Urol.* 1991; 146:1040–5.

7. Handley R, Carr TW, Travis D et al. Deferred treatment for prostate cancer. *Br. J. Urol.* 1988; 62:249–53.
8. Adolfsson J, Carstensen J, Lowhagen T. Deferred treatment in clinically localized prostate carcinoma. *Br. J. Urol.* 1992; 69:183–7.
9. Lu-Yao GL, Yao SL. Population-based study of long-term survival in patients with clinically localised prostate cancer. *Lancet.* 1997; 349:906–10.
10. Waaler G, Stenwig AE. Prognosis of localised prostatic cancer managed by 'Watch and Wait' policy. *Br. J. Urol.* 1993; 72:214–9.
11. Whitmore WF, Warner IA, Thompson IM. Expectant management of localized prostatic cancer. *Cancer* 1991; 67:1091–6.
12. George NJR. Natural history of localised prostatic cancer managed by conservative therapy alone. *Lancet* 1988; 1:494–6.
13. Aus G, Hugosson I, Norlen L. Long-term survival and mortality in prostate cancer treated with noncurative intent. *J. Urol.* 1995; 154:460–5.
14. Sandblom G, Dufmats M, Varenhorst E. Long-term survival in a Swedish population-based cohort of men with prostate cancer. *Urology* 2000; 56:442–7.
15. Madsen PO, Graversen PH, Gasser TC Treatment of localized prostatic cancer. Radical prostatectomy versus placebo. A 15-year follow-up. *Scand. J. Urol. Nephrol.* Suppl. 1988; 110:95–100.
16. Albertsen PC, Hanley IA, Gleason DF et al. Competing risk analysis of men aged 55 to 74 years at diagnosis managed conservatively for clinically localized prostate cancer. *JAMA* 1998; 280:975–80.
17. Albertsen PC. Prostate cancer mortality trends in Oregon and Connecticut. Presented at the *Whistler International Conference on Prostate Cancer*, March 2001.
18. Wilt TJ, Brawer MK. The Prostate Cancer Intervention Versus Observation Trial (PIVOT). *Oncology* 1997; 1133–43.
19. Chodak GW. The management of localized prostate cancer. *J. Urol.* 1994; 152:1766.
20. Sakr WA, Haas GP, Cassin BF et al. The frequency of carcinoma and intraepithelial neoplasia of the prostate in young male patients. *J. Urol.* 1993; 150:379–85.
21. Pound CR, Partin AW, Eisenberger MA et al. Natural history of progression after PSA elevation following radical prostatectomy. *JAMA* 1999; 281:1591–7.
22. Kattan MW, Zelefsky MJ, Kupelian PA et al. Pretreatment nomogram for predicting the outcome of three-dimensional conformal radiotherapy in prostate cancer. *J. Clin. Oncol.* 2000; 18:3352–9.
23. Zietman AL, Thakral H, Wilson L et al. Conservative management of prostate cancer in the prostate specific antigen era: the incidence and time course of subsequent therapy. *J. Urol.* 2001; 166:1702–6.
24. Holmberg L, Bill-Axelson A, Helgesen F et al. A randomized trial comparing radical prostatectomy with watchful waiting in early prostate cancer. *NEJM* 2002; 347:781–9.

Surgery

Is Surgery still Necessary for Prostate Cancer?

Chapter 26

Ehab A. El-Gabry and Leonard G. Gomella
Department of Urology, Kimmel Cancer Center, Thomas Jefferson University, Philadelphia, PA, USA

INTRODUCTION

Before attempting to answer whether or not surgery is necessary for prostate cancer, let us first review the facts. In 2002, prostate cancer will have killed approximately 30 000 men; it is the most common cancer diagnosed in American men and the second cause of cancer-specific death. A new diagnosis of prostate cancer is made every 3 minutes and a man dies of the disease every 15 minutes.[1] Owing to an aging population, this death rate may increase in the future. These facts provide compelling evidence that prostate cancer is a potentially lethal disease.

To reduce the number of deaths from prostate cancer, an effective treatment must be employed for localized disease. A current list of potentially effective treatment modalities for localized prostate cancer is shown in **Table 26.1**. As for any organ-confined malignancy, if curative treatment is intended, elimination of the entire organ is mandatory. Leaving the original cell of origin, with its inherent genetic alterations, is likely to result in an increase in disease recurrence over the long term. Accordingly, of all the available treatment options for prostate cancer, radical prostatectomy has the greatest likelihood of curing organ-confined disease over the long term.[2] This line of discussion should not imply that all men with localized prostate cancer should undergo radical prostatectomy.

Multiple factors should be considered before planning a treatment strategy for prostate cancer. These factors are presented in **Table 26.2**. In general, owing to the slow natural history of the disease, patients with a life expectancy of less than 10–15 years might not benefit from the long-term advantage of surgery. Surgery should be reserved for patients with a high likelihood of organ-confined disease. Thus, prediction of the pathological stage using nomograms that consider prognostic factors, such as clinical stage, Gleason grade and serum prostate specific antigen (PSA) provide important information for proper treatment choice.[3] However, playing the percentages can be sometimes problematic. Although it goes without saying that, in a patient with extensive lymph node metastasis, surgery would be unwise, in a patient with 40% chance of organ-confined disease versus a 60% chance of capsular penetration, the decision to deny surgery is difficult.

Table 26.1. Current treatment modalities for clinically localized prostate cancer.

Watchful waiting
Monotherapies
Retropubic radical prostatectomy
Perineal radical prostatectomy
Laparoscopic prostatectomy
External beam radiation
Brachytherapy
Cryotherapies
Multimodality therapies
Neoadjuvant and advuvant hormonal therapy (surgical and radiation)
Brachytherapy combined with external beam radiation

Table 26.2. Factors influencing treatment strategies.

Patient-related factors
Age
Comorbidity
Patient preference/physician preference
Disease-related factors
PSA
Gleason score
Clinical staging

The morbidity and the expected life quality associated with each form of treatment is another important factor to be considered. In most cases, the relative importance of this factor on influencing treatment decisions is decided by every patient in a subjective fashion. Finally, the salvage options of treatment failure should be considered.

In this chapter, we will compare radical prostatectomy (RP) to current treatment modalities in terms of efficacy, complications, quality of life and salvage of treatment failure. In addition, retropubic versus perineal versus laparoscopic prostatectomy approaches would be discussed.

RADICAL PROSTATECTOMY VERSUS SURVEILLANCE

Various studies have questioned the appropriateness of aggressive treatment for localized disease and have suggested surveillance instead.[4–9] Advocates for surveillance argue that early radical treatment for PC has not been proven to alter the natural history of the disease and thus treatment-associated morbidities, especially in younger patients, could not be justified.[10] An often quoted Swedish study that suggests an advantage to surveillance has reported a 15-year cancer-specific survival of up to 81%.[11] However, careful interpretation of the study reveals the following limitations: a preponderance of older patients (mean age of 70–72 years), and the majority of the reported cases had favorable histologies in the form of moderate and well-differentiated disease. Accordingly, this study cannot be applied to younger, healthier populations or patients with unfavorable histology. In contrast, another Swedish study has reported 63% mortality rates due to prostate cancer among patients who were managed expectantly with a follow-up exceeding 10 years.[12]

In a study by Albersten, a significant survival advantage for surgery over surveillance was noted in treating poorly differentiated disease (Gleason score 8–10).[13] Moreover, a watchful waiting strategy is not free of complications. Local progression from prostate cancer can result in ureteral obstruction or bladder outlet obstruction. In a review by Aus and associates, 17% of patients who were managed expectantly underwent procedures to relieve upper tract obstruction and 41 had at least one channel transurethral resection of the prostate (TURP).[12] Metastatic prostate cancer can cause severe bone pain, pathologic fracture, spinal cord compression and death.

Presently, the Scandinavian Prostate Cancer Group and the US Prostate Cancer Intervention Versus Observation Trial (PIVOT) are two large prospective randomized controlled studies comparing radical prostatectomy and watchful waiting as treatment modalities for localized prostate cancer.[14] Until the results of these studies are available, prostate cancer remains the second leading cause of cancer death in men with a 50% suggested death rate in men who are diagnosed with a moderately well-differentiated disease and a life expectancy of 15 years or more.[15] We believe that younger males would be best served by surgery, since they are at risk of their disease for many years. Watchful waiting should only be reserved for older patients with low Gleason scores.

RADICAL PROSTATECTOMY VERSUS EXTERNAL BEAM RADIATION

Although early reports comparing the two modalities demonstrated a clear advantage for RP over external beam radiation therapy (EBRT),[16,17] the studies were shown to have several flaws. More recent retrospective studies have evaluated the treatment outcomes of EBRT and RP in comparably staged prostate cancer patients. When stratified by preoperative biopsy grade, T stage and serum PSA, the rates of biochemical relapse-free survival and cause-specific survival for the two modalities were similar at 5 and 7 years post-treatment.[18–23] However, since EBRT is a treatment that does not remove the cancer-harboring organ (the prostate), it is more likely to fail at >10 years than a modality that eliminates the source of the cancer (surgery). Accordingly, 7 years treatment outcomes may not be long enough to demonstrate any difference. Limited data have been presented that the addition of hormonal therapy may impact on the outcomes of high-risk prostate cancer treated by radiation. At present, there is no multimodality comparison available between radical prostatectomy and this group.

Although the absence of a randomized trial that directly compares the outcomes in patients with localized prostate cancer treated by RP versus EBRT, several observations can be made from available studies. Data suggest an advantage for surgery in treating clinically confined disease. When outcomes of surgery versus radiation were compared in high-risk cases (PSA >10, Gleason ≥7) surgery was significantly proven to be better when negative margins were achieved, where in low-risk cases (PSA ≤10, Gleason score ≤6), positive margins after surgery were an adverse prognostic factor.[18] In this era, where PSA-based early detection has led to the emergence of a new patient population with an increased proportion of organ-confined disease, surgery should be the best potentially curative option.

In addition, several studies suggested that a substantial percentage of patients with clinically localized prostate cancer would still harbor persistent cancer within the prostate after radiation therapy. Freiha et al. reported a 27% (39 out of 146 patients) positive biopsy rate 18 months or more following completion of radiation therapy, of which 19% subsequently developed metastases.[24] Scardino et al. reported a post-irradiation positive biopsy rate of 22% with stage B1, with local recurrence developing in 58% of the patients with a positive biopsy by 5 years and in 82% by 10 years.[25] The abovementioned studies further validate the argument for the importance of total surgical removal of the prostate gland. The challenge is whether this can be done with reasonable morbidity.

Radiation therapy has been promoted partly because of a perception that it does not affect potency. In a recent study, Mantz et al. noted an overall EBRT patient potency preservation of 71.3% versus 66.2% for patients who underwent nerve-sparing radical prostatectomy (NSRP). However, in their study EBRT was shown to offer higher potency preservation rates than NSRP for patients above 70 years of age.[26] In another study, sexuality was moderately or severely impaired in 71% of the RP group and 50% of the EBRT group.[27] The above data clearly demonstrate that impotence is a common sequela of both EBRT and RP for localized prostate cancer. Consequently, treatment decisions should not be based on potency consideration.

Although various studies have shown that the quality of life (QOL) after RP and EBRT is comparable, several surveys have demonstrated that radical prostatectomy can be performed with a high degree of patient satisfaction.[28,29] This can be partly attributed to the patient uncertainty about cure with EBRT. While the definition of disease-free survival (DFS) after RP is straightforward (an undetectable serum PSA level), the DFS after EBRT is often more difficult to assess.[19] This was translated into a statistical advantage for RP in terms of patients' worries about cure.[30]

Although salvage radiation therapy can be safely administrated after surgery failure, salvage RP after radiation failure is associated with dramatic side effects. Potency is virtually impossible to regain, total urinary incontinence rates are reported as high as 60%, with many patients requiring artificial sphincters, severe rectal injuries are seen in up to 15%, some of which required colostomy.[31–33]

In summary, we believe that in younger healthy males with >10 years life expectancy, and high likehood of organ-confined disease, surgery should be recommended.

RADICAL PROSTATECTOMY VERSUS BRACHYTHERAPY

In the recent years, advances in brachytherapy technology has resulted in a renewed interest in this treatment option. The current indications for brachytherapy as a monotherapy for PC as recommended by Ragde et al.[34] are presented in **Table 26.3**. Proponents of seed implantation argue that the 5- and 10-year follow-up data in low-risk patients demonstrate PSA levels that were comparable to reported RP series, and better than some ERBT published data.[34] Since brachytherapy is indicated for a highly favorable population, an argument can be made that this select group of patients have influenced these results. When the Johns Hopkins group applied the stringent criteria of Ragde's series to their RP series, a biochemical no evidence of disease (NED) rate of almost 98% at 7 years was achieved versus only 79% for brachytherapy.[35] In addition, progression curves showed a plateau with long-term follow-up for men with Gleason 5–6 tumors who underwent surgery, whereas a progressive downward trend was observed in the data of Ragde et al. Moreover, when used to treat tumors of Gleason scores of 7 or higher, brachytherapy was proven to be inadequate.[36] Since needle biopsy specimens have a 35% potential chance of being undergraded,[37] a question may be raised about the effectiveness of brachytherapy in such a setting.

Table 26.3. Brachytherapy optimum inclusion criteria.

Criteria
Clinical stage (T1a–T2a)
Gleason sum (2–6)
PSA (≤10 ng/ml)
Prostate volume <60 cm^3
Urine flow rate >15 cm^3/second
AUA symptom score <10–12

Again, there is a widespread perception that there is no impairment in sexual function with brachytherapy. After reviewing 66 references published over the last 10 years, Peneau observed an impotence rate of 25% associated with brachytherapy and progressive decrease in sexual potency with time.[38] In a study comparing general and disease-specific health-related quality of life in men undergoing brachytherapy to those undergoing radical prostatectomy and age-matched healthy controls, sexual function and bother were equivalent in RP and brachytherapy groups, and both were worse than in healthy controls.[39] The same study has shown that general health-related quality of life did not differ greatly among the three groups. Urinary leakage was better in the brachytherapy group than in the prostatectomy group; however, both were worse than controls. The brachytherapy group had significantly worse AUA symptom index scores and worse bowel function than RP group. Patients who underwent combined brachytherapy and ERBT performed worse in all general and disease-specific health-related quality of life domains compared to those who had brachytherapy as a monotherapy. In contrast to common beliefs, brachytherapy may not offer a better quality of life over other treatments.

Lastly, although early literature suggested a benefit for brachytherapy over RP in terms of cost effectiveness,[40] more recent studies demonstrated that the costs of brachytherapy are substantially higher than RP.[41,42] Although treatment expense should not be the primary consideration in management strategies, its role may become more important in choosing the proper treatment modality in the face of doubt. Although we believe that brachytherapy is an acceptable approach, in the young and healthy patient, surgery offers a more reliable treatment option.

RADICAL PROSTATECOMY VERSUS CRYOSURGERY

In the early 1990s, prostate cryosurgery experienced a revival as a result of technical advancement in transrectal ultrasound and improved knowledge of cryobiology. However, years of experience have shown cryosurgery to be significantly inferior to radical prostatectomy and radiation therapy. Shinohara et al. reported PSA failure of up to 48% of patients with PSA nadirs between 0.1 and 0.4 ng/ml at a follow-up of <2 years.[43] Cohen et al. reported a 60% biochemical NED rate 21 months or more after treatment.[44] These results are roughly 30% less than what would be expected with RP. In addition, when compared to surgery, the complication rate of cryosurgery appears unacceptably high. Impotence is almost a universal complication of the procedure. Approximately 10% of cryosurgery-treated patients experience urethral sloughing. Incontinence rates up to 15% have been reported. Even worse, the complication of rectourethral fistula may occur in up to 3% of cases.[43,44]

With a success rate of almost 30% less than RP and a higher complication rate, cryosurgery should not be considered as a viable option for the vast majority of patients with newly diagnosed prostate cancer. It is obvious that at this point of time, RP offers patients with localized disease a significantly superior and more durable treatment outcome with less morbidity.

RADICAL PROSTATECTOMY APPROACHES

Currently, there are three different approaches that can be used to perform radical proctectomy: retropubic, perineal and laparoscopic. Approximately 90% of procedures are performed by the retropubic route, with 8–9% done perineally and only 1–2% laparoscopically. A frequently asked question is which approach is better? The literature is most complete on the retropubic and perineal approaches. Limited data are available on the laparoscopic approach, since this has only been utilized at highly selected centers since 1998.

Studies comparing the retropubic versus perineal approach have demonstrated comparable rates of cure, positive margins, PSA failures and tumor progression.[45–48] The retropubic approach has been reported to have a higher incidence for positive apical margins, while the perineal approach has a higher incidence of positive anterior margins. However, the impact of these findings on the clinical outcome remains unclear. Various studies have reported the advantages and disadvantages of both routes.[49,50] The reported advantages of the retropubic approach are: retropubic anatomy is more familiar to urologists, preservation of the neurovascular bundle is performed more easily, rectal injuries are fewer and pelvic lymphadenectomy can be performed in the same setting. The reported advantages of the perineal approach are: less postoperative ileus and length of hospital stay, and decreased likelihood of receiving perioperative blood products. In summary, both approaches are fairly comparable in advantages and disadvantages; the choice of the approach is dependent upon the surgeon's familiarity with the procedure, and the patient's own preference.

The preliminary results of laparoscopic prostatectomy are comparable to the standard approaches discussed above.[51] However, until the long follow-up data are available and the learning curve improves, the widespread use of this procedure should still be approached cautiously.

CONCLUSION

Radical prostatectomy remains a mainstay of prostate cancer treatment. In organ-confined disease it offers the only potentially curative option. Currently available data suggest that, in the young and healthy patient, the procedure is associated with negligible mortality and acceptable morbidity, while offering the highest disease-free survival rates.

REFERENCES

1. Walsh PC. Prostate cancer kills: strategy to reduce deaths. *Urology* 1994; 44:463–6.
2. El-Gabry EA, Strup SE, Gomella LG. Deciding on radical prostatectomy: the physician's perspective. *Semin. Urol. Oncol.* 2000; 18:205–13.
3. Partin AW, Yoo J, Carter HB et al. The use of prostate specific antigen, clinical stage and Gleason score to predict pathological stage in men with localized prostate cancer [see comments]. *J. Urol.* 1993; 150:110–14.
4. Adolfsson J, Carstensen J, Lowhagen T. Deferred treatment in clinically localised prostatic carcinoma. *Br. J. Urol.* 1992; 69:183–7.
5. Moskovitz B, Nitecki S, Richter Levin D. Cancer of the prostate: is there a need for aggressive treatment? *Urol. Int.* 1987; 42:49–52.
6. George NJ. Natural history of localised prostatic cancer managed by conservative therapy alone. *Lancet* 1988; 1:494–7.
7. Handley R, Carr TW, Travis D et al. Deferred treatment for prostate cancer. *Br. J. Urol.* 1988; 62:249–53.
8. Goodman CM, Busuttil A, Chisholm GD. Age, and size and grade of tumour predict prognosis in incidentally diagnosed carcinoma of the prostate. *Br. J. Urol.* 1988; 62:576–80.
9. Whitmore WF Jr, Warner JA, Thompson IM Jr. Expectant management of localized prostatic cancer. *Cancer* 1991; 67:1091–6.
10. Drachenberg DE. Treatment of prostate cancer: watchful waiting, radical prostatectomy, and cryoablation. *Semin. Surg. Oncol.* 2000; 18:37–44.
11. Johansson JE, Holmberg L, Johansson S et al. Fifteen-year survival in prostate cancer. A prospective, population-based study in Sweden [see comments] [published erratum appears in *JAMA* 1997; 278:206]. *JAMA* 1997; 277:467–71.
12. Aus G, Hugosson J, Norlen L. Long-term survival and mortality in prostate cancer treated with noncurative intent [see comments]. *J. Urol.* 1995; 154:460–5.
13. Albertsen PC, Fryback DG, Storer BE et al. Long-term survival among men with conservatively treated localized prostate cancer [see comments]. *JAMA* 1995; 274:626–31.
14. Wilt TJ, Brawer MK. The Prostate Cancer Intervention Versus Observation Trial (PIVOT). *Oncology (Hunting.)* 1997; 11:1133–9; discussion 1139–40, 1143.
15. Walsh PC. The natural history of localized prostate cancer: a guide to therapy. In: Walsh PC (ed.) *Campbell's Urology*, pp. 2539–46. Philadelphia: W.B. Saunders, 1997.
16. Paulson DF, Lin GH, Hinshaw W et al. Radical surgery versus radiotherapy for adenocarcinoma of the prostate. *J. Urol.* 1982; 128:502–4.
17. Stamey TA, Ferrari MK, Schmid HP. The value of serial prostate specific antigen determinations 5 years after radiotherapy: steeply increasing values characterize 80% of patients [see comments]. *J. Urol.* 1993; 150:1856–9.
18. Kupelian P, Katcher J, Levin H et al. External beam radiotherapy versus radical prostatectomy for clinical stage T1-2 prostate cancer: therapeutic implications of stratification by pretreatment PSA levels and biopsy Gleason scores [see comments]. *Cancer J. Sci. Am.* 1997; 3:78–87.
19. Zietman AL, Coen JJ, Shipley WU et al. Radical radiation therapy in the management of prostatic adenocarcinoma: the initial prostate specific antigen value as a predictor of treatment outcome. *J. Urol.* 1994; 151:640–5.

20. Zagars GK, Pollack A. Radiation therapy for T1 and T2 prostate cancer: prostate-specific antigen and disease outcome. *Urology* 1995; 45:476–83.
21. Zagars GK. Prostate specific antigen as an outcome variable for T1 and T2 prostate cancer treated by radiation therapy [see comments]. *J. Urol.* 1994; 152:1786–91.
22. Zagars GK. Serum PSA as a tumor marker for patients undergoing definitive radiation therapy. *Urol. Clin. North Am.* 1993; 20:737–47.
23. Martinez AA, Gonzalez JA, Chung AK et al. A comparison of external beam radiation therapy versus radical prostatectomy for patients with low risk prostate carcinoma diagnosed, staged, and treated at a single institution. *Cancer* 2000; 88:425–32.
24. Freiha FS, Bagshaw MA. Carcinoma of the prostate: results of post-irradiation biopsy. *Prostate* 1984; 5:19–25.
25. Scardino PT, Frankel JM, Wheeler TM et al. The prognostic significance of post-irradiation biopsy results in patients with prostatic cancer. *J. Urol.* 1986; 135:510–16.
26. Mantz CA, Nautiyal J, Awan A et al. Potency preservation following conformal radiotherapy for localized prostate cancer: impact of neoadjuvant androgen blockade, treatment technique, and patient-related factors [see comments]. *Cancer J. Sci. Am.* 1999; 5:230–6.
27. Lilleby W, Fossa SD, Waehre HR et al. Long-term morbidity and quality of life in patients with localized prostate cancer undergoing definitive radiotherapy or radical prostatectomy. *Int. J. Radiat. Oncol. Biol. Phys.* 1999; 43:735–43.
28. Klein EA, Grass JA, Calabrese DA et al. Maintaining quality of care and patient satisfaction with radical prostatectomy in the era of cost containment. *Urology* 1996; 48:269–76.
29. Fowler FJ Jr, Barry MJ, Lu-Yao G et al. Patient-reported complications and follow-up treatment after radical prostatectomy. The National Medicare Experience: 1988–1990 (updated June 1993). *Urology* 1993; 42:622–9.
30. Fowler FJ Jr, Barry MJ, Lu-Yao G et al. Outcomes of external-beam radiation therapy for prostate cancer: a study of Medicare beneficiaries in three surveillance, epidemiology, and end results areas. *J. Clin. Oncol.* 1996; 14:2258–65.
31. Rogers E, Ohori M, Kassabian VS et al. Salvage radical prostatectomy: outcome measured by serum prostate specific antigen levels [see comments]. *J. Urol.* 1995; 153:104–10.
32. Lerner SE, Blute ML, Zincke H. Critical evaluation of salvage surgery for radio-recurrent/resistant prostate cancer. *J. Urol.* 1995; 154:1103–9.
33. Pontes JE, Montie J, Klein E et al. Salvage surgery for radiation failure in prostate cancer. *Cancer* 1993; 71:976–80.
34. Ragde H, Korb L. Brachytherapy for clinically localized prostate cancer. *Semin. Surg .Oncol.* 2000; 18:45–51.
35. Polascik TJ, Pound CR, Deweese TL et al. Comparision of radical prostatectomy and iodine 125 interstitial radiotherapy for the treatment of clinically localized prostate cancer: a 7-year biochemical (PSA) progression analysis. *Urology* 1998; 51:884–90.
36. D'Amico AV, Whittington R, Malkowicz SB et al. Biochemical outcome after radical prostatectomy, external beam radiation therapy, or interstitial radiation therapy for clinically localized prostate cancer [see comments]. *JAMA* 1998; 280:969–74.
37. Steinberg DM, Sauvageot J, Piantadosi S et al. Correlation of prostate needle biopsy and radical prostatectomy Gleason grade in academic and community settings. *Am. J. Surg. Pathol.* 1997; 21:566–76.
38. Peneau M. [Interstitial radiotherapy and prostate cancer. Analysis of the literature. Subcommittee of Prostate Cancer of C.C.A.F.U.]. *Prog. Urol.* 1999; 9:440–51.
39. Brandeis JM, Litwin MS, Burnison CM et al. Quality of life outcomes after brachytherapy for early stage prostate cancer. *J. Urol.* 2000; 163:851–7.
40. Blasko JC, Ragde H, Luse RW et al. Should brachytherapy be considered a therapeutic option in localized prostate cancer? *Urol. Clin. North Am.* 1996; 23:633–50.
41. Kohan AD, Armenakas NA, Fracchia JA. The perioperative charge equivalence of interstitial brachytherapy and radical prostatectomy with 1-year followup. *J. Urol.* 2000; 163:511–14.
42. Ciezki JP, Klein EA, Angermeier KW et al. Cost comparison of radical prostatectomy and transperineal brachytherapy for localized prostate cancer. *Urology* 2000; 55:68–72.
43. Shinohara K, Rhee B, Presti JC Jr et al. Cryosurgical ablation of prostate cancer: patterns of cancer recurrence. *J. Urol.* 1997; 158:2206–9; discussion 2209–10.
44. Cohen JK, Miller RJ, Rooker GM et al. Cryosurgical ablation of the prostate: two-year prostate-specific antigen and biopsy results. *Urology* 1996; 47:395–401.
45. Frazier HA, Robertson JE, Paulson DF. Radical prostatectomy: the pros and cons of the perineal versus retropubic approach. *J. Urol.* 1992; 147:888–90.
46. Oesterling JE. Radical prostatectomy: the retropubic approach [editorial]. *Urology* 1996; 47:460–2.
47. Sullivan LD, Weir MJ, Kinahan JF et al. A comparison of the relative merits of radical perineal and radical retropubic prostatectomy. *Br. J. Urol. Int.* 85:95–100.
48. Weldon VE, Tavel FR, Neuwirth H et al. Patterns of positive specimen margins and detectable prostate specific antigen after radical perineal prostatectomy [see comments] [published erratum appears in *J. Urol.* 1995; 154:538]. *J. Urol.* 1995; 153:1565–9.
49. Resnick MI. Radical prostatectomy: the perineal approach [editorial]. *Urology* 1996; 47:457–9.
50. Walther PJ. Radical perineal vs. retropubic prostatectomy: a review of optimal application and technical considerations in the utilization of these exposures. *Eur. Urol.* 1993; 24:34–8.
51. Guillonneau B, Vallancien G. Laparoscopic radical prostatectomy: the Montsouris experience. *J. Urol.* 2000; 163:418–22.

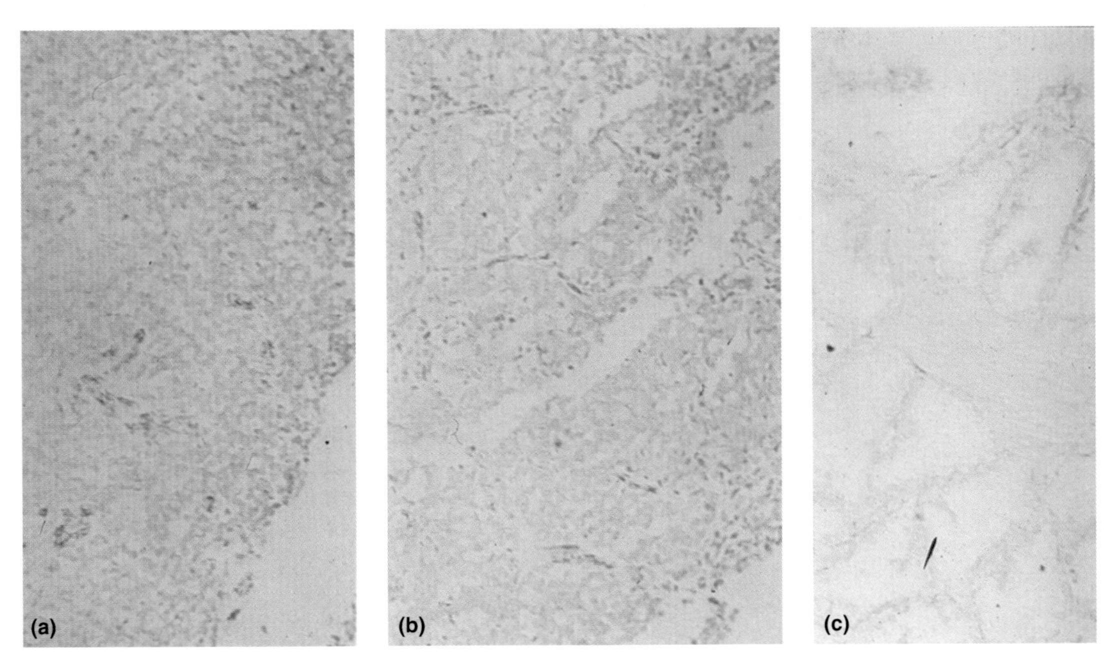

Color plate 1. Less microvessel density was demonstrated in prostate tissue of lower Gleason score (a, darker stained endothelial cells) than in prostate cancer tissue of higher Gleason score (b). Last panel (c) demonstrates ablation of vascularity after radiation and/or hormonal therapy, one of the mechanisms involved in cell death. Reproduced from Mydlo et al. *Eur. Urol.* 1998; 34:426–32, with permission from Karger, Basel. *(See Fig. 7.1.)*

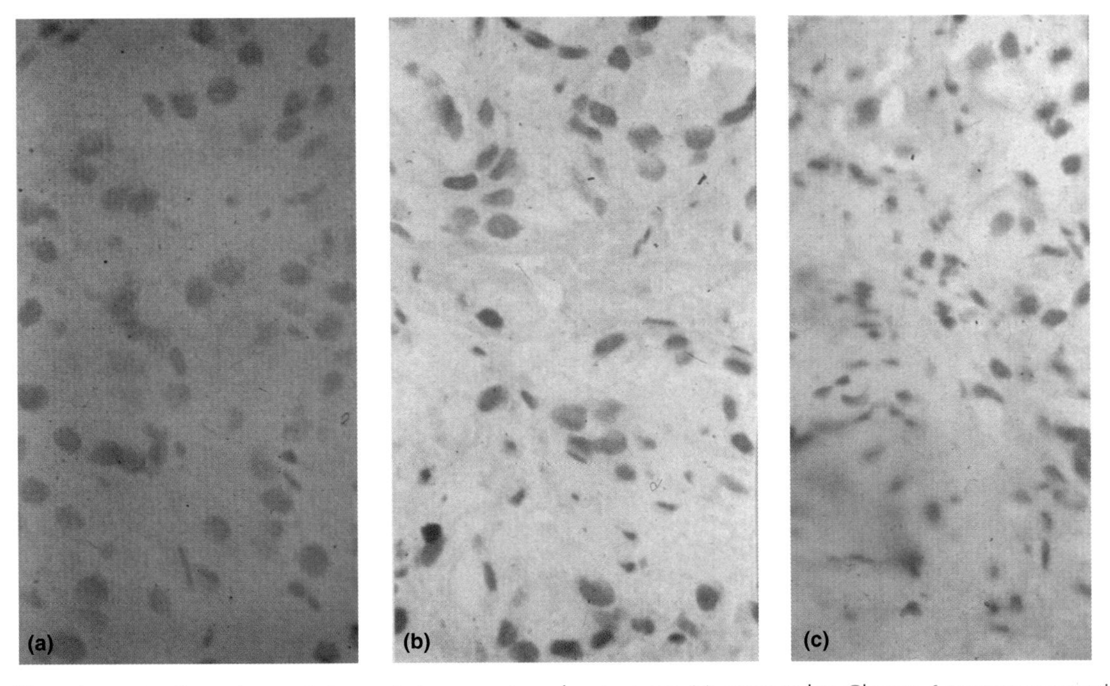

Color plate 2. Normal prostate tissue does not demonstrate expression of mutant p53 (a) compared to Gleason 6 prostate cancer tissue (b, brown avitin-biotin stained cells). Last panel (c) demonstrates the persistence of expression of mutant p53 even after radiation therapy, suggesting that these tumors are resistant to such therapy. Reproduced from Mydlo et al. *Eur. Urol.* 1998; 34:426–32, with permission from Karger, Basel. *(See Fig. 7.2.)*

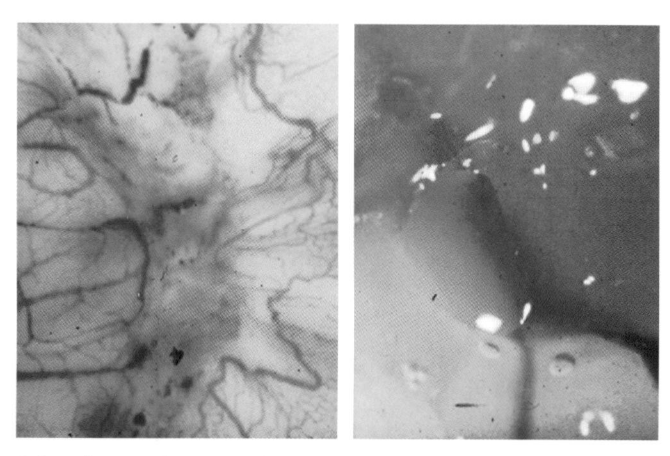

Color plate 3. Greater angiogenic response, as measured by neovascularity on the chorioallantoic membrane (CAM) assay, is demonstrated from purified adipose tissue FGF-2 (bFGF, left) than from benign or cancerous prostate tissue FGF-2 (right). Reprinted from *Journal of Urology*, 1988; 140:1575–9 with permission from Lippincott, Williams and Wilkins. *(See Fig. 19.1.)*

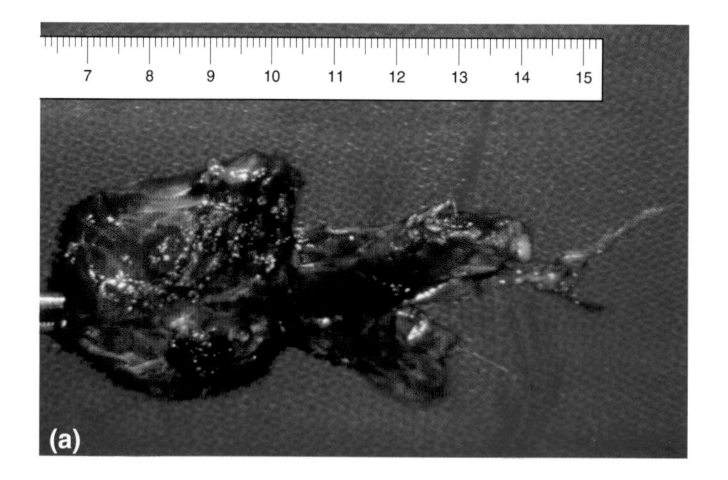

Color plate 4. (a) There were greater positive margins on the pathology specimens, as well as a greater microvessel density (MVD), as demonstrated by the darker stained endothelial cells, (b) in prostate cancer tissue from more obese patients than more lean patients. These were stratified for PSA and Gleason score. *(See Fig. 19.3.)*

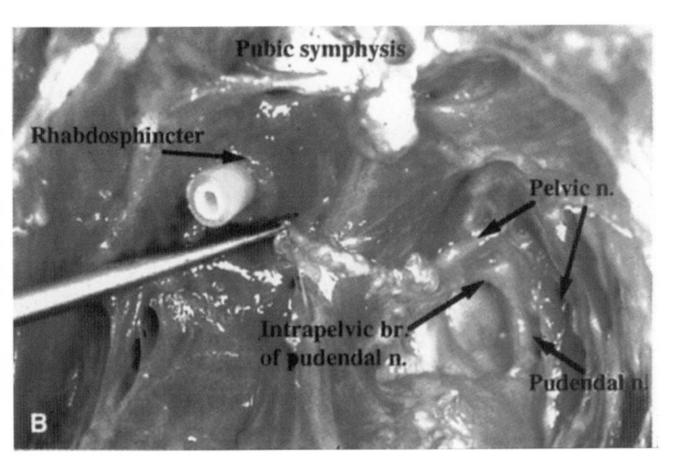

Color plate 5. Relation of the pelvic nerve and the intrapelvic branch of the pudendal nerve to the prostatic apex. The prostate has been removed. The forceps point to the pelvic nerve and the intrapelvic branch of the pudendal nerve as they enter the dorsolateral aspect of the rhabdosphincter. Reprinted from Hollaburgh et al., Copyright 1997, with permission from Elsevier Science. *(See Fig. 30.2.)*

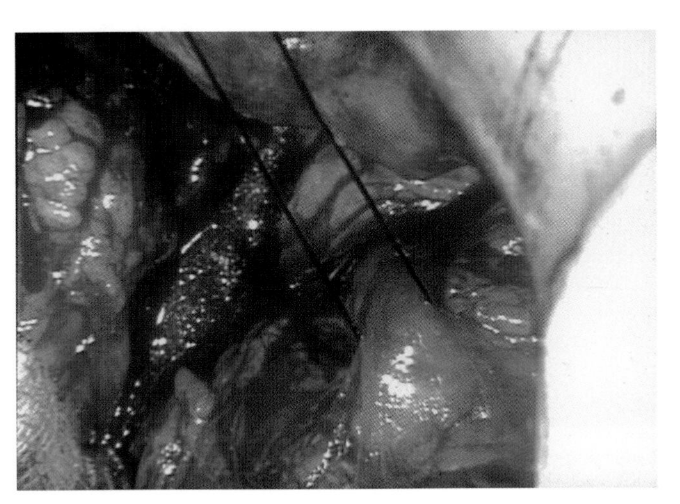

Color plate 6. Isolation of the cavernous nerve proximal and lateral to the base of the prostate. A suture is used to mark the cavernous nerve for grafting. Reprinted from *Urologic Clinics of North America* 2001; 28:843 with permission from W.B. Saunders Co. *(See Fig. 30.3.)*

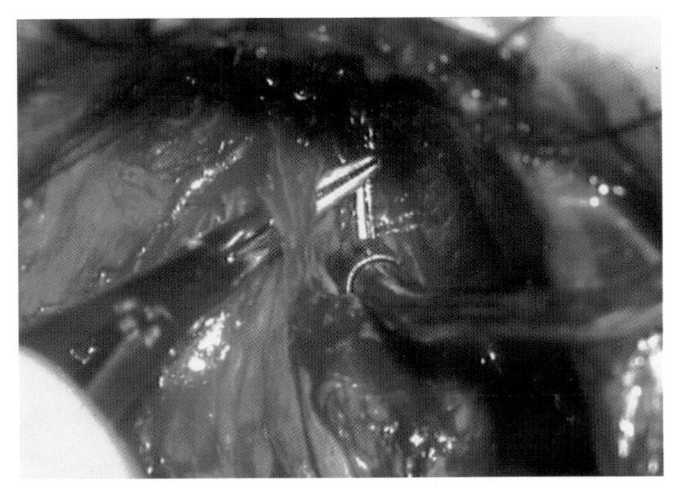

Color plate 7. Isolation of the cavernous nerve distal to the prostatic apex. Reprinted from *Urologic Clinics of North America* 2001; 28:844 with permission from W.B. Saunders Co. *(See Fig. 30.4.)*

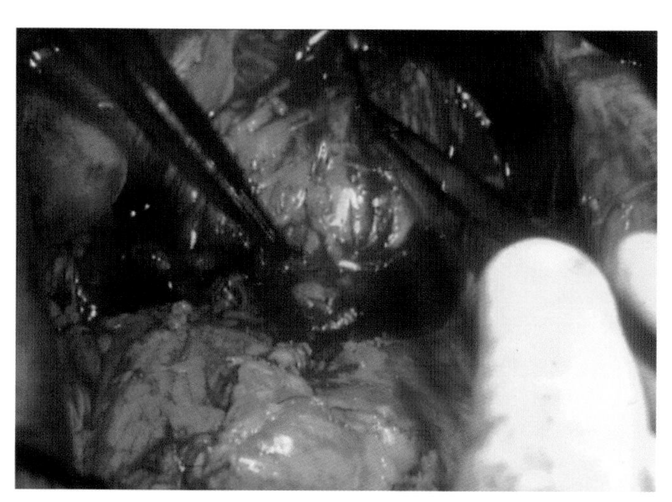

Color plate 8. Transected ends of the cavernous nerve. Both the proximal and distal ends of the excised cavernous nerve have been marked with silk ties. Denonvilliers' fascia and the left NVB have been excised exposing perirectal fat. The right NVB has been spared. Reprinted from *Urologic Clinics of North America* 2001; 28:845 with permission from W.B. Saunders Co. *(See Fig. 30.5.)*

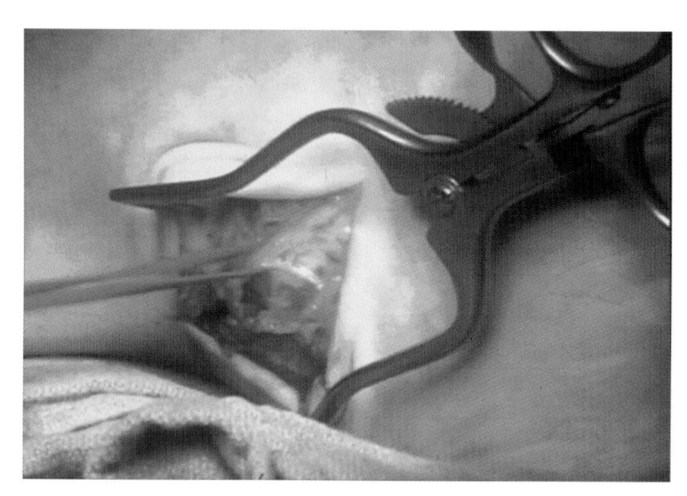

Color plate 9. Sural nerve harvest. The sural nerve is approached through an incision placed immediately inferior to the lateral malleolus of the fibula. Reprinted from *Urologic Clinics of North America* 2001; 28:845 with permission from W.B. Saunders Co. *(See Fig. 30.6.)*

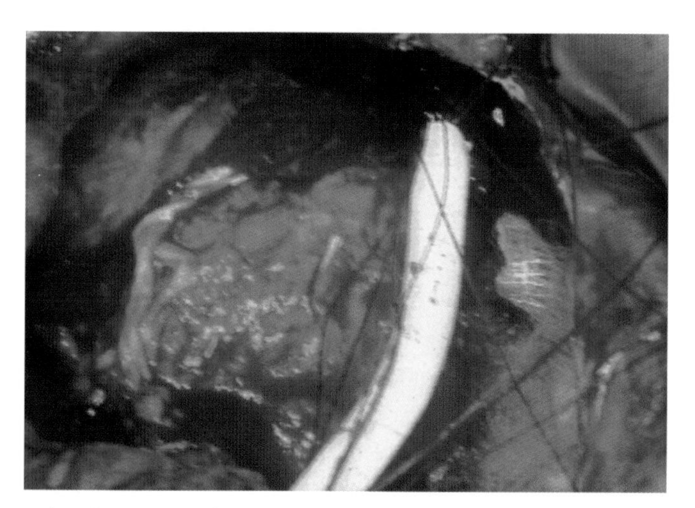

Color plate 10. Sural interposition graft *in situ*. A sural nerve segment has been used to bridge the cut ends of the excised left neurovascular bundle. The vesicourethral anastomotic sutures have not yet been tied. Reprinted from *Urologic Clinics of North America* 2001; 28:846 with permission from W.B. Saunders Co. *(See Fig. 30.7.)*

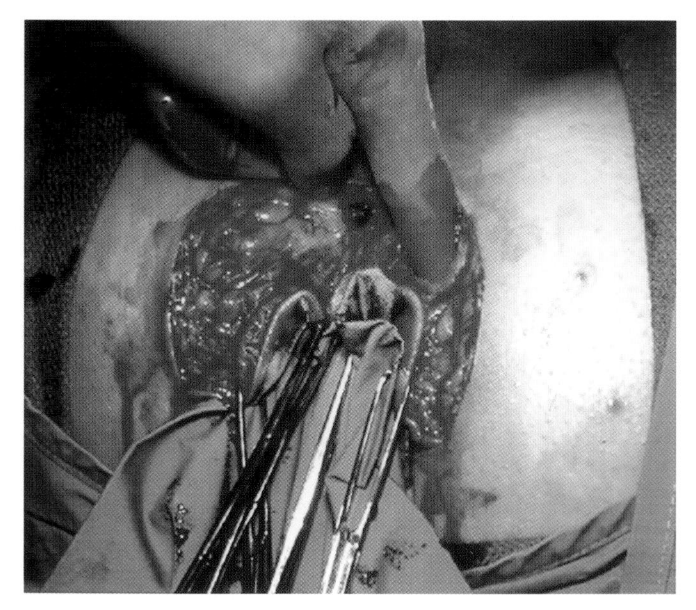

Color plate 11. Surgical dissection of the ishiorectal fossa in radical perineal prostatectomy. *(See Fig. 33.1.)*

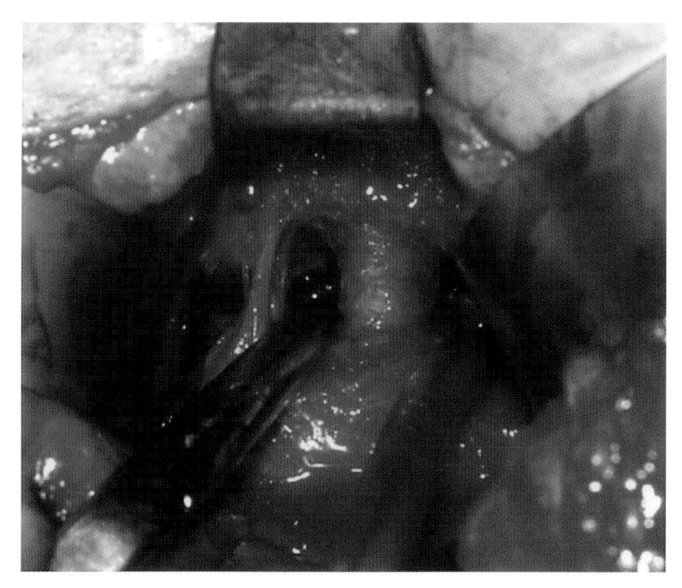

Color plate 12. Dissection of the apex of the prostate. *(See Fig. 33.2.)*

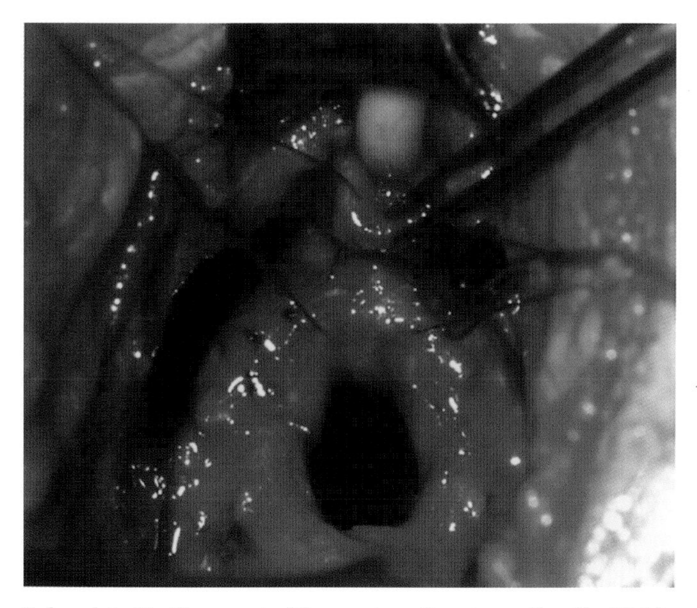

Color plate 13. Placement of the anastomotic sutures. *(See Fig. 33.3.)*

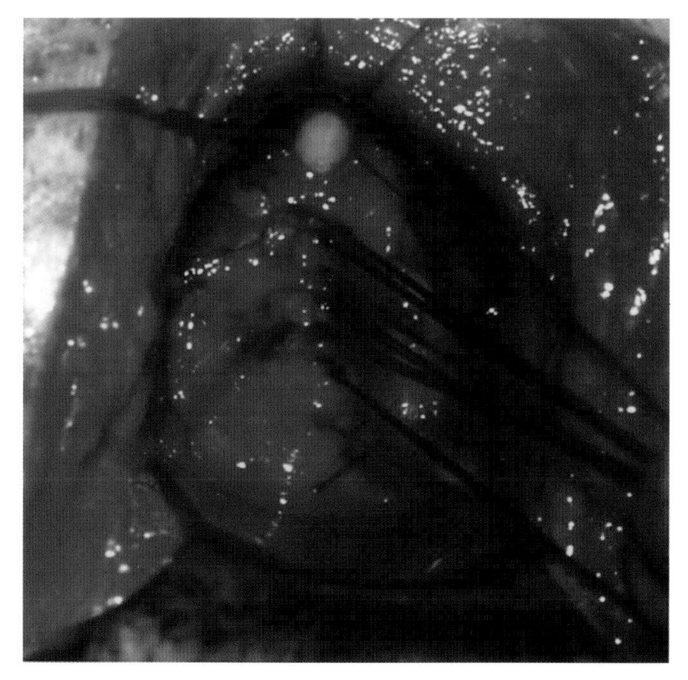

Color plate 14. Reconstruction of the bladder neck in a racket handle fashion. *(See Fig. 33.4.)*

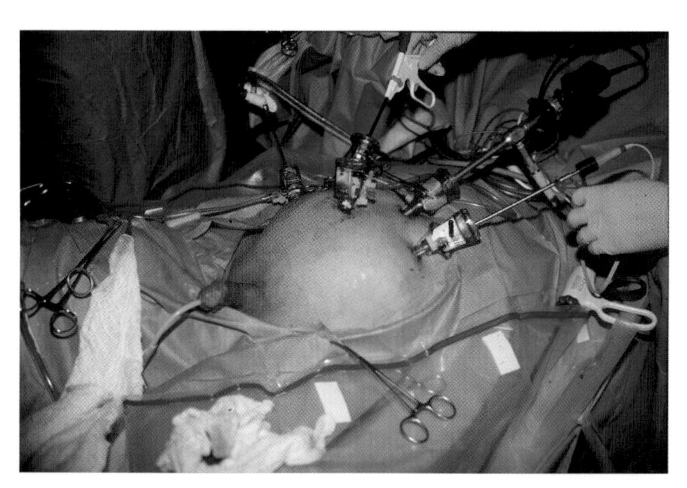

Color plate 15. Port placements are depicted with the surgeon operating through two 12-mm ports on either side of the umbilical port. The umbilical 12-mm port is used for the 0° laparoscope. Two other 5-mm ports are positioned: one midway between the umbilicus and the pubic symphysis, and one between the right lateral 12-mm port and the anterior superior iliac spine. *(See Fig. 37.1.)*

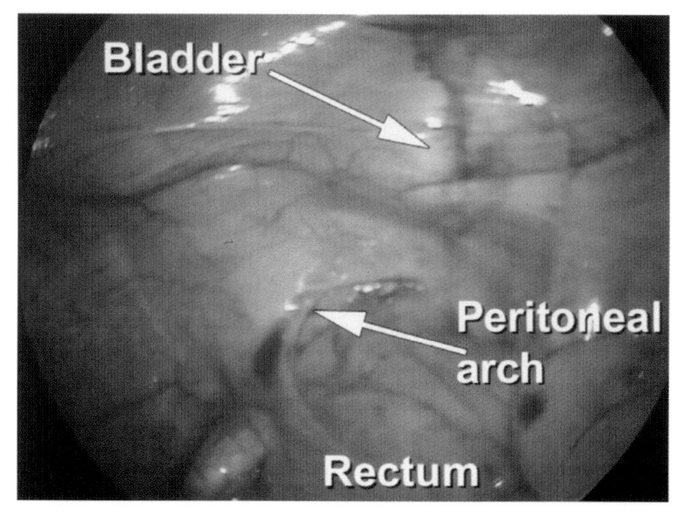

Color plate 16. Seminal vesicle dissection. The peritoneum is incised on the lower peritoneal arch in the Pouch of Douglas, to dissect the seminal vesicles and the vasa deferentia. *(See Fig. 37.2.)*

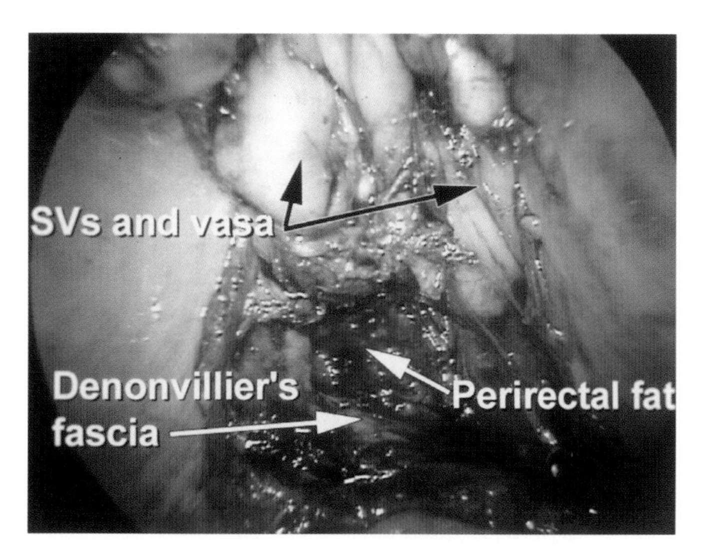

Color plate 17. Incision of Denonvillier's fascia. After division of the vasa and dissection of the seminal vesicles (SVs), the Denonvillier's fascia is transversely incised exposing the perirectal fat. *(See Fig. 37.3.)*

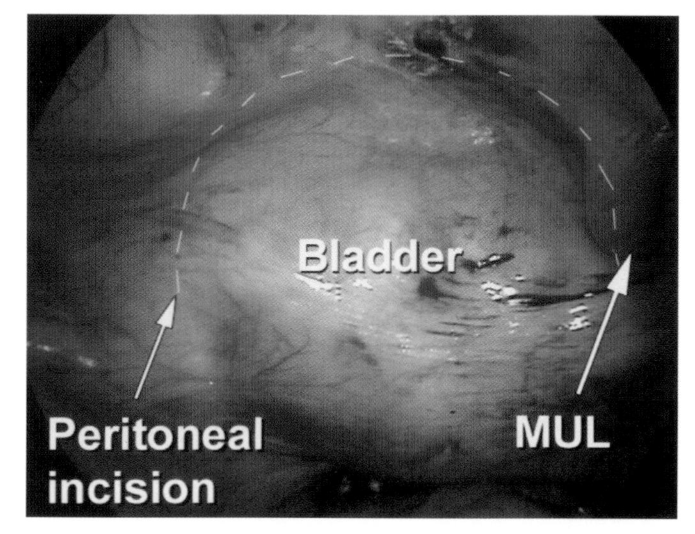

Color plate 18. Retropubic dissection. An inverted U-shaped peritoneal incision is made from one medial umbilical ligament (MUL) to the other, to drop the bladder and enter the retropubic space. *(See Fig. 37.4.)*

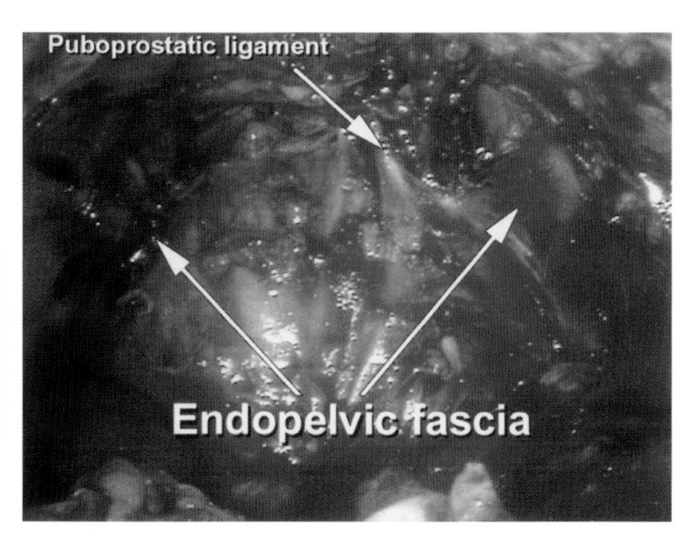

Color plate 19. Endopelvic fascia dissection. The endopelvic fascia is dissected and the puboprostatic ligaments exposed. *(See Fig. 37.5.)*

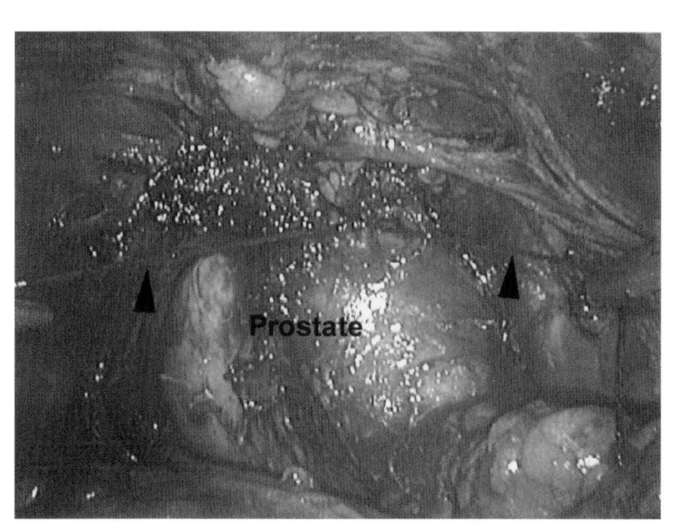

Color plate 20. Ligation of the dorsal venous complex. After the endopelvic fascia is incised and the puboprostatic ligaments divided, the dorsal venous complex is ligated. *(See Fig. 37.6.)*

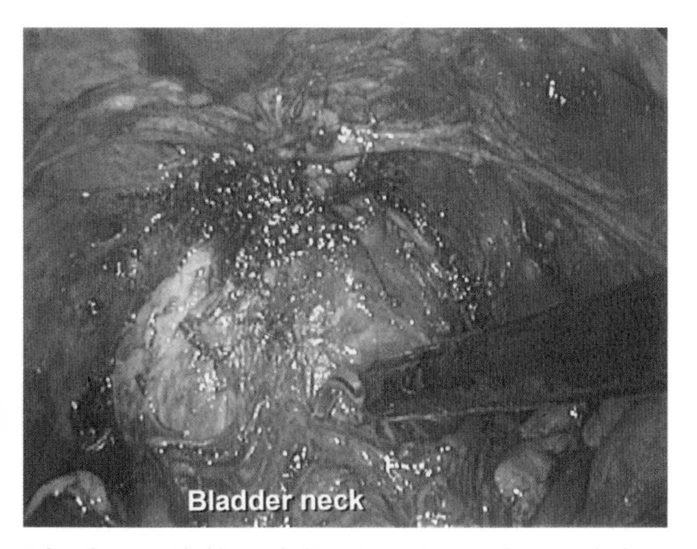

Color plate 21. Bladder neck dissection. Dissection between the base of the bladder and the bladder neck. *(See Fig. 37.7.)*

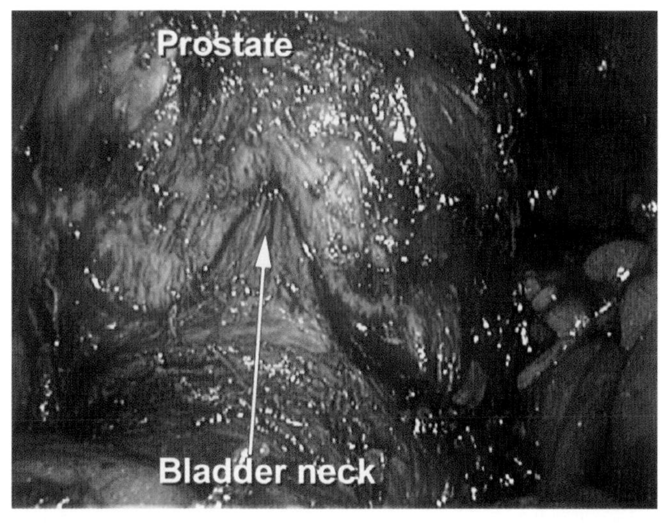

Color plate 22. Bladder neck dissection. The anterior bladder neck dissection is complete, exposing the fibers of the bladder neck. *(See Fig. 37.8.)*

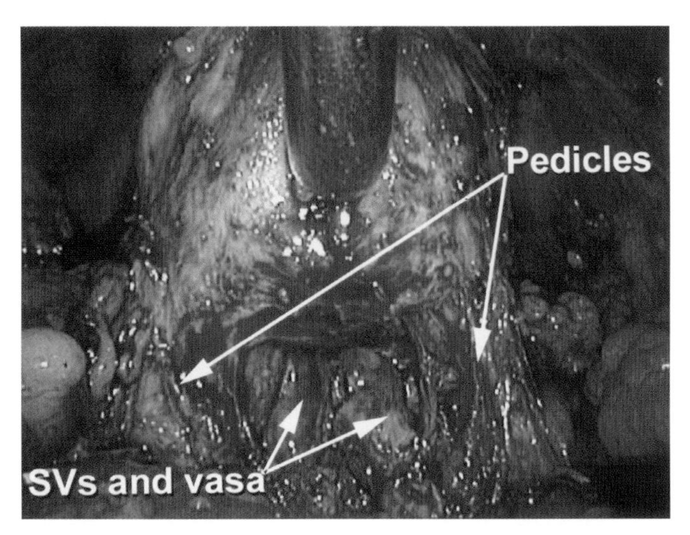

Color plate 23. Posterior bladder neck dissection. The posterior bladder neck dissection is complete exposing previously dissected seminal vesicles (SVs). *(See Fig. 37.9.)*

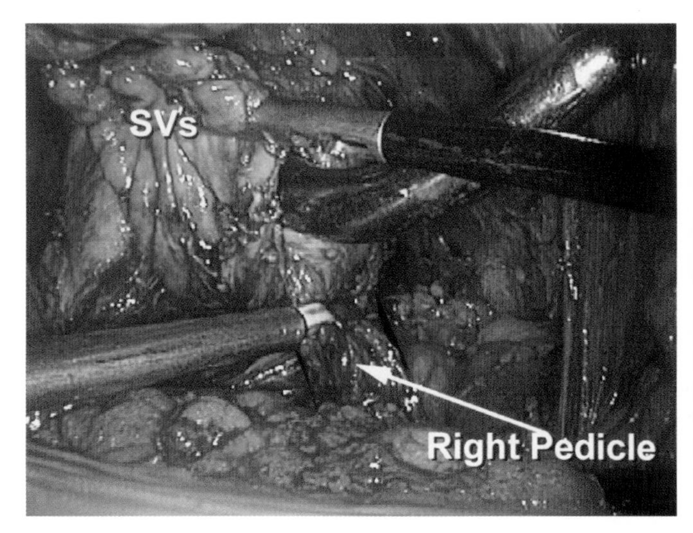

Color plate 24. The seminal vesicles (SVs) are held up, exposing the prostatic pedicles. The pedicles are divided close to the prostate, preserving the neurovascular bundles. *(See Fig. 37.10.)*

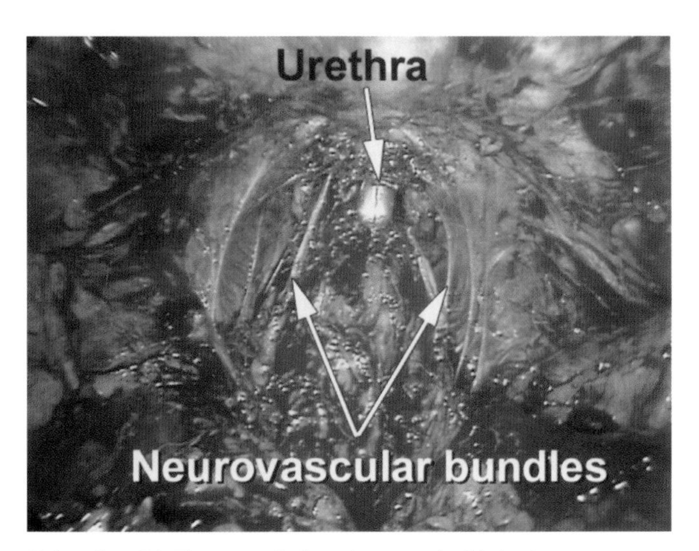

Color plate 25. The prostatic fossa is exposed with the intact neurovascular bundles on both sides. *(See Fig. 37.11.)*

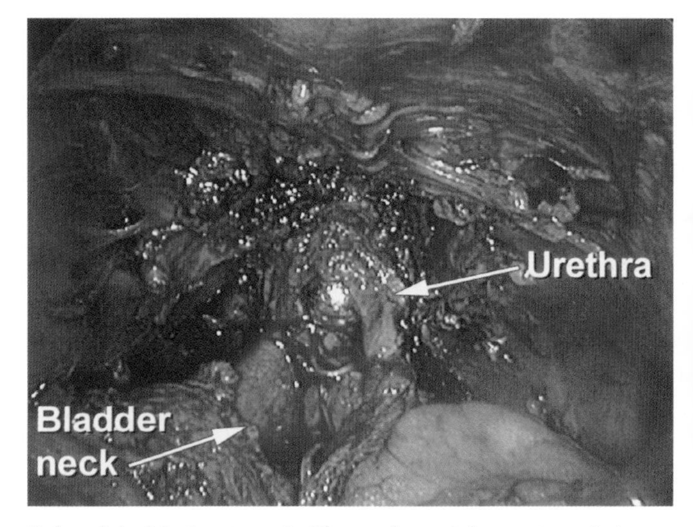

Color plate 26. Anastomosis. The urethrovesicle anastomosis is performed with interrupted 2/0 polyglactin suture. A metal urethral bougie facilitates the anastomosis. *(See Fig. 37.12.)*

Indications for Pelvic Lymphadenectomy

Richard E. Link and Ronald A. Morton

Scott Department of Urology, Baylor College of Medicine, Houston, TX, USA

INTRODUCTION

The radical prostatectomy remains one of the most frequently chosen treatment options for men with clinically localized prostate cancer who are in good health and who are otherwise good surgical candidates. Oncologic control with radical prostatectomy is excellent for early-stage disease with contemporary series reporting biochemical recurrence rates of only 12–15% at 5 years for clinical stage T1c disease.[1,2] Yet radical prostatectomy has a significant degree of associated long-term morbidity in the form of incontinence and erectile dysfunction. Since radical prostatectomy plays little if any role in the management of metastatic prostate cancer, there has been significant interest in refining preoperative staging to identify patients with nodal metastasis who should be spared extirpative surgery. For this reason, it became common practice during the 1980s to perform bilateral pelvic lymphadenectomy prior to proceeding with radical prostatectomy.

The complete pelvic lymphadenectomy involves the clearance of all nodal and fibrofatty tissue between the external and internal iliac arteries from the bifurcation of the common iliac artery, proximally, to the circumflex iliac artery and endopelvic fascia, distally. It can be performed either as the first step in a retropubic radical prostatectomy through a classic 9 cm lower midline incision or as a separate procedure either laparoscopically or through a 6 cm 'mini-laparotomy' incision.[3–11] Evaluation of the nodal packets is performed either by simple gross inspection or with frozen section histopathology, in accordance with the surgeon's preference. When performed as part of a combined procedure with a radical prostatectomy, if the nodes are deemed positive for metastatic disease, customarily the radical prostatectomy is aborted.

Bilateral pelvic lymphadenectomy adds additional operating time, cost and complication risk to the radical prostatectomy procedure. When performed independently, it requires a separate anesthetic and exposes the patient to all the cardiopulmonary, thromboembolic and wound complications associated with any pelvic surgical procedure. For these reasons, several authors have attempted to identify subsets of prostate cancer patients in whom pelvic lymphadenectomy can be omitted with acceptable risk.[12–18]

The role for pelvic lymphadenectomy in the management of prostate cancer has evolved significantly over the past 10 years. Advances in our understanding of preoperative staging, histopathology, surgical technique and risk have driven this process. In this chapter we will attempt systematically to review the literature on this topic and clarify the indications for pelvic lymphadenectomy as they stand today.

PREOPERATIVE STAGING

The decision of whether to perform pelvic lymphadenectomy in an individual patient can be a difficult one and is dependent on the accuracy of currently available preoperative staging for prostate cancer. If prostate cancer could be staged preoperatively with 100% specificity and sensitivity, then pelvic lymphadenectomy would be unnecessary. Unfortunately, this is not the case. Preoperative staging involves the synthesis of data from prostate specific antigen (PSA) values, biopsy pathology, physical examination and, in some circumstances, imaging studies. By interpreting these data within the framework of our experience with the natural history of prostate cancer, one can derive a statistical prediction for risk of nodal metastasis.

In 1980, Kramer et al. reported that no patient with Gleason sum 2–4 cancer, irrespective of clinical stage, had nodal disease on pelvic lymphadenectomy.[19] In subsequent studies, however, Gleason sum was not found to predict clinical stage reliably.[20,21] It has been reported that clinical stage better predicted nodal metastases than did histologic grade. Yet digital rectal examination, alone, is associated with significant clinical understaging as demonstrated by the high rate (22–63%) of unsuspected periprostatic extension at radical prostatectomy.[22–24] Likewise, although preoperative PSA correlates with disease progression, it does not adequately predict specific aspects of pathologic stage (nodal status, capsular penetration, seminal vesicle involvement) in individual patients.[25–29] For example, the rate of organ-confined prostate cancer ranges from 53% to 67% for men with PSA between 4 and 10 ng/ml and from 31% to 56% for PSA between 10 and 20 ng/ml.[23,27,30]

Several groups have attempted to derive models for predicting nodal disease based on a combination of multiple factors.[12,15,17,27] Partin et al. analyzed data from 703 patients and generated a nomogram for predicting nodal metastases based on three factors (clinical stage, preoperative PSA and tumor biopsy grade).[27] This model was subsequently validated in a larger multicenter study of 4133 men and accurately predicted nodal metastases in 82.9% of patients.[25] Using a similar concept, Bluestein et al. performed a retrospective review of 1632 patients who underwent pelvic lymphadenectomy at the Mayo Clinic for staging of prostate cancer.[15] These authors proposed and tested a model based on multivariate logistic regression analysis that incorporated similar factors to predict nodal metastasis. Using this method, they determined that 29% of their patients with clinical stages T1a to T2c disease could have been spared pelvic lymphadenectomy with only a 3% rate of missed nodal metastases. Bishoff et al. reported similar results the following year. They demonstrated that 20–63% of their prostate cancer patients could be spared pelvic lymphadenectomy when accepting a 2–10% risk of missed nodal metastasis.[14] These results are encouraging and suggest that many patients can be spared pelvic lymphadenectomy solely by analyzing preoperative PSA, Gleason grade and clinical stage, without incurring an unacceptable risk of missing regional metastasis. The exact level of 'acceptable' risk tolerable by an individual patient and surgeon, of course, may vary significantly.

Preoperative imaging studies

Modern axial imaging plays an important role in the preoperative staging of several urologic malignancies, most notably bladder, renal and testes neoplasms. However, despite some early favorable results, radiographic imaging has not proven to be of much value in the preoperative staging of nodal metastasis in prostate cancer. Suarez et al. preoperatively screened 94 patients with computed tomography (CT) scans who subsequently underwent pelvic lymphadenectomy.[31] Eighteen patients in this series had positive nodal metastases despite negative CT scans and neither patient with a suspicious preoperative CT had nodal disease. In a larger series, 409 prostate cancer patients with negative preoperative CT scans underwent pelvic lymphadenectomy and radical prostatectomy, and 3.7% were found to harbor nodal metastases at operation.[32] Likewise, neither magnetic resonance imaging (MRI) nor bipedal lymphangiography has demonstrated greater sensitivity than CT for detection of nodal metastases.[33–36]

It is clear that standard axial imaging does not have the sensitivity to identify micrometastatic prostate cancer in regional lymph nodes. One alternative approach to this problem may be the detection of metastatic cells using a radiolabeled monoclonal antibody to prostate specific membrane antigen (^{111}In-capromab penditide, the ProstaScint® scan, Cytogen Corporation, Princeton, NJ).[37–40] In a prospective study, 198 patients with clinical T2 or T3 disease were scanned with this agent prior to surgical lymphadenectomy. For this high-risk population, the monoclonal antibody scan demonstrates a positive predictive value of almost 67%.[40] Whether this technology will be valuable in staging low-risk patients with clinical T1 disease remains to be investigated. Of concern, there are a number of circumstances that can lead to false-positive results, and interpretation of the study is somewhat operator dependent.[41–43]

ACCURACY OF FROZEN-SECTION HISTOPATHOLOGY AFTER PELVIC LYMPHADENECTOMY

It is clear that, under all but the most obvious circumstances, preoperative studies cannot accurately predict pelvic lymph node involvement in men diagnosed with adenocarcinoma of the prostate. Consequently, bilateral pelvic lymphadenectomy is often performed as the initial surgical maneuver at the start of a radical retropubic prostatectomy. In this setting, some surgeons proceed with prostatectomy, if the excised nodal packets are grossly normal and others await analysis by frozen-section histopathology before proceeding. The false-negative rate for frozen-section analysis as compared to permanent sections has been reported to range from 7% to 27.5%.[44–48] When only nodes with micrometastatic disease are examined, the rate of false-negative frozen section histopathology may be higher than 30%.[45] For this reason, the use of frozen sections should be individualized to a given clinical situation and may be more appropriate for patients with adverse prognostic factors (PSA >10; T2 or T3 disease; higher Gleason grade) and a correspondingly higher risk for nodal spread.

MORBIDITY OF PELVIC LYMPHADENECTOMY

Several confounding factors complicate any evaluation of the morbidity of pelvic lymphadenectomy. First, the extent of the lymphatic dissection has differed between surgeons and over time as the technique has evolved. Second, pelvic lymphadenectomy can be performed in a number of ways: (1) through a traditional lower midline incision; (2) a mini-lap incision; or (3) laparoscopically. These factors influence the overall complication rate, for example, the incidence of intraoperative neural and vascular injury, the rate of lymphocele formation and the extent of postoperative lymphedema. Ordinarily, pelvic lymphadenectomy is performed immediately before definitive treatment for prostate cancer, such as radical prostatectomy, or radioactive seed placement under the same anesthetic or external beam radiation within the perioperative period. For this reason, it can be difficult to differentiate complications attributable to the pelvic lymphadenectomy procedure from those related to the therapy and/or the adjuvant therapy. In **Table 27.1** we review the complication rates for 26 published pelvic lymphadenectomy series. The data are separated both by technique and whether or not lymphadenectomy was accompanied by a therapeutic intervention. Keeping this in mind, we review complication rates for 25 published series of pelvic lymphadenectomy in **Table 27.1**.[9,11,49–70] Note that only five of these series include solely patients undergoing open pelvic lymphadenectomy as an independent procedure.[49,56,62,63,67]

Several complications can be directly linked to the lymphadenectomy procedure itself. Injury of the obturator nerve, obturator blood vessels, iliac vein and/or artery, can all occur during pelvic lymphadenectomy and would not be anticipated during any of the therapeutic interventions that are commonly associated with prostate cancer interventions. Delayed

Table 27.1. Complication rates for published series of pelvic lymphadenectomy. Rates for individual complications are expressed as a percentage.

									Early postoperative																	
					Intraoperative				Pelvic or retroperitoneal			Wound				Cardiovascular					Late postoperative					
	Year	No. Pt	Total	%	Death	Neural inury	Vascular injury	Visceral injury	Abscess	Hematoma	Lymphocele	Infection	Hematoma	Seroma	Dehiscence	Arrythmias	Myocardial infarction	Pulmonary embolus	Cerebrovascular accident	Thrombophlebitis	Pneumonia	Gastrointestinal	Epididymo-orchitis	Genital/extremity edema	Incisional hernia	Lymphocutaneous fistula
LYMPHADENECTOMY ONLY (OPEN)																										
Freiha and Salzman	1977	65	18	28	0.0	0.0	0.0	0.0	0.0	0.0	0.0	6.2	0.0	0.0	3.1	0.0	0.0	0.0	0.0	0.0	7.7	4.6	0.0	NS	NS	NS
McCullough et al.	1977	30	6	20	0.0	3.3	0.0	0.0	0.0	3.3	13.3	0.0	0.0	0.0	0.0	0.0	0.0	0.0	0.0	0.0	0.0	0.0	0.0	NS	NS	NS
Babcock and Grayhack	1979	73	25	34	0.0	NS	NS	NS	0.0	0.0	1.4	0.0	0.0	0.0	0.0	0.0	0.0	1.4	0.0	1.4	0.0	1.4	0.0	NS	NS	NS
Lieskovsky et al.	1980	17	9	53	0.0	NS	NS	NS	0.0	0.0	0.0	5.9	0.0	0.0	0.0	0.0	0.0	0.0	0.0	5.9	0.0	0.0	0.0	41.2	0.0	0.0
Paul et al.	1983	150	77	51	0.0	1.3	1.3	0.7	0.7	1.3	0.7	5.3	2.7	0.7	0.0	0.7	0.7	0.7	0.0	0.0	0.0	4.7	4.0	5.3	2.7	0.7
		335	135	40																						
LYMPHADENECTOMY (open) WITH RP, SEEDS OR XRT																										
Mclaughlin et al.	1976	60	17	28	0.0	1.7	0.0	0.0	3.3	1.7	6.7	0.0	0.0	0.0	0.0	0.0	0.0	3.3	0.0	0.0	0.0	0.0	0.0	8.3	0.0	0.0
McCullough et al.	1977	30	16	53	0.0	0.0	0.0	0.0	3.3	0.0	6.7	3.3	0.0	0.0	0.0	0.0	0.0	6.7	0.0	6.7	3.3	6.7	0.0	NS	NS	NS
Fowler et al.	1979	300	135	45	0.7	2.0	0.0	0.0	4.7	1.3	2.7*	2.7*	1.7	0.0	0.0	0.0	0.0	3.7	0.3	3.0	0.3	0.7	0.0	3.3	1.0	1.3
Herr	1979	75	17	23	1.3	0.0	NS	0.0	1.3	0.0	2.7	2.7	0.0	0.0	0.0	0.0	1.3	2.7	0.0	4.0	0.0	0.0	0.0	1.3	0.0	0.0
Brendler et al.	1980	125	9	7	0.8	NS	NS	NS	0.0	0.0	0.0	2.4	0.0	0.0	0.0	0.8	0.8	0.8	0.0	0.8	1.6	0.0	0.0	0.0	0.0	0.0
Grossman et al.	1980	91	9	10	0.0	0.0	0.0	0.0	0.0	0.0	1.1	1.1	0.0	0.0	0.0	0.0	0.0	1.1	0.0	2.2	0.0	0.0	0.0	NS	NS	NS
Lieskovsky et al.	1980	65	27	42	0.0	NS	NS	NS	0.0	0.0	3.1	3.1	1.5	0.0	0.0	0.0	0.0	4.6	0.0	3.1	0.0	0.0	0.0	12.3	0.0	0.0
Cumes et al.	1981	36	11	31	0.0	NS	NS	8.3	8.3	0.0	0.0	8.3	0.0	0.0	0.0	2.8	0.0	0.0	0.0	0.0	0.0	5.6	0.0	5.6	0.0	0.0
Sogani et al.**	1981	187	NS	NS	NS	NS	NS	NS	NS	NS	4.8	NS	NS	NS	NS	NS	NS	NS	NS	NS	NS	NS	NS	NS	NS	NS
Igel et al.	1987	692	240	35	0.6	NS	NS	1.6	0.1	1.2	0.9	1.0	0.7	0.0	NS	0.0	0.0	2.7	0.0	1.2	0.0	NS	NS	NS	NS	NS
Donohue et al.	1990	284	56	20	0.4	2.5	1.1	1.8	1.1	0.7	4.2	2.5	0.4	0.0	1.1	0.0	0.4	1.8	0.0	2.1	0.7	0.7	0.0	0.0	0.0	0.4
McDowell et al.	1990	217	48	22	0.0	0.5	0.0	0.0	0.0	0.9	4.6	0.9	1.4	2.8	0.0	0.5	0.0	0.9	0.0	1.4	0.9	3.7	0.0	0.5	0.9	0.0
Campbell et al.	1995	245	10	4	0.0	0.4	0.4	0.0	0.0	0.0	1.6	0.0	0.0	0.0	0.0	0.0	0.0	0.0	0.0	1.6	0.0	NS	NS	NS	NS	NS
		2407	595	27																						
LYMPHADENECTOMY (LAPAROSCOPIC)																										
Kavoussi et al.	1993	372	55	15	0.0	0.5	2.4	1.1	0.5	0.8	1.3	0.5	0.3	0.0	0.3	0.0	0.0	0.0	0.0	1.3	0.0	1.9	0.0	0.8	0.0	0.0
Schuessler et al.	1993	147	NS	31	0.0	NS	NS	1.4	NS	3.4	2.7	NS	NS	NS	NS	NS	NS	NS	NS	NS	NS	NS	NS	NS	NS	NS
Lang et al.	1994	100	9	9	0.0	1.0	0.0	1.0	1.0	1.0	1.0	0.0	0.0	0.0	0.0	0.0	0.0	0.0	0.0	2.0	0.0	1.0	0.0	NS	NS	NS
Parra et al.	1994	96	16	17	0.0	1.0	3.1	0.0	0.0	0.0	4.2	0.0	0.0	0.0	0.0	0.0	0.0	0.0	1.0	0.0	0.0	0.0	0.0	0.0	0.0	0.0
Rukstalis et al.	1994	73	11	15	0.0	0.0	0.0	0.0	1.4	0.0	0.0	0.0	4.1	0.0	0.0	0.0	0.0	0.0	0.0	2.0	0.0	1.0	0.0	0.0	0.0	0.0
Raboy et al.	1997	125	7	6	0.8	0.0	0.0	0.0	0.0	0.0	2.4	0.0	0.0	0.0	0.0	0.0	0.0	0.8	0.0	0.0	0.0	0.0	0.0	NS	NS	NS
Stone et al.	1997	189	24	27	0.0	1.6	0.0	0.0	1.1	0.0	0.0	0.0	0.0	0.0	0.0	0.0	0.0	0.0	0.0	0.0	0.0	0.0	0.0	7.9	0.0	0.0
Freid et al.**	1998	111	NS	NS	NS	NS	NS	NS	NS	NS	3.5	NS	NS	NS	NS	NS	NS	NS	NS	NS	NS	NS	NS	NS	NS	NS
		1213	122	13																						

** – Focus of study only on rate of lymphocele development.
NS – Not specified in published results. Note that some studies do not track late postoperative complications and others focus only on rate of lymphocele formation.

complications associated with the pelvic lymphadenectomy, independent of cancer treatment, would include lymphocele, lymphocutaneous fistulae and lower extremity or genital edema. A lymphocyst or lymphocele is a lymph-filled space, without a distinct epithelial lining, which forms in the retroperitoneum in response to a lymphatic leak.[70–73] It was first described in the gynecologic literature as a complication occurring in 13–50% of patients following pelvic lymphadenectomy during radical hysterectomy.[73–75] In contemporary radical prostatectomy series the reported frequency of symptomatic lymphocele has been much lower, generally less than 5%, when compared to previous reports. It is quite likely that the rate of asymptomatic lymphocele formation is much higher. For example, in a series of 23 patients who had scheduled CT scans after laparoscopic pelvic lymphadenectomy, asymptomatic lymphoceles were identified in 30.5%.[55] Symptomatic lymphoceles can be unilateral or bilateral with 80–90% observed within 3 weeks of pelvic lymphadenectomy.[70] In patients presenting with a lymphocele, physical examination commonly demonstrates an ovoid, smooth, non-tender mass in the iliac fossa parallel to the inguinal ligament. Clinical suspicion can be confirmed by axial imaging with CT or by pelvic ultrasound.[76,77] Symptoms are related either to compression of adjacent structures, such as the bladder, ureter, colon or iliac vessels, or to the effects of inflammation. Conservative measures, such as catheter drainage with or without sclerotherapy, are indicated for the initial management of symptomatic lymphocele.[71,78,79] If conservative measures prove ineffective, then more invasive steps, specifically, laparoscopic or open marsupialization are the preferred treatment options.[71,80–83] At this time, we favor laparoscopic management of pelvic lymphocele.

In addition to morbidity directly related to the surgical procedure, pelvic lymphadenectomy performed as an independent procedure subjects the patient to all the anesthetic, cardiovascular, pulmonary, gastrointestinal, infectious and wound complications associated with any surgical procedure. Mortality across all studies has been less than 1.5%, even though many of these patients underwent radical prostatectomy at the time of pelvic lymphadnectomy. For isolated pelvic lymphadenectomy, Paul et al. reviewed the rate of complications in a series of 150 patients who underwent an open procedure.[67] Wound infection (5.3%) and prolonged ileus (>5 days; 4.6%) were the most commonly observed morbidities. Major cardiopulmonary events occurred in less than 2% of patients. Even lower complication rates have been reported in contemporary series of laparoscopic pelvic lymphadenectomy.[8,60]

RECOMMENDATIONS

Pelvic lymphadenectomy is not without cost and can substantially increase the cost of providing care to men undergoing curative procedures for prostate cancer. It has been reported that pelvic lymphadenectomy, in the setting of a radical prostatectomy, can cost anywhere from $1600 to $3200 dollars.[84] If all men in the USA that undergo radical prostatectomy on a yearly basis also had a bilateral pelvic lymph node dissection, the cost would be approximately $213 000 000.[84,85] Given the limited resources available to allow for the provision of health care, one could make a cogent argument for limiting pelvic lymphadenectomy to those patients that are likely to benefit clinically from the information that can be derived.

However, there are a number of other concerns that influence a physicians' decision to proceed with pelvic lymphadenectomy. Some surgeons regard the staging information obtained to be invaluable when counseling patients about their prostate cancer. Since the information cannot reliably be obtained radiographically, most prefer to perform a pelvic lymphadenectomy in conjunction with radical prostatectomy. The staging information obtained may also be used to select patients for adjuvant treatments or inclusion in clinical trials.[86,87] Given the wide variety of reasons for performing pelvic lymphadenectomy in conjunction with prostate cancer treatment, it is difficult to establish strict guidelines for indications. However, the use of nomograms and other clinical prediction instruments allows one to develop strategies that can aid in the decision-making process regarding performing pelvic lymphadenectomy. If one agrees that a 5% risk for missing prostate cancer metastatic to lymph nodes is at the limit of acceptability, then broad guidelines can be established. Using the Partin nomogram, one can establish criteria for performing bilateral pelvic lymphadenectomy at the time of radical prostatectomy.[25,27] It should be noted that these guidelines only apply to patients treated by radical prostatectomy and are not necessarily relevant for patients undergoing other forms of therapy. In **Table 27.2** we have outlined clinical parameters that can be used to select patients for whom pelvic lymphadenectomy is appropriate at the time of radical prostatectomy. The PSA, clinical stage and tumor grade cut-offs used are derived from the Partin tables, which have been validated on over 4000 men who have undergone radical prostatectomy for clinically localized prostate cancer. An alternative approach to **Table 27.2** would be linear regression and development of a mathematical model to predict lymph node metastases.[88–91] Using a cohort of 1960 men who underwent radical prostatectomy for clinically localized prostate cancer, we have developed an equation that predicts whether or not a pelvic lymphadenectomy is indicated based on available clinical parameters. If one uses the same 5% rate for acceptable false-negative results, then the following equation predicts the need for performing a lymph node dissection:

Table 27.2. Biopsy Gleason scores that justify bilateral pelvic lymphadenectomy at the time of radical prostatectomy

	Serum PSA (ng/ml)			
Clinical stage	0.0–4.0	4.1–10.0	10.1–20.0	>20.0
T1a	none	none	none	ND
T1b	6–10	6–10	5–10	ND
T1c	none	8–10	7–10	All
T2a	7–10	7–10	6–10	6–10
T2b	6–10	5–10	5–10	All
T2c	5–10	All	All	All
T3a	All	All	All	All

Recommendations based on clinical stage and preoperative PSA. The values are based on accepting a less than 5% risk for missing nodal disease. Predictions for risk are based on published nomograms.

$$gx = -6.860 + 1.112 \times \ln(PSA) + 1.182 \times PS1 + 2.298 \times PS2 + 2.069 \times GL1 + 3.741 \times GL2 - 0.620 \times GL1 \times \ln(PSA) - 1.053 \times GL2 \times \ln(PSA)$$

where $PS1 = 1$, if clinical stage is T2 or less, 0 otherwise; $PS2 = 1$, if clinical stage is T3 or greater, 0 otherwise; $GL1 = 1$, if biopsy Gleason score is ≤ 7, 0 otherwise; and $GL2 = 1$, if biopsy Gleason score is >7, 0 otherwise.

The calculated value is $\exp(gx)/[1+\exp(gx)]$. If the calculated value is greater than 0.0148, then there is a high likelihood of positive lymph nodes and, although the false-positive rate is around 60%, a lymph node dissection is clearly indicated. If the calculated value is less than 0.0148, then the probability of a positive lymph node dissection is exceedingly low, and the cost and risk of lymphadenectomy are not justified. Although somewhat cumbersome, this equation lends itself to development of a simple calculator that can be written for any of a number of handheld devices.[91,92] This nomogram is based strictly upon positive node rate. In a more elaborate decision-analysis framework, the value of knowing that the pelvic lymph nodes are negative for both the patient and physician would be incorporated into the decision-making process.[90] Moreover, the morbidity associated with pelvic lymphadenectomy will also have to be quantified and considered in such a decision-making process. Thus, a decision-analysis process that takes into account financial factors, prognostic value, mental health status and surgical risk will ultimately provide the most informative predictor of the need for performing a pelvic lymphadenectomy in patients being treated for prostate cancer.

REFERENCES

1. Han M, Partin AW, Pound CR et al. Long-term biochemical disease-free and cancer-specific survival following anatomic radical retropubic prostatectomy. The 15-year Johns Hopkins experience. *Urol. Clin. North Am.* 2001; 28:555–65.
2. Morton RA, Steiner MS, Walsh PC. Cancer control following anatomical radical prostatectomy: an interim report. *J. Urol.* 1991; 145:1197–200.

3. Arai Y, Ishitoya S, Okuba K et al. Mini-laparotomy staging pelvic lymph node dissection for localized prostate cancer. *Int. J. Urol.* 1995; 2:121–3.
4. Brant LA, Brant WO, Brown MH et al. A new minimally invasive open pelvic lymphadenectomy surgical technique for the staging of prostate cancer. *Urology* 1996; 47:416–21.
5. Idom CB Jr, Steiner MS. Minilaparotomy staging pelvic lymphadenectomy follow-up: a safe alternative to standard and laparoscopic pelvic lymphadenectomy. *World J. Urol.* 1998; 16:396–9.
6. Parra RO, Andrus C, Boullier J. Staging laparoscopic pelvic lymph node dissection: comparison of results with open pelvic lymphadenectomy. *J. Urol.* 1992; 147:875–8.
7. Steiner MS, Marshall FF. Mini-laparotomy staging pelvic lymphadenectomy (minilap). Alternative to standard and laparoscopic pelvic lymphadenectomy. *Urology* 1993; 41:201–6.
8. Stone NN, Stock RG. Laparoscopic pelvic lymph node dissection in the staging of prostate cancer. *Mt Sinai J. Med.* 1999; 66:26–30.
9. Stone NN, Stock RG, Unger, P. Laparoscopic pelvic lymph node dissection for prostate cancer: comparison of the extended and modified techniques. *J. Urol.* 1997; 158:1891–4.
10. Kerbl K, Clayman RV, Petros JA et al. Staging pelvic lymphadenectomy for prostate cancer: a comparison of laparoscopic and open techniques. *J. Urol.* 1993; 150:396–8.
11. Schuessler WW, Pharand D, Vancaillie TG. Laparoscopic standard pelvic node dissection for carcinoma of the prostate: is it accurate? *J. Urol.* 1993;150:898–901.
12. Alagiri M, Colton MD, Seidmon EJ et al. The staging pelvic lymphadenectomy: implications as an adjunctive procedure for clinically localized prostate cancer. *Br. J. Urol.* 1997; 80:243–6.
13. Bangma CH, Hop WC, Schroder FH. Eliminating the need for peroperative frozen section analysis of pelvic lymph nodes during radical prostatectomy. *Br. J. Urol.* 1995; 76:595–9.
14. Bishoff JT, Reyes A, Thompson IM et al. Pelvic lymphadenectomy can be omitted in selected patients with carcinoma of the prostate: development of a system of patient selection. *Urology*, 1995; 45:270–4.
15. Bluestein DL, Bostwick DG, Bergstralh EJ et al. Eliminating the need for bilateral pelvic lymphadenectomy in select patients with prostate cancer. *J. Urol.* 1994; 151:1315–20.
16. El Galley RE, Keane TE, Petros JA et al. Evaluation of staging lymphadenectomy in prostate cancer. *Urology* 1998; 52:663–7.
17. Hoenig DM, Chi S, Porter C et al. Risk of nodal metastases at laparoscopic pelvic lymphadenectomy using PSA, Gleason score, and clinical stage in men with localized prostate cancer. *J. Endourol.* 1997; 11:263–5.
18. Parra RO, Isorna S, Perez MG et al. Radical perineal prostatectomy without pelvic lymphadenectomy: selection criteria and early results. *J. Urol.* 1996; 155:612–15.
19. Kramer SA, Spahr J, Brendler CB et al. Experience with Gleason's histopathologic grading in prostatic cancer. *J. Urol.* 1980; 124:223–5.
20. Zincke H, Farrow GM, Myers RP et al. Relationship between grade and stage of adenocarcinoma of the prostate and regional pelvic lymph node metastases. *J. Urol.* 1982; 128:498–501.
21. Sagalowsky AI, Milam H, Reveley LR et al. Prediction of lymphatic metastases by Gleason histologic grading in prostatic cancer. *J. Urol.* 1982; 128:951–2.
22. McNeal JE, Villers AA, Redwine EA et al. Capsular penetration in prostate cancer. Significance for natural history and treatment. *Am. J. Surg. Pathol.* 1990; 14:240–7.
23. Perrotti M, Pantuck A, Rabbani F et al. Review of staging modalities in clinically localized prostate cancer. *Urology* 1999; 54:208–14.
24. Rosen MA, Goldstone L, Lapin S et al. Frequency and location of extracapsular extension and positive surgical margins in radical prostatectomy specimens. *J. Urol.* 1992; 148:331–7.
25. Partin AW, Kattan MW, Subong EN et al. Combination of prostate-specific antigen, clinical stage, and Gleason score to predict pathological stage of localized prostate cancer. A multi-institutional update. *JAMA* 1997; 277:1445–51.
26. Partin AW, Carter HB. The use of prostate-specific antigen and free/total prostate-specific antigen in the diagnosis of localized prostate cancer. *Urol. Clin. North Am.* 1996; 23:531–40.
27. Partin AW, Yoo J, Carter HB et al. The use of prostate specific antigen, clinical stage and Gleason score to predict pathological stage in men with localized prostate cancer. *J. Urol.* 1993; 150:110–14.
28. Stamey TA, Johnstone I M, McNeal JE et al. Preoperative serum prostate specific antigen levels between 2 and 22 ng/ml correlate poorly with post-radical prostatectomy cancer morphology: prostate specific antigen cure rates appear constant between 2 and 9 ng/ml. *J. Urol.* 2002; 167:103–11.
29. Stamey TA, Kabalin JN, McNeal JE et al. Prostate specific antigen in the diagnosis and treatment of adenocarcinoma of the prostate. II. Radical prostatectomy treated patients. *J. Urol.* 1989; 141:1076–83.
30. Narayan P, Gajendran V, Taylor SP et al. The role of transrectal ultrasound-guided biopsy-based staging, preoperative serum prostate-specific antigen, and biopsy Gleason score in prediction of final pathologic diagnosis in prostate cancer. *Urology* 1995; 46:205–12.
31. Suarez P, Mondes L, Bernardo N et al. [Correlation between computed axial tomography and ileum obturating lymphadenectomy in localized adenocarcinoma of the prostate]. *Arch. Esp. Urol.* 1997; 50:131–3.
32. Levran Z, Gonzalez JA, Diokno AC et al. Are pelvic computed tomography, bone scan and pelvic lymphadenectomy necessary in the staging of prostatic cancer? *Br. J. Urol.* 1995; 75:778–81.
33. Perrotti M, Kaufman RP Jr, Jennings TA et al. Endo-rectal coil magnetic resonance imaging in clinically localized prostate cancer: is it accurate? *J. Urol.* 1996; 156:106–9.
34. Nicolas V, Beese M, Keulers A et al. [MR tomography in prostatic carcinoma: comparison of conventional and endorectal MRT]. *Rofo Fortschr. Geb. Rontgenstr. Neuen Bildgeb. Verfahr.* 1994; 161:319–26.
35. Rorvik J, Halvorsen OJ, Albrektsen G et al. Use of pelvic surface coil MR imaging for assessment of clinically localized prostate cancer with histopathological correlation. *Clin. Radiol.* 1999; 54:164–9.
36. Rorvik J, Halvorsen OJ, Albrektsen G et al. Lymphangiography combined with biopsy and computer tomography to detect lymph node metastases in localized prostate cancer. *Scand. J. Urol. Nephrol.* 1998; 32:116–19.
37. Babaian RJ, Sayer J, Podoloff DA et al. Radioimmunoscintigraphy of pelvic lymph nodes with 111indium-labeled monoclonal antibody CYT-356. *J. Urol.* 1994; 152:1952–5.
38. Hinkle GH, Burgers JK, Olsen JO et al. Prostate cancer abdominal metastases detected with indium-111 capromab pendetide. *J. Nucl. Med.* 1998; 39:650–2.
39. Kahn D, Williams RD, Manyak MJ et al. 111Indium-capromab pendetide in the evaluation of patients with residual or recurrent prostate cancer after radical prostatectomy. The ProstaScint Study Group. *J. Urol.* 1998; 159:2041–6.
40. Polascik TJ, Manyak MJ, Haseman MK et al. Comparison of clinical staging algorithms and 111indium-capromab pendetide immunoscintigraphy in the prediction of lymph node involvement in high risk prostate carcinoma patients. *Cancer* 1999; 85:1586–92.
41. Valliappan S, Joyce JM, Myers DT. Possible false-positive metastatic prostate cancer on an In-111 capromab pendetide scan as a result of a pelvic kidney. *Clin. Nucl. Med.* 1999; 24:984–5.
42. Scott DL, Halkar RK, Fischer A et al. False-positive 111indium capromab pendetide scan due to benign myelolipoma. *J. Urol.* 2001; 165:910–11.
43. Zucker RJ Bradley YC. Indium-111 capromab pendetide (ProstaScint) uptake in a meningioma. *Clin. Nucl. Med* 2001; 26:568–9.
44. Catalona WJ, Stein AJ. Accuracy of frozen section detection of lymph node metastases in prostatic carcinoma. *J. Urol.* 1982; 127:460–1.
45. Epstein JI, Oesterling JE, Eggleston JC et al. Frozen section detection of lymph node metastases in prostatic carcinoma: accuracy in grossly uninvolved pelvic lymphadenectomy specimens. *J. Urol.* 1986; 136:1234–7.
46. Fowler JE Jr, Torgerson L, McLeod DG et al. Radical prostatectomy with pelvic lymphadenectomy: observations on the

accuracy of staging with lymph node frozen sections. *J. Urol.* 1981; 126:618–19.
47. Kramolowsky EV, Narayana AS, Platz CE et al. The frozen section in lymphadenectomy for carcinoma of the prostate. *J. Urol.* 1984; 131:899–900.
48. Sadlowski RW, Donahue DJ, Richman AV et al. Accuracy of frozen section diagnosis in pelvic lymph node staging biopsies for adenocarcinoma of the prostate. *J. Urol.* 1983; 129:324–6.
49. Babcock JR, Grayhack JT. Morbidity of pelvic lymphadenectomy. *Urology* 1979; 13:483–6.
50. Brendler CB, Cleeve LK, Anderson EE et al. Staging pelvic lymphadenectomy for carcinoma of the prostate risk versus benefit. *J. Urol.* 1980; 124:849–50.
51. Campbell SC, Klein EA, Levin HS et al. Open pelvic lymph node dissection for prostate cancer: a reassessment. *Urology* 1995; 46:352–5.
52. Cumes DM, Goffinet DR, Martinez A et al. Complication of 125 iodine implantation and pelvic lymphadenectomy for prostatic cancer with special reference to patients who had failed external beam therapy as their initial mode of therapy. *J. Urol.* 1981; 126:620–2.
53. Donohue RE, Mani JH, Whitesel JA et al. Intraoperative and early complications of staging pelvic lymph node dissection in prostatic adenocarcinoma. *Urology* 1990; 35:223–7.
54. Fowler JE Jr, Barzell W, Hilaris BS et al. Complications of 125iodine implantation and pelvic lymphadenectomy in the treatment of prostatic cancer. *J. Urol.* 1979; 121:447–51.
55. Freid RM, Siegel D, Smith AD et al. Lymphoceles after laparoscopic pelvic node dissection. *Urology* 1998; 51:131–4.
56. Freiha FS, Salzman J. Surgical staging of prostatic cancer: transperitoneal versus extraperitoneal lymphadenectomy. *J. Urol.* 1977; 118:616–17.
57. Grossman IC, Carpiniello V, Greenberg SH et al. Staging pelvic lymphadenectomy for carcinoma of the prostate review of 91 cases. *J. Urol.* 1980; 124:632–4.
58. Herr HW. Complications of pelvic lymphadenectomy and retropubic prostatic 125I implantation. *Urology* 1979; 14:226–9.
59. Igel TC, Barrett DM, Segura JW et al. Perioperative and postoperative complications from bilateral pelvic lymphadenectomy and radical retropubic prostatectomy. *J. Urol.* 1987; 137:1189–91.
60. Kavoussi LR, Sosa E, Chandhoke P et al. Complications of laparoscopic pelvic lymph node dissection. *J. Urol.* 1993; 149:322–5.
61. Lang GS, Ruckle HC, Hadley HR et al. One hundred consecutive laparoscopic pelvic lymph node dissections: comparing complications of the first 50 cases to the second 50 cases. *Urology* 1994; 44:221–5.
62. Lieskovsky G, Skinner DG, Weisenburger T. Pelvic lymphadenectomy in the management of carcinoma of the prostate. *J. Urol.* 1980; 124:635–8.
63. McCullough DL, McLaughlin AP, Gittes RF. Morbidity of pelvic lymphadenectomy and radical prostatectomy for prostatic cancer. *J. Urol.* 1977; 117:206–7.
64. McDowell GC, Johnson JW, Tenney DM et al. Pelvic lymphadenectomy for staging clinically localized prostate cancer. Indications, complications, and results in 217 cases. *Urology* 1990; 35:476–82.
65. McLaughlin AP, Saltzstein SL, McCullough DL et al. Prostatic carcinoma: incidence and location of unsuspected lymphatic metastases. *J. Urol.* 1976; 115:89–94.
66. Parra RO, Hagood PG, Boullier JA et al. Complications of laparoscopic urological surgery: experience at St. Louis University. *J. Urol.* 1994; 151:681–4.
67. Paul DB, Loening SA, Narayana AS et al. Morbidity from pelvic lymphadenectomy in staging carcinoma of the prostate. *J. Urol.* 1983; 129:1141–4.
68. Raboy A, Adler H, Albert P. Extraperitoneal endoscopic pelvic lymph node dissection: a review of 125 patients. *J. Urol.* 1997; 158:2202–4.
69. Rukstalis DB, Gerber GS, Vogelzang NJ et al. Laparoscopic pelvic lymph node dissection: a review of 103 consecutive cases. *J. Urol.* 1994; 151:670–4.
70. Sogani PC, Watson RC, Whitmore WF Jr. Lymphocele after pelvic lymphadenectomy for urologic cancer. *Urology* 1981; 17:39–43.
71. Glass LL, Cockett AT. Lymphoceles: diagnosis and management in urologic patients. *Urology* 1998; 51:135–40.
72. Kim JK, Jeong YY, Kim YH et al. Postoperative pelvic lymphocele: treatment with simple percutaneous catheter drainage. *Radiology* 1999; 212:390–4.
73. Scholz HS, Petru E, Benedicic C et al. Fibrin application for preventing lymphocysts after retroperitoneal lymphadenectomy in patients with gynecologic malignancies. *Gynecol. Oncol.* 2002; 84:43–6.
74. Conte M, Panici PB, Guariglia L et al. Pelvic lymphocele following radical para-aortic and pelvic lymphadenectomy for cervical carcinoma: incidence rate and percutaneous management. *Obstet. Gynecol.* 1990; 76:268–71.
75. Querleu D, Leblanc E, Castelain B. Laparoscopic pelvic lymphadenectomy in the staging of early carcinoma of the cervix. *Am. J. Obstet. Gynecol.* 1991; 164:579–81.
76. Chow CC, Daly BD, Burney TL et al. Complications after laparoscopic pelvic lymphadenectomy: CT diagnosis. *Am. J. Roentgenol.* 1994; 163:353–6.
77. Spring DB, Schroeder D, Babu S et al. Ultrasonic evaluation of lymphocele formation after staging lymphadenectomy for prostatic carcinoma. *Radiology* 1981; 141:479–83.
78. Akhan O, Cekirge S, Ozmen M et al. Percutaneous transcatheter ethanol sclerotherapy of postoperative pelvic lymphoceles. *Cardiovasc. Intervent. Radiol.* 1992; 15:224–7.
79. Teiche PE, Pauer W, Schmid N. Use of talcum in sclerotherapy of pelvic lymphoceles. *Tech. Urol.* 1999; 5:52–3.
80. Albala DM, Kevwitch MK, Waters WB. Treatment of persistent lymphatic drainage after laparoscopic pelvic lymph node dissection and radical retropubic prostatectomy. *J. Endourol.* 1993; 7:337–40.
81. Fallick ML, Long JP. Laparoscopic marsupialization of lymphocele after laparoscopic lymph node dissection. *J. Endourol.* 1996; 10:533–4.
82. McDougall EM, Clayman RV. Endoscopic management of persistent lymphocele following laparoscopic pelvic lymphadenectomy. *Urology* 1994; 43:404–7.
83. Molnar BG, Magos AL, Walker PG. Laparoscopic excision and marsupialisation of bilateral pelvic lymphocysts following extended hysterectomy and pelvic lymphadenectomy for endometrial carcinoma. *Br. J. Obstet. Gynaecol.* 1997; 104:263–6.
84. Link RE, Morton RA. Indications for pelvic lymphadenectomy in prostate cancer. *Urol. Clin. North Am.* 2001; 28:491–8.
85. Moul JW. Prostate specific antigen only progression of prostate cancer. *J. Urol.* 2000; 163:1632–42.
86. Partin AW, Piantadosi S, Sanda MG et al. Selection of men at high risk for disease recurrence for experimental adjuvant therapy following radical prostatectomy. *Urology* 1995; 45:831–8.
87. Messing EM, Manola J, Sarosdy M et al. Immediate hormonal therapy compared with observation after radical prostatectomy and pelvic lymphadenectomy in men with node-positive prostate cancer. *N. Engl. J. Med.* 1999; 341:1781–8.
88. Harrison LE, Karpeh MS, Brennan MF. Extended lymphadenectomy is associated with a survival benefit for node-negative gastric cancer. *J. Gastrointest. Surg.* 1998; 2:126–31.
89. Dillioglugil O, Leibman BD, Kattan MW et al. Hazard rates for progression after radical prostatectomy for clinically localized prostate cancer. *Urology* 1997; 50:93–9.
90. Kattan M. Statistical prediction models, artificial neural networks, and the sophism 'I am a patient, not a statistic'. *J. Clin. Oncol.* 2002; 20:885–7.
91. Kattan MW, Eastham JA, Stapleton AM et al. A preoperative nomogram for disease recurrence following radical prostatectomy for prostate cancer. *J. Natl Cancer Inst.* 1998; 90:766–71.
92. Kattan MW, Wheeler TM, Scardino PT. Postoperative nomogram for disease recurrence after radical prostatectomy for prostate cancer. *J. Clin. Oncol.* 1999; 17:1499–507.

The Surgical Anatomy of the Prostate

S. Bruce Malkowicz
Division of Urology, Hospital of the University of Pennsylvania, Philadelphia, PA, USA

INTRODUCTION

The surgical anatomy of the prostate gland is challenging and complex, owing to the significant variation of gland architecture among patients and the constraints imposed by the body habitus of the patients. The pioneering work by Walsh and associates during the past two decades has allowed the rational application of anatomic findings to radical surgery for disease control of prostate cancer. This fundamental basis for understanding prostate surgical anatomy has been augmented by the work of other outstanding investigators, yet it is generally felt that prostate anatomy may still not be optimally understood.

The principal goals of prostate cancer surgery are disease control, preservation of urinary control and preservation of erectile function. The contemporary description of the anatomy of the prostate in conjuction with the conscientious application of surgery has allowed for an iterative process, which allows for continual refinement of our understanding of anatomy, modification of technique and improvement of measurable outcomes. The aim of this review is to describe the basic general anatomy of the prostate gland with an emphasis on recent findings that facilitate the desired goals of cancer surgery.

EMBRYOLOGY AND ZONAL ANATOMY

The prostate begins to develop at approximately 8 weeks from the pelvic urethra. Epithelial outgrowths continue to grow with accompanying lumenal development. This is further modified by the effects of androgens and stromal–epithelial growth factor interactions.[1,2] The prostate is an ovoid structure of fibromuscular and glandular components. It encases the urethra, which angulates at its midpoint toward the bladder neck. The gland is covered by a 0.5 mm capsule of smooth muscle and connective tissue. Benign glandular elements may travel through its thickness. The zonal anatomy described by McNeal[3–5] does not exactly correspond to the clinical impression of two lateral lobes separated by a central sulcus, but provides greater understanding of pertinent surgical findings when one considers that the morphology of the adult gland is in great part the result of the distortions caused by hypertrophy of the transition zone tissue. It is also useful to realize that the anterior fibromuscular stroma includes the preprostatic sphincter and the detrusor apron, and not just the fibrous anterior sheath particular to the prostate.[4,5]

VASCULAR SUPPLY

The principal arterial supply to the prostate is provided by branches of the inferior vesical artery, which is in turn supplied by the hypogastric artery. The capsular artery and its branches run posteriolateral to the prostate in parallel with the cavernosal nerves. The branches of the capsular artery enter the prostate at a right angle and can be sources of bleeding during dissection, while the artery itself terminates in the pelvic diaphragm. Occasionally an aberrant pudendal artery can be seen coursing lateral to the prostate. The effort to preserve this during surgery may aid in preserving erectile function postoperatively.[6]

The urethral arteries enter the prostate at the junction of the bladder neck and prostate from four quadrants (at 11, 1, 5 and 7 o'clock) and then run parallel with the urethra. The blood supply of the seminal vesicle and vas deferens comes from the vesiculodeferential artery, which in turn is a branch of the superior vesicular artery. Of technical concern is the consistent, significant blood supply encountered at the tip of the seminal vesicle, which should be controlled in all cases to avoid unnecessary and annoying blood loss during this portion of the operation.

The venous supply of the penis and prostate can present a considerable challenge and be the source of significant morbidity, if difficulty is encountered with these structures during dissection. The dorsal venous complex emerges from the pubic arch, and develops anterior and posterior components. The anterior vessels splay over the surface of the prostate, and can be variable in their size and density. Bunching of these vessels can prevent significant back bleeding during dissection.[7–10] The posterior divisions of these veins run along neurovascular bundles running posterior lateral to the prostate.

LYMPHATIC DRAINAGE

Lymphatic drainage from the prostate is primarily to the obturator and interior iliac lymph nodes. Drainage is also present to the external iliac, presacral and paraortic nodes, which can occasionally be the site of a 'skip metastasis' in an otherwise node-negative specimen. The internal lymphatics of the prostate are discrete.

INNERVATION OF THE PROSTATE

Autonomic innervation of the prostate is provided by the cavernous nerves, which follow the course of the capsular arteries. Afferent nerves travel through the prostate to the pelvic plexuses with further transmission directed to thoracolumbar and pelvic spinal segments. The prostatic plexus consists of multiple ganglia found at the prostate and bladder base.

The careful identification and mapping of the neurovascular bundles has provided a major advance in reducing the morbidity of radical prostatectomy.[11] With the preservation of the nerve bundles according to appropriate anatomic guidelines, cancer control may be maintained without unnecessarily compromising erectile function.[12,13] The relationship of the nerve bundles to the rectum has been described as fixed, while the size of the prostate can have a significant impact on the relative position of the nerves to the gland (lateral versus posteriolateral).

Branches extend from the nerves to the capsule and are in association with the capsular vessels. This arrangement may provide the egress for perineural spread of tumor and is a consideration in higher volume disease.[14,15] As a general practical matter, these branches can tether the neurovascular bundle to the prostate surface and should be individually addressed when prominent.[16] The neural investment of the urethra will be discussed in the description of continence mechanisms.

FASCIAL ANATOMY

The retropubic approach to the prostate consists of the initial identification and incision of several facial layers starting with the midline incision of the endopelvic fascia to avoid the exposure of the underside of the rectus muscles and damage to the epigastric vessesls (**Fig. 28.1**). Exposure of the pelvis reveals the parietal fascia of the pelvis, which condenses at the sulci on either side of the bladder to form the tendinous arch of the pelvis.[6,7]

The component of this fascia that sweeps over the bladder covers a robust venous supply of interconnecting vessels. Aggressive entry through this fascial plane can result in brisk venous hemorrhage. Initial entry through this fascial plane is best accomplished laterally to avoid the condensation of veins that can be encountered close to the prostate. Bunching of the prostatovesical veins over the prostate with a curved Babcock clamp often demonstrates a 'free space' that provides a lateral avascular plane of access along the prostate during a nerve sparing procedure. When the sacrifice of the bundle is a planned portion of the operation, this incision is made lateral to the groove between the prostate and the rectum, which is occupied by the neurovascular structures.

The most prominent fascial structures encountered during surgery are the puboprostatic ligaments.[17,18] Upon careful dissection and evaluation, they are actually pubovesical ligaments, which appear to terminate in the prostate secondary to

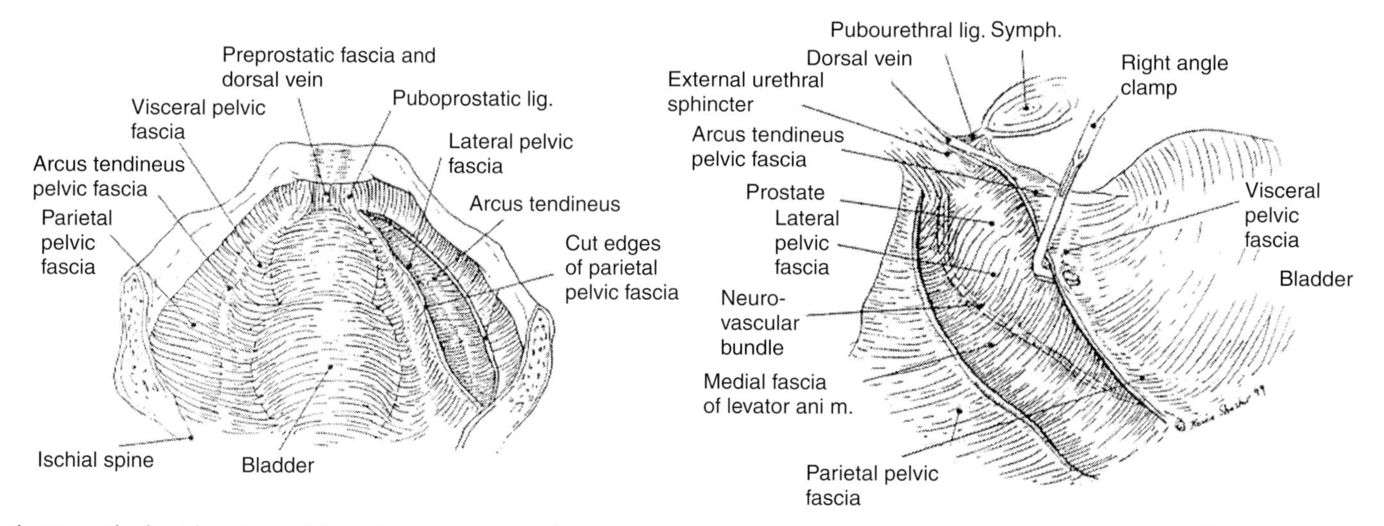

Fig. 28.1. The fascial anatomy of the male pelvis (reprinted from Steiner MS. *Urology* 2000; 55:429 with permission from Elsevier).[7]

prostatic growth during maturation.[19] These structures must be divided at some point during the performance of a retropubic radical prostatectomy and are generally incised at a more proximal site near the prostate rather than a distal location the pubic insertion. In this manner more support is provided for the structures constituting the continence mechanism of the urethra. Additionally, distal dissection is more likely to encounter the dorsal venous vascular complex with ensuing unnecessary bleeding and possible shortening of the urethra.

Denonvilliers' fascia is the prominent covering of the posterior prostate and the seminal vesicles. This rectoprostatic fascia provides a significant barrier between the prostate and rectum, and is generally more adherent to the prostate. Cephalad, it can appear in several layers over the seminal vesicles. Distally, it terminates in a fibrous plate just below the urethra described as the median fibrous raphe, which has a distal extension to the central tendon of the perineum.[19]

THE URETHRA AND ASSOCIATE CONTINENCE STRUCTURES

In most descriptions, male continence mechanisms are divided into two continence zones: the internal or proximal complex, consisting of the intraprostatic urethra and bladder neck; the external complex, consisting of the urethra distal to the verumontanum; the puborectalis aspect of the levator ani; and the additional fascial support structures previously mentioned.[20–22] The urethral component of the external complex is a sophisticated arrangement of elastic fibers, longitudinal smooth muscle and the external rhabdosphincter.[23,24]

The rhabdosphincter is a component of the urethra that is composed of slow-twitch and fast-twitch fibers that initially give an 'omega' shape appearance, suggesting muscle deficiency, particularly at the base.[19,25,26] Current studies suggest, however, that this is a function of lateral thickening of the muscle with similar dorsal and ventral distributions of muscle fibers.[20,21,24] Some of these fibers, however, are not circumferential and do project laterally to flanking fibrous bands, which provide support for the urethra.[7] An associated yet separate structure involved with continence is the puboperineales, which by their sling-like arrangement about the urethra and the fast-twitch component of muscle fibers, work as a voluntary fast-stopping mechanism for continence.[22]

The innervation for this continence region is somewhat complex. Dual innervation of the rhabdosphincter has been suggested by several investigators.[25,27] Recent work by Steiner and associates using fresh cadaver dissections has provided greater insight regarding the neuroanatomy of the urethra and suggests some surgical modifications that may aid in optimizing continence in surgical patients.[28,29] In general, the autonomic component of the urethra is derived from the pelvic nerve and inferior hypogastric plexus. The somatic component is provided by the pudendal nerve. A perineal component, the perineal nerve, provides somatic innervation for the urinary rhabdosphincter. Additionally, as the pudendal nerve heads toward the ischioanal fossa, a terminal branch of the intrapelvic component of the pudendal nerve can be identified, which innervates the rhabdosphincter at the 5 and 7 o'clock positions (**Fig. 28.2**). Thus, the somatic innervation of the rhabdosphincter is from two components of the pudendal nerve.

The smooth muscle component of the proximal continence mechanism is innervated by intrapelvic branches of the inferior hypogastric plexus that are distinct from the neurovascular bundle cavernosal nerves and enter the inferiolateral aspect of the rhabdosphincter.[28,29] These autonomic fibers also enter at the 5 and 7 o'clock positions along with the intrapelvic branches of the pudendal nerves, and course through the rhabdosphincter to supply the muscle component of the urethra (**Fig. 28.3**). The intrapelvic component of the pudendal nerve was found to emanate from the pudendal up to 2 cm proximal to the rhabdosphincter or branch of the pudendal at the rhabdosphincter. More recent histochemical studies have also supported the concept of dual (mixed) innervation.[30]

The practical implication of these findings bears on surgical technique. If a significant proportion of innervation to vital continence mechanisms can be demonstrated at the posteriolateral portion of the rhabdosphincter, it would be prudent to avoid spreading and dissection of the urethra in that location during the apical portion of the operation. An effort to avoid disruption or traction on the pelvic floor structures during surgery is further rationalized by these data. The 5 and 7 o'clock entry of this dual innervation also suggests that suture placement should be modified to more lateral and medial locations.[7]

The continence mechanism is often obscured anatomically by the variations in the anatomy of the prostate apex and the effect of benign enlargement in general.[31] The impact of variable prostate anatomy on the optimal performance of a radical prostatectomy has been elegantly investigated and described by Myers in several studies and the following descriptions are drawn primarily from his work.[19,23,32]

What is generally thought of as only the anterior fibromuscular stroma contains not only the commisure, but also a portion of the preprostatic sphincter and the collection of longitudinal smooth muscle as well as prominent veins referred to as the detrusor apron.[9] The apron is an extension of the longitudinal smooth muscle of the bladder, while the preprostatic sphincter is an extension of the circular smooth muscle fibers of the bladder neck. The maturation of the prostate displaces these structures. With the development of benign prostatic hyperplasia, one also sees the enlargement of the avascular plane beneath the apron.[23] This can facilitate vascular control in larger glands, but also suggests that passage of a vascular clamp should be avoided in smaller glands, where the margin for damage to the urethra is greater.[32]

In addition to structural displacement by general enlargement, there is significant variation in the morphology of the gland at the apex. This is best demonstrated during surgery by downward pressure at the vesicoprostatic junction, which then accentuates the presenting apical anatomy.[32] While multiple variations may be encountered, the issue is conceptually simplified by the presence or absence of an anterior apical notch (**Fig. 28.4**).

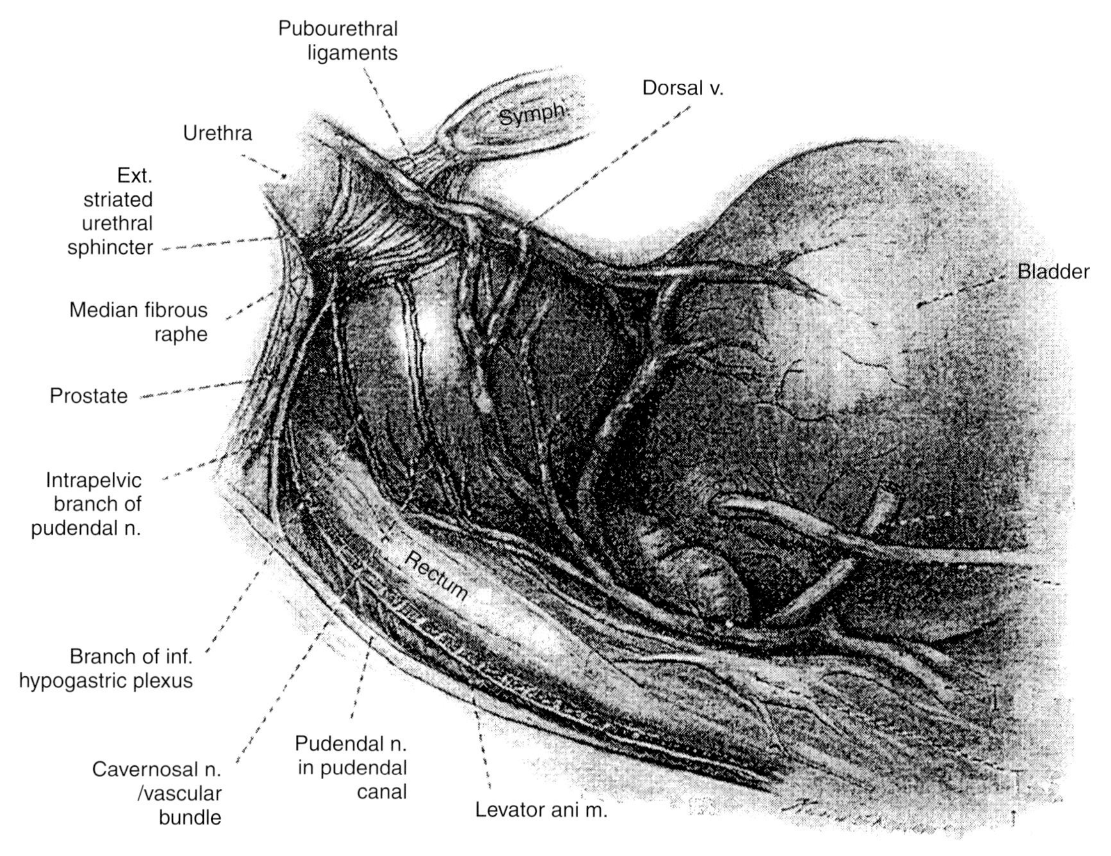

Fig. 28.2. Diagram of pudendal nerve anatomy demonstrating an intrapelvic branch innervating the rhabdosphincter (reprinted from Steiner MS. *Urology* 2000; 55:428 with permission from Elsevier).[7]

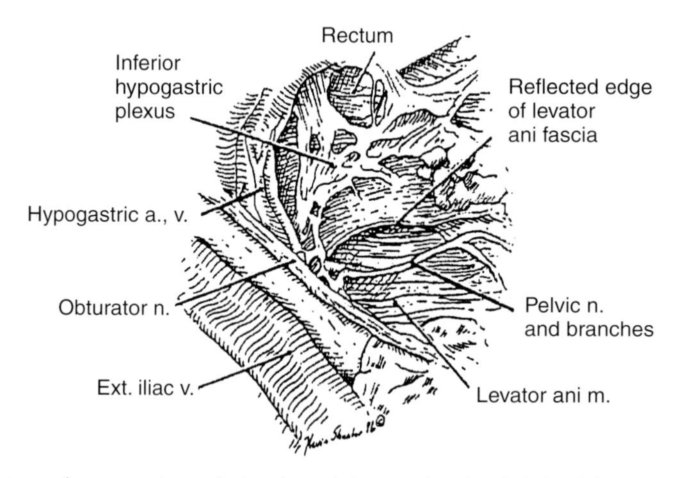

Fig. 28.3. Diagram of pelvic nerve anatomy demonstrating a distinct branch innervating the rhabdosphincter.

If a significant portion of the periprostatic urethra is enveloped in lateral lobe hypertrophy, division at the apex can sacrifice several millimeters of continence-preserving urethra. In the ideal situation, the apex falls off sharply at the urethral level of the verumontanum, providing optimal length without compromising cancer control. Prostate hypertrophy may be so great, however, as to envelope the periprostatic urethra for several millimeters beyond the verumontanum, and any attempt to recover this length would result in an unacceptable compromise of the prostatic margin.[32]

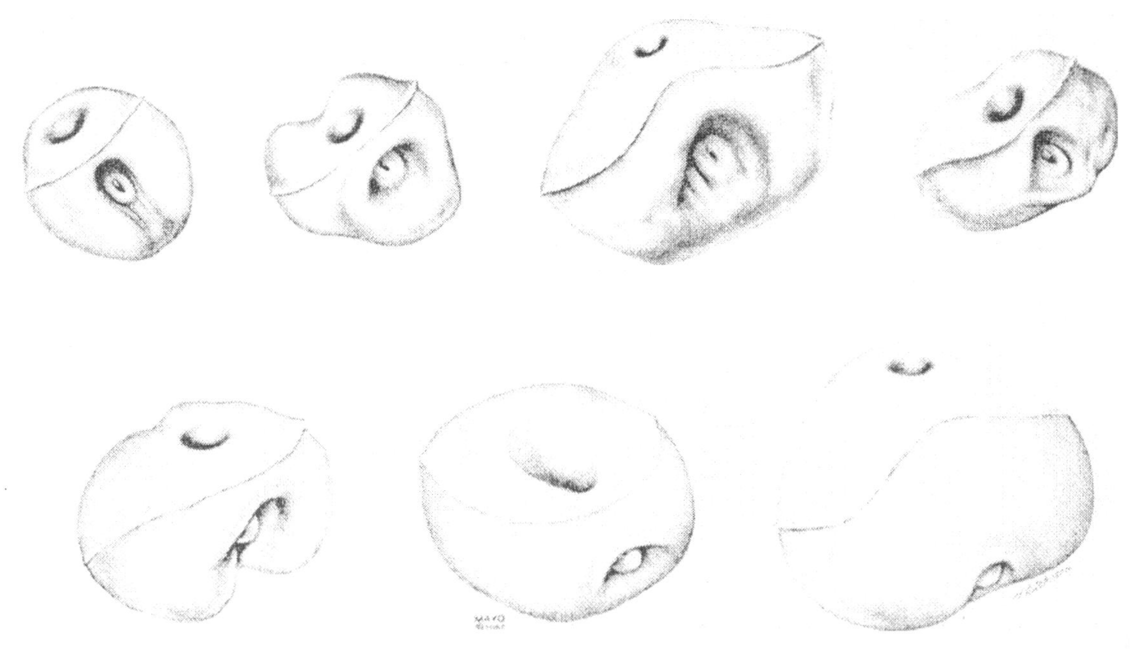

Fig. 28.4. Multiple variations of the prostate apex secondary to the effect of benign hypertrophy (reprinted from Myers RP, Goellner JR, Cahill DR. *J. Urol.* 1987; 138:544 with permission from Lippincott, Williams and Wilkins).[32]

SYNTHESIS

The past decade has seen the incorporation of radical prostatectomy into the therapeutic arsenal of most urologists. It is performed in a region of anatomic constraints, yet the careful delineation of anatomic relationships and morphologic variability allows surgeons the ability to optimize outcomes with regard to cancer control, maintenance of urinary continence and restoration of sexual function. Recent reviews emphasize certain anatomic observations that aid in the execution of this procedure:[7,19,33]

- Multiple micropedicles connect the neurovascular bundles along the lateral portion of the prostate. These can be significant near the urethra and must be dealt with carefully along the entire length of the prostate to avoid attrition of nerve fibers.
- The prostate is covered by a detrusor apron, which should not be bunched or divided beyond the prostate–urethral junction.
- The puboprostatic ligaments, in reality, are pubovesical ligaments that at some point need to be divided.
- The urethra is sphincteric from the verumontanum to the penile bulb. It is the principal source of continence post-prostatectomy, thus preservation of the urethra and associated structures is of paramount importance.
- In this regard, all attempts are made to preserve the circumferential musculature of the rhabdosphincter and the fascial support structures are left intact as much as possible (puboprostatic ligaments superiorly and median fibrous raphe posteriorly). This is accomplished by not passing clamps through or exerting traction on these structures during division.
- The dual innervation of the urethral sphincter is also best preserved by avoiding the intrapelvic branch of the pudendal nerve and awareness of the 5 and 7 o'clock entry of the somatic and autonomic innervation of the external sphincter. Awareness of anatomic variation and attention to detail can optimize this procedure for cancer control, and minimize potential long-term morbidity.

REFERENCES

1. Cunha G. Androgenic effects upon prostatic epithelium are mediated via trophic influences from stroma. *Prog. Clin. Biol. Res.* 1984; 145:81–93.
2. Kellokumpa-Lehtin P, Pelliniemi LJ. Hormonal regulation and differentiation of human fetal prostate and Leydig cells in utero. *Folia Histochem. Cytobiol.* 1988; 26:113–24.
3. McNeal JE. The prostate and prostatic urethra: a morphologic synthesis. *J. Urol.* 1972; 107:1008.
4. McNeal JE. Normal and pathologic anatomy of the prostate. *Urology* 1981; 17(Suppl. 3):11–24.
5. McNeal JE. Anatomy of the prostate gland. *Am. J. Surg. Path.* 1988; 12:619–33.
6. Brooks JD, Chao WM, Kerr J. Male pelvic anatomy reconstructed from the Visible Human data set. *J. Urol.* 1998; 159:868.
7. Steiner MS. Continence-preserving anatomic radical retropubic prostatectomy. *Urology* 2000; 55:427–35.
8. Reiner WG, Walsh PC. An anatomical approach to the surgical management of the dorsal vein and Santorini's plexus during radical retropubic prostatectomy. *J. Urol.* 1979; 121:198.
9. Myers RP. Improving the exposure of the prostate in radical retropubic prostatectomy: longitudinal bunching of the deep venous plexus. *J. Urol.* 1989; 142:1282.
10. Myers RP. Anatomical variation of the superficial preprostatic veins with respect to radical retropubic prostatectomy. *J. Urol.* 1991; 145:992.
11. Walsh PC, Donker PJ. Impotence following radical prostatectomy: insight into etiology and prevention. *J. Urol.* 1982; 128:492.

12. Walsh PC, Lepor H, Eggleston JC. Radical prostatectomy with preservation of sexual function: anatomical and pathological considerations. *Prostate* 1983; 4:473–85.
13. Walsh PC. Anatomic radical retropubic prostatectomy. In: Walsh PC, Retlk AB, Vaughan J et al. (eds) *Campbell's Urology*, pp. 2565–88. Philadelphia, PA: W.B. Saunders, 1997.
14. Villers A, McNeal JE, Freiha FS et al. Invasion of Denovilliers fascia in radical prostatectomy specimens. *J. Urol.* 1993; 149:793–799.
15. Villers A, McNeal JE, Redwine EA et al. The role of perineural space invasion in the local spread of prostatic adenocarcinoma. *J. Urol.* 1989; 142:763–70.
16. Walsh PC. Anatomic radical prostatectomy: evolution of the surgical technique. *J. Urol.* 1998; 160:2418–27.
17. Steiner MS. The puboprostatic ligament and the male urethral suspensory mechanism: an anatomic study. *Urology* 1994; 44:530–41.
18. Lowe BA. Preservation of the anterior urethral ligamentous attachments in maintaining post-prostatectomy urinary continence: a comparative study. *J. Urol.* 1997; 158:2137–41.
19. Myers RP. Radical prostatectomy: pertinent surgical anatomy. *Atlas Urol. Clin. North Am.* 1994; 22:1–23.
20. Dorschner W, Stolzenburg JU, Leutert G. A new theory of micturition and urinary continence based on histomorphological studies. 1. The musculus detrusor vesicae: occlusive function or support of mictruition? *Urol. Int.* 1994; 52:61–72.
21. Gosling JA, Dixon JS, Critchley HO et al. A comparative study of the human external sphincter and peri-urethral levator ani muscles. *Br. J. Urol.* 1981; 53:35–43.
22. Myers RP, Cahill DR, Kay PA et al. Puboperineales: muscular boundaries of the male urogenital hiatus in 3D from magnetic resonance imaging. *J. Urol.* 2000; 164:1412–19.
23. Myers RP. Male urethral spincteric anatomy and radical prostatectomy. *Urol. Clin. North Am.* 1991; 18:211–27.
24. Narayan P, Konety B, Aslam K et al. Neuroanatomy of the external urethral sphincter: implications for urinary continence preservation during radical prostate surgery. *J. Urol.* 1995; 153:337–48.
25. Gosling JA, Dixon JS. The structure and innervation of smooth muscle in the wall of the bladder neck and proximal urethra. *J. Urol.* 1975; 47:549–58.
26. Burnett AL, Mostwin JL. In situ anatomical study of the male urethral spincteric complex: relevance to continence preservation following major pelvic surgery. *J. Urol.* 1998; 160:1301–6.
27. Elbadawi A, Schenk EA. A new theory of the innervation of bladder musculature. 2 Innervation of the vesicourethral junction and external sphincter. *J. Urol.* 1974; 111:613–15.
28. Hollabaugh Jr RS, Dmochowski RR, Steiner MS. Neuroanatomy of the male rhabdosphincter. *Urology* 1997; 49:426–34.
29. Hollabaugh RS, Dmochowski RR, Kneib TG et al. Preservation of putative continence nerves during radical retropubic prostatectomy leads to more rapid return of urinary continence. *Urology* 1998; 51:960–7.
30. Zvara P, Carrier S, Kour NW et al. The detailed neuroanatomy of the human striated urethral sphincter. *Br. J. Urol.* 1994; 74:182–7.
31. Turner-Warwick R. The sphincter mechanisms: their relation to prostatic enlargement and its treatment. In: Hinman JF (ed.) *Benign Prostatic Hypertrophy*, p. 809. New York: Springer-Verlag, 1983.
32. Myers RP, Goellner JR, Cahill DR. Prostate shape, external striated urethral sphincter and radical prostatectomy: the apical dissection. *J. Urol.* 1987; 138:543–50.
33. Kaye KW, Creed KE, Wilson GJ et al. Urinary continence after radical retropubic prostatectomy. Analysis and synthesis of contributing factors: a unified concept. *J. Urol.* 1997; 80:444–501.

Radical Retropubic Prostatectomy

Chapter 29

John W. Davis and Paul F. Schellhammer

Department of Urology, Eastern Virginia School of Medicine, Norfolk, VA, USA

HISTORY

Localized prostate cancer can be treated by one or more of several treatment options including observation, androgen deprivation, chemotherapy, cryotherapy, brachytherapy, external beam radiation therapy and surgical removal of the gland. Randomized trials comparing modern treatment strategies are not available; therefore, patients and treating physicians must evaluate survival and quality of life results from individual therapies and make their best judgment as to how to proceed. Radiotherapeutic techniques have been employed in prostate cancer treatment for many decades; however, the long-term results are difficult to predict for currently treated patients. Techniques have continually changed in terms of dosimetry, number and definition of fields of treatment, use of neoadjuvant hormonal ablation, and imaging technique for planning. Radical retropubic prostatectomy has several attractive aspects: 10–20-year survival results are available from multiple institutions, surgical removal of the primary organ is a time-tested principle of oncology, and the pathologic staging provides valuable information to the patient as risk of failure and need for further treatment to minimize failure. Nomograms are now available for each of the three major therpeutic options: radiation therapy, brachytherapy and radical prostatectomy. They will help to identify expectations for each modality.[1–3]

Since the description of the anatomic approach by Walsh, the radical retropubic prostatectomy has become one of the topics in urology about which most is written. Nearly every minute step of this operation has been described and illustrated in detail. Countless additional papers have challenged old concepts and introduced newer techniques designed to solve ongoing problems with improving cancer control and postoperative quality of life. Additional studies have introduced techniques to decrease blood loss and transfusions, and the use of clinical pathways designed to decrease length of hospital stay and overall costs.

No single text chapter can capture all of the concepts and opinions related to this operation. Ultimately, each surgeon with a serious interest in prostate cancer decides how to perform the operation based on his or her initial training and subsequent experience. Subsequent modifications to a surgeon's technique must be carefully planned based upon the nature of his or her surgical practice, a critical interpretation of the literature, and ongoing review his or her personal outcomes.

In this chapter, we set reasonable goals for discussion. After establishing historical relevance, the technique of radical retropubic prostatectomy is presented based upon over a decade of experience and critical self-review at Eastern Virginia Medical School in Norfolk, Virginia. Next, we review some interesting technical, pathologic and quality of life information that may help the younger surgeon develop his or her technique. At the time of this writing, traditional radical prostatectomy is being challenged on many fronts by laparoscopic radical prostatectomy, brachytherapy and conformal external beam radiation therapy dose escalation with or without neoadjuvant/adjuvant hormonal deprivation. Despite a lack of proof from randomized clinical trials, traditional urologic teaching is that radical prostatectomy offers the best chance of cure when disease is organ confined. Efforts continue to substantiate those differences through the clinical trial process: the ongoing PIVOT trial, which is approaching 1000 patients randomized to radical prostatectomy or expectant management, and The American College of Surgeons Clinical Trial group is sponsoring a trial which will randomize 2000 men with Gleason scores ≤6, and prostate specific antigen (PSA) <10 to radical prostatectomy or brachytherapy (SPIRIT: Surgical Prostatectomy Interstitial Radiation Intervention Trial). The Scandinavian Prostate Clinical Trial group has enrolled >700 patients to radical prostatectomy versus expectant management, and analysis will be available in 2003.

PREOPERATIVE COUNSELING

New patients seen in the clinic need individual face-to-face time for counseling, preferably in conjunction with their spouse or significant other. A complete history and physical, medical and family history, etc. is completed with special attention to comorbid factors, prior cancers, smoking habits, etc. that may affect their longevity. A life expectancy of 10 or more years for a surgical candidate is an accepted figure. Patients over age 70 are occasionally considered for surgery, if highly motivated and with evidence of family longevity. Patients with comorbid factors often require consultation from medical colleagues for assessment on optimization for surgery.

Theoretically, one could spend an entire office day talking about prostate cancer with a handful of patients. Unfortunately, this is not a financially viable option in today's health care market. In addition, it is our experience that localized prostate cancer is the most frequent second or more opinion visit in a urologic oncologist's practice. Therefore, streamlined patient information is a must, so that face-to-face time can be efficient. We have found it useful to produce a videotape of our oncologists discussing the treatment options for localized prostate cancer. Handouts and suggested print and Internet-based reading are also helpful. Our view of the urologist role is to provide unbiased interpretation of the information concerning the pros and cons of prostate cancer treatment options.[4] We routinely recommend that our patients discuss their treatment options with a radiation oncologist, and reassure them that a 2–3-month delay in treatment while seeking the necessary medical opinions and reading is beneficial to their long-term satisfaction with treatment and is unlikely to decrease their chances of a cure, given the extremely slow growth of the tumor. There is no clear evidence to suggest that beginning androgen deprivation while the undecided patient reviews treatment options is beneficial.

It is our practice and recommendation that review of prostate biopsy slides with a designated genitourinary pathologist is a critical step in quality assurance, and maximizing accuracy of preoperative planning.[5] Given that radiation therapy and brachytherapy are competitive treatment alternatives, urologists performing radical prostatectomy have a duty to review and adjust their technique periodically to maximize their results. As Gleason scoring has a learning curve and subjective component, re-review by a genitourinary pathologist may change the score in a significant number of cases. As Gleason score is a major determinant in predicting pathologic stage, relying on outside pathologist scoring allows potentially significant inconsistency when interpreting one's pathologic organ-confined and margin-free rates. Furthermore, a significant change in Gleason score may alter one's recommendations regarding watchful waiting, brachytherapy, neoadjuvant androgen deprivation or the appropriateness of nerve-sparing surgery. Finally, in rare circumstances, the presence of tumor cannot be confirmed and one of the mimics of cancer is diagnosed.

We do not obtain whole body nuclear bone scanning or computed tomography (CT) of the abdomen and pelvis for Gleason 7 or less, PSA < 10, and stage T2a or less. Use of the prostascint scan has waned recently owing to the frequently equivocal results, but may be considered in higher risk cases – usually in the circumstance where external beam radiation therapy (XRT) with the long-term neoadjuvant hormone ablation is the primary treatment consideration.

The criteria of PSA < 10, PSA < 7 and stage $\leq$ T2a have been used to eliminate the dissection of pelvic lymph nodes in conjunction with radical perineal prostatectomy.[6] We have started to adopt this approach with the radical retropubic operation as well. Our patients usually predonate 2 units of autologous blood for their procedure, although there is increasing evidence that this is not necessary. Other strategies have been described such as using only banked blood if one's transfusion rates are low, hemodilution and preoperative erythropoetin.[7]

Informed consent is carried out, quoting a $<1\%$ operative mortality and a $<5\%$ risk of major/minor complications.[8] Quality of life literature is cited, including our own experience with radical prostatectomy, brachytherapy and XRT.[9]

A common issue for many men is the appropriateness of nerve preservation, while also controlling cancer. Positive surgical margins increase the risk of PSA recurrence and need for additional treatments,[10] and the neurovascular bundle is a common site of spread of disease. Many factors affect return of potency following radical prostatectomy including age, number of bundles preserved and presurgical potency.[11] In addition, preservation of both nerves (and for some patients under 55 years of age unilateral nerve sparing) is required for response to Sildenafil.[12,13] It has been suggested that nerves may be preserved in the majority of men with potentially curable disease, and that intraoperative findings can guide the decision based upon induration in the lateral pelvic fascia, adherence of the neurovascular bundle to the prostate while it is being released, and inadequate tissue covering the posterolateral surface of the prostate once the prostate had been removed, leading to secondary wide excision of the neurovascular bundle.[14] Other authors have shown poor sensitivity of intraoperative findings.[15]

In our practice, we do not see a high volume of men self-selecting for nerve sparing, and have used the procedure less frequently. In the past 5 years, the distribution has been 50% non-nerve sparing, 27% unilateral nerve sparing and 23% bilateral nerve sparing. In general, nerve sparing is not performed in men with preoperative poor sexual function, men who stress that survival is the overwhelming first priority, apical induration or positive-biopsy perineural infiltration,[16,17] and in the majority of men over age 70. Nevertheless, these are only guidelines. Multiple preoperative factors affect the chance of non-organ-confined disease and positive margins, including PSA, Gleason score, stage, perineural invasion, number/percentage positive cores, length of core involvement and volume of Gleason 4/5 disease. Graefen et al. have performed a multivariate analysis of these factors, and found that the risk of extracapsular extension was significantly affected by the presence of Gleason 4 in biopsy cores, the percentage of positive cores and PSA.[18] Ultimately, the risks and benefits of nerve preservation must be discussed with the patient and the decision individualized. Patients seeking multiple urologic opinions, as well as urology residents and fellows working with different experienced surgeons, will note varied attitudes and recommendations regarding nerve sparing.

TECHNICAL DISCUSSIONS

Positioning

The patient is positioned with the lower back supported by a gell-role or elevated kidney rest, and the table is flexed to open the pelvis. We position the patient in the low lithotomy position with Allen stirrups. This position allows access to the rectum in the rare event of rectal injury, allows for perineal pressure to expose the urethra and a position for a second assistant. Other surgeons utilize the supine position with the legs in frog-leg split or split with spreaderbars.

We perform flexible cystoscopy, once the patient is positioned, to look for a median lobe, check the position of the ureteral orifices and rule out other bladder pathology. The yield is low and recent reports have suggested this step is unnecessary. However, the additional 2 minutes may provide useful information.

Incision and exposure

A low-midline incision is made from just below the umbilicus to above the pubic bone. After cauterizing through the fat, the fascia is divided in the midline and the rectus muscles bluntly spread laterally – the pyrimidalis muscle is divided at the inferior aspect of the incision until the pubic bone is exposed. The thin transversalis fascia is divided over the bladder and peritoneum, at which time entry into the extraperitoneal space of Retzius is easily accomplished. The surgeon's hand or a sponge stick can then be used to sweep the bladder off the pelvic sidewalls. The vas deferens is identified over the external iliac vein and the peritoneum underneath is swept superiorly. If more exposure is necessary, the posterior sheath can be divided a short distance at both lateral edges near the abdominal wall to allow more cephalad retraction of the peritoneum. A retractor system of choice is installed for pelvic lymph node dissections such that the abdominal wall/rectus muscle is retracted laterally beyond the pelvic side wall, the bladder is retracted medially and the peritoneal pocket over the iliac vessels is retracted superior to expose the vein.

LYMPH NODE DISSECTION

Pelvic lymph node dissection for prostate cancer is a diagnostic/staging procedure with no evidence of inherent therapeutic value (but which may be useful in tumor reduction in conjunction with other treatments). Dissection borders are the medial aspect of the external iliac vein laterally, the obturator nerve inferiorly, the vessel bifurcation superior and the node of Cloquet distally. The obturator nerve is always preserved to maintain medial abduction of the thigh. Obturator vessels are preserved when possible, and dissection below the nerve is unnecessary and likely to cause bleeding. Frozen sections are not routinely obtained. Recent data from the Eastern Cooperative oncology Group (ECOG),[19] and the historical series from Mayo[20] suggest a benefit to radical prostatectomy and androgen deprivation, and removal of the primary tumor may provide an advantage to response to chemotherapy. However, some surgeons advocate curtailment of surgery in the face of grossly positive nodes.

PROSTATECTOMY

After completing the lymph node dissection, the prostate dissection begins by readjusting the exposure. The bladder is retracted cephalad by hand, a malleable retractor or a notched malleable retractor. Superficial fat and the superficial dorsal vein are dissected off the apex with cautery. A medial retraction on the prostate exposes the edges of the endopelvic fascia, which must often be cleaned of fat and levator muscles with a peanut dissector. The lateral edges of endopelvic fascia are then divided with scissors proximally to the bladder neck and distally towards the puboprostatic ligaments. Dissection near the ligaments must be carefully visualized and often branches of the dorsal vein run lateral to the ligament.

We find it helpful to define the proper planes for neurovascular bundle preservation or division[21] (**Fig. 29.1**). Using medial retraction with a sponge stick, the neurovascular bundle running posterior–lateral to the prostate is visualized. Often there are large veins running through the complex with a thin overlying layer of fascia. For nerve preservation, the plane of thin lateral prostatic fascia above the neurovascular bundle is sharply and carefully divided. The amount of dissection possible at this stage varies, as some men have extremely large veins in this plane that will bleed if aggressive dissection is carried out. If bleeding is encountered, suture ligation with a small, superficially placed, resorbable suture (i.e. 4–0 Vicryl) is preferred over cautery. If non-nerve-sparing dissection is

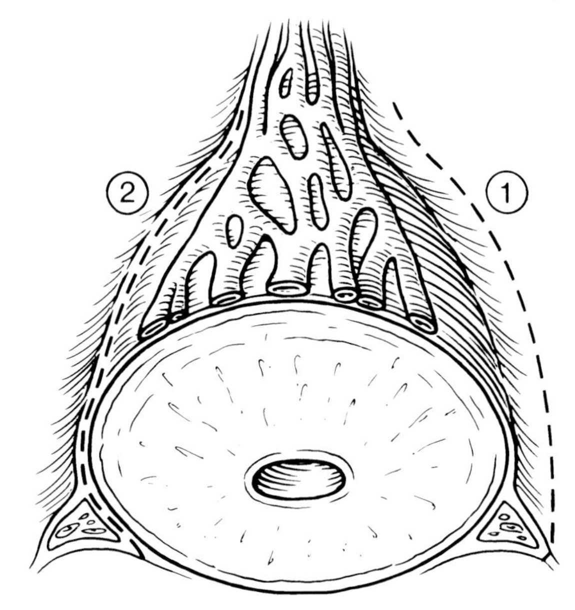

Fig. 29.1. Prior to division of the dorsal vein complex, the lines of incision are prospectively defined: laterally over the ventral rectal fascia in the non-nerve-sparing procedure (1) or medially through the visceral prostatic fascia in the nerve-sparing procedure (2).

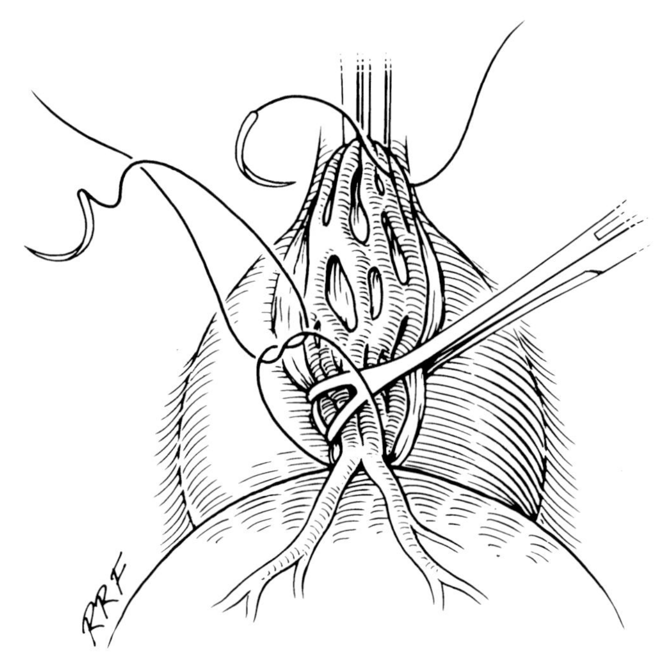

Fig. 29.2. A Babcock clamp is used to gather the visceral edges of the endopelvic fascia and a suture ligature is placed to control back bleeding.

planned, a much wider plane of sharp dissection is carried out that exposes the underlying perirectal fat. Proper dissection of these planes at this step of the operation helps direct the proper plane of lateral pedicle dissection and division once the urethra is divided, such that the amount of periprostatic tissue dissected with the specimen is maximized.[22] Next, the bunching of the proximal dorsal vein is accomplished. We find it easiest to draw up the leaves of the cut endopelvic fascia with a Babcock or Allis clamp and sew under this layer with a 2–0 Vicryl CT-1 needle (straightened) (**Fig. 29.2**)

DORSAL VEIN/APICAL DISSECTION

In the past, the puboprostatic ligaments were divided to allow easy palpation of the urethra, and passage of a clamp between the urethra and dorsal vein complex for control and division. However, many authors have reported the advantage of leaving the puboprostatics mostly intact for early return of urinary continence. By sweeping the levator muscles off of the apex of the prostate and urethra, the catheter can be palpated. A short distance of puboprostatic ligament is cut so that the apex and complex can be palpated. For a right-handed surgeon standing on the patient's left, the left index finger palpates the apex of the prostate and the urethral cathether to guide a right-angled clamp to the plane between the urethra and the dorsal vein complex. The McDougal clamp is popular and a Stile clamp works well in our experience as well. The clamp should be placed only a short distance past the apex, as placement too distally will often stir up venous bleeding that is deep and difficult to control until the prostate is removed. Generally, mild pressure is applied, and the clamp will 'pop' with finger tip dissection through the plane to the left side.

Numerous techniques for dorsal vein control have been described.[23] In our experience, bringing sutures around the complex with a clamp and attempting to tie distally usually results in these sutures being cut or dislodged during division of the complex. Thus, oversewing of the complex is required. More successful techniques may include bringing the back end of a suture around the complex and then suture ligating distally, or starting a suture ligating anteriorly, dividing the complex, and then quickly oversewing. A secure method for ligature is illustrated in **Fig. 29.3**. We have also had personal success with using a Debakey hypogastric clamp to occlude the distal aspect of the complex, dividing the complex, and then suture ligating around the clamp (**Fig. 29.4**). Regardless, there is usually a vein or two that escapes these maneuvers and requires additional suture ligature. Placement of sutures is optimized by placing one's body parallel to the patient, facing the head, and sewing underhand from patient left to right. It is imperative that the dorsal vein complex be completely controlled before proceeding, as dissection of the urethra and posterior apex will otherwise be difficult to visualize. Sutures should be in the 2–0 range with Monocryl or Vicryl, and needles such as UR-5, UR-6, CT-1 or CT-2, depending on surgeon preference. As a final word of caution, suture ligatures should be kept in the narrow region of the complex and anterior to the urethra. Wide placement of surtures that incorporate lateral levator muscles may avulse veins running in these structures and may adversely affect continence.

In unusual circumstances, the surgeon will be faced with a narrow deep pelvis, in addition to a notched-like 'hump' growing from the pubic ramus. This notch may completely obscure proper vision and palpation of the urethra and dorsal vein complex. Proceeding with apical dissection under such circumstances may lead to bleeding that is extremely difficult to control. When this anatomy is encountered, it may be safer to call for an orthopedic tray and use osteotomes, and a mallet to carve this notch out of the way. The pubic ramus is then smoothened with a file and bone wax.

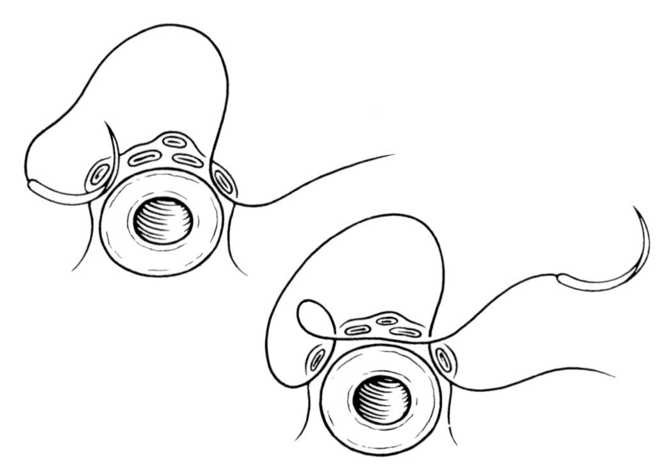

Fig. 29.3. A suture is used to secure the dorsal vein complex. It is placed to include vessels that fall lateral to the urethra.

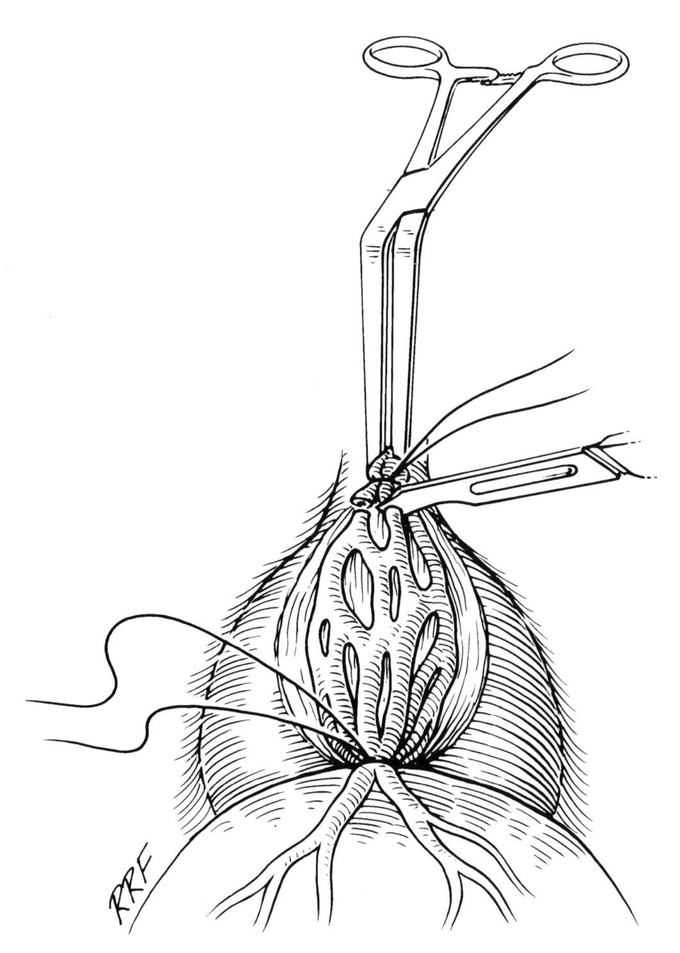

Fig. 29.4. A DeBakey hypogastric vascular clamp is applied distal to the suture prior to division to the urethra with knife and/or scissor. This minimizes blood loss.

After control of the dorsal vein complex, the urethra is easily palpable. A thin clamp is used to gently separate the lateral leaves of the rhabdosphincter. At this point, many surgeons have described passing a right-angle clamp underneath the urethra to assist with division. Recent anatomic descriptions of the innervation of the rhapdosphincter, however, suggest that manipulation in this area should be minimized.[24] We have found it just as effective to simply cut the anterior urethra with a knife without a clamp underneath. Once the anterior urethra is divided and the catheter visualized, the balloon is deflated and the catheter withdrawn, and the posterior urethra is divided. Sharp dissection is continued through the posterior Denonvilliers' fascia such that the proper plane between the fascia and perirectal fat is established. Pulling cephalad with the cut end of the foley before establishing this plane may lead to starting the posterior apical dissection above Denonvilliers' and possibly increasing the rate of positive margins at the apex.[25] Using an allis clamp to move the prostate (**Fig. 29.5**), we sharply dissect past Denonvilliers' first, establish the plane immediately anterior to the anterior longitudinal fibers of the rectum, which are covered with a fibrofatty layer, then reinsert the foley catheter through the prostatic urethra to provide cephalad traction.

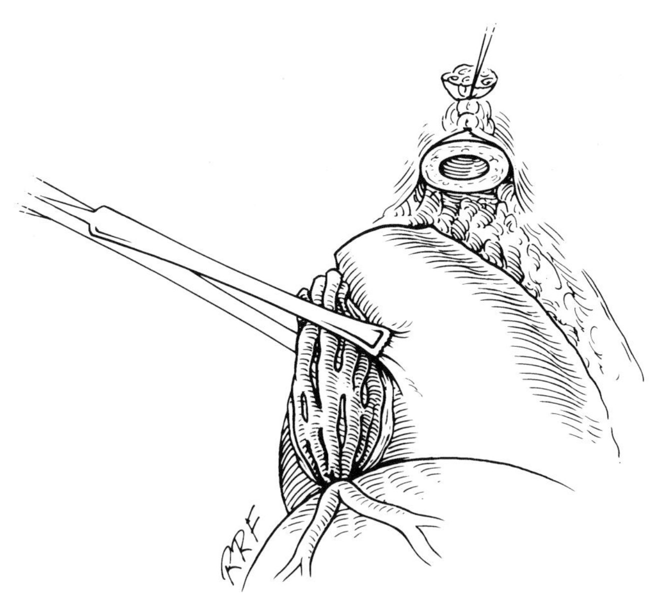

Fig. 29.5. An Allis clamp permits movement of the prostate without the slippage often encountered with a sponge stick traction. In this figure, note the incision in the ventral rectal fascia, which keeps the neurovascular bundle and surrounding soft tissue with the specimen to minimize the chance of a positive margin.

LATERAL PEDICLES

Once the proper plane is established beneath Denonvilliers' fascia, an index finger can gentle separate the prostate from the rectum, and the lateral pedicles are clearly defined. As previously described, we define the plane of nerve-sparing versus non-nerve-sparing dissection at the beginning of the dissection. Using these previously defined planes, a right-angle clamp is used to dissect the lateral pedicles, which are controlled with sutures or small clips. Cautery needs to be avoided for nerve-sparing dissections to avoid damage. Cephalad traction is provided by pulling on the foley catheter. Once the lateral pedicles are divided, the neurovascular bundle is visible posterior–lateral, if spared. Venous bleeding may occur and is controlled with superficially placed fine sutures. **Figure 29.6** summarizes these maneuvers in the sagital plane.

BLADDER NECK DISSECTION

Once the lateral pedicles are divided, the bladder neck can be approached anteriorly. Cautery is used to divide the detrussor fibers distal to the foley balloon. Once the bladder is entered, the balloon is deflated and the end of the foley brought out of the prostate for upward traction to expose the posterior bladder neck. We use a needle point bovie to divide the posterior mucosa that will be used for the anastomosis sharply. If a large median lobe is present, it must be pulled out of the bladder with the prostate and the mucosa may be dissected from the lobe to create a posterior lip followed by the division of the posterior bladder neck made just proximal to the lobe.

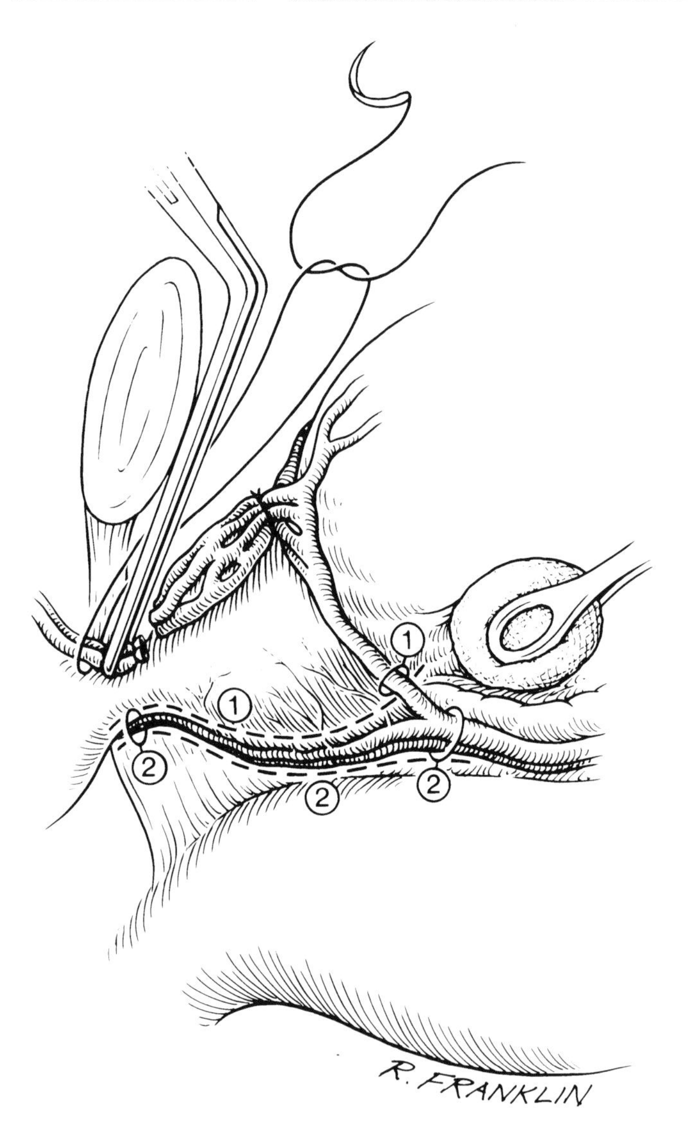

Fig. 29.6. The sagittal view demonstrates the placement of the sutures to prevent back bleeding from the dorsal vein complex and the placement of the hypogastric clamp before the division in the dorsal vein complex. Upward and cephalad pressure on the bladder with a sponge stick will flatten and straighten the visceral prostatic fascia.

Cautery is then used to finish the division of the posterior detrussor fibers. With gentle posterior spreading, the vasa deferentia are identified and ligated with clips. A tonsil clamp is then used to identify the remaining lateral detrussor fibers, which can be clamped and suture ligated. The seminal vesicles are then dissected out, being sure to clip the vasculature at the tips. Evidence that complete dissection of the seminal vesicle tip is unnecessary has been presented.[26] Advantages are avoidance of bleeding and minimizing neural damage.[27] The specimen is then inspected and sent for pathologic examination. Separate sections are sent from the bladder neck, apex and, occasionally, the pedicle for more specific margin definition.

Alternatively, the bladder neck can be approached posteriorly by dividing the posterior Denonviellier's fascia near the seminal vesicles, which are then dissected out along with the vasa deferentia. The remaining bladder neck is then divided. This approach is necessary, if bladder neck preservation is planned, as detrussor fibers can be dissected off circumferentially until all that is left is prostatic urethra, which is divided, leaving a much smaller cystotomy to close. Bladder neck preservation has been advocated as enhancing the return of continence, keeping dissection away from the ureteral orifices and simplifying reconstruction. However, a recent randomized trial showed higher rates of positive margins only at the bladder neck with preservation and no difference in baseline, 2- and 6-month continence.[28] A retrospective study of 222 bladder-neck-sparing cases showed increased positive margins in pT3a cases compared to standard bladder neck resection.[29]

BLADDER NECK RECONSTRUCTION

To ensure a seal between the urethra and bladder neck, bladder mucosa is everted with interrupted, fine absorbable sutures. Indigo carmine may be given intravenously to help visualize the ureteral orifices, especially if a large median lobe dissection has distorted the trigone. A 2–0 Vicryl suture is used to close the posterior bladder neck in running or interrupted fashion, such that the bladder opening is approximately 28F in size. Care must be taken posteriorly to stay superficial and not incorporate the ureteral orifices in the closure. If the orifices are close, 5F feeding tubes can be placed temporarily as guides during reconstruction.

ANASTOMOSIS

As we position the patient in the low lithotomy position, a sponge stick is placed to the perineum and cephalad pressure is applied by the assistant to expose the urethral stump. If the patient is supine, the Greenwald sound with retractor blades may be necessary to expose the stump. Six 2–0 Monocryl sutures on UR-6 needles are inserted, avoiding the 5 and 7 o'clock positions in nerve-sparing cases. One or both of the anterior sutures incorporate the dorsal venous complex for additional strength (**Fig. 29.7**). Sutures are placed through the corresponding positions on the bladder with knots on the outside. We organize our sutures by placing hemostats on the posterior sutures: Kelly's on the mid-sutures, and tonsils and the anterior sutures. A 20F 5 cm^3 soft catheter is placed into the bladder. The table is then flattened, the bladder brought down to the pelvis and the sutures tied from anterior to posterior. A catheter-tipped syringe irrigates the bladder with 100–150 cm^3 saline to check for leaks.

We place two Jackson Pratt drains to the pelvis, close the fascia with 0-Nylon running, and close the skin with a subcuticular Monocryl, as patients will be going home in 3 days, and an additional office visit for staple removal is saved.

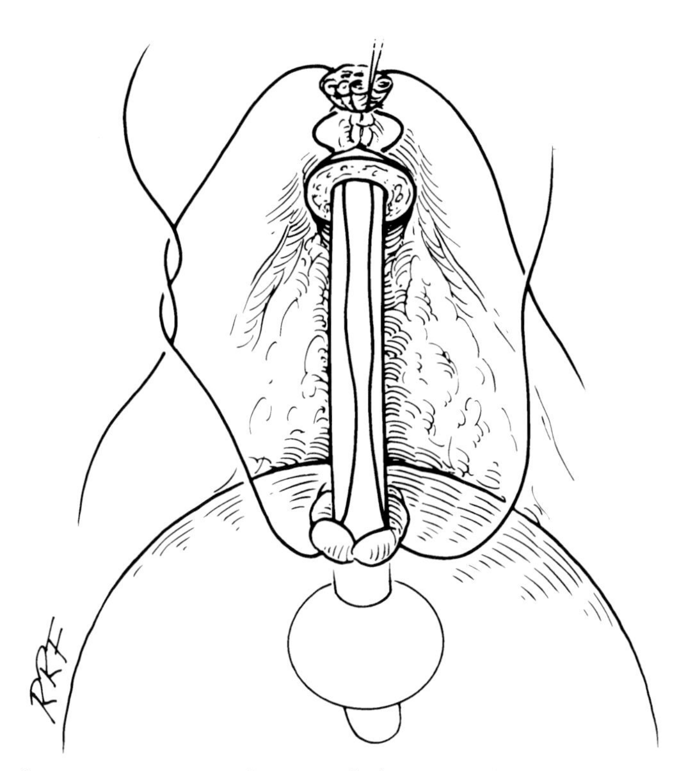

Fig. 29.7. Construction of vesicourethral anastomosis.

HOSPITAL ROUTINE

On postoperative day one, pain is controlled with an epidural, and sips of liquid are begun, and hemoglobin and creatinine are checked. On postoperative day 2, the epidural is removed, and oral analgesic begun, and the diet is advanced as tolerated Catheter teaching is begun and one drain is removed in the afternoon after ambulation, if the output remains low. On postoperative day 3, the second drain is removed, and the patient is discharged on oral pain medications, a stool softener and macrodantin 50 mg per day for suppression of bacturia, while the catheter is in place.

In the event of prolonged drain output, the creatinine level is checked. If it is more than twice the serum level, the drain is pulled back and placed to gravity until drainage resolves. Mild catheter traction may also help. If drainage is lymphatic, 12 cm^3 of 1/4% betadine solution is instilled through the drain and then aspirated to sclerose the lymphatics.[30,31] In either situation, patients may be discharged with the drain in place, cut short to near the skin and drained with an ostomy device for further evaluation in the office.

Postoperatively, patients return for cathether removal in 2–3 weeks without a cystogram. Early catheter removal closer to 1 week has been advocated in conjunction with a cystogram showing no or very minimal leak.[32–34] Patients are encouraged to perform frequent Kegal exercises to enhance urinary control.[35]

NEW ISSUES IN THE RADICAL RETROPUBIC PROSTATECTOMY

Without a doubt, experienced prostate surgeons will identify numerous points of our technique that they perform differently. The point to emphasize for surgeons in training is that there are many ways to perform this operation that will be successful, so long as the principles of oncologic surgery are followed, while trying to minimize the side effects of incontinence and, when indicated, impotency. Again, we emphasize that careful ongoing review of one's personal pathologic and quality of life results is needed. Walsh has even recently described keeping video coverage of each case so that outcomes can be correlated with the original surgical technique.

However, several technical modificiations have been introduced that are aimed at making a significant impact on quality of life outcomes.

Cavernosal mapping

Cavernous nerve stimulation has recently been described. The Cavermap Surgical Aid (Uromed Corp., Boston, MA) is a system composed of a control unit, a probe handle with disposable tip containing eight stimulation electrodes in a 1.2 cm linear array,[36] and a penile tumescence sensor. The device detects a 0.5% increase in penile girth as a positive response. Use of the device consists of establishing an initial response by a negative control (anterior bladder neck, for example) followed by a positive response at the posteriorlateral apex. The device is then used to map the course of the neurovascular bundle testing proximal and distal to proposed dissection planes (allow 45–60 seconds for detumescence). Following nerve-sparing prostatectomy, a positive response should be present at the bundle at the level of the bladder. Klotz et al.[37] performed a randomized trial comparing Cavermap nerve-sparing versus conventional nerve-sparing surgery. In both groups, Cavermap was used at the end of dissection to document response. End-points were potency by questionnaire and Rigiscan at 1 year. Cavermap prolonged the operation by a mean of 17 minutes and increased blood loss by 25% (but not transfusion rates). A bilateral response was associated with recovery in 68%, while no response was associated with 0% recovery. At 6 months, erections in the Cavermap were seen in 60% compared to 45% in conventional nerve sparing ($P = 0.10$), and Rigiscan results were improved ($P = 0.24$).

While Cavermap in this trial showed improved results, Klotz[36] has postulated that the Cavermap may improve the results in the control group, as the same surgeons were performing Cavermap versus non-Cavermap dissections by random assignment. Thus, Cavermap may not be an ongoing necessity for optimized outcomes, but may assist surgeons with less experience with the operation. Furthermore, nerve-sparing with Cavermap may lead to improved responses to Sildenafil, as well as assistance with sural nerve grafting.[36] Other authors, however, have reported their experience with Cavermap to be suboptimal.[38–40] Technology may provide more sensitive and specific monitoring in the future.

Sural nerve grafts

The decision to resect a neurovascular bundle is a significant challenge that requires individual counseling with the patient and careful review of one's results with positive margins. In terms of extracapsular, seminal vesicle and lymph node disease, all the urologist can do is obtain CT and bone scans (when clinically appropriate), and estimate the risks of these adverse findings with nomograms. Surgical margins, however, are under the control of the surgeon, and the posterior–lateral margin is a common area of spread of prostate cancer. Thus, non-nerve-sparing surgery is often indicated when prognostic factors such as Gleason score, PSA, palpable disease and high number/percentage of positive biopsy cores are present. The patient's age and potency status are also relevant factors. At the extremes the decisions are easy – the 68-year-old with erectile dysfunction should receive non-nerve-sparing, while the 52-year-old potent male with a small volume of Gleason 6 at the base or mid-prostate with a T1c lesion, PSA 4.5 should be offered at least a unilateral and probably a bilateral nerve-sparing approach.

Many patients, however, fall in the middle and have Gleason 4+3; 3+4 disease, apical disease, bilateral palpable disease, numerous bilateral positive cores, perineural involvement, and/or PSA >10. For the younger, fully potent patients, the decision is complicated by the simultaneous need for disease-free survival beyond 20–30 years and the desire to retain potency without the need for a vacuum erection pump, intracaverosal treatment or a prosthesis.

In an effort to offer the urologist more surgical options, Kim et al. have published preliminary data on resection of neurovascular bundles with interposition of sural nerve grafts. Their first report of nine patients treated with bilateral grafts showed improving spontaneous erections at 4–5 months.[41] They have subsequently reported their technique on sural nerve harvesting,[42] more details of the technique[43] and their 1-year follow-up reporting unassisted intercourse in 4/12 patients, and an an additional 5/12 with a '40–60%' recovery of erections. The greatest return of function occurred at 14–18 months after surgery.[44]

These preliminary reports are promising, as patients with bilateral resection of neurovascular bundles rarely achieve spontaneous erections. Results with unilateral grafts will be more difficult to interpret, as erectile activity is preserved in some of these patients without grafting. Many authors, however, have reported that unilateral nerve-preservation results in a more than 50% reduction in erectile function, and thus there is potential for unilateral grafts to show a statistically significant benefit in a randomized trial.

The topic of sural nerve grafts has been debated in a recent publication.[45] Both sides of the issue are well presented. While nerve grafts show efficacy, there is no expectation that they will ever be as effective as careful preservation of neurovascular bundles when cancer control can be accomplished. In theory, neurovascular bundle resection is only a benefit when there is extracapsular disease in that bundle that would have resulted in a positive margin, if nerve sparing was performed. In addition, it can be argued that a benefit is only present when the only site of a positive margin is at the neurovascular bundle. Thus, the surgeon is still left with the initial dilemma of when to resect a neurovascular bundle and, ultimately, the nature of one's patient population and referral patterns will determine the efficacy of nerve grafts. If a urologist establishes a practice known for bilateral nerve sparing, there will be a natural self-selection towards younger patients with low-risk disease, and nerve grafts are unlikely to show a benefit. On the other hand, if a urologist establishes a practice known for sural nerve grafting, there will be a natural self-selection towards higher risk disease where nerve grafting will show a benefit. In some European centers, PSA screening is not often recommended by primary care physicians, and the population of men undergoing radical prostatectomy have more advanced disease than those in the USA and nerve grafting may prove beneficial (Stefan Loening, personal communication). More data will be forthcoming from several centers including the Baylor College of Medicine, the Memorial Sloan–Kettering Cancer Center and the University of Texas MD Anderson Cancer Center. In addition, an experienced laparoscopic radical prostatectomy center has begun sural nerve grafting (Ingolf Türk, Charite Hospital, Berlin, personal communication).

REFERENCES

1. Kattan MW, Potters L, Blasko JC et al. Pretreatment nomogram for predicting freedom from recurrence after permanent prostate brachytherapy in prostate cancer. *Urology* 2001; 58:393–9.
2. Kattan MW, Zelefsky MJ, Kupelian PA et al. Pretreatment nomogram for predicting the outcome of three-dimensional conformal radiotherapy in prostate cancer. *J. Clin. Oncol.* 2000; 18:3352–9.
3. Kattan MW, Wheeler TM, Scardino PT. Postoperative nomogram for disease recurrence after radical prostatectomy for prostate cancer. *J. Clin. Oncol.* 1999; 17:1499–507.
4. Schellhammer PF. Patient counseling regarding treatment of T1/2 (A/B) prostate cancer. *Am. J. Clin. Oncol.* 1995; 18:532–7.
5. Murphy WM, Rivera-Ramirez I, Luciani LG et al. Second opinion of anatomical pathology: a complex issue not easily reduced to matters of right and wrong. *J. Urol.* 2001; 165:1957–9.
6. Bluestein DL, Bostwick DG, Bergstralh EL et al. Eliminating the need for bilateral pelvic lymphadenectomy in select patients with prostate cancer. *J. Urol.* 1994; 151:1315–20.
7. Nieder AM, Rosemblum N, Lepor H. Comparison of two different doses of preoperative recombinant erythropoietin in men undergoing radical retropubic prostatectomy. *Urology* 2001; 57:737–41.
8. Lepor H, Neider AM, Ferrandino MN. Intraoperative and postoperative complications of radical retropubic prostatectomy in a consecutive series of 1,000 cases. *J. Urol.* 2001; 166:1729–33.
9. Davis JW, Kuban DA, Lynch DF et al. Quality of life after treatment for localized prostate cancer: differences based on treatment modality. *J. Urol.* 2001; 166:947–52.
10. Grossfeld GD, Chang JJ, Broering JM et al. Impact of positive surgical margins on prostate cancer recurrence and the use of secondary cancer treatment: data from the CaPSURE database. *J. Urol.* 2000; 163:1171–7.
11. Rabani F, Stapleton AM, Kattan MW et al. Factors predicting recovery of erections after radical prostatectomy. *J. Urol.* 2000; 164:1929–34.
12. Zagaja GP, Mhoon DA, Aikens JE et al. Sildenafil in the treatment of erectile dysfunction after radical prostatectomy. *Urology* 2000; 56:631–4.

13. Zippe CD, Jhaveri FM, Klein EA et al. Role of Viagra after radical prostatectomy. *Urology* 2000; 55:241–5.
14. Walsh PC. Nerve grafts are rarely necessary and are unlikely to improve sexual function in men undergoing anatomic radical prostatectomy. *Urology* 2001; 57:1020–4.
15. Vaidya A, Hawke C, Tiguert R et al. Intraoperative T staging in radical retropubic prostatectomy: is it reliable? *Urology* 2001; 57:949–54.
16. Bastacky SI, Walsh PC, Epstein JI. Relationship between perineural tumor invasion on needle biopsy and radical prostatectomy capsular penetration in clinical stage B adenocarcinoma of the prostate. *Am. J. Surg. Pathol.* 1993; 17:336–41.
17. D'Amico AV, Wu Y, Chen MH et al. Perineural invasion as a predictor of biochemical outcome following radical prostatectomy for select men with clinically localized prostate cancer. *J. Urol.* 2001; 165:126–9.
18. Graefen M, Haese A, Pichlmeier U et al. A validated strategy for side specific prediction of organ confined prostate cancer: a tool to select for nerve sparing radical prostatectomy. *J. Urol.* 2001; 165:857–63.
19. Messing EM, Manola J, Sarosdy M et al. Immediate hormonal therapy compared with observation after radical prostatectomy and pelvic lymphadenectomy in men with node-positive prostate cancer. *N. Engl. J. Med.* 1999; 341:1781–8.
20. Ghavamian R, Bergstralh EJ, Blute ML et al. Radical retropubic prostatectomy plus orchiectomy versus orchiectomy alone for pTxN+ prostate cancer: a matched comparison. *J. Urol.* 1999; 161:1223–7.
21. Freiha, FS. Radical retropubic postatectomy. In: Crawford ED, Das S (eds) *Cancer of the Prostate*, pp. 189–223. New York: Marcel Dekker, 1993.
22. Stephenson RA, Middleton RG, Abott TM. Wide excision (nonnerve sparing) radical retropubic prostatectomy using an initial perirectal dissection. *J. Urol.* 1997; 157:251–5, 1997.
23. Lynch DF, Schellhammer PF. Techniques for management of the dorsal venous complex in radical retropubic prostatectomy. *Atlas Urol. Clin. North Am* . 1994; 2:81–94.
24. Steiner MS. Anatomic basis for the continence-preserving radical retropubic prostatectomy. *Semin. Urol. Oncol.* 2000; 18:9–18.
25. Villers A, McNeal JE, Freiha FS et al. Invasion of Denonvilliers' fascia in radical prostatectomy specimens. *J. Urol.* 1993; 149:793–8.
26. John H, Hauri D. Seminal vesicle sparing radical prostatectomy: effect on urinary continence. *J. Urol.* 2001; 165(no. 5, Suppl.): abstract 1598.
27. John H, Hauri D. Seminal vesicle-sparing radical prostatectomy: a novel concept to restore early urinary continence. *Urology* 2000; 55:820–4.
28. Srougi M, Nesrallah LJ, Kauffmann JR et al. Urinary continence and pathological outcome after bladder neck preservation during radical retropubic prostatectomy: a randomized prospective trial. *J. Urol.* 2001; 165:815–18.
29. Marcovich R, Wojno KJ, Wei JT et al. Bladder neck-sparing modification of radical prostatectomy adversely affects surgical margins in pathologic T3a prostate cancer. *Urology* 2000; 55:904–8.
30. Gilliland JD, Spies JB, Brown SB et al. Lymphoceles: percutaneous treatment with povidone–iodine sclerosis. *Radiology* 1989; 171:227–9.
31. Cohan RH, Saeed M, Schwab J et al. Povidone–iodine sclerosis of pelvic lymphoceles: a prospective study. *Urol. Radiol.* 1988; 10:203–6.
32. De Marco RT, Bihrle R, Foster RS. Early catheter removal following radical retropubic prostatectomy. *Semin. Urol. Oncol.* 2000; 18:57–9.
33. Lepor H, Nieder AM, Fraiman MC. Early removal of urinary catheter after radical retropubic prostatectomy is both feasible and desirable. *Urology* 2001; 58:425–9.
34. Souto CA, Teloken C, Souto JC et al. Experience with early catheter removal after radical retropubic prostatectomy. *J. Urol.* 2000; 163:865–6.
35. Van Kampen M, De Weerdt W, Van Poppel H et al. Effect of pelvic-floor re-education on duration and degree of incontinence after radical prostatectomy: a randomized controlled trial. *Lancet* 2000; 355:98–102.
36. Klotz L. Neurostimulation during radical prostatectomy: improving nerve-sparing techniques. *Semin. Urol. Oncol.* 2000; 18:46–50.
37. Klotz LH, Jewett MAJ, Heaton J et al. A phase 3 trial of intraoperative cavernous nerve mapping during radical prostatectomy. *J. Urol.* 1999; 161:592.
38. Walsh PC, Marschke P, Catalona WJ et al. Efficacy of first-generation Cavermap to verify location and function of cavernous nerves during radical prostatectomy: a multi-institutional evaluation by experienced surgeons. *Urology* 2001; 57:491–4.
39. Holzbeierlein J, Peterson M, Smith JA Jr. Variability of results of cavernous nerve stimulation during radical prostatectomy. *J. Urol.* 2001; 165:108–10.
40. Kim HL, Stoffel DS, Mhoon DA et al. A positive cavermap response poorly predicts recovery of potency after radical prostatectomy. *Urology* 2000; 56:561–4.
41. Kim ED, Scardino PT, Hampel O et al. Interposition of sural nerve restores function of cavernous nerves resected during radical prostatectomy. *J. Urol.* 1999; 161:188–92.
42. Kim ED, Seo JT. Minimally invasive technique for sural nerve harvesting: technical description and follow-up. *Urology* 2001; 57:921–4.
43. Kim ED, Scardino PT, Hampel O et al. Interposition of sural nerve restores function of cavernous nerves resected during radical prostatectomy. *J. Urol.* 2001; 161:188–92.
44. Kim ED, Nath R, Kadmon D et al. Bilateral nerve graft during radical retropubic prostatectomy: 1-year follow-up. *J. Urol.* 2001; 165:1950–6.
45. Kim ED, Scardino PT, Kadmon D et al. Interposition sural nerve grafting during radical retropubic prostatectomy. *Urology* 2001; 57:211–16.

Sural Nerve Interposition Graft During Radical Prostatectomy

Eduardo I. Canto and Kevin M. Slawin

Scott Department of Urology, Baylor College of Medicine, Texas Medical Center, Houston, TX, USA

INTRODUCTION

Because synthesis of proteins in neurons is limited to the cell body, sectioning of a nerve axon results in the death of the distal axonal segment. Upon axotomy, synaptic transmission is rapidly lost and the distal axonal segment begins to degenerate in a process that occurs over a 1–2-month period. This process, termed Wallerian degeneration, is named after Augustus Waller, a British physician who described it in the 19th century. The essence of Wallerian degeneration is that, in the segment distal to the site of injury, the myelin sheath (if present) and axon degenerate completely, but the perineural sheaths and Schwann cells remain.[1–3]

Peripheral nerve axons, unlike their central nervous system counterparts, are capable of regenerating after transection. In order for nerve conduction to be re-established after transection, the proximal axon must grow down the remaining perineural sheath all the way to the target organ. This process is mediated by Schwann cells, and occurs even when the distal nerve segment is devascularized.[4] Within 24 hours after injury, Schwann cells begin to proliferate and form solid cellular columns. The Schwann-cell columns guide sprouting axons. Only the axons that enter the Schwann-cell columns persist.[1–3] The process of peripheral nerve regeneration after complete transection is, therefore, dependent on the correct tension-free alignment of the cut ends. If there is loss of a long segment of nerve, a conduit must be provided to guide the sprouting axons to the distal cut end of the nerve. Although a number of synthetic and tissue conduits have been used for this purpose, the most commonly used material consists of an autologous nerve segment harvested at the time of definitive repair. Unlike synthetic or other tissue conduits, an autologous nerve graft provides the regenerating axon with a microenvironment similar to the one found in the distal nerve segment.

Interposition nerve grafting using a segment of the sural nerve has been performed for over 100 years. It is considered a standard surgical technique for the reconstruction of peripheral nerve injuries involving significant segmental loss. The facial nerve, brachial plexus, median and ulnar nerves, sciatic nerve, and numerous other motor and sensory nerves are successfully grafted on a routine basis following injury.[1–3] When the cavernous nerve is interrupted by resection at the time of non-nerve-sparing radical retropubic prostatectomy (RRP), Wallerian degeneration occurs in the distal segment. Interposition nerve grafting, in this situation, can provide a physical conduit that bridges the iatrogenic gap, guiding the regenerating autonomic nerve axons back to their original target. If meticulous grafting has been performed, regenerating fibers will cross both of the suture-line interfaces and establish connections to the end-organ receptors, allowing the nerve function to resume. Successful regeneration, therefore, involves the extension of axonal growth cones through the interposed graft segment and down the distal *in situ* segment. A typical cavernous nerve graft measures 6.5 cm in length, and the distal *in situ* segment to the most distal smooth muscle cells of the corpora cavernosa is about 8–10 cm. The rate of peripheral nerve regeneration is, on average, about 1–2 mm per day, and is probably slower in middle-aged adults.[1–3] The predicted time to recovery of function, therefore, is more than 6 months.

The cavernosal nerves are composed almost exclusively of parasympathetic and sympathetic fibers from the pelvic plexus.[5] Although most peripheral nerve injuries repaired with nerve grafts involve damage to short segments of somatic nerves, the ability of autonomic nerves to regenerate after transection was demonstrated in the 1950s. Vagophrenic anastomosis was shown to result in diaphragmatic contractions, and penile erections were reported to occur after cross-anastomosis of the lumbar and sacral roots.[6,7] Although the results after bridging large gaps are not as good as those involving short gaps, grafts of up to

14 cm in length have been performed successfully.[8] Graft repair of autonomic nerves may be aided by the fact that restoration of autonomic function does not seem to require the same precise and accurate reconstitution following reinnervation as that required for the restoration of highly specialized somatic functions. Furthermore, regeneration of autonomic postganglionic fibers may be faster than that of somatic nerves.[9]

VASCULAR SUPPLY AND INNERVATION OF THE CORPORA CAVERNOSA

The arterial supply to the corpora cavernosa is usually from the internal pudendal artery via the common penile artery. The three branches of the common penile artery are the cavernosal artery, the dorsal penile artery and the bulbourethral artery. The cavernosal artery, during its course within the corpus cavernosum, gives off helicine arteries that feed the trabecular erectile tissue and sinusoids. The dorsal artery of the penis is responsible for engorgement of the glans penis. Accessory arteries arising on the obturator, vesical, femoral or external iliac arteries may exist. Only on rare occasions are branches of the obturator or vesical arteries the predominant supply to the corpora cavernosa, making vasculogenic erectile dysfunction after RRP uncommon.[5,10]

The penis receives innervation from both autonomic and somatic pathways. Somatosensory perception from the penile skin, urethra, corpora cavernosa and glans penis is mediated by the dorsal penile nerve. It reaches the sacral spinal cord via the pudendal nerve. Somatomotor penile innervation is also mediated by branches of the pudendal nerve, and is responsible for contraction of the ischiocavernosus muscle during the rigid erection phase and the rhythmic contraction of the bulbocavernosus muscle during ejaculation.[5]

Penile autonomic innervation is responsible for regulating the tone of the cavernous smooth musculature and the intracavernous arteriolar smooth muscle. Relaxation of these muscles results in tumescence. While the penis is in the flaccid state, blood flow is limited by the basal tone of these smooth muscles. This tonic contraction is mediated by sympathetic stimulus from the cavernous nerves. The sympathetic pathway originates from the T10 to L2 spinal segments and travels through the sympathetic chain ganglia to the hypogastric plexus via the splanchnic nerves. The hypogastric plexus divides into two hypogastric nerves at the level of the aortic bifurcation. These two nerves enter the pelvis medial to the internal iliac vessels and deep to the endopelvic fascia. Both the hypogastric nerves and the pelvic continuation of the sympathetic trunks send branches to the pelvic plexus. In humans, the sympathetic fibers that contribute to the cavernosal nerves and regulate erection originate mostly from the tenth to twelfth thoracic spinal cord segments. The parasympathetic neurons of the cavernosal nerves originate from the intermediolateral cell column of the sacral spinal cord and exit through the sacral foramina reaching the pelvic plexus via the pelvic splanchnic nerves (nervi erigentes). Parasympathetic stimulus results in relaxation of the intracavernous arteriolar smooth muscle and engorgement of the corpora cavernosa.[5]

The pelvic plexus lies in the retroperitoneum on the anterolateral aspect of the rectum, approximately 5–11 cm from the anal verge, or at the level of the seminal vesicles. The most caudal aspect of the pelvic plexus may extend to the so-called lateral pedicle of the prostate. The branches of the inferior vesical artery that feed the bladder neck and prostate traverse the pelvic plexus at this level. The cavernous nerves leave the most caudal aspect of the pelvic plexus at the level of the base of the prostate and travel within the leaves of the lateral pelvic fascia on the dorsolateral aspect of the prostate.[10,11] As the cavernous nerves reach the apex of the prostate, 9–12 mm cranial to the genitourinary diaphragm, there is further branching. The smaller branches travel along the anterolateral aspect of the urethra in close approximation to it while the larger of these branches travel 4–7 mm lateral to the membranous urethra.

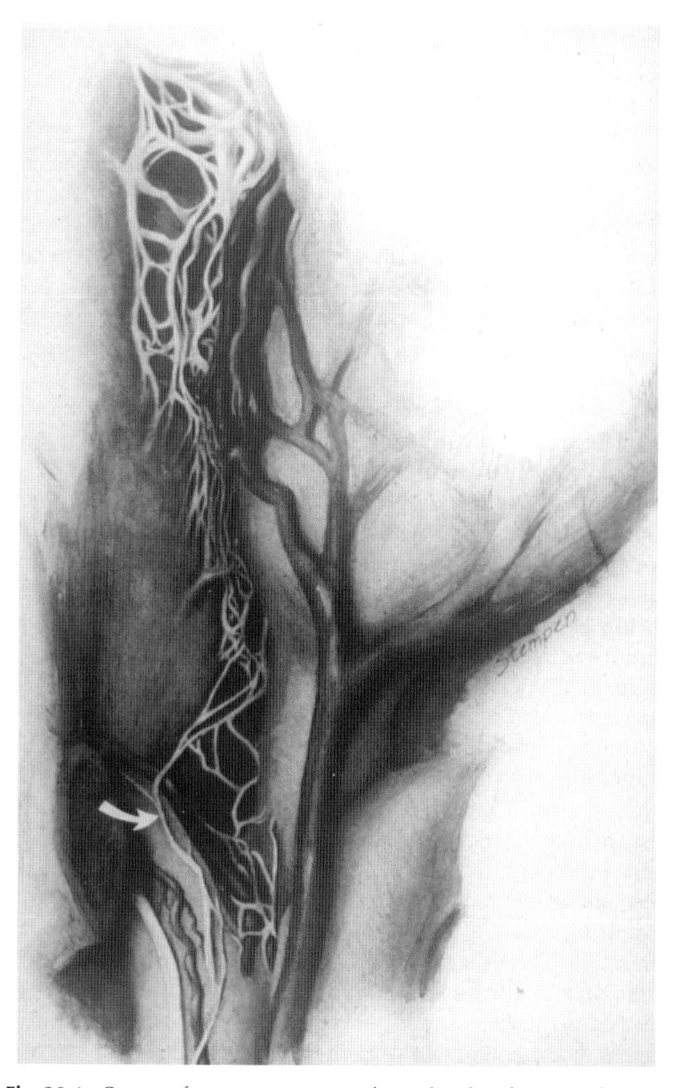

Fig. 30.1. Course of cavernous nerves. The endopelvic fascia on the right has been removed exposing the cavernous nerves. Reprinted from Paick et al.,[13] Copyright 1993, with permission from Elsevier Science.

These lateral branches traverse the genitourinary diaphragm at an oblique angle and turn medially at the distal margin of the diaphragm entering the corpora cavernosa mostly through the hilum of the penis. Fibers that do not penetrate the tunica albuginea either connect to or merge into the dorsal nerve of the penis (**Fig. 30.1**). The dorsal nerve of the penis, therefore, carries not only somatic, but also autonomic nerve fibers.[10,12,13]

The principle behind nerve-sparing RRP, as described by Walsh, is to preserve the cavernous nerves by separating the lateral pelvic fascia from the prostatic surgical capsule in order to leave the cavernous nerves *in situ* once the specimen is removed. When the vascular pedicle that lies at the lateral aspect of the base of the prostate is dissected during a nerve-sparing RRP, care must be taken to avoid damaging the most caudal tip of the pelvic plexus. Because the cavernous nerves travel along the anterolateral aspect of the membranous urethra immediately distal to the apex of the prostate, the cavernous nerves are also at risk of damage during dissection of the prostatic apex, transection of the urethra or placement of the urethral anastomotic sutures. Along its course, the cavernous nerve gives off small branches that innervate the rectum, prostate and urethra. Therefore, damage to the cavernous nerves during RRP may not only affect potency but also can affect urinary continence.

INNERVATION OF THE MALE URINARY SPHINCTER

In the absence of an intact bladder neck, as in the case of post-RRP patients, urinary continence is dependent on the sphincter mechanism found at the level of the membranous urethra (from the prostatic apex to the perineal membrane). This sphincter mechanism consists of the prostatomembranous urethra, the periurethral musculature or rhabdosphincter, and the extrinsic paraurethral musculature and connective tissue structures of the pelvis. All three intrinsic layers of the prostatomembranous urethra contribute to the generation of closing pressure. These include the pseudostratified columnar epithelium of the urethral lumen, the submucosal connective tissue and the fibroelastic tissue enmeshed with longitudinal smooth muscle of the urethral wall.[5] The rhabdosphincter is a concentric muscular structure in the shape of a cylinder composed mostly of type I, slow-twitch striated muscle. It surrounds the prostatomembranous urethra. Its fibers extend from the base of the bladder to the perineal membrane ventrally and from the prostate to the perineal membrane dorsally. The muscle thickness is similar ventrally and dorsally, but is greatest laterally. The rhabdosphincter attaches to the subpubic arch via the paired puboprostatic ligaments. It fuses posteriorly with the perineal membrane via the median fibrous raphe that extends to the central tendon of the perineum. Contraction of the rhabdosphincter results in elevation of the urethra against the undersurface of the pubis and occlusion of the urethral lumen.[12,14] In contrast to the rhabdosphincter, the main pelvic muscle, the levator ani, consists of fast-twitch, type II striated muscle fibers designed for forceful, rapid and short-lived contractions.[14]

The sphincter mechanism at the membranous urethra is innervated by both autonomic and somatic nerves.[15–18] Branches of the cavernous nerves distal to the apex of the prostate provide mostly sympathetic innervation to the intrinsic smooth muscle fibers of the membranous urethra and probably to the rhabdosphincter. In addition, a pelvic autonomic nerve that originates from the pelvic plexus and travels beneath the muscle fascia of the levator ani sends branches to the dorsolateral aspect of the rhabdosphincter at the level of the prostatic apex from beneath the levator ani's superior fascia.[17,18] The pudendal nerve provides somatic motor innervation of the pelvic floor and rhabdosphincter. Both extrapelvic and intrapelvic branches of the pudendal nerve innervate the rhabdosphincter. Extrapelvic somatic innervation of the rhabdosphincter is mediated by branches of the perineal nerve. The pudendal nerve also gives an intrapelvic branch to the posterolateral aspect of the rhabdosphincter before leaving the pelvis through the ischioanal fossa. This branch travels alongside the autonomic pelvic nerve.[16–19] What probably is a sensory nerve branch to the rhabdosphincter from the dorsal nerve of the penis has been identified by some, but not all, investigators.[18,20]

Post-prostatectomy incontinence, in most cases, is a result of sphincter dysfunction.[21] Dysfunction of the membranous urethral sphincter complex can be the result of direct injury to the rhabdosphincter or to the nerves that innervate it. The pudendal nerve and its extrapelvic branches are unlikely to be injured during RRP. However, dissection of the apex of the prostate may result in injury to the putative sensory branch from the dorsal nerve of the penis, the sphincter branches of the pelvic autonomic nerve and the intrapelvic branch of the pudendal nerve (**Fig. 30.2**). For example, the distance between the sensory branch arising from the dorsal nerve of the penis and the prostatic apex is only 0.3 to 1.3 cm.[18,20]

The specimen from a nerve-sparing RRP should include only the prostate, Denonvilliers' fascia and the seminal vesicles.

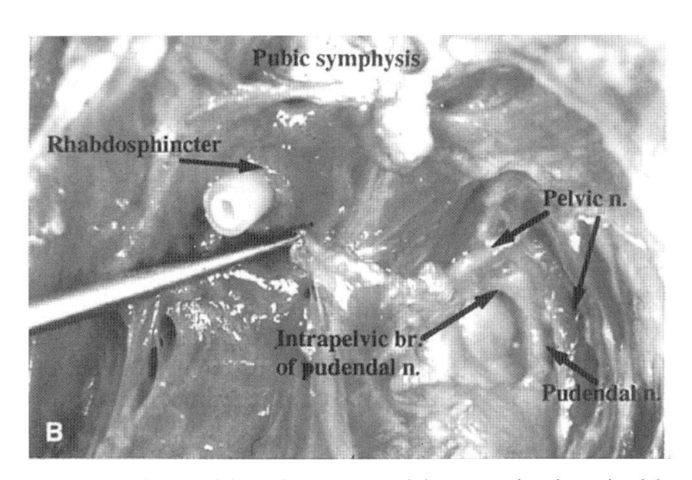

Fig. 30.2. Relation of the pelvic nerve and the intrapelvic branch of the pudendal nerve to the prostatic apex. The prostate has been removed. The forceps point to the pelvic nerve and the intrapelvic branch of the pudendal nerve as they enter the dorsolateral aspect of the rhabdosphincter. Reprinted from Hollaburgh et al.,[17] Copyright 1997, with permission from Elsevier Science. (*See Color plate 5.*)

The lateral pelvic fascia and cavernous nerves should not be part of the specimen. Because the dissection around the prostate, including apex and base, is done between the lateral pelvic fascia and the prostatic capsule, the risk of damage to the cavernous nerves, the pelvic autonomic nerve to the rhabdosphincter, and the branch of the pudendal nerve that travels to the rhabdosphincter from within the pelvis is reduced. Intentional non-nerve-sparing RRP with wide resection of one or both neurovascular bundles (NVB), on the other hand, guarantees resection of the cavernous nerves and risks damage or transection of the autonomic pelvic nerve and the pudendal branch that travels with it from within the pelvis to the rhabdosphincter. Non-nerve-sparing RRP can, therefore, damage both potency and continence mechanisms.

RATIONALE FOR INTERPOSITION SURAL NERVE GRAFT

Given our current knowledge of the anatomy, function and regenerative ability of the pelvic autonomic nerves, it is not unreasonable to think that one could improve the functional outcome for patients with locally advanced prostate cancer who have undergone either unilateral or bilateral wide resection of the neurovascular bundle(s) (leaving the nerve intact on the prostatectomy specimen) by interposing a nerve graft to the free ends of the resected cavernous nerve(s). In patients with pathologic stage T3 disease, this approach would have the potential to improve cancer control by reducing the incidence of positive surgical margins as compared with nerve-sparing surgery and to reduce the incidence of impotence, and possibly incontinence, as compared with non-nerve-sparing surgery without nerve graft. Four key conditions determine whether patients can benefit from this approach. First, cavernous nerve function must be restored on the grafted side once the neuroregenerative process is complete. Second, since most NVB resections are unilateral, patients who undergo unilateral NVB resection must have reduced potency and/or continence compared to that in patients who undergo bilateral nerve-sparing surgery. Third, the surgeon must be able to predict the presence and location of extracapsular extension (ECE) of the carcinoma in order to avoid inappropriately resecting potentially salvageable NVBs. Finally, wide resection of the NVB must improve cancer control in the presence of ECE along the resected NVB.

The ability of the cavernous nerve to regenerate after complete transection was demonstrated in rodents as early as 1992. In these studies, cavernous nerve function was restored by interposition nerve grafting with the genitofemoral nerve to repair defects created in the cavernous nerve by excision of short segments.[22,23] Walsh was the first to perform interposition nerve grafting of excised cavernous nerves in humans at the time of RRP. He used a genitofemoral nerve segment for interposition grafting. However, he did not see a clear benefit and the procedure was abandoned.[24] Conclusive evidence that the human cavernous nerves could regenerate after wide local excision with interposition graft repair was first reported by an interdisciplinary group from Baylor College of Medicine, led by Peter Scardino.[25] The first autologous unilateral interposition nerve graft using the sural nerve was performed in January 1997. The first bilateral nerve graft was performed in March of 1997. As far as we know, these were the first successful efforts, in humans, to restore erectile function after bilateral cavernous nerve resection with repair by interposition nerve grafting at the time of RRP. Analysis of erectile function in the first 12 patients with at least 12 months of follow-up after bilateral nerve grafts revealed that four had spontaneous erections sufficient for unassisted intercourse, three had no spontaneous erections and the remaining five had partial erections.[26] Of the five men with partial erections, two reported erections sufficient for intercourse with sildenafil (Viagra) treatment. None of the patients in a control group of men who had undergone bilateral nerve resection without nerve graft had erections. A multidisciplinary team involving specialists in erectile dysfunction, prostate cancer surgery and microsurgical nerve reconstruction contributed to the development of this novel technique. The sural nerve was selected for interposition grafting because it has a larger caliber than the genitofemoral nerve. The smaller caliber of the genitofemoral nerve may explain the failure of earlier attempts at cavernous nerve reconstruction.

Unilateral NVB resection significantly decreases postoperative potency, especially in men older than 50. High-volume academic centers have reported decreases in potency rates of about 50% in patients who undergo unilateral NVB resection.[27–30] Nevertheless, a distinction should be made between 'non-nerve-sparing' surgery and purposeful, wide resection of the NVB. For example, it has been reported that potency rates after radical prostatectomy performed in a community setting fall within a very narrow range, irrespective of the reported degree of nerve-sparing performed as part of the surgery. Of men who had unilateral or bilateral nerve-sparing surgery in one study, 41% and 44%, respectively, were potent. Surprisingly, 33% of men who had bilateral non-nerve-sparing procedures were also potent.[31] Given the experience at most academic centers, it is unlikely that patients who were included in this study as undergoing non-nerve-sparing surgery underwent wide resection of the NVB. In our experience, unilateral, intentional wide resection of one NVB decreases postoperative potency. Few, if any, patients that undergo intentional wide resection of both NVB are potent postoperatively.

Tumor involvement of the NVB cannot yet be predicted with 100% certainty. Nevertheless, the presence of ECE can be predicted relatively accurately by integrating a number of parameters including the prostate specific antigen (PSA), the % free PSA, results of the digital rectal examination (DRE), transrectal ultrasound findings, and the extent and Gleason score of the cancer in the biopsy specimens. Currently, at our institution, in patients for whom the decision to perform nerve-sparing RRP appears equivocal, a 12-core biopsy is performed, including six sextant-directed cores, and six laterally directed cores at the base, mid and apical prostate, if one has not already been obtained at diagnosis. We have recently shown that this 12-core extended, regional, transrectal ultrasound-guided

Table 30.1. Rate of positive surgical margins as a function of final pathological stage and degree of nerve sparing in consecutive patients who underwent either bilateral nerve-sparing or unilateral non-nerve-sparing surgery. Reprinted with permission of MedReviews, LLC, from Slawin KM, Canto EI et al. Sural nerve interposition grafting during radical prostatectomy. *Reviews in Urology* 2002; 4:19.

	Number of patients with positive surgical margins (%)	
Pathologic Stages	Unilateral nerve sparing	Bilateral nerve sparing
All	15/149 (10.1)	16/201 (8.0)
pT2	3/70 (4.3)	12/173 (6.9)
pT3a	5/61 (8.2)	4/27 (14.8)
Focal	1/33 (3.0)	2/14 (14.3)
Established	3/25 (12.0)	2/12 (16.7)

biopsy has a high correlation with ECE in the sextant in which there is a positive laterally directed biopsy. In a recent study at our institution involving 76 patients (23.7% of which had ECE), the presence of cancer on an individual laterally directed biopsy core was a significant predictor of ECE independent of DRE, preoperative PSA and transrectal ultrasound results ($P=0.0046$). However, the presence of cancer on an individual sextant biopsy core was not predictive of ECE ($P=0.0525$).[32] Integrating PSA, % free PSA, DRE, 12-core biopsy data and intraoperative findings, we have been relatively successful in predicting the presence and location of ECE. An analysis of a recent cohort of 149 patients who underwent unilateral nerve-sparing surgery with contralateral NVB resection by a single surgeon (KMS) revealed that 79 (53%) had ECE. Data regarding the site of ECE were available for 66 of these patients. Of the 66, 61 (92%) had ECE on the same side as the resected NVB. Despite these encouraging results, we are currently developing nomograms designed to improve our ability to predict ECE at the NVB preoperatively.

Locally advanced prostate cancer is, to a certain extent, curable with the appropriate surgical approach. Data from Baylor College of Medicine show that, in patients with specimen-confined pathologic stage T3 disease, long-term PSA progression-free survival is greater than 70%, indicating that the majority of these patients can achieve cancer cure, if their disease is adequately excised.[33] The most common site of ECE of prostate cancer is the posterolateral region, a finding that is likely to be related not only to the fact that most cancers originate in this area of the peripheral zone, but also to the biological behavior of prostate cancer.[34] Recent studies have found a complex interaction between nerve growth factors and prostate cancer.[35,36] This may be the molecular explanation for Villers' histopathologic finding that prostate cancer traverses the prostatic capsule preferentially through the perineural spaces along perforating branches of the neurovascular bundle.[37] Coincidentally, these branches are found most commonly along the posterolateral border of the prostate at the prostatic base and apex. Clinically, perineural invasion is associated with a higher rate of both ECE and prostate cancer progression.[38–40]

Because of the high likelihood that the ECE is in the posterolateral aspect of the prostate in T3 specimens, one would expect that NVB resection on the side of the ECE would reduce the positive margin rate. This is our experience, as revealed by an analysis of the influence of nerve sparing on the positive surgical margin rate (+SM) by one surgeon (KMS). In patients with pT3a disease, the +SM rate was 14.8% in patients undergoing bilateral nerve-sparing surgery, but only 8.2% in patients that had the NVB widely resected on one side. The patients who had unilateral NVB resection had a lower positive margin rate, despite a higher rate of adverse factors prompting resection of the NVB. For example, 73% of patients who had unilateral NVB resection had a positive DRE, and 69% had a Gleason score ≥ 7, compared with 56% and 42%, respectively, in patients who had bilateral nerve-sparing surgery (**Table 30.1**). Others have also shown that the presence of established extracapsular disease in the region of the NVB increases the risk of a positive surgical margin if nerve-sparing RRP is performed.[38] Many authors have shown that positive surgical margins have a negative impact on PSA progression-free survival after radical prostatectomy.[41,42] Indeed, positive surgical margins are an independent predictor of PSA progression-free survival even in multivariate analyses that control for other parameters of tumor aggressiveness, such as PSA, Gleason score, pathologic stage, etc.[33,43–45] Nevertheless, longer follow-up is needed to prove that NVB resection improves cancer-free survival in patients with non-metastatic, locally advanced disease.

OPERATIVE TECHNIQUE

Sural nerve interposition graft (SNG) is not a substitute for nerve-sparing surgery. Ideally, an NVB is resected only when there is extracapsular extension associated with it, with the goal of improving the odds for cancer cure. Bilateral nerve-sparing surgery is preferred whenever the surgeon believes that he can achieve a negative surgical margin. At our institution, of the 500 most recent prostatectomies performed for typical indications by a single surgeon (KMS), almost 60% of patients had bilateral nerve-sparing surgery. On the other hand, bilateral NVB excision is rarely performed; only 7% of patients had both NVBs resected at RRP. However, 34% of patients, comprising the majority of patients considered candidates for sural nerve grafting, had unilateral NVB resection at RRP.

SNG at the time of RRP requires only minor modifications to our nerve-sparing approach to RRP. We have found the CaverMap Surgical Aide device (Alliant Medical Technologies, Norwood, MA) to be a useful tool, particularly in corroborating isolation of the cavernous nerves lateral and proximal to the base of the prostate. Because the nerves are not accompanied by a discrete group of blood vessels at this location, their identification is not as straightforward as it is at the dorsolateral aspect of the prostate. Below, we describe the technique for performing a unilateral nerve graft with contralateral nerve sparing. This is the procedure most commonly performed.

Proper patient positioning is essential to any surgery. We have not found a need to change our patient positioning to allow for SNG. We place patients on the operative table with the breakpoint of the table between the patient's umbilicus and symphysis pubis. Unless there are any contraindications, e.g. significant venous stasis disease, we use the right leg for sural nerve harvest. This allows for the radical prostatectomy and nerve harvest to be performed simultaneously without the need to change the position of the instrument table or the technician assisting the RRP team. In almost all cases, only one sural nerve needs to be harvested, even for bilateral sural nerve grafting. Up to 30 cm can be harvested from one leg. When a unilateral nerve graft is performed, we usually divide the harvested nerve segment into two parts. These are then placed in parallel to increase the total diameter of the cavernous nerve encompassed by the graft. Because the cavernous nerve at the level of the prostate is not a discrete nerve, but rather a nerve plexus, the larger the diameter of the interposition graft, the higher the number of regenerating axons that will grow through the graft.

After flexing the operative table to approximately 150°, we apply Trendelenburg rotation until the lower abdomen is horizontal. In preparation for use of the CaverMap Surgical Aide device, the mercury-containing tumescence sensor that measures changes in penile girth is placed around the base of the penis and the device is grounded to the retractor. The pelvic lymph node dissection is carried out through a midline incision that extends only 8–9 cm from the pubis. With the use of the Omni-tract retractor (Minnesota Scientific, Inc., Minneapolis, MN), we have not found a need to extend this incision in order to perform the RRP or the SNG, except in extremely overweight patients.

Once the endopelvic fascia has been opened and the dorsal venous complex has been transected and ligated, we palpate the prostate, paying particular attention to the dorsolateral aspect from apex to base.[46] Information gathered at this point should be incorporated with knowledge of preoperative PSA, % free PSA, results of DRE and transrectal ultrasound, Gleason score, cancer length in the biopsy specimen, and location of positive biopsies to provide the best means of predicting the likelihood and location of ECE. This information should indicate whether a high likelihood of ECE at either NVB warrants wide excision and possible SNG.

CAVERMAP SURGICAL AIDE-ASSISTED CONFIRMATION OF THE CAVERNOSAL NERVES

The CaverMap Surgical Aide device consists of a probe that delivers a biphasic electrical current pulse via a small metal tip and a mercury-filled tumescence sensor that measures changes in penile girth. During nerve stimulation, the device delivers increasing amounts of the biphasic current via the probe tip (from 8 to 20 mA, current increases every 20 seconds). Changes in penile girth, as measured by the mercury-filled tumescence sensor, are reported as a percentage change in the device's electronic display. Both increases and decreases in penile girth are measured. We use the CaverMap Surgical Aide device to confirm, functionally, that the cavernous nerve has been identified at two locations, where it approaches the prostate (proximal and lateral to the prostatic base) and distal to the prostatic apex. It is unclear whether the routine use of the CaverMap Surgical Aide device during nerve-sparing RRP improves postoperative erectile function.[47–50] Furthermore, data assessing the impact on outcomes of the use of the CaverMap Surgical Aide device during non-nerve-sparing RRP with SNG are inconclusive. Nevertheless, the CaverMap Surgical Aide device is able to provide accurate identification, intraoperatively, of the NVBs with high sensitivity and specificity. By analyzing a total of 324 intraoperative CaverMap Surgical Aide responses, we determined that the optimal cut-off for change in penile girth for distinguishing between a significant (positive) and non-significant (negative) response is 1%. When this optimal cut-off was used, the CaverMap Surgical Aide was able to identify the NVBs with >90% sensitivity and >80% specificity.[51] Others have also found that a 1% change in penile girth as cut-off yields high specificity.[47]

In order to facilitate the nerve-grafting procedure, we identify the nerve prior to proceeding with any periprostatic dissection. The proximal end of the graft will be anastomosed to the cavernous nerve at a location 1 cm proximal and lateral to the base of the prostate. At this location, the vascular bundle runs anteromedial to the cavernous nerves. In order to isolate the nerve, we incise the pelvic fascia that covers the perivesical fat at either side of the visualized, essentially avascular, nerve bundle, and loop it with a silk suture (**Fig. 30.3**). We use the CaverMap Surgical Aide device at this point to confirm, functionally, that the nerves to the corpora cavernosa have, indeed, been isolated. We place the probe directly on top of the isolated bundle with minimal traction on the suture to pull the nerve bundle off the soft tissue surrounding it. The start button on the CaverMap Surgical Aide probe initiates the calibration process. The device first calibrates to a stable 'zero' baseline, ensuring that there is adequate contact between the tissue and the metal strips on the undersurface of the probe. It progressively increases the current output and constantly monitors the penile girth via the mercury-filled tumescence sensor at the base of the penis. If the device reaches a current of 20 mA without producing a greater than 1% change in penile girth, we dissect further in an attempt to

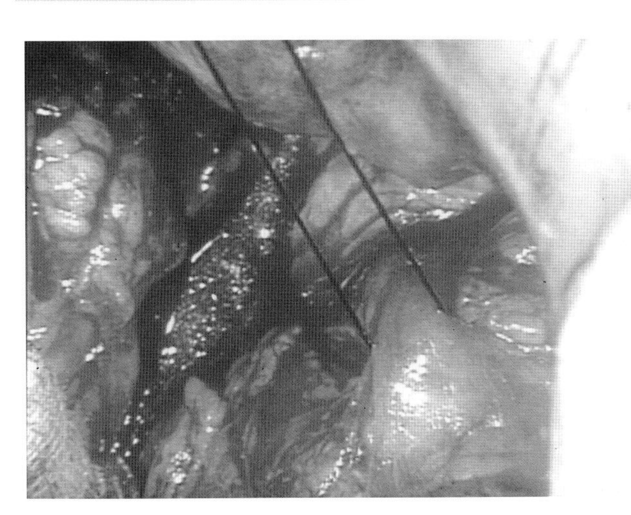

Fig. 30.3. Isolation of the cavernous nerve proximal and lateral to the base of the prostate. A suture is used to mark the cavernous nerve for grafting. Reprinted from *Urologic Clinics of North America* 2001; 28:843 with permission from W.B. Saunders Co. (*See Color plate 6.*)

isolate a bundle that produces a better signal. We then tie the silk suture to mark the location of the isolated nerve. Comfortable that we have isolated the cavernous nerve, we complete the dissection lateral to the NVB by incising the pelvic fascia until perirectal fat lobules are seen bulging through the incision. We then extend it both proximally and distally to the apex of the prostate. The proximal NVB eventually will be divided just distal to the silk suture.

On the side where the nerve is to be spared, we dissect the NVB off of the prostate prior to transecting the urethra and performing the apical dissection. The function of the spared NVB can also be assessed using the CaverMap Surgical Aide device by placing it over the posterolateral aspect of the prostate, where the proximal NVB is typically localized. We divide the lateral pelvic fascia along the ventral edge of the NVB from mid-prostate to apex and we mobilize the NVB posterolaterally off the prostate. To avoid entrapment of the NVBs in the vesicourethral anastomosis, we bluntly widen the plane between the urethra and the NVBs with a Kittner dissector. Finally, we sharply incise Denonvilliers' fascia between the bundle and the prostate, exposing perirectal fat.

Having performed almost the entire nerve-sparing procedure on one side and identified and marked the limits of the dissection on the side where the nerve is to be sacrificed, we proceed to transect the urethra, rectourethralis, and the anterior and posterior layers of Denonvilliers' fascia. In order to complete the wide resection of the NVB on the side of the tumor, we approach the apex of the prostate from lateral to medial, starting at the incised pelvic fascia. This maneuver puts at risk both the intrapelvic branch of the pudendal nerve and the branches of the autonomic pelvic nerve that travel with it, as well as the lateral branches of the cavernous nerves that travel across the genitourinary membrane. In order to minimize the potential for damage to these nerves, we begin the dissection only a few millimeters lateral to the edge of the prostate and limit to a minimum any deep dissection. The perirectal fat marks the deep limit of the dissection. Prior to transecting the distal NVB, we place a suture to mark the future distal anastomotic site for the SNG (**Fig. 30.4**). By carrying the dissection from lateral to medial, we ensure that we maintain the proper plane at the level of the posterior edge of Denonvilliers' fascia. We then completely mobilize the prostate off the rectum in the plane between the rectum and the posterior layer of Denonvilliers' fascia. After removing the prostate, we place the final anastomotic suture at 6 o'clock. At this point, we often stimulate the preserved NVB to confirm that it is functional (**Fig. 30.5**).

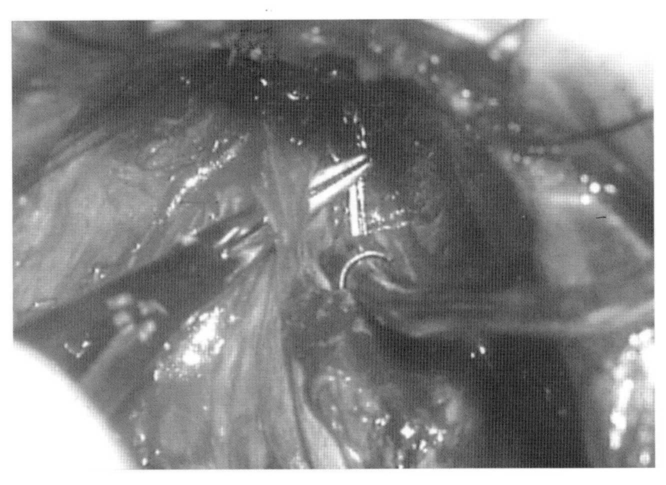

Fig. 30.4. Isolation of the cavernous nerve distal to the prostatic apex. Reprinted from *Urologic Clinics of North America* 2001; 28:844 with permission from W.B. Saunders Co. (*See Color plate 7.*)

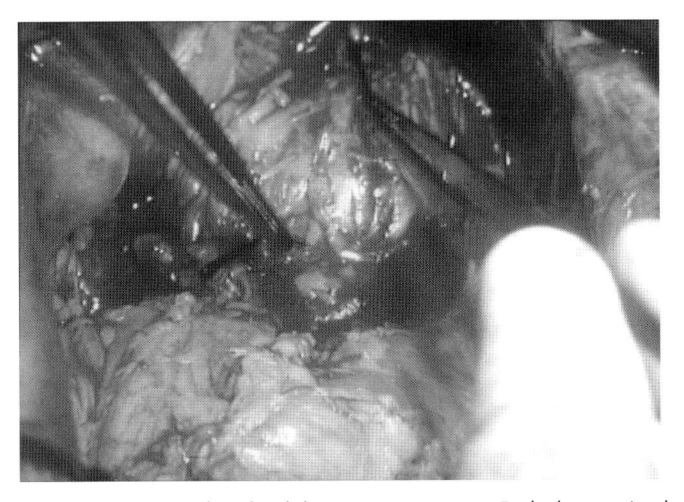

Fig. 30.5. Transected ends of the cavernous nerve. Both the proximal and distal ends of the excised cavernous nerve have been marked with silk ties. Denonvilliers' fascia and the left NVB have been excised exposing perirectal fat. The right NVB has been spared. Reprinted from *Urologic Clinics of North America* 2001; 28:845 with permission from W.B. Saunders Co. (*See Color plate 8.*)

SURAL NERVE HARVEST

Once the prostate has been removed, the urologic surgeon reconstructs the bladder neck while the plastic surgeon simultaneously performs the sural nerve harvest. The sural nerve starts in the popliteal fossa and travels between the two heads of the gastrocnemius muscle. Once it enters the foot, it supplies the skin on the lateral margin of the foot and the fifth digit. Although there is some variability in the origin and branching of the sural nerve, in general, it is readily found deep to the medial surface of the lesser saphenous vein. The plastic surgeon approaches the sural nerve through an incision placed immediately inferior to the lateral malleolus of the fibula. Once the sural nerve is isolated, the plastic surgeon surrounds it with a vessel loop and under gentle axial traction palpates for the proximal nerve in the posterior calf (**Fig. 30.6**). He makes a second transverse incision over the area where the nerve is palpated, approximately two-thirds of the distance down the calf, and he enters the superficial fascia of the leg with tenotomy scissors. After separating the nerve from the adjoining lesser saphenous vein, the plastic surgeon clamps and transects the sural nerve at the proximal end of the dissection. In order to prevent the development of a neuroma, the plastic surgeon buries the proximal stump within the gastrocnemius muscle. Care must be taken not to bury the stump too deeply in order to avoid injuring the tibial nerve. The pain of a neuroma can be debilitating, especially if reflex sympathetic dystrophy and causalgia develop. However, none of our patients has developed such a problem.

The plastic surgeon then brings the harvested sural nerve into the distal incision. Branching of the nerve frequently exists, and each of these branches must be identified and severed so that the nerve is delivered intact. A third incision is rarely required for bilateral grafts.

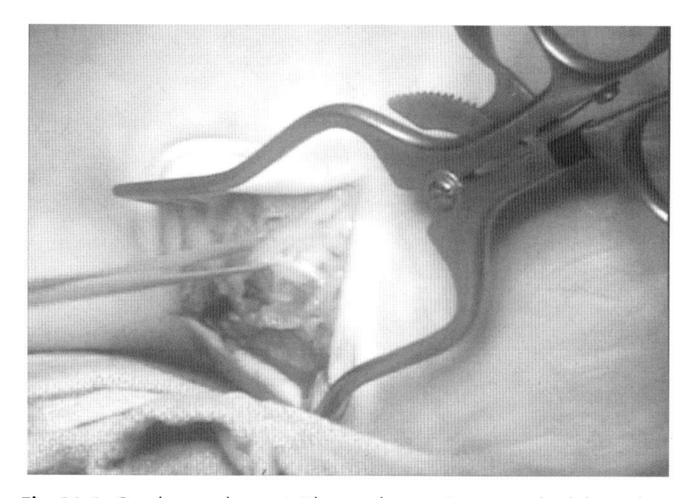

Fig. 30.6. Sural nerve harvest. The sural nerve is approached through an incision placed immediately inferior to the lateral malleolus of the fibula. Reprinted from *Urologic Clinics of North America* 2001; 28:845 with permission from W.B. Saunders Co. (*See Color plate 9.*)

INTERPOSITION SURAL NERVE GRAFTING

While the plastic surgeon is harvesting the sural nerve, we finish the bladder neck reconstruction and place the anastomotic sutures on the bladder. The sural nerve grafting is performed immediately prior to tying down the vesicourethral anastomosis. To facilitate the exposure for nerve grafting, we tag the anastomotic sutures and move them to the side opposite the graft site. In order to ensure a bloodless field, we place several 4×4 sponges along the contralateral NVB. In order to compensate for graft shrinkage and to provide a tension-free repair, the plastic surgeon cuts the harvested segment of sural nerve to a length that is approximately 10% longer than the gap to be bridged. When unilateral nerve sparing is performed, as in most cases, the harvested segment, about 20 cm, is usually long enough to fashion a 'double-barrel' graft composed of two parallel segments of equal length.

Prior to placing the nerve graft, we usually stimulate the distal transected end of the cavernous nerve using a smaller tipped probe for the CaverMap Surgical Aide. We remove the suture used to mark the nerve ends and the plastic surgeon trims the nerve ends as needed. Immediately after removal of the prostate, the deep pelvis is an inhospitable environment for performing a sural nerve graft. Not only is it difficult to perform microsurgery using long instruments but the field invariably is bloodier than that encountered in areas of the body where nerve grafting is usually performed. In order to expedite the anastomosis, the plastic surgeon uses 2 mm sections of plastic tubing and microsurgical clips rather than knots to secure each anastomotic suture. This procedure is further facilitated by melting the free end of a 5 cm, 7–0 Prolene suture in order to create an expansion in the diameter of the suture. Using 4× magnification, the plastic surgeon places the anastomotic sutures. Because the tail of the suture now has an expansion, upward traction on the suture brings the graft and the *in situ* cut edge into contact in an end-to-end fashion. In order to secure the suture in place, the plastic surgeon applies a metallic vascular clip to the suture end opposite the one with the expansion, at the point where it exits the nerve epineurium. Since the 7–0 Prolene is too small to be held by a clip directly, the plastic surgeon guides a 2 mm by 1 mm section of silicone tubing over the needle down the free suture segment, and the clip is applied to this tube, thereby securing the 7–0 suture. The use of microclips rather than microsurgical knots to secure the end of the suture saves time, reduces nerve handling and creates a satisfactory microsurgical reconstruction in a non-microsurgical environment. The cavernous nerve is a plexus of nerve fibers rather than a discrete nerve. The use of a relatively large caliber nerve and two parallel graft segments maximizes the amount of cavernous nerve plexus that can make end-to-end contact with the graft (**Fig. 30.7**).

Once the SNG is in place, we take the table off flexion and tie down the vesicourethral anastomotic sutures. Care must be taken not to disturb the nerve graft during completion of the anastomosis. We tie the 6 o'clock anastomotic suture from the

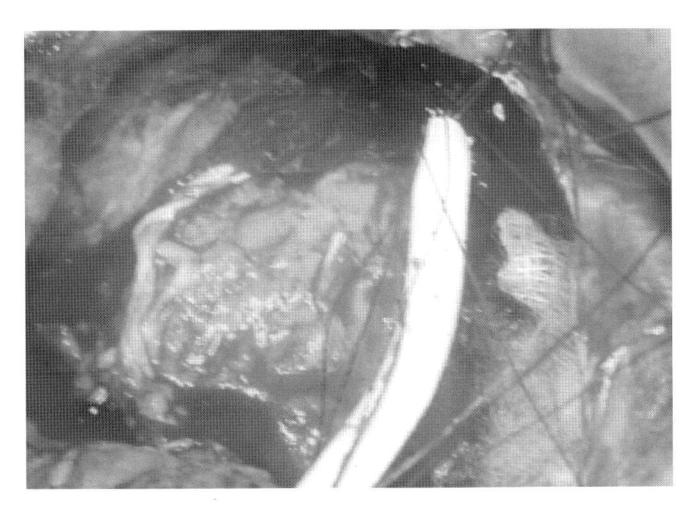

Fig. 30.7. Sural interposition graft *in situ*. A sural nerve segment has been used to bridge the cut ends of the excised left neurovascular bundle. The vesicourethral anastomotic sutures have not yet been tied. Reprinted from *Urologic Clinics of North America* 2001; 28:846 with permission from W.B. Saunders Co. (*See Color plate 10.*)

side opposite the nerve graft. We place only one Jackson–Pratt drain on the side opposite the nerve graft.

Despite the potential complications associated with sural nerve harvest, including formation of hematoma, neuroma, infection, pain, reflex sympathetic dystrophy and delayed ambulation, patients who have received an SNG typically have been able to follow our routine postoperative care pathway. They are ambulatory by postoperative day 1, and are routinely discharged home on postoperative day 2 or 3. Except for one minor leg wound infection, there have been no complications directly attributable to SNG. For the initial 4 weeks after surgery, we ask patients to wrap the ankle of the donor leg with an Ace® elastic bandage and to elevate it when not ambulating. These measures prevent edema and minimize discomfort.

We advise patients to use sildenafil beginning 6 weeks after surgery and to use a vacuum erection device daily as soon as continence has been restored (by 2–3 months after surgery). If patients find that erections are still inadequate for intercourse, we offer intracavernous injection therapy. Since the rate of nerve regeneration has been estimated at 1–2 mm per day in adults, patients are counseled that improved erections as a result of SNG function will not be evident until well after the sixth month post-RRP.[1–3,52]

RECOVERY OF POTENCY AND CONTINENCE AFTER SURAL NERVE GRAFT

We showed that the cavernous nerve can regenerate after resection and interposition grafting with our initial series of bilateral SNG. The indications for bilateral non-nerve-sparing surgery, however, are rare. Almost all of the nerve graft procedures that we perform are unilateral. What are the benefits of unilateral SNG as compared to unilateral NVB resection without nerve graft? Although not randomized, our data suggest that unilateral SNG improves potency and may also improve continence.

A recent analysis of 93 potent patients who underwent unilateral nerve-sparing RRP (between November 1994 and November 1999) with ($n=51$) or without ($n=42$) contralateral SNG showed that 32 of the 51 men who had unilateral SNG recovered erectile function compared with 7 of the 42 men who did not have SNG. The average time to recovery of erectile function was significantly shorter in the SNG group [mean time to potency recovery 13.7 months, 95% confidence interval (CI) 11.4–15.9 months] than in the non-SNG group (mean time to potency recovery 65.9 months, 95% CI 57.7–74.1 months). Because men who opted for SNG were appreciably younger, we performed both age-adjusted and subgroup analyses. After adjustment for age, the Kaplan–Meier probability of recovery of erectile function remained significantly higher in the SNG group than in the non-SNG group at 6, 12, 18 and 24 months ($P < 0.05$). The age-adjusted hazard rate when the SNG group was compared with the non-SNG group was 4.04 ($P=0.003$), indicating the Kaplan–Meier probability of erectile function recovery in the grafted patients was four-fold higher than that in the non-grafted patients after adjustment for age. When the analysis was restricted to a subgroup of men aged 60 years or younger ($n = 51$) the positive association between the probability of erectile function recovery and SNG persisted (**Table 30.2**). The data for this analysis were obtained from the erectile function domain score of the International Index of Erectile Function Questionnaire, or a modified 'recall' version of it. Only men with preoperative erectile function domain scores, without any therapy, of greater than 25/30 were

Table 30.2. Probability and rate of erectile function recovery after unilateral neurovascular bundle resection with or without unilateral sural nerve interposition graft (SNG) in patients 60 years of age or younger ($n=51$).

	Cumulative Kaplan–Meier probability of erectile function recovery (95% confidence interval)				Erectile function recovery hazard rate	
	6 months	12 months	18 months	24 months	Unadjusted	Age adjusted
SNG $n=33$	7.9% (0–16.5)	43.6% (27.6–59.7)	74.4% (59.0–89.8)	85.8% (71.2–100.0)	3.13 $P=0.02$	3.43 $P=0.02$
Non-SNG $n=18$	0%	15.4% (0–35.0)	38.5% (12.0–64.9)	38.5% (12.0–64.9)		

included in the analysis. Postoperative potency was defined as an erectile function domain score of greater than 17/30 without any therapy. Median follow-up for the SNG group was 10 months with a mean of 10.5 and range of 0–24 months. Mean and median follow-up for the non-SNG group was 36 months with a range of 0–78 months. Mean and median ages in the non-SNG group were 64 and 65 years, respectively, with a range of 47–77 years. Mean and median ages were 56 years in the SNG group, with a range of 44–68 years.

We have also analyzed postoperative continence after both nerve-sparing and non-nerve-sparing RRP, and analyzed the impact of nerve grafting in this setting. Here, too, the data were obtained by a self-administered prostate cancer-specific, health-related quality-of-life questionnaire. Of the 377 questionnaires administered to consecutive patients who underwent RRP by a single surgeon (KMS), 261 were satisfactorily completed. Three questions specific to urinary control were analyzed. Patients who had undergone bilateral nerve-sparing RRP were compared with those who had undergone unilateral nerve resection without SNG using the Kaplan–Meier method and Multivariate Cox Regression analysis. The data show an improved time to and level of urinary control with degree of nerve sparing ($P < 0.0158$). Patients who had undergone unilateral nerve resection without SNG were then compared with those who had undergone unilateral nerve resection with SNG. Those who received an SNG demonstrated a significant improvement in time to and level of urinary control ($P < 0.0001$). SNG status remained an independent predictor of urinary control when the analysis was controlled for age (hazards ratio >7, 95% CI 3.3–15.1) ($P < 0.0001$). The median age at the time of RRP was 60.3 years (range 40–77 years) with a median follow-up of 20 months (range 3–54 months).[53]

Although the results presented here are preliminary, they show a significant improvement in both potency and continence with the use of sural nerve interposition graft in patients who undergo unilateral or bilateral NVB resection. Wide resection of the NVB followed by SNG in patients with T3 disease and ECE at the posterolateral margin is a logical extension of our current knowledge of pelvic neuroanatomy. Autonomic nerves to both the corpora cavernosa and the external urethral sphincter complex travel through the cavernous nerves. Potency has been shown to be reduced by about 50% in patients who undergo unilateral wide resection of the NVB.[27–30] We have shown, as proof of principle in bilaterally resected NVBs at the time of RRP, that SNG can restore erectile function.[25,26] Improvement in erectile function after unilateral sural nerve graft in patients who undergo wide resection of one NVB is, therefore, not unexpected. Other groups have also shown improved potency with SNG.[54]

Our data also suggest that SNG can improve postoperative continence. The effect of NVB resection on postoperative continence is controversial. In our hands, NVB resection results in reduced postoperative continence as measured by responses to questions about urinary control in a prostate cancer-specific, health-related quality-of-life questionnaire. Although it is possible that damage to nerves other than the cavernous nerves contributes to the observed postoperative decrease in urinary control, the improvement noted in patients with SNG suggests that autonomic innervation of the distal urethral sphincter can contribute significantly to post-RRP urinary control. The effects of SNG on postoperative potency and continence will need to be evaluated in a multi-institutional, controlled randomized trial.

THE FUTURE OF RRP

The retropubic approach to radical prostatectomy has been improved significantly since Millin's initial report in 1947.[55] Detailed studies of human pelvic anatomy over the last two decades facilitated improvements in technique that have resulted in decreased morbidity and mortality from RRP. Furthermore, the advent of PSA-based prostate cancer screening has allowed for the diagnosis of prostate cancer at much earlier stages. This, in turn, has improved our ability to cure prostate cancer with RRP.

Although clinically relevant discoveries regarding pelvic anatomy will continue to be made, and new serum markers undoubtedly will improve our ability to detect clinically relevant prostate cancer at its earliest stages, advances in minimally invasive, robotic surgery are likely to revolutionize the way RRP is performed. RRP, in its current form, is essentially a minimally invasive procedure. It is performed completely extraperitoneally; it can be done through an incision 8–9 cm long, and there is minimal immediate postoperative morbidity associated with the retropubic surgical approach. Nevertheless, robotic devices have the potential to improve on the surgeon's technical ability. They already offer a range of motion greater than that of current laparoscopic instruments, and will eventually be capable of more accurate and precise movements than those of human hands alone.

Given the number of tumor markers under study, it is very likely that the concentration of one or a combination of these will be found to correlate better with disease stage than PSA and/or % free PSA. Tumor markers that correlate with pathologic stage in combination with improved ultrasound-guided prostate biopsy schemes are likely to make significant contributions to our ability to predict pathologic stage preoperatively. Improvements in our ability to predict pathologic stage preoperatively and intraoperatively have the potential to both improve cancer cure rates and reduce postoperative morbidity by allowing the surgeon to perform wide resection of the NVB via either an open or a minimally invasive technique, only when it will benefit the patient.

REFERENCES

1. Lundborg G, Danielsen N. Injury degeneration and reconstruction. In: Gelberman RH (ed.) *Operative Nerve Repair and Reconstruction*, p. 844. Philadelphia: J.B. Lippincott Co., 1991.
2. Mackinnon SE, Dellon AL. *Surgery of the Peripheral Nerve*. New York: Thieme Medical Publishers, 1988.

3. Sutherland S. *Nerve Grafting and Related Methods of Nerve Repair*, pp. 467–77. Edinburgh: Churchill Livingstone, 1991.
4. Kline DG, Hackett ER, Davis GD et al. Effects of mobilization on the blood supply and regeneration of injured nerves. *J. Surg. Res.* 1972; 12:254–66.
5. Brooks JD. Anatomy of the lower urinary tract and male genitalia. In: Walsh PC, Retik AB, Vaughan ED et al. (eds) *Campbell's Urology*, 7th edn, pp. 89–128. Philadelphia: W.B. Saunders, 1998.
6. Brown JO, Satinsky VP. Functional restoration of the paralyzed diaphragm following the cross-union of the vagus and phrenic nerves. *Am. J. Med. Sci.* 1951; 222:613–16.
7. Freeman LW. Observations on spinal nerve root transplantation in the male Guinea baboon. *Ann. Surg.* 1952; 136:206–8.
8. Seddon HJ. The use of autogenous grafts for the repair of large gaps in peripheral nerves. *Br. J. Surg.* 1947; 35:151–67.
9. Neuman C, Grunefest H, Berry C et al. Return of function of sweat glands after cutting or crushing sympathetic nerves. *Proc. Soc. Exp. Biol. Med.* 1943; 54:27–8.
10. Benoit G, Droupy S, Quillard J et al. Supra and infralevator neurovascular pathways to the penile corpora cavernosa. *J. Anat.* 1999; 195:605–15.
11. Lepor H, Gregerman M, Crosby R et al. Precise localization of the autonomic nerves from the pelvic plexus to the corpora cavernosa: a detailed anatomical study of the adult male pelvis. *J. Urol.* 1985; 133:207–12.
12. Burnett AL, Mostwin JL. In situ anatomical study of the male urethral sphincteric complex: relevance to continence preservation following major pelvic surgery. *J. Urol.* 1998; 160:1301–6.
13. Paick JS, Donatucci CF, Lue TF. Anatomy of cavernous nerves distal to prostate: microdissection study in adult male cadavers. *Urology* 1993; 42:145–9.
14. Myers RP. Male urethral sphincteric anatomy and radical prostatectomy. *Urol. Clin. North Am.* 1991; 18:211–27.
15. Kumagai A, Koyanagi T, Takahashi Y. The innervation of the external urethral sphincter; an ultrastructural study in male human subjects. *Urol. Res.* 1987; 15:39–43.
16. Juenemann KP, Lue TF, Schmidt RA et al. Clinical significance of sacral and pudendal nerve anatomy. *J. Urol.* 1988; 139:74–80.
17. Hollabaugh RS Jr, Dmochowski RR, Steiner MS. Neuroanatomy of the male rhabdosphincter. *Urology* 1997; 49:426–34.
18. Zvara P, Carrier S, Kour NW et al. The detailed neuroanatomy of the human striated urethral sphincter. *Br. J. Urol.* 1994; 74:182–7.
19. Tanagho EA, Schmidt RA, de Araujo CG. Urinary striated sphincter: what is its nerve supply? *Urology* 1982; 20:415–17.
20. Narayan P, Konety B, Aslam K et al. Neuroanatomy of the external urethral sphincter: implications for urinary continence preservation during radical prostate surgery. *J. Urol.* 1995; 153:337–41.
21. Kleinhans B, Gerharz E, Melekos M et al. Changes of urodynamic findings after radical retropubic prostatectomy. *Eur. Urol.* 1999; 35:217–21.
22. Quinlan DM, Nelson RJ, Walsh PC. Cavernous nerve grafts restore erectile function in denervated rats. *J. Urol.* 1991; 145:380–3.
23. Ball RA, Richie JP, Vickers MA. Microsurgical nerve graft repair of the ablated cavernosal nerves in the rat. *J. Surg. Res.* 1992; 53:280–6.
24. Walsh PC. When is it necessary to excise the nerves and are nerve grafts likely to improve sexual function in men undergoing anatomic radical prostatectomy? News Highlights, Brady Urological Institute Web Site accessed April 11, 2001; http://prostate.urol.jhu.edu/highlights/nerve_grafts.html.
25. Kim ED, Scardino PT, Hampel O et al. Interposition of sural nerve restores function of cavernous nerves resected during radical prostatectomy. *J. Urol.* 1999; 161:188–92.
26. Kim ED, Nath R, Kadmon D et al. Bilateral nerve graft during radical retropubic prostatectomy: 1-year followup. *J. Urol.* 2001; 165:1950–6.
27. Catalona WJ, Carvalhal GF, Mager DE et al. Potency, continence and complication rates in 1,870 consecutive radical retropubic prostatectomies. *J. Urol.* 1999; 162:433–8.
28. Rabbani F, Stapleton AM, Kattan MW et al. Factors predicting recovery of erections after radical prostatectomy. *J. Urol.* 2000; 164:1929–34.
29. Geary ES, Dendinger TE, Freiha FS et al. Nerve sparing radical prostatectomy: a different view. *J. Urol.* 1995; 154:145–9.
30. Quinlan DM, Epstein JI, Carter BS et al. Sexual function following radical prostatectomy: influence of preservation of neurovascular bundles. *J. Urol.* 1991; 145:998–1002.
31. Stanford JL, Feng Z, Hamilton AS et al. Urinary and sexual function after radical prostatectomy for clinically localized prostate cancer: the Prostate Cancer Outcomes Study. *JAMA* 2000; 283:354–60.
32. Gore JL, Shariat S, Ayala G et al. Individual laterally-directed biopsy cores predict the site of extracapsular extension in patients undergoing radical prostatectomy. *96th Annual Meeting of the American Urological Association*, Anaheim, CA, 2001.
33. Hull GW, Rabbani F, Abbas F et al. Cancer control with radical prostatectomy alone in 1000 consecutive patients. *J. Urol.* 2002; 167:528–34.
34. Rosen MA, Goldstone L, Lapin S et al. Frequency and location of extracapsular extension and positive surgical margins in radical prostatectomy specimens. *J. Urol.* 1992; 148:331–7.
35. Paul AB, Grant ES, Habib FK. The expression and localization of beta-nerve growth factor (beta-NGF) in benign and malignant human prostate tissue: relationship to neuroendocrine differentiation. *Br. J. Cancer* 1996; 74:1990–6.
36. Walch ET, Marchetti D. Role of neurotrophins and neurotrophins receptors in the in vitro invasion and heparanase production of human prostate cancer cells. *Clin. Exp. Metast.* 1999; 17:307–14.
37. Villers A, McNeal JE, Redwine EA et al. The role of perineural space invasion in the local spread of prostatic adenocarcinoma. *J. Urol.* 1989; 142:763–8.
38. D'Amico AV, Wu Y, Chen MH et al. Perineural invasion as a predictor of biochemical outcome following radical prostatectomy for select men with clinically localized prostate cancer. *J. Urol.* 2001; 165:126–9.
39. de la Taille A, Katz A, Bagiella E et al. Perineural invasion on prostate needle biopsy: an independent predictor of final pathologic stage. *Urology* 1999; 54:1039–43.
40. Epstein JI. The role of perineural invasion and other biopsy characteristics as prognostic markers for localized prostate cancer. *Semin. Urol. Oncol.* 1998; 16:124–8.
41. Epstein JI, Partin AW, Sauvageot J et al. Prediction of progression following radical prostatectomy. A multivariate analysis of 721 men with long-term follow-up. *Am. J. Surg. Pathol.* 1996; 20:286–92.
42. Ohori M, Wheeler TM, Kattan MW et al. Prognostic significance of positive surgical margins in radical prostatectomy specimens. *J. Urol.* 1995; 154:1818–24.
43. Demsar J, Zupan B, Kattan MW et al. Naive Bayesian-based nomogram for prediction of prostate cancer recurrence. *Stud. Health Technol. Inform.* 1999; 68:436–41.
44. Kattan MW, Wheeler TM, Scardino PT. Postoperative nomogram for disease recurrence after radical prostatectomy for prostate cancer. *J. Clin. Oncol.* 1999; 17:1499–507.
45. Shariat SF, Shalev M, Menesses-Diaz A et al. Preoperative plasma levels of transforming growth factor beta(1) (TGF-beta(1)) strongly predict progression in patients undergoing radical prostatectomy. *J. Clin. Oncol.* 2001; 19:2856–64.
46. Stapleton AMF, Scardino PT. Nerve-sparing radical retropubic prostatectomy. In: McConnell JD, Scardino PT (eds) *Atlas of Clinical Urology*. Philadelphia: Current Medicine, 1999.
47. Kim HL, Stoffel DS, Mhoon DA et al. A positive cavermap response poorly predicts recovery of potency after radical prostatectomy. *Urology* 2000; 56:561–4.

48. Klotz L. Intraoperative cavernous nerve stimulation during nerve sparing radical prostatectomy: how and when? *Curr. Opin. Urol.* 2000; 10:239–43.
49. Klotz L. Neurostimulation during radical prostatectomy: improving nerve-sparing techniques. *Semin. Urol. Oncol.* 2000; 18:46–50.
50. Klotz L, Heaton J, Jewett M et al. A randomized phase 3 study of intraoperative cavernous nerve stimulation with penile tumescence monitoring to improve nerve sparing during radical prostatectomy. *J. Urol.* 2000; 164:1573–8.
51. Karakiewicz PI, Shariat S, Gore JL et al. The cavermap surgical aid has a high specificity in identifying the neurovascular bundle (NVB) during radical prostatectomy (RP). *96th Annual Meeting of the American Urological Association*, Anaheim, CA 2001.
52. Millesi H. Healing of nerves. *Clin. Plast. Surg.* 1977; 4:459–73.
53. Singh H, Shariat S, Canto EI et al. Impact of interposition sural nerve graft on urinary control in patients undergoing radical prostatectomy. *97th Annual Meeting of the American Urological Association*, Orlando, FL, 2002.
54. Wood CG, Change D, Kroll S et al. A phase I study of autologous sural nerve grafting to preserve potency after radical prostatectomy. *Annual Meeting of the Society of Surgical Oncology*, Washington, DC, 2001.
55. Millin T. *Retropubic Urinary Surgery*. London: Livingstone, 1947.

Radical Prostatectomy: The Retropubic Antegrade Approach

Bulent Akduman and E. David Crawford

Department of Urology, University of Colorado Health Science Center, Denver, CO, USA

INTRODUCTION

Prostate cancer is the most common solid tumor in men. Approximately 198 000 new cases will be diagnosed in 2001, with about 20% of these patients presenting with metastatic disease at the time of the diagnosis.[1] Early diagnosis is paramount to the successful management of prostate cancer. The widespread public awareness and early detection efforts for prostate cancer have resulted in the majority of patients having their conditions diagnosed at a stage when treatment with curative intent can be considered.

Patients diagnosed with a clinically localized prostate cancer have a variety of management choices, including radical prostatectomy, brachytherapy, external beam radiation therapy as well as conservative management such as watchful waiting. Treatment decision can be done based on the nature of each cancer. The effectiveness of treatment, the life expectancy of the patient and the quality of life issues should be discussed with active participation of the patient. Radical retropubic prostatectomy remains the time-honored therapy for locally confined prostate cancer. Although the retropubic approach to simple prostatectomy was first described by Millin[2] in 1947, it was not widely accepted to radical prostatectomy because of high morbidity and mortality rates. The advantages of retropubic radical prostatectomy are a lower rate of positive surgical margins, more consistent preservation of neurovascular bundles and fewer rectal injuries. In addition, a staging pelvic lymphadenectomy can be done in the same session. Technical refinements in radical retropubic prostatectomy have resulted in lower rates of urinary incontinence and erectile dysfunction, less blood loss and shorter hospital stays, and lower rates of positive surgical margins.[3–7]

A modification of Campbell's technique[8] of radical prostatectomy has been performed in our center.[8] The technique emphasizes dissection of the prostate from the base to the apex. Removal of all malignant tissue with acceptable tumor-free margins, and tissue dissection performed in the proper sequence to minimize iatrogenic lymphatic and hematogenous dissemination are provided with this technique. There is less time-consuming hemorrhage early in the procedure because sealing the dorsal vein of the penis and periurethral venous plexus is one of the final steps. In addition, we routinely use a new device called the LigaSure™ Vessel sealing system to dissect and seal the vessels. The LigaSure™ Vessel Sealing System works by applying a precise amount of bipolar energy and pressure to denature the elastin and collagen in vessel walls, resulting in permanent occlusion. Since the system uses the body's own collagen to re-form the tissue, it resists dislodgment and leaves no foreign material behind. With the use of this technique, placement of the urethral sutures under direct vision at the time of anastomosis also decreases the chance of incontinence. Our clinical results demonstrate that mean operative time, blood loss and hospital stay were lower when using the modified technique rather than the conventional approach. Decreased operative time offers a potential advantage in terms of cardiac and pulmonary morbidity, especially in elderly patients.

PREOPERATIVE ASSESSMENT

Radical prostatectomy is considered as a curative treatment. It is highly important for preoperative staging to exclude any evidence of locally advanced or distant metastatic disease. The presence of any residual disease after radical prostatectomy denotes treatment failure. The likelihood of post-radical prostatectomy failure is associated with increased pathologic stage[9]. On the other hand, the clinical staging of prostate cancer is limited using conventional preoperative diagnostic methods.

The likelihood of extracapsular disease is increased in patients with high prostate specific antigen (PSA) levels and Gleason scores. It was demonstrated that the chances of having organ-confined disease in patients with a PSA level of <10 ng/ml is 70–80%. The chance decreases to 50% for men with PSA >10 ng/ml and to approximately 25% for men with PSA >50 ng/ml.[10] However, PSA values vary widely within a given stage and overlap between different stages; the predictive value of PSA in determining pathologic stage is weak.[11]

Digital rectal examination understages organ-confined disease. It was shown that only 52% of patients with cT2 disease and 19% of patients with cT3 disease had organ-confined tumors.[10] Although there are reports of cure using radical prostatectomy for men with cT3 disease, the presence of a bulky extraprostatic tumor and the high risk for metastatic disease limits the use of radical prostatectomy in these patients.[12] However, patients with cT1 or cT2 tumors that are upstaged to pT3 after the surgery can experience durable cancer remissions.[13,14]

It is well known that tumor grade is strongly related to stage and clinical prognosis. It was demonstrated that tumors with a Gleason score of 6 on prostate needle biopsy had a 24% risk of capsular penetration and a 29% risk of positive surgical margins. This increased to a 62% risk of capsular penetration and a 48% risk of positive surgical margins for cancers with a Gleason score of 7, and to 85% and 59%, respectively, for cancers with a Gleason score of 8–10.[15]

Perineural invasion on needle biopsy and increased tumor volume are associated with capsular penetration and extracapsular extension.[16,17] The odds of having extraprostatic disease are increased when all three biopsy cores from a single side of the prostate are positive for cancer on sextant biopsy.

Patient-related factors are also important in decision making for radical prostatectomy. Advancing age and poor health status limit the indications of radical prostatectomy as a treatment option. Risk for incontinence and impotence, hospitalization and recovery period as well as fear of being under anesthesia should be discussed in detail before the surgery.

ANATOMIC CONSIDERATIONS

Knowledge of the anatomic features of the prostate is imperative for the success of the operation. The pelvic fascia reflects medially off the obturator internus muscle at the arcus tendineus. One leaf of the fascia sweeps medially and thickens to form a sheath covering the anterior and lateral surfaces of the prostate. This condensation is known as the endopelvic fascia. The endopelvic fascia fuses with Denonvilliers' fascia at the posterolateral margin of the prostate. The anatomic importance of the endopelvic fascia lies not only in its support of the prostate but also in the fact that it contains, beneath its lateral reflections, much of the vascular and lymphatic supply of the gland. The anterior and lateral reflections of the endopelvic fascia must be divided in the course of a radical prostatectomy.

The puboprostatic ligaments represent endopelvic fascia's anterior reflection from the pubic bone, which supports the gland and blends with the prostatic capsule. The puboprostatic ligaments are avascular structures. These ligaments are attached to the inferior border of the pubis, lateral to the cartilaginous symphysis, and on the prostatic side they blend with the lateral leaves of the endopelvic fascia at the prostatovesical junction. Sharp division of the puboprostatic ligaments resulted in their retraction because of the smooth muscle component.

Denonvilliers' facsia is compressed between the posterior surface of the seminal vesicles, the prostate, the bladder and the anterior rectal wall. Overlying the ventral rectal surface and intermingled with the external longitudinal musculature is a layer of areolar and fatty rectal fascia known as the posterior layer of Denonvilliers' fascia. The remnant of the peritoneal cul-de-sac in the fetus is located anterior to the rectal fascia. This layer is known as the anterior layer of Denonvilliers' fascia. This anterior layer splits and one leaf passes caudally over the posterior surface of the seminal vesicles, vasa and prostatic capsule. It then extends to meet the anterior layer at the apex of the prostate. The second leaf extends over the ventral surface of the seminal vesicles to the bladder base, where it reflects superiorly over the fundus of the bladder and then fuses with the first leaf. The preferred plane in an attempt to separate the prostate from the anterior rectal wall is between the anterior and posterior layers of Denonvilliers' fascia.

The inferior vesical artery provides small branches to the inferior, posterior extent of the seminal vesicles and the base of the bladder and prostate. It terminates in two large groups of prostatic vessels: the urethral and capsular groups. The urethral vessels enter the prostate at the posterolateral vesicoprostatic junction, providing arterial supply to the vesical neck and periurethral portion of the gland. The capsular branches run along the pelvic side wall in the lateral pelvic fascia, traveling posterolateral to the prostate providing branches that course ventrally and dorsally to supply the outer portion of the prostate. The capsular vessels provide the macroscopic landmark that aids in the identification of the microscopic branches of the pelvic plexus that innervate the corpora cavernosa.

The venous drainage of the prostate is into the plexus of Santorini. The deep dorsal vein leaves the penis under Buck's fascia between the corpora cavernosa and penetrates the urogenital diaphragm, dividing into three major branches: the superficial, and the right and left lateral venous plexuses. The superficial branch, which travels between the puboprostatic ligaments, is a centrally located vein overlying the bladder neck and prostate. The lateral plexuses traverse posterolaterally and communicate freely with the pudendal, obturator and vesical plexuses. These plexuses interconnect with other venous systems to form the inferior vesical vein, which empties into the internal iliac vein.

The autonomic innervation of the pelvic organs and external genitalia arises from the pelvic plexus, which is formed by parasympathetic visceral efferent preganglionic fibers, nervi erigentes, which arise from the sacral center (S2–S4) and sympathetic fibers from the thoracolumbar center (T11–L2). The pelvic plexus provides visceral branches that innervate the ureter, bladder, prostate, seminal vesicles, rectum, membranous urethra and corpus cavernosa. Branches that contain somatic motor axons travel through the pelvic plexus to supply the levator ani, coccygeus and striated urethral musculature. The nerves innervating the prostate travel outside the capsule of the prostate and Denonvilliers' fascia until they perforate the capsule where they enter the prostate. The branches to the membranous urethra and corpora cavernosa also travel outside the prostate dorsolaterally in the lateral pelvic fascia between the prostate and rectum. At the level of the membranous urethra they travel at 3 and 9 o'clock. After piercing the urogenital diaphragm, they pass behind the dorsal penile artery and dorsal penile nerve before entering the corpora cavernosa.

SURGICAL PROCEDURE

Patients are admitted on an outpatient basis. They are advised to refrain from consuming anything by mouth starting from midnight before the procedure. Routine bowel preparations are not performed. A broad-spectrum antibiotic is given preoperatively and for 48 hours after the procedure. Radical prostatectomy candidates often have a good performance status and the incidence of hypercoagulability due to malignancy is also likely to be low. The benefit of preventive measures for pulmonary embolus and deep venous thrombosis, such as sequential compression devices, has yet to be shown. Again, the efficacy of minidose heparin has yet to be established and it appears to be associated with an increase in the incidence of prolonged lymphatic drainage or lymphocele formation.[18] For this reason, we routinely do not take these kinds of measures to prevent pulmonary embolus and deep venous thrombosis, but instead concentrate on early ambulation.

Patients are placed in the supine position on the operative table. The kidney rest under the lumbar spine area is elevated, providing slight hyperextension. The patient is draped in the usual manner with a sterile towel placed under the penis. A sterile 22-French catheter is inserted and connected to a gravity drainage system.

A midline incision is made beginning at the symphysis pubis and extending superiorly to the level below the umbilicus, continuing through the subcutaneous layer into the retropubic space of Retzius. Pelvic lymphadenectomy is reserved for patients with high risk for lymph node involvement (high Gleason score, high PSA level and poor clinical stage).

The first phase of our technique is division of the bladder neck from the prostate. The index finger identifies the balloon of the Foley catheter; the line between the bladder neck and the prostate is marked with cautery. The laparoscopic LigaSure device is used to dissect and seal the tissue between the bladder neck and prostate (**Fig. 31.1**). The dissection is extended laterally. A Babcock clamp is applied to the distal end of the balloon. With the use of a Kitner dissector, this dissection is carried down about a half centimeter to identify a clear urethral tube. A surgical knife is used to incise the upper portion of the urethra and bladder neck. The Foley catheter is deflated, advanced out and pulled over the pubic bone, and secured to the other end of the catheter with a clamp. Using cautery, a U-shaped incision is made in the lower portion of the urethra and underlying muscles, which is a continuation of the trigone.

The second phase is division of the vasa and dissection of the seminal vesicles. After the vasa is identified and dissected with a right-angle clamp, the LigaSure device is applied to seal and divide the vasa (**Fig. 31.2**). A Babcock clamp is placed on the vasa and pulled cephalad. The vessels around the seminal vesicle are sealed and divided using the LigaSure device (**Fig. 31.3**).

The third phase is incision of Denonvilliers' fascia and lateral pedicles. The posterior leaf of the anterior layers of Denonvilliers' fascia is sharply divided using scissors. The posterior side of the prostate is dissected using the LigaSure device and blunt dissection. After the lateral pedicles are

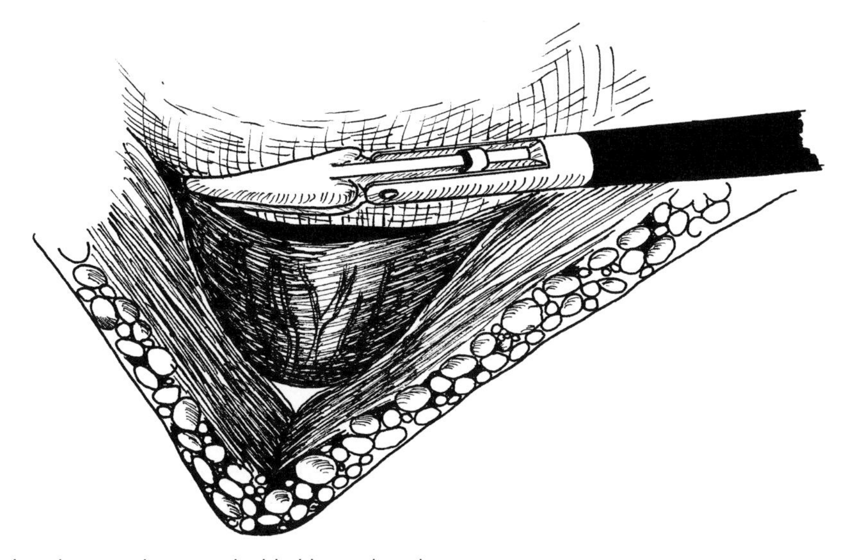

Fig. 31.1. Dissection and sealing the tissue between the bladder neck and prostate.

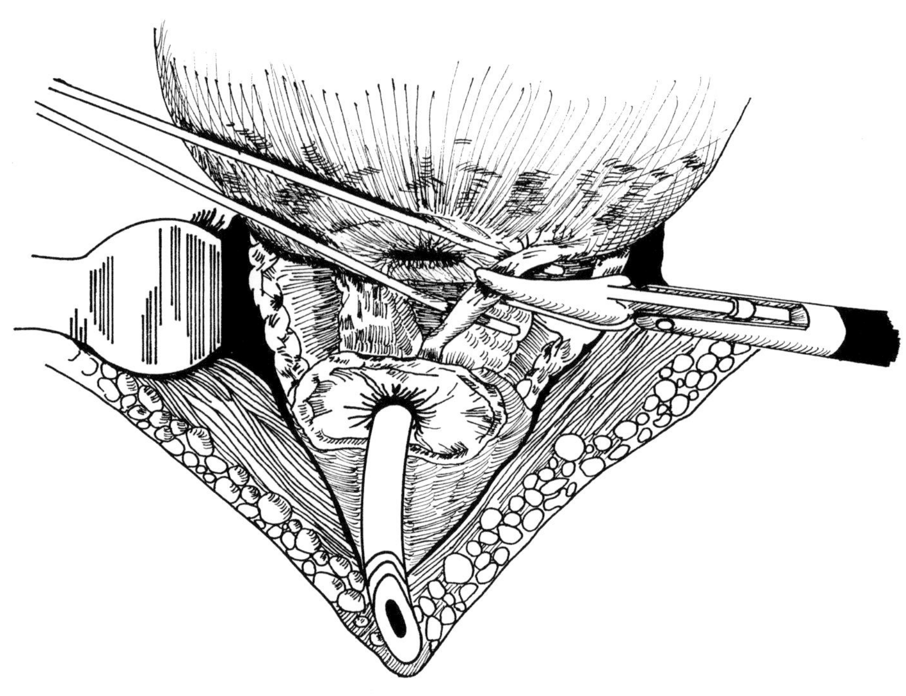

Fig. 31.2. Sealing and division of the vasa.

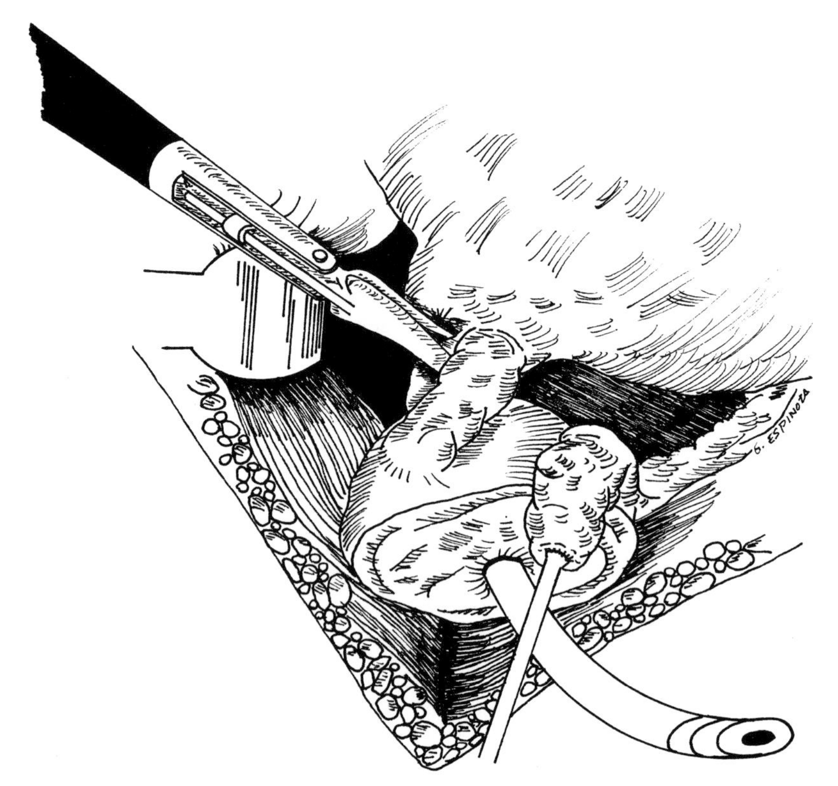

Fig. 31.3. Sealing and division of the vessels around the seminal vesicle.

identified, they are dissected and sealed using the LigaSure device (**Fig. 31.4**).

The fourth phase is division of the puboprostatic ligaments and dorsal vein complex. Endopelvic fascia, which remains towards the apex, is dissected and sealed with LigaSure (**Fig. 31.5**). After the puboprostatic ligament is identified, it is cut sharply where it attaches to the pubic bone. The underlying dorsal vein complex is sealed with the LigaSure device in

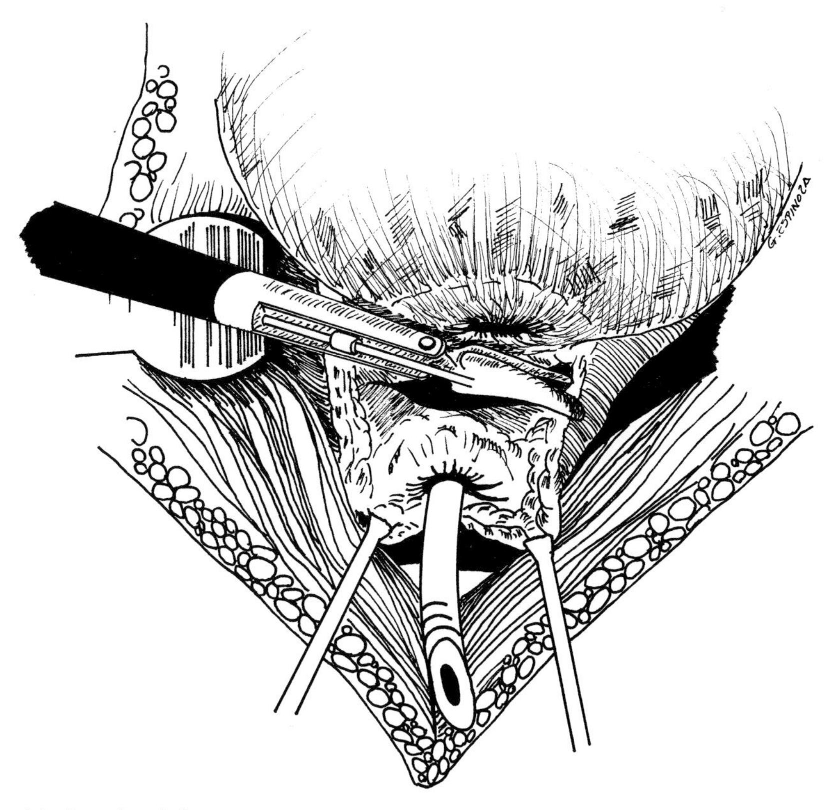

Fig. 31.4. Dissection and sealing of the lateral pedicles.

Fig. 31.5. Dissection and sealing of the endopelvic fascia.

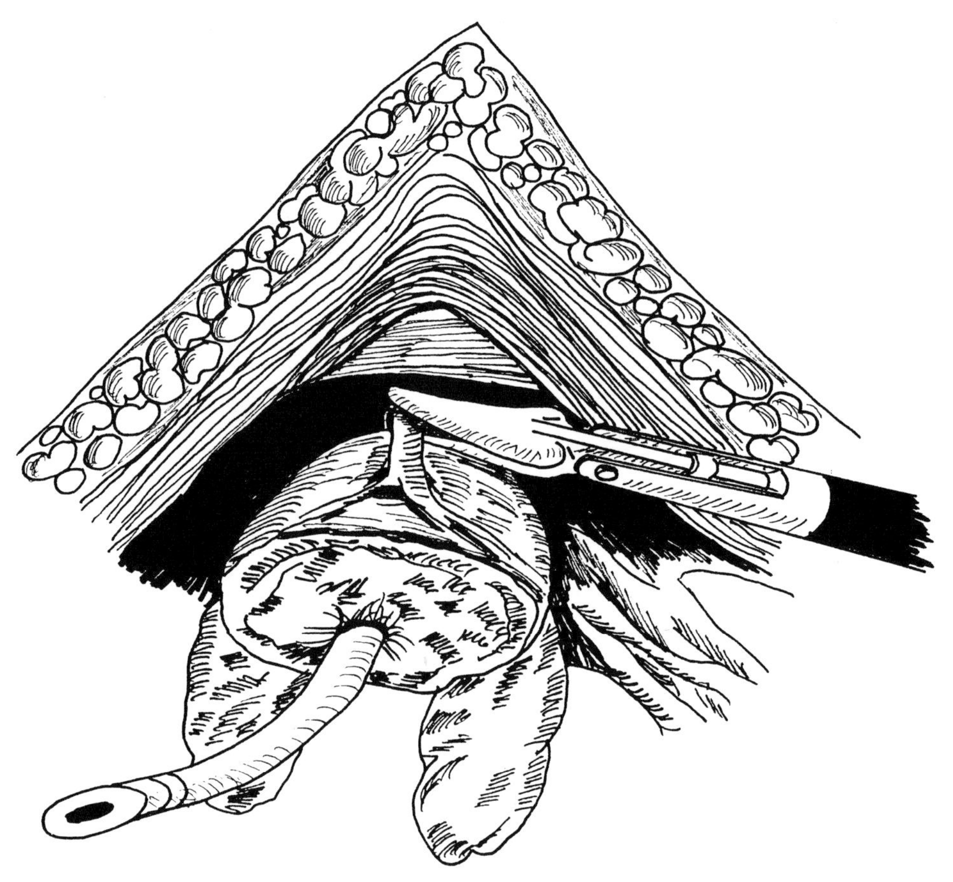

Fig. 31.6. Sealing of the dorsal vein complex.

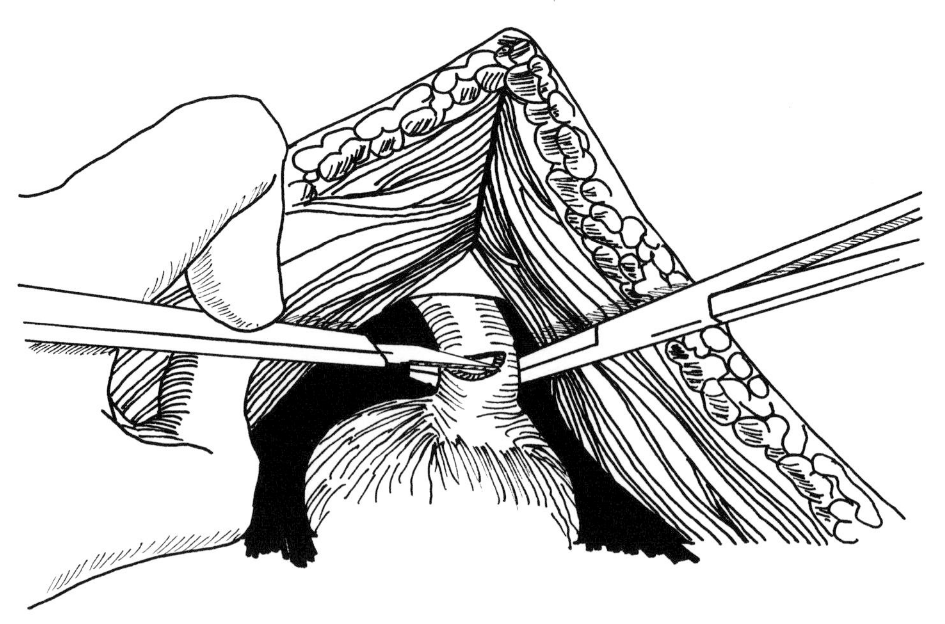

Fig. 31.7. Dissection of the anterior urethra.

a majority of the cases (**Fig. 31.6**). The dissection is carried down towards the apex of the prostate using scissors. After the apex of the prostate is palpated, the anterior urethra is cut sharply with scissors (**Fig. 31.7**).

The final phase is closure. After the urethra has been transected, 2–0 colored monocryl sutures are placed at the 12, 2, 4, 8, 10, and 6 o'clock positions prior to cutting the posterior side of the urethra (**Fig. 31.8**). After the 4, 6, and 8 o'clock

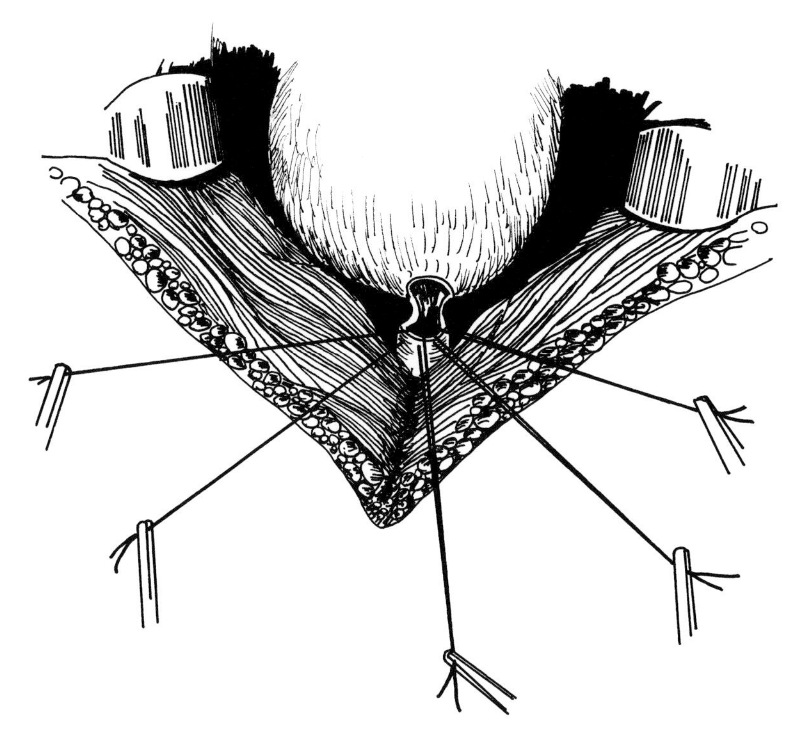

Fig. 31.8. Closure phase of the procedure.

sutures are placed on the bladder neck, a 22F Foley catheter is passed through the urethra into the bladder. Finally, the 10, 2, and 12 o'clock sutures are placed on the bladder neck.

A 0.5-inch Penrose drain is placed near the anastomotic site and brought through a separate stab wound; the surgical incision is closed. If the rectus muscles have been divided from their tendinous insertion in the symphysis pubis, they are approximated. The rectus fascia is closed with figure-of-eight absorbable sutures and the subcutaneous tissues are then reapproximated. The skin is closed in the usual fashion.

POSTOPERATIVE MANAGEMENT

It is important to monitor the patients' vital signs and urinary output. Urine should be clear and blood clots, which may obstruct the drainage of the urethral catheter, should be avoided. We keep very mild traction on the urethral catheter for hours and then release it. In high-risk patients, central venous pressure lines should be inserted before the surgical procedure and this group of patients may need to be observed in the surgical intensive care unit postoperatively. Intravenous fluids are administered until the patients can tolerate fluid intake by mouth.

Patients should be ambulating on the first postoperative day, and striving to increase their ambulation and eating a regular diet so that discharge can be anticipated on the first or second postoperative day. The urethral catheter is left in place for 1–2 weeks, and the drain is advanced and usually removed by the first or second postoperative day. However, in the rare event of significant urine leakage, it is left in place until this abates. Patients are provided with postoperative instructions, including information about Kegel exercises, at hospital discharge so they can begin the exercises as soon as urethral catheter is removed.

COMPLICATIONS

Complications of the radical prostatectomy can be grouped as intraoperative complications, early postoperative complications and late postoperative complications. Hemorrhage is the most common intraoperative complication during radical prostatectomy. A more thorough understanding of the anatomy of the dorsal vein complex and the periprostatic fascia has allowed development of techniques to control the vessels that are responsible for hemorrhage.[19,20] The key steps to reduce blood loss are complete control of the dorsal vein complex and anterior periprostatic veins, identification and control of the small branches from the neurovascular bundles to the prostate posterolaterally, and dissection of the seminal vesicles and vas with control of many small vessels between the base of the bladder and the seminal vesicles.[5] In our series, estimated blood loss is significantly reduced with the use of LigaSure TM Vessel

Sealing System. Mean blood loss is 240 ml with this technique. Because most patients will not require transfusion for a blood loss of less than 1000 ml, we no longer require autologous blood donation. Our series is now of 550 consecutive patients without any blood transfusions.

Rectal injury, ureteral injury and obturator nerve injury are less common intraoperative complications. In fact, in over 1800 cases, there has not been a rectal injury. Rectal injury occurs rarely when developing the plane between the rectum and Denonvilliers' fascia. Patients who have undergone salvage prostatectomies after radiation therapy are at greater risk of rectal injury. When it occurs, the rectum should be closed in two layers after the edges of the incision have been freshened. Interposing omentum between the rectal injury and vesicourethral anastomosis to redsuce the risk of vesicourethral fistula is recommended. At the end of the procedure, the anal sphincter should be dilated. Ureteral injuries and obturator nerve injuries are extremely rare. If ureteral injury occurs, reimplantation of the ureter should be undertaken. Obturator nerve injury should be managed by reanastomosis using fine non-absorbable sutures.

Pulmonary embolism or deep venous thrombosis, urine leak, wound infection, and loss of catheter can be seen in the early postoperative period. Pulmonary embolism and deep venous thrombosis occur in approximately 2% of patients after radical prostatectomy.[21] Shorter anesthesia time, early postoperative ambulation and shorter hospital stay are likely to reduce thromboembolic complications. The benefit of preventive measures, such as sequential compression devices and minidose heparin, has yet to be established. Clot retention, catheter plugging or disruption of the anastomosis can be a cause of leak in the early postoperative period. When it occurs, a combination of catheter drainage with retropubic drainage should allow closure of the leak in a short period of time. If the patient has significant hematuria, periodic gentle irrigation of the catheter should be done to maintain drainage. Even complete disruption heals, if the anastomosis is stented with a catheter. In early postoperative period, consideration should be given to surgical revision of the anastomosis along with drainage of hematoma. The likelihood of bladder neck contracture is high with significant disruption of the anastomosis. Significant urinary extravasation, bleeding or excessive subcutaneous fat may be at greater risk for wound infection. Relatively loose approximation of the skin or the use of suprafascial subcutaneous suction drain in high-risk patients for wound infection allow drainage of serous and serosanguinous fluid, thereby limiting the risk of fluid collection. When wound infection occurs, it is treated by opening the incision and allowing healing by secondary intent. Urethral catheter can fall out following the surgery due to catheter malfunction or injury. In the very early postoperative period, the use of a flexible cytoscope with passage of a floopy guidewire and then Council-tip catheter is the safest course of action. In the late postoperative period (>7 days), consideration should be given to leaving the catheter out as long as the patient is not in urinary retention. In the face of retention or severe hematuria, placement of an open or closed suprapubic cystostomy rather than catheter change should be considered.

Incontinence, impotence and bladder neck contracture are the late postoperative complications. Preservation of external striated sphincter is of paramount importance in attempting to minimize the risk of incontinence. Bladder neck and intrinsic urethral mechanism are responsible passively to maintain the continence. The rate of total continence in contemporary surgical series is between 80% and 95%.[22–24] In our series, the rate of stress incontinence is 3–5% and total incontinence is around 1%. Patients' age, diabetes, chronic hypertension, cardiovascular disease, multiple medication and decreased preoperative erectile dysfunction influence the risk of erectile dysfunction after the surgery. In addition to patient characteristics contributing to postoperative erectile dysfunction, a number of technical factors may contribute to this outcome as well. Preservation of cavernosal nerves, excessive traction, manipulation or local cauterization of the neurovascular bundles is responsible for the risk of erectile dysfunction after the surgery. Bladder neck contracture is usually caused by poor mucosa-to-mucosa apposition of the bladder to the urethra at the time of the anastomosis. It has been reported to occur in 0.5–15% of cases.[22,25] A major complaint is a dribbling stream but at times patients can have overflow incontinence. Urethral dilatation is usually effective in the management of bladder neck contracture.

REFERENCES

1. Greenlee RT, Hill-Harmon MB, Murray T et al. Estimated new cancer cases by gender, US, 2001. *CA Cancer J. Clin.* 2001; 51:15–36.
2. Millin T. *Retropubic Urinary Surgery*. Baltimore, MD: Williams & Wilkins, 1947.
3. Catalona WJ. Surgical management of prostate cancer: contemporary results with anatomic radical prostatectomy. *Cancer* 1995; 75:1903.
4. Walsh PC. Technique of vesicourethral anastomisis may influence recovery of sexual function following radical prostatectomy. *Atlas Urol. Clin. North Am.* 1994; 2:59–64.
5. Goad JR, Scardino PT. Modifications in the technique of radical retropubic prostatectomy to minimize blood loss. *Atlas Urol. Clin. North Am.* 1994; 20:65.
6. Leibman BD, Dillioglugil O, Abbas F et al. Impact of a clinical pathway for radical retropubic prostatectomy. *Urology* 1998; 52:94–9.
7. Wieder JA, Soloway MS. Incidence, etiology, location, prevention and treatment of positive surgical margins after radical prostatectomy for prostate cancer. *J. Urol.* 1998; 160:299–315.
8. Campbell EW. Total prostatectomy with preliminary ligation of the vascular pedicles. *J. Urol.* 1959; 81:464.
9. Catalona WJ, Smith DS. Cancer recurrence and survival rates after anatomic radical retropubic prostatectomy for prostate cancer: intermediate-term results. *J. Urol.* 1998; 160:2428–34.
10. Partin AW, Yoo J, Carter HB et al. The use of prostate specific antigen, clinical stage and Gleason score to predict pathological stage in men with localized prostate cancer. *J. Urol.* 1993; 150:110–14.
11. Oesterling JE, Martin SK, Berstralh EJ et al. The use of prostate specific antigen in staging patients with newly diagnosed prostate cancer. *JAMA* 1993; 269:57–60.
12. Ohori M, Wheeler TM, Dunn JK et al. Pathologic features and prognosis of prostate cancer detectable with current diagnostic tests. *J. Urol.* 1994; 152:1712–20.

13. Van den Ouden D, Hop WC, Schroder FH. Progression in a survival of patients with locally advanced prostate cancer (T3) treated with radical prostatectomy as monotherapy. *J. Urol.* 1998; 160:1392–7.
14. Catalona WJ, Smith DS. Five-year tumor recurrence rates after anatomic radical retropubic prostatectomy for prostate cancer. *J. Urol.* 1994; 152:1837–42.
15. Oesterling JE, Brendler CB, Epstein JI et al. Correlation of clinical stage, serum prostatic acid phosphatase, and preoperative Gleason grade with final pathological stage in 275 patients with clinically localized adenocarcinoma of the prostate. *J. Urol.* 1987; 138:92–8.
16. Epstein JI. The role of perineural invasion and other biopsy characteristics as prognostic markers for localized prostate cancer. *Semin. Urol. Oncol.* 1998; 16:124–8.
17. Mc Neal JE. Cancer volume and site of origin of adenocarcinoma of the prostate: relationship to local and distant spread. *Human Pathol.* 1992; 23:258–66.
18. Catalona WJ, Kadmon D, Crane DB. Effect of minidose heparin on lymphocele formation following extraperitoneal pelvic lymphadenectomy. *J. Urol.* 1980; 123:890–2.
19. Reiner WG, Walsh PC. An anatomical approach to the surgical management of the dorsal vein and Santorini's plexus during radical retropubic surgery. *J. Urol.* 1979; 121:198–200.
20. Hrebinko RL, O'Donnel WF. Control of the deep dorsal venous complex in radical retropubic prostatectomy. *J. Urol.* 1993; 149:799–800.
21. Cisek LJ, Walsh PC. Thromboembolic complications following radical retropubic prostatectomy. *Urology* 1993; 42:406–8.
22. Catalona WJ, Carvalhal GF, Mager DE et al. Potency, continence, complication rates in 1870 consecutive radical retropubic prostatectomy. *J. Urol.* 1999; 162:433–8.
23. Zincke H, Oesterling JE, Blute ML et al. Long-term (15 years) results after radical prostatectomy for clinically localized (stage Tc or lower) prostate cancer. *J. Urol.* 1994; 152: 1850–7.
24. Walsh PC, Partin AW, Epstein JI. Cancer control and quality of life following anatomical radical retropubic prostatectomy: results at 10 years. *J. Urol.* 1994; 152:1831–6.
25. Lerner SE, Blute ML, Lieber MM et al. Morbidity of contemporary radical retropubic prostatectomy for localized prostate cancer. *Oncology* 1995; 9:379–82.

Mini-lap Radical Retropubic Prostatectomy

Fray F. Marshall
Department of Urology, Emory University School of Medicine, Atlanta, GA, USA

INTRODUCTION

The American Cancer Society estimates that 643 000 new cases of cancer will be diagnosed in the year 2001, and prostate cancer leads all other cancers with an estimated 198 100 new cases.[1] The SEER data reveals a lifetime risk of 16.03% for developing prostate cancer,[2] and with prostate specific antigen (PSA) screening and heightened awareness, more men present with clinically localized disease in the modern era.[3] From the time of diagnosis, patients are confronted with several treatment strategies, including radical prostatectomy, and recent research shows that approximately 30% with newly diagnosed prostate cancer will elect to undergo surgery.

Initially, the considerable morbidity associated with radical prostatectomy made it unattractive to many men but, since the development of the anatomical approach to radical prostatectomy first described by Walsh in 1983, outcomes have improved. Many subsequent refinements to radical prostatectomy have been described. Coupling early detection with focused dissection of the puboprostatic ligaments, rhabdosphincter, neurovascular bundle and bladder neck, prostatectomy has led to improved cancer control, further reductions in morbidity and increased patient acceptance.[4,5]

A further modification of the technique, the mini-laparotomy radical retropubic prostatectomy, reduces the incisional morbidity associated with standard open radical retropubic prostatectomy without compromising surgical cure. It originates from the initial application of pelvic lymphadenectomy in men with prostate cancer in which both cost and feasibility proved superior in relation to laparoscopy.[6–9] Extending the concept of the mini-lap incision, it too became apparent that radical prostatectomy would be possible. Hand-held retractors have been replaced by a fixed retractor system and only one surgical assistant is necessary. Mini-lap prostatectomy employs standard technique and requires minimal additional training, making it an attractive minimally invasive operative approach.

TECHNIQUE OF MINI-LAPAROTOMY RADICAL RETROPUBIC PROSTATECTOMY

Bilateral lower extremity sequential compression devices are placed and preoperative antibiotics are administered before induction of anesthesia. Following regional or general anesthesia, the doctor positions the patient so that the umbilicus lies just below the break in the table. Current preference is given to general anesthesia over epidural because of the slower effect of regional anesthesia and potentiation of postoperative ileus. The table is then flexed maximally to extend the distance from the umbilicus to the pubis. Trendelenburg positioning, so that the lower abdomen is parallel to the floor, further optimizes exposure to the pelvic organs.

A 7–8-cm vertical incision is made extending distally to the pubic bone. The rectus fascia is then divided in the midline, and the rectus muscles are separated. The transversalis fascia is then divided and the retropubic space is bluntly developed. We use the Omni-Tract system with a set of specially designed blades (**Fig. 32.1**), including a Y-shaped blade (Marshall), that allows the entire procedure to be performed with a single operative assistant. Bilateral pelvic lymphadenectomies are performed in 10–15 minutes as described previously.[6] Briefly, the retropubic space is bluntly developed proximally to the origin of the hypogastric artery and vein. Specially designed curved Mayo blades that allow a slightly larger working radius through a smaller incision are used to move the contralateral rectus muscle laterally. Short and long Harrington blades are used to retract the bladder medially and peritoneal contents superiorly. The lymphadenectomy starts by incision of the tissue along the medial aspect of the external iliac vein. The dissection continues to the lateral pelvic side wall, proximally to the bifurcation of the common iliac vein, and distally to Coopers's ligament, where a large right-angle clip is placed on to the nodal package. In sweeping the nodal package off the pelvic side wall, the surgeon identifies the obturator nerve and vessels

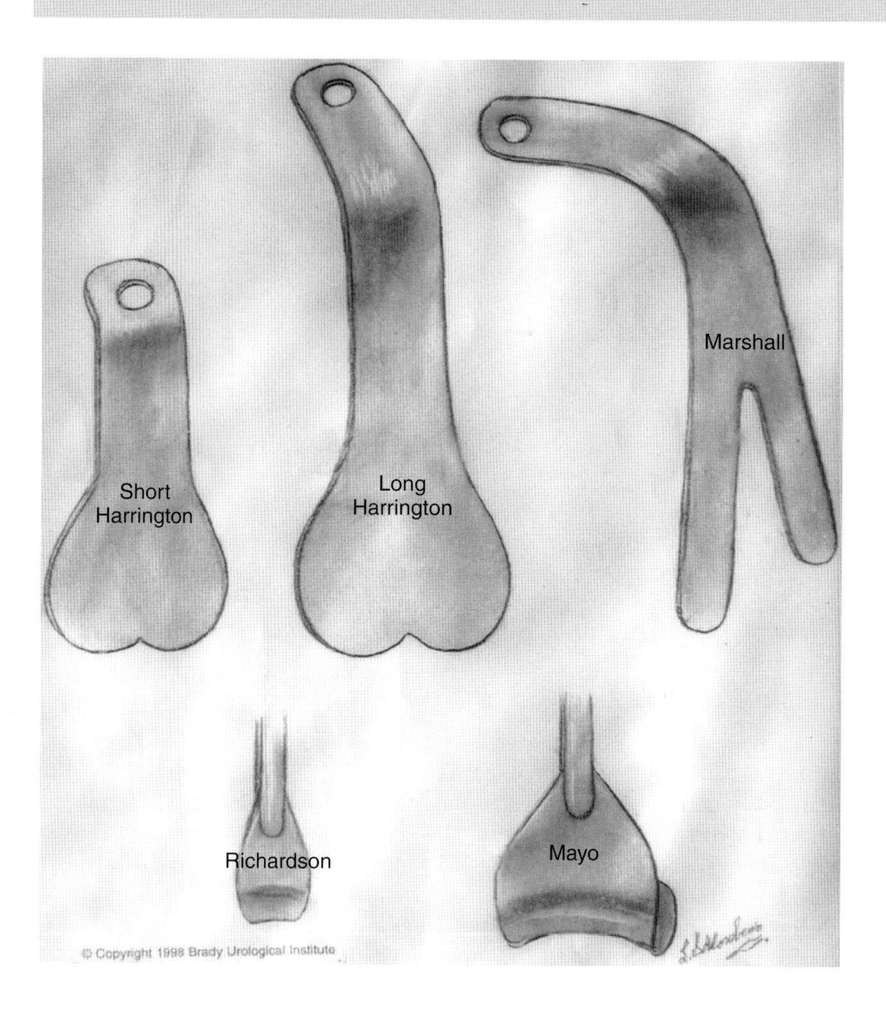

Fig. 32.1.

on the lateral aspect of the nodal package. The nodes are resected and a large clip is placed at the base of the nodal package. The obturator fossa is dissected free from all remaining lymphatic and adipose tissue. Pelvic lymphadenectomy is primarily for staging purposes. Frozen sections are not obtained unless there are clinical suspicions or elevated PSA or high Gleason's grade.

The radical prostatectomy is commenced with the Mayo blades retracting the rectus muscles laterally and the specifically designed Marshall Y blade retracting the bladder proximally (**Fig. 32.2**). The surgeon clears the anterior surface of the prostate, removing fibroadipose tissue and securing the superficial branch of the dorsal vein of the penis with either ligature or cautery. The endopelvic fascia is then divided lateral to the prostate and extends down to the levator ani muscle fibers. The incision in the endopelvic fascia extends superiorly to the puboprostatic ligaments and dorsally along the lateral edge of the prostate. The puboprostatic ligaments are generally preserved but may be partially divided near their insertion on the prostate, taking care not to enter the dorsal venous complex and to preserve the portion of the puboprostatic ligament that provides support to the proximal urethra. Preservation of these fibers is likely to be important in the preservation of continence. A figure of eight 3–0 monofilament, absorbable suture is placed on the veins of the anterior aspect of the prostate to secure the divided endopelvic fascia and prevent venous back bleeding (**Fig. 32.3a**). The dorsal venous complex is sharply exposed above the urethra. A right-angle clamp is no longer used to encircle the dorsal venous complex at this time. We instead use a 2–0 monofilament, absorbable suture placed at the level of the pubis just underneath the dorsal venous complex in the form of a ligature (**Fig. 32.3b**). Care is taken to manipulate the tissue judiciously in order to prevent injury to the urethra or the external urethral sphincter. The assistant retracts the prostate in a cephalad direction with a sponge stick, and the surgeon divides the dorsal venous complex with a long-handled scalpel until the anterior aspect of the urethra is seen. The division of the dorsal venous complex should be at an angle so that the prostate is not entered at its apex (**Fig. 32.4**). The surgeon obtains further control of bleeding from the dorsal venous complex with the same 2–0 monofilament, absorbable suture, again taking care to avoid incorporating any significant external sphincter fibers.

The neurovascular bundles can be identified at this point posterolateral to the urethra. Dissection starts at the lateral aspect of the prostate by first releasing the fascia overlying the bundles. The assistant simplifies this maneuver, if the prostate is rolled to the contralateral side with a sponge stick. The

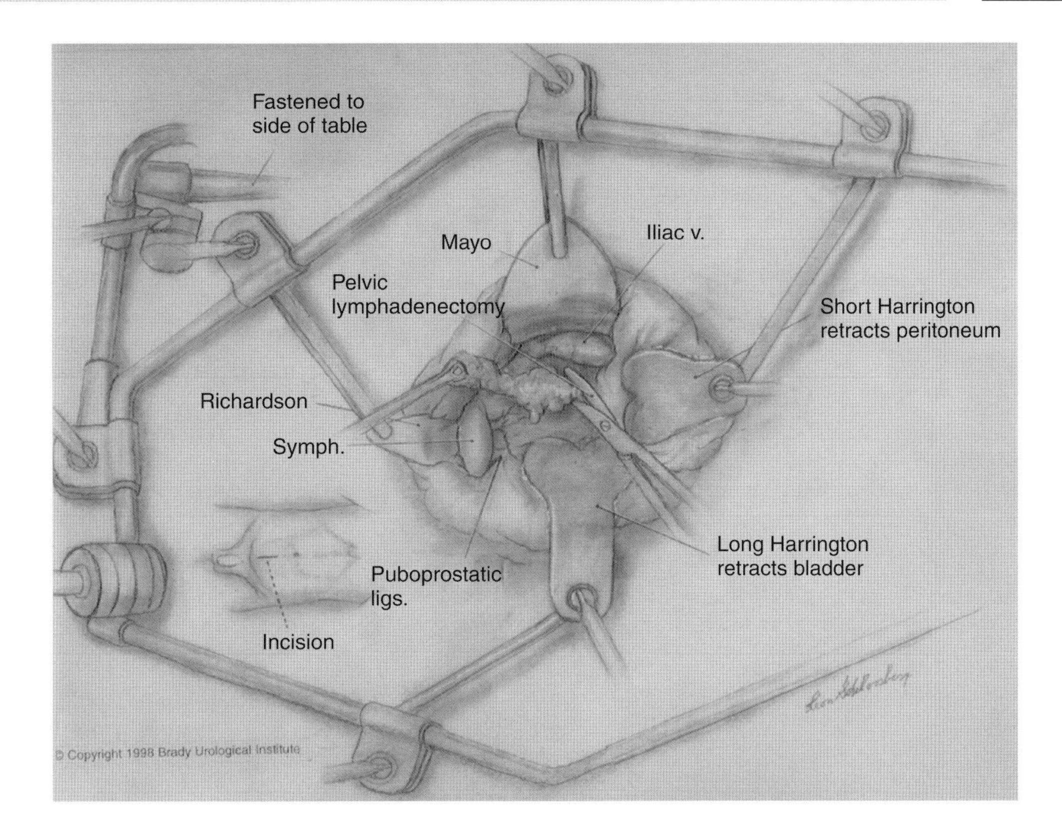

Fig. 32.2.

neurovascular bundles themselves may then be dissected from the lateral edge of the prostate, and any small perforating vessels may be clipped. Electrocautery in this area is to be strictly avoided. Apically, the neurovascular bundles may be further separated from the prostate after division of the urethra.

The urethra is sharply divided anteriorly at right angles to the urethra leaving an approximately 5–10 mm distal stump. The surgeon grasps the urethral catheter and retracts superiolaterally in order to facilitate division of the remaining circumference of the urethra. Only a strip of intact urethra remains for retraction at the 6 o'clock position. Four 3–0 absorbable sutures are placed at the 8, 10, 2 and 4 o'clock positions. The 8 and 10 o'clock sutures are easiest, if passed from outside to in, whereas the 2 and 4 o'clock sutures may be passed from inside to out and later reversed with a round French eye needle. Following deflation of the balloon, the catheter is withdrawn so that only its tip remains visible at the distal urethral stump. A 3–0 absorbable suture is then placed at the 6 o'clock position (**Fig. 32.5**), and the remainder of the urethra divided. The apex of the prostate is mobilized with preservation of the neurovascular bundles, if indicated.

The surgeon then develops the plane between the prostate and the rectum in the midline. A catheter is reinserted into the divided prostatic urethra and used for superior traction after initial mobilization of the prostate. The lateral vascular pedicles are divided after ligation with surgical clips. Once the prostate has been mobilized laterally and superiorly, Denonvilliers' fascia is divided transversely and the seminal vesicles are identified and dissected free. The vasculature to the seminal vesicles is then clipped and the vasa are clipped and divided. The entire prostate, seminal vesicles and vasa can now be rotated superiorly and the plane between the seminal vesicles and bladder is dissected posteriorly with right-angle scissors (**Fig. 32.6**). Care is taken to avoid the lateral venous plexus. Indigo carmine is given intravenously. Meanwhile, the bladder neck is sharply isolated and spared. Ultimately, the bladder neck is divided anteriorly and the specimen removed (**Fig. 32.7**). The ureteral orifices can be identified with the efflux of indigo carmine dye. Most often it is possible to preserve the bladder neck intact. If the neck requires reconstruction, interrupted 2–0 absorbable sutures evert the mucosa and reduce the lumen for eventual closure. Eversion of the vesical neck is done with interrupted 6–0 absorpable sutures. The surgeon completes the vesicourethral anastomosis over an 18F Foley catheter with the previously placed 3–0 absorbable sutures in the distal urethra (**Fig. 32.8**). The urethral catheter is left for at least 2 weeks. A single closed suction drain is brought through a separate stab wound and left for 2–3 days. The fascia is closed with a running #1 suture and the skin closed with staples.

OUTCOMES

Previously published articles detailing the mini-lap radical retropubic prostatectomy (RRP) from a series of 522 patients confirm cancer control and morbidity are similar to the results obtained through open standard RRP.[10] A summary of the

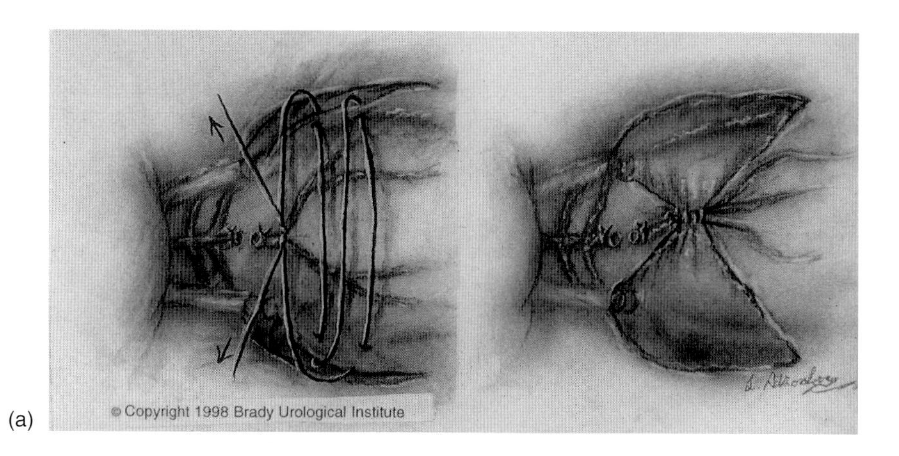

(a)

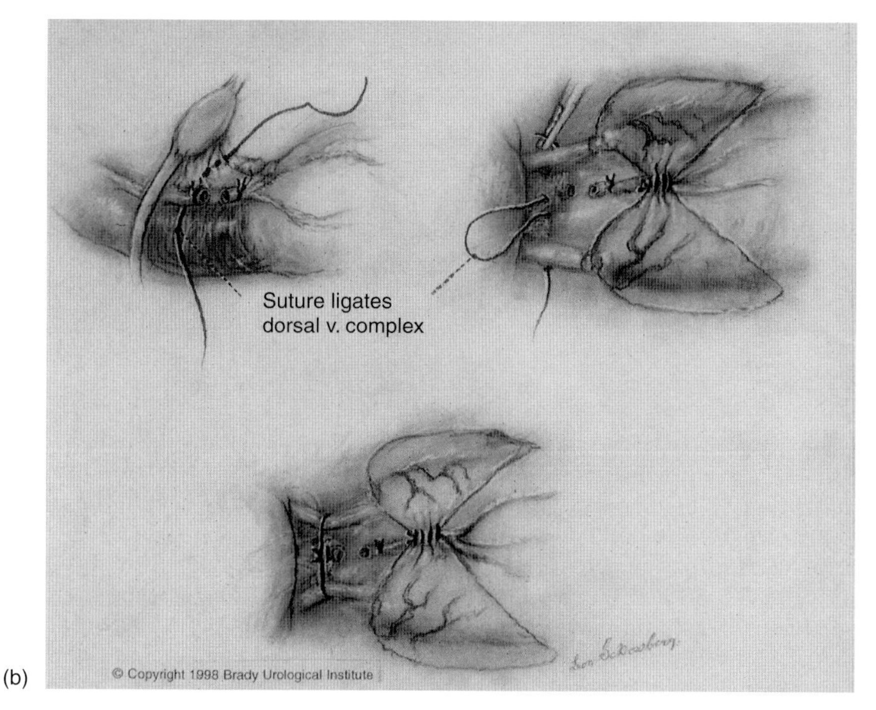

(b)

Fig. 32.3.

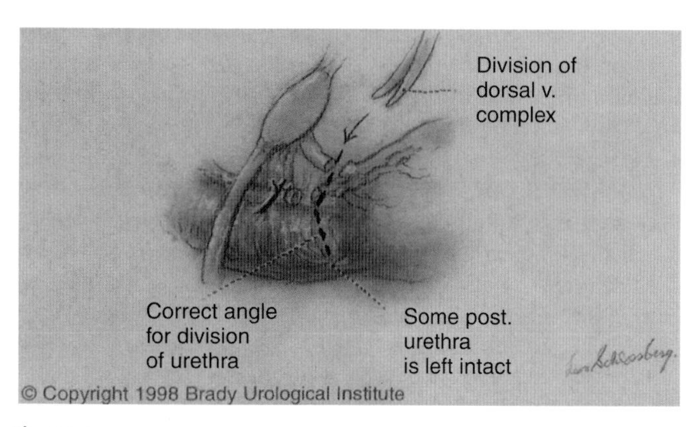

Fig. 32.4.

review shows that the average patient age was 60.5 years, the mean preoperative PSA was 8.4 ng/ml and 37% clinical stage T1c. In follow-up, no perioperative mortalities have occurred, 83% have a PSA level <0.2 with a mean follow-up of 2.6 years and 85% of patients have reported continence.

From February 1999 to June 2000, 125 consecutive patients at Emory University Hospital underwent a nerve-sparing RRP with bladder neck sparing and bilateral pelvic lymphadenectomy. A recent review of these patients reaffirmed the earlier results. Multiple parameters were examined by medical record or telephone interview. At the time of operation, the average age was 57 years (range 43–72 years), mean serum PSA was 7.0 mg/dl (1.0–31 mg/dl), mean height was 70.9 inches (62–79 inches) and mean weight was 87.9 kg (63.8–130.0 kg). Clinical staging revealed 2 T1a (1.6%), 79 T1c (63.2%), 37 T2a (29.6%) and 7 T2b (5.6%) patients. Preoperative assessment noted 100% with continence, 76.1% were potent, 16.2% utilized erectile aids and 7.7% were impotent. Comorbidity was present in a significant majority. Prevalence was as follows: hypertension, 34.4%; tobacco usage, 15.2%; alcohol consumption, 37.6%; diabetes, 8.0%; renal insufficiency, 3.2%; hypothyroidism,

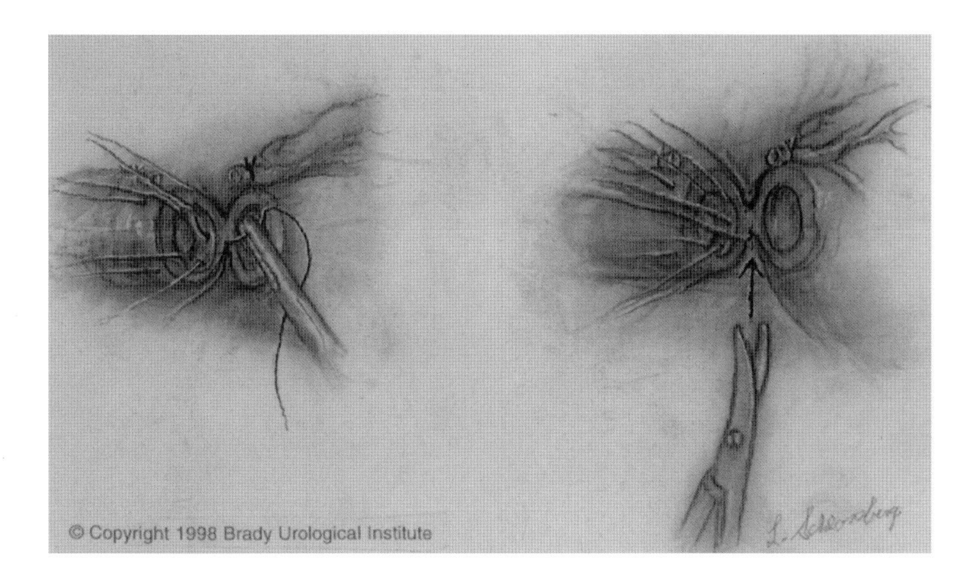

Fig. 32.5.

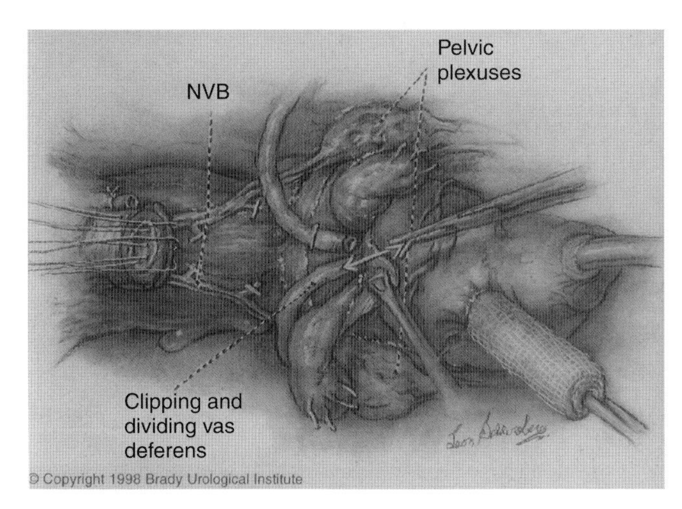

Fig. 32.6.

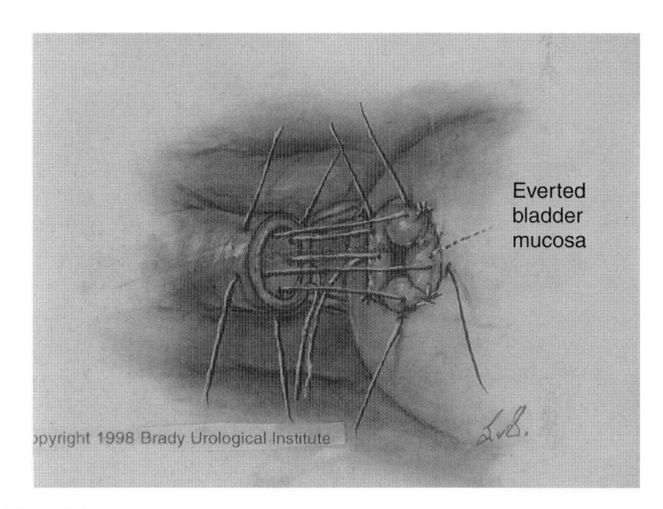

Fig. 32.8.

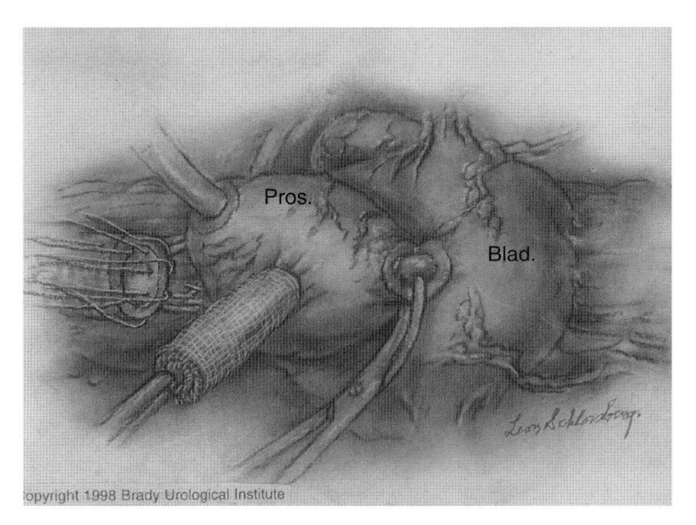

Fig. 32.7.

5.6%; myocardial infarction, 4.8%; angioplasty or bypass grafting, 6.4%.

Average operation time was 162.9 minutes (124–227 minutes) with a mean blood loss of 1051 ml and median stay was 3 days. Final pathologic staging showed 1 pT10, 1 pT1a, 30 pT2a, 68 pT2b, 32 pT3a, 2 pT3b and 1 pT4a cancers. A 2.4% (3 out of 125) allogenic transfusion rate and margin positivity rate of 13.6% (17 out of 125) was found. Regarding bladder neck margins, pathologic examination found 90 out of 93 were negative (fibromuscular tissue), 1 out of 93 showed adenocarcinoma and 2 out of 93 noted residual benign prostatic tissue. At a mean 21.8 months (15–31 months) postoperative follow-up, PSA recurrence equaled 8.5%, continence 98.3% (0–1 pads/day) and potency (intercourse) 58.4% for the entire population. No mortality has occurred.

DISCUSSION

Mini-lap RRP limits surgical morbidity without sacrificing outcomes in a urologic practice population with significant comorbidity. Moreover, with costs and manpower being an ever present issue in surgery, a second surgical assistant has become a rare luxury. The retractor, with specially designed blades, functions as a stationary robot with multiple degrees of freedom. It replaces the second operative assistant by providing fixed deep and superficial retraction. Such a system allows the surgeon ample exposure to perform the dissection through a 7–8 cm incision while reducing the incisional morbidity. Patients tolerate the small mini-lap incision well with minimal analgesia requirements after discharge. In fact, many of our patients find the catheter more bothersome that the incision.

Other operative modifications, including sparing of the puboprostatic ligaments, passage of a suture ligature rather than a right angle for securing the dorsal venous complex and sharp resection at the apex allows for more precise dissection. These maneuvers act to preserve the external sphincter with its fixation intact, contributing to improved continence rates. Preservation of the external sphincter has been shown in several series to be significant in preservation of continence.[11,12] It was rare for patients to note full continence initially more than 5 years ago, but it occurs more frequently now.[13] Although the role bladder neck preservation plays in continence remains controversial, many believe it provides a speedier return of continence[14] and reduces the incidence of bladder neck contracture.[15] Dissection of the bladder neck is performed more carefully now, aided by palpating the catheter and defining the natural plane of dissection between the prostate and bladder. Some may question whether the small incision limits the surgeon's ability to perform the operation effectively, but our resultant rates for continence, impotence and margin positive specimens follow historic norms. In addition, the technical challenge of bladder neck sparing is executed without undue difficulty or surgical compromise as only 1 of 93 patients where the bladder neck was spared exhibited residual positive adenocarcinoma on bladder neck biopsy.

Technology has similarly evolved with the advancements in anatomic dissection. Most notably, several studies to date have documented the feasibility of laparoscopic prostatectomy, but prolonged surgical times and technical skills training are necessary. Even the most able laparoscopic surgeons report mean operative times ranging from 4 to 6 hours with a steep learning curve.[16,17] Some experts estimate 80 cases or more may be needed to reach a plateau with regard to operative time and complication rates.[18] Five ports are typically used for the dissection with a 3–5 cm extension of the umbilical port for specimen extraction. Nascent results appear promising and equal those found in open prostatectomy in relation to continence, potency and margin status, but long-term oncologic results are pending.

Regarding laparoscopy, do these outcomes confer superiority and, if so, can these techniques be utilized by urologists at large? Adding the port incisions with the extension for specimen extraction approaches the totality of the mini-lap incision. Some proponents offer that blood loss is better with laparoscopic technique, but collaborative care pathways reducing preoperative autologous blood donation along with improved anatomic definition has limited transfusion to as low as 1% in some open series.[19] Patients generally leave the hospital on postoperative day 2 or 3. Equivalent outcomes do not cross the necessary threshold to embrace the technology when satisfactory alternatives exist.

Technically familiar to most urologists, mini-lap prostatectomy requires no additional training, utilizes standard equipment, is minimally invasive and extraperitoneal, can be performed under regional anesthesia and is performed reliably in 2–3 hours. With decreased postoperative morbidity provided by a shortened incision and faster operative times, a mini-lap should be competitive in terms of morbidity and cost in relation to other less invasive laparoscopic techniques. In addition to cost, these factors favorably contrast with laparoscopic series and make mini-lap prostatectomy a more attractive minimally invasive operative approach. We believe mini-lap prostatectomy compares favorably with standard open RRP and laparoscopic RRP.

REFERENCES

1. Ries LAG, Eisner MP, Kosary CL et al. *SEER Cancer Statistics Review, 1973–1997*. Bethesda, MD: National Cancer Institute, 2000.
2. *SEER Cancer Statistics Review (1973–1998)*. Bethesda, MD: National Cancer Institute, 1998.
3. Mettlin C, Murphy GP, Menck HR. Changes in patterns of prostate cancer care in the United States: results of American College of Surgeons Commission of Cancer studies, 1974–1993. *Prostate* 1997; 32:221–6.
4. Walsh PC, Donker PJ. Impotence following radical prostatecomy: insight into etiology and prevention. *J. Urol.* 1982; 128:492–7.
5. Steiner MS, Morton RA, Walsh PC. Impact of anatomical radical prostectomy on urinary incontinence. *J. Urol.* 1991; 145:512–15.
6. Steiner MS, Marshall FF. Mini-laparotomy staging pelvic lymphadenectomy (minilap): alternative to standard and laparoscopic pelvic lymphadenectomy. *Urology* 1993; 41:201–6.
7. Troxel S, Winfield HN. Comparative financial analysis of laparoscopic versus open pelvic lymph node dissection for men with cancer of the prostate. *J. Urol.* 1994; 151:675–80.
8. St Lezin M, Cherrie R, Cattolica EV. Comparison of laparoscopic and minilaparotomy pelvic lymphadenectomy for prostate cancer staging in a community practice. *Urology* 1997; 49:60–4.
9. Brant LA, Brant WO, Brown MH et al. A new minimally invasive open pelvic lymphadenectomy surgical technique for the staging of prostate cancer. *Urology* 1996; 47:416–21.
10. Marshall FF, Chan D, Partin AW et al. Minilaparotomy radical retropubic prostectomy: technique and results. *J. Urol.* 1998; 160:2440–5.
11. Kaye KW, Creed KE, Wilson GJ et al. Urinary continence after radical retropubic prostatectomy. Analysis and synthesis of contributing factors: a unified concept. *Br. J. Urol.* 1997; 80:444–51.
12. Eastham JA, Kattan MW, Rogers E et al. Risk factors for urinary incontinence after radical prostatectomy. *J. Urol.* 1996; 156:1707–13.
13. Kielb S, Dunn RL, Rashid MG et al. Assessment of early continence recovery after radical prostatectomy: patient reported symptoms and impairment. *J. Urol.* 2001; 166:958–61.
14. Klein E. Early continence after radical prostatectomy. *J. Urol.* 1992; 148:92–5.

15. Shelfo SW, Obek C, Soloway MS. Update on bladder neck preservation during radical retropubic prostatectomy: impact on pathologic outcome, anastomotic strictures and continence. *Urology* 1998; 51:73–8.
16. Guillonneau B, Vallancien G. Laparoscopic radical prostatectomy: The Montsouris experience. *J. Urol.* 2000; 163:418–22.
17. Abbou CC, Salomon P, Hoznek A et al. Laparoscopic radical prostatectomy: preliminary results. *Urology* 2000; 55:630–4.
18. Schulam PG, Link RE. Laparoscopic radical prostatectomy. *World J. Urol.* 2000; 18:278–82.
19. Holzbeierlein JM, Smith JA. Radical prostatectomy and collaborative care pathways. *Semin. Urol. Oncol.* 2000; 18:60–5.

Radical Perineal Prostatectomy

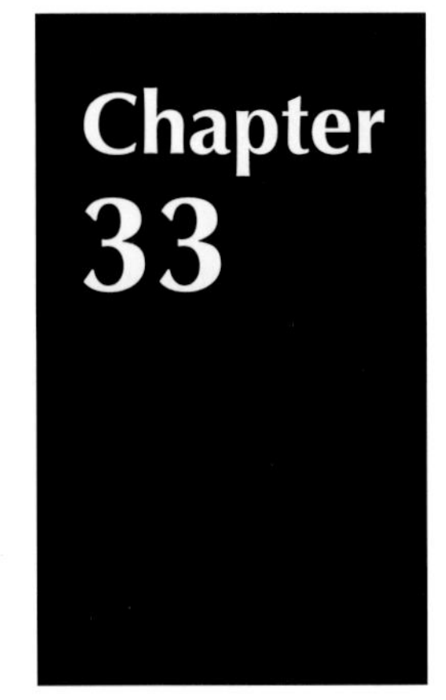

Philipp Dahm and David F. Paulson*

Duke University Medical Center, Division of Urology, Department of Surgery, Durham, NC, USA

INTRODUCTION

Radical prostatectomy remains the treatment of choice for clinically organ-confined prostate cancer. Radical retropubic prostatectomy (RRP) is the surgical approach most urologists are familiar with. Meanwhile radical perineal prostatectomy (RPP) offers some distinct advantages over the retropubic approach and constitutes an attractive option for treating prostate cancer in many patients. Its long-term efficacy in treating clinically organ-confined disease has been demonstrated in large patient series with over 25 years of follow-up.[1–4] RPP has the distinct advantage of allowing prostatic dissection in a relatively avascular surgical field, providing good exposure for reconstruction of the urethrovesical anastomosis, and allowing dependent postoperative drainage of the prostatic fossa. Modifications of the classic RPP include the nerve-sparing approach to preserve potency[5,6] as well as the wide-field dissection RPP for suspected locally advanced disease.[7] The inability to perform a simultaneous staging pelvic node dissection may have been the main drawback of the perineal approach. This argument, however, has lost much of its significance owing to the recent trend of earlier prostate cancer detection and the availability of prostate cancer specific antigen (PSA) based normograms to predict the likelihood of nodal involvement.[8–10] These features make RPP an attractive treatment choice for patients with clinically organ-confined prostate cancer.

EVOLUTION OF RPP

The development of the perineal approach to the prostate may be traced back to the 'blind' perineal lithotomies performed to treat bladder stones through a median perineal incision as early as 400 BC. Theodor Billroth is credited with the first planned perineal prostatectomy for malignancy in 1867, when he enucleated a prostate cancer through a median perineal incision.[7] Meanwhile, the origin of modern perineal prostatectomy for prostate cancer dates back to Hugh Hampton Young. In 1902 he developed a prostatic retractor, which he used for enucleating adenomatous tissue through the bladder to relieve outlet obstruction. In so doing he discovered incidental carcinomas in three prostatectomy specimens. After studying the spread of prostate cancer within the prostate and along the surrounding structures, he performed the first radical perineal prostatectomy in April of 1904. Young's radical prostatectomy included removal of the entire prostate, Denonvillier's fascia, both seminal vesicles, the ampullae of the vas deferens and the vesical neck with portions of the trigone. John Dees subsequently modified Young's classic technique by removing less of the bladder neck, thus improving continence.[11] Vest described a modification of the wound closure with placement of sutures from the vesical neck through the apex of the perineal wound. These sutures were designed to align the urethrovesical junction and relieve potential tension on the anastomosis. More recently, authors have described a potency-sparing modification of RPP[5,6] as well as a wide-field dissection incorporating the periprostatic fascia with the adjacent neurovascular bundles.[7]

INDICATIONS FOR RPP

RPP is mainly performed for clinically organ-confined prostate cancer. Patients subjected to this procedure should have a projected life expectancy of at least 10–15 years. RPP in the setting of suspected extracapsular extension or as a salvage procedure following failed local radiation treatment is rarely performed for select cases only. While RPP is technically feasible in the vast majority of patients, certain features deserve consideration. These are listed below.

- Positioning of the patient in the exaggerated lithotomy position requires a sufficient degree of mobility for hip flexion. This could represent a potential problem in elderly patients; however, this may be readily assessed at the time

*To whom correspondence should be sent.

of the preoperative clinic visit. A history of hip ankylosis, spinal stenosis or vertebral fractures is a relative contraindication for RPP.

- RPP does not allow simultaneous bilateral pelvic node dissection (BPLND) staging. Patients with a PSA ≥20 ng/ml and/or a component of Gleason grade 4 or 5 are recommended to undergo prior BPLND through a mini-laparotomy incision or laparoscopically immediately before prostatectomy under the same anesthesia. The patient is repositioned while the frozen sections of the lymph nodes are processed. Both approaches add little extra morbidity to the procedure. There is some consensus that radical prostatectomy should be aborted in the setting of metastatic disease in the regional lymph nodes.
- Massively obese patients with a 'large-barrel abdomen' with large amounts of intrabdominal fat may require high ventilation pressures in the exaggerated lithotomy position (>40 mmHg), which may preclude the operation for anesthesiology reasons. Meanwhile, obese patient with a large pannus are not at risk and are excellent candidates, since RPP avoids dissection of deep layers of subcutaneous fat.
- The surgical removal of very large prostates glands (>100 g) may be technically challenging and preclude *en bloc* removal of the prostate. In our experience, this has not compromised the outcome in terms of tumor control; however, a more detailed investigation has yet to be done.

PATIENT PREPARATION

RPP candidates are routinely admitted to the hospital on the day of surgery after complete mechanical and antibiotic bowel preparation at home. Optimal patient preparation begins with a clear liquid diet 2 days prior to surgery. The day before surgery, patients are started on 4 liters of polyethylene glycol solution, followed by three doses each of erythromycin base and neomycin at 2-hour intervals. They are encouraged to drink at least 3 liters of clear liquids until midnight from which time they remain NPO (nothing per mouth) until the morning. Upon presentation to preoperative holding they are started on maintenance intravenous fluids in order to keep them well hydrated. Preoperative intravenous antibiotics for prophylaxis typically consist of cefazolin and ciprofloxacin.

SURGICAL TECHNIQUE OF CLASSIC PERINEAL PROSTATECTOMY

Positioning

The patient is placed in an exaggerated lithotomy position. The sacrum is brought to the edge of the table with the buttocks extended approximately 15 cm over the table. A bolster of folded towels is placed below the sacrum to further elevate the pelvis. Correct positioning will place the perineum parallel to the floor and provide optimal exposure.

Access to the prostate

The Lowsley retractor is placed through the urethra into the bladder. Correct placement of the retractor may be confirmed by performing a digital rectal examination using an additional glove. Movement of the retractor allows the palpating finger to reassess the relative position, size and mobility of the prostate. A skin incision is made 1.5 cm above the anal verge and extended posterolaterally on either side, medial to the ischial tuberosities. The incision is carried out to just below the anus on either side. Using cutting cautery for hemostasis, a defect is created in the superficial perineal fascia on either side of the central tendon. This space is then extended posteriorly towards the floor, thereby exposing the ischiorectal fossa (**Fig. 33.1**). Two Allis clamps are then placed on the anal verge to put the central tendon on gentle stretch. The surgeon then passes a finger anterior to the rectum and beneath the central tendon of the perineum. The central tendon is then divided alongside the upper skin edge using electrocautery. After division of the central tendon, the rectal sphincter will be seen as an arch overlying the rectum with the anterior rectal fascia evident as glistening fascial structure. The anterior retractor is then placed and used to retract the anal sphincter anteriorly. The rectourethralis muscle is then identified as a band in the midline. With an examining finger placed in the rectum to identify its course, the anterior fascial surface of the rectum can be identified in its relationship to the prostate. Metzenbaum scissors are then used to divide the rectourethralis in the midline and displace its fibers to either side. At this point, the prostate is separated from the posterior surface of the rectum in the midline by blunt dissection. A moist gauze is placed over the rectum to protect it from inadvertent injury and displacement is maintained by a posterior weighted retractor.

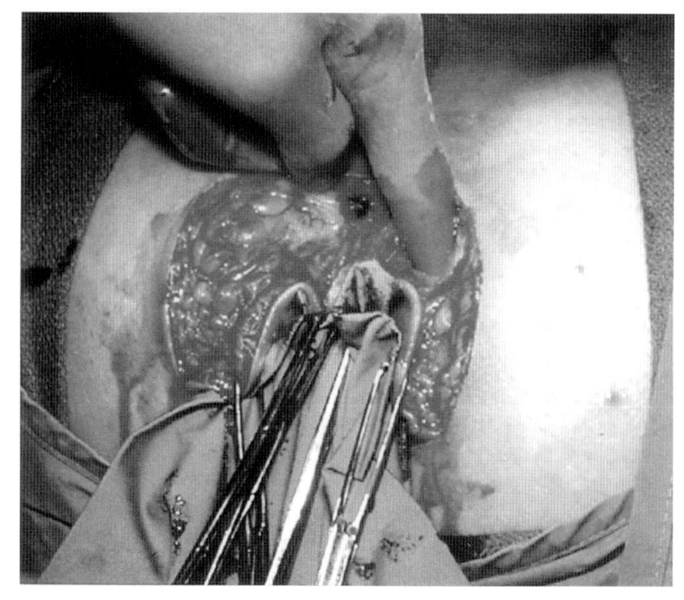

Fig. 33.1. Surgical dissection of the ishiorectal fossa in radical perineal prostatectomy. (*See Color plate 11.*)

Dissection of the prostate

Classic perineal prostatectomy then proceeds with a transverse incision of Denonvillier's fascia just below the apex of the prostate. The posterior layer of Denonvillier's fascia is dissected away from the prostatic surface. A combination of sharp and blunt dissection is then used to develop the membranous urethra distal to the prostatic apex but beneath the puboprostatic dorsal venous complex (**Fig. 33.2**). A right-angled clamp is then passed behind the urethra, the Lowsley retractor is removed and the urethra is sharply divided using a scalpel on a long handle. At this point the urethral stump can be tagged.

Alternatively the two anterior anastomotic sutures may be placed in the urethra. A Young retractor or straight Lowsley is placed through the prostatic urethra into the bladder and the wings are opened. With pressure on the retractor to displace the prostate posteriorly, and alternating sharp and blunt dissection, a plane can be developed on either side of the midline beneath the puboprostatic dorsal venous complex. The investing fibers in the midline, which secure the apical part of the prostate anteriorly are divided sharply. As the dissection is carried out further, the bladder neck is encountered. It is important to dissect free the anterior bladder neck adequately so that a tension-free vesicourethral anastomosis can be achieved later. The prostatovesical junction is then identified and the prostate is sharply cut away from the circular detrusor fibers, thereby sparing the bladder neck whenever possible. After sharp entry into the bladder between 10 and 2 o'clock, the Young retractor is withdrawn and a Foley catheter is passed through the prostatic urethra and brought out superiorly through the line of incision between the prostate and the bladder. Both ends are grasped with a Kelly clamp and the Foley catheter is used to manipulate the specimen during the remainder of the case. Traction on this catheter further defines the plane of cleavage between the bladder neck and the prostate, which should be incised until the prostate is attached only between 5 and 7 o'clock at the posterior bladder neck. The trigonal fibers are then divided distal to the ureteral orifices. Once the ampullae of the vas deferens are identified, they are isolated with a right-angle clamp, and the patient side controlled with right-angled clip. The ampullae are then divided on the specimen side. The specimen is now held in place by the vascular supply of the prostate posterolaterally at 5 and 7 o'clock, and by the seminal vesicles. The vascular supply of the prostate must be identified and controlled using either surgical clips or absorbable ligatures. At this point, only the seminal vesicles and the investing fascia fix the prostate posteriorly. The investing fascia should be incised and swept off the seminal vesicles bluntly. The vascular supply of the seminal vesicles, entering at the tip, is identified and controlled surgical clips and the specimen removed. Our current practice is to send bladder neck margins routinely for frozen sections.

Reconstruction

Once hemostasis is established, the bladder neck is reconstructed with the intent to establish continuity between the reconstructed bladder neck and the membranous urethra. Intravenous indigo-carmine administered by the anesthesiologist helps to identify the ureteral orifices on either side. Reconstruction is performed uniformly using slowly absorbable, monofilament suture material. A running 4–0 suture is used on either side of the bladder neck to evert the mucosa. Vest sutures of 0-suture material are placed in a mattress fashion at 10 o'clock and 2 o'clock in the open bladder neck. These sutures serve to align the anastomosis and take off any potential tension. Two 2–0 sutures are placed in the membranous urethra and brought to the anterior bladder neck at 10 and 2 o'clock within the Vest sutures later to be tied down as anastomotic sutures (**Fig. 33.3**). The bladder neck is then approximated with interrupted 0-sutures in a racket handle fashion

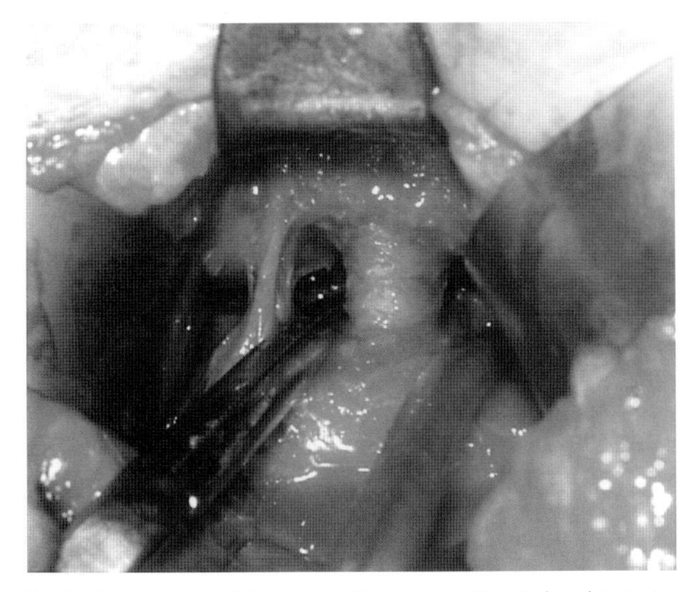

Fig. 33.2. Dissection of the apex of the prostate. (*See Color plate 12.*)

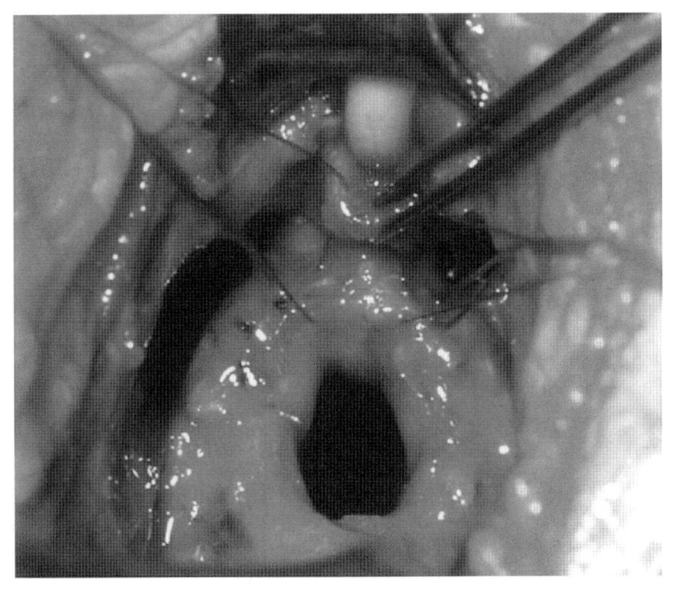

Fig. 33.3. Placement of the anastomotic sutures. (*See Color plate 13.*)

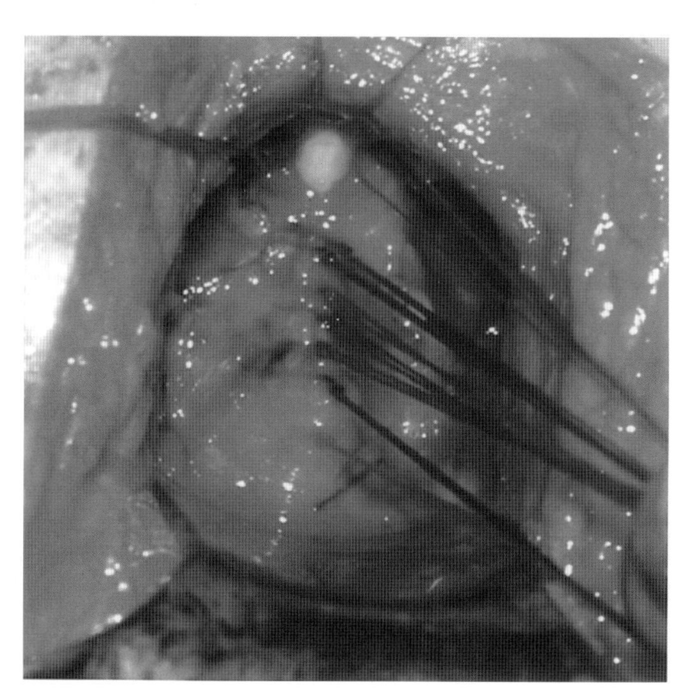

Fig. 33.4. Reconstruction of the bladder neck in a racket handle fashion. (*See Color plate 14.*)

from posterior to anterior (**Fig. 33.4**). The last two sutures are left long to serve as posterior Vest sutures. An 18 French Foley catheter is passed through the urethra into the bladder, which should fit snugly once the approximation has been completed. Two further 2–0 sutures are then placed in the membranous urethra at 4 and 6 o'clock, and in the equivalent position in the reconstructed bladder neck. The four 2–0 sutures, which have been placed at 10, 2, 4 and 6 o'clock are then tied down under gentle traction of the inflated Foley catheter and after releasing the posterior weighted retractor. A watertight anastomosis is thereby created under direct vision, which may be tested by irrigating the Foley catheter.

Closure

The four double-length sutures of 0-suture material are then brought out through the perineal body, but beneath the skin, paralleling the urethra. The incision is then copiously irrigated and the rectum inspected for any evidence of injury. A Penrose drain is placed anterior to the rectal surface and secured with a stitch. The rectourethralis, fibers of the levator ani and central tendon are then brought back together in the midline using absorbable suture to obliterate any dead space and restoring anatomy. Meticulous reconstruction of the perineum is felt to be important to prevent problems of fecal incontinence reported by some authors.[12] The skin margins are then closed with interrupted 2–0 chromic sutures leaving the ends approximately 5 cm long so that they do not produce any discomfort in the postoperative period. A compression dressing is then applied to the perineum before taking the patient out of lithotomy position. The catheter is well secured both to the foreskin and penis as well as the patient's leg to prevent inadvertent pulling on the catheter and potential disruption of the anastomosis.

MODIFICATIONS OF SURGICAL TECHNIQUE

Nerve-sparing radical perineal prostatectomy

Nerve-sparing in radical perineal prostatectomy is performed with equal efficacy as in the retropubic approach. It does, however, require an intimate understanding of the course of the neurovascular bundles as viewed from the perineum to prevent unintended injury. The procedure differs in that initially a vertical instead of a transverse incision of the posterior layer of Denonvillier's fascia is made. Reflection of this layer over the apex of the prostate gland and laterally, allows establishing a plane between the prostate and the neurovascular bundles. Posterolateral dissection of the prostate is kept close to the prostatic substance and small clips rather than electrocautery are used to achieve hemostasis. Finally, care is taken not to inadvertently injure the neurovascular bundles, which course in close proximity, when dissecting out the tips of the seminal vesicles.

Wide-field dissection radical perineal prostatectomy

In patients considered at high risk for extracapsular disease, the procedure may be extended to encompass the lateral pelvic fascia with the removed prostate. Wide-field radical perineal prostatectomy always results in sacrifice of the neurovascular bundles because the margin of resection always includes the periprostatic fascia. The prostate is exposed as described. However, after the rectum is displaced posteriorly, a surgical plane is developed between the posterior layers of Denonvillier's fascia – the anterior rectal surface posteriorly and the levator musculature anteriorly. The prostate remains surrounded by the periprostatic lateral pelvic fascia. All fibrovascular pedicles are divided as far distant from the prostate as possible. The seminal vesicles are dissected as described, but clipped and dissected at such a level to include the neurovascular bundle at that level. Additionally, wider margins of the bladder neck may be taken to achieve negative margins followed by reconstruction as described.

POSTOPERATIVE PATIENT CARE

Immediate postoperative care

From the operating room, the patient is taken to the recovery room where he is typically monitored for approximately 2 hours to ensure appropriate recovery from general anesthesia, adequate urinary drainage and good pain control. Serum electrolytes and a hematocrit are routinely obtained postoperatively but rarely affect patient management in a patient with normal preoperative values. Since the average blood loss in radical perineal prostatectomy is low and averages at less than 500 ml, postoperative transfusions are a rare event. In the setting of gross hematuria, the catheter may be gently irrigated by the nursing staff. Typically, the small doses of fentanyl or morphine administered in the recovery room are the only intravenous narcotics the patients require during their postoperative hospital course. Oral analgesics are started in the recovery room. Patients will routinely tolerate a clear liquid diet the evening of surgery and advance to their regular preoperative diet on the morning of postoperative day 1.

Further postoperative care

The surgical dressing is removed on postoperative day 1 and replaced by a dressing gauze. Problems with delayed postoperative bleeding are an exceedingly rare event. Direct compression of the surgical area by the patient's own body weight is likely to be an important factor. The Penrose drain is left in place until postoperative day 2 or the time of the patient's first bowel movement. At that time the patient is encouraged to either take a antiseptic sitzbath or alternatively to clean himself using a hand-held shower. The patient is encouraged to ambulate the hallways on postoperative day 1 and to spend as much time out of bed as possible. It is unusual for patients to require intravenous narcotics over the further hospital course. Criteria for patient discharge are the ability to tolerate a regular diet and ambulate without assistance as well as good pain control with oral analgesics. These criteria are normally met towards the end of postoperative day 1 and patients are typically discharged home then or on the morning of postoperative day 2. While outpatient RPP has been described to be safe in select patients,[13] the safety and overall benefit of such an approach remains to be determined.

COMPLICATIONS AND THEIR MANAGEMENT

Rectal injury

Rectal injury is a well-recognized risk of radical prostate surgery. It is a potentially concerning, but rare complication of the perineal approach, with an incidence of 1–3% in the hands of experienced surgeons.[7] An increased risk of rectal injury in RPP compared to RRP relates to the intimate relationship of the plane of initial dissection to the rectum. They most commonly occur at the time of dissection of the rectourethralis muscle or during placement of the posterior weighted retractor. With strict adherence to the correct surgical technique and careful placement of the retractors, however, injury appears avoidable in most cases. If rectal injury does occur, prompt recognition and appropriate management may minimize subsequent morbidity to the patient. This should consist in copious irrigation of the surgical field followed by primary closure of the defect in several layers using an absorbable monofilament. A diverting colostomy is rarely necessary, unless gross fecal soiling is present, which is not encountered if the patient has been adequately prepared using both a mechanical and antibiotic bowel preparation. Postoperative management includes the use of broad-spectrum antibiotics and maintaining patients on a clear liquid diet for 3 days. We recommend anal dilatation on a twice a day basis to evacuate air and prevent any potential pressure build-up at the site of injury. Placement of a rectal tube is unnecessary. The outlined management usually results in uncomplicated healing of the rectal injury with no adverse consequences to the patient.

Ureteral injury

Problems in reconstruction of the bladder neck may occur, if the posterior margin of the incision has been carried so close to the ureteral orifices that drainage may be compromised. In this case, the ureteral orifices can be stented using open-ended ureteral catheters followed by bladder neck closure from 6 to 12 o'clock of the detrusor musculature only, not incorporating the bladder mucosa. These are typically left in place for 5 to 7 days. Alternatively the ureteral orifice may be reimplanted through the perineal incision. The limited exposure, however, makes this maneuver technically challenging. While it has been necessary in less than a dozen patients in the senior author's professional experience of over 2000 RPPs, it has resulted in unobstructed ureteral drainage as documented by postoperative intravenous pyelogram (IVP) in all cases. Should primary ureteral reimplantation not be feasible, patients may alternatively be managed by inserting a percutaneous nephrostomy tube on postoperative day 1. This delay will allow for some degree of dilatation of the collecting system to occur, which facilitates tube placement. Ureteral reimplantation may then be performed in a delayed fashion using an abdominal approach ideally at 6–12 weeks postoperatively.

Lower extremity neuropraxia

Lower extremity neuropraxia is a unique problem associated with the exaggerated lithotomy position. While it is relatively common in the immediate postoperative setting occurring in up to 20% of patients, symptoms are usually transient and consist of sensory deficits below the knee only.[14] These symptoms invariably resolve completely, in most cases prior to the patient's discharge from the hospital on postoperative day 2. The etiology of these injuries is mostly attributable to a stretch injury to the sciatic or common peroneal nerve, and less often due to direct compression injury to the sensory nerves of ankle of foot. The recognition of these mechanisms of injury, often resulting from a lack of awareness by inexperienced operating room personnel has resulted in a significant reduction of this problem at our institution. While we consider postoperative lower extremity neuropraxia more of a nuisance than a true complication, it may result in significant postoperative patient's anxiety. Besides making all efforts to minimize its incidence, which involves patient positioning by the senior surgeon himself, we routinely discuss this issue with the patient preoperatively. Management is expectant and conservative. Further diagnostic work-up is only indicated if neurological symptoms are progressive, which suggests a different etiology, such as spinal stenosis or a herniated disk.

Urinary problems

Potential urinary problems in the early postoperative period include the inability to void after removal of the catheter or dependent urinary drainage from the perineal incision. In the hands of the senior author, both problems are relatively rare with an incidence of less than 5%. Catheters are removed on postoperative days 12–14 without the need for a cystogram. Urinary control is evident immediately in approximately 60% of patients, while continence is regained over the course of the next weeks in the remainder. In those unable to void due to residual edema at the site of the anastomosis, the catheter is replaced and left for one additional week. Alternatively a suprapubic tube could be placed.

Extravasation of urine may result from leakage at the bladder neck or distal to it. If the patient leaks from the perineum only when voiding, the leak is distal to the reconstructed bladder neck. The patients should be reassured and instructed to void only while seated on the toilet or under the shower. These leaks close with time and continence is maintained. If the patient drains continuously through the perineal incision, the catheter is replaced and left in place for one additional week after the time leakage stops.

COMPARISON WITH RETROPUBIC APPROACH

Relatively few studies in the literature have directly compared the relative merits of RPP versus RRP and no prospective, randomized trial has been published thus far. Frazier et al. published the first comprehensive comparison of the two approaches in 1992 based on the results from our institution.[5] This study retrospectively compared the clinical outcome of 122 patients undergoing RPP and 51 patients who underwent RRP in a 2-year period of time from 1988 to 1989. The selection of the approach was based on the preference of the attending surgeon. Variables analyzed included operative time, blood loss, transfusion requirements, rate of positive margins and length of stay, as well as complication rates. Comparison of operative time was limited to those RPP patients who also underwent a staging lymph node dissection and included the time to reposition and redrape. The authors found no difference in the incidence of positive margins, operative time, length of stay in the hospital, long-term complication rates and clinical outcome. Meanwhile there was a significant difference in estimated intraoperative blood loss (EBL) and number of perioperative transfusions. The median EBL was 565 ml for the RPP group versus 2000 ml for the RPP group ($P < 0.001$). The median transfusion requirements in the RPP patients was 0 units, while the RRP patients received a median of 3 units of packed red blood cells perioperatively. A smaller, but otherwise comparable study from a different institution made similar observations and came to the same conclusions.[15] The two techniques are comparable in their ability to control organ-confined prostate cancer and are associated with a low morbidity rate. However, the perineal approach is associated with a significantly lower blood loss, resulting in fewer transfusions. While recent series by experienced RRP surgeons have reported lower blood losses and transfusion requirements for this approach, a similar trend has been observed in RPP patients. Currently, the median EBL for the last 100 consecutive RPP patients has been less than half a liter and none of these patients has required a blood transfusion. Recognizing the advantage in transfusion requirements appears important for a generation of surgeons less familiar with the specific advantages of the perineal approach. Blood transfusions constitute a significant cost factor but, more importantly, may be a point of major concern to patients worried about the risk of contracting hepatitis or the human immunodeficiency virus associated with heterologous blood transfusions.

CLINICAL OUTCOME AFTER RPP

Continence

Modern RPP achieves excellent continence rates. The percentage of patients that achieve total urinary continence within 1 year postoperatively has been reported to range between 92% and 96%, depending on the definition used.[4,5,16] Allowing for minimal leakage with strenuous exercise and use of one minipad per day, we quote our patients a continence rate of 98%. Continence following RPP is typically achieved early in the postoperative course, which has been suggested to constitute an advantage over RRP.[12] Weldon et al. reported a return of continence with no routine pad use in 23% of patients at 1 month, 56% by 3 months and 90% by 6 months.[16]

Our own experience has been very similar. In an ongoing prospective quality of life study, 92% of patients reported use of no or a single pad at 6 months after RPP. By this point in time the patient's subjective quality of life in regards to urinary function and bother returned to baseline.[17] Patients may, however, expect a further improvement of their continence up to 1 year following surgery.

Potency

All forms of treatment for prostate cancer affect erectile function. Appropriately selected RPP candidates (PSA $\leq$10 ng/ml and Gleason score $\leq$6) may be offered a nerve-sparing approach. Nerve-sparing RPP is equally effective in preserving potency as the retropubic approach and may be performed unilaterally or bilaterally.[6] It has been our practice to offer bilateral nerve-sparing only to those patients with a single positive biopsy core. Most patients undergo a unilateral nerve-sparing RPP with deliberate sacrifice of the neurovascular bundle on that side on which side biopsies were positive, suggesting the location of the bulk of the tumor. Frazier et al. found 77% of patients to be potent 1 year after nerve-sparing RPP.[5] Weldon et al. analyzed the time course to recovery of potency and found that potency returned in 50% of patients after 1 year and in 70% after 2 years. A further important observation was the strong inverse correlation of the rate of successful potency sparing and increasing age. Potency returned in all patients less than 50 years of age, but only 29% of patients age 70 or older.[16] Our experience confirms these observations. It is most important to communicate realistic expectations to patients preoperatively regarding the likelihood of regaining potency as well as the extended time this may take. Appropriately selected patients with realistic expectations meanwhile have excellent outcomes after a nerve-sparing approach with high satisfaction ratings regarding their perceived quality of life in terms of sexual function.

Disease control

The long-term efficacy of RPP in achieving disease control in patients with clinically organ-confined prostate cancer has been documented in large patient series with extended follow-up.[2] We have used the term ‘cancer-associated death’ to define outcome in terms of survival in this patient population: When a patient dies of any cause with a biologically active malignancy, as witnessed by a rising PSA level ⩾0.5, survival is considered a function of the underlying tumor biology and the event considered a ‘cancer-associated death’. Studies from our institution using these end-points have demonstrated the profound impact of the biology of the tumor on the time to recurrence and long-term survival. They further suggest that many current series of alternative curative therapies may lack follow-up of sufficient duration to permit comparison with the results of radical surgery. While the use of different staging systems and end-points of outcome cloud a direct comparison between RPP and RRP, the disease control achieved by RPP and RRP may be considered equivalent.

Patients with organ-confined disease in the RPP specimen experience long-term survival with 10- and 15-year cancer-associated survival rates of 92.9% and 85.5%, respectively. The median time to cancer-associated death in a series of 1091 patients with pathologically organ-confined disease and a median follow-up of 5 years was not reached (**Fig. 33.5**). This was true across all Gleason score categories (Gleason sums 2–4, 5–6, 7, 8–10) including patients with high-grade disease (Gleason sum 8–10) (**Fig. 33.6**). These data suggest that low-volume, high-grade disease may be successfully treated by radical surgery as previously suggested by Ohori et al.[18] These findings argue for the potential benefit of prostate cancer screening programs for earlier disease detection.

Patients with specimen-confined and margin-positive disease experienced significantly lower, nevertheless extended long-term survival rates (**Table 33.1**). The outcome of patients with specimen-confined disease was not as good as that of patients with organ-confined disease, but significantly better that the outcome of patients with positive margins (**Fig. 33.5**). These findings suggest that wide-field dissection of the prostate with sacrifice of the neurovascular bundles may benefit patients suspected to extracapsular extension.

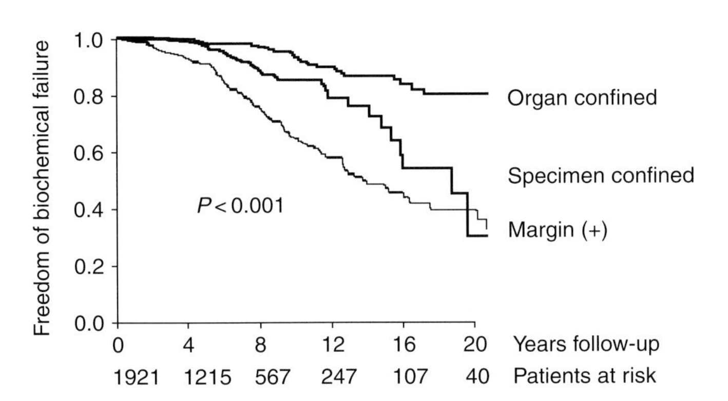

Fig. 33.5. Cancer-associated survival of patients undergoing radical perineal prostatectomy grouped by pathological stage.

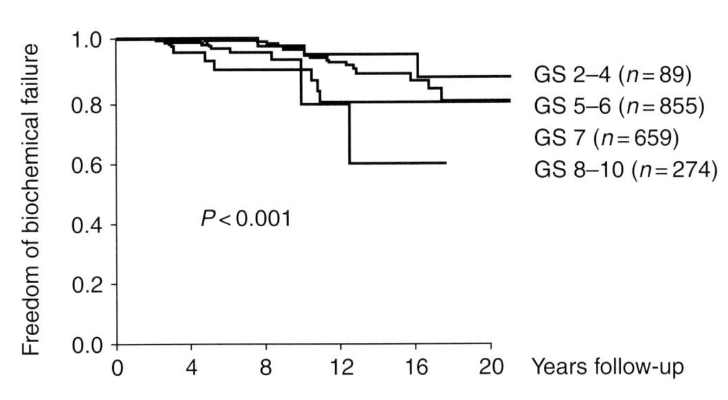

Fig. 33.6. Cancer-associated survival of patients undergoing radical perineal prostatectomy with pathologically organ-confined disease grouped by Gleason sum (GS) categories.

Table 33.1. Cancer-associated survival of patients ($n=1921$) following radical perineal prostatectomy. Patients are grouped by pathological stage.

Pathological stage	Patients	5-year	10-year	15-year
Organ confined	1120	98.5±0.5	92.9±1.6	85.5±2.7
Specimen confined	386	96.7±1.1	85.3±2.7	69.2±6.6
Margin positive	408	90.7±1.6	64.5±3.5	47.8±4.5

Margins

The incidence of positive margins may be considered both a function of the pathological stage as well as the appropriate surgical techniques. Analysis of the positive margin rates at our institution over the last 10 years has shown a dramatic decrease in the rate of positive margins, which coincides with the increasing proportion of patients diagnosed by PSA screening (**Fig. 33.7**). Others have reported similar trends in RRP series.[18] Most studies have not demonstrated any difference in the rate of positive margins of patients undergoing RPP versus RRP.[5,19] In a retrospective review of 122 patients who underwent RPP and 51 patients who underwent RRP at our institution, overall positive margin rates were comparable (29% and 31%, respectively).[5] Furthermore, there was no difference in the incidence of positive margins at different locations nor was there a difference in the incidence of capsular perforations. Findings of a single report[20] suggesting a high rate of surgically induced positive margins in patients undergoing RPP (43% vs. 29%) compared to RRP have not been substantiated by others and may, in fact, be a reflection of

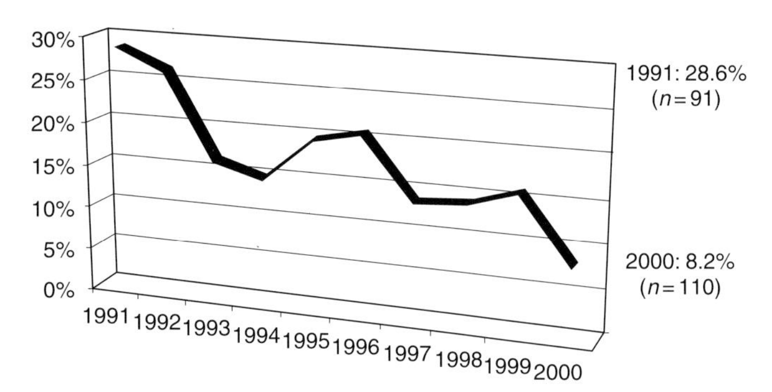

Fig. 33.7. Positive margin rate in radical perineal prostatectomy from 1991 to 2000 ($n=1267$).

a surgeon's insufficient familiarity with the perineal approach rather than a shortcoming of RPP.

CONCLUSION

Radical perineal prostatectomy constitutes an excellent option for treating clinically organ-confined prostate cancer. The key advantages include long-term disease control, low morbidity and fast patient recovery. These features translate into direct and indirect cost savings, factors that may gain increasing importance in the future. It appears likely that the resurgence of the perineal approach will continue, making it important for contemporary urologists to familiarize themselves with this procedure.

REFERENCES

1. Belt E, Schroeder FH. Total perineal prostatectomy for carcinoma of the prostate. *J. Urol.* 1972; 107:91–6.
2. Iselin CE, Robertson JE, Paulson DF. Radical perineal prostatectomy: oncological outcome during a 20-year period. *J. Urol.* 1999; 161:163–8.
3. Paulson DF, Robertson JE, Daubert LM et al. Radical prostatectomy in stage A prostatic adenocarcinoma. *J. Urol.* 1988; 140:535–9.
4. Gibbons RP. Total prostatectomy for clinically localized prostate cancer: long-term surgical results and current morbidity. *NCI Monographs* 1988; 123–6.
5. Frazier HA, Robertson JE, Paulson DF. Radical prostatectomy: the pros and cons of the perineal versus retropubic approach. *J. Urol.* 1992; 147:888–90.
6. Weldon VE, Tavel FR. Potency-sparing radical perineal prostatectomy: anatomy, surgical technique and initial results. *J. Urol.* 1988; 140:559–62.
7. Thrasher JB, Paulson DF. Reappraisal of radical perineal prostatectomy. *Eur. Urol.* 1992; 22:1–8.
8. Gingrich JR, Paulson DF. The impact of PSA on prostate cancer management. Can we abandon routine staging pelvic lymphadenectomy? *Surg. Oncol. Clinics North Am.* 1995; 4:335–44.
9. Partin AW, Kattan MW, Subong EN et al. Combination of prostate-specific antigen, clinical stage, and Gleason score to predict pathological stage of localized prostate cancer. A multi-institutional update. *JAMA* 1997; 277:1445–51.
10. Parra RO, Isorna S, Perez MG et al. Radical perineal prostatectomy without pelvic lymphadenectomy: selection criteria and early results. *J. Urol.* 1996; 155:612–5.
11. Dees JE. Radical perineal prostatectomy for carcinoma. *J. Urol.* 1970; 104:160–2.
12. Bishoff JT, Motley G, Optenberg SA et al. Incidence of fecal and urinary incontinence following radical perineal and retropubic prostatectomy in a national population. *J. Urol.* 1998; 160:454–8.
13. Ruiz-Deya G, Davis R, Srivastav SK et al. Outpatient radical prostatectomy: impact of standard perineal approach on patient outcome. *J. Urol.* 2001; 166:581–6.
14. Price DT, Vieweg J, Roland F et al. Transient lower extremity neurapraxia associated with radical perineal prostatectomy: a complication of the exaggerated lithotomy position. *J. Urol.* 1998; 160:1376–8.
15. Haab F, Boccon-Gibod L, Delmas V et al. Perineal versus retropubic radical prostatectomy for T1, T2 prostate cancer. *Br. J. Urol.* 1994; 74:626–9.
16. Weldon VE, Tavel FR, Neuwirth H. Continence, potency and morbidity after radical perineal prostatectomy. *J. Urol.* 1997; 158:1470–5.
17. Lance RS, Freidrichs PA, Kane C et al. A comparison of radical retropubic with perineal prostatectomy for localized prostate cancer within the Uniformed Services Urology Research Group. *Br. J. Urol. Int.* 2001; 87:61.
18. Han M, Partin AW, Pound CR et al. Long-term biochemical disease-free survival following radical retropubic prostatectomy. *Urol. Clinics North Am.* 2001; 28:555–65.
19. Scolieri MJ, Resnick MI. The technique of radical perineal prostatectomy. *Urol. Clinics North Am.* 2001; 28:521–33.
20. Boccon-Gibod L, Ravery V, Vordos D et al. Radical prostatectomy for prostate cancer: the perineal approach increases the risk of surgically induced positive margins and capsular incisions. *J. Urol.* 1998; 160:1383–5.

Stress Incontinence After Radical Prostatectomy

Eric S. Rovner and Alan J. Wein
Division of Urology, Department of Surgery, University of Pennsylvania School of Medicine, Hospital of the University of Pennsylvania, Philadelphia, PA, USA

INTRODUCTION

The immediate and primary goal of radical surgery for the treatment of localized prostate cancer is to remove all involved cancerous tissue. This can be successfully achieved in the majority of well-staged patients. The secondary surgical goals remain the preservation of normal genitourinary function, including potency and urinary continence. Urinary incontinence following radical prostatectomy (RP) is a devastating complication. It is often a life-altering event for the patient with significant quality of life implications, including psychological, economic and social effects.[1] Despite a well-meaning, thoughtful and complete preoperative discussion of this potential adverse outcome, patients are often angry, frustrated, confused and not uncommonly litigious when confronted with the reality of persistent urinary incontinence (UI) months or years after their surgery. Although the spontaneous remission rate of UI following RP is quite high, persistent UI is a source of constant dissatisfaction and irritation for both the patient and surgeon alike. This unfortunate condition may often cast considerable negative overtones on an otherwise successful extirpative surgery for prostate cancer even in the setting of an apparently cancer-free, and thus cured, patient.

Conceptually, in the absence of a fistula, urinary incontinence following RP may only result from abnormalities of the urinary bladder or bladder outlet. Bladder conditions that may be responsible for post-prostatectomy incontinence (PPI) include bladder overactivity or underactivity (impaired contractility). Although these are important and common causes of urinary incontinence following RP, this chapter will primarily discuss PPI only as it results from decreased outlet resistance. Decreased outlet resistance, when responsible for urinary incontinence, results in the condition known as stress, or sphincteric, incontinence. In the absence of prostate surgery or neurological illness, and excluding those with bladder overactivity and/or those with overflow incontinence owing to bladder outlet obstruction or impaired contractility, the prevalence of UI in adult males is extraordinarily low.

EPIDEMIOLOGY

The incidence of UI following RP varies considerably depending on the definition utilized as well as the method of data acquisition. This has been a major impediment to research in the field and has contributed to the wide range of reported prevalence estimates. Even among experts, there is a clear lack of consensus of a definition of UI following RP. Urinary incontinence has been defined by the International Continence Society (ICS) as the objective involuntary loss of urine per urethra which is a social or hygienic problem.[2] However, this definition has rarely been utilized in the literature when accumulating data in post-prostatectomy patients. Often the data reported in the literature are either from retrospective chart review or, in some instances, from postal questionnaires. This precludes strict use of the ICS definition, since the patient is not present to demonstrate the presence or absence of UI objectively and, furthermore, the degree to which it is necessary to have a 'social and hygienic problem' is not quantified as part of the definition. Clinically, UI can be defined quite narrowly or broadly. The use of any pads or protective undergarments is usually a good indicator of incontinence, but some postoperative patients with little or no incontinence use pads only in select circumstances and relate that the pads are never wet. Alternatively use of a very broad definition of continence (i.e. 'any wetness at any time') would be very inclusive and results in an alarmingly high prevalence of UI. Reported incidence also clearly varies with the methodology of data collection.[3] Data can be obtained by direct physician interview, telephone interview, third-party interview, chart review, written patient questionnaire and via semiquantitative objective methods, such as pad testing, voiding diaries and urodynamics. Use of any of these methodologies introduces significant limitations, either scientific or practical, as well as creating sources of intrinsic error. Thus, when reviewing the UI literature, it is important to understand the methodology in order to interpret the information properly. Other factors that could influence the reported prevalence rate of PPI include the time interval from surgery to data collection, owing to the spontaneous resolution rate of PPI with time,[4] and the experience of the operating surgeon.[5]

The historical incidence of UI after radical prostatectomy varies from 2.5% to 87%.[6,7] These data include patients with both bladder- and sphincter-related incontinence. Some authors report incontinence as being present in patients with very mild stress incontinence, and others only report those with gravitational or total incontinence. The incidence of total incontinence varies from 0 to 12.5%.[6,8] Several recent series are reviewed in **Table 34.1**.

PATHOPHYSIOLOGY

In the normal male there exist two separate continence zones or sphincters: the proximal urethral sphincter (PUS) and the distal urethral sphincter (DUS). The relative contributions of each sphincteric zone in the maintenance of urinary continence in the normal male is unknown. Under certain clinical circumstances, such as pelvic fracture, neurological disease or surgical resection of the bladder neck, either the PUS or DUS can be incompetent with complete urinary continence being maintained by the other sphincteric region. PPI owing to sphincteric incontinence implies a loss of urethral resistance at both of these sphincter regions.

The PUS is located in the region of the bladder neck and prostatic urethra proximal to the level of the verumontanum. This is sometimes referred to as the 'internal sphincter' and is composed of primarily the smooth muscle of the bladder neck. A true 'anatomic' muscular sphincter at this level cannot

Table 34.1. Prevalence of urinary incontinence following radical retropubic prostatectomy: recent selected series.

Author	Institution	Year	Number of patients	Follow-up	Data procurement	Definition of urinary incontinence	% Continent	% Total incontinent
Steiner et al.	Johns Hopkins	1991	593	>1 year	Patient reported	No pads	92	0
Murphy et al.	Multi-institutional	1994	1796	<1 year (38.6%), remainder >1 year	Medical record and chart review	Complete control or occasional leakage but no pad use	81.1	3.6
Geary et al.	Stanford	1995	458	>1 year	Patient interview	No pads	80.1	5.2
Eastham et al.	Baylor	1996	581	>2 years	Mailed questionnaire or patient interview	No use or occasional use of pads	91	4
Goluboff et al.	Columbia	1998	615	3.3 years	Mailed questionnaire	No pads	91.8	0.8
Catalona et al.	Washington U.	1999	1325	>1.5 years	Questionnaire or patient interview	No requirement for protection of outer garments	92	n/a
Walsh et al.	Johns Hopkins	2000	64	1.5 years	Vailidated QOL instrument	No pads	93	n/a
Stanford et al.	Multi-institutional	2000	1291	>1.5 years	Mailed questionnaire	Total control or occasional leakage	91.6	1.6
Kao et al.	Multi-institutional	2000	1013	>6 months	Mailed questionnaire	Any problem at all with dripping or leaking urine	65.6	n/a

be identified. Instead, the smooth muscle at the level of the bladder neck maintains a baseline tone creating sufficient urethral resistance to prevent the egress of urine into the proximal urethra. This baseline tone at the level of the bladder neck increases in response to bladder filling by complex neurologic spinal reflex arcs thought to be at least partially mediated by the sympathetic nervous system. This produces a continuing advantage of urethral outlet resistance over the bladder expulsive forces and thus the maintenance of urinary continence. The smooth muscle of the bladder neck surrounds the transitional cell lining of the bladder neck, which comprises the 'mucosal seal' mechanism at this level. The PUS is mainly innervated by the autonomic nervous system via branches of the pelvic plexus.

Normally the bladder neck and proximal urethra remain closed in the normal male on a static cystogram such that no contrast enters the proximal urethra. During radical prostatectomy, the PUS as well as the proximal portion of the DUS is removed, including the verumontanum and the prostatic apex. Therefore, continence in the post-prostatectomy patient is dependent on the preservation of the distal portion of the DUS. However, O'Donnell et al. and others have demonstrated that the contrast on a postoperative cystogram in the continent patient lies at the level of the vesicourethral anastomosis, at least 1 cm proximal to the DUS[9,10] (**Fig. 34.1**). Thus, at least some degree of passive continence is maintained at this level in some patients. Nevertheless, the function of the bladder neck in the continent post-RP patient, the role of bladder neck sparing radical prostatectomy procedures or bladder neck reconstruction and/or tubularization in the preservation of urinary continence following RP is controversial. Some authors have advocated extensive bladder neck reconstructive procedures, including complex tubularization procedures,[11,12] while others have suggested that painstaking preservation of the bladder neck can impact on postoperative continence.[13] However, radiographic or urodynamic evidence confirming this concept is, at present, lacking.

The segment of the male urethra from the level of the verumontanum distally, through the membranous urethra, is the region of the distal urethral sphincteric mechanism or DUS. The DUS or 'external sphincter' has both smooth and striated muscle components as well as an important contribution from the spongy vascular tissue of the urethra. The smooth muscle component is embedded within the urethral wall and is considered a continuation of the smooth muscle of the detrusor by most authors.[14] The striated component has both intrinsic and extrinsic components. The intrinsic or intramural component (also referred to in the literature as the 'rhabdosphincter') lies within the urethral wall. This contains 'slow-twitch' muscle fibers capable of sustaining tone over the urethral lumen for long periods of time, providing passive urinary continence. The extrinsic or extramural component of the DUS consists of the striated muscle located between the leaves of the urogenital diaphragm outside and separate from the urethral wall. This component, composed of primarily 'fast-twitch' striated muscle is probably related to the levator ani musculature of the pelvis and augments the activity of the

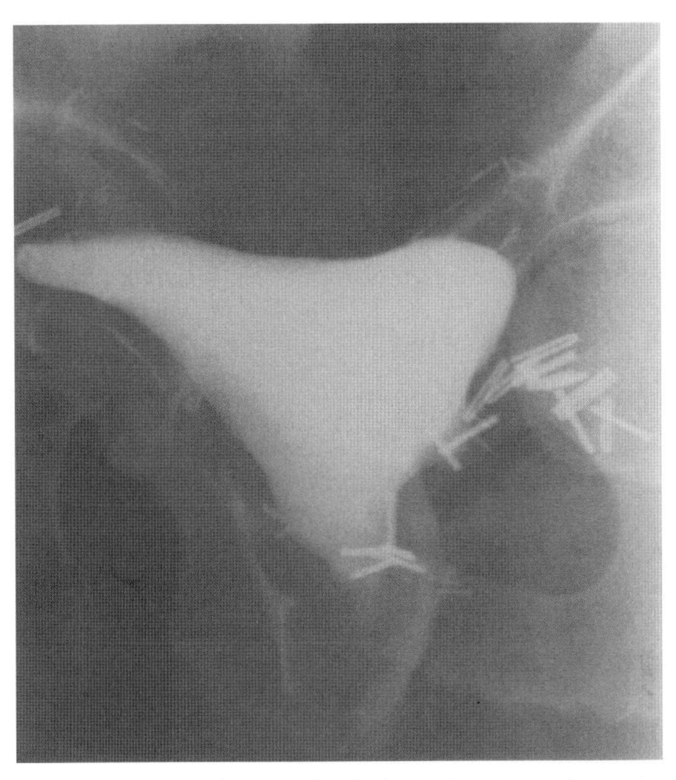

Fig. 34.1. Cystogram demonstrating the level of passive continence following radical prostatectomy. This patient had mild symptoms of stress incontinence.

intrinsic component of the DUS during sudden increases in abdominal pressure (e.g. coughing, sneezing, etc.).[15] The activity of the extrinsic component of the DUS is preserved in most patients with sphincteric PPI and can be demonstrated even in patients with florid incontinence by asking the patient to interrupt the urinary stream volitionally during voiding. Successfully stopping the urinary stream by forceful contraction of the pelvic floor indicates preserved function of the extrinsic portion of the DUS but does not imply passive continence. Innervation of the DUS is primarily via the pudendal nerve (S2–S4) with variable contributions from the autonomic nervous system. As in the female, an important contribution to the maintenance of continence in the male is the compressible spongy vascular tissue of the urethral wall, and the mucosal lining and secretions into the urethral lumen. Like the female, the supple spongy tissue of the urethral wall is coapted and gently compressed by the aforementioned muscular sphincters creating a 'mucosal seal' at the level of the sphincters. This 'mucosal seal' acts in the same fashion as a washer in a faucet in order to prevent leakage. When damaged or deficient, the loss of this contribution to outlet resistance usually results in incontinence. This usually cannot be compensated for by increased sphincteric resistance or external compression (slings or artificial sphincters) at the level of the damaged segment. Slings or artificial compression devices will only be effective if placed at another location along the urethral lumen where the mucosal seal is still intact.

RISK FACTORS AND PREVENTION

As noted previously, sphincteric incontinence in the male is exceedingly rare, except in the setting of lower urinary tract surgery such as prostatectomy. An important well-defined risk factor for UI following prostatectomy includes pre-existing bladder or sphincter dysfunction.[16] The reported incidence of bladder dysfunction as a sole or contributing cause of UI following prostatectomy is highly variable, but has been reported to be as high as 95%.[17] Notably, this study and others combined patients with PPI resulting from surgery for both benign and malignant disease. Leach and colleagues looked at 210 patients with PPI, 159 following RP and 51 following transurethral resection of the prostate (TURP) or open prostatectomy, for benign disease.[18] Almost 20% of patients had detrusor dysfunction as the sole urodynamic cause for PPI, while an additional 40% had combined sphincteric and detrusor dysfunction. Other authors have reported detrusor dysfunction as the sole or contributing cause of PPI in 38–78% of patients.[19–21] The prevalence of detrusor dysfunction in the literature as a contributing factor to PPI may be quite high owing to the mixed population of patients reporting PPI in these series. Those with PPI following prostatectomy for benign disease might be expected to have a higher incidence of detrusor dysfunction owing to their symptomatic presentation with obstructive or irritative symptoms, some of which might be due to pre-existing bladder dysfunction, as opposed to those undergoing RP, many of whom may have been asymptomatic. In a series of 74 patients with PPI following RP, Chao and Mayo reported only 4% with detrusor dysfunction alone, and an additional 39% with combined detrusor and sphincteric dysfunction.[22] Groutz and colleagues studied 83 patients with PPI following RP using sophisticated video urodynamics evaluation, pad tests and voiding diaries.[23] They concluded that detrusor dysfunction was present in 33.7% of patients, and was the only urodynamic abnormality in 3.6%. However, in only 7.2% (6) of patients was detrusor dysfunction considered to be the primary cause of PPI. Our experience would agree.

Several studies have attempted to identify factors that might be associated with PPI. Advanced age[24,25] and the lack of surgical expertise[5] seem to correlate with an increased risk of persistent PPI. Eastham and colleagues conducted a univariate and multivariate analysis of 514 patients undergoing RP at one institution.[26] In a univariate analysis, prostate size, blood loss and prior history of TURP were considered adverse risk factors. In the multivariate analysis, risk factors for PPI included advanced age, resection of one or both neurovascular bundles, development of a postoperative anastomotic stricture and, finally, surgical technique. Conversely, in a review of over 700 patients at Johns Hopkins, no correlation was noted between PPI and the preservation of one or both neurovascular bundles. These authors concluded that intrinsic anatomical factors were responsible for PPI.[27] Others have also reported no difference in continence following resection of one or both neurovascular bundles.[28,29] The impact of prior TURP on postoperative continence following RP is controversial. Historically, several authors reported an increased incidence of urinary incontinence in those undergoing RP following TURP.[30–32] It is possible that these series included patients with significant pre-existing detrusor dysfunction (see discussion earlier). Other historical series do not report this relationship.[26,33] More recent series do not support an increased risk of PPI in patients who have had prior TURP.[29,34]

Surgical techniques for radical prostatectomy have evolved over the last several decades as a better understanding of the pertinent pelvic neurovascular anatomy has emerged.[35] Modifications in surgical technique have allowed better preservation of erectile function, lower intraoperative blood loss and shorter hospital stays. However, whether any of these technical modifications have improved overall continence rates is unclear. In several instances, a number of other modifications to RP have been proposed with the goal of reducing or eliminating postoperative sphincteric incontinence. Careful preservation of the bladder neck fibers during dissection of the prostate has been suggested as providing a faster return to continence postoperatively[36,37] and improving continence overall.[13] Owing to concerns in preserving cancer-free surgical margins during RP, bladder neck preservation is not always possible. Lowe reported that preservation of the bladder neck did not result in improved overall continence but, as compared to those without bladder neck preservation, the time to achieve maximal continence was shorter.[38] Furthermore, it is unproven whether the additional time invested in this sometimes tedious dissection can provide a surgically reconstructed and functionally continent bladder neck as compared to other approaches.[39] When compared to those patients without bladder neck preservation at RP, Srougi and colleagues could find no difference in overall continence at 6 months.[40] Others have also found no difference in continence with bladder neck sparing procedures.[41] It is generally accepted that preservation of adequate functional urethral length during the apical prostatic dissection is important. The surgeon must be cognizant of the variety of the anatomic configurations of the prostate (e.g. 'croissant' vs. 'donut')[42,43] in order to maximize functional urethral length and attain adequate surgical margins. Maintaining a long urethral stump[26] with minimal manipulation during the dissection[44] may be additional important factors in preserving sphincteric function. Among the myriad of permutations of RP, minimal dissection distal at the apex of the prostate in the region of striated sphincter,[45] preservation of the puboprostatic ligaments,[46,47] anastomotic inclusion of the posterior layer of Denonvilliers fascia,[48] sparing a portion of the seminal vesicals[49] and limiting the number of anastomotic sutures or precise placement of the anastomotic sutures[29] have all been proposed as potentially providing improved postoperative continence overall or time to achieve maximum continence. Finally, reconstruction and tubularization of the bladder neck has been proposed by several authors as a method of improving continence. Seamans and Benson reported a 97% continence rate following bladder neck tubularization in 29 patients at follow-up of 6 months.[48–50] Continence was achieved in seven patients in less than 24 hours following catheter removal and in 27 patients by 3 months postoperatively. Steiner et al. compared a group of 69 patients undergoing bladder neck reconstruction and tubularization to 45 patients undergoing a standard RP.[12] At 6 months, continence was complete in 87% of patients with reconstructed bladder neck vs. 47% of controls.

CLINICAL EVALUATION OF PPI

The initial evaluation of the patient with UI following RP centers on a thorough history and physical examination. Symptoms of urinary frequency, urgency and obstruction are assessed. Obstructive symptoms and poor urinary stream may suggest impaired bladder contractility or, more commonly, a concomitant urethral stricture or postoperative anastomotic contracture. These symptoms warrant a post-void residual urine check and perhaps non-invasive uroflowmetry. The patient is queried regarding the sensation of bladder filling. No sensation of bladder filling or a total lack of urinary urgency implies gravitational incontinence or an undiagnosed neurological condition. Incontinence is further characterized as dribbling, episodic or gravitational, predictable or unpredictable. Episodic, predictable incontinence implies mild to moderate sphincteric incontinence. Gravitational incontinence generally implies severe sphincteric incontinence. Its relationship to physical activities is assessed, as this gives an indirect measure of the degree of sphincteric compromise. Timing of incontinent episodes and the use of day and/or night-time pads are important diagnostic variables. The complete absence of incontinence during sleep strongly implies sphincteric incontinence as the etiology of PPI. Many patients with even severe gravitational daytime incontinence due to impaired sphincteric function may have little or no leakage while recumbent as the bladder, in a dependent position relative to the urethra, and in the absence of poor compliance or involuntary contractions, accommodates well to increasing volumes of urine. The number and type of absorbent products should also be quantified. Some qualitative measure of the degree of wetness of the pads (i.e. soaked, wet, damp, almost dry, etc.) is sometimes helpful in assessing the degree of wetness. Other pertinent points to be covered include neurological history, the presence of concomitant fecal incontinence, and a history of prior pelvic surgery, such as rectal surgery or pelvic radiation therapy. The elapsed time from RP and the last postoperative PSA measurement are noted, as these might have significant short- and long-term therapeutic implications. Finally, the number, type, intensity and success of prior interventions are assessed.

The physical examination should focus on neurological integrity of the pelvis and perineum, as well as the objective demonstration of PPI. Although wet pads are virtually diagnostic of PPI, the direct observation of Valsalva or cough-induced leakage per urethra is highly suggestive of sphincteric incontinence.

A pad test and voiding diary provides objective determinations of voiding frequency, voided volumes, fluid intake, the number and severity of incontinence episodes as well as objectively quantifies the volume of leakage.

Cystourethroscopy should be performed to examine for urethral stricture, bladder neck contracture, and other abnormalities of the bladder and urethra. Direct visualization of incomplete coaptation of the DUS is seen in some patients with sphincteric incontinence. Anatomic configuration of the vesicourethral anastomosis and scarring at the bladder neck should be assessed, as it may have implications for future therapeutic interventions.

A careful urodynamic study, with or without fluoroscopy (video urodynamics) lends confirmation to the proposed diagnosis that is based initially on the history and physical examination. The role of the urodynamic evaluation is to assess and define the abnormal lower urinary tract physiology responsible for PPI. A non-invasive uroflowmetry and a catheterized postvoid residual determination begin the study. Abnormalities in either of these two studies might suggest impaired bladder contractility or outlet obstruction as primary or contributing factors to PPI. A supranormal flow rate with a normal pattern suggests sphincteric incontinence. A filling cystometrogram and multichannel pressure-flow urodynamics will evaluate the bladder and urethra for relative contributions to PPI. Filling cystometry evaluates for involuntary bladder contractions, bladder compliance, bladder capacity, bladder sensation and, during provocative maneuvers, such as the Valsalva maneuver, sphincteric function. In patients undergoing video urodynamics, any contrast located beyond the DUS following a Valsalva maneuver or cough is abnormal and, in the absence of detrusor activity, is diagnostic of sphincteric incontinence. A Valsalva leak point pressure, in essence, provides a measurement of sphincteric resistance.[51] In some patients, a Valsalva leak point pressure may be helpful in determining appropriate therapy by providing prognostic information regarding some types of treatment.[52] The voiding portion of the study will provide information regarding bladder contractility, bladder outlet obstruction and bladder emptying. The pressure-flow portion of the urodynamic tracing integrates the contractile function of the detrusor as a function of total outlet resistance. High pressure and low flow indicates obstruction. Low-pressure voiding does not necessarily imply impaired bladder contractility. In the male with significant sphincteric compromise, voiding pressures may be very low owing to negligible outlet resistance. Electromyography of the pelvic floor is probably unnecessary in patients undergoing urodynamics for the evaluation of PPI unless neurological illness is known or suspected.

TREATMENT

Once the diagnosis of sphincteric incontinence has been established as the etiology of PPI, and any coexistent bladder dysfunction has been addressed and controlled, treatment is directed toward increasing outlet resistance. Options consist of behavioral measures, pharmacological treatment and surgery. Nevertheless, it is important to provide interim support and measures to control PPI while the diagnostic evaluation is proceeding and therapy is being planned. Secondary complications of the skin, such as fungal infection, due to PPI are not uncommon and are usually easily prevented. This is a source of considerable patient irritation and a few minutes in the office reviewing the judicious use of skin barrier products and skin protectants, as well as advice regarding absorbent products, such as liners, pads, diapers and reusable washable undergarments, can be time well spent. The intermittent and appropriate application of penile clamps, condom catheters, as well as information regarding urine collection devices, should also be provided and reviewed. In combination with some confident reassurance that the vast majority of patients with PPI have either spontaneous resolution of their incontinence or that it can be successfully treated, these simple measures can alleviate considerable patient anxiety and frustration.

Behavioral modification and pelvic floor exercise

Pelvic floor exercise is an integral component of a behavioral modification program (see **Table 34.2**) and is often utilized to treat PPI in the immediate postoperative period. This program consists of: (1) patient education regarding the function of the lower urinary tract; (2) fluid and dietary management; (3) timed voiding, and bladder training; (4) pelvic floor exercises; and (5) a voiding log or diary. For the patient with PPI, the aim of behavioral therapy is to help regain bladder control by increasing the effective capacity of the bladder and improving outlet resistance, thereby reducing the symptoms of UI. This type of program can be used for both sphincteric and bladder-related causes of PPI. Keeping a record in the form of a frequency/volume chart or voiding log plays a central role. Dietary items, such as coffee, tea and alcohol, may precipitate symptoms, and this will become obvious upon review of the voiding log, if this information has been included. The initiation of pelvic floor exercises, which can improve outlet resistance and inhibit the micturition reflex, may result in a gradual increase in bladder capacity. These exercises are taught in such a way that patients can use this physiologic mechanism to inhibit an impending or beginning bladder contraction. A properly timed pelvic floor exercise (voluntary contraction of the striated muscle of the pelvic floor) can increase outlet resistance instantaneously to prevent the flow of urine through the urethra. Female patients with stress incontinence can be instructed to volitionally contract their pelvic floor during those physical activities that result in UI, such as coughing, sneezing or rising to a standing position.[53] It is unclear whether this is useful in men with PPI. Often patients are initially unable to 'find' and voluntarily contract their pelvic floor muscles effectively. This may be especially true in the postoperative patient. Biofeedback and/or electrical stimulation may be quite beneficial as adjunctive measures in enabling patients to locate and utilize their pelvic floor muscles in an effective manner.

Behavioral modification programs have been utilized with significant success in treating females with stress incontinence, overactive bladder and mixed incontinence. Relatively few studies have addressed their application to PPI. It is important to understand that behavioral therapy methods are not standardized, and studies vary considerably in their approach to treatment and treatment protocols. Meaglia et al. reported on the use of behavioral training augmented with a strong verbal support group in 24 patients with PPI.[54] Results were better in those with post-TURP incontinence (74% improvement) as compared to those with post-RP incontinence (33% improvement rate). Burgio et al. reported on behavioral modification and pelvic floor exercise in 20 patients with post-RP incontinence.[55] Patients were instructed in relaxation of their abdominal musculature and intermittent interruption of their normal urinary stream while voiding. A total of 78.3% of those with stress incontinence symptoms improved. Van Kampen and colleagues reported on a randomized, controlled study of the effect of pelvic floor exercise on overall continence in 102 patients following RP.[56] After 3 months, 88% of patients in the pelvic floor re-education program had regained continence

Table 34.2. Key elements of a behavioral modification program.

1. Patient education in lower urinary tract function
   A primer on how the lower urinary tract fills, stores and empties urine.
2. Dietary and fluid management
   Fluid moderation during the day (unless medically contraindicated) and restriction of fluids in the evening/night-time to alleviate nocturia. Certain items, such as coffee, tea and alcohol, can act as bladder irritants and, to some degree, create diuresis. Reduction or elimination of these items can improve symptoms.
3. Timed, prompted or scheduled voiding
   Regularly emptying the bladder on a schedule will prevent the accumulation of large amounts of urine in the bladder. With the use of judicious fluid managment and pelvic floor exercise, the voiding interval can be gradually increased.
4. Pelvic floor exercises
   Should be performed properly and regularly emphasizing both endurance training and 'quick flicks'. Biofeedback techniques and electrical stimulation are adjunctive instructional measures.
5. Voiding diary
   Frequency–volume chart reminds the patient of the need to void regularly. In combination with timed voiding, pelvic floor exercise and fluid management, the voiding diary monitors success of the program.

as compared to 56% of controls. At one year, the pelvic floor re-education arm maintained a 14% improvement over the control group. Moore and colleagues randomized 63 post-RP patients to standard therapy, intensive pelvic floor rehabilitation or pelvic floor rehabilitation and electrical stimulation at 8 weeks following surgery.[57] In contrast to Van Kampen et al., they found that there was no difference in continence rates or degree of incontinence between the three groups at 8 months, although all three groups had considerable improvement. Franke and colleagues looked at whether a structured program of pelvic floor exercise initiated 6 weeks following RP would improve postoperative continence. A total of 30 patients were randomized to five sessions of postoperative biofeedback or no directed therapy. They found no difference in overall continence, pad test results or voiding diaries between those who received the additional training and those who did not.[58] Finally, Bales et al.[59] randomized 100 patients to preoperative biofeedback and pelvic floor training or no preoperative training. They found that preoperative pelvic floor instruction and training using biofeedback did not influence overall continence rates at 6 months following RP as compared to simple verbal and written instructions. Overall, it does appear that pelvic floor exercises and behavioral modification provide some short-term relief from PPI symptoms, and perhaps shorten the interval to achieving maximum continence. However, whether these therapies are effective in the long term, above and beyond the spontaneous improvement rate of PPI, and whether these therapies can be effective in patients who remain with significant incontinence more than 1 year from RP is unclear. Overall, behavioral therapy can be utilized by any health care professional. It is simple, inexpensive (as long as not overburdened with 'bells and whistles') and probably effective in the short term in most patients with PPI. Importantly, it is essentially free of adverse effects. In the incontinent patient with a perceived postoperative adverse event, this is probably not insignificant. It does, however, require patient motivation and a time commitment.

Periurethral injectable agents

Teflon paste, silicone, fat and collagen have all been utilized as injectable agents to provide improved continence following prostatectomy. Osther and Rohl treated 25 patients with transperineal or transurethral injection of Teflon paste with 'good to moderate' results in 24% of patients with minimal morbidity.[60] The retreatment rate, however, was 72%. The patients with the worst results had poor bladder compliance. Additionally, patients that failed the initial injection were usually no better on subsequent treatment. Deane et al. reported a 60% reapplication rate in their Teflon series, noting that patients requiring repeated injections did not improve, which they felt was probably due to local scarring from prior surgery.[61] Despite some initally encouraging results, most of their patients failed within 3 months. Politano reviewed his greater than 20 year experience with Teflon paste for all types of post-prostatectomy incontinence.[62] Of the over 700 patients treated, those with post-RP incontinence (67% cured or improved) did not fare as well over those with post-TURP incontinence (88% cured or improved) or those with UI following open prostatectomy (74% cured or improved). Although the results with injectable Teflon were promising, subsequent reports of complications, including particle migration, granulomatous reaction and local irritation, have prevented widespread usage.[63]

Initially reported by Shortliffe et al. in 1989, gluteraldehyde cross-linked collagen has demonstrated some efficacy as a relatively non-invasive treatment modality for UI following prostatectomy.[64] Cummings et al. reported on 19 men who had received collagen for PPI: 11 out of 19 were considered improved at a mean follow-up of 10.4 months.[65] Failure correlated with a history of bladder neck contracture or severe incontinence. Similar results were obtained by Elsergany and Ghoneim, who reported on transurethral collagen injection in 35 patients.[66] A total of 20% of patients were reportedly dry and an additional 31% significantly improved. A history of radiation, urethral stricture or bladder overactivity correlated with failure. Smith, Appell and colleagues looked at transurethral collagen injection in 62 patients with PPI.[67] Fifty-four patients had UI following RP and eight had UI following TURP. Successful patients had undergone a mean of four injection sessions with a total of 20 ml of collagen material injected. At a mean of 29 months from the last injection, 8.1% of patients were dry and an additional 38.7% achieved 'social continence'. Post-TURP patients had a better response as compared to post-RP patients with 62.5% versus 35.7% considered successful respectively. Adverse prognostic factors included a history of severe incontinence or a history of an incision for bladder neck contracture. Aboseif et al. reported on the results of transurethral injection of collagen under local anesthesia in 88 men with incontinence due to intrinsic sphincteric deficiency.[68] Results were reported after a mean of 3.5 sessions per patient and a mean total of 25 ml of collagen material injected per patient. Forty-two patients (47.7%) were considered 'nearly' dry, an additional 19 (21.6%) had substantial improvement but still required 1–3 pads per day; 13 patients reported no improvement.

Longer term results with collagen appear less satisfactory. Faerber reported on 68 men with post-prostatectomy incontinence followed for a mean of 38 months.[69] These patients underwent an average of five injections per patient (range 3–15) with an average of 36 ml (range 8–125 ml) of collagen material injected per patient. A total of 10% were considered cured (no pads), an additional 10% were significantly improved (>50% reduction in pad use), but 67% of patients had little or no improvement and 13% had worsening of their incontinence. An antegrade approach to collagen injection was initially reported by Klutke et al. in 1995.[70] Using suprapubic transvesical percutaneous access, 20 patients were injected with collagen at the bladder neck. At mean follow-up of 8.5 months, 25% of patients were dry and 45% of patients were considered significantly improved.[71] Subsequently, at a follow-up of 28 months, 10% of patients were still considered dry with 35% improved.[72] Appel and colleagues also reported on the antegrade approach to collagen injection in a series of 24 men with PPI, all of whom had previously received collagen through a retrograde transurethral approach.[73] A mean of 7.1 ml of material was injected. A total of 18 out of 24 (75%) were considered dry at 6 months with only 9 out of 24 (37.5%) remaining dry at 1 year.

Although generally considered safe, complications have been reported with collagen in up to 20% of patients,[74] including *de novo* detrusor instability, periurethral abscess formation, urinary retention, hematuria and a systemic delayed hypersensitivity reaction.[74–78] Potential adverse prognostic factors for the treatment of PPI with collagen appear to be a low Valsalva leak point pressure, prior history of transurethral incision of a bladder neck contracture, severe incontinence, radiation therapy and detrusor overactivity.[52,65–67,79]

Urethral compression procedures

Historically, applying fixed compression of the urethra had been found to have modest short-term success with poor long-term results. In 1919, Young described an operation compressing the urethra by narrowing the membranous urethra and approximating the transverse perineal and levator muscles.[80] Later, Kaufman reported on fixed compression of the urethra by reorienting the penile crura.[81] Providing fixed urethral resistance using a prosthetic device has existed in a number of forms. Kaufman and Raz reported a 70% success rate using a perineally implanted silicone gel prosthesis in 86 patients.[82–84] Long-term follow-up for this procedure did not demonstrate good durability in patients with post-RP incontinence.[85] Yarbrough and colleagues reported cure of postprostatectomy incontinence in 16 of 22 patients using a silicone gel 'pillow'.[86] After adding a Marlex strap for fixation, six out of seven patients were cured. There were no reported complications. However, owing to complications or poor long-term efficacy, many of these rather ingenious procedures have been abandoned.

Recently, new interest in fixed urethral resistance has emerged. Schaeffer and colleagues have described their results with a bulbourethral sling using vascular graft material.[87] A total of 64 patients with sphincteric incontinence following RP underwent a combined perineal and abdominal approach that placed vascular graft bolsters beneath the bulbar urethra. The bulbourethral sling could be 'tightened' postoperatively, if necessary. A combination of chart review, telephone interview and patient interview were used for data collection. At a mean follow-up of 18 months, 56% were dry and an additional 8% were significantly improved; 27% of patients required a revision of the procedure or 'tightening'; and 6% of patients had erosion and 3% had infection of the prosthesis. Adjuvant radiation therapy was felt to be associated with a significant risk of failure from the procedure. In a follow-up study, a questionnaire was mailed to 61 of the patients who had undergone the bulbourethral sling.[88] Of the 66 patients who had undergone the procedure, one had died and four had complications requiring removal of the bolsters and thus were not included in the analysis. At a median follow-up of 9.6 months, 41% of patients considered themselves cured, 53% were not using pads, and 85% were using two pads per day or less. However, persistent perineal numbness or pain was present in 52% of patients. Madjar and colleagues have attempted to provide fixed urethral resistance with a perineally placed bone-anchored sling.[89] Sixteen men with PPI underwent the procedure. At a mean follow-up of 12.2 months, 12 of 14 with stress incontinence were cured, two had more than 50% reduction in pad use and two patients with mixed incontinence were improved.

Artificial urinary sphincter

The 'gold standard' therapy for sphincteric incontinence in the male is the AMS 800 (American Medical Systems, Minnetonka, MN) artificial urinary sphincter prosthesis (**Fig. 34.2**). Of the available options, this device offers the greatest chance of cure from sphincteric incontinence. Previous models of the artificial urinary sphincter had considerable technical problems, including a high rate of urethral erosion.[90] The introduction of the narrow-backed urethral cuff in 1987 has resulted in fewer complications, including a decreased urethral atrophy rate and fewer urethral erosions without compromising mechanical reliability[91–93] (**Fig. 34.3**). Acceptable social continence can be expected in more than 90% of patients.[94–96] Although perfect continence is not achieved in the majority of cases, significant improvement in quality of life is likely following implantation of the AMS 800.[97,98] Long-term reliability rates are good. Montague reported a reoperation rate of only 12% at a mean follow-up of more than 7 years in 113 patients.[99] Other

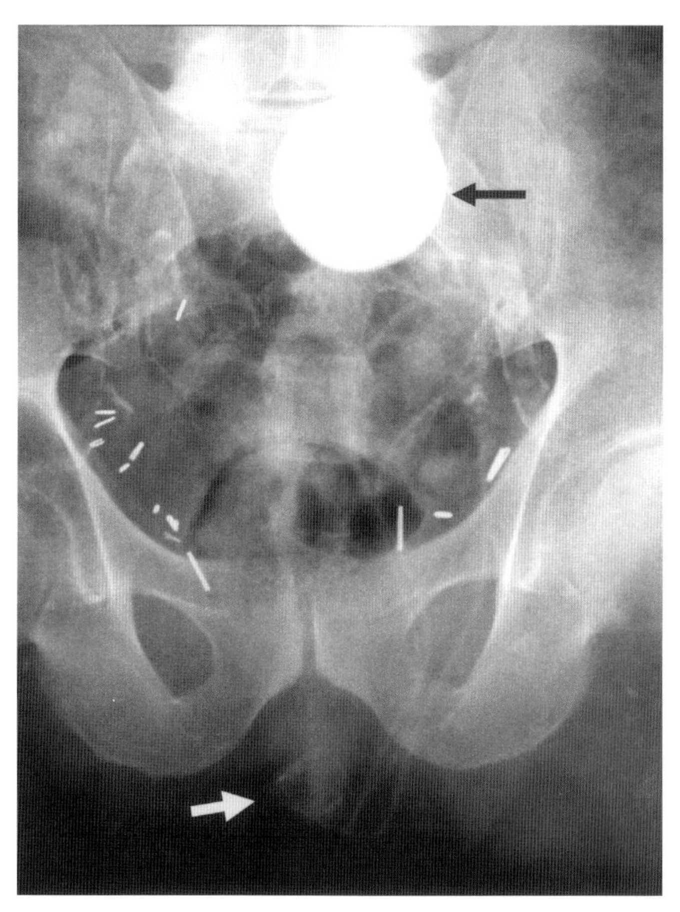

Fig. 34.2. Plain anterioposterior X-ray demonstrating the location of the reservior (black arrow) and cuff (white arrow) of an AMS 800 artificial sphincter.

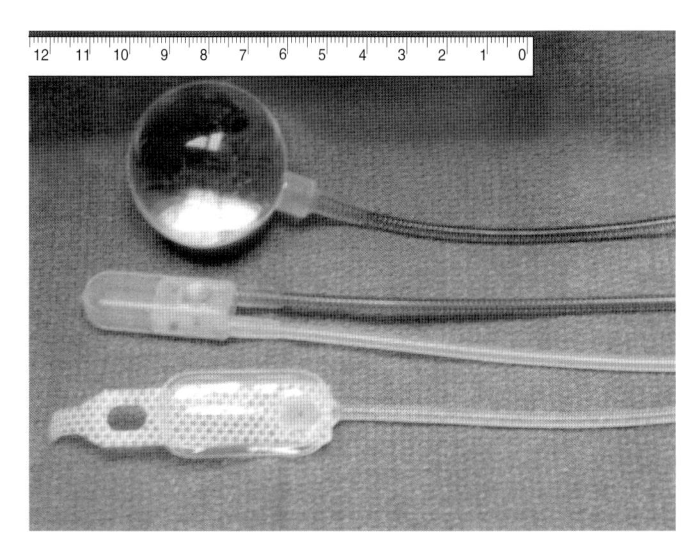

Fig. 34.3. Components of the AMS artificial urinary sphincter. At the top is the reservoir, in the middle the control pump and the lower component is the urethral cuff.

long-term series report overall revision rates for all causes of failure (infection, erosion, atrophy, etc.) as between 21% and 33%.[91,94,97,100–102] Urethral atrophy is the most common reason for revision of the artificial sphincter[103] but the dreaded complications of erosion or infection can occur in up to 1–12% of patients[95,101] (**Fig. 34.4**).

The most significant adverse factor in implanting the AMS 800 prosthesis is a prior history of external beam radiotherapy.[95,103–105] These patients are at greater risk for urethral erosion and urethral atrophy. Nevertheless, successful implantation can be achieved in greater than 90% of these patients.[103,106] Routine deactivation of the device at night has been recommended to decrease the reoperation rate[95] but controlled studies supporting this recommendation are lacking. Urethral catheterization, especially traumatic urethral catheterization following AMS 800 implantation is also associated with urethral cuff erosion.[105] Interestingly, prior intraurethral collagen injection does not appear to decrease the success or long-term durability of the AMS 800.[107]

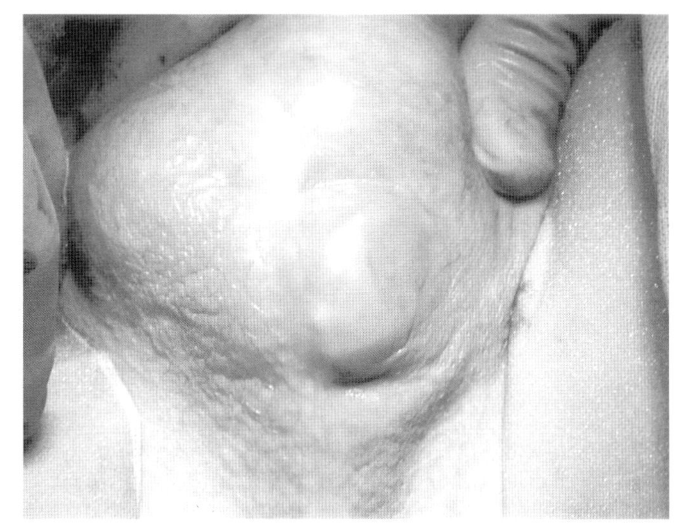

Fig. 34.4. Infection of the AMS 800 sphincter characterized by erythema, edema and swelling of the scrotum and perineum.

Recurrent incontinence following implantation of the AMS 800 may have several etiologies and a careful evaluation is necessary. Bladder factors, including new onset detrusor overactivity should always be considered and can be diagnosed by a well-performed urodynamic evaluation. Urethral erosion and/or mechanical malfunction, including fluid leak, may result in the sudden reappearance of incontinence. Cystoscopy, voiding cystourethrography (**Fig. 34.5**) or video urodynamics may be helpful in assessing for these possibilities. Urethral atrophy may present with slowly increasing incontinence. Perfusion sphincterometry is a simple low-cost

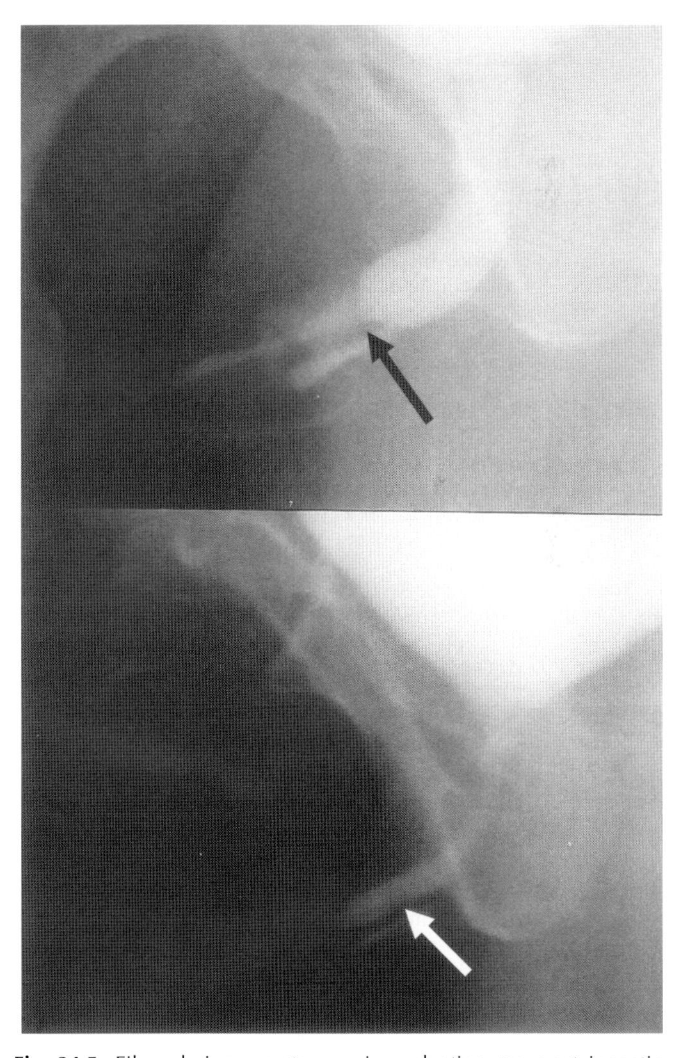

Fig. 34.5. Films during a cystogram in evaluation recurrent incontinence following AMS 800 implantation. No extravasation of contrast is seen to suggest urethral erosion. In the upper image, in the activated state, the column of contrast in the urethra stops at the level of the urethral cuff (black arrow). There appears to be contrast distal to the cuff, which represents the filling catheter. In the lower image, when the device is deactivated, contrast is seen to flow through the cuff (white arrow). Urodynamic evaluation revealed an overactive bladder as the cause of the recurrent incontinence.

method that can be utilized to assess for cuff atrophy.[108] The options for treatment for urethral atrophy include placement of a tandem cuff,[109,110] downsizing the cuff, relocating the cuff more proximally[111] or changing to a higher pressure reservoir.

REFERENCES

1. Herr HW. Quality of life of incontinent men after radical prostatectomy. *J. Urol.* 1994; 151:652–4.
2. Abrams P, Blaivas JG, Stanton SL et al. The standardisation of terminology of lower urinary tract function. The International Continence Society Committee on Standardisation of Terminology. *Scand. J. Urol. Nephrol. Suppl.* 1988; 114:5–19.
3. Wei JT, Montie JE. Comparison of patients, and physicians, rating of urinary incontinence following radical prostatectomy. *Semin. Urol. Oncol.* 2000; 18:76–80.
4. Hammerer P, Huland H. Urodynamic evaluation of changes in urinary control after radical retropubic prostatectomy. *J. Urol.* 1997; 157:233–6.
5. Albertsen P. Risk of complication and re-admission associated with radical prostatectomy in community practice: a Medicare claims analysis. *J. Urol.* 1997; 157:93.
6. Walsh PC, Jewett HJ. Radical surgery for prostatic cancer. *Cancer* 1980; 45:1906–11.
7. Rudy DC, Woodside JR, Crawford ED. Urodynamic evaluation of incontinence in patients undergoing modified Campbell radical retropubic prostatectomy: a prospective study. *J. Urol.* 1984; 132:708–12.
8. Veenema RJ, Gursel EO, Lattimer JK. Radical retropubic prostatectomy for cancer: a 20-year experience. *J. Urol.* 117:330–1.
9. O'Donnell PD, Brookover T, Hewett M et al. Continence level following radical prostatectomy. *Urology* 1990; 36:511–12.
10. Caine M, Edwards D. The peripheral control of micturition: a cine-radiographic study. *Br. J. Urol.* 1958; 30:34–6.
11. Presti JC Jr, Schmidt RA, Narayan PA et al. Pathophysiology of urinary incontinence after radical prostatectomy. *J. Urol.* 1990; 143:975–8.
12. Steiner MS, Burnett AL, Brooks JD et al. Tubularized neourethra following radical retropubic prostatectomy. *J. Urol.* 1993; 150:407–9.
13. Gaker DL, Gaker LB, Stewart JF et al. Radical prostatectomy with preservation of urinary continence. *J. Urol.* 1996; 156:445–9.
14. Elbadawi A. Neuromorphologic basis of vesicourethral function. I. Histochemistry, ultrastructure, and function of intrinsic nerves of the bladder and urethra. *Neurourol. Urodyn.* 1982; 1:3–50.
15. Haab F, Yamaguchi R, Leach GE. Postprostatectomy incontinence. *Urol. Clinics North Am.* 1996; 23:447–57.
16. Aboseif SR, Konety B, Schmidt RA et al. Preoperative urodynamic evaluation: does it predict the degree of urinary continence after radical retropubic prostatectomy? *Urol. Int.* 1994; 53:68–73.
17. Goluboff ET, Chang DT, Olsson CA et al. Urodynamics and the etiology of post-prostatectomy urinary incontinence: the initial Columbia experience. *J. Urol.* 1995; 153:1034–7.
18. Leach GE, Trockman B, Wong A et al. Post-prostatectomy incontinence: urodynamic findings and treatment outcomes. *J. Urol.* 1996; 155:1256–9.
19. Fitzpatrick JM, Gardiner RA, Worth PH. The evaluation of 68 patients with post-prostatectomy incontinence. *Br. J. Urol.* 1979; 51:552–5.
20. Khan Z, Mieza M, Starer P et al. Post-prostatectomy incontinence. A urodynamic and fluoroscopic point of view. *Urology* 1991; 38:483–8.
21. Mayo ME, Ansell JS. Urodynamic assessment of incontinence after prostatectomy. *J. Urol.* 1979; 122:60–1.
22. Chao R, Mayo ME. Incontinence after radical prostatectomy: detrusor or sphincter causes. [see comments]. *J. Urol.* 1995; 154:16–18.
23. Groutz A, Blaivas JG, Chaikin DC et al. The pathophysiology of post-radical prostatectomy incontinence: a clinical and video urodynamic study. *J. Urol.* 2000; 163:1767–70.
24. Kerr LA, Zincke H. Radical retropubic prostatectomy for prostate cancer in the elderly and the young: complications and prognosis. *Eur. Urol.* 25:305–11.
25. Catalona WJ, Carvalhal GF, Mager DE et al. Potency, continence and complication rates in 1,870 consecutive radical retropubic prostatectomies. *J. Urol.* 1999; 162:433–8.
26. Eastham JA, Kattan MW, Rogers E et al. Risk factors for urinary incontinence after radical prostatectomy. *J. Urol.* 1996; 156:1707–13.
27. Walsh PC, Partin AW, Epstein JI. Cancer control and quality of life following anatomical radical retropubic prostatectomy: results at 10 years. *J. Urol.* 1994; 152:1831–6.
28. Catalona WJ, Basler JW. Return of erections and urinary continence following nerve sparing radical retropubic prostatectomy. *J. Urol.* 1993; 150:905–7.
29. Steiner MS, Morton RA, Walsh PC. Impact of anatomical radical prostatectomy on urinary continence. *J. Urol.* 1991; 145:512–14.
30. Bass RB, Barrett DM. Radical retropubic prostatectomy after transurethral prostatic resection. *J. Urol.* 1980; 124:495–7.
31. Elder JS, Gibbons RP, Correa RJ Jr et al. Morbidity of radical perineal prostatectomy following transurethral resection of the prostate. *J. Urol.* 1984; 132:55–7.
32. Nichols RT, Barry JM, Hodges CV. The morbidity of radical prostatectomy for multifocal stage I prostatic adenocarcinoma. *J. Urol.* 117:83–4.
33. Lindner A, deKernion JB, Smith RB et al. Risk of urinary incontinence following radical prostatectomy. *J. Urol.* 129: 1007–8.
34. Ramon J, Rossignol G, Leandri P et al. Morbidity of radical retropubic prostatectomy following previous prostate resection. *J. Surg. Oncol.* 1994; 55:14–19.
35. Walsh PC. Anatomic radical prostatectomy: evolution of the surgical technique. *J. Urol.* 1998; 160:2418–24.
36. Licht MR, Klein EA, Tuason L, Levin H. Impact of bladder neck preservation during radical prostatectomy on continence and cancer control. *Urology* 1994; 44:883–7.
37. Braslis KG, Petsch M, Lim A et al. Bladder neck preservation following radical prostatectomy: continence and margins. *Eur. Urol.* 1995; 28:202–8.
38. Lowe BA. Comparison of bladder neck preservation to bladder neck resection in maintaining postrostatectomy urinary continence. *Urology* 1996; 48:889–93.
39. Poon M, Ruckle H, Bamshad BR et al. Radical retropubic prostatectomy: bladder neck preservation versus reconstruction. *J. Urol.* 2000; 163:194–8.
40. Srougi M, Nesrallah LJ, Kauffmann JR et al. Urinary continence and pathological outcome after bladder neck preservation during radical retropubic prostatectomy: a randomized prospective trial. *J. Urol.* 2001; 165:815–18.
41. Wei JT, Dunn RL, Marcovich R et al. Prospective assessment of patient reported urinary continence after radical prostatectomy. *J. Urol.* 2000; 164:744–8.
42. Myers RP. Male urethral sphincteric anatomy and radical prostatectomy. *Urol. Clinics North Am.* 1991; 18:211–27.
43. Myers RP, Goellner JR, Cahill DR. Prostate shape, external striated urethral sphincter and radical prostatectomy: the apical dissection. *J. Urol.* 1987; 138:543–50.
44. Hollabaugh RS Jr, Dmochowski RR, Kneib TG et al. Preservation of putative continence nerves during radical retropubic prostatectomy leads to more rapid return of urinary continence. *Urology* 1998; 51:960–7.

45. Kaye KW, Creed KE, Wilson GJ et al. Urinary continence after radical retropubic prostatectomy. Analysis and synthesis of contributing factors: a unified concept. *Br. J. Urol.* 1997; 80:444–501.
46. Jarow JP. Puboprostatic ligament sparing radical retropubic prostatectomy. *Semin. Urol. Oncol.* 2000; 18:28–32.
47. Lowe BA. Preservation of the anterior urethral ligamentous attachments in maintaining post-prostatectomy urinary continence: a comparative study. *J. Urol.* 1997; 158:2137–41.
48. Klein EA. Modified apical dissection for early continence after radical prostatectomy. *Prostate* 1993; 22:217–23.
49. John H, Hauri D. Seminal vesicle-sparing radical prostatectomy: a novel concept to restore early urinary continence. *Urology* 2000; 55:820–4.
50. Seaman EK, Benson MC. Improved continence with tubularized bladder neck reconstruction following radical retropubic prostatectomy. *Urology* 1996; 47:532–5.
51. Gudziak MR, McGuire EJ, Gormley EA. Urodynamic assessment of urethral sphincter function in post-prostatectomy incontinence. *J. Urol.* 1996; 156:1131–4.
52. Sanchez-Ortiz RF, Broderick GA, Chaikin DC et al. Collagen injection therapy for post-radical retropubic prostatectomy incontinence: role of Valsalva leak point pressure. *J. Urol.* 1997; 158:2132–6.
53. Miller JM, Ashton-Miller JA, DeLancey JO. A pelvic muscle precontraction can reduce cough-related urine loss in selected women with mild SUI. *J. Am. Geriat. Soci.* 1998; 46:870–4.
54. Meaglia JP, Joseph AC, Chang M et al. Post-prostatectomy urinary incontinence: response to behavioral training. *J. Urol.* 1990; 144:674–6.
55. Burgio KL, Stutzman RE, Engel BT. Behavioral training for post-prostatectomy urinary incontinence. *J. Urol.* 1989; 141:303–6.
56. Van Kampen M, De Weerdt W, Van Poppel H et al. Effect of pelvic-floor re-education on duration and degree of incontinence after radical prostatectomy: a randomised controlled trial. *Lancet* 2000; 355:98–102.
57. Moore KN, Griffiths D, Hughton A. Urinary incontinence after radical prostatectomy: a randomized controlled trial comparing pelvic muscle exercises with or without electrical stimulation. *Br. J. Urol. Int.* 1999; 83:57–65.
58. Franke JJ, Gilbert WB, Grier J et al. Early post-prostatectomy pelvic floor biofeedback. *J. Urol.* 2000; 163:191–3.
59. Bales GT, Gerber GS, Minor TX et al. Effect of preoperative biofeedback/pelvic floor training on continence in men undergoing radical prostatectomy. *Urology* 2000; 56:627–30.
60. Osther PJ, Rohl HF. Teflon injections in post-prostatectomy incontinence. *Scand. J. Urol. Nephrol.* 1988; 22:171–4.
61. Deane AM, English P, Hehir M et al. Teflon injection in stress incontinence. *Br. J. Urol.* 1985; 57:78–80.
62. Politano VA. Transurethral polytef injection for post-prostatectomy urinary incontinence. *Br. J. Urol.* 1992; 69:26–8.
63. Malizia AA Jr, Reiman HM, Myers RP et al. Migration and granulomatous reaction after periurethral injection of polytef (Teflon). *JAMA* 1984; 251:3277–81.
64. Shortliffe LM, Freiha FS, Kessler R et al. Treatment of urinary incontinence by the periurethral implantation of glutaraldehyde cross-linked collagen. *J. Urol.* 1989; 141:538–41.
65. Cummings JM, Boullier JA, Parra RO. Transurethral collagen injections in the therapy of post-radical prostatectomy stress incontinence. *J. Urol.* 1996; 155:1011–13.
66. Elsergany R, Ghoniem GM. Collagen injection for intrinsic sphincteric deficiency in men: a reasonable option in selected patients. *J. Urol.* 1998; 159:1504–6.
67. Smith DN, Appell RA, Rackley RR et al. Collagen injection therapy for post-prostatectomy incontinence. *J. Urol.* 1998; 160:364–7.
68. Aboseif SR, O'Connell HE, Usui A et al. Collagen injection for intrinsic sphincteric deficiency in men. *J. Urol.* 1996; 155:10–13.
69. Faerber GJ, Richardson TD. Long-term results of transurethral collagen injection in men with intrinsic sphincter deficiency. *J. Endourol.* 11:273–7.
70. Klutke CG, Nadler RB, Andriole GL. Surgeon's workshop: antegrade collagen injection: new technique for postprostatectomy stress incontinence. *J. Endourol.* 1995; 9:513–15.
71. Klutke CG, Nadler RB, Tiemann D et al. Early results with antegrade collagen injection for post-radical prostatectomy stress urinary incontinence. *J. Urol.* 1996; 156:1703–6.
72. Klutke JJ, Subir C, Andriole G et al. Long-term results after antegrade collagen injection for stress urinary incontinence following radical retropubic prostatectomy. *Urology* 1999; 53:974–7.
73. Appell RA, Vasavada SP, Rackley RR et al. Percutaneous antegrade collagen injection therapy for urinary incontinence following radical prostatectomy. *Urology* 1996; 48:769–72.
74. Stothers L, Goldenberg SL, Leone EF. Complications of periurethral collagen injection for stress urinary incontinence. *J. Urol.* 159:806–7.
75. Bernier PA, Zimmern PE, Saboorian MH et al. Female outlet obstruction after repeated collagen injections. *Urology.* 1997; 50:618–21.
76. McLennan MT, Bent AE. Suburethral abscess: a complication of periurethral collagen injection therapy. *Obstet. Gynecol.* 1998; 92:650–2.
77. Stothers L, Goldenberg SL. Delayed hypersensitivity and systemic arthralgia following transurethral collagen injection for stress urinary incontinence. *J. Urol.* 1998; 159:1507–9.
78. Sweat SD, Lightner DJ. Complications of sterile abscess formation and pulmonary embolism following periurethral bulking agents. *J. Urol.* 1999; 161:93–6.
79. Martins FE, Bennett CJ, Dunn M et al. Adverse prognostic features of collagen injection therapy for urinary incontinence following radical retropubic prostatectomy. *J. Urol.* 1997; 158:1745–9.
80. Young HH. An operation for the cure of incontinence of urine. *Surg. Gynecol. Obstet.* 1919; 28:84–5.
81. Kaufman JJ. Surgical treatment of post-prostatectomy incontinence: use of the penile crura to compress the bulbous urethra. *Trans. Am. Associ. Genito-Urin. Surg.* 1971; 63:113–20.
82. Kaufman JJ. Urethral compression operations for the treatment of post-prostatectomy incontinence. *J. Urol.* 1973; 110:93–6.
83. Kaufman JJ. Treatment of post-prostatectomy urinary incontinence using a silicone gel prosthesis. *Br. J. Urol.* 1973; 45:646–53.
84. Kaufman JJ, Raz S. Passive urethral compression with a silicone gel prosthesis for the treatment of male urinary incontinence. *Mayo Clin. Proc.* 1976; 51:373.
85. Graham SD, Carson CC III, Anderson EE. Long-term results with the Kaufman prosthesis. *J. Urol.* 1982; 128:328–30.
86. Yarbrough WJ, Semerdjian HS, Miller HC. George Washington University technique for surgical correction of post-prostatectomy incontinence. *J. Urol.* 1975; 113:47–9.
87. Schaeffer AJ, Clemens JQ, Ferrari M et al. The male bulbourethral sling procedure for post-radical prostatectomy incontinence. *J. Urol.* 1998; 159:1510–15.
88. Clemens JQ, Bushman W, Schaeffer AJ. Questionnaire based results of the bulbourethral sling procedure. *J. Urol.* 1999; 162:1972–6.
89. Madjar S, Jacoby K, Giberti C et al. Bone anchored sling for the treatment of post-prostatectomy incontinence. *J. Urol.* 2001; 165:72–6.
90. Bruskewitz R, Raz S, Smith RB et al. AMS 742 sphincter: UCLA experience. *J. Urol.* 1980; 124:812–14.
91. Elliott DS, Barrett DM. Mayo Clinic long-term analysis of the functional durability of the AMS 800 artificial urinary sphincter: a review of 323 cases. *J. Urol.* 1998; 159:1206–08.
92. Leo ME, Barrett DM. Success of the narrow-backed cuff design of the AMS800 artificial urinary sphincter: analysis of 144 patients. *J. Urol.* 1993; 150:1412–14.

93. Light JK, Reynolds JC. Impact of the new cuff design on reliability of the AS800 artificial urinary sphincter. *J. Urol.* 1992; 147:609–11.
94. Leibovich BC, Barrett DM. Use of the artificial urinary sphincter in men and women. *World J. Urol.* 1997; 15:316–19.
95. Marks JL, Light JK. Management of urinary incontinence after prostatectomy with the artificial urinary sphincter. *J. Urol.* 1989; 142:302–4.
96. Fishman IJ, Shabsigh R, Scott FB. Experience with the artificial urinary sphincter model AS800 in 148 patients. *J. Urol.* 1989; 141:307–10.
97. Haab F, Trockman BA, Zimmern PE et al. Quality of life and continence assessment of the artificial urinary sphincter in men with minimum 3.5 years of followup. *J. Urol.* 1997; 158:435–9.
98. Litwiller SE, Kim KB, Fone PD et al. Post-prostatectomy incontinence and the artificial urinary sphincter: a long-term study of patient satisfaction and criteria for success. *J. Urol.* 1996; 156:1975–80.
99. Montague DK, Angermeier KW, Paolone DR. Long-term continence and patient satisfaction after artificial sphincter implantation for urinary incontinence after prostatectomy. *J. Urol.* 2001; 166:547–9.
100. Klijn AJ, Hop WC, Mickisch G et al. The artificial urinary sphincter in men incontinent after radical prostatectomy: 5 year actuarial adequate function rates. *Br. J. Urol.* 1998; 82:530–3.
101. Mottet N, Boyer C, Chartier-Kastler E et al. Artificial urinary sphincter AMS 800 for urinary incontinence after radical prostatectomy: the French experience. *Urol. Int.* 1998; 60(Suppl. 2):25–9.
102. Bosch JL, Klijn AJ, Schroder FH et al. The artificial urinary sphincter in 86 patients with intrinsic sphincter deficiency: satisfactory actuarial adequate function rates. *Eur. Urol.* 2000; 38:156–60.
103. Martins FE, Boyd SD. Artificial urinary sphincter in patients following major pelvic surgery and/or radiotherapy: are they less favorable candidates? *J. Urol.* 1995; 153:1188–93.
104. Manunta A, Guille F, Patard JJ et al. Artificial sphincter insertion after radiotherapy: is it worthwhile? *Br. J. Urol. Int.* 2000; 85:490–2.
105. Martins FE, Boyd SD. Post-operative risk factors associated with artificial urinary sphincter infection-erosion. *Br. J. Urol.* 1995; 75:354–8.
106. Perez LM, Webster GD. Successful outcome of artificial urinary sphincters in men with post-prostatectomy urinary incontinence despite adverse implantation features. *J. Urol.* 1992; 148:1166–70.
107. Gomes CM, Broderick GA, Sanchez-Ortiz RF et al. Artificial urinary sphincter for post-prostatectomy incontinence: impact of prior collagen injection on cost and clinical outcome. *J. Urol.* 2000; 163:87–90.
108. Choe JM, Battino BS, Bell TE. Retrograde perfusion sphincterometry with a flexible cystoscope: method of troubleshooting the AMS 800. *Urology* 2000; 56:317–19.
109. Kowalczyk JJ, Spicer DL, Mulcahy JJ. Long-term experience with the double-cuff AMS 800 artificial urinary sphincter. *Urology* 1996; 47:895–7.
110. Brito CG, Mulcahy JJ, Mitchell ME et sal. Use of a double cuff AMS800 urinary sphincter for severe stress incontinence. *J. Urol.* 1993; 149:283–5.
111. Couillard DR, Vapnek JM, Stone AR. Proximal artificial sphincter cuff repositioning for urethral atrophy incontinence. *Urology* 1995; 45:653–6.

Prognostic Significance of Positive Surgical Margins

Vivek Narain and David P. Wood
Department of Urology, Wayne State University School of Medicine and Karmanos Cancer Institute, Detroit, MI, USA

INTRODUCTION

Radical prostatectomy is commonly performed on patients with prostate cancer presumed to be confined to the prostate. The current success of early detection efforts has resulted in an increased number of men who are candidates for radical prostatectomy. However, the finding of a positive surgical margin in a radical retropubic prostatectomy specimen is an all too common occurrence in patients with clinically localized prostate cancer. The practice of analyzing radical prostatectomy specimens by thin sections and analysis for tumor cells at the inked margin of resection has led to a high incidence of positive surgical margins in radical prostatectomy specimens.[1]

The reasons for a high incidence of positive surgical margins have been detailed in pathological analysis performed at several institutions with the anatomical observation that the majority of prostate cancers arise in the peripheral zone of the prostate and that early microscopic capsular penetration frequently is present.[2,3] Combined with the fact that little periprostatic tissue exists in the pelvis to allow for wider surgical excision, this results in a high frequency of surgical margin involvement. Furthermore, Villers has identified the superior pedicle of the neurovascular bundle that supplies the prostate base as the possible pathway of the intraprostatic to periprostatic spread of prostate cancer.[4] Positive surgical margins may also occur in an otherwise organ-confined cancer because of an inadvertent incision through the capsule.[5] In some series, incision into the cancer was observed in 61–73% of the sites of positive surgical margin, and is particularly prone to occur at the apex.[6,7]

DEFINITION OF A POSITIVE SURGICAL MARGIN

Because the technique of specimen processing impacts on the incidence of positive surgical margins, a standard technique using 2–3 mm sections has been described by True in the hope of improving the likelihood of detecting positive margins. A positive surgical margin is defined as the presence of tumor cells at the inked surgical margin of resection. The margin extent is classified as focal, extensive or equivocal. A focal margin is defined as a margin present in only one step section and involving one gland in that section. Involvement greater than this is classified as an extensive margin. An equivocal margin is used to describe tumor cells in contact with ink in the absence of periprostatic tissue. This phenomenon occurs at the site of hemostatic clips and sharp dissection close to the capsule.[9]

The location of the positive surgical margin is classified as apical, anterior, bladder neck, posterolateral or posterior. Bladder neck or apical margins are identified as being present if tumor is in contact with ink in the respective zones. The posterior surface represents the concave aspect of the gland in contact with the rectal surface. The anterior surface describes the corresponding width of the gland anteriorly. Remaining lateral aspects of the gland are designated as posterolateral, lying between these two surfaces. The posterolateral segment includes the area next to the neurovascular bundles.

Extracapsular perforation is defined as the presence of neoplastic cells in contact with periprostatic fat, connective tissue, or into adjacent bladder or skeletal muscle, which may be associated with positive surgical margin. If the positive surgical margin occurs in an area without periprostatic tissue, this is designated a positive surgical margin.

The positive margin rate is highly dependent on the technique the pathologist uses to process and evaluate the radical prostatectomy specimen. For example, different methods of processing the apical region may result in varying incidences of positive surgical margins.[10] Voges et al. made the distinction between positive surgical margins due to incision into the prostate versus those due to capsular penetration.[11] They found that positive apical margins arising from small volume tumor (less than 4 cm^3) were more frequently the result of incision into the prostate, and those arising from large volume tumors (greater than 12 cm^3) were mostly due to capsular penetration. Capsular incision exposing benign tissue has not been associated with biochemical failure.[12]

FREQUENCY AND LOCATION OF POSITIVE SURGICAL MARGINS

Researchers using different surgical approaches and techniques for radical prostatectomy have reported positive margins with variable frequency at different sites. Overall, the incidence of positive surgical margins in radical prostatectomy series ranges from 11% to 62%.[6,11,13–20] Wieder et al. reported an overall rate of positive surgical margin after radical retropubic prostatectomy to be 28%, ranging from 10% to 59% by pathologic stage.[21] The recent decline in positive surgical margins has been attributed to stage migration and improved surgical techniques.

Several studies have evaluated the anatomic sites of margin positivity in patients with organ-confined prostate cancer who have received no adjuvant therapy.[5–7,22–27] About 65–80% of positive surgical margins involves one anatomic site, whereas about 20–35% of cases involve at least two or more anatomic sites involved. The rate of positive surgical margins also increases if concomitant extraprostatic involvement is present. The most common location of a positive surgical margin is the apex, which accounts for 7–62% of all positive margin sites and occurs in 42–46% of patients with a solitary positive margin. Positive margins at the apex that result from incision into the prostate (that is, in the absence of capsular perforation) are reported in 15–87% of apical positive margins[6,7,9] (**Table 35.2**). Attempts to preserve membranous urethral length or the neurovascular bundles may compromise the extent of cancer excision and leave a positive apical margin. In some cases, it is unclear if a positive apical margin is a true tumor extension or an artifact of the pathologic processing when the dorsal venous complex is incised because the anterior/apical periprostatic tissue spreads, exposing the apex of the prostate. If tumors are present in this exposed apical tissue, a false-positive margin may be identified.

Involvement of the posterior margin (rectal surface) is the second most commonly involved site (15–40%), reflecting direct invasion with extracapsular perforation by posterior peripheral zone tumors. They occur in locations throughout the length of the prostate, from apex to base. This highlights the importance of clean, sharp dissection through Denonvilliers' fascia with either an antegrade or retrograde approach to avoid stripping the capsule away from the prostate and exposing tumor. A posterolateral margin accounts for about 16–40% of all positive margins and 21% of all solitary margins. The posterior prostate margin is a frequent site of extracapsular perforation in the region of the neurovascular bundles throughout the length of the prostate.[28] Positive margins in this location have also been reported as a result of preservation of the neurovascular bundle. Rosen et al. showed that over half of the posterolateral positive margins resulted from incision into the prostatic capsule at the site of attempted nerve sparing.[5] Stamey et al. reported that one-third of the posterolateral margin in their series was through the same mechanism.[7] This maneuver may breach the prostatic capsule and create a positive margin in the presence of organ-confined disease.

An anterior positive surgical margin is present in about 2.5–15% cases but involvement of the anterior surface as the only positive margin is uncommon (0.5–6%). Weldon et al. noted a 25% incidence of anterior positive margins with perineal prostatectomy, frequently from tearing the anterior capsule when the puboprostatic ligaments are not sharply divided.[10] Equivocal positive margins at this site do not occur commonly because of the limited amount of anterior dissection in this area in the presence of mostly peripheral zone tumors.

Only 1.4–19% of patients have a positive bladder neck margin. Isolated positive bladder neck margins are uncommon, even via a perineal approach.[10] Gomez et al. reported that the incidence of positive bladder neck margins is not increased when utilizing a bladder neck preservation technique to improve urinary incontinence.[27]

PREOPERATIVE PREDICTORS OF POSITIVE MARGIN STATUS

The preoperative prediction of risk for residual disease is highly dependent on accurate staging and grading techniques. For example, Van Den Ouden et al. showed that the incidence of positive surgical margins ranged from 15% (T0) to 47% (T3) disease.[20] Similarly, they reported that high-grade lesions are associated with a significantly greater frequency of residual disease than that of low-grade lesions.

Digital rectal examination (DRE), computed tomography (CT) scan and endorectal magnetic resonance imaging (MRI) are too insensitive to identify or predict accurately either extracapsular or margin positive disease. Although the preoperative serum prostate specific antigen (PSA) level roughly correlates with the presence of positive surgical margins, for patients with a PSA between 4 and 10, the PSA level lacks discrimination.

Tigrani et al. reported that patients with three or more positive sextant biopsies are at a higher risk (37% vs. 14%) of positive surgical margins than those with fewer than three.[29] Univariate analysis revealed patients with bilateral positive biopsies at the base and midgland had a higher frequency of a positive margin than those who had zero or one positive biopsy in those regions. Interestingly, patients with bilateral positive biopsies in the apex did not have a statistically higher frequency of positive margins. In their study, multivariable logistic regression using PSA, primary Gleason grade, Gleason score and the number of positive sextant biopsy revealed that only the number of positive sextant biopsies was statistically significant predictor of margin status.

There is a well-established correlation between positive surgical margins and tumor volume[6,7,30,31] as well as between higher tumor volume and the number of positive margin sites.[5–7,9,13] Preoperative parameters of serum PSA, clinical stage and biopsy Gleason grade have all been demonstrated to be associated with a higher rate of positive surgical margins, in part through their correlation with tumor volume. Shah et al. reported that percent tumor volumes greater than 26% on prostatectomy specimen had a significantly higher likelihood for positive apical soft tissue margins.[19] Watson et al. reported that when 15% of the gland was involved with tumor, 54% of cases had a positive margin at more than one site.[9] However, the ability to preoperative predict tumor volume preoperatively remains problematic.

High-tumor grade is also associated with greater tumor volume and positive surgical margins.[6,13,27] Tifilli et al. reported that the specimen Gleason score was a strong predictor of a positive surgical margin.[32] Of patients with specimen Gleason score 6 or less, 7, and 8 or more, positive surgical margins were present in 31%, 47.6%, and 67.8% respectively. Shah et al. also reported that men who had positive surgical margins had significantly higher Gleason scores compared to those with negative surgical margins.[19]

PROGNOSTIC SIGNIFICANCE OF POSITIVE SURGICAL MARGINS

The clinical significance of positive surgical margins has been questioned because biochemical or clinical disease progression does not develop in the majority of these patients. Between 25% and 60% of men with positive surgical margins will have PSA biochemical progression within 5 years from the date of surgery, with 20% cases experiencing local recurrence[10,33–37] (**Table 35.1**).

Table 35.1. Clinical or biochemical relapse in patients with positive surgical margins following radical prostatectomy for clinically organ-confined prostate cancer.

Study	Clinical/biochemical relapse	Minimum follow-up
Epstein[30,39]	38%	5 year
Blute[41]	22%	5 year
Catalona[42]	26%	5 year
Kupelian[43]	63%	5 year
Watson[9]	25%	23 months
Boccon-Gibod[12]	37%	25 months

In general, the presence of a positive margin is a poor prognositic feature, although not invariably so. Intuitively, a positive resection margin often implies residual disease; however, the high rate of biochemical disease-free progression demonstrates the limited specificity of this pathological finding for residual localized disease. Whether positive surgical margins are an independent prognostic factor for progression and decrease cancer-specific and overall survival is controversial.

Paulson reported that the status of the surgical margins was the most important prognostic feature of patients treated with radical perineal prostatectomy, if those with lymph node metastases were excluded.[35] They noted that 60% of patients with positive surgical margins had clinical recurrence during a 10-year period, compared with only 19% of those with organ-confined disease during the same interval. If the surgical margins were positive, 40% of the patients died of prostate cancer at 13.5 years compared to 10%, if the margins were negative. In a separate article the same authors reported a 12% risk of recurrence in patients with organ-confined cancer compared with a 60% risk in patients with positive surgical margins with or without extraprostatic extension or seminal vesicle invasion.[38] Similarly, Epstein found positive surgical margins to be an independent predictor of progression. In their series, 38% of patients experienced clinical or biochemical recurrence with a minimum of 5-year follow-up.[30,39] In a separate paper, they reported on a population of patients with clinical stage T2 prostate cancer and capsular penetration in whom the frequency of positive margins was assessed.[40] Among 196 radical prostatectomy specimens 93 had focal capsular penetration. In this group there was no difference in progression among patients with positive or negative margins. However, among 103 patients with established capsular penetration, there was a progressive increase of failure in patients with positive margins and high-grade tumor as compared to those with negative margins and low-grade tumor. Blute et al. found that positive margin was not associated with death or clinical recurrence, but was for the combined end-point of clinical recurrence and PSA failure.[41] The 5-year survival rate was 89% for those without positive margins versus 78% for those with positive surgical margins. Multivariate analysis of 2334 patients in their study revealed that positive surgical margins were a significant predictor of clinical recurrence and PSA failure after controlling for Gleason score, preoperative PSA and DNA ploidy.

Although principles of surgery suggest that wider excision leading to negative surgical margins is an important prognostic indicator, Shekarriz et al. found the number and location of the positive margins did not have a significant impact of disease-free survival.[23] Ohari et al. reported that positive surgical margins were a significant predictor of progression by univariate analysis but did not retain independent prognostic value by multivariate analysis. In their study, only extraprostatic extension, seminal vesicle invasion and Gleason score were significant predictors of PSA progression.[31] They also reported that the interaction between positive surgical margins and other variables, such as extracapsular extension and seminal vesicle invasion, were not significant, suggesting that the effect of positive surgical margin does not depend on extracapsular extension or seminal vesicle invasion.[16] Similarly, Partin et al. did not find the status of surgical margins to be a significant predictor of cancer progression.[44]

IMPACT OF MARGIN LOCATION AND EXTENT OF MARGIN INVOLVEMENT

The impact of margin location on tumor progression and disease-free survival is unclear. The site and degree of margin involvement may have important implications regarding the risk of residual cancer and prognosis. Watson et al. reported that patients with tumor involvement at more than one margin site (in general reflecting larger tumor volume) had a 48% progression rate compared to a 29% progression rate for those with a single positive margin.[9] They reported a 0% progression in men with a solitary anterior positive margin. Solitary apical, posterior and posterolateral positive margins progressed in 25%, 29% and 45% of the cases respectively. However, in a multivariate analysis, the site of margin involvement was not a significant predictor of progression.

Blute et al., however, reported that the impact of positive margin status on recurrence-free survival appears to be site specific.[24] When adjusted for PSA and grade, patients with a positive margin at the prostate base had a worse prognosis. Five-year progression-free survival using a combined clinical and PSA failure end-point for those patients with a positive margin at the prostate base was 56% versus 86% for those with a positive margin elsewhere ($P < 0.001$). Patients with positive surgical margins at the apex/urethra, anterior/posterior prostate, or multiple anatomic sites of positivity as long as the base was clear had slightly decreased (79%, 78%, 82%) 5-year PSA failure rates compared to negative margin status (86% clinical and PSA failure-free survival). Also, multiple positive margins were not associated with higher clinical recurrence in their study. Van den Ouden reported that disease-free survival was significantly less for patients with apical compared with posterolateral positive margins.[20]

The presence of equivocal or focally positive margins has been found to have a variable impact on progression. Watson et al. did not find the extent of margin involvement to be a significant factor for progression.[9] Ohari et al. identified a group with equivocal positive margins whose progression rate was similar to patients with negative surgical margins.[16] Conversely, Epstein et al. noted that progression in patients with equivocal positive margins occurred with similar frequency as patients with extensive positive margins.[33] An exception was an identifiable subgroup in that series with incision into the capsule at the apex, with no increased risk of progression. They also documented that focal margin involvement had a significantly higher incidence of both local recurrence and disease progression, compared to progression in margin-negative patients, although the rate was less than for extensive margin involvement.

Reports analyzing the extent of positive margins (the amount of tumor in contact with ink) and the impact of disease recurrence have been complicated by inconsistent definitions of focal and extensive surgical margins. Epstein et al. defined focal positive margins as 'involved sites were limited and present in only one or two areas'.[33] However, Watson et al. defined it as only one gland in contact with ink in only one section.[9]

PROGNOSTIC IMPLICATION OF POSITIVE APICAL MARGIN

The intimate anatomical relationship of the prostate to surrounding structures limits the ability to obtain wide surgical margins. Furthermore, the prostate lacks a true anatomical capsule, and normal condensation of fibromuscular tissue present in the posterolateral margin is not present at the apex.[45] Stamey et al. also noted the absence of a defined capsule at the apex more than anywhere else in the prostate gland.[3] McNeal suggested that cancer in the apex of the gland may represent invasion outside of the prostate because it is difficult to identify the capsule at the apex.[2] Since the apex of the prostate does not have a well-defined capsular boundary, some investigators suggest that cancer in the apex of the prostate gland is equivalent to extraprostatic extension or a positive surgical margin.[2,3]

In addition to an absent well-defined capsule at this location, the method of pathological processing and surgical technique may contribute to the high incidence of a positive apical margin. Cephalad retraction with the Foley catheter may macerate the apical component of the surgical specimen, creating false-positive margins.

The importance of prostatic apex dissection in anatomical radical prostatectomy, as it relates to potency and urinary continence, is well established. The proximity of the rhabdosphincter must be preserved to minimize incontinence after prostatectomy. The anatomical location of the apex of the prostate deep in the pelvis, and the proximity with critical structures responsible for preservation of continence and sexual potency make the proper dissection of this area challenging. Before the anatomic approach, the high incidence of incontinence after radical prostatectomy had led patients and physicians to consider seriously alternatives to surgery even for the most appropriate candidates. Attempts to preserve the neurovascular bundles or maximal urethral length may compromise the extent of cancer excision, resulting in positive surgical margins at the apex.

Table 35.2. Location and frequency of positive surgical margin.

Sites of positive margin	Overall frequency of all positive margin sites	Frequency of solitary positive margin only
Apex/urethral	7–62%	42–46%
Posteriolateral	16–40%	21%
Anterior	2.5–15%	0.5–6%
Bladder neck	1.4–19%	0%

Fesseha et al. reported that the finding of a positive apical margin in the absence of positive surgical margin or extraprostatic extension elsewhere in the prostate has the same prognosis as organ-confined disease.[46] Similarly, Ohari et al. identified in their series 23 patients with a positive margin in otherwise organ-confined disease, 15 (67%) of whom had an apical margin positive.[16] There was no difference in failure rate in this group when compared to cases with organ-confined disease and negative margins.

POSITIVE SURGICAL MARGINS WITH EXTRAPROSTATIC EXTENSION OR SEMINAL VESICLE INVOLVEMENT

The incidence of extracapsular disease ranges from 28% to 48%.[3,13,33] The chance of finding a positive margin is more frequent with tumors of large volume, high Gleason grade and extracapsular penetration. The interaction between extraprostatic extension and positive seminal vesicles, Gleason score and positive surgical margin is complex. Tefilli et al. found that Gleason score, positive seminal vesicles and positive margins were significant predictors of PSA failure after radical prostatectomy.[32] D'Ámico et al. noted that most positive margins (61%) also had extraprostatic extension.[36] Most agree that seminal vesicle invasion predominates over other factors in predicting PSA failure after radical prostatectomy.

Many investigators have reported that a positive surgical margin with established extraprostatic extension is associated with an increased likelihood of local or distant recurrence.[16,31,39] Ohari et al. reported that positive surgical margins in association with established extraprostatic extension has an adverse impact on prognosis only in the absence of a high-grade cancer (Gleason 7 or greater), seminal vesicle invasion or lymph node metastasis, which are much stronger predictors of prognosis.[16] Partin et al. showed that, after 3.5 years of follow-up from radical prostatectomy, the failure rate of patients with positive margins and those with extraprostatic extension and negative surgical margins were same.[47] However, their patients with high-grade tumors and established capsular penetration behaved like patients with positive margins.

Seminal vesicle invasion is often seen with positive surgical margins. Between 32% and 50% of patients with positive surgical margins have seminal vesicle invasion.[6,9,13,20] Seminal vesicle invasion carries alone an ominous prognosis and it will often determine the long-term outcome of these patients, if present with a positive surgical margin.[16,20,48] Teffili et al. evaluated 93 patients with seminal vesicle invasion and found that patients with seminal vesicle invasion who have a negative surgical margin have a more favorable outcome than if the margins were positive.[47] Specifically, at a median follow-up of 3.6 years, patients with negative surgical margins had a 65% chance of being free of biochemical recurrence versus a 30% chance of patients with positive margins. Lymph node metastasis has been reported in as many as 24% of patients with positive margins.[6,20]

ATTEMPTS TO DECREASE POSITIVE SURGICAL MARGINS

Because a significant number of cases reported by Stamey et al.[6] had an iatrogenic positive apical margin, some authors have proposed a perineal surgical approach for more accurate dissection of the apex. Weldon et al. reported 7% solitary positive apical margin using perineal prostatectomy.[10] The lower incidence of positive apical margin using a perineal approach suggests improved exposure at the apex may be an advantage. Boccon-Gibod et al. performed a retrospective non-randomized study to compare 48 perineal with 46 retropubic prostatectomy specimens of identical pathological stage, and concluded that, as far as apical margins were concerned, the perineal approach was a much safer than the retropubic approach.[12] However, they also reported that the incidence of positive surgical margins in organ-confined tumors was higher in the perineal than retropubic group (43% vs. 29%). Furthermore, they documented that the incidence of capsular incisions exposing benign glandular tissue was significantly higher in the perineal than retropubic group (90% vs. 37%) irrespective of positive surgical margins. They concluded that, although overall the perineal and retropubic approaches to radical prostatectomy for clinically (T1, T2) localized prostate cancer are similar as far as positive margins and biochemical failure rates are concerned, the retropubic probably is superior to the perineal approach for control of organ-confined cancer.

Ohari reported a steady decline with time in the frequency of positive surgical margin (24% to 8%) at their institution, partly due to improved surgical techniques during radical retropubic prostatectomy.[16] Since positive surgical margins occurred more frequently posterolaterally in their series, in the area of neurovascular bundle, they started resecting all or part of the neurovascular bundle more often. They further modified their technique to approach the neurovascular bundle laterally, preserving the bundles more often while allowing a wide dissection around the apex of the prostate, especially posteriorly.

Stephenson et al. reported their experience in 53 non-nerve-sparing radical retropubic prostatectomies performed with attention paid to extending the margin of attached periprostatic tissues.[49] They accomplished this primarily by initial perirectal release of periprostatic tissues at the level of longitudinal rectal fibers posterior and lateral to the prostate to ensure that the maximal quantities of periprostatic tissue will remain with the prostate specimen and will not be attenuated or sheared away at subsequent stages of the procedure. Using this technique, they noted a positive surgical margin rate in only 13% and capsular penetration in 89% of cases.

The role of nerve-sparing surgery remains controversial with regard to its association with positive surgical margins. Although postoperative erectile dysfunction is multifactorial, bilateral excision of neurovascular bundles to reduce post-operative positive surgical margins compromises postoperative potency. Several reports have suggested nerve-sparing techniques do not significantly increase the rate of positive surgical margins.[6,33] Partin et al. reported that, although wider excision of the neurovascular bundles in an attempt to obtain a negative margin may delay recurrence initially, most patients with established capsular penetration ultimately failed radical prostatectomy despite wide excision of periprostatic soft tissue by the end of 43 months.[48] In men with any apical or near apical nodule, Stamey et al. recommended wide excision of the adjacent periprostatic tissue, including each neurovascular bundle.[5]

In an attempt to improve urinary continence, several investigators have advocated dissecting the prostate away from the bladder so as to preserve the fibers of the bladder neck.[50] One problem with such bladder neck-sparing procedures is that tumor may be left in the unresected tissue of the bladder neck. Gomez et al. noted that 6% of patients will have a positive margin at the bladder neck in radical prostatectomy specimen when a bladder neck sparing technique is used.[27] Wood et al. performed circumferential bladder neck biopsies *in situ* after bladder neck-sparing radical prostatectomy and found 12% of patients had a positive surgical margin.[25] In their study, the posterior bladder was the most common site of bladder neck involvement accounting for 70% of the positive finding. They recommended frozen section analysis of biopsy specimen from the posterior bladder neck at the 4, 6 and 8 o'clock positions to identify residual benign tissue or prostate cancer before the vesicourethral anastomosis is performed. Licht et al. noted a positive bladder neck margin of 6.8% in 206 patients with bladder neck preservation, which was associated in all with higher grade tumor, more advanced local stage and other positive margin sites.[51] However, local recurrence and PSA failure rates were independent of bladder neck preservation or resection.

Since prostate cancer is sensitive to hormonal manipulation, surgeons have attempted to overcome the problem of positive surgical margins by advocating neo-adjuvant hormonal treatment for locally advanced tumors.[52] The development of gonadotropin-releasing hormone agonists and antiandrogens provided the possibility of reversible androgen blockade, which was given as 3-month pretreatment before radical prostatectomy. These agents made induction therapy possible without the cardiovascular risks associated with diethylstilbestrol or the disadvantages of orchiectomy. Many investigators have reported downstaging and even downgrading of the tumor using this approach with gonadotropin-releasing hormone analog, with or without and antiandrogen in randomized trials.

Both non-randomized and randomized studies have reported a decrease in positive surgical margins after hormonal pretreatment. Soloway et al. noted that the rate of positive surgical margins was six fold less in patients treated with hormonal therapy than in those who underwent only radical prostatectomy for clinical T2b disease.[53] Goldenberg et al. reported the results of the Canadian Urologic Oncology Group in patients with stage T1–T2 prostate cancer and similarly found that neoadjuvant hormonal therapy decreased the rate of positive surgical margins from 65% in the surgery group to 28% in the hormonal therapy group.[54] However, Aus reported that, even though early progression was delayed by approximately a year in the pretreatment group, at a median follow-up of 3 years, there was no difference in progression-free survival between the two treatment arms.[55]

IMPACT OF RACE ON POSITIVE SURGICAL MARGINS

African American men have been found to have a higher positive surgical margin rates compared to Caucasian men in several studies[23,56] but not others.[57] Shekarriz et al. investigated the differences between African American men and Caucasian men in terms of the site and multifocality of the positive surgical margins, and if race was an independent prognostic factor for disease-free survival in patients with positive surgical margins.[23] They found no significant difference in the frequency of multifocality of the positive surgical margins between the two races. Even though the positive surgical margins in African American men were located significantly more often at the base than their Caucasian counterparts, the location of the positive surgical margins did not impact on the disease-free survival between the two groups. In those with multifocal positive surgical margins, African American males had a worse disease-free survival compared to Caucasian males. They concluded that the prognostic value of multifocal positive margins appears to be race dependent. The same group also reported that extracapsular extension and positive surgical margins are significant predictors of outcome among African American men and White men with non-organ-confined prostate cancer but not in organ-confined cases.[58]

CONCLUSION

Positive surgical margins correlate well with preoperative serum PSA, clinical stage, biopsy and specimen Gleason score and tumor volume. The overall rate of positive surgical margin is around 28%, with apex or urethral margin being the most common sites of involvement. However, owing to the recent stage migration and improved surgical techniques, there has been a recent decline in positive surgical margin rates.

The presence of a positive margin is generally a poor prognostic sign, with biochemical or clinical recurrence noted in as many as 63% of patients within 5 years from the date of surgery. Positive surgical margin in the presence of seminal vesicle invasion also carries a poor prognosis compared to patients with a negative surgical margin status.

REFERENCE

1. Sakr WA, Wheeler TM, Blute M. Staging and reporting of prostate cancer – sampling of the radical prostatectomy specimen. *Cancer* 1996; 78:366–8.
2. McNeal JE, Villers AA, Redwine EA et al. Capsular penetration in prostate cancer: significance for natural history and treatment. *Am. J. Surg. Path.* 1990; 14:240.

3. Stamey TA. Evaluation of radical prostatectomy capsular margins of resection: the significance of margins designated as negative, closely approaching, and positive. *Am. J. Surg. Path.* 14:626.
4. Villers A. Extracapsular tumor extension in prostatic cancer: pathways of spread and implications for radical prostatectomy. In: Stamey TA (ed.) *Monographs in Urology*, Vol. 15, no. 4. Montverde, FL: Medical Directions Publishing, 1995.
5. Rosen MA, Goldstone L, Lapin S et al. Frequency and location of extracapsular extension and positive surgical margins in radical prostatectomy specimens. *J. Urol.* 1992; 148:331–7.
6. Stamey TA, Villers AA, McNeal JE et al. Positive surgical margins at radical prostatectomy: importance of the apical dissection. *J. Urol.* 1990; 143:1166–73.
7. Acerman DA, Barry JM, Wicklund RA et al. Analysis of risk factors associated with prostate cancer extension to the surgical margin and pelvic node metastasis at radical prostatectomy. *J. Urol.* 1993; 150:1845–50.
8. True LD. Surgical pathology examination of the prostate gland: practice survey by American Society of Clinical Pathologists. *Am. J. Clin. Pathol.* 1994; 102:572–9.
9. Watson RB, Civantos F, Soloway M. Positive surgical margins with radical prostatectomy: detailed pathological analysis and prognosis. *Urology* 1996; 48:80–8.
10. Weldon VE, Tavel FR, Neuwirth H et al. Patterns of positive specimen margins and detectable prostate specific antigen after radical perineal prostatectomy. *J. Urol.* 1995; 153:1565–9.
11. Voges GE, McNeal JE, Redwine EA. Morphologic analysis of surgical margins with positive findings in prostatectomy for adenocarcinoma of the prostate. *Cancer* 1992; 69:520–6.
12. Boccon-Gibod L, Ravery V, Vordos D et al. Radical prostatectomy for prostate cancer: the perineal approach increases the risk of surgically induced positive surgical margins and capsular incisions. *J. Urol.* 1998; 160:1383–5.
13. Jones EC. Resection margin status in radical retropubic prostatectomy specimens: relationship to type of operation, tumor size, tumor grade and local tumor extension. *J. Urol.* 1990; 144:89–93.
14. Catalona WJ, Smith DS. 5 year tumor recurrence rates after anatomical radical retropubic prostatectomy for prostate cancer. *J. Urol.* 1994; 152:1837–42.
15. Walsh PC, Partin AW, Epstein JI. Cancer control and quality of life following anatomical radical retropubic prostatectomy: results at 10 years. *J. Urol.* 1994; 152:1831–6.
16. Ohari M, Wheeler TM, Kattan MW et al. Prognostic significance of positive surgical margins in radical prostatectomy specimens. *J. Urol.* 1995; 154:1818–24.
17. Alasikafi NF, Brendler CB. Surgical modification of radical retropubic prostatectomy to decrease incidence of positive surgical margins. *J. Urol.* 1998; 159:1281–5.
18. Obek C, Sadek S, Lai S. Positive surgical margins with radical retropubic prostatectomy: anatomic site specific pathologic analysis and impact on prognosis. *Urology* 1999; 54:682–8.
19. Shah O, Malameed J, Lepor H. Analysis of apical soft tissue margins during radical retropubic prostatectomy. *J. Urol.* 2001; 165:1943–9.
20. Van den Ouden D, Bentvelsen FM, Boeve ER et al. Positive surgical margins after radical prostatectomy correlation with local recurrence and distant progression. *Br. J. Urol.* 1993; 72:489–94.
21. Weider JA, Soloway MS. Incidence, etiology, location, prevention and treatment of positive surgical margins after radical prostatectomy. *J. Urol.* 1998; 60:299–315.
22. Watson RB, Civantos F, Soloway MS. Positive surgical margins with radical prostatectomy: detailed pathological analysis and prognosis. *Urology* 1996; 48:80–8.
23. Shekarriz B, Tiguert R, Upadhayay J. Impact of location and multifocality of positive surgical margins on disease-free survival following radical prostatectomy: a comparison between African-American and white men. *Urology* 2000; 55:899–903.
24. Blute ML, Bostwick DG, Bergstralh EJ. Anatomic site-specific positive margins in organ-confined prostate cancer and its impact on outcome after radical prostatectomy. *Urology* 1997; 50:733–9.
25. Wood DP, Peretsman SJ, Seay TM. Incidence of benign and malignant prostate tissue in biopsies of the bladder neck after a radical prostatectomy. *J. Urol.* 1995; 154:1443–6.
26. Catalona WJ, Briggs SW. Nerve sparing radial prostatectomy: evaluation of results after 250 patients. *J. Urol.* 1990; 143:538–44.
27. Gomez CA, Soloway MS, Civantos F et al. Bladder neck preservation and its impact on positive surgical margins during radical prostatectomy. *Urology* 1993; 42:689–94.
28. McNeal JE, Villers AA, Redwine EA et al. Capsular penetration in prostate cancer: significance for natural history and treatment. *Am. J. Surg. Path.* 1990; 14:240.
29. Tigrani VS, Bhargava V, Shinohara K et al. Number of positive systematic sextant biopsies predicts surgical margin status at radical prostatectomy. *Urology* 1999; 54:689–93.
30. Epstein JI, Carmichael MJ, Partin AW et al. Is tumor volume an independent predictor of progression following radical prostatectomy? A multivariate analysis of 185 clinical stage B adenocarcinoma of the prostate with 5 years of follow up. *J. Urol.* 1993; 149:1478–1.
31. Ohari M, Abbas F, Wheeler TM et al. Pathological features and prognostic significance of prostate cancer in the apical section determined by whole mount histology. *J. Urol.* 1999; 161:500–4.
32. Tefilli MV, Gheiler EL, Tiguert R et al. Should Gleason score 7 prostate cancer be considered a unique grade category? *Urology* 1999; 53:372–7.
33. Epstein JI, Pizov G, Walsh PC. Correlation of pathologic findings with progression after radical retropubic prostatectomy. *Cancer* 1993; 71:3582–93.
34. Walsh PC, Partin AW, Epstein JI. Cancer control and quality of life following anatomical radical retropubic prostatectomy: results at 10 years. *J. Urol.* 1994; 152:1831–6.
35. Paulson DF. Impact of radical prostatectomy in the management of clinically localized disease. *J. Urol.* 1994; 152:1826–30.
36. D'Amico AV, Whittington R, Malkowicz SB et al. A multivariate analysis of clinical and pathological factors that predict for prostate specific antigen failure after radical prostatectomy for prostate cancer. *J. Urol.* 1995; 154:131–8.
37. Smith RC, Brender CG, Partin AW. Extended follow up of the influence of whole excision of the neurovascular bundle on prognosis in men with clinically localized prostate cancer (abstract 654). *J. Urol.* 1995; 153:392A.
38. Paulson DF, Moul JW, Walther PJ. Radical prostatectomy for clinical stage T1-T2N0M prostate adenocarcinoma: long term results. *J. Urol.* 144:1180–1.
39. Epstein JI. Incidence and significance of positive surgical margins in radical prostatectomy specimens. *Urol. Clin. North Am.* 1996; 23:651–63.
40. Epstein JI. Evaluation of radical prostatectomy capsular margins of resection: the significance of margins designated as negative, closely approaching, and positive. *Am. J. Surg. Path.* 1990; 14:240.
41. Blute ML, Bostwick DG, Seay TM et al. Pathologic staging of prostate cancer: impact of margin status. *Cancer* 1998; 82:902.
42. Catalona WJ, Stein AJ. Staging errors in clinically localized prostate cancer. *J. Urol.* 1982; 127:452–6.
43. Kupelian P, Katcher J, Levin H. Correlation of clinical and pathologic factors with rising prostate-specific antigen profiles after radical prostatectomy alone for clinically localized prostate cancer. *Urology* 1996; 48:249–60.
44. Partin AW, Pound CR, Clemens JQ et al. Serum PSA after anatomic radical prostatectomy. *Urol. Clin. North Am.* 1993; 20:713–25.
45. Ayala AG, Ro JY, Babaian R. The prostatic capsule: does it exist? Its importance in the staging and treatment of prostatic carcinoma. *Am. J. Surg. Path.* 1989; 13:21.
46. Fesseha T, Sakr W, Grignon D et al. Prognostic implications of a positive apical margin in radical prostatectomy specimens. *J. Urol.* 1997; 158:2176–9.

47. Partin AW, Borland RN, Epstein JI et al. Influence of wide excision of the neurovascular bundles on prognosis in men with clinically localized prostate cancer with established capsular penetration. *J. Urol.* 1993; 150:142–8.
48. Tefilli MV, Gheiler EL, Tiguert R et al. Prognostic indicators in patients with seminal vesicle involvement following radical prostatectomy for clinically localized prostate cancer. *J. Urol.* 1998; 160:802–6.
49. Stephenson RA, Middleton RG, Abbott TM. Wide excision (nonnerve sparing) radical retropubic prostatectomy using an initial perirectal dissection. *J. Urol.* 1997; 157:251–5.
50. Latif A. Preservation of bladder neck fibers in radical prostatectomy. *Urology* 1993; 41:566–7.
51. Licht MR, Klein EA, Tuason L et al. Impact of bladder neck preservation during radical prostatectomy on continence and cancer control. *Urology* 1994; 44:883–7.
52. Fair WR. Why not neo-adjuvant therapy for prostatic carcinoma? *Prog. Clin. Biol. Res.* 1991; 370:305.
53. Soloway MS, Sharifi R, Wajsman Z et al. for the Lupron Depot Neoadjuvant Prostate Cancer Study Group. Randomized prospective study comparing radial prostatectomy alone versus radical prostatectomy preceded by androgen blockade in clinical stage B2(T2bNxM0) prostate cancer. *J. Urol.* 1995; 154:424–8.
54. Goldenberg SL, Klotz LH, Srigley J et al. Randomized, prospective controlled study comparing radical prostatectomy alone and neoadjuvant androgen withdrawal in the treatment of localized prostate cancer. *J. Urol.* 1996; 156:873–7.
55. Aus G, Abrahamsson P, Ahlgren G et al. Hormonal treatment before radical prostatectomy: a 3 year follow up. *J. Urol.* 1998; 159:2013–17.
56. Powell IJ, Heilbrun LK, Sakr W. The predictive value of race as a clinical prognostic factor among patients with clinically localized prostate cancer: a multivariate analysis of positive surgical margins. *Urology* 1997; 49:726–31.
57. Witte MN, Kattan MW, Albani J et al. Race is not an independent predictor of positive surgical margins after radical prostatectomy. *Urology* 1999; 54:869–74.
58. Powell IJ, Banerjee M, Novallo M et al. Prostate cancer biochemical recurrence stage for stage is more frequent among African-American than White men with locally advanced but not organ confined disease. *Urology* 2000; 55:246–51.

Bloodless Surgery and Radical Retropubic Prostatectomy

Chapter 36

Ciril J. Godec

Department of Urology, Long Island College Hospital, Brooklyn, New York, USA

INTRODUCTION

The term bloodless is really a misnomer. Today bloodless surgery means performing surgery without the use of blood products. Only a few years ago many experienced urologists considered blood transfusion far less risky than almost any major surgical procedure and thus had a cavalier approach to transfusion. The old concept, still embraced by many urologists, that allogeneic blood is a safe and effective therapeutic modality, has been challenged by numerous reports of transfusion reactions, disease transmission and immunomodulation related to blood elements.

Radical retropubic prostatectomy (RRP) traditionally has been associated with significant blood loss. In coping with this blood loss the urological surgeon has three options: meticulous surgery with precise knowledge of surgical anatomy; augmentation of the oxygen-carrying capacity of red blood cells; and transfusion to compensate for blood loss.

Transfusion can be either allogeneic (homologous), i.e. ABO compatible from another human, or autologous, which is the patient's own blood. Autologous blood can be prepared as a predonation 2–3 weeks before surgery; as normovolemic hemodilution, where the anesthesiologist immediately before surgery removes blood from the patient, replacing the volume with crystalloids or colloids, and then at the end of the procedure reinfuses the initially removed blood into the patient; or via cell saver, where lost blood is collected, filtered and then reinfused back into the patient.

SURGICAL TECHNIQUES AND BLOOD LOSS

Meticulous surgery is by far the best option to minimize blood loss. The urological surgeon should be thoroughly familiar with the details of the vascular and neurological anatomy of the area. Historically the surgical anatomy of the area has been poorly understood. As late as 1998, Patrick Walsh, the pioneer of the modern era of prostate cancer surgery, declared 'It is humbling to realize that even today the basic anatomy may not be known or understood'.[1]

Beneventi and Noback were the first researchers to perform an anatomical study of the venous system of the prostate.[2] Through routine postmortem examinations they documented the interconnecting branches of the deep dorsal vein of the penis. Ten years later Weyrauch described that these interconnecting venous plexuses have free anastamoses with the pubic, pudendal, deep epigastric and obturator veins.[3] Confusion in the nomenclature of the pelvic vascular complex clouds our knowledge of the vascular anatomy of this area.[4] Even the presence or absence of the valvular system in the venous system in the pelvis is controversial. Some authors consider the venous system valve-free,[5] yet others have found valves present on venographic studies.[6] Reiner and Walsh were the first to use the knowledge of the venous anatomy in the operating room.[7] No longer was the supraurethral complex transected without the control of bleeding. Their technique, in which the puboprostatic ligaments (PPL) were transected near the pubic bone, significantly minimized blood loss, as transection at this level usually does not cause bleeding. Albers et al. did not find any vessels of significance inside PPL on histological examinations.[8] In our own study we found veins immediately under PPL in some patients and, because we cannot predict the specific anatomy of a particular patient, we ligate the supraurethral complex *en bloc*, including PPL, in all patients. The PPL on microscopic sections contain collagen and some smooth muscle fibers, which retract after transection, pulling the urethra behind the pubic bone and thus contributing to the continence mechanisms after surgery.

Blood loss during RRP was significant, especially in the early era before Walsh made his pioneering contribution to this surgery. Significant bleeding requiring blood transfusion is still the major intraoperative complication during RRP.[9] The main source of bleeding still is the division of supraurethral complex. Bleeding during the pelvic lymph node dissection is minimal but can be significant if the branches of the hypogastric vessels are transected. If the opening of the endopelvic fascia is too close to the prostate, the lateral part of Santorini's plexus can be injured. Most urological surgeons today perform the nerve-sparing RRP, which is associated with a higher blood loss than conventional surgery without preservation of the neurovascular bundle.[10] Although blood loss is diminishing, today it still is significant **(Table 36.1).**

With increased experience with this procedure, blood loss is diminishing.[17,19] Nevertheless, some surgeons still report a significant blood loss despite an otherwise overall excellent outcome.[11]

My interest in bloodless surgery was triggered because our hospital is located in the vicinity of the Watch Tower, which is the world Headquarters of Jehovah's Witnesses. Up to 10 years ago, I was routinely refusing to perform RRP on the Witness patients – as was everybody else in the USA. With careful surgical technique and some modifications of RRP, we can now safely operate Witness patients.

Minimizing the blood loss should start before the surgery. The patient awaiting RRP should stop all medications which could potentially increase bleeding during surgery – aspirin, non-steroidal anti-inflammatory drugs, coumadin, heparin and thrombolithic agents. Preoperative phlebotomies should be restricted and the use of pediatric blood samples encouraged. Strict adherence to the sound surgical principles of bleeding control is of paramount importance. The use of locally acting agents that promote clotting is helpful – topical thrombin, fibrin glue and collagen. Sometimes fibrinolytic drugs are advisable, such as tranexemic acid, aprotinin, aminocaproic acid and desmopressin acetate.

The best tool to prevent transfusion is to minimize bleeding by precise and anatomic surgery. Most bleeding during the RRP is venous. Thus temporary compression is the best tool to prevent bleeding. We utilize the standard Walsh[9] technique with some modifications.[21] During ligation of the dorsal venous complex, we ligate the PPL together with the dorsal venous complex. After incision of the endopelvic fascia, we extend the incision to the lateral prostatic fascia, which gives us access to the neurovascular pedicles, which we carefully dissect posteriorly towards the rectum. During dissection of the lateral pedicles, we ligate, clip or ligature every bleeding vessel immediately. Speedy control of every bleeding vessel is probably the major contributing factor enabling us to minimize blood loss. During dissection of the base of the prostate from the bladder neck, we dissect it in a mushroom fashion along an avascular plane,[22] which exists between the prostate base and the bladder neck.

Table 36.1. Blood loss during RRP.

Authors	Blood loss (ml)
Catalona et al.[11]	1500
Keetch et al.[12]	1500
Hammerer et al.[13]	1200
Igel et al.[14]	1018
Hautman et al.[15]	900
Lerner et al.[16]	844
Zincke et al.[17]	600
Koch et al.[18]	579
Leandri et al.[19]	530
Guilloneau[20]	290

RRP today remains one of the bloodiest surgical procedures in urology. Despite the general familiarity with the surgical anatomy of the prostate, blood loss remains significant.[23] The temporary clamping of the internal hypogastric arteries diminished blood loss for some authors.[24] They used the bulldog clamps on hypogastric arteries in order to decrease blood perfusion of the prostate during surgery. Some other researchers in the field could not document any difference in blood loss with the bulldog clamp placement.[25] Although the operating field becomes clearer when bulldog clamps are applied, a statistically significant decrease in blood loss did not result. Kavoussi et al. reported that the blood loss in the group with hypogastric artery occlusion was 1420 ml (range 200–2500 ml) and in the group without occlusion was 1605 ml (range 250–3800 ml).[25] There was no statistically significant difference between the two groups. They concluded that 'hemorrhage remains a potential problem even with the meticulous care being exercised during the dissection'.

Rainwater and Segura compared two groups of patients, one with non-ligation and the other with ligation of the deep dorsal venous complex.[26] The group with ligation was further subdivided into nerve-sparing and non-nerve-sparing groups. The non-ligation group had a mean loss of 1262 ml (range 150–7500 ml), the ligated group a loss of 1020 ml (range 100–4320 ml), a statistically significant difference. Although the nerve-sparing procedure usually leads to a higher blood loss, the Rainwater study found a surprisingly lower blood loss in the group with the nerve-sparing procedure than that with a non-nerve-sparing procedure. Walsh in the editorial comment to the same article reported an average blood loss of 1400 ml without temporary occlusion of the hypogastric arteries and 1175 ml with occlusion of arteries.

Hrebinko and O'Donnell reported a mean blood loss of 982 ml (range 400–2200 ml) with ligation of the deep dorsal vein complex.[27] They developed a space between the urethra and the deep dorsal venous complex large enough to insert an index finger up to the proximal interphalangeal joint. They used a high coagulating current (60–80 W, which is much higher than normal). In most of their cases, the PPL were not ligated separately, they were ligated *en bloc* with the dorsal vein complex.

Another attempt to minimize bleeding was Berger and Ireton's combined infrapubic and retropubic ligation of the dorsal vein of the penis.[28] With this method, they ligate the dorsal vein of the penis before they divide the PPL. The novelty behind this method is prevention of bleeding by inadvertent injury of the dorsal vein of the penis while transecting the PPL.

Koch et al. studied the effect of sequential compression devices on intraoperative blood loss during radical prostatectomy and found no significant difference in either the mean

(1241 ml without and 930 ml with sequential compression devices, $P = 0.164$) or median (885 ml and 800 ml) blood loss.[29] Their conclusion was that sequential compression devices might be better indicated for preventing pulmonary embolus than for diminishing blood loss. Other authors came to different conclusions. Strup et al. reported that the use of sequential compression devices was associated with increased blood loss during radical pelvic surgery.[30] The literature regarding blood loss with use of sequential compression devices is inconclusive. Most reports recommend the use of this technique more for the prevention of deep venous thrombosis, which could be a potential complication of radical pelvic surgery, than for minimizing blood loss.

As urological surgeons, we should relearn the Halsteadian principles of gentle tissue handling, anatomic dissection and pre-emptive control of blood vessels within the supraurethral complex and along the lateral prostatic pedicles, as this provides still the best approach to minimizing blood loss. The latest surgical techniques such as Ligasure, harmonic scalpel, gamma knife and argon beam coagulator can all be utilized if the urologist is well versed in their use.

Laparoscopic technique to perform radical prostatectomy was first utilized 10 years ago by Schuessler et al.[31] It took another 6 years for the authors to publish an additional eight cases.[32] In the second report, the authors did not encourage the use of this technique and concluded 'laparoscopy is not an efficacious surgical alternative to open prostatectomy for malignancy'. Nevertheless, European urologists continued to explore this modality and subsequently developed a feasible, reproducible and teachable technique.[20,33,34] Although we do not know the long-term outcome of this procedure, especially regarding cancer control, we do know that in most series this technique has a much lower blood loss than open surgery.

In the Mountsouris experience, the transfusion rate was 4.7% and steadily decreased with experience.[20] For the first 50 patients, the transfusion rate was 14% but fell to 3.3% for the next 330 patients, and then to only 2.5% for the last 200 patients. In this series the reduction in transfusion rate parallels their blood loss; their overall average blood loss was 380 ml but for the last 350 patients was 290 ml.

Most bleeding involved in RRP is venous in nature. The pneumoperitonium used with the laparoscopic approach, with its pressure of 12 mmHg or less, contributes to diminished bleeding, especially during the dissection of lateral pedicles, where most venous oozing occurs during open surgery. Prompt coagulation with bipolar current during laparoscopy is necessary so that all bleeding vessels are promptly occluded, because bleeding interferes with clear vision and could make this procedure risky and dangerous. Very likely laparoscopic radical prostatectomy is here to stay, unless 'early' long-term results show that cancer control is significantly less than with open surgery.

ANESTHESIA AND BLEEDING

Does the type of anesthesia have some impact on the blood loss during radical prostatectomy? Many authors have explored this topic and again results are inconclusive. Peters and Walsh, in comparing the use of regional anesthetic techniques (spinal or epidural) to general anesthesia, found that regional anesthesia may slightly reduce the need for blood replacement, but this reduction was not statistically significant.[24] In a more recent study, Stevens et al. demonstrated that the use of combined thoracic epidural anesthesia and light general anesthesia led to significantly less (mean 34%) intraoperative blood loss compared with patients receiving general anesthesia.[35] Their explanation was twofold. First, patients on epidural anesthesia breathe spontaneously and may have lower venous pressure, which may diminish intraoperative bleeding. Second, patients on epidural anesthesia may have more intense sympathetic blockade of the splanchnic circulation, which can lead to lower venous pressure and thus decreased blood loss. Shir et al. divided their patients into three groups: epidural, combined epidural and general, and general anesthesia.[36] The mean blood loss in the epidural group (1490 ml) was significantly less than in both the combined (1810 ml) and general (1940 ml) groups. However, their conclusion was different: in their opinion, the epidural anesthesia did not reduce bleeding but general anesthesia increased blood loss. This study was small and did not take into consideration the duration of surgery and initial hematocrit, both of which can have a significant impact on blood loss and the requirement for transfusion.

The preference of the urologist usually determines anesthesia. Sometimes the patient's history is important in the choice of anesthesia. A history of coagulopathy or any type of hematologic disorder would prevent the use of regional anesthesia. Severe osteoarthritis or severe spinal disorder would also dictate against the use of epidural anesthesia.

BLOOD TRANSFUSION

Allogeneic transfusion is the most frequent form of blood replacement. Patients receiving blood transfusion face approximately three in 10 000 risk of contracting a serious or fatal transfusion-related disease.[37] Infectious complications are perceived to be the most dangerous risk. In the public's view, human immunodeficiency virus (HIV) represents the most feared complication of blood transfusion; however, transfusion-acquired HIV accounts for less than 20 cases per year.[38] The great majority (90%) of post-transfusion hepatitis is caused by the hepatitis C virus, and only a small proportion (2%) by the hepatitis B virus. The others are due to a variety of viruses, such as Barr virus or hepatitis A virus.[39]

There are also several non-infectious complications of allogeneic transfusion. The most significant are immunosuppression, alloimmunization and graft versus host diseases. The most significant for our discussion is immunosuppression, which can have both positive and negative consequences. Increased risk of reactivation of latent viruses, cancer recurrence and metastases on one side and improved survival in allograft transplant on the other side imply that transfusion triggers some significant immunomodulatory reactions. Allogeneic transfusion increases humoral immunity but at same time decreases cell-mediated immunity. The beneficial effect is exploited in transplant surgery where immunosuppression can improve allograft survival.[40] Already in 1973, Opelz et al. in their seminal paper documented improved renal transplant survival.[41] Peters et al. reported on the reduced recurrence of Crohn's disease in patients receiving multiple blood transfusions.[42]

The main negative impact of transfusion immunosuppression is cancer recurrence and subsequent decreased survival of patients who received transfusion. Bordin et al. thought that the immunosupressive effect of allogeneic transfusion is related to the exposure to leukocytes.[43] Some studies supported the suggestion that leukocyte reduction might lead to a lower rate of postoperative infections[44] but some found no difference in cancer recurrence or rate of infection in transfused versus non-transfused patients.[45] One group of researchers reported that the operative blood loss and not the type of blood (allogeneic versus autologous) was related to decreased recurrence-free survival after RRP.[46] Their conclusion suggested that the operative events leading to transfusion are more significant than immunological effects of transfusion. The operative blood loss was the most significant predictor of disease progression.

DECISION-MAKING PROCESS FOR TRANSFUSION

The main purpose of red blood cell transfusion is the augmentation of oxygen delivery and prevention of tissue hypoxia.[47] The decision for transfusion should be made only after exploring other options to increase oxygen delivery to the patient and when the patient's cardiopulmonary mechanisms are insufficient to compensate for blood loss. The urologist should understand the physiology of anemia and its consequences in order to make a comprehensive transfusion decision. The heart provides the primary response to anemia by increasing cardiac output through increased stroke volume. In patients with coronary artery disease, the heart is not able to provide for increased body oxygen delivery. If the demand for oxygen continues, the ischemic heart converts to an anaerobic metabolism, which can lead to myocardial infarction.

Transfusion decisions should be individualized and made jointly between the urologist and the anesthesiologist. The laboratory values of hemoglobin are less critical than the careful assessment of the patient preoperatively. It is true that some patients bleed more than others due to their specific physiology, but it is also true that some surgeons have higher blood loss than others due to their experience and their level of surgical skill.

ALTERNATIVES TO ALLOGENEIC TRANSFUSION

The principal advantage of autologous transfusion is the avoidance of complications associated with allogeneic blood transfusion. In clinical practice today, we utilize three forms of autologous transfusions: predonation of autologous blood; acute normovolemic hemodilution; and reinfusion of blood via cell saver.

Preoperative blood donation is still the standard of care in many institutions in the USA today. The first autologous transfusion was described in 1874.[48] With this technique, patients get their own blood, which is the safest form of transfusion. With predonation, patients are reducing their exposure to allogeneic blood. During RRP without predonation, the patient allogeneic transfusion rate is 60–70%,[17] but only 5–20% of patients who predonated blood required allogeneic transfusion. Nevertheless, predonation recently has undergone some critical reappraisal. In many centers, blood loss during RRP is steadily decreasing. Improved surgical technique has further contributed to diminished blood loss to the point that autologous predonation has become unnecessary.[49] Also reappraisal of the trigger points for transfusion has diminished the overall need for blood during RRP. This surgery is usually performed in patients whose life expectancy is at least 10 years and significant comorbidity is infrequent. For sicker patients, we use alternative treatments, such as external beam radiation, seed implant or cryosurgery. One recent study showed no difference in homologous blood transfusion based on preoperative autologous donation status.[50] In their report, the recent rate of homologous transfusion was less than 3% overall with no difference between donors and nondonors. The American Society of Anesthesiologists Task force on Blood Component Therapy suggested that transfusion is usually indicated when hemoglobin is less than 6 g/ml.[51]

Autologous blood transfusion in urology is steadily decreasing. Goad et al. identified three phases regarding the use of allogeneic transfusion for RRP.[49] In the initial phase between 1983 and 1988, autologous blood donation was used preoperatively in only 13%. The allogeneic transfusion rate was 62% for non-donors and 46% for donors. In the second phase from 1989 to 1991, autologous donation increased to 81% and the allogeneic transfusion decreased to 37% for non-donors and to 7% for donors. During the third phase from 1991–1992, the homologous transfusion rate further decreased to 11% for non-donors and to 4% for donors. The transfusion triggers were lowered to less than 7 mg/dl for hemodynamically stable patients.

Similar results in the use of allogeneic transfusion was reported by Toy et al.[52] who found for patients who underwent RRP in a period from 1987 to 1991 rates of 66% for non-donors and 20% for donors. Koch and Smith reported even further decrease in the use of allogeneic transfusion.[18] Their average intraoperative blood loss was only 579 ml and their allogeneic transfusion rate of only 2.4%. Their experience showed that 98% of patients can undergo RRP without transfusion. The discharge hematocrit in their patients was 33%. None of their patients had any ischemic cardiovascular or cerebrovascular accident in the postoperative period. Their conclusion was that the elimination of autologous blood donation was a cost-effective measure in their hands. Shekarriz et al. recently recommended against the use of autologous blood donation because the need for allogeneic transfusion was less than 1%.[10]

Autologous transfusion is much less risky than allogeneic but it is not risk-free. The process of donation itself can be complicated for some patients. In one report, the incidence of vasovagal reaction was 2–5%.[53] Incorrect labeling of donated blood can lead to clerical error not only for allogeneic but also for autologous blood. Infection of stored blood is another possible risk. Collecting and storing of autologous blood is more expensive than for allogeneic blood. Autologous blood does not need to undergo the same degree of screening and testing and most of the unused blood is discarded.

Many of the reported series on blood loss are based on a single surgeon or a single institution, and broad extrapolations are not always applicable to the general urological community. Even in the most experienced hands, blood loss can be high. In those cases, two or three units of autologous blood are not sufficient and allogeneic transfusion will be necessary. In the near future, at least for urological oncology surgery, autologous transfusion will no longer be necessary.

In acute normovolemic hemodilution (ANH) blood is removed immediately before surgery and replaced by colloids or crystalloids. Although the patient's hematocrit is lowered, more colloids and crystalloids and smaller amounts of red blood cells are lost during the surgery.[54] This method represents a cost-effective alternative to predonation of autologous blood.[55] When acute normovolemic hemodilution is used, the blood never leaves the operating room, thereby completely eliminating potential clerical error; it is effective, simple and convenient. It is the only transfusion technique that provides fresh, whole blood for immediate use in the operating room. It probably is the most frequently underused form of autologous transfusion because it is perceived by some urologists as a technique that prolongs operating time, requiring additional personnel and costly monitoring.[56] During the initial stage of RRP, there is no significant blood loss, especially if the patient is first undergoing pelvic lymphadenectomy. This gives the anesthesiologist the time to complete hemodilution after induction of anesthesia. Atallah et al. documented that hemodilution could be performed in awake or anesthetized patients without jeopardizing hemodynamic stability.[57] In a study by Monk et al. the authors found that the central venous pressure remained normal throughout the hemodilution.[55] There was no perioperative cardiovascular morbidity. ANH was well tolerated even in elderly patients. The duration of hospital stay was shortened. ANH offers extra convenience for patients, as they do not need to come to the hospital for autologous predonation.

Cell savers have been used in surgical procedures since 1970. In cell saver the blood salvaged during surgery is passed through a filtration and centrifuge system, which separates functional red blood cells, which are then reinfused into the patient. Until recently, this technique has been avoided in oncological surgery for fear of reinfusing malignant cells. Many reports indicate that reinfusion with cell saver is safe and does not lead to increased rate of malignancy.[58,59] Recently introduced leukocyte depletion filters eliminate residual malignant cells together with leukocyte-associated infectious viruses.[60]

Blood can be collected even postoperatively. Significant bleeding postoperatively is rare, especially bleeding requiring transfusion. A small hematoma can be treated expectantly. A larger one could cause some anatomic deformity in the anastomotic area between bladder and urethra, and should be explored as in any other surgery. Hedican and Walsh reported on seven patients with delayed postoperative bleeding in their series of 1350 patients.[61] Four were treated surgically and did well. The remaining three were treated conservatively by draining their pelvic hematoma through urethral anastomosis; all three developed bladder neck contracture and urinary incontinence persisted in two patients.

USE OF PHARMACOLOGICAL AGENTS

If indicated, the use of locally acting agents (fibrin glue, collagen, topical thrombin) that promote clotting mechanisms might help to stop bleeding.[62] In rare instances and if the urologist is familiar with the agents, antifibrinolytic drugs may reduce blood loss. In his meta-analysis, Fremes et al. documented that aprotinin, desmopressin, epsilon aminocaproic acid and tranexemic acid are the most frequently used agents in the prevention of postoperative bleeding.[63] The concern that the prophylactic use of hemostatic drugs could create additional morbidity by creating hypercoagulable conditions is not identified in this meta-analysis. Routine use of these agents is not advisable, but for patients who have increased risk of bleeding, their use might be clinically beneficial.

Epoetin alfa is today the most frequently used in radical prostatectomy patients. There are many reports about the administration of Epoetin alfa in patients undergoing major elective surgery. Atabek et al. reported on two groups of patients, one treated with Epoetin alfa and intravenous iron, and the other with intravenous iron alone.[64] All were Jehovah's Witness patients with severe postsurgical anemia (hematocrit less than 25%). Hematocrit 1 week later was significantly higher in the Epogen group (19.3% vs. 12.5%; P 0.0005). Hematocrit continued to rise during the second week of Epogen use. Some other researchers documented similar results with Epogen treatment on patients with severe anemia.[65] Monk et al. were the first to report on the use of Epoetin alfa in RRP patients.[66] In their study, the Epoetin alfa increased the mean preoperative value of hematocrit from 43% to 47%. Chun et al. divided their patients into two groups, one with Epoetin alfa and the other with autologous blood donation.[67] They compared the use of allogeneic blood transfusion and the cost. The use of allogeneic blood was identical in the two groups: the cost for Epoetin group was $540 and for autologous blood group was $657. Their conclusion was that Epoetin alfa offers patients greater convenience and less time commitment. One theoretical concern of Epoetin therapy is hypercoagulability, which could lead to an increased risk of stroke. In a study by Erikssen et al., the researchers monitored 2014 patients and documented an increased risk of cardiovascular mortality only when hematocrit values were above 53%.[68] Epoetin alfa significantly reduces the exposure to allogeneic blood and the associated risk.

CONCLUSION

Radical retropubic prostatectomy used to be, and for some urologists still is, one of the bloodiest surgeries performed in urology today. A general urologist whose practice includes urological oncology performs RRP more frequently than any other open surgery. The blood loss and transfusion practice vary enormously from urologist to urologist. Our efforts should be twofold: minimizing blood loss, and avoiding transfusion. By steadily improving our surgical technique and surgical skills to the extent humanly possible in every surgery, we might have the most significant impact on the use of transfusion. We should know anatomical details; every bleeding vessel should be immediately controlled. Less bleeding means better visibility during surgery, open or laparoscopic, and very likely better cancer surgery with a lower rate of incontinence and impotence. We should familiarize ourselves with the latest technological innovations, which can help us to diminish blood loss. We should be open-minded and be able to discern a mere technological gimmick from a clinically useful new tool.

Transfusion in essence is a blood transplant. We should treat is as such. We should be constantly aware of the risks of transfusion and we should discuss them honestly with our patients. We should respect our patient's decision if they refuse blood for religious or any other reason.

Even today, few urologists would agree to perform RRP if sufficient blood components were not available. It is comforting to know that transfusion is available when we operate. Sometimes, however, we will not have this option and the patient still wants his cancer out. Many urologists today operate with a very low blood loss, and some with minimal or close to zero transfusion rate. All other options should be explored before the decision to transfuse is made. Very likely, if the FDA would have to approve blood today as a new medicine, it probably would not pass the scrutiny of their standards.

REFERENCES

1. Walsh PC. Anatomic radical prostatectomy: evolution of the surgical technique. *J. Urol.* 1998; 160:2418–24.
2. Beneventi F, Noback GJ. Distribution of the blood vessels of the prostate gland and urinary bladder: application to retropubic prostatectomy. *J. Urol.* 1949; 62:663–71.
3. Weyrauch HM. *Surgery of the Prostate*, pp. 29–32. Philadelphia: W. B. Saunders Co. 1959.
4. Clegg EJ. Arterial supply of the human prostate and seminal vesicles. *J. Anat.* 1955; 80:209–16.
5. Abeshouse BS, Ruben ME. Prostatic and periprostatic phlebography. *J. Urol.* 1952; 68:640–6.
6. Fitzparick TJ. Venography of the deep dorsal venous and valvular systems. *J. Urol.* 1974; 111:518–20.
7. Reiner WG, Walsh PC. An anatomical approach to the surgical management of the dorsal vein and Santorini's plexus during radical retropubic surgery. *J. Urol.* 1979; 121:198–200.
8. Albers DD, Faulkner KK, Cheatham WN et al. Surgical anatomy of the pubovesical (puboprostatic) ligaments. *J. Urol.* 1973; 109:388–92.
9. Walsh PC. Anatomic radical retropubic prostatectomy. In: Walsh PC, Retik AB, Vaughan ED et al. (eds) *Campbell's Urology*, pp. 2565–88 Philadelphia: W.B. Saunders, 1998.
10. Shekarriz B, Upadhyay J, Wood DP. Intraoperative, perioperative, and long-term complication of radical prostatectomy. *Urol. Clinics North Am.* 2001; 28:639–53.
11. Catalona WJ, Carvalhal GF, Mager DE et al. Potency, continence and complication rate in 1,870 consecutive radical retropubic prostatectomies. *J. Urol.* 1999; 162:433–8.
12. Keetch DW, Andriole GL, Catalona WJ. *Complications of Radical Retropubic Prostatectomy, Lesson* 6, pp.46–51 Howtow, TX: AUA Update Series, 1994.
13. Hammerer P, Hubner D, Gonnerman D et al. [Perioperative and postoperative complications of pelvic lymphadenectomy and radical prostatectomy in 320 consecutive patients]. *Urologe A* 1995; 34:334–42.
14. Igel TC, Barret DM, Segura JW et al. Perioperative and postoperative complications from bilateral pelvic lymphadenectomy and radical retropubic prostatectomy. *J. Urol.* 1987; 137:1189–91.
15. Hautman RE, Sauter TW, Wanderoth UK. Radical retropubic prostatectomy: morbidity and urinary continence in 418 consecutive cases. *Urology* 1994; 43:47–51.
16. Lerner SE, Blute ML, Lieber MM et al. Morbidity of contemporary radical prostatectomy for localized prostate cancer. *Oncology (Huntingt.)* 1995; 9:379–82.
17. Zincke H, Bergstrahl EJ, Blute ML et al. Radical prostatectomy for clinically localized prostate cancer: long-term of 1143 patients from a single institution. *J. Clin. Oncol.* 1994; 12:2254–62.
18. Koch OK, Smith Jr JA. Blood loss during radical retropubic prostatectomy: is preoperative autologous blood donation indicated? *J. Urol.* 1996; 156:1077–9.
19. Leandri P, Rossignol G, Gautier JR et al. Radical retropubic prostatectomy: morbidity and quality of life. Experience with 620 consecutive cases. *J. Urol.* 1992; 147:883–7.
20. Guillonneau B, Valencien G. Laparoscopic radical prostatectomy: the Mountsouris technique. *J. Urol.* 2000; 163:418–22.
21. Godec C, Plauker M, Rudberg A et al. Radical retropubic prostatectomy with minimal blood loss. *Curr. Surg. Techn. Urol.* 1995; 8, issue 4.
22. Meyer PP. Practical surgical anatomy for radical prostatectomy. *Urol. Clinics North Am.* 2001; 28:473–90.
23. Walsh PC. Editorial comment. *J. Urol.* 1991; 146:65.
24. Peters CA, Walsh PC. Blood transfusion and anesthetic practices in radical retropubic prostatectomy. *J. Urol.* 1985; 134:81–3.
25. Kavoussi LR, Myers JA, Catalona WJ. Effect of temporary occlusion of hypogastric arteries on blood loss during radical retropubic prostatectomy. *J. Urol.* 1991; 146:362–5.
26. Rainwater LM, Segura JW. Technical consideration in radical retropubic prostatectomy: blood loss after ligation of dorsal venous complex. *J. Urol.* 1990; 143:1163–5.
27. Hrebinko RL, O'Donnell WF. Control of the deep dorsal venous complex in radical retropubic prostatectomy. *J. Urol.* 1993; 149:799–800.
28. Berger RE, Ireton R. Combined infrapubic and retropubic ligation of the dorsal vein of the penis during radical retropubic surgery. *J. Urol.* 1983; 130:1107–9.
29. Koch MO, Brandell DL, Smith A Jr. The effect of sequential compression devices on the intraoperative blood loss during radical prostatectomy. *J. Urol.* 1994; 152:1178–9.
30. Strup SE, Gudziak M, Mulholland SG, Gomella LG. The effect of intermittent pneumatic compression devices on intraoperative blood loss during radical prostatectomy and radical cystectomy. *J. Urol.* 1993; 1560:1176–8.
31. Schluesser W, Kavoussi LR, Clayman R et al. Laparoscopic radical prostatectomy: initial case report. *J. Urol.* 1992; 4:246A.
32. Schluesser W, Schulam P, Clayman R et al. Laparoscopic radical prostatectomy: initial short-tem experience. *Urology* 1997; 50:854–7.
33. Abbou CC, Salomon L, Hoznek A et al. Laparoscopic radical prostatectomy: preliminary results. *Urology* 2000; 55:630–4.
34. Rassweiler J, Siemon O, Abdel Salam Y et al. Laparoscopic radical prostatectomy: the Heilbron experience. *J. Endourol.* 1999; 13(Suppl. 1):A46.

35. Stevens RA, Mikat-Stevens M, Flanigan R et al. Does the choice of anesthetic technique affect the recovery of bowel function after radical prostatectomy? *Urology* 1998; 52:213–18.
36. Shir Y, Raja SN, Frank SM et al. Intraoperative blood loss during radical retropubic prostatectomy: epidural versus general anesthesia. *Urology* 1994; 45:993–9.
37. Dodd RY. The risk of transfusion-transmitted infection. *N. Engl. J. Med.* 1992; 327:419–21.
38. Selik RM, Ward JW, Buehler JW. Trends in transfusion-associated acquired immune deficiency syndrome in the United States, 1982 through 1991. *Transfusion* 1993; 33:890–3.
39. Zuckerman AJ. The new GB viruses. *Lancet* 1995; 345:890.
40. Lagaaij EL, Henneman IP, Ruigrok M et al. Effect of one-HLA-DR-antigen-matched and completely HLA-DR-mismatched blood transfusion on survival on heart and kidney allografts. *N. Engl. J. Med.* 1989; 321:701–5.
41. Opelz G, Sengar DPS, Mickey MR et al. Effect of blood transfusion on subsequent kidney transplants. *Transplant Proc.* 1973; 5:253–9.
42. Peters WR, Fry RD, Fleshman JW et al. Multiple blood transfusion reduce the recurrence rate of Crohn's disease. *Dis. Colon Rectum* 1989; 32:749–53.
43. Bordin JO, Heddle NM, Blachman MA. Biologic effects of leukocytes present in transfused cellular blood products. *Blood* 1994; 84:1703–21.
44. Heiss MM, Mempel W, Delanoff C et al. Blood transfusion-modulated tumor recurrence: first results of a randomized study of autologous versus allogeneic blood transfusion in colorectal cancer surgery. *J. Clin. Oncol.* 1994; 12:1859–67.
45. Busch ORC, Hop WCJ, Hoynck van Papendrecht MAW et al. Blood transfusion and prognosis in colorectal cancer. *N. Engl. J. Med.* 1993; 328:1372–6.
46. Oefelein MG, Colangelo LA, Rademaker AW et al. Intraoperative blood loss and prognosis in prostate cancer patients undergoing radical retropubic prostatectomy. *J. Urol.* 1995; 154:442–7.
47. American College of Physicians. Practice strategy for elective red blood cell transfusion. *Ann. Intern. Med.* 1992; 116:403–6.
48. Highmore W. Practical remarks on overlooked source of blood supply for transfusion in post-partum hemorrhage. *Lancet* 1874; I:89.
49. Goad JR, Eastham JA, Fitzgerald KB et al. Radical retropubic prostatectomy: limited benefit of autologous blood donation. *J. Urol.* 1995; 154:2103–9.
50. Goldschlag B, Afzal N, Carter HB et al. Is preoperative donation of autologous blood rational for radical retropubic prostatectomy? *J. Urol.* 2000; 164:1968–72.
51. Practice guidelines for blood component therapy; a report by the American Society for Anesthesiologists Task force on Blood Component Therapy. *Anesthesiology* 1996; 84:732–47.
52. Toy PT, Menozzi D, Strauss RG et al. Efficacy of preoperative donation of blood for autologous use in radical prostatectomy. *Transfusion* 1993; 33:721–4.
53. Goh M, Kleer CG, Kielczewski P et al. Autologous blood donation prior to anatomical radical retropubic prostatectomy: is it necessary? *Urology* 1997; 49:569–74.
54. Goodnough LT, Grishaber JE, Monk TG et al. Acute preoperative hemodilution in patients undergoing radical prostatectomy: a case study analysis of efficacy. *Anesth. Analg.* 1994; 78:932–7.
55. Monk TG, Goodnough LT, Birkmeyer JD et al. Acute normovolemic hemodilution is a cost-effective alternative to preoperative autologous blood donation in patients undergoing radical retropubic prostatectomy. *Transfusion* 1995; 35:559–65.
56. Stehling L, Zauder HL. Controversies in transfusion medicine. Perioperative hemodilution. *Transfusion* 1994; 34:265–8.
57. Atallah MM, Abdelbaky SM, Said MM. Does timing of hemodilution influence the stress response and overall outcome? *Anesth. Analg.* 1993; 76:113–17.
58. Fong J, Gurewitsch ED, Kump L et al. Clearance of fetal products and subsequent immunoreactivity of blood salvaged at cesarean delivery. *Obstet. Gynecol.* 1999; 93:968–72.
59. Rebarber A, Lonser R, Jackson S et al. The safety of intraoperative autologous blood collection and autotransfusion during cesarean section. *Am. J. Obstet. Gynecol.* 1998; 179:715–20.
60. Karczewski DM, Lema MJ, Glaves D. The efficiency of an autotransfusion system for tumor cell removal from blood salvaged during cancer surgery. *Anesthes. Analg.* 1994; 78:1131–5.
61. Hedican SP, Walsh PC. Postoperative bleeding following radical retropubic prostatectomy. *J. Urol.* 1994; 152:1181–3.
62. Kram HB, Nathan RC, Stafford FJ et al. Fibrin glue achieves hemostasis in patients with coagulation disorders. *Arch. Surg.* 1989; 124:384–7.
63. Fremes S, Wong B, Lee E et al. Metaanalysis of prophylacitc drug treatment in the prevention of postoperative bleeding. *Ann. Thorac. Surg.* 1994; 58:1580–8.
64. Atabek U, Alvarez R, Pello MJ et al. Erythropoetin accelerates hematocrit recovery in postsurgical anemia. *Am. Surg.* 1995; 61:74–7.
65. Ford PA, Henry DH. Using r-HuEPO in patients unwilling to accept blood transfusions. *Erythropoiesis* 1996; 7:63.
66. Monk TG, Goodnough LT, Andriole GL et al. Preoperative recombinant human erythropoietin therapy improves outcome with acute normovolemic hemodilution. *Blood* 1994; 84:466a (abstract).
67. Chun TY, Martin S, Lapor H. Preoperative recombinant human erythropetin injection versus preoperative autologous blood donation in patients undergoing radical retropubic prostatectomy. *Urology* 1997; 50:727–32.
68. Erikssen G, Thaulow E, Sandvik L et al. Hematocrit: a predictor of cardiovascular mortality? *J. Intern. Med.* 1993; 234:493–9.

Laparoscopic Radical Prostatectomy

Chapter 37

Chandru P. Sundaram and Gerald L. Andriole

Division of Urologic Surgery, Washington University School of Medicine, St Louis, MO, USA

INTRODUCTION

Recent advances in laparoscopic equipment and suturing skills has resulted in laparoscopic radical prostatectomy (LRP) becoming an acceptable option for the treatment of localized prostate cancer. Schuessler and colleagues first performed this operation in 1991.[1] Between 1991 and 1995, nine laparoscopic radical prostatectomies were performed. However, the surgery was difficult with long operating times. It was, therefore, felt that the laparoscopic approach for the treatment of prostate cancer offered no advantage over open surgery. In 1998, Guillonneau and colleagues reported their initial experience with the surgery with early results of the transperitoneal approach comparable to contemporary series of open radical prostatectomy.[2] Since then, several centers have performed the LRP in increasing numbers. The surgery is now established as a feasible option with acceptable early oncological results. This chapter will discuss the operative technique involved with a transperitoneal laparoscopic radical prostatectomy followed by a review of the recent published clinical series.

OPERATIVE TECHNIQUE

The operating technique described in this chapter is based on the work of Bertrand Guillonneau and Guy Vallencian in Paris.[3] Details of the technique have been modified based on our initial experience at Washington University. We will discuss the difficulties that we faced, to help shorten the learning curve of others attempting to start performing these operations.

PATIENT SELECTION

There is no difference between open and laparoscopic prostatectomy as far as patient selection is concerned. However, during a surgeon's initial experience with this operation, it is recommended that patients be selected with early cancers that do not require laparoscopic pelvic lymph node dissection. This is suggested to limit the operating time, which is likely to be prolonged during the early patients. In the first 20 patients we have selected patients with prostate specific antigen (PSA) less than 10 and a Gleason score of 6 or below.

Patients that should be avoided during the learning curve include those with the following: obesity; history of radiotherapy to the prostate; transurethral resection of the prostate; and previous bladder and prostate surgery, and laparoscopic inguinal hernia repair. Neoadjuvant hormonal treatment can make surgery difficult due to periprostatic adhesions and fibrosis. The nerve-sparing technique is difficult during a surgeon's early experience. We would, therefore, recommend that patients with preoperative erectile dysfunction be chosen and that the nerve-sparing technique should not be attempted initially. Patients with prostates that are larger than 80 g and less than 20 g would also be relatively difficult initially. As experience is gained, indications for laparoscopic prostatectomy should include all patients who are candidates for radical prostatectomy. A large median lobe can make bladder neck preservation difficult and would necessitate bladder neck reconstruction.

PREOPERATIVE PREPARATION

Patients are requested to take two bottles of magnesium citrate at home the day before surgery as well as a bisacodyl suppository the night before the surgery. Aspirin and other non-steroidal analgesics are stopped a week before surgery. Blood loss during surgery with the laparoscopic approach is less than 500 ml typically and, therefore, preparation for autologous blood transfusion is not required. Blood of the patient is typed and screened.

PATIENT POSITIONING

The patient is positioned supine with his legs on spreader bars. Pneumatic intermittent compression boots are used as a prophylaxis against deep vein thrombosis. The abdomen is prepared with antibiotic solution from the xiphisternum to the perineum including the genitals. Both arms are placed beside the patient. The patient is secured to the table with adhesive tape. Elastic adhesive tape is used across the chest, and to secure the thighs and the lower extremities. Foam pads are used to pad the patient in all bony prominences to minimize injury as a result of the positioning. Shoulder support is not required if the strapping is securely performed. A steep 40° Trendelenburg position is achieved before the patient is draped in order to ensure that the patient does not move on the table. A 20F Foley catheter is inserted per urethra. Low-molecular-weight subcutaneous heparin can be administered, especially in patients who are at risk for postoperative deep vein thrombosis. The anus is exposed during patient draping for insertion of a surgeon's finger or a rectal bougie, should this be required during the surgery to facilitate dissection in the perirectal region.

The voice-activated AESOP (Computer Motion Inc., Santa Barbara, CA) camera-holding robot, if used, is secured to the operating table adjacent to the right shoulder of the patient. The surgeon stands on an elevated platform adjacent to the left shoulder of the patient facing the right toe of the patient. The assistant's position is on the right of the patient just caudal to the position of the AESOP camera holder. Should the robot not be used, a second assistant stands on the right of the patient to hold the camera. The nurse is positioned beside the left lower extremity of the patient. We use two ceiling-mounted video monitors placed at the foot end of the patient just above each lower extremity at the eye level of the operating surgeon to minimize surgeon fatigue. One video monitor between the legs can also be used.

PORT PLACEMENT

A total of five laparoscopic ports are used: three 12-mm ports and two 5-mm ports. Pneumoperitoneum is established with a Veress needle at the umbilicus. A 12-mm port is then inserted at that site. The second 12-mm port is inserted between the umbilicus and the left anterior superior iliac spine. The third 12-mm port is inserted at the lateral border of the right rectus abdominis muscle two fingerbreadths below the umbilicus. The fourth 5-mm port is inserted between the third port and the right anterior superior iliac spine. The fifth 5-mm port is inserted between the umbilicus and the pubic symphysis in the midline (**Fig. 37.1**).

In tall patients, the lateral 12-mm ports are moved about 2–3 cm caudally to allow access to the urethra during the urethrovesical anastomosis. The other alternative for port placement includes a fan configuration where two ports are placed on each side between the umbilicus and the anterior superior iliac spine. The surgeon operates through the two ports on either side of the umbilicus. With port placement in the fan configuration, the surgeon operates through the two ports on the left side and the assistant uses the two ports on the right side.

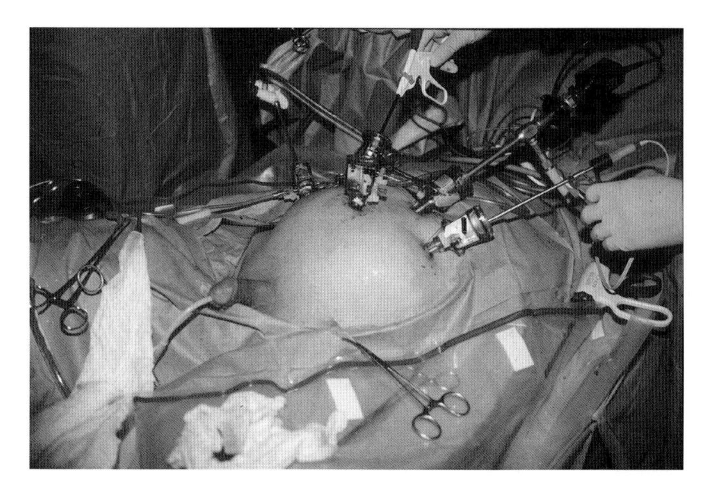

Fig. 37.1. Port placements are depicted with the surgeon operating through two 12-mm ports on either side of the umbilical port. The umbilical 12-mm port is used for the 0° laparoscope. Two other 5-mm ports are positioned: one midway between the umbilicus and the pubic symphysis, and one between the right lateral 12-mm port and the anterior superior iliac spine. (*See Color plate 15.*)

SEMINAL VESICLE DISSECTION

After the ports are placed, the patient is placed in a 40° steep Trendelenburg position. It is ensured that the bladder is empty by using the suction irrigator to evacuate the urine within the Foley catheter, since the Foley catheter does not adequately drain the bladder in a steep Trendelenburg position. The sigmoid colon is retracted superiorly by the assistant using the right lateral port. The suction irrigation is used through the lower midline port by the assistant to help with suction during the dissection. There are two peritoneal arches that are evident anteriorly in the rectovesical pouch of Douglas. The lower peritoneal arch is the landmark for dissection of the seminal vesicles (**Fig. 37.2**). A transverse incision is made at this arch. The vas deferens is held up and dissected towards the prostate. The vas deferens is coagulated with bipolar coagulation or clipped and divided. The assistant then holds the vas deferens anteriorly, exposing the seminal vesicles on either side. After both vasa deferentia are dissected, the seminal vesicles are dissected from the base to the apex, taking care to control the artery to the seminal vesicle with bipolar coagulation before division. After the seminal vesicles and vasa deferentia are dissected on both sides, an incision is made in the Denonvilliers' fascia about 2 mm posterior to the base of the seminal vesicles. Visualization of the perirectal fat identifies the right plane during this dissection (**Fig. 37.3**). Dissection can then be carried out in that plane towards the prostatic apex.

Should the vas deferens or the seminal vesicles not be readily apparent after incision of the peritoneum in the pouch of Douglas, the vas deferens is identified more laterally along the

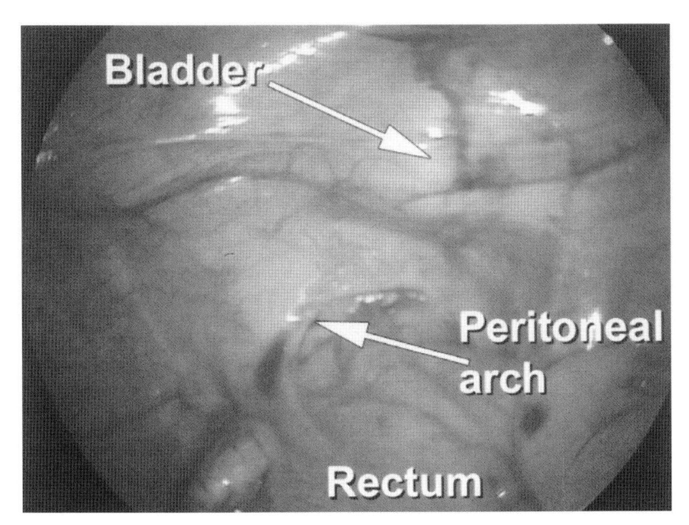

Fig. 37.2. Seminal vesicle dissection. The peritoneum is incised on the lower peritoneal arch in the Pouch of Douglas, to dissect the seminal vesicles and the vasa deferentia. (*See Color plate 16.*)

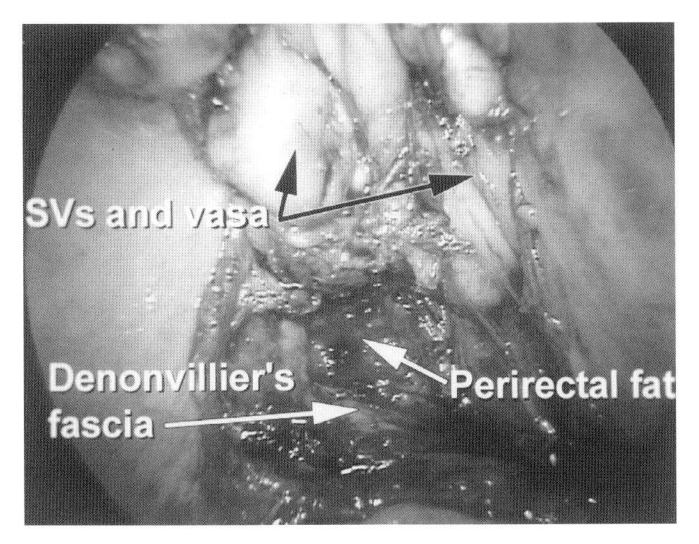

Fig. 37.3. Incision of Denonvillier's fascia. After division of the vasa and dissection of the seminal vesicles (SVs), the Denonvillier's fascia is transversely incised exposing the perirectal fat. (*See Color plate 17.*)

lateral pelvic wall and followed posteriorly towards the prostate. During the dissection of the seminal vesicles, it is essential to remain close to the seminal vesicles in order to prevent damage to the neurovascular bundle and the distal ureter.

During the incision of the Denonvilliers' fascia, if the plane is not readily apparent, an assistant's finger in the rectum or rectal bougie would help identify the rectal wall and avoid rectal injury. Should injury to the rectal wall be apparent, it can be primarily closed in two layers after thorough irrigation of the pelvis provided there is no gross fecal contamination.

RETROPUBIC DISSECTION

The bladder is distended with 150 ml of saline via the urethral Foley catheter. This helps identify the margin of the bladder

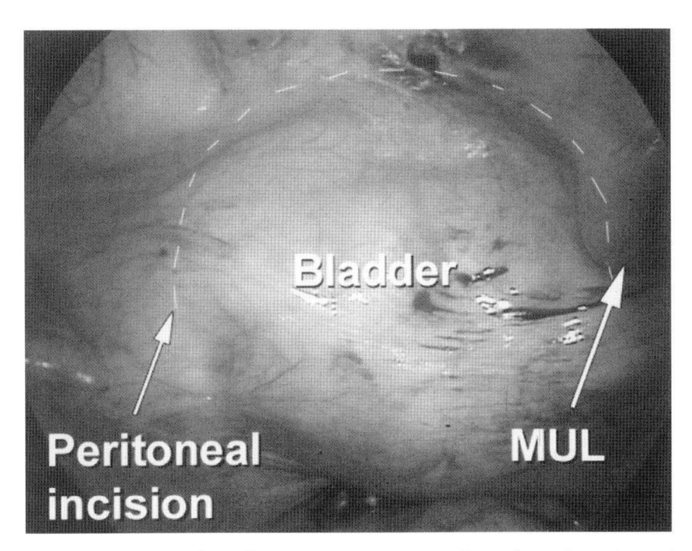

Fig. 37.4. Retropubic dissection. An inverted U-shaped peritoneal incision is made from one medial umbilical ligament (MUL) to the other, to drop the bladder and enter the retropubic space. (*See Color plate 18.*)

that is visible through the peritoneum. An inverted U-shaped incision is made from one medial umbilical ligament to the other (**Fig. 37.4**). Dissection is begun laterally just medial to the medial umbilical ligament on each side until the loose retropubic areolar tissue is identified and the pubic bone is felt. After this dissection is continued medially on both sides, the bladder is held up by the urachus in the midline. The urachus is then divided after bipolar coagulation. In some patients the access to the retropubic space is narrow and deep. In these situations, the medial umbilical ligament on both sides can be divided after bipolar coagulation.

The peritoneal incision in the midline before dissection of the bladder should be as high as possible on the anterior abdominal wall in order to prevent inadvertent injury to the dome of the bladder. This had occurred on one occasion in the authors' series. The bladder injury was repaired in two layers laparoscopically with no further complications in that patient. Dissection in this plane is usually bloodless and bleeding could suggest injury to the bladder wall.

DISSECTION OF ENDOPELVIC FASCIA AND DIVISION OF PUBOPROSTATIC LIGAMENTS

Before the dissection is begun, the bladder is emptied via the Foley catheter using the laparoscopic suction irrigator. The endopelvic fascia on both sides is exposed using blunt dissection with a laparoscopic Kittner. The superficial dorsal venous complex is coagulated and divided. The puboprostatic ligaments are likewise exposed (**Fig. 37.5**). The endopelvic fascia is then incised just lateral to the prostatic surface along the lateral pelvic wall. The endopelvic fascia is incised with the endoshears. The small veins identified during this dissection can be controlled with bipolar electrocoagulation. The puboprostatic ligaments are divided close to the posterior surface of the

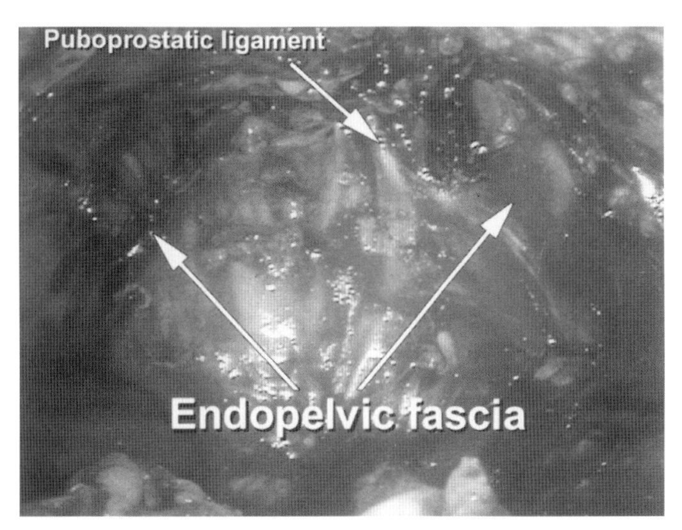

Fig. 37.5. Endopelvic fascia dissection. The endopelvic fascia is dissected and the puboprostatic ligaments exposed. (*See Color plate 19.*)

pubic bone. The lateral surface of the prostate on both sides are exposed and dissected until the posterior edge of the prostate is visualized on both sides.

LIGATION OF THE DORSAL VENOUS COMPLEX

The Foley catheter is replaced with a 22-French curved metal bougie. The metal bougie helps place the dorsal venous complex on stretch. This is also done by cephalad retraction of the prostate. The apex is visualized and the dorsal venous complex ligated (**Fig. 37.6**) with a figure-of-eight stitch with 0 polyglactin on a 36 mm CT-1 needle (Ethicon Inc., Somerville, NJ). The suture is applied by holding the needle with the right hand backhanded with the curve of the needle parallel to the curve of the pubic bone. A second back-bleeding stitch can be applied on the anterior surface of the prostate to identify the base of the prostate to help with bladder neck dissection. The authors, however, have found that this back-bleeding stitch is not required and the vessels can be controlled during the bladder neck dissection. The dorsal venous complex after ligation is not divided at this stage of the operation.

BLADDER NECK DISSECTION

Bladder neck dissection is now performed by the authors using the Harmonic Scalpel (Ethicon Endo-Surgery, Cincinnati, OH), although the endoshears with bipolar coagulation may also be used. The plane of dissection between the base of the prostate and the bladder neck is identified by the margin of the perivesical fatty tissue (**Fig. 37.7**). Also the distinction of the floppy bladder wall and the solid prostatic surface can be visualized by gently tapping on the anterior surface of the bladder wall with a laparoscopic instrument. The relative movement, between the prostate and the bladder wall when the metal bougie is moved, also provides a visual queue to identification of the plane of dissection. The dissection of the bladder neck is then performed. Dissection can be carried out on both sides of the midline exposing the vertical fibers of the bladder neck in the midline (**Fig. 37.8**). With tactile feedback afforded by the metal bougie, the anterior bladder neck is incised in the midline with the endoshears. Coagulation is not used during this maneuver for the potential for coagulating the urethra in the presence of the metal bougie within the urethra. After the anterior bladder neck is divided, the metal bougie is brought out through this opening and the base of prostate is rotated anteriorly. The posterior wall of the bladder neck is divided and is then held with a laparoscopic grasper. The plane between the posterior bladder neck and the base of the prostate is identified. It is essential to proceed vertically in order to avoid

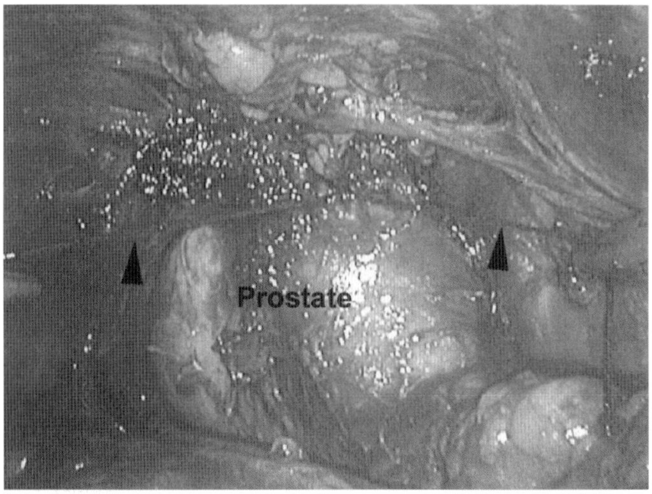

Fig. 37.6. Ligation of the dorsal venous complex. After the endopelvic fascia is incised and the puboprostatic ligaments divided, the dorsal venous complex is ligated. (*See Color plate 20.*)

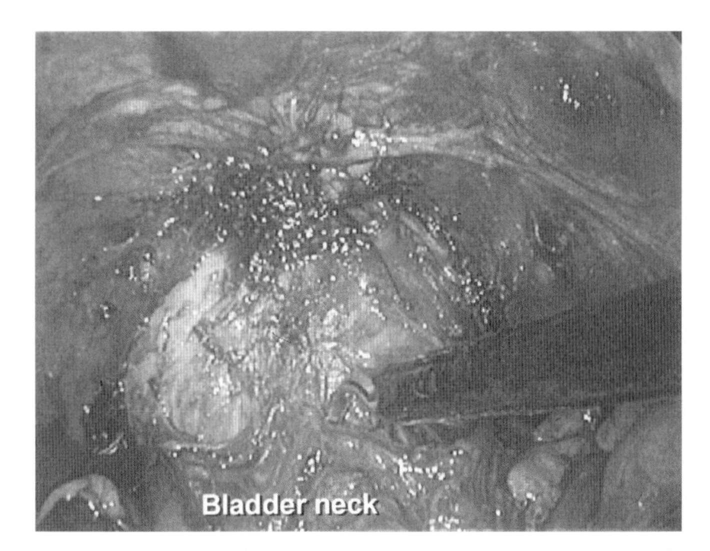

Fig. 37.7. Bladder neck dissection. Dissection between the base of the bladder and the bladder neck. (*See Color plate 21.*)

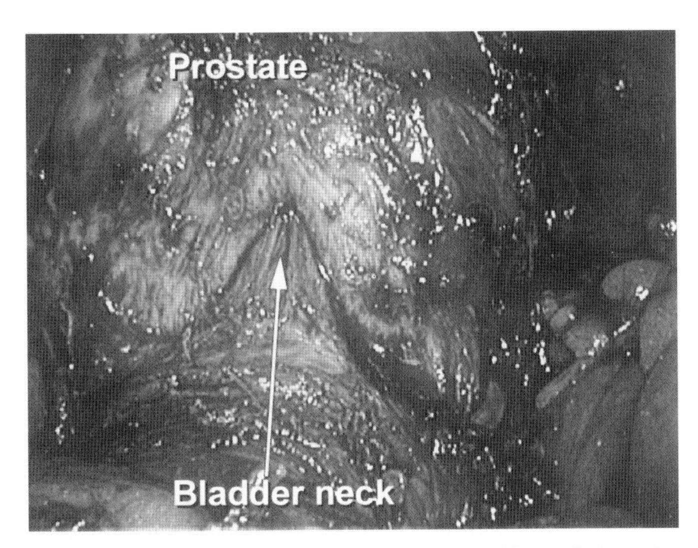

Fig. 37.8. Bladder neck dissection. The anterior bladder neck dissection is complete, exposing the fibers of the bladder neck. (*See Color plate 22.*)

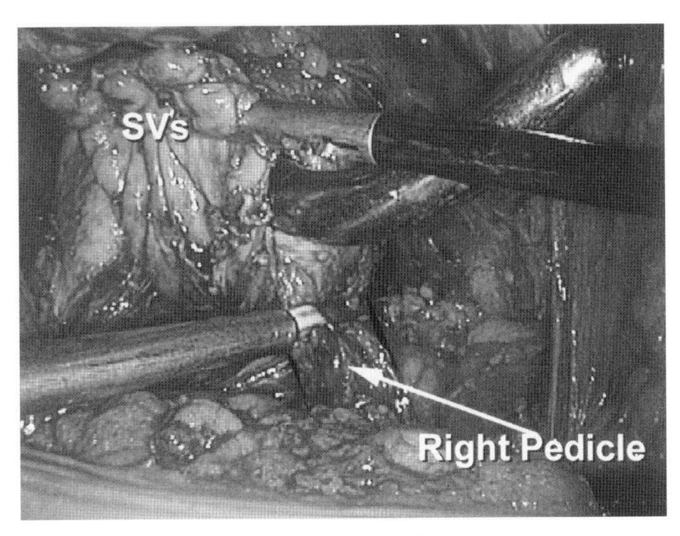

Fig. 37.10. The seminal vesicles (SVs) are held up, exposing the prostatic pedicles. The pedicles are divided close to the prostate, preserving the neurovascular bundles. (*See Color plate 24.*)

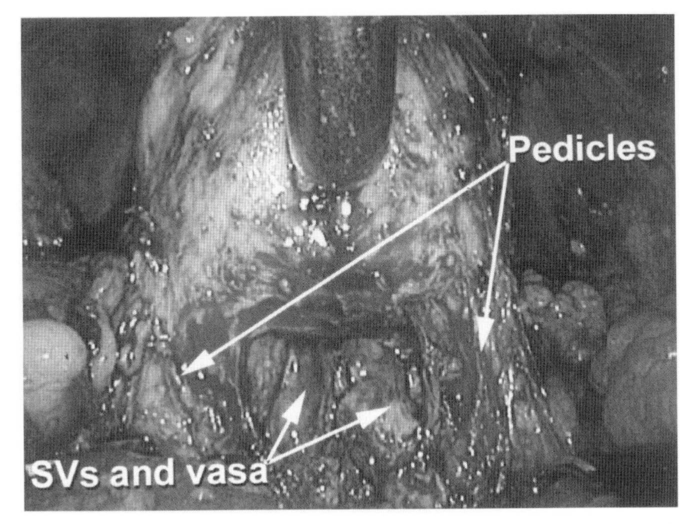

Fig. 37.9. Posterior bladder neck dissection. The posterior bladder neck dissection is complete exposing previously dissected seminal vesicles (SVs). (*See Color plate 23.*)

dissection into the prostate. In the presence of a large median lobe, the median lobe is retracted anteriorly to identify the posterior bladder neck.

Dissection in the proper plane posteriorly will result in exposure of the Denonvilliers' fascia that was previously opened during dissection of the seminal vesicle (**Fig. 37.9**). The seminal vesicles and the vasa deferentia are then visualized through this opening and held up with a locking atraumatic grasper by the assistant. The metal bougie at this stage can be removed.

DIVISION OF THE PROSTATIC PEDICLES

The prostatic pedicles are put on stretch by holding the seminal vesicles anteriorly (**Fig. 37.10**). The prostatic pedicles are divided in a medial to lateral direction. The neurovascular bundles that run along the posterolateral surface of the prostate are spared during control of the lateral pedicles. The lateral pelvic fascia on the lateral aspect of the prostate is carefully incised and the neurovascular bundle dissected off the prostatic capsule. Completion of the division of the prostatic pedicles is accomplished without injury to the neurovascular bundles (**Fig. 37.11**).

We now use the harmonic scalpel for the prostatic pedicles, since the harmonic scalpel has minimal spread of heat and thereby minimizes damage of the neighboring neurovascular bundles. The nerve-sparing transection of the prostatic pedicles on the opposite side is similarly performed before the base of the prostate is free. Minor hemorrhage that occurs during the nerve-sparing technique may be ignored and usually stops spontaneously during the remaining prostatic dissection. Increasing the intraperitoneal pressure to 20 mm Hg for a few minutes also helps control bleeding. The nerve-sparing technique can add up to an hour of operating time to this procedure and is not possible during the early experience of a surgeon.

Guillonneau and associates have used the narrow-tipped bipolar forceps successfully for the nerve-sparing technique. Gill and associates at the Cleveland Clinic have used the articulating endo-GIA stapler (U.S. Surgical, Norwalk, CT) with the 2.5 mm staple width, gray color staple load for the non-nerve-sparing technique. For the nerve-sparing procedure, they have described the use of the Hem-o-lok clip (Weck Systems, Research Triangle Park, NC), which is a polymer-ligating clip with a locking tip.

The remnants of the Denonvilliers' fascia are then divided to free the posterior aspect of the prostate completely up to its apex. The prostate is now attached only in the apical region.

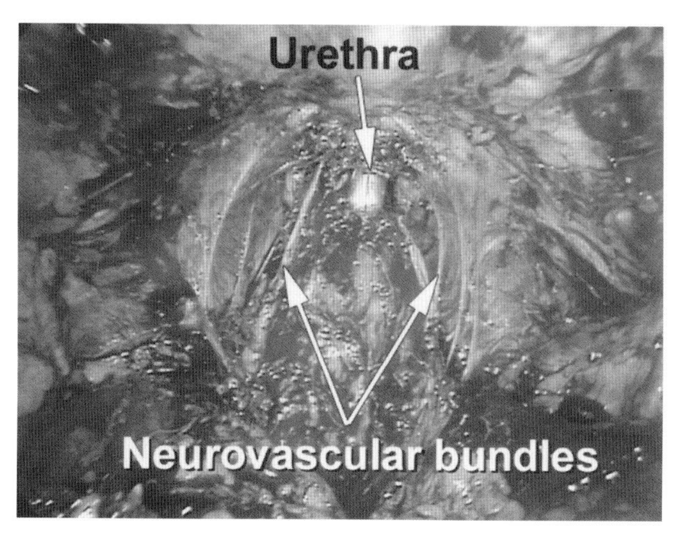

Fig. 37.11. The prostatic fossa is exposed with the intact neurovascular bundles on both sides. (*See Color plate 25.*)

DIVISION OF THE DORSAL VENOUS COMPLEX AND URETHRA

The deep dorsal venous complex that was previously ligated is now divided using the endoshears or the harmonic scalpel. The metal bougie within the prostate helps displace the prostate posteriorly and place the dorsal venous complex on stretch. The curve of the prostatic apex is followed during this procedure to reach the prostatic apex and expose the anterior urethra. During the division of the dorsal venous complex, the previously placed stitch may become dislodged. Bleeding from the complex, however, is not significant and is usually controlled with a combination of increasing the intra-abdominal pressure and precise bipolar coagulation of the bleeding vessels. Occasionally a figure-of-eight hemostatic stitch needs to be applied. 2–0 polyglactin on a 36 mm CT-1 needle or 26 mm SH needle (Ethicon Inc., Somerville, NJ) is usually used for suture ligature of this bleeding, should it be required. Using the endoshears or a laparoscopic knife, the anterior wall of the urethra is divided and the metal bougie within the urethra exposed. The metal bougie is then delivered through this opening and the posterior urethral wall transected. The prostate is then retracted to one side and the rectourethralis is divided, safeguarding the neurovascular bundles. The same procedure is repeated on the opposite side. The prostate along with the seminal vesicles is then entirely freed.

BLADDER NECK RECONSTRUCTION

Bladder neck reconstruction is normally not required, since the bladder neck is preserved during bladder neck dissection. The diameter of the lumen of the bladder neck usually corresponds to the lumen of the urethral stump when bladder neck preservation has been successful. When the patient has had a previous transurethral resection of the prostate (TURP) or if the patient has a large median lobe, bladder neck preservation may not be entirely successful. Under these circumstances, the bladder neck is reconstructed with a racket-handle technique. At our center we reconstruct the bladder anteriorly rather than posteriorly as is done traditionally with open surgery. We do not evert the bladder neck mucosa and rely on mucosal apposition during precise urethrovesical anastomosis.

URETHROVESICAL ANASTOMOSIS

We use Ethicon needle drivers during the procedure and use the two ports on either side of the umbilicus. The authors use interrupted 2–0 polyglactin sutures on a 26-mm SH needle with intracorporeal knot tying. The 17.45-mm RB1 needle can also be used (Ethicon Inc., Somerville, NJ). The first suture is placed inside out on the urethra and outside in on the bladder with a right-hand forehand approach at the 5 o' clock position. The second stitch is a 6 o'clock stitch, which is placed right-hand forehand inside out on the urethra and left-hand forehand outside in on the bladder and tied within the lumen. The third stitch is a 7 o'clock stitch right-hand forehand inside out on the urethra and left-hand forehand outside in on the bladder. These sutures are tied within the lumen of the anastomosis and have not caused problems with intraluminal calcification. The metal sound within the urethra helps guide the needles through the full thickness of the urethra. A perineal sponge stick is used to exert pressure in the perineum to help visualize the urethral stump clearly. A total of 6–12 interrupted sutures are placed, depending on the size of the bladder neck and the urethra to create a watertight anastomosis. All other sutures are placed with extraluminal knot tying. After the three posterior sutures are placed, the lateral sutures of both sides are placed: the right-hand forehand outside in on the bladder and left-hand backhand inside out on the urethra on the right side. The left side stitches are placed left-hand forehand outside in on the bladder and right-hand backhand

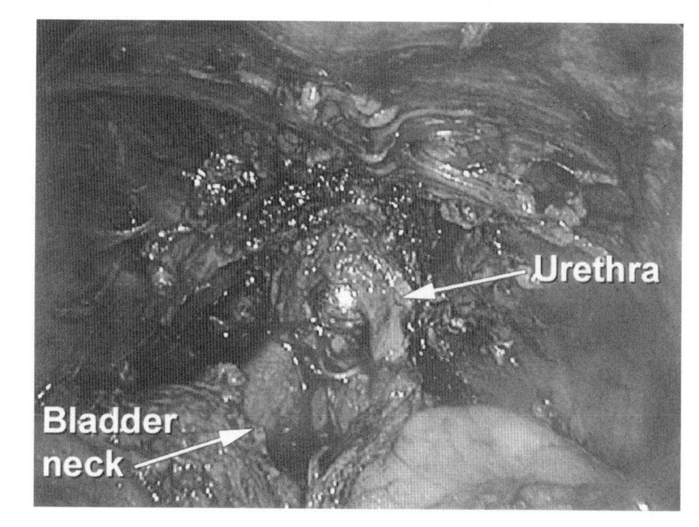

Fig. 37.12. Anastomosis. The urethrovesicle anastomosis is performed with interrupted 2/0 polyglactin suture. A metal urethral bougie facilitates the anastomosis. (*See Color plate 26.*)

inside out on the urethra. The anterior stitches are placed at the 1 and 11 o'clock positions: right-hand forehand outside in on the urethra and right-hand forehand inside out on the bladder for the 1 o'clock position. The 11 o'clock position is also similarly inserted. These last two stitches are inserted and not tied until the Foley catheter is confirmed under visual guidance to be within the bladder through the anastomosis. Ten milliliters of water is instilled into the balloon of the 18F Foley catheter, and the two last stitches are tied (**Fig. 37.12**).

The urethrovesical anastomosis, during initial experience, is the most time-consuming and difficult part of the operation, especially for those surgeons who are not facile at intracorporeal suturing. However, with experience, this is predictable and precise. Practice in a pelvic trainer results in considerable and significant decrease in the operating time for this part of the surgery. After the anastomosis is complete, the Foley catheter is irrigated to ensure that the anastomosis is watertight.

A total of three knots are tied for each of these sutures that are inserted, the first one being a surgeon's knot to ensure that these sutures do not slip and the anastomosis is secure. The suture length is about 7 inches for each interrupted stitch. However, as experience is gained, a 9-inch stitch can be used for two or three interrupted stitches. A self-righting needle driver must not be used for the laparoscopic prostatectomy, since the needle would need to be adjusted to various different angles for each of the sutures, depending on its position in the urethrovesical anastomosis. In our experience, there were four anastomotic leaks in our first eight patients, but all of these were treated conservatively with continued urethral catheter drainage. All patients thereafter did not have an anastomotic leak.

DRAIN INSERTION

A 10-French Blake drain is inserted through a 5-mm port site and placed in the region of the urethrovesical anastomosis. The 12-mm ports are closed with 0 polyglactin with a Carter-Thomason needle device. The intra-abdominal pressure is decreased to 5 mm Hg to confirm that there is no bleeding. The prostate that was previously placed in an Endocatch bag (US Surgical Corp, Norwalk, CT) is brought out through the umbilical incision by extension of incision vertically. Care is taken not to exert undue traction on the Endocatch bag, since it is not strong, and can rupture and result in the prostate dropping into the abdominal cavity.

POSTOPERATIVE RECOVERY

At our institution, patients typically are discharged home on the second postoperative day. The patient is on a clear liquid diet on the first postoperative day and on a regular diet on the second postoperative day. Compression stockings are used during the hospitalization. The Foley catheter is typically removed in about 1 week at our center. In Guillonneau's series, with the last 140 patients the catheter has been removed at a mean of 4.2 days. With increasing experience, the integrity of the anastomosis is secure and the duration of catheterization decreases considerably.[4]

EXTRAPERITONEAL APPROACH

The extraperitoneal approach has been described and may be less invasive than the transperitoneal approach. The benefit of the extraperitoneal approach is the preservation of the peritoneal integrity and the elimination of the potential for intraperitoneal injury. There is, however, less operating space than the transperitoneal approach, and the seminal vesicle dissection would need to be done at a later stage of the operation as with the open approach. Early results suggest that this approach may be comparable to the transperitoneal approach.[5]

RESULTS

Table 37.1. Operative parameters from several large series.

Authors	Number of patients	Mean operating time (minutes)	Estimated blood loss (ml)	Hospital stay (days)	Catheter indwell time (days)	Positive margins (%)	Continence (%)	Potency
Guillonneau et al.[4]	240	232	370	5.2 (in last 140 patients)	4.2 (140 patients)	13.75	84 (in 127 patients at 6 months)	45% (at 3 months in selected patients)
Bollens et al.[5]	50	330	708	N/A	7.8	20	85	75
Abbou et al.[6]	43	255	N/A	4.5	4	27.9	84	23
Turk et al.[7]	145	265	185	N/A	5.5	23.4	92 at 9 months	N/A
Zippe et al.[8]	50	324	225	1.6	9 (in 60% of patients)	20	76 at 6 months	Incomplete data

LEARNING CURVE

Laparoscopic radical prostatectomy is a technically challenging surgery and requires advanced laparoscopic skills. The learning curve was about 40 patients for the Montsouris team. This was because they were developing the technique during the early experience. Since the surgical technique has now been established, the learning curve should be shorter. Furthermore, as urologists at several centers become proficient at the surgery, the learning curve can be further reduced. This can be achieved by an experienced surgeon assisting a novice surgeon. During the first ten patients, the anastomosis can be challenging because no other urologic laparoscopic surgery requires complex intracorporeal suturing. Practice before surgery on the pelvic trainer is essential during the learning curve to minimize operating time and to achieve a watertight anastomosis. Operating times are prolonged during the early patients and was 8.5 hours during our first 20 patients and 6.0 hours during our last five patients in an experience of 25 patients. The average time in the first 50 patients was between 4.2 hours and 5.5 hours.[5,6,8] With even more experience, the mean time drops to about 3 hours for the transperitoneal approach.[3] Complications with 1228 LRPs performed by 13 surgeons in six European centers included: conversion to open surgery in 26 (2%), rectal or bowel perforations in 15 (1.2%), ureteral injuries in 12 (1%), anastomotic leak in 69 (5.6%) and thromboembolic complications in two patients. Major bleeding from the epigastric artery occurred in three patients. Twenty-three patients (1.9%) were reoperated for portsite hernias in ten, ureteral injuries in five, bleeding in five, and anastomotic leak in three.[9] Operative parameters are compared in **Table 37.1**.

ADVANTAGES OF THE LAPAROSCOPIC APPROACH

Decreased blood loss and possibly shorter duration of catheterization seem to be the obvious advantages. Blood loss is less partly due to the tamponading effect of the pneumoperitoneum. In most laparoscopic series the estimated blood loss was less than 500 ml compared to an about 1000 ml blood loss after open surgery.[10] The urinary continence following surgery appears satisfactory. Erectile function may not completely recover for 1–2 years following surgery. The efficacy of the nerve-sparing technique in the LRP has, therefore, not been adequately studied. Magnification and the better visualization that is provided during laparoscopy can theoretically result in more accurate dissection. This can result in a superior nerve-sparing technique. The positive margin rate of 13.75–27.9% is comparable to the series by Weider and Soloway,[11] who reported 23% positive margins in T1c disease. The positive margin rate during our early experience with LRP was, however, due to violation of the capsule. The bladder neck-sparing technique has been employed with the LRP. Bladder neck-sparing has not been implicated to contribute to increased incidence of positive margins in the vast majority of patients.[12] Furthermore, the bladder neck can be widely excised and reconstructed during LRP, should it be required.

The LRP is a minimally invasive option with encouraging early results for the treatment of localized prostate cancer. There has been no randomized or prospective matched study comparing the open and laparoscopic approaches. Furthermore, the LRP series have less than 5 years of follow-up; PSA progression rates over the long term are, therefore, unknown. The LRP should, therefore, be studied prospectively in several large centers before it becomes an accepted option for the treatment of localized prostate cancer.

REFERENCES

1. Schuessler WW, Schulam PG, Clayman RV et al. Laparoscopic radical prostatectomy: initial short-term experience. *Urology* 1997; 50:854–7.
2. Guillonneau B, Cathelineau X, Barret E et al. Laparoscopic radical prostatectomy. Preliminary evaluation after 28 interventions. *Presse Med.* 1998; 27:1570–4.
3. Guillonneau B, Vallancien G. Laparoscopic radical prostatectomy: the Montsouris technique. *J. Urol.* 2000; 163:1643–9.
4. Guillonneau B, Rozet F, Barret E et al. Laparoscopic radical prostatectomy: assessment after 240 procedures. *Urol. Clin. North Am.* 2001; 28:189–202.
5. Bollens R, Vanden Bossche M, Roumeguere T et al. Laparoscopic radical prostatectomy: analysis of the first series of extraperitoneal approach. *J. Urol.* 165(Suppl.):Abstract 1354.
6. Abbou C, Salomon L, Hoznek A et al. Laparoscopic radical prostatectomy: preliminary results. *Urology* 2000; 55:630–4.
7. Tuerk I, Deger S, Winkelmann B, Loening SA. Laparoscopic radical prostatectomy – the Berlin Experience. *J. Urol.* 2001; 165(Suppl.):Abstract 1340.
8. Zippe CD, Meraney AM, Sung GT et al. Laparoscopic radical prostatectomy in the USA. Cleveland Clinic experience series of 50 patients. *J. Urol.* 2001; 165(Suppl.):Abstract 1341.
9. Sulser T, Guillonneau B, Vallancien G et al. Complications and initial experience with 1228 laparoscopic radical prostatectomies at 6 European centers. *J. Urol.* 2001; 165(Suppl.):Abstract 615.
10. Walsh PC. The status of radical prostatectomy in the United States in 1993: where do we go from here? *J. Urol.* 1994; 152:1816.
11. Wieder JA, Soloway MS. Incidence, etiology, location, prevention and treatment of positive surgical margins after radical prostatectomy for prostate cancer. *J. Urol.* 1998; 160:299–315.
12. Soloway MS, Neulander E. Bladder-neck preservation during radical retropubic prostatectomy. *Semin. Urol. Oncol.* 2000; 18:51–6.

Pitfalls of Laparoscopic Radical Prostatectomy

Chapter 38

András Hoznek and Clément-Claude Abbou
Service d'Urologie CHU Henri Mondor, Créteil, France

INTRODUCTION

Laparoscopy has become an integral part of urologic surgery. Its indications have been gradually extended to the most advanced oncologic and reconstructive procedures. Within this frame, radical prostatectomy is of major interest considering the incidence and clinical significance of the disease. This procedure comprises several steps of challenging dissection where the preservation of delicate nerve and muscle structures needs to be reconciled with safe tumor excision. The intervention ends with vesicourethral anastomosis, which is considered to be the most difficult reconstructive procedure in urologic laparoscopy.

Each step of the surgical procedure involves distinct risks as the anatomy and pathologic process of each individual patient has an infinite variability. Knowledge of these risks is the key to their prevention. Herein, based on our single center experience of more than 300 wholly standardized procedures, we give a comprehensive review of the pitfalls of laparoscopic radical prostatectomy.

SURGICAL TECHNIQUE

Patient selection and preoperative preparation

The body habitus of the patient does usually not significantly influence the degree of difficulty of the surgery. However, in obese patients achieving access may be more difficult. The thickness of the abdominal wall diminishes the available length of the trocars and consequently the ability to reach deeply situated structures with the instruments. Obese patients frequently have a large amount of perivesical fat and the bladder may be difficult to retract during the vesicourethral anastomosis. It often obstructs the view on the bladder neck and urethra when using a 0° laparoscope. The 30° laparoscope may be a valuable help in these circumstances.

Previous abdominal surgery sometimes leads to parietal or intestinal adhesions, which may increase risks while creating access.

Periprostatic fibrosis increases the difficulty of dissecting the prostate, particularly on its posterior aspect. This situation is often met after neoadjuvant hormone therapy, previous prostatitis, repeated transrectal biopsies, previous transurethral resection of the prostate (TURP) or retropubic prostatectomy.

Patients should remain on a low-residue diet for 3–4 days before surgery. This facilitates the retraction of the bowel during surgical resection.

Patient positioning and creating access

As the procedure lasts for several hours under general anesthesia, positioning of the patient should be made very meticulously to avoid decubitus lesions or neuromuscular injury. The patient is placed supine in a steep Trendelenburg position. Spreader bars enable the legs to be opened to 30°. Silicon pads are used to protect pressure points. We do not use shoulder holders because there is a risk of nerve compression.

After preparing the skin from the xyphoid process to midthigh, drapes are placed in a sterile fashion and an 18 French three-way catheter is inserted into the bladder.

We use five trocars. First, we perform a hemi-circumferential incision at the lower margin of the umbilicus. An open Hasson technique is used to gain access to the abdominal cavity under direct vision. In obese patients, care should be taken not to cleave and insufflate the preperitoneal space otherwise the peritoneum falls towards the bowels, thus significantly reducing the working space. The insertion of secondary trocars is almost impossible in such circumstances.

Other sites of primary access can be used in the presence of a surgical scar around the umbilicus. Then, a 12-mm trocar with a foam grip is inserted, the abdomen is insufflated. The abdominal cavity is inspected thoroughly. The Trendelenburg position is exaggerated to 30°, which displaces the bowels cephalad by gravity. In case of adhesions, an intact site is chosen for the first trocar puncture. Throughout the whole procedure, the insufflation pressure is maintained below 10–12 mmHg. However, during puncture with secondary trocars, it is safer to increase CO_2 pressure to 20 mmHg. This avoids the abdominal wall coming into close contact with the bowels and vascular structures, thus diminishing the risk of trocar injury. In the presence of parietal bowel adhesions, they can be easily detached with monopolar rotating-tip coagulating scissors before the insertion of all other trocars. The monopolar scissors should be used very carefully because they can cause thermal injury to the bowels. During laparoscopy, we use both monopolar and bipolar electrocautery. The advantage of monopolar scissor is that the same instrument can be used to cut and coagulate. In contrast, bipolar electrocautery can only be used for hemostatasis. The monopolar instrument is handled by the primary surgeon and the assistant is responsible for activating the bipolar one.

Several factors should be taken into account when choosing the site for the trocars. Trocars should be sufficiently distanced from each other to prevent collision between instruments and to offer an optimal angle between the needle-holders during the suturing step. Given the length of the instruments, there is a limit above which trocars cannot be placed, otherwise instruments would not reach structures below the pubic symphysis. With these considerations in mind, we elaborated a five-trocar configuration (**Fig. 38.1**).

The use of large trocars is another source of complication. Postoperative hernias occur only when trocars in the region of 10–12 mm are used. Therefore, the number of large trocars used should be kept to minimum. In our practice, we use only two 12-mm trocars: one for the laparoscope, and a second near to the lateral margin of the rectus sheath on the surgeon's side. The latter is indispensable for inserting the needles and large instruments such as vascular clip appliers.

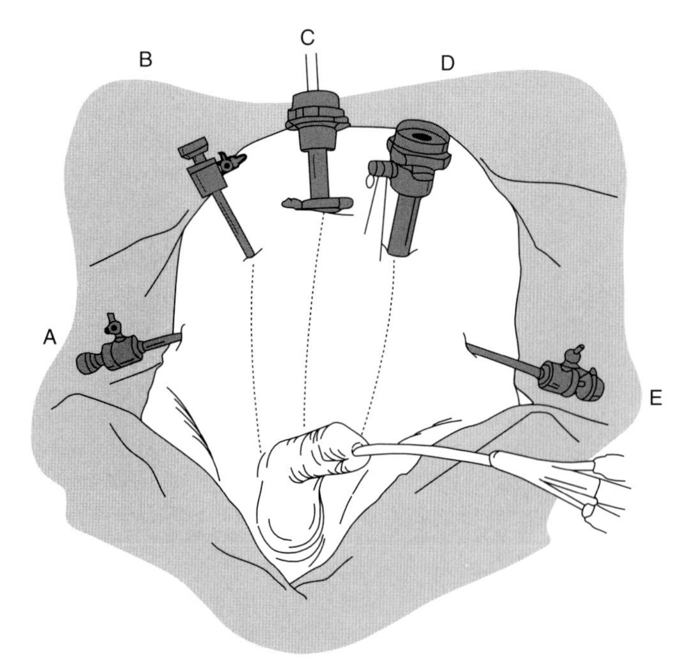

Fig. 38.1. Position of the trocars. (A) 5-mm trocar for the assistant. (B) 5-mm trocar for the assistant. This trocar is used by the primary surgeon during the vesicourethral anastomosis. (C) 12-mm trocar for the camera. (D) 12-mm trocar for the primary surgeon. This is used by the assistant during the vesicourethralis anastomosis. (E) 5-mm trocar for the primary surgeon.

Parietal bleeding rarely occurs during trocar insertion. The typical site of bleeding is the two trocars near the lateral margin of the rectus sheath. This bleeding is due to injury of the branches of the inferior epigastric artery and can be managed with the use of a balloon trocar that compresses the vessels. The drawback is the impossibility of pushing this trocar deep into the abdomen, thus reducing the available length of the instruments. However, this method is usually satisfactory; at the end of the procedure, each trocar is removed under direct vision to verify that there is no residual bleeding.

Posterior dissection

The pouch of Douglas is widely exposed and insufflation pressure is diminished to 12 mmHg. In order to achieve better visualization of the pouch, it is helpful to retract the sigmoid colon cephalad and to the left of the abdominal wall using a straight transcutaneous needle. This is passed under vision through an appendix epiploicae and then again through the abdominal wall. The sigmoid is lifted with the suture and fixed with a straight clamp. The transcutaenous needle should be inserted laterally and above all the trocars otherwise the fixed bowel would get in the way of the instruments during their introduction.

The peritoneum is incised over the vasa deferentia and the seminal vesicles. The vas deferens is usually easily identified because it deforms the surface of the overlying peritoneum. There is, however, a risk of confusion with the ureter and, therefore, the laparoscopic anatomy of these two structures must be well understood. When inspecting the anterior aspect of the Douglas pouch, two transverse inverted U-shaped peritoneal folds can be observed. The more superficial fold corresponds to the ureters and the deeper to the distal portion of the vasa deferentia. The transverse peritoneotomy should be done on this deeper fold. The dissection goes on along the posterior surface of the seminal vesicles downwards and the Denonvillier's fascia is incised. It is important not to dissect the vasa and seminal vesicles free of the bladder base at this step otherwise these structures will hang and obstruct the view when dealing with the prostatorectal cleavage. The latter is performed at both sides of the rectum until the levator ani muscles are reached. The vasa deferentia are dissected and sectioned only when this is done (**Fig. 38.2**). During the dissection of the seminal vesicles, it should be remembered that the cavernous nerves lie close to the tip of the seminal vesicles. Two large arteries are typically identified supplying each seminal vesicle from the lateral side. These are divided immediately adjacent to the seminal vesicles after being controlled with surgical clips. When nerve sparing has been included in the goals of the surgery, we prefer to use hemoclips rather than any kind of thermal energy to achieve hemostasis of these vessels. The vas deferentia are clipped and divided.

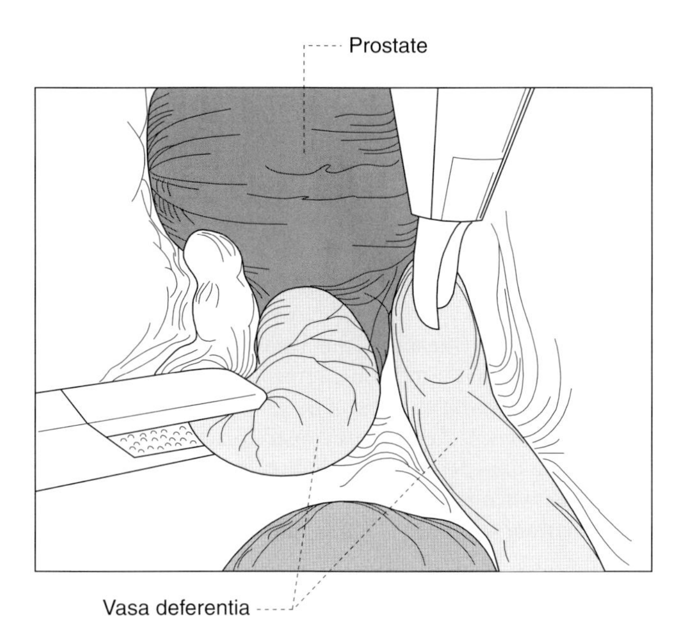

Fig. 38.2. Retroprostatic dissection.

Anterior attachments to these structures are dissected using blunt and monopolar scissor dissection up to the location of the bladder neck.

Again, previous TURP, multiple biopsies, prostatitis and neoadjuvant hormonal therapy are the main causes of bothersome fibrosis during this dissection.

Dissection of retzius space

The dissection of the bladder is much facilitated if it is previously filled with 200 ml of saline, which results in a slight downward traction due to gravity. The upper limit of the bladder in men is particularly high and the incision of the anterior parietal peritoneum should be performed as close to the umbilicus as possible in the midline. The optic is retracted until the two umbilical arteries and urachus can be identified. There is an avascular plane between the bladder and the abdominal wall. If bleeding occurs, it means that one is too close to the bladder. The peritoneum should be incised laterally to facilitate the passage of instruments to the deep part of the pelvis. The symphysis pubis is rapidly reached, and care should be taken not to injure the small vessels that run perpendicular to the rami of the pubic bone. These are collaterals between the external iliac vein and the obturator vein (accessory obturator vein). Their bleeding is difficult to stop; however, if it occurs, bipolar forceps represent the most effective remedy.

The endopelvic fascia is incised on either side of the prostate and incisions carried towards the apex. The levator muscle attachments are peeled off the prostate bluntly. The puboprostatic ligaments are divided sharply for a few millimeters to aid in the apical dissection.

Ligature of the dorsal venous complex

The margin between the urethra and dorsal vein complex is easily identified due to the superb light conditions and magnification of the laparoscope (**Fig. 38.3**). A 2–0 Vicryl stitch is introduced and secured around the superficial tissue at the base of the prostate and a long tail is left to be used for retraction.

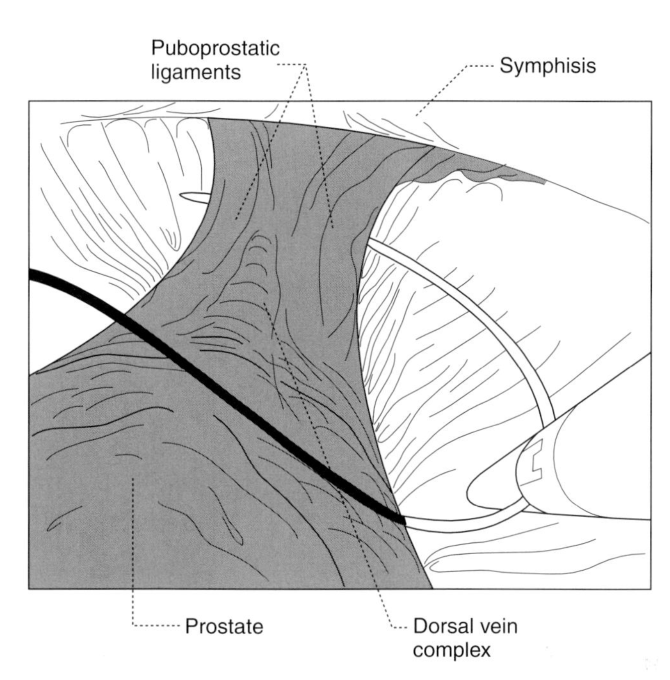

Fig. 38.3. Ligature of the dorsal vein complex.

The assistant grasps the stitch and retracts posteriorly to put stretch on the apex. A second 2–0 Vicryl stitch is used to place a figure-of-eight stitch around the Santorini plexus. We prefer intracorporeal knotting. It is more precise than extracorporeal knotting because the knot can be brought farther under the symphysis. The dorsal vein complex is not divided at this step. Postponing the section of the vein helps to achieve hemostasis because of intravenous coagulation.

Transection of the bladder neck

We begin the transection at the anterior aspect of the bladder neck a few millimeters cephalad to the distal suture of the dorsal venous complex. Because of the lack of manual palpation, the identification of the bladder neck differs from open surgery. The recognition of the landmarks is facilitated by a couple of rules (**Fig. 38.4**). The consistency of the prostate and the bladder are totally different. The prostate is a solid gland, whereas the bladder is mobile when palpated with endoscopic instruments. In addition, the prostate is covered laterally by a plain fascial layer, while the bladder is surrounded by fatty tissue. Although bladder neck-sparing is perfectly feasible by laparoscopy, its utility is questionable because of the risk of positive surgical margins at this site. We prefer to section the bladder at a safe distance and perform a racket handle before the vesicourethral anastomosis.

To facilitate the transection of the bladder neck, the assistant retracts anteriorly on the prostate base stitch to elevate the bladder neck. The section is carried out between the muscular fibers of the detrusor and the prostatic capsule. The anterior

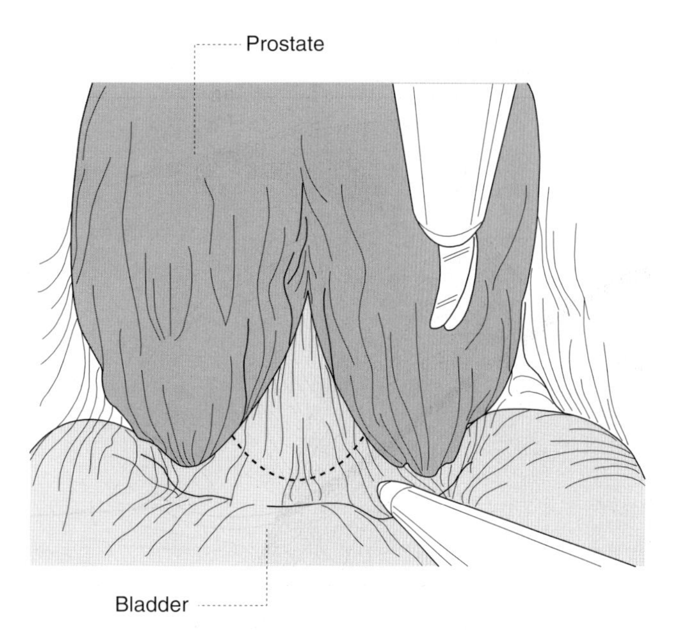

Fig. 38.4. Dissection of the bladder neck.

bladder neck is opened sharply, the catheter balloon is deflated and the catheter delivered through the opening. The assistant now grasps the catheter with a toothed grasper. The catheter is pulled at the urethral meatus and is secured with a Kelly clamp, allowing the assistant to elevate the bladder neck with anterior traction. The posterior bladder neck is incised, as well as the Denonvillier's fascia. The surgeon soon encounters an empty space, indicating that the primary posterior dissection has been reached. The seminal vesicles and vasa deferentia are apparent and delivered anteriorly.

The assistant now grasps the left seminal vesicle and retracts towards the right, while he places the suction tip behind the prostate and retracts posteriorly putting the left prostatic pedicle on stretch. The surgeon is able to view the prostate laterally and posteriorly, and must progress slowly with clips and sharp dissection until the posterolateral aspect of the prostate is reached.

During the transection of the bladder neck, there are two special circumstances of which one should be aware. The first occurs after previous TURP. In this situation, the outlines of the prostate are blurred and the dissection is rendered more difficult because of periprostatic fibrosis. Special care should be taken when incising the posterior margin of the bladder neck because of the proximity of the ureters. At the beginning of the laparoscopic experience, the distance of the incision line from the ureteral orifices tends to be overestimated due to the magnified vision and the incision may occur too close to the orifices. This leads to two kind of hazardous situations during the step of the vesicourethral anastomosis. If the suture is placed too superficially, there is a risk of a tear, resulting in anastomotic leakage. On the other hand, if the suture is placed too deep, it can result in ureteral occlusion. The best way to avoid this situation is to place bilateral double-J stents preoperatively. However, this lengthens operative time.

The second difficulty is met in the presence of an enlarged median lobe. In this case, the section of the posterior margin of the bladder neck should begin with a submucosal incision, tangentially to the bladder trigone, contouring the bulging median lobe. The danger during this maneuver is to enter the cleavage plane of benign prostatic hyperplasia (BPH), i.e. below the surgical capsule of the prostate, thus obtaining positive margins.

Apical dissection

Next, the assistant releases the seminal vesicle and again grasps the proximal stitch on the prostate retracting posteriorly to help expose the apex. The dorsal venous complex is transected between the two previously placed ligatures. This section of the dorsal vein is perpendicular to its axis, but then the plane between the vein and the urethra is developed in an oblique manner caudally. This is important to avoid positive apical margins. Occasionally, the dorsal venous complex produces collaterals to the lateral aspect of the prostate that may bleed. The bipolar forceps allow hemostasis to be achieved. If nerve sparing is to be undertaken, the neurovascular bundles are freed on each side of the urethra (**Figs 38.5** and **38.6**). The bundles have perpendicular vascular branches to the prostate. To avoid thermal injury, we prefer to use hemoclips before sectioning these vessels.

Section of the lateral pedicles

The periprostatic fascia has two layers: the outer layer, called levator fascia, and an inner layer, called prostatic fascia. The neurovascular bundles run between these two layers (**Fig. 38.5**). At this point they are in proximity with the Denonvillier's fascia. If nerve-sparing is to be undertaken, the two layers of

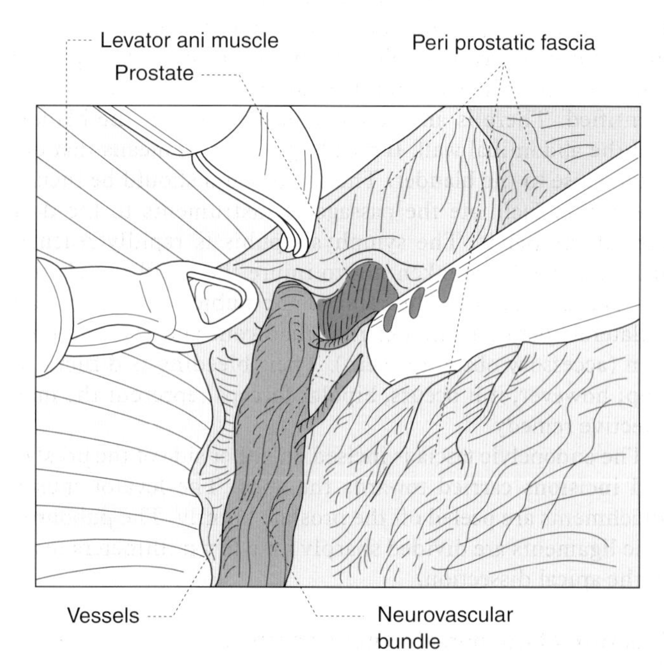

Fig. 38.5. Nerve sparing. After incision of the levator fascia, the neurovascular bundle is released of the prostate.

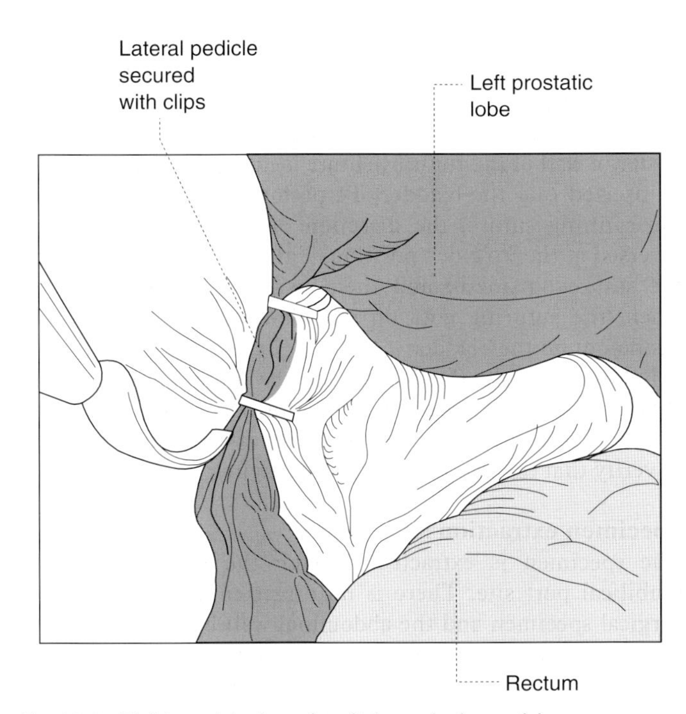

Fig. 38.6. Division of the lateral pedicles at the base of the prostate.

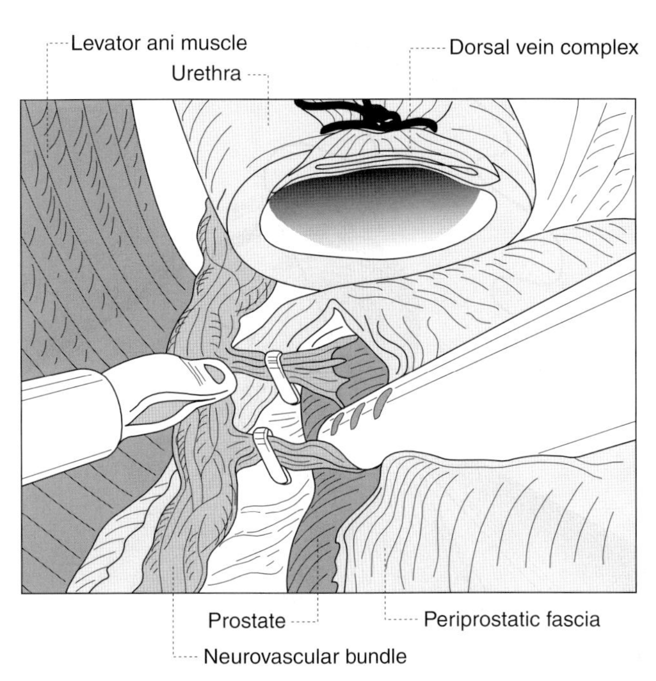

Fig. 38.7. Arterial branches originating in the neurovascular bundles and running towards the prostate are clipped and sectioned.

the periprostatic fascia should be separated and the obturator fascia detached until a subtle groove appears at the lateral aspect of the prostate, indicating the border of the bundles. The bundles can thus be detached from the prostate. The assistant grasps the left seminal vesicle and retracts towards the right, putting the left prostatic pedicle on stretch. Once the pedicle has been divided in contact with the prostate, the surgeon incises the periprostatic fascia and extends this incision to the apex. The neurovascular bundle is identified and allowed to fall laterally from the prostate by clipping and dividing the small branches leading to the prostate as they are encountered. The right side pedicle and proximal neurovascular bundle preservation are performed in a similar fashion.

During this maneuver, care should be taken not to overstretch the nerves; an elongation of 10% can lead to nerve damage. Again, we use hemoclips before sectioning the vascular branches running towards the prostate. The separation of the bundles is complete when the perirectal fatty tissue appears medially.

Then, the urethra is divided anteriorly and the catheter withdrawn until the posterior urethral mucosa is seen, which is then sharply divided. The assistant, grasping the proximal prostate stitch, retracts the prostate to each side in an exaggerated manner, allowing the suction tip to be positioned under the rectourethralis and above the rectum, which permits the surgeon to complete the prostatectomy (**Figs 38.7** and **38.8**). A watertight endocatch is introduced and the prostate placed within. The specimen sac is placed in the upper abdomen.

Vesicourethral reconstruction

A tennis-racket reconstruction is usually performed. We prefer not to preserve the bladder neck to avoid positive margins. If the bladder neck cannot be brought down to the urethra without tension, the solution is to incise sagitally the anterior aspect of the bladder and perform a tennis-racket closure posteriorly.

We routinely perform two hemicircumferential[1] and more recently a single circumferential running suture for the anastomosis with 3–0 Vicryl. The use of a 5/8 tapered needle

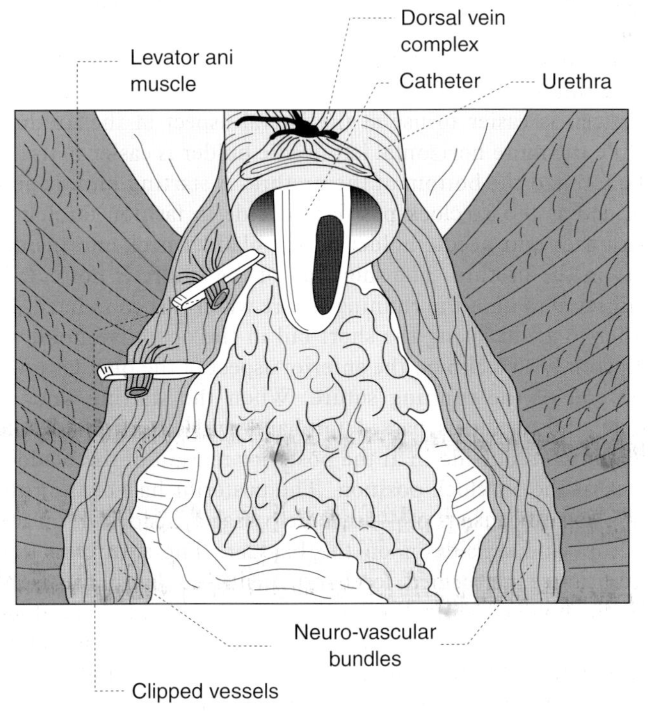

Fig. 38.8. Operative view after excision of the prostate.

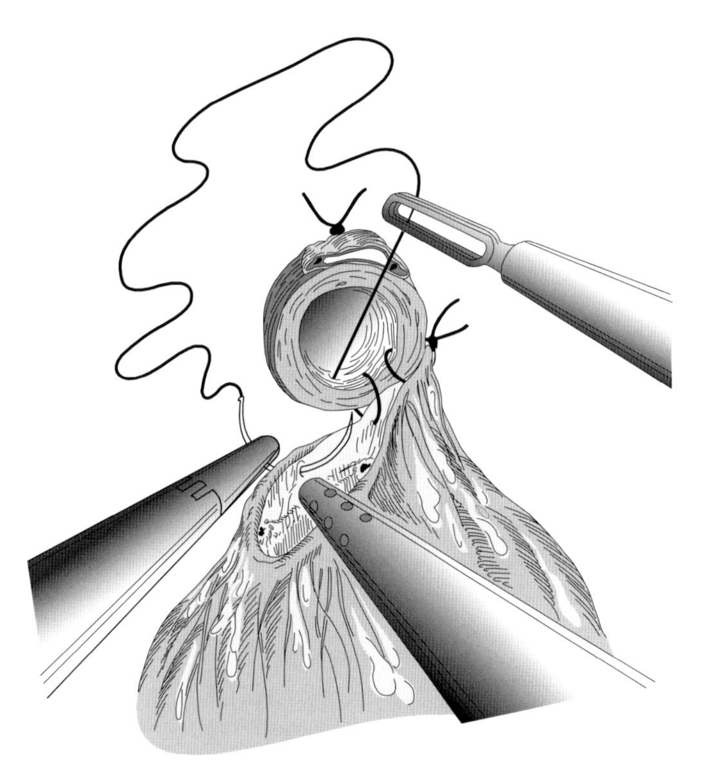

Fig. 38.9. Urethrovesical anastomosis: the circumferential running suture begins at the 3 o'clock position and is completed in a clockwise direction.

facilitates the passage of the suture. The length of the suture should be approximately 20 cm. If the suture is too long, it is not comfortable to use and, if too short, leads to difficulties during knot-tying when finishing the anastomosis.

It is more comfortable to use two needle holders. The right-hand needle holder is inserted through the right paramedian port and the left needle holder through the left lateral port. By doing so, the angle between the needle holders is close to 90°. As a rule of thumb, the right needle holder, which is more vertical, is easier to use at the lateral aspect of the urethra, while the more horizontal left needle holder is easier to use at the top and the bottom of the urethra. A starting knot is performed at the 3 o'clock position outside-in fashion forehand, with the right needle holder on the bladder neck and inside-out on the urethra. The short tail of the suture is not cut; it will serve to knot the running suture when it is completed. In order to avoid losing the short tail in the operative field, we attach it with a clip on the anterior aspect of the bladder until the end of the running suture. When the starting knot is done, the needle is passed forehand with the right needle holder from outside-in at the posterior margin of the bladder neck at the 5 o'clock position. The posterior half of the running suture consists of 4–5 needle passages inside-out on the urethra and outside-in on the bladder. The sutures of the urethra are performed forehand, inside-out with the right needle holder, while the sutures on the bladder are done outside-in forehand with the left needle holder (**Fig. 38.9**). Meanwhile, the assistant helps to retract the bladder neck cranially with the suction-irrigation passed through the left para-median port. In obese patients, it may be difficult to obtain an optimal view on the urethra and bladder neck because of the excessive amount of perivesical fat, which obstructs the view of the laparoscope. In such cases, it is better to use the 30° laparoscope directed downwards. When the posterior half of the running suture is done, the Foley catheter is inserted into the bladder. To perform the anterior half of the running suture, the direction of the needle passage is reversed at the 9 o'clock position. This is achieved by passing the suture outside-in and then inside-out on the bladder. Then, the suturing goes on outside-in on the urethra and inside-out on the bladder.

Before inflating the balloon of the Foley catheter, care should be taken to ensure that it is not caught in the running suture. The bladder is then filled with 200 ml saline in order to verify that the anastomosis is watertight.

Specimen extraction and closure

The specimen is extracted through the slightly enlarged umbilical port site. There is no direct contact between the surgical specimen and the abdominal wall because of the use of the watertight endoscopic bag. A small suction drain is inserted through the left lateral 5-mm port site and placed in the Retzius space near the anastomosis. The fascia is carefully closed at the umbilical port site. The other ports are only closed with intradermic sutures.

POSTOPERATIVE CARE

The suction drain can be removed when the amount of evacuated liquid is less than 50 ml. In the typical patient, this occurs on the first postoperative day. If this is not the case and a significant quantity of drainage fluid is observed, the creatinine level of the drainage fluid should be taken. If this analysis reveals lymphorrhea and not urine leakage, the drain can be removed because the lymph is reabsorbed on the peritoneal surface.

At postoperative days 2–4 (days 3 or 4 when done over the weekend) a gravitational cystography is performed in all patients.[2] The bladder is filled with 250 ml of water-soluble contrast, the balloon of the catheter is deflated and the patient is asked to perform a Valsalva maneuver. With contrast still running, the catheter is slowly pulled and the anastomotic area carefully imaged during Valsalva as well.

When no anastomotic leak is identified, the catheter is immediately removed. If a leak is observed, the catheter should be left in place and another cystography performed 6 days later.

COMPLICATIONS

During open surgery, the main injury to the patient results from the large incision. In contrast, after laparoscopic radical prostatectomy, the typical patient can ambulate within the first 48 hours after surgery and resume normal activity much faster. If this is not the case, a thorough postoperative evaluation should be performed.

Since laparoscopic radical prostatectomy uses a transperitoneal approach in contrast to traditional retropubic prostatectomy, unusual complications may occur. Rectal injuries may occur during the initial rectoprostatic dissection or, more frequently, during the division of the lateral pedicles or the apex of the prostate. Most of the time, these injuries are minimal; they are immediately recognized and repaired, thus avoiding the necessity of temporary colostomy.

Since the peritoneal cavity is opened, postoperative paralytic ileus and abdominal pain is encountered in patients with more pronounced intraoperative bleeding or postoperative urine leak due to the irritating effect of blood or urine at the peritoneal surface. These complications usually resolve spontaneously within a few days. As it has been already mentioned, the ureters are at risk of injury during the posterior dissection if they are mistaken for the vas deferens or during the section of the posterior bladder neck, especially in patients with previous TURP. The accurate identification of the anatomic landmarks and the insertion of two double-J stents provide a guarantee to avoid these problems. Urinary leakage does not have a significant impact on postoperative course, provided it is timely recognized. Therefore, we routinely perform a retrograde urethrocystogram before taking out the urinary catheter. If leakage occurs, the catheter should be kept in place for a few more days.

CONCLUSIONS

Laparoscopic radical prostatectomy is now a well-standardized procedure, which has acquired its maturity. It has become the procedure of choice in a growing number of centers, where it is routinely used and taught. The task that remains is to compare long-term results of laparoscopic radical prostatectomy with its traditional counterparts in order to establish its real place in the urologic armamentarium. Future development in instrumentation may alleviate the technical drawbacks and facilitate acquisition of the necessary surgical skill.

REFERENCES

1. Hoznek A, Salomon L, Rabii R et al. Vesicourethral anastomosis during laparoscopic radical prostatectomy: the running suture method. *J. Endourol.* 2000; 14:749–53.
2. Nadu A, Salomon L, Hoznek A et al. Early removal of the catheter after laparoscopic radical prostatectomy. *J. Urol.* 2001; 166:1662–4.

Robotic Retropubic Radical Prostatectomy

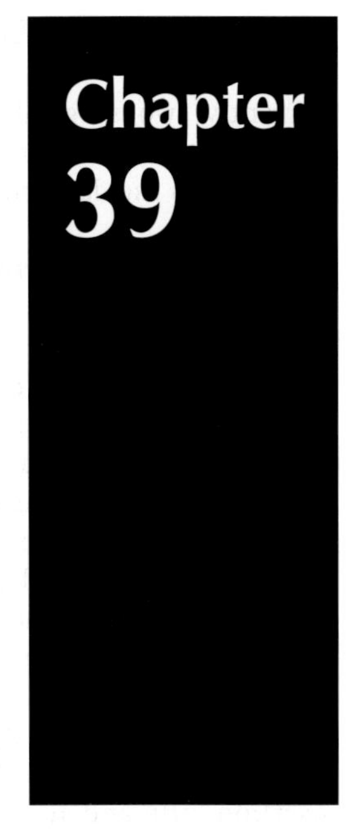

Anoop M. Meraney
Department of Urology, The Cleveland Clinic Foundation, Cleveland, OH, USA

Inderbir S. Gill*
Department of Urology, The Cleveland Clinic Foundation, Cleveland, OH, USA

INTRODUCTION

Minimal access surgery has been widely implemented for advanced urological procedures, including radical retropubic prostatectomy. Laparoscopic radical prostatectomy is now routinely performed at several eminent centers worldwide. More recently, minimally invasive robotic systems have been developed in order to execute laparoscopic procedures. In this chapter we describe the current status of robotic radical prostatectomy with an emphasis on the benefits and drawbacks of robotic surgery.

Laparoscopy is associated with numerous established benefits compared to open surgery. However, laparoscopy is also associated with certain constraints, which include: (1) lack of stereovision, which in turn results in difficulty with depth perception and an obtunded hand–eye coordination; (2) impaired tactile feedback; (3) fixed position of the trocars and long shafts of laparoscopic instruments that cause the instruments to work as levers intracorporeally, resulting in mirrored hand movements; (4) exaggeration of amplitude of motion and hand tremor; and (5) a restricted range of motion of the instruments. Also, the surgeon depends completely on the skill and coordination of the assistant for control of the laparoscope. As such, performance of advanced laparoscopy involves acquisition of specialized skills and application of unique surgical techniques. Robotic systems for minimally invasive surgery are designed to overcome these constraints associated with conventional laparoscopy and provide improved surgical precision.

A robotic system is essentially a programmable automated task performance system controlled by microprocessors. Robotic systems for minimally invasive surgery consist of three robotic arms and a control station, the surgeon console. One of the three robotic arms is dedicated for control of the laparoscope, and the other two arms are designed to operate various interchangeable laparoscopic instruments. The surgeon is seated at the surgeon console from where he manipulates the robotic arms to execute surgical tasks effectively.

HISTORICAL BACKGROUND

Initial application of robotics in urology was for prostate biopsies[1] and transurethral resection of the prostate (TURP).[2] The Green Telemanipulator surgical system (SRI International, Menlo Park, CA) was a robot designed to perform surgical tasks in the battlefield.[3] The initial experimental laparoscopic telesurgical procedure was reported by Schurr and colleagues.[4] They utilized the ARTEMIS (Advanced Robotic Telemanipulator for Minimally Invasive Surgery) system for a robotic laparoscopic cholecystectomy in a porcine model. The da Vinci surgical system (Intuitive surgical Inc., Mountain View, CA) has been developed from its prototype, the Green telemanipulator, through the combined efforts of Stanford Research Institute, Massachusetts Institute of Technology and International Business Machines Corporation. While the prototype was capable of only four degrees of freedom, the more advanced and modern version is capable of seven degrees of freedom of motion. Another telerobotic system capable of performing minimally invasive procedures is the Zeus Microsurgical Robotic System (Computer Motion, Goleta, LA) and the Aesop arm voice-controlled laparoscopic manipulator. A newer version of Zeus system, the Z2P, provides three dimensional (3D) vision, similar to the da Vinci robotic system.

*To whom correspondence should be addressed.

TECHNICAL SPECIFICATIONS

Currently two surgical systems are commercially available and widely utilized for laparoscopic telesurgical procedures: the Z2P and the da Vinci robotic system.

Design of working arms of the robotic system

For a surgical procedure to be performed laparoscopically, the three robotic arms are strategically placed around the operative field. Each working arm of the robot is comprised of several joints and links capable of delivering rotational and translational movements. The robotic arms in the da Vinci system are mounted on a patient-side cart, which is wheeled into position beside the operating room table. Robotic arms of the Zeus system are independent units, each of which are individually mounted on the side rail of the operating room table.

The distal appendage in the da Vinci system incorporates a unique 'Endowrist', which provides increased dexterity. The end effector in the da Vinci robotic system delivers a total of seven degrees of freedom of motion including yaw, pitch, insertion, grip and an added three degrees resulting from the Endowrist. The Zeus robotic system is capable of delivering six degrees of freedom of motion. In comparison, during conventional laparoscopy only four degrees of motion are possible.

Optics

Stereoscopic vision in the da Vinci system is facilitated by the use of a two-channel endoscope. Individual three-chip cameras sample the two channels. The output from each of the cameras is displayed on individual cathode ray tubes. The Z2P robot provides stereovision using passive eyewear. A shutter glass on the screen polarizes the image into images for the left and right eyes, resulting in 3D visual perception. Further, the design of the surgeon console of the da Vinci system combined with stereovision provides an impression that the surgeon is actually immersed within the operative field, thereby providing visual characteristics similar to open surgery.[5]

Features incorporated to improve surgical precision

During laparoscopy, the surgeon's hand movements are exaggerated at the tips of the instruments. The elongated shafts of the laparoscopic instruments, and their fixed position within the trocar, results in a lever action. This causes increased amplitude of motion at the tips of the instruments. The master–slave manipulators are equipped with motion scaling technology; whereby the movement of the surgeons' hands can be scaled down to provide added precision. The scale can be adjusted in order to tailor to the specific procedure being performed and the preferences of the individual surgeon. For example, a scale of 1:4 translates into 1 mm motion in the body for 4 mm of motion by the surgeon controlling the console. Furthermore, the system is also equipped with a motion filter, which is capable of eliminating unintentional hand tremor.

The da Vinci robot incorporates haptic feedback technology that permits detection of 0.6 N of force, which is equivalent to 4 mm tissue deflection. However, the current feedback system has poor sensitivity compared to conventional laparoscopy.

Safety features

Both the da Vinci and Zeus robotic systems have a safety clutch incorporated into the system, which is operated by pressing a foot pedal. The clutch prevents unattended movement of the robotic arms. The Aesop arm has a built-in safety feature by which the laparoscope is automatically released from the clasp of the robotic arm when a pressure of more than 5 lb is sensed. This feature safeguards against any inadvertent injury.

SURGICAL TECHNIQUE

The procedure is performed under general anesthesia and endotracheal intubation. The patient is positioned supine on the operative table, with arms adducted, and legs abducted and placed in Allen stirrups. A Trendelenburg position is employed in order to displace small bowel loops from the pelvis into the upper abdomen. The procedure described below is performed utilizing the da Vinci robotic system. Other groups have performed the procedure utilizing the Zeus robotic system.

The operative field is prepared and draped in the usual fashion, and a 22F Foley catheter is inserted and maintained sterile within the operative field. A 1.5 cm incision is created in the inferior umbilical crease and the peritoneum is accessed utilizing the open technique. A 12-mm cannula is inserted and pneumoperitoneum is established. A 12-mm 0-degree endoscope is employed and all subsequent ports are introduced under laparoscopic guidance. Two 8-mm trocars are inserted pararectally, bilaterally. Two additional 10-mm ports are introduced 2–3 fingerbreadths superomedial to the right and left anterior superior iliac spine.

The robotic arms are then wheeled in adjacent to the operating table. The two robotic manipulators are connected to the two-pararectal ports and the camera is set up. The assistant utilizes the lateral ports during the procedure. Various robotic instruments are utilized for the procedure, including the hook electrode, scissors, graspers and needle drivers. The procedure is performed utilizing the master–slave manipulator by the surgeon seated at the control console. The assistant standing besides the patient operates the clip applicator and suction device through the lateral ports.

Operative steps

Initially, the sigmoid colon is retracted cephalad by utilizing a suture, which is passed through the appendices epiploicae and suspended from the anterior abdominal wall.

Dissection of the seminal vesicles and the vas deferens

The posterior parietal peritoneum overlying the seminal vesicles is incised. The vasa are identified and divided bilaterally. Anterior traction on the vasa by the assistant facilitates the subsequent identification and dissection of the seminal vesicles, and the Denonvillier's fascia. Since the artery to the vesicles lies in close proximity to the cavernous nerves, the use of electrocautery is eliminated and the arteries are controlled utilizing hemoclips bilaterally. The Denonvillier's fascia is incised at this time and a plane of dissection is developed between the prostate anteriorly and the rectum posteriorly.

Entering the prevesical space
The bladder is distended with 200 ml of saline. An inverted U-shaped peritoneal incision is created starting laterally medial to the medial umbilical ligament of one side, and is extended anteriorly and medially towards the anterior abdominal wall encompassing the urachus, and then over to the contralateral side. A combination of blunt and sharp dissection is employed to enter the space of Retzius.

Dissection of the endopelvic fascia and control of the dorsal vein
The endopelvic fascia are identified, dissected and incised bilaterally. The Foley catheter is replaced with a metal sound, and a 2–0 vicryl suture on a 37-mm needle is utilized to control the dorsal vein utilizing robotic intracorporeal suturing and knot-tying techniques. The suture is anchored anteriorly to the periosteum of the pubis as a urethral suspension stitch. A suture is placed on the anterior surface of the prostate to prevent back-bleeding from the dorsal vein following its transection.

Dissection of the bladder neck and control of the lateral pedicles
The next step involves precise identification of the bladder neck, which is facilitated by tactile identification of the tip of the metal sound just beyond the base of the prostate. The anterior bladder neck followed by the posterior bladder neck is incised, and the previously dissected seminal vesicles and vasa are delivered anteriorly. Lateral traction on the vesicles by the assistant facilitates dissection of the pedicles. Next, the lateral pedicles are controlled with endoclips and dissection is continued anteriorly along the lateral surface of the gland in order to dissect the neurovascular bundles bilaterally.

Apical dissection and urethrovesical anastomosis
The dorsal vein is divided at this time followed by dissection of the prostatic apex. The urethra is identified and divided distal to the apex of the gland. This is followed by division of the fibers of the rectourethralis muscle. The surgical specimen is then placed in an Endocatch bag and placed in the right upper quadrant of the abdomen.

The urethrovesical anastomosis is performed utilizing two running 2–0 vicryl sutures. One suture begins at 5 o'clock position and is run in a clockwise manner up to the 11 o'clock position, while the second row is run in an anticlockwise manner. Given that the robotic system permits greater surgical dexterity than can be achieved with conventional laparoscopy, suturing with the robot is easier to perform and is more precise.

At the completion of the anastomosis a 10-mm JP drain is placed in the pelvis through one of the lateral ports and the specimen is extracted through a circumumbilical extension of the umbilical port site.

RESULTS

Binder and colleagues reported a series of ten robotic-assisted laparoscopic radical prostatectomies.[6] Patients selected for the procedure had an average age of 60.5 years (range 57–69 years), clinical stages T1b–T2b and the average prostate specific antigen (PSA) level was 6.4 ng/ml (range 0.5–22.4 ng/ml). In one patient, the procedure was converted to open for control of hemorrhage. Median operative time for nine robotic cases was 9 hours (range 8.75–11 hours). The catheter was removed after a median time of 18 days (range 5–23 days). Bilateral pelvic lymphadenectomy was routinely performed in all cases. Lymph nodes were positive for cancer in one patient. Pathologic tumor stages were pT2a (2), pT2b (4), pT3a (2) and pT3b (2). Three patients had positive surgical margins. Pasticier and colleagues reported their experience with robotic-assisted radical prostatectomy in five patients.[7] Average age of the patients was 58 years, average PSA level was 12, and Gleason score was 12. A nerve-sparing radical prostatectomy procedure was performed in all cases. Average time for installation of the robot was 93 minutes (range 76–149 minutes). Average operative time was 222 minutes (range 150–381 minutes), average blood loss was 800 ml (range 700–1600 ml), and average length of hospital stay was 5.5 days. The average catheter time was 6.5 days. One patient had a urine leak postoperatively. Four of the five patients were continent in the immediate postoperative period. One of five patients had a positive surgical margin. The authors commented that the urethrovesical anastomosis was easier to perform with the robotic system compared to conventional laparoscopic radical prostatectomy.

Rassweiler and colleagues reported their experience with seven robotic laparoscopic radical prostatectomies.[8] The average age of the patients was 64.2 years (range 57–71 years) and the average preoperative PSA was 8.4 ng/ml (range 2.4–12 ng/ml) and Gleason score 6. Preoperative staging included T1/c (1), T2a (4), T2b (1) and T3 (1) tumors. The mean operative time was 351 minutes. Pelvic lymphadenectomy was performed in all cases. Pathological staging revealed pT2a (1), pT2b (1) and pT3a (4) tumors. There were no positive surgical margins. Postoperatively catheters were removed after an average of 7.3 days.

COMMENTS

Minimally invasive robotic procedures are gaining widespread application in general surgery,[9] gynecology,[10] cardiac surgery,[11] orthopedics[12] and urology. Currently, over 70 da Vinci robotic systems have been installed at various medical centers in the USA, Europe and in Japan. In urology, robotics has been utilized to perform procedures, including pyeloplasty,[13] nephrectomy, adrenalectomy[14] and radical prostatectomy.

Laparoscopic radical prostatectomy is a technically challenging and strenuous procedure for the surgeon. The ergonomic design of the surgeon console makes the procedure more comfortable to perform robotically. The surgeon comfortably rests his head on either side of the view port. The 3D immersive vision of the robotic system simulates open surgery. The magnified visual field promotes increased accuracy during dissection and reconstruction. Robotic control of the laparoscope results in performance of procedures with fewer assistants. Furthermore, an advantage of the robotic arm is that it does not fatigue with prolonged operative time. The 'Endowrist' technology of the da Vinci robotic system facilitates increased freedom of motion. This enables improved surgical dexterity, especially for fine dissection and for the performance of intracorporeal suturing.[15] Moreover, the utilization of filter tremors and motion scaling software eliminates unintended movements and thereby further improves precision. The improved quality of vision and increased surgical dexterity result in superior surgical precision and accuracy. Although specialized training is required to operate the system, it is relatively easy to learn for any experienced surgeon. Surgeons without any prior experience with conventional laparoscopy have reported successful performance of advanced laparoscopic procedures utilizing the system.

Advances in telecommunication technology have enabled rapid transfer of audio and visual data. State-of-the-art telecommunications networks integrated with robotic technology have enabled the performance of robotic surgery in a telepresent manner. High-speed optical cables and integrated ISDN lines are being utilized in order to telementor surgical procedures across continents.[16] The Socrates telecollaboration system facilitates state-of-the-art combined audio and video conferencing. It facilitates shared control of the robotic laparoscope during telementoring sessions.

The costs for the robotic systems range from $750 000–1 000 000. This prohibitive cost has deterred more widespread implementation of robotic technology. A disadvantage of the robotic system is its large size, which makes its cumbersome to handle. Also, the installation of the system is time consuming and may take up to 90 minutes to set up, prior to the procedure. The design of the arms often makes it uncomfortable for the assistant during the case, and may obscure direct visual communication between the surgeon and the assistant. Only a limited selection of instruments is available for use with the robotic arms. Currently, robotic clip applicators, suction devices and retractors are not yet available. Another drawback of the robotic system is that, during the procedures, instruments need to be manually exchanged by the scrub nurse or assistant. During conventional laparoscopy with two-dimensional (2D) vision, a sense of depth perception is acquired with experience. This 'learned 3D' vision is similar to the depth perception perceived while watching a 2D television or movie. Surgeons accustomed to 2D visual images need to adapt to the 3D vision technology of the robotic system. A clear disadvantage of the robotic system is that tactile feedback is suboptimal. This makes it difficult to gauge the tension to be applied on sutures during intracorporeal knot tying. This may result in fraying of sutures. Although the robotic camera arm can be a great asset, it deprives the assistant of an invaluable learning experience; this is a potential disadvantage for budding laparoscopists.

Robotic technology is constantly evolving, which will ultimately lead to further improvement of the minimally invasive robotic systems. Potential avenues of improvement include: (1) reduction in size; (2) a wider array of robotic instruments; (3) introduction of more robotic arms to assist in complex procedures; and (4) incorporation of artificial intelligence into robotics and the development of 'smart' robotic systems.

Currently robotic radical prostatectomy is only being performed at select centers. It has proven to be feasible. With improved robotic designs and advances in technology, it remains to be seen whether this technique will result in improved short- and long-term results for patients with prostatic cancer.

REFERENCES

1. Pisani E, Montanari E, Deiana G et al. Robotized prostate biopsy. *Minim. Invas. Ther.* 1995; 4:289–91.
2. Davies BL, Hibberd RD, Coptcoat MJ et al. A surgeon robot prostatectomy: a laboratory evaluation. *J. Med. Eng. Technol.* 1989; 13:273.
3. Green PE, Piantanida TA, Hill JW et al. Telepresence: dexterious procedures in a virtual opening field. *Am. Surg.* 1991; 57:192.
4. Schurr MO, Buess G, Rininsland H et al. The Artemis manipulator system for endoscopic surgery. *Endoskopie Heute.* 1996; 9:245–51.
5. Sung GT, Gill IS. Robotic laparoscopic surgery: a comparison of the Da Vinci and Zeus systems. *Urology* 2001; 58:893–8.
6. Binder J, Kramer W. Robotically-assisted laparoscopic radical prostatectomy. *Br. J. Urol. Int.* 2001; 87:408–10.
7. Pasticier G, Rietbergen JB, Guillonneau B et al. Robotically-assisted laparoscopic radical prostatectomy: feasibility study in men. *Eur. Urol.* 2001; 40:70–4.
8. Rassweiler J, Binder J, Frede T. Robotic and telesurgery: will they change our future. *Curr. Opin. Urol.* 2001; 11:309–20.
9. Cardiere GB, Himpens J, Vertruyen M et al. The world's first obesity surgery performed by a surgeon at a distance. *Obes. Surg.* 1999; 9:206–9.
10. Falcone T, Goldberg J, Garcia-Ruiz A. Full robotic assistance for laparoscopic tubal anastomosis: a case report. *J. Laparoendosc. Adv. Surg.* 1999; 9:107–12.
11. Shennib H, Bastawisy A, Mcloughlin J et al. Robotic computer-assisted telemanipulation enhances coronary artery bypass. *J. Thorac. Cardiovasc. Surg.* 1999; 117:310–13.
12. Paul HA, Bargner WL, Mittelstadt B et al. Development of a surgical robot for cementless hip arthroplasty. *Clin. Orthop.* 1992; 285:57–66.
13. Sung GT, Gill IS, Hsu THS. Robotic-assisted laparoscopic pyeloplasty: a pilot study. *Urology* 1999; 53:1099–103.
14. Sung GT, Gill IS, Meraney AM et al. Telepresent robotic laparoscopic nephrectomy and adrenalectomy: the initial feasibility study (Abstract no.33). *J. Urol.* 2000; 163:8.
15. Falk V, McLoughlin J, Guthart G et al. Dexterity enhancement in endoscopic surgery by a computer-controlled mechanical wrist. *Minim. Invas. Ther. Allied Technol.* 1999; 8:235–42.
16. Janetschek G, Bartsch G, Kavoussi LR. Transcontinental interactive laparoscopic telesurgery between the United States and Europe. *J. Urol.* 1998; 160:1413.

Radiation Therapy

Does Radiation Therapy Really Work for Prostate Cancer?

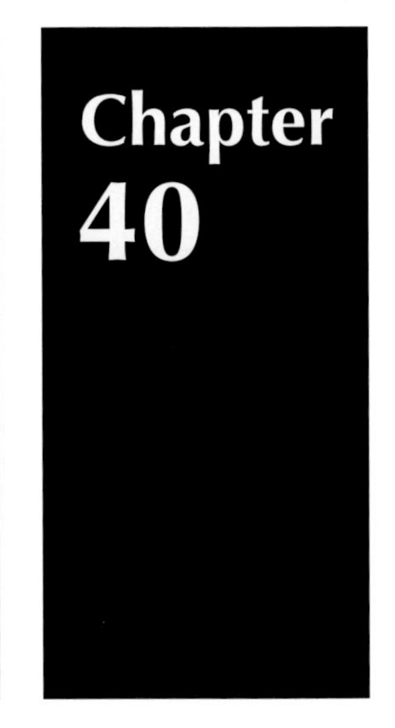

Steven J. DiBiase and Stephen C. Jacobs

Department of Radiation Oncology and Department of Surgery, University of Maryland School of Medicine, Baltimore, MD, USA

INTRODUCTION

It is estimated that 198 100 men will have been diagnosed with prostate cancer in 2001 and that 31 500 deaths will have occurred. Widespread screening with prostate specific antigen (PSA) has led to increased detection of prostate cancer at an early stage, when the tumor is confined to the prostate and, therefore, curable. The improvement in outcome in men with prostate cancer over the last 30 years is illustrated by the trend in 5-year cancer survival rates, which increased from 68% in 1974–1976 to 75% in 1980–1982, and to 94% in 1989–1996.[1]

However, the most effective therapy for an individual patient with early-stage prostate cancer is not clear. Management options for localized prostate cancer include surgery, radiation therapy (external beam or brachytherapy, with or without hormone therapy), or observation, also termed watchful waiting. For men with apparently organ-confined prostate cancer, many issues must be considered in making the choice between these treatments.

Radiation therapy is an effective treatment option for men with early-stage prostate cancer; it is used for definitive therapy in about one-third of men with clinically localized disease.[2] Technological advances in the delivery of radiation therapy, coupled with earlier diagnosis, have led to steadily improving outcomes with radiation over the past two decades.[3] Newer techniques to deliver ionizing radiation include conformal treatment planning for external beam radiation (EBRT), brachytherapy and proton beam therapy. Modern PSA-based series show equivalent biochemical cure rates with radical prostatectomy (RP) and radiation therapy, when men are stratified equally for serum PSA levels at diagnosis, tumor stage and Gleason score.[4,5]

Significant technological advances have led to the development of several different radiotherapeutic approaches over the last decade.[6–8] The choice of treatment offered to a patient is often dependent upon institutional custom, individual physician judgment, patient preference and the availability of resources.[9]

The use of radiation therapy for clinically localized prostate cancer will be reviewed here. An emphasis will be taken to define the curative role of radiation therapy with an objective, evidence-based approach on biochemical control rates and morbidity.

ANATOMIC CONSIDERATIONS

The goal of radiotherapy is the accurate delivery of a therapeutic dose of ionizing radiation to the tumor while minimizing the dose received by normal surrounding tissues.[10] The prostate gland is a midline pelvic structure that lies in close proximity to the rectum and bladder. Treatment planning for the radiotherapeutic management of prostate cancer must take into account the volume and distribution of both tumor tissue and these normal structures.[11] The major sequelae following radiotherapy involve the gastrointestinal (GI) and genitourinary (GU) tract.

Definitive radiation therapy is usually delivered to the entire prostate gland because of the multifocal nature of prostate cancer and our inability to localize all malignant areas accurately within the gland by non-invasive means. Radiation treatment fields are tailored in individual patients based upon the estimated risk of seminal vesicle and regional lymph node spread.[12,13] As the clinical stage, serum PSA concentration and Gleason score increase, the risk of seminal vesicle and pelvic lymph node involvement increases.[14,15] The lymphatic drainage of the prostate gland is accomplished by a periprostatic network that drains into both the external and internal iliac lymph nodes.[16] In men at high risk for pelvic nodal spread, whole pelvis radiation therapy fields, which encompass the external and internal iliac lymph node regions, are utilized.[6]

DEFINING CLINICAL OUTCOME FROM RADIATION THERAPY

Although radiotherapy is an established treatment option for the treatment of clinically localized disease, the best measure of total tumor eradication is uncertain. Historically, the success of radiotherapy in the treatment of prostate cancer was defined by the absence of local progression on physical examination, positive prostatic biopsies, distant metastases or cancer-related death.[17,18] However, with the advent of the serum PSA testing, the assessment of clinical outcome has changed. The pretreatment serum PSA level is of significant prognostic importance in risk stratification for prediction of treatment outcome, and it is commonly used as an end-point for assessing treatment response after radiotherapy (termed biochemical recurrence or failure). Men who have not had a rise in serum PSA following treatment are considered biochemically without evidence of disease (bNED).

Risk of biochemical failure and outcome stratification

The interpretation of outcome data for radiotherapy in early-stage prostate cancer requires stratification for known prognostic factors and knowledge of the treatment delivered. Information derived from predictive models can be used to define risk groups for either biochemical recurrence or death following treatment:[19–21]

- Low risk – 80–90% 5-year PSA failure-free survival rate; 1992 AJCC clinical stage T1c or T2a, and PSA ≤10 ng/ml and biopsy Gleason score of 6 or less.
- Intermediate risk – approximately 50% 5-year PSA failure-free survival rate; 1992 AJCC clinical stage T2b, or PSA between 10 and 20 ng/ml or biopsy Gleason score of 7.
- High risk – 25–33% 5-year PSA failure-free survival rate; 1992 AJCC clinical stage T2c disease, or PSA >20 ng/ml, or biopsy Gleason score of 8 or higher.

As an example, in a multi-institutional pooled analysis that included 1765 patients with T1/2 tumors, recursive partitioning analysis of the initial PSA level, palpation stage and Gleason score was used to derive four separate prognostic groups:[20]

- Group 1 – serum PSA level <9.2 ng/ml.
- Group 2 – serum PSA level between 9.2 and 19.7 ng/ml.
- Group 3 – serum PSA level at least 19.7 ng/ml, with a Gleason score of 2–6.
- Group 4 – serum PSA level of at least 19.7 ng/ml, with a Gleason score of 7–10.

The estimated rates of survival free of biochemical failure at 5 years were 81%, 69%, 47% and 29% for groups 1–4, respectively. Further, the information derived from the Partin model permits the pretreatment stratification of patients by clinical tumor (T) stage, Gleason score and preoperative serum PSA level into defined risk groups for PSA failure following either RP or EBRT.[14,21]

Percentage of positive biopsies

The fraction of prostate biopsies that contain tumor may be an additional independent predictor of biochemical outcome in men with low-, intermediate- and high-risk clinically localized prostate cancer undergoing radical prostatectomy. The percentage of positive biopsies is also a clinically significant factor in men undergoing radiation therapy for clinically localized disease. In one report, the biochemical outcome of men with intermediate-risk disease (see earlier) who had <34% positive biopsies approached that of men with low-risk disease, while for those with >50% positive biopsies, biochemical outcome was comparable to men with high-risk disease.[22]

Clinical significance of biochemical failure

Although biochemical failure is accepted as a legitimate end-point for defining treatment outcome, its clinical significance remains unclear. Following definitive local treatment, overt metastatic disease may not become evident for many years, even with biochemical failure and the relationship to cancer-related mortality is questionable. These issues are illustrated by the following two studies in men undergoing RP for clinically localized disease.

- In one series, men with a rising serum PSA following radical prostatectomy had a survival rate that was similar to men without biochemical failure (88% and 93%, respectively).[23]
- In a second report, only 34% of all men with a biochemical recurrence following RP developed metastatic disease at an average of 8 years; the median time to death after the development of metastases was 5 years.[24] Men with rising serum PSA within 2 years were less likely to be free of clinically evident metastatic disease at 7 years compared with those who failed after 2 years, both for men with Gleason score 5–7 disease (47% vs. 77%), and Gleason 8–10 disease (21% vs. 47%).

The correlation between biochemical failure, clinical failure and cause-specific survival depends upon the definition of biochemical failure. The definition of a biochemical failure following radiation is more complicated than following RP. The mean time to the nadir of the serum PSA (nPSA) typically occurs around 18 months, although it may be prolonged; it was 33 months in one series.[25] The rate of PSA decline following therapy does not appear to correlate with the risk of further relapse. PSA levels can fluctuate significantly after radiation therapy, especially in men undergoing prostate brachytherapy. Over 25% of such patients will experience a transient rise in their PSA at 12–24 months without subsequent biochemical progression. Further, it is unreasonable to expect PSA levels to fall to undetectable levels following a course of radiation therapy, since some prostatic glandular tissue remains.

Despite these issues, nadir serum PSA is a strong indicator of treatment success following radiation therapy.[26,27] Although lower nadir values have been associated with improved outcomes, no absolute level has been determined to reflect treatment success from treatment failure. The American Society for Therapeutic Radiology and Oncology (ASTRO) Consensus Panel recommended that nPSA be considered a prognostic factor similar to pretreatment variables, such as serum PSA, Gleason score and T stage.[28]

Biochemical control rates after radiation therapy or surgery are durable beyond 5 years of follow-up. Of 302 patients available for follow-up beyond 5 years and who had bNED control, Shipley et al. noted that only 5% relapsed from their bNED status from the fifth to eighth year.[20] This rate of loss of bNED status beyond 5 years is similar to those that have been reported following radical prostatectomy.[29] Furthermore, in one series of 446 men treated with conventional radiotherapy at the Massachusetts General Hospital, over 73% of men who had a non-rising serum PSA level <2 ng/ml at 5 years following EBRT remained biochemically disease-free at 10 years after treatment. The lower the PSA level at 5 years, the more durable the response.[30]

Relative stability of the nadir value is important, since a rising value is likely to indicate recurrence. The rate of PSA decline following therapy does not correlate with the risk of further relapse. To standardize the use of serum PSA testing for outcome assessment following radiotherapy, ASTRO organized a consensus panel to define biochemical failure and outline its role in assessment of treatment response following radiation.[28] The panel agreed upon four guidelines, as listed below.

- A PSA recurrence is defined as three consecutive increases in PSA after radiation therapy or a single rise so great as to trigger the initiation of hormone therapy. For censored data analysis, the date of failure should be the midpoint between the postirradiation nadir PSA and the first of the three consecutive rises. It was recommended that series be presented with a minimum period of observation of 24 months. Furthermore, it was recommended that PSA determinations be obtained at 3- or 4-month intervals during the first 2 years after the completion of radiation therapy, and every 6 months thereafter.
- PSA recurrence is an appropriate early end-point for clinical trials; however, it may not be a justification to initiate additional treatment.
- No level of PSA failure has, as yet, been shown to be a surrogate for clinical progression or survival.
- Although the PSA nadir is a strong prognostic factor, no absolute level is a valid cut-point for separating successful and unsuccessful treatments. nPSA is similar in prognostic value to pretreatment prognostic variables.

Using this definition of biochemical failure following radiotherapy in a retrospective review that included 568 men who underwent prostate radiotherapy for clinically localized disease, the 5-year actuarial rates of distant metastases-free survival were significantly worse for those with a biochemical failure (74% vs. 99%, respectively).[31]

RTOG prognostic model to predict long-term survival following radiation

Based upon data from 1557 men enrolled in four different Radiation Therapy Oncology Group (RTOG) trials for clinically localized prostate cancer, the RTOG developed a model based upon these factors to predict disease-specific survival following definitive radiotherapy for prostate cancer.[32] Disease-specific survival was defined as death from prostate cancer or the complications of treatment, or death from other causes, if active prostate cancer was present. Four subgroups were identified that had similar disease-specific survival as listed below.

- Group 1 – Gleason score 2–6, T1–2 Nx.
- Group 2 – Gleason score 2–6, T3 Nx; or Gleason score 2–6, N+; or Gleason score 7, T1–2 Nx.
- Group 3 – Gleason score 7, T3 Nx; or Gleason score 7, N+; or Gleason score 8–10, T1–2 Nx.
- Group 4 – Gleason score 8–10, T3 Nx; or Gleason score 8–10, N+.

The 5-, 10- and 15-year disease-specific survival were significantly different for these groups. Pretreatment serum PSA was not included in this model.

Role of postradiation prostate biopsy to assess for local failure

Some men who undergo definitive radiation treatment for organ-confined prostate cancer will fail locally without distant metastases. Some of these men can be potentially cured with salvage prostatectomy. Most will be identified by a progressive rise in their serum PSA above the nadir value following radiation therapy, without clinical evidence of metastatic disease. In patients considering salvage surgery, prostate biopsy is recommended.[33]

In men with stable serum PSA levels following definitive radiation, the benefit of routine postradiation biopsies is controversial. The long doubling time of many prostate tumors, coupled with radiobiologic data indicating that cell death following radiation is a postmitotic event, suggests that the time course of disappearance of viable cancer from the prostate is prolonged. As a result, false-positive biopsies may be due to delayed tumor regression, and indeterminate biopsies (usually showing radiation effect in viable tumor cells) are of uncertain significance.[34] In one series, 30% of indeterminate biopsies showed eventual clearance of tumor at a mean time of 30 months following radiotherapy.[34]

Despite these caveats, the persistence of viable tumor beyond 18 months may have some clinical significance, especially in men with high pretreatment serum PSA levels. In one series of 498 men treated with conventional radiation over a 10-year period who had postradiation biopsies following the completion of treatment, a positive biopsy between 24 and 36 months following treatment was an independent predictor of outcome in both univariate and multivariate analysis.[34] However, positive rebiopsy results do not appear to add significant clinical information to that provided by sequential serum PSA measurements.[33] As a result, rebiopsy is not routine following radiation therapy, and is only indicated if the PSA is rising and salvage surgery or brachytherapy are being considered.

Comparing outcome of radiation and radical prostatectomy

Because a randomized trial comparing radical prostatectomy to radiation for men with clinically localized prostate cancer has not been performed in the modern era, risk stratification must be used to compare contemporaneously treated patients. As an example, in a recent series from the Cleveland Clinic, the biochemical outcomes of 2127 men treated with either external radiation therapy or radical prostatectomy between 1987 and 1998 were compared.[35] When patients were stratified by serum PSA, Gleason score and stage, there was no difference in biochemical relapse-free survival at 5 years.[35]

However, late recurrences following radiation therapy can occur, since the tumor may not be completely destroyed. As a result, 10 years or more after treatment, the outcome with radiotherapy may not be as favorable as with surgery.[36] However, the available data on this issue using modern radiotherapy techniques are limited.

CONVENTIONAL EXTERNAL BEAM RADIATION THERAPY

Conventional EBRT utilizes treatment planning methods in which the prostate and other target tissues are identified by the anatomy of surrounding structures (contrast-defined bladder and rectum).[17] Typically, a fluoroscopic simulation is performed to design the appropriate treatment portals. After 1980, simulation techniques have commonly included information obtained from computed tomography (CT) scans to localize and plan the appropriate target volumes. Much of the published literature detailing long-term results from EBRT is based upon the older, non-CT-based methods of treatment planning.[18]

Treatment results

Multiple contemporary series have reported rates of biochemical failure-free survival following conventional EBRT (usually at doses <70 Gy) for clinically localized prostate cancer, stratified by pretreatment serum PSA levels.[15,17,20,21,32,37] In general, as the pretreatment serum PSA levels increase, rates of biochemical control decrease[20] as shown in **Table 40.1**.

Outcome is also influenced by pretreatment T stage and histologic grade. In one series of 1044 men with stage T1–T4 prostate cancer treated with conventional technique, the rates of biochemical control for men with T1–2 disease were 60% and 40% at 5 and 10 years, respectively.[38] When subdivided by grade, the 10-year survival for well, moderate and poorly differentiated tumors was 53%, 42%, and 20%, respectively.

Table 40.1. Pre-treatment PSA vs biochemical control.

Serum PSA (ng/ml)	bNED
<10	60–100%
10–20	40–75%
20–30	20–60%
>30	10–35%

Importance of dose

Improved outcomes have been reported with the advent of modern techniques to escalate the radiation dose safely to the prostate gland.[3,39–41] The following illustrate the range of findings in contemporary trials performed in the PSA era.

- In a randomized phase III trial comparing 70 versus 78 Gy in men with localized prostate cancer, made possible with the use of a three-dimensional conformal radiation therapy boost, higher doses were associated with a significant improvement in freedom from biochemical and/or clinical failure at 5 years in men with a pretreatment PSA of >10 ng/ml (75% vs. 48%).[42] Although there were no differences in acute bowel or bladder toxicity between the two groups, the incidence of late rectal complications was higher when 25% of the rectum received 70 Gy or more.[43]
- In another series that included 1041 consecutive patients with clinically localized prostate cancer receiving conventional EBRT, the likelihood of biochemical relapse-free survival (bRFS) at 5 and 8 years was significantly better for those receiving ⩾72 Gy compared to <72 Gy (87% vs. 55%, and 87% vs. 51%, respectively).[44] The 8-year clinical relapse rates were also significantly better for doses ⩾72 Gy (17% vs. 5%).
- Several institutions are reporting improved disease-free survival with higher than conventional radiation doses, often administered utilizing conformal techniques (see below).[3,13,37,40,41] As an example, in an analysis of 1465 men treated on four randomized radiation therapy trials conducted in the RTOG for localized prostate cancer, radiation doses >66 Gy were associated with a 29% lower relative risk of death from prostate cancer.[41] In a second report in which conformal or intensity-modulated techniques were used to achieve higher radiation doses, men with favorable risk disease who received ⩾75.6 Gy had a significantly better 5-year PSA relapse-free survival compared to men receiving 64.8–70.2 Gy (90% vs. 77%)[45] (see 'Intensity-modulated radiation therapy' later). The corresponding rates for those with intermediate or unfavorable risk cases were 70% versus 50%, and 47% versus 21%, respectively.

Complications of external beam radiation therapy

The morbidity associated with EBRT is very low with modern techniques. Treatment-related sequelae during or after EBRT can be grouped into four categories: GI, GU, sexual and other. In order to judge treatment-related morbidity following EBRT, the RTOG developed the commonly used physician-based acute and late morbidity scales.[46] In evaluating toxicity, it should be noted that patient-reported scores usually exceed those reported by physicians.[47]

Gastrointestinal complications

Acute intestinal toxicity during radiation therapy manifests as enteritis due to the effects of radiation on the small and large intestine. The severity is proportional to the amount of bowel in the radiation field.

After radiation is completed, acute symptoms usually return to baseline within 3–8 weeks. In patients with T1–T2 tumors, long-term (late) side effects persist in a low percentage of patients, manifesting as persistent diarrhea, tenesmus, rectal/anal strictures or hematochezia.[48] The reported incidence of radiation proctitis ranges from 2% to 39%, depending upon the definition used, and the dose and field of radiation therapy employed. In a review of the experience of multi-institutional RTOG randomized trials 7506 and 7706, the use of non-conformal EBRT was associated with a 3.3% incidence of grade 3–5 late GI toxicity.[49]

Urinary complications

Up to one-half of patients experience some degree of urinary frequency, dysuria and urgency during EBRT. These acute symptoms typically resolve within 3 weeks after the completion of radiation. Late side effects are uncommon in modern series; one multi-institutional review reported a urinary incontinence rate of 0.3% after EBRT.[48]

Long-term GU toxicity includes urethral stricture, cystitis, hematuria and bladder contracture. In a review of two large RTOG randomized trials, the incidence of grade 3 or higher GU toxicity was 8%, with half of these attributed to urethral stricture that could be managed with a dilation procedure as an outpatient.[49]

Sexual dysfunction

Impotence is reported in 20–50% of men following EBRT, depending upon the definition of potency and the time frame of assessment.[47] More contemporary series have revealed that 60–70% of men who are potent prior to radiation retain potency following treatment.[50,51] However, deterioration of potency with age, intercurrent diseases, such as hypertension, cardiovascular disease and diabetes, and the use of medications all compromise erectile function in this population of older men with prostate cancer.

Comparison of sequelae from radiation therapy and radical prostatectomy

There are no randomized trials comparing the incidence and severity of complications from radiation versus radical prostatectomy. However, there are several large prospective series of men treated with either modality for clinically localized disease that provide some valuable information regarding adverse effects.

In one prospective single-institution cohort study of 279 men with early-stage prostate cancer who underwent either radical prostatectomy or external beam radiation, patient-reported complications were as follows.[47]

- Irritative bladder symptoms were reported at 3 and 12 months following radiation and surgery by a similar number of patients.[47] In contrast, substantial urinary incontinence and the need for absorptive pads at 12 months were reported less frequently after radiation (2% vs. 11% and 5% vs. 35%, respectively). Incontinence following radiation only developed in men older than 65.
- Irritative bowel complaints, such as bowel urgency, were more common following radiation (28% vs. 8% at 3 months and 19% vs. 6% at 12 months).
- Sexual dysfunction, which was nearly universal in men 3 months postoperatively, improved at 12 months, especially in men younger than age 65. In contrast, sexual dysfunction following radiation was less common, but increased in incidence between 3 and 12 months.

These important differences in urinary, bowel and sexual function continue to be evident 2 years following treatment with either radiation or radical prostatectomy, as illustrated in a subset of 1591 men aged 55–74 who were enrolled on the Prostate Cancer Outcomes Study.[52] Almost 2 years following treatment, patients undergoing radical prostatectomy had a more than twofold higher risk of urinary incontinence compared with those treated with EBRT (9.6% vs. 3.5%) and were more likely to be impotent (80% vs. 62%). On the other hand, men receiving radiation therapy reported greater declines in bowel function (diarrhea, bowel urgency and painful hemorrhoids).

THREE-DIMENSIONAL CONFORMAL RADIATION THERAPY

Three-dimensional conformal radiation therapy (3D-CRT) refers to the delivery of radiation to a three-dimensional volume using appropriate imaging studies and computer software.[53,54] The development of sophisticated imaging modalities, particularly helical CT, in conjunction with advanced treatment planning software for 3D-CRT, permits a more precise delivery of radiation to the prostate by better delineation and immobilization of the gland. This decreases the treatment margins and minimizes the volume of normal tissue receiving a clinically significant radiation dose.[9]

Axial CT images of the pelvis are usually used to identify the prostate, rectum and bladder. Treatment planning computers are then used to calculate the dose in three dimensions. The radiation beam arrangements are planned to the prostate target volume (the gross tumor volume, GTV) plus a margin of grossly normal surrounding tissue, termed the clinical target volume (CTV).[55] The planning target volume (PTV) is the final treatment volume that encompasses the GTV, CTV, and any error that results from daily set up and prostatic motion.[12] Either multileaf collimation or cerrobend blocks are used to shape the treatment portal. Dose–volume histograms, which are the visual representation of the amount of dose received by particular target volume, permit the radiation oncologist to assign a specific radiation dose to a particular volume of tissue, maximizing the potential to deliver the highest doses to the areas at highest risk.[5,6]

Treatment results

There is a growing body of evidence to support the use of 3D-CRT over conventional external beam irradiation in the management of localized prostate cancer. Contemporary single arm studies suggest an improved therapeutic ratio for 3D-CRT over conventional EBRT[6] as listed below.

- In one study, baseline serum PSA data were available for 170 consecutive men with clinically localized prostate cancer who were treated conformally, and 90 who received conventional EBRT at a single institution.[56] Among men undergoing irradiation of the prostate only, the 1-year post-treatment serum PSA values were ≤1.5 ng/ml in 76% and 55% of conformally versus conventionally treated men. The corresponding values among those receiving pelvic plus prostatic irradiation were 56% and 38%, respectively.
- Similar data were noted in a second non-randomized comparison that included 146 men undergoing 3D-CRT and 131 treated conventionally for clinical stage T1c or T2 disease. The use of conformal treatment was associated with a significantly higher bNED rate (94% vs. 56%) overall, and in men with Gleason score 5–7 tumors, a higher 5-year bNED rate (96% vs. 53%).[57]

The importance of risk stratification in the interpretation of therapeutic results from 3D-CRT was illustrated in at least two series[3,40] as follows.

- In one retrospective series that included 743 men receiving 3D-CRT, the probability of biochemical control at 5 years, defined as a serum PSA less than 1 ng/ml, for the favorable, intermediate and poor-risk groups was 85%, 65% and 35%, respectively.
- In a second series, 163 men with T1–3, Gleason score 7 prostate cancer received 3D-CRT (76 Gy) at a single institution.[58] The 5-year biochemical control rate for all patients was 66%; stratified by pretreatment serum PSA levels, 5-year bNED rates were 83%, 65% and 21% for serum PSA levels 0–9.9, 10–19.9 and ≥20 ng/ml, respectively.

The subgroup that appears to benefit most from 3D-CRT are men with pretreatment serum PSA levels >10 ng/ml, in whom the use of 3D-CRT results in an almost 30% improvement in bNED control at 5 years compared with conventional EBRT.[43,45] This was illustrated in one retrospective series that included 232 men treated with 3D-CRT at doses from 63 to 79 Gy.[40] For men with pretreatment serum PSA levels 10–20 ng/ml, bNED rates for those treated to 70 and 76 Gy were 35% versus 75%, respectively; for those with serum PSA values >20 ng/ml, the corresponding vales were 10% and 32%, respectively.

Although most contemporary series rely on the reporting of biochemical failure as a clinical end-point for comparison of treatment results, overall survival is high following 3D-CRT, even in men who have biochemical failure. As an example, in one series of 718 men who received 3D-CRT for clinically localized disease, biochemical failure developed in 154 patients (21%).[59] However, the 5-year overall and cause-specific survival after biochemical failure were 58% and 73%.

Complications of 3D-CRT

Despite the delivery of higher doses of radiation, the use of 3D-CRT has been associated with less acute and late GI and GU toxicity compared with EBRT.[60–64]

GI toxicity

In particular, the risk of late radiation proctitis appears to be significantly reduced with the use of 3D-CRT compared with EBRT.

- In one series of 721 men with clinically localized prostate cancer who were treated with 3D-CRT to a median dose of 68.4 Gy (range 59.4–80.4), 537 had clinical stage T1–2 tumors.[64] Although grade 1 or 2 chronic rectal morbidity was observed in 11%, the actuarial risk of a grade 3 or 4 complication was 3% at both 3 and 5 years.
- One trial directly compared the toxicity of conformal and conventional radiation in 225 men with clinically localized prostate cancer who were randomly assigned to receive 64 Gy administered either by conformal or conventional techniques.[61] The primary end-point was the development of late radiation complications >3 months after treatment, as measured by the RTOG score.[65] Conformal treatment was associated with significantly less grade 1 (37% vs. 56%) or grade 2 (5% vs. 15%) proctitis or bleeding.
- The incidence of chronic GI complications following 3D-CRT, especially late rectal bleeding, appears to be dose related. In one study, the 3-year incidence of grade 2 rectal bleeding requiring medication or 1–2 laser coagulations was 11% in patients receiving <73 Gy versus 22% in those receiving >73 Gy. More severe grade 3 or higher GI toxicity was seen in 1.7% and 7% of patients receiving less than and greater than 73 Gy, respectively.[40]

GU toxicity

Acute GU toxicity is common during treatment but late toxicity is uncommon. In one series of 198 patients with clinically localized prostate cancer treated with 3D-CRT to a median dose of 73.8 Gy (range 66–79.2 Gy), grade 1 and 2 acute GU toxicity was noted in 40% and 33% of men, respectively.[66] There were no episodes of grade 3 acute toxicity (e.g. hourly nocturia, gross hematuria, need for narcotic analgesics or catheterization for acute urinary retention).

The incidence of chronic treatment-related GU injury appears to be less with conformal than with conventional techniques. As an example, in one series of 616 men treated with definitive prostate irradiation, the incidence of late GU injury was reduced from 4.6% with conventional techniques to 0.2% with conformal RT, despite higher doses in the 3D-CRT group (almost one-half received >74Gy).[60]

Intensity-modulated radiation therapy

Intensity-modulated radiation therapy (IMRT) is an advanced form of 3D-CRT that can create a dose distribution around a complex and irregular target volume.[67,68] In contrast to conventional or conformal radiation, in which a constant dose rate is administered to a defined field, IMRT delivers non-uniform beam intensities to the target volume. The intensity of the beam is changed during IMRT either by fluctuating the opening of the radiation beam (collimator) with a fixed gantry position ('step and shoot') or by changing the beam opening during an arc.[67–69]

Similar to 3D-CRT, the possible benefit of IMRT is the ability to escalate doses safely to the prostate gland while reducing the complication rate from irradiation of surrounding normal tissue, especially the rectum. As clinical experience grows using IMRT, its optimal use for the treatment of prostate cancer will become evident.[13,67]

In one series from Memorial Sloan–Kettering, 171 men with localized prostate cancer were treated to 81 Gy using IMRT and 61 men achieved this dose by 3D-CRT.[70] The use of IMRT permitted improved coverage of the CTV by the prescription dose, and less rectal toxicity when compared to the 3D-CRT technique. Acute and late urinary toxicity were not different between the two groups.[70] However, the combined rates of acute grade 1 and 2 rectal toxicity (45% vs. 61%), and the incidence of late grade 2 to 3 rectal bleeding (3% vs. 15%) were significantly lower in the IMRT-treated patients.

An overall reduction in irradiated bowel with IMRT as compared to 3D-CRT has also been shown by others. In a study in which the prostate target field, small bowel, colon, rectum and bladder were outlined on CT planning scans of ten men with prostate cancer, the mean percentage volume of small bowel and colon receiving >45 Gy during radiation of the prostate and pelvic lymph nodes for conventional techniques, 3D-CRT and IMRT was 21%, 18% and 5%, respectively.[68] The rectal volume irradiated to >45 Gy was reduced from 50% (3D-CRT) to 6% (IMRT) and the bladder from 52% to 7%. In a second report that included 100 men with localized prostate cancer, 83% of men treated with IMRT had no GI complaints and 27% had no genitourinary complaints during treatment.[67]

HORMONE THERAPY WITH RADIATION

The rationale for neoadjuvant hormone therapy before definitive local therapy rests upon the apoptotic effect (programmed cell death) of castration in prostate cancer cells. Medical androgen ablation therapy decreases circulating levels of serum testosterone, causing a decrease in tumor size before definitive local therapy is administered. In patients who receive EBRT, for example, the use of neoadjuvant therapy may permit smaller treatment fields that encompass less normal tissue volume, thereby minimizing treatment-related toxicity,[71] although data to support this relationship is lacking.

Randomized trials of neoadjuvant or postradiation hormone therapy plus radiation versus radiation alone in patients with locally advanced prostate cancer have shown benefit in disease-free survival, time to progression disease[72,73] and overall survival.[72,74,75]

- In one trial sponsored by the EORTC, 415 men with poor-risk clinical stage T1–4, N0 disease were randomly assigned to external beam radiation alone or radiation plus goserelin for 3 years, starting on the first day of radiation therapy.[74] Men in the combined treatment group also received cyproterone acetate (150 mg per day) for 1 month, starting 1 week before the first injection of goserelin. After a median follow-up of 45 months, combined hormone ablation and radiation was associated with a significant improvement in both 5-year relapse-free (85% vs. 48%) and overall survival (79% vs. 62%).
- RTOG trial 86-10 randomly assigned 471 men with T2–T4 tumors that were greater than 25 cm^3 to radiation therapy with or without 4 months of the LHRH antagonist goserelin, plus the antiandrogen flutamide.[72] Hormone therapy was administered as a short course 2 months before and 2 months during radiation therapy. The addition of antiandrogen therapy significantly improved the 8-year rates of local control (42% vs. 30%), disease-free survival (33% vs. 21%), biochemical disease-free survival (24% vs. 10%) and cause-specific mortality (23% vs. 31%). In subset analysis, the beneficial effect was preferentially seen in men with Gleason score 2–6 tumors. In that population, there was also a beneficial impact on survival (70% vs. 52%).
- The optimal duration of hormone therapy was addressed in a second RTOG trial (92-02) that enrolled 1520 men with T2c–T4 prostate cancer.[75] All patients received 4 months of hormone suppression with goserelin and flutamide 2 months before and 2 months during EBRT, then they were randomized to receive no further therapy or 24 months of additional goserelin. In a preliminary report, after 4.8 years of follow-up, the use of use long-term hormone suppression was associated with a significant improvement in 5-year disease-free survival (54% vs. 34%), and a reduction in local recurrence and distant metastases. In men with Gleason score 8–10 tumors, long-term therapy was also associated with a survival advantage at 5 years (80% vs. 70%).
- For men with clinically localized disease, the stratification of men by their pretreatment risk of disease-related death may permit the identification of subgroups of men with prostate cancer for whom longer duration of hormone therapy is warranted. In a meta-analysis that included 2200 men enrolled on five randomized RTOG trials for clinically localized prostate cancer, men were stratified into four prognostic risk groups based upon primary tumor, nodal status and Gleason score.[32] Patients in risk group 2, who had bulky or T3 disease, or Gleason score 7, T1–2 tumors appeared to have a disease-free survival benefit at 8 years with the addition of 4 months of hormone therapy. In contrast, for men in risk groups 3 and 4 (predominantly Gleason scores 8–10), long-term hormonal therapy was associated with a significant improvement in overall survival by 20% at 8 years.[76]

CONCLUSION

In the last several years, advances in the delivery of radiation therapy have been significant, and now more emphatically provide an effective alternative to radical prostatectomy for the curative management of localized disease. When men are stratified based upon their known risk factors (serum PSA, Gleason score and clinical stage), comparable cure rates are achieved with modern dose selection and treatment technique. As a result of improvements in outcomes, emphasis should be placed on morbidity and quality of life issues of comparable treatments as important end-points in the future to guide patient treatment choices.

REFERENCES

1. Greenlee RT, Hill-Harmon MB, Murray T et al. Cancer statistics, 2001. *CA Cancer J. Clin.* 2001; 51:15–30.
2. Mettlin CJ, Murphy GP, Cunningham MP et al. The National Cancer Data Base report on race, age, and region variations in prostate cancer treatment. *Cancer* 1997; 80:1261–6.
3. Zelefsky MJ, Leibel SA, Gaudin PB et al. Dose escalation with three-dimensional conformal radiation therapy affects the outcome in prostate cancer. *Int. J. Radiat. Oncol. Biol. Phys.* 1998; 41:491–500.
4. Fiveash JB, Hanks G, Roach M et al. 3D conformal radiation therapy (3DCRT) for high grade prostate cancer: a multi-institutional review. *Int. J. Radiat. Oncol. Biol. Phys.* 2000; 47:335–42.
5. Zelefsky MJ, Leibel SA, Kutcher GJ et al. Three-dimensional conformal radiotherapy and dose escalation: where do we stand? *Semin. Radiat. Oncol.* 1998; 8:107–14.
6. King CR, DiPetrillo TA, Wazer DE. Optimal radiotherapy for prostate cancer: predictions for conventional external beam, IMRT, and brachytherapy from radiobiologic models. *Int. J. Radiat. Oncol. Biol. Phys.* 2000; 46:165–72.
7. Sharkey J, Chovnick SD, Behar RJ et al. Evolution of techniques for ultrasound-guided palladium 103 brachytherapy in 950 patients with prostate cancer. *Tech. Urol.* 2000; 6:128–34.
8. Mohan DS, Kupelian PA, Willoughby TR. Short-course intensity-modulated radiotherapy for localized prostate cancer with daily transabdominal ultrasound localization of the prostate gland. *Int. J. Radiat. Oncol. Biol. Phys.* 2000; 46:575–80.
9. Lanciano R, Thomas G, Eifel PJ. 20 years of progress in radiation oncology: prostate cancer. *Semin. Radiat. Oncol.* 1997; 7:121–6.
10. Pollack JM. Radiation therapy options in the treatment of prostate cancer. *Cancer Invest.* 2000; 18:66–77.
11. Roach M III, Pickett B, Weil M et al. The 'critical volume tolerance method' for estimating the limits of dose escalation during three-dimensional conformal radiotherapy for prostate cancer. *Int. J. Radiat. Oncol. Biol. Phys.* 1996; 35:1019–25.
12. Michalski JM, Purdy JA, Winter K et al. Preliminary report of toxicity following 3D radiation therapy for prostate cancer on 3DOG/RTOG 9406. *Int. J. Radiat. Oncol. Biol. Phys.* 2000; 46:391–402.
13. Zelefsky M. IMRT offers superior dose distribution compared to 3D-CRT. *Radiother. Oncol.* 2000; 55:241–9.
14. Partin AW, Kattan MW et al. Combination of prostate-specific antigen, clinical stage, and Gleason score to predict pathological stage of localized prostate cancer. A multi-institutional update. *JAMA* 1997; 277:1445–51.
15. D'Amico A, Whittington R, Malkowicz B. A multivariate analysis of clinical and pathologic factors that predict for prostate specific antigen failure after radical prostatectomy for prostate cancer. *J. Urol.* 1995; 154:131–8.
16. Raboy A, Adler H, Albert P. Extraperitoneal endoscopic pelvic lymph node dissection: a review of 125 patients. *J. Urol.* 1997; 158:2202–5.
17. Perez CA, Hanks GE, Leibel SA et al. Localized carcinoma of the prostate (stages T1B, T1C, T2, and T3). Review of management with external beam radiation therapy. *Cancer* 1993; 72:3156–73.
18. Hanks GE, Hanlon A, Schultheiss T et al. Early prostate cancer: the national results of radiation treatment from the Patterns of Care and Radiation Therapy Oncology Group studies with prospects for improvement with conformal radiation and adjuvant androgen deprivation. *J. Urol.* 1994; 152:1775–80.
19. D'Amico AV, Whittington R, Malkowicz SB et al. Prostate specific antigen outcome based on the extent of extracapsular extension and margin status in patients with seminal vesicle negative prostate carcinoma of Gleason score < or = 7. *Cancer* 2000; 88:2110–15.
20. Shipley WU, Thames HD, Sandler HM et al. Radiation therapy for clinically localized prostate cancer. *JAMA* 1999; 281:1598–604.
21. D'Amico AR, Whittington R, Malkowicz B et al. Pretreatment nomogram for prostate-specific antigen recurrence after radical prostatectomy or external beam radiation therapy for clinically localized prostate cancer. *J. Clin. Oncol.* 1999; 17:168–72.
22. D'Amico A, Schultz D, Silver B et al. The clinical utility of the percent of positive prostate biopsies in predicting biochemical outcome following external beam radiation therapy for patients with clinically localized prostate cancer. *Int. J. Radiat. Oncol. Biol. Phys.* 2001; 49:679–85.
23. Jhaveri FM, Zippe C, Klein EA et al. Biochemical failure does not predict overall survival after radical prostatectomy for localized prostate cancer: 10-year results. *Urology* 1999; 54:884–90.
24. Pound CR, Partin AW, Eisenberger MA et al. Natural history of progression after PSA elevation following radical prostatectomy. *JAMA* 1999; 281:1591–7.
25. Crook J, Choan E, Perry G et al. Serum prostate-specific antigen profile following radiotherapy for prostate cancer: implications for patterns of failure and definition of cure. *Urology* 1998; 51:566–72.
26. Critz FA, Levinson AK, Williams WH et al. The PSA nadir that indicates potential cure after radiotherapy for prostate cancer. *Urology* 1997; 49:322–6.
27. Aref I, Eapen L, Agboola O et al. The relationship between biochemical failure and time to nadir in patients treated with external beam therapy for T1–T3 prostate carcinoma. *Radiother. Oncol.* 1998; 48:203–7.
28. Consensus statement: guidelines for PSA following radiation therapy. American Society for Therapeutic Radiology and Oncology Consensus Panel. *Int. J. Radiat. Oncol. Biol. Phys.* 1997; 37:1035–40.
29. Pound CR, Partin AW, Epstein JI et al. Prostatic specific antigen after anatomic radical retropubic prostatectomy: patterns of recurrence and cancer control. *Urol. Clin. North Am.* 1997; 24:395–406.
30. Yock TI, Thakral AL, Zeitman AL et al. The durability of PSA failure-free survival from 5 years to 13 years following radiotherapy in patients with localized prostate cancer. *Int. J. Radiat. Oncol. Biol. Phys.* 2000; 48(Suppl.):227–8.
31. Horwitz E, Vicini F, Ziaja EL et al. The correlation between the ASTRO Consensus Panel definition of biochemical failure and clinical outcome for patients with prostate cancer treated with external beam irradiation. *Int. J. Radiat. Oncol. Biol. Phys.* 1998; 41:267–72.
32. Roach M, Lu J, Pilepich MV et al. Four prognostic groups predict long-term survival from prostate cancer following radiotherapy alone on Radiation Therapy Oncology Group clinical trials. *Int. J. Radiat. Oncol. Biol. Phys.* 2000; 47:609–15.

33. Cox JD, Gallagher MJ, Hammond EH et al. Consensus statements on radiation therapy of prostate cancer: guidelines for prostate re-biopsy after radiation and for radiation therapy with rising prostate-specific antigen levels after radical prostatectomy. American Society for Therapeutic Radiology and Oncology Consensus Panel. *J. Clin. Oncol.* 1999; 17:1155–62.
34. Crook J, Malone S, Perry G et al. Postradiotherapy prostate biopsies: what do they really mean? Results for 498 patients. *Int. J. Radiat. Oncol. Biol. Phys.* 2000; 48:355–67.
35. Kupelian PA, Katcher J, Levin H et al. External beam radiotherapy versus radical prostatectomy for clinical stage T1–T2 prostate cancer: therapeutic implications of stratification by pretreatment PSA levels and biopsy Gleason scores. *Cancer J. Sci. Am.* 1997; 3:78–87.
36. Goluboff ET, Benson MC. External beam radiation therapy does not offer long-term control of prostate cancer. *Urol. Clin. North Am.* 1996; 23:617–21.
37. Hanks GE, Hanlon AL, Pinover WH et al. Survival advantage for prostate cancer patients treated with high-dose three-dimensional conformal radiotherapy. *Cancer J. Sci. Am.* 1999; 5:152–8.
38. Zietman AL, Coen JJ, Dallow K et al. The treatment of prostate cancer by conventional radiation therapy: an analysis of long-term outcome. *Int. J. Radiat. Oncol. Biol. Phys.* 1995; 32:287–92.
39. Hanks G, Hanlon A, Pinover W et al. Dose selection for prostate cancer patients based on dose comparison and dose response studies. *Int. J. Radiat. Oncol. Biol. Phys.* 2000; 46:823–32.
40. Hanks GE, Hanlon AL, Schultheiss TE et al. Dose escalation with 3D conformal treatment: five year outcomes, treatment optimization, and future directions. *Int. J. Radiat. Oncol. Biol. Phys.* 1998; 41:501–10.
41. Valicenti RK, Lu J, Pilepich MV et al. Survival advantage from higher-dose radiation therapy for clinically localized prostate cancer treated on the radiation therapy oncology group trials. *J. Clin. Oncol.* 2000; 18:2740–6.
42. Pollack A, Zagars G, Smith LG et al. Preliminary results of a randomized radiotherapy dose-escalation study comparing 70 Gy with 78 Gy for prostate cancer. *J. Clin. Oncol.* 2000; 18:3904–11.
43. Storey MR, Pollack A, Zagars G et al. Complications from radiotherapy dose escalation in prostate cancer: preliminary results of a randomized trial. *Int. J. Radiat. Oncol. Biol. Phys.* 2000; 48:635–42.
44. Kupelian PA, Mohan DS, Lyons J et al. Higher than standard radiation doses (>=72 Gy) with or without androgen deprivation in the treatment of localized prostate cancer. *Int. J. Radiat. Oncol. Biol. Phys.* 2000; 46:567–74.
45. Zelefsky M, Fuks Z, Hunt M et al. High dose radiation delivered by intensity modulated conformal radiotherapy improves the outcome of localized prostate cancer. *J. Urol.* 2001; 166:876–81.
46. Cox JD, Stetz J, Pajak TF. Toxicity criteria of the Radiation Therapy Oncology Group (RTOG) and the European Organization for Research and Treatment of Cancer (EORTC). *Int. J. Radiat. Oncol. Biol. Phys.* 1995; 31:1341–6.
47. Talcott JA, Rieker P, Clark JA et al. Patient-reported symptoms after primary therapy for early prostate cancer: results of a prospective cohort study. *J. Clin. Oncol.* 1998; 16:275–83.
48. Shipley JW, Zietman AL, Hanks G. Treatment related sequelae following external beam radiation for prostate cancer – a review with an update in patients with stages T1 and T2 tumors. *J. Urol.* 1994; 152:1799–802.
49. Lawton CA, Won M, Pilepich MV. Long-term treatment sequelae following external beam irradiation for adenocarcinoma of the prostate: analysis of RTOG studies 7506 and 7706. *Int. J. Radiat. Oncol. Biol. Phys.* 1991; 21:935–9.
50. Hamilton AS, Stanford JL, Gilliland FD et al. Health outcomes after external beam radiation therapy for clinically localized prostate cancer: results from the prostate cancer outcomes study. *J. Clin. Oncol.* 2001; 19:2517–26.
51. Mantz C, Song P, Farhangi E et al. Potency probability following conformal megavoltage radiotherapy using conventional doses for localized prostate cancer. *Int. J. Radiat. Oncol. Biol. Phys.* 1997; 37:551–7.
52. Potosky A, Legler J, Albertsen P et al. Health outcomes after prostatectomy or radiotherapy for prostate cancer: results from the prostate cancer outcomes study. *J. Natl Cancer Inst.* 2000; 92:1582–92.
53. Hanks GE, Lee WR, Hanlon AL et al. Conformal technique dose escalation for prostate cancer: biochemical evidence of improved cancer control with higher doses in patients with pretreatment prostate-specific antigen > or =10 ng/ml. *Int. J. Radiat. Oncol. Biol. Phys.* 1996; 35:861–8.
54. Hartford AC, Niemierko A, Adams JA et al. Conformal irradiation of the prostate: estimating long-term rectal bleeding risk using dose-volume histograms. *Int. J. Radiat. Oncol. Biol. Phys.* 1996; 36:721–30.
55. Bedford JL, Khoo VS, Webb S et al. Optimization of coplanar six-field techniques for conformal radiotherapy of the prostate. *Int. J. Radiat. Oncol. Biol. Phys.* 2000; 46:231–8.
56. Corn B, Hanks G, Schultheiss TC. Conformal treatment of prostate cancer with improved targeting: superior prostate-specific antigen response compared to standard treatment. *Int. J. Radiat. Oncol. Biol. Phys.* 1995; 32:325–30.
57. Perez C, Michalski J, Purdy JA et al. Three-dimensional conformal therapy or standard irradiation in localized carcinoma of the prostate: preliminary results of a nonrandomized comparison. *Int. J. Radiat. Oncol. Biol. Phys.* 2000; 47:629–37.
58. Anderson PR, Hanlon A, Horwitz E et al. Outcome and predictive factors for patients with Gleason score 7 prostate carcinoma treated with three-dimensional conformal external beam radiation therapy. *Cancer* 2000; 89:2565–9.
59. Sandler HM, Dunn RL, McLaughlin PW et al. Overall survival after prostate-specific-antigen-detected recurrence following conformal radiation therapy. *Int. J. Radiat. Oncol. Biol. Phys.* 2000; 48:629–33.
60. Schultheiss TE, Lee WR, Hunt MA et al. Late GI and GU complications in the treatment of prostate cancer. *Int. J. Radiat. Oncol. Biol. Phys.* 1997; 37:3–11.
61. Dearnaley DP, Khoo VS, Norman AR et al. Comparison of radiation side-effects of conformal and conventional radiotherapy in prostate cancer: a randomized trial. *Lancet* 1999; 353:267–72.
62. Hanks GE, Schultheiss TE, Hanlon AL et al. Optimization of conformal radiation treatment of prostate cancer: report of a dose escalation study. *Int. J. Radiat. Oncol. Biol. Phys.* 1997; 37:543–50.
63. Nguyen LN, Pollack A, Zagars G. Late effects after radiotherapy for prostate cancer in a randomized dose-response study: results of a self assessment questionnaire. *Urology* 1998; 51:991–7.
64. Sandler HM, McLaughlin PW, Ten Haken RK et al. Three dimensional conformal radiotherapy for the treatment of prostate cancer: low risk of chronic rectal morbidity observed in a large series of patients. *Int. J. Radiat. Oncol. Biol. Phys.* 1995; 33:797–801.
65. Rubin P. Late effects of normal tissues (LENT) consensus conference, including RTOG/EORTC SOMA scales. *Int. J. Radiat. Oncol. Biol. Phys.* 1995; 31:1035–46.
66. Chou RH, Wilder RB, Ji M et al. Acute toxicity of three-dimensional conformal radiotherapy in prostate cancer patients eligible for implant monotherapy. *Int. J. Radiat. Oncol. Biol. Phys.* 2000; 47:115–19.
67. Teh BS, Woo SY, Butler EB. Intensity modulated radiation therapy (IMRT): a new promising technology in radiation oncology. *Oncologist* 1999; 4:433–42.
68. Nutting CM, Convery DJ, Cosgrove VP et al. Reduction of small and large bowel irradiation using an optimized intensity-modulated pelvic radiotherapy technique in patients with prostate cancer. *Int. J. Radiat. Oncol. Biol. Phys.* 2000; 48:649–56.

69. De Neve W, Claus F, Van Houtte P et al. Intensity modulated radiotherapy with dynamic multileaf collimator. Technique and clinical experience. *Cancer Radiother*. 1999; 3:378–92.
70. Zelefsky M, Fuks Z, Happersett L et al. Clinical experience with intensity modulated radiation therapy (IMRT) in prostate cancer. *Radiother. Oncol*. 2000; 55:241–9.
71. Forman JD, Kumar R, Haas G et al. Neoadjuvant hormonal downsizing of localized carcinoma of the prostate: effects on the volume of normal tissue irradiation. *Cancer Invest*. 1995; 13:8–15.
72. Pilepich MV, Winter K, John MJ et al. Phase III radiation therapy oncology group (RTOG) trial 86-10 of androgen deprivation adjuvant to definitive radiotherapy in locally advance carcinoma of the prostate. *Int. J. Radiat. Oncol. Biol. Phys*. 2001; 50:1243–52.
73. Lawton CA, Winter K, Byhardt R et al. Androgen suppression plus radiation versus radiation alone for patients with D1 (pN+) adenocarcinoma of the prostate (results based on a national prospective randomized trial, RTOG 85–31). *Int. J. Radiat. Oncol. Biol. Phys*. 1997; 38:931–9.
74. Bolla M, Gonzalez D, Warde P et al. Improved survival in patients with locally advanced prostate cancer treated with radiotherapy and goserelin. *N. Engl. J. Med*. 1997; 337:295–300.
75. Hanks G, Lu J, Machtay M et al. RTOG protocol 9202: a phase III trial of the use of long term total androgen suppression following neoadjuvant hormonal cytoreduction and radiotherapy in locally advanced carcinoma of the prostate. *Int. J. Radiat. Oncol. Biol. Phys*. 2000; 48:112S.
76. Roach M, Lu J, Pilepich MV et al. Predicting long-term survival, and the need for hormonal therapy: a meta-analysis of RTOG prostate cancer trials. *Int. J. Radiat. Oncol. Biol. Phys*. 2000; 47:617–27.

Is One Form of Radiation Therapy Better Over Another?

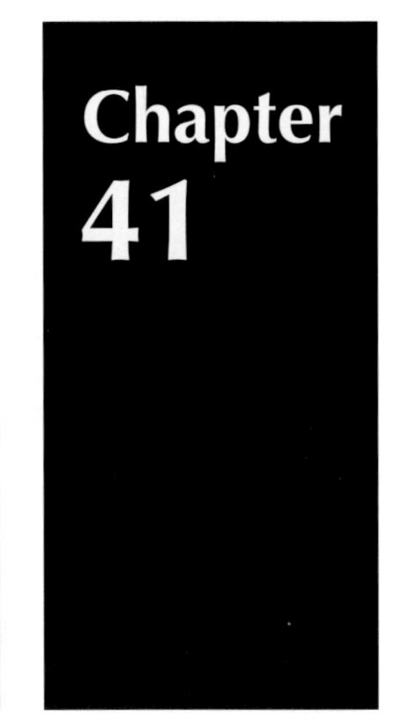

Richard K. Valicenti

Department of Radiation Oncology, Thomas Jefferson University, Philadelphia, PA, USA

INTRODUCTION

The optimal radiotherapeutic approach to treat clinically localized prostate cancer has yet to be established. Current modern radiotherapeutic options involve either the primary use of external beam radiation therapy or the delivery of interstitial brachytherapy. The state-of-the-art techniques consist of three-dimensional conformal radiation therapy (3D-CRT), intensity modulated radiation therapy (IMRT), or image-guided transperineal brachytherapy (IGTPB). Since there are no prospective data comparing directly these radiotherapeutic approaches, the optimal form of therapy at times depends on imprecise criteria relating to relative treatment efficacy, side effects, patient selection and perceived quality of life. In light of this absence of randomized trials, prostate cancer patients are faced with confusion in deciding on treatment for their disease.

Comparisons of outcomes have many limitations and are difficult to carry out successfully. Despite recent reports suggesting excellent results for IGTPB and 3D-CRT, there is a need to compare treatment outcomes in patients with similar pretreatment risk characteristics and the same definition of biochemical relapse. Even if these conditions are met, it is possible only to address approximately the relative benefits of external beam radiation therapy and IGTPB.

In the following review, comparisons will focus on several important series from single institutions and address comparisons between different institutions. The latter series have as an advantage that the same definition of failure was used across treatment groups.

COMPARISON OF EXTERNAL BEAM RADIATION THERAPY WITH TRANSPERINEAL BRACHYTHERAPY

Several recent single-institution series suggest that appropriately selected patients have similar control rates at 5 years between external beam radiation therapy and IGTPB.[1–4] Although these studies were non-randomized, they were restricted to patients with complete pretreatment staging information (including prostate specific antigen; PSA) so that stratification according to risk groups was possible. This strategy is useful to compare outcomes among different forms of therapy but is by no means a substitute to well-designed prospective randomized trials.[5]

There are several purported models used in risk appropriation for men with prostate cancer. For patients with clinically localized prostate cancer, the serum PSA, tumor Gleason score and tumor stage are widely available, and are useful in combination to stratify outcome. Such popular strategies lead to comparisons of different treatments and to assess relative therapeutic benefit. Several authors have reported the results of external beam radiation therapy versus IGTPB according to prognostic groups.[5,6]

In one particular analysis, the outcomes of the major forms of local therapy were classified according to three risk groups. The risk groups were determined by using a combination of prognostic factors, such as the PSA value, Gleason score and clinical stage.[5] The three risk groups were defined as follows.

1. Low-risk group: approximately 80% 10-year PSA failure-free survival, 1992 AJCC clinical stage T1c, 2a, Gleason score $\leqslant 6$ and PSA $\leqslant 10$.
2. Intermediate-risk group: approximately 50% 10-year PSA failure-free survival, 1992 AJCC clinical stage T2b, or PSA >10 and $\leqslant 20$ ng/ml or biopsy Gleason score 7.
3. High-risk group: approximately 33% 10-year PSA failure-free survival, 1992 AJCC stage T2c disease, or PSA >20 ng/ml or biopsy Gleason score $\geqslant 8$.

In this analysis, PSA failure was defined according to the ASTRO 1996 consensus statement.[7]

The risk stratification schemes at best yield limited information regarding the relative efficacy of IGTPB and 3D-CRT.

It is important to keep in mind that no conclusions can be drawn from this non-randomized retrospective comparison. The results of this study evaluated outcome among the three risk groups. High-risk patients as defined as above treated using a radical prostatectomy (RP) or radiation therapy (RT) did significantly better ($P \leq 0.01$) than those managed with implant with or without androgen deprivation. Specifically, high-risk patients managed with implant with or without androgen deprivation had at least a 2.4–3.3-fold increased risk of PSA failure compared with those treated with 3D-CRT. Intermediate-risk patients managed with implant alone had a 3.9-fold increased risk of PSA failure compared with those patients managed with 3D-CRT. Patients with biopsy Gleason scores of 7 did not have significantly different PSA failure-free survival when managed with implant and neoadjuvant androgen deprivation. In the low-risk groups, freedom from biochemical progression were not significantly different.

Potential pitfalls of retrospective analyses generally involve patient selection, physician bias, non-uniform methods of treatment and variation of follow-up time. Other criticisms of the D'Amico study were that patients were treated with neoadjuvant hormonal therapy, varying radiation doses and duration of follow-up depended on type of treatment. Nevertheless, the data from this study would suggest that, in appropriately selected patients, the efficacy of IGTPB appears to be similar to 3D-CRT. Since this study was a retrospective evaluation, it is impossible to determine conclusively the relative benefits of these two forms of radiotherapeutic delivery.

In another retrospective comparison, low-risk patients were uniformly treated with similar radiotherapeutic approaches.[6] Zelefsky et al. reported on a group of 282 prostate cancer patients receiving either IGTPB ($n = 137$) or 3D-CRT ($n = 145$). Again, in this study, the 3D-CRT patients had significantly longer follow-up times (36 months) compared to the IGTPB patients (24 months). There was also the use of neoadjuvant hormonal therapy to reduce the volume of rectum or bladder exposed to high doses of therapy. Unlike the D'Amico study, this analysis also reported on the incidence of late toxicity.

The authors defined biochemical relapse according to the ASTRO consensus definition. Eleven patients (8%) in the 3D-CRT group and 12 patients (8%) in the IGTPB group developed biochemical failure. The 5-year freedom from biochemical relapse rates for 3D-CRT and the IGTPB groups were 88% and 82%, respectively ($P = 0.09$). These results were consistent with reports from other institutions.[8,9]

ACUTE AND LATE TOXICITY

Comparison of morbidity between 3D-CRT and IGTPB has been limited because of the lack of consistent toxicity scoring guidelines and stratification criteria. There has also been a lack of the use of actuarial methods to report toxicity for patients treated with IGTPB. In addition, various implantation techniques, source distribution patterns, various definitions of reference dose and beam orientation have contributed to a wide range of morbidities reported after 3D-CRT or IGTPB. Nevertheless, investigators have made historical qualitative comparisons among the various radiotherapeutic modalities for prostate cancer.

Urinary toxicity

Mild to moderate urinary side effects are the most common sequelae after 3D-CRT of IGTPB, typically radiation-induced urethritis or prostatitis. In 3D-CRT patients, 90% may have no or only mild (grade 1) acute genitourinary (GU) toxicity that may require no therapeutic intervention. As many as 20% may have grade 2 urinary side effects that may require medications for relief. This is in sharp contrast to IGTPB patients where no more than 60% have minimal or no GU toxicity. It is well documented that there is a low incidence (<5%) of both grade 3 or higher urinary toxicity (urinary retention requiring catherization) or rectal toxicity (bleeding)[10,11] for patients treated on a phase I/II dose escalation study, RTOG 9406. These data compared favorably to historical controls.

The extent of less severe urinary side effects have been reported by the group at Memorial Sloan–Kettering.[6,11] Whereas acute urinary toxicity after 3D-CRT persists typically no longer than 6 weeks after completion of treatment, moderate grade 2 urinary symptoms may persist for at least 6–12 months after IGTPB. Zelefsky et al. found that protracted grade 2 urinary symptoms were more prevalent among patients treated with IGTPB compared with 3D-CRT.[5] Grade 2 urinary toxicity that persisted for more than 1 year after IGTPB was observed in 45 patients (31%). In these patients, the median duration of grade 2 urinary symptoms was 23 months (range 12–70 months). In contrast, grade 2 urinary symptoms resolved within 4–6 weeks after completion of 3D-CRT. The 5-year actuarial likelihood of late grade 2 urinary toxicity for the 3D-CRT patients was 8%.

Of note, patients undergoing IGTPB are at higher risk of developing late grade 3 than 3D-CRT patients. Late grade 3 urinary toxicity (urethral strictures) develops in 1–2% in 3D-CRT patients as compared to 5–10% in men undergoing IGTPB. The increased risk of developing such urinary symptoms after IGTPB is related to higher doses ultimately delivered to the urethra, which can reach as high as 1.5–2.0 times more than the prescription dose. The acute urinary symptoms and late urinary morbidity may correlate with the central target doses and the proximity of the seed placement to the urethra. Patients often develop this condition after external beam irradiation if there was a prior transurethral resection of prostate. Other potential factors may include a dose above 70.2 Gy.[1]

Unlike patients undergoing 3D-CRT, in IGTPB patients, the late urethral toxicity is closely correlated with the urethral dose from the implant and not always explainable by prior surgical intervention. A higher incidence of grade 2 GU side effects with IGTPB compared with 3D-CRT results from the higher urethral doses with implants, which is on average 150% of the prescription doses. Other possible reasons for a higher incidence of grade 2 or higher GU toxicity may relate to technical factors, such as the use of higher activity seeds or a non-peripheral-based seed implant approach. This justifies the evaluation of novel computerized optimization programs for treatment planning to constrain the urethral dose and reduce the number of seed placements near the urethra. Whether the loading pattern and seed activity has an impact of the incidence of late toxicity remains to be seen.

Sexual function

Reduction in disease-specific mortality is traditionally the primary measure in evaluating the benefits of treatment for clinically localized prostate cancer. However, if the reduction in death proves small or indeterminate among favorable patients treated with either 3D-CRT or various IGTPB treatment modalities, other outcome variables may play an important role in the selection of primary therapy. Such variables may include a patient's desire to maintain optimal function in physical, psychological and social domains. Often, potential detriments in sexual function and associated quality of life (QOL) need to be weighted in the balance of a man's management decision for his prostate cancer.

Three-dimensional conformal radiation therapy is able to cause sexual impotence by disruption of the arteriolar system supplying the corporal muscles.[12] The mechanism of sexual dysfunction after transperineal brachytherapy (TPB) has been less well evaluated.

Secondly, overall satisfaction with sex life and sexual function may change at different rates over time depending on the radiotherapeutic modality. It may well be that men receiving IGTPB recover at slower rates than those treated with 3D-CRT, since the mechanism of impotency may be neurovascular injury rather than solely microvascular angiopathy. In addition, the radiotherapeutic doses delivered over several months in palladium or iodine brachytherapy as opposed to 8 weeks in external beam potentially prolongs the tissue effect. Another factor is whether the selection of isotope and the actual delivered dose by postimplant dosimetry had any bearing on the rates of potency cited. This is particularly important since data indicate that the use of palladium and higher radiation doses may negatively affect sexual function.[13]

Recently, several studies have been published addressing sexual function and overall quality of sex life after radiation therapy for prostate cancer.[14] Some of these studies did not use validated questionnaires to analyze outcome. Others restricted their analysis to only one type of primary radiotherapeutic treatment, i.e. conventional external beam radiation therapy (EBRT), 3D-CRT or IGTPB.[15–20]

In 3D-CRT treatment planning, there is greater use of more fields, shaped blocks and multileaf collimation, and computer planning systems. These factors may allow one to reduce the dose to adjacent structures, such as the penile bulb, and thus reduce the incidence of sexual dysfunction. In retrospective studies with known potency status before treatment, the rates of dysfunction differ depending of radiotherapeutic technique. Postradiation erectile dysfunction varied from 17% to 84% and, with 3D-CRT, from 27% to 49%.[15–20]

QUALITY OF LIFE ISSUES

Currently, there is limited number of reports of health-related quality of life after prostate cancer treatment comparing IGTPB with 3D-CRT.[21,22] In one cross-sectional survey study, Davis et al. studied patients treated at a single community medical center between 1995 and 1999.[22] Totals of 269, 142 and 222 men who underwent radical prostatectomy, Pd103 implants and conventional external beam radiation therapy were studied using validated quality of life survey instruments. In this study, men who received conventional external beam radiation therapy reported lower bowel function and more bother with their bowels. All treatment groups reported decreased sexual function. Urinary bother scores were unaffected in all three groups. Irritative urinary symptoms were significant in men who received Pd103 brachytherapy and who were surveyed less than 1 year after treatment. In addition, satisfaction with treatment was equivalent among the treatment groups.

Unfortunately, global prospective longitudinal QOL study to assess the changes in QOL for patients treated with IGTPB and 3D-CRT is lacking. Important information from QOL assessments will allow patients and physicians to select optimal treatment that best suits their needs and lifestyle.

SELECTION OF PATIENTS FOR TREATMENT

There is an urgent need to select patients appropriately for radical local treatment and to determine the optimal radiotherapeutic technique. By evaluating tumor stage, Gleason score, prostatic specific antigen level and perhaps percent positive core biopsy, it is possible to determine which patients are at risk for micrometastatic disease and should not undergo radiation therapy as monotherapy. In addition, patients who have a greater than 15% risk of extracapsular disease should not undergo IGTPB alone, and should be considered for combined hormonal therapy and external beam radiation therapy. This approach is justified given recent prospective data.[23,24] In selecting patients for 3D-CRT, a critical issue is whether pelvic irradiation or androgen deprivation should be a component of the overall treatment plan.

The ideal patients for permanent interstitial implantation are those with low risk features who have PSA levels ≤ 10 ng/ml and Gleason scores <7. The prostate gland size should preferably be less than 50 cm^3. In addition, a history of transurethral resection of the prostate (TURP) may increase the risk of urinary morbidity after IGTPB. Patients with pre-existing urinary obstructive symptoms are more likely to experience acute urinary morbidity after IGTPB and should be recommended to undergo 3D-CRT. However, patients with bilateral hip replacements may have a relative contraindication for external beam radiation therapy and may be more suitable for IGTPB. In such patients, computer tomography (CT)-based treatment planning is technically difficult because of the artifact caused by the prostheses making it difficult to visualize the target volume. Other relative contraindications for 3D-CRT include small bowel in close proximity to a high-dose region of 3D-CRT or a history of inflammatory bowel disease.

REFERENCES

1. Zelefsky MJ, Cowen D, Fuks Z et al. Long term tolerance of high dose three-dimensional conformal radiotherapy in patients with localized prostate carcinoma. *Cancer* 1999; 85:2460–8.
2. Wallner K, Roy J, Zelefsky M et al. Short-term freedom from disease progression after I-125 prostate implantation. *Int. J. Radiat. Biol. Phys*. 1994; 30:405–9.
3. Hanks GE, Lee WR, Hanlon AL et al. Conformal technique dose escalation for prostate cancer: biochemical evidence of improved cancer control with higher doses in patients with pretreatment prostate-specific antigen >10 ng/ml. *Int. J. Radiat. Oncol. Biol. Phys*. 1996; 35:862–8.
4. Ragde H, Blasko JC, Grimm PD et al. Brachytherapy for clinically localized prostate cancer: results at 7- and 8-year follow-up. *Semin. Surg. Oncol*. 1997; 13:438–43.
5. D'Amico AV, Whittengton R, Malkowicz B et al. Biochemical outcome after radical prostatectomy, external beam radiation therapy, or interstitial radiation therapy for clinically localized prostate cancer. *JAMA* 1998; 280:969–74.
6. Zelefsky MJ, Wallner KE, Ling CC et al. Comparison of the 5-year outcome and morbidity of three-dimensional conformal radiotherapy versus transperineal permanent Iodine-125 implantation for early-stage prostatic cancer. *J. Clin. Oncol*. 1999; 17:517–22.
7. American Society for Therapeutic Radiology and Oncology Consensus Panel. Guidelines for PSA following radiation therapy. *Int. J. Radiat. Oncol. Biol. Phys*. 1997; 37:1035–41.
8. Kupelian P, Katcher J, Levin H et al. External beam radiotherapy versus radical prostatectomy for clinical stage T1–T2 prostate cancer: therapeutic implications of stratification by PSA levels and biopsy Gleason scores. *Cancer J. Sci. Am*. 1997; 3:78–87.
9. Stock RG, Stone NN. The effect of prognostic factors on therapeutic outcome following transperineal prostate brachytherapy. *Semin. Surg. Oncol*. 1997; 13:454–60.
10. Michalski JM, Purdy JA, Winter K et al. Preliminary report of toxicity following 3D radiation therapy for prostate cancer on 3DOG?RTOG 9406. *Int. J. Radiat. Oncol. Biol. Phys*. 2000; 46:391–402.
11. Arterberry VE, Wallner K, Roy J et al. Short-term morbidity from CT-planned transperineal I-125 prostate implants. *Int. J. Radiat. Oncol. Biol. Phys*. 1993; 25:661–7.
12. Zelefsky MJ, Eid JF. Elucidating the etiology of erectile dysfunction after definitive therapy for prostate cancer. *Int. J. Radiat. Oncol. Biol. Phys*. 1998; 40:129–33.
13. Stock RG, Stone NN, Iannuzzi C. Sexual potency following interactive untrasound-guided brachytherapy for prostate cancer. *Int. J. Radiat. Oncol. Biol. Phys*. 1996; 35:267.
14. Talcott JA, Rieker P, Clark JA et al. Patients-reported symptoms after primary therapy for early prostate cancer: results of a prospective cohort study. *J. Clin. Oncol*. 1998; 16:275–83.
15. Roach M, Chinn DM, Holland J et al. A pilot survey of sexual function and quality of life following 3D conformal radiotherapy for clinically localized prostate cancer. *Int. J. Radiat. Oncol. Biol. Phys*. 1996; 35:869–74.
16. Chen CT, Valicenti RK, Lu JD et al. Does hormonal therapy influence sexual function in men receiving 3D conformal radiation therapy for prostate cancer? *Int. J. Radiat. Oncol. Biol. Phys*. 2001; 50:591–5.
17. Fulmer BR, Petroni GR, Bissonette EA et al. Prospective assessment of voiding and sexual function after treatment for localized prostate cancer: comparison of radical prostatectomy to hormonal-brachytherapy with and without external beam radiotherapy. *Cancer* 2001 (in press).
18. Krupski T, Petroni GR, Bissonette EA et al. Quality-of-life comparison of radical prostatectomy and interstitial brachytherapy in the treatment of clinically localized prostate cancer. *Urology* 2000; 55:736–42.
19. Stock RG, Kao J, Stone NN. Penile erectile function after permanent radioactive seed implantation for treatment of prostate cancer. *J. Urol*. 2001; 165:436–9.
20. Wilder RB, Chou RH, Ryu JK et al. Potency preservation after three-dimensional conformal radiotherapy for prostate cancer. *Am. J. Clin. Oncol*. 2000; 23:330–33.
21. Litwin MS, Hays RD, Fink A et al. Quality-of-life outcome in men treated for localized prostate cancer. *JAMA* 1995; 273:129–35.
22. Davis JW, Kuban DA, Lynch DF et al. Quality of life after treatment for localized prostate cancer: differences based on treatment modality. *J. Urol*. 2001; 166:947–52.
23. Laverdiere J, Gomez JL, Cusan L et al. Beneficial effect of combination hormonal therapy administered prior and following external beam radiation therapy in localized prostate cancer. *Int. J. Radiat. Oncol. Biol. Phys*. 1997; 37:247.
24. Pilepich MV, Sause WT, Shipley WU et al. Androgen deprivation with radiation therapy compared with radiation therapy alone for locally advanced prostatic carcinoma: a randomized comparative trail of the radiation therapy oncology group. *Urology* 1995; 45:616–23.

Radiation Therapy for Early-stage Prostate Cancer – Could It Parallel Prostatectomy?

Chapter 42

Glen Gejerman
Department of Radiation Oncology, Hackensack University Medical Center, Hackensack, NJ, USA

Neil Sherman
Department of Urology, University of Medicine and Dentistry of New Jersey, NJ, USA

RADIATION THERAPY: EQUIVALENT OPTION OR INFERIOR THERAPY?

One of the most vexing controversies in urologic oncology has been determining the appropriate management for patients with early-stage prostate cancer. Despite, or perhaps because of, rapid technological advances in the diagnosis staging and treatment of localized prostate cancer, the optimal curative treatment for this disease has been disputed. Owing to the lack of a well-designed randomized prospective trial with long-term biochemical, disease-specific and overall survival data, the patient diagnosed with organ-confined disease is subject to the heated debate on the comparative merits of the two definitive therapies – radical prostatectomy and external beam radiotherapy. In an attempt to resolve this dilemma, the American Urological Association (AUA) convened the Prostate Cancer Clinical Guidelines Panel in 1989.[1] To estimate survival rates for patients choosing between radical prostatectomy, external beam radiotherapy, brachytherapy or surveillance, a comprehensive survey and analysis of published outcomes data were performed. After detailed review of 1453 articles, practice policy recommendations were based on 165 papers considered sufficient for outcomes data extraction. Because the heterogeneous clinical and pathological characteristics among the series did not allow valid comparisons of the treatments, no consensus regarding superiority of any one modality could be made. The panel presented treatment alternatives as options and proposed either radical prostatectomy or radiation therapy for those patients with a relatively long life expectancy.

Despite these published guidelines, many urologists have concluded that radiation therapy is an inadequate treatment for early-stage prostate cancer. This steadfast conviction, unencumbered by conclusive data, initially emerged in 1982 when Paulson et al.[2] reported the results of a phase 3 randomized trial comparing radical prostatectomy to radiation therapy in 97 patients with clinical stage T1–2N0M0 disease. With time to first evidence of treatment failure as the end-point, 85% of patients who were treated surgically were free of disease at 5 years as opposed to only 59% of those treated by radiation therapy. The authors concluded that surgery was more effective than radiation therapy in establishing disease control. Further skepticism regarding definitive radiotherapy arose when Schelhammer et al.[3] reported disappointing biochemical control rates after treatment with a significant proportion of patients who were clinically without evidence of disease showing disease progression by prostate specific antigen (PSA) evaluation. The most disparaging data were reported by Stamey et al.,[4] who showed that only 20% of irradiated patients had a stable PSA at a median follow-up of 7.7 years. The fact that 80% of these patients had precipitously increasing PSA levels 5 years after therapy led the authors to suggest that ‘irradiation actually accelerates prostate cancer growth rate’.[5] The latter study resulted in emotive and sometimes vitriolic editorials proclaiming ‘these data should serve to refute and demonstrate the fallacy of the observation made frequently by radiation oncologists that radiation and surgery provide equivalent overall 10 year survival rates’.[6]

Despite the above denouncement of alchemy to those who would equate the efficacy of radiotherapy to that of the ‘gold standard radical prostatectomy’, a critical review of the studies cited above demonstrates that serious flaws in their design execution and interpretation render the data inconclusive.

The numerous irregularities in the Paulson study[2] have been well summarized by Hanks.[7] An unusual method of randomization allocated treatment to each participating institution with four envelopes – two with surgery and two with external beam irradiation. Thus, investigators knew the randomization for at least 25% and sometimes 50% of enrolled patients allowing for selection bias. Additionally, 16 of 106 patients did not receive their assigned treatment, and two patients who had positive resection margins and local recurrence were censored from the surgical arm allowing for further bias. The end-point selected (time to first failure) was an inappropriate gauge of the two treatment modalities because, while the patients received local treatment, local control was not evaluated. Finally, the rate of metastasis in the radiation arm was more consistent with stage T3–4 disease and not stage T1–2. The disease-free survival of 59% is essentially identical to the 58% 5-year disease-free survival reported by the Uro-oncology group in T3–4 patients treated with external beam irradiation.[8] Zietman and Shipley[9] put the Stamey data[4] in proper context by noting that these patients represented an outdated experience – reflecting the spectrum of clinical stages referred for radiation therapy prior to the widespread use PSA screening. Of the 113 men reported on, 50 had stage T3–4 or node-positive disease. Of the remaining 63, 49 had T2 disease and the proportion of those with T2b/T2c is unknown. The nodal status was also unknown and only 26% of patients had well-differentiated tumors. Thus, the percentage of patients who could have anticipated cure after external radiotherapy (i.e. those with early-stage low-grade disease) was only a small segment of the overall population studied. Zagars and Pollack[10] addressed the issue of whether unsuccessful radiotherapy leads to an accelerated growth rate. They estimated the PSA doubling time in 75 of 461 T1/T2 patients who had a biochemical failure after radiotherapy and found that the PSA velocity increased proportionately with the pretreatment PSA and grade. They concluded that the dramatic PSA doubling time that Stamey[4] described was due to the fact that recurrent cancers reflect the more aggressive end of the disease spectrum and are expected to manifest more biological aggression than the pretreatment clonogens from which they emerge.

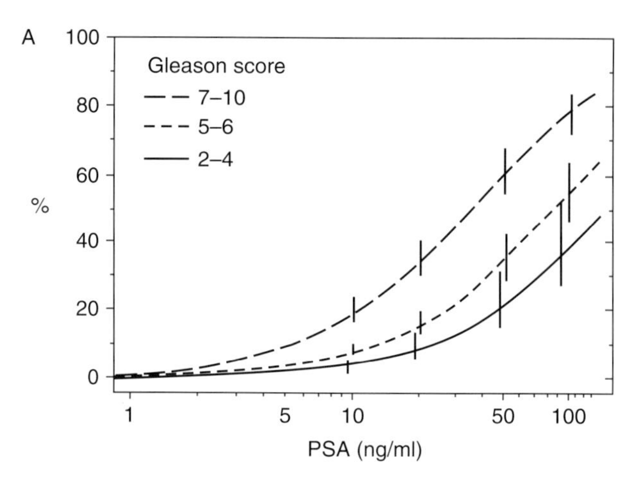

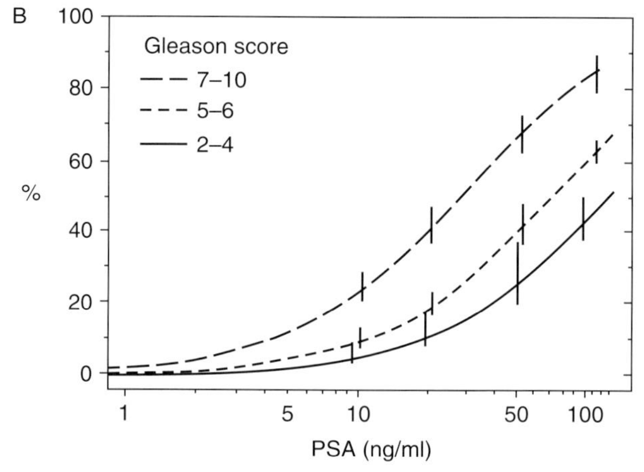

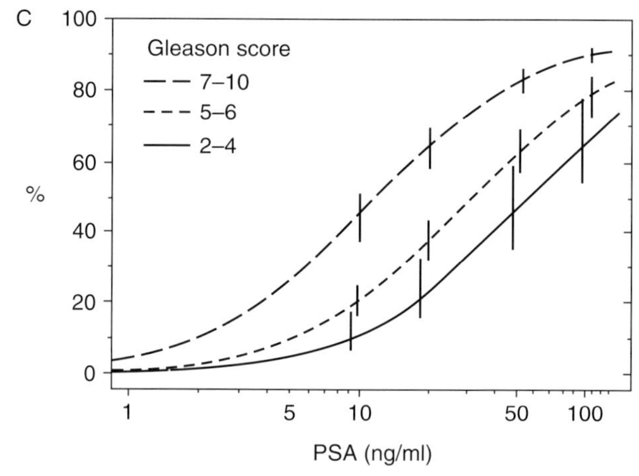

Fig. 42.1. Risk of relapse within 5 years according to pretherapy PSA value, Gleason score and clinical stage. Vertical lines represent the 95% confidence interval for PSA values of 10, 20, 50 and 100 ng/ml. (A) stage T1a–2a; (B) stage T2b–c; (C) stage T3–4. *Cancer* Vol. 79, No. 2, 1997, page 42. Copyright © 1997 American Cancer Society. Reprinted by permission of Wiley-Liss, Inc., a subsidiary of John Wiley & Sons, Inc.

RETROSPECTIVE STUDIES

The most significant deficiency of these and other series published prior to the PSA era is the lack of proper risk stratification. While the studies separated patients into different groups based on anatomic staging and this did provide some measure of prognostic gradient, significant heterogeneity existed within each stage. The spectrum of PSA values and tumor grade encountered in otherwise homogeneously staged patients reflects the different potential for aggressive biological behavior. Numerous studies have demonstrated that the pretreatment PSA,[11–15] Gleason score[16,17] and the combination of PSA with other clinical factors[18] are significant predictors of post-treatment outcome (**Fig. 42.1**). In several multivariate analyses, pretreatment PSA was found to be the single most significant variable in predicting biochemical disease-free survival.[10,12,13,19–23] Most of these studies found a marginal or less significant contribution from the tumor grade.[10,13,20] Initial PSA, Gleason score and clinical stage have been correlated with overall tumor burden and the risk of extracapsular spread, seminal vesicle invasion and lymph node metastases.[24–27] Owing to the prognostic importance of these pretreatment factors, no meaningful comparison of efficacy can be made between different treatment modalities without proper patient stratification. Most surgical series are comprised of young men with early-stage-low grade-low PSA disease. Kuban et al.[11] analyzed the pretreatment characteristics of prostatectomy and external beam patients from a single institution and found significant differences in the distribution of favorable versus unfavorable prognostic variables. The discrepancies were even more striking when these subsets were compared to a surgical series from Johns Hopkins.[14] Nearly four times as many Hopkins surgical patients had PSA <4 compared to those treated with external beam and PSA >20 was found in 35% of those irradiated compared with 6% of the surgical patients. Similarly, Zietman et al.[13] reported PSA values greater than 15 in 33% of patients treated with definitive radiotherapy compared with only 15% in the Hopkins study. It is apparent that only studies that include the initial PSA, Gleason score and clinical stage should be considered when comparing the curative potential of external beam therapy to radical prostatectomy.

Retrospective studies reporting outcomes after definitive radiotherapy or surgery that include pretreatment characteristics are summarized in **Table 42.1**. Zagars and Pollack[10] reported the outcome for 461 patients with T1 and T2 prostate cancer treated with a median of 65 Gy. The overall freedom from relapse was 70% and was significantly associated with pretreatment PSA. The 5-year PSA-free survival was 91% for those with a pretreatment PSA <4, 69% PSA = 4.1–10, 62% PSA 10.1–20 and 38% with PSA >20. The pretreatment PSA was the most important prognosticator ($P < 0.0001$), the PSA nadir was a highly significant variable ($P = 0.0008$), and tumor grade was marginally significant ($P < 0.057$). Similarly Perez et al.[28] found a close correlation between chemical disease-free survival, and both the pretreatment PSA and the postirradiation PSA nadir. Their 6-year chemical disease-free survival rates according to pretreatment PSA level for T1 tumors were: 100% PSA ≤4, 80% PSA 4.1–20 and 50% PSA >20. For T2 tumors the rates were 91% PSA ≤4, 81% PSA 4.1–10, 55% PSA 10.1–20, 63% PSA 20.1–40 and 46% PSA >40. Zietman et al.[13] analyzed the 4-year biochemical-free survival rate in 161 patients treated with 68.4–72 Gy. The likelihood of being free of disease was a function of the initial PSA: 81% PSA ≤4, 43% PSA 4.1–10, 31% PSA 10.1–20 and 6% PSA 20.1–50. Shipley et al.[22] evaluated a combined data set of 1814 men treated at six institutions and multivariate analysis demonstrated that the pretreatment PSA level was a highly significant independent predictor of PSA-free survival ($P < 0.001$). The PSA-free survival rates 5 years after radiotherapy were 81% for initial PSA <10, 68% for initial PSA 10–<20, 51% for initial PSA 20–< 30, and 31% for initial PSA ≥30. These and other radiation series[12,18,29] note worse outcomes with higher pretreatment PSA levels; however, the 5-year PSA-free survival of 81–96% for initial PSA <4 and 67–93% for initial PSA 4.1–10 is not significantly different than that reported in surgical series.[14,15,17,20,30–32] Thus, a balanced analysis of retrospective studies with stratification based on initial PSA demonstrates that radiotherapy is equivalent to prostatectomy in achieving PSA-free survival for patients in the low- to intermediate-risk categories.

There is a prevailing selection process in this country so that younger men with earlier stages of disease are culled for radical prostatectomy while older men with more advanced stage of disease are referred for radiotherapy.[11,13] Therefore, when analyzing biochemical-free survival data from a single institution, surgery will often appear to be more effective than radiation therapy. When patients in both treatment groups are properly stratified, however, the bias is removed and the two modalities produce similar PSA-free survival rates (**Table 42.2**). D'Amico et al.[20] reported an overall 4-year PSA-free survival of 64% for 757 patients treated with prostatectomy versus 47% for 867 patients treated with radiotherapy ($P = 0.96$). Multivariable analysis demonstrated that the pretreatment PSA, Gleason score and clinical stage (T1/T2 vs. T3/T4) were significant variables in predicting time to PSA failure. A statistically significant increase in T3/T4, PSA >20 ng/ml and Gleason score 7 was noted in the radiation group ($P < 0.0001$). Using the results of the regression analysis, patients were divided into low risk (PSA = 4–10, Gleason ≤4), intermediate risk (PSA 4–10, Gleason 5–7 or PSA 10.1–20 and Gleason ≤7) and high risk (PSA 4.1–20 and Gleason ≥8) categories. The 2-year PSA-free survival for patients treated by prostatectomy versus radiation was 98% versus 92% ($P = 0.45$), 77% versus 81% ($P = 0.86$) and 51% versus 53% for patients at low, intermediate and high risk for post-treatment biochemical failure. Thus, patients with prognostically favorable disease fared equally well with surgery or radiation, while those with prognostically unfavorable disease fared equally poorly. Kupelian et al.[21] also documented that patients receiving radiotherapy had more aggressive pretreatment characteristics than their surgical counterparts. Clinical stage T2b–c was encountered in 42% of radiotherapy as opposed to 33% of the prostatectomy patients ($P = 0.018$). The median presenting PSA was 12.1 ng/ml in the radiotherapy group versus 8 ng/ml in the prostatectomy patients. PSA levels greater than 10 ng/ml were noted in 58% of the radiotherapy patients as opposed to 39% of those undergoing surgery ($P < 0.001$). A multivariate analysis was performed and only the initial PSA and Gleason score were independent predictors of PSA-free survival. After the patients were divided into a low-risk group (PSA ≤10.1 and Gleason score ≤6) and a high-risk one (PSA >10 and Gleason score ≥7) the 5-year relapse-free survival was 81% for the low-risk group and 34% for the high-risk group. Forty-eight per cent of patients treated with surgery were low risk versus 33% in the radiotherapy group. For low-risk patients treated with surgery, the 5-year relapse-free survival was 80% versus 81% for those treated with radiotherapy. For high-risk patients treated with radiotherapy, the 5-year relapse-free survival was 26% versus 21% for prostatectomy patients with a positive margin and 62% for negative margins. The authors concluded that radiotherapy and prostatectomy achieved similar results for patients with low-risk features (PSA ≤10.1 and Gleason score ≤6). The outcome in the high-risk group was better with surgery, if negative margins were achieved. Keyser et al.[19] subsequently

Table 42.1. Retrospective studies with pretreatment patient characteristics.

Author (reference)	Modality	Study interval	Patient number	Definition of biochemical recurrence	PSA distribution	Gleason distribution	Follow-up	Biochemical control							
								PSA <4	PSA 4.1–10	PSA 10.1–20	PSA >20	Gleason 2–4	Gleason 5–6	Gleason 7	Gleason 8–10
Babaian (17)	RP	1987–1993	265	Postop. PSA ⩾0.1	<4: 23%; 4.1–10: 49%; 10.1–20: 23%; >20: 4.7%	At least 7 in 86.4%	Min. 48 mos	90%	84%	76%	92%*	NA	89%	87%	66%
Walsh (30)	RP	1982–1991	955	Postop. PSA ⩾0.2	<4: 42%; 4.1–10: 36%; 10.1–20: 16%; >20: 6%	2–4: 10.5%; 5–6: 70% 7: 15.5%; 8–10: 4%	Avg. 53 mos	92%	83%	56%	45%	98%	90%	62%	46%
Pound (15)	RP	1982–1995	1623	Postop. PSA ⩾0.2	<4: 29%; 4.1–10: 46%; 10.1–20: 19%; >20: 6%	2–4: 4%; 5–6: 54%; 7: 33%; 8–10: 9%	Avg. 60 mos	94%	82%	72%	54%	100%	95%	66%	41%
Catalona (31)	RP	1983–1993	925	Postop. PSA ⩾0.6	<4: 15%; 4.1–10: 52%; >10: 33%	2–4: 21%; 5–7: 66%; 8–10: 13%	Avg. 28 mos	97%	97%	83%†	NA	91%	90%‡	NA	74%
Pisansky (18)	RT	1987–1993	500	2 or more consecutive PSA values that rose >1 ng/ml from post-Tx nadir without subsequent spontaneous decline	<4: 15%; 4–10: 33%; 10–20: 29%; >20: 24%	2–4: 79%; 5–6: 50%; 7: 25%; 8–10: 10%	Median 43 mos	96%	93%	78%	48%	94%	85%	67%	44%
Brachman (29)	RT	1988–1995	1527	ASTRO definition	<4: 9%; 4–10: 37%; 10.1–20: 32%; >20: 22%	2–4: 28%; 5–6: 46%; 7: 18%; 8–10: 8%	Median 41 mos	87%	76%	53%	49%	78%	72%	51%	52%
Zagars (68)	RT	1987–1993	707	2 or more consecutive elevations above the nadir	<4: 20%; 4–10: 33.5%; 10–20: 28%; >20: 18.5%	2–4: 26%; 5–6: 41% 7: 21%; 8–10: 11%	Mean 32.4 mos	89%	57%	53%	33%	67%	66%	52%	45%

Zietman (13)	RT	1988–1992	161	PSA>1.0 ng/ml after 2 years of f/u or a 10% incremental rise before 2 years of f/u	<4: 19%; 4–10: 27%; 10.1–20: 18%; >20: 36%	2–4: 23%; 5–6: 47%; 7–10: 21%	Median 32 mos	81%	43%	31%	6%	73%	NA	40%	NA
Shipley (22)	RT	1988–1955	1765	ASTRO definition	<4: 14%; 4–10: 36%; 10.1–20: 26%; >20: 24%	2–4: 22%; 5–6: 42%; 7–10: 26%; ? 10%	Median 41 mos	81%	81%	68%	40%	75%	73%	7–10 =53%	
Lee (67)	RT	1986–1993	500	<1.5 ng/ml	<4: 10%; 4–10: 32%; 10.1–20: 30%; >20: 28%	2–4: 28%; 5–6: 47%; 7: 21%; 8–10: 6%	Median 20 mos	85%	81%	59%	34%	NA	NA	NA	NA
Zagars (12)	RT	1987–1991	269	2 or more consecutive PSA increases	NA	2–4: 42%; 5–6: 35%; 7: 15%; 8–10: 6%	Median 30 mos	86%	67%	10–30 = 45%	>30 = 20%	80%	60%	57%	30%
Zagars and Pollack (10)	RT	1987–1993	461	2 or more consecutive PSA increases	<4: 25%; 4–10: 37%; 10.1–20: 26%; >20: 12%	2–4: 30%; 5–6: 39%; 7: 21%; 8–10: 8%	Median 31 mos	91%	69%	62%	38%	78%	70%	65%	47%
Kuban (11)	RT	1975–1989	652	'An abnormal' PSA	NA	2–4: 29%; 5–7: 41%; 8–10: 29%	Median 78 mos	69%	57%	56%	20%	74%	Gleason 5–7 =	48%	33%

*Very few patients.

†PSA >10.

‡Gleason 5–7.

RP, radical prostatectomy; RT, radiation therapy; postop., postoperative; f/u, follow-up; mos, months; Min., minimum; Avg., average.

Table 42.2. Surgery versus radiation therapy: retrospective series with stratification into risk groups.

Author (reference)	Modality	Study interval	Patient number	Definition of biochemical recurrence	PSA distribution	Gleason distribution	Follow-up	Biochemical control: Low-risk group	Intermediate risk	High risk
Kupelian (21)	RP	1987–1993	298	>0.2 ng/ml	<4: 18%; 4–10: 43%; 10.1–20: 23%; >20: 16%	2–5: 33%; 6: 38%; 7: 18%; 8–10: 11%	Median 42 mos	85%		35%
Kupelian (21)	RT	1987–1993	253	2 or more consecutive elevations above the nadir	<4: 12%; 4–10: 30%; 10.1–20: 30%; >20: 28%	2–5: 29%; 6: 40%; 7: 17%; 8–10: 14%	Median 42 mos	81%		35%
D'Amico (20)	RP	1989–1995	757	2 serial detectable PSA after nadir	<4: 11%; 4–10: 53%; 10.1–20: 24%; >20: 12%	2–4: 23%; 5–6: 54% 7: 13%; 8–10: 10%	Median 26.4 mos	98%	77%	51%
D'Amico (20)	RT	1989–1995	867	2 or more serial elevations above the nadir or PSA >1.0 ng/ml	<4: 16%; 4–10: 36%; 10.1–20: 23%; >20: 25%	2–4: 14%; 5–6: 46% 7: 28%; 8–10: 13%	Median 30 mos	92%	81%	53%
Martinez (23)	RP	1987–1994	157	>0.2 ng/ml	<4: 25%; 4–10: 75%	2–4: 31%; 5–6: 69%	Median 66 mos	67%		
Martinez (23)	RT	1987–1994	225	ASTRO definition	<4: 30%; 4–10: 70%	2–4: 34%; 5–6: 66%	Median 66 mos	69%		

RP, radical prostatectomy; RT, radiation therapy; mos, months.

reported a larger experience from the same institution and confirmed that, for patients with clinical stage T1–2 and PSA ≤10, there is no difference in biochemical failure rates between radiation and surgery. Martinez et al.[23] found similar biochemical control rates for patients at low risk (PSA ≤10 and Gleason score ≤6) treated with surgery or radiotherapy. For radical prostatectomy patients, the 7-year PSA-free survival was 69% versus 67% for patients treated with radiotherapy. The authors concluded that low-risk prostate cancer patients treated with either definitive modality could expect similar long-term rates of cure.

CASE FOR DOSE ESCALATION

Despite the apparent equivalence between radiotherapy and prostatectomy seen in retrospective studies, the problem of persistent local disease remains disconcerting. As many as 46% of patients with early-stage disease will have positive surgical margins at prostatectomy and local recurrence rates as high as 35% have been reported.[33] The incidence of positive biopsies after radiotherapy for all stages of disease has ranged from 19% to 65%[34–37] but is confounded by difficulty in distinguishing recurrence from postradiation histopathology. One study analyzing serial biopsies found that 31% of patients with positive biopsies 1 year after radiotherapy had negative biopsies 12 months later but found a high rate of local treatment failure associated with any positive biopsy within 36 months of completing radiotherapy.[38] Others have questioned the significance of a positive biopsy after radiotherapy.[39] Nevertheless, the fact that high-risk patients treated with radiotherapy have similar PSA-free survival rates as those prostatectomy patients with positive surgical margins[21] indicates that conventional radiotherapy does not completely eliminate intraprostatic tumor. Persistence of cancer within the

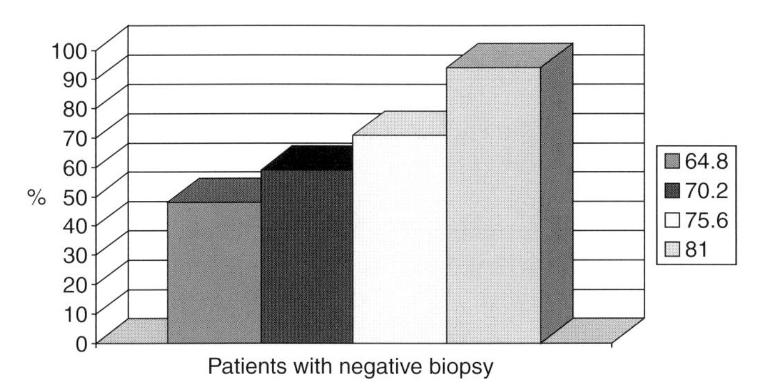

Fig. 42.2. Effect of radiation dose on local control.

prostate gland after conventional radiotherapy may be related to the resistance of clonogenic cells to inadequate radiation doses.

The amount of radiation dose delivered to the prostate gland has been shown to affect local control and long term outcome.[40–56] Hanks[42] analyzed the effect of dose on infield recurrence in 1516 patients from three national Patterns of Care outcome trials. As the dose increased from less than 60 Gy to 70 Gy, the rate of local failure decreased from 25% to 14% in stage C patients and 23% to 14% in stage B patients. Similarly, Perez et al.[40] found improved local control rates for higher doses in a range from 60 Gy to over 70 Gy. Shipley et al.[44] reported the results of a dose escalation trial in patients with T3–T4 disease treated with a combination of photons and a perineal proton boost to 75.6 Gy (cobalt gray equivalent) or 67.2 Gy with standard radiotherapy. Subgroup analysis demonstrated that, for patients with poorly differentiated tumors, high-dose therapy resulted in fewer local failures. Roach et al.[46] demonstrated a similar dose response in patients with poorly differentiated histology. The 3-year PSA-free survival for patients who received higher than 71 Gy was 83% compared with 0% for those treated with lower doses. Histologic evidence of the relationship between radiation dose and tumor control was demonstrated by Zelefsky et al.,[55] who rebiopsied 136 patients after they received from 64.8 to 81 Gy in a phase I dose escalation trial. The rate of positive biopsies decreased linearly as the radiation dose was increased in 5.4 Gy increments (**Fig. 42.2**). The positive biopsy rate was only 7% in those patients who received 81 Gy as compared with 48% after 75.6 Gy, 45% after 70.2 Gy and 57% after 64.8 Gy. Data suggesting that a local recurrence may increase the chance for systemic dissemination[57,58] underscore the importance of eradicating local disease. The goal of improved local control combined with recent technological advances provides the impetus for dose escalation with conformal therapy.

CONFORMAL RADIOTHERAPY

The introduction of three-dimensional (3D) imaging modalities, such as computed tomography (CT) and magnetic resonance imaging scans, coupled with the increased availability of sophisticated computer systems has helped move radiation oncology into the era of 3D conformal therapy (3D-CRT). 3D-CRT is a method of irradiating a target volume with multiple X-ray beams of uniform intensity whose direction and boundaries are individually determined based on 3D anatomical information. The radiation beam is geometrically shaped to conform to the 'beam's eye view' of the tumor volume. Beam's eye view is an approach to treatment planning that uses beams that are shaped to match the cross-section of the target volume as seen along the axis of the treatment beam from various angles. Using graphic display tools in the treatment planning software, the target volume coverage is scrutinized and the amount of normal tissue that is exposed to radiation is determined. The proper number and orientation of radiation beams is then chosen to minimize the traversal of normal tissues. 3D-CRT thereby overcomes the major limitations of conventional two-dimensional radiotherapy – uncertainty in the location and shape of the target and its relationship with neighboring normal tissue. Unless one uses 3D anatomic information derived from modern imaging techniques, the true spatial extension of disease and the real volume of normal contiguous structures exposed to radiation dose can not be appreciated. For irregularly shaped organs that are not well separated from critical structures (like the prostate gland) the addition of beam intensity modulation to the geometric shaping inherent in 3D-CRT will allow an extra degree of freedom to enhance the degree of conformality.[59] Intensity-modulated radiation therapy is a form of 3D-CRT in which beams of varying intensity are applied across a target volume. In order to maximize uniform dose throughout the tumor volume while minimizing dose to the surrounding structures, a non-uniform beam profile is created for each treatment portal. Conformal radiotherapy treatment planning is much more complex than the conventional technique of evaluating isodose distributions through one or more cross-sections. Sophisticated algorithms that seek to maximize tumor dose subject to a limitation of the maximum dose allowed to critical structures require advanced computer and graphic resources. This type of inverse planning can be considered the mirror image of CT scan image reconstruction.[60] The goal in CT image reconstruction is to produce an image based on the underlying density distribution in transverse slices from X-ray projections of a large number of angles. The quest in conformal radiotherapy planning is: given the planned dose maximum in the tumor and the desired dose minimum in critical structures, calculate the beam profiles for a number of angles that yield this distribution.

Conformal radiotherapy allows for a more accurate delineation of the tumor that ensures more consistent dosimetric coverage. Corn et al.[61] demonstrated that, even when similar doses are delivered, conformal therapy achieves better PSA-free survival than conventional irradiation. The authors concluded that conformal radiotherapy is inherently more accurate and allows for a more precise deposition of dose within the target. Employing conformal techniques restricts the high-dose distribution to the prostate gland, while minimizing dose to the bladder and rectum. Soffen et al.[62] demonstrated a 14% reduction in the volume of bladder and rectum exposed to doses greater than 55 Gy when using conformal techniques. This translated into a significant reduction in acute toxicity with 34% grade 2 genitourinary and gastrointestinal morbidity for conformal therapy compared with 57% for conventional radiotherapy.[63] Perez et al.[64] reported the incidence of moderate diarrhea was 3–6% for patients treated with conformal therapy as opposed to 9–21% for those treated with standard therapy. Similarly, in a randomized study,[65] a significantly lower incidence of proctitis and rectal bleeding was found in men treated with conformal therapy versus standard techniques (43% vs. 71%). These reductions in acute and chronic genitourinary and gastrointestinal toxicity have permitted the escalation of radiation dose to levels that until recently were considered unsafe.

DOSE ESCALATION WITH CONFORMAL RADIOTHERAPY

Pollack et al.[49,52] described the impact of dose escalation on 1127 patients with stage T1–T4 prostate cancer treated consecutively from 1987 through 1997 at the M.D. Anderson Cancer Center. Despite the retrospective nature of this analysis, the patients were treated in a fairly uniform fashion. Androgen suppression was not used, none of the patients received elective lymph node irradiation and all patients were initially treated to a dose of 46 Gy with a conventional technique. The same arrangement was used for the boost in 982 patients and 145 patients received a boost with conformal therapy after the initial 46 Gy. The median radiotherapy dose increased from 64 Gy (1987–1989) through 66 Gy (1990–1991), 68 Gy (1992–1993) and 70 Gy (1994–1995). After 1993, as part of a prospective trial, patients were randomized to either 70 Gy or 78 Gy. Outcomes were analyzed based on three radiation dose groups consisting of ⩽67 Gy (n = 500), >67–77 Gy (n = 495), and >77 Gy (n = 132). Actuarial 4-year PSA-free survival rates for the entire cohort were 54%, 71% and 77% in the low-, intermediate- and high-dose levels. A highly significant difference in PSA-free survival was found when the dose was increased from ⩽67 Gy to >67–77 Gy ($P < 0.0001$). Increasing the dose to the intermediate level resulted in a dramatic improvement in control rates seen in all patient prognostic groups and sustained at 7 years. In contrast, only patients with initial PSA >10 showed improvement in the PSA-free survival when increasing the dose beyond 77 Gy. These patients had a 4-year PSA-free survival rate of 61% when treated with >67–77 Gy compared with 91% when the dose was escalated above 77 Gy. The authors confirmed these results in a subsequent randomized trial,[66] demonstrating a substantial improvement in PSA-free survival when the prescribed dose was increased from 70 Gy to 78 Gy. An analysis was performed to determine which subgroup of patients benefited from the 8 Gy increment. A dose–response was described for those with an initial PSA >10 so that the 5-year PSA-free survival rate was 75% in patients receiving 78 Gy versus 48% in those treated with 70 Gy. In contrast, in patients with PSA ⩽10, no apparent improvement was found and the 5-year PSA-free survival was approximately 80% at both dose levels.

A retrospective analysis from the Mayo Clinic, however, demonstrated an improved outcome for all patient subsets treated with dose escalation.[53] The outcomes of 1041 patients treated between 1986 and 1999 were examined based on doses above or below 72 Gy. The treatment technique was reasonably consistent over the 12-year interval. A conventional technique was used in 562 patients and 479 patients were treated with conformal radiotherapy using the M.D. Anderson technique described above. After 1994, androgen deprivation was used more frequently so that 58% of patients treated to ⩾72 Gy were suppressed as opposed to 6% of those receiving less than 72 Gy. While a similar proportion of patients had initial PSA ⩽10 in the <72 Gy and the ⩾72 Gy groups, overall the tumors were more advanced in the higher dose cohort. All patient subsets benefited from radiation dose escalation beyond 72 Gy. The 5-year PSA-free survival improved from 73% to 96% in patients with presenting PSA ⩽10 (P = 0.004) and from 38% to 79% in those with PSA >10 ($P < 0.001$). The improved outcome was durable so that the 8-year PSA-free survival rate was 87% for those receiving ⩾72 Gy versus 51% for those treated with lower doses. A dose of 78 Gy was delivered to 288 patients and accordant with the M.D. Anderson prospective trial, a dose–response was seen even when comparing survival to those patients treated to the 72–75 Gy range.

Hanks et al.[45,48,50,51] have reported their extensive experience with conformally delivered dose escalation from the Fox Chase Cancer Center. Analysis of 618 consecutive patients treated with 3D-CRT from 1989 to 1997 demonstrated a dose–response for all patients except those patients with PSA ⩽10 and favorable pretreatment characteristics and those with PSA ⩾20 with unfavorable risk factors. A 5 Gy dose increment from 73 Gy to 78 Gy was shown to improve the 5-year PSA-free survival by 15–45% in various patient subgroups. Patients were divided by PSA levels as well as favorable features (stage T1–T2a, Gleason score ⩽6, and no perineural invasion) and unfavorable ones (stage T2b–T3, Gleason score 7–10, and/or perineural invasion). Increasing the median delivered dose from 73 Gy to 78 Gy improved the 5-year PSA-free survival from 72% to 88% for patients with PSA <10 with unfavorable features, 71% to 86% for those with PSA 10–19.9 and favorable features, 56% to 78% for those with PSA 10–19.9 and unfavorable features, and 20% to 63% for those with PSA ⩾20 and favorable features. In patients with PSA ⩾20 and unfavorable features metastasis dominated the outcome and no dose–response was seen. In a matched-pair analysis of low dose (<74 Gy) versus high dose (>74 Gy), the end-points of PSA-free survival, freedom from distant metastases and cause-specific survival were all shown to have a statistically significant advantage with the high-dose treatment.[50]

In a recent update of the Memorial Sloan–Kettering dose-escalation program, Zelefsky et al.[56] described a dose–response for all patient prognostic categories and found dose to be the most important variable affecting PSA-free survival. The dose was escalated from 64.8 Gy to 75.6 Gy in 810 patients treated with 3D-CRT and from 81 Gy to 86.4 Gy in 229 treated with intensity-modulated therapy. Patients were classified by

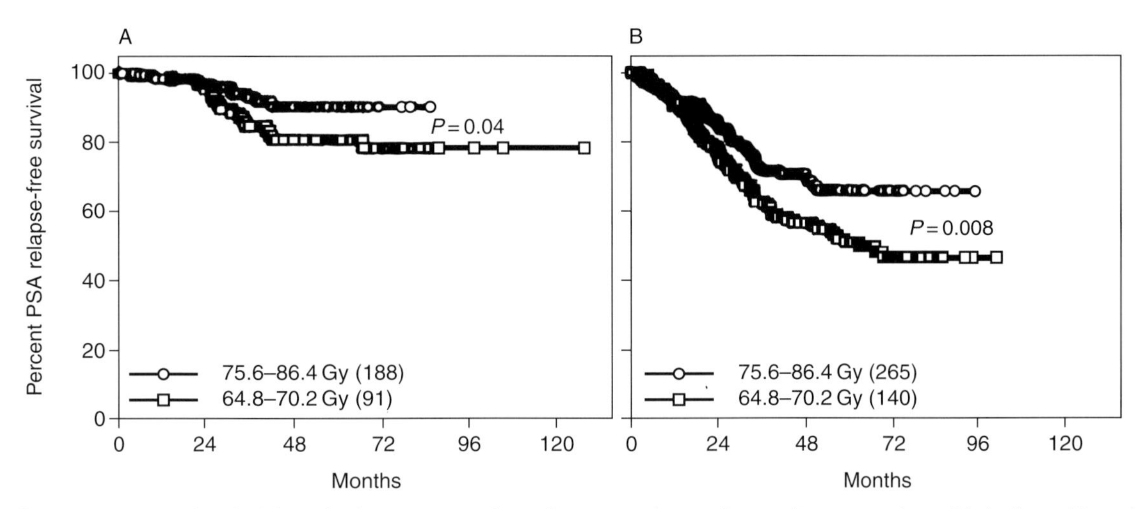

Fig. 42.3. Kaplan–Meier actuarial probability of achieving PSA relapse-free survival according to dose: (A) in favorable high- and low-dose groups at 3–5 years; and (B) in intermediate high and low-risk groups at 3–5 years. Reprinted with permission from Zelefsky, Fuks, Hunt et al. High dose radiation delivered by intensity modulated conformal radiotherapy improves the outcome of localized prostate cancer. *J. Urol.* 2001; 166:876–81.

pretreatment parameters, such that those with PSA <10, clinical stage T1–2, and Gleason score ≤6 were considered low risk. An increase in the value of 1 and 2 or more of the variables classified the patient into an intermediate- and high-risk category. The overall 5-year PSA-free survival was 85% for patients in the low-risk, 58% for those in the intermediate-risk and 38% for those in the high-risk group classified by clinical stage, PSA and Gleason score (**Figs 42.3** and **42.4**). Increasing the dose from 64.8–70.2 Gy to ≥75.6 Gy improved the 5-year PSA-free survival from 77% to 90% in the low-risk group and 50% to 70% in the intermediate one. Patients in the high-risk group who received 81 Gy had a 67% PSA-free survival compared with 43% for those receiving 75.6 Gy and 21% for those treated with 64.8–70.2 Gy.

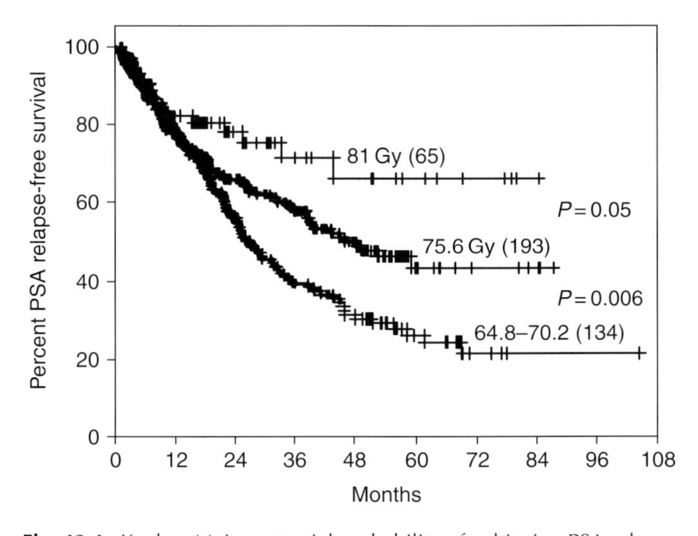

Fig. 42.4. Kaplan–Meier actuarial probability of achieving PSA relapse-free survival in unfavorable prognostic risk subgroups according to three dose levels. Reprinted with permission from Zelefsky, Fuks, Hunt et al. High dose radiation delivered by intensity modulated conformal radiotherapy improves the outcome of localized prostate cancer. *J. Urol.* 2001; 166:876–81.

CONCLUSION

In summary, reports of post-radiotherapy persistent local disease and high failure rates in the pre-PSA era led some to conclude that radiotherapy is an inadequate treatment for prostate cancer. None the less, a thorough review of the surgical and radiation series that include information on initial PSA, Gleason score and clinical stage show equivalent biochemical control rates with proper stratification. Patients with early-stage low to intermediate risk disease fare equally well with either definitive radiotherapy or prostatectomy. Improvements in anatomic imaging, treatment planning software and dose-delivery systems make it possible to maximize the radiation dose to the prostate gland while minimizing dose to the contiguous bladder and rectum. Conformal radiotherapy has allowed dose escalation to the prostate target volume with a resultant improvement in tumor eradication. The 5–7-year PSA-free survival rates are equivalent to those achieved by radical prostatectomy.

ACKNOWLEDGMENTS

The authors wish to thank Dr Bhadrasain Vikram and Dr Eduard Mullokandov for their review of this chapter and their insightful suggestions.

REFERENCES

1. Middleton RG, Thompson IM, Austenfeld MS et al. Prostate cancer clinical guidelines panel summary report on the management of clinically localized prostate cancer. *J. Urol.* 1995; 154:2144–8.
2. Paulson DF, Lin GH, Hinshaw W et al. The Uro-Oncology Research Group. Radical surgery versus radiotherapy for adenocarcinoma of the prostate. *J. Urol.* 1982; 128:502–4.
3. Schellhammer PF, El-Mahdi AM, Wright GL Jr et al. Prostate-specific antigen to determine progression-free survival after radiation therapy for localized carcinoma of prostate. *Urology* 1993; 42:13–20.

4. Stamey TA, Ferrari MK, Schmid HP. The value of serial prostate specific antigen determinations 5 years after radiotherapy: steeply increasing values characterize 80% of patients. *J. Urol.* 1993; 150:1856–9.
5. Stamey TA, Ferrari MK, Schmid HP. Editorial: The value of serial prostate specific antigen determinations 5 years after radiotherapy: steeply increasing values characterize 80% of patients. *J. Urol.* 1994; 152:1565–6.
6. Brendler CB. Editorial: Prostate cancer. *J. Urol.* 1993; 150:1865–6.
7. Hanks GE. More on the Uro-oncology research group report of radical surgery vs. radiotherapy for adenocarcinoma of the prostate. *Int. J. Radiat. Oncol. Biol. Phys.* 1988; 14:1053.
8. Paulson DF, Cline WA Jr, Koefoot RB Jr et al. Radical surgery versus radiotherapy for adenocarcinoma of the prostate. *J. Urol.* 1982; 128:502–4.
9. Zietman AL, Shipley WU. Editorial: The value of serial prostate specific antigen determinations 5 years after radiotherapy: steeply increasing values characterize 80% of patients. *J. Urol.* 1994; 152:1564–5.
10. Zagars GK, Pollack AP. Radiation therapy for T1 and T2 prostate cancer: prostate-specific antigen and disease outcome. *Urology* 1995; 45:476–83.
11. Kuban DA, El-Mahdi AM, Schellhammer PF. Prostate-specific antigen for pretreatment prediction and post treatment evaluation of outcome after definitive irradiation for prostate cancer. *Int. J. Radiat. Oncol. Biol. Phys.* 1995; 12:307–16.
12. Zagars GK. Prostate specific antigen as an outcome variable for T1 and T2 prostate cancer treated by radiation therapy. *J. Urol.* 1994; 152:1786–91.
13. Zietman AL, Coen JJ, Shipley WU et al. Radical radiation therapy in the management of prostatic adenocarcinoma: the initial prostate specific antigen value as a predictor of treatment outcome. *J. Urol.* 1994; 151:640–5.
14. Partin AW, Pound CR, Clemens JQ et al. Serum PSA after anatomic radical prostatectomy. *Urol. Clin. North Am.* 1993; 20:713–24.
15. Pound CR, Partin AW, Epstein JI et al. Prostate-specific antigen after anatomic radical retropubic prostatectomy. *Urol. Clin. North Am.* 1997; 24:395–406.
16. Roach M, Lu J, Pilepich MV et al. Four prognostic groups predict long-term survival from prostate cancer following radiotherapy alone on radiation therapy oncology group clinical trials. *Int. J. Radiat. Oncol. Biol. Phys.* 2000; 47:609–15.
17. Babaian RJ, Troncosco P, Bhadkamkar VA et al. Analysis of clinicopathologic factors predicting outcome after radical prostatectomy. *Cancer* 2000; 91:1414–22.
18. Pisansky TM, Kahn MJ, Rasp GM et al. A multiple prognostic index predictive of disease outcome after irradiation for clinically localized prostate carcinoma. *Cancer* 1997; 79:337–44.
19. Keyser DK, Kupelian PA, Zippe CD et al. Stage T1–2 prostate cancer with pretreatment prostate-specific antigen level ≤10 ng/ml: radiation therapy or surgery? *Int. J. Radiat. Oncol. Biol. Phys.* 1997; 38:723–9.
20. D'Amico AV, Whittington R, Kaplan I et al. Equivalent biochemical failure-free survival after external beam radiation therapy or radical prostatectomy in patients with a pretreatment prostate specific antigen of >4–20 ng/ml. *Int. J. Radiat. Oncol. Biol. Phys.* 1997; 37:1053–8.
21. Kupelian P, Katcher J, Levin H et al. External beam radiotherapy versus radical prostatectomy for clinical stage T1–2 prostate cancer: therapeutic implications of stratification by pretreatment PSA levels and biopsy Gleason scores. *Cancer J. Sci. Am.* 1997; 3:78–87.
22. Shipley WU, Thames HD, Sandler HM et al. Radiation therapy for clinically localized prostate cancer: a multi-institutional pooled analysis. *JAMA* 1999; 281:1598–604.
23. Martinez AA, Gonzalez JA, Chung AK et al. A comparison of external beam radiation therapy versus radical prostatectomy for patients with low risk prostate carcinoma diagnosed staged and treated at a single institution. *Cancer* 2000; 88:425–32.
24. Stamey TA, Kabalin JN, McNeal JE et al. Prostate specific antigen in the diagnosis and treatment of adenocarcinoma of the prostate. *J. Urol.* 1989; 141:1076.
25. Partin AW, Kattan MW, Subong EN et al. Combination of prostate specific antigen, clinical stage, and Gleason score to predict pathological stage of localized prostate cancer. A multi-institutional update. *JAMA* 1997; 277:1445–51.
26. Ziada AM, Lisle TC, Snow PB et al. Impact of differed variables on the outcome of patients with clinically confined prostate carcinoma: prediction of pathologic stage and biochemical failure using an artificial neural network. *Cancer* (Suppl.) 2001; 91:1653–60.
27. Han M, Snow PB, Brandt JM et al. Evaluation of artificial neural networks for the prediction of pathologic stage in prostate carcinoma. *Cancer* (Suppl.) 91:1661–6.
28. Perez CA, Michalski JM, Lockett MA. Chemical disease-free survival in localized carcinoma of prostate treated with external beam irradiation: comparison of American society of therapeutic radiology and oncology consensus or 1 ng/ml as endpoint. *Int. J. Radiat. Oncol. Biol. Phys.* 2001; 49:1287–96.
29. Brachman DG, Thomas T, Hilbe J et al. Failure-free survival following brachytherapy alone or external beam irradiation alone for T1–2 prostate tumors in 2222 patients: results from a single practice. *Int. J. Radiat. Oncol. Biol. Phys.* 2000; 48:111–17.
30. Walsh PC, Partin AW, Epstein JI. Cancer control and quality of life following anatomical radical retropubic prostatectomy: results at 10 years. *J. Urol.* 1994; 152:1831–6.
31. Catalona WJ, Smith DS. 5-year tumor recurrence rates after anatomical radical retropubic prostatectomy for prostate cancer. *J. Urol.* 1994; 152:1837–42.
32. Schneider SB, Schweitzer VG, Parker RG et al. The prognostic value of PSA levels in radiation therapy of patients with carcinoma of the prostate: the UCLA Experience 1988–1992. *Am. J. Clin. Oncol.* 1996; 19:65–72.
33. Zietman AL, Shipley WU, Willett CG. Residual disease after radical surgery or radiation therapy for prostate cancer: clinical significance and therapeutic implications. *Cancer* 1993; 71:959–69.
34. Kabalin JN, Hodge KK, McNeal JE et al. Identification of residual cancer in the prostate following radiation therapy: role of transrectal ultrasound guided biopsy and prostate specific antigen. *J. Urol.* 1989; 142:326–31.
35. Miller EB, Ladaga LE, El-Mahdi AM et al. Reevaluation of prostate biopsy after definitive radiation therapy. *Urology* 1993; 41:311–16.
36. Babaian RJ, Kojima M, Saitoh M et al. Detection of residual prostate cancer after external radiotherapy. *Cancer* 1995; 75:2153–8.
37. Crook J, Malone S, Perry G et al. Postradiotherapy prostate biopsies: what do they really mean? Results for 498 patients. *Int. J. Radiat. Oncol. Biol. Phys.* 2000; 48:355–67.
38. Scardino PT, Wheeler TM. Local control of prostate cancer with radiotherapy: frequency and prognostic significance of positive results of postirradiation prostate biopsy. *Natl Cancer Inst. Monograph* 1988; 7:95–103.
39. Cox JD, Kline RW. Do prostatic biopsies 12 months or more after external irradiation for adenocarcinoma stage III predict long-term survival. *Int. J. Radiat. Oncol. Biol. Phys.* 1983; 9:299–303.
40. Perez C, Walz BJ, Zivnuska FR et al. Irradiation of carcinoma of the prostate localized to the pelvis: analysis of tumor response and prognosis. *Int. J. Radiat. Oncol. Biol. Phys.* 1980; 6:555–63.
41. Stock RG, Stone NN, Tabert A et al. A dose–response study for I-125 prostate implants. *Int. J. Radiat. Oncol. Biol. Phys.* 1998; 41:101–8.

42. Hanks GE. External-beam radiation therapy for clinically localized prostate cancer: patterns of care studies in the United States. *Natl Cancer Inst. Monograph* 1988; 7:75–84.
43. Leibel SA, Zelefsky MJ, Kutcher GJ et al. Three-dimensional conformal radiation therapy in localized carcinoma of the prostate: interim report of a phase 1 dose-escalation study. *J. Urol.* 1994; 152:1792–8.
44. Shipley WU, Verhey LJ, Munzenrider JE et al. Advanced prostate cancer: the results of a randomized comparative trial of high dose irradiation of boosting with conformal protons compared with conventional dose irradiation using photons alone. *Int. J. Radiat. Oncol. Biol. Phys.* 1995; 32:3–12.
45. Hanks GE, Lee WR, Hanlon AL et al. Conformal technique dose escalation for prostate cancer: biochemical evidence of improved cancer control with higher doses in pateints with pretreatment prostate-specific antigen ≥10 ng/ml. *Int. J. Radiat. Oncol. Biol. Phys.* 1996; 35:861–8.
46. Roach M, Meehan S, Kroll S et al. Radiotherapy for high grade clinically localized adenocarcinoma of the prostate. *J. Urol.* 1996; 156:1719–23.
47. Forman JD, Duclos M, Shamsa F et al. Hyperfractionated conformal radiotherapy in locally advanced prostate cancer: results of a dose escalation study. *Int. J. Radiat. Oncol. Biol. Phys.* 1996; 34:655–62.
48. Pinover WH, Hanlon AL, Horwitz EM et al. Defining the appropriate radiation dose for pretreatment PSA <10 ng/ml prostate cancer. *Int. J. Radiat. Oncol. Biol. Phys.* 2000; 47:649–54.
49. Pollack A, Zagars GK. External beam radiotherapy dose response of prostate cancer. *Int. J. Radiat. Oncol. Biol. Phys.* 1997; 39:1011–18.
50. Hanks GE, Hanlon AL, Pinover WH et al. Survival advantage for prostate cancer patients treated with high-dose three-dimensional conformal radiotherapy. *Cancer J. Sci. Am.* 1999; 5:152–8.
51. Hanks GE, Hanlon AL, Pinover WH et al. Dose selection for prostate cancer patients based on dose comparison and dose response studies. *Int. J. Radiol. Oncol. Biol. Phys.* 2000; 46:823–32.
52. Pollack A, Smith LG, von Eschenbach AC. External beam radiotherapy dose response characteristics of 1127 men with prostate cancer treated in the PSA era. *Int. J. Radiat. Oncol. Biol. Phys.* 2000; 46:507–12.
53. Kupelian PA, Mohan DS, Lyons J et al. Higher than standard radiation doses (≥72 Gy) with or without androgen deprivation in the treatment of localized prostate cancer. *Int. J. Radiat. Oncol. Biol. Phys.* 2000; 46:567–74.
54. Fukunaga-Johnson N, Sandler HM, McLaughlin PW et al. Results of 3D conformal radiotherapy in the treatment of localized prostate cancer. *Int. J. Radiat. Oncol. Biol. Phys.* 1997; 38:311–17.
55. Zelefsky MJ, Leibel SA, Gaudin PB et al. Dose escalation with three-dimensional conformal radiation therapy affects the outcome in prostate cancer. *Int. J. Radiat. Oncol. Biol. Phys.* 1998; 41:491–500.
56. Zelefsky MJ, Fuks Z, Hunt M et al. High dose radiation delivered by intensity modulated conformal radiotherapy improves the outcome of localized prostate cancer. *J. Urol.* 2001; 166:876–81.
57. Fuks Z, Leibel SA, Wallner KE et al. The effect of local control on metastatic dissemination in carcinoma of the prostate: long-term results in patients treated with 125-I implantation. *Int. J. Radiat. Oncol. Biol. Phys.* 1991; 21:537–47.
58. Zagars GK, von Eschenbach AC, Ayala AG et al. The influence of local control on metastatic dissemination of prostate cancer treated by external beam megavoltage radiation therapy. *Cancer* 1991; 68:2370–7.
59. Verhey LJ. Comparison of three-dimensional conformal radiation therapy and intensity-modulated radiation therapy systems. *Semin. Radiat. Oncol.* 1999; 9:78–98.
60. Bortfeld T. Optimized planning using physical objectives and constraints. *Semin. Radiat. Oncol.* 1999; 9:20–34.
61. Corn BW, Hanks GE, Schultheiss TE et al. Conformal treatment of prostate cancer with improved targeting: superior prostate-specific antigen response compared to standard treatment. *Int. J. Radiat. Oncol. Biol. Phys.* 1995; 32:325–30.
62. Soffen EM, Hanks GE, Hunt MA et al. Conformal static field radiation therapy treatment of early prostate cancer vs. non-conformal techniques: a reduction in acute morbidity. *Int. J. Radiat. Oncol. Biol. Phys.* 1992; 24:485–8.
63. Hanks GE, Schultheiss TE, Hunt M et al. Factors influencing incidence of acute grade 2 morbidity in conformal and standard radiation treatment of prostate cancer. *Int. J. Radiat. Oncol. Biol. Phys.* 1995; 31:25–9.
64. Perez C, Michalski J, Drzymala et al. 3D conformal therapy and potential for IMRT in localized carcinoma of the prostate. In: *Theory and Practice of Intensity Modulated Radiation Therapy*, pp. 199–217. Madison, WI: Advanced Medical Publishing, 1997.
65. Dearnaley DP, Khoo VS, Norman AR et al. Comparison of radiation side effects of conformal and conventional radiotherapy in prostate cancer: a randomized trial. *Lancet* 1999; 353:267–71.
66. Pollack A, Zagars GK, Smith LG et al. Preliminary results of a randomized radiotherapy does-escalation study comparing 70 Gy with 78Gy for prostate cancer. *J. Clin. Oncol.* 2000; 18:3904–11.
67. Lee WRL, Hanks GE, Schultheiss TE et al. Localized prostate cancer treated by external beam radiotherapy alone: serum prostate specific antigen driven outcome analysis. *J. Clin. Oncol.* 1995; 13:464–8.
68. Zagars GK, Pollack A, Kavadi VS et al. Prostate specific antigen and radiation therapy for clinically localized prostate cancer. *Int. J. Radiat. Onc. Biol. Phys.* 1995; 32:293–306.

Brachytherapy for Prostate Cancer

Steven J. DiBiase and Stephen C. Jacobs

Department of Radiation Oncology and Department of Surgery, University of Maryland School of Medicine, Baltimore, MD, USA

INTRODUCTION

For clinically localized prostate cancer, the optimal treatment remains controversial. Patients are offered radical prostatectomy (RP), external beam radiation therapy (EBRT) or prostate brachytherapy (PB) as definitive management. Although no prospective randomized studies exist comparing these treatment options, there have been several retrospective studies that have demonstrated equivalent outcomes for localized disease.[1–3] With improved outcomes, PB has experienced a resurgence in popularity in the last several years. In 1995, about 4% of men with localized prostate cancer received PB; it is estimated that 40–50% of men may undergo this procedure by 2006, if current trends continue.[4] Modern transrectal ultrasound imaging and computer-based dosimetry systems have contributed to improved recurrence-free survival.[5,6] Numerous series have revealed 5-year recurrence-free survival rates that compare favorably to EBRT and RP in favorable-risk patients.[1,7,8] Because of such favorable outcomes, a recent health policy brief from the American Urological Association projects that PB will soon surpass RP as the treatment of choice for early-stage prostate cancers.[4]

Prostate brachytherapy can be delivered through the placement of permanent radioactive sources ('seeds') or through a temporary radioactive source delivered through a catheter, typically referred to as high dose rate (HDR) therapy. Although the initial experience with PB entailed a free-hand approach performed during an open surgical procedure, today, the technique has evolved into a closed, ultrasound-guided, interstitial transperineal approach. Current techniques for PB are associated with excellent biochemical control in patients with localized prostate cancer. None the less, treatment protocols and techniques for PB vary from center to center. Controversy exists regarding the selection of patients for PB alone or in combination with external beam irradiation and hormonal therapy. This review will take an evidence-based approach to the efficacy and toxicity of PB in the treatment of localized disease. In addition, the indications and relative contraindications for PB are detailed and new directions in PB likely to have further effects on the outcome of treated patients are highlighted.

HISTORIC PERSPECTIVE

Permanent interstitial implantation for prostate cancer was first reported in 1910 after a radium source was inserted through a urethral catheter.[9] This treatment approach gradually fell into disfavor because of radiation safety issues using this isotope and the complications that were observed using this crude approach. In the 1960s, initial experience began to surmount at several centers in the USA using a 'free-hand' retropubic approach for seed placement within the prostate gland.[10–12] This procedure was commonly performed in conjunction with a pelvic lymph node dissection. Investigators at Baylor University reported on 510 patients treated between 1965 and 1980 using ^{198}Au combined with EBRT. In this series, Carlton and Scardino reported acceptable 15-year survival in node-negative, low-volume disease patients.[10–12] At the same time, investigators at Memorial Sloan–Kettering Cancer Center (MSKCC) reported on 1078 patients treated with permanent ^{125}I. With a median follow-up of 11 years, Whitmore and Zelefsky reported overall local recurrence-free survival rates at 5, 10 and 15 years to be 59%, 36% and 21%, respectively.[10–12] Schellhammer similarly reported poor local control rates at 15 years using a similar retropubic approach.[10–12] It was concluded from such series that retropubic implantation was not as effective as other approaches for the eradication of localized prostate cancer.

MODERN TECHNIQUES

With the development of transrectal ultrasound and pelvic computed tomography (CT) imaging, the precision of radioactive source placement and subsequent outcomes has improved. Ultrasound-guided implantation was first described by Holm et al. in 1983.[13] This opened the door for a large amount of clinical experience subsequently to develop over the ensuing years. Transrectal ultrasound permitted the transperineal placement of sources by visualizing axial images of the prostate gland. Initial experience was reported from investigators in Seattle, where a computerized pre-plan approach was implemented using ultrasound guidance.[5] Thereafter, numerous institutions began implant programs throughout the 1990s and transperineal guidance was the preferred approach. Investigators from MSKCC described a transperineal approach using CT guidance with encouraging tumor control rates and acceptable morbidity.[14]

There are different approaches to the placement of radioactive sources into the prostate gland. One approach entails the use of preloaded needles that have seeds positioned in a manner that was determined by either an ultrasound or CT scan prior to the date of the radioactive seed placement.[15] Alternatively, a Mick applicator is used to inject seeds separately through the transperineal needles.[16] Both delivery methods have been shown to produce comparable results and advantages are described by users of both methods.

Two radioactive isotopes are used during most PB procedures. Commonly, iodine 125 (^{125}I) and palladium 103 (^{103}Pd) are used in permanent interstitial placement. ^{125}I has a half-life of 60 days and ^{103}Pd has a half-life of 17 days. Although physician preference varies in their use across the country, no objective evidence supports the use of one isotope over the other in terms of treatment efficacy or toxicity.[17]

Over the last several years, various institutions have reported on the use of real-time intraoperative planning in PB.[7,18,19] Faster treatment planning computers have allowed investigators to develop a seed-loading pattern in the operating room prior to the actual placement of seeds in the prostate gland. Typically, axial ultrasound images of the prostate are acquired into a treatment planning computer using a transrectal ultrasound. Then, a loading pattern is determined and used by the treating physician. Proponents of intraoperative planning argue that this method of seed placement improves the outcome of seed placement, since the plan is custom-made to the patient's position and anatomy on the actual treatment day. However, randomized evidence to support such a position is not available.

Brachytherapy is considered to have several advantages over EBRT for prostate cancer[3] as listed below.

- The-low energy radioactive sources [iodine-125 (^{125}I) or palladium-103 (^{103}Pd)] have limited tissue penetration. This allows a sharp drop-off of the radiation dose at the edge of the gland, limiting the dose received to normal tissues and minimizing potential treatment-related complications.
- Prostate gland movement, which can significantly affect the accuracy of EBRT, is less of a problem during implantation using real-time ultrasound imaging.
- A single outpatient treatment for placement of the implant is convenient and less time-consuming as compared to conventional EBRT, which is typically administered over 7 weeks.
- The precision and conformation of brachytherapy permits the administration of high radiation doses with a low risk of rectal morbidity.

Typically, PB is delivered through a transperineal approach via needles or catheters that are placed under the guidance of various imaging modalities, most commonly transrectal ultrasound (TRUS).[6,20,21] The radiation dose is either delivered with permanent, low dose rate (LDR) implantable radioactive sources or, less commonly, by the temporary placement of high dose rate sources through afterloading catheters, permitting a temporary but high-dose exposure of the prostate gland to the radiation source.

High dose rate techniques

HDR techniques are performed at a limited number of centers; they are usually combined with EBRT (see 'High dose rate brachytherapy' below). Following CT-based treatment planning, several high-dose fractions, ranging from 4 to 6 Gy each, are administered over an interval of 24–36 hours using isotopes such as iridium-192 (^{192}Ir). Treatment is followed by EBRT to the prostate and periprostatic tissues to a dose of 45–50.4 Gy using conventional techniques.

Low dose rate techniques

In the more commonly used LDR method using permanent implants, typically 75–125 radioactive seeds measuring about 5 mm in length are placed strategically within the prostate gland under the guidance of TRUS and a computerized treatment plan. Free hand placement of the seeds is associated with inhomogeneous dose distribution and a high rate of local failure, owing to subtherapeutic dosing of other areas. The specified dose of radiation is emitted over a 4–10-month period of time (depending upon the specific isotope).[19] Approximately 60% of all permanent implants use ^{125}I and 40% use ^{103}Pd.[17,22]

Although PB is often enthusiastically offered to patients, and its use for definitive radiation treatment for localized disease is increasing,[4] much remains to be learned about PB with respect to quality assurance, long-term outcome and the impact of additional treatments, such as androgen ablation (see later).

RISK GROUP STRATIFICATION

The development of risk-group stratification has increased our understanding of localized prostate cancer. Previous investigators have stratified localized prostate cancer patients into low risk, intermediate risk or high risk for biochemical failure after radiotherapy based on the patient's pretreatment prostate specific antigen level (PSA), Gleason score and stage.[1,23–25] Surgical series have also confirmed the importance of such risk groups in determining the risk of biochemical failure after radical prostatectomy.[26] Such pretreatment selection factors have commonly been employed to determine eligibility for PB as monotherapy, typically restricting patients to a PSA <10, Gleason score of 6 or less, and stage T1–T2a.[22,27,28] Because of the risk of extracapsular extension with higher PSA, stage and Gleason score, the use of PB as monotherapy has often been discouraged.[29] Higher risk patients are commonly offered additional therapy, including external beam radiotherapy and hormonal deprivation therapy, in order to eradicate extracapsular spread.[30]

With the growth in popularity of PB, many patients that do not have traditional implant alone criteria have been offered PB as monotherapy. Unfortunately, there is a paucity of data on the outcome of such patients who received PB as monotherapy for intermediate and high-risk disease.

BIOCHEMICAL OUTCOME AFTER PROSTATE BRACHYTHERAPY

Treatment results

There are no randomized trials that compare PB to either radical prostatectomy or EBRT. Several retrospective series have compared treatment results in men who received PB with or without EBRT with those in men treated with other modalities.

Brachytherapy alone

In general, compared to other treatment modalities, the use of PB appears to be equivalent to other treatment modalities for men with low-risk disease, but may be associated with an inferior outcome in men with higher Gleason grade or higher pretreatment serum PSA levels.[1,5,7,31]

- One non-randomized series compared 5-year biochemical failure-free survival for men with clinical T1–T2 prostate cancer who were treated with either PB (n=695) or EBRT (n=1527) at a single institution.[32] The use of EBRT was associated with a superior outcome for men with Gleason 8–10 tumors (52% vs. 28%), while the outcome for lower grade lesions was similar between PB and EBRT. Men with pretreatment serum PSA values 10–20 ng/ml also had a significantly superior outcome with EBRT (70% vs. 53%), while outcomes were similar for those with initial PSA ranges of 0–4 ng/ml, or 4–10 ng/ml and >20 ng/ml.
- Similar results were noted in a second retrospective cohort study of 1872 men with clinically localized prostate cancer who were treated with PB, EBRT or RP.[1] There was no difference in the rate of biochemical failure among PB, RP and EBRT in men with low-risk tumors (stage T1c and T2a, serum PSA ≤10 ng/ml and Gleason score ≤6). However, the study was insufficiently powered to detect significant differences in outcome among men with Gleason score 2–4 and 5–6 tumors.

Men with high-risk (stage T2c or serum PSA >20 ng/ml or Gleason score ≥8) or intermediate-risk tumors (stage T2b or Gleason score of 7 or serum PSA >10 ng/ml), who were treated with PB (with or without neoadjuvant hormone therapy), did significantly worse than men who underwent RP (relative risk of PSA failure 3.0 and 3.1, respectively). Of note, the median follow-up in this trial was relatively short, which compromises its usefulness at this time.

Brachytherapy plus EBRT

Men who have clinically localized prostate cancer with adverse features tend not to fare as well with PB monotherapy. Potential reasons for this include the inability to deliver adequate doses to the periprostatic tissues in patients at higher risk for extraprostatic disease. Combining EBRT or three-dimensional conformal external radiation therapy (3D-CRT) with a brachytherapy boost in such patients may permit the delivery of higher radiation doses to achieve optimal tumor control.

- In one report, 54 men were treated to 45 Gy with conventional EBRT followed by PB with ^{125}I.[33] With a median follow-up of almost 10 years, 75% maintained a serum PSA level <0.4 ng/ml. Only 24% and 41% of these men had Gleason scores >7 and PSA levels >10 ng/ml, respectively.
- In a second series, 689 men with clinically localized prostate cancer received PB with ^{125}I, followed 3 weeks later by EBRT (45 Gy).[34] Pretreatment Gleason score was <7 in 76% and serum PSA <10 ng/ml in 73% of the men. PSA relapse was defined as a nadir >0.2 ng/ml, or a progressive rise from that nadir. With a median follow-up of 4 years, the biochemical no evidence of disease (bNED) rates for men with pretreatment serum PSA levels of 0–4 ng/ml (n=50), 4–10 ng/ml (n=451), 10–20 ng/ml (n=144), and >20 ng/ml (n=44) were 94%, 93%, 75% and 69%, respectively.
- A third report suggests no significant benefit to the addition of EBRT to brachytherapy in 403 men receiving PB monotherapy and 213 men receiving EBRT (45 Gy) plus PB.[35] With a median follow-up of 58 months, the bNED outcomes by risk group for monotherapy versus combined therapy were: low risk, 94% vs. 87%; intermediate risk, 84% vs. 85%; and high risk, 54% vs. 62%. These differences did not reach statistical significance for any risk group.

TOXICITY

Complications of prostate brachytherapy

Complications following PB are generally limited to the genitourinary (GU) and gastrointestinal (GI) tracts.

Early complications

The most common early symptoms following PB are related to the GU tract. Most radiation-induced symptoms do not appear for several days following the implant; they include urinary frequency, urgency and dysuria.[14] In one series, 90% of 310 men receiving PB for localized disease had grade 1 or 2 urinary symptoms during the first 12 months following therapy.[36] In a second report of 92 men with T1 or T2 prostate cancer undergoing PB, 46% had urinary symptoms that were substantial enough at 1 month following implantation to require medication, and 14% had persistent urinary symptoms ≤ grade 2 after 2 years.[14] Additionally, obstructive uropathy requiring urinary catheter placement has been reported in 5–22% of patients.[14,30,37]

Men with very large prostate glands or a prior transurethral resection are at a higher risk for side effects after interstitial therapy. Prostatic inflammation and swelling can occur acutely in these men, and urinary retention can be severe enough to require catheterization. The risk of urinary retention in large prostates after HDR therapy is not such a problem. Transurethral resection to improve flow after brachytherapy is a relative risk for urinary incontinence, as smaller TURP defects are less problematic than larger ones.

Late complications

Late complications of PB include irritative voiding symptoms, persistent urinary retention, rectal urgency, bowel frequency, rectal bleeding or ulceration, and prostatorectal fistulas.[38–41] In one series, 19 of 109 patients (19%) developed persistent, bright-red rectal bleeding, from 1 to 28 months following ^{125}I implantation.[42] Urinary retention requiring intermittent or long-term urinary catheterization is uncommon, occurring in fewer than 5%.[43] In a second report of 105 men treated an average of 5.2 years before survey, bowel symptoms were uncommon (4–9%) unless external beam irradiation had also been administered.[38]

Compared to 3D-CRT, PB is associated with both more acute and chronic urinary morbidity. This was illustrated in one series that compared outcomes in men with favorable risk prostate cancer who received either 3D-CRT ($n=137$) or transperineal ^{125}I implantation ($n=145$).[7] In the PB group, 31% had grade 2 urinary symptoms persisting after 1 year, and the median duration of grade 2 urinary symptoms was 23 months (range 12–70 months). In contrast, acute grade 2 urinary symptoms resolved within 4–6 weeks after completion of 3D-CRT, and at 5 years, only 8% still had ≤ grade 2 urinary toxicity. The 5-year likelihood of post-treatment erectile dysfunction among patients who were initially potent before therapy was similar between the two groups, as was the 5-year likelihood of grade 2 late rectal toxicity (6% and 11%, respectively, for 3D-CRT and PB).

The impact of brachytherapy on potency was evaluated in several reports[38,44,45] as follows.

- In one series, 416 men underwent radioactive seed implantation for T1–2 prostate cancer.[44] In the 313 men who were potent prior to treatment, potency was retained in 79% at 3 years and 59% at 6 years.
- Similar results were noted in a second review of 482 men who were potent before treatment.[45] At 5 years, potency was preserved overall in 53%; 76% of those treated with PB monotherapy maintained potency.
- In a third report surveying patient-reported symptoms at an average of 5.2 years following treatment, complete impotence (51%) and erections inadequate for penetration without manual assistance (72%) were seen in men who had full erections at baseline.[38]

High dose rate brachytherapy

HDR brachytherapy is usually administered in conjunction with EBRT. In one series that included 171 men, of whom one-third had either T3 or high-grade disease, 50 Gy of EBRT was followed by two 15 Gy boosts of ^{192}Ir.[46] The 5-year bNED rate was 79%. Late toxic effects included chronic proctitis in 16% and chronic cystitis in 11%.

In a second series, 104 men received three or four fractions of ^{192}Ir followed by 50.4 Gy of EBRT.[47] The majority had T1–2 disease, and the median pretreatment serum PSA level was 8.1 ng/ml. The 5-year bNED rate was 84% for those men with a pretreatment PSA level <20 ng/ml. The only severe late toxicity was the development of a urethral stricture in 8%.

ROLE OF ANDROGEN DEPRIVATION

Neoadjuvant hormone therapy with brachytherapy

The role of androgen deprivation therapy is incompletely studied in men undergoing PB. However, one retrospective report included 163 men, with clinically localized prostate cancer and an estimated prostate gland volume of 60 g, who received neoadjuvant androgen deprivation therapy for an average of 3.4 months prior to PB as monotherapy or in combination with EBRT.[48] The outcome in these men was compared to that of 263 matched men, treated at the same institution, who did not receive androgen deprivation. The use of neoadjuvant hormone therapy was not associated with any improvement in outcome.

However, in another study involving 296 patients who received brachytherapy, neoadjuvant hormonal therapy (3 months prior to, and 3 months after the implant) significantly increased the percentage of negative biopsies (3.5% vs. 14%). These results were most pronounced for men presenting with PSA >10 ng/ml, clinical stage ≥T2b, or Gleason score ≥7 (i.e. high-risk patients).[49]

INDICATIONS AND CONTRAINDICATIONS

Various selection criteria are used in the recommendation for PB. Because of anatomic constraints, larger prostate volumes are a relative contraindication for permanent PB. As the volume of the prostate gland increases, the risk of pubic arch interference increases. Many investigators use a volume cut-off volume around 50 cm^3 in order to select appropriate candidates for PB. However, various intraoperative techniques are possible to gain access to the prostate gland's periphery in a tightly constrained pelvis. One option is to increase the hip flexion to a maximal dorsal lithotomy position. Alternatively, the ultrasound probe and needle trajectory can be angled more anteriorly to get under the pubic bone. In addition to technical difficulties, some investigators have reported worse urinary toxicity in men with larger glands.

Some investigators argue that a history of a TURP is a relative contraindication to PB. Although initial reports suggested an increased incidence of urinary incontinence in men with a prior history of a TURP, recent reports do not substantiate. Different rates of toxicity between investigators are likely secondary to different degrees of a TURP defect in reported patients. The authors of this chapter judge each patient's case separately, avoiding PB in men with very large TURP defects.

Although inflammatory bowel disease is a contraindication to those men undergoing EBRT for prostate cancer, owing to the significant rate of GI toxicity, bowel disease has not been shown to influence the toxicity of PB. Wallner et al.[14] reported on six patients with inflammatory bowel disease undergoing PB and did not find an increased incidence of GI toxicity.

FUTURE DIRECTIONS

Because of our inability to localize prostate cancer during prostate brachytherapy with current imaging, radioactive seeds are placed throughout the entire prostate gland in order to eradicate this malignancy. Recent advances in magnetic resonance spectroscopic imaging (MRSI) have allowed the biochemical visualization of cancerous regions inside the prostate gland. Because prostate cancer is associated with low levels of citrate and zinc as compared to the normal glandular tissue, MRSI provides a metabolic mapping of the cancer inside the gland. Although recent reports suggest an improved outcome in prostate cancer by escalating the radiation dose, corresponding increased morbidity has resulted. As such, MRSI provides a possible avenue to improve PB by providing a selective means to escalate the radiation dose to the tumor-bearing region(s) and potentially improve the therapeutic ratio.

Investigators at the University of Maryland recently reported on the use of MRSI in PB. In a series of 15 patients, MRSI allowed selective dose escalation of an additional 30% of the prescription dose to areas within the gland determined to harbor cancer. Post-implant dosimetry on axial CT images confirmed the additional dose to citrate-deficient regions and it revealed acceptable doses to the rectum and urethra as compared to conventionally treated patients. Longer follow-up will be required to assess the efficacy of such an approach.

CONCLUSION

Prostate brachytherapy is a curative treatment option for localized prostate cancer. Modern techniques have a significantly improved outcome and have limited morbidity from this treatment approach. As follow-up approaches 10 years for several large series in the literature, biochemical cure rates appear similar to radical prostatectomy and external beam radiotherapy. Furthermore, biological imaging devices offer additional promise to the treating physicians by allowing the therapeutic ratio to be improved by concentrating the high radiation dose to areas within the gland with the highest tumor burden. Over the next decade, technologic advances coupled with an increased understanding of PB will likely continue to improve our outcomes in this disease.

REFERENCES

1. D'Amico A. Biochemical outcome after radical prostatectomy, external beam radiotherapy, or interstitial radiation therapy for clinically localized prostate cancer. *JAMA* 1998; 280:969–74.
2. Martinez A, Gonzalez JA, Chung AK et al. A comparison of external beam radiation therapy versus radical prostatectomy for patient with low risk prostate carcinoma diagnosed, stated, and treated at a single institution. *Cancer* 2000; 88:425–32.
3. Stokes SH. Comparison of biochemical disease-free survival of patients with localized carcinoma of the prostate undergoing radical prostatectomy, transperineal ultrasound-guided radioactive seed implantation, or definitive external beam irradiation. *Int. J. Radiat. Oncol. Biol. Phys.* 2000; 47:129–36.
4. Hudson R. Brachytherapy treatments increasing among Medicare population. *Health Policy Brief of the American Urologic Association* 1999; IX:1.
5. Blasko J. Pd-103 brachytherapy for prostate carcinoma. *Int. J. Radiat. Oncol. Biol. Phys.* 2000; 46:839–50.
6. Ragde H, Korb L. Brachytherapy for clinically localized prostate cancer. *Semin. Surg. Oncol.* 2000; 18:45–51.
7. Zelefsky MJ, Wallner KE, Ling CC et al. Comparison of the 5-year outcome and morbidity of three-dimensional conformal radiotherapy versus transperineal permanent iodine-125 implantation for early-stage prostatic cancer. *J. Clin. Oncol.* 1999; 17:517–22.
8. Brachman DG, Thomas T, Hilbe J et al. Failure-free survival following brachytherapy alone or external beam irradiation alone for T1–2 prostate tumors in 2222 patients: results from a single practice. *Int. J. Radiat. Oncol. Biol. Phys.* 2000; 48:111–17.
9. Paschkis R, Tittinger W. Radiumbehandlung eines Prostata-Sarkoms. *Wein Klin. Woschr.* 1910; 23:1715–16.
10. Carlton CE, Scardino PT. Long-term results after combined radioactive gold seed implantation and external beam radiotherapy for localized prostatic cancer. In: Coffey DS, Dorr FA, Karr JP (eds) *A Multidisciplinary Analysis of Controversies in the Management of Prostate Cancer*, pp. 109–21. New York: Plenum Press 1988.
11. Schellhammer PF, Moriarty R, Bostwick D et al. Fifteen-year minimum follow-up of a prostate brachytherapy series: comparing the past with the present. *Urology* 2000; 56:436–9.
12. Zelefsky MJ, Whitmore WF Jr. Long-term results of retropubic permanent 125iodine implantation of the prostate for clinically localized prostatic cancer. *J. Urol.* 1997; 158:23–30.
13. Holm H, Juul N, Pederson JF et al. Transperineal iodine-125 seed implantation in prostatic cancer guided by transrectal ultrasonography. *J. Urol.* 1983; 130:283–6.
14. Wallner K, Roy J, Harrison L. Tumor control and morbidity following transperineal iodine 125 implantation for stage T1/T2 prostatic carcinoma. *J. Clin. Oncol.* 1996; 14:449–53.
15. Sylvester J, Blasko JC, Grimm P et al. Interstitial implantation techniques in prostate cancer. *J. Surg. Oncol.* 2000; 66:65–75.
16. Zelefsky MJ, Hollister T, Raben A et al. Five-year biochemical outcome and toxicity with transperineal CT-planned permanent I-125 prostate implantation for patients with localized prostate cancer. *Int. J. Radiat. Oncol. Biol. Phys.* 47:1261–6.
17. Gelblum DY, Potters L, Ashley R et al. Urinary morbidity following ultrasound-guided transperineal prostate seed implantation. *Int. J. Radiat. Oncol. Biol. Phys.* 1999; 45:59–67.
18. D'Amico AV, Cormack R, Tempany CM et al. Real-time magnetic resonance image-guided interstitial brachytherapy in the

treatment of select patients with clinically localized prostate cancer. *Int. J. Radiat. Oncol. Biol. Phys.* 1998; 42:507–15.
19. Nag S, Bice W, DeWyngaert K et al. The American Brachytherapy Society recommendations for permanent prostate brachytherapy postimplant dosimetric analysis. *Int. J. Radiat. Oncol. Biol. Phys.* 2000; 46:221–30.
20. Sharkey J, Chovnick SD, Behar RJ et al. Evolution of techniques for ultrasound-guided palladium 103 brachytherapy in 950 patients with prostate cancer. *Tech. Urol.* 2000; 6:128–34.
21. Nag S. Brachytherapy for prostate cancer: summary of American Brachytherapy Society recommendations. *Semin. Urol. Oncol.* 2000; 18:133–6.
22. Merrick GS, Butler WM, Dorsey AT et al. Prostatic conformal brachytherapy: 125I/103Pd postoperative dosimetric analysis. *Radiat. Oncol. Invest.* 1997; 5:305–13.
23. D'Amico A, Whittington R, Malkowicz B. A multivariate analysis of clinical and pathologic factors that predict for prostate specific antigen failure after radical prostatectomy for prostate cancer. *J. Urol.* 1995; 154:131–8.
24. Hanks G, Hanlon A, Pinover W et al. Dose selection for prostate cancer patients based on dose comparison and dose response studies. *Int. J. Radiat. Oncol. Biol. Phys.* 2000; 46:823–32.
25. Zelefsky MJ, Lyass O, Fuks Z et al. Predictors of improved outcome for patients with localized prostate cancer treated with neoadjuvant androgen ablation therapy and three-dimensional conformal radiotherapy. *J. Clin. Oncol.* 1998; 16:3380–5.
26. Walsh PC. Radical prostatectomy for localized prostate cancer provides durable cancer control with excellent quality of life: a structured debate. *J. Urol.* 2000; 163:1802–7.
27. Stock RG, Stone NN. The effect of prognostic factors on therapeutic outcome following transperineal prostate brachytherapy. *Semin. Surg. Oncol.* 1997; 13:454–60.
28. Nag S, Beyer D, Friedland J et al. American Brachytherapy Society (ABS) recommendations for transperineal permanent brachytherapy of prostate cancer. *Int. J. Radiat. Oncol. Biol. Phys.* 1999; 44:789–99.
29. Partin AW, Kattan MW, Subong EN. Combination of prostate-specific antigen, clinical stage, and Gleason score to predict pathological stage of localized prostate cancer. A multi-institutional update. *JAMA* 1997; 277:1445–51.
30. Dattoli M, Wallner K, Sorace R et al. 103Pd brachytherapy and external beam irradiation for clinically localized, high-risk prostatic carcinoma. *Int. J. Radiat. Oncol. Biol. Phys.* 1996; 35:875–81.
31. Beyer DC, Priestley JB Jr. Biochemical disease-free survival following 125I prostate implantation. *Int. J. Radiat. Oncol. Biol. Phys.* 1997; 37:559–63.
32. Beyer DC, Brachman DG. Failure free survival following brachytherapy alone for prostate cancer: comparison with external beam radiotherapy. *Radiother. Oncol.* 2000; 57:263–7.
33. Ragde H, Elgamal AA, Snow PB et al. Ten-year disease free survival after transperineal sonography-guided iodine-125 brachytherapy with or without 45-gray external beam irradiation in the treatment of patients with clinically localized, low to high Gleason grade prostate carcinoma. *Cancer* 1998; 83:989–1001.
34. Critz FA, Williams WH, Levinson AK et al. Simultaneous irradiation for prostate cancer: intermediate results with modern techniques. *J. Urol.* 164:738–41.
35. Blasko JC, Grimm PD, Sylvester JE et al. The role of external beam radiotherapy with I-125/Pd-103 brachytherapy for prostate carcinoma. *Radiother. Oncol.* 2000; 57:273–8.
36. Grimm P, Blasko J, Ragde H et al. Does brachytherapy have a role in the treatment of prostate cancer? *Hematol. Oncol. Clin. North Am.* 1996; 10:653–73.
37. Kaye KW, Olson DJ, Payne JT. Detailed preliminary analysis of 125iodine implantation for localized prostate cancer using a percutaneous approach. *J. Urol.* 1995; 153:1020–5.
38. Talcott JA, Clark JA, Stark PC et al. Long-term treatment related complications of brachytherapy for early prostate cancer: a survey of patients previously treated. *J. Urol.* 2001; 166:494–9.
39. Gelblum DY, Potters L. Rectal complications associated with transperineal interstitial brachytherapy for prostate cancer. *Int. J. Radiat. Oncol. Biol. Phys.* 2000; 48:119–24.
40. Kleinberg L, Wallner K, Roy J. et al. Treatment-related symptoms during the first year following transperineal implantation. *Int. J. Radiat. Oncol. Biol. Phys.* 1994; 28:985–90.
41. Theodorescu D, Gillenwater JY, Koutrouvelis PG. Prostatourethral-rectal fistula after prostate brachytherapy. *Cancer* 2000; 89:2085–91.
42. Hu K, Wallner K. Clinical course of rectal bleeding following I-125 prostate brachytherapy. *Int. J. Radiat. Oncol. Biol. Phys.* 1998; 41:263–72.
43. Ragde H, Grado GL, Nadir BS et al. Modern prostate brachytherapy. *CA Cancer J. Clin.* 2000; 50:380–94.
44. Stock R, Kao J, Stone NN. Penile erectile function after permanent radioactive seed implantation for treatment of prostate cancer. *J. Urol.* 2001; 165:436–9.
45. Potters L, Torre T, Fearn PA et al. Potency after permanent prostate brachytherapy for localized prostate cancer. *Int. J. Radiat. Oncol. Biol. Phys.* 2001; 50:1235–42.
46. Kovacs G, Galalae R, Loch T et al. Prostate preservation by combined external beam and HDR brachytherapy in nodal negative prostate cancer. *Strahlenther. Onkol.* 1999; 175:87–92.
47. Mate TP, Gottesman JE, Hatton J et al. High dose-rate afterloading 192Iridium prostate brachytherapy: feasibility report. *Int. J. Radiat. Oncol. Biol. Phys.* 1998; 41:525–33.
48. Potters L, Torre T Ashley R et al. Examining the role of neoadjuvant androgen deprivation in patients undergoing prostate brachytherapy. *J. Clin. Oncol.* 2000; 18:1187–92.
49. Stone NN, Stock R Unger P. Effects of neoadjuvant hormonal therapy. *Mol. Urol.* 2000; 4:163–8.

Hormonal Therapy/ Chemotherapy

Osteoporosis and Prostate Cancer

Barry Stein
Department of Urology, Rhode Island Hospital, and Division of Urology, Brown Medical School, Providence, RI, USA

Seetharaman Ashok
Division of Urology, Brown Medical School, Providence, RI, USA

BACKGROUND

Osteoporosis in the male is a newly appreciated problem. Long associated with the aging woman, osteoporosis is now known to affect more than 2 000 000 American males over the age of 50 at this time, with another pool of 3 100 000 men who are at risk for developing it.[1–5] About one out of every eight men over the age of 50 will at some point have an osteoporosis-related fracture.[6,7] Every year about 100 000 men will suffer an osteoporosis-related hip fracture, one-third of which will die within the year. In addition tens of thousands of men will have a fracture of the wrist, spine or rib. Physicians who treat men need to be increasingly sensitive to this problem. Thirty-six per cent of the osteoporosis in men is due to low androgen levels, which can occur as a result of hypogonadism that is either congenital, as part of the aging process, or due to acute androgen deprivation, such as in the treatment of advanced carcinoma of the prostate.[8] In the latter case, given the usually advanced age of the patient, the acute loss of androgens is occurring in addition to the bone density loss of aging.[7] This can lead to a higher incidence of osteoporosis and its complications. Considering that hormone therapy is initiated earlier today owing to the PSA monitoring of prostate cancer patients, a longer duration of androgen deprivation therapy can be anticipated, and the incidence of osteoporosis may yet increase.

WHAT IS OSTEOPOROSIS?

Osteoporosis has been defined by the NIH Consensus Panel 2000 as being a skeletal disorder characterized by compromised bone strength, which predisposes to an increased risk of bone fracture.[3] Overall bone strength is comprised of both bone density and the quality of the bone itself.[5] The bone density can be measured and quantified as grams of mineral per volume of bone. The bone quality in turn is determined by the bone architecture, turnover, damage and mineralization. Peak bone mass is the amount of bone at the end of skeletal growth, and bone mass will then decrease throughout life.[5] Peak bone mass in men tends to be greater than that in women because men have larger bones and current techniques for measurement do not correct completely for this size difference.[4] Although the term 'osteoporosis' implies loss of bone density, this may not be the case. For example, men who are hypogonadotropic from birth may never have developed optimal bone density in the first place, so a low bone density is not related in them to loss of bone, but lack of development. Androgens are necessary for the maintenance of bone mass in adult males, but estrogens are also necessary. These are produced in men by the aromatase conversion of testosterone.

Both men and women will thus lose bone mineral density as a natural consequence of the aging process; however, men will lose less bone density with aging.[5] Overall, men will lose about 15–45% of trabecular bone and 5–15% of cortical bone, as compared to women who will lose 30–50% and 25–30%, respectively.[9] This loss of trabecular bone in both men and women explains the increase in fractures seen and why, although hip fractures occur in both sexes, it is twice as common in women.[5,9] Additional risk factors include lifestyle decisions, such as smoking and excessive alcohol consumption, inactivity and low calcium intake.

At this time, there is no accurate measurement of overall bone strength, so we use the measurement of bone mineral density as its proxy. Bone mineral density accounts for 70–85% of bone strength, and the measurement of bone mineral density correlates with the load-bearing capacity of the skeleton.[5] DEXA (dual-energy X-ray absorption) scans measure bone density and, as such, are our best gauge of bone strength. This technology utilizes X-rays of two different energy levels so as to distinguish between bone and the surrounding soft tissue structures. A two-dimensional image is produced, which can be used to measure bone density.[3] A number of prospective studies have been performed, most of which but not all in women, show that DEXA scans can in fact predict the future risk of fracture. Each decrease in bone mineral density of 1 standard deviation (SD) in the hip, translates into a 1.5–3.0 increased risk of fracture.[5]

In reading DEXA scans, the measurement of the patient's bone mineral density is compared to one of two cohorts of men, whose results are already in the computer's data bank. The *z*-score compares the patient's DEXA results to an age- and ideally race-matched cohort of men, and will read out how that man compares to this cohort by both per cent of normal and by SD readings. In reality, the cohort is based on the Caucasian male, as insufficient data exist at this time relative to other races. The *t*-score compares the patient's bone mineral density to young healthy men between the ages of 30 and 45. This represents the optimum age for peak bone mineral density, and the *t*-score is the more important measurement for the determination of bone density loss. Thus all results presented in this chapter will focus on the *t*-score for determination of the bone density status. In interpreting DEXA scans, a bone mineral density that is between 1 and 2.5 SD less than the cohort is read as osteopenia, while a bone mineral density loss of >2.5 SD is defined as osteoporosis.[2] When a DEXA scan reading is >2.5 SD less than the cohort of young healthy men, and the patient has a fracture, this is termed severe osteoporosis.

THE SCOPE OF THE PROBLEM IN PROSTATE CANCER

We became aware of the problems of bone mineral density loss when, within the space of 1 year, three of the author's patients suffered hip fractures. We then began to evaluate our patients who were being treated with androgen deprivation therapy (ADT) with DEXA scans. Of our 75 patients studied to date, 70 were on leuprolide therapy and five had undergone bilateral orchiectomy.[10] All patients underwent an in-office screening DEXA scan utilizing the fourth digit of the non-dominant hand. When this proved abnormal, a full table body scan of the hip and spine was performed for comparison. The patients' ages ranged from 46 to 98, with a mean of 76.4. The patients' duration of ADT ranged from just beginning treatment to 13 years on therapy, with an average of 3.35 years. Of the entire series, only 34 (45.4%) had a normal bone mineral density. Of the 41 patients with abnormal DEXA scans, 25 (61%) had osteopenia, while 16 (39%, or 21.3% of the entire group of 75 men) had osteoporosis.

When we examined the men by age, we found that the men over the age of 70 had a greater risk of loss of bone mineral density than men less than 70. Overall, 68.4% of the men less than 70 years of age had a normal bone mineral density, compared to only 37.5% of the men over 70. Of the 25 men with osteopenia, 19 (76%) were older than 70, while all of the 16 men with osteoporosis were over 70. Duration of therapy also appeared to be important. We examined the 34 men who were on treatment for less than 2 years and compared their DEXA results to the men on ADT for more than 2 years. Of the less than 2 year group, 58.8% had a normal DEXA scan, while only 34% of the men on treatment for more than 2 years had a normal DEXA scan. We further examined the 34 men on therapy for less than 2 years, and found that those with normal bone mineral density had an average age of 72.4 years, while those with abnormal DEXA scans, had an average age of 81.3. We examined the men by the number of years on therapy and as the duration of therapy became longer, the bone mineral density decreased. Thus, from our own work, we conclude that the older the man is and the longer he is on ADT, the likelier he is to experience a loss in bone mineral density. Other interesting studies have confirmed our results.

Smith and associates initially reported on 41 men with locally advanced prostate cancer, without metastatic disease, and who had not yet undergone ADT.[11] These men then underwent DEXA scan of the hip and spine. The mean age in this series was 68. Of this group of men, 66% had normal bone densities, while 29% had osteopenia and 5% had osteoporosis. This compares very favorably to our findings in men on therapy for less than 2 years.

Stoch et al. studied three groups of men: group 1 consisted of controls solicited via a newpaper advertisement; group 2 were men with cancer of the prostate, but not on ADT; and group 3 were men on ADT for at least 6 months, with a mean of 41 months of therapy.[12] The men underwent evaluation of bone mineral density by a variety of techniques including finger, spine and hip DEXA scans. They found that the normal rate of bone loss due to aging is 0.5–1.0% per year, but that LHRH analog therapy was associated with more than a decade increase in this loss. They also reported an incidence of osteoporotic fractures similar to other groups.

Kiratli et al. studied 36 patients from baseline to 10 years after initiation of ADT therapy for bone mineral density with DEXA scans.[13] They also found an increasing trend for loss of bone mineral density over time, more pronounced in the hip. Surgical castration appeared to be more likely than LHRH analogs to lead to a loss in bone density. They evaluated intermittent LHRH analog therapy and concluded that after 6 years there was a trend toward less bone mineral density loss with this therapy; however, their numbers were quite small.

Daniell performed two studies on osteoporosis and ADT. In the first paper, he reviewed the records of 235 men with prostate cancer, and from this culled the names of 17 men who had undergone orchiectomy between 1983 and 1990, and were still alive in 1995.[14] He then performed DEXA scans of the femoral neck and compared the results to 23 controls. He found ten osteoporotic fractures in the larger group, eight of which were found in the 17 orchiectomy patients. Of the 16 men who survived for >60 months, six had osteoporotic fractures and reduced bone mineral density on DEXA scans. The incidence continued to increase over time (**Fig. 44.1**). In a follow-up study, Daniell evaluated 26 men prior to orchiectomy or LHRH

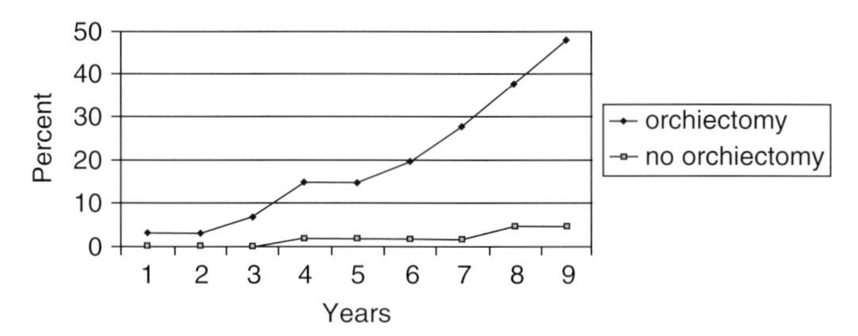

Fig. 44.1. Osteoporotic fracture incidence.

analog therapy and followed them for 6–42 months, comparing them to 12 controls.[15] They found that bone mineral density in the ADT patients fell about 4% per year for years 1–2, and 2% per year every year thereafter. The loss continued at a pace of 1.4–2.6% per year for years 3–8. Both orchiectomy and LHRH analog therapy were likely to cause this loss.

Hatano et al. studied 218 men treated for 6 months or more with LHRH analogs for prostate cancer.[16] Fourteen of these patients (6%) developed a bone fracture during their treatment. The bone density of the lumbar vertebrae was measured by quantitative computed tomography (QCT) scan. The average age was of the men was 78 years and the duration of therapy ranged from 11 to 46 months (average 28 months). The bone density was significantly lower in the patients with a fracture than in those without. Of the patients with a fracture, 12 out of 14 had a good recovery but two patients (one lumbar and one hip fracture) cannot walk unaided. In the patients with a fracture, the duration of therapy was longer (28 vs. 18 months, $P < 0.05$).

Preston et al. evaluated 38 patients with prostate cancer who were treated with LHRH analog or orchiectomy with or without antiandrogen therapy and 38 aged-matched controls not on androgen deprivation therapy were compared.[17] Bone mineral density was measured using DEXA scans of the forearm, femoral neck, trochanter, total hip and spine, every 6 months up to 24 months. The mean age of the treated patients was 73.7 years and the duration of therapy was 29.6 months. The bone mineral density changed significantly between the groups over time at all sites except the lumbar spine, with the treated patients having a greater loss of bone mineral density.

Wei and associates studied 32 men with prostate cancer who were about to begin androgen deprivation therapy or who had been on therapy for more than 1 year.[18] Bone mineral density was measured by DEXA scan of the spine, hip and forearm. Of the eight men not yet on androgen deprivation therapy, 63% had osteopenia or osteoporosis. Of the 24 men on treatment for more than 1 year, 88% were abnormal.

Eriksson et al. compared two groups of men on hormone therapy for prostate cancer: group 1 (11 patients) were treated with orchiectomy alone, while group 2 (16 patients) underwent orchiectomy plus estrogen therapy (IM or PO).[1] They then measured bone mineral density of the femoral neck, trochanter and Ward's Triangle. There was a decrease in bone mineral density in the orchiectomy-only patients, which was not seen in the orchiectomy plus estrogen patients. Statistical significance was achieved only in the forearm.

Townsend and associates reviewed the records of 224 prostate cancer patients treated with LHRH analogs in order to determine the incidence of fractures. The duration of treatment was from 1 to 96 months.[19] In all 22 fractures (9%) were found, with 5% of the fractures attributed to osteoporosis. However, DEXA scans were not performed and a great number of the 36% attributed to trauma may have also had underlying bone density weakness, with fractures precipitated by otherwise a minor trauma.

RECOMMENDED EVALUATION AND TREATMENT

Given the high incidence of bone mineral density loss in men with prostate cancer undergoing hormone therapy, we recommend baseline evaluation with a DEXA scan (**Fig. 44.2**). If the baseline scan is normal, no further evaluation is necessary at that time and a follow-up scan should be performed in 1–2 years. If the scan is abnormal, then treatment should be discussed with the patient, explaining the risks and benefits of treatment.

Men with abnormal scans should first be provided with counseling on nutrition and lifestyle issues. They should be instructed to eat a balanced, healthy diet, especially high in calcium content. If appropriate, they should stop smoking, moderate alcohol consumption and begin a regimen of physical exercise. Exposure to sunlight is also suggested, providing that they do not have skin cancer.

Our initial medical treatment for osteopenia is bisphosphonate therapy at osteopenic doses, vitamin D and calcium. Currently, we use alendronate 35 mg once a week, and vitamin D 400 IU or more, and calcium carbonate or citrate 1000 mg daily. The vitamin D and calcium are often available as a combination marketed specifically for osteoporosis. The initial treatment for osteoporosis is identical except for increasing the dose of bisphosphonate, in this case to alendronate 70 mg weekly.

According to the NIH Consensus Conference 2000, both alendronate and risendronate are bisphosphonates, and have been shown to reduce the risk of vertebral fractures by 30–50% in randomized, clinical trials, although the majority of such trials involve female osteoporosis.[3] Orwoll et al. reported on a randomized double-blind trial in men with osteoporosis, evaluating alendronate 10 mg daily ($n = 146$) versus placebo ($n = 95$), with all men receiving calcium carbonate 500 mg and vitamin

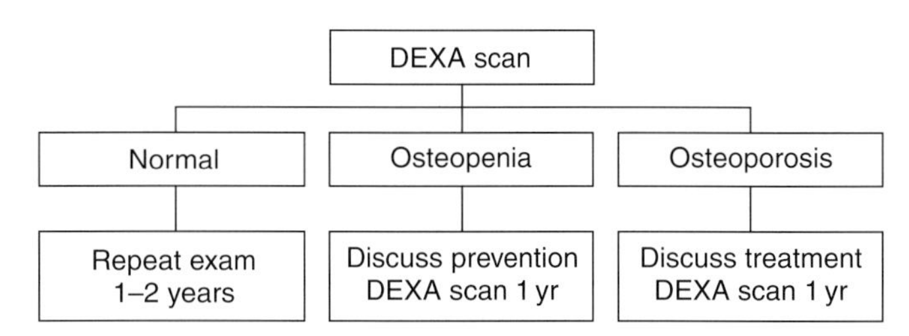

Fig. 44.2. Recommended management based on DEXA scan.

D 400 IU daily.[8] All men underwent DEXA scans of the lumbar spine, hip and total body up to 24 months. In the placebo group, bone density remained unchanged, while in the alendronate arm the bone density increased, particularly in the lumbar spine. These changes were not related to testosterone or estradiol levels. The incidence of vertebral fractures in the placebo group was 7.1%, while in the alendronate group it was only 0.8% ($P=0.02$). Frediani and associates also performed a placebo-controlled study of alendronate 10 mg daily ($n=30$) versus placebo ($n=30$), with all men receiving calcium 500 mg daily. These men were followed up to 24 months with DEXA scans of the hip and spine. The men on placebo in this study had a loss of bone mineral density of 2.8–3.6%, while those on alendronate had a bone mineral density gain of 3.4–6.3%.

Smith et al. recently published a new series of 43 men treated either with leuprolide alone (22) or leuprolide plus intravenous pamidronate (a bisphosphonate) in an attempt to prevent bone loss.[20] All of the men were placed on vitamin D and calcium supplements. The patients were evaluated up to 48 weeks with repeat DEXA scans. The authors found that, in the patients on leuprolide and supplemental therapy alone, the bone mineral density decreased by 3.3% in the spine, 2.1% in the trocanter and 1.8% in the total hip. By contrast, those patients whose treatment included bisphosphonate therapy experienced no change in their bone mineral density.

Diamond and associates studied 12 men with disseminated prostate cancer who were treated with LHRH analogs and an androgen antagonist for up to 12 months with DEXA scans of the femoral neck, and QCT scans of the lumbar spine.[21] The lumbar spine QCT decreased by 6.6%, and the femoral neck DEXA scan decreased by 6.5%. The patients were given cyclic etidronate (bisphosphonate therapy) for the second 6 months of this study, and they found that the bone density of the spine then increased by 7.8%.

CONCLUSION

Osteoporosis is a major health threat to the aging male population, but especially to men with prostate cancer on ADT. This can be diagnosed easily with DEXA scans and can be successfully treated. The men at highest risk are those >70 years of age, and on ADT for >2 years. Successful treatment can be undertaken with bisphosphonates, vitamin D and calcium. This diagnosis and treatment in men is just as important as diagnosing women with osteoporosis.

REFERENCES

1. Eriksson S, Eriksson A, Stege R et al. Bone mineral density in patients with prostatic cancer treated with orchidectomy and with estrogens. *Calcif. Tissue Int.* 1995; 57:97–9.
2. World Health Organization. Assessment of fracture risk and its application to screening for postmenopausal osteoporosis. *WHO Tech. Rep. Ser.*, 1994.
3. National Institutes of Health Consensus Development Conference Statement. *Osteoporosis Prevention, Diagnosis, and Therapy*. Bethesda, MA: NIH.
4. Seeman E. Osteoporosis in men: epidemiology, pathophysiology, and treatment possibilities. *Am. J. Med.* 1993; 95(Suppl. 5A):22–8.
5. Delmas PD, Chapurlat RD. Osteoporosis. In: DeGroot LJ, Jameson JL (eds) *Endocrinology*, 4th edn, pp. 1244–57. Philadelphia, PA: W.B. Saunders, 2001.
6. Zmuda JM, Cauley JA, Glynn NW et al. Posterior–anterior and lateral dual-energy x-ray absorptiometry for the assessment of vertebral osteoporosis and bone loss among older men. *J. Bone Mineral Res.* 2000; 15:1417–23.
7. Orwoll ES, Bevan L, Phipps KR. Determinants of bone mineral density in older men. *Osteoporos. Int.* 2000; 11:815–21.
8. Orwoll E, Ettinger M, Weiss S et al. Alendronate for the treatment of osteoporosis in men. *N. Engl. J. Med.* 2000; 343:604–10.
9. Francis RM. Male osteoporosis. *Rheumatology* 2000; 39:1055–7.
10. Stein BS. Unpublished data, 2001.
11. Smith MR, McGovern FJ, Fallon MA et al. Low bone mineral density in hormone-naïve men with prostate carcinoma. *Cancer* 2001; 91:2238–45.
12. Stoch SA, Parker RA, Chen L et al. Bone loss in men with prostate cancer treated with gonadotropin-releasing hormone agonists. *J. Clin. Endocrinol. Metab.* 2001; 86:2787–91.
13. Kiratli BJ, Srinivas S, Perkash I et al. Progressive decrease in bone density over 10 years of androgen deprivation therapy in patients with prostate cancer. *Urology* 2001; 57:127–32.
14. Daniell HW. Osteoporosis after orchiectomy for prostate cancer. *J. Urol.* 1997; 157:439–44.
15. Daniell HW, Dunn SR, Ferguson DW et al. Progressive osteoporosis during androgen deprivation therapy for prostate cancer. *J. Urol.* 2000; 163:181–6.
16. Hatano T, Oishi Y, Furuta A et al. Incidence of bone fracture in patients receiving luteinizing hormone-releasing hormone agonists for prostate cancer. *Br. J. Urol. Int.* 2000; 86:449–52.
17. Preston DM, Torrens DM, Javier HI et al. Bone mineral density changes in men receiving androgen deprivation for advanced prostate cancer. *J. Urol.* 2000; 163(Suppl. 4):262.
18. Wei JT, Gross M, Jaffe CA et al. Androgen deprivation therapy for prostate cancer results in significant loss of bone density. *Urology* 1999; 54:607–11.
19. Townsend MF, Sanders WH, Northway RO et al. Bone fractures associated with luteinizing hormone-releasing agonists used in the treatment of prostate carcinoma. *Cancer* 1997; 79:545–50.
20. Smith MR, McGovern FJ, Zietman AL et al. Pamidronate to prevent bone loss during androgen-deprivation therapy for prostate cancer. *N. Engl. J. Med.* 2001; 345:948–55.
21. Diamond T, Campbell J, Bryant C et al. The effect of combined androgen blockade on bone turnover and bone mineral densities in men treated for prostate cancer. *Cancer* 1998; 83:1561–6.

Neoadjuvant Hormonal Treatment Prior to Curative Treatment in Prostate Cancer

Michael R. van Balken and Frans M.J. Debruyne

Department of Urology, University Medical Center Nijmegen, Nijmegen, The Netherlands

NEOADJUVANT HORMONAL TREATMENT PRIOR TO RADICAL PROSTATECTOMY

Radical prostatectomy is widely performed in patients with early, locally confined prostate cancer and may lead to cure of the disease, provided that all malignant cells are removed.[1,2] Unfortunately, none of the currently used diagnostic tools are able to reliably identify patients with true locally confined disease. As a result, it has been shown in numerous studies that clinical staging regularly underestimates pathological stage: about 30% of patients with prostate cancer defined as cT1 or cT2 are found to have T3 tumors.[3,4] Positive surgical margins can be found in 10–20% of cT1 patients and in 30–60% of cT2 patients, leading to an adverse prognosis.[1,5] Given the well-known androgen sensitivity of prostate cancer, neoadjuvant hormonal treatment has been explored as a way to increase the rate of organ-confined disease and ultimately potentially improve disease progression and survival.[6] This idea is not new: the first descriptions of using hormonal treatment to shrink the prostate, thus making it more suitable for operative removal, date back to 1941.[7] With the availability of reversible luteinizing hormone-releasing hormone (LHRH) analogs and non-steroidal antiandrogens, there is a renewed interest in preoperative endocrine manipulation of prostate cancer in an attempt to improve long-term results.[6] However, in contrast to external beam radiation, in which neoadjuvant hormonal treatment seems to have earned its place, the role of neoadjuvant treatment prior to radical prostatectomy remains controversial. For the following we pay tribute to the data of the European Study Group on Neoadjuvant Treatment of Prostate Cancer, as put forward by Schulman et al.,[6] and to the recent review of van Poppel.[1]

Effects of neoadjuvant hormone treatment prior to radical prostatectomy

Positive effects of neoadjuvant hormone treatment[1]

Recorded effects that might be of benefit are:

- downstaging of the tumor from:
  - cT3 to pT2[8,9]
  - pT3 to pT2[10]
  - cT1 or cT2 to complete tumor regression;[11,12]
- downsizing of the prostate gland;[13]
- reduction of tumor density;[14]
- fewer positive surgical margins;[15,16]
- reduction of the incidence of prostatic intraepithelial neoplasia (PIN);[17,18]
- reduction of lymph node involvement;[9]
- lower proliferative activity;[13,19]
- reduction of circulating cancer cells during surgery;[4]
- a delay in early and late prostate specific antigen (PSA) progression;[9,20]
- facilitating surgical procedures and improving postoperative urinary incontinence.[21]

Negative effects of neoadjuvant hormone treatment[1]

Recorded effects that might be disadvantagous are:

- less extensive resection by the surgeon because of downstaging/decreasing prostate volume;[22]
- more difficult surgery because in smaller prostates it is more difficult to recognize the apex;[23]
- more difficult surgery because of periprostatic fibrosis and seminal vesicle adherence, and therefore a higher risk for extensive surgical bleeding;[20,24]

- more difficult pathological evaluation, especially of possible positive margins because of pycnotic changes;[25]
- the costs and side effects of hormonal treatment;[26]
- development of hormone resistance, if therapy continues over a prolonged period;[13]
- patient stress because of operation delay.[1]

Results of the European study group on neoadjuvant treatment of prostate cancer

Patients and methods used in the study[6]

From October 1991 to December 1995, 487 patients were randomized in a prospective randomized multicenter trial.[27] For inclusion, patients had to have T2–3NxM0 prostate carcinoma, confirmed by histology. PSA had to be below 100. Patients were randomized for direct radical retropubic prostatectomy ('control group') or neoadjuvant hormonal therapy (NHT) using the LHRH analog goserelin (3.6 mg subcutaneously/month) and flutamide (250 mg trice daily) for a period of 3 months followed by radical retropubic prostatectomy ('NHT group'). At randomization, patients were stratified for clinical T category and pathological grade of the tumor for each participating center. Parameters investigated were: serial PSA levels, pretreatment and post-treatment prostate volumes (TRUS), clinical stage before and after NHT and ease of surgery. All pathologists used the same standardized prostatectomy step-section protocol and all surgical specimens were classified according to the TNM classification (1987 version). Clinical downstaging was defined as a clinical stage after NHT lower than the clinical stage at baseline, and pathological downstaging was defined as a pathological stage after NHT lower than the clinical stage at baseline. Only in pT4, pN2, cM+ patients or in patients with postoperative PSA levels >1.0 ng/ml on two subsequent occasions were urologists allowed to start postsurgical treatment. The end-points used were the abovementioned postoperative PSA levels >1.0 ng/ml on two subsequent occasions, distant metastases or local recurrence (biopsy proven).

Of the 487 randomized patients, 21 were excluded because they were not eligible (cM1 disease in eight, cT1 in five, concurrent malignancy in two, a PSA of 105 at screening in one and postponed surgery because of concurrent disease in five). No data were available in 30 patients and 34 patients were lost to follow-up before surgery. Therefore, a total of 402 patients was used for evaluation: 192 in the NHT group, and 210 in the control group (direct radical prostatectomy). The distribution of patients according to stage and grade is presented in **Table 45.1**.

Early postoperative effects[6]

In the NHT group, mean prostate volume (measured with TRUS) decreased from 37.7 cm^3 at baseline to 26.8 cm^3 after 3 months of NHT. Mean PSA level likewise decreased from 20.5 ng/ml to 0.8 ng/ml after neoadjuvant treatment. In 33% of patients PSA became undetectable. No statistical significant differences could be found between the NHT group and the control group with regard to mean duration of surgery (163 and 159 minutes, $P=0.25$), mean blood loss (1082 and 1150 ml, $P=0.95$) and mean hospital stay 16.2 and 16.8 days, $P=0.51$). More important, no differences could be found in the number of dissections described as good, moderate or poor (82 and 102, 78 and 65, and 27 and 35, respectively; $P=0.15$). In two patients a prostate could no longer be identified. Details of the final histopathological stage after surgery compared with the initial clinical stage at diagnosis for the NHT and the control group are presented in **Tables 45.2** and **45.3**.

Clinical downstaging was observed in 57 of 192 patients (30%), pathological downstaging in 30 of 192 patients (15%) in the NHT group. Although pathological downstaging was also seen in the control group (15 of 210 patients, 7%), the difference was statistically significant ($P=0.002$). The differences

Table 45.1. Distribution of patients according to initial stage and grade.

	Grade 1	Grade 2	Grade 3
NHT T2 (n=105)	38	61	6
Control T2 (n=115)	37	69	9
NHT T3 (n=87)	7	60	20
Control T3 (n=95)	12	62	21
Total NHT (n=192)	45	121	26
Total control (n=210)	49	131	30

Table 45.2. Final histopathological stage after surgery compared with the initial stage at the diagnosis: NHT group.

NHT	cT2 N0 M0 (n=105)	cT3 N0 M0 (n=87)
pTx N0	0	1 (1%)
pT0 N0	4 (4%)	1 (1%)
pT1 N0	4 (4%)	0
pT2 N0	66 (62%)	20 (23%)
pT3 N0	23 (22%)	39 (45%)
pT4 N0	3 (3%)	3 (3%)
pTx N+	5 (5%)	23 (27%)

Table 45.3. Final histopathological stage after surgery compared with the initial stage at the diagnosis: control group.

Control	cT2 N0 M0 (n=115)	cT3 N0 M0 (n=95)
pTx N0	–	1
pT0 N0	1	0
pT1 N0	2 (2%)	0
pT2 N0	46 (40%)	11 (12%)
pT3 N0	46 (40%)	48 (51%)
pT4 N0	3 (3%)	4 (4%)
pTx N+	17 (15%)	31 (33%)

Table 45.4. Effects of 3 months NHT on positive margins after radical prostatectomy.

	Negative Margins	Positive Margins	
NHT T2	90 (87%)	14 (13%)	
Control T2	71 (63%)	41 (37%)	$P<0.01$
NHT T3	48 (58%)	35 (42%)	
Control T3	35 (39%)	55 (61%)	$P=0.01$
Total NHT	138 (74%)	49 (26%)	
Total control	106 (52%)	96 (48%)	$P<0.01$

in positive margins can be seen in **Table 45.4**. Especially of interest is the difference in pathologically confirmed lymph node positive disease: in the NHT group, 28 of 192 patients (15%) had positive lymph nodes compared with 48 of 210 patients in the control group (23%), $P=0.038$.

PSA progression[6]

PSA progression was diagnosed if a patient had a PSA on two subsequent occasions of >1 ng/ml (34 patients in the NHT group, 39 in the control group), if the patient was found to have pT4 or pN2 disease (13 patients in the NHT group, 26 in the control group), if metastases were observed before a PSA of >1 ng/ml was detected (three patients in the NHT group, two in the control group) or if local recurrence was histologically confirmed before a PSA of >1 ng/ml was detected (one patient in the control group). These differences were not statistically significant. Also substratification of patients in groups with cT2 only, cT3 only, pathologically organ-confined disease or PSA levels below 10 or 20 revealed no statistical differences in time to PSA progression. The overall time to PSA progression is illustrated in **Fig. 45.1**.

Local recurrence rate and distant metastases[6]

A total of 18 of 189 patients in the NHT group (10%) and 33 of 209 (16%) patients in the control group developed local recurrence. When looking in cT2 patients only, there was a statistical difference ($P=0.03$), as 3 of 102 (3%) NHT patients and 12 of 114 control patients (11%) had a local recurrence. Only for cT3 patients was there no such difference. With regard to distant metastases, no advantage of NHT could be found as neither evaluation of the overall groups nor evaluation of cT2 or cT3 patients revealed statistically significant differences (7% and 6%, 6% and 5%, and 8% and 6%, respectively). For details of the time to local recurrence and distant metastases, see **Figs 45.2** and **45.3**.

Death related to prostate cancer was observed in three patients in the NHT and in five patients in the control group.

Discussion and conclusion

In accordance with other studies, the abovementioned data of the European Study Group on Neoadjuvant Treatment of Prostate Cancer[6] revealed further evidence that neoadjuvant hormonal treatment prior to radical prostatectomy lowers the rate of positive surgical margins, decreases the likelihood of positive nodes, clinically downstages the tumor in one-third of patients but, ultimately, did not decrease the number of patients with PSA progression in a statistically significant manner.[9,19,28–31] The same can be stated for local recurrence and distant metastases, although, when looking at cT2 patients only, a statistical significant difference could be found in the time to local recurrence ($P=0.03$) in favor of the NHT group. Nevertheless, in the same group, this difference was not seen in the time to PSA progression.

Clinical downstaging is reported in many studies but is not usually in accordance to the pathological examination.[32,33] Also in the European study, clinical downstaging was observed in about one-third of all patients but pathological confirmation was not in accordance. Immunohistochemical staining of tumor cells to improve pathological evaluation has not yet been proved to be beneficial.[34] Especially in cT3 patients and in patients with locally advanced disease and/or poorly differentiated tumors, favorable pathological changes after 3 months are only rarely seen.[35]

Bearing these last observations in mind, it might be that the duration of NHT is not optimal. In their report of 2001,

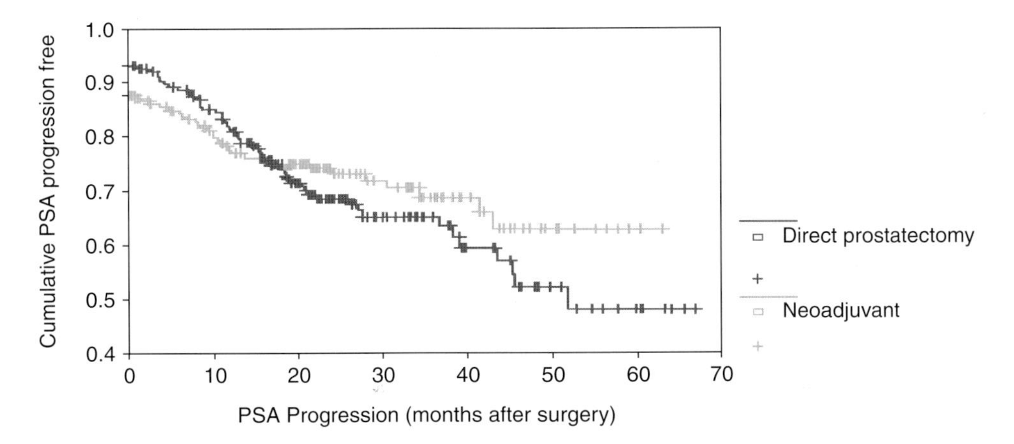

Fig. 45.1. Time to biochemical progression (logrank $P=0.18$).

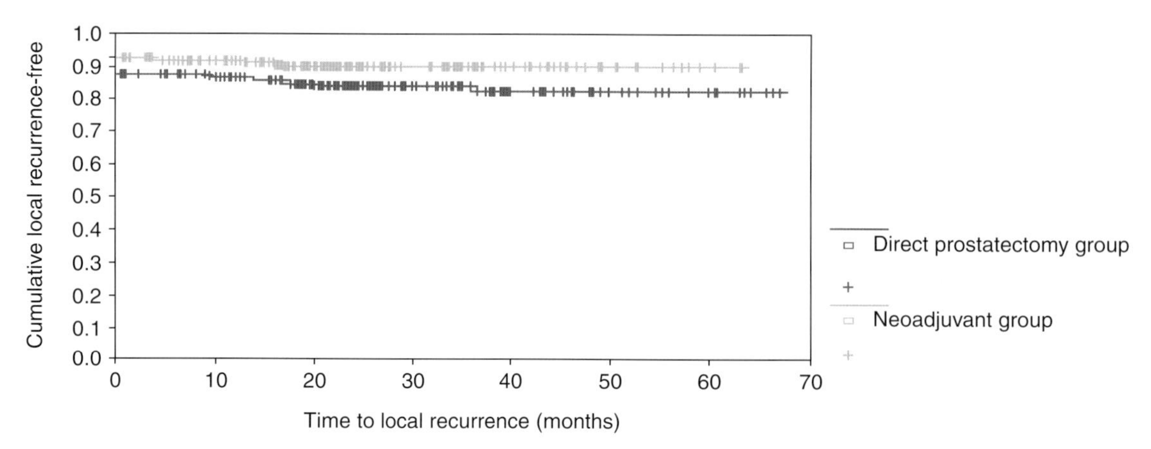

Fig. 45.2. Time to local recurrence (logrank $P=0.07$).

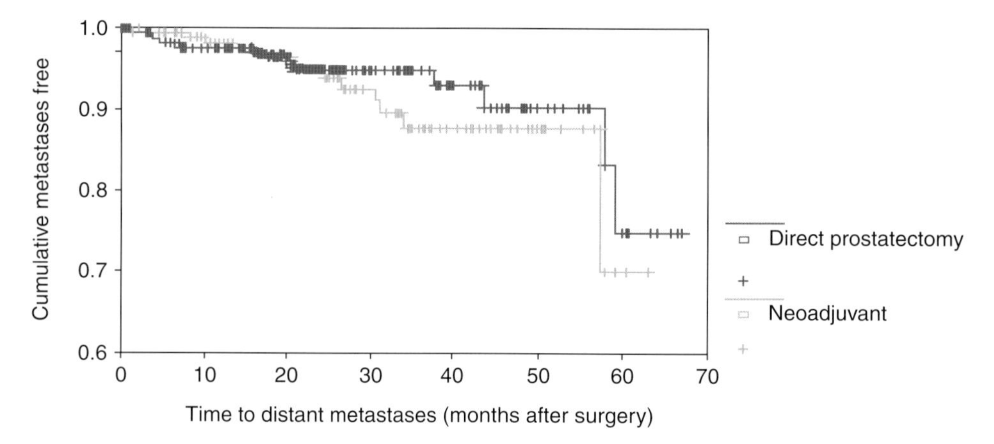

Fig. 45.3. Time to distant metastasis (logrank $P=0.37$).

Meyer et al. state that, after examining 756 men prospectively between 1991 and 1998 (516 men radical prostatectomy alone, 129 men radical prostatectomy after ≤92 days of neoadjuvant hormonal treatment and 111 men radical prostatectomy after ≥93 days of neoadjuvant hormonal treatment), a minimum of 3 months of neoadjuvant therapy is needed to obtain a real but delayed effect on the risk of PSA failure. This effect was not observed before 3 years of follow-up, but was still present after 8 years.[36] Gleave et al.[37] compared 3 months of NHT with 8 months of NHT, reporting significant changes in preoperative PSA nadir, preoperative prostate volume, positive margins, pT0 staging after treatment and positive lymph nodes in favor of the 8 months treatment group. However, this group also had evidently more adverse events ($P < 0.0001$).

The extensively described study of the European Study Group on Neoadjuvant Treatment of Prostate Cancer reveals no proof for the disadvantages mentioned by van Poppel[1] of less extensive or more difficult surgery, as no differences were found in blood loss, operation time or the surgeons' evaluation of the difficulty of the dissection. The other disadvantages mentioned still stand: patient stress owing to operation delay will only increase if NHT over 3 months is advocated; the pathological evaluation of the surgical specimen remains difficult in the absence of better immunohistochemical markers; and, most importantly, hormonal treatment is expensive and has evident side effects. As up to now no real effects have been observed on time to PSA progression, or better, on disease-specific or overall survival, the use of NHT outside study protocols should not be recommended. Further research should be performed, especially on the substratification of patient groups that may benefit the most from NHT. The study of the European Study Group on Neoadjuvant Treatment of Prostate Cancer at least suggests that cT2 patients, with proven better local control, might be the main target for neoadjuvant hormonal treatment prior to radical prostatectomy.

NEOADJUVANT HORMONAL TREATMENT PRIOR TO RADIATION THERAPY

Introduction

Various strategies to improve treatment results with radiotherapy for prostate cancer have been suggested, like increasing the radiation dose to the primary tumor and reducing tumor volume before radiotherapy.[28] In search for such improvements, based on early results of hormonal treatment prior to surgery, neoadjuvant hormonal treatment was also tried in patients with localized prostate cancer undergoing radiation therapy. Nowadays, hormonal treatment before radiotherapy has well earned its place, although there are still several questions to be solved.

Effects of neoadjuvant hormone treatment prior to radiation therapy

Neoadjuvant androgen deprivation therapy can be advantageous in many ways, as is partly outlined earlier. With regard to radiation therapy, one of the main advantages is the reduction of tumor size and thereby a smaller volume of tissue to be irradiated.[38–40] This can be beneficial in reducing acute and long-term complications associated with a confined tumor dose and may reduce the number of clonogens that need to be eradicated by radiation.[28] Neoadjuvant hormonal treatment in combination with radiotherapy may have additional effects in:

- enhancing tumor cell death by common cell death mechanisms (e.g. apoptosis);[41]
- removing tumor cells from the active cycling phase to the resting phase, reducing the rate of accelerated repopulation during radiation therapy;[28]
- inhibiting angiogenesis with tumor bulk reduction as a result, thereby improving the oxygenation of the remaining clonogenic cells. Radiation effects, therefore can increase.[28]

Basic studies are now performed to further clarify the mechanisms of action.[42]

Results of the Radiation Therapy Oncology Group

Patients and methods used in the study[43]

The Radiation Therapy Oncology Group (RTOG) has conducted a number of trials aimed at the evaluation of androgen deprivation adjuvant to definitive radiotherapy. RTOG 86–10 was the first phase III trial testing the value of hormonal treatment in a neoadjuvant format. Preliminary analysis undertaken in 1994 was reported in 1995.[44] A long-term update with mature follow-up data has only recently been published.[43] Patients were randomized to receive either radiotherapy alone, or goserelin and flutamide for 2 months before radiotherapy and during the radiotherapy course. The prostatic target volume was to receive a boost to a total prescribed dose of 65–70 Gy, the daily given doses being 1.8–2 Gy 4–5 times a week. End-points of the study were locoregional control (primary end-point), and disease-free and overall survival (secondary end-points).[43] Eligible patients were patients with bulky primary tumors, clinical stage T2–4. A total of 471 patients were included from 1987 through 1991. Fifteen patients were excluded because of no follow-up ($n=4$), too small a tumor ($n=5$), refusal of treatment ($n=3$), lung or bone metastases ($n=2$) or benign disease ($n=1$). Of 226 analysable patients receiving neoadjuvant hormonal treatment, 186 patients completed both goserelin and flutamide therapy as planned per the study guidelines. Adverse reactions were the main reason to stop hormonal treatment in the remaining patients in this group. The control arm consisted of 230 analyzable patients. The median follow-up reached 6.7 years for all patients and 8.6 years for live patients.

Results[43]

A highly significant improvement was seen in local control as well as in the reduction in disease progression in patients with neoadjuvant hormonal treatment. Subset analyses indicate that the beneficial effect appears preferentially in patients with a lower Gleason score (2–6). In this group ($n=129$) all parameters measured – local failure, distant metastases, no (biochemical) evidence of disease, survival and cause-specific failure – were significantly better in the neoadjuvant-treated group. Local failure at 8 years was 21% in the neoadjuvant group and 46% in the control group ($P=0.005$). For distant metastases and survival, these figures were 13% versus 34% ($P=0.006$) and 70% versus 52% ($P=0.015$), respectively. In patients with a higher Gleason score [7 ($n=176$) or 8–10 ($n=124$)] only, no biochemical evidence of disease reached statistical significance.

Other studies on neoadjuvant hormonal treatment prior to radiotherapy

Besides the abovementioned RTOG trial, randomized trials comparing radiotherapy with or without neoadjuvant androgen deprivation are difficult to find. In 1997, a study by Laverdiere was published[45] describing 120 patients with T2b to T3 tumors of the prostate randomized to radiation therapy alone, 3 months of neoadjuvant hormonal treatment with an LHRH agonist plus flutamide before radiotherapy, or 3 months of neoadjuvant hormonal treatment with an LHRH agonist plus flutamide before radiotherapy with the hormonal treatment continued during and after radiotherapy. In 68 patients the prostate was biopsied 2 years after radiotherapy, revealing positive biopsies in 68.8% of the patients with radiotherapy alone, 28.6% in those treated neoadjuvantly and 5.6% in patients in whom hormonal treatment was continued during and after radiation therapy.

A 12-week course of the steroidal antiandrogen cyproterone acetate has been used in a randomized multicenter trial in 208 men with stage B2–C disease, which significantly increased the proportion of patients free of clinical and biochemical evidence of disease after 18–36 months (70.6 vs. 48.7% and 47.4 vs. 21.5%, respectively).[46] However, follow-up is short and positive results using antiandrogens alone were not found in the study by Wilson et al.[47]

On an experimental level, the study of Zietman[42] is worthwhile mentioning: in Shionogi mice, using tumor control as an end-point, a dose–response curve was constructed. Tumor control dose for 50% of the tumor (TCD 50) fell from 89 Gy to 60.3 Gy, when orchiectomy was performed within 1–2 days before radiotherapy. When tumors were allowed to regress maximally, the TCD 50 dropped even further, to 42.1.

Discussion and conclusion

As can be found in the studies by RTOG[43] and Laverdiere[45], there seems to be a benefit for patients treated neoadjuvantly with androgen deprivation prior to radiotherapy, with regard to the incidence of local failures, disease progression and reduction in disease-specific survival. It should be noted that these effects were only profound in the subgroup with low Gleason scores,[2–6] while patients with Gleason scores 7–10 hardly improved compared to radiotherapy alone. It needs to be considered that, in these patients, different strategies for hormonal treatment in combination with radiotherapy are to be found.

One of the issues to be addressed in future research is the optimal length of neoadjuvant therapy before radiotherapy. Up till now, most patients were treated for 2 or 3 months, but studies are being performed using hormone therapy for much longer periods of time prior to radiation therapy.[48] Besides, continuing androgen therapy during or even after radiotherapy may further increase the beneficial effects. The type of androgen deprivation used seems to be less disputable as the best results were obtained in trials using LHRH agonists in combination with antiandrogens, whereas the scarce studies on antiandrogens alone present conflicting outcomes.

A major point of criticism of the studies performed on neoadjuvant treatment is the lack of a control arm with hormonal treatment only, as the effects may theoretically be the result of hormone alone instead of its combination with radiation. This issue is currently being addressed in a National Cancer Institute of Canada Clinical Trials Group/Canadian Uro-Oncology Group (CUOG) study in which patients are randomized to hormonal therapy alone or hormonal treatment and radiotherapy.[28]

Finally, as in neoadjuvant treatment prior to radical prostatectomy, there is further need for substratification of patient groups that may benefit the most from NHT. Research concerning this problem should not only be restricted to clinical parameters, but should also involve basic research into, for example, genetic changes such as p53 mutation and bcl-2 expression, which are both considered to increase radioresistance.[49,50]

REFERENCES

1. Van Poppel H. Neoadjuvant hormone therapy and radical prostatectomy: the jury is still out. *Eur. Urol.* 2001; 39(Suppl. 1):10–14.
2. Walsh PC, Jewett HJ. Radical surgery for prostatic cancer. *Cancer* 1980; 45(7 Suppl.):1906–11.
3. Iversen P, Adolfsson J, Johansson JE. Localised prostate cancer: a Scandinavian view. *Monogr. Urol.* 1994; 15:93–112.
4. Fair WR, Scher HI. Neoadjuvant hormonal therapy plus surgery for prostate cancer. The MSKCC experience. *Surg. Oncol. Clin. North Am.* 1997; 6:831–46.
5. Watson RB, Soloway MS. Neoadjuvant hormonal treatment before radical prostatectomy. *Semin. Urol. Oncol.* 1996; 14(2 Suppl. 2): 48–55.
6. Schulman CC, Debruyne FM, Forster G et al. 4-Year follow-up results of a European prospective randomized study on neoadjuvant hormonal therapy prior to radical prostatectomy in T2–3N0M0 prostate cancer. European Study Group on Neoadjuvant Treatment of Prostate Cancer. *Eur. Urol.* 2000; 38:706–13.
7. Huggings C, Hodges CV. Studies on prostatic cancer: the effect of castration, of estrogen and of androgen injection on serum phospatases in metastases in metastatic carcinoma of the prostate. *Cancer Res.* 1941; 1:293–7.
8. Bergstralh EJ, Amling CL, Martin S et al. Longterm outcome following preoperative androgen deprivation therapy for clinical stage T3 prostate cancer. *J. Urol.* 1997; 157:332.
9. Aus G, Abrahamsson PA, Ahlgren G et al. Hormonal treatment before radical prostatectomy: a 3-year followup. *J. Urol.* 1998; 159:2013–16.
10. Montironi R, Magi-Galluzzi C, Muzzonigro G et al. Effects of combination endocrine treatment on normal prostate, prostatic intraepithelial neoplasia, and prostatic adenocarcinoma. *J. Clin. Pathol.* 1994; 47:906–13.
11. Villavicencio MH, Chechile TG, Salinas DD et al. Prognostic factors in patients with prostate cancer treated with neoadjuvant hormonal therapy and radical prostatectomy. *Arch. Esp. Urol.* 1996; 49:797–806.
12. Kollermann MW, Pantel K, Enzmann T et al. Supersensitive PSA-monitored neoadjuvant hormone treatment of clinically localized prostate cancer: effects on positive margins, tumor detection and epithelial cells in bone marrow. *Eur. Urol.* 1998; 34:318–24.
13. van der Kwast TH, Tetu B, Candas B et al. Prolonged neoadjuvant combined androgen blockade leads to a further reduction of prostatic tumor volume: three versus six months of endocrine therapy. *Urology* 1999; 53:523–9.
14. Hellstrom M, Ranepall P, Wester K et al. Effect of androgen deprivation on epithelial and mesenchymal tissue components in localized prostate cancer. *Br. J. Urol.* 1997; 79:421–6.
15. Montironi R, Diamanti L, Santinelli A et al. Effect of total androgen ablation on pathological stage and resection limit status of prostate cancer. Initial results of the Italian PROSIT study. *Pathol. Res. Pract.* 1999; 195:201–8.
16. Bonney WW. Neoadjuvant androgen ablation in localised prostatic cancer: a meta analysis of randomised trials. *J. Urol.* 1997; 157:392.
17. Van Poppel H, De Ridder D, Elgamal AA et al. Neoadjuvant hormonal therapy before radical prostatectomy decreases the number of positive surgical margins in stage T2 prostate cancer: interim results of a prospective randomized trial. The Belgian Uro-Oncological Study Group. *J. Urol.* 1995; 154:429–34.
18. Fradet Y. The role of neoadjuvant androgen deprivation prior to radical prostatectomy. *Urol. Clin. North Am.* 1996; 23:575–85.
19. Labrie F, Dupont A, Cusan L et al. Downstaging of localized prostate cancer by neoadjuvant therapy with flutamide and lupron: the first controlled and randomized trial. *Clin. Invest. Med.* 1993; 16:499–509.
20. Meyer F, Moore L, Bairati I et al. Neoadjuvant hormonal therapy before radical prostatectomy and risk of prostate specific antigen failure. *J. Urol.* 1999; 162:2024–8.
21. Schulman CC, Sassine AM. Neoadjuvant hormonal deprivation before radical prostatectomy. *Eur. Urol.* 1993; 24:450–5.
22. Labrie F, Cusan L, Gomez J et al. Down-staging of early stage prostate cancer before radical prostatectomy: the first randomised trial of neoadjuvant combination therapy with flutamide and a luteinising hormone-releasing hormone agonist. *Urology* 1994; 44:29–37.
23. Van Poppel H, Ameye F, Oyen R, Van d V, Baert L. Neoadjuvant hormonotherapy does not facilitate radical prostatectomy. *Acta Urol. Belg.* 1992; 60:73–82.
24. Soloway MS, Sharifi R, Wajsman Z et al. Randomized prospective study comparing radical prostatectomy alone versus radical prostatectomy preceded by androgen blockade in clinical stage B2 (T2bNxM0) prostate cancer. The Lupron Depot Neoadjuvant Prostate Cancer Study Group. *J. Urol.* 1995; 154:424–8.
25. Van de Voorde WM, Elgamal AA, Van Poppel HP et al. Morphologic and immunohistochemical changes in prostate cancer after preoperative hormonal therapy. A comparative study of radical prostatectomies. *Cancer* 1994; 74:3164–75.
26. Tyrrell CJ, Kaisary AV, Iversen P et al. A randomised comparison of 'Casodex' (bicalutamide) 150 mg monotherapy versus castration in the treatment of metastatic and locally advanced prostate cancer. *Eur. Urol.* 1998; 33:447–56.

27. Witjes WP, Schulman CC, Debruyne FM. Preliminary results of a prospective randomized study comparing radical prostatectomy versus radical prostatectomy associated with neoadjuvant hormonal combination therapy in T2–3 N0 M0 prostatic carcinoma. The European Study Group on Neoadjuvant Treatment of Prostate Cancer. *Urology* 1997; 49(3A Suppl.):65–9.
28. Lee HH, Warde P, Jewett MA. Neoadjuvant hormonal therapy in carcinoma of the prostate. *Br. J. Urol. Int.* 1999; 83:438–48.
29. Schulman CC, Wildschutz T, Zlotta AR. Neoadjuvant hormonal treatment prior to radical prostatectomy: facts and open questions. *Eur. Urol.* 1997; 32(Suppl. 3):41–7.
30. Witjes WP, Oosterhof GO, Schaafsma HE et al. Current status of neoadjuvant therapy in localized prostate cancer. *Prostate* 1995; 27:297–303.
31. Goldenberg SL, Klotz LH, Srigley J et al. Randomized, prospective, controlled study comparing radical prostatectomy alone and neoadjuvant androgen withdrawal in the treatment of localized prostate cancer. Canadian Urologic Oncology Group. *J. Urol.* 1996; 156:873–7.
32. Macfarlane MT, Abi-Aad A, Stein A et al. Neoadjuvant hormonal deprivation in patients with locally advanced prostate cancer. *J. Urol.* 1993; 150:132–4.
33. Narayan P, Lowe BA, Carroll PR et al. Neoadjuvant hormonal therapy and radical prostatectomy for clinical stage C carcinoma of the prostate. *Br. J. Urol.* 1994; 73:544–8.
34. Gleave ME, Goldenberg SL, Jones EC et al. Longer duration of neoadjuvant hormonal therapy prior to radical prostatectomy in clinically confined prostate cancer: biochemical and pathological effects. *Mol. Urol.* 1997; 1:199–204.
35. Srougi M, Kaufmann JR, Nesrallah A, Leite KR. High Gleason score predicts poor pathologic outcome after neoadjuvant androgen deprivation for locally advanced prostate cancer. *Mol. Urol.* 2001; 2:195–9.
36. Meyer F, Bairati I, Bedard C et al. Duration of neoadjuvant androgen deprivation therapy before radical prostatectomy and disease-free survival in men with prostate cancer. *Urology* 2001; 58(2 Suppl. 1):71–7.
37. Gleave ME, Goldenberg SL, Chin JL et al. Randomized comparative study of 3 versus 8-month neoadjuvant hormonal therapy before radical prostatectomy: biochemical and pathological effects. *J. Urol.* 2001; 166:500–6.
38. Lilleby W, Fossa SD, Knutsen BH et al. Computed tomography/magnetic resonance based volume changes of the primary tumour in patients with prostate cancer with or without androgen deprivation. *Radiother. Oncol.* 2000; 57:195–200.
39. Shearer RJ, Davies JH, Gelister JS et al. Hormonal cytoreduction and radiotherapy for carcinoma of the prostate. *Br. J. Urol.* 1992; 69:521–4.
40. Forman JD, Kumar R, Haas G et al. Neoadjuvant hormonal downsizing of localized carcinoma of the prostate: effects on the volume of normal tissue irradiation. *Cancer Invest.* 1995; 13:8–15.
41. Widmark A, Damber JE, Bergh A et al. Estramustine potentiates the effects of irradiation on the Dunning (R3327) rat prostatic adenocarcinoma. *Prostate* 1994; 24:79–83.
42. Zietman AL, Nakfoor BM, Prince EA et al. The effect of androgen deprivation and radiation therapy on an androgen-sensitive murine tumor: an in vitro and in vivo study. *Cancer J. Sci. Am.* 1997; 3:31–6.
43. Pilepich MV, Winter K, John MJ et al. Phase III radiation therapy oncology group (RTOG) trial 86-10 of androgen deprivation adjuvant to definitive radiotherapy in locally advanced carcinoma of the prostate. *Int. J. Radiat. Oncol. Biol. Phys.* 2001; 50:1243–52.
44. Pilepich MV, Krall JM, al Sarraf M et al. Androgen deprivation with radiation therapy compared with radiation therapy alone for locally advanced prostatic carcinoma: a randomized comparative trial of the Radiation Therapy Oncology Group. *Urology* 1995; 45:616–23.
45. Laverdiere J, Gomez JL, Cusan L et al. Beneficial effect of combination hormonal therapy administered prior and following external beam radiation therapy in localized prostate cancer. *Int. J. Radiat. Oncol. Biol. Phys.* 1997; 37:247–52.
46. Porter AT, Elhilali M, Manji M et al. A phase III randomized trial to evaluate the efficacy of neoadjuvant therapy prior to curative radiotherapy in locally advanced prostate cancer patients. *Pros. Am. Soc. Clin. Oncol.* 1997; 16:315.
47. Wilson KS, Ludgate CM, Wilson AG et al. Neoadjuvant hormone therapy and radical radiotherapy for localized prostate cancer: poorer biochemical outcome using flutamide alone. *Can. J. Urol.* 2000; 7:1099–103.
48. Ludgate CM, Lim JT, Wilson AG et al. Neoadjuvant hormone therapy and external beam radiation for localized prostate cancer: Vancouver Island Cancer Centre experience. *Can. J. Urol.* 2000; 7:937–43.
49. Heidenberg HB, Sesterhenn IA, Gaddipati JP et al. Alteration of the tumor suppressor gene p53 in a high fraction of hormone refractory prostate cancer. *J. Urol.* 1995; 154:414–21.
50. McDonnell TJ, Deane N, Platt FM et al. bcl-2-immunoglobulin transgenic mice demonstrate extended B cell survival and follicular lymphoproliferation. *Cell* 1989; 57:79–88.

Androgen-independent Prostate Cancer: The Evolving Role of Chemotherapy

Beth A. Hellerstedt and Kenneth J. Pienta
Medical Center, University of Michigan, Ann Arbor, MI, USA

INTRODUCTION

Androgen-independent prostate cancer (AIPC) is currently an incurable disease. Only a decade ago, the risks associated with available chemotherapy far outweighed the meager response rates seen with treatment of AIPC. Recent trials, however, suggest that AIPC can be responsive to novel chemotherapeutic agents. New approaches, including combination chemotherapy and biological therapy, are providing promise for longer survival and improved quality of life.

One of the most important advances in the field of prostate cancer treatment is the use of prostate specific antigen (PSA) as a reliable tumor marker. As prostate cancer metastasizes most frequently to bone, accurate measurement of disease burden and response to treatment is difficult. A number of phase II studies have investigated the correlation between PSA and overall survival.[1–4] Importantly, a recent randomized, placebo-controlled phase III study conducted to assess the effect of suramin on AIPC was also designed to determine whether post-treatment declines in PSA were associated with clinical measures of improvement.[5] A total of 460 patients with AIPC were randomized to receive suramin plus hydrocortisone or placebo plus hydrocortisone. In this group of patients, a post-treatment decrease in PSA of $\geqslant$50% lasting $\geqslant$28 days was associated with significantly longer durations of median overall survival, median overall pain-free survival and median time to pain progression. This represents the first phase III data specifically designed to study the correlation between PSA decline and overall survival/clinical end-points. The 1999 consensus statement by the National Cancer Institute states that a PSA decrease of $\geqslant$50% for at least two measurements 6 weeks apart is considered a partial response to chemotherapy in trials for AIPC.[6]

NEW PROGNOSTIC AND RISK-STRATIFICATION FACTORS

Although the median duration of survival in men with AIPC is approximately 12 months, the heterogeneity of response to chemotherapy suggests the existence of subgroups of responsive and non-responsive disease. Therefore, prediction of response for the individual patient is impossible. Most of the risk stratification or prognostic factors in AIPC are associated with tumor burden (high PSA, alkaline phosphatase) and patient status (performance scores, weight loss). New methods of disease evaluation may be useful in tailoring therapy to both risk and response.

Reverse transcriptase polymerase chain reaction (RT-PCR) of peripheral blood appears to be emerging as a method of risk stratification for all stages of prostate cancer. The original assays were often limited by lack of both sensitivity and specificity. Three large trials suggested that detecting circulating tumor cells by analysis of peripheral blood with RT-PCR predicts biochemical relapse after radical prostatectomy.[7–9] Another trial did not find such a correlation.[10]

New assays and new targets are providing more promise. CALGB 9480, a prospective, randomized trial of three doses of suramin for treatment of AIPC, was also designed to evaluate peripheral blood for the presence of PSA mRNA using RT-PCR.[11] Peripheral blood samples were acquired in 193 patients, and deemed adequate for measurement in 156 patients. Forty-eight per cent of samples were positive by RT-PCR. The median survival for patients with a negative RT-PCR was 18 months, compared with 13 months for patients with a positive study ($P = 0.04$). In a multivariate analysis, RT-PCR was an independent predictor of survival, adjusting for alkaline phosphatase and performance status. RT-PCR was still of borderline significance when measurable disease was included in the model, suggesting that a positive test may confer information distinct from tumor burden.

If the significant survival benefit associated with a negative RT-PCR reflects properties of the tumor cell itself, molecular characterization of the differences between circulating and non-circulating cells could provide further therapeutic targets. In the future, technologies such as RT-PCR could assist with stratification of risk–benefit ratios for all interventions, from prostatectomy to salvage chemotherapy.

CHEMOTHERAPY

The use of chemotherapy in AIPC has undergone a revolution in the past decade. Once labeled 'chemotherapy resistant', new agents have shown great promise in both palliative and response endpoints.

While the new agents have been tested extensively in phase II trials, only a few phase III trials have been conducted using chemotherapy in AIPC in the last decade. A summary of these trials is shown in **Table 46.1**.

Table 46.1. Phase III trials using chemotherapy in the treatment of AIPC.

Regimen	Number of patients	Conclusions
Prednisone (P) vs. mitoxantrone and prednisone (MP)[47]	161	Primary end-point: palliative response • 38% of MP patients met criteria for primary response, compared to 21% of patients in P group • Crossover allowed • Time to disease progression longer in MP group, but overall survival no different
Hydrocortisone (H) vs. mitoxantrone and hydrocortisone (M+H)[48]	242	Primary end-point: 50% increase in survival duration, assuming 12-month survival duration in patients receiving only H • Appropriately powered, with a sensitivity analysis demonstrating the improbability that a small improvement in survival would be missed • No difference between the two groups in survival duration • Modest benefits in time to disease progression and time to treatment failure in M+H group (3.7 months vs. 2.3 months) • In post-hoc analysis, patients with ⩾50% decline in PSA had median survival of 20 months, regardless of which treatment was received; 38% of patients in the M+H group achieved this result, compared to 22% in the H group
Vinblastine vs. estramustine and vinblastine (EV)[49]	201	Primary end-point: overall survival (OS) • No difference in OS seen between the two groups, but only powered to detect a 50% or greater increase • 25% of patients in each arm discontinued treatment before completing one 8-week treatment cycle • Superiority of EV in time to progression and sustained PSA response
Placebo plus hydrocortisone vs. suramin plus hydrocortisone (S+H)[50]	460	Primary end-point: reduction of pain and opioid requirements • S+H showed a greater overall mean reduction in combined pain scores and opioid use, and a minor positive effect in time to disease progression and the proportion of patients achieving a ⩾50% reduction in PSA • Overall survival was no different between the groups. The median overall survival was <10 months, suggesting a patient population with advanced disease
Mitoxantrone and prednisone vs. estramustine and taxotere	Ongoing	Southwest Oncology Group study 9916 • Currently accruing patients. Accrual is slow because phase II data on taxanes render it difficult for patients and physicians to accept randomization to a regimen considered to be palliative

Table 46.2. Phase II trials using chemotherapy in AIPC.

Regimen	Response rate – measurable disease	Response rate – PSA	Reference
Estramustine, etoposide	42–57%	22–86%	53–57
Estramustine, etoposide, cisplatin		61%	58
Etoposide, carboplatin	7%	54%	59
Estramustine, vinblastine	14–40%	46–54%	60–62
Estramustine, IV vinorelbine	0%	30–38%	63, 64
Estramustine, etoposide, vinorelbine		56%	65
Paclitaxel (weekly)	0%	75%	66
Paclitaxel (weekly), high-dose oral estramustine	33%	50%	67
Paclitaxel (weekly), estramustine	27–40%	58–71%	68–70
Paclitaxel (q21 days), estramustine	44%	53%	71
Paclitaxel, estramustine, carboplatin (TEC)	45%	67%	72
Paclitaxel, estramustine (IV), carboplatin (Hi-TEC)	61%	63%	73
Paclitaxel, estramustine, etoposide (TEE)		65%	74
Paclitaxel, estramustine, etoposide, carboplatin (TEEC)	58%	65%	75
Docetaxel (weekly)		58%	76
Docetaxel (q21 days)	28%	38–46%	77, 78
Docetaxel, vinorelbine, G-CSF	66%	63%	79
Docetaxel (weekly), estramustine	11%	71%	80, 81
Docetaxel (q21 days), estramustine	6–25%	39–82%	82–85
Estramustine, docetaxel, carboplatin (ETP)	Discontinued due to thrombocytopenia		86
Docetaxel, estramustine, dexamethasone, prednisone		75%	87
Docetaxel (q21 days), estramustine, hydrocortisone	50%	68%	88
Cyclophosphamide, vincristine, dexamethasone (CVD)	29%	29%	89
Cyclophosphamide, UFT (uracil-tegafour), estramustine	67%	62%	90
Cyclophosphamide, estramustine		44%	91
Epirubicin, cisplatin	14%	32%	92
Epirubicin, mitomycin C, 5-FU (5-fluorouracil)		47%	93

Despite a dearth of phase III data, chemotherapy has been gaining momentum in a multitude of phase II trials, employing both single-agent and combination regimens.[12] The microtubule inhibitors have been the targeted group of drugs, based on *in vitro* data.[13,14] Practitioners face significant difficulty in interpreting this wealth of information. As no single formulation has emerged as 'standard', the basic rules of treatment apply: assessing the patient's status and tailoring the choice of regimen accordingly. The phase II data listed in **Table 46.2** contains many references. As is often the case with phase II data, many of the studies contain small numbers of patients, and inclusion criteria vary widely between groups. Often, only a handful of patients are evaluable for measurable response, which must evoke caution when evaluating this data point.

Overall, the regimens listed in **Table 46.2** all have some degree of efficacy. As expected, combination regimens have higher toxicities, particularly myelosuppression. Estramustine therapy is associated with a growing literature on thromboembolic phenomenon, and new evidence exists implicating that risks may be higher in men who are not receiving androgen ablation.[15] Several studies included heavily pretreated patients, suggesting that even these patients can tolerate and respond to newer active drugs. An illustration of a stepwise approach to treatment is provided in **Fig. 46.1** Although a few regimens are cited as examples, to date there are no phase III data to indicate that one regimen demonstrates a survival advantage over another.

DECREASING ESTRAMUSTINE TOXICITY

Estramustine is a conjugate of 17β-estradiol with the alkylating agent nor-nitrogen mustard. Originally thought to act exclusively

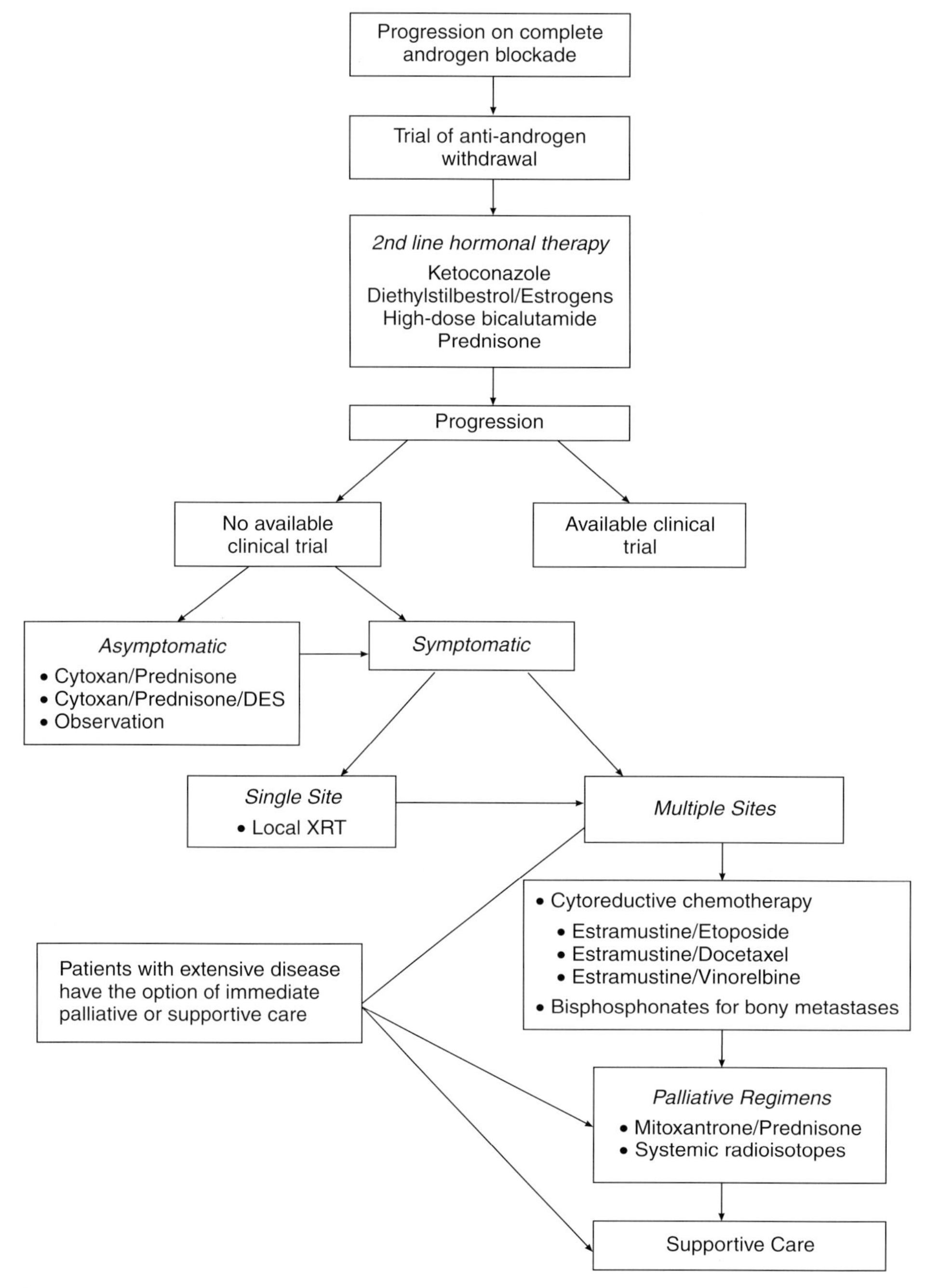

Fig. 46.1. An example of a stepwise approach to therapy in AIPC.

as an alkylator or an estrogen, its cytotoxic effects are now thought to be secondary to binding of tubulin and microtubule associated proteins, promoting tubule disassembly and interruption of mitosis.[16–19] *In vitro* studies suggested synergy of estramustine with a variety of other chemotherapeutic agents, including the taxanes.[20] Although estramustine is now used extensively in combination with other microtubule inhibitors, practitioners had all but abandoned the drug in the early 1990s. Significant toxicities, including gastrointestinal, cardiac and thrombotic, often precluded its use. As physicians at that time were unaware of the need to avoid calcium-containing foods and antacids when taking the drug, efficacy may have been underestimated secondary to poor absorption. Many case reports highlighted other morbid syndromes attributed to estramustine, including hypocalcemia,[21] hemolytic–uremic syndrome,[22] secondary myelodysplastic syndrome,[23] and acute myelogenous leukemia.[24] Prophylactic measures, such as the addition of aspirin or low-dose warfarin to decrease thrombotic events, have not translated into decreased incidence of thrombosis.[25] However, the ability of estramustine to potentiate the cytotoxic effects of taxanes *in vitro* has established it as a mainstay of current therapy.

The incremental benefit of estramustine remains an issue. Although more widely used in combination with estramustine, taxanes have been used effectively as single agents in the treatment of AIPC. Friedland et al. conducted a study of 21 patients with AIPC using docetaxel at a dose of 75 mg/m^2 q 21 days.[26] Seven patients had a decrease in PSA by >50%, and six patients had radiographic response on bone scan or computed tomography (CT) scan. Picus et al. conducted a similar study in 35 patients.[27] Sixteen patients had a >50% decline in PSA. Seven patients had a measurable disease response. The response was maintained for a median of 9 months and overall survival was 27 months. In the phase III trial of estramustine and vinblastine versus vinblastine alone, no difference was seen in overall survival.[28] However, the study was underpowered, and the combination arm showed significant improvements in time to progression and sustained PSA response.

Given the myriad of doses and schedules used in clinical practice, the precise role of estramustine becomes even more difficult to define. Knowledge about the ideal dosing of the drug is limited. Conventional dosing of 4–9 capsules in divided doses may not be ideal when the drug is used in combination. Also, many regimens continue estramustine for weeks, which may exacerbate toxicity without improving efficacy. Three trials using estramustine and etoposide highlight dose disparity (**Table 46.3**). Similar studies using estramustine and q 21 days docetaxel also highlight schedule disparity (**Table 46.4**). Seventy kilograms is used as a representative weight in regimens requiring weight-based dosing.

Given these confusing results, the necessity of determining the appropriate dose and schedule of estramustine remains paramount. Two small studies reported in abstract form attempt to address this question.[29,30] In the first cohort, consisting of an 8-week cycle, estramustine is dosed at 420 mg three times per day (TID) on days 1–4, with docetaxel administered on day 3 weekly for 6 weeks (5040 mg of estramustine/8 weeks). In the second cohort, consisting of a 6-week cycle, estramustine is dosed at 420 mg X4 on days 1 and 2, and 280 mg X5 on day 3, with docetaxel administered on day 2 weekly X2 (4700 mg estramustine/6 weeks). Eighteen patients were enrolled, 12 of whom had received prior chemotherapy. Fourteen patients (78%) had a PSA response. Four patients with measurable disease had objective responses. Twelve patients had symptomatic improvement or an increase in performance status. Although only 18 patients were studied, the comparable results in terms of PSA and measurable disease response support the concept of decreasing estramustine dose.

Table 46.3. Dosing regimens of estramustine and etoposide.

Estramustine dose	Etoposide dose	Response rates	Toxicity	Reference
15 mg/kg per day in four divided doses, 3 weeks on, 1 week off (22 050 mg/cycle)	50 mg/m^2 per day in two divided doses, 3 weeks on, 1 week off	PSA: 55% Measurable: 50%	Diarrhea 12%, nausea 29%, thromboembolic events (TE) 10%, cardiac 3%	55
10 mg/kg per day in four divided doses, 3 weeks on, 1 week off (14 700 mg/cycle)	50 mg/m^2 per day in two divided doses, 3 weeks on, 1 week off	PSA: 39% Measurable: 53%	Diarrhea 3%, nausea 24%, TE 6%	54
140 mg TID, 3 weeks on, 1 week off (8820 mg/cycle)	50 mg/m^2 per day in two divided doses, 3 weeks on, 1 week off	PSA: 58% Measurable: 45%	Diarrhea 22%, nausea 63%, TE 4%	56

Table 46.4. Dosing regimens of estramustine and docetaxel.

Estramustine dose	Docetaxel dose	Response rates	Toxicity	Reference
280 mg q6 hours × 5 (1400 mg/cycle)	70 mg/m^2	PSA: 39% Measurable: 25%	Diarrhea 11%, thromboembolic events (TE) 0%	82
14 mg/kg per day continuously (20580 mg/cycle)	40–80 mg/m^2	PSA: 82% Measurable: 16%	Diarrhea 59%, nausea 100%, TE 12%	85
280 mg TID × 5 days (4200 mg/cycle)	40–80 mg/m^2	PSA: 63% Measurable: 28%	Diarrhea 15%, nausea 29%, vomiting 12%, hypocalcemia 59%, TE 8.8%	94
10 mg/kg per day × 5 days (3500 mg/cycle)	70 mg/m^2	PSA: 68% Measurable: 50%	Diarrhea 34%, nausea 34%, vomiting 15%, hypocalcemia 41%, cardiac 4%, TE 9%	88

Table 46.5. Selected chemotherapy regimens.

Estramustine and etoposide[54–56]
- Estramustine 6–15 mg/kg per day PO in three divided doses on days 1–21
- Etoposide 50 mg/m^2 PO on days 1–21
- 28-day treatment cycle

Paclitaxel and estramustine[71]
- Paclitaxel 120 mg/m^2, by 96-hour continuous infusion on days 1–4
- Estramustine 600 mg/m^2 per day PO, continuously
- 21-day treatment cycle

Mitoxantrone and prednisone[49]
- Mitoxantrone 12 mg/m^2 on day 1
- Prednisone 5 mg PO BID, continuously
- 21-day treatment cycle

Estramustine, docetaxel, hydrocortisone[88]
- Estramustine 10 mg/kg per day PO, days 1–5
- Docetaxel 70 mg/m^2 on day 2
- Hydrocortisone 40 mg PO qd, continuously
- 21-day treatment cycle

Cytoxan, prednisone (+/– diethylstilbestrol; DES)[31]
- Cytoxan 100 mg PO qd, days 1–20
- Prednisone 10 mg PO qd, continuously
- DES 1 mg PO qd, continuously
- 30-day treatment cycle

Estramustine and vinorelbine[64]
- Estramustine 140 mg PO TID, days 1–3 and 8–10
- Vinorelbine 22.5 mg/m^2 IV, day 2 and day 9
- 28-day treatment cycle

Note: Many physicians also prescribe low-dose Coumadin (2 mg/day) when using regimens containing estramustine or diethylstilbestrol.

COMBINATION THERAPY

Many types of hormonal therapies have been studied in combination with chemotherapy. Although the studies are preliminary, combination therapy deserves further exploration. **Table 46.5** provides doses and schedules of some of the most commonly used regimens.

Pienta et al. treated 58 patients with diethylstilbestrol, prednisone and oral cyclophosphamide. The combination was well tolerated, with no episodes of significant myelosuppression and minimal nausea. Three of seven patients (43%) had partial responses in soft tissue lesions and 21 of 58 patients (36%) had a PSA decline of ⩾50% from pretreatment levels.[31]

Sella et al. conducted a phase II trial of high-dose ketoconazole with weekly adriamycin in 39 patients.[32] A PSA response of ⩾50% was demonstrated in 55% of evaluable patients and 7 of 12 patients with soft tissue metastasis had partial responses. Unfortunately, toxicities were severe: 17 patients required hospitalization for complications and two patients with a history of coronary disease experienced sudden cardiac death. This study did not prophylactically use hydrocortisone and 24 patients developed adrenal insufficiency.

Another phase II trial evaluated 103 patients who were treated with 2–3 cycles of ketoconazole and doxorubicin with alternating estramustine and vinblastine (KAVE). Patients who responded to chemotherapy were then randomly assigned to weekly doxorubicin with or without strontium. The strontium arm was associated with improved overall survival when compared to chemotherapy alone (27.7 months vs. 16.8 months; P=0.0014).[33]

BISPHOSPHONATES

New data are emerging regarding potential benefits of bisphosphoriates in prostate cancer. Bone metastases can be found in 8% of Caucasian men and 14% of African American men at initial diagnosis.[34] Although prostate cancer metastases are considered to be osteoblastic, new evidence suggests that the inciting event is bone resorption.[35] Increased levels of markers associated with bone resorption, such as pyridinium cross-links, pyridionoline and deoxypyridinoline, have been found in patients with bone lesions.[36]

Small studies and case series regarding the efficacy of bisphosphonates in pain control for bone metastases were first reported in the late 1980s. A number of follow-up studies using first-generation oral bisphosphonates produced disappointing results, so the entire class of drug was largely abandoned. However, in the late 1990s, *in vitro* data prompted another wave of studies, using intravenous, second-generation bisphosphonates in higher doses. These studies showed clinical benefit in conjunction with decreases in markers of bone resorption. Recently, *in vitro* data suggest that bisphosphonates can inhibit adhesion of tumor cells to the bony matrix.[37] A follow-up study demonstrated the ability of bisphosphonates to both prevent tumor cell invasion into the bony matrix, and to inhibit matrix metalloproteinases, known mediators of uncontrolled cell growth.[38] **Table 46.6** provides a summary of some published studies to date.

A recent large study evaluated the efficacy of clodronate in 85 patients with AIPC and painful osseous metastases.[39] Clodronate was given intravenously for 8 days, followed by an oral daily maintenance dose. A significant decrease was noted in the mean visual analog pain score (7.9 to 2.5, $P < 0.001$) in 75% of patients. A total of 19 patients required no analgesics and 45 had a significant decrease in required dose. The improvement in pain was paralleled by an improvement in performance status, from a Karnofsky score of 45% to 70%.

The newest bisphosphonate, zoledronic acid, has been evaluated in a randomized placebo-controlled trial of 422 patients

Table 46.6. Bisphosphonates in the treatment of AIPC.

Study	Number of patients	Bisphosphonate	Response
Adami (1985)[95]	17	IV clodronate	Sustained pain relief and improvement in performance status in 16 patients
Smith (1989)[96]	57	PO etidronate	Randomized, double-blind, placebo-controlled study. No difference between etidronate and placebo
Clarke (1992)[97]	42	IV pamidronate	Decrease in markers of bone destruction
Kymala (1993)[98]	57	IV clodronate with PO maintenance	Used in combination with estramustine. IV clodronate used for first 5 days, then PO clodronate versus placebo for 3 months. Only 10% improvement from placebo in the clodronate group
Lipton (1994)[99]	58	IV pamidronate	Decreased bone pain, but no healing of bone lesions observed
Vinholes (1997)[100]	52	IV pamidronate	Randomized, double-blind placebo-controlled study. Pamidronate group demonstrated less pain, sustained QOL and decreased bone resorption markers
Adami (1997)[101]	12	PO alendronate	11 patients reported decreased bone pain
Pelger (1998)[102]	28	IV olpadronate with or without PO maintenance	76% of patients reported decreased bone pain and analgesic use, sustained only in those patients on maintenance. Clinical results correlated with markers of bone resorption
Coleman (1999)[103]	16	Ibandronate	Randomized, double-blind, placebo-controlled dose-finding study in 110 total patients, 16 with prostate cancer. Showed potent effects on bone resorption, no efficacy data reported

with AIPC (Saad F, personal communication). Significantly fewer patients in the zoledronic acid group experienced skeletal-related events and pathologic fractures. The time to first skeletal event, vertebral fracture and pathologic fracture were significantly longer in the treatment group. Pain control at 3 months and 9 months was also significantly improved. This study is the first large, randomized, placebo-controlled trial to demonstrate a protective effect of bisphosphonates in patients with AIPC. These data provide support for bisphosphonates as part of standard therapy for patients with bone metastases.

As androgen ablation also places men at higher risk of osteoporosis, bisphosphonates may be useful in prevention of secondary fractures. Even in a study of intermittent androgen blockade, evaluation of bone mineral density in the lumbar spine revealed osteopenia in 46% and osteoporosis in 20% of patients, with similar values seen for the hip.[40] A similar study found that 50% of men on androgen blockade for at least 12 months had asymptomatic vertebral fractures.[41] A phase III, randomized, double-blind, placebo-controlled trial of oral clodronate given to 311 patients with androgen-sensitive prostate cancer has provided preliminary data that administration of clodronate delayed the time to the development of symptomatic bone progression by 6 months.[40] As the recent zoledronic acid trial in AIPC patients supports the use of bisphosphonates, a similar trial should be undertaken in men with androgen-sensitive disease.

NOVEL THERAPIES

Following recent trends for other cancers, the treatment of AIPC is looking outward from chemotherapy. The following is a sample of newly described targets and therapies that highlight translational approaches, but is in no way an exhaustive list.

Statins

Recent developments in the understanding of bone mineralization may be applicable to prevention and treatment of metastatic disease. The HMG CoA reductase inhibitors ('statins') have recently been shown to have *in vitro* effects on osteoblast differentiation and bone formation.[43] Effects seen at low concentrations of drug included enhanced alkaline phosphatase activity and mineralization, and were manifest in a dose- and time-dependent fashion. *In vivo* data from rodents also show that simvastatin stimulates bone formation.[44] Previous studies of osteoblast activity have centered on growth factors, such as transforming growth factor-β (TGFβ) and fibroblast growth factors (FGF), which stimulate osteoblast proliferation. Unfortunately, these molecules also inhibit osteoblast differentiation[45] and also cause proliferation of cells in the overlying subcutaneous tissue.[42]

In vitro studies have demonstrated an effect of statins on osteoblast differentiation via the induction of bone morphogenetic protein-2 (BMP-2). Mundy et al. reported that lovastatin, simvastatin, mevastatin and fluvastatin all increased new bone formation in murine calvarial bones in organ culture, and after *in vivo* injection.[42] An increase in osteoblast numbers and differentiation was seen. To determine whether this effect could be reproduced with systemic administration, ovariectomized rats were given oral doses of statins. Treated rats had an increase in trabecular bone volume between 39% and 94%. Although much of the increased volume was attributed to osteoblast activity, it was noted that treatment also reduced numbers of viable osteoclasts. The current statins are targeted to the liver, although some data exist that postmenopausal women taking statins have a greater bone mineral density and decreased risk of hip fracture.[46] Effects were seen at doses considered to be similar to those used for cholesterol reduction, so whether more bone-targeted preparations are necessary to promote meaningful osteoblastic activity is unknown.

Antiangiogenesis agents

Major steps have been taken in the delineation of tumor-associated angiogenesis. *In vitro* data are accumulating to evaluate the role of the major ligands of the angiogenesis pathway, VEG-F (vascular endothelial growth factor) and bFGF (basic fibroblast growth factor) in prostate cancer cell lines.

Thalidomide is a glutamic acid derivative thought to have antiangiogenic activity. A randomized, phase II trial of 63 patients utilized low-dose (200 mg daily) and high-dose (up to 1200 mg daily) thalidomide in AIPC patients.[47] Prior cytotoxic treatment was allowed. The high-dose arm was terminated early as none of the 13 patients enrolled had a ⩾50% reduction in PSA. The low-dose arm was then expanded to 50 patients. Nine patients (14%) had a ⩾50% decline in PSA. Four patients (6%) had a PSA decline of ⩾50% that was sustained for >150 days. No complete or partial responses were seen in patients with measurable disease on CT scan or bone scan. A total of 560 adverse events were reported. The most common complaints were fatigue, constipation and peripheral neuropathy. Median survival for all 63 patients is 15.8 months.

This study, while appearing to provide only a minor role for thalidomide in AIPC, must be viewed in the appropriate context. First, the mechanism of antiangiogenic agents could be to reduce the potential for metastatic growth. In comparison to chemotherapy, their role may be considered cytostatic as opposed to cytotoxic. Thalidomide and other antiangiogenesis agents may be best suited for patients with a low disease burden. In this study, however, patients had markedly advanced disease and were heavily pretreated. Additionally, *in vitro* data have suggested that thalidomide may up-regulate the secretion of PSA in certain prostate cancer cell lines.[48] If this holds true *in vivo*, even small declines in PSA may correlate to a significant amount of tumoricidal activity. Further studies in this area are eagerly awaited.

CONCLUSIONS

Clinical research is developing all fields of oncology, but the realm of androgen-independent prostate cancer may be on the cusp of a revolution. Such action is needed for a disease that affects thousands of men annually. However, only thoughtful clinical trials will provide the answers sought by patients and practitioners alike.

REFERENCES

1. Kelly WK, Scher HI, Mazumdar M. Prostate-specific antigen as a measure of disease outcome in metastatic hormone-refractory prostate cancer. *J. Clin. Oncol.* 1993; 11:607–15.
2. Smith DC, Dunn RL, Strawderman MS et al. Change in serum prostate-specific antigen as a marker of response to cytotoxic therapy for hormone-refractory prostate cancer. *J. Clin. Oncol.* 1998; 16:1835–43.
3. Scher HI, Kelly WK, Zhang ZF et al. Post-therapy serum prostate-specific antigen level and survival in patients with androgen-independent prostate cancer. *J. Natl Cancer Inst.* 1999; 91:244–51.
4. Sridhara R, Eisenberger M, Sinibaldi V et al. Evaluation of prostate specific antigen as a surrogate marker for response of hormone refractory prostate cancer to suramin therapy. *J. Clin. Oncol.* 1995; 13:2944–53.
5. Small EJ, McMillan A, Meyer M et al. Serum prostate-specific antigen decline as a marker of clinical outcome in hormone-refractory prostate cancer patients: association with progression-free survival, pain endpoints, and survival. *J. Clin. Oncol.* 2001; 19:1304–11.
6. Bubley GJ, Carducci M, Dahut W et al. Eligibility and response guidelines for phase II clinical trials in androgen-independent prostate cancer: recommendations from the Prostate-Specific Antigen Working Group. *J. Clin. Oncol.* 1999; 17; 3461–7.
7. Katz AE, de Vries GM, Begg MD et al. Enhanced reverse transcriptase-polymerase chain reaction for prostate specific antigen as an indicator of true pathologic stage patients with prostate cancer. *Cancer* 1995; 75:1642–8.
8. De la Taille A, Olsson CA, Buttyan R et al. Blood-based reverse transcriptase polymerase chain reaction assays for prostatic specific antigen: long-term follow-up confirms the potential utility of this assay in identifying patients more likely to have biochemical recurrence (rising PSA) following radical prostatectomy. *Int. J. Cancer* 1999; 20:360–4.
9. Mejean A, Vona G, Nalpas B et al. Detection of circulating prostate-derived cells in patients with prostate adenocarcinoma is an independent risk factor for tumor recurrence. *J. Urol.* 2000; 163:2022–9.
10. Okegawa T, Nutahara K, Higashihara E. Preoperative nested reverse transcription-polymerase chain reaction for prostate specific membrane antigen predicts biochemical recurrence after radical prostatectomy. *Br. J. Urol. Int.* 1999; 84:112–17.
11. Kantoff PW, Halabi S, Farmer DA et al. Prognostic significance of reverse transcriptase polymerase chain reaction for prostate-specific antigen in men with hormone-refractory prostate cancer. *J. Clin. Oncol.* 2001; 19:3025–8.
12. Smith DC. Chemotherapy for hormone refractory prostate cancer. *Urol. Clin. North Am.* 1999; 26:323–31.
13. Hartley-Asp B, Kruse E. Nuclear protein matrix as a target for estramustine-induced cell death. *Prostate* 1986; 9:387–95.
14. Tew KD, Stearns ME. Hormone-independent, non-alkylating mechanism of cytotoxicity for estramustine. *Urol. Res.* 1987; 15:155–60.
15. Cooney KA, Fardig J, Olson K et al. Vascular events secondary to estramustine: does androgen ablation influence risk? *Proc. Am. Soc. Clin. Oncol.* 2001; 20:2440a.

16. Speicher LA, Laing N, Barone LR et al. Interaction of an estramustine photoaffinity analogue with cytoskeletal proteins in prostate carcinoma cells. *Mol. Pharmacol.* 1994; 46:866–72.
17. Stearns ME, Wang M, Tew KD et al. Estramustine binds a MAP-1-like protein to inhibit microtubule assembly in vitro and disrupt microtubule organization in DU 145 cells. *J. Cell Biol.* 1988; 107:2647–56.
18. Hartley-Asp B, Gunnarsson PO. Growth and cell survival following treatment with estramustine, nor-nitrogen mustard, estradiol and testosterone of a human prostatic cancer cell line (DU 145). *J. Urol.* 1982; 127:818–22.
19. Eklöv, Nilsson S, Larson A et al. Evidence for a non-estrogenic cytostatic effect of estramustine on human prostatic carcinoma cells in vivo. *Prostate* 1992; 20:43–50.
20. Kreis W, Budman DR, Calabro A. Unique synergism or antagonism of combinations of chemotherapeutic and hormonal agents in human prostate cancer cell lines. *Br. J. Urol.* 1997; 79:196–202.
21. Vatan R, Le Bougeant P, Constans J et al. Hypocalcemia in a patient treated with estramustine. *Presse Med.* 1999; 28:1070–1.
22. Tassinari D, Sartori S, Panzini I et al. Hemolytic–uremic syndrome during therapy with estramustine phosphate for advanced prostatic cancer. *Oncology* 1999; 56:112–13.
23. Ando M, Tamayose K, Sugimoto K et al. Secondary myelodysplastic syndrome after treatment of prostate cancer with oral estramustine. *Am. J. Hematol.* 2001; 67:274–5.
24. Munshi HG, Pienta KJ, Smith DC. Chemotherapy in patients with prostate specific antigen-only disease after primary therapy for prostate carcinoma: a phase II trial of oral estramustine and oral etoposide. *Cancer* 2001; 91:2175–80.
25. Kelly WK, Curley T, Slovin S et al. Paclitaxel, estramustine phosphate and carboplatin in patients with advanced prostate cancer. *J. Clin. Oncol.* 2001; 19:44–53.
26. Friedland D, Cohen J, Miller R Jr et al. A phase II trial of docetaxel (Taxotere) in hormone-refractory prostate cancer: correlation of antitumor effect to phosphorylation of bcl-2. *Semin Oncol.* 1999; 26(5 Suppl. 17):19–23.
27. Picus J, Schultz M. Docetaxel (taxotere) as monotherapy in the treatment of hormone-refractory prostate cancer: preliminary results. *Semin Oncol.* 1999; 26(5 Suppl. 17):14–18.
28. Hudes G, Einhorn L, Ross E et al. Vinblastine versus vinblastine plus oral estramustine phosphate for patients with hormone-refractory prostate cancer: a Hoosier Oncology Group and Fox Chase Network phase III trial. *J. Clin. Oncol.* 1999; 17:3160–6.
29. Natale RB, Zaretsky S. Phase I/II trial of estramustine (E) with taxotere (T) or vinorelbine (V) in patients with metastatic hormone-refractory prostate cancer (HRPC). *Proc. Am. Soc. Clin. Oncol.* 1999; 18:338a.
30. Natale RB, Zaretsky S. Phase I/II trial of estramustine (E) and taxotere (T) in patients with hormone-refractory prostate cancer. *Proc. Am. Soc. Clin. Oncol.* 1999; 18:348a.
31. Pienta KJ, Esper PS, Smith DC. The oral regimen of Cytoxan, prednisone and diethylstilbestrol is an active, non-toxic treatment for patients with hormone-refractory prostate cancer. *Proc. Am. Soc. Clin. Oncol.* 1997; 16:1104a.
32. Sella A, Kilbourn R, Amato R et al. Phase II study of ketoconazole combined with weekly doxorubicin in patients with androgen-independent prostate cancer. *J. Clin. Oncol.* 1994; 12:683–6.
33. Tu S, Millikan RE, Mengistu B et al. Bone-targeted therapy for advanced androgen-independent carcinoma of the prostate: a randomized phase II trial. *Lancet* 2001; 357:336–41.
34. Carlin BI, Andriole, GL. The natural history, skeletal complications, and management of bone metastases in patients with prostate carcinoma. *Cancer* 2000; 88(Suppl. 12):2989–94.
35. Chaarhon SA, Chapuy MC, Delvin EE et al. Histomorphometric analysis of sclerotic bone metastases from prostatic carcinoma with special reference to osteomalacia. *Cancer* 1983; 51:918–24.
36. Ikeda I, Miura T, Kondo I. Pyridinium cross-links as urinary markers of bone metastases in patients with prostate cancer. *Br. J. Urol.* 1996; 77:102–6.
37. Boissier S, Magnetto S, Frappart L et al. Bisphosphonates inhibit prostate and breast carcinoma cell adhesion to unmineralized and mineralized bone extracellular matrices. *Cancer Res.* 1997; 57:3890–4.
38. Boissier S, Ferreras M, Peyruchaud O et al. Bisphosphonates inhibit breast and prostate carcinoma cell invasion, an early event in the formation of bone metastases. *Cancer Res.* 2000; 60:2949–54.
39. Heidenreich A, Hofmann R, Engelmann UH. The use of bisphosphonate for the palliative treatment of painful bone metastasis due to hormone refractory prostate cancer. *J. Urol.* 2001; 165:136–40.
40. Malone S, Donker R, Perry G et al. Long term side effect of intermittent androgen suppression therapy in prostate cancer: results of a phase II study. *Proc. Am. Soc. Clin. Oncol.* 2001; 20:2390a.
41. Modi S, Wood L, Siminoski K et al. A comparison of the prevalence of osteoporosis and vertebral fractures in men with prostate cancer on various androgen deprivation therapies: preliminary report. *Proc. Am. Soc. Clin. Oncol.* 2001; 20:2420a.
42. Dearnaley DP, Sydes MR. Preliminary results that oral clodronate delays symptomatic progression of bone metastases from prostate cancer: first results of the MRC Pr05 Trial. *Proc. Am. Soc. Clin. Oncol.* 2001; 20:693a.
43. Maeda T, Matsunuma A, Kawane T et al. Simvastatin promotes osteoblast differentiation and mineralization in MC3T3-E1 cells. *Biochem. Biophys. Res. Commun.* 2001; 280:874–7.
44. Mundy G, Garrett R, Harris S et al. Stimulation of bone formation in vitro and in rodents by statins. *Science* 1999; 286:1946–9.
45. Harris SE, Bonewald LF, Harris MA et al. Effects of transforming growth factor beta on bone nodule formation and expression of bone morphogenetic protein 2, osteocalcin, osteopontin, alkaline phosphatase, and type I collagen mRNA in long-term cultures of fetal rat calvarial osteoblasts. *J. Bone Miner. Res.* 1994; 9:855–63.
46. Bauer DC, Sklarin PM, Stone KL et al. Biochemical markers of bone turnover and prediction of hip bone loss in older women: the study of osteoporotic fractures. *J. Bone Miner. Res.* 1999; 14:1404–10.
47. Figg WD, Dahut W, Duray P et al. A randomized phase II trial of thalidomide, an angiogenesis inhibitor, in patients with androgen-independent prostate cancer. *Clin. Cancer Res.* 2001; 7:1888–93.
48. Dixon SC, Kruger EA, Bauer KS et al. Thalidomide up-regulates prostate-specific antigen secretion from LNCaP cells. *Cancer Chemother. Pharmacol.* 1999; 43S:78–84.
49. Tannock IF, Osoba D, Stockler MR et al. Mitoxantrone plus prednisone or prednisone alone for symptomatic hormone-resistant prostate cancer: a Canadian randomized trial with palliative endpoints. *J. Clin. Oncol.* 1996; 14:1756–64.
50. Kantoff PW, Halabi S, Conaway M et al. Hydrocortisone with or without mitoxantrone in men with hormone-refractory prostate cancer: results of the Cancer and Leukemia Group B 9182 study. *J. Clin. Oncol.* 1999; 17:2505–13.
51. Hudes G, Einhorn L, Ross E et al. Vinblastine versus vinblastine plus oral estramustine phosphate for patients with hormone-refractory prostate cancer: a Hoosier Oncology Group and Fox Chase Network phase III trial. *J. Clin. Oncol.* 1999; 17:3160–6.
52. Small EJ, Meyer M, Marshall ME et al. Suramin therapy for patients with symptomatic hormone-refractory prostate cancer: results of a randomized phase III trial comparing suramin plus hydrocortisone to placebo plus hydrocortisone. *J. Clin. Oncol.* 2000; 18:1440–50.
53. Pienta KJ, Fisher EI, Eisenberger MA et al. A phase II trial of estramustine and etoposide in hormone refractory prostate cancer: a Southwest Oncology Group trial (SWOG 9407). *Prostate* 2001; 46:257–61.
54. Pienta KJ, Redman BG, Bandekar R et al. A phase II trial of oral estramustine and oral etoposide in hormone refractory prostate cancer. *Urology* 1997; 50:401–6, discussion 406–7.

55. Pienta KJ, Redman B, Hussain M et al. Phase II evaluation of oral estramustine and oral etoposide in hormone refractory adenocarcinoma of the prostate. *J. Clin. Oncol.* 1994; 12:2005–12.
56. Dimopoulos MA, Panopoulos C, Bamia C et al. Oral estramustine and oral etoposide for hormone-refractory prostate cancer. *Urology* 1997; 50:754–8.
57. Cruciani G. Phase II oral estramustine and oral etoposide in hormone refractory adenocarcinoma of the prostate: Instituto Oncologico Romagnolo, Gruppo Onco-Urologico. *Proc. Am. Soc. Clin. Oncol.* 1998; 17:329a.
58. Frank SJ, Amsterdam A, Kelly WK et al. Platinum-based chemotherapy for patients with poorly differentiated hormone refractory prostate cancers (HRPC): response and pathologic cor-relations. *Proc. Am. Soc. Clin. Oncol.* 1995; 14:601a.
59. Olver IN, Keefe D, Myers M. Ambulatory carboplatin and oral etoposide for metastatic hormone resistant prostate cancer. *Proc. Am. Soc. Clin. Oncol.* 2001; 20:2399a.
60. Amato RJ, Ellerhorst J, Bui C et al. Estramustine and vinblastine for patients with progressive androgen-independent adenocarcinoma of the prostate. *Urol. Oncol.* 1995; 1:168–73.
61. Hudes GR, Greenberg R, Krigel RL et al. Phase II study of estraustine and vinblastine, two microtubule inhibitors, in hormone-refractory prostate cancer. *J. Clin. Oncol.* 1992; 10:1754–61.
62. Seidman AD, Scher HI, Petrylak D et al. Estramustine and vinblastine: use of prostate specific antigen as a clinical trial end point for hormone refractory prostatic cancer. *J. Urol.* 1992; 47:931–4.
63. Carles J, Domenech M, Gelabert-Mas A et al. Phase II study of estramustine and vinorelbine in hormone-refractory prostate carcinoma patients. *Acta Oncol.* 1998; 37:187–91.
64. Pienta KJ, Olson KB. Treatment of metastatic hormone refractory prostate cancer with estramustine (emcyt) and vinorelbine. *Proc. Am. Soc. Clin. Oncol.* 2001; 20:2421a.
65. Colleoni M, Graiff C, Vicario G et al. Phase II study of estramustine, oral etoposide, and vinorelbine in hormone-refractory prostate cancer. *Am. J. Clin. Oncol.* 1997; 29:383–6.
66. Frassoldati A, Nicolini M, Brausi M et al. Weekly paclitaxel in hormone-refractory prostate cancer. *Proc. Am. Soc. Clin. Oncol.* 2001; 20:2396a.
67. Ferrari AC, Chachoua A, Singh H et al. A phase I/II study of weekly paclitaxel and 3 days of high dose oral estramustine in patients with hormone-refractory prostate carcinoma. *Cancer* 2001; 91:2039–45.
68. Hudes GR, Manola J, Conroy J et al. Phase II study of weekly paclitaxel (P) by 1-hour infusion plus reduced-dose oral estramustine (EMP) in metastatic, hormone-refractory prostate carcinoma (HRPC): a trial of the Eastern Cooperative Oncology Group. *Proc. Am. Soc. Clin. Oncol.* 2001; 20:697a.
69. Vaishampayan UN, Fontana JA, Hussain M. An active regimen of weekly estramustine and paclitaxel in metastatic, androgen independent prostate cancer. *Proc. Am. Soc. Clin. Oncol.* 2001; 20:748a.
70. Athanasiadis A, Tsavdaridis D, Athanassiades I, et al. Paclitaxel (P) and estramustine phosphate (EP) in patients with hormone-refractory prostate cancer (HRPC)-a phase II study. *Proc. Am. Soc. Clin. Oncol.* 2001; 20:755a.
71. Hudes GR, Nathan F, Khater C et al. Phase II trial of 96-hour paclitaxel plus oral estramustine phosphate in metastatic, hormone-refractory prostate cancer. *J. Clin. Oncol.* 1997; 15:3156–63.
72. Kelly WK, Curley T, Slovin S et al. Paclitaxel, estramustine phosphate and carboplatin in patients with advanced prostate cancer. *J. Clin. Oncol.* 2001; 19:44–53.
73. Solit DB, Kelly WK, Fallon M et al. Phase I/II study of intravenous (IV) estramustine (EMP), paclitaxel and carboplatin (Hi-TEC) in patients with castrate, metastatic prostate cancer. *Proc. Am. Soc. Clin. Oncol.* 2001; 20:735a.
74. Smith DC, Esper P, Strawderman M et al. Phase II trial of oral estramustine, oral etoposide, and intravenous paclitaxel in hormone-refractory prostate cancer. *J. Clin. Oncol.* 1999; 17:1664–71.
75. Chay CH, Smith DC, Fardig J et al. Phase II trial of paclitaxel, estramustine, etoposide and carboplatin (TEEC) in the treatment of hormone refractory prostate cancer. *Proc. Am. Soc. Clin. Oncol.* 2001; 20:2435a.
76. Gravis G, Bladou F, Salem N et al. Efficacy, quality of life (QOL) and tolerance with weekly docetaxel (D) in metastatic hormone refractory prostate cancer (HRPC) patients. *Proc. Am. Soc. Clin. Oncol.* 2001; 20:2433a.
77. Friedland D, Cohen J, Miller R Jr et al. A phase II trial of docetaxel (Taxotere) in hormone-refractory prostate cancer: correlation of antitumor effect to phosphorylation of Bcl-2. *Semin. Oncol.* 1999; 26(5 Suppl. 17):19–23.
78. Picus J, Schultz M. Docetaxel (Taxotere) as monotherapy in the treatment of hormone-refractory prostate cancer: preliminary results. *Semin. Oncol.* 1999; 26(5 Suppl. 17):14–18.
79. Baranwal A, Amjad M, Naidu S et al. A phase II trial with docetaxel, vinorelbine and G-CSF in patients with hormone refractory prostate cancer. *Proc. Am. Soc. Clin. Oncol.* 2001; 20:2410a.
80. Kosty MP, Ferreira A, Bryntesen T. Weekly docetaxel and low-dose estramustine phosphate in hormone refractory prostate cancer: a phase II study. *Proc. Am. Soc. Clin. Oncol.* 2001; 20:2360a.
81. Rajasenan KK, Friedland DM, Lembersky BC et al. Weekly docetaxel alone or with estramustine has significant activity in patients with hormone-refractory prostate cancer previously treated with conventionally dosed docetaxel. *Proc. Am. Soc. Clin. Oncol.* 2001; 20:2434a.
82. Sinibaldi VJ, Carducci M, Laufer M et al. Preliminary evaluation of a short course of estramustine phosphate and docetaxel (Taxotere) in the treatment of hormone-refractory prostate cancer. *Semin. Oncol.* 1999; 26(5 Suppl. 17):45–8.
83. Kreis W, Budman D. Daily oral estramustine and intermittent intravenous docetaxel as chemotherapeutic treatment for metastatic, hormone-refractory prostate cancer. *Semin. Oncol.* 1999; 26(5 Suppl. 17):34–8.
84. Petrylak DP, Macarthur R, O'Connor J et al. Phase I/II studies of docetaxel (Taxotere) with estramustine in men with hormone-refractory prostate cancer. *Semin. Oncol.* 1999; 26(5 Suppl. 17): 28–33.
85. Kreis W, Budman DR, Fetten J et al. Phase I trial of the combination of daily estramustine phosphate and intermittent docetaxel in patients with metastatic, hormone-refractory prostate carcinoma. *Ann. Oncol.* 1999; 10:33–8.
86. George DJ, Jacobson JO, Prisby J et al. A phase I study of estramustine, docetaxel and carboplatin (ETP) in patients with hormone refractory prostate cancer (HRPC). *Proc. Am. Soc. Clin. Oncol.* 2001; 20:2371a.
87. Bracarda S, Pinaglia D, DeAngelis V et al. Phase II study of docetaxel (D), estramustine (E), dexamethasone (DEX) and low dose prednisone (P) in hormone-refractory prostate cancer (HRPC). *Proc. Am. Soc. Clin. Oncol.* 2001; 20:2398a.
88. Savarese DM, Halabi S, Hars V et al. Phase II study of docetaxel, estramustine and low-dose hydrocortisone in men with hormone-refractory prostate cancer: a final report of CALGB 9780. *J. Clin. Oncol.* 2001; 19:2509–16.
89. Assikis VJ, Tu S, Papandreou C et al. Phase II study of cyclophosphamide (CTX), vincristine (VCR), dexamethasone (DEX) for patients with metastatic androgen independent adenocarcinoma of the prostate (AIPCa). *Proc. Am. Soc. Clin. Oncol.* 2001; 20:2380a.
90. Nishimura K, Nonomura N, Tokizane T et al. Oral combination chemotherapy for hormone-refractory prostate cancer on an outpatient basis. *Proc. Am. Soc. Clin. Oncol.* 2001; 20:2385a.

91. Bracarda S, Tonato M, Rosi P et al. Oral estramustine and cyclophosphamide in patients with metastatic hormone refractory prostate carcinoma: a phase II study. *Cancer* 2000; 88:1438–44.
92. Huan SD, Stewart DJ, Aitken SE et al. Combination of epirubicin and cisplatin in hormone-refractory prostate cancer. *Am. J. Clin. Oncol.* 1999; 22:471–4.
93. Recchia F, Sica G, De Filippis S et al. Phase II study of epirubicin, mitomycin C, and 5-fluorouracil in hormone-refractory prostatic carcinoma. *Am. J. Clin. Oncol.* 2001; 24:232–6.
94. Petrylak DP, Macarthur RB, O'Connor J et al. Phase I trial of docetaxel with estramustine in androgen-independent prostate cancer. *J. Clin. Oncol.* 1999; 17:958–67.
95. Adami S, Salvagno G, Guarrera G et al. Dichloro-methylene-diphosphonate in patients with prostate carcinoma metastatic to the skeleton. *J. Urol.* 1985; 134:1152–4.
96. Smith JA. Palliation of painful bony metastases from prostate cancer using sodium etidronate: results of a randomized, prospective, double-blind, placebo-controlled study. *J. Urol.* 1989; 141:85–7.
97. Clarke NW, McClure J, George NJR. Disodium pamidronate identifies differential osteoclastic bone resorption in metastatic prostate cancer. *Br. J. Urol.* 1992; 69:64–70.
98. Kymala T, Tammela TLJ, Risteli L et al. Evaluation of the effect of oral clodronate on skeletal metastases with type 1 collagen metabolites. A controlled trial of the Finnish Prostate Group. *Eur. J. Cancer* 1993; 29A:821–5.
99. Lipton A, Glover D, Harvey H et al. Pamidronate in the treatment of bone metastases: results of two dose-ranging trials in patients with breast or prostate cancer. *Ann. Oncol.* 1994; 5(Suppl. 7):S31–5.
100. Vinholes JJF, Purohit OP, Abbey ME et al. Relationships between biochemical and symptomatic response in a double-blind randomized trial of pamidronate for metastatic bone disease. *Ann. Oncol.* 1997; 8:1243–50.
101. Adami S. Bisphosphonates in prostate carcinoma. *Cancer* 1997; 80:1674–9.
102. Pelger RCM, Hamdy NAT, Zwinderman AH et al. Effects of the bisphosphonate olpadronate in patients with carcinoma of the prostate metastatic to the skeleton. *Bone* 1998; 22:403–8.
103. Coleman RE, Purohit OP, Black C et al. Double-blind, randomized, placebo-controlled, dose-finding study of oral ibandronate in patients with metastatic bone disease. *Ann. Oncol.* 1999; 10:311–16.

Monotherapy with Antiandrogens in the Treatment of Prostate Cancer

Don W.W. Newling
VU Medical Center, Department of Urology, Amsterdam, The Netherlands

INTRODUCTION

The development and normal functioning of the adult prostate gland is heavily dependent upon the presence of androgens. Although between 70% and 90% of testosterone is produced by the testicles, under certain circumstances the contribution by the adrenal glands can also be of importance. For the last 70 years it has been known that removal of male hormones by surgical or medical castration, or inhibition of their activities by the exhibition and administration of female hormones can drastically influence the development of benign and malignant prostate tissue.[1]

Androgen activity is governed by the presence and expression of the androgen receptor and, regardless of the sort of androgens it is exposed to, blockage of this receptor will materially alter the activity of the prostate cell. Because all malignant prostate cells will ultimately become androgen independent, the efficacy of antiandrogen treatment is of limited palliative use. Debate still continues over whether a percentage of androgen-independent cells are present when the cancer is still very small or whether that independence is achieved by changes occurring in the androgen receptor or its expression during the course of hormonal therapy.

Androgen receptor antagonists, antiandrogens, were developed in the late 1970s and are structurally of two different classes, steroidal and non-steroidal. Because they can inhibit androgen activity regardless of its source, as monotherapy the receptor antagonists have always been of interest. In addition, steroidal androgen receptor antagonists will inhibit the production of luteinizing hormone from the pituitary and, therefore, induce a degree of medical castration as well as interrupting the androgen pathway at a cellular level.

In the late 1980s, owing largely to the work of Labrie,[2] the concept of removing the testicular source of androgens by surgical castration or the use the recently developed LHRH agonists, combined with the administration of an antiandrogen, the so-called maximal androgen blockade, became appealing. This form of therapy has, however, not been proven to be more effective in the majority of patients than medical or surgical castration alone.[3] More recently, differences have been noted, not only between the steroidal and non-steroidal antiandrogens but also among the non-steroidal antiandrogens themselves, showing that they probably have different modes of action and, therefore, may ultimately be found to have different roles to play in the treatment of advanced prostate cancer.

THE ANDROGEN RECEPTOR

The androgen receptor is a member of the family of nuclear receptor transcription factors.[4] The protein has a DNA-binding domain, a ligand-binding domain and a number of transactivation domains. The androgen receptor gene is situated on the X chromosome. Normally the androgen receptor is situated in the cytoplasm but in the presence of androgens migrates into the nucleus. Testosterone is converted by 5α-reductase in the prostate cell and as 5α-dehydrotestosterone attaches to the androgen receptor, a dimer of these molecules then makes its way to the nucleas and attaches via the DNA-binding segment to the nuclear DNA. The androgen receptor's antagonists compete with 5-dehydrotestosterone for the ligand-binding domain of the receptor.

When the activated receptor arrives on the DNA of the prostate cell, it sets in motion a series of transcription events, which stimulate genes in the cell cycle to give rise to cell proliferation, suppresses apoptosis and also causes the production of a number of proteins of different enzymatic functions, notably prostate specific antigen (PSA), the most important marker of prostatic cellular activity. In the proliferative pathway the sequence of events involves the production of proteins, which in turn stimulate genes to produce a variety of growth factors, such as epidermal growth factor (EGF), insulin-like growth factor and fibroblast growth factor (FGF), which difuse out of the cell and, by an autocrine or paracrine mechanism via their own specific receptors, stimulate further cellular proliferation. Some of the androgen target genes are the cyclin-dependent kinases CDK-2, CDK-4 and also BCL-2, the most important androgen-regulated gene in apoptosis.[5,6]

The development of androgen-independent prostate cancer may involve growth factor pathways that completely bypass the androgen receptor and act independently, such as EGF or insulin growth factor pathways, or the development of changes in the receptor itself, leading to overexpression or mutation. Such a mutation has been clearly defined in the LINCaP prostate cancer cell line, where threonine to alanine substitution, at position 877, has a dramatic effect on ligand specificity, enabling the cell to proliferate in response, not only to testosterone and dehydrotestosterone, but also to antiandrogens.[7]

That such a mutation occurs clinically has been shown in the clinical manifestation of the antiandrogen withdrawal phenomenon, where patients who have been treated with maximal androgen blockade (MAB) and who have progressed can respond to withdrawal of the antiandrogen. This phenomenon has been shown with all the antiandrogens, although initially reported with flutamide.[8]

ANDROGEN-DEPENDENT AND ANDROGEN-INDEPENDENT PROSTATE GROWTH

The development of androgen-independent proliferation in prostate cancer is a gradual process and there are a number of steps along the way where the careful use of different modilities of endocrine therapy can cause repeated responses of the malignant cells.

Table 47.1 shows the development of the completely endocrine independent cell occurring in four stages. The hormone-naive patient, who has been diagnosed as suffering from prostate cancer that is no longer to be cured by radical prostatectomy or radiotherapy, will respond to the use of monotherapy with androgen receptor antagonists, either steroidal or non-steroidal, or medical or surgical castration. Within a period of 1.5–3 years the cells will show resistance to this monotherapeutic option and require further hormone treatment. In patients who have received monotherapy with an antiandrogen, there will still be normal levels of testosterone in the blood, but in some instances the levels may be slightly elevated. Withdrawal of the patient's own androgens by means of medical or surgical castration will give rise to further response in approximately half of the patients. In patients who have been medically or surgically castrated, there will be a response in approximately 30% to the administration of an androgen receptor antagonist, which will block the androgens, or androgen precursors, produced by the adrenal glands. At this point, the patient's own testosterone will be at castrate level. Although the tumor is probably now androgen resistant, it will still respond to other endocrine manipulations, such as the administration of corticosteroids or, in the case of MAB, to the withdrawal of an antiandrogen. In addition, at this stage, many patients will respond to estrogens, probably as a result of their cytotoxic effects. Only after further proliferation of the cancer cells is there talk of a truely endocrine-independent tumor.[9]

Table 47.1. Methodological classification of prostate cancer based on hormone sensitivity.

Category		Tumor factors	Host factor
1. Androgen dependent	Endocrine naive: 0 no prior hormone therapy	Antitumor effect: 1. Androgens withdrawal 2. Antiandrogens are administered	Physiological levels of androgens in the blood
2. Androgen dependent	Endocrine sensitive: 1. Relapse after neoadjuvant therapy 2. Intermittent therapy-planned discontinuation of hormones 3. Relapse on antiandrogens alone	Decrease in proliferation if: 1. Androgens are withdrawn 2. Antiandrogens are administered (except group 3)	Non-castrate levels of androgens in the blood
3. Androgen independent	Endocrine sensitive	Decrease in proliferation in response to: 1. Adrenal androgen blockade 2. Corticosteroids 3. Withdrawal of agents that bind steroid hormone receptors 4. Other hormone manipulations	Castrate levels of testosterone
4. Hormone independent	Androgen independent and endocrine insensitive	Insensitive to all hormonal manipulation(s)	Castrate levels of testosterone

ANTIANDROGEN MONOTHERAPY

Of the four commercially available antiandrogens, only three have been exhaustively investigated and used in monotherapy. Nilutamide has only been used prospectively in one small phase II trial and some further information has accrued from a phase IV study, where 80 patients received effective monotherapy with the compound. The remaining three, cyproterone, flutamide and bicalutamide, have all been used as monotherapy in both phase II and phase III trials. Only one, bicalutamide, has undergone exhaustive dose ranging studies. The others have been used with a degree of empiricism. The second steroidal antiandrogen, WIN 1178, although used experimentally, has not found a place clinically as yet.[10]

In Japan, chlormadinone as a progestogen, and, therefore, a steroidal antiandrogen, has been used extensively, although it has seldom been used elsewhere.

CYPROTERONE ACETATE

Since the late 1970s the steroidal androgen receptor antagonist and antiprogestational agent, cyproterone acetate (CPA), has been available. Jacobi et al. and later Tunn showed cyproterone acetate in a dose of 300 mg weekly by intramuscular injection is as effective as diethylstilbestrol in a dose of 3 mg/day.[11,12]

In 1976, the European Organization for Research and Treatment in Cancer (EORTC) started a randomized phase III study of cyproterone acetate, 250 mg/day versus diethylstilbestrol (DES) 3 mg/day, and medroxyprogesterone acetate in a low dose of 200 mg/weekly after 8 weeks at a starting dose 500 mg intramuscularly × 3 weekly. A total of 217 patients were recruited in this study and the end-points were response, progression and survival. Responses in both localized and metastatic disease were measured but their inaccuracy led to subsequent dropping of this parameter of hormonal activity from future studies. Nevertheless, the response rates were similar between cyproterone acetate and diethylstilbestrol, both being superior to medroxyprogesterone acetate. Similarly, the time to progression of the first two and cancer-specific survival did not appear to be statistically significant different. Medroxyprogesterone acetate remained the weakest of the three and this may well have been a reflection of the dose of the compound used. Being a steroid, CPA exhibited many of the same side effects as diethylstilbestrol but to a lesser extent.

Nevertheless, significant cardiovascular events occured in around 8% of patients and there were cardiovascular deaths. Gynecomastia was less of a problem than with DES, a loss of libido and loss of potency occurred in approximately 70%.[13] Subsequently, the EORTC Genitourinary (GU) Group carried out a further trial of CPA monotherapy versus flutamide in patients presenting with asymptomatic, metastatic prostate cancer. The final therapeutic analysis of this trial consisting of 310 patients is not yet available. What has been found to be a dramatic feature of the study, has been the side-effect profile. The side-effect profile of flutamide, notably with diarrhea in 23%, nausea, loss of appetite and liver function disturbances was considerably worse than that of cyproterone acetate. In addition, in a small number of patients who were sexually active at the beginning of the trial, the same percentage remain sexually active on both arms, showing that loss of potency and libido is not inevitable with a steroidal androgen-receptor antagonist.[14]

As monotherapy, CPA has also found a role in the treatment of hot flushes caused by medical or surgical castration. In doses as low as 50 mg/day patients have reported relief of this distressing symptom in over 50% of cases.[15] CPA, along with all the other antiandrogens, is also used when starting therapy with an LHRH agonist to inhibit the testosterone search, which can occur in the first 4 weeks of treatment.

NON-STEROIDAL ANTIANDROGENS

Flutamide

Flutamide was originally used in a number of phase II trials, notably that by Whitmore and Sogani[16], in patients with advanced metastatic prostate cancer. It showed a subjective response in over 70% and an objective response of just under 50%. It was also noted that there were some good subjective responses where patients had been previously surgically castrated. Unlike the steroidal antiandrogens, the non-steroidal anti-androgens appear to exert no cardiovascular side effects. Flutamide does, however, cause gastrointestinal side effects in approximately a quarter of patients, most noticably diarrhea in between 17% and 23% of patients plus nausea, anorexia and depression in a further 15%. In addition, hepatotoxicity has been shown in between 3% and 8% of cases. Two different dosage schedules have been employed in phase II studies and one small phase III study: 250 mg 3×/day and 1500 mg/day. The side effects are noticebly worse in the higher dosages and there appeared to be no additional clinical benefit. True dose finding studies have not been carried with this compound.[17]

Three important phase III studies have been carried out using flutamide 250 mg in one arm. Boccon-Gibod carried out a study of 104 patients, half of whom underwent surgical castration and the other half received flutamide monotherapy. There was no therapeutic difference in the two arms. Flutamide did, however, give rise to more side effects than were experienced in the castration arm.[18] In a large Italian study with over 420 patients, Pavone-Macaluso compared flutamide with MAB. Again, there appeared to be no difference in the time to progression and cancer-related survival. Once more, the side-effect profile of flutamide led to withdrawal of patients because of diarrhea, and there was some hepatic toxicity.[19] The last important trial was mentioned above and comprises that of the EORTC GU Group with 310 patients randomized between flutamide and cyproterone acetate. The final therapeutic analyses is still awaited.

Nilutamide

This non-steroidal antiandrogen has been investigated in one phase II study[20] comprising only 26 patients. Further information on its possible role as monotherapy has also been obtained in a recent phase IV study, a postmarketing surveillance study, carried out by Schasfort et al. wherein 80 patients for one reason or another used the compound as monotherapy. It has one great advantage over flutamide in having a long half-life, which allows a once-daily dosage.[21]

It has an entirely different side-effect profile compared to other non-steroidal antiandrogens. In around a quarter of the patients there is a mild alcohol intolerance. In 15%, in various combined trials, there appears to be visual disturbances, amounting to a failure of light–dark adaptation in approximately 5%, making it extremely dangerous for patients on this compound to drive at night. Interstitial pneumonitis with pulmonary fibrosis has also been described in a small percentage of patients subjected to nilutamide therapy. That the compound is an effective antiandrogen has been shown in a large phase III study, when the compound was used as part of MAB.[22] In this study with over 400 patients, the combined therapy arm appeared superior to orchiectomy alone. At the present time, there are no ongoing studies with nilutamide monotherapy.

As with monotherapy with all non-steroidal antiandrogens, there is a risk of gynecomastia. This is caused by the normal levels of circulating testosterone being unable to attach to receptors in the normal way and being converted by aromitase to oestrogens.

Bicalutamide

Of the non-steroidal antiandrogens, bicalutamide has been shown experimentally to have the greatest affinity for the androgen receptor. It, therefore, would theoretically appear to be the most powerful of this group of compounds.[23] It is also the compound with the longest half-life and, therefore, can be given in a once-daily dosage. The side-effect profile is favorable compared to the other two non-steroidal antiandrogens. Gynecomastia is the only serious side effect and that may be preventable by pretreatment radiotherapy to the breast tissue or the use of an antioestrogen or aromatase inhibitors. Apart from the pharmacological side effects anticipated from androgen blockade, there are no significant cardiovascular or gastrointestinal side effects reported.

In extensive dose-ranging studies, series of phase II and phase III studies confirmed the dose of 150 mg/day as the appropriate one for monotherapy with this compound. Even in doses as high as 600 mg/day, there were no additional dose-limiting side effects.[24]

In two large phase III studies, comparing 50 mg Casodex/day with castration, Casodex was found to be inferior in terms of the time to progression and survival. In further phase III studies, comparing Casodex 150 mg/day and castration, in patients with non-metastatic disease ($n = 480$), Casodex was found to be equivalent to castration with fewer side effects and in patients with small-volume metastatic disease and PSA levels <400, there was also equal efficacy.[25]

In the further development of monotherapy with bicalutamide, a large worldwide study with over 8000 patients has recently been completed, where patients following radical prostatectomy or radiatherapy, or being managed by a policy of surveillance, were randomized without evidence of disease progression to receive placebo or bicalutamide. In an interim analysis, it is already clear that those who have been treated with bicalutamide have developed fewer signs of progression and, most noticebly, the development of metastases than those who received no androgen antagonist. Although only a handful of events has occurred, the difference being approximately 100% in the two arms was so dramatic that it was felt necessary to publish the results and, therefore, give collaborators a change to offer patients therapy earlier than was planned.[26]

The early results of a phase II study of bicalutamide in advanced prostate cancer patients who have failed conventional hormonal manipulation have recently been published. Although showing no definitive objective response, there has been subjective improvement in serious symptoms, such as pain and lethargy, with an improvement in the quality of life of patients in the first year of therapy.[27]

The exact mechanism of this response has not been identified but it may be a reflection of the fact that of the androgen receptor antagonist, bicalutamide appears to be one that maintains an antagonist activity longer than the others. In other words, if mutation occurs at the androgen receptor, it does not mean that bicalutamide will automatically act as an agonist.

Table 47.2. Side effects of androgen-receptor antagonists.

Cardiovascular	DES > CPA > MPA >>> NSAA
Gynecomastia	Bical = Flu = Nil > DES > MPA = CPA
Liver function disturbance	Flu > DES > CPA = Flu > Nil = Bical
Sexual function	DES > MPA = CPA = Flu > Nil = Bical
Hot flushes	DES >> MPA = CPA > Flu = Nil = Bical
Alcohol intolerance	DES = CPA > Flu > Bical
Visual problems	Nil > MPA >> the others
Pulmonal fibrosis	Nil >>> the others

Bical, bicalutamide; CPA, cyproterone acetate; DES, diethylstilbestrol; Flu, flutamide; MPA, medroxyprogesterone acetate; Nil, nilutamide; NSAA, non-steroidal antiandrogens.

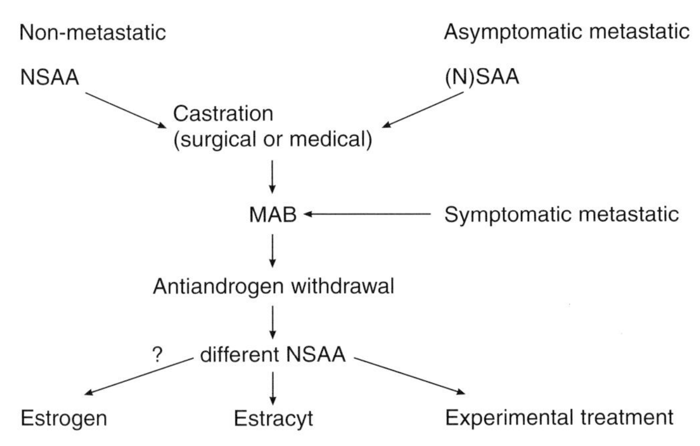

Fig. 47.1. A progressive step-up form of antiandrogen treatment.

CONCLUSION

Androgen receptor antagonists have an established place in the treatment of prostate cancer. With the development of less toxic compounds, such as bicalutamide and the demonstration that early treatment with hormone therapy may confer a benefit, patients are likely to be exposed to these agents for longer periods of time. **Table 47.2** compares the toxicities of the androgen receptor antagonists and explains why the suggestion has been made that bicalutamide, being the least toxic, is the most suitable for early therapy.

With the gradual development of hormone-independent disease, a progressive step-up form of antiandrogen treatment might be the most appropriate. Such a scheme can be seen in **Fig. 47.1**.

The EORTC has recently launched a study comparing the so-called step-up therapy to MAB, the hope being that, with the progressive step-up of therapy, the development of antiandrogen resistance can be further delayed.

In the future, the use of androgen receptor antagonists will be explored as neoadjuvant therapy for radiotherapy, and adjuvant therapy following conformal beam or brachy therapy. As scientists have developed the ability to predict patients at high risk of developing prostate cancer and particularly those with premalignant histological changes in the prostate, the use of relatively non-toxic antiandrogen therapies as preventative modalities may also come into the picture. The suggestion to try monotherapy with an androgen-receptor antagonist, in an intermittent fashion, has been made but as yet no results from clinical trials are available.

REFERENCES

1. Huggins C, Hodge CV. Studies on prostate cancer: the effect of castration, the administration of oestrogen and of androgen injection on serum phosphatase in metastatic carcinoma of the prostate. *Cancer Res*. 1941; 1:203–97.
2. Labrie F, Dupont E, Balancier E. Complete androgen blockade for the treatment of prostate cancer. In: Vita WT de, Helman S, Rosenberg SE (eds) *Important Advances in Oncology*, pp. 193–200 Philadelphia: Lippincott, 1985.
3. Prostate Cancer Trialists Collaborative Group. MAB in advanced prostate cancer – An overview of the randomized trials. *Lancet* 2000; 355:1491–8.
4. Beto M, Herlich P, Schulz G. Steroid hormone receptors – many actors in search of plot. *Cell* 1995; 83:851–7.
5. Chen Y, Robles AI, Martinez LA et al. Expression of G1 cycline, cycline dependent kinase inhibitors in androgen induced prostatic proliferation in castrated rats. *Cell Growth differentiation* 1996; 7:1571–8.
6. MacDonnell TJ, Troncoscol P, Brisbay SM et al. Expression the proto-oncogene BCL-2 in the prostate and its association with emergence of androgen independent prostate cancer. *Cancer Res*. 1992; 52:6940–44.
7. Veldscholte J, Ris-Stalpers C, Kuiper G et al. A mutation of the ligand binding domain of the androgen receptor of human LINCaP cell affects steroid binding characteristics in response to anti-androgens. *Biochem. Biophys. Res. Commun.* 1999; 173:534–40.
8. Kelly WK, Scher HI. Prostate specific antigen decline after anti-androgen withdrawal, the flutamide withdrawal syndrome. *J. Urol*. 1993; 149:607–11.
9. Newling DWW, Fosså SD, Andersson L et al. Assessment of hormone refractory prostate cancer. *Urology* 1997; 49:46–53.
10. Kolvenbag, G, Blackledge GRP. Worldwide activity and safety of bicalutamide – a summary review. *Urology* 1996; 47(Suppl.): 70–9.
11. Jacobi GH, Altwein JE, Kurth KH et al. The treatment of advanced prostate cancer with parenteral cyproterone acetate – a phase III randomised trial. *Br. J. Urol*. 1980; 52:2008–213.
12. Tunn UW, Weiglein W, Saborovsky J. Clinical experience with cyproterone acetate in a randomised and in an open trial. In: Murphy JP, Khoury S, Tuss R et al. (eds) *Prostate Cancer, Part A. Research: Endocrine Treatment and Histopathology*, pp. 365–8. New York: Alan R. Liss, 1987.
13. Pavone-Macaluso M, De Voogt HJ, Vigiano G et al. Comparison of diethylstilbestrol, cyproterone acetate, medroxy progesterone acetate in the treatment of advanced prostate cancer. Final analysis of a randomized phase III trial of the EORTC. *J. Urol*. 1986; 136:624–30.
14. Schröder FH, Colette L, de Reijke TM, et al. and members of the EORTC GU-Group. Prostate cancer treated by anti-androgens – is sexual function preserved? *Br. J. Cancer* 2000; 82:283–90.
15. Radlmaier A, Bormacher K, Neuman F. Hot flushes, mechanism and prevention. In: Schröder FH. (ed.) *EORTC GU-Group Monograph 8. Treatment of Prostatic Cancer, Facts and Controversies*, pp. 131–40. New York: Wiley-Liss, 1990.
16. Sogani PC, Vagaiwala MR, Whitmore WF Jr. Experience with flutamide in patients with advanced prostatic cancer without prior endocrine therapy. *Cancer* 1984; 54:744–50.
17. Neri RO, Florance K, Koziol P et al. A biological profile of a non-steroidal anti-androgen Sch 13521. *Endocrinology* 1972; 91:427–37.
18. Boccod-Gibod L. Are non-steroidal antiandrogens appropriate as monotherapy in advanced prostate? *J. Urol*. 1998; 33:139–46.
19. Pavone-Macaluso M. Flutamide monotherapy vs combined androgen blockade in advanced prostate cancer. Interim report of an Italian multicenter, randomized study. *SIU 23rd Congress*, 1994, Sydney, Australia (abstract)
20. Dijkman JA, Janknegt J, De Reijke T et al. Long-term efficacy and safety of nilutamide + castration in advanced prostate cancer and the significance of early PSA normalisation. Int. Anandrone Study Group. *J. Urol*. 1997; 158:160–3.
21. Decensi AU, Boccardo F, Guaneri D et al. Monotherapy with nilutamide, a pure non-steroidal anti-androgen in untreated patients with metastatic carcinoma of the prostate. *J. Urol*. 1991; 146:377–81.

22. Furr BJA, Tacke H. Preclinical development of bicalutamide. *Urology* 1996; 47(Suppl. 1A):13–25.
23. Schasfort EMC, v.d. Beek C, Newling DWW. Safety and efficacy of a non steroidal anti-androgen based on results of a post-marketing surveillance of nilutamide. *Prostate Cancer Prostat. Dis*. 2001; 4:112–17.
24. Kolvenbag GJ, Blackledge GRP. Worldwide activity and safety of bicalutamide. Summary review. *Urology* 1996; 47(Suppl.): 70–9.
25. Iversen T, Turrell CA, Kaisery AV et al. Casodex, bicalutamide 150 mgms monotherapy compared with castration in patients with previously untreated metastatic prostate cancer. Results from two multi center randomised trials with median follow-up of four years. *Urology* 1998; 51:389–96.
26. Wirth M, Tyrrell C, Wallace M et al. Bicalutamide (Casodex) 150 mgms as immediate therapy in patients with locally advanced prostate cancer, significantly reduces the risk of disease progression. *Urology* 2001; 58:146–50.
27. Kucuk O, Fisher I, Moispour M et al. A phase II trial of bicalutamide in patients with advanced prostate cancer in whom conventional hormonal therapy failed – a South West Oncology Group study, SWOG 9235. *Urology* 2001; 58:53–8.

Cryoablation

Is Cryoablation Here to Stay?

Chapter 48

Daniel B. Rukstalis
Division of Surgery, MCP Hahnemann University, Philadelphia, PA, USA

INTRODUCTION

The clinical longevity of any cancer treatment appears to be related to many traditional and diverse factors, such as therapeutic efficacy, treatment-related toxicity and ease of application. Additionally, there is most often a sound rationale for the therapy extrapolated from results of experimentation with animals coupled with repeatable clinical observations. Finally, patient preferences, defined as utilities for various outcome states, and quality of life issues have also gained prominence in determining the medical community's enthusiasm for any particular treatment. Therefore, it is necessary to examine each of the diverse factors mentioned above in order to determine the long-term viability of percutaneous cryoablation as a treatment option for clinically localized prostate cancer.

It has long been understood that rapid cooling to a lethal temperature can destroy living cells. One has only to remember the frostbitten extremities of the Arctic explorers of old to recognize the effect of extreme cold on human tissues. However, the injury caused by those unfocused cold temperatures was unpredictable and often not uniformly lethal. The premeditated destruction of human tissue requires the ability to apply the cold-inducing agent in a focused manner, reaching lethal temperatures in a predictable and controllable fashion. This was initially accomplished in the 1960s by several clinical investigators using liquid nitrogen to destroy animal and human bone, liver and prostate.

In particular, Loening and colleagues used automated cryosurgical equipment via an open perineal approach for prostatic cryoablation with successful eradication of localized prostate cancer in many patients.[1] The ensuing 40 years since those early experiences have seen many technical advances such that the targeted prostatic cryoablation procedure of today is an accurate and reasonable therapy for clinically localized prostate cancer. The remainder of this chapter will examine each of the pertinent factors that influence the clinical usefulness of this minimally invasive technique.

EXPERIMENTAL RATIONALE

At its most basic level, cryoablation of human tissue relies on the fact that very cold temperatures delivered in a rapid fashion can result in lethal injury to individual cells. In the 1960s it was determined that a temperature of –20 °C held for 1 minute could lead to cellular necrosis.[2] Further experiments with cells grown in tissue culture identified the basic features of the cryosurgical approach, namely rapid freezing followed by a slow thaw process repeated over a number of cycles. This technique resulted in cellular damage through several mechanisms and formed the foundation for a clinical procedure for targeted tissue destruction.

Rapid freezing of tissue results in ice-crystal formation in the extracellular compartment that propagates through vascular channels, as tissue grows colder.[3] Cells are damaged either by the mechanical forces of growing ice or by dehydration as the extracellular compartment becomes hyperosmolar. Experimental evidence from tumor cells grown in suspension suggests that the rate of cooling (more or less than 50 °C/minute) influences the ratio of intracellular ice formation to cellular dehydration. Rapid rates of cooling result in the formation of intracellular ice that is highly lethal to cells. Slower rates of cooling allow cells to dehydrate with a more variable cytotoxic effect.[4] During the thaw phase, ice crystals continue to injure cell membranes while water returns to the already damaged cells leading to further metabolic injury. This injury appears to stimulate the cellular equivalent of self-destruction called apoptosis.[5] Lethal temperatures, presently considered to be below –40 °C, increase cell death through the formation of intracellular ice that disrupts cell membranes.

Although ice-crystal formation and cellular dehydration appear to play a prominent role in cold-induced tissue injury, it is also likely that vascular stasis with subsequent cellular anoxia contributes to tissue destruction. Vasoconstriction develops at temperatures of –20 °C with a halt to blood flow as freezing progresses. Blood flow resumes as the tissue warms with a resulting hyperemia and edema that ultimately results in endothelial cell damage, platelet aggregation and thrombosis of most small vessels within hours of thawing.[6]

The degree of tissue destruction is determined by individual cell damage during the freezing process. This damage can be maximized by cooling to temperatures below −40 °C, the temperature at which intracellular ice is created, in a fashion as rapid as possible. Other influential factors include a prolonged hold time at the lowest end temperature, a slow thaw period and multiple freeze–thaw cycles. Importantly, different tumor types appear to respond variably to these factors. The prostate cancer cell line AT-1 is highly insensitive to the rate of cooling, while the prostate cancer cell line ND-1 demonstrates a high degree of intracellular ice formation and cell death at low cooling rates.[7] Clinically, the cryosurgeon has the most control over the coldest temperature achieved and the number of freeze–thaw cycles. Larson and co-investigators examined the prostates of six men treated with cryoablation followed by radical prostatectomy to evaluate the tissue effects of temperature and number of freeze–thaw cycles.[8] Coagulative necrosis was maximal with a double freeze–thaw process at a critical temperature of −41.4 °C. These investigators emphasized the importance of the double freeze technique coupled with careful intraoperative temperature monitoring.

The *in vitro* findings of cold-induced cell death or apoptosis stimulated interest in animal models of cryoablation. Mathematical models have suggested that lethal ice is created approximately 1 cm from a 3-mm cryoprobe cooled to −160 °C and several investigators have attempted to confirm this finding using *in vivo* models. Additionally, investigators examined the relationship of the size of the iceball created within tissue to the ultimate size of the cryolesion on histologic sectioning. These experiments have confirmed the ability of cold injury to cause reproducible necrosis in many tissues including liver and prostate.[9,10] Successful prostatic cryoablation necessitates the destruction of a volume of tissue beyond the capacity of a single cryoprobe given the thermal characteristics described above. Therefore, multiple probes must be positioned into the prostate in such an array that all prostatic tissue can be included within the lethal iceball. The physical parameters influencing cryoprobe placement in the prostate are illustrated in **Fig. 48.1**.

Finally, *in vivo* experiments have suggested that cryodestruction of living tissue elicits an immunological response by the host that may itself eradicate cancer cells.[11] In particular, experiments with MRMT-1 mammary adenocarcinoma cells and R3327 prostate adenocarcinoma cells in rats demonstrated the appearance of tumor immunity by 7–10 weeks following cryoablation that is associated with a decrease in the growth of metastatic lesions. However, this phenomenon has not been detected in all model systems. Cryoablation of the Dunning AT-1 cell line, an androgen-insensitive cell line, in Copenhagen rats was associated with a minimal increase in T-cell counts without effect on local tumor growth rate.[12] Although the potential for a cryoimmunologic effect against residual prostate cancer cells is exciting, cryoablation must still be scientifically considered as a local therapy alone.

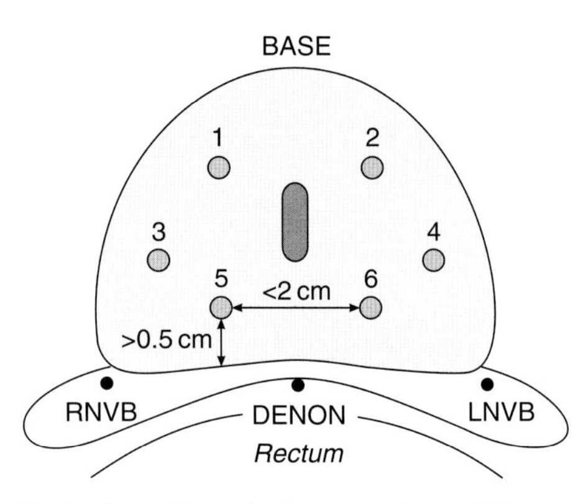

Fig. 48.1. Optimal position of six cryoprobes within the transverse section of the prostate. Probes must be no more than 2 cm apart. Additionally, the posterior probes must be no closer than 0.5 cm to the posterior prostatic capsule to optimize destruction of all prostatic tissue.

Early clinical experience with prostate cryoablation employed an open perineal approach during which the freezing process was controlled visually with minimal ability to protect adjacent structures, such as the bladder and rectum. However, this problem was overcome by Onik and co-workers with the incorporation of real-time ultrasound imaging in a dog model, thereby providing a method for monitoring and controlling the freezing process in humans.[13] Additionally, transrectal ultrasound facilitates the accurate placement of multiple cryoprobes, such that all prostatic tissue is within 1 cm of a heat sink. Subsequently, this same group established the clinical feasibility of ultrasound-guided prostate cryoablation in 1993.[14]

EVOLUTION OF CRYOSURGICAL TECHNIQUE

It is reasonable to assume that a simple surgical technique has a higher chance of general acceptance by physicians and their patients than a more complicated procedure, when given equivalent outcomes. Therefore, the percutaneous placement of 4–8 cryoprobes into the prostate under transrectal ultrasound guidance represents a potentially attractive therapy. However, the amount of equipment currently needed for a targeted cryoablation can appear intimidating to the individual accustomed to removing a prostate with a scalpel or placing needles through a rigid template for the implantation of radioactive seeds. The most simple early prostate cryoablation techniques involved an open perineal incision with placement of probes under direct visual and digital control with little other equipment necessary outside of the machine for pumping liquid nitrogen through the probes. Although conceptually simple, this procedure was complicated by injury to adjacent organs, such as the bladder and rectum, as well as urethral occlusion due to necrosis and fibrosis. It was apparent that a more conformal approach was required and the procedure fell out of favor.

It was not the surgeons that reinvigorated prostate cryoablation but the radiologists. Advanced imaging technology, such as transrectal ultrasound, provided a margin of safety for the conceptually straightforward destruction of prostate tissue with cryoprobes. As already acknowledged, Onik and co-workers determined the ultrasound appearance of the iceball in 1988 and the clinical feasibility of transrectal ultrasound for guiding cryoprobe placement in 1993. Real-time ultrasound facilitated visual monitoring of the iceball as it developed and provided an opportunity for tailoring the freezing process to avoid injury to the rectum, bladder or external sphincter. Clinically significant tissue destruction was assumed to take place when the entirety of the prostate was engulfed in the iceball with little ability to predict the actual lowest temperatures achieved within the cancer. Tissue destruction did indeed take place with many clinically localized cancers eradicated by this procedure. Additionally, the incidence of rectal injury was significantly reduced.

However, several hurdles remained. One such hurdle was the circumstance that cryodestruction of the urethra frequently resulted in bladder outlet obstruction due to sloughing of the now necrotic urethral lumen. Men required prolonged catheter drainage or secondary transurethral surgical procedures to resolve this complication. The development of a double lumen urethral warming catheter by Cohen and co-workers solved the problem of urethral necrosis and led to a substantive reduction in the number of secondary procedures.[15,16]

Despite these accomplishments, cryoablation of the prostate needed further modifications in order to achieve the efficacy for cancer eradication that would justify it as a reasonable choice for men with clinically localized prostate cancer. Basic research into the thermal history of prostate freezing had demonstrated that the leading edge of the iceball was visible on ultrasound but lethal ice existed only some distance internal to the leading edge. Applicable temperature information was not provided to the cryosurgeon by transrectal ultrasound monitoring. Rather, important thermal information was obtained through the placement of thermosensors that recorded the temperature at specific points outside the prostatic capsule facilitating more accurate tissue destruction with improved safety.[17] Together these clinical advances have shaped the current prostate cryoablation procedure into a safe and conformal technique for the targeted destruction of prostate cancer.

Current and future technology has potential for even greater efficacy with improved toxicity profiles. Traditional liquid nitrogen machines have given way to argon gas-based units, which appear capable of the rapid freezing and warming of tissue, thereby reducing the risk of freezing adjacent organs. Computer-guided systems now provide improved intraoperative treatment planning with a potential for increased tissue destruction.[18] An illustration of a computer-guided treatment plan is provided in **Fig. 48.2**. Finally, the application of magnetic resonance imaging holds promise for even greater therapeutic accuracy in the not to distant future.[19] Indeed, targeted prostatic cryoablation may evolve into a very focused technique for destruction of a limited amount of prostate tissue coupled with chemotherapeutic agents designed to increase the prostate cancer cell's sensitivity to cold-induced cell death.

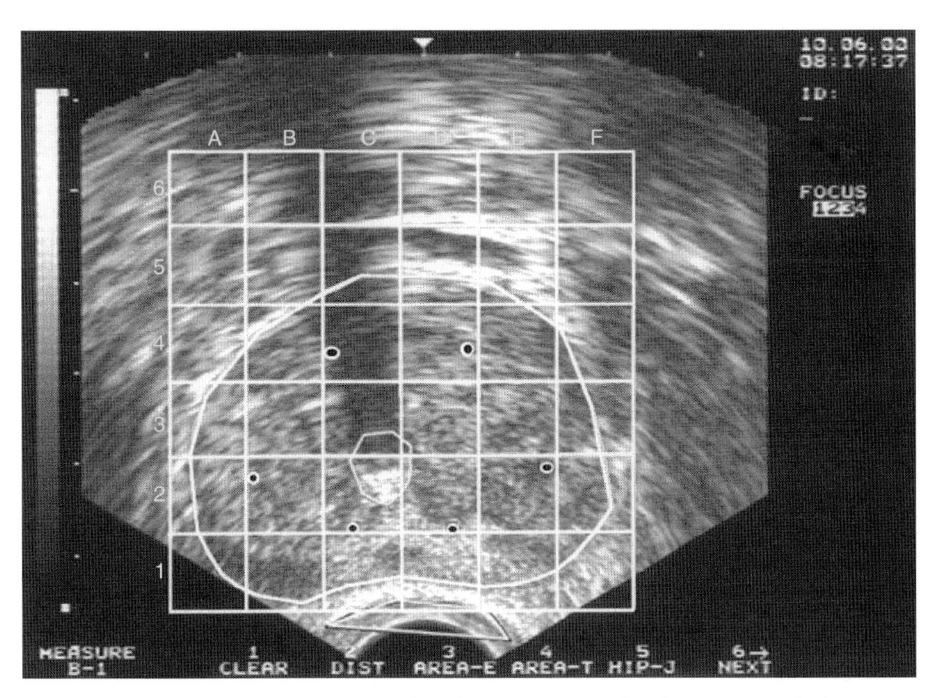

Fig. 48.2. A transverse ultrasound image of the prostate with an overlay of a computer-simulated treatment grid. The computer has calculated the optimal position for the planned six cryoprobes needed to treat the entire prostate adequately. This provides real-time intraoperative treatment planning.

THERAPEUTIC EFFICACY

It appears that prostate cryoablation has a sound experimental foundation and has evolved into a straightforward image-guided percutaneous technique. However, general acceptance of this treatment option requires demonstration of clinically meaningful outcomes both in terms of disease eradication and toxicity. Much published information is now available from retrospective single-institution reports and pooled analysis of combined patient series. However, the enthusiastic reader should still view these results with caution, since definitive outcome measures have not been established. The overall patient follow-up is not of sufficient length to examine disease-specific survival and often the non-experimental study design limits the generalization of data to the community at large.[20] Despite these admitted limitations, a review of the literature suggests that prostate cryoablation as initial therapy for clinically localized disease compares favorably to other local modalities.

Surrogate end-points for disease-free survival include the survival free of biochemical failure, nadir prostate specific antigen (PSA) level achieved and results of postprocedure prostate biopsies. The transrectal prostate biopsy arguably represents the most accurate end-point for local disease ablation. **Table 48.1** depicts the results of prostate biopsies following cryoablation and selected radiation therapy series.[21,22] Cohen et al. reported in 1996 on a series of 383 men treated with 448 cryosurgical procedures.[23] A total of 364 men underwent postprocedure ultrasound-guided prostate biopsy with 11 cores obtained at each setting. Overall, 18% (21 of 114) of the men had evidence of residual cancer at 21 or more months after cryoablation. These investigators provided a further analysis of this series in 1997 following an examination of a total of 2077 biopsy cores.[24] In this report, 14.5% of the men demonstrated residual adenocarcinoma, while 80% of the cores had no evidence of unaltered glandular elements. Subsequently, several other groups have published the results from their patient series with a positive biopsy incidence of 2.4–38%.[25,26] The clinical results evaluated in these reports were predominantly achieved with the use of a 5-cryoprobe configuration (**Fig. 48.1**).

Table 48.1. Results of prostate biopsy following local therapy.

Cryoablation		
Long (2001)[32]	T1–2a, PSA <10, GG <7	12%
	>T2a, PSA >10, GG >6	24%
Cohen (1996)[23]	PSA <10	18%
	PSA >10	24.1%
Koppie (1999)[26]	Overall	38%
Lee (1999)[25]	Overall	3.6%
Radiation Therapy		
Stone (2000)[22]	Brachytherapy	11%
Crook (1995)[21]	External beam	30.5%

GG = Gleason grade

Importantly, the technique has continued to evolve and the current approach is to place six or more cryoprobes within the prostate, an advance that is likely to improve the destruction of all viable prostatic tissue.

The natural history of residual prostate cancer found on follow-up prostate biopsy after cryoablation is uncertain. However, it is safe to assume that the majority of these men will experience a clinically identifiable progression, if the remaining cancer is of significant volume. Although some percentage of cases will spontaneously convert to negative status on repeat biopsy, or achieve stable disease, the majority of men with a positive postcryoablation biopsy will require another procedure for complete local disease eradication. The therapeutic options for these men include all modalities originally available for treatment, including repeat cryoablation, brachytherapy, external beam radiation therapy, radical prostatectomy, androgen ablation or observation.

Serum PSA levels have been found to correlate with outcome following radical prostatectomy and radiation therapy. However, the most accurate parameter for post-therapy PSA changes, such as PSA nadir achieved or the pattern of increase, for the calculation of PSA progression-free survival is uncertain. In particular, after radical prostatectomy, the calculation of PSA progression-free survival depends upon the PSA nadir cut-point employed for analysis with higher rates of progression seen with lower cut-points.[27] A range of nadir PSA values from 0.2 to 0.4 ng/ml has been suggested following radical prostatectomy for predicting progression-free survival. Additionally, debate continues regarding the most appropriate measurement for PSA progression following radiation therapy. The ASTRO Consensus Panel recommended that PSA failure is considered to occur midway between the PSA nadir and the first of three consecutive rises in the serum PSA following radiation therapy.[28] Still others believe that the PSA nadir achieved following radiation therapy is more prognostic for subsequent failure with recommendations for values below 0.5 ng/ml.[29] Unfortunately, this debate limits our ability to evaluate prostate cryoablation as a therapy using short-term PSA follow-up information since the literature employs a variety of PSA criteria. Additionally, it further complicates the comparison of cryoablation to the more established modalities outside of prospective randomized trials, which have not been performed to date.

That being said, clinical investigators have attempted to evaluate the therapeutic outcome of prostate cryoablation with parameters such as PSA nadir, percentage of men below 0.5 or 1.0 ng/ml, and two successive elevations in serum PSA following cryosurgery. It appears that between 49% and 73% of cases will manifest an undetectable PSA (<0.2) following cryoablation.[30,31] The most robust analysis of treatment outcomes was presented by Long in 2001 using a pooled analysis of 975 patients treated at five institutions between 1993 and 1998.[32] The 5-year actuarial PSA progression-free survival was 52% for a PSA cut-point of 0.5 ng/ml and 63%, if the PSA threshold was increased to 1.0 ng/ml. Patients in the lowest risk group demonstrated a 76% 5-year PSA progression free-survival, which the authors believed compared well to results for brachytherapy

Table 48.2. Serum PSA results after *in situ* therapy.

	PSA nadir	
Cryoablation		
Koppie (1999)[26]	<0.1 ng/ml	49%
	<0.4 ng/ml	21%
	>0.5 ng/ml	30%
Gould (1999)[31]	<0.1 ng/ml	66.7%
	<0.2 ng/ml	96%
Long (2001)[32]	<0.5 ng/ml	60%
	<1.0 ng/ml	76%
Radiation therapy		
Grimm (2001)[35]	<0.2 ng/ml	81%
Shipley (1999)[33]	<0.5 ng/ml	43.9%
	0.5–0.9 ng/ml	26%
Ragde (1999)*[34]	ASCO definition	66%

*Data reported as 10-year PSA progression free survival.

or external beam radiation therapy. An abbreviated compilation of PSA results for each of the *in situ* therapies, prostate cryoablation, external beam radiation therapy and brachytherapy, is presented in **Table 48.2**.[33–35] In general, prostate cryoablation performed for a clinically localized prostate cancer reliably eradicates all glandular elements in at least 50% of the cases with undetectable serum PSA levels achieved in the majority of men.[26,31–35]

TREATMENT COMPLICATIONS

The gradual understanding of the protracted natural history of localized prostate adenocarcinoma by the urologic community has resulted in incorporation of quality of life calculations into the decision tree for treatment of this disease. In fact, it sometimes seems that the most successful treatment is now first and foremost one that causes the least toxicity rather than one that eradicates the most cancers. The development and implementation of validated health-related quality of life questionnaires by Litwin and others have provided a mechanism for the rationale evaluation of patient preferences for particular outcomes following treatment for localized prostate cancer.[36] In general, men are significantly bothered by urinary and sexual dysfunction following radical prostatectomy and radiation therapy. It can be argued that this bother is the driving force for the creation of putatively less invasive strategies such as prostate cryoablation. If this is the case, then percutaneous cryoablation must demonstrate an improved, and not simply similar, toxicity profile relative to the established therapies. A compendium of adverse events following cryoablation is provided in **Table 48.3**.

Table 48.3. Adverse events following prostate cryoablation.

Major complications	Minor complications*
Urinary incontinence	Urinary tract infections
Erectile dysfunction	Perineal pain
Urethral sloughing with obstruction	Penile/scrotal edema
Urinary retention	Epididymitis
Urethrorectal fistula	Hyperactive detrusor
	Hematuria
	Prostatitis

*Minor complications are self-limiting without the need for another intervention other than appropriate medical therapy.

A review of the evolution of the surgical technique for prostatic cryoablation has been provided earlier in this chapter during which advances such as transrectal ultrasound and the development of a urethral warming catheter have been acknowledged. These advances have successfully altered the toxicity profile of the procedure such that this section will focus, data permitting, on the most recent results of modern temperature-controlled percutaneous cryoablation. A review of 223 men treated with prostate cryoablation between 1993 and 1994, at a time when temperature monitoring was experimental and the urethral warming catheter was not yet approved by the Food and Drug Administration (FDA), was published in 1999 by Badalament.[37] Urinary incontinence requiring absorbent pads developed in 4.3% (9 of 208) but only one man required multiple pads for significant leakage of urine. Endoscopic procedures were performed in 10% (22 of 223) of the cases for bladder outlet obstruction due to bladder neck fibrosis or sloughing of necrotic urethral tissue. Erectile dysfunction was common following cryoablation with only 15% of men able to achieve an erection sufficient for unaided vaginal penetration. Finally, only one patient developed a urethrorectal fistula in this series.

Several investigators have documented the low incidence of urinary incontinence following cryoablation as primary therapy for localized prostate cancer. This is a potentially attractive situation for men with prostate cancer, since it appears that urinary function is the most influential outcome in determining satisfaction with therapy.[38] Urinary incontinence that required the use of an absorbent pad has been seen in 1–7.5% of men. This compares well to the results following radical prostatectomy, external beam radiation therapy and brachytherapy, which exhibit rates of 4–57%, 0–9% and 0–18%, respectively.[39–41]

Sexual dysfunction following definitive therapy for a localized prostate cancer represents another major domain for patient dissatisfaction. Each treatment modality for prostate cancer has undergone modifications in an effort to reduce the incidence of post-treatment erectile dysfunction. Nerve-sparing techniques have been developed for the radical prostatectomy, while brachytherapy, in one sense, represents a modification of radiation therapy designed to reduce this complication. Erectile dysfunction is a common toxicity of prostatic cryoablation even with transrectal ultrasound and temperature monitoring. The incidence of this adverse event ranges from 40% to 100%. In fact, preservation of erectile function may imply that an inadequate freeze was obtained since the lethal iceball must include the posterior prostatic capsule. A prospective study of 15 sexually active men demonstrated that 9 of 15 men reported persistent erectile dysfunction at 6 months following the procedure. Importantly, these men exhibited a poor response to intracavernosal injections of vasoactive agents, suggesting a postoperative vascular injury.[42] Although modifications of the cryoablation procedure may be possible, secondary erectile dysfunction must be considered a common event.

Although urinary and sexual function are most commonly adversely impacted upon by cryoablation, the development of a urethrorectal fistula is perhaps the most concerning adverse event. Improved visualization of the iceball coupled with more accurate temperature monitoring in the region of the anterior rectal wall has reduced the incidence of this adverse outcome. Contemporary patient series referenced in this chapter report an incidence of 0–2.4% in previously untreated men. Despite the very low incidence, this complication has become identified with the cryoablation procedure and can serve to limit patient and physician enthusiasm. However, a urethrorectal fistula may develop following any of the available treatment modalities for localized prostate cancer and, as such, requires that the treating physician monitors the patient closely for this outcome irrespective of the treatment approach.

CONCLUSIONS

Percutaneous cryoablation of the prostate represents a contemporary treatment for localized prostate cancer. A sound scientific rationale has led to the evolution of a sophisticated surgical technique that exhibits reasonable rates of disease eradication with acceptable toxicity. The question as to whether or not cryoablation remains a viable treatment strategy will be answered by treating physicians and patients alike. The fact that physicians overwhelmingly recommend therapies that they themselves deliver suggests that there is a significant educational requirement before urologists embrace this procedure.[43] Additionally, physicians must now consider the costs of any treatment when deciding on treatment validity. Prostatic cryoablation appears to be cost effective when compared to the radical prostatectomy with overall lower hospital costs.[44] Furthermore, the federal government has judged the outcome data for cryoablation favorably and the procedure was assigned a current procedure terminology (CPT) code for reimbursement in January 2001.

Men with prostate cancer appear to prefer clinical outcomes in relation to their overall health state, suggesting that men in better health place a higher value on length of life.[45] The information available from serum PSA determinations and prostate biopsy results suggest that cryoablation closely approximates the efficacy of more established treatments. Men in less good health appear to value quality of life outcomes that may be maximized with the minimally invasive cryoablation procedure more highly. It, therefore, appears that ultrasound-guided percutaneous cryoablation of the prostate with temperature monitoring will remain a viable treatment choice for men with localized prostate cancer.

REFERENCES

1. Loening S, Hawtrey C, Bonney W et al. Cryotherapy of prostate, cancer. *Prostate* 1980; 1:279–86.
2. Gage AA, Baust J. Mechanisms of tissue injury in cryosurgery. *Cryobiology* 1998; 37:171–86.
3. Rubinsky B, Lee CY, Bastacky J et al. The process of freezing and the mechanism of damage during hepatic cryosurgery. *Cryobiology* 1990; 27:85–97.
4. Bischof JC, Smith D, Pazhzyannur PV et al. Cryosurgery of dunning AT-1 rat prostate tumor: thermal, biophysical, and viability response at the cellular and tissue level. *Cryobiology* 1997; 34:42–69.
5. Clarke DM, Hollistei WR, Baust J et al. Cryosurgical modeling: sequence of freezing and cytotoxic agent application affects cell death. *Mol. Urol.* 1999; 3:25–31.
6. Whittaker D. Vascular response in oral mucosa following cryosurgery. *J. Periodont. Res.* 1977; 12:55–63.
7. Smith DJ, Fahssi WM, Swanlund DJ et al. A parametric study of freezing injury in AT-1 rat prostate tumor cells. *Cryobiology* 1999; 39:13–28.
8. Larson TR, Robertson DW, Corica A et al. *In vivo* interstitial temperature mapping of the human prostate during cryosurgery with correlation to histopathologic outcomes. *Urology* 2000; 55:547–52.
9. Reddy KP, Ablin RJ. Immunologic and morphologic effects of cryosurgery of the monkey (macaque) prostate. *Res. Exp. Med. (Berl.)* 1979; 175:123–38.
10. Onik G, Porterfield B, Rubinsky B et al. Percutaneous transperineal prostate cryosurgery using transrectal ultrasound guidance: animal model. *Urology* 1991; 37:277–81.
11. Ablin RJ. Cryoimmunotherapy. *Br. Med. J.* 1972; 3:476.
12. Hoffmann NE, Coad JE, Huot CS et al. Investigation of the mechanism and the effect of cryoimmunology in the Copenhagen rat. *Cryobiology* 2001; 42:59–68.
13. Onik G, Cobb C, Cohen J et al. US characteristics of frozen prostate. *Radiology* 1988; 168:629–31.
14. Onik GM, Cohen JK, Reyes GD et al. Transrectal ultrasound-guided percutaneous radical cryosurgical ablation of the prostate. *Cancer* 1993; 72:1291–9.
15. Cohen JK, Miller RJ. Thermal protection of urethra during cryosurgery of prostate. *Cryobiology* 1994; 31:313–16.
16. Cohen JK, Miller RJ, Shuman BA. Urethral warming catheter for use during cryoablation of the prostate. *Urology* 1995; 45:861–4.
17. Lee F, Bahn DK, Mc Hugh TA et al. US-guided percutaneous cryoablation of prostate cancer. *Radiology* 1994; 192:769–76.
18. Jankun M, Kelly TJ, Zaim A et al. Computer model for cryosurgery of the prostate. *Comput. Aided Surg.* 1999; 4:193–9.
19. Wong TZ, Sluerman SG, Fielding JR et al. Open-configuration MR imaging, intervention, and surgery of the urinary tract. *Urol. Clin. North. Am.* 1998; 25:113–22.
20. Albertsen PC, Hanley JA, Murphy-Setzko M. Statistical considerations when assessing outcomes following treatment for prostate cancer. *J. Urol.* 1999; 162:439–44.
21. Crook J, Malone S, Perry G et al. Postradiotherapy prostate biopsies: what do they really mean? Results for 498 patients. *Int. J. Radiat. Oncol. Biol. Phys.* 2000; 48:355–67.
22. Stone NN, Stock RG, Unger P et al. Biopsy results after real-time ultrasound-guided transperineal implants for stage T1-T2 prostate cancer. *J. Endourol.* 2000; 14:375–80.
23. Cohen JK, Miller RJ, Rooker GM, et al. Cryosurgical ablation of the prostate: two-year prostate-specific antigen and biopsy results. *Urology* 1996; 47:395–401.

24. Shuman BA, Cohen JK, Miller RJ et al. Histological presence of viable prostatic glands on routine biopsy following cryosurgical ablation of the prostate. *J. Urol.* 1997; 157:552–5.
25. Lee F, Bahn DK, Badalament RA et al. Cryosurgery for prostate cancer: improved glandular ablation by use of 6 to 8 cryoprobes. *Urology* 1999; 54:135–40.
26. Koppie TM, Shinohaia K, Grossfeld GD et al. The efficacy of cryosurgical ablation of prostate cancer: the University of California, San Francisco experience. *J. Urol.* 1999; 162:427–32.
27. Amling CL, Bergstralh EJ, Blute ML, et al. Defining prostate specific antigen progression after radical prostatectomy: what is the most appropriate cut point? *J. Urol.* 2001; 165:1146–51.
28. D'Amico AV, Schuitz D, Loffredo M et al. Biochemical outcome following external beam radiation therapy with or without androgen suppression therapy for clinically localized prostate cancer. *JAMA* 2000; 284:1280–3.
29. Critz FA, Levison AK, Williams WH et al. Prostate specific antigen nadir achieved by men apparently cured of prostate cancer by radiotherapy. *J. Urol.* 1999; 161:1199–203; discussion 1203–5.
30. Bahn DK, Lee F, Solomon MH et al. Prostate cancer: US-guided percutaneous cryoablation. Work in progress. *Radiology* 1995; 194:551–6.
31. Gould RS. Total cryosurgery of the prostate versus standard cryosurgery versus radical prostatectomy: comparison of early results and the role of transurethral resection in cryosurgery. *J. Urol.* 1999; 162:1653–7.
32. Long JP, Bahn D, Lee F et al. Five-year retrospective, multi-institutional pooled analysis of cancer-related outcomes after cryosurgical ablation of the prostate. *Urology* 2001; 57:518–23.
33. Shipley WU, Thames HD, Sandler HM et al. Radiation therapy for clinically localized prostate cancer: a multi-institutional pooled analysis. *JAMA* 1999; 281:1598–604.
34. Ragde H, Korb LJ, Elgamal AA et al. Modern prostate brachytherapy. Prostate specific antigen results in 219 patients with up to 12 years of observed follow-up. *Cancer* 2000; 89:135–41.
35. Grimm PD, Blasko JC, Sylvester JE et al. 10-year biochemical (prostate-specific antigen) control of prostate cancer with (125) I brachytherapy. *Int. J. Radiat. Oncol. Biol. Phys.* 2001; 51:31–40.
36. Litwin MS, Hays RD, Fink A et al. Quality-of-life outcomes in men treated for localized prostate cancer. *JAMA* 1995; 273:129–35.
37. Badalament RA, Bahn DK, Kim H et al. Patient-reported complications after cryoablation therapy for prostate cancer. *Urology* 1999; 54:295–300.
38. Carvalhal GF, Smith DS, Ramos C et al. Correlates of dissatisfaction with treatment in patients with prostate cancer diagnosed through screening. *J. Urol.* 1999; 162:113–8.
39. Litwin MS, Melmed GY, Nakazon T. Life after radical prostatectomy: a longitudinal study. *J. Urol.* 2001; 166:587–92.
40. McCammon KA, Kolm P, Man B et al. Comparative quality of life analysis after radical prostatectomy or external beam radiation for localized prostate cancer. *Urology* 1999; 54:509–16.
41. Talcott JA, Clark JA, Stark PC et al. Long-term treatment related complications of brachytherapy for early prostate cancer: a survery of patients previously treated. *J. Urol.* 2001; 166:494–9.
42. Aboseif S, Shinohara K, Bonrakchanyavat S et al. The effect of cryosurgical ablation of the prostate on erectile function. *Br. J. Urol.* 1997; 80:918–22.
43. Fowler FJ, Mc Naughton Collins M, Albertsen PC et al. Comparison of recommendations by urologists and radiation oncologists for treatment of clinically localized prostate cancer. *JAMA* 2000; 283:3217–22.
44. Benoit RM, Cohen JK, Miller RJ Jr. Comparison of the hospital costs for radical prostatectomy and cryosurgical ablation of the prostate. *Urology* 1998; 52:820–4.
45. Saigal CS, Gornbein J, Nerse R et al. Predictors of utilities for health states in early stage prostate cancer. *J. Urol.* 2001; 166:942–6.

Salvage Cryoablation of the Prostate

Chapter 49

Aaron E. Katz
Department of Urology, College of Physicians & Surgeons of Columbia University, New York, USA

INTRODUCTION

Recent scientific and technological advances have challenged the traditional treatment options for patients with localized prostate cancer. Many of these advances contain operating skills, which can be learned with relative ease, as the technical components within the procedure are well known to the experienced surgeon. Over the past decade, there has been a strong movement within urologic oncology towards minimally invasive diagnostic procedures as well as therapies. Two prime examples of this for prostate cancer are laparascopic radical prostatectomy and brachytherapy. The goals of minimally invasive therapies for a malignancy of a solid organ are to eradicate the local disease, shorten hospital stay, limit post-operative morbidities, quicken return to daily functions and work, and to reduce the overall cost of the procedure. Although some of these therapies are relatively new, they are gaining popularity rather quickly, and several worldwide experiences have demonstrated that they are quite effective at achieving most or all of these goals.

Recently, there has been increased interest amongst urologists in cryosurgical ablation for locally advanced or radiation-recurrent prostate cancer. Cryosurgery is defined as the *in situ* freezing of a tissue. The goals of cryosurgery for prostate cancer are to ablate the entire gland, rendering the patient free of disease, while preventing the freezing to surrounding structures, such as the bladder, rectum and external striated sphincter. Currently, the procedure is performed percutaneously, introducing small cryoprobes though the perineal skin into the gland and with the aid of a computer-guided system. The placement of the probes is performed using ultrasound guidance.

Modern techniques of cryosurgery include an argon-based system to generate lethal ice. This system allows the urologist to monitor the iceball formation visually by ultrasound and thermally by temperature sensors. These devices have dramatically reduced the morbidity of the procedure and have allowed the urologists to determine the end-points of cryosugery. Once the outer freezing edge of the iceball extends beyond the posterior capsule of the gland and into Denonvillier's fascia, the freezing process can be stopped, thereby preventing damage to the anterior rectal wall. This has resulted in a significant reduction in rectal fistulas. Most current series report a 0–2% incidence of rectal fistulas.[1–3]

Renewed interest in cryosurgery has focused on the potential role in patients with radiation recurrent cancers. Owing to modifications in the technical aspects of delivery of lethal ice, the procedure has been associated with a decrease in morbidity.[1] In this chapter we will review the latest information with respect to the techniques and outcomes for patients undergoing salvage cryoablation.

HISTORY OF CRYOSURGERY

Cryosurgery has been used for the treatment of many benign and malignant conditions. The use of freezing techniques in the treatment of carcinoma began in London during the 1850s when 'iced' (–18° to –22 °C) saline solutions were utilized to treat advanced carcinoma of the breast, uterus and cervix. In 1907 cryosurgery was utilized in dermatological patients, with the reported cure rate for skin cancer equal to that obtained by surgical excision. In recent years, cryosurgery has been utilized for the treatment of a wide variety of diseases, including liver, prostate and kidney malignancies. Initially, cryosurgical ablation of the prostate in humans was performed using a transurethral approach.[4] However, several problems became apparent: gross sloughing, severe bladder outlet obstruction and prolonged hospitalizations. Modifications of the technique involved the open perineal exposure of the posterior prostate surface with direct application of the cryoprobes.[5] This approach produced much less urethral sloughing than the transurethral approach. However, effective monitoring of the iceball in both techniques was inadequate or absent.

This caused freezing of the surrounding structures, fistulas, and urinary and fecal incontinence in a significant number of patients undergoing this operation.

After the development of transrectal ultrasound imaging of the prostate (TRUS), Onik et al.[6] described a modification of the technique by placing percutaneous cryoprobes into the prostate under TRUS. The TRUS accurately monitored the propagation of the iceball and prevented damage of the surrounding structures. Current cryoablation of the prostate continues to be performed using the TRUS technique, and the formation of ice at the tips of the small cryoprobes, use of thermosouple devices and a urethral warming device has led to a dramatic decline in the morbidity of cryosurgery.[1]

MECHANISMS OF TISSUE INJURY IN CRYOSURGERY

To understand the mechanisms of tissue injury in cryosurgery, we should understand the freeze–thaw cycle, which is the critical component of cryosurgery. All cryobiological studies showed that rapid cooling is more destructive.[7] For this reason in cryosurgery, the cooling rate is always as high as possible in order to produce the lethal intracellular ice.

Mechanisms of tissue injury during freezing cycles

Direct cell injury

As the temperature falls, the function and structure of the cells, including their constituent proteins and lipids, are stressed, cell metabolism progressively fails and death results. Rapid cooling does not allow time for water to leave the cells leading to ice-crystal formation, which primarily starts in the extracellular space, creating a hyperosmotic extracellular environment, which in turn withdraws water from the cells. As the freezing process continues, the cells shrink and ice-crystal expansion leads to rupturing of the cell membranes.

Vascular stasis

The effect of circulatory stagnation after cold injury is already known from investigations on frostbite. Cellular anoxia is commonly considered to be the main mechanism of injury in cryosurgery. The initial response to the cooling of tissue is vasoconstriction and a decrease in the flow of blood and, finally, the circulation ceases.

Mechanisms of tissue injury during thawing cycles

Recrystallization

During thawing, ice crystals fuse to form large crystals, which are disruptive to cell membranes. Furthermore, as the ice melts, the extracellular environment becomes hypotonic and withdraws water from the surrounding tissue, which may also rupture the cell membrane. The damaged tissue is subject to continued metabolic derangements during this time.

Vascular stasis

As the tissue thaws and the tissue temperature climbs over 0 °C, the circulation returns, now with vasodilatation. The hyperemic response is brief and increased vascular permeability occurs within a few minutes. Edema develops and progresses over few hours. Changes in the capillary endothelial cells that progress to defects in the endothelial cell junctions have been observed about 2 hours after thawing.

Histopathological changes after cryosurgery

1. Histopathologically: cryoablated tissues start to show features of irreversible cell death at 1 hour. Electron microscopy reveals chromatin condensation, loss of nuclear membrane, partial fragmentation, and thrombi in capillaries and cytoplasmic vacuolization of membranes. Light microscopy reveals signs of acute inflammation, vascular congestion, well-defined areas of interstitial hemorrhage and early coagulation necrosis.
2. Granulation tissue, basal cell hyperplasia, cell swelling, focal hemosiderin deposits and thickening of small nerves was also noted.[8]
3. Marked fibrosis with hyalinization, squamous metaplasia of regenerating duct epithelia and basal cell hyperplasia of the gland acini. Scattered normal ducts with flattened epithelia; patent or virtual lumina were occasionally seen. Coagulative necrosis could also be observed in some biopsy fragments obtained 12 months after cryotherapy.

SALVAGE CRYOSURGICAL ABLATION OF THE PROSTATE

Radiation therapy for localized prostate cancer is a main form of therapy for patients with newly diagnosed and localized prostate cancer. It has been estimated that nearly one-third of newly diagnosed prostate cancer patients will choose one form of radiation therapy as their primary treatment.[9] Despite modifications of delivering radiation to the gland, such as intensity modulation, three-dimensional (3D) conformal and computer-guided seed implantation, a number of these patients will have a rise in the serum prostate specific antigen (PSA) value in the years after radiation. According to the recent literature, the number of biochemical failures has ranged from 20% to 66%.[10] One of the criticisms in the past is that many of the investigators in the past have differed with regards to their definition of biochemical failure. One of the advances with regards to following patients came out of the 1997 ASTRO Conference.[11] At that meeting, a biochemical failure was defined as three consecutive PSA rises separated by 3–4 month intervals. What has not been established is the number of patients with recurrent prostate cancer. Since PSA itself is not a cancer-specific protein, there are several reasons for the PSA rise in the radiated patient besides cancer.

Following radiation therapy, serum PSA levels should slowly start to decline. If a patient should demonstrate a change in rectal examination or have a rise in serum PSA values, a restaging prostate biopsy should be performed. We have followed an algorithm, which is outlined in **Fig. 49.1**. Patients with recurrent prostate cancer should undergo a restaging evaluation to rule out metastasis. Currently, patients with

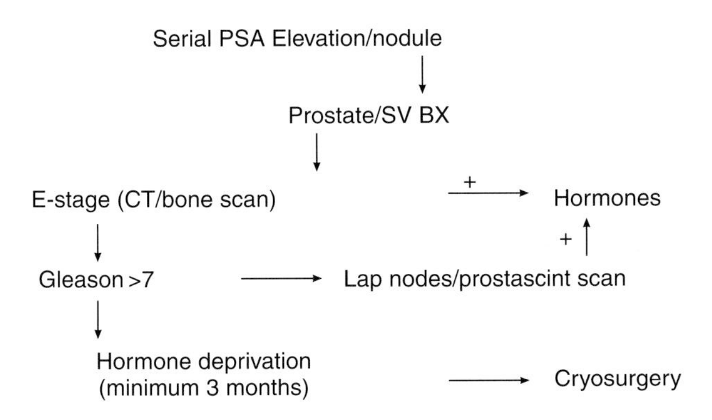

Fig. 49.1. Algorithm used to follow patients after radiation with a rising PSA value of change in rectal examination.

clinically localized recurrent disease have limited therapeutic options. Salvage radical prostatectomy is a very technically challenging procedure and has been associated with high comorbidities and a prolonged hospital stay. Recently, new thoughts have been given towards cytotoxic chemotherapy, although this is not curative and should be used only as a palliative treatment in late stages. Additional radiation therapy is not acceptable, as these tumors are clearly radioresistant and further radiation therapy will put the patients at a higher risk for radiation-induced complications. It is well known that radiation may lead to periurethral fibrosis and functional impairment of the external sphincter, and further local radiation therapy may result in a high rate of urinary incontinence. For salvage cryosurgery to become a reasonable therapeutic option, an acceptable complication rate and a prolonged disease-free survival rate is clearly necessary.

Patient evaluation

Currently, there are no defined guidelines for urologists to follow that would enable them to select patients properly for salvage cryoablation. The optimal candidates for the procedure would be those patients that are candidates for localized therapy. For patients with a locally recurrent cancer following radiation, cryosurgery should be considered when there is absence of clinical evidence of metastatic deposits. Salvage cryosurgery can be performed under spinal anesthesia or general anesthesia. Patients with a prior history of transurethral resection of the prostate (TURP) should be exluded from cryosurgery, especially if there is a large TUR defect present, as these patients are at increased risk for sloughing and urinary incontinence.

If a prostate biopsy reveals recurrent disease in the gland, a metastatic survey should be performed. It has been our practice to obtain a computed tomography (CT) scan of the abdomen and pelvis as well as a bone scan. There may be a role in performing an open or laparascopic biopsy of the pelvic lymph nodes. If we look at the lymph node positivity in these patients from the salvage radical prostatectomy series, it appears that between 20% and 40% of these patients will have prostate cancer metastases to the lymph nodes. We would caution novice laparascopic surgeons that dissection of these nodes could be technically difficult as there can be adhence to the pelvic side wall and external iliac vessels. The role of the prostatic scan is unknown for clinically staging these patients but can be helpful in select cases for patients that do not want to undergo a staging pelvic node dissection. Exclusion criteria included patients with prostate cancer invading the seminal vesicles or lymph node diagnosed by biopsy under TRUS guidance.

Role of hormonal therapy

The uses of neoadjuvant hormonal therapy prior to cryosurgery is still controversial. We have routinely used hormonal therapy for 3 months prior to cryosurgery for two reasons:

1. Androgen deprivation will lead to decrease in the prostate gland size that might increase the efficacy of cryosurgery. Reducing prostate volume may also increase the space between the posterior prostate capsule and the anterior rectal wall. This increased space will allow for greater depth of penetration of the iceball beyond the capsule, increasing cell death in this area.
2. Despite negative metastatic evaluation, a significant number of patients may have microscopic disease in their peripheral circulation, especially those patients with high-risk disease or radiation-recurrent cancers. Hormonal therapy has the additional advantage of clearing these hematogenous micrometastasis, although up to 40% of distant metastasize do not respond to hormonal therapy.

Preoperative care

Patients are given a Fleets enema on the morning of the procedure and Flagyl 500 mg IV at the start of the procedure.

Operative technique

We have used the technique reported by Onik et al. and Pisters et al. The CRYOCare system (Endocare Inc, Irvine, CA) was utilized, and made use of argon and helium gases to freeze and thaw the tissue, respectively. Cryosurgical procedures were performed under spinal anesthesia.

Briefly, under flexible cystoscopic guidance, a 10F Cook suprapubic catheter is placed and the bladder was left distended. A urethral warming catheter was then inserted prior to freezing the tissue (CMSI, Maryland). It was warmed to 38 °C and remained in place for 2 hours after the completion of the procedure. Prior to the start of the freezing cycle, six cryoprobes were placed into the prostate under TRUS guidance (two anteriorly, two posteriomedially and two posteriolaterally). Simultaneously, thermocouple devices, or temperature monitor probes, were placed adjacent to each of the two neurovascular bundles, the apex, Denonvillier's space and in the external sphincter. Freezing is initiated by activating the two anterior probes, which was followed by starting the two posterior probes. A double freeze–thaw technique was used in all cases. The outer edge of the iceball had a hyperechoic appearance and could be seen readily on TRUS.

Table 49.1. Patient population.

	Patient characteristics
Number of patients	43
Mean age (range) (years)	69.4 (48.1–83.6)
Mean precryosurgery serum PSA (range) (ng/ml)	7.07 (0.6–50)
Mean precryosurgery Gleason sum (range)	7.3 (4–9)
Mean interval between radiation therapy and cryosurgery (range) (years)	3.9 (0.3–10.7)
Mean follow-up (range)	21.9 (1.2–54)
PSA nadir	
≤0.1 ng/ml	26 (60%)
>0.1 and ≤4 ng/ml	16 (37%)
>4 and ≤10 ng/ml	1 (3%)

Freezing was completed when the following criteria were met:

1. The temperatures were less than –40 °C at each of the neurovascular bundles.
2. The apical temperature was less than –10 °C.
3. All of the prostatic tissue appeared to have frozen as visualized by ultrasound.

Postoperative care

Patients are usually discharged home the following morning with oral antibiotics (Ciprofloxacin 500 mg BID) for 5 days. The suprapubic tube was left open and patients were instructed to clamp the tube on postoperative day number four.

Results of salvage cryosurgery

Columbia University experience

The mean patient age of our patients was 69.4 years with mean serum PSA of 7.07 ng/ml. **Table 49.1** summarizes patient characteristics. Of our 43 patients, 19 (44%) had a precryotherapy serum PSA less than 4.0 ng/ml, 18 (42%) between 4 and 10 ng/ml and 6 (14%) above 10 ng/ml. Gleason sum on presurgical procedure needle prostate biopsies was between two and four in one case (2%), between five and seven in 23 cases (53%) and between eight and ten in 19 cases (45%).

During the follow-up, six patients had urinary problems including two with urgency resolved with anticholinergic drug and four patients (9%) with stress incontinence requiring one or more pad/day. Of these four patients, two required artificial sphincter for severe incontinence.

In our experience in a series of 43 patients, we found urinary incontinence to be present in only 7.9% of the patients (**Table 49.2**). This reduction in incontinence rates over previously published reports is likely to be due to our ability to monitor the temperature within the external sphincter. When the temperature readings within the sphincter reach values below 0 °C, the iceball can be thawed, thus preventing damage to the area and allowing the patient to remain continent. Moreover in our series, we did not experience any rectal fistula. This may be attributed to improvements in ultrasound technology and the placement of a temperature monitor probe in Denonvillier's space. The position of the devices needs to be confirmed and rechecked before and after each freeze–thaw cycle.

In addition to thermocouples, we believe that continuous warming of the urethra has helped lower the morbidity of this procedure in the modern era of cryosurgery. The urethral warming system was approved by the Food and Drug Administration (FDA), in an attempt to maintain the integrity of the urethral mucosa and thus prevent urethral sloughing during the freezing cycle. In our study, we left the warmer in the urethra during the freezing and for an additional 2 hours postoperatively. This innovation affected a major improvement in incidence of urethral sloughing and obstruction. The potential down

Table 49.2. Incidence of complications for overall patients.

	Overall complication rate	
	Number/Total	%
Major complications		
Incontinence	4/43	7.9%
Obstruction	2/43	5%
Urethral dilatation	2/43	5%
Rectal fistula	0/43	0%
Hydronephrosis	0/43	0%
Renal failure	0/43	0%
Death	0/43	0%
Minor complications		
Rectal pain	11/43	26%
Perineum swelling – scrotal edema	5/43	12%
Lower tract infection	4/43	9%
Hematuria	2/43	5%

Table 49.3. Salvage cryosurgery: follow-up.

Study	Number of Patients	Follow-up (months)	Biochemical recurrence	Positive prostate biopsy
Pisters (1997)[12]	150	13.5 (1.2–32.2)		25/110 (23%)
Lee (1997)[13]	43			At 3 months 3/20 (15%) At 1 year 7/20 (35%)
Miller (1996)[14]	33	17.1 (4.1–34.3)		9/33 (27.3%)
Present study	43	21.9 (1.2–54)	At 6 months: 79% BRFS At 12 months: 66% BRFS	3/8 (37%)

BRFS, biochemical recurrence-free survival (Kaplan–Meier curve analysis).

side of a urethral warming system is that periurethral cancers can be left behind. Although this is theoretically possible, it is our belief that the benefits of a warming system outweigh the risks, especially in the radiated patient.

We continue to have nearly one-third of our patients experience rectal discomfort following salvage cryosurgery. It is clear that this side effect is also higher in the previously radiated group of patients that undergo cryosurgery than in those patients that ever received prior radiation. Most of the patients are managed conservatively with warm baths and anti-inflammatory agents, and can persist for up to 6 weeks. In those patients with persistent pain, we have found the use of nitroglycerin suppositories to be helpful. It is possible that these patients are experiencing some form of ischemia in the area of the anterior rectal wall. The use of these suppositories may restore some blood to an area that has been previously radiated.

Biochemical recurrence

With a mean follow-up of 21.9 months, 26 patients (60%) reached a serum PSA nadir less than 0.1 ng/ml (undetectable), 16 (37%) had a PSA less than 4 ng/ml, and one (3%) had a PSA level less than 10 ng/ml (**Table 49.3**). Biochemical recurrence-free survival estimated by Kaplan–Meier curve analysis was 79% at 6 months, and 66% at 12 months (**Table 49.3**). Eight of our patients had needle prostate biopsy: three of them had a local recurrence (37%).

We found that the biochemical recurrence-free survivals differed between the group of patients who reached a PSA nadir less than 0.1 ng/ml (undetectable) versus patients who did not reached a undetectable serum PSA (73% vs. 30%, logrank = 7.11, $P = 0.0076$).

Multivariate analysis was performed in order to determine the most independent prognostic factors of PSA recurrence after cryotherapy (**Table 49.4**). In this analysis, we included preoperative parameters (serum PSA, Gleason sum on needle biopsy, and the interval between radiation therapy and cryotherapy), peroperative parameter (number of freeze) and postoperative parameter (value of the PSA nadir). The value of the PSA nadir >0.1 ng/ml was an independent predictor of PSA recurrence after cryotherapy.

Table 49.4. Multivariate analysis to define the independent predictors of PSA recurrence after salvage cryosurgery.

	Risk ratios	*P* value
Precryosurgery serum PSA	0.60 (0.17–2.13)	0.43
Precryosurgery Gleason sum	0.999 (0.35–2.82)	0.99
Interval between radiation therapy and cryotherapy	0.991 (0.96–1.02)	0.51
Number of freeze	0.67 (0.26–1.76)	0.41
PSA nadir >0.1 ng/ml	2.98 (1.25–7.10)	0.01

Canadian experience

A group of cryosurgeons in Western Ontario, Canada, recently published their long-term experience with salvage cryoablation.[15] In their study, 125 patients with biopsy-proven radiation recurrent disease underwent cryoablation-based using an argon-based system. The study included routine biopsies on these patients after cryosurgery. The overall negative biopsy rate was 97%. This included biopsy data from over 700 cores of tissue with a median follow-up of 18.6 months. Urinary incontinence was found in 6.7% of the cases, and they had 3% of rectal fistulas. The results of salvage cryosurgery studies are summarized in **Table 49.5**.[23–29]

Compared to previously published studies of different treatment modalities for patients with recurrent prostate cancer after radiation therapy, we have favorable complication rates (**Table 49.6**). In our study the biochemical recurrence-free survival calculated from Kaplan–Meier curves was 86% at 12 months and 74% at 24 months (**Fig. 49.2**).

SUMMARY

There is sufficient evidence both from the laboratory and clinical studies that cryoablation destroys cancer cells. Despite advances in radiation oncology, there will be a percentage of those

Table 49.5. Salvage cryosurgery for radiorecurrent prostate cancer: comparisons of the most frequent complications reported in the literature.

Study	Number of patients	Incontinence	Impotence	Obstruction/ Retention	Rectal fistula	Urethral strictures	Pain	Hydronephrosis
Pisters (1997)[12]	150	60%*	72%‡	43%	1%	0%	N/A	0%
Lee (1997)[13]	46	8.7%†	N/A	N/A	8.7%	N/A	N/A	N/A
Miller (1996)[14]	33	10.3%*	N/A	9%	0%	5.1%	N/A	0%
Katz (2002)[15]	43	9%*	N/A	4%	0%	0%	24%	0%

N/A, not available.
*Incontinence = need of ≥1 pad/day.
†Incontinence = any urinary troubling.
‡Preoperative potency not available.

Table 49.6. Comparison of complications of different treatment options for radioresistant prostate cancer patients.

Author	Type	Number of patients	INC	Obstruction	Rectal injury	US	BR
Pontes et al. (1993)[17]	RP	43	30%	2.3%	9%	N/A	60%
Ahlering et al. (1992)[18]	RP	11	64%	N/A	N/A	N/A	N/A
Rogers et al. (1995)[19]	RP	40	58%	N/A	15%	2.5%	75%
Lee et al. (1997)[15]	Cryo	46	9%	N/A	8.7%	N/A	53%
Amling et al. (1999)[20]	RP	108	23%	N/A	6%	N/A	74%
Miller et al. (1996)[14]	Cryo	33	9%	4%	0%	5.1%	40%
Present	Cryo	38	7.9%	0%	0%	0%	74%

RP, radical prostatectomy; Cryo, cryosurgery; N/A, not available; INC, incontinence; US, urethral sloughing; BR, biochemical recurrence.

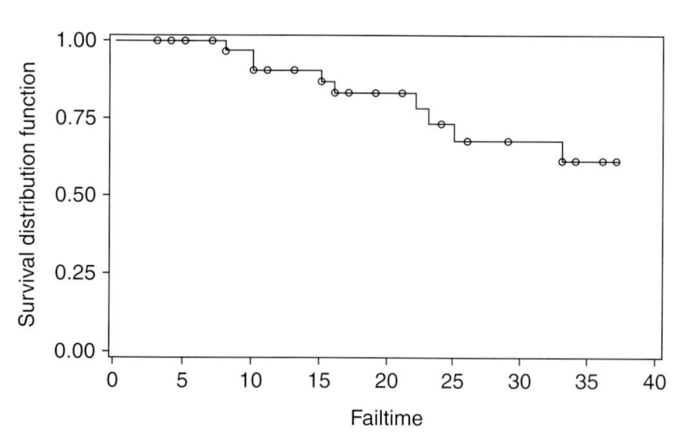

Fig. 49.2.

patients that will recur locally. Current data on salvage cryotherapy indicate that this therapy can impact local tumor control and possibly prevent progression of the disease (**Fig. 49.2**). Our data support the role of cryotherapy as a viable option in the management of patients who have biopsy-proven local failure following radiation therapy for prostate cancer. Further refinements in technique and equipment may enhance cryosurgical results.

REFERENCES

1. Saliken JC, Donnelly BJ, Rewcastle JC. The evolution and state of modern technology for prostate cryosurgery. *Urology* 2002; 60 (Suppl 2A):26–33.
2. Carson CC 3d, Zincke H, Utz DC et al. Radical prostatectomy after radiotherapy for prostatic cancer. *J. Urol.* 1980; 124:237–9.
3. Cespedes RD, Pisters LL, von Eschenbach AC et al. Long-term followup of incontinence and obstruction after salvage cryosurgical ablation of the prostate: results in 143 patients. *J. Urol.* 1997; 157:237–40.
4. Soanes WA, Gonder MJ. Use of cryosurgery in prostatic cancer. *J. Urol.* 1968; 99:793–7.
5. Flocks RH, Nelson CM, Boatman DL. Perineal cryosurgery for prostatic carcinoma. *J. Urol.* 1972; 108:933–5.
6. Onik GM, Cohen JK, Reyes GD et al. Transrectal ultrasound-guided percutaneous radical cryosurgical ablation of the prostate. *Cancer* 1993; 72:1291–9.
7. Hoffman NE, Bischoff JC. The cryobiology of cryosurgical injury. *Urology* 2002; 60(Suppl 2A):40–9.

8. Falconieri G, Lugnani F, Zanconati F, Signoretto D, Di Bonito L. Histopathology of the frozen prostate. The microscopic bases of prostatic carcinoma cryoablation. *Pathol. Res. Pract.* 1996; 192: 579–87.
9. Brawer MK. The management of radiation failure in prostate cancer. *Rev. Urol.* 2002; 4(Suppl 2):S1.
10. Brawer MK. Radiation therapy failure in prostate cancer patients: risk and methods of detection. *Rev. Urol.* 2002; 4(Suppl 2):S2–S11.
11. Panel AsfTRaOC. Consensus statement: guidelines for PSA following radiation therapy. *Int. J. Radiat. Oncol. Biol. Phys.* 1997; 37:1035–41.
12. Pisters LL, von Eschenbach AC, Scott SM et al. The efficacy and complications of salvage cryotherapy of the prostate. *J. Urol.* 1997; 157:921–5.
13. Lee F, Bahn tic K, McHugh TA et al. Cryosurgery of prostate cancer. Use of adjuvant hormonal therapy and temperature monitoring – a one year follow-up. *Anticancer Res.* 1997; 17:1511–15.
14. Miller RJ Jr, Cohen JK, Shuman B et al. Percutaneous, transperineal cryosurgery of the prostate as salvage therapy for post radiation recurrence of adenocarcinoma. *Cancer* 1996; 77:1510–14.
15. Chin JL, Pautler SE, Mouraviev V, Touma N, Moore K, Downey DB. Results of salvage cryoablation of the prostate after radiation: identifying predictors of treatment failure and complications. *J. Urol.* 2001; 165:1937–42.
16. Katz AE, Ghafar MA. Selection of salvage cryotherapy patients. *Rev. Urol.* 2002; 4(Suppl 2):S18–S23.
17. Pontes JE, Montie J, Klein E, Huben R. Salvage surgery for radiation failure in prostate cancer. *Cancer* 1993; 71(Suppl 3): 976–80.
18. Ahlering TE, Lieskovsky G, Skinner DG. Salvage surgery plus androgen deprivation for radioresistant prostatic adenocarcinoma. *J. Urol.* 1992; 147(3 Pt 2): 900–2.
19. Rogers E, Ohori M, Kassabian VS et al. Salvage radical prostatectomy: outcome measured by serum prostate specific antigen levels. *J. Urol.* 1995; 153:104–10.
20. Amling CL, Lerner SE, Martin SK, Slezak JM, Blute ML, Zincke H. Deoxyribonucleic acid ploidy and serum prostate specific antigen predict outcome following salvage prostatectomy for radiation refractory prostate cancer. *J. Urol.* 1999; 161(3):857–62.

Patients' Experiences

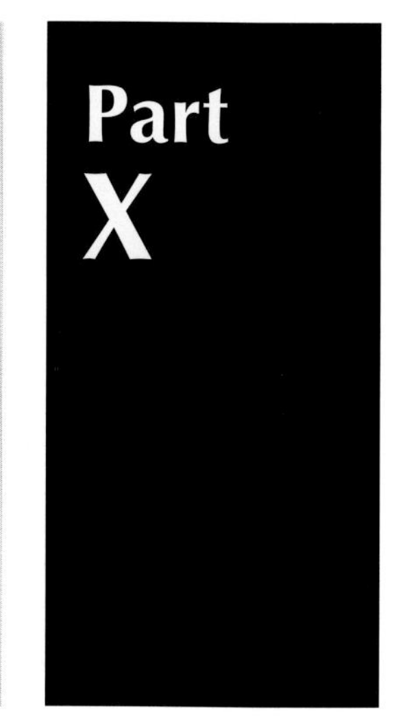

Quality of Life: Impact of Prostate Cancer and its Treatment

John W. Davis and Paul F. Schellhammer

Department of Urology and The Virginia Prostate Center of the Eastern Virginia Medical School and Sentara Cancer Institute Norfolk, VA, USA

INTRODUCTION

The primary end-points of cancer research are survival and quality of life. For many diseases, quality of life refers to the acute toxicity of treatment, such as incisional pain from surgery, or emesis and allopecia from chemotherapy. If survival is significantly enhanced, then acute toxicity is well accepted. The decision to treat may be further simplified by the absence of effective competing treatment modalities. Thus, for localized kidney cancer, the decision for nephrectomy is often simple, as is the decision to treat leukemia with chemotherapy.

Prostate cancer presents a unique challenge for quality of life assessment. The disease can be treated by numerous modalities including watchful waiting, chemotherapy, hormonal therapy, external beam radiation therapy, brachytherapy, radical perineal prostatectomy, laparoscopic prostatectomy and radical retropubic prostatectomy. Randomized comparisons between modern curative therapies are not available and, therefore, conclusions regarding survival and quality of life differences are affected by selection bias. As recently as 10 years ago, quality of life studies were limited to simplified questionnaires or reviews of office charts that reported on impotence and incontinence rates after radical prostatectomy compared to impotence and bowel side effects of radiation therapy. In short, radical prostatectomy was found to be associated with urinary incontinence and impotence, while radiation therapy was associated with bowel irritation and a lesser degree of impotence.

Modern quality of life research, however, has become far more sophisticated with the use of validated survey instruments that target general and disease-specific quality of life, and the emphasis on patient completion of questionnaires over physician interviews. In addition, several of the recent trends in prostate cancer listed below will affect quality of life compared to the results from 5 to 10 years ago.

1. Prostate specific antigen (PSA) screening has resulted in a significant down-staging of disease. At Eastern Virginia Medical school, the distribution of prostate cancers at diagnosis from 1980 to 1985 were 17% T1a/b, 38% T2, 35% T3 and 10% TxM1. From 1996 to 1999, the distribution was 2% T1a/b, 47% T1C, 38% T2, 5% T3 and 8%TxM1. Thus, PSA-detected-only cancers have significantly increased, while locally advanced cancers have decreased. Reports from Tyrol, Austria, indicate that screening significantly decreases overall mortality.[1] The impact on quality of life is that more patients may be cured by monotherapy [radical prostatectomy (RP), external beam radiotherapy (XRT) or seeds], and fewer patients will require the increased morbidity of multimodal therapies for locally advanced disease, such as XRT with 3 months neoadjuvant and 2–3 years of adjuvant androgen ablation. While the combination of androgen ablation and XRT has been shown to increase progression-free and overall survival,[2,3] the quality of life effects are just beginning to be understood. Nejat et al.[4] demonstrated that patients treated with 2 years of androgen ablation were not able to recover normal serum testosterone levels. Davis et al.[5] showed that patients treated with brachytherapy and neoadjuvant androgen deprivation for 3–5 months required a median of 11 months to normalize serum testosterone levels. More data are needed to understand fully the impact of adjuvant AD strategies on quality of life.
2. Several large surgical series have been reported demonstrating far fewer complications and better return of continence and sexual function[6–11] compared to reports from the Medicare database.[12] There are many possible explanations for this discrepancy, but it is certainly logical to conclude that quality of life will be favorably impacted by well-trained surgeons who perform the procedure on a regular basis.

3. Conventional XRT can be associated with bowel side effects. In part, driven by high PSA failure rates for conventional XRT monotherapy, centers of excellence have increased their dosage significantly to 72–80 Gy using conformal beam imaging, with or without intensity-modulated radiotherapy (IMRT). Conformal beam allows more precise delivery of the dose with less dosage given to the surrounding bowel and bladder. Early reports indicate better outcomes and fewer bowel and bladder complications compared to conventional techniques.
4. Brachytherapy has quickly emerged as a popular treatment option and may soon represent 50% of all localized treatment selections. Early reports indicate a favorable morbidity profile, but validated quality of life reports are beginning to emerge demonstrating different but significant side effects compared to standard surgical and external beam radiation options.
5. Laparoscopic radical prostatectomy is emerging as a treatment option after the technical success, low blood loss and equivalent operative times reported by Guillonneau and Vallancien.[13] Early operative, transfusion, pathologic, continence and potency rates have been reported, but no validated quality of life study has been published.

For these reasons as well as the availability of a number of validated survey instruments available, quality of life research has drawn increasing interest in the past few years, and has become a standard part of prostate cancer clinical trial design.

QUALITY OF LIFE RESEARCH MADE STRAIGHTFORWARD

Survival studies are fairly straightforward to understand, i.e. Kaplan–Meier curves are presented with 5 or more years follow-up and subgroups such as stage, Gleason, PSA, etc. are presented. By comparison, quality of life (QOL) studies can be tedious to read – numerous scales, comparisons, statistical and validation data, etc. Furthermore, QOL assessments face the difficult task of quantifying the qualitative. In essence, there are those who write QOL studies and those who read them. In this chapter, we introduce the fundamental concepts and terminology of QOL research. Most of the terms used are present in the methods section of a study, but word limitations generally prohibit the opportunity to elaborate or give background information. We will proceed by asking and answering common questions aimed at those who have not studied QOL research methods.

How does one develop a QOL instrument?

The process of developing a QOL instrument is a tedious and multidisciplinary effort that incorporates the goals of the clinician and the psychometric properties of survey research. During the development phase, a common technique is to devise discussion groups of clinicians (often urologists and radiation oncologists), nurses, patients and patients' spouses. Four key concepts must be addressed in developing the ideal set of questions, responses and scoring methods:

1. The questions must be easily read and interpreted at a basic education level.
2. The questions must be answered consistently.
3. The question should accurately reflect the targeted research goals.
4. The overall volume of questions should strike an optimum balance.

The later goal is a particular challenge, as researchers would like to gather as much information as possible about general and disease-specific QOL; however, as the number of questions grows, the likelihood increases that patients tire and either fail to complete the survey or give random answers. Non-responders and incomplete data are a serious problem, and studies with <70% response rates are at risk for biased results.

There are several methods of constructing questions and responses. The most common are termed categorical, which are a set of mutually exclusive, collectively exhaustive responses, or Likert-like, which are levels of agreement to disagreement with the question.[14] A less often used method is visual analog responses, which are a line anchored at each end with words representing a range of quality of life, and the patient makes a mark somewhere in between. Complex question patterns such as 'if this response, then go to question 5, otherwise go to question 10,' etc. should be avoided.[15]

Once questions are devised, they are tested among patients with and without disease to determine several psychometric properties. The results from these validation studies are often quoted and/or referenced in the methods section of QOL studies to give the reader an indication as to how well the questions perform in patients, and hence, how likely the results and conclusions of the study are valid. Commonly used terms are defined as follows.

1. Test–retest reliability: the measure of how stable responses are over time. Commonly, the questions are administered once as a baseline, and again 1 month later. The time interval is arbitrary, but must not be too long as to reflect valid changes in quality of life over time.[14]
2. Internal consistency: the measure of how similar an individual's responses are in several items.[16] The term Cronbach's coefficient alpha is the statistical measure of internal consistency.[17] Reliability by both test–retest and internal consistency should be greater than 0.70.[14]
3. Validity: the expression of how well the instrument measures its intended quality of life attribute. Several subtypes have been described based upon whether validity is assessed by a lay person, expert or correlation with 'gold standard' instrument that measures the same or similar variables.[14]
4. Administration: patient-completed questionnaires risk incomplete responses, which reduce validity of assessment of some domains (i.e. several domains may be skipped). Phone

interview and responses reduces or eliminates this problem, but requires well-trained interviewers and is costly.

How does one choose a QOL instrument?

The Southwest Oncology Group (SWOG) has published guidelines for measuring quality of life in clinical trials: (1) quality of life assessment should include measures of physical function, emotions, symptoms, global QOL, social functioning and protocol specific measures; (2) data should be collected by patient-completed questionnaires using categorical responses; and (3) instruments should be established with published psychometric properties.[18,19] Once these guidelines are met, the investigator can consider several qualities of an instrument. As stated, the total length of time required to complete the survey(s) must be reasonable – no more than 20 minutes. One must also consider the context in which a survey was developed. For example, the American Urological Association (AUA) symptom index was developed and validated in benign prostatic hyperplasia (BPH) populations and may not be as sensitive to urinary changes after brachytherapy. The UCLA Prostate Cancer Index (PCI) urinary function questions are purely incontinence questions and, therefore, irritative symptoms from brachytherapy may be missed. Some scales such as the SF-36 have subscales, whereas others only have a single final summary score. The UCLA PCI and others have both disease-specific function and bother scores, whereas the EORTC QLQ-C and FACT-P do not separate function from bother. Thus, investigators must carefully assess what questions they want to answer and select the best available instrument. If foreign language is a barrier, then one must look for instruments that have been not only translated but also cross-culturally validated to ensure that reliability and validity are still present in the new language.

What are the most commonly used QOL instruments?

Rand 36-item health survey (SF-36)[20]

The SF-36 is a 36-item questionnaire with eight domains targeting physical and emotional issues. Each of the eight scales are scored from 0 to 100 with higher scores indicating higher function. The SF-36 has been shown to be reliable and valid.[21] Differences of 6.5 to 8.3 points between comparison groups are considered clinically meaningful.[22,23]

FACT-G and FACT-P

The Functional Assessment of Cancer Therapy has a general scale, which includes six areas: physical, social–family, relationship with doctor, emotional, functional and miscellaneous. No distinction is made between function and bother. The scale has been validated[24] and translated into numerous languages. The FACT-P prostate-specific scale adds 12 items covering areas such as weight, appetite, pain, and general assessment of urinary, sexual and bowel function. Validity studies have shown that the scale is sensitive to changes in disease stage, performance status and change in PSA.[25]

EORTC

The EORTC QLQ-C30 is a 30-item instrument designed to capture cancer-specific quality of life across several malignancies, such as prostate, lung, head and neck, breast and ovarian, and to be a readily available tool for clinical trials.[26,27] The instrument is copyrighted and according to their website, www.eortc.be/home/qol, it has been translated and validated into 38 languages and used in more than 1500 studies worldwide. The QLQ-C30 addresses various domains common to cancer treatments. The general scales include physical, role function, emotions, cognitive and social function. In addition, a global health scale, symptom scales (fatigue, nausea/vomiting, pain), and single items covering dyspnea, insomnia, appetite loss, constipation, diarrhea and financial difficulties.

Supplemental disease-specific models are available for head and neck, lung, and breast, and prostate and bladder modules are under the final stages of development. The prostate module has 11 questions: four for urinary symptoms, three for sexual function, and single items for hot flashes, weight loss, weight gain and sexual satisfaction.

TAG Life/Family scales

These scales are a pair of non-disease-specific QOL measures developed and validated for use in patients enrolled in CaPSURE – a national observational cohort study of men with prostate cancer.[28] TAG Life is a six-item domain addressing the effect cancer may have on the patient's ability to walk, sleep, perform normal work, recreational and social activities, overall mood and enjoyment of life. The TAG Family is a two-item domain addressing the effect cancer may have on relations with the patient's spouse/significant other and other family members. Both scales are scored 0–100 with higher scores indicating better outcomes.

UCLA Prostate Cancer Index (PCI)

The UCLA PCI measures disease-specific HRQOL with a six domain survey: bowel function, bowel bother, sexual function, sexual bother, urinary function and urinary bother. The functional domains are multi-item scales focusing on incontinence, proctitis and erectile dysfunction. The bother scales are one-item scales assessing how much the patient is bothered by each dysfunction. The six scales are scored 0–100 with higher scores indicating higher function. For bother domains, the scoring may seem contradictory, so it is important to emphasize that higher bother scores indicate better outcomes, i.e. less bother. The survey's test–retest reliability has been documented.[29,30] Differences of ten points are considered clinically significant by its authors;[31] however, this cut-off is a clinical impression yet to be validated.

AUASI

AUA symptom index measures obstructive and irritative urinary symptoms. The seven items cover frequency, nocturia, weak stream, hesitancy, intermittency, incomplete emptying and urgency. It has a test–retest reliability $r=0.92$, and internal consistency alpha$=0.86$.[32]

EPIC

The EPIC survey[33] aims to improve upon the UCLA PCI survey by validating additional questions targeted at treatment-related urinary side effects from brachytherapy. In the UCLA PCI survey, the function questions only address urinary incontinence. To account for brachytherapy-related side effects, many QOL studies using the UCLA PCI have added the AUA symptom index score (AUASI), which asks questions relating to urinary frequency, hesitancy and irritation. However, the AUASI was designed and validated for benign prostate hypertrophy patients, and not cancer patients treated by brachytherapy. Such patients are known to complain of hematuria, painful urination, and possibly painful or bloody stools after treatment. Thus, the use of the UCLA PCI and AUASI may artificially favor brachytherapy, because it mostly captures the incontinence symptoms of prostatectomy, but not as much of the irritable and painful urinary symptoms of brachytherapy. In addition, many brachytherapy and XRT patients are treated with several months of neoadjuvant hormone ablation. Therefore, the EPIC survey has added a domain of hormone-related questions targeting depression, hot flashes, etc.

Overall, the EPIC survey is a more 'prostate specific' survey, as it adds questions related to urinary function, function-specific bother questions and the hormone domain. To keep the survey length reasonable, the EPIC utilizes the SF-12 to cover general QOL, instead of the SF-36. The loss of questions and multiple domains going from the SF-36 to the SF-12 may concern some, however, prior studies such as Davis et al.[34] and Brandeis et al.[31] have shown that there are relatively few side-effects detected among healthy men being treated for localized prostate cancer.

What is a clinically meaningful difference and how is it determined?

As with clinical trials assessing survival end-points, QOL instruments need guidelines as to how much of a difference in scores is clinically meaningful. In clinical trials, this is often a judgment, i.e. a 10% advantage in 5-year survival with adjuvant chemotherapy may or may not be clinically meaningful depending upon the toxicity. In QOL instruments, the commonly used SF-36 assessment of general QOL has a recommendation of 6–8 points being clinically significant.[22,23,35] For the UCLA PCI and EPIC, a ten-point difference is recommended by the authors[36] (both the SF-36 and UCLA PCI contain several domains scored on a 100-point scale). Guyatt et al.[37] have suggested a 0.3–0.5 standard deviation as representing a clinically meaningful change. Litwin[36] has used this concept in a longitudinal study to determine what level of return of a QOL function represents a fair 'return to baseline'. For many instruments, however, this concept has not been addressed.

Who best determines quality of life?

After treatment for prostate cancer, it is standard for the treating physician to follow the patient's PSA, and inquire as to QOL changes. Thus, physician-obtained QOL data are nearly universally available, and any reported series of radical prostatectomy or radiation therapy will include an estimate of urinary incontinence, erectile dysfunction, bowel toxicity, etc. However, as the science of survey research began to be applied to cancer patients, it became clear that physician estimates and patient estimates of QOL often disagree.[38] Litwin et al.[39] utilized the CAPSURE database to assess differences in physician and prostate cancer patients' estimates of QOL using the SF-36 and UCLA PCI. Substantial differences were seen in assessment of physical, urinary, bowel, sexual function, fatigue and pain, with urologists generally underestimating quality of life. This trend has been noted by other studies,[40–42] and explains why most modern QOL studies distribute questionnaires with return envelopes addressed to research coordinators, so as to remove the physician from the process as much as possible.

How do spouses perform as estimators of quality of life? Kornblith et al.[43] studied a large sample of patients and spouses with validated instruments and found that spouses reported greater psychological distress but less sexual dysfunction than patients. Sneeuw et al.,[44] noting poor response rates in studies of metastatic prostate cancer, compared patients and spouses assessments of QOL for D2 prostate cancer. Using the EORTC survey, they noted that proxies rated patients with more impairment in physical and role function, sleep disturbance, weight loss and sexual satisfaction. There was a low level of agreement for sexual function. However, when they looked at the severity of disagreement, there was 60% exact agreement and most of the disagreements were off by only one response category. In their estimate, their proxies rated QOL more accurately than the urologists in the Litwin study,[39] and proposed that spouses may be a reasonable method of collecting QOL data.

What are the commonly utilized study designs?

As with any clinical question, the randomized clinical trial would be the ideal method of determining the magnitude of differences in QOL between surgery versus radiation techniques. Such a trial would utilize validated general and disease-specific quality of life surveys, and would measure QOL at baseline, and post-treatment at 3, 12, and 24 months (intervals could certainly be changed if acute toxicity vs. longer toxicity were the primary question). While any of the abovementioned surveys could be appropriate, in our opinion, the UCLA PCI has the advantage of having several published studies with which to compare. However, if brachytherapy is involved in the study, the EPIC may be a more appropriate for the reasons listed above. Unfortunately, no such randomized trials exist for survival end-points, much less quality of life.

However, the American College of Surgeons has recently approved a trial comparing prostatectomy to brachytherapy for low-risk disease – the SPIRIT Trial (Surgical Prostatectomy Interstitial Radiation Trial) – and this protocol will include QOL assessment using the EPIC survey.

In the absence of randomized data, the two most common methods utilized are the cross-sectional study and the longitudinal study. A cross-section study has the advantage of being an inexpensive and quick method of collecting data. Scores are often compared to published, age-matched values of men without prostate cancer. A longitudinal study (with pretreatment values, if possible) is a more sensitive method, however, as trends in recovery or deterioration can be identified. With pretreatment values, patients serve as their own controls.

On this note, we should point out two recently published large observational studies that will be referred to in the following discussions.

1. Prostate Cancer Outcomes Study (PCOS). The PCOS is an National Cancer Institute-initiated study that was launched in 1994 as a longitudinal, community-based study of prostate cancer.[45] After patients were diagnosed and entered into the study, patients were asked to recall their pretreatment QOL, and were then studied at 6, 12 and 24 months after treatment, using the SF-36 and validated disease-specific instruments. The study primarily included men treated with RP or XRT, but also included and separately reported men treated with hormonal therapy. The study's strength lies in its large size, longitudinal design and multicenter community-based design, which makes the results more generalizable to the population at large, in contrast to data from single academic medical centers. Its potential confounders include the low response rate of 62% (with further loss to follow-up over time), and the use of recall bias. The use of recall for pretreatment assessment has been criticized,[46] and Litwin[47] found poor accuracy in recall among men treated for prostate cancer (JCO'99). However, the PCOS authors performed their own validation study and found no significant recall bias.[48]
2. Cancer of the Prostate Strategic Urological Research Endeavor (CaPSURE). CaPSURE is a large national observational database of prostate cancer patients treated by academic and community urologists, starting in 1995, totally 27 sites and 4061 patients.[28] It includes all stages and a variety of treatments. Follow-up is completed at 3-month intervals using the SF-36 and UCLA PCI.

PROSTATE CANCER QUALITY OF LIFE IN 2002 – WHAT HAVE WE LEARNED?

Radical prostatectomy

Radical prostatectomy is historically associated with significant urinary incontinence and impotence. Recent improvements in the anatomic approach to the procedure and nerve preservation have improved but not eliminated these side effects. During the first year after surgery, continence and potency (with nerve sparing) improves, but the degree of improvement and final outcome varies widely in the literature. Talcott et al.[49] reported a prospective cohort study and found that, at 12 months, impotence was 75% and incontinence was 35%. In another study,[50] they noted high impotency rates among bilateral nerve-sparing patients, and essentially no benefit for unilateral nerve sparing. In a study of Medicare outcomes, Fowler et al.[12] reported that 23% of patients post-RP reported daily dripping more than a few drops, and only 11% had erections sufficient for intercourse. However, Walsh[6] reported a longitudinal cohort study of 64 men including baseline values, and at 18 months continence was 93% and potency was 86%. They attributed their results to better surgical technique.

In the PCOS report on radical prostatectomy,[51] the findings at 18 months included incontinence in 8.4% and impotence in 59.9%. Age, nerve sparing and race affected sexual function and, beyond 18 months, 41.9% reported their sexual function was a moderate to large problem. Litwin et al.[36] reported a single-institution longitudinal study that included baseline measurements. As with the PCOS, response rates dropped from 90% over time to 65% overall. In the SF-36, 60% reached baseline by 3 months, and 90% by 12 months with a mean 4.5 month recovery. The domains role physical, role emotional and social function showed the greatest improvement. In some domains, as social and emotional, the scores improved from baseline, indicating that patients adjust to their diagnosis once it is treated and time passes. Urinary function improved 21% at 3 months, 56% at 12 months, and 63% at 30 months, with over 80% eventually recovering urinary bother. Sexual function recovered 33% at 1 year and 40% at 2 years, with 60% bother recovery at 11 months. Age, marital status and income were predictors of sexual function recovery; however, nerve sparing was not factored.

Radiation therapy

Radiation therapy is traditionally associated with bowel side effects and erectile dysfunction. As noted previously, the technique has evolved tremendously with conformal beam planning, dose escalation and frequent use of neoadjuvant/adjuvant androgen deprivation. Thus, the era studied and technique used will significantly affect QOL results.

In the PCOS study,[52] bowel function decreased at 6 months but improved to baseline by 24 months. Urinary function and bother were essentially not affected by treatment. Sexual function declined over time, such that by 24 months, 43% of all men with pretreatment potency were no longer potent. Bother assessment also declined over time. Other studies have also documented this pattern of sexual function decline.[49,53–55] However, reports of impotence after conformal beam radiation therapy have been lower (13–29%)[56–58] and et al. reported decreased bowel complications (34% to 10%, $P < 0.04$) for moderate to major changes in bowel function. The importance of technique were also demonstrated by Hanlon et al.[59] in a cross-sectional study of three-dimensional (3D) conformal radiation therapy with or without whole pelvis boost. Patients treated with pelvic boost reported more rectal urgency, use of pads for bowel incontinence and lower overall satisfaction.

Brachytherapy with or without XRT

Quality of life data from brachytherapy is just beginning to emerge. Cross-sectional data from Davis et al.[34] showed that brachytherapy is associated with erectile dysfunction and irritation/obstruction urinary side effects that improve in the first 12 months after treatment, and a decrease in erectile function. Scores from the SF-36 were no different than age-matched controls. Sanchez-Ortiz et al.[60] also used the SF-36/UCLA PCI and found that among pretreatment potent men, 51% developed erectile dysfunction; however, this finding did not correlate with androgen deprivation, SF-36 scores or satisfaction rates. Brandeis et al.[31] also performed a cross-sectional study and found the same pattern of SF-36, sexual function and urinary function changes, but also found that patients who had combination brachytherapy/XRT had even lower sexual and urinary function scores. Krupski et al.[61] also performed a cross-sectional study using the FACT G, AUASI and a separate sexual function inventory. Brachytherapy/XRT patients reported more urinary irritation and sexual function loss compared to brachyetherapy alone.

Lee et al.[62] performed one of the first longitudinal studies with pretreatment data to include modern brachytherapy. Using the FACT-G, FACT-P and International Prostate Symptom Score (IPSS), they found that FACT G and P scales declined at 1 and 3 months, but returned to baseline by 12 months. The IPSS scores doubled for 1 and 3 months, but returned to baseline by 12 months.

Comparative studies for localized disease

As we have established the general patterns of QOL after the three major treatments for localized disease, the question remains: can we properly compare the modalities to each other? As stated, no randomized trials are available and, therefore, retrospective and observational cohort studies must be interpreted, understanding the inherent weaknesses in these study designs. In addition, there are subtle points of technique that one can look for in the methods section that may shed light on the validity of the conclusions.

1. Pretreatment data. In many cases, the decision to analyze quality of life came after treatment or after suitable QOL scales were available. Therefore, authors must either omit pretreatment data, compare to published norms or ask patients to recall their pretreatment function. As mentioned above, this issue is controversial and it is not established which method is preferred.
2. Selection bias. As all QOL comparisons are non-controlled, there will be selection biases that affect age, comorbidity and probably pretreatment erectile/urinary function. Thus, these QOL scores must be statistically adjusted, and there is considerable variation in technique. Some studies omit comorbidity adjustment, some perform a simple comorbidity assessment or others a complex adjustment (presence and severity of comorbidity).
3. Time since treatment. For cross-sectional analysis, time since treatment must be controlled. For longitudinal, the follow-up times must be defined.
4. Response rate and missing data. While important for all QOL studies, for comparative treatment modality studies, additional selection bias may be entered through differences in response rates or frequency of incomplete responses. In general, a response rate of over 70% is desired and the treatment of missing data should be included in the methods, as this often leads to the exclusion of additional patients from the final analysis. For unknown reasons, minority groups have lower response rates.[34,63] Telephone calls to remind patients to consider participating in the study are common and may improve response rates, but telephone interviews to obtain data for part or all of a study is controversial, as patients may respond differently to an interviewer and have questions explained that otherwise would not occur in interpreting the written questions alone.[46]

The PCOS trial has published a comparative study of RP versus XRT with 24 months of follow-up.[63] SF-36 scores were mostly equivalent, although XRT scored lower for general health. They confirmed other studies demonstrating greater decreases in sexual and urinary function after RP, but more bowel side effect from XRT. Urinary bother mirrored loss of urinary function; however, sexual bother differed more by age than by sexual function – older patients reporting less bother.

The CaPSURE longitudinal study of RP versus XRT study also reports the same patterns of change, with XRT reporting declining sexual function from 1 to 2 years, with RP improving, such that the two groups were roughly equal by 2 years.[64] Urinary function improved after the initial loss for RP, but remains lower than XRT at 2 years, and the bowel side effects of XRT improve to near RP scores by 2 years.

Madalinska et al.[65] also performed a longitudinal study in the context of their clinical trial of prostate cancer screening in the Netherlands. They compared RP versus XRT using the SF-36 and UCLA PCI, also comparing patients who were diagnosed from screening clinics versus clinically detected. In contrast to the PCOS and CaPSURE studies that suffered from declining participation over time, the Madalinska study achieved a 91% baseline response rate, which held at 87% at 12 months. Despite theoretic selection biases, the baseline assessment of urinary and sexual function was not different between the groups. The same pattern of sexual, urinary and bowel function were noted, and the paper presented data to support the novel concept that screen-detected versus clinically detected cancers had the same quality of life.

Studies that include brachytherapy include Krupski et al.[61] who compared RP versus brachytherapy with a cross-sectional analysis of the first year after treatment. Using the FACT-G, AUA and a separate urinary/sexual inventory, they found that RP and brachytherapy had the same FACT-G scores. Sexual and urinary scores were not significantly different. However, patients treated by brachytherapy/XRT showed more urinary symptoms and declining sexual function.

Lee et al.,[62] as mentioned, have performed one of the few longitudinal studies that include brachytherapy, RP and XRT. As stated, using the FACT-G and IPSS surveys, all significant QOL changes were detected in the first 3 months of treatment and most had recovered to baseline by 12 months.

In the cross-sectional study by Davis et al.[34] with 5 years of follow-up, brachytherapy patients had no incontinence or bowel dysfunction compared to age-matched norms. Urinary irritation was demonstrated with elevated AUA scores, which were mostly seen in patients treated within a year of the survey.

Of interest, urinary bother scores were not significantly different between RP and brachytherapy, thus supporting the notion that incontinence is not the only side effect of treatment important to patients. In terms of sexual function, brachytherapy suffered sexual dysfunction that was more severe than XRT, but less severe than RP. However, both nerve-sparing RP patients and brachytherapy patients using erectile aids showed scores approaching or equal to age-matched controls.

Since validating the EPIC survey, the Michigan group[66] has published their comprehensive cross-sectional analysis of brachytherapy, XRT and RP compared to age-matched controls. They included the SF-36, FACT-G and P, AUASI, and EPIC, and achieved a 72–79% response rate – a remarkable achievement for such a large set of questions. As with the previously cited studies, no significant differences were seen in the general health measures. However, with the EPIC survey's additional urinary obstruction/irritation, and hormonal domain, brachytherapy appeared just as morbid as RP and XRT. As with other studies, RP was associated with worse urinary incontinence and sexual function, and XRT with worse bowel and sexual function. By contrast, brachytherapy was associated with urinary irritation and obstruction, worse bowel function, and worse sexual function compared to controls. In the bother domains, brachytherapy reported significant bother for urinary, sexual and bowel; XRT reported significant sexual and bowel bother; and RP reported only sexual bother. Furthermore, in an analysis of long-term morbidity (>1 year after treatment), no significant difference was seen between RP and brachytherapy for urinary incontinence. In the hormonal domains, XRT and brachytherapy patients treated with neoadjuvant/adjuvant androgen deprivation reported lower function and bother scores. A unique finding of this study was that patients who had experienced a PSA recurrence reported significantly lower sexual and hormonal scores, and marginally lower general scores. Overall, in the opinion of these authors, brachytherapy was less favorable than RP or XRT in several domains, and showed no advantage in any domain studied over a year after treatment.

As mentioned, this study was retrospective, single-institutional and cross-sectional, and may be biased by these factors. The next step, of course, will be longitudinal studies with pretreatment data. In the more distant future, the SPIRIT trial will study RP versus brachytherapy using the EPIC survey.

Satisfaction rates

A curious yet constant feature of all QOL studies we have reviewed, is that patients report very high satisfaction with their treatment – regardless of differences in quality of life. Papers reported thus far have a 70–90% range of satisfaction with the original treatment. In our study,[34] only biochemical failure was associated with a decrease in satisfaction rates, and watchful waiting was a popular second choice, rather than an alternate curative treatment. These findings indicate that, if men are appropriately counseled, they are likely to be satisfied with their result. Thus, randomized trials with longitudinal data may provide more accurate descriptions of quality of life after localized treatment, but such data may or may not affect men's treatment decision, unless survival data proves conclusive advantages for one modality over another.

Watchful waiting

Few data are available on the results of watchful waiting for men with prostate cancer. Common sense dictates that this group of men will not suffer the immediate toxicity of treatment, and should maintain quality of life consistent with age-matched men without prostate cancer. However, in a combined series from Massachusetts General Hospital and Eastern Virginia Medical School, men frequently proceeded to definitive treatment within the first decade: 57% at 5 years, and 74% at 7 years.[67] The authors noted that slight increases in PSA often triggered treatment, but that physicians often perceived the patients initiated treatment, while patients perceived just the opposite. In addition, Jonler et al.[68] studied quality of life in a group of 71 patients from Denmark who elected watchful waiting. During follow-up, 31% had required transurethral resection of the prostate (TURP), 8% underwent XRT and 44% hormonal deprivation. They found high rates of incontinence (21% using pads, 37% leaked daily), and significant reduction in sexual function. Nevertheless, the satisfaction rate was 85%. More data are needed regarding emotional or other effects of watchful waiting. In the meantime, available evidence suggests that watchful waiting often represents delayed therapy, and that QOL benefits may only be temporary.

Salvage local therapy

Salvage radical prostatectomy is well known to be a morbid procedure, with high incontinence and impotence rates. Salvage radiation therapy may also add morbidity to radical prostatectomy. Impotence is generally the rule, but continence is often preserved if XRT is delivered after enough time has passed for the patient to regain postoperative continence. However, rigorous study of salvage treatments using validated instruments is extremely limited. Tefilli et al.[69] used the FACT-G and FACT-P instruments to study their series of patients with locally recurrent prostate cancer who underwent salvage prostatectomy versus salvage XRT. As one would expect, salvage prostatectomy was associated with significantly lower scores for physical well-being, and urinary continence compared to salvage XRT.

Perrotte et al.[70] studied the MD Anderson series of salvage cryotherapy using a modified UCLA PCI and SF36, and AUASI. They were able to demonstrate significantly better incontinence and pain scores with a urethral warmer catheter compared to no warming catheter, and higher impotence with a double freeze–thaw cycle. Overall satisfaction rates were 33%, and they saw no significant advantage over salvage prostatectomy.

Metastatic disease

Metastatic prostate cancer includes a range of disease burden. For many men treated by curative surgery or radiation who subsequently experience PSA progression, quality of life is mostly affected by their primary treatment. As mentioned, the studies from Eastern Virginia Medical School and Michigan demonstrated satisfaction rate and QOL changes for patients with a rising PSA after definitive local therapy.[5,66] As Pound et al.[71] have demonstrated, men may persist with an elevated PSA for a median 8 years after radical prostatectomy before metastatic disease is documented and subsequently survive a median 5 more years.

For most men with metastatic lymph or bony disease, however, quality of life is primarily affected by androgen deprivation in the form of LHRH agonist or surgical castration (with or without complete androgen blockade with an oral non-steroidal antitestosterone agent such as flutamide or bicalutamide). Herr[72] has reported a study of men with locally advanced or failed local therapy who elected immediate versus delayed hormonal therapy. Using the EORTC scale, they found that men on androgen therapy, as expected, experienced more fatigue, loss of energy, emotional distress and overall lower quality of life than men on deferred therapy. Combined androgen therapy showed greater adverse effects than androgen monotherapy. Thus, validated general QOL scales are sensitive to androgen deprivation therapies.

In a cross-sectional study of patients with metastatic disease using the SF-36, EORTC and a disease-specific module, patients in remission (hormone sensitive) showed similar scores to age-matched norms, whereas those with progressing disease showed lower scores for bodily pain, vitality, social function and mental health.[73] However, no differences were seen in sexual function, sexual satisfaction, hot flashes or diarrhea.

A common dilemma for patients needing androgen deprivation therapy is whether to undergo surgical versus medical castration. While costs, trips to the doctor's office and body image are common features that direct the decision, two recent studies have looked at QOL issues. In the PCOS observational study,[74] a large group of patients were identified who were treated with primary androgen ablation for localized, locally advanced and metastatic disease by either LHRH agonist or orchiectomy. This study was non-randomized and several baseline variables were different, making comparisons difficult. Nevertheless, significant declines in sexual function and interest were found with either therapy. Of interest, stage and other prognostic factors did not affect quality of life and satisfaction rates were 90% in each group. Litwin et al.[75] performed a longitudinal study using the SF-36 and UCLA PCI. The study is small (47 combined androgen ablation vs. 16 orchiectomy) but achieved an impressive 84% response rate. They found no differences in SF-36 scores between the groups; social function, emotional well-being and pain were the highest scores, while energy/fatigue, and general health perceptions were the lowest. In the 12 months after diagnosis, they noted improvement in SF-36 scores. Sexual function was low in both groups, but sexual bother much less affected, indicating that patients adjust well to their diagnosis.

Finally, a critical difference between surgical and medical castration is the irreversible nature of the former. Do patients regret this choice? Clark et al.[76] recently addressed this issue with selected SF-36 items, newly formulated questions and patient focus groups aimed at identifying regretful patients. As one may guess, regretful patients were more common in surgical patients (43% vs 36%, $P=0.030$), and regretful men scored poorer on several generic and prostate specific scales. Survey and comments from focus groups indicated that regretful men rated their communication with their physician as poor.

CONCLUSIONS

Quality of life is an important field of prostate cancer research. While treatment modalities continue to improve the ability to cure localized disease and minimize morbidity, QOL survey instruments are improving our ability to obtain objective documentation of outcomes. The science of survey research is well established and has been incorporated into urologic research. Several survey instruments are available to help understand the general and disease-specific consequences of cancer treatment. The discrete concepts of function versus bother are also well established in prostate cancer quality of life.

Several large observational studies have recently been published that confirm early conceptions of surgery versus radiation therapy, and now brachytherapy data are beginning to be published. Future studies will include more comparisons of multiple modalities, using validated instruments and longitudinal data with pretreatment assessment. Such studies will help men make informed treatment decisions, although current data indicate that men are highly satisfied with their treatment and, by inference, well counseled by their physicians.

An as yet unquantified effect of QOL assessment also requires consideration. When patients are presented with a list of problems to consider, they may see the selections describing the most morbid outcomes as evidence of their future and, understandably, become depressed.

REFERENCES

1. Bartsch G, Horninger W, Klocker H et al. Prostate cancer mortality after introduction of prostate-specific antigen mass screening in the Federal State of Tyrol, Austria. *Urology* 2001; 58:417–24.
2. Bolla M, Gonzalez D, Warde P et al. Improved survival in patients with locally advanced prostate cancer treated with radiotherapy and goserelin. *N. Engl. J. Med.* 1997; 337:295–300.
3. Hanks G, Lu J, Machtay M et al. RTOG Protocol 92–02: A phase III trial of the use of long term total androgen suppression following neoadjuvant hormonal cytoreduction and radiotherapy in locally advanced carcinoma of the prostate. *Proc. Am. Soc. Ther. Rad. Oncol. (ASTRO)*, Boston, MA. *Int. J. Radiat. Oncol. Biol. Phys.* 2000; 48:112 (Abstract #4).
4. Nejat RJ, Rashid HH, Bagiella E et al. A prospective analysis of time to normalization of serum testosterone after withdrawal of androgen deprivation therapy. *J. Urol.* 2000; 164:1891.
5. Davis JW, Clements MA, Brassil D et al. Recovery of serum testosterone after brachytherapy with neoadjuvant androgen deprivation for localized prostate cancer. *J. Urol.* 2001; 165:327 (#1344).

6. Walsh PC, Marschke P, Ricker D et al. Patient-reported urinary continence and sexual function after anatomic radical prostatectomy. *Urology* 2000; 55:58.
7. Catalona WJ, Carvalhal GF, Mager DE et al. Potency, continence and complication rates in 1,870 consecutive radical retropubic prostatectomies. *J. Urol.* 1999; 162:433–8.
8. Eastham JA, Kattan MW, Rogers E et al. Risk factors for urinary incontinence after radical prostatectomy. *J. Urol.* 1996; 156:1707–13.
9. Rabbani F, Stapleton AM, Kattan MW et al. Factors predicting recovery of erections after radical prostatectomy. *J. Urol.* 2000; 164:1929–34.
10. Lepor H, Nieder AM, Ferrandino MN. Intraoperative and postoperative complications of radical retropubic prostatectomy in a consecutive series of 1,000 cases. *J. Urol.* 2001; 166:1729–33.
11. Lerner SE, Blute ML, Lieber, MM et al. Morbidity of contemporary radical retropubic prostatectomy for localized prostate cancer. *Oncology* 1995; 9:379–82.
12. Fowler FJ, Barry MJ, Lu-Yao G et al. Patient-reported complications and follow-up treatment after radical prostatectomy. The National Medicare Experience: 1988–1990 (updated June 1993). *Urology* 1993; 42:622–9.
13. Guillonneau B, Vallancien G. Laparoscopic radical prostatectomy: the Montsouris experience. *J. Urol.* 2000; 163:418–22.
14. Penson DF, Litwin MS. *Health-related Quality of Life in Patients with Urologic Cancers*. AUA Update Series XVI (5). Houston, TX: American Urological Association, 1997.
15. Litwin MS. Analyzing health-related quality of life. In: Vogelzang et al. (eds) *Comprehensive Textbook of Genitourinary Oncology*, pp. 16–22. Philadelphia: Lippincott Williams and Wilkins, 2000.
16. Tulsk DA. An introduction to test theory. *Oncology* 1990; 4:43–8.
17. Cronbach LJ. Coefficient alpha and the internal structure of tests. *Psychometrika* 1951; 16:297–334.
18. Moinpour Cm, Feigl P, Metch B et al. *J. Natl Cancer Inst.* 1989; 81:485–95.
19. Moinpour CM. Quality of life assessment in Southwest Oncology Group trials. *Oncology* 1990; 4:79–89.
20. Hays RD, Sherbourne CD. The MOS 36-item Health Survey 1.0. *Health Econ.* 1993; 2:217–27.
21. Ware JE Jr, Sherbourne CD. The MOS 36-item short form health survey (SF-36). I. Conceptual framework and item selection. *Med. Care.* 1993; 30:473–83.
22. Jaeschke R, Singer J, Guyatt GH. Measurement of health status. Ascertaining the minimal clinically important difference. *Controlled Clin Trials.* 1989; 10:407–15.
23. Juniper EF, Guyatt GH, Willan A, Griffith LE. Determining a minimal important change in a disease-specific Quality of Life Questionnaire. *J. Clin. Epidemiol.* 1994; 47:81–7.
24. Cella DF, Tulsky DS, Gray G et al. The Functional Assessment of Cancer therapy Scale: development and validation of the general measure. *J. Clin. Oncol.* 1993; 11:570–9.
25. Esper P, Mo F, Chodak G et al. Measuring quality of life in men with prostate cancer using the Functional Assessment of Cancer Therapy – Prostate Instrument. *Urology* 1997; 50:920–8.
26. Aaronson NK, Ahmedzai S, Bergman B et al. The European Organization for Research and the treatment of Cancer QLQ-C30: a quality of life instrument for use in international clinical trials in oncology. *J. Natl Cancer Inst.* 1993; 85:365–76.
27. Osoba D, Aaronson N, Zee B et al. Modification of the EORTC QLQ-C30 (version 2.0) based on content validity and reliability testing in large samples of patients with cancer. *Qual. Life Res.* 1997; 6:103.
28. Lubeck DP, Litwin MS, Henning JM et al. The CaPSURE database: a methodology for clinical practice and research in prostate cancer. CaPSURE Research Panel. Cancer of the Prostate Strategic Urologic Research Endeavor. *Urology*. 1996; 48:773–7.
29. Litwin MS, Hays RD, Fink A et al. Quality-of-life outcomes in men treated for localized prostate cancer. *JAMA* 1995; 273:129–35.
30. Lubeck DP, Litwin MS, Henning JM et al. Measurement of health-related quality of life in men with prostate cancer: the CaPSURE database. *Qual. Life Res.* 1997; 6:385–92.
31. Brandeis JM, Litwin MS, Burnison CM et al. Quality of life outcomes after brachytherapy for early state prostate cancer. *J. Urol.* 2000; 163:851–7.
32. Barry MJ, Fowler FJ Jr, O'Leary MP et al. The American Urological Association symptom index for benign prostatic hyperplasia. *J. Urol.* 1992; 148:1549.
33. Wei JT, Dunn RL, Litwin MS et al. Development and validation of the Expanded Prostate cancer Index Composite (EPIC) for comprehensive assessment of health-related quality of life in men with prostate cancer. *Urology* 2000; 56:899–905.
34. Davis JW, Kuban DA, Lynch DF et al. Quality of life after treatment for localized prostate cancer: differences based on treatment modality. *J. Urol.* 2001; 166:947–52.
35. Ware JE, Kisinski M, Keller SK. *SF-36 Physical and Mental Health Summary Scales: A User's Manual*. Boston: The Health Institute, New England Medical Center, 1994.
36. Litwin MS, Melmed GY, Nakazon T. Life after radical prostatectomy: a longitudinal study. *J. Urol.* 2001; 166:587–92.
37. Guyatt GH, Bombardier C, Tugwell PX. Measuring disease-specific quality of life in clinical trials. *Can. Med. Assoc. J.* 1986; 134:889.
38. Slevin ML, Plant H, Lynch D et al. Who should measure quality of life, the doctor of the patient? *Br. J. Cancer* 1988; 57:109–12.
39. Litwin MS, Lubeck DP, Henning JM et al. Differences in urologist and patient assessments of health related quality of life in men with prostate cancer: results of the CaPSURE database. *J. Urol.* 1998; 159:1988–92.
40. McCammon KA, Kolm P, Main B et al. Comparative quality-of-life analysis after radical prostatectomy or external beam radiation for localized prostate cancer. *Urology* 1999; 54:509–16.
41. Calais da Silva F. Quality of life in prostatic carcinoma. *Eur. Urol.* 1993; 24(Suppl. 2):113.
42. Fossa SD, Aaronson NK, Newling D et al. Quality of life and treatment of hormone resistant metastatic prostatic cancer. The EORTC Genito-Urinary Group. *Eur. J. Cancer* 1990; 26:1133.
43. Kornblith AB, Herr HW, Ofman US et al. Quality of lifepatients with prostate cancer and their spouses. The value of a data base in clinical care. *Cancer* 1994; 73:2791.
44. Sneeuw KCA, Albertsen PC, Aaronson NK. Comparison of patient and spouse assessments of health related quality of life in men with metastatic prostate cancer. *J. Urol.* 2001; 165:478–82.
45. Potosky AL, Harlan LC, Stanford JL et al. Prostate cancer practice patterns and quality of life: the Prostate Cancer Outcomes Study. *J. Natl Cancer Inst.* 1999; 91:1719–24.
46. Dalkin BL. Letter re: Health outcomes after prostatectomy or radiotherapy for prostate cancer: results from the Prostate Cancer Outcomes Study. *J. Natl Cancer Inst.* 2001; 93:401.
47. Litwin MS, McGuigan KA. Accuracy of recall in health-related quality-of-life assessment among men treated for prostate cancer. *J. Clin. Oncol.* 1999; 17:2882–8.
48. Legler J, Potosky AL, Gilliland FD et al. Validation of retrospective recall of disease-targeted function. Results from the Prostate Cancer Outcomes Study. *Med. Care* 2000; 38:847–57.
49. Talcott JA, Rieker P, Clark JA et al. Patient-reported symptoms after primary therapy for early prostate cancer: results of a prospective cohort study. *J. Clin. Oncol.* 1998; 16:275.
50. Talcott JA, Rieker P, Propert KJ et al. Patient-reported impotence and incontinence after nerve-sparing radical prostatectomy. *J. Natl Cancer Inst.* 1997; 89:1117–23.

51. Stanford JL, Feng Z, Hamilton AS et al. Urinary and sexual function after radical prostatectomy for clinically localized prostate cancer, the Prostate Cancer Outcomes Study. *JAMA* 2000; 283:354–60.
52. Hamilton AS, Stanford JL, Gilliland FD et al. Health outcomes after external-beam radiation therapy for clinically localized prostate cancer: results from the Prostate Cancer Outcomes Study. *J. Clin. Oncol.* 2001; 19:2517–26.
53. Beard CJ, Propert KJ, Rieker PP et al. Complications after treatment with external-beam irradiation in early-stage prostate cancer patients: a prospective multi-institutional outcomes study. *J. Clin. Oncol.* 1997; 15:223–9.
54. Lubeck DP, Litwin MS, Henning JM et al. Changes in health-related quality of life in the first year after treatment for prostate cancer: Results from CaPSURE. *Urology* 1999; 53:180–6.
55. Turner SL, Adams K, Bull CA et al. Sexual dysfunction after radical radiation therapy for prostate cancer: a prospective evaluation. *Urology* 1999; 54:124–9.
56. Reddy SM, Ruby J, Wallace M et al. Patient self-assessment of complications and quality of life after conformal neutron and photon irradiation for localized prostate cancer. *Radiat. Oncol. Invest.* 1997; 5:252–6.
57. Mantz CA, Nautiyal J, Awan A et al. Potency preservation following conformal radiotherapy for localized prostate cancer: impact of neoadjuvant androgen blockade, treatment technique, and patient-related factors. *Cancer J. Sci. Am.* 1999; 5:230–6.
58. Nguyen LN, Pollack A, Zagars GK. Late effects after radiotherapy for prostate cancer in a randomized dose-response study: results of a self-assessment questionnaire. *Urology* 1998; 51:991–7.
59. Hanlon AL, Bruner DW, Peter R et al. Quality of life study in prostate cancer patients treated with three-dimensional conformal radiation therapy: comparing late bowel and bladder quality of life symptoms to that of the normal population. *Int. J. Radiat. Oncol. Biol. Phys.* 2001; 49:51–9.
60. Sanchez-Ortiz RF, Broderick GA, Rovner ES et al. Erectile function and quality of life after interstitial radiation therapy for prostate cancer. *Int. J. Impot. Res.* 2000; 12(Suppl. 3):S18–24.
61. Krupski T, Petroni GR, Bissonette EA et al. Quality-of-life comparison of radical prostatectomy and interstitial brachytherapy in the treatment of clinically localized prostate cancer. *Urology* 2000; 55:736–42.
62. Lee WR, Hall MC, McQuellon RP et al. A prospective quality-of-life study in men with clinically localized prostate carcinoma treated with radical prostatectomy, external beam radiotherapy, or interstitial brachytherapy. *Int. J. Radiat. Oncol. Biol. Phys.* 2001; 51:614–23.
63. Potosky AL, Legler J, Albertsen PC et al. Health outcomes after prostatectomy or radiotherapy for prostate cancer: results from the prostate cancer outcomes study. *J. Natl Cancer Inst.* 2000; 92:1582–92.
64. Litwin MS, Flanders SC, Pasta DJ et al. Sexual function and bother after radical prostatectomy or radiation for prostate cancer: multivariate quality-of-life analysis from CaPSURE. Cancer of the Prostate Strategic Urologic Research Endeavor. *Urology* 1999; 54:503–8.
65. Madalinska JB, Essink-Bot ML, de Koning HJ et al. Health-related quality-of-life effects of radical prostatectomy and primary radiotherapy for screen-detected or clinically diagnosed localized prostate cancer. *J. Clin. Oncol.* 2001; 19:1619–28.
66. Wei JT, Dunn RL, Sandler HM et al. Comprehensive comparison of health-related quality of life after contemporary therapies for localized prostate cancer. *J. Clin. Oncol.* 2002; 20:557–66.
67. Zietman AL, Thakral H, Wilson L et al. Conservative management of prostate cancer in the prostate specific antigen era: the incidence and time course of subsequent therapy. *J. Urol.* 2001; 166:1702–6.
68. Jonler M, Nielsen OS, Wolf H. Urinary symptoms, potency, and quality of life in patients with localized prostate cancer followed up with deferred treatment. *Urology* 1998; 52:1055.
69. Tefilli MV, Gheiler EL, Tiguert R et al. Quality of life in patients undergoing salvage procedures for locally recurrent prostate cancer. *J. Surg. Oncol.* 1998; 69:156–61.
70. Perrotte P, Litwin MS, McGuire EJ et al. Quality of life after salvage cryotherapy: the impact of treatment parameters. *J. Urol.* 1999; 162:398–402.
71. Pound CR, Partin AW, Eisenberger MA et al. Natural history of progression after PSA elevation following radical prostatectomy. *JAMA* 1999; 281:1591–7.
72. Herr HW, O'Sullivan M. Quality of life of asymptomatic men with nonmetastatic prostate cancer on androgen deprivation therapy. *J. Urol.* 2000; 163:1743–6.
73. Albertsen P, Aaronson N, Muller M et al. Health related quality of life among patients with metastatic prostate cancer. *Urology* 1997; 49:207–17.
74. Potosky AL, Knopf K, Clegg LX et al. Quality-of-life outcomes after primary androgen deprivation therapy: results from the Prostate Cancer Outcomes Study. *J. Clin. Oncol.* 2001; 19:3750–7.
75. Litwin MS, Shpall AI, Dorey F et al. Quality-of-life outcomes in long-term survivors of advanced prostate cancer. *Am. J. Clin. Oncol.* 1998; 21:327–32.
76. Clark JA, Wray NP, Ashton CM. Living with treatment decisions: regrets and quality of life among men treated for metastatic prostate cancer. *J. Clin. Oncol.* 2001; 19:72–80.

Sexual Aspects of Prostate Cancer Treatment

T. Casey McCullough, Phillip C. Ginsberg and Richard C. Harkaway

Department of Surgery, Division of Urology, Albert Einstein Medical Center, Philadelphia, PA, USA

INTRODUCTION

Prostate cancer is the most common tumor diagnosed among men in the USA. In 2001 an estimated 180 000 new cases will have been diagnosed and 32 000 deaths will have occurred secondary to this disease.[1] The patient with prostate cancer has numerous treatment options, which include radical retropubic prostatectomy, radical perineal prostatectomy, laparoscopic radical prostatectomy, external beam radiation therapy, brachytherapy, cryosurgery, hormonal therapy, and expectant or deferred therapy. Each of these treatments has associated risks and side effects, which must be carefully explained to patients to facilitate their decision-making process.

Erectile dysfunction (ED) is one of the most common and troubling complications following treatment for prostate cancer. It is defined as the inability to have an erection sufficient for vaginal penetration and orgasm.[2] Because prostate cancer is a slow-growing neoplasm, and because more men are being diagnosed earlier in their lives, urologists are faced with a therapeutic dilemma. How can the prostate be successfully extirpated from the pelvis without compromising the delicate nerves that are responsible for erection? In 1982, Walsh and Donker introduced the anatomic radical retropubic prostatectomy, which preserved erections by avoiding injury to the cavernous neurovascular bundles.[3] This finding greatly improved the quality of life of men with localized prostatic carcinoma, since the morbidity of erectile dysfunction following radical prostatectomy could be minimized. Still, potency rates vary widely in the literature, from 20% to greater than 80% after radical prostatectomy.[1,2,4–17] This wide discrepancy stems from differences in study design, study size and patient selection.

Sexual dysfunction is important to men since many are sexually active well into their ninth decade.[18] Prostatic surgery has long been associated with erectile dysfunction but patients today expect this procedure to be performed with low morbidity. Therefore, therapy for sexual dysfunction is a prime issue before considering any kind of prostate surgery. Numerous treatments for erectile dysfunction secondary to radical prostatectomy, radiation therapy, cryosurgery and hormonal therapy are available today.

RADICAL RETROPUBIC PROSTATECTOMY

Nerve sparing versus non-nerve sparing

Most men presenting with prostate cancer will have localized disease and, therefore, a long disease-free survival after treatment for their cancer.[19] Successful treatment of prostate cancer not only depends on the length of time patients are cancer free, but also a patient's sexual function, which comprises a major component of their quality of life.[19] It is recognized that nerve-sparing procedures provide the patient with a better quality of life and quicker return of erections.[2,16] Quinlan et al. reported that there was a twofold increase in the risk of impotence when one neurovascular bundle was widely excised.[2] This implies a neurogenic component to the physiologic mechanism of erection. Similarly, Geary et al. found that the potency rates increased with the number of neurovascular bundles spared.[16] The number of intact neurovascular bundles was the most significant factor affecting postoperative potency.

Potency rates

Several studies have demonstrated that younger patients have a greater chance of retaining their potency after radical prostatectomy.[2,11,16,20,21] Younger patients typically have smaller tumor burdens, are more sexually active preoperatively and, therefore, are more likely to undergo a nerve-sparing procedure.[16] One study demonstrated that men younger than 70 years old who had a bilateral nerve-sparing procedure had a threefold increase in their odds of retaining their potency.[20] Preserving potency revolves around sparing the nerves of varing sizes that comprise the neurovascular bundle, not just the large nerves in the area.[15]

The pathological stage of the tumor also plays a role in predicting postoperative potency. Eggleston and Walsh found that with more advanced disease the potency rate decreased because, with extraprostatic extension of tumor, the neurovascular bundles had to be more widely excised.[14] Quinlan et al. reported that the relative risk of impotence increased by a factor of two, if the tumor had invaded through the prostatic capsule or the seminal vesicles.[2] The pathological stage dictates what type of nerve-sparing procedure can be performed. Prostate cancer that extends through the capsule produces a desmoplastic response, which can be seen intraoperatively.[14,15,22,23] The surgeon can then decide at that time whether to sacrifice the neurovascular bundles. Thus sacrificing the neurovascular bundles secondary to the fibrosis seen during the procedure would ultimately increase the rate of impotence as discussed above.

Neurostimulation/caverMap

During radical prostatectomy it is possible to evaluate the functional capacity of the neurovascular bundles with the CaverMap Surgical Aid. This device stimulates the cavernous nerves, which will produce a 1% or greater change in the circumference of the penile shaft if the bundles are intact.[23] Several studies have attempted to correlate a positive CaverMap response with recovery of potency postoperatively.[23,24] Klotz and Herschorn found 94% of patients who had a positive CaverMap response intraoperatively had the ability to have erections after surgery.[24] However, only 31% reported recovery with the ability to achieve erections consistently sufficient for penetration.[24] This pilot study was also limited by small sample size.

Kim et al. reported that a positive CaverMap response poorly predicted recovery of potency after radical prostatectomy.[23] A positive CaverMap response was noted in 77% of men but only 18% reported regaining sufficient potency for sexual intercourse.[23] None of the patients who had both neurovascular bundles excised could achieve erections postoperatively. This group of patients also had a negative CaverMap response, which validates the neurogenic component to erectile dysfunction. Because the high CaverMap response rate did not correlate with the low potency rate postoperatively, it is hypothesized that erectile function is most likely a multifactorial process.

Etiology of erectile dysfunction after radical retropubic prostatectomy

Penile erection is a complex phenomenon that includes coordinated interaction of the nervous, arterial, venous and sinusoidal systems.[13] While the neurogenic component to erectile dysfunction after radical prostatectomy is secondary to disruption of the cavernosal nerves,[1] the arterial component is thought to arise from the disruption of accessory pudendal arteries or atherosclerosis.[25] Breza et al. demonstrated that an accessory pudendal artery supplied additional blood to the penis in seven out of ten cadavers, and because of its close proximity to the prostate and bladder, could easily be compromised during radical pelvic surgery.[26] Aboseif et al. showed a 40% incidence of erectile dysfunction after nerve-sparing radical prostatectomy. These patients had a minimal response after surgery to an intracavernosal vasoactive agent despite an adequate response preoperatively.[13] This suggests a vascular etiology of postsurgical impotence, which was confirmed with reduced diameter and velocity of penile blood flow via pulsed Doppler ultrasonography.[13,27] Polascik and Walsh conferred that arterial insufficiency is a major factor contributing to impotence along with neurovascular bundle disruption.[28] However, they identified accessory internal pudendal arteries in only 4% of the patients and could not demonstrate improved sexual function when the accessory pudendal arteries were preserved than when they were sacrificed.[28]

Intraoperative damage to the neurovascular bundles, which initiate erections, is the main cause of erectile dysfunction after prostatectomy.[1] These neurovascular bundles, comprised of cavernous nerves, release nitric oxide in response to sexual stimuli, which initiates a molecular cascade resulting in increased levels of cyclic guanosine monophosphate (cGMP). This cGMP potentiates corporal vasodilatation and smooth muscle relaxation, which ultimately produces an erection.[9] Erectile dysfunction occurs immediately after radical prostatectomy secondary to disruption of the neurovascular bundles, and the return of sexual function may take up to 18 months in some individuals.[1,20] This varied length of time of return of sexual function may be due to patient age. Younger patients are more likely to have excessive erectile capacity and greater neurovascular regenerative capabilities.[21]

Another etiology of post-radical prostatecomy erectile dysfunction is the physiologic development of erectile dysfunction in the aging man. The risk of erectile dysfunction is estimated at 26/1000 men annually and increases with age, lower educational status, diabetes, heart disease and hypertension.[29] It is not clear as to how much of the erectile dysfunction is secondary to the prostate surgery verus the patients' comorbid conditions.

Helgason et al. found that men with prostate cancer had a significantly greater risk of impaired function in all aspects of sexuality than those without prostate cancer.[18] They found the prevalence of physiological impotence to be 31% in the random population, while 71% in the group with prostate cancer. Medications contributing to impotence were diuretics, hydrogen blockers and warfarin.[18]

Miller et al. examined whether changes in serum hormone levels influenced erectile dysfunction after radical prostatectomy.[30] They found a statistically significant increase in serum testosterone, free testosterone, estradiol, luteinizing hormone (LH) and follicle-stimulating hormone (FSH) as well as a decrease in dihydrotestosterone (DHT).[30] This implies erectile dysfunction after radical prostatectomy can not be explained by androgen deficiency.

Sexual function after radical prostatectomy

Prostate cancer and its treatment have been associated with decreased sexual desire and function.[18] One important component of sexual function, orgasm, is often overlooked. Orgasmic pleasure is accompanied by contractions of muscles in the genital tract and pelvic floor, with stimuli reaching the brain via the somatic pudendal nerve.[31] Orgasm is compromised after radical prostatectomy as there are no contractions of the prostate and seminal vesicles, and the ejaculate is decreased or absent.[31] The orgasm experienced after surgery is often less intense and may be associated with an involuntary loss of urine. Decreased sexual desire, erection, orgasm and ejaculation are distressing factors affecting patients after radical surgery and must, therefore, be addressed early in the course of treatment.

Treatment of erectile dysfunction

Since March of 1998, when the use of sildenafil citrate (Viagra) was officially approved in the USA, it has had an immense impact on the treatment of erectile dysfunction. Pharmacologically, sildenafil potentiates the vasodilatory effects of cyclic guanosine monophosphate by inhibiting type 5 phosphodiesterase, which would normally break cGMP into guanosine monophosphate. Several studies have examined how well sildenafil restores erectile function post-radical prostatectomy.[8,9,10]

It seems the factors that significantly influence the return of erections after surgery are the same factors that influence potency after the use of sildenafil. Earlier it was stated that the number of neurovascular bundles spared during surgery and the patient's age were significant factors in predicting potency post-radical prostatectomy. There was a quite uniform response to sildenafil among those patients who underwent bilateral nerve-sparing procedures. Zippe et al. reported an 80% success rate of achieving erections sufficient for vaginal intercourse after a bilateral procedure with 100 mg of sildenafil.[8] They found the maximum number of doses required for a positive response was three and the quality of erections was excellent with a mean duration of vaginal intercourse of 6.92 minutes.[8]

Zagaja et al. found a similar relationship between the number of neurovascular bundles spared and postoperative potency after the use of sildenafil. Patients in this study received either 50 mg or 100 mg of sildenafil and had response rates of 80% for men younger than 55 years, 45% in men aged 56–65, and 33% in men aged 66 or greater.[9] These percentages represent only those individuals in whom both neurovascular bundles were preserved. Similarly, Feng et al. reported that 71% of bilateral nerve-sparing procedures had a positive response to sildenafil.[10]

When one neurovascular bundle was excised, the response to sildenafil was more disappointing. Zippe and Zagaja both reported potency rates of 0% when just one neurovascular bundle was excised despite therapeutic doses of sildenafil.[8,9] Feng et al. interestingly found that 80% of patients receiving unilateral nerve-sparing surgery responded to sildenafil, with erections sufficient for satisfactory sexual intercourse.[10] The potency rates with sildenafil after non-nerve-sparing radical prostatectomy are uniformly very low with rates between 0% and 6%.[8,9,10]

These findings suggest that nerve preservation is essential to the recovery of potency after radical prostatectomy. The greater the number of neurovascular bundles, the greater amount of nitric oxide can be produced to stimulate cyclic guanosine monophosphate to induce corporal vasodilatation. Sildenafil's enhanced relaxant effect on the penile smooth muscle vasculature prolongs erections. When one neurovascular bundle is excised, there is less nitric oxide present to initiate this cascade. If both cavernosal nerves are spared, then why do older patients have inferior potency rates compared to their younger peers? Potency is a multifactorial process and thus vascular factors must play a more significant role as patients' age. Older patients are more likely to have comorbid conditions like diabetes, atherosclerosis, and cardiomyopathies that compromise not only peripheral vascular compliance but also cardiac output.

The return of potency after radical prostatectomy varies with time and can range from 4 to 40 months.[4,8,9,16] One study found that no patient regained sexual function with the aid of sildenafil sooner than 9 months postoperatively, and this was attributed to neuropraxia.[9] Stimulating the penile vasculature soon after surgery would prevent penile fibrosis, loss of elasticity and possibly increase the positive response rate to sildenafil.[8,9] A study is currently underway evaluating whether nightly sildenafil expedites the return of erections.

Intracavernous injections of alprostadil (MUSE), a powerful vasoactive agent and PGE1 analog, have been used after radical prostatectomy to assist in the recovery of spontaneous erections. Alprostadil when introduced into the urethra is absorbed into the local circulation and causes relaxation of the trabecular smooth muscle in the corpora.[32] Montorsi et al. found that 67% of patients recovered erections after surgery with 3 weekly injections of alprostadil after just 3 months.[12] Soon after prostatectomy the corpora cavernosa are in a state of relative hypoxia as decreased nitric oxide is available to dilate the penile vasculature. This hypoxia causes tissue fibrosis leading to decreased penile elasticity and erectile dysfunction.[12,33,34] Since sildenafil appears to be ineffective early after radical prostatectomy, it is reasonable to use alprostadil in the interim to promote increased corporal oxygenation and prevent fibrosis.[10,12]

Alprostadil is also an effective alternative in those patients who fail treatment with sildenafil. Harkaway et al. found 50% of sildenafil non-responders after radical prostatectomy reported improved erectile function with alprostadil.[32] Alprostadil differs from sildenafil in that it is a direct-acting agent that promotes the inflow and retention of blood, and is not dependant on sexual stimulation or intact penile innervation to function. With these two agents' differing mechanisms of action, it is apparent that combination therapy may benefit.[35]

The poor rate of recovery of potency after non-nerve-sparing radical prostatectomy have prompted some researchers to investigate new options to treat erectile dysfunction.

One such innovation involves using autologous nerve grafts to replace the resected cavernous nerves. Kim et al. performed bilateral sural nerve grafts to restore sexual function after the neurovascular bundles had been widely excised.[36] The nerve graft provides a conduit where regenerating nerve fibers can follow a structured framework to bridge the gap between the transected cavernosal nerves.[36] Although this study was limited in size, the authors found the return of spontaneous, unassisted erectile activity satisfactory for penetration and intercourse in two of their patients. This is significant, since men who otherwise have minimal chance for recovery of erections may benefit from autologous nerve grafts.

Vacuum erection devices are effective after radical prostatectomy because the injury is neurogenic. The erection produced is not a natural erection, however, as a low-flow ischemic erection is produced. Vacuum erection appliances should be used in patients with preoperative erectile dysfunction, but not those patients who were potent before surgery and are trying to regain erectile function. In preoperative impotent patients who failed sildenafil, vacuum erection devices and injections, the insertion of a penile prosthesis is an acceptable alternative to restore sexual functioning.

Centrally acting agents, such as melanotan II, apomorphine and oxytocin, represent a new class of agents to treat erectile dysfunction. Melanotan II in particular works on the hypothalamus to induce sexual desire, although its mechanism of action is unknown.[37] Wessells et al. also found that melanotan II initiated erections in half of their administrations of the drug, which is given by subcutaneous injection.[37] While none of the patients in this study had undergone radical prostatectomy, the results are encouraging in that centrally acting agents could work synergistically with established agents, like sildenafil, to augment the erectile response after nerve-sparing surgery.

Cancer control and quality of life

Localized prostate cancer is multifocal in origin having an average of seven separate cancers, and when treated with radical prostatectomy a long disease-free survival is expected.[38] Patients with prostate cancer evaluating different treatment options must weigh the consequences and alternatives associated with those options. The patient's quality of life after radical surgery must be evaluated, both in terms of what the patient desires from the surgery and what realistically the surgeon expects to occur – and these two viewpoints should be compared to prevent any surprises postoperatively. Quality of life refers to the psychosocial, emotional and physical outcomes of treatment as perceived by the patient.[39] Sexual function remains one of the critical components of quality of life and can be objectively evaluated by quality of life studies, such as the RAND/UCLA Prostate Cancer Index with reasonable accuracy.[19,40–43]

Some studies reveal that nerve-sparing radical prostatectomy patients have a significantly better quality of life in terms of sexual function.[19,20] Gralnek et al. found that nerve-sparing patients with spontaneous erections sufficient for intercourse had significantly better sexual function and bother scores than those men without spontaneous erectile activity.[19] This correlates with what was discussed earlier in that, as the number of neurovascualar bundles spared increases, so does the potency rate.[2,11,16,20] Gralnek also noted that the use of erectile aids, such as vacuum devices and medications, failed to increase the sexual quality of life among patients who underwent nerve-sparing surgery to the same level of those patients who regained their erectile function.[19] Therefore, while the current treatments for erectile dysfunction do aid in the return of potency in some individuals, further research in this area is needed.

The main purpose of radical prostatectomy is to eradicate the cancer in its entirety and preserving sexual function is of secondary importance. It is certainly possible with locally confined disease that the two objectives stated above be completed simultaneously. There is no reason to believe that by performing a nerve-sparing procedure that cancer control is compromised. Walsh and others have noted that the neurovascular bundles lie outside the capsule and fascia of the prostate, and that organ-confined adenocarcinoma can be successfully removed without disrupting the erectile mechanism.[2,14,15,17,44] Interestingly, even when irreversible erectile dysfunction occurs postoperatively, most men report that they are satisfied with their treatment choice and would choose radical prostatectomy again. Satisfaction rates after radical prostatectomy have ranged from 74% to 91% and most of those respondants would choose to have radical prostatectomy again even though the risk for impotence exists.[4,6,19,45,46] This is reassuring since being free of disease remains the main determinant of patient satisfaction. In those patients where a nerve-sparing procedure is unlikely to be performed, patients are willing to sacrifice some quality of life for the opportunity to be disease free.

LAPAROSCOPIC AND PERINEAL PROSTATECTOMY

Minimally invasive surgery has been popularized recently as patients generally return to work faster, have decreased postoperative pain and have shortened hospital stays. The laparoscopic radical prostatectomy, introduced in 1997 by Schuessler et al.,[47] has not produced any advantage over retropubic prostatectomy in terms of return of sexual function. Schuessler initially reported laparoscopic prostatectomy on nine patients with preservation of potency in 50% of those patients potent preoperatively.[47] Guillonneau in his laparoscopic series reported a postoperative erection rate at 45% after only 3 months of follow-up.[48] Salomon et al. found 25% of their patients postoperatively able to engage in sexual intercourse.[49] While these initial potency rates fare well within the range seen for retropubic prostatectomy, the increased surgical time, steep learning curve, increased cost of intrumentation and similar surgical outcome make laparoscopic prostatectomy a procedure in evolution.[47–49]

Radical perineal prostatectomy, first introduced in 1901 by Proust and then in 1905 by Young,[50] attempts to minimize postoperative morbidity while still offering excellent cancer control. In terms of sexual dysfunction after perineal prostatectomy, Harris et al. report 73% preservation of spontaneous erectile activity but 36% able to achieve vaginal penetration.[51] Further research is needed to confirm the potency rates after radical perineal prostatectomy.

EXTERNAL BEAM RADIATION

External beam radiation treatment represents the other major modality to treat prostate cancer. Just as radical prostatectomy has evolved, so too has radiation treatment. Today patients desiring radiation therapy have options between external beam radiotherapy without computed tomography, three-dimensional conformal radiotherapy, conformal therapy with dose escalation, proton beam therapy and transperineal interstitial seed implantation.[52]

Potency rates

Potency rates following external beam radiation therapy vary widely, from 23% to 73% depending on the type of radiation delivered and length of follow-up.[53–59] Some authors have noted that potency was retained in a greater percentage of patients who had a higher level of sexual functioning prior to radiation treatment.[53,60–62] This observation correlates with the findings from the radical prostatectomy data, which also show that preoperative sexual functioning is a significant predictor of postoperative impotence.

Three-dimensional conformal radiotherapy (3D-CRT) differs from conventional external beam radiotherapy in its treatment planning and delivery technique. The former obtains images of the patient's internal anatomy by computed tomography scans in the treatment position, and then delivers a more precise radiation dose to spare more normal tissue. The goal is to irradiate at higher doses with less morbidity. Studies have been contradictory concerning the ability of 3D-CRT to offer patients higher potency rates versus traditional therapy. Beard et al. found 20% complete impotence in the conformal group as compared to 44% in the whole pelvis radiation group.[52] This suggests that larger radiation fields negatively affect post-treatment erectile function. Other data suggest that there is no difference in the risk of impotence between conformal therapy and conventional radiotherapy.[58,63] One study by Zelefsky et al. found a higher incidence of erectile dysfunction with higher doses of 3D-CRT.[64] At 5 years, 68% of patients who received ⩾75.6 Gy were impotent compared to 52% who received lower doses.[64]

It appears that potency rates gradually decrease as the number of years post-treatment increases. Wilder et al. found that the potency preservation rate decreased with each year after treatment from 100%, to 83%, to 63% at 3 years after treatment.[65] Several other studies have reported similar results.[57,59,66,67] The question remains at what point in time do patients after 3D-CRT become impotent? The median time to impotence after conformal radiotherapy varies between 14 and 19 months.[64,66] This implies that studies that report potency rates after radiotherapy to the prostate at or before 1 year are reporting falsely elevated figures.

Four to eight years after traditional radiotherapy to the prostate sexual function appears to stabilize. Previously it was discussed that the rate of impotence increases as a function of time after radiotherapy. How does this rate change over extended periods of time and does it always increase? Fransson et al. reported that sexual function at 4 years predicts what sexual function will be at 8 years after radiotherapy.[68]

Etiology of erectile dysfunction after external beam radiotherapy

The causes of impotence are arteriogenic, neurogenic and pyschogenic in nature. After radiotherapy the most likely cause of erectile dysfunction is arteriogenic.[65,69,70] Goldstein et al. postulated that postradiation impotence is secondary to damage to the internal pudendal and penile arteries.[70] Radiation causes proliferation of the intima of vessels and is prothrombotic, favoring atherosclerotic plaque formation.[71] The histologic changes after radiation therapy were initially demonstrated in animal studies, where damaged vessels became sites of fibroblastic proliferation and lipid deposition.[72–74] These changes would certainly impede blood flow to the penis and cause erectile dysfunction. It is not surprising that a higher percentage of patients after radiotherapy present with erectile dysfunction, if they have a history of hypertension, atherosclerosis and tobacco use.[70]

In contrast to the factors affecting erectile dysfunction after radical prostatectomy, younger age did not seem to play a significant role after radiotherapy in some studies.[58,60,64] Instead, preradiotherapy sexual dysfunction, vascular disease and diabetes were significantly associated with the development of impotence.[61,66] Thus, normal potency status prior to radiation treatment increased the chances of retaining potency.[59] Zelefsky et al. also found no correlation between prior transurethral resection of the prostate or neoadjuvant androgen deprivation in predicting erectile dysfunction after radiotherapy.[64] Likewise, Formenti et al. reported that moderate dose adjuvant radiotherapy, 45–54 Gy, did not affect sexual potency after radical prostatectomy.[75,76]

Treatment of erectile dysfunction after radiotherapy

There are fewer data outlining the treatment of erectile dysfunction after radiotherapy as compared with surgery. Radiation treatment for prostate cancer used to be reserved for the elderly, who were not likely to care about postradiation impotence. This trend is changing as more and more younger patients are electing for radiotherapy as initial treatment for their prostate cancer – and, as a result, potency is a major concern.[67]

Sildenafil, similar to retropubic prostatectomy patients, remains the first-line agent to treat erectile dysfunction after radiotherapy to the prostate. Zelefsky et al. reported significant improvement in the firmness of erection, durability of erection and increased frequency of sexual activity after sildenafil use.[77] Patients who had at least partial erections prior to radiation treatment were most likely to respond to sildenafil and at a significant rate of 90%. Patients with poor pretreatment erectile activity were not likely to respond to sildenafil after treatment. The 74% response rate to sildenafil after radiotherapy that Zelefsky reported is improved over the 43% response rate after radical prostatectomy.[15] This is not surprising given the fact that erectile dysfunction after radiation therapy is secondary to arteriogenic causes, while after radical prostatectomy it is thought to arise from cavernosal disruption. Sildenafil is more effective at treating the vascular causes of impotence compared to the neurogenic causes.

Kedia et al. reported similar findings. The majority of patients after radiotherapy (71%) were able to achieve an erection sufficient for vaginal penetration.[67] Sildenafil had a greater effect on those patients who were sexually active prior to radiation treatment and had to be titrated up to the 100 mg dose in 80% of patients. Weber et al. found that the response rate to sildenafil increased progressively throughout the study period.[78] From 1 to 5 weeks postoperatively, the response rate increased from 40%, 57%, 66%, 69% and 74%. This suggests that using sildenafil after radiation should not be discontinued prematurely, as many more patients will respond to the medication if given the chance to do so.

Intracavernosal agents have been used after radiotherapy to treat erectile dysfunction. Pierce et al. reported an initial study where phentolamine and papaverine, and later prostaglandin E1 (PGE1) injected intracavernosally resulted in satisfactory erections in men after radiotherapy.[79] Agents like alprostadil and PGE1 can be used after sildenafil failure or to augment the response to sildenafil in those subjects who require additional treatment.

Quality of life after radiotherapy

Patients who choose external beam radiation treatment for their prostate cancer seek cure above all else. The goal of any cancer treatment is to eradicate the tumor cells, with quality of life issues taking a close second. Physicians must help patients weigh the benefits of cancer treatment with the consequences, like erectile dysfunction. In addition to not being able to achieve an erection sufficient for intercourse, patients complained about decreased orgasmic pleasure and reduced ejaculate volume in one series by Helgason et al.[80] Erectile dysfunction closely parallels the quality of life patients experience and all factors that comprise erectile dysfunction need to be properly addressed to patients beforehand, like orgasmic pleasure and the ability to ejaculate.

Another aspect that affects patient's lives postradiation is the subjective feeling of being cancer free. Fowler et al. discovered that radiation patients were less likely to say they were 'cancer free' and worried more frequently about their prostate cancer than surgical patients.[81] Because the prostate is irradiated and not physically removed from the pelvis, it is understandable why patients would feel this way. Even so, most patients who choose radiation therapy are certain they made the right choice and would choose radiation therapy in the future, if given the chance. Jonler et al. reported 81% satisfaction with radiation treatment and 97% would choose radiation again.[82] Similarly, another study found 84% of patients feeling no discomfort after radiation, 91% satisfied with their choice, and 97% willing to choose radiation treatment again.[56]

BRACHYTHERAPY

Brachytherapy, another treatment option for patients with prostate cancer, is the implantation of radioactive isotopes into a body cavity or tissue.[83] This mode of therapy is improved today with better isotopes (palladium-103), computed tomography and transrectal ultrasound. Potency rates after interstitial brachytherapy have generally been better than those for either radical prostatectomy or external beam radiation therapy. Brachytherapy potency rates have been reported between 47% and 97% in several studies.[63,84–90] Factors that had negative impacts on postbrachytherapy potency were pretreatment erectile dysfunction, high implant dose (D90 ≥160 Gy for I-125 and ≥100 Gy for Pd-103), and short duration hormonal therapy of 5–6 months.[89] No difference in postbrachytherapy erectile dysfunction was noted between iodine-125 or palladium-103.[91]

The etiology of impotence after brachytherapy is thought to arise from either excessive radiation exposure to the neurovascular bundles[92,93] or vascular etiologies.[90] DiBiase et al. reported that the patients who developed early postimplant impotence had doses to the neurovascular bundles that far exceeded the average values.[93]

Sildenafil has proved successful in treating erectile dysfunction after interstitial brachytherapy as it has for radical prostatectomy and radiotherapy. Merrick et al. reported an 80% success rate in achieving erections sufficient for vaginal penetration after sildenafil use in brachytherapy patients.[90] It was stated earlier that patients after bilateral nerve-sparing prostatectomy had a greater response to sildenafil. Here, too, brachytherapy patients have bilateral nerve integrity, which supports the vascular cause of impotence in these patients.

In terms of quality of life after brachytherapy, it compared well with radical prostatectomy. When brachytherapy was combined with external beam radiation therapy, sexual dysfunction did not improve over time.[94] The dual therapy adversely affected patients' lives and reduced the frequency of satisfactory erections. Brachytherapy as monotherapy had similar results to radical prostatectomy in terms of sexual dysfunction.[94]

CRYOSURGERY

Cryosurgery, first performed on humans in the mid-1960s, was largely abandoned until recently due to a high complication rate. Recently, there has been a trend toward implementing minimally invasive techniques to treat prostate cancer. Cryosurgery is one of these minimally invasive procedures, and is defined as the *in situ* reduction of tumor by the application of subzero temperatures.[95]

Potency after cryosurgery has been poor. Rates of erectile dysfunction after this procedure have ranged from 100% to 10% depending on the length of follow-up.[91,95–99] Immediately after cryosurgery, impotence is almost universal as the neurovascular bundles thaw. As time progresses after surgery, there is evidence that erectile function returns in some patients. Bahn et al. reported impotence rates of 90% after 6 months, but only 41% after 1 year.[100] Other studies confirmed that the impotence is usually transient with 86% of patients recovering erectile function to some degree.[95,97,99] Immediate postoperative impotence is caused by the iceball extending into the lateral periprostatic tissue, which shelters the neurovascular bundles.[98] The neurovascular bundles are, therefore, destroyed by the extreme cold temperature. In addition, patients have noted decreased sensitivity and parasthesia of the glans penis and general penile numbness 3 months after cryosurgery.[91,98] This may be secondary to damage to the pudendal nerve during the freezing process. These parasthesias over the penis may point to a neurogenic cause of erectile dysfunction after cryosurgery. However, a decreased peak velocity of blood flow within the cavernosal arteries was discovered after cryosurgery in another study – implying a vascular cause.[101]

Cryosurgery has the potential to become a low-morbidity treatment option for elderly men with organ-confined disease or local recurrence after radiotherapy. Patients should not be told that cryosurgery has decreased rates of impotence as compared to radical prostatectomy and external beam radiation, and that long-term results are unknown. Erectile function improves with time after cryosurgery and patients may benefit from sildenafil.

HORMONAL THERAPY

Patients with high-grade or metastatic prostate carcinoma are frequently medically or surgically castrated to inhibit the deleterious effects of testosterone. As a result, androgen deficiency can lead to a loss of libido and decreased sexual performance.[102] Physicians have many options to induce androgen blockade, such as low-dose estrogens, luteinizing hormone-releasing hormone (LHRH) agonists, antiandrogens and orchiectomy.

The antiandrogens, like flutamide and bicalutamide, competitively inhibit the binding of testosterone and DHT to the androgen receptor.[103,104] The goal of hormonal therapy is to provide maximal androgen blockade while minimizing sexual side effects. Several studies have evaluated the use of these agents and their affect on potency.

Migliari et al. showed that monotherapy with bicalutamide did not induce significant changes in nocturnal penile tumescence recordings or libido, sexual desire and frequency of sexual intercourse.[103] This differs with studies on other antiandrogens, like medroxyprogesterone acetate and cyproterone acetate, where the latter decreased sexual drive. This suggests that pure antiandrogen therapy does not impair sexual function.[103,105] Schroder et al. reported that potency preservation with antiandrogen monotherapy using either flutamide or cyproterone acetate was not maintained.[106] A loss of sexual function was found in roughly 80% of patients with either antiandrogen after 2 years of observation. They did note, however, that the loss of sexual function with both antiandrogens was gradual with 10–20% of men retaining sexual activity after 2–6 years of treatment.

LHRH agonists, like leuprolide and goserelin, decrease gonadotropin release, which suppresses testicular androgen production. LHRH agonists decrease sexual desire, sexual intercourse, and frequency, duration and rigidity of nocturnal erections when serum testosterone levels fall to castration levels.[107] These agents are potent inhibitors of the male sexual response and must be used judiciously in younger patients who wish to preserve sexual function.

When bicalutamide was compared with flutamide plus goserelin in prostate cancer patients, fewer patients in the bicalutamide group had erectile dysfunction, which ensured a better quality of life.[102] The bicalutamide group of patients enjoyed a higher emotional well-being, vitality and social functioning.

Finasteride and flutamide combined as androgen ablative therapy for advanced prostate adenocarcinoma have been studied. Finasteride inhibits 5α-reductase, decreases the conversion of testosterone to 5α-dihydrotestosterone and provides a more complete intraprostatic androgen blockade.[104] Combined therapy with these agents decreased serum prostate specific antigen (PSA) greater than either agent alone and accounted for an overall potency rate of 82% – with potency defined as the ability to maintain an erection satisfactory for sexual intercourse.[104] Combined hormonal therapy has the promise of delivering sufficient androgen blockade while maintaining adequate erectile dysfunction. Greenstein et al. found that surgical castration with orchiectomy reduced serum testosterone to a level not statistically different than that with hormonal castration.[108] Orchiectomy is associated with decreased potency and sexual activity.

CONCLUSIONS

Sexual dysfunction after any form of prostatic surgical or hormonal intervention can adversely affect the quality of life of patients. While cancer control is the primary focus, one must be attuned to the patient's sexual functioning after treatment. While sildenafil is the first-line agent for erectile dysfunction, other agents deserve mention. Pending sildenafil failure, intraurethral agents, vacuum tumescence devices, and penile implantations and sexual rehabilitation are all viable options to treat erectile dysfunction in patients after prostate cancer. Vacuum tumescence devices can be combined with other hormones, like sildenafil, or used alone with excellent results of up to 94% potency.[109]

It is unlikely that most men with prostate cancer have some form of erectile dysfunction at the time of their diagnosis. They may have other health problems that could interfere with their erectile ability, like cardiovascular disease, diabetes mellitus or multiple medications.

Sexual function after radical prostatectomy hinges on several variables. Younger age and the number of neurovascular bundles spared correlated with increased potency postsurgery.

Some studies included pathological grade in the likelihood of maintaining erectile function, but this is likely secondary to larger tumors requiring wider excision of the neurovascular bundles to achieve adequate cancer control. Taking frozen sections of the lateral prostatic pedicle, using CaverMap, and autologous nerve grafting may all aid to increase the sexual ability of patients postoperatively.

Radiation therapy is more likely to preserve potency if patients are able to achieve erections sufficient for vaginal penetration preradiotherapy. Smokers and patients with vascular disease are more likely to experience problems with erections after radiation treatment. Radiation therapy does offer increased potency rates when compared to hormonal therapy.[110]

With hormonal treatment, LHRH agonists and orchiectomy both produce adequate androgen ablation with similar rates of erectile dysfunction. LHRH agonists alone produced fewer sexual side effects than combined maximum androgen blockade. The newer antiandrogens are less likely to produce erectile dysfunction. Hormonal therapy in general has higher rates of sexual dysfunction than radical prostatectomy or radiation treatment due to its central effect of decreasing sexual desire.

Finally, patients who fail initial measures to restore normal sexual function should not be deferred from seeking additional help from sexual rehabilitation. This evaluation should assess not only the quality of erections, but also how the patient's sexual dysfunction affects his life in general. Participation of the patient's sexual partner is key, with both individuals requiring proper counseling to address the physical and psychological issues involved with the sexual dysfunction in their relationship.

REFERENCES

1. Siegel T, Moul JW, Spevak M et al. The development of erectile dysfunction in men treated for prostate cancer. *J. Urol.* 2001; 165:430–5.
2. Quinlan DM, Epstein JI, Carter BS et al. Sexual function following radical prostatectomy: influence of preservation of neurovascular bundles. *J. Urol.* 1991; 145:998–1002.
3. Walsh PC, Donker PJ. Impotence following radical prostatectomy: insight into etiology and prevention. *J. Urol.* 1982; 128:492–7.
4. Stanford JL, Feng Z, Hamilton AS et al. Urinary and sexual function after radical prostatectomy for clinically localized prostate cancer: the prostate cancer outcomes study. *JAMA* 2000; 283:354–60.
5. Drago JR, Badalament RA, Nesbitt JA. Radical prostatectomy 1972–1987 single institutional experience: comparison of standard radical prostatectomy and nerve sparing technique. *Urology* 1990; 35:377–80.
6. Gaylis FD, Friedel WE, Armas OA. Radical retropubic prostatectomy outcomes at a community hospital. *J. Urol.* 1998; 159:167–71.
7. Ramon J, Rossignol G, Leandri P et al. Morbidity of radical retropubic prostatectomy following previous prostate resection. *J. Surg. Oncol.* 1994; 55:14–19.
8. Zippe CD, Kedia AW, Kedia K et al. Treatment of erectile dysfunction after radical prostatectomy with sildenafil citrate (Viagra). *Urology* 1998; 52:963–6.
9. Zagaja GP, Mhoon DA, Aikens JE et al. Sildenafil in the treatment of erectile dysfunction after radical prostatectomy. *Urology* 2000; 56:631–4.
10. Feng MI, Huang S, Kaptein J et al. Effect of sildenafil citrate on post-radical prostatectomy erectile dysfunction. *J. Urol.* 2000; 164:1935–8.
11. Rabbani F, Stapleton AM, Kattan MW et al. Factors predicting recovery of erections after radical prostatectomy. *J. Urol.* 2000; 164:1929–34.
12. Montorsi F, Guazzoni G, Strambi LF et al. Recovery of spontaneous erectile function after nerve-sparing radical retropubic prostatectomy with and without early intracavernous injections of alprostadil: results of a prospective, randomized trial. *J. Urol.* 1997; 158:1408–10.
13. Aboseif S, Shinohara K, Breza J et al. Role of penile vascular injury in erectile dysfunction after radical prostatectomy. *Br. J. Urol.* 1994; 73:75–82.
14. Eggleston JC, Walsh PC. Radical prostatectomy with preservation of sexual dysfunction: pathological findings in the first 100 cases. *J. Urol.* 1985; 134:1146–8.
15. Walsh PC, Epstein JI, Lowe FC. Potency following radical prostatectomy with wide unilateral excision of the neurvascular bundle. *J. Urol.* 1987; 138:823–7.
16. Geary ES, Dendinger TE, Freiha FS et al. Nerve sparing radical prostatectomy: a different view. *J. Urol.* 1995; 154:145–9.
17. Walsh PC. Radical prostatectomy for localized prostate cancer provides durable cancer control with excellent quality of life: a structured debate. *J. Urol.* 2000; 163:1802–7.
18. Helgason AR, Adolfsson J, Dickman P et al. Factors associated with waning sexual function among elderly men and prostate cancer patients. *J. Urol.* 1997; 158:155–9.
19. Gralnek D, Wessells H, Cui H et al. Differences in sexual function and quality of life after nerve sparing and nonnerve sparing radical retropubic prostatectomy. *J. Urol.* 2000; 163:1166–70.
20. Catalona WJ, Carvalhal GF, Mager DE et al. Potency, continence, and complication rates in 1,870 consecutive radical retropubic prostatectomies. *J. Urol.* 1999; 162:433–8.
21. Catalona WJ, Basler JW. Return of erections and urinary continence following nerve sparing radical retropubic prostatectomy. *J. Urol.* 1993; 150:905–7.
22. Epstein JI. Evaluation of radical prostatectomy capsular margins of resection. The significance of margins designated as negative, closely approaching, and positive. *Am. J. Surg. Path.* 1990; 14:626–30.
23. Kim HL, Stoffel DS, Mhoon DA et al. A positive cavermap response poorly predicts recovery of potency after radical prostatectomy. *Urology* 2000; 56:561–4.
24. Klotz L, Herschorn S. Early experience with intraoperative cavernous nerve stimulation with penile tumescence monitoring to improve nerve sparing during radical prostatectomy. *Urology* 1998; 52:537–42.
25. Mulhall JP, Graydon RJ. The hemodynamics of erectile dysfunction following nerve-sparing radical retropubic prostatectomy. *Int. J. Impotence Res.* 1996; 8:91–4.
26. Breza J, Aboseif SR, Orvis BR et al. Detailed anatomy of penile neurovascular structures surgical significance. *J. Urol.* 1989; 141:437–43.
27. Kim ED, Blackburn D, McVary KT. Post-radical prostatectomy penile blood flow: assessment with color flow Doppler ultrasound. *J. Urol.* 1994; 152:2276–9.
28. Polascik TJ, Walsh PC. Radical retropubic prostatectomy: the influence of accessory pudendal arteries on the recovery of sexual function. *J. Urol.* 1995; 153:150–2.
29. Johannes CB, Araujo AB, Feldman HA et al. Incidence of erectile dysfunction in men 40–69 years old: longitudinal results

from the Massachusetts Male Aging Study. *J. Urol.* 2000; 163:460–8.
30. Miller LR, Partin AW, Chan DW et al. Influence of radical prostatectomy on serum hormone levels. *J. Urol.* 1998; 160:449–53.
31. Koeman M, Van Driel V, Schultz WCM et al. Orgasm after radical prostatectomy. *Br. J. Urol.* 1996; 77:861–4.
32. Jaffe JS, Antell M, Ginsberg PC et al. Use of intraurethral alprostadil (MUSE) in a group of sildenafil citrate non-responders. *J. Urol.* 2002; 167:282–3.
33. Sattar AA, Salpigides G, Vanderhaeghen JJ et al. Eavernous oxygen tension and smooth muscle fibers: relation and function. *J. Urol.* 1995; 154:1736–9.
34. Christ GJ, Lerner SE, Kim DC et al. Endothelin-1 as a putative modulator of erectile dysfunction: characteristics of contraction of isolated corporal tissue strips. *J. Urol.* 1995; 153:1998–2001.
35. Mydlo JH, Volpe MA, Macchia RJ. Initial results utilizing combination therapy in patients with a suboptimal response to either alprostadil or sildenafil monotherapy. *Eur. Urol.* 2000; 38:30–4.
36. Kim ED, Scardino PT, Hampel O et al. Interposition of sural nerve restores function of cavernous nerves resected during radical prostatectomy. *J. Urol.* 1999; 161:188–92.
37. Wessells H, Gralnek D, Dorr R et al. Effect of an alpha-melanocyte stimulating hormone analog on penile erection and sexual desire in men with organic erectile dysfunction. *Urology* 2000; 56:641–6.
38. Bastacky SI, Wojno KJ, Walsh PC et al. Pathologic features of hereditary prostate cancer. *J. Urol.* 1995; 153:987–91.
39. Bland KI. Quality of life management for cancer patients. *CA Cancer J. Clin.* 1997; 47:194–6.
40. Hays RD, Sherbourne CD, Mazel RM. The RAND 36-item health survey. *Health Econ.* 1993; 2:217–21.
41. Ware JE, Sherbourne CD. The MOS 36-item short-form health survey, conceptual framework and item selection. *Med. Care* 1992; 30:473–5.
42. Schag CA, Ganz PA, Heinrich RL. Cancer rehabilitation evaluation system-short form (CARES-SF). *Cancer* 1991; 68:1406–9.
43. Tulsky DS, Cella DF, Bonomi A et al. Development and validation of new quality of life measures for patients with cancer. *Proc. Soc. Behav. Med.* 1990; 11:45–50.
44. Catalona WJ, Dresner SM. Nerve-sparing radical prostatectomy: extraprostatic tumor extension and preservation of erectile dysfunction. *J. Urol.* 1985; 134:1149–53.
45. Jonler M, Messing EM, Rhodes PR et al. Sequelae of radical prostatectomy. *Br. J. Urol.* 1994; 75:48–53.
46. Fowler FJ, Barry MJ, Lu-Yao G et al. Effect of radical prostatectomy for prostate cancer on patient quality of life: results from a medicare survey. *Urology* 1995; 45:1007–13.
47. Schuessler WW, Schulam PG, Clayman RV et al. Laparoscopic radical prostatectomy: initial short-term experience. *Urology* 1997; 50:854–7.
48. Guillonneau B, Vallancien G. Laparoscopic radical prostatectomy: the mountsouris experience. *J. Urol.* 2000; 163:418–23.
49. Salomon JF, Hoznek A, Bellot J et al. Laparoscopic radical prostatectomy: preliminary results. *Eur. Urol.* 2000; 37:615–20.
50. Young HH. The early diagnosis and radical cure of carcinoma of the prostate: being a study of 40 cases and presentations of radical operation which was carried out in 4 cases. *Johns Hopkins Hosp. Bull.* 1905; 16:315–7.
51. Harris MJ, Thompson IM. The anatomic radical perineal prostatectomy: a contemporary and anatomic approach. *Urology* 1996; 48:762–8.
52. Beard CJ, Propert KJ, Rieker PP et al. Complications after treatment with external-beam irradiation in early-stage prostate cancer patients: a prospective multiinstitutional outcomes study. *J. Clin. Oncol.* 1997; 15:223–9.
53. Banker FL. The preservation of potency after external beam irradiation for prostate cancer. *Int. J. Radiat. Oncol.* 1988; 15:219–20.
54. Chinn DM, Holland J, Crownover RL et al. Potency following high-dose three-dimensional conformal radiotherapy and the impact of prior major urologic surgical procedures in patients treated for prostate cancer. *Int. J. Radiat. Oncol.* 1995; 33:15–22.
55. Dattoli M, Wallner K, Sorace R et al. 103Pd brachytherapy and external beam irradiation for clinically localized, high risk prostatic carcinoma. *Int. J. Radiat. Oncol.* 1996; 35:875–9.
56. Reddy SM, Ruby J, Wallace M et al. Patient self-assessment of complications and quality of life after conformal neutron and photon irradiation for localized prostate cancer. *Radiat. Oncol. Invest.* 1997; 5:252–6.
57. Al-Abany M, Steineck G, Agren CAK et al. Improving the preservation of erectile function after external beam radiation therapy for prostate cancer. *Radiother. Oncol.* 2000; 57:201–6.
58. Nguyen LN, Pollack A, Zagars GK. Late effects after radiotherapy for prostate cancer in a randomized dose–response study: results of a self-assessment questionnaire. *Urology* 1998; 51:991–7.
59. Turner SL, Adams K, Bull CA et al. Sexual dysfunction after radical radiation therapy for prostate cancer: a prospective evaluation. *Urology* 1999; 54:124–9.
60. Hart KB, Duclos M, Shamsa F et al. Potency following conformal neutron/photon irradiation for localized prostate cancer. *Int. J. Radiat. Oncol.* 1996; 35:881–4.
61. Shipley WU, Zietman AL, Hanks GE et al. Treatment related sequelae following external beam radiation for prostate cancer: a review with an update in patients with stages T1 and T2 tumor. *J. Urol.* 1994; 152:1799–805.
62. Beckendorf V, Hay M, Rozan R et al. Changes in sexual function after radiotherapy treatment of prostate cancer. *Br. J. Urol.* 1996; 77:118–23.
63. Zelefsky MJ, Wallner KE, Ling, CC et al. Comparison of the 5-year outcome and morbidity of three-dimensional conformal radiotherapy versus transperineal permanent iodine-125 implantation for early-stage prostatic cancer. *J. Clin. Oncol.* 1999; 17:517–22.
64. Zelefsky MJ, Cowen D, Fuks Z et al. Long term tolerance of high dose three-dimensional conformal radiotherapy in patients with localized prostate carcinoma. *Cancer* 1999; 85:2460–8.
65. Wilder RB, Chou RH, Ryu JK et al. Potency preservation after three-dimensional conformal radiotherapy for prostate cancer: preliminary results. *Am. J. Clin. Oncol.* 2000; 23:330–3.
66. Mantz CA, Song P, Farhangi E et al. Potency probability following conformal megavoltage radiotherapy using conventional doses for localized prostate cancer. *Int. J. Radiat. Oncol.* 1997; 37:551–7.
67. Kedia S, Zippe CD, Agarwal A et al. Treatment of erectile dysfunction with sildenafil citrate (VIAGRA) after radiation therapy for prostate cancer. *Urology* 1999; 54:308–12.
68. Fransson P, and Widmark A. Late side effects unchanged 4–8 years after radiotherapy for prostate carcinoma. *Cancer* 1999; 85:678–88.
69. Mittal B. A study of penile circulation before and after radiation in patients with prostate cancer and its effect on impotence. *Int. J. Radiat. Oncol.* 1985; 11:1121–5.
70. Goldstein I, Feldman MI, Deckers PJ et al. Radiation-associated impotence. *JAMA* 251:903–10.
71. Gold H. Production of atherosclerosis in the rat: effect of x-ray and high fat diet. *Arch. Pathol. Lab. Med.* 1961; 71:268–73.
72. Warren S. Effect of radiation on normal tissues. *Arch. Pathol. Lab. Med.* 1942; 34:1070–9.
73. Lindsay S, Kohn HI, Dakin RL et al. Aortic arteriosclerosis in the dog after localized aortic X-irradiation. *Circ. Res.* 1962; 10:51–60.
74. Konings AWT, Hardonk MJ, Wieringa, RA et al. Initial events in radiation-induced atheromatosis: activation of lysosomal enzymes. *Strahlentherapie* 1975; 150:444–8.
75. Formenti SC, Lieskovsky G, Skinner D et al. Update on impact of moderate dose of adjuvant radiation on urinary continence

and sexual potency in prostate cancer patients treated with nerve-sparing prostatectomy. *Urology* 2000; 56:453–8.
76. Formenti SC, Lieskovsky G, Simoneau AR et al. Impact of moderate dose of postoperative radiation on urinary continence and potency in patients with prostate cancer treated with nerve sparing prostatectomy. *J. Urol.* 1996; 155:616–19.
77. Zelefsky MJ, McKee AB, Lee H et al. Efficacy of oral sildenafil in patients with erectile dysfunction after radiotherapy for carcinoma of the prostate. *Urology* 1999; 53:775–8.
78. Weber DC, Bieri S, Kurtz JM et al. Prospective pilot study of sildenafil for treatment of postradiotherapy erectile dysfunction in patients with prostate cancer. *J. Clin. Oncol.* 1999; 17:3444–9.
79. Pierce LJ, Whittington R, Hanno PM et al. Pharmacologic erection with intracavernosal injection for men with sexual dysfunction following irradiation: a preliminary report. *Int. J. Radiat. Oncol.* 1991; 21:1311–14.
80. Helgason AR, Fredrikson M, Adolfsson J et al. Decreased sexual capacity after external radiation therapy for prostate cancer impairs quality of life. *Int. J. Radiat. Oncol.* 1995; 32:33–9.
81. Fowler FJ, Barry MJ, Lu-Yao G et al. Outcomes of external-beam radiation for prostate cancer: a study of medicare beneficiaries in three surveillance, epidemiology, and end results areas. *J. Clin. Oncol.* 1996; 14:2258–65.
82. Jonler M, Ritter MA, Brinkmann R et al. Sequelae of definitive radiation therapy for prostate cancer localized to the pelvis. *Urology* 1994; 44:876–82.
83. Sylvester J, Blasko JC, Grimm P et al. Interstitial implantation techniques in prostate cancer. *J. Surg. Oncol.* 1997; 66:65–75.
84. Stock RG, Stone NN, Iannuzzi C. Sexual potency following interactive ultrasound-guided brachytherapy for prostate cancer. *Int. J. Radiat. Oncol.* 1996; 35:267–72.
85. Benoit RM, Naslund MJ, Cohen JK. A comparison of complications between ultrasound-guided prostate brachytherapy and open prostate brachytherapy. *Int. J. Radiat. Oncol.* 2000; 47:909–13.
86. Herr HW. Preservation of sexual potency in prostatic cancer patients after 125-I implantation. *J. Am. Gerlatr. Soc.* 1979; 27:17–19.
87. Stone NN, Stock RG. Brachytherapy for prostate cancer: real-time three-dimensional interactive seed implantation. *Techni. Urol.* 1995; 1:72–80.
88. Sanchez-Ortiz RF, Broderick GA, Rovner ES et al. Erectile function and quality of life after interstitial radiation therapy for prostate cancer. *Int. J. Impotence Res.* 2000; 3:S18–24.
89. Stock RG, Kao J, Stone NN. Penile erectile function after permanent radioactive seed implantation for treatment of prostate cancer. *J. Urol.* 2001; 165:436–9.
90. Merrick GS, Butler WM, Lief JH et al. Efficacy of sildenafil citrate in prostate brachytherapy patients with erectile dysfunction. *Urology* 1999; 53:1112–16.
91. Chaikin DC, Broderick GA, Malloy TR et al. Erectile dysfunction following minimally invasive treatments for prostate cancer. *Urology* 1996; 48:100–4.
92. Merrick GS, Butler WM, Dorsey AT et al. A comparison of radiation dose to the neurovascular bundles in men with and without prostate brachytherapy-induced erectile dysfunction. *Int. J. Radiat. Oncol.* 2000; 48:1069–74.
93. DiBiase SJ, Wallner K, Tralins K et al. Brachytherapy radiation doses to the neurovascular bundles. *Int. J. Radiat. Oncol.* 2000; 46:1301–7.
94. Krupski T, Petroni GR, Bissonette EA et al. Quality-of-life comparison of radical prostatectomy and interstitial brachytherapy in the treatment of clinically localized prostate cancer. *Urology* 2000; 55:736–42.
95. Robinson JW, Saliken JC, Donnelly BJ et al. Quality-of-life outcomes for men treated with cryosurgery for localized prostate carcinoma. *Cancer* 1999; 86:1793–801.
96. Mack D, Jungwirth A, Adam U et al. Long-term follow-up after open perineal cryotherapy in patients with locally confined prostate cancer. *Eur. Urol.* 1997; 32:129–32.
97. Saliken JC, Donnelly BJ, Brasher P et al. Outcome and safety of transrectal US-guided percutaneous cryotherapy for localized prostate cancer. *J. Vasc. Intervent. Rad.* 1999; 10:199–208.
98. Shinohara K, Connolly JA, Presti JC et al. Cryosurgical treatment of localized prostate cancer (stages T1 to T4): preliminary results. *J. Urol.* 1996; 156:115–21.
99. Wieder J, Schmidt JD, Casola G et al. Transrectal ultrasound-guided transperineal cryoablation in the treatment of prostate carcinoma: preliminary results. *J. Urol.* 1995; 154:435–41.
100. Bahn DK, Lee F, Solomon MH et al. Prostate cancer: US-guided percutaneous cryoablation. *Radiology* 1995; 194:551–55.
101. Aboseif S, Shinohara K, Borirakchanyavat S et al. The effect of cryosurgical ablation of the prostate on erectile function. *Br. J. Urol.* 1997; 80:918–22.
102. Boccardo F, Rubagotti A, Barichello M et al. Bicalutamide monotherapy versus flutamide plus goserelin in prostate cancer patients: results of an Italian prostate cancer project study. *J. Clin. Oncol.* 1999; 17:2027–38.
103. Migliari R, Muscas G, Usai E. Effect of casodex on sleep-related erections in patients with advanced prostate cancer. *J. Urol.* 1992; 148:338–41.
104. Brufsky A, Fontaine-Rothe P, Berlane K et al. Finasteride and flutamide as potency-sparing androgen-ablative therapy for advanced adenocarcinoma of the prostate. *Urology* 1997; 49:913–20.
105. Tyrell CJ. Tolerability and quality of life aspects with the anti-androgen casodex as monotherapy for prostate cancer. *Eur. Urol.* 1994; 26:15–9.
106. Schroder FH, Collette L, de Reijke TM et al. Prostate cancer treated by anti-androgens: is sexual function preserved? *Br. J. Cancer* 2000; 82:283–90.
107. Marumo K, Baba S, Murai M. Erectile function and nocturnal penile tumescence in patients with prostate cancer undergoing luteinizing hormone-releasing hormone agonist therapy. *Int. J. Urol.* 1999; 6:19–23.
108. Greenstein A, Plymate SR, Katz PG. Visually stimulated erection in castrated men. *J. Urol.* 1995; 153:650–2.
109. Korenman SG, Viosca SP. Use of a vaccum tumescence device in the management of impotence in men with a history of penile implant or severe pelvic disease. *J. Am. Geriatr. Soc.* 1992; 40:61–4.
110. Bergman B, Damber JE, Littbrand B et al. Sexual function in prostatic cancer patients treated with radiotherapy, orchiectomy, or oestrogens. *Br. J. Urol.* 1984; 56:64–9.

Prostate Cancer: A Patient's Perspective

Chapter 52

Thomas Anderson
Office of Community Relations, Temple University, Philadelphia, PA, USA

I am an African American male who was diagnosed with prostate cancer in January 1995. I have been a survivor for the past 8 years and hope to survive and see many, many more years in good health.

I would like to acknowledge my loving wife, Ruth Anderson, my son, Sean Anderson, my mother, Lillian Buckins, my brother Robert, who is also a prostate cancer survivor, and other members of my family and friends for their prayers and the support I received while going through my ordeal with prostate cancer. Believe me, without their support and concern, I would not be alive today to write about my experiences and perspective as a patient who was diagnosed and survived prostate cancer.

BEFORE BEING DIAGNOSED

I was in basically good health for a man of my age. My cholesterol count was under 180; blood pressure level was normal, I did not and do not have any cardiovascular problems, no allergies, no respiratory problems, etc. Like most men who were active athletically in their young adult years, I knew I was in good health and did not get a regular physical each year, as I felt I did not need a physical each year, as we all should do after age 40, and especially if there is a history of cancer or other diseases in your family history.

In 1980, my wife became aware of prostate cancer having an impact on the longevity of African–American males lives after age 50. I was in my mid-forties at that time and aware of her concern. I promised her that I would start having regular physical examinations. I made an appointment and went to see my medical physician, who administered the different tests for a complete physical, blood pressure, cholesterol and general body functions. This examination found me to be in good physical health.

I sought out a urologist/oncologist, a physician with training and experience in the evaluation of cancer. The urologist gave me a digital rectal examination (DRE) and a blood test called prostate specific antigen (PSA) screening to test an enzyme normally produced by prostate cells, both normal and cancerous. I was found to have a normal PSA of 0.5, which gave me a baseline or level of my PSA count that I could use as reference when having my next PSA screening.

BEING DIAGNOSED

Each year in September, which is my birth month, I started having my physical and urological examinations. In 1984, when I had my yearly urological examination, my urologist thought he felt a growth on my prostate from the DRE. He recommended that I have an ultrasound and a biopsy so that he could visualize the entire prostate gland with ultrasound waves and remove pieces of prostate tissue for microscopic analysis to ascertain if any cancer was present. Fortunately, he did not find any cancerous cells, but asked me to make sure I had my DRE and PSA examination each year to keep a check on it.

In 1986, I left my original urologist because I felt as though I was not getting any personal attention and I was just another patient. This urologist was a department chairperson, with a large practice and most times I was seen by one of his medical residents. After researching and gathering information on other urologists, I ended up at my employer's hospital, Temple University. I was able to get an appointment with the department chairperson, whom I found to be just what the doctor ordered, a doctor with a personality and the personal touch.

In 1990, my PSA moved up to 2.5, which was still in the so-called normal range up to 4.0. If the range was above 4.0, you needed further diagnosis. My urologist asked me not to be concerned and to keep having my examinations yearly in order to keep track of my PSA level.

In September of 1994, I went to an American Cancer Society's prostate cancer screening to have my DRE and PSA test examination. The PSA screening test came back at 4.5 and I was told to see my urologist for further analysis. Because of a stringent work schedule, which required my immediate attention, I was unable to go to see my urologist immediately. In January 1995, I made the appointment and went to see my urologist. This was 4 months after I was given the notification that my PSA had moved to a level of 4.5, which was two points higher from the previous year's PSA reading. I gave him the results of the screening that was taken in September 1994. He immediately gave me a DRE and had my blood drawn for a PSA screening, and within 10 days he received the PSA results. The PSA count had moved from a 4.5 level to a 6.0. He recommended that I come in for an ultrasound and biopsy examination procedure, which I consented to have done.

In consultation with my urologist, he informed me that I would have some discomfort during the procedure. During the procedure, it was stated to me that there were some areas around my prostate in question and a biopsy would also be performed. Over the next 10 days, I was hoping that the test would come back with good results. My urologist called me with the results, and informed me that he had good news and bad news. I asked for the bad news first. They found cancer in my prostate gland. The good news was that the cancer was located inside the prostate gland and in a well-defined grouping.

DECISION AND CHOICE OF TREATMENT

My urologist asked me to come in for consultation and to bring my wife with me. I was told that I had 6 months to make a decision and choice of treatment. At this point, I was in a state of shock, not knowing how to tell my wife that I had been diagnosed with prostate cancer. It took me a month to get enough courage to break the good and bad news to her. When I told her, she became very upset, but I assured her that there was enough time to make a decision.

I made an appointment for my wife and I to see my urologist. During the consultation, he told my wife that this was a time for support and loving care, and that I had a good chance to recover because of my physical condition. He talked to us about several optional treatments and procedures. He recommended that I get a second opinion, if I did not feel comfortable with his diagnosis. I had great confidence in my urologist and the medical laboratories at Temple University Hospital, so I did not seek a second opinion. He gave me all the information about choice of treatments so I could make a decision.

His first choice was surgery, performing a radical prostatectomy to remove the entire prostate gland and adjacent glands. He assured me that he could perform and leave the nerves intact that are responsible for penile erections (nerve-sparing technique). He recommended that I make an appointment to see a radiation oncologist for consultation about their choice of treatments. The radiation oncologist talked to me about several specialized treatments: external beam radiation to destroy the cancer cells in the affected areas; radioactive seed implants, placing little rice-size 'pellets' in the prostrate gland to destroy the cancer cells; and cryosurgery, controlled freezing of the prostate gland to destroy cancer cells.

This was quite a bit of information to receive and make a decision on which method to choose. I started reading up by going to the medical sections of bookstores and public libraries. In addition, I consulted with some of my associates and friends who knew individuals who had been treated for prostate cancer. After much thought and further consultation with my wife and son, I decided to have my urologist perform the radical surgery operation to remove my prostate using the nerve-sparing technique method.

SURGERY AND RECOVERY

After several presurgical examinations, I went into the hospital on June 29, 1995 for surgery. At 7 am on June 30, 1995, I was rolled into the operating room for surgery. I was in the operating room for 4 hours and in recovery for 2 hours. Under medication, I was returned to my room where I would be for the next 5 days. My urologist and his medical staff came to see me that evening, and stated that everything went very well and I should be on my way to recovery. The 5-day stay in the hospital gave me the opportunity to adjust to the morphine pump for pain, which I had for 2 days, intravenous fluids, surgical clamps, a catheter/bag and support stockings to protect from blood in my legs after surgery.

Before leaving the hospital on the fifth day, I was given instructions on how to do my walking exercises, the amount of physical activity to be taken and care of my Foley catheter. The Foley catheter had to be worn for a minimum of 2 weeks. I returned to my urologist on the third week to have the surgical clamps and Foley catheter removed. Everything went fine and I returned home for the next 6 weeks to recuperate from the surgery. After the sixth week, I returned to my urologist to review the pathology report and discuss long-term follow-up. At that time I was told to return in 3 months for my first of many periodic PSA and DRE screening tests, which would be routinely performed for the rest of my life. My first PSA after my prostatectomy was 0.1 and it has stayed at this level for the past 8 years.

SUPPORT GROUP AWARENESS

After my ordeal of surviving prostate cancer, I discovered that there were many men, especially African American men, who knew someone that needed advice and education about prostate cancer and diet. I was asked by an associate who was a breast cancer survivor if I would be interested in becoming a volunteer for the American Cancer Society, to help organize a Man-To-Man prostate cancer support group for the Southeast Region of Pennsylvania Division to include Philadelphia, Pennsylvania, where I live and work. In July of 1996, a year after my surgery, I joined the American Cancer Society. Since that time, I have assisted in organizing eight annual prostate cancer conferences. These conferences have given over 1200 men who are survivors, who may have prostate cancer or who are in need of information, an exceptional opportunity to hear and question leading authorities on prostate cancer. The initial goals of the conferences are to: increase understanding of prostate cancer and treatment choices; identify ways to manage side effects of the disease and treatment; enhance better communication between patients and family members, friends and healthcare providers; provide a forum to meet with various experts and prostate cancer patients who share their concerns.

In 2000, our Southeast Region Philadelphia area American Cancer Society made a commitment to promote a national program for prostate cancer in the African American community. This program is called 'Let's Talk About It'. It is a community-based prostate awareness and education program for African American men that was developed collaboratively by the American Cancer Society and 100 Black Men of America, Inc. I have made a personal commitment to the 100 Black Men and the American Cancer Society program, because prostate cancer is the most common form of cancer found in African American men. It occurs more often in African American men than any other racial or ethnic group. They are twice as likely to die of the disease than any other cultural group of men.

I underwent a radical prostatectomy. I did not have to awaken from that operation; I have had a stable PSA reading of 0.1 since the operation with reasonable good health. I am a survivor, because I finally listened to my wife, started receiving a yearly medical examination, and lot of prayers. I'm on a 'Mission' to tell the story of my survival from prostate cancer and hopefully save many lives.

Governmental Policies

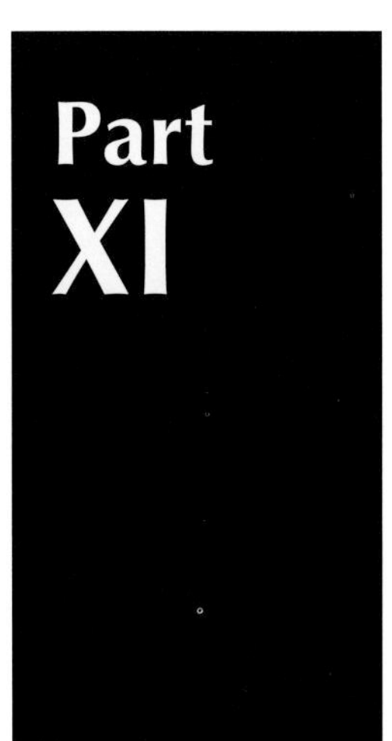

Prostate Cancer Economics

Lars Ellison
Johns Hopkins Hospital, Baltimore, MD, USA

Carl Olsson
Columbia University College of Physicians and Surgeons

INTRODUCTION

Medical care and research is a delicate balance between competing illnesses and finite resources. Since its introduction in 1937, the National Cancer Institute (NCI) has served as the central agency within the National Institutes of Health for the coordination of cancer research. The majority of the NCI's 65-year history has been marked by development of a research agenda based on assessment of disease risk. Increasingly, however, large patient and physician constituencies have become powerful voices raising awareness on Capital Hill of discrepancies in health care funding.

In this chapter, we will attempt to provide structure to the debates that have changed prostate cancer research funding patterns over the past 10 years. We begin with a review of the literature concerning costs associated with screening, diagnosis and the various treatment strategies. Second, we will follow the political and public relations campaigns that opened the now public dialogue on this subject. Finally, we will look at additional information that adds substance to the claims of equity in funding.

COST OF TREATMENT

Total United States health care expenditures are projected to exceed $1.5 trillion in 2002.[1] The Medicare budget is estimated to be $243 billion for that same year.[1] Prostate cancer is a major economic industry in the USA. Estimates of expenditures beyond screening and diagnosis suggest the Medicare program alone paid more than $1.8 billion for treatment of prostate cancer in 1997.[2] Prostate cancer has high prevalence and incidence rates, in conjunction with long expected survival; therefore, the costs of treatment for an individual often accrue for many years. The two major contributors to the cost of treatment are radical prostatectomy and luteinizing hormone-releasing hormone (LHRH) agonists.[2] Radiation therapy, either as external beam or brachytherapy, is an increasingly important source of expenditure as well.

Analysis of the financial impact of a health care intervention can be measured either by examining cost data or charge data. The cost of an intervention is related to the line-item expense of delivering the intervention. Cost itemization accounts for the institutional and physician outlay required to provide a given service. Wholesale price of goods is relatively stable across different geographic regions. For this reason, cost is accepted as the gold standard for the estimation of the economic impact of an intervention on the health care system. Charge data reports either the amount billed for a service (physician or hospital perspective) or the amount paid (insurers perspective). Charge data reflects either the 'assigned' value of the intervention by a physician or hospital, or the 'accepted' value as reflected in payment by either insurers or the patient. From administrative data sources, it is often difficult to differentiate between the dollar amounts billed versus the dollar amount paid. In addition, the amount paid for services is dependent on the coding practices of the physician as well as the regional reimbursement schedule. Therefore, the subjective nature of charge data makes it a poor proxy for health care dollar accounting. Unfortunately, charge data are easily accessed and are often used in medical economic analysis.

Radical prostatectomy remains the primary initial form of treatment for men less than 70 years of age. Rates of radical prostatectomy rose rapidly in the early 1990s in response to a rapid rise in the incidence of the disease.[3] By 1994, aggressive screening with prostate specific antigen (PSA) had tapped the prevalent pool of undetected cases within the population. In addition, more rational age-specific screening strategies were adopted. These two phenomena led to an overall drop in the incidence rate. Rates of radical prostatectomy dropped in step with the declining incidence. At the same time, a number of studies began to demonstrate little survival benefit for men with less than a 10-year life expectancy.[4–7] While overall rates of radical prostatectomy have declined, rates for men less than 65 years old have steadily risen since 1990.[3]

Significant changes in preoperative, perioperative and post-operative management have occurred over the past 10 years.

The practice of preoperative admission is now all but historical. Radical prostatectomy has been examined from a number of different perspectives using both charge and cost data. Several studies have documented the value of critical pathways for postoperative care. Kock reported critical pathways successfully reduced operating time, length of stay and 'unnecessary medications'.[8] As a result, hospital charges were reduced by 44% ($13 783 to $7741). Litwin reported a 12% reduction in hospital costs ($7916 to $6934, respectively) and a 28% reduction in hospital charge ($17 005 to $13 524) between the year 1993 and 1994 with the introduction of a critical pathway at UCLA.[9]

Wagner analyzed charge data from a single institution in 1996–1997. The mean hospital charge associated with radical prostatectomy was $15 100.[10] Using Medicare data, Brandeis reported charge data for the years 1993 to 1996.[11] All charges associated with prostate cancer were totaled for the fiscal quarter in which the diagnosis was made, through two fiscal quarters after the delivery of treatment. The mean charge associated with the group undergoing radical prostatectomy was $19 019. Ellison reported the volume of procedures done at a hospital was inversely related to total hospital charges.[12] Hospitals that performed less than 25 cases per year charged more than institutions that performed greater than 54 cases per year ($15 600 vs. $13 500, respectively).

Radiation therapy has accounted for an increasing percentage of initial treatment. This has become an increasing popular strategy in part as a result of dissemination of patient reported outcomes data after radical prostatectomy. Because the treatment is 'non-invasive' the perception remains that this is a less expensive approach. Several studies have examined this issue. External beam radiation therapy is delivered in fractionated doses over a period of 4–8 weeks. Tailored dosimetry remains dynamic throughout the course in response to patient tolerance and side-effect symptomatology. For this reason cost analysis is difficult to obtain.

Brandeis compared Medicare charges for brachytherapy and external beam radiation for the years 1993–1996.[11] Combined therapy was found to generate the highest total charges. When modalities were used in isolation the mean charge for brachytherapy was $15 301 versus $15 937 for external beam therapy.

In contrast, Ceizki examined cost data at a single institution comparing brachytherapy to radical prostatectomy.[13] Brachytherapy costs were 85–105% higher than radical prostatectomy. Wagner calculated single institution hospital charges for brachytherapy.[10] The mean hospital charge associated with brachytherapy was $21 000. The single most important line item for brachytherapy was the cost of interstitial seeds. Kohan found little difference in total hospital charges when comparing brachytherapy to radical prostatectomy ($13 904 vs. $13 886, respectively).[14]

Hormone ablation therapy is the second largest Medicare expenditure for the treatment of prostate cancer. Between 1993 and 1997 the cost of this drug to the Medicare program rose 120%, from $330 million to over $760 million.[2] In 1997, this single medication accounted for 64% of all Medicare reimbursments to urologists for urologic care. The efficacy of medical hormone ablation for the control of PSA failure has not been shown to be superior to orchiectomy alone. In addition, there are no data to suggest that medical hormone ablation substantially effects survival or quality of life relative to surgical castration. Decision analytic modeling has been used to determine thresholds of efficacy and costs per quality adjusted life year (QALY). Bennett used NCI 0036 clinical trials results as baseline assumptions, and determined that flutamide increased survival among patients with minimal disease by 2.1 quality-adjusted months at a cost of $41 000 per QALY.[7] For patients with severe disease, their survival improved by 2.6 quality-adjusted months at a cost of $52 700 per QALY.

The primary concern surrounding the above data is not with the associated reported values, but the fact that cost data for radiation therapy and hormone therapy are incomplete. Within the accounting structure for 'cost' are the subcategories of direct and indirect costs. Direct costs include the value of all the goods, services and other resources that are consumed in the provision of care. Indirect costs include those values associated with the ongoing care of a patient that are not related to health care. Within this category are such items as cost of transportation and lost productivity as well as morbidity and mortality costs. Different treatments generate costs in both the direct and indirect categories. The absolute cost, at the bottom line in either category will differ substantially between treatments. For example, travel costs and impact on productivity will differ significantly between radical prostatectomy and external beam radiation therapy for the first 8–10 weeks after initiation of treatment.

Great efforts have been made to analyze the financial impact of the diagnosis and treatment of prostate cancer. Many of these efforts have fallen far short of their intentions. From a pure accounting standpoint, it remains impossible to identify the true 'bottom line'. This is as true for the patient receiving his bill at the end of a hospitalization as it is for a large health maintenance organization (HMO) at the end of the fiscal year.

NCI RESEARCH FUNDING HISTORY

Prostate cancer research receives funding from a large number of private and public sources. The National Institutes of Health (NIH) is the federal government's umbrella organization through which the majority of appropriations are funneled. The NCI serves as central agency within the NIH for the oversight of intramural and extramural government-funded research. Certainly there are other organizations (the Department of Defense, the Centers for Disease Control, for example) within the federal government that play an important role in this regard. Our focus is the NCI and a review of the timeline of change in prostate cancer funding as well as a broad outline of the current research portfolio.

Federal funding of cancer research has been in place for over 50 years. The National Cancer Institute act of 1937 (public law 244)[15] outlined a multilevel approach for federal involvement in cancer research, diagnosis, epidemiology and treatment: first, establishment of the National Cancer Institute within the Public Health Service; second, direction of the Surgeon General to 'promote the coordination of researches conducted by the Institute and similar researches conducted by other agencies, organizations, and individuals'; third, to establish the National Cancer Advisory Council; and fourth, 'to purchase radium; to make such radium available ... for the study of the cause, prevention, or methods of diagnosis or treatment of cancer, or for the treatment of cancer'. Between 1938 and 1968, the NCI received a cumulative total of $1.69 billion in federal funds;[16] however, by the FY2002, the annual budget of the NCI was $5.03 billion.[17]

Cancer research became a major federal initiative under the Nixon administration. The National Cancer Act of 1971 (PL 92-218)[18] greatly expanded the role and independence of the NCI. Among the provisions were efforts to extend to the Director the 'coordination of all the activities of the NIH relating to cancer'. The Director was to present a budget directly to the President and Congress after review and comment (*without change*) by the Secretary of Health, the Director of the NIH and the National Cancer Advisory Board. In addition to the annual budgets of 1972–74, $70 million was allocated for development of 15 new cancer training and research centers.

As a check to the unprecedented latitude of the director of the NCI, PL 92-218 created the 'Presidents Cancer Panel'. This panel consisted of three individuals appointed by the president who served as a liaison to and watchdog for the president. As stipulated, the Panel was to meet with the Director not less than 12 times per year, and was to report progress and problems directly to the President. The members of this panel were clearly afforded a significant amount of influence over the priorities of the NCI at any given time. The current members are: Harold Freeman, a surgeon and Chief Executive Officer of North General Hospital in New York City; Maureen Wilson, a Ph.D. basic science researcher involved with tumor virology; and Frances Visco, a lawyer and president of the National Breast Cancer Coalition.

The Community Mental Health Center Extension Act and Biomedical Research and Research Training Amendments of 1978 [PL 95-622][19] modified the 1971 National Cancer Act. At the core of the changes was a consolidation of the duties of the director and the scope of the NCI mandate for control of research. In addition, the intense scrutiny of the Presidents Cancer Council was reduced. Most importantly, language was added which expanded the research agenda to include programs on prevention as well as the impact of environmental and occupational exposure in carcinogenesis.

The Health Research Extension Act of 1985 [PL 99-158][20] and the Health Omnibus Programs Extension of 1988 [PL 100-607][21] further expanded the scope of the NCI to include research on continuing care of cancer patients and their families and rehabilitation research.

Significant changes for urology came with the NIH Revitalization Amendments of 1993 [PL 103-43].[22] Within these amendments were mandates for intensified and expanded research programs in breast/women's cancers and prostate cancer. However, also included was an unprecedented and egregious example of special interest lobbying. The specific language within the amendments was '(Sec. 1911) Requires studies: (1) on environmental and other potential risk factors contributing to the incidence of breast cancer in specified counties in New York State and two other northeastern U.S. counties listed in a specified report'.[22] In other words, a case-control study of elevated breast cancer rates on Long Island was mandated as a result of pressure from women in that area who believed they suffered excess cancer incidence due to environmental exposures. This marked the opening of the door to aggressive lobbying from disease-specific special interest groups. To its credit, the American Urological Association responded in kind, and has been a powerful voice for increased appropriations for prostate cancer research.

Since the 105th Congress, a number of bipartisan pieces of legislation have come to committee and floor vote. Of these are included the Prostate Cancer Research Commitment Resolution of 1999 (S Res-92),[23] which expressed the need for increases in funding for research within the NIH and the Department of Defense. The House Resolution Raising Public Awareness of Prostate Cancer (H Res-211)[24] encouraged the dissemination of the importance of prostate cancer screening. The Prostate Cancer Research and Prevention Act introduced by Senator Bill Frist (R-TN) suggested revision and extension of the prostate cancer preventive health program.[25] Finally, a bill that generated a great deal of debate, the Stamp Out Prostate Cancer Act was passed in 1999, and was based on the prior and analogous The Stamp Out Breast Cancer Act (Public Law 105-41) of 1997.[26–28]

FUNDING COMPARISONS

There is a long and contentious history associated with the allocation of federal funds for cancer research. Much of the controversy arose in the early 1990s. With impending budgetary constraint, advocacy groups increasingly lobbied congressional offices and NIH for more research on their specific disease of interest. These efforts succeeded for certain diseases (e.g. AIDS and breast cancer) at the expense of others. Limited budget resources added to the intensity of fierce lobbying for disease-specific mandates in the NIH budget. As a result of the continuing controversy over disease funding, congressional hearings were held in 1997 addressing the mechanisms for establishing research priorities at NIH.

As an example of the impact of disproportional funding related to disease incidence, the following describes federal funding for AIDS research for 1998 and 1999:

> For FY1998, NIH is allocating a total of $1.61 billion for AIDS research, or 12% of the $13.6 billion total NIH budget. Funding for research on AIDS in FY1998 is second only to the National Cancer Institute ($2.3 billion). Government-wide AIDS spending is estimated at $9.67 billion in FY1999.[29]

The allocation for AIDS research is startling in its magnitude. Furthermore, if examined in light of disease prevalence and mortality, the level of funding is even more disproportionate.

Overall, there have been a total of 566 000 reported cases of AIDS and 340 000 AIDS-related deaths in the USA between the years 1980 and 1996.[30] In contrast, there were 1 359 000 new cancer cases and 554 000 cancer-related deaths in the USA in 1996 alone. There are twice as many incident cases of cancer and 60% more cancer deaths in the single year 1996 than occurred as a result of AIDS during the entire epidemic from 1980 to 1996.

PROSTATE CANCER FUNDING

Prostate cancer research funding by the NCI has significantly increased since 1998. In the early 1990s, the annual allocation grew from $13 million in 1990 to $71 million in 1996.[17] Even with these increases, the 1996 funding level for prostate cancer ranked among the lowest four cancers by incidence. However, with mandates from Congress to expand funding of prostate (and breast cancer), the levels continued to rise. Allocations rose to $135.7 million in 1999 and $203.2 million in 2000. By 2000, prostate cancer funding was second only to breast cancer ($438.7 million). The expected level for prostate cancer funding within the NCI for FY2003 is $340 million. As demonstrated above, the use of absolute allocations as the measure of funding parity, while interesting, do not account for differences in disease impact on society. While federal support of prostate cancer research rose dramatically in the late 1990s, there remain disparities in allocations when examined in relation to various epidemiological measures.

Incidence rates may be used as a weight to adjust absolute levels of allocation. In other words, we could look at dollars per incident case of cancer. In 2000, there were 180 400 new cases of prostate cancer in the USA. This was second only to breast cancer, with 184 200 new cases.[17] The rise in absolute appropriations might at first glance lead one to believe the current distribution of funds is reasonable. However, if these levels are adjusted for incidence, we find that prostate cancer ranks eleventh behind Hodgkin's disease and just above lung and bronchus cancer (**Fig. 53.1**).

Alternatively, annual disease-specific mortality could be the measure upon which allocations are adjusted. In 2000, the prostate cancer-related death rate ranked fourth behind lung, colon and breast cancer at 31 900 deaths/year. When adjusting appropriations based on mortality rate, prostate cancer ranks sixth behind leukemia (**Fig. 53.2**). What is more, while breast ranks second after this adjustment, colon cancer ranks tenth and lung and bronchus cancer ranks fourteenth.

Clearly the decision process influencing allocation of funds is driven by a number of 'hard' and 'soft' variables. Among the hard variables, health-services researchers have examined, as above, the ability of incidence and mortality as well as years of life lost and prevalence to predict appropriations. No single crude epidemiological measure sufficiently describes the relationship. Presumably a complex model weighting each variable would approximate the process. Gross examined NIH budget requests for 1996 in relation to incidence, prevalence, mortality and disability-adjusted life-years (DALYs).[31] The DALY is a downward adjustment of the value of a year of life with a specific disability relative to no disability. Of the four measures, the authors found a significant relationship between NIH funding and DALYs. However, the prediction model accompanying this analysis accounted for only 62% of the variability of funding by disease. Therefore, the interplay of incidence, prevalence and mortality, as well as the efforts of the numerous lobbying groups on Capital Hill also influenced the process.

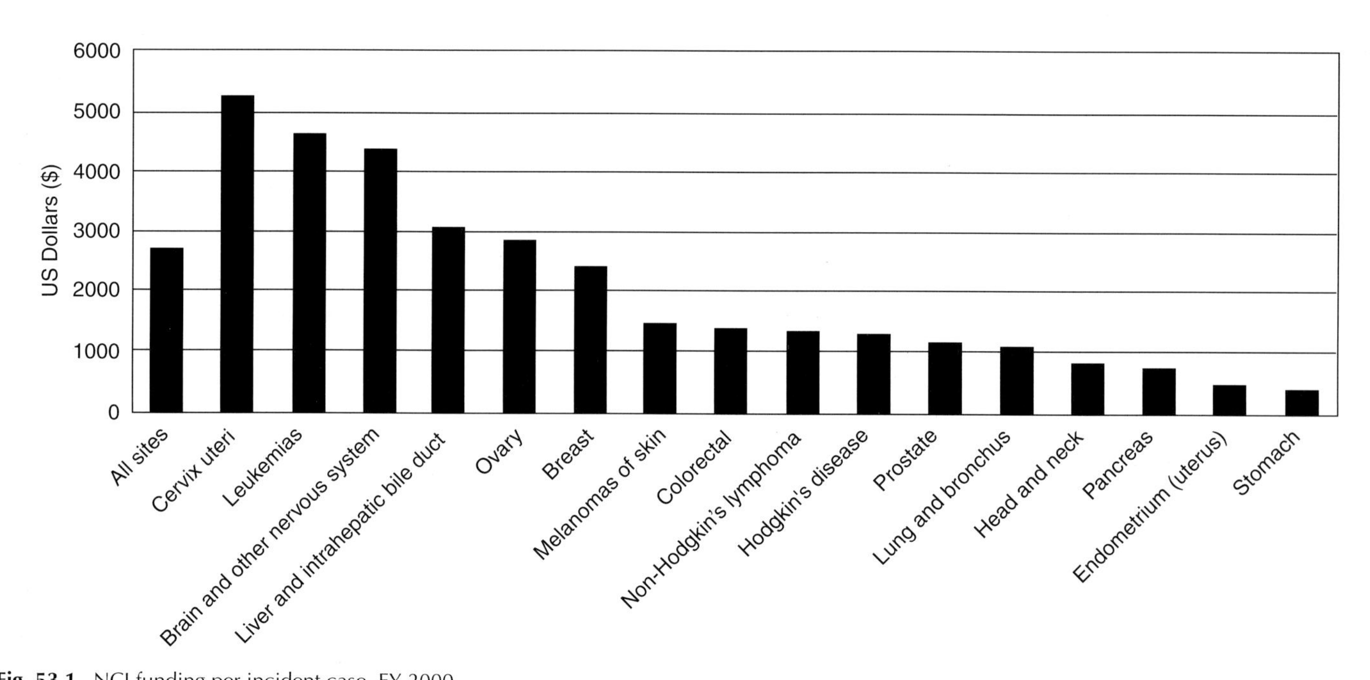

Fig. 53.1. NCI funding per incident case, FY 2000.

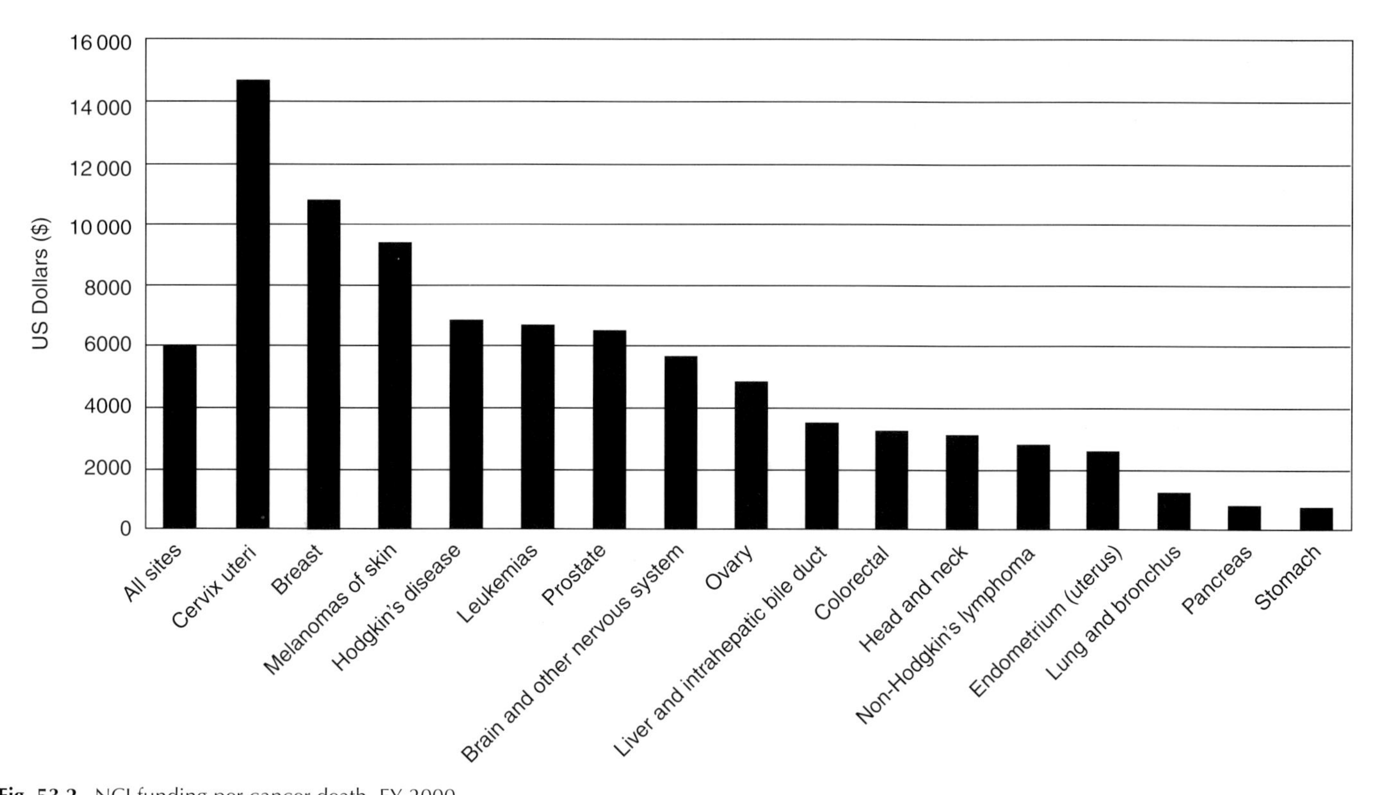

Fig. 53.2. NCI funding per cancer death, FY 2000.

While there has never been a full disclosure of the methods used, the NCI/NIH has publicly acknowledged the above debates.[32] In response, they identify several constraints within the process that limit the flexibility for and speed of change. First, for any given year, 75% of the budget is allocated to continuation of multiyear projects. Second, they identify the need for rapid response to unforeseen public health needs (i.e. Ebola virus and smallpox investigations). Finally, it is suggested that many research activities, while assigned to a given disease, will generate findings that translate to and inform other disease processes.

What is the value of weighting allocations? A bar for assisting to determine the level of federal funding for research for a given disease is non-existent. Superficially there appears to be a rank order to determining funding levels. However, when the system fails to demonstrate logical consistency, it can only follow that external forces have distorted the process to their own ends. In this case, it is clear that the special interest groups that have come to play such an important role on Capital Hill have negatively impacted the distribution of federal funds. The above examples, using incidence and mortality as weighting systems, allow for simple calculation of funding levels based on known and accepted epidemiologic measures.

Politics are intertwined with health care policy and funding. While this is not a new phenomenon, the level of interaction is unprecedented and growing. The 'War on Cancer' has been waged for close to 30 years, and significant strides have been made. In the past 10 years the determination of funding allocation has been adversely influenced by special interest groups. Given that lobbying has *de facto* been legislated (PL 103-43),[22] to its credit the American Urological Association (AUA) has been effective in advancing prostate cancer as a significant agenda item within the NCI.

While prostate cancer has benefited from funding trends in the past decade, there remain casualties of the funding process in urological cancer: kidney cancer and bladder cancer. Together kidney and bladder accounted for nearly 87 000 new cancer cases and 25 000 cancer deaths in the year 2000.[17] Compared to lymphomas or leukemias, kidney and bladder cancer have higher rates of incidence and mortality. The funding levels for these sites are not included as a line item in the published annual budget submitted to Congress. However, the NCI does report its ongoing research portfolio by cancer site. As a crude comparison, there are currently 788 funded studies of lymphoma and 984 funded studies of leukemia. The NCI currently reports 232 funded studies of kidney cancer and 337 funded studies of bladder cancer. As a comparison, there are 2253 current funded studies of prostate cancer. Based on average study costs, the estimated budget allocation for either site is estimated to be approximately $20 million. With recent studies suggesting a rise in the incidence of kidney cancer, it would certainly seem prudent for the NCI to identify this as an area of particular interest, as well as address the apparent deficiency in bladder cancer research funding.

It can never be assumed that the security of future funding for prostate cancer is assured. However, because of the high incidence and mortality rates associated with prostate cancer, urologists and their patients may be in a unique position proactively to help define the future process of cancer research funding. As one of the major recipients of NCI support, urology could play a leadership role in this regard. By calling for a process of benchmarking allocation by a combination of epidemiological measures, urology could maintain their current funding levels for prostate cancer and improve funding for bladder and kidney cancer, while at the same time leveling the playing field for other medical conditions that rely on NCI-sponsored funding.

CONCLUSION

In this chapter we have attempted to provide structure to the debates that have changed prostate cancer research funding patterns over the past 10 years. The AUA has served as an effective advocate for promoting the urologic research agenda at the federal level. This aside, the legislative legacy has created a funding system that forces competition between political and public relations campaigns, the result of which is a distorted research agenda. The underlying epidemiologic measures of cancer (e.g. prevalence, incidence and disease-specific mortality) are key determinates for evaluating the significance of a disease within a population. So, too, should they guide us in determining the levels of funding necessary for setting a national research agenda. Hopefully, urology as a specialty can help guide the process of redefining the federal research agenda and the funding process.

REFERENCES

1. *National Health Expenditure Projections 2000–2010*. Baltimore, MD: Health Care Financing Administration, 2001.
2. Holtgrewe HL. The economics of prostate cancer. In: Murphy G, Khoury S, Partin AW et al. (eds) *Prostate Cancer*, pp. 497–514. Plymouth, UK: Health Publication Ltd, 2000.
3. Ellison LM, Heaney JA, Birkmeyer JD. Trends in the use of radical prostatectomy for treatment of prostate cancer. *Eff. Clin. Pract.* 1999; 2:228–33.
4. Launois R. Cost-effectiveness analysis of strategies for screening prostatic cancer. *Dev. Health Econ. Public Policy* 1992; 1:81–108.
5. Littrup PJ, Goodman AC, Mettlin CJ. The benefit and cost of prostate cancer early detection. The Investigators of the American Cancer Society – National Prostate Cancer Detection Project. *CA Cancer J. Clin.* 1993; 43:134–49.
6. Abramson N, Cotton S, Eckels R et al. Voluntary screening program for prostate cancer: detection rate and cost. *South Med. J.* 1994; 87:785–8.
7. Bennett CL, Matchar D, McCrory D et al. Cost-effective models for flutamide for prostate carcinoma patients: are they helpful to policy makers? *Cancer* 1996; 77:1854–61.
8. Koch MO, Smith JA Jr. Influence of patient age and co-morbidity on outcome of a collaborative care pathway after radical prostatectomy and cystoprostatectomy. *J. Urol.* 1996; 155:1681–4.
9. Litwin MS, Smith RB, Thind A et al. Cost-efficient radical prostatectomy with a clinical care path. *J. Urol.* 1996; 155:989–93.
10. Wagner TT 3rd, Young D, Bahnson RR. Charge and length of hospital stay analysis of radical retropubic prostatectomy and transperineal prostate brachytherapy. *J. Urol.* 1999; 161:1216–18.
11. Brandeis J, Pashos CL, Henning JM et al. A nationwide charge comparison of the principal treatments for early stage prostate carcinoma. *Cancer* 2000; 89:1792–9.
12. Ellison LM, Heaney JA, Birkmeyer JD. The effect of hospital volume on mortality and resource use after radical prostatectomy. *J. Urol.* 2000; 163:867–9.
13. Ciezki JP, Klein EA, Angermeier KW et al. Cost comparison of radical prostatectomy and transperineal brachytherapy for localized prostate cancer. *Urology* 2000; 55:68–72.
14. Kohan AD, Armenakas NA, Fracchia JA. The perioperative charge equivalence of interstitial brachytherapy and radical prostatectomy with 1-year followup. *J. Urol.* 2000; 163:511–14.
15. National Cancer Institute Act of 1937, 1937, 244, 75th Congress, 8/5/1937.
16. National Cancer Institute. *Appropriations of The NCI 1938–1998*. Bethesda, MD: National Cancer Institute, 1998.
17. *The 2002 NCI Budget Request*. Bethesda, MD: National Cancer Institute, 2001.
18. National Cancer Act of 1971, 1971, S 1828, PL 92–218, 92nd Congress, 12/23/1971.
19. The Community Mental Health Center Extension Act, 1978, S 2450, PL 96-622, 95th Congress, 11/09/1978.
20. Health Research Extension Act of 1985, 1985, HR 2409, PL 99-158, 99th Congress, 11/20/1985.
21. Health Omnibus Extension of 1988, 1988, S 2889, PL 100-607, 100th Congress, 11/04/1988.
22. National Institutes of Health Revitalization Act of 1993, 1993, S 1, PL 103-43, 103rd Congress, 6/10/1993.
23. Prostate Cancer Research Commitment Resolution of 1999, 1999, S Res-92, 106th Congress, 10/26/1999.
24. Expressing the sense of the House of Representatives regarding the importance of raising public awareness of prostate cancer, and of regular testing and examinations in the fight against prostate cancer, 1999, H Res-211, 106th Congress, 6/22/1999, h4677–82.
25. Prostate Cancer Research and Prevention Act, 2000, S 1243, Senate, 106th Congress.
26. Stamp Out Prostate Cancer Act of 1999, 1999, 106th Congress.
27. Stamp Out Breast Cancer Act, 1997, PL 105-41, 105th Congress.
28. Woloshin S, Schwartz LM. The U.S. Postal Service and cancer screening – stamps of approval? *N. Engl. J. Med.* 1999; 340:884–7.
29. Johnson Ja. *AIDS Funding for Federal Government Programs: fy1981–fy1999*. Congressional Research Service Science, Technology, and Medicine Division Library of Congress, 1998.
30. Kirschstein R. *Disease-specific Estimates of Direct and Indirect costs of Illness and NIH Support*. Washington, DC: Department of Health and Human Services, 2000.
31. Gross CP, Anderson GF, Powe NR. The relation between funding by the National Institutes of Health and the burden of disease. *N. Engl. J. Med.* 1999; 340:1881–7.
32. Setting WGoP. *Setting Research Priorities at the National Institutes of Health*. Bethesda, MD: National Institutes of Health, 1997. (http://www.nih.gov/news/ResPriority/priority.htm)

New Horizons for Prostate Cancer

Part XII

Molecular Therapeutics in Prostate Cancer

Brian Nicholson and Dan Theodorescu

Department of Urology, University of Virginia Health Science Center, Charlottesville, VA, USA

INTRODUCTION

The purpose of this chapter is to provide information on the molecular basis of prostate cancer biology and to identify some of the targets for therapy, and highlight some potential strategies for molecular treatment. Here we give a synopsis of what we have learned regarding molecular biology of cancer in general, and the directions research might take in the future in order to impact prostate cancer specifically. This work is certainly not encyclopedic in nature and we apologize in advance to colleagues whose work we were no able to include. Hope lies in learning to utilize some of these molecular workings for better prevention, diagnosis and treatment of the most common solid organ cancer in men.

Prostate cancer is a formidable disease and at current rates of diagnosis will affect one in six men living in the USA.[1] Many of these men are diagnosed at an early stage of the disease and can be effectively treated by surgery or radiation. However, a significant fraction of men are diagnosed with later stage disease or progress despite early curative therapeutic attempts. Unfortunately, many of these men succumb to prostate cancer, as management options are limited and not always successful. Through an understanding of the molecular processes that occur in the development and progression of prostate cancer, novel therapies will arise that will provide longer survival, better quality of life and a chance for cure in men afflicted with this disease.

THE LIFE OF THE CELL

Cell growth

Normal cells abide by an internal clock: they progress through a sequence of events known as the cell cycle. Different cells have varying time on this clock and the body regulates which cells divide into new (daughter) cells and how long it takes them to do so. It is in this manner that the body replaces worn out cells or makes more cells when needed, for example, the production of liver cells in response to injury and the turnover of gastrointestinal epithelial lining cells. Cancer cells, in contrast, escape the body's regulatory system and multiply despite being unwanted and unnecessary. Cells have a period when duplication of DNA occurs (synthesis phase), and a period when chromosomal copies are segregated to opposite ends of the cell (mitosis), permitting subsequent cellular division. These two phases are separated by checkpoints wherein cells determine that there are no mistakes in the replication procedure. If errors are found, they are either repaired or, if the mistake is irreparable, the cell is programmed to stop replication and will die, thus preventing the production of faulty progeny. Therefore, it is a balance between mitosis and programmed cell death (apoptosis) that gives the body a relatively constant number of healthy cells throughout life (**Figs 54.1, 54.2**).

The cell is usually in a quiescent state, during which time it performs the functions for which it was created; this is termed the G_0 phase. Growth factors and steroid hormones can stimulate the cell to enter into the beginning of the growth cycle. The initial cell-cycle phase is termed the G_1 phase (first gap phase) and represents the presynthesis period, where the cell accumulates the building blocks necessary for replication. There is a checkpoint between the G_1 phase and the synthesis phase (S phase) known as the G_1/S checkpoint. Once preparations are completed, the S phase begins and each chromosome

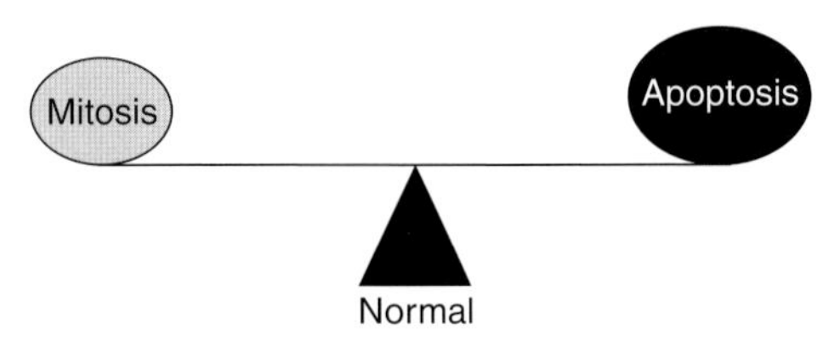

Fig. 54.1. Normal cells demonstrate a balance between mitosis and apoptosis giving the body a relatively constant number of healthy cells throughout life. This balance is dependent upon proper regulation of control mechanisms.

is copied. Following the complete replication of the DNA content, there is a second gap period or G_2 phase during which time the cell prepares for mitosis. When all is in order, confirmed at the G_2/M checkpoint, mitosis occurs in the M phase. The cell divides into two daughter cells; the nuclei reorganize, cytoskeletal components are rebuilt, and cell membranes are sealed, thus completing the cell cycle.

Checkpoint regulators play a large role in preventing the replication of faulty cells and thus the formation of cancer. Cells utilize a number of brakes at these checkpoints and the master brake is the retinoblastoma (Rb) protein. Rb is present in all cells and works in the nucleus at the G_1/S checkpoint to halt progression. When the cell is ready to proceed, Rb is modified by phosphorylation via the regulatory protein cyclin-dependent kinase (CDK). CDK, as the name implies, requires the protein cyclin for its activation. Phosphorylation of Rb removes the brake by inactivation and allows the cell to enter the S phase. Some tumor suppressor genes also act as brakes, the best known being p53. CDK itself can be inactivated by cyclin-dependent kinase inhibitors (CDKI), which in turn are induced by the protein p53. When a cell recognizes damage to its DNA, the p53 protein is activated, thus indirectly preventing the Rb brake from removal (**Fig. 54.3**).

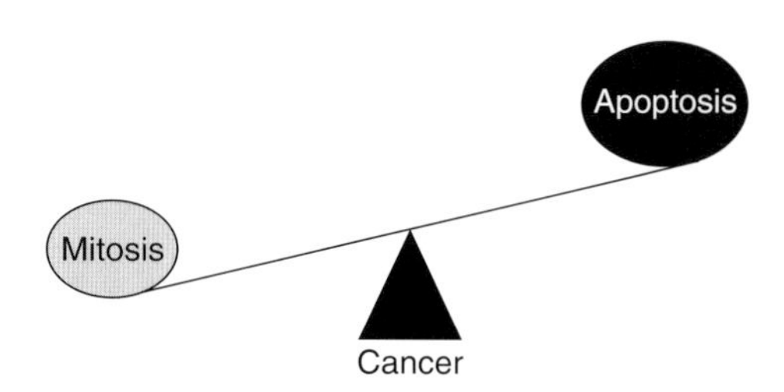

Fig. 54.2. Cancer cells demonstrate an imbalance with mitosis outweighing apoptosis. The improper regulation of cell cycle control and growth factors, activation of oncogenes or inactivation of tumor suppressor genes promote cancer growth by increasing the rate of cell replication or decreasing the rate of cell death.

Programmed cell death (apoptosis)

The decision for a cell to enter programmed cell death, or apoptosis, can be initiated by the detection of irreparable DNA damage. The mechanisms by which cells identify damaged DNA is incompletely described, but may involve the addition of a special polymer to the ends of broken DNA on to which poly-ADP chains are then added. This poly-ADP string sets off a signal that begins the cell death pathway. In addition to DNA damage, cells can be disrupted from the extracellular

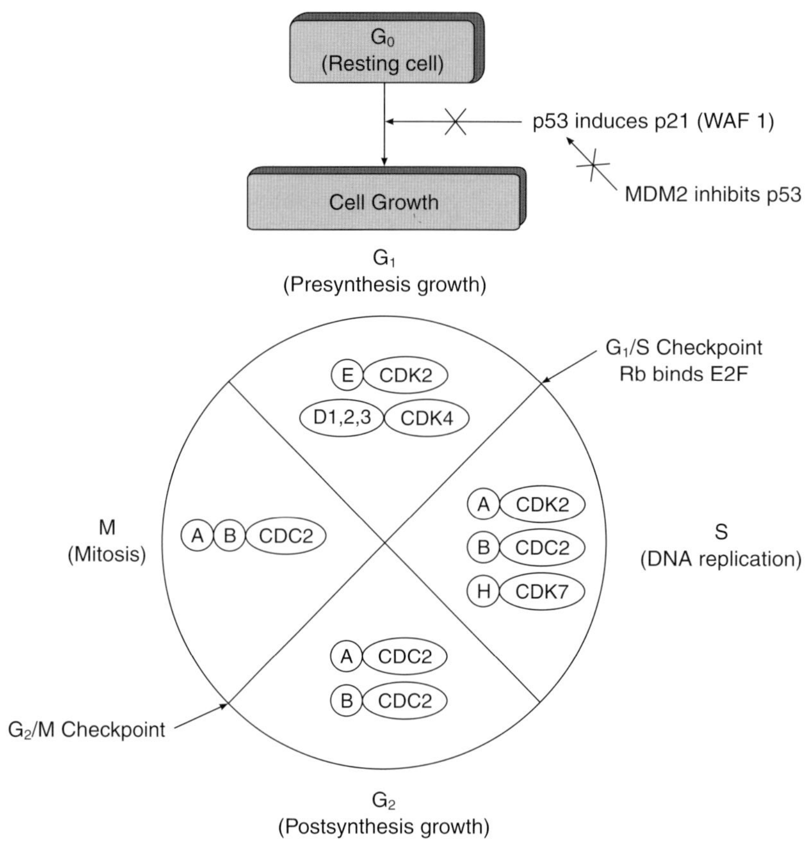

Fig. 54.3. Regulation of the cell cycle depends largely on cyclins (A, B, D, E and H) that are produced and degraded during specific phases of the cell cycle. Cyclins activate cyclin-dependent kinases (CDK2, CDK4, CDK7 and CDC2) that remain at a relatively constant level throughout the cell cycle, resulting in phosphorylation of key regulatory proteins. For example, cyclin–CDK complexes in G_1-phase phosphorylate Rb protein releasing the transcription factor E2F, which then induces proteins required to enter S phase.

Table 54.1. Factors affecting cellular growth and death.

Growth-promoting factors	Growth-inhibiting factors	Death-promoting factors	Death-inhibiting factors
Androgen	Retinoblastoma protein (Rb)	Tumor necrosis factor (TNF)	Bcl-2
Fibroblast growth factor 2	p53	Fas (CD95)	
Fibroblast growth factor 7	p21	Transforming growth factor-β (TGFβ)	
Epithelial growth factor	p16		

matrix anchorage or the internal cytoskeleton may be damaged, which also set a course for cell destruction. Furthermore, there are antagonists to growth factors such as tumor necrosis factor (TNF) and Fas-ligand that can signal the start of apoptosis. Negative growth factors such as transforming growth factor-β (TGFβ) are also known to activate cell death in epithelial cells.

The absence of 'survival factors' may also induce cell death. A notable example in prostate cancer is how the absence of androgen can induce cell death. Similarly, loss of epidermal growth factor (EGF) or fibroblast growth factors (FGF2 and FGF7) will promote cell death. Interestingly, vascular endothelial factor (VEGF), a potent angiogenic factor in many developmental and malignant processes, including prostate cancer, is also a 'survival factor' for tumor endothelial cells.[2] Loss of this factor not only prevents new angiogenesis but also destroys many pre-existing tumor endothelial vessels by this mechanism, resulting in necrosis of the tumor cells previously supplied by these vascular arcades.

Just as there are brakes for cell growth, there are also brakes for cell death. It is these regulators that are manipulated by cancer cells to prevent apoptosis allowing for unchecked tumor growth. Manipulation of these factors, in order to force the cell into a suicidal death, is under active investigation. Moreover, the survival factor B-cell chronic lymphocytic leukemia/lymphoma 2 gene (Bcl-2) was originally identified in follicular cell lymphoma and inhibits apoptosis.[3] The Bcl-2 binding protein bax forms heterodimer with Bcl-2 and is required to remove the brake and allow apoptosis to proceed[4] (**Table 54.1**).

Once positive signals are activated, or negative signals removed, a series of events take place that result in death of the cell. Early events include the induction of the protease interleukin-converting enzyme (ICE), which activates one cascade of events, and the activation of several genes including the receptor for TGFβ and a nuclease that degrades DNA. Nuclear fragmentation ensues, which is the irreversible step, followed by destruction of the nucleus and finally by phagocytosis of the fragmented cell by macrophages and other scavengers.

The DNA itself has a mitotic clock. Normal cells accomplish approximately 50 doublings before they are no longer able to divide. This is due to a quirk in the replication of the DNA; DNA polymerase requires a template for attachment in order to replicate chromosomes. The lagging strand of bidirectional replication is incompletely copied and the cell loses a small portion of the chromosome each division. Therefore, the end of chromosomes consists of a repeating segment of DNA that is not used for encoding messages, called telomeres. The telomeres are designed to be clipped and may be measured experimentally to indicate the age of the cell, akin to rings in a tree trunk. When the telomeres become too short, the cell assumes a pathway to death. Immortal cells, including cancer cells and stem cells, such as bone marrow precursors, intestinal cells and spermatogonia, avoid this problem by expressing an enzyme called telomerase. New telomere length is obtained through the action of telomerase, which synthesizes the DNA repeating segment on its own template. In this way, they avoid a major pathway to apoptosis. As one would expect, telomerase activity is one of the best markers available to delineate cancer cells from differentiated epithelial cells of the same organ, as been shown for prostate cancer cells.[5]

MECHANISMS OF CANCER DEVELOPMENT AND GROWTH OF THE PRIMARY TUMOR

DNA changes

Mutation

If only one of the cells in any given organ fails the checkpoint mechanisms and divides with faulty DNA, the birth of cancer may ensue. One major way DNA is damaged is by reactive oxygen species, such as superoxide, hydrogen peroxide and hydroxyl radicals. These are formed as byproducts of mitochondrial aerobic oxidation incompletely reducing oxygen to water, as well as by carcinogens, ionizing radiation and ultraviolet light. These byproducts are termed free radicals because they lack the ability to be paired; they have the propensity to attack proteins, lipids and DNA within the cell. Specifically, the nucleotide guanine is converted to 8-oxoguanine, which leads to mutations in the DNA as 8-oxoguanine preferentially mispairs with adenine. Cells attempt to repair this error by excising the 8-oxoguanine and replacing it with the proper guanine, excreting 8-oxoguanine in the urine. Since reactive oxygen species are always being created, the cell has a number of enzymes that protect against damage from these species as well as damage from other harmful agents that attack DNA. Antioxidants act as scavengers for reactive oxygen species.

These include the lipid-soluble vitamin E, selenium, lycopene and others, and have been reviewed in Chapter 55.

Methylation

In contrast to changing the composition of the DNA code (mutation), modulation of the DNA in the form of methylation can occur. The pattern of DNA methylation can affect the expression of genes by altering protein interaction with DNA. The state of methylation of the nucleotide cytosine is inheritable and is termed DNA imprinting. Genes have a promoter region that regulates their activity and contains repetitive units containing cytosine termed CpG islands. These islands change the local three-dimensional structure of the DNA containing the gene, causing them to remain in an inactive state. Clearly, if heavy methylation of these CpG islands within the promoters of tumor suppressor genes occurs, as reported in nearly one-half of tumor suppressor genes in cancers, protection against the development of tumors is diminished.[6] The most common genetic change in prostate cancer is hypermethylation of the CpG islands of glutathione-S-transferase π, one of the major enzymes protecting against reactive oxygen species damage.[7] Through this link, one can see how methylation abnormalities can enhance mutation risk.

Prostate cancer genetics

Multiple insults are required to form cancer, as cells have back-up mechanisms of protection. As proposed by Alfred Knudson, an individual inherits two copies of genes and both copies must be affected to induce cancer.[8] While most prostate cancers appear to be sporadic, 10% are inherited. The first reported prostate cancer susceptibility gene is found on chromosome 1 (HPC1).[9] Families that carry the HPC1 gene have multiple affected members and prostate cancer tends to be diagnosed at an earlier age (less than 65). There are other genes that contribute to familial prostate cancer and further characterizations of these abnormalities are active areas of research. Further evidence for a genetic predisposition towards prostate cancer is the difference between racial groups. For example, Black men have a greater incidence of the disease that also tends to present earlier than in White men. However, unlike previously believed, these men do not have more aggressive cancers than their Caucasian counterparts.[10] Sporadic, or non-inherited, prostate cancer then represents 90% of the men with this disease. These men tend to harbor genetic abnormalities, and chromosomes 7, 8, 13 and 17 have all been described to have changes in prostate cancer. Some of the mutated genes include Rb, bcl-2, p53, androgen receptor, PTEN, p16, p27 and ras. For example, the mutation of the tumor suppressor gene p53 is very common in advanced disease, but not frequent in localized prostate cancer and thus serves as a good predictor of tumor behavior.[11,12]

Signal transduction

There is a tremendous amount of communication that occurs between cells within an organ and between different organs. A signal that operates within the cell in which it was made is termed an 'intracrine' factor. Autocrine factors are secreted by a cell and, in turn, signal the very same cell. Cells also signal neighboring cells via a 'paracrine' mechanism and signals can be transported through the circulation to affect distant cells in an endocrine fashion (the latter signals being known as hormones). Nerves can transport signals ('neurocrine') and immune cells can secrete 'cytokines' as signals. Moreover, cells can communicate directly by utilizing pores formed by cell adhesion molecules (CAMs) connected to the cytoskeleton, akin to adjoining hotel rooms. Additionally, cells make contact with extracellular matrix (ECM) by binding integrins that also attach to the cytoskeleton. In fact, what a cell touches largely determines what a cell does, from the embryonic stages of organogenesis to cancer cells trying to attach in foreign sites in the process of 'metastasis'. This is an active area of research and finding ways to modulate cell signaling is a major effort.

As discussed, growth factors affect the regulation of cell growth and death. This is accomplished via interactions with receptors on the cell surface, which leads to cellular events or a cascade of cellular events that will eventually transmit the signal to the nucleus and affect the expression of genes. At least five families of growth factors influence prostate cancer in an autocrine or paracrine fashion: TGFβ, EGFs, FGFs, insulin-like growth factors (IGFs) and platelet-derived growth factors (PDGFs). Additionally, hormones acting in an endocrine manner, such as androgen, can also alter the growth characteristics of prostate cancer.

The TGFβ family contains at least five forms, with the $TGF\beta_1$ form being the only one inappropriately expressed in prostate cancer and benign prostatic hyperplasia (BPH).[13] The usual function of $TGF\beta_1$ is inhibition of normal epithelial cell growth, but it has been shown to be stimulatory in several epithelial cell lines as a function of tumor progression.[14,15] In addition, it can stimulate growth of prostatic supporting stromal cells under some conditions. $TGF\beta_1$ can also encourage tumor progression by promoting angiogenesis and inhibiting the immune response. Prostate cancer is often insensitive to the growth inhibition normally affected by $TGF\beta_1$. Of the FGF family, the FGF-1 and FGF-2 subtypes are able to promote angiogenesis and can be mitogenic in animal models. FGF-2 has been demonstrated to increase expression in highly malignant experimental prostate cancer cell lines as opposed to less aggressive prostate cancer cells.[16] The EGF family includes EGF, TGFα and amphiregulin, all of which are expressed in both prostate cancer and normal prostatic tissue. EGF and TGFα, are mitogens and both interact with the EGF receptor. Significant epithelial changes occur in response to TGFα, including dysplastic changes in the epithelium resembling prostatic epithelial neoplasia (PIN). The role of the IGF axis in prostate carcinogenesis is gaining attention. Elevated plasma levels of IGF-1 predict an increased prostate cancer risk of up to sixfold, though direct evidence that IGF-1 causes the initiation of prostate cancer is lacking. Ongoing work to delineate the role of the IGF axis in prostate cancer development and progression should provide important information on these growth factors and their binding proteins (IGFBPs). Generally, IGFs promote prostate cancer growth while IGFBPs are largely inhibitory.[17] IGFBPs are overexpressed in prostate cancer cells, with the exception of IGFBP-3, while IGFs are usually down-regulated. IGFs along with the IGF receptors and IGFBPs are regulated by androgens, and the IGF axis is involved in the evolution towards androgen independence. Although IGFBPs are predominately inhibitory to prostate cancer cell growth, the enzyme prostatic specific antigen (PSA) is capable of cleaving IGFBP-3, resulting in mitotic activity.[18]

ANDROGEN REGULATION

Androgen regulates both the development of the prostate and many aspects of prostate cancer growth. Eunuchs do not get prostate cancer, therefore, demonstrating that this hormone is necessary for the development of this disease. Androgens have been shown to increase the level of oxygen free-radical formation in androgen-dependent LNCaP cells but not androgen-independent DU145 cells by altering the balance between pro-oxidant and antioxidant states. In large part this is secondary to increased mitochondrial activity in response to mitogenic stimulus and can be inhibited by antioxidants such as ascorbic acid.[19]

Prostate cancers are initially composed of a majority of cells that require androgen for proper growth and thus lack of androgen can cause these cells to die. The Nobel Prize was awarded to Charles Huggins in 1966 for this discovery, which led to present day hormonal therapy. However, the larger and more advanced the cancer is prior to chemical or surgical castration, the lower the response rate, namely the number of cells that are going to die and thus tumor shrinkage in response to androgen withdrawal. In addition, after castration, the cancer continues to grow, albeit more slowly, and during this time often becomes androgen independent (AI), no longer requiring androgens for proliferation.

The androgen receptor (AR) is encoded on the X chromosome and is responsible for binding androgen and taking it to the nucleus, where it affects a cell response. The action of androgen on target cells is mediated through the androgen receptor (AR), which exists in the cytoplasm in an inactive hypophosphorylated form in complexes with heat shock proteins (hsp). Androgen enters the cell and binds the ligand-binding domain of AR, resulting in dissociation with heat shock proteins and phosphorylation of AR. These effects result in conformational changes in AR producing ligand-activated ARs that form homodimers and are translocated into the nucleus. In the nucleus, AR dimers associate with the androgen-response elements (AREs) in the promoters of hundreds of target genes thereby regulating transcription.[20] A number of coactivator proteins, such as TIF-2 and GRIP-1, and corepressor proteins bind the AR-DNA complex and influence gene transcription. The AR is up-regulated after androgens are withdrawn by both surgical and pharmacologic castration. Complicating the story is the ability of AR to respond to peptide growth factors even in the absence of androgen. IGF-I, EGF and keratinocyte growth factor (KGF) have been demonstrated to activate AR in this manner.[21] During tumor progression and development of the AI phenotype, spontaneously occurring mutations in the AR can lead to promiscuous activation of this receptor by progesterone or antiandrogen medications used in the initial treatment of this cancer. This may allow estrogens or antiandrogen therapy to activate the AR, perhaps explaining the improvement seen in some patients after withdrawing antiandrogen therapy.[22] The frequency of androgen receptor mutations in AI prostate cancer is unknown, although an increase in the X-chromosome region containing AR has been demonstrated in some men unsuccessfully treated with hormonal therapy. Thus, it is possible that, instead of becoming insensitive to androgen, prostate cancer may become supersensitive or even respond to non-androgen steroid hormones. In addition to mutations in the AR, androgen-independent prostate cancers demonstrate a heterogeneous loss of AR expression in 20–30% of tumors. Regulation of AR occurs at least in part through DNA methylation of a specific region of the minimal promoter, similar to silencing due to methylation in many other genes.[23,24] The androgen receptor, therefore, certainly plays an important role in the subset of prostate cancers that are hormone refractory.

Vascular endothelial growth factor is one of the major regulators of angiogenesis in prostate cancer and is down-regulated by antiandrogen therapy. In both *in vitro* and *in vivo* mouse xenograft human prostate cancer models, androgen withdrawal results in decreased VEGF expression and reversal of neovascularization in xenografts. This was found to be an early event in the tumor response to therapy.[25] Furthermore, complete androgen blockade before radical prostatectomy is demonstrated to down-regulate the expression of VEGF and decrease vascularization in tumor specimens, except in areas with neuroendocrine (NE) features.[26] These data suggest that the acute response to androgen withdrawal resulting in decreased vascularization could inhibit primary tumor growth and prevent the formation of micrometastasis.

PROGRESSION OF DISEASE

Two histologic lesions are considered malignant precursors of invasive prostate cancer. Prostatic intraepithelial neoplasia is the histologic abnormality most commonly associated with prostate cancer and is proposed as a precursor of invasive prostate cancer. This lesion is the prostate equivalent of 'carcinoma *in situ*' in other cancers. PIN is segregated into high-grade (HGPIN) and low-grade (LGPIN) classifications, and it is the HGPIN variety that is considered to be the most likely precursor to invasive carcinoma.[27] Autopsy studies indicate a relationship between HGPIN and prostate carcinoma with HGPIN development occurring as early as the fourth decade of life and 5–10 years prior to clinically detectable cancer.[28] In addition, HGPIN is frequently found in conjunction with carcinoma in radical prostatectomy specimens and reported to be present in up to 86% of specimens in one study.[29] Furthermore, the present of HGPIN alone on prostate needle biopsy portends a higher likelihood of adenocarcinoma and is an indication for repeat biopsy.[30] What is still unclear is the exact risk of harboring invasive cancer in such a situation. As early as 1954 Franks proposed that epithelial hyperplasia was a precursor to prostatic carcinoma.[31] Ten years later, McNeal described PIN ('atypical hyperplasia') and published the classic paper associating PIN and adenocarcinoma in 1986.[32,33] Prostate carcinoma and PIN share multifocality and zonal similarities, and a majority of carcinomas have adjacent HGPIN within 2 mm.[34] Atypical adenomatous hyperplasia (AAH) has characteristics that are intermediate between BPH and low-grade carcinoma, but has far less evidence for acting as a precursor than HGPIN.[35]

METASTASIS

Metastatic sites

Metastasis is the battle line in prostate cancer, as localized prostate cancer can often be effectively treated while metastatic prostate cancer is currently incurable. Prostate cancer grows at varying rates, but most men diagnosed with metastatic prostate cancer die over a period of months to years.[36] Therefore, the forefront of prostate cancer research is dedicated to preventing the development or curing patients with metastasis. Understanding the changes that occur at the molecular level during the progression to the metastatic phenotype is a necessary precursor to the development of novel therapeutic approaches. Prostate most frequently metastasizes to lymph nodes (pelvic and abdominal) and to bone (60–70% of patients suffering metastasis having bony involvement) with the spine, pelvis, sternum, ribs and femurs being the most common skeletal sites.[37] Lymphatics originating from the prostate travel to the pelvis (iliac lymph nodes) upwards retroperitoneally along the great vessels towards the heart.

The first report of prostate cancer bone metastasis was made by Thompson in 1854 and is responsible for much of the morbidity associated with this condition.[38] Batson first described the venous plexus that drains the prostate and proposed a mechanical mechanism of prostate cancer seeding to the spine.[39] In contrast, Paget proposed the 'seed-and-soil' theory for metastasis, which highlights the importance of host factors at the metastatic site ('soil') that determines the preferential growth of metastatic prostate cancer cells ('seed') to sites such as bone.[40] The former hypothesis is no longer thought to be valid.

Development of metastasis

The steps in the progression to metastasis include angiogenesis, cell attachment, invasion (basement membrane degradation), migration to a suitable environment and proliferation.[41] Each of these steps is regulated and is, therefore, a potential target for molecular treatment. Interfering with any or all of these steps should impede the ability of cancer cells to metastasize.

As tumors grow, they require new blood vessels from which they obtain oxygen and nutrients, without which they are unable to grow beyond 2 mm.[42] Furthermore, the addition of new vessels increases the odds for tumor cells reaching the circulation through which they may metastasize. In particular, prostate cancer has a negative correlation between progression and the degree of angiogenesis independent of Gleason score.[43] Angiogenesis can be divided into several steps: proliferation of endothelial cells; breakdown of ECM; migration of endothelial cells toward the chemotactic angiogenic stimulus (such as VEGF); and, finally, tube formation followed by blood circulation through the lumen.[44] Studies suggest that tumor cells release soluble factors that induce an angiogenic response.[45] Angiogenic factors include VEGF, members of the FGF family, angiogenin, TNFα, IGF-I, hepatocyte growth factor (HGF) and others. Members of the TGFβ family have been shown to have both angiogenic and antiangiogenic effects depending on the systems studied, probably by affecting different tumor and host populations. There are other inhibitors of angiogenesis and modification or removal of these factors is a future therapeutic possibility. These inhibitors include interferon-α (IFNα), platelet factor-4, thrombospondin, angiostatin, pigment epithelium-derived factor and tissue inhibitors of metalloproteinases (TIMPs). Angiogenesis is a highly regulated event balancing proangiogenic and antiangiogenic factors.

Cells within organs are generally attached to a foundation (ECM) and cancer cells must disassociate with the ECM in order to escape the primary tumor. Integrins are membrane proteins that bind a variety of ECM molecules including laminin, fibronectin, vitronectin and collagens. Furthermore, integrins have a crucial role in the attachment of tumor cells to ECM.[46] The vitronectin receptor ($\alpha_v\beta_3$ integrin) is involved in the bone attachment mechanism of osteoclasts and is important in tumor metastasis to bone.[47] Cadherins are proteins that anchor cells to one another. For example, epithelial cadherins (E-cadherins) link to E-cadherins on adjacent cells through interactions with the catenin family of proteins (such as β-catenin). E-cadherin function is often lost during progression of many cancers, including prostate cancer.[48] Additionally, mutations in β-catenin occur in prostate cancer, or the promoter for E-cadherin can be hypermethylated, and E-cadherin can be disrupted by the matrix metalloproteinase (MMP) family member stromelysin-1.[49] Other types of cell adhesion actually promote metastasis; these include immunoglobulins and vascular cell adhesion molecule-1, which as the name implies is involved in cell attachment to vascular endothelium.

The ECM forms a barrier through which cancer cells must traverse to escape the primary tumor. Invasion involves proteolysis of the ECM, pseudopodial extension and cell migration.[50] The matrix metalloproteinases are a zinc binding family of proteins that disrupt ECM components; they are secreted in a proenzyme form and must be activated in the extracellular spaces. There are three classes of these molecules: interstitial collagenases, stromelysins and gelatinases (type IV collagenases). Metalloproteinases inhibition occurs by the action of TIMPs. Tumor cell motility is stimulated by hepatocyte growth factor/scatter factor (HGF/SF), IGF-II and autotaxin (ATX). Additionally, ECM proteins vitronectin, fibronectin, laminin, type I collagen, type IV collagen and thrombospondin promote motility by chemotaxis or through interactions with integrin receptors.[51] Host-secreted factors, sometimes called homing factors, cause tumor cells to move towards the organs in which they are produced and include IGF-I, interleukin-8 (IL-8) and histamine.[41]

Tumor establishment at the metastatic site occurs under the influence of paracrine and autocrine growth factors. However, as tumors progress and become increasingly malignant, they become decreasingly dependent on exogenous factors for growth. In the case of bone metastasis, bone marrow-derived growth factors include TGF-β, IGF-I and IGF-II. Furthermore, osteoblastic lesions arising from prostate cancer have increased growth in response to basic fibroblast growth factor (bFGF) secreted by osteoblasts.[52] IL-6 is another stimulatory factor secreted by osteoblasts and prostate cancer cells may respond to this interleukin as they express the IL-6 receptor.[53]

MOLECULAR TARGETS IN THE TREATMENT OF PROSTATE CANCER

Targeting cellular processes: rationally based therapeutics

As mentioned above, androgen withdrawal is the only effective treatment modality for advanced prostate cancer and will provide an objective response in the majority of patients. Unfortunately, progression to androgen independent disease, often causing death, occurs in many of these cases within a few years.[54] Hormone refractory prostate cancer (HRPC), therefore, is the main cause of the demise of patients with advanced disease. Until very recently, strategies that involved cytotoxic agents were the mainstay of investigative efforts in prostate cancer research. Despite the tremendous efforts to find an effective combination of chemotherapeutic agents to combat prostate cancer, results have been disappointing. Although the primary goal in the management of prostate cancer should be prevention of clinically significant disease, by alterations in lifestyle and diet, for example, the likelihood of significantly affecting these changes is low. Therefore, research efforts are needed to aim at all phases of prostate cancer, from prevention to treatment of advanced localized disease to management of metastatic disease. The promise of molecular therapeutics resides with the potential of increased efficacy and decreased morbidity for all stages of prostate cancer by virtue of increased specificity.

A paradigm for translation of basic science research and therapeutic development may be found in the discoveries leading to the effective use of imatinib mesylate (Gleevec™) in leukemia.[55] After 30 years of basic science research, imatinib mesylate was the first drug approved by the Food and Drug Administration (FDA) that directly inhibits a protein known to cause cancer.[56] The FDA approved imatinib mesylate in a very quick two and one-half years (a process which often lasts 10 years) primarily because the drug was extremely effective in phase I and phase II trials, but also because the basic science research provided a clear rationale for this treatment. Genetic studies identified a translocation event in chronic myeloid leukemia (CML) patients resulting in the Philadelphia chromosome (Ph+). Biochemical studies then demonstrated that in Ph+ patients, production of a Bcr-Abl protein-tyrosine kinase fusion protein causes CML. This abnormality in the constitutively active Bcr-Abl protein-tyrosine kinase was targeted for therapeutic intervention and led to the discovery that the protein-tyrosine kinase inhibitor protein imatinib mesylate can constrain Bcr-Abl *in vitro*, and leads to inhibition of tumor growth and induction of apoptosis *in vivo*.[57] In phase I trials, 53 of 54 patients treated with a dose of 300 mg or greater had a complete hematologic response.[58] Additionally, in advanced disease, a 55% response rate (19% complete response) was observed in CML patients with myeloid blast crisis and a 70% response rate (20% complete response) in acute lymphocytic leukemia (ALL) patients harboring the Philadelphia chromosome combined with patients suffering a lymphoid blast crisis.[59] The rational stepwise research approach resulting in imatinib mesylate approval serves as a model to cancer investigators and accentuates the value of careful characterization of tumor mechanisms before therapeutic investigation.

Chemoprevention

As mentioned above, generation of free radicals has significant potentially detrimental effects on cellular processes that may lead to carcinogenesis and tumor progression. The epidemiology of prostate cancer is incredibly different in Asian countries as compared to Western Europe and the USA. Additionally, men who migrate from Asia to the USA acquire an epidemiologic profile similar to other Americans.[60] Although the reason for these changes is not completely delineated, dietary factors, particularly the ‘Western diet’ high in saturated fats, have been touted as a positive risk factor of prostate cancer incidence and mortality.[61] A putative protective effect of soy proteins, particularly the isoflavones, has been postulated.[62] Furthermore, selenium also has been postulated to be a protective agent, although more studies are needed to confirm this hypothesis.[63] A comprehensive discussion of this subject is presented in Chapter 55.

The antioxidant lycopene, present in tomatoes, was shown in the Health Professionals Follow-up Study (HPFS) to lower the risk of developing prostate cancer.[64] Oxidative stress, in combination with faulty cellular defenses, has been suggested from both epidemiologic and molecular biology studies. The ability to inhibit free-radical damage by modulation of oxidative mechanisms has a promising outlook as a prostate cancer chemoprevention strategy. The aforementioned Phase 2 enzyme GSTP 1 is inactivated in prostate cancer and is not expressed in PIN as well. Replacing the function of this enzyme by gene therapy or agents that induce a large amount of other Phase 2 enzymes may be an effective method of stopping prostatic carcinogenesis or progression of prostate cancer.[65]

Hormonal activity

Dihydrotestosterone (DHT), the 5α-reduced product of testosterone, binds to androgen receptors with 2.5-fold higher affinity than testosterone itself, and serves as the major regulator of prostatic tumor growth. Androgen-mediated cancer proliferation occurs by direct effects of androgens, indirectly by growth factors stimulated by androgens, and by a combination of these mechanisms.[66] Approximately 7000 mg of testosterone is secreted daily by the testes of which only 7% is converted into DHT in peripheral tissues.[66] The testes produce 95% of androgens in men, while the adrenal gland accounts for the remaining 5% of hormone.[67] Testosterone, as well as precursors to androgens, such as androstenedione, dehydroepiandrosterone (DHEA) and DHEA sulfate, originate in the adrenal gland and are likewise converted to DHT in peripheral tissues. Approximately 40% of prostatic DHT originates from steroids of adrenal origin.[68] Androgen-dependent prostate cancer contains androgen receptors that bind DHT and transmit proliferative signals to the nucleus.[69]

The responsiveness of the prostate gland and most prostate cancers to androgen indicates the importance of this hormone, as well as the ability to manipulate its action, in prostate cancer treatment. Many abnormalities in AR expression occur during prostate cancer progression and the AR gene is amplified in 30% of hormone refractory prostate cancers and multiple copies of chromosome X are found in 20% of such tumors.[70,71] Additionally, mutations of the androgen receptor occur, but the frequency in primary prostate cancer is controversial.[72] An early study found a 30% incidence of androgen receptor mutations in primary prostate cancers, while others have found a much lower frequency ranging from 0% to 5%. However, all investigators find mutations in metastatic disease ranging from 21% to 50%.[73] The frequency and type of mutations appear to be influenced by selective pressure exerted by antiandrogens.[74] As previously discussed, these mutations may be the reason we see the flutamide withdrawal response.[75] Drawing a parallel with breast cancer, in which estrogens are known to play an important role and the antiestrogen tamoxifen has been shown to decrease risk, the Prostate Cancer Prevention Trial (PCPT) is studying 18 000 men to determine if the 5α-reductase inhibitor finasteride will effect the development of clinically significant prostate cancer.[76,77] In time, the PCPT should provide insight regarding the efficacy of finasteride on preventing invasive prostate cancer.

In contrast to breast cancer, where estrogen receptor (ER) and progesterone receptor (PR) expressions are lost in hormone refractory disease, AR mRNA is upregulated *in vitro* in androgen-independent prostate cancer cell lines.[78] Furthermore, *in vivo* studies have demonstrated high levels of AR expression as well as increased expression of androgen-regulated genes in castrate versus hormonally intact human prostate cancer xenografts.[79] The maintenance of androgen-regulated genes in the absence of androgen could occur via an AR-independent mechanism; however, the importance of androgen-independent activation of AR itself is becoming increasingly recognized. In the absence of androgen, cytosolic AR can be phosphorylated and activated by alternative kinase pathways. For example, the protein kinase A (PKA) activator, forskolin, was shown to activate AR *in vitro* in the absence of androgen; this effect could be blocked by a PKA inhibitor protein and partially blocked by the competitive inhibitors flutamide and bicalutamide. The authors demonstrated that AR activation in this manner is dependent on a functional AR DNA-binding domain by mutational studies.[80] Other studies have shown androgen-independent activation of AR involving mitogen-activated protein (MAP) kinase, HER-2/neu receptor tyrosine kinase and cyclin-dependent kinases.[81–84] A separate mechanism of androgen-independent AR activation occurs through direct binding of growth factors to cytosolic AR. As previously mentioned, IGF-1, KGF and EGF all are capable of activating the AR in the absence of androgen.[21] FGF-1 and FGF-2 however, were shown to lack this ability *in vitro*.[85] Interestingly, blockade of the EGF receptor stimulated pathway with the specific inhibitor of PKA (H89) in DU145 cells was found to inhibit not only the action of EGF on the MAP kinase system, but also IGF-1 activation of MAP kinase as well as the interaction between the kinase pathways PKA and MAP kinase.[86] PKA pathway inhibition with H89 in DU145 cells has also been shown to abolish the neuropeptide calcitonin, which is secreted in neuroendocrine variants of androgen-independent tumors, mediated activation of MAP kinase.[87] Similarly, the epidermal growth factor receptor (HER)-2/neu inhibitor tyrphostin AG825, a cell-permeable tyrosine kinase inhibitor, preferentially induced apoptosis in androgen-independent C4-2 cells but not androgen-dependent LNCaP cells.[88] This complex and convergent activation of AR, along with mutations in AR that allow activation by antiandrogens, likely plays an important role in androgen-independent mitogenic stimulation by AR in hormone refractory disease. Strategies inhibiting AR activation in the absence of androgen could be utilized to reduce autonomous tumor growth. A monoclonal antibody against AR has been developed (F52.24.4) against the C-terminal portion of the DNA binding domain.[89] It may be possible that disruption of AR interaction with DNA by such an antibody could inhibit the penultimate step in AR-responsive genes. Similarly, any strategy to knockout AR in hormone-independent prostate cancers may prove to alleviate the mitogenic stimulus by AR in these cancers. Otherwise, delineating the relative importance of the aforementioned mechanisms of AR activation could narrow the approach to designing treatment alternatives.

Angiogenesis

Agents that have the ability to impede a tumor's ability to form new blood vessels will decrease the size to which a tumor may grow and perhaps inhibit a tumor's ability to metastasize by decreasing the number of vessels to which it has access. The ability of a tumor to grow beyond 2 mm in diameter depends on both tumor cell proliferation and inducing the growth of new capillary blood vessels from the host, a process called angiogenesis. This occurs via secretion of soluble factors such as VEGF. This may permit a tumor to expand by as much as 1000–16 000 times its primary volume in weeks or months.[45] Thus, no matter how strong a growth stimulus a cancer cell has, further tumor growth does not occur unless a tumor becomes vascularized.[90]

During angiogenesis, endothelial cells move from a resting state to one of rapid growth when exposed to diffusible factors secreted by tumor cells. VEGF was the first selective angiogenic growth factor to be purified, and is still a pre-eminent molecule in this area. Many human tumor biopsies exhibit enhanced expression of VEGF by malignant cells and VEGF receptor in adjacent endothelial cells. Abrogation of VEGF function with monoclonal anti-VEGF antibodies results in complete suppression of prostate cancer-induced angiogenesis, and prevents tumor growth beyond the initial prevascular growth phase.[91] Tissue staining has demonstrated that human prostate cancer is positive for VEGF, while BPH and normal prostate cells displayed little VEGF staining and vascularity.[92] Other studies have demonstrated that castration inhibits prostate cancer VEGF production, but had no effect on other angiogenic factors.[93] Since surgical or chemical castration is a mainstay of prostate cancer therapy, this finding would suggest that VEGF plays an important role in this process. Additionally, increased VEGF expression has been related to neuroendocrine differentiation in prostate cancer, a known poor prognostic factor for survival.[94,95] Taken together, these data suggest that the prostate tumor growth advantage conferred by VEGF expression appears to be a consequence of stimulation of angiogenesis. One therapeutic design might include utilizing antisense-VEGF cDNA with gene therapy to disrupt angiogenesis in prostate cancer.

The various VEGF forms bind to two tyrosine-kinase receptors, VEGFR-1 (flt-1) and VEGFR-2 (KDR/flk-1), which are expressed almost exclusively in endothelial cells. Endothelial cells express in addition the neuropilin-1 and neuropilin-2 coreceptors, which bind selectively to the 165 amino acid form of VEGF (VEGF165). Fetal liver kinase 1 (Flk-1) receptor tyrosine kinase associates with VEGF as a high-affinity ligand and is suggested to have a major role in angiogenesis.[96] Analysis of VEGF and Flk-1 receptor expression in benign prostate glands, PIN and prostatic carcinomas of different Gleason scores, was performed on 21 radical prostatectomy specimens. In all benign glands, VEGF and Flk-1 expression were confined almost exclusively to the basal cell layer; PIN labeling was no longer confined to the basal cell layer, but also was seen in all neoplastic secretory cells. All carcinomas stained positive for both markers and there was a trend for increasing labeling intensity with increasing cellular dedifferentiation.[97] The drug SU5416 decreases Flk-1 phosphorylation and inhibits vascular endothelial growth factor (VEGF)-driven neovascularization. Phase I and phase II trials of SU5416 in patients with intermediate to hormone-refractory prostate cancer are currently active. In addition, a small-molecular-weight inhibitor of KDR and Flt-1 and compatible with chronic oral administration was recently developed.[98] ZD4190, a substituted 4-anilinoquinazoline, is a potent inhibitor of VEGF-stimulated human vascular endothelial cells (HUVEC) proliferation *in vitro*. Chronic once-daily oral dosing of ZD4190 to mice bearing established human tumor xenografts (breast, lung, prostate and ovarian) elicited significant antitumor activity and at doses that would not be expected to have any direct antiproliferative effect on tumor cells. Prolonged tumor cytostasis was further demonstrated in a PC-3 xenograft model with 10 weeks of ZD4190 dosing, and upon withdrawal of therapy, tumor growth resumed after a short delay consistent with its purported effect on angiogenesis.

The fungus *Aspergillus fumigatus* secretes an antibiotic fumagillin and, along with its synthetic analog TNP-470, inhibits endothelial cell growth *in vitro* but is not toxic.[99] TNP-470 has been found to be effective *in vivo* against renal tumors, rhabdomyosarcomas and hepatomas.[100–102] Additionally, PC-3 cell xenograft tumor growth is inhibited by TNP-470, which is also synergistic when used with cisplatin despite the fact that PC-3 cells in monolayer culture were insensitive to this agent.[103] Studies of the transcriptional activation of androgen receptor and PSA in prostate cancer cells *in vitro* revealed a 1.2-fold and a 1.4 induction with TNP-470, respectively, indicating that using PSA as an end-point in clinical trials might be misleading.[104] A phase I trial of TNP-470 in 33 patients with metastatic and androgen-independent prostate cancer has been completed. Dose escalation was performed and the dose-limiting toxic effect was a characteristic neuropsychiatric symptom complex that resolved after discontinuation of the drug. The authors report no definite antitumor effect in their trial.[105] Recently, a CKD-731 analog has been developed and is reported to have 1000-fold more inhibition of endothelial cell growth than TNP-470.[106] Fumagillin analogs show promise and may provide the specificity needed to suppress endothelial mitogens without being prohibitively toxic.

Platelet factor 4 (PF4) is a chemokine derived from the precursor β-thromboglobulin. It is produced in megakaryocytes and platelets, and released from *alpha* granules in activated platelets. PF4 release from platelets results in its rapid binding to endothelial cells, where it is then released by heparin in a time-dependent manner. Immunological functions of PF4 include chemotaxis for monocytes and neutrophils, promoting neutrophil attachment to endothelial cells, and activating neutrophils causing degranulation. PF4 also has procoagulation properties: neutralizing the anticoagulatory activity of heparin sulfate in the extracellular matrix of endothelial cells, inhibiting local antithrombin III activity, and accelerating the formation of blood clots after injury.[107] An additional potential therapeutic action of this molecule is its ability to inhibit collagenase *in vitro*, though no *in vivo* confirmation of this effect has been reported.[108] A recombinant form of human PF4 (rHuPF4) inhibited blood vessel proliferation in the chicken chorioallantoic membrane in a dose-dependent manner and *in vitro* studies suggested that the angiostatic effect was due to specific inhibition of growth factor-stimulated endothelial cell proliferation. This antiangiogenic effect could be abrogated by adding heparin to the assay.[109] Further studies demonstrated that rHuPF4 inhibited the migration of human endothelial cells *in vitro* and suppressed tumor growth in murine melanoma and human colon carcinoma cell lines *in vivo*.[110] This group then designed another PF4 analog (rPF4-241) that lacked affinity for heparin, but retained antitumor properties and inhibited angiogenesis in the chicken chorioallantoic membrane. Daily intralesional injections of rPF4-241 significantly inhibited the growth of both murine melanoma and human colon carcinoma tumors in mice, but showed no effect on these cell lines *in vitro*, suggesting that the effect is due to inhibition of angiogenesis and not secondary to direct tumor toxicity.[111] The use of platelet factor 4 and its analog has not yet been reported in prostate cancer studies but this molecule is attractive due to its multiplicity of actions, namely antiangiogenesis, procoagulation and immune cell modulation.

Thalidomide was marketed in Europe as a sedative, but was withdrawn 30 years ago because it has potent teratogenic effects that cause stunted limb growth (dysmelia) in humans. *In vitro* data suggested that thalidomide has antiangiogenic activity induced by basic fibroblast growth factor in a rabbit cornea assay.[112] A report on a randomized phase II study of thalidomide in patients with androgen-independent prostate cancer has recently been released. A total of 63 patients were enrolled in the study; 50 patients were on the low-dose arm and received a dose of 200 mg/day, while 13 patients were on the high-dose arm and received an initial dose of 200 mg/day that escalated to 1200 mg/day. A serum PSA level decline of greater than or equal to 50% was noted in 18% of patients on the low-dose arm, but in none of the patients on the high-dose arm. Also, a total of 27% of all patients had a decline in PSA of greater than or equal to 40%, often associated with an improvement of clinical symptoms. Only four patients were maintained for greater than 150 days and the most prevalent complications were constipation, fatigue and neurological disorders. The authors note that the decline in PSA in these patients may be particularly important as preclinical studies showed thalidomide increasing PSA levels.[113]

Endostatin was discovered as an angiogenesis inhibitor produced by hemangioendothelioma, and was determined to be a 20 kDa C-terminal fragment of collagen XVIII. Endostatin was demonstrated to inhibit endothelial proliferation specifically and was found to be a potent inhibitor of angiogenesis and tumor growth. Primary tumors treated with endostatin regressed to dormant microscopic lesions similar to those found in the angiostatin-treated tumors,[114] with immunohistochemistry revealing high proliferation balanced by apoptosis in tumor cells and blocked angiogenesis without apparent toxicity.[115] A transgenic mouse model developed by insertion of an SV40 early-region transforming sequence under the regulatory control of a rat prostatic steroid-binding promoter was used to evaluate the effects of endostatin treatment on spontaneous prostate cancer tumorigenesis. The SV40 Tag functionally inactivates p53 and Rb through the direct binding to these proteins and appears to interfere with cell-cycle regulation. Adenomas develop in about one-third of animals between 6 and 8 months of age and approximately 40% of male mice develop invasive prostate adenocarcinomas by 9 months of age. Mouse endostatin expressed in yeast was administered to mice 7 weeks prior to the expected visibility of tumors. While the authors do not report a decrease in tumor burden as seen with mammary adenocarcinomas in transgenic females with this model, they did prolong their survival time for an additional 74 days.[116] In human patients with prostate cancer, a single nucleotide polymorphism (D104N) may have impaired the function of endostatin in 13 men heterozygous for the polymorphism D104N and 13 men homozygous for the allele men diagnosed with prostate cancer. Serum enzyme-linked immunosorbent assay (ELISA) analysis demonstrated endostatin levels were similar both in carriers and non-carriers of this mutation. The results of statistical analysis predict that individuals heterozygous for N104 have a 2.5 times greater chance of developing prostate cancer when compared with men containing two wild-type endostatin alleles. Based on sequence comparison and structural modeling, this polymorphism in endostatin may inhibit the ability to interact with other molecules.[117]

Interestingly, angiostatin, an internal fragment of plasminogen, is a potent inhibitor of angiogenesis, which selectively inhibits endothelial cell proliferation.[118] When given systemically, angiostatin potently inhibits tumor growth and can maintain metastatic and primary tumors in a dormant state defined by a balance of proliferation and apoptosis of the tumor cells. Angiostatin was identified while studying the phenomenon of inhibition of tumor growth by tumor mass. In the original animal model, a primary tumor almost completely suppresses the growth of its remote metastases. However, after tumor removal, the previously dormant metastases neovascularize and grow. Hence when the primary tumor is present, metastatic growth is suppressed by a circulating angiogenesis inhibitor. Serum and urine from tumor-bearing mice, but not from controls, specifically inhibit endothelial cell proliferation. The activity copurifies with a 38 kDa plasminogen fragment, which was named angiostatin. Human angiostatin, obtained from a limited proteolytic digest of human plasminogen, has similar activities. Systemic administration of angiostatin, but not intact plasminogen, potently blocks neovascularization and growth of metastases and primary tumors.

Supplementing agents of endogenous origin, such as plasminogen, which subsequently is cleaved into angiostatin by proteolysis by tumors, and endostatin may prove useful to reduce primary tumor growth and the establishment of metastasis that requires neovascularization.[115,119,120] It is important to determine appropriate end-points for antiangiogenesis trials as this mode of therapy may inhibit tumor growth with variable degrees of apoptosis. It is, therefore, probable that effective use of these therapeutic interventions would best be used in combination with other treatment modalities.

Invasion

Proximity to blood vessels is paramount to a tumor's ability to reach the circulation, the step to metastasis is attachment and invasion of cells into the vasculature. Attachment of epithelial cells involves several junctional structures including desmosomes and tight junctions. These contacts are mediated by calcium-dependent interactions with the cadherin cell-adhesion molecule family, the classic cadherin being E-cadherin that binds to the cytoskeleton via catenins (e.g. β-catenin). Disruption of the cadherin–catenin complex decreases cell–cell adhesion and low levels of E-cadherin have been associated with a more aggressive phenotype of prostate cancer. Replacing E-cadherin in a rat model deficient in this protein has been shown to decrease the invasiveness of cancer cells.[121] In contrast, CD44 is a protein involved in cell adhesion to the extracellular matrix protein hyaluronic acid (HA) and high cell surface expression correlates with poor outcomes. Forced expression of CD44 variants transform a rat pancreatic carcinoma cell line from non-metastatic to metastatic.[122] Additionally, blocking CD44 with antibodies can inhibit metastatic formation.[123]

Integrins are proteins that interact with the extracellular matrix to initiate signal transduction pathways. This interaction with the extracellular matrix results in focal adhesions, which forms complexes with the cytoskeleton. Focal adhesion kinase (FAK) can autophosphorylate resulting in activation of the mitogen-activated protein kinase pathway, which has been linked to the induction of cell migration.[124] PTEN is a tumor suppressor gene encoding a protein tyrosine phosphatase. PTEN interacts with FAK and is sometimes mutated in prostate cancer; loss of PTEN function results in alterations in the FAK pathway and leads to an invasive phenotype.[125] Inhibition of an integrin-linked kinase-directed PTEN-mutant prostate cancer cell lines towards apoptosis in one report.[126] The integrins are comprised of α- and β-subunits in heterodimers. Normal basal cells of the prostate contain multiple combinations of these subunits to bind ECM proteins, such as laminin receptors ($\alpha3\beta1$ and $\alpha6\beta1$), and an $\alpha6\beta4$ dimer that forms hemidesmosomes that are down-regulated in PIN and prostate cancer specimens.[127] Expression of the integrins $\alpha2$, $\alpha4$, $\alpha5$, αv and $\beta4$ is lost in carcinoma; however, the laminin receptors $\alpha3\beta1$ and $\alpha6\beta1$ are retained even in invasive prostate carcinoma. The predominate laminin receptor is $\alpha6\beta1$ and tumor cells with high levels of $\alpha6$ integrin are more invasive when injected into immuno-deficient mice using a diaphragm invasion assay indicating that $\alpha6$ integrin may confer an invasive phenotype. Effective strategies to combat invasiveness could include inhibiting the presence or function of $\alpha6\beta1$ integrin, blocking the expression or function of laminin, or preventing the loss of $\beta4$ integrin.[128]

Protease inhibitors that inhibit basement membrane proteases are a class of molecules that may impede the tumor cell ability to penetrate the vasculature as the basement membrane forms a barrier. Matrix metalloproteinases are proteins that break down basement membranes, while the tissue inhibitors of matrix metalloproteinases are being studied in many types of cancer as invasion inhibitors. There is some evidence that TIMP-3 may have the additional property of causing apoptosis in some cells.[129] Immunohistochemical studies in human prostate cancer tissues show a correlation between increased levels of MMP-2 and MMP-9, and absence of TIMP-1 and TIMP-2 in higher Gleason sum tumors (8–10) as compared to lower Gleason sum specimens. Additionally, TIMP-1 and TIMP-2 expression was high in organ-confined disease while absent in locally advanced cancers.[130] Several MMP inhibitors, including derivatives of doxycycline and tetracycline, have been shown to be inhibitory to prostate cancer metastasis in model systems. For instance, the tetracycline derivative CMT-3 inhibited both tumor growth and metastasis in a rat model.[131] IL-10 treatment of PC-3ML cell tumors in the severe combined immunodeficiency (SCID) mouse model was an effective inhibitor of spinal metastasis and increased tumor-free survival rates. IL-10 treatment of the PC-3 ML cells and the SCID mice reduced the number of spinal metastases from 70% seen in the natural progression of the model to 5% of the mice. Additionally, following discontinuation of IL-10 treatment after 30 days, the mice remained tumor free and mouse survival rates increased dramatically, from less than 30% in untreated mice to about 85% in IL-10-treated mice. To further delineate the mechanism behind these findings, the authors measured expression of MMPs and TIMPs by ELISA assay in IL-10 treated PC-3ML cells. IL-10 treatment of the PC-3 ML cells down-regulated MMP-2 and MMP-9 while up-regulating TIMP-1, but not TIMP-2, expression. IL-10-treated mice exhibited similar changes in MMP-2, MMP-9 and TIMP-1 expression. Lastly, IL-10 receptor antibodies blocked the IL-10 effects on PC-3ML cells.[132] Alendronate, a potent bisphosphonate compound, has been shown to inhibit $TGF\beta_1$-induced MMP-2 secretion in PC-3ML cells, while TIMP-2 secretion was unaffected. The relative imbalance between the molar stoichiometry of TIMP-2 to MMP-2 resulted in decreased collagen solubilization.[133] Several well-tolerated, orally active MMP inhibitors (MMPIs) have been generated that demonstrate efficacy in mouse cancer models. Marimastat (BB-2516) was the first matrix metalloproteinase inhibitor to have entered clinical trials in the field of oncology and has completed phase I and phase II trials in prostate and colon cancer patients.[134] Marimastat was generally well tolerated in phase I trials and phase II trials used serum PSA as a marker in patients with prostate cancer. The authors reported a 58% response rate (no increase in serum PSA over the course of the study plus partial response defined as 0–25% increase in serum PSA per 4 weeks) using doses of greater than 50 mg twice daily.[135] Other MMPIs have been developed, are in various stages of preclinical and clinical trials, and include inhibitor batimastat (BB-94), Bay 12-9566 and prinomastat (Ag3340).

The urokinase-type plasminogen activator (uPA) likely plays a key role in tissue degradation in both normal and cancerous tissues. Increased expression of uPA has been reported in many cancers, including prostate, and gene amplification has been identified in a portion of hormone refractory prostate cancers and may play a role in the high expression of uPA. Prostate cancer cell lines that contain this gene amplification (i.e. PC-3) are more sensitive to the urokinase inhibitor amiloride as compared to prostate cancer cell lines that lack uPA gene amplification (e.g. LNCaP).[136] Overexpression of uPA by the rat prostate-cancer cell line Dunning R3227, Mat-LyLu, results in increased tumor metastasis to several sites. Histological examination of skeletal lesions has shown them to be primarily osteoblastic. A selective inhibitor of uPA enzymatic activity, 4-iodobenzo(b)thiophene-2-carboxamidine (B-428) used in this model resulted in a marked decrease in primary tumor volume and weight as well as in the development of tumor metastases when compared with controls.[137] In a similar study, a mutant recombinant murine uPA, which retains receptor binding but not proteolytic activity, was made by polymerase chain reaction (PCR) mutagenesis and transfected into the highly metastatic rat Dunning MAT-LyLu prostate cancer cell line. A clone stably expressing uPA was injected into Copenhagen rats, and tumors found in these animals were significantly smaller with fewer metastases than in control animals. Additionally, mean microvessel density in transfected tumors was fourfold lower than that in animals with tumors derived from the control tumor cell line.[138] These studies demonstrate that uPA-specific inhibitors can decrease primary tumor volume and invasiveness as well as metastasis in a model of prostate cancer. To determine the effect bone cells have on prostate cancer cell expression of basement membrane degrading proteins, serum-free conditioned medium harvested from osteoblast cultures was used to stimulate the *in vitro* chemotaxis of prostate cancer cells and invasion of a reconstituted basement membrane (Matrigel). This enhanced invasive activity was due to osteoblast cell-conditioned media stimulated secretion of uPA and matrix MMP-9. Additionally, inhibition of these matrix-degrading proteases by neutralizing antibodies or by inhibitors of their catalytic activity reduced Matrigel invasion, thus demonstrating that factors produced during osteogenesis by bone cells stimulates prostate cancer cell chemotaxis and matrix proteases expression, thus representing potential targets for alternative therapies deterring the progression of prostate cancer metastasis to bone.[139]

Cell–cell interactions and metastasis

Cell–cell interactions

Interactions between prostate cells and stromal cells are important in every aspect of prostate regulation. Beginning with development of the prostate *in utero*, continuing with post-pubertal growth and differentiation, through the development of BPH and prostate cancer later in life, stromal–epithelial interactions are a driving force determining how a prostate behaves. Furthermore, these interactions can accelerate local prostate cancer growth, stimulate distant metastatic tumors and are involved in the development of hormone independence.[140,141] Interactions may occur by multiple means involving communication directly or indirectly. The extracellular matrix forms connections with cells providing one route of cross-talk, otherwise cells communicate directly between one another and by paracrine factors interacting with cellular receptors.

To modulate the integrin receptor-mediated communication with the ECM, a number of molecules may be used. Integrin-specific antibodies, such as integrin $\alpha v\beta 3$ antagonist antibodies has been shown to cause tumor regression, with induction of apoptosis of angiogenic blood vessels.[142] Cell surface peptides attached to chemotherapeutic agents may home to tumor blood vessels; a peptide with an αv integrin binding motif can target tumors and, when attached to toxic agents, can act with some specificity.[143,144] Based on three-dimensional structures of cell-surface receptor molecules, designer molecules can be synthesized to recognize these receptors specifically. Laminin-like peptides designed to inhibit degradation of the $\beta 1$ chain of laminin can promote cell attachment and, although not shown to inhibit metastasis *in vivo*, demonstrate an alternate strategy for fighting metastatic spread of tumors.[145] Chemokines, or proinflammatory mediators that control leukocyte migration and up-regulation of adhesion receptors, may be antagonized by synthetic inhibitors again with the goal of stopping metastasis.[146]

Implantation at metastatic sites

In contrast to the mechanisms for invasion into the vasculature, extravasation into organs does not seem to rely heavily on cellular attachment schemes. As previously noted, the milieu of potential metastatic sites plays a major role in the ability of a cancer cell to multiply, utilize angiogenesis, and avoid the immune system to survive and form metastases. Bone is the best-studied system because of the proclivity of prostate cancer to metastasize skeletally. Prostate cancer cells have osteomimetic properties, which are likely to support metastasis within the bone environment and reciprocal interactions between prostate cancer, and bone stromal growth factors leading to gene expression of osteopontin (OPN), osteocalcin (OC), and bone sialoprotein (BSP) may occur. Furthermore, prostate cancer metastases in the bone are frequently osteoblastic and likely due to the secretion of soluble factors by prostate cancer cells, which stimulate bone production.[147] In the mouse model, the prostate cancer cell line PC-3 localized preferentially to human bone implanted in the hindlegs, specifically to the reconstituted bone marrow cavity. PC-3 cells found in the human bone stained strongly for parathyroid hormone-related protein (PTHrP), TNFα and IL-6, which is consistent with osteoclast recruitment and activity.[148] PC-3 tumors found in the bone have been found to be osteolytic in nature, consistent with the recruitment of osteoclasts.[149] Osteocalcin is expressed in some prostate cancer specimens by reverse transcription PCR and immunohistochemical staining. Expression in transiently transfected prostate cancer cell lines show up-regulation in androgen-independent lines as compared to androgen-dependent lines. Gene delivery with Ad-OC-TK (OC promoter-driven herpes-simplex virus thymidine kinase) was shown to be effective at destroying prostate-cancer cell lines *in vitro* and prostate tumor xenografts *in vivo* in both subcutaneous and bone sites.[150] Characterization of the OC promoter in PC-3 cells shows activation by transcription factors (Runx2, JunD/Fra-2 and Sp-1) that are responsible for the high OC promoter activity in PC3 cells.[151] Despite the observed tendency for prostate cancer metastasis to bone, and some understanding of the osteomimetic properties seen in advanced prostate cancer, characterization of the processes responsible for the prostate cancer–bone interactions requires further development to implicate targets fully for therapy. However, interventions to disrupt these interactions by gene therapy or small designer molecules may provide effective treatment to prevent or treat bony metastasis.

Growth factors, the cell cycle and apoptosis

Growth factors that interact with receptors can be targeted with receptor-specific antibodies attached to therapeutic molecules, such as toxins and radioactive isotopes.[152] This strategy has been used in phase II trials in breast cancer. Miyake et al. have developed antisense oligodeoxynucleotides (ODN) against some genes up-regulated in prostate cancer after androgen withdrawal and in progression to androgen independence. These genes, which have antiapoptotic or mitogenic activity, are felt to confer resistance to androgen withdrawal and cytotoxic chemotherapy. The authors find a delay in the progression to androgen independence by enhancing apoptotic cell death induced by androgen ablation with

ODNs, as well as an additive or synergistic effect with ODNs and chemotherapy in prostate cancer models.[153]

Suramin is an anthelmintic drug that has been used in clinical trials in the treatment of patients with hormone-refractory prostate cancer since the late 1980s. It has been shown to have some efficacy and is commonly employed in combination with multiple treatment modalities, for example, synergistic action has been seen with hydrocortisone, doxorubicin and TNFα.[154] The mechanism of action of suramin is incompletely elucidated though evidence exists to suggest antihormonal and direct antiproliferative effects. *In vitro* inhibition of the growth of PC-3 cells by suramin may be caused, at least in part, by growth factor antagonism (including, but not exclusively bFGF) by the drug.[155–157] A recent Southwest Oncology Group Study evaluated the feasibility of administering a combination of suramin and hydrocortisone in addition to androgen deprivation in 62 patients (59 assessed after the first cycle) with newly diagnosed metastatic prostate cancer. Suramin was administered on a 78-day fixed dosing schedule (one cycle), and treatment were repeated every 6 months for a total of four cycles. Thirty-two (54%) of 59 patients received a second cycle, 13 (22%) of 59 patients received a third cycle, and only five patients (8%) received a fourth cycle. There was one therapy-related death; grade 4 toxicities were noted in 14 patients during first and second courses, and neurotoxicity of grade 3 or higher was observed in 16 patients during the first and second cycles. Overall, only 54% of the patients demonstrated acceptable limits of toxicity. The authors conclude that suramin plus hydrocortisone and androgen deprivation has limited applicability in the treatment of patients with newly diagnosed metastatic prostate cancer. Although response was not an end-point for this study, of the 57 patients who were assessable for response, no patient had a complete or partial response: 30 patients (53%) had stable disease, and 9 patients (16%) had progressive disease. Of the 59 patients, 40 (68%) failed therapy and 34 (58%) patients were dead, including one early death.[158]

The epidermal growth factor receptor interacts with TGFα secreted by androgen-independent prostate cancer cells to stimulate growth by an autonomous feedback loop. Phosphorylation and activation of EGFR in androgen-independent PC-3 and DU145 cells was decreased with anti-EGFR antibody resulting in inhibited proliferation. Additional studies showed that anti-EGFR enhanced the sensitivity of PC3 cells to the cytotoxic and cytostatic effects of TNFα.[159] Examination of the mechanisms involved in EGFR stimulated proliferation demonstrated that treatment with mAb225 (anti-EGFR) induced G_1 cell-cycle arrest. This was accompanied by a marked decrease in cyclin-dependent kinase 2, cyclin A and cyclin E-associated histone H1 kinase activities, as well as a sustained increase in cell-cycle inhibitor p27KIP1 by both transcriptional and translational activation.[160] Utilizing an *in vivo* system, treatment with C225 (anti-EGFR) alone or in combination with doxorubicin significantly inhibited tumor progression of well-established DU145 and PC-3 xenografts in nude mice.[161] Using XenoMouse technology, ABX-EGF, a human IgG2 monoclonal antibody, has recently been developed. ABX-EGF binds specifically to human EGFR with high affinity and blocks the binding of both EGF and TGFα; a phase I trial has been initiated in patients with advanced prostate cancer.[162]

Small bioactive peptides offer another promising approach to the problem of HRPC. The G-protein coupled peptides bombesin/gastrin-releasing peptide (GRP), endothelin-1 (ET-1) and neurotensin have been shown to have significant effects in prostate cancer.[163] Bombesin/GRP secretion by neuroendocrine-type prostate cancer cells may be partially responsible for progression, androgen independence and hence a poor prognosis.[164] Synthetic bombesin/GRP receptor antagonists have been shown to inhibit the growth of androgen-independent prostate cancer cell lines and are undergoing clinical trials.[165] Somatostatin analogs interact with receptors in prostate cancer and one analog significantly reduced androgen-independent prostate tumor growth in mice and had some effect of stabilizing disease in men with advanced prostate cancer.[166]

Cell-cycle inhibitors and agents that disrupt cell division are another strategy for potential utilization against prostate cancer. Most of these agents are in the preclinical stage of investigation. They include inhibitors of microtubules, which are critical for segregation of chromosomes during mitosis, tyrosine kinase inhibitors, cyclin-dependent kinase inhibitors, DNA synthesis inhibitors and thymidylate synthase inhibitors. Because tumor cells generally replicate more quickly than normal somatic cells, these agents could provide some specificity towards cancer, with a goal to restore normal life span to cancerous cells.

Estramustine phosphate, a nitrogen mustard derivative of 17 β-estradiol, has been used for treatment of prostate cancer patients since the 1960s. Mechanisms of action include inhibiting microtubule assembly and promoting disassembly of polymerized microtubules by interacting with microtubule-associated binding proteins, in addition to decreasing serum testosterone levels.[167,168] Cytotoxic effects derive mainly from microtubule polymer inhibition required in the formation of the spindle pole apparatus during mitosis. Multiple agents that inhibit microtubule polymers have been used together and colchicine with estramustine phosphate is cytotoxic *in vitro* in PC-3 cells.[169] Clinically, estramustine phosphate has been used in combination with another microtubule inhibitor, vinblastine, without additive side affects and modest results were observed.[170] One video microscopy study suggests that estramustine phosphate mostly stabilizes microtubules in an attenuated state, perhaps explaining its additive effect with more direct microtubule depolymerizing agents, such as taxol and vinblastine.[171] Estramustine phosphate in combination with chemotherapeutics and other microtubule polymer inhibitors have been evaluated in many clinical trials. The results have been modest but encouraging (PSA decreases greater than 50% in 25–75%, which is usually predictive of symptomatic improvement) and estramustine continues to be a reasonable treatment option in hormone-refractory advanced prostate cancer.

Tyrosine kinase (TK) receptors are the major form of growth factor receptors in cells, regulating cell proliferation, cell differentiation and signaling processes. Tyrosine kinase inhibitor, RG-13022 (tyrphostin), was demonstrated to inhibit TGFα phosphorylation of the EGFR and proliferation in LNCaP and PC-3 cells as well as stimulation by EGF itself. Inhibition of androgen-stimulated growth was also seen, suggesting that androgen-induced regulation involves TK pathways.[172] A similar study using an EGF-R selective tyrosine kinase inhibitor, ZM252868, inhibited basal growth in DU145 cells in addition to EGF- and TGFα-stimulated growth. Interestingly, only TGFα-stimulated PC-3 cell growth was inhibited, and the distribution of EGFR by immunohistochemistry varied between DU145 and PC-3 cells, with EGFR being predominately located on the cell membrane and in the cytoplasm, respectively.[173] The high-affinity tyrosine kinase-linked receptor for nerve growth factor, trkA, has been implicated in prostatic cancer growth; the trk tyrosine kinase inhibitor, CEP-751 (KT6587), was demonstrated to inhibit prostatic cancer growth in nine different animal models. Inhibition was independent of the tumor growth rate, androgen sensitivity, metastatic ability or state of tumor differentiation. CEP-751 was found to be selective for cancerous versus normal prostate cells, affected the growth of only a limited number of non-prostate tumors, and induced cell death in a cell cycle-independent fashion implying that it could be used in both slowly and quickly dividing tumors.[174]

Bcl-2 has been the target of several ODN studies; in one such report, LNCaP cells *in vitro* treated with Antisense Bcl-2 (ODN) treatment reduced Bcl-2 messenger RNA and protein levels by >90% in a sequence-specific and dose-dependent manner, and Bcl-2 mRNA levels returned to pretreatment levels by 48 hours after discontinuing treatment. Athymic male mice bearing subcutaneous LNCaP tumors were castrated and injected with Bcl-2 ODN or controls; LNCaP tumor growth and serum PSA levels were 90% lower in mice treated with antisense Bcl-2 ODN compared with mismatch or reverse polarity ODN controls.[175] Antisense Bcl-2 oligodeoxynucleotides after castration have been shown to decrease the progression to androgen independence in the mouse model and androgen-independent LNCaP tumor regression in mice occurred with a novel method of administering paclitaxel with antisense Bcl-2 oligodeoxynucleotide.[176,177]

NOVEL THERAPEUTIC DELIVERY SYSTEMS IN THE TREATMENT OF PROSTATE CANCER

Gene therapy involves delivering recombinant genetic material, in the form of DNA or RNA to combat disease. Major strategies involve modifying gene expression to correct deficiencies or block inappropriate gene expression, inducing 'suicide' genes to promote cancer cell death and modulating the interactions of the immune system with cancer cells. This is performed either by removing tissue and genetically altering it *ex vivo* or delivering the genetic material *in vivo*. *In vivo* gene therapy has been limited largely by inefficient delivery systems and improvements in gene vectors, and delivery will allow practical use of clinical gene therapy for a variety of conditions. The ideal delivery system would be non-toxic to normal cells, deliver the genetic information efficiently with specificity to the targeted system and be inexpensive yet easily administered.

Tumor cell vaccines are a typical *ex vivo* gene therapy strategy for cancer and rely on the ability of cancer-specific antigens to elicit an immune response. An approach is to harvest tumor cells from the patient, genetically modify the cells (usually with retroviral transfection) so that they can stimulate the immune system, irradiate the cells so that they are non-tumorigenic, and reinject the cells into the patient. The first tumor vaccine trial for prostate cancer began in 1994 when granulocyte-macrophage colony-stimulating factor (GM-CSF) was transfected using retrovirus into a patient's prostate cancer cells *in vitro*. The cells were then injected subcutaneously and found to be safe in phase I trials.[178] Data from this trial indicate vaccination activated new T-cell and B-cell immune responses against PCA antigens. T-cell responses, evaluated by assessing delayed-type hypersensitivity (DTH) reactions against untransduced autologous tumor cells, were evident in two of eight patients before vaccination and in seven of eight patients after treatment. These data are the first to suggest that both T-cell and B-cell immune responses to prostate cancer can be generated by treatment with irradiated, GM-CSF gene-transduced prostate cancer vaccines. A limitation in the trial by Simons et al. was generating vaccine cells in culture. This group has embarked on a new trial to estimate the efficacy of using the *ex vivo*, GM-CSF-transduced, prostate cancer cell lines PC-3 and LNCaP as a vaccine. They reported results on 21 patients treated and one had a partial PSA response of greater than 7 months in duration, 14 had stable disease and six patients underwent progression. By 3 months PSA velocity or slope decreased in 71% of cases. Additionally, numerous new postvaccination IgG antibodies were identified in these patients, indicating that immune tolerance to prostate cancer-associated antigens may be broken.[179]

The first report to demonstrate anticancer activity of gene therapy in human prostate cancer, by Herman and colleagues, used intratumoral injection of replication-deficient adenovirus carrying the thymidine kinase gene from the herpes simplex virus, followed by parenteral administration of the prodrug gancyclovir. They report promising results with some patients with local recurrence after radiotherapy obtaining a greater than 50% decrease in PSA lasting 6 weeks to 1 year.[180] These data support the usefulness of cytoreductive gene therapy, by 'suicide' genes, or by increasing tumor cell sensitivity to chemotherapeutics. Replication-deficient adenovirus containing the herpes simplex virus thymidine kinase (HSV-tk) gene ('suicide' gene therapy) has also been used in men with localized prostate cancer. Multiple and/or repeat intraprostatic injections was followed by intravenous ganciclovir or oral valaciclovir 14 days after injection. A total of 52 patients were treated, with a total of 76 gene therapy cycles; toxic events were recorded in 16 of 29 patients (55.2%) who were given multiple viral injections into the prostate, seven of 20 (35%) who received two cycles of 'suicide' gene therapy and three of four (75%) who received a third course of gene therapy. All toxic events were mild (grades 1 to 2) and resolved completely once the therapy course was terminated. Mean follow-up was 12.8 months and preliminary results for 28 patients indicated a mean decrease of 44% in serum PSA in 43% of patients.[181]

A current study is evaluating the effect of p53 replacement therapy utilizing intraprostatic Ad5CMVp53 injection in men with localized prostate cancer prior to radical prostatectomy. Patients are followed for tumor volume response by transrectal ultrasound and magnetic resonance imaging (MRI) and injections are continued every 2 weeks until tumor shrinkage abates, after which time they undergo radical prostatectomy. Results are not available at this time as this trial is still open for enrollment.[182] Logothetis et al. reported initial results using a similar gene replacement strategy of adenovirus containing wild-type p53 (AdCMVp53) driven by the CMV promoter before radical prostatectomy in 17 patients with locally advanced prostate cancer. AdCMVp53 was administered via a transperineal route under ultrasound guidance. In this phase I–II study, three patients who completed a repeat course of therapy had a greater than 25% decrease in tumor size, as measured by endorectal coil MRI after the initial treatment course and there was no grade 3 or 4 toxicity in the 14 evaluable patients. Further results on efficacy due to the apparent radiological responses, correlations with gene expression, pathological findings at prostatectomy and surgical outcomes are pending.[183]

Investigators have recently completed a phase I trial in 24 patients with locally advanced prostate cancer using gene-based immunotherapy. A functional DNA–lipid complex encoding the IL-2 gene was administered intraprostatically, under transrectal ultrasound guidance, into the hypoechoic tumor lesion. Patients were stratified into two groups: group 1 included patients who underwent radical prostatectomy after the completion of the treatment regimen; and group 2 consisted of patients who had failed a prior therapy. IL-2 gene therapy was well tolerated, with no grade 3 or 4 toxic reactions occurring. Post-treatment prostate specimens were attained and compared with the transrectal biopsies performed prior to therapy. Evidence of systemic immune activation after IL-2 gene therapy included an increase in the intensity of T-cell infiltration seen on immunohistochemistry of tissue samples from the injected tumor sites, and increased proliferation rates of peripheral blood lymphocytes that were cocultured with patient serum collected after treatment. Transient decreases in serum PSA were seen in 16 of 24 patients (67%) on day 1. Fourteen of the patients persisted in this decrease to day 8 (58%). However, in eight patients the PSA level rose after therapy. More patients (9–10) in group 2 responded to the IL-2 gene injections and six of the nine also had lower than baseline PSA levels at week 10 after treatment.[184] This group has also reported on novel chimeric PSA enhancers that exhibit increased activity in prostate cancer gene therapy vectors. They addressed the problem of low transcriptionally active native PSA enhancer and promoter, which otherwise confers prostate-specific expression when inserted into adenovirus vectors. The investigators exploited the mechanism by which androgen receptor molecules bind cooperatively to androgen response elements in the PSA enhancer core, and act synergistically with AR bound to the proximal PSA promoter to regulate transcriptional output. They generated chimeric enhancer constructs by inserting four tandem copies of the proximal AREI element, duplicating the enhancer core, or by removing intervening sequences between the enhancer and promoter. They obtained chimeric PSA enhancer constructs, which were highly androgen inducible and retained a high degree of tissue specificity even in an adenoviral vector (Ad-PSE-BC-luc). They also demonstrated that augmented activity of the chimeric constructs *in vivo* correlated with their ability to recruit AR and critical coactivators *in vitro*. Furthermore, systemic administration of Ad-PSE-BC-luc into SCID mice harboring the LAPC-9 human prostate cancer xenografts showed that this prostate-specific vector retained tissue discriminatory capability compared with a comparable cytomegalovirus (CMV) promoter driven vector.[185]

Kwon and colleagues have utilized regulation of T-lymphocyte proliferation by the stimulatory HLA-B7 family with T-cell surface inhibitory receptor CTLA4 in a transgenic mouse model of metastatic prostate cancer growth after tumor resection. They modified their tumor vaccine with the addition of anti-CTLA4 antibodies and showed metastatic relapse dropped from 97.4% in controls to 44% in animals treated immediately after tumor resection.[186] Additionally, anti-CTLA4 antibody treatment can delay tumor growth in a transgenic mouse model.[187]

In vivo gene therapies for prostate cancer utilize vectors to deliver genetic information to tissues. Viruses are commonly used and types include adenovirus, retrovirus, adeno-associated virus, herpes virus and poxvirus. Non-viral vectors include liposomes, DNA-coated gold particles, plasmid DNA and polymer DNA complexes. Overcoming the obstacle for gene therapy vector inefficiency is being approached with development of less immunogenic viral vectors and 'stealth liposomes,' which evade the first-pass destruction by the reticular endothelial system. Furthermore, developing more specific therapies by targeting specific cell receptors or using prostate-specific gene promoters (e.g. prostate-specific membrane antigen and human kallekrein II) may allow for lower effective doses and perhaps decreasing the side effects commonly seen with viral vectors.

Correction of genetic alterations of tumor suppressor genes and cancer-promoting oncogenes can target a multitude of genes in the pathways to cancer development or progression. Restoration of one of the tumor-suppressor genes (Rb, p53, p21, and p16) have been shown to alter malignant phenotype, but would have to be delivered to every cancer cell to eradicate the cancer. The retinoblastoma gene has been shown to be mutated in human prostate cancer specimens, and gene therapy has shown some effectiveness in model systems of neuroendocrine and lung tumors.[188,189] Mutations in p53 are reported at a wide range of rates; however, p53 gene therapy suppressed tumorigenesis and induced apoptosis *in vitro*.[190] Furthermore, *in vivo* studies have shown some benefit of p53 gene therapy in prostate tumor models.[191] Similarly, p21 gene therapy has been shown to suppress growth in prostate tumor cells *in vitro* and *in vivo*.[192,193]

A concept that is being expanded is restoration of genes that confer increased chemosensitivity or radiosensitivity to prostate cancer tumors. One study has looked at proliferation of tumor cells after incubation with various combinations of p53 adenovirus (p53 Ad) and chemotherapeutic drugs. The authors found p53 Ad combined with cisplatin, doxorubicin, 5-fluorouracil, methotrexate or etoposide inhibited cell proliferation more effectively than chemotherapy alone in multiple human tumor cell lines, including DU-145 prostate cells. Human tumor xenografts in SCID mice dosed with intraperitoneal or intratumoral p53 Ad with or without chemotherapeutic drugs showed greater anticancer efficacy with combination therapy in four human tumor xenograft models *in vivo*.[194] Another study evaluated gene therapy enhancement of prodrugs and radiation therapy. Prostate tumor cells (PC-3) were transduced with adenovirus containing a fusion gene encoding the *E. coli* cytosine deaminase and herpes simplex virus type-1 thymidine kinase under the control of a cytomegalovirus promoter. PC-3 cells expressing this protein were sensitized to killing by the normally innocuous prodrugs 5-fluorocytosine and ganciclovir. In addition, radiation-induced killing was enhanced in virally infected cells in the presence of the prodrugs.[195] Additionally, some evidence exists that gene therapy may be cooperative with androgen deprivation. Castration was combined with HSV-tk plus GCV using an androgen-sensitive mouse prostate cancer cell line, which led to markedly enhanced tumor growth suppression in both subcutaneous and orthotopic models compared with either treatment alone. An enhanced survival was also observed, in which combination-treated animals lived twice as long as controls in the subcutaneous model and over 50% longer than controls in the orthotopic model.[196] A very interesting study utilized a constructed recombinant adenovirus containing *E. coli* cytosine deaminase and herpes simplex virus type 1 thymidine kinase fusion gene under the control of a human inducible heat shock protein 70 promotional sequence. Heating at 41°C for 1 hour induced strong expression of the fusion gene product. Heat-induced expression of the CD-TK protein significantly reduced the survival of PC-3 cells in the presence of both 5-fluorocytosine and ganciclovir prodrugs. These data represent a novel form of gene therapy involving the transduction and regulation of a double suicide gene in tumor cells and may provide a unique application for hyperthermia in cancer therapy.[197] In the future, perhaps hsp promoters could be used in combination with thermal energy modalities.

Gene transfer trials sponsored by the National Institutes of Health are listed on the World Wide Web at http://www4.od.nih.gov/oba/rac/clinicaltrial.htm and currently list 40 trials.

CONCLUSIONS

The observation of phenotypic differences between normal cells and cancer cells led to basic science experiments to discover why these changes occur. A cascade of discoveries brought about some answers and many more questions. Utilization of the rapidly growing wealth of basic science knowledge leads to therapeutic intervention, first at the level of the test tube or Petri dish, then to animal experimentation and finally in clinical trials. The outlook for prostate cancer patients is promising. Better methods of screening and diagnosis, as well as recognition of tumor markers, are improving the survival of patients afflicted with prostate cancer. New treatments for advanced prostate cancer are being evaluated at an increasing rate and significant advances for the therapy of prostate cancer should come sooner rather than later. It is likely that a combination of the strategies reviewed in this chapter will provide the most effective weapon in the battle against prostate cancer.

REFERENCES

1. Greenlee RT, Murray T, Bolden S et al. Cancer statistics, 2000. *CA Cancer J. Clin.* 2000; 50:7–33.
2. Benjamin LE, Golijanin D, Itin A et al. Selective ablation of immature blood vessels in established human tumors follows vascular endothelial growth factor withdrawal. *J. Clin. Invest.* 1999; 103:159–65.
3. Bissonnette RP, Echeverri F, Mahboubi A et al. Apoptotic cell death induced by c-myc is inhibited by bcl-2. *Nature* 1992; 359:552–4.
4. Oltvai ZN, Milliman CL, Korsmeyer SJ. Bcl-2 heterodimerizes in vivo with a conserved homolog, Bax, that accelerates programmed cell death. Cell 1993; 74:609–19.
5. Sommerfeld HJ, Meeker AK, Piatyszek MA et al. Telomerase activity: a prevalent marker of malignant human prostate tissue. *Cancer Res.* 1996; 56:218–22.
6. Rountree MR, Bachman KE, Herman JG et al. DNA methylation, chromatin inheritance, and cancer. *Oncogene* 2001; 20:3156–65.
7. Lee WH, Morton RA, Epstein JI et al. Cytidine methylation of regulatory sequences near the pi-class glutathione S-transferase gene accompanies human prostatic carcinogenesis. *Proc. Natl Acad. Sci. USA* 1994; 91:11733–7.
8. Knudson AG Jr. Mutation and cancer: statistical study of retinoblastoma. *Proc. Natl Acad. Sci. USA* 1971; 68:820–3.
9. Smith JR, Freije D, Carpten JD et al. Major susceptibility locus for prostate cancer on chromosome 1 suggested by a genome-wide search. *Science* 1996; 274:1371–4.
10. Merrill RM, Lyon JL. Explaining the difference in prostate cancer mortality rates between white and black men in the United States. *Urology* 2000; 55:730–5.
11. Krupski T, Petroni GR, Frierson HF, Jr. et al. Microvessel density, p53, retinoblastoma, and chromogranin A immunohistochemistry as predictors of disease-specific survival following radical prostatectomy for carcinoma of the prostate. *Urology* 2000; 55:743–9.
12. Theodorescu D, Broder SR, Boyd JC et al. p53, bcl-2 and retinoblastoma proteins as long-term prognostic markers in localized carcinoma of the prostate. *J. Urol.* 1997; 158:131–7.
13. Truong LD, Kadmon D, McCune BK et al. Association of transforming growth factor-beta 1 with prostate cancer: an immunohistochemical study. *Hum. Pathol.* 1993; 24:4–9.
14. Huang F, Newman E, Theodorescu D et al. Transforming growth factor beta 1 (TGF beta 1) is an autocrine positive regulator of colon carcinoma U9 cells in vivo as shown by

transfection of a TGF beta 1 antisense expression plasmid. *Cell Growth Differ*. 1995; 6:1635–42.
15. Theodorescu D, Caltabiano M, Greig R et al. Reduction of TGF-beta activity abrogates growth promoting tumor cell–cell interactions in vivo. *J. Cell Physiol*. 1991; 148:380–90.
16. Nakamoto T, Chang CS, Li AK et al. Basic fibroblast growth factor in human prostate cancer cells. *Cancer Res*. 1992; 52:571–7.
17. Grimberg A, Cohen P. Role of insulin-like growth factors and their binding proteins in growth control and carcinogenesis. *J. Cell Physiol*. 2000; 183:1–9.
18. Cohen P, Graves HC, Peehl DM et al. Prostate-specific antigen (PSA) is an insulin-like growth factor binding protein-3 protease found in seminal plasma. *J. Clin. Endocrinol. Metab*. 1992; 75:1046–53.
19. Ripple MO, Henry WF, Rago RP et al. Prooxidant-antioxidant shift induced by androgen treatment of human prostate carcinoma cells. *J. Natl Cancer Inst*. 1997; 89:40–8.
20. Koivisto P, Kolmer M, Visakorpi T et al. Androgen receptor gene and hormonal therapy failure of prostate cancer. *Am. J. Pathol*. 1998; 152:1–9.
21. Culig Z, Hobisch A, Cronauer MV et al. Androgen receptor activation in prostatic tumor cell lines by insulin- like growth factor-I, keratinocyte growth factor, and epidermal growth factor. *Cancer Res*. 1994; 54:5474–8.
22. Sartor O, Cooper M, Weinberger M et al. Surprising activity of flutamide withdrawal, when combined with aminoglutethimide, in treatment of 'hormone-refractory' prostate cancer. *J. Natl Cancer Inst*. 1994; 86:222–7.
23. Jarrard DF, Kinoshita H, Shi Y et al. Methylation of the androgen receptor promoter CpG island is associated with loss of androgen receptor expression in prostate cancer cells. *Cancer Res*. 1998; 58:5310–14.
24. Kinoshita H, Shi Y, Sandefur C et al. Methylation of the androgen receptor minimal promoter silences transcription in human prostate cancer. *Cancer Res*. 2000; 60:3623–30.
25. Stewart RJ, Panigrahy D, Flynn E et al. Vascular endothelial growth factor expression and tumor angiogenesis are regulated by androgens in hormone responsive human prostate carcinoma: evidence for androgen dependent destabilization of vascular endothelial growth factor transcripts. *J. Urol*. 2001; 165:688–93.
26. Mazzucchelli R, Montironi R, Santinelli A et al. Vascular endothelial growth factor expression and capillary architecture in high-grade PIN and prostate cancer in untreated and androgen-ablated patients. *Prostate* 2000; 45:72–9.
27. Graham SD Jr, Bostwick DG, Hoisaeter A et al. Report of the Committee on Staging and Pathology. *Cancer* 1992; 70:359–61.
28. Sakr WA, Grignon DJ, Crissman JD et al. High grade prostatic intraepithelial neoplasia (HGPIN) and prostatic adenocarcinoma between the ages of 20–69: an autopsy study of 249 cases. *In Vivo* 1994; 8:439–43.
29. Qian J, Wollan P, Bostwick DG. The extent and multicentricity of high-grade prostatic intraepithelial neoplasia in clinically localized prostatic adenocarcinoma. *Hum. Pathol*. 1997; 28:143–8.
30. Davidson D, Bostwick DG, Qian J et al. Prostatic intraepithelial neoplasia is a risk factor for adenocarcinoma: predictive accuracy in needle biopsies. *J. Urol*. 1995; 154:1295–9.
31. Franks LM. Atrophy and hyperplasia in the prostate proper. *J. Pathol. Bacteriol*. 1954; 98:617–21.
32. McNeal JE. Morphogenesis of prostatic carcinoma. *Cancer* 1965; 18:1659–66.
33. McNeal JE, Bostwick DG. Intraductal dysplasia: a premalignant lesion of the prostate. *Hum. Pathol*. 1986; 17:64–71.
34. Qian J, Bostwick DG. The extent and zonal location of prostatic intraepithelial neoplasia and atypical adenomatous hyperplasia: relationship with carcinoma in radical prostatectomy specimens. *Pathol. Res. Pract*. 1995; 191:860–7.
35. Helpap B, Bonkhoff H, Cockett A et al. Relationship between atypical adenomatous hyperplasia (AAH), prostatic intraepithelial neoplasia (PIN) and prostatic adenocarcinoma. *Pathologica* 1997; 89:288–300.
36. Plesnicar S. The course of metastatic disease originating from carcinoma of the prostate. *Clin. Exp. Metastasis* 1985; 3:103–10.
37. Mintz ER, Smith GG. Autopsy findings in 100 cases of prostate cancer. *N. Engl J. Med*. 1934; 211:479–87.
38. Thompson H. *Trans. Path. Soc. Lond*. 1854; 5:204.
39. Batson OV. The function of the vertebral veins and their role in the spread of metastasis. *Ann. Surg*. 1942; 112:138–49.
40. Paget S. The distribution of secondary growths in cancer of the breast. *Lancet* 1889; 1:571–3.
41. Woodhouse EC, Chuaqui RF, Liotta LA. General mechanisms of metastasis. *Cancer* 1997; 80:1529–37.
42. Folkman J, Klagsbrun M. Angiogenic factors. *Science* 1987; 235:442–7.
43. Silberman MA, Partin AW et al. Tumor angiogenesis correlates with progression after radical prostatectomy but not with pathologic stage in Gleason sum 5 to 7 adenocarcinoma of the prostate. *Cancer* 1997; 79:772–9.
44. Denijn M, Ruiter, DJ. The possible role of angiogenesis in the metastatic potential of human melanoma. Clinicopathological aspects. *Melanoma Res*. 1993; 3:5–14.
45. Gimbrone MA Jr, Leapman SB, Cotran RS et al. Tumor dormancy in vivo by prevention of neovascularization. *J. Exp. Med*. 1972; 136:261–76.
46. Goldbrunner RH, Haugland HK, Klein CE et al. ECM dependent and integrin mediated tumor cell migration of human glioma and melanoma cell lines under serum-free conditions. *Anticancer Res*. 1996; 16:3679–87.
47. Zheng DQ, Woodard AS, Fornaro M et al. Prostatic carcinoma cell migration via alpha(v)beta3 integrin is modulated by a focal adhesion kinase pathway. *Cancer Res*. 1999; 59:1655-64.
48. Umbas R, Isaacs WB, Bringuier PP et al. Decreased E-cadherin expression is associated with poor prognosis in patients with prostate cancer. *Cancer Res*. 1994; 54:3929–33.
49. Lochter A, Galosy S, Muschler J et al. Matrix metalloproteinase stromelysin-1 triggers a cascade of molecular alterations that leads to stable epithelial-to-mesenchymal conversion and a premalignant phenotype in mammary epithelial cells. *J. Cell Biol*. 1997; 139:1861–72.
50. Stetler-Stevenson WG, Aznavoorian S, Liotta LA. Tumor cell interactions with the extracellular matrix during invasion and metastasis. *Annu. Rev. Cell Biol*. 1993; 9:541–73.
51. Leavesley DI, Ferguson GD, Wayner EA et al. Requirement of the integrin beta 3 subunit for carcinoma cell spreading or migration on vitronectin and fibrinogen. *J. Cell Biol*. 1992; 117:1101–7.
52. Gleave M, Hsieh JT, Gao CA et al. Acceleration of human prostate cancer growth in vivo by factors produced by prostate and bone fibroblasts. *Cancer Res*. 1991; 51:3753–61.
53. Siegall CB, Schwab G, Nordan RP et al. Expression of the interleukin 6 receptor and interleukin 6 in prostate carcinoma cells. *Cancer Res*. 1990; 50:7786–8.
54. Denis L and Murphy GP. Overview of phase III trials on combined androgen treatment in patients with metastatic prostate cancer. *Cancer* 1993; 72:3888–95.
55. Garber K. STI571 revolution: can the newer targeted drugs measure up? *J. Natl Cancer Inst*. 2001; 93:970–3.
56. Arnold K. After 30 years of laboratory work, a quick approval for STI571. *J. Natl Cancer Inst*. 2001; 93:972.
57. Buchdunger E, Zimmermann J, Mett H et al. Inhibition of the Abl protein-tyrosine kinase in vitro and in vivo by a 2-phenylaminopyrimidine derivative. *Cancer Res*. 1996; 56:100–4.
58. Druker BJ, Talpaz M, Resta DJ et al. Efficacy and safety of a specific inhibitor of the BCR-ABL tyrosine kinase in chronic myeloid leukemia. *N. Engl J. Med*. 2001; 344:1031–7.

59. Druker BJ, Sawyers CL, Kantarjian H et al. Activity of a specific inhibitor of the BCR-ABL tyrosine kinase in the blast crisis of chronic myeloid leukemia and acute lymphoblastic leukemia with the Philadelphia chromosome. *N. Engl J. Med.* 2001; 344:1038–42.
60. Whittemore AS, Kolonel LN, Wu AH et al. Prostate cancer in relation to diet, physical activity, and body size in blacks, whites, and Asians in the United States and Canada. *J. Natl Cancer Inst.* 1995; 87:652–61.
61. Kolonel LN, Yoshizawa CN, Hankin JH. Diet and prostatic cancer: a case-control study in Hawaii. *Am. J. Epidemiol.* 1988; 127:999–1012.
62. Blumenfeld AJ, Fleshner N, Casselman B et al. Nutritional aspects of prostate cancer: a review. *Can. J. Urol.* 2000; 7:927–35; discussion 936.
63. Clark LC, Dalkin B, Krongrad A et al. Decreased incidence of prostate cancer with selenium supplementation: results of a double-blind cancer prevention trial. *Br. J. Urol.* 1998; 81:7 30–4.
64. Giovannucci E, Ascherio A, Rimm EB et al. Intake of carotenoids and retinol in relation to risk of prostate cancer. *J. Natl Cancer Inst.* 1995; 87:1767–76.
65. Wang JS, Shen X, He X et al. Protective alterations in phase 1 and 2 metabolism of aflatoxin B1 by oltipraz in residents of Qidong, People's Republic of China. *J. Natl Cancer Inst.* 1999; 91:347–54.
66. Denis LJ, Griffiths K. Endocrine treatment in prostate cancer. *Semin. Surg. Oncol.* 2000; 18:52–74.
67. Sanford EJ, Paulson DF, Rohner TJ Jr et al. The effects of castration on adrenal testosterone secretion in men with prostatic carcinoma. *J. Urol.* 1977; 118:1019–21.
68. Geller J, Albert JD, Nachtsheim DA et al. Comparison of prostatic cancer tissue dihydrotestosterone levels at the time of relapse following orchiectomy or estrogen therapy. *J. Urol.* 1984; 132:693–6.
69. Newmark JR, Hardy DO, Tonb DC et al. Androgen receptor gene mutations in human prostate cancer. *Proc. Natl Acad. Sci. USA* 1992; 89:6319–23.
70. Visakorpi T, Hyytinen E, Koivisto P et al. In vivo amplification of the androgen receptor gene and progression of human prostate cancer. *Nature Genet.* 1995; 9:401–6.
71. Koivisto P, Kononen J, Palmberg C et al. Androgen receptor gene amplification: a possible molecular mechanism for androgen deprivation therapy failure in prostate cancer. *Cancer Res.* 1997; 57:314–19.
72. Taplin ME, Bubley GJ, Shuster TD et al. Mutation of the androgen-receptor gene in metastatic androgen-independent prostate cancer. *N. Engl. J. Med.* 1995; 332:1393–8.
73. Avila DM, Zoppi S, McPhaul MJ. The androgen receptor (AR) in syndromes of androgen insensitivity and in prostate cancer. *J. Steroid Biochem. Mol. Biol.* 2001; 76:135–42.
74. Taplin ME, Bubley GJ, Ko YJ et al. Selection for androgen receptor mutations in prostate cancers treated with androgen antagonist. *Cancer Res.* 1999; 59:2511–15.
75. Richie JP. Anti-androgens and other hormonal therapies for prostate cancer. *Urology* 1999; 54:15–18.
76. Fisher B, Costantino JP, Wickerham DL. et al. Tamoxifen for prevention of breast cancer: report of the National Surgical Adjuvant Breast and Bowel Project P-1 Study. *J. Natl Cancer Inst.* 1998; 90:1371–88.
77. Thompson IM, Coltman CA Jr, Crowley J. Chemoprevention of prostate cancer: the Prostate Cancer Prevention Trial. *Prostate* 1997; 33:217–21.
78. Dai JL, Maiorino CA, Gkonos PJ Burnstein KL. Androgenic up-regulation of androgen receptor cDNA expression in androgen-independent prostate cancer cells. *Steroids* 1996; 61:531–9.
79. Gregory CW, Hamil KG, Kim D et al. Androgen receptor expression in androgen-independent prostate cancer is associated with increased expression of androgen-regulated genes. *Cancer Res.* 1998; 58:5718–24.
80. Nazareth LV, Weigel NL. Activation of the human androgen receptor through a protein kinase A signaling pathway. *J. Biol. Chem.* 1996; 271:19900–7.
81. Abreu-Martin MT, Chari A, Palladino AA et al. Mitogen-activated protein kinase kinase kinase 1 activates androgen receptor-dependent transcription and apoptosis in prostate cancer. *Mol. Cell Biol.* 1999; 19:5143–54.
82. Craft N, Shostak Y, Carey M et al. A mechanism for hormone-independent prostate cancer through modulation of androgen receptor signaling by the HER-2/neu tyrosine kinase. *Nature Med.* 1999; 5:280–5.
83. Yeh S, Lin HK, Kang HY et al. From HER2/Neu signal cascade to androgen receptor and its coactivators: a novel pathway by induction of androgen target genes through MAP kinase in prostate cancer cells. *Proc. Natl Acad. Sci. USA* 1999; 96:5458–63.
84. Gregory CW, Johnson RT Jr, Presnell SC et al. Androgen receptor regulation of G1 cyclin and cyclin-dependent kinase function in the CWR22 human prostate cancer xenograft. *J. Androl.* 2001; 22:537–48.
85. Shain SA. Neither fibroblast growth factor-1 nor fibroblast growth factor-2 is an androgen receptor coactivator in androgen-resistant prostate cancer. *Mol. Urol.* 2001; 5:121–30.
86. Putz T, Culig Z, Eder IE et al. Epidermal growth factor (EGF) receptor blockade inhibits the action of EGF, insulin-like growth factor I, and a protein kinase A activator on the mitogen-activated protein kinase pathway in prostate cancer cell lines. *Cancer Res.* 1999; 59:227–33.
87. Segawa N, Nakamura M, Nakamura Y et al. Phosphorylation of mitogen-activated protein kinase is inhibited by calcitonin in DU145 prostate cancer cells. *Cancer Res.* 2001; 61:6060–3.
88. Murillo H, Schmidt LJ, Tindall DJ. Tyrphostin AG825 triggers p38 mitogen-activated protein kinase-dependent apoptosis in androgen-independent prostate cancer cells C4 and C4-2. *Cancer Res.* 2001; 61:7408–12.
89. Veldscholte J, Berrevoets CA, Zegers ND et al. Hormone-induced dissociation of the androgen receptor-heat-shock protein complex: use of a new monoclonal antibody to distinguish transformed from nontransformed receptors. *Biochemistry* 1992; 31:7422–30.
90. Folkman J. The role of angiogenesis in tumor growth. *Semin. Cancer Biol.* 1992; 3:65–71.
91. Borgstrom P, Bourdon MA, Hillan KJ et al. Neutralizing anti-vascular endothelial growth factor antibody completely inhibits angiogenesis and growth of human prostate carcinoma micro tumors in vivo. *Prostate* 1998; 35:1–10.
92. Ferrer FA, Miller LJ, Andrawis RI et al. Angiogenesis and prostate cancer: in vivo and in vitro expression of angiogenesis factors by prostate cancer cells. *Urology* 1998; 51:161–7.
93. Joseph IB, Isaacs JT. Potentiation of the antiangiogenic ability of linomide by androgen ablation involves down-regulation of vascular endothelial growth factor in human androgen-responsive prostatic cancers. *Cancer Res.* 1997; 57:1054–7.
94. Harper ME, Glynne-Jones E, Goddard L et al. Vascular endothelial growth factor (VEGF) expression in prostatic tumours and its relationship to neuroendocrine cells. *Br. J. Cancer* 1996; 74:910–16.
95. Theodorescu D, Broder SR, Boyd JC et al. Cathepsin D and chromogranin A as predictors of long term disease specific survival after radical prostatectomy for localized carcinoma of the prostate. *Cancer* 1997; 80:2109–19.
96. Millauer B, Wizigmann-Voos S, Schnurch H et al. High affinity VEGF binding and developmental expression suggest Flk-1 as a major regulator of vasculogenesis and angiogenesis. *Cell* 1993; 72:835–46.
97. Kollermann J, Helpap B. Expression of vascular endothelial growth factor (VEGF) and VEGF receptor Flk-1 in benign,

premalignant, and malignant prostate tissue. *Am. J. Clin. Pathol.* 2001; 116:115–21.
98. Wedge SR, Ogilvie DJ, Dukes M et al. ZD4190: an orally active inhibitor of vascular endothelial growth factor signaling with broad-spectrum antitumor efficacy. *Cancer Res.* 2000; 60: 970–5.
99. Ingber D, Fujita T, Kishimoto S et al. Synthetic analogues of fumagillin that inhibit angiogenesis and suppress tumour growth. *Nature* 1990; 348:555–7.
100. Morita T, Shinohara N, Tokue A. Antitumour effect of a synthetic analogue of fumagillin on murine renal carcinoma. *Br. J. Urol.* 1994; 74:416–21.
101. Kalebic T, Tsokos M, Helman LJ. Suppression of rhabdomyosarcoma growth by fumagillin analog TNP-470. *Int. J. Cancer* 1996; 68:596–9.
102. Yoshida T, Kaneko Y, Tsukamoto A et al. Suppression of hepatoma growth and angiogenesis by a fumagillin derivative TNP470: possible involvement of nitric oxide synthase. *Cancer Res.* 1998; 58:3751–6.
103. Yamaoka M, Yamamoto T, Ikeyama S et al. Angiogenesis inhibitor TNP-470 (AGM-1470) potently inhibits the tumor growth of hormone-independent human breast and prostate carcinoma cell lines. *Cancer Res.* 1993; 53:5233–6.
104. Horti J, Dixon SC, Logothetis CJ et al. Increased transcriptional activity of prostate-specific antigen in the presence of TNP-470, an angiogenesis inhibitor. *Br. J. Cancer* 1999; 79:1588–93.
105. Logothetis CJ, Wu KK, Finn LD et al. Phase I trial of the angiogenesis inhibitor TNP-470 for progressive androgen-independent prostate cancer. *Clin. Cancer Res.* 2001; 7:1198–203.
106. Han CK, Ahn SK, Choi NS et al. Design and synthesis of highly potent fumagillin analogues from homology modeling for a human MetAP-2. *Bioorg. Med. Chem. Lett.* 2000; 10:39–43.
107. Zucker MB, Katz IR. Platelet factor 4: production, structure, and physiologic and immunologic action. *Proc. Soc. Exp. Biol. Med.* 1991; 198:693–702.
108. Hiti-Harper J, Wohl H, Harper E. Platelet factor 4: an inhibitor of collagenase. *Science* 1978; 199:991–2.
109. Maione TE, Gray GS, Petro J et al. Inhibition of angiogenesis by recombinant human platelet factor-4 and related peptides. *Science* 1990; 247:77–9.
110. Sharpe RJ, Byers HR, Scott CF et al. Growth inhibition of murine melanoma and human colon carcinoma by recombinant human platelet factor 4. *J. Natl Cancer Inst.* 1990; 82:848–53.
111. Maione TE, Gray GS, Hunt AJ et al. Inhibition of tumor growth in mice by an analogue of platelet factor 4 that lacks affinity for heparin and retains potent angiostatic activity. *Cancer Res.* 1991; 51:2077–83.
112. D'Amato RJ, Loughnan MS, Flynn E et al. Thalidomide is an inhibitor of angiogenesis. *Proc. Natl Acad. Sci. USA* 1994; 91:4082–5.
113. Figg WD, Dahut W, Duray P et al. A randomized phase II trial of thalidomide, an angiogenesis inhibitor, in patients with androgen-independent prostate cancer. *Clin. Cancer Res.* 2001; 7:1888–93.
114. O'Reilly MS, Holmgren L, Chen C et al. Angiostatin induces and sustains dormancy of human primary tumors in mice. *Nature Med.* 1996; 2:689–92.
115. O'Reilly MS, Boehm T, Shing Y et al. Endostatin: an endogenous inhibitor of angiogenesis and tumor growth. *Cell* 1997; 88:277–85.
116. Yokoyama Y, Green JE, Sukhatme VP et al. Effect of endostatinon spontaneous tumorigenesis of mammary adenocarcinoma in a transgenic mouse model. *Cancer Res.* 2000; 60:4362–5.
117. Iughetti P, Suzuki O, Godoi PH et al. Polymorphism in endostatin, an angiogenesis inhibitor, predisposes for the development of prostatic adenocarcinoma. *Cancer Res.* 2001; 61:7375–8.
118. O'Reilly, MS. Angiostatin: an endogenous inhibitor of angiogenesis and of tumor growth. Exs 1997; 79:273–94.
119. Gately S, Twardowski P, Stack MS et al. Human prostate carcinoma cells express enzymatic activity that converts human plasminogen to the angiogenesis inhibitor, angiostatin. *Cancer Res.* 1996; 56:4887–90.
120. Gately S, Twardowski P, Stack MS et al. The mechanism of cancer-mediated conversion of plasminogen to the angiogenesis inhibitor angiostatin. *Proc. Natl Acad. Sci. USA* 1997; 94:10868–72.
121. Luo J, Lubaroff DM, Hendrix MJ. Suppression of prostate cancer invasive potential and matrix metalloproteinase activity by E-cadherin transfection. *Cancer Res.* 1999; 59:3552–6.
122. Gunthert U, Hofmann M, Rudy W et al. A new variant of glycoprotein CD44 confers metastatic potential to rat carcinoma cells. *Cell* 1991; 65:13–24.
123. Seiter S, Arch R, Reber S et al. Prevention of tumor metastasis formation by anti-variant CD44. *J. Exp. Med.* 1993; 177:443–55.
124. Klemke RL, Cai S, Giannini AL et al. Regulation of cell motility by mitogen-activated protein kinase. *J. Cell Biol.* 1997; 137:481–92.
125. Tamura M, Gu J, Tran H et al. PTEN gene and integrin signaling in cancer. *J. Natl Cancer Inst.* 1999; 91:1820–8.
126. Persad S, Attwell S, Gray V et al. Inhibition of integrin-linked kinase (ILK) suppresses activation of protein kinase B/Akt and induces cell cycle arrest and apoptosis of PTEN-mutant prostate cancer cells. *Proc. Natl Acad. Sci. USA* 2000; 97:3207–12.
127. Allen MV, Smith GJ, Juliano R et al. Downregulation of the beta4 integrin subunit in prostatic carcinoma and prostatic intraepithelial neoplasia. *Hum. Pathol.* 1998; 29:311–18.
128. Cress AE, Rabinovitz I, Zhu W et al. The alpha 6 beta 1 and alpha 6 beta 4 integrins in human prostate cancer progression. *Cancer Metast. Rev.* 1995; 14:219–28.
129. Baker AH, George SJ, Zaltsman AB et al. Inhibition of invasion and induction of apoptotic cell death of cancer cell lines by overexpression of TIMP-3. *Br. J. Cancer* 1999; 79:1347–55.
130. Wood M, Fudge K, Mohler JL et al. In situ hybridization studies of metalloproteinases 2 and 9 and TIMP-1 and TIMP-2 expression in human prostate cancer. *Clin. Exp. Metast.* 1997; 15:246–58.
131. Lokeshwar BL. MMP inhibition in prostate cancer. *Ann. N. York Acad. Sci.* 1999; 878:271–89.
132. Stearns ME, Fudge K, Garcia F et al. IL-10 inhibition of human prostate PC-3 ML cell metastases in SCID mice: IL-10 stimulation of TIMP-1 and inhibition of MMP-2/MMP-9 expression. *Invasion Metast.* 1997; 17:62–74.
133. Stearns ME. Alendronate blocks TGF-beta1 stimulated collagen 1 degradation by human prostate PC-3 ML cells. *Clini. Exp. Metast.* 1998; 16:332–9.
134. Nemunaitis J, Poole C, Primrose J et al. Combined analysis of studies of the effects of the matrix metalloproteinase inhibitor marimastat on serum tumor markers in advanced cancer: selection of a biologically active and tolerable dose for longer-term studies. *Clin. Cancer Res.* 1998; 4:1101–9.
135. Steward WP. Marimastat (BB2516): current status of development. *Cancer Chemothera. Pharmacol.* 1999; 43(Suppl.):S56–60.
136. Helenius MA, Saramaki OR, Linja MJ et al. Amplification of urokinase gene in prostate cancer. *Cancer Res.* 2001; 61:5340–4.
137. Rabbani SA, Harakidas P, Davidson DJ et al. Prevention of prostate-cancer metastasis in vivo by a novel synthetic inhibitor of urokinase-type plasminogen activator (uPA). *Int. J. Cancer* 1995; 63:840–5.
138. Evans CP, Elfman F, Parangi S et al. Inhibition of prostate cancer neovascularization and growth by urokinase-plasminogen activator receptor blockade. *Cancer Res.* 1997; 57:3594–9.

139. Festuccia C, Giunciuglio D, Guerra F et al. Osteoblasts modulate secretion of urokinase-type plasminogen activator (uPA) and matrix metalloproteinase-9 (MMP-9) in human prostate cancer cells promoting migration and matrigel invasion. *Oncol. Res.* 1999; 11:17–31.
140. Camps JL, Chang SM, Hsu TC et al. Fibroblast-mediated acceleration of human epithelial tumor growth in vivo. *Proc. Natl Acad. Sci. USA* 1990; 87:75–9.
141. Gleave ME, Hsieh JT, von Eschenbach AC et al. Prostate and bone fibroblasts induce human prostate cancer growth in vivo: implications for bidirectional tumor-stromal cell interaction in prostate carcinoma growth and metastasis. *J. Urol.* 1992; 147:1151–9.
142. Brooks PC, Montgomery AM, Rosenfeld M et al. Integrin alpha v beta 3 antagonists promote tumor regression by inducing apoptosis of angiogenic blood vessels. *Cell* 1994; 79:1157–64.
143. Pasqualini R, Koivunen E, Ruoslahti E. Alpha v integrins as receptors for tumor targeting by circulating ligands. *Nature Biotechnol.* 1997; 15:542–6.
144. Arap W, Pasqualini R, Ruoslahti E. Cancer treatment by targeted drug delivery to tumor vasculature in a mouse model. *Science* 1998; 279:377–80.
145. Zhao M, Kleinman HK, Mokotoff M. Synthetic laminin-like peptides and pseudopeptides as potential antimetastatic agents. *J. Med. Chem.* 1994; 37:3383–8.
146. Saunders J, Tarby CM. Opportunities for novel therapeutic agents acting at chemokine receptors. *Drug Discov. Today* 1999; 4:80–92.
147. Charhon SA, Chapuy MC, Delvin EE et al. Histomorphometric analysis of sclerotic bone metastases from prostatic carcinoma special reference to osteomalacia. *Cancer* 1983; 51:918–24.
148. Tsingotjidou AS, Zotalis G, Jackson KR et al. Development of an animal model for prostate cancer cell metastasis to adult human bone. *Anticancer Res.* 2001; 21:971–8.
149. Yonou H, Yokose T, Kamijo T et al. Establishment of a novel species- and tissue-specific metastasis model of human prostate cancer in humanized non-obese diabetic/severe combined immunodeficient mice engrafted with human adult lung and bone. *Cancer Res.* 2001; 61:2177–82.
150. Koeneman KS, Kao C, Ko SC et al. Osteocalcin-directed gene therapy for prostate-cancer bone metastasis. *World J. Urol.* 2000; 18:102–10.
151. Yeung F, Law WK, Yeh CH et al. Regulation of human osteocalcin (hOC) promoter in hormone-independent human prostate cancer cells. *J. Biol. Chem.* 2001; 29:29.
152. Baselga J, Norton L, Albanell J et al. Recombinant humanized anti-HER2 antibody (Herceptin) enhances the antitumor activity of paclitaxel and doxorubicin against HER2/neu overexpressing human breast cancer xenografts. *Cancer Res.* 1998; 58:2825–31.
153. Miyake H, Hara I, Kamidono S et al. Novel therapeutic strategy for advanced prostate cancer using antisense oligodeoxynucleotides targeting anti-apoptotic genes upregulated after androgen withdrawal to delay androgen-independent progression and enhance chemosensitivity. *Int. J. Urol.* 2001; 8:337–49.
154. Fruehauf JP, Myers CE, Sinha BK. Synergistic activity of suramin with tumor necrosis factor alpha and doxorubicin on human prostate cancer cell lines. *J. Natl Cancer Inst.* 1990; 82:1206–9.
155. Pienta KJ, Isaacs WB, Vindivich D et al. The effects of basic fibroblast growth factor and suramin on cell motility and growth of rat prostate cancer cells. *J. Urol.* 1991; 145:199–202.
156. La Rocca RV, Danesi R, Cooper MR et al. Effect of suramin on human prostate cancer cells in vitro. *J. Urol.* 1991; 145:393–8.
157. Ewing MW, Liu SC, Gnarra JR et al. Effect of suramin on the mitogenic response of the human prostate carcinoma cell line PC-3. *Cancer* 1993; 71:1151–8.
158. Hussain M, Fisher EI, Petrylak DP et al. Androgen deprivation and four courses of fixed-schedule suramin treatment in patients with newly diagnosed metastatic prostate cancer: a Southwest Oncology Group Study. *J. Clin. Oncol.* 2000; 18:1043–9.
159. Fong CJ, Sherwood ER, Mendelsohn J et al. Epidermal growth factor receptor monoclonal antibody inhibits constitutive receptor phosphorylation, reduces autonomous growth, and sensitizes androgen-independent prostatic carcinoma cells to tumor necrosis factor alpha. *Cancer Res.* 1992; 52:5887–92.
160. Peng D, Fan Z, Lu Y et al. Anti-epidermal growth factor receptor monoclonal antibody 225 up-regulates p27KIP1 and induces G1 arrest in prostatic cancer cell line DU145. *Cancer Res.* 1996; 56:3666–9.
161. Prewett M, Rockwell P, Rockwell RF et al. I. The biologic effects of C225, a chimeric monoclonal antibody to the EGFR, on human prostate carcinoma. *J. Immunother. Emphasis Tumor Immunol.* 1996; 19:419–27.
162. Yang XD, Jia XC, Corvalan JR et al. Development of ABX-EGF, a fully human anti-EGF receptor monoclonal antibody, for cancer therapy. *Crit. Rev. Oncol. Hematol.* 2001; 38:17–23.
163. Nelson JB, Carducci MA. Small bioactive peptides and cell surface peptidases in androgen-independent prostate cancer. *Cancer Invest.* 2000; 18:87–96.
164. Aprikian AG, Han K, Guy L et al. Neuroendocrine differentiation and the bombesin/gastrin-releasing peptide family of neuropeptides in the progression of human prostate cancer. *Prostate* Suppl. 1998; 8:52–61.
165. Jungwirth A, Galvan G, Pinski J et al. Luteinizing hormone-releasing hormone antagonist Cetrorelix (SB-75) and bombesin antagonist RC-3940-II inhibit the growth of androgen-independent PC-3 prostate cancer in nude mice. *Prostate* 1997; 32:164–72.
166. Pinski J, Schally AV, Halmos G et al. Effect of somatostatin analog RC-160 and bombesin/gastrin releasing peptide antagonist RC-3095 on growth of PC-3 human prostate-cancer xenografts in nude mice. *Int. J. Cancer* 1993; 55:963–7.
167. Wallin M, Deinum J, Friden B. Interaction of estramustine phosphate with microtubule-associated proteins. *FEBS Lett.* 1985; 179:289–93.
168. Karr JP, Wajsman Z, Kirdani RY et al. Effects of diethylstilbestrol and estramustine phosphate on serum sex hormone binding globulin and testosterone levels in prostate cancer patients. *J. Urol.* 1980; 124:232–6.
169. Fakih M, Yagoda A, Replogle T et al. Inhibition of prostate cancer growth by estramustine and colchicine. *Prostate* 1995; 26:310–15.
170. Hudes GR, Greenberg R, Krigel RL et al. Phase II study of estramustine and vinblastine, two microtubule inhibitors, in hormone-refractory prostate cancer. *J. Clin. Oncol.* 1992; 10:1754–61.
171. Panda D, Miller HP, Islam K et al. Stabilization of microtubule dynamics by estramustine by binding to a novel site in tubulin: a possible mechanistic basis for its antitumor action. *Proc. Natl Acad. Sci. USA* 1997; 94:10560–4.
172. Kondapaka BS, Reddy KB. Tyrosine kinase inhibitor as a novel signal transduction and antiproliferative agent: prostate cancer. *Mol. Cell Endocrinol.* 1996; 117:53–58.
173. Jones HE, Dutkowski CM, Barrow D et al. EGF-R selective tyrosine kinase inhibitor reveals variable growth responses in prostate carcinoma cell lines PC-3 and DU-145. *Int. J. Cancer* 1997; 71:1010–18.
174. Dionne CA, Camoratto AM, Jani JP et al. Cell cycle-independent death of prostate adenocarcinoma is induced by the trk tyrosine kinase inhibitor CEP-751 (KT6587). *Clin. Cancer Res.* 1998; 4:1887–98.
175. Gleave M, Tolcher A, Miyake H et al. Progression to androgen independence is delayed by adjuvant treatment with antisense Bcl-2 oligodeoxynucleotides after castration in the

LNCaP prostate tumor model. *Clin. Cancer Res.* 1999; 5:2891–8.
176. Miyake H, Tolcher A, Gleave ME. Antisense Bcl-2 oligodeoxynucleotides inhibit progression to androgen-independence after castration in the Shionogi tumor model. *Cancer Res.* 1999; 59:4030–4.
177. Leung S, Miyake H, Zellweger T et al. Synergistic chemosensitization and inhibition of progression to androgen independence by antisense Bcl-2 oligodeoxynucleotide and paclitaxel in the LNCaP prostate tumor model. *Int. J. Cancer* 2001; 91:846–50.
178. Simons JW, Mikhak B, Chang JF et al. Induction of immunity to prostate cancer antigens: results of a clinical trial of vaccination with irradiated autologous prostate tumor cells engineered to secrete granulocyte-macrophage colony-stimulating factor using ex vivo gene transfer. *Cancer Res.* 1999; 59:5160–8.
179. Simons JW, Mikhak B, Chang JF. Clinical activity and broken immunologic tolerance from ex vivo GM-CSF gene transduced prostate cancer vaccines. *J. Urol.* Suppl. 1999; 161:abstract 188.
180. Herman JR, Adler HL, Aguilar-Cordova E et al. In situ gene therapy for adenocarcinoma of the prostate: a phase I clinical trial. *Hum. Gene Ther.* 1999; 10:1239–49.
181. Shalev M, Kadmon D, Teh BS et al. Suicide gene therapy toxicity after multiple and repeat injections in patients with localized prostate cancer. *J. Urol.* 2000; 163:1747–50.
182. Sweeney P. and Pisters LL. Ad5CMVp53 gene therapy for locally advanced prostate cancer – where do we stand? *World J. Urol.* 2000; 18:121–4.
183. Logothetis CJ, Hossan E, Pettaway CA. AD-p53 intraprostatic gene therapy preceding radical prostatectomy (RP): an in vivo model for targeted therapy development. *J. Urol.* Suppl. 1999; 161:abstract 1142.
184. Belldegrun A, Tso CL, Zisman A et al. Interleukin 2 gene therapy for prostate cancer: phase I clinical trial and basic biology. *Hum. Gene Ther.* 2001; 12:883–92.
185. Wu L, Matherly J, Smallwood A et al. PSA enhancers exhibit augmented activity in prostate cancer gene therapy vectors. *Gene Ther.* 2001; 8:1416–26.
186. Kwon ED, Foster BA, Hurwitz AA et al. Elimination of residual metastatic prostate cancer after surgery and adjunctive cytotoxic T lymphocyte-associated antigen 4 (CTLA-4) blockade immunotherapy. *Proc. Natl Acad. Sci. USA* 1999; 96:15074–9.
187. Hurwitz AA, Foster BA, Kwon ED et al. Combination immunotherapy of primary prostate cancer in a transgenic mouse model using CTLA-4 blockade. *Cancer Res.* 2000; 60:2444–8.
188. Kubota Y, Fujinami K, Uemura H et al. Retinoblastoma gene mutations in primary human prostate cancer. *Prostate* 1995; 27:314–20.
189. Nikitin AY, Juarez-Perez MI, Li S et al. RB-mediated suppression of spontaneous multiple neuroendocrine neoplasia and lung metastases in Rb+/– mice. *Proc. Natl Acad. Sci. USA* 1999; 96:3916–21.
190. Yang C, Cirielli C, Capogrossi MC et al. Adenovirus-mediated wild-type p53 expression induces apoptosis and suppresses tumorigenesis of prostatic tumor cells. *Cancer Res.* 1995; 55:4210–13.
191. Eastham JA, Grafton W, Martin CM et al. Suppression of primary tumor growth and the progression to metastasis with p53 adenovirus in human prostate cancer. *J. Urol.* 2000; 164:814–19.
192. Eastham JA, Hall SJ, Sehgal I et al. In vivo gene therapy with p53 or p21 adenovirus for prostate cancer. *Cancer Res.* 1995; 55:5151–5.
193. Steiner MS, Zhang Y, Farooq F et al. Adenoviral vector containing wild-type p16 suppresses prostate cancer growth and prolongs survival by inducing cell senescence. *Cancer Gene Ther.* 2000; 7:360–72.
194. Gurnani M, Lipari P, Dell J et al. Adenovirus-mediated p53 gene therapy has greater efficacy when combined with chemotherapy against human head and neck, ovarian, prostate, and breast cancer. *Cancer Chemother. Pharmacol.* 1999; 44:143–51.
195. Blackburn RV, Galoforo SS, Corry PM et al. Adenoviral transduction of a cytosine deaminase/thymidine kinase fusion gene into prostate carcinoma cells enhances prodrug and radiation sensitivity. *Int. J. Cancer* 1999; 82:293–7.
196. Hall SJ, Mutchnik SE, Yang G et al. Cooperative therapeutic effects of androgen ablation and adenovirus-mediated herpes simplex virus thymidine kinase gene and ganciclovir therapy in experimental prostate cancer. *Cancer Gene Ther.* 1999; 6:54–63.
197. Blackburn RV, Galoforo SS, Corry PM et al. Adenoviral-mediated transfer of a heat-inducible double suicide gene into prostate carcinoma cells. *Cancer Res* 1998; 58:1358–62.

Antioxidants and Phytotherapy

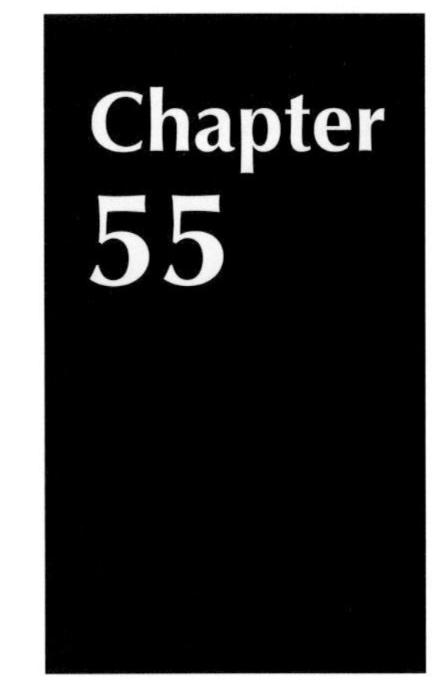

Charles E. Myers and Dan Theodorescu
Department of Urology, University of Virginia, Charlottesville, VA, USA

INTRODUCTION

Patients with chronic diseases are increasingly seeking alternative medicine therapies. Thirty-four per cent of Americans surveyed by phone in 1991 and 42% in 1997 reported using at least one type of alternative medicine. In addition, in 1991 one-third of alternative medicine users consulted a provider of alternative medicine, while in 1997 this figure rose to 46%.[1,2] Extrapolation to the US population suggested that the 629 million visits to providers of alternative medicine exceeded the 386 million visits to all US primary care physicians in 1997.[2] Of the $13.7 billion spent on alternative medicine in 1990, three-quarters were paid out of pocket,[1] while of the $21.2 billion spent in 1997, $12.2 billion was paid out of pocket.[2] Increasing health insurance coverage of alternative therapies also supports increasing patients usage.[3] Because of such consumer demand, the National Institutes of Health (NIH) created the National Center for Complementary and Alternative Medicine (NCCAM). This center is dedicated to exploring complementary and alternative healing practices in the context of rigorous science, training complementary and alternative medicine (CAM) researchers and disseminating authoritative information (http://nccam.nih.gov). Complementary and alternative healthcare and medical practices are those healthcare and medical practices that are not currently an integral part of conventional medicine.* The list of practices that are considered CAM changes continually as CAM practices and therapies that are proven safe and effective become accepted as 'mainstream' healthcare practices. Today, CAM practices may be grouped within five major domains:† (1) alternative medical systems; (2) mind–body interventions; (3) biologically-based treatments; (4) manipulative and body-based methods; and (5) energy therapies. The individual systems and treatments comprising these categories are too numerous to list in this document. Thus, only limited examples are provided within each.

The 'biologically-based treatments' category of CAM includes natural and biologically-based practices, interventions and products, many of which overlap with conventional medicine's use of dietary supplements. Included are herbal, special dietary, orthomolecular and individual biological therapies. Herbal therapies employ individual or mixtures of herbs for therapeutic value. A herb is a plant or plant part that produces and contains chemical substances that act upon the body. Orthomolecular therapies aim to treat disease with varying concentrations of chemicals, such as selenium, and megadoses of vitamins. Biological therapies include, for example, the use of laetrile and shark cartilage to treat cancer, and bee pollen to treat autoimmune and inflammatory diseases. In the current chapter, we will discuss several biologically-based therapies relevant to prostate cancer and highlight the relevant biology and epidemiology pertaining to these approaches.

USE OF COMPLEMENTARY AND ALTERNATIVE MEDICINE IN PROSTATE CANCER PATIENTS

The use of alternative medicine therapies for cancer patients varies from 7% to 64% in published studies with an average prevalence of 31.4% [Ernst, 1998 #2103].[7] Interestingly, it appears that up to 72% of those patients using alternative therapies do not inform their doctors.[1] Many patients hold high expectations for their alternative medicines. Most patients who utilize complementary therapy expect it to improve the quality of their life, over 70% expect it to boost their immune system and 62% believe that it will prolong

*The term conventional medicine refers to medicine as practiced by holders of MD (medical doctor) or DO (doctor of osteopathy) degrees, some of whom may also practice complementary and alternative medicine. Other terms for conventional medicine are allopathy, Western, regular and mainstream medicine, and biomedicine (source: http://nccam.nih.gov).
†These are the categories within which NCCAM has chosen to group the numerous CAM practices; others employ different broad groupings (source: http://nccam.nih.gov).

their life. As many as 37% of CAM users anticipate that the therapy will cure their disorder.[4] Lazar and O'Connor reported several factors influencing the use of complementary or alternative medicine including: the attempt to control side effects; the need to be involved and to have some control over their therapy; the desire to avoid toxicities of conventional medicines; and the diagnosis of a disease for which proper treatment is lacking.[5] Those that choose not to use complementary medicine (CM) do so for a variety of reasons including satisfying results from conventional medicine and confidence in medical professionals.[6]

We have previously demonstrated that up to 43% of clinically localized prostate cancer patients treated with curative intent were found to use at least one form of CM. In this study, vitamins and herbal therapies were found to be among the most prevalent forms of complementary therapies. It was also demonstrated that higher pretreatment Gleason scores were associated with increased usage of CM.[7] Other studies have shown a positive correlation with increased socioeconomic status and education.[8] These data led us to carry out another study to query as to the specific complementary therapies being used, the time frame of this usage, and the major motivational aspects and sources of information involved in this decision.

In this study, information from 238 patients with localized prostate cancer treated with curative intent by radical surgery or radiation, including brachytherapy, was included. A total of 37% of patients acknowledged using some type of complementary therapy. Our data indicate similar overall use of CM among the treatment groups. The most common CM type was vitamins, utilized by 35% of patients, followed by herbal products and dietary changes used by 12% of respondents. Vitamin E and lycopene were the most commonly used single items in each group, respectively, while the most common dietary change was decreased fat intake. A total of 43% of patients began using CM before starting conventional treatment, while 32% began after initiation of such treatment. The majority of patients who utilized CM indicated they would never discontinue these therapies. The most common reason for using CM was the patient's impression that it made them feel better and, secondarily, that it helped cure their cancer. Physicians were listed as the most common source of information regarding CM. Surprisingly, almost twice as many patients identified physicians as being advocates of CM than critics of this type of care. Many patients felt that their urologist or radiation oncologist was neutral or chose not to discuss the topic of CM with patients. However, for those physicians that did discuss this topic with patients, more patients felt that their urologist or radiation oncologist encouraged the use of complementary therapies than discouraged their use.

Since alternative therapies may alter results of standard cancer therapies,[9] produce side effects and confound results of clinical trials, new patient evaluations should include questions concerning any herbal preparations or practice of any alternative methods of cancer treatment. The potential confounding effects due to alternative medicine use are particularly relevant in a disease such as prostate cancer, whose status is assessed by biochemical markers (i.e. prostate specific antigen – PSA). The most compelling finding is that several of these botanical therapies have shown striking clinical effects on human prostate cancer[9] as well as in experimental *in vitro* studies. This will be discussed in more detail below.

DIETARY SUPPLEMENT HEALTH AND EDUCATION ACT OF 1994

The past few years have witnessed a dramatic increase in the use of herbal products as well as other methods associated with alternative and complementary medicine. What has fueled this shift toward herbal and nutritional supplements? Many patients view herbal products as offering a gentler, more natural form of therapy with fewer side effects. While this may be true for some supplements, the side effects of many herbal products have simply not been publicized adequately. In addition, the use of herbal products also puts the patient in control of his treatment, since most do not require prescriptions.

As the American public turned their attention to natural products, the political and legal environment also began to change. One key event was the passage by Congress of a major new law governing the marketing and sale of these products. This law, called the Dietary Supplement Health and Education Act of 1994 (Public Law No:103-417)[10] (full text of legislation can be found at http://thomas.loc.gov), represents the result of a debate dating back to the early 1960s, between the Food and Drug Administration (FDA) and parties seeking to provide Americans with less restricted access to supplements.

Key aspects of the Act represent a different philosophy than that which had dominated the FDA:

1. 'There is a link between the ingestion of certain nutrients or dietary supplements and the prevention of chronic diseases such as cancer, heart disease, and osteoporosis'.
2. 'Healthful diets may mitigate the need for expensive medical procedures, such as coronary bypass surgery or angioplasty'.
3. 'Preventive health measures, including education, good nutrition, and appropriate use of safe nutritional supplements will limit the incidence of chronic diseases, and reduce long-term health expenditures'.
4. 'There is a growing need for emphasis on the dissemination of information linking nutrition and long-term health'.
5. 'Consumers should be empowered to make choices about preventative health care programs based on data from scientific studies on health benefits related to particular dietary supplements'.
6. 'Studies indicate that consumers are placing increasing reliance on the use of nontraditional health care providers to avoid the excessive costs of traditional medical services and to obtain more holistic considerations of their needs'.

7. 'Dietary supplements are safe within a broad range of intake, and safety problems with supplements are relatively rare'.
8. 'Legislative action that protects the right of access of consumers to safe dietary supplements is necessary in order to promote wellness'.

This Act declared dietary supplements were neither foods nor drugs. Instead, they created a new category of food by specifically defining dietary supplements to include dietary ingredients, such as vitamins, minerals, herbs or other botanicals, amino acids or other dietary supplements. A dietary supplement must be intended for ingestion, such as a tablet, capsule, powder, softgel, gelcap or liquid form. The Act also specifically states that customers can be provided with publications on supplements in connection with the sale of the dietary supplements. These publications must not be false or misleading, not promote a specific brand of supplement, and provide a balanced view of the available scientific literature. In the store, it must be physically separated from the supplement. The Act allowed supplements to carry claims that they help preserve general well-being. Additionally, they can claim that they help preserve the structure or function of parts of the body. However, they can not state that the supplements are effective treatment of disease, because this would be a drug claim. This section of the Act has been the subject of discussion and litigation between the FDA and representatives of the supplement industry.

The Act specifically empowers the FDA to oversee quality control of supplements and they are given the power to remove unsafe supplements from the market. In practice, this process has not worked very well. The plain fact is that the quality of supplements on the market is very variable. Studies have shown that supplements from some manufacturers can contain 25% or less of the active ingredient than stated on the label (www.consumerlabs.com). It also appears that supplements on the market are not adequately monitored for hazardous contaminants (www.consumerlabs.com). Therefore, it would appear that there is a need for increased supervision of the supplement industry to ensure Americans of the potency and safety of the supplements they purchase. While we wait for these changes, there are some promising trends. Several major pharmaceutical firms have purchased supplement companies and this promises improvement in quality control and standardization. Additionally, several major supplement manufacturers are funding clinical trials that specifically document the value and side effects of their herbal extracts.

In summary, there is no doubt that Americans now have much greater access to nutritional or herbal supplements as a result of this Act. Additionally, Americans have much greater access to literature on nutritional and herbal supplements. However, these supplements range radically in their value. Among these supplements are therapeutic agents with a strong scientific basis and proven clinical value. Most of these lack FDA approval because no company has been willing to take them through the expensive process the FDA mandates. In contrast, other herbal products have no scientific basis, no sound clinical trial documentation of efficacy and may pose serious health risks.

ORTHOMOLECULAR THERAPIES

Antioxidants

Antioxidants represent one of the most common supplement classes ingested by prostate cancer patients. Widely used examples include vitamin C, vitamin E, selenium and plant polyphenols, such as those from green tea, grape seed and pine bark. There is a growing body of scientific and clinical work supporting the use of these compounds by patients with cancer and other diseases.

Antioxidant defense network

Many biochemical reactions essential for life carry with them the threat of oxidative damage to the tissues of your body. For example, the process by which mitochondria produce the ATP that drives your muscles and other tissues creates side products, such as hydrogen peroxide, that can damage tissues by oxidation. Exposure to sunlight can enhance oxidative damage to skin cells. Chemicals that naturally occur in food can cause oxidative damage. We are able to eat these foods safely only because our bodies do such a great job defending us against oxidative damage. An example of this are fava beans, which can cause severe injury if consumed by people who lack the normal defenses against oxidative damage.

We now know that the body has a comprehensive antioxidant defense network in which each component part has a role to play.[11,12] What are the components of this antioxidant defense network? One group of enzymes function to destroy or detoxify common oxidants. For example, hydrogen peroxide is one oxidant commonly formed as a byproduct as the tissues in the body perform their daily tasks. There are several enzymes capable of detoxifying hydrogen peroxide. The enzyme, catalase, reacts two hydrogen peroxide molecules together to form oxygen and water. A second family of enzymes, called glutathione peroxidases, reduces hydrogen peroxide to water. Most of the known glutathione peroxidases require selenium. A second group of components are vitamins that act as antioxidants. Vitamin C and vitamin E are prominent members of this group. Vitamin C is soluble in water and acts as an antioxidant in the water phase of the cell. Vitamin E is soluble in body fat and other lipids, but not in water. For this reason, vitamin E acts as an antioxidant for body lipids.

A third group of components are the dietary antioxidants. It is now apparent that most vegetables and grains contain antioxidants. Tomatoes contain the red pigment, lycopene. Green tea contains antioxidant polyphenols. Onions and garlic have large amounts of sulfur-containing chemicals that are strong antioxidants. Fruits, such as blueberries, strawberries and raspberries, are another rich source of antioxidants. It now appears that these plant antioxidants act to bolster the effectiveness of other members of the antioxidant network. The final component of the antioxidant defense network are a group of proteins that sequester iron and copper. This is important because free iron and copper can stimulate the conversion of peroxides into free radicals that rapidly react with and destroy normal tissues. Under normal conditions, this system is so effective that free iron and copper do not exist in body fluids or tissues. When sequestration of iron fails, it causes hemochromatosis and hemosiderosis. When sequestration of copper fails, it causes Wilson's disease.

When all four components of the antioxidant defense network are functioning optimally, the body can effectively handle attack by a wide range of oxidants without sustaining serious injury. Oxidative damage can include injury to DNA, the genetic material in a cell. This genetic damage can foster the development of cancer or promote the progression of cancer from a slow-growing local problem to one that grows and spreads rapidly. Thus, a fully functioning antioxidant network can lower the risk that cancer will develop. However, this network operates so that various components overlap in their function, so that a deficiency in one component can be compensated by the others. For example, a low level of vitamin E may not be of great consequence if selenium, vitamin C and dietary antioxidants are present at optimal levels. This is an important point to keep in mind when reading the results of clinical trials: any antioxidant can appear to have no impact if the subjects are taking in large amounts other antioxidants and/or are not exposed to significant oxidant stress. Conversely, if the people in the trial are all subject to some oxidative stress, antioxidants may prove more effective than they would be in normal subjects. For example, cigarette smoking subjects the body to increased oxidative stress and smokers may have a greater need for antioxidants than non-smokers. This will prove important when we discuss some of the clinical trials involving antioxidants and prostate cancer.

Oxidants and the development of prostate cancer

Why should antioxidants alter the risk of dying from prostate cancer? Several papers provide a possible answer to this question.[13,14] These authors have shown that prostate cancer cells exposed to testosterone produce hydrogen peroxide and other oxidants. This led them to propose that testosterone exposure results in the production of oxidants that cause genetic damage. Genetic damage can play a role in the progression as well as the genesis of cancer. The implication of this line of research is that strengthening the antioxidant network may lessen the risk of developing prostate cancer. It may also slow the progression of this disease from a slow-growing cancer limited to the prostate gland to one that has spread throughout the body and become resistant to all therapies. There are several antioxidants that may alter the natural history of prostate cancer.

Selenium

Over the past few years, there has been a dramatic increase in our knowledge about selenium and its effects on living organisms. A search of the Library of Medicine's online database, PubMed, show more than 1000 scientific articles published on selenium since January 1998. One major factor behind this surge in interest about the health effects of selenium has been the publication, in 1996, of a large randomized controlled clinical trial that demonstrated a marked reduction in cancer deaths associated with increased selenium intake.[15,16] This trial was initiated in 1983 as a randomized controlled trial testing the impact of supplementation with 200 μg of selenium yeast per day on the risk of skin cancer. This clinical trial enrolled 1300 individuals residing in the Mid-Atlantic Coastal Plain from Virginia to South Carolina, an area long known to have low soil and water selenium levels. After 10 years, the overall cancer death rate was 50% lower in the subjects on selenium as compared to the control group. The cancers responsible for this difference were carcinomas of the prostate, colon and lung; prostate cancer deaths were decreased by 64%, colon cancer by 40% and lung cancer by 30%. The four major causes of cancer death are those of the lung, prostate, colon and breast. Thus, selenium dramatically reduced the death rate of three of the four most common causes of cancer death in the USA. Furthermore, there were no cases of selenium toxicity among the people in this trial who took selenium supplements for years.

Selenium and prostate cancer Since the publication of this randomized controlled trial, there has been one major study designed to expand on the validity of these findings to address one criticism, namely that serum selenium levels used were just a snapshot of selenium levels at one point in time, while the development of prostate cancer takes many years and is more likely to be influenced by the average selenium level during that time period. Since hair and nails are formed at the rate of about one millimeter per day and thus are a better estimate of average selenium level, a recent study has examined the selenium content of nail clippings from close to 34 000 men and found that the risk of developing metastatic prostate cancer decreased as the selenium content of the nail clippings increased.[17] When men with the highest and lowest nail selenium content were compared, a 65% reduction in the risk of metastatic prostate cancer was found in the former group. What is the next step in this line of investigation? The National Cancer Institute (NCI) has just funded a large randomized controlled trial in which men will be randomized between control, vitamin E, selenium and vitamin E plus selenium.

Selenium and the immune system in cancer There are a number of studies that suggest adequate selenium levels are necessary for proper function of the immune system. The most convincing study is a randomized double-blind controlled clinical trial in patients with head and neck cancer.[18] Cytotoxic T cells and natural killer cells are the most important parts of the immune system in terms of resistance to cancer. Cytotoxic T-cell number and natural killer cell function is usually depressed in head and neck cancer patients. Patients in this study all had untreated squamous carcinoma of the head and neck, and were randomized to placebo or selenium at a dose of 200 μg per day for 8 weeks. Plasma selenium levels were measured at the start and during the trial to ensure patients took the selenium as directed. During the 8 weeks, 88% of the patients took the selenium dose assigned. At the start of the trial the average selenium levels in the two groups were 91.3 and 94.4 μg/l of blood plasma. After 8 weeks of selenium supplementation, the average selenium level in the men on this mineral increased from 91.3 μg/l to 105.3 μg/l.

Despite this minor increase in plasma selenium, there were major changes in the immune system. In the patients on placebo, cytotoxic T cells killed only 7% of the tumor cells. After 8 weeks of selenium supplementation, the T cells were able to kill 78% of the tumor cell targets. This is a rather remarkable shift given the short duration of treatment and small increase in selenium levels. To place these findings in perspective, in the Clarke study, prolonged administration of the same dose of selenium, 200 μg per day, eventually increases the selenium levels to close to 200 μg/l rather than the 105.3 μg/l after eight weeks reported in this paper on head and neck cancer.

Selenium and heavy metals Metals such as arsenic, lead, mercury, cadmium and thallium are all quite poisonous. Furthermore, industrial uses of these metals have led to large-scale environmental contamination. One of these, cadmium, has been specifically implicated as a cause of prostate cancer. Selenium alters the toxicity of heavy metal ions in several ways.[19,20] It is important to note that selenium can bind these metal ions into complexes that are insoluble in water. To a certain extent, selenium can bind to these metal ions in the gastrointestinal tract, diminishing the absorption of both selenium and the toxic metals. Once selenium has been absorbed, it can form complexes with these metal ions, lessening or delaying the toxicity of these metal ions and potentially reducing the risk of cadmium induced prostate cancer.

Vitamin E

Vitamin E is soluble in lipids but not in water. As a result, it largely acts to prevent oxidative damage to the lipids in a cell. In laboratory studies, vitamin E has been shown to act in concert with antioxidants present in the cytosol, such as vitamin C, glutathione and selenium. The first evidence that vitamin E might alter the progression of prostate cancer came as an unexpected result of a clinical trial designed to test the role of this vitamin in the prevention of lung cancer in smokers. In this trial, close to 29 133 male cigarette smokers were randomized to placebo, β-carotene, vitamin E or β-carotene plus vitamin E.[21] At the point where men had been on the trial for 5–8 years, death rates from prostate cancer were 40% less in the men on vitamin E alone compared to placebo. In contrast, the men on β-carotene experienced a 30% increase in mortality from this malignancy. The group on β-carotene plus vitamin E was not significantly different from the control group.

Vitamin E is not a single chemical substance but a name given to a family of fat-soluble antioxidants. These compounds differ in their content in foods as well as their ability to act as antioxidants. While α-tocopherol is widely available and the form most commonly used in laboratory and clinical studies, it is not the most active antioxidant. Recently, several groups have reported that γ-tocopherol is more active against prostate cancer cells than is the α form of this vitamin.[22] In contrast, other groups have found γ-tocopherol no more active than α-tocopherol as an anticancer agent.[23–25] Instead, they have found the most active tocopherol is the δ-form. Additionally, tocotrienols are considerably more active than the corresponding tocopherols, with the most active preparation currently available commercially being concentrated palm oil tocotrienols. Finally, vitamin E succinate appears to exert anticancer activity equal to or greater than any of the naturally occurring tocopherols or tocotrienols.

There is now evidence that vitamin E kills prostate cancer cells by a unique mechanism. All cells in the body have the capacity to commit suicide. There are several means by which this suicide can be triggered. One means of accomplishing this is by activating certain proteins, called 'death receptors' on the surface of cells. These death receptor proteins are designed so that when they are activated, the cell bearing them rapidly destroys itself. One of the most common of these proteins is called Fas. When vitamin E-like chemicals kill prostate cancer cells, Fas rapidly appears on the surface of the prostate cancer cells and appears to undergo activation, leading to cell death. These findings suggest that vitamin E does more than prevent genetic damage caused by oxidants released as a result of testosterone exposure. In fact, it may well be that vitamin E-like molecules may actually be orally active chemotherapeutic agents with low toxicity. This concept has led at least one group to synthesize analogs of vitamin E chemically with the hope of increasing the anticancer activity of this family of compounds.

Vitamin A

Vitamin A and its derivatives seem to have a protective effect against various cancers. However, epidemiological data on prostate cancer and vitamin A intake are conflicting. Some studies demonstrate a significant trend of increased prostatic cancer risk associated with decreasing serum retinol levels,[26] while others fail to show a protective value.[27] Some epidemiological studies have even shown an increased risk of prostate cancer with supplemental vitamin A intake.[28] A possible explanation for this discrepancy may be the source of dietary vitamin A. For example, in Asia and other areas with a low incidence of prostate cancer, vitamin A is largely obtained from vegetables, while the main source of vitamin A in the USA is fat. The positive correlation between vitamin A and prostate cancer may be due to higher dietary fat intake.

In a recent placebo-controlled, randomized intervention trial, β-carotene caused an increase in prostate cancer incidence.[21] As described above the 'Alpha-Tocopherol Beta-Carotene Cancer Prevention Study' included 29 133 eligible male cigarette smokers randomly assigned to receive 20 mg β-carotene, 50 mg tocopherol, β-carotene and tocopherol or placebo daily for 5–8 years.[21] β-Carotene treatment did not result in a decrease in cancer at any of the major sites but rather an increase at several sites, notably the lungs, prostate and stomach. Lung cancer increased by 18% (474 vs. 402 cases, 95% confidence interval 3–36, $P=0.01$), prostate cancer by 25% (138 vs. 112 cases) and stomach cancer by 25% (70 vs. 56). Interestingly, the increase in cancer incidence was most pronounced in men consuming alcohol. Another possibility is that the dose–response curve of chemopreventive agents may be bell shaped with excess administration reducing the efficacy or even promoting tumors with agents, such as alcohol shifting the dose–response curve. The biphasic effect of retinoic acid on LNCaP (a popular human prostate cancer cell line used in experimental laboratory studies) growth should be considered when designing *in vivo* studies to determine the impact of retinoic acid on solid prostatic tumor growth. In addition, the fact that retinoic acid increases PSA secretion may complicate the interpretation of serum PSA levels used for outcome assessment.[29]

Vitamin D

Vitamin D is a general term given to antirachitic substances synthesized in the body in response to ultraviolet radiation and found in foods activated by ultraviolet radiation. Calcitriol (1,25-D3) is the active form of vitamin D. Because the body can synthesize vitamin D, it has been reclassified as a hormone.[30] In 1990 it was first suggested that vitamin D_3 deficiency might increase the risk of prostate cancer.[31] Subsequent studies showed an inverse association between ultraviolet solar exposure and mortality rates for prostate cancer[32] and a direct association between vitamin D levels and prostate cancer.[33] Unfortunately, other similar studies were not able to demonstrate this association.[34]

Prostate cell division is influenced by two steroid hormones, testosterone and vitamin D, the action of each being mediated by its respective receptor. The first study examined genetic polymorphisms in the androgen receptor (AR) and the vitamin D receptor (VDR), in a case-control pilot study of prostate cancer where 57 non-Hispanic White case patients with prostate cancer and 169 non-Hispanic White control subjects were genotyped for a previously described microsatellite (CAG repeats) in the AR gene and for a newly discovered poly-A microsatellite in the 3′-untranslated region (3′UTR) of the VDR gene.[35] Both the AR and the VDR polymorphisms were associated, individually with prostate cancer. Adjusted odds ratios (ORs) [95% confidence intervals (CIs)] for prostate cancer were 2.10 (95% CI = 1.11–3.99) for individuals carrying an AR CAG allele with fewer than 20 repeats versus an allele with 20 or more repeats and 4.61 (95% CI = 1.34–15.82) for individuals carrying at least one long (A18–A22) VDR poly-A allele versus two short (A14–A17) poly-A alleles. For both the AR and VDR genes, the at-risk genotypes were more strongly associated with advanced disease than with localized disease.

Another study tested the hypothesis that vitamin D receptor gene polymorphisms are associated with prostate cancer risk using a case-control study of 108 men undergoing radical prostatectomy and 170 male urology clinic controls with no history of cancer.[36] Among the Caucasian control group, 22% were homozygous for the presence of a TaqI RFLP at codon 352 (genotype tt), but only 8% of cases had this genotype ($P<0.01$). A similar trend was seen among the small number of African Americans in this study (13% for controls, 8% for cases), although the difference was not statistically significant. Race-adjusted combined analysis suggests that men who are homozygous for the t allele (shown to correlate with higher serum levels of the active form of vitamin D) have one-third the risk of developing prostate cancer requiring prostatectomy compared to men who are heterozygotes or homozygous for the T allele (odds ratio MH = 0.34; 95% CI 0.16–0.76; $P<0.01$).

Substantial experimental data indicate that 1,25-dihydroxyvitamin D_3 (calcitriol) has potent antiproliferative effects on human prostate cancer cells.[37] Because of this and the genetic alterations of the VDR outlined above, investigators carried out an open-label, non-randomized pilot trial to determine whether calcitriol therapy is safe and efficacious for early recurrent prostate cancer. Their hypothesis was that calcitriol therapy slows the rate of rise of PSA compared with the pretreatment rate. After primary treatment with radiation or surgery, cancer recurrence was indicated by rising serum PSA levels documented on at least three occasions. Seven subjects completed 6–15 months of calcitriol therapy, starting with 0.5 μg calcitriol daily and slowly increasing to a maximum dose of 2.5 μg daily depending on individual calciuric and calcemic responses. Each subject served as his own control, comparing the rate of PSA rise before and after calcitriol treatment. As determined by multiple regression analysis, the rate of PSA rise during versus before calcitriol therapy significantly decreased in six of seven patients, while in the remaining man a deceleration in the rate of PSA rise did not reach statistical significance.

In conclusion, it would appear that variation of the VDR gene is associated with prostate cancer susceptibility and therapeutic exploitation of vitamin D biology is a potentially fruitful approach in prostate cancer. These results support recent ecological, population and *in vitro* studies, suggesting that vitamin D is an important determinant of prostate cancer risk and, if confirmed, suggest strategies for chemoprevention/treatment of this common cancer.

Vitamin C

Vitamin C is the major circulating water-soluble antioxidant, which acts as a free-radical scavenger, resulting in inhibition of malignant transformation *in vitro*[38] and a decrease in carcinogen-induced chromosome damage.[39] Despite the popularity of vitamin C as a cancer preventive, to date no consistent relationship between vitamin C levels and prostate cancer has been demonstrated.[40] However, treatment of androgen independent (DU145) and dependent (LNCaP) human prostate cancer cell lines, while growing in culture dishes with vitamin C, resulted in a dose- and time-dependent decrease in cell survival and cell division. Vitamin C induced these changes through the production of hydrogen peroxide, which damages cells by generating free radicals. These results suggest that ascorbic acid could be a potent anticancer agent for prostate cancer cells.[41]

HERBAL THERAPIES EMPLOYING INDIVIDUAL OR MIXTURES OF HERBS

Monophenols and polyphenols in green tea

Plant phenols are one of the most potent groups of antioxidants. They can occur as monophenols or as clusters of monophenols, called polyphenols. The major polyphenols from green tea are epigallocatechin-3-gallate (EGCG), epigallocatechin and epicatechin-3 gallate. Together, these make up 30–40% of the solids extracted from green tea leaves during brewing. However, polyphenols are found in a wide range of plant products that also have health benefits and may make a welcome alternative to green tea. Green tea intake has been associated with a decreased risk of cancers of the prostate, colon, pancreas, skin and other organs.[42–44] Seemingly, green tea prevents the damage caused by a large number of cancer-causing chemicals and even excessive sun exposure. It appears that green tea can not only prevent cancer, but is able to stop the growth or even kill human cancer cells. This ability was documented in human breast, lung, colon, pancreatic and prostate cancers in tissue culture and in animal models.

The contents of green tea have been carefully examined to determine the chemical in the tea responsible for the anticancer activity. There is a growing consensus that most of the useful activity is caused by a polyphenol called epigallocatechin gallate or EGCG. This compound caused the rapid shrinkage of human prostate cancers growing in mouse xenograft models. However, there is no known mechanism by which an antioxidant can cause cancer cells to stop growing or cause rapid shrinkage of a tumor mass. For this reason, it is very likely EGCG exerts its anticancer activity by a mechanism independent of its antioxidant activity.

EGCG and other green tea polyphenols are sensitive to light and air. EGCG and other tea polyphenols are most stable at a pH of less than 5.5 and when the tea is brewed at 180 °F or less. The amount of EGCG found in green tea appears to be effective as a cancer prevention agent. The amount needed to stop the growth of cancer cells or to cause the cancer to shrink rapidly in the mice is much larger – projecting from the animal experiments a dose equivalent to 10 or more cups per day would be necessary. Human clinical trials have progressed to phase I, where a dose of green tea extract of 1000 mg per day (equivalent to 10–12 cups per day) was well tolerated. At present, we are unaware of any published Phase II or III clinical trials in prostate cancer.

Saw palmetto

Saw palmetto is a compound derived from the fruit extract of the American dwarf palm tree that is native to the Atlantic Coastal Plain from North Carolina south through Northern Florida. The common herbal preparation represents the extracted lipids and sterols. Recent clinical trials testing saw palmetto in the treatment of benign prostatic hyperplasia (BPH) were reviewed in the *Journal of the American Medical Association* in 1998.[45] After critically evaluating the quality of the clinical trials and the consistency of the results, the authors concluded that saw palmetto was an effective treatment for BPH. This conclusion has been confirmed in two subsequent extensive literature reviews.[46,47]

One problem with these reviews are that they do not take into account that various commercially available saw palmetto products vary in their potency and quality. There are many different ways to extract these lipids and sterols, yielding products with differences in their biologic effects. Also, the different brands and preparations available with varying amount of extract per capsule. Consumer Labs, Inc. has tested 27 of the products on the market and found that only 17 contain an adequate amount of the specific fatty acids and sterols needed to produce a therapeutic effect. A list of the manufactures that passed this test is available at their website, www.consumerlabs.com.

The only sure way of knowing whether a given saw palmetto extract is active is for the manufacturer to sponsor clinical trials that document its activity. Clinical trials are expensive and most herbal products companies have been reluctant to participate in such stringent tests of the value of their products. The economic justification was that the consumer did not demand this investment and the cost of such trials eroded the profitability their business. A few companies have been fore-sighted enough to take part in clinical testing of their herbal extracts. In those few situations, we have solid evidence of the value of a herbal product as well as sound information about the effective dose and possible side effects.

The biochemical basis for the activity of saw palmetto against benign enlargement of the prostate almost certainly includes inhibition of the formation of dihydrotestosterone (DHT) within the prostate gland. Interestingly, saw palmetto inhibits formation of DHT in the prostate gland, but not in many other tissues and may not alter blood levels of this hormone.[48,49] The action of testosterone and dihydrotestosterone on the prostate gland is markedly stimulated by the hormone, prolactin. Saw palmetto is reported to block the response of prostate cells to prolactin.[50]

Saw palmetto has also been reported to inhibit the enzyme, 5-lipoxygenase, the enzyme that converts arachidonic acid to the eicosanoid, 5-HETE. This eicosanoid is known to stimulate the growth of prostate cancer cells.[51]

In BPH, the increased size of the gland can arise from an increase in the number of cells lining the prostate ducts as well as an increase in the number of stromal cells in the space between the ducts. With this background, it is interesting to note that saw palmetto has been shown to cause the death of both the stromal cells and the cells lining the prostate ducts.[52,53] This has been followed by one study showing that saw palmetto was able to slow the growth of prostate cancer cells and, at high enough concentrations, even kill these cancer cells. To date, there are no clinical trials testing saw palmetto in the treatment of prostate cancer, so it is difficult to know whether this observation is clinically relevant.

In summary, specific saw palmetto extracts appear to have useful activity in the treatment of BPH. This product appears to alter prostate biology through a number of mechanisms, some of which might also be relevant to the treatment of prostate cancer. This is especially true if these are considered as part of a multiagent regimen such as PC-SPES when saw palmetto is combined with other agents. Unfortunately, clinical trial documentation of the activity of saw palmetto extract against prostate cancer is completely lacking.

Black cohosh root

The alcoholic extract of black cohosh root is widely used by women in Europe and now the USA as a treatment for menopausal symptoms.[54–56] Virtually all of the clinical studies have been done with one preparation, Remifemin, originally developed in Europe and now marketed in the USA. Many men with prostate cancer on hormonal therapy are using Remifemin or other black cohosh root preparations as treatment for their hot flashes and other symptoms of male menopause.

Black cohosh was one of the medicinal plants widely used by various American Indian tribes and adapted by early European settlers as a treatment for rheumatism. The use of black cohosh for menopausal symptoms became widespread during the 1800s in both America and Europe. At present, virtually all significant clinical and laboratory investigation on black cohosh have been performed by European investigators. Numerous clinical trials, including randomized controlled trials, have compared Remifemin with placebo and a variety of estrogen preparations commonly used as hormone replacement therapy for women. It is now well documented that this herbal preparation effectively reduces hot flashes in women. Remifemin also appears to be effective in countering the depression, insomnia and other psychiatric complications of female menopause. In addition, this black cohosh extract appears to prevent bone loss in animals after surgical removal of the ovaries, raising the possibility that it might also ameliorate the development of osteoporosis. This possibility has yet to be tested in humans. Remifemin also appears to act on the female genitalia, where it has been reported to cause estrogen-like effects on the vaginal mucosa.

The production of estrogen by the ovary is stimulated by luteinizing hormone (LH). When the brain senses that sufficient estrogen is present, it decreases the production of LH, shutting down the production of additional estrogen.[57] During menopause in women, LH production increases because the brain senses the absence of estrogen. The magnitude of this LH rise has been reported to correlate with the severity of menopausal symptoms. A majority of the studies indicate that a dose of Remifemin sufficient to reverse menopausal symptoms also blocks the release of LH.

As with most herbal preparations, Remifemin is composed of a mixture of different chemicals. Fractions have been identified that can bind to the estrogen receptor and appear to mimic the actions of estrogen. Another fraction can block LH release even though it has no estrogenic activity. These results suggest that it would be possible to prepare black cohosh extracts that selectively suppress the response of the brain to menopause, such as hot flashes, depression and insomnia, that lack any activity at the estrogen receptor.

The most common side effect of Remifemin is transient gastric distress seen in approximately 7% of women. In fact, it seems to be as well or better tolerated than the commonly used estrogenic medications. In particular, it appears to be much less likely to cause uterine bleeding. In standard test systems, it also does not cause mutations or stimulate the development of cancers. This is important because naturally occurring estrogens may promote the development of cancers of the breast and uterus.

What about the use of Remifemin and other black cohosh extracts in men? There are no clinical trials in which Remifemin or other black cohosh extracts were administered to men. In men already on hormonal therapy, the estrogenic activity of this supplement may cause breast enlargement and other estrogen-dependent side effects. We have no information about the impact of Remifemin and other black cohosh preparations on prostate cancer growth and spread, and it is impossible to predict its impact on this disease. Therefore, its use in men is questionable at best until we have a better understanding of its impact on the physiology of the human male and its effects on prostate cancer.

Lycopene

It has been long known that intake of fruits, vegetables and grains are associated with a decrease in the risk of many cancers, but the specific components of these foods responsible for this reduced risk remain a matter of controversy. Lycopene is a member of a broad group of plant pigments, called carotenoids, that includes β-carotene (orange color of carrots) and lutein (yellow color in many vegetables). Carotenoids, especially β-carotene, have long been thought to play an important role in cancer prevention.[58] Lycopene is the red pigment found in tomatoes, pink grapefruit, watermelon and apricots, and is related to β-carotene and other carotinoids in structure.

Evidence strongly indicates that intake of tomato products and lycopene offer protection against cancers of the stomach, lung and prostate. Lycopene also appears to reduce the risk of cancers of the oral cavity, esophagus, breast, pancreas, cervix and colon. What is more interesting is that none of the 72 studies reviewed in preparation of this chapter report a link between lycopene and an increase in the risk of any cancer. Furthermore, there are no reports of toxicity, regardless of dose.

In a recent review, Giovannucci discussed study after study in which the intake of lycopene, but not other carotenoids, such as β-carotene, correlates with protection from cancer.[59] This is consistent with a recent randomized controlled trial that found supplemental β-carotene actually increased the risk of death from prostate cancer. This is an important point, because many dietitians and alternative health practitioners are still recommending β-carotene to men with prostate cancer. It is important to stress that there is no evidence to support this practice.

Lycopene is well absorbed from cooked tomato products, such as tomato sauce or tomato paste, but not from fresh tomatoes. Additionally, lycopene absorption is enhanced by the addition of oil.[60] Why should lycopene reduce the risk of cancer? Carotenoids can act as antioxidants and as precursors to Vitamin A. Among the common carotenoids, lycopene is the most effective antioxidant, and this property may be important given the activity of vitamin E and selenium against this cancer.

Several observations instill confidence in the importance of lycopene. First, the studies that report a reduced risk of cancer associated with lycopene involved the USA, Italy, Netherlands, Spain, Sweden, Australia, Iran, China and Japan, indicating the protective effect persists despite widely different dietary patterns, lifestyles and racial background. Second, the protective effect extends to a wide range of human cancers, each of which has its own unique biology and is caused by different mechanisms. Third, in every country and with every cancer examined, lycopene intake never correlated with a significant increase in the risk of cancer.

What is lacking? Final proof of lycopene's importance will require a randomized controlled clinical trial in which large numbers of people take a placebo or a defined amount of lycopene over a prolonged period of time, and the impact on cancer frequency and cancer deaths measured. Since tomato products are often used in prepared foods in ways that are not obvious, this study will be difficult to conduct.

PC-SPES

PC-SPES is a combination of seven Chinese medicinal herbs [Reishi (*Ganoderma lucidum*) spores, Balkal skullcap (*Scutellaria baicalenesis*) root, Rabdosia (*Rabdosia rubescens*) root, Dyer's wood (*Isatis indigotica*) root, mum (*Dendranthema morifolium*) flower, san-qi ginseng (*Panax notoginseng*) bark and licorice (*Glycyrrhiza glabra*) root] with the addition of one North American herb, saw palmetto. This herbal product appeared a few years ago and is widely used by patients as a treatment for their prostate cancer. The story of this herbal product illustrates the positive and negative aspects of the current regulatory environment with regard to supplements. This product did not proceed through the standard process by which prescription drugs are approved by the FDA. Shortly after this product was introduced, many of us became aware that in some patients this product induced a significant decrease in tumor size. Physicians initially had no information about appropriate dosing, side effects and antitumor activity of this preparation.

Physicians specializing in prostate cancer have gathered experience with the use of this herbal product, largely because patients decided on their own to try it. Reports about PC-SPES then began to appear in medical literature. In one of the first clinical and experimental studies to evaluate the effect of PC-SPES in humans, DiPaola and associates[61] tested the estrogenic activity of PC-SPES in yeast and mice, and in men with prostate cancer. In this study, they measured the estrogenic activity of PC-SPES with transcriptional-activation assays in yeast and a biologic assay in mice. The clinical activity of PC-SPES was evaluated in eight patients with hormone-sensitive prostate cancer by measuring serum prostate-specific antigen and testosterone concentrations during and after treatment. In complementary yeast assays, a 1:200 dilution of an ethanol extract of PC-SPES had estrogenic activity similar to that of 1 nM estradiol, and in ovariectomized CD-1 mice, the herbal mixture increased uterine weights substantially. In six out of six men with hormone-sensitive prostate cancer, PC-SPES decreased serum testosterone concentrations and in eight out of eight patients it decreased serum concentrations of PSA. All eight patients had breast tenderness and loss of libido, and one had venous thrombosis. High-performance liquid chromatography, gas chromatography and mass spectrometry showed that PC-SPES contains estrogenic organic compounds that are distinct from diethylstilbestrol, estrone and estradiol.

This exciting study was followed by several other recent studies which examined both *in vitro* and clinical effects of PC-SPES. At UCSF,[62] 33 patients with androgen-dependent prostate cancer (ADPCa) and 37 patients with androgen-independent prostate cancer (AIPCa) were treated with PC-SPES (nine capsules daily). All ADPCa patients experienced a PSA decline of >80%, with a median duration of 57+ weeks. No patient developed PSA progression. Almost all of these patients had declines of testosterone to the anorchid range. Nineteen (54%) of 35 AIPCa patients had a PSA decline of >50%, a commonly used surrogate index for a robust response in these clinical situations.[62] Median time to PSA progression was 4 months. Of 25 patients with positive bone scans, two had improvement, seven had stable disease, 11 had progressive disease and five did not have a repeat bone scan because of PSA progression. Severe toxicities included thromboembolic events and allergic reactions. Other frequent toxicities included gynecomastia/gynecodynia, leg cramps and grade 1 or 2 diarrhea.

The study at Columbia University evaluated 69 patients with prostate cancer treated with PC-SPES (three capsules daily).[63] Serum PSA responses and side effects were evaluated. Of the patients with prostate cancer, 82% had decreased serum PSA 2 months, 78% 6 months and 88% 12 months after treatment with PC-SPES. Side effects in the treated patient population included nipple tenderness in 42% and phlebitis requiring heparinization in 2%. In a recent study from the Dana–Farber Cancer Institute, 23 patients with AIPCa were treated with PC-SPES (six capsules daily). With a median follow-up of 8 months, 20 patients experienced a post-therapy decline in PSA. Twelve patients (52%) had a >50% decline in PSA. The median duration of the PSA response was 2.5 months. Toxicity was mild and included nipple tenderness, nausea and diarrhea. In univariate analyses, older patients and those with a longer duration of initial androgen ablation therapy were more likely to respond to PC-SPES.[64]

In recent *in vitro* studies from Columbia University,[63] PC-SPES was evaluated for its ability to induce apoptosis on human prostate cancer cell lines LNCaP (hormone sensitive), PC3 (hormone independent) and DU145 (hormone independent). The effect of oral PC-SPES on growth of PC3 tumors present in male immunodeficient mice was studied. All of the cultured prostate cancer cell lines had a significant dose-dependent induction of apoptosis following exposure to PC-SPES. Immunodeficient mice xenografted with the PC3 cell line had reduced tumor volume compared with sham-treated controls when they were treated with a PC-SPES extract from the time of tumor cell implantation but not when the treatment was begun 1 week after tumor cell implantation.

Soy and genistein

Several reviews have suggested that soy food consumption may contribute to the relatively low rates of breast, colon and prostate cancers in countries such as China and Japan.[65,66] Soy contains isoflavones, such as genistein and daidzein, which are thought to account for the health benefits of this legume. You will see these compounds alternatively referred to by their specific chemical name, such as genistein, as isoflavones or as phytoestrogens. The latter term arises from the fact that these isoflavones mimic some of the biological effects of the female sex hormone, estrogen. In the laboratory, genistein has well-documented activity against prostate cancer.[67–71] High concentrations of genistein and other soy isoflavones cause the death of prostate cancer cells. At somewhat lower concentrations, these isoflavones arrest the growth of prostate cancer and block tumor cell invasiveness. Additionally, there are now several studies in which soy products or isolated soy isoflavones have been demonstrated to slow the growth of human prostate cancer cells in animal models.[68,72,73]

The mechanism by which soy isoflavones might slow the growth and/or spread of prostate cancer is unclear. Genistein, at concentrations obtainable in humans, can block the action of both the epidermal growth factor and Her-2/neu receptors. This means that genistein can theoretically block one of the major mechanisms by which prostate cancer cells become hormone-refractory. While these isoflavones act as estrogenic compounds in laboratory assays, men on soy-rich diets have only minor alterations in sex hormone levels. Finally, soy isoflavones appear to block tumor angiogenesis in laboratory models by inhibiting the growth of proliferating endothelial cells.[74]

There is also controversy on the best soy product to use in order to obtain high blood levels of genistein and other soy isoflavones. Genistein is absorbed much more effectively from fermented soy products such as miso, natto and tempeh than it is from soy beans, tofu or soy milk. Additionally, soy phytoestrogen or genistein tablets or capsules currently on the market would easily permit the ingestion of several grams of soy isoflavones per day. An additional complication is that blood levels of genistein may underestimate the levels in the prostate. When genistein levels are measured in prostatic fluid, the concentration is 5–10 times higher than in the blood.

Despite these findings, use of soy isoflavones in the treatment or prevention of prostate cancer in men is questionable. While there are quite a few epidemiology studies that show a correlation between high soy intake and a reduced risk of prostate cancer, no randomized controlled clinical trial showing that these soy products prolong the life of men with this disease has yet been published.[75,76] We even lack clinical trials that show soy products arrest or slow the growth of prostate cancer in men. We do not even know the dose and schedule for the administration of soy isoflavones most likely to affect human prostate cancer. On the other hand, there appear to be other health benefits associated with the use of soy protein as a substitute for animal proteins, prompting the FDA to allow firms marketing soy protein to claim that these products have a favorable impact on the course of coronary heart disease.[77]

SUMMARY AND CONCLUSIONS

The 'Dietary Supplement Health and Education Act of 1994' guarantees Americans ready access to herbal products. While selected herbal products have impressive therapeutic activity in tissue culture and animal models, with few exceptions clinical trial documentation of activity is less than impressive. In fact, the common pattern is for initially promising laboratory findings to lead to widespread commercial availability without any intervening clinical investigation. An additional problem of great concern is that the commercially available herbal products vary widely in their quality. Nevertheless, the increasing number of clinical trials carried out with these materials combined with the advent of modern molecular biological drug-screening techniques promises to revolutionize drug discovery and lead to effective anticancer compounds in the next 5–10 years.

REFERENCES

1. Eisenberg DM, Kessler RC, Foster C et al. Unconventional medicine in the United States. Prevalence, costs, and patterns of use. *N. Engl. J. Med.* 1993; 328:246–52.
2. Eisenberg DM, Davis RB, Ettner SL et al. Trends in alternative medicine use in the United States, 1990–1997: results of a follow-up national survey. *JAMA*. 1998; 280:1569–75.
3. Cassileth BR, Chapman CC. Alternative and complementary cancer therapies. *Cancer* 1996; 77:1026–34.
4. Richardson MA, Sanders T, Palmer JL et al. Complementary/alternative medicine use in a comprehensive cancer center and the implications for oncology. *J. Clin. Oncol.* 2000; 18:2505–14.
5. Lazar JS, O'Connor BB. Talking with patients about their use of alternative therapies. *Prim. Care* 1997; 24:699–714.
6. Bold J, Leis A. Unconventional therapy use among children with cancer in Saskatchewan. *J. Pediatr. Oncol. Nurs.* 2001; 18:16–25.
7. Lippert MC, McClain R, Boyd JC et al. Alternative medicine use in patients with localized prostate carcinoma treated with curative intent. *Cancer* 1999; 86:2642–8.
8. Eisenberg DM, Kessler RC, Foster C et al. Unconventional medicine in the United States. Prevalence, costs, and patterns of use. *N. Engl. J. Med.* 1993; 328:246–52.
9. DiPaola RS, Zhang H, Lambert GH et al. Clinical and biologic activity of an estrogenic herbal combination (PC-SPES) in prostate cancer. *N. Engl. J. Med.* 1998; 339:785–91.
10. Bass IS, Young AL. *Dietary Supplement Health and Education Act: a legislative history and analysis*. Washington, DC: The Food and Drug Law Institute, 1996.
11. Clinton S. The dietary antioxidant network and prostate carcinoma. *Cancer* 1999; 86:1629–31.
12. Halliwell B. Antioxidant defense mechanisms: from the beginning to the end (of the beginning). *Free Radical Res.* 1999; 31:261–72.
13. Ripple MO, Henry WF, Rago RP et al. Prooxidant-antioxidant shift induced by androgen treatment of human prostate carcinoma cells. *J. Natl Cancer Inst.* 1997; 89:40–8.

14. Ripple GH, Wilding, G. Drug development in prostate cancer. *Semin. Oncol.* 1999; 26:217–26.
15. Clark LC, Combs GF Jr, Turnbull BW et al. Effects of selenium supplementation for cancer prevention in patients with carcinoma of the skin. A randomized controlled trial. Nutritional Prevention of Cancer Study Group [published erratum appears in *JAMA* 1997; 277:1520]. *JAMA* 1996; 276:1957–63.
16. Clark LC, Dalkin B, Krongrad A et al. Decreased incidence of prostate cancer with selenium supplementation: results of a double-blind cancer prevention trial. *Br. J. Urol.* 1998; 81:730–4.
17. Yoshizawa K, Willett WC, Morris SJ et al. Study of prediagnostic selenium level in toenails and the risk of advanced prostate cancer. *J. Natl Cancer Inst.* 1998; 90:1219–24.
18. Kiremidjian-Schumacher L, Roy M, Glickman R et al. Selenium and immunocompetence in patients with head and neck cancer. *Biol. Trace Elem. Res.* 2000; 73:97–111.
19. Zikic RV, Stajn AS, Ognjanovic BI et al. The effect of cadmium and selenium on the antioxidant enzyme activities in rat heart. *J. Environ. Pathol. Toxicol. Oncol.* 1998; 17:259–64.
20. Whanger PD. Selenium in the treatment of heavy metal poisoning and chemical carcinogenesis. *J. Trace Elem. Electrolytes Health Dis.* 1992; 6:209–21.
21. Heinonen OP, Albanes D, Virtamo J et al. Prostate cancer and supplementation with alpha-tocopherol and beta-carotene: incidence and mortality in a controlled trial. *J. Natl Cancer Inst.* 1998; 90:440–6.
22. Moyad MA, Brumfield SK, Pienta KJ. Vitamin E, alpha- and gamma-tocopherol, and prostate cancer. *Semin. Urol. Oncol.* 1999; 17:85–90.
23. Israel K, Yu W, Sanders BG et al. Vitamin E succinate induces apoptosis in human prostate cancer cells: role for Fas in vitamin E succinate-triggered apoptosis. *Nutr. Cancer* 2000; 36:90–100.
24. Gunawardena K, Murray DK, Meikle AW. Vitamin E and other antioxidants inhibit human prostate cancer cells through apoptosis. *Prostate* 2000; 44:287–95.
25. McIntyre BS, Briski KP, Gapor A et al. Antiproliferative and apoptotic effects of tocopherols and tocotrienols on preneoplastic and neoplastic mouse mammary epithelial cells. *Proc. Soc. Exp. Biol. Med.* 2000; 224:292–301.
26. Hayes RB, Bogdanovicz JF, Schroeder FH et al. Serum retinol and prostate cancer. *Cancer* 1988; 62:2021–6.
27. Eichholzer M, Stahelin HB, Gey KF et al. Prediction of male cancer mortality by plasma levels of interacting vitamins: 17-year follow-up of the prospective Basel study. *Int. J. Cancer* 1996; 66:145–50.
28. Hsing AW, McLaughlin JK, Schuman LM et al. Diet, tobacco use, and fatal prostate cancer: results from the Lutheran Brotherhood Cohort Study. *Cancer Res.* 1990; 50:6836–40.
29. Fong CJ, Sutkowski DM, Braun EJ et al. Effect of retinoic acid on the proliferation and secretory activity of androgen-responsive prostatic carcinoma cells. *J. Urol.* 1993; 149:1190–4.
30. Bouillon R, Okamura WH, Norman AW. Structure–function relationships in the vitamin D endocrine system. *Endocr. Rev.* 1995; 16:200–57.
31. Schwartz GG, Hulka BS. Is vitamin D deficiency a risk factor for prostate cancer? (Hypothesis). *Anticancer Res.* 1990; 10:1307–11.
32. Hanchette CL, Schwartz GG. Geographic patterns of prostate cancer mortality. Evidence for a protective effect of ultraviolet radiation. *Cancer* 1992; 70:2861–9.
33. Corder EH, Guess HA, Hulka BS et al. Vitamin D and prostate cancer: a prediagnostic study with stored sera. *Cancer Epidemiol. Biomarkers Prev.* 1993; 2:467–72.
34. Braun MM, Helzlsouer KJ, Hollis BW et al. Prostate cancer and prediagnostic levels of serum vitamin D metabolites (Maryland, United States). *Cancer Causes Control* 1995; 6:235–9.
35. Ingles SA, Ross RK, Yu MC et al. Association of prostate cancer risk with genetic polymorphisms in vitamin D receptor and androgen receptor. *J. Natl Cancer Inst.* 1997; 89:166–70.
36. Taylor JA, Hirvonen A, Watson M et al. Association of prostate cancer with vitamin D receptor gene polymorphism. *Cancer Res.* 1996; 56:4108–10.
37. Peehl DM, Skowronski RJ, Leung GK et al. Antiproliferative effects of 1,25-dihydroxyvitamin D3 on primary cultures of human prostatic cells. *Cancer Res.* 1994; 54:805–10.
38. Sies H, Stahl W, Sundquist AR. Antioxidant functions of vitamins. Vitamins E and C, beta-carotene, and other carotenoids. *Ann N. York Acad. Sci.* 1992; 669:7–20.
39. Pohl H, Reidy JA. Vitamin C intake influences the bleomycin-induced chromosome damage assay: implications for detection of cancer susceptibility and chromosome breakage syndromes. *Mutat. Res.* 1989; 224:247–52.
40. Fair WR, Fleshner NE, Heston W. Cancer of the prostate: a nutritional disease? *Urology* 1997; 50:840–8.
41. Maramag C, Menon M, Balaji KC et al. Effect of vitamin C on prostate cancer cells in vitro: effect on cell number, viability, and DNA synthesis. *Prostate* 1997; 32:188–95.
42. Ahmad N, Feyes DK, Nieminen AL et al. Green tea constituent epigallocatechin-3-gallate and induction of apoptosis and cell cycle arrest in human carcinoma cells. *J. Natl Cancer Inst.* 1997; 89:1881–6.
43. Liao S, Umekita Y, Guo J et al. Growth inhibition and regression of human prostate and breast tumors in athymic mice by tea epigallocatechin gallate. *Cancer Lett.* 1995; 96:239–43.
44. Valcic S, Timmermann BN, Alberts DS et al. Inhibitory effect of six green tea catechins and caffeine on the growth of four selected human tumor cell lines. *Anticancer Drugs* 1996; 7:461–8.
45. Wilt TJ, Ishani A, Stark G et al. Saw palmetto extracts for treatment of benign prostatic hyperplasia: a systematic review [published erratum appears in *JAMA* 1999; 281:515]. *JAMA* 1998; 280:1604–9.
46. Wilt T, Ishani A, Stark G et al. Serenoa repens for benign prostatic hyperplasia. *Cochrane Database Syst. Rev.* 2000; 2.
47. Boyle P, Robertson C, Lowe F, Roehrborn C. Meta-analysis of clinical trials of permixon in the treatment of symptomatic benign prostatic hyperplasia. *Urology* 2000; 55:533–9.
48. Bayne CW, Ross M, Donnelly F et al. The selectivity and specificity of the actions of the lipido-sterolic extract of Serenoa repens (Permixon) on the prostate. *J. Urol.* 2000; 164:876–81.
49. Di Silverio F, Monti S, Sciarra A et al. Effects of long-term treatment with Serenoa repens (Permixon) on the concentrations and regional distribution of androgens and epidermal growth factor in benign prostatic hyperplasia. *Prostate* 1998; 37:77–83.
50. Van Coppenolle F, Le Bourhis X, Carpentier F et al. Pharmacological effects of the lipidosterolic extract of Serenoa repens (Permixon) on rat prostate hyperplasia induced by hyperprolactinemia: comparison with finasteride. *Prostate* 2000; 43:49–58.
51. Paubert-Braquet M, Mencia Huerta JM, Cousse H et al. Effect of the lipidic lipidosterolic extract of Serenoa repens (Permixon) on the ionophore A23187-stimulated production of leukotriene B4 (LTB4) from human polymorphonuclear neutrophils. *Prostaglandins Leukot. Essent. Fatty Acids* 1997; 57:299–304.
52. Ravenna L, Di Silverio F, Russo MA et al. Effects of the lipidosterolic extract of Serenoa repens (Permixon) on human prostatic cell lines. *Prostate* 1996; 29:219–30.
53. Vacherot F, Azzouz M, Gil-Diez-De-Medina S et al. Induction of apoptosis and inhibition of cell proliferation by the lipidosterolic extract of serenoa repens (LSESr, Permixon(R)) in benign prostatic hyperplasia. *Prostate* 2000; 45:259–66.
54. Hardy ML. Herbs of special interest to women. *J. Am. Pharm. Assoc. (Wash.)* 2000; 40:234–42; quiz 327–9.
55. Wade C, Kronenberg F, Kelly A et al. Hormone-modulating herbs: implications for women's health. *J. Am. Med. Womens Assoc.* 1999; 54:181–3.
56. Liske E. Therapeutic efficacy and safety of Cimicifuga racemosa for gynecologic disorders. *Adv. Ther.* 1998; 15:45–53.

57. Duker EM, Kopanski L, Jarry H et al. Effects of extracts from Cimicifuga racemosa on gonadotropin release in menopausal women and ovariectomized rats. *Planta Med.* 1991; 57:420–4.
58. Agarwal S, Rao AV. Tomato lycopene and its role in human health and chronic diseases. *Cmaj.* 2000; 163:739–44.
59. Giovannucci E. Tomatoes, tomato-based products, lycopene, and cancer: review of the epidemiologic literature, *J. Natl Cancer Inst.* 1999; 91:317–31.
60. Lee A, Thurnham DI, Chopra M. Consumption of tomato products with olive oil but not sunflower oil increases the antioxidant activity of plasma. *Free Radic. Biol. Med.* 2000; 29:1051–5.
61. DiPaola RS, Zhang H, Lambert GH et al. Clinical and biologic activity of an estrogenic herbal combination (PC-SPES) in prostate cancer. *N. Engl. J. Med.* 1998; 339:785–91.
62. Small EJ, Frohlich MW, Bok R et al. Prospective trial of the herbal supplement PC-SPES in patients with progressive prostate cancer. *J. Clin. Oncol.* 2000; 18:3595–603.
63. de la Taille A, Buttyan R, Hayek O et al. Herbal therapy PC-SPES: in vitro effects and evaluation of its efficacy in 69 patients with prostate cancer. *J. Urol.* 2000; 164:1229–34.
64. Oh WK, George DJ, Hackmann K et al. Activity of the herbal combination, PC-SPES, in the treatment of patients with androgen-independent prostate cancer. *Urology* 2001; 57:122–6.
65. Messina, M. Soy, soy phytoestrogens (isoflavones), and breast cancer. *Am. J. Clin. Nutr.* 1999; 70:574–5.
66. Messina MJ, Persky V, Setchell KD et al. Soy intake and cancer risk: a review of the in vitro and in vivo data. *Nutr. Cancer* 1994; 21:113–31.
67. Bylund A, Zhang JX, Bergh A et al. Rye bran and soy protein delay growth and increase apoptosis of human LNCaP prostate adenocarcinoma in nude mice. *Prostate* 2000; 42:304–14.
68. Aronson WJ, Tymchuk CN, Elashoff RM et al. Decreased growth of human prostate LNCaP tumors in SCID mice fed a low-fat, soy protein diet with isoflavones. *Nutr. Cancer* 1999; 35:130–6.
69. Shen JC, Klein RD, Wei Q et al. Low-dose genistein induces cyclin-dependent kinase inhibitors and G(1) cell-cycle arrest in human prostate cancer cells. *Mol. Carcinog.* 2000; 29:92–102.
70. Yang Z, Liu S, Chen X et al. Induction of apoptotic cell death and in vivo growth inhibition of human cancer cells by a saturated branched-chain fatty acid, 13- methyltetradecanoic acid. *Cancer Res.* 2000; 60:505–9.
71. Zhou JR, Gugger ET, Tanaka T et al. Soybean phytochemicals inhibit the growth of transplantable human prostate carcinoma and tumor angiogenesis in mice. *J. Nutr.* 1999; 129:1628–35.
72. Pollard M, Wolter W. Prevention of spontaneous prostate-related cancer in Lobund-Wistar rats by a soy protein isolate/isoflavone diet. *Prostate* 2000; 45:101–5.
73. Bylund A, Zhang JX, Bergh A et al. Rye bran and soy protein delay growth and increase apoptosis of human LNCaP prostate adenocarcinoma in nude mice. *Prostate* 2000; 42:304–14.
74. Koroma BM, de Juan E Jr. Inhibition of protein tyrosine phosphorylation in endothelial cells: relationship to antiproliferative action of genistein. *Biochem. Soc. Trans.* 1997; 25:35–40.
75. Habito RC, Montalto J, Leslie E et al. Effects of replacing meat with soyabean in the diet on sex hormone concentrations in healthy adult males. *Br. J. Nutr.* 2000; 84:557–63.
76. Jacobsen BK, Knutsen SF, Fraser GE. Does high soy milk intake reduce prostate cancer incidence? The Adventist Health Study (United States). *Cancer Causes Control* 1998; 9:553–7.
77. Stein K. FDA approves health claim labeling for foods containing soy protein. *J. Am. Diet. Assoc.* 2000; 100:292.

What is the Newest Technology in Treating Prostate Cancer? Transrectal High-intensity Focused Ultrasound

Chapter 56

Stephan Madersbacher and Michael Marberger
Department of Urology, University of Vienna, Austria

PHYSICAL PRINCIPLE

As an ultrasound wave propagates through a medium, it is progressively adsorbed and the energy is converted to heat in any medium that is not ideally viscoelastic, such as biological tissue.[1–4] In diagnostic ultrasound systems, the amount of heat generated is extremely small and has not been shown to have any harmful effect. For therapeutic purposes, the intensities exceed those of diagnostic systems by the factor 10^3–10^4, which induce tissue temperatures in the focal area of >80 °C; this technique is known as high-intensity focused ultrasound (HIFU).[1–4] The sharp heat impulse leads to a coagulative necrosis of all cellular elements within this focal area.[1–4] As the HIFU beam is emitted in a focused way, thermal damage to intervening tissue and areas in the vicinity of the focal zone can be avoided. Hence, HIFU is the only means currently available permitting contact- and irradiation-free in-depth tissue ablation in any solid organ accessible for ultrasound.[1–4]

The thermal effect of HIFU on tissue is dependent on a number of factors, such as: (1) the ultrasonic site intensity throughout the tissues; (2) the absorption coefficients of the tissues; (3) the temperature rise throughout the exposed tissues [for short time intervals (<0.1 seconds), the temperature rise is proportional to ultrasonic intensity; as time increases, temperature rises are modified by thermal conduction and simple proportionality is no longer valid]; (4) the damage integral on tissue strictures, which estimates the effects of temperature elevation [tissue changes occur if the heat induction exceeds the threshold level of protein degradation of 45–47 °C].[3]

ROUTE OF HIFU APPLICATION FOR PROSTATIC TISSUE ABLATION

To ablate prostatic tissue by HIFU, a transabdominal, perineal or transrectal approach is feasible; for several reasons, the transrectal one is clearly preferable.[3,4] First of all, the prostate is shielded by the bony pelvis making is difficult for it to be approached by HIFU transabdominally. Second, the close proximity of a transrectal HIFU transducer to the prostate allows the use of shorter HIFU-beam focal lengths and lower energy levels.[3,4] Consequently, the mechanical energy to intervening tissue is reduced, thus limiting thermal damage to these structures (particularly the rectum). Finally, ultrasound-guided imaging is clearly facilitated transrectally as higher ultrasound frequencies can be used. In fact, solid histological and clinical data for prostate tissue ablation are only available for transrectal devices.

TRANSRECTAL HIFU DEVICES

Transrectal devices incorporate an imaging and therapy transducer on a single probe sheath (**Figs 56.1** and **56.2**)[4]. For air-free coupling of the HIFU beam to tissue (i.e. the rectal wall), the probe is covered by a condom, which is filled, after rectal insertion, with degassed water. In one device (Sonablate,™ Focus Surgery Inc., USA) the same 4.0 MHz transducer is used for imaging and therapy (**Figs 56.1** and **56.2**). The focal length is dependent on the crystal used; presently, different

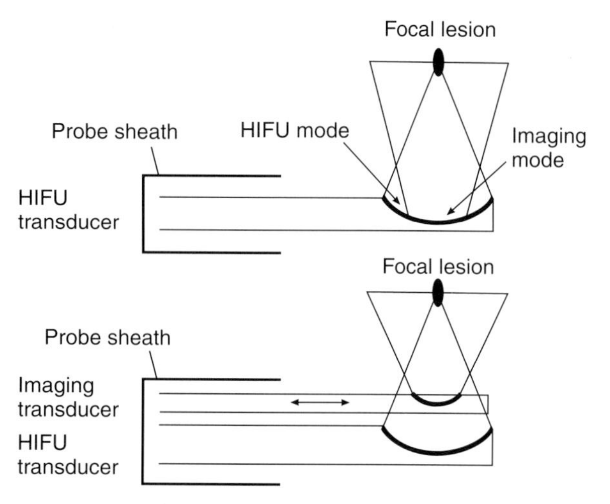

Fig. 56.1. Transrectal HIFU systems. Top: the same transducer operating at 4.0 MHz is used for imaging and therapy (Sonablate,™ Focus Surgery Inc., USA). Bottom: in this device, two separate transducers for imaging and therapy are incorporated in a transrectal probe sheath (Ablatherm,™ EDAP-TMS, France). Adapted from Madersbacher et al.[4]

focal lengths between 2.5 and 4.5 cm are available.[4] The site intensity can be varied from 1260 to 2200 W/cm^2. A 4-second interval of therapy (=power on) is followed by 12 seconds power off, which is used for obtaining an image update and moving the transducer electronically to the next treatment location. The second transrectal device (Ablatherm,™ EDAP-TMS, France) uses two separate transducers, one for imaging and operating at 7.5 MHz, and one for the HIFU therapy at 2.25–3.0 MHz (**Figs 56.1** and **56.2**).[5–8] Site intensities in this system have been reported to range between 700 and 2200 W/cm^2.

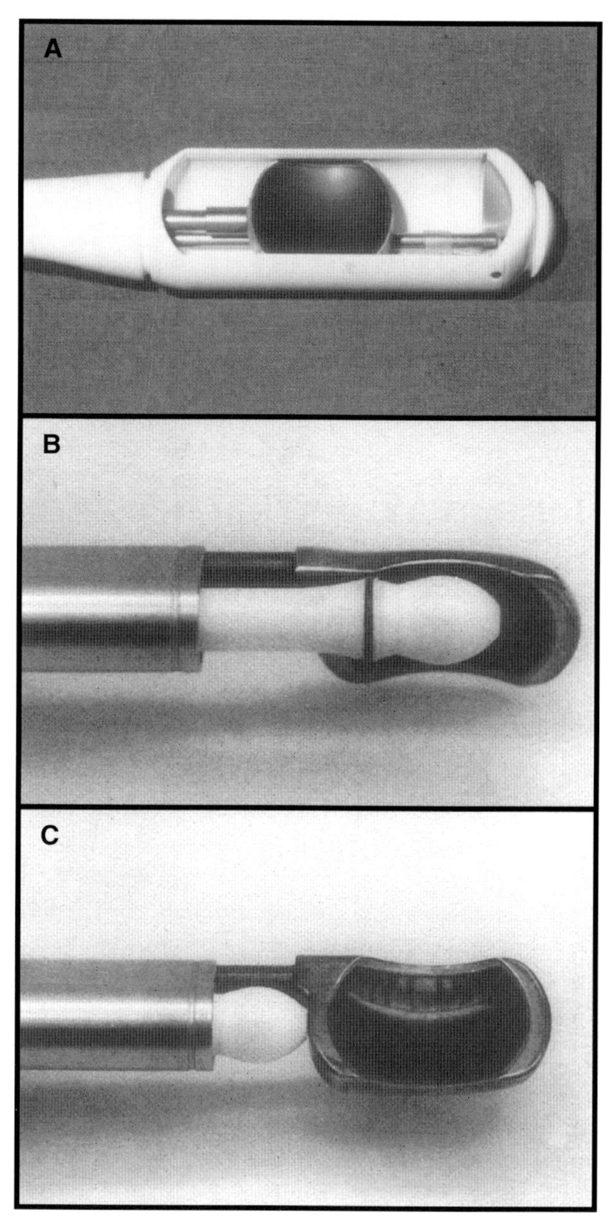

Fig. 56.2. Transrectal HIFU probes. (A) Sonablate™ (Focus Surgery Inc., USA). (B,C) Ablatherm™ (EDAP-TMS, France). Once imaging is complete (B), the imaging transducer is retracted (C) and HIFU therapy is initiated.

EFFECT OF HIFU ON EXPERIMENTAL TUMORS

The antineoplastic effect of HIFU has been documented in several experimental settings.[5] Fry and Johnson implanted hamster medulloblastoma cells in rats, which were subsequently treated by HIFU (900 W/cm^2, 1 MHz, 7 seconds).[9] The tumor cure rate was 29% in rats treated with HIFU and 40% in rates with a combination of HIFU and chemotherapy.[9] Moore et al. and Yang et al. have studied the effect of HIFU on the Morris 3924-A hepatoma implanted into rats.[10,11] The tumor volume in treated animals was substantially smaller than in the untreated control group. However, although the entire tumor was included into the target zone, no tumor was completely destroyed. In 1992, Chapelon et al. reported on the effect of HIFU on the Dunning R3327 prostatic adenocarcinoma implanted in Copenhagen rats.[12] In study 2 of this series, 25 rats implanted with the AT2 subline were treated with an acoustic intensity of 820 W/cm^2. Complete tumor necrosis was achieved in 24 (96%) of cases and 16 (64%) appeared to be cured, whereas all rats in the control group died of progressive tumor growth within 60 days of tumor implantation.[12] Similar data have been reported by Bataille et al. using the same cell line (**Fig. 56.3a**).[13]

These data demonstrate that HIFU is capable of destroying tumor cells in experimental settings.[5,14] It needs to be emphasized that none of the results are 100% perfect. In all series, local recurrences were seen or viable cells were identified with the target zone. The reason for this is not fully understood; most likely the efficacy of each HIFU shot varies as the penetration of the ultrasound beam into tissue can be reduced under certain circumstances, such as tissue (micro)cavitation.

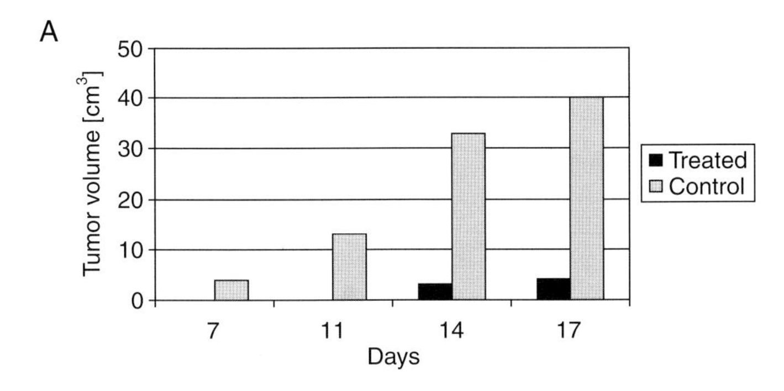

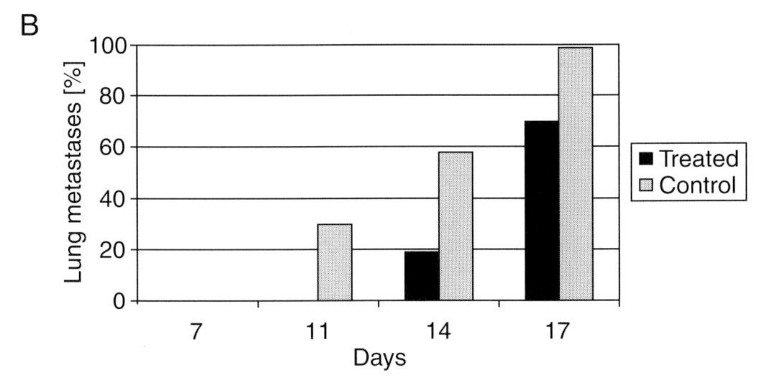

Fig. 56.3. Impact of transcutaneous HIFU treatment ('treatment') compared to sham-treated animals ('control') on tumor volume (A) and the development of lung metastases (B) in the MAT LyLu Dunning R-3327 cell line. Transcutaneous HIFU treatment resulted in a significantly smaller tumor volume at all time points ($P<0.01$) (A) and a lower incidence of lung metastases (B). Adapted from Bataille et al.[13]

EFFECT OF HIFU ON THE CANINE PROSTATE

Foster et al. studied the histological impact of HIFU on 26 canine prostates.[15] Animals were sacrificed from 2 hours up to 12 weeks following treatment. An intraprostatic coagulative necrosis was consistently present. In all animals, surrounding and intervening structures, such as the rectal wall and prostate capsule, were intact. Gelet et al. have conducted a similar study on 37 beagle dogs.[5] An intraprostatic coagulative necrosis was present in the majority of animals, which subsequently formed to a cystic cavity after 4 weeks. Both studies demonstrate that transrectal HIFU is capable of inducing precise and predictable intraprostatic tissue ablation. If technical parameters are set appropriately, thermal damage to intervening structures can be avoided.

A canine study has also addressed the issue of increasing the volume of necrosis, which is an important point for the indication localized prostate cancer.[16] With a modified treatment protocol involving several HIFU-treatment zones and transducers, the authors attempted to ablate the entire canine prostate.[16] On histology, a massive coagulative necrosis comprising approximately 90% of the gland was present. Long-term histological analysis demonstrated only a small rim of prostatic tissue lining a huge intraprostatic cavity.[16] These studies demonstrate the possibility of performing a subtotal prostatic ablation with the present HIFU technology. The major disadvantage of this approach is the long treatment time (4–8 hours). In the near future new HIFU transducers will be available, which will reduce treatment time by 30–50%.

RISK OF METASTASIS FOLLOWING HIFU TREATMENT

The application of HIFU for oncological indications, such as prostate cancer, raises the important issue of whether HIFU has an impact on the development of distant metastases. Chapelon et al. determined this issue in an experimental prostate cancer model. In the control population, 28% of animals developed distant metastases, while this percentage dropped to 16% in HIFU-treated animals.[12] In parallel, Oosterhof et al. have noted no difference in the metastatic rate of HIFU-treated and control animals, and Bataille et al. reported on a lower incidence of lymph node and lung metastases following HIFU-therapy (**Fig. 56.3b**).[13,17] Based on these data, it can be concluded that HIFU applied to cancer tissue does not accelerate the development of distant metastases; in contrast, numerous studies indicated that HIFU treatment reduces the rate of metastases.[12,13,17]

EFFECT OF HIFU ON THE HUMAN PROSTATE

To determine the histological impact of transrectal HIFU on the human prostate, 54 prostates were subjected to HIFU treatment *in vivo* prior to radical retropubic prostatectomy (RPE) at the authors's institution.[3,18–20] HIFU therapy was performed using transducers with focal lengths of 3.0, 3.5 and 4.0 cm, and the site intensity varied from 1260 to 2000 W/cm^2. All specimens were analyzed by whole-mount histological sections with volumetric analysis of the area of necrosis.[3,18–20] Mapping of thermal lesions was possible in all specimens (**Fig. 56.4**).[3,18–20] Transrectal HIFU consistently induced sharply delineated intraprostatic coagulative necrosis within the target area, whereas alterations of periprostatic structures were absent.[3,18–20] The cross-sectional area of necrosis increased significantly depending on HIFU-beam focal length and site intensity.[20] HIFU beam transmission and therapeutic effect were comparable in benign and malignant prostatic tissue.[20] Subsequently, HIFU was applied to unilateral histologically proven T2a/T2b prostate cancer ($n=10$) in an attempt to destroy all cancer prior to RPE.[20] Prostate cancer was always correctly targeted. In seven individuals, prostate cancer was partially (mean 53%; range 38–77%) destroyed; in the remaining three cases, the entire tumor was ablated. These data demonstrate that transrectal HIFU is capable of destroying prostate cancer without contact and without irradiation.[20] As all HIFU-treatments were performed prior to radical prostatectomy under the same anesthesia, treatment time had to be kept short. This fact explains why, despite adequate targeting, cancer was only partially destroyed in the majority of patients.

To obtain a more detailed insight into the effect of HIFU on prostatic cells, our group studied the heat shock protein 27

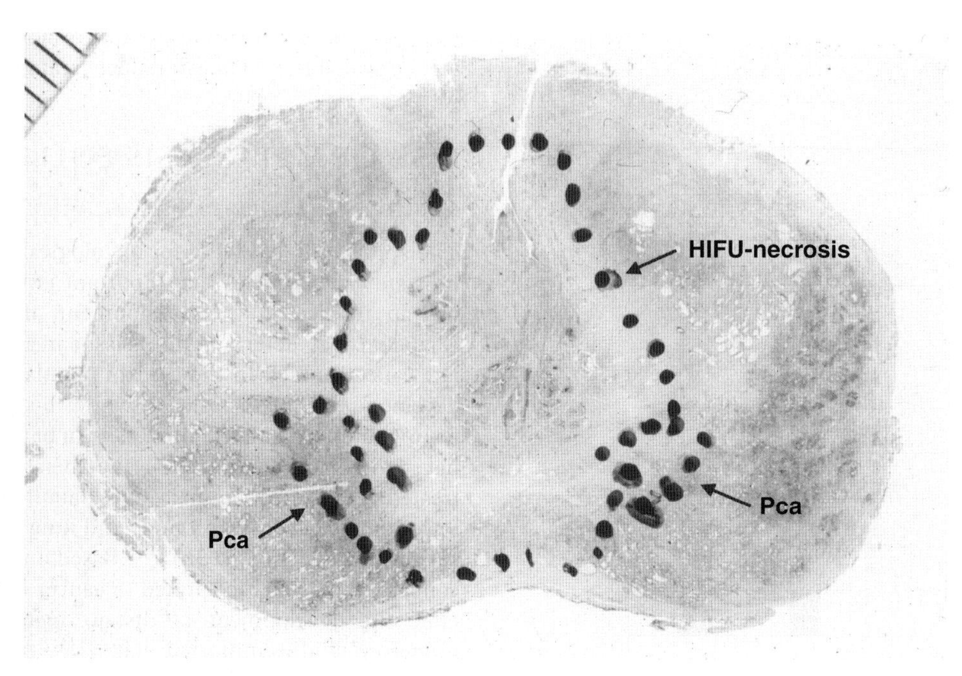

Fig. 56.4. Macroscopic view of a radical prostatectomy specimen removed 3 hours after transrectal HIFU treatment. A targeted transrectal HIFU induced a necrotic area in the periurethral region. The posterior prostate capsule was intact. Pca, prostate cancer.

expression of normal and malignant prostatic cells *in vitro* and *in vivo*.[21] It was shown that benign and malignant human prostatic cells respond to heat by an increased expression of heat shock protein 27 *in vivo*.

A group in Nijmegen performed HIFU-treatments 1–2 weeks prior to radical prostatectomy in nine patients.[22,23] The lobe in which the prostate cancer was confirmed histologically was treated. Histological analysis revealed a very sharp delineation between treated and untreated areas. On the dorsal border, however, incomplete destruction of tissue was noted and, in two cases, a small residual tumor was seen. Most epithelial glands in the center of the HIFU lesions revealed signs of necrosis. However, glands without apparently necrotic features were also located in the HIFU lesion, raising the question of whether lethal destruction had occurred. This epithelium reacted with antibodies to pancytokeratin, PSA and Ki67, but did not express cytokeratin 8, which is indicative of severe cell damage. HIFU treatment was incomplete at the ventral, lateral and dorsal sides of the prostate lobe treated. These human histological analyses demonstrate that HIFU-treatment induces a spectrum of morphological changes ranging from apparent light microscopic necrosis to more subtle ultrastructural cell damage. However, HIFU treatment does not affect the whole of the area treated, but leaves vital tissue at the ventral, lateral and dorsal sides of the prostate.

CLINICAL DATA

To date, three institutions have reported on their experience with transrectal HIFU as a minimally invasive therapy for localized prostate cancer.[6–8,24–29] Gelet et al. initiated a phase II clinical trial in 1992 and reported in 1999 on 50 patients with localized prostate cancer unsuitable for radical prostatectomy who underwent a total of 113 HIFU treatment sessions.[7] Median follow-up was 24 months (range 3–46 months). Negative post-HIFU biopsies and a prostate specific antigen (PSA) nadir of less than 4.0 ng/ml were seen in 28 patients (56%), 6% had negative biopsies but a serum PSA exceeding 4.0 ng/ml, yet 38% had residual cancer on the follow-up biopsies.[7] The complication rate was as high as 50% with the first prototype; this number declined to 17% (urinary retention, urinary tract infection, bladder neck sclerosis, urinary incontinence) in those treated with a modified HIFU system.

The Munich group treated 184 patients with a total of 232 HIFU sessions; 80% were treated as a primary therapy for localized prostate cancer, 10% for local recurrences and 10% for local debulking.[27–29] Mean operating time was 90 minutes. Follow-up biopsies were negative in 80%, and a PSA nadir of less than 4.0 ng/ml was observed in 97%, including 61% with a PSA nadir of <0.5 ng/ml (**Table 56.1**). All patients received a

Table 56.1. Outcome following HIFU treatment of localized prostate cancer ($n = 184$). Modified from Thüroff et al.[29]

PSA nadir following HIFU (ng/ml)	
<0.1	32%
0.1–0.5	29%
0.51–4.0	36%
>4.0	3%

Table 56.2. Outcome following HIFU therapy for localized prostate cancer is dependent on the extent of treatment. Modified from Beerlage et al.[24]

Outcome	Selective (%)	Global (%)
Biopsy negative PSA < 4.0 ng/ml	25	60
Biopsy negative PSA > 4.0 ng/ml	3	8
Biopsy positive PSA < 4.0 ng/ml	37	26
Biopsy positive PSA > 4.0 ng/ml	35	6

suprapubic tube (average 29 days), and 33% needed a transurethral debris resection averaging 7 g. It was clearly shown that, to be successful, treatment of the entire prostate is required (**Table 56.2**). Gelet et al. analysed outcome predictors in 102 patients.[8] As expected, patients with lower baseline PSA (<10 ng/ml), less than five positive preoperative biopsies and lower Gleason score (<7) had a significantly better outcome.[8]

After primary HIFU treatment, no severe side effects (fistula, second- or third-grade urinary incontinence, rectal mucosal burns) were observed in more recent series, once the learning curve and technical problems had been overcome.[6–8,24–29] The most frequent side effects were urinary tract infection and stress urinary tract infection, by far the most frequent auxiliary treatment was a transurethral resection of the prostate (TURP) for removing necrotic/obstructive tissue, which was necessary in 30% of patients in the most recent series due to an increase in HIFU dosage.

Kiel et al. treated 62 patients with a total of 73 HIFU sessions; follow-up biopsy data were reported on 48 patients.[26] Median follow-up was 15 months (range 5–29 months). Negative follow-up with a serum PSA of <4.0 ng/ml was seen in 69%.

FUTURE TECHNICAL IMPROVEMENTS

Various technical innovations are expected to facilitate clinical application of HIFU in the future. The most important step would be the availability of phased-array transducers with variable focal lengths. Once these transducers are at hand, it is possible to mark the target zone in two planes and the respective area will be ablated precisely. In addition, longer focal lengths are required to enable ablation of the ventral part of the prostate. The availability of varying focal sizes and shorter duty cycles would also help to cut down treatment time. Incorporation of magnetic resonance imaging (MRI) techniques permits better visualization of target tissue, and monitoring of temperature changes and treatment effect.[30] In experimental settings, this approach yielded encouraging data.[30] Hence, the objective to be envisaged for clinical HIFU is non-invasive, transcutaneous, ultrasound- or MRI-guided anticancer treatment for both curative and palliative purposes. Recent data also indicate that HIFU may be a powerful instrument for *in vivo* transfection of plasmid DNA and, therefore, might play an important role in future gene-therapy protocols involving organs accessible for ultrasound.[31]

CONCLUSION

Experimental, canine and human histological data have shown the capability of transrectal HIFU to ablate prostate cancer and, therefore, render this technique highly attractive as a minimally invasive treatment for localized prostate cancer. Although several hundred patients have been treated with this approach predominantly in two centers (Munich and Lyon) in the past 8 years, this technique has not gained widespread acceptance, presumably owing to the sophisticated and expensive equipment and operation room set-up, the lack of long-term data and because reliable ablation of the entire prostate (a prerequisite of good long-term data) is difficult to achieve with the current technology.

Preliminary clinical data for the indication localized prostate cancer, however, are promising yet have to be interpreted with caution. In parallel to radiotherapy, long-term tumor control can only be expected if the PSA nadir following HIFU treatment is below 0.5 ng/ml (**Table 56.1**). This critical value is currently reached in only 50–70% of HIFU-treated patients, indicating incomplete prostate and/or prostate cancer ablation (**Table 56.1**). Owing to the multifocality of prostate cancer, the entire prostate needs to be treated to enable local tumor control (**Table 56.1**). This goal cannot be reliable achieved with the current HIFU technology. To ablate the entire prostate, several treatment zones and different HIFU transducers have to be placed, as it is not possible to cover the entire prostate with one treatment zone. If these treatment zones are overlapped incompletely, prostate tissue and tumor cells might escape treatment. Furthermore, it is not known whether it is possible to ablate malignant tissue at or beyond the prostate capsule. Finally, one has to be aware that developing and optimizing a new technology is expensive and, in particular, time consuming, as a multitude of different power levels, treatment patterns, duty cycles and transducer designs are possible.

REFERENCES

1. Ter Haar G. Focused ultrasound therapy. *Curr. Opin. Urol.* 1994; 4:89–92.
2. Foster RS, Bihrle R. High-intensity focused ultrasound. *Curr. Opin. Urol.* 1995; 5:10–12.
3. Madersbacher S, Marberger M. Therapeutic applications of ultrasound in urology. In: Marberger M (ed.) *Application of Newer Forms of Therapeutic Energy in Urology*, pp. 115–36. Oxford: Isis Medical Media Ltd, 1995.
4. Madersbacher S, Marberger M. High-intensity focused ultrasound for prostatic tissue ablation. *Curr. Opin. Urol.* 1996; 6:28–32.
5. Gelet A, Chapelon JY. Effects of high-intensity focused ultrasound on malignant cells and tissues. In: Marberger M (ed.) *Application*

of Newer Forms of Therapeutic Energy in Urology, pp. 107–14. Oxford: Isis Medical Media Ltd, 1995.

6. Gelet A, Chapelon JY, Bouvier R et al. Treatment of prostate cancer with transrectal focused ultrasound: early clinical experience. *Eur. Urol.* 1996; 29:174–83.
7. Gelet A, Chapelon JY, Bouvier R et al. Local control of prostate cancer by transrectal high intensity focused ultrasound therapy: preliminary results: *J. Urol.* 1999; 161:156–62.
8. Gelet A, Chapelon JY, Bouvier R et al. Transrectal high intensity focused ultrasound for the treatment of localized prostate cancer: factors influencing outcome. *Eur. Urol.* 2001; 40:124–9.
9. Fry FJ, Johnson LK. Tumor irradiation with intense ultrasound. *Ultrasound Med. Biol.* 1978; 4:337–41.
10. Moore WE, Lopez RM, Matthews DE et al. Evaluation of high-intensity therapeutic ultrasound irradiation in the treatment of experimental hepatoma. *J. Pediatr. Surg.* 1989; 24:30–3.
11. Yang R, Sanghvi NT, Rescorla FJ et al. Liver cancer ablation with extracorporeal high-intensity focused ultrasound. *Eur. Urol.* 1993; 23(Suppl. 1):17–22.
12. Chapelon JY, Margonari J, Vernier F et al. In vivo effects of high-intensity focused ultrasound on prostatic adenocarcinoma Dunning R3327. *Cancer Res.* 1992; 52:6353–7.
13. Bataille N, Vallancien G, Chopin D. Antitumoral local effect and metastatic risk of focused extracorporeal pyrotherapy on Dunning R-3327 tumors. *Eur. Urol.* 1996; 29:72–7.
14. Hill CR, ter Haar G. High intensity focused ultrasound – potential for cancer treatment. *Br. J. Radiol.* 1995; 68:1296–303.
15. Foster RS, Bihrle R, Sanghvi N et al. Production of prostatic lesions in canines using transrectally administered high-intensity focused ultrasound. *Eur. Urol.* 1993; 23:330–6.
16. Kincaide LE, Sanghvi NT, Cummings O et al. Noninvasive ultrasound subtotal ablation of the prostate in dogs. *Am. J. Vet. Res.* 1996; 57:1225–7.
17. Oosterhof GO, Cornel EB, Smits GA et al. Influence of high-intensity focused ultrasound on the development of metastases. *Eur. Urol.* 1997; 32:91–5.
18. Susani M, Madersbacher S, Kratzik C et al. Morphology of tissue destruction induced by focused ultrasound. *Eur. Urol.* 1993; 23(Suppl. 1):34–8.
19. Madersbacher S, Kratzik C, Susani M et al. Tissue ablation in benign prostatic hyperplasia with high intensity focused ultrasound. *J. Urol.* 1994; 152:1956–61.
20. Madersbacher S, Pedevilla M, Vingers L et al. Effect of high-intensity focused ultrasound on human prostate cancer in vivo. *Cancer Res.* 1995; 55:3346–51.
21. Madersbacher S, Gröbl M, Kramer G et al. Regulation of heat shock protein 27 expression of prostatic cells in response to heat treatment. *Prostate* 1998; 37:174–81.
22. Beerlage HP, van Leenders GJ, Oosterhof GO et al. High-intensity focused ultrasound (HIFU) followed after one to two weeks by radical prostatectomy: results of a prospective study. *Prostate* 1999; 39:41–6.
23. Van Leenders GJ, Beerlage HP, Ruijter ET et al. Histopathological changes associated with high intensity focused ultrasound (HIFU) treatment for localised adenocarcinoma of the prostate. *J. Clin. Pathol.* 2000; 53:391–4.
24. Beerlage HP, Thuroff S, Debruyne FM et al. Transrectal high-intensity focused ultrasound using the Ablatherm device in the treatment of localized prostate cancer. *Urology* 1999; 54:273–7.
25. Chapelon JY, Ribault M, Birer A et al. Treatment of localised prostate cancer with transrectal high intensity focused ultrasound. *Eur. J. Ultrasound* 1999; 9:31–8.
26. Kiel HJ, Wieland WF, Rossler W. Local control of prostate cancer by transrectal HIFU-therapy. *Arch. Ital. Urol. Androl.* 2000; 72:313–19.
27. Chaussy C, Thüroff S. High-intensity focused ultrasound in localized prostate cancer. *J. Endourol.* 2000; 14:293–9.
28. Chaussy C, Thüroff S. Results and side effects of high intensity focused ultrasound in localized prostate cancer. *J. Endourol.* 2001; 15:437–40.
29. Thüroff S, Chaussy C. Therapy of localized prostate cancer with high intensity focused ultrasound. Results and side effects *Urologe [A]* 2001; 40:191–4.
30. Sokka SD, Hynynen KH. The feasibility of MRI-guided whole prostate ablation with a linear aperiodic intracavitary ultrasound phased array. *Phys. Med. Biol.* 2000; 45:3373–83.
31. Huber PE, Pfisterer P. In vitro and in vivo transfection of plasmid DNA in the Dunning prostate tumor R3327-AT1 is enhanced by focused ultrasound. *Gene Ther.* 2000; 7:1516–26.

New Markers for Prostate Cancer Detection: What is on the Horizon?

Chapter 57

Bob Djavan, Mesut Remzi and Michael Marberger
Department of Urology, University of Vienna, Vienna, Austria

WHY DO WE NEED NEW TUMOR MARKERS FOR PROSTATE CANCER DETECTION?

Although utilization of serum prostate specific antigen (PSA) testing for early detection of prostate cancer is generally accepted, the specificity of the cut-off levels currently used is low. As the threshold value of PSA used in cancer detection is lowered, more curable cancers are detected. However, as the cut-offs for PSA testing are lowered, correspondingly more unnecessary negative, potentially morbid and costly biopsies will be required, unless the specificity of cancer detection is increased. Studies have demonstrated that 70–80% of men with a PSA of 2.5–10 ng/ml who had biopsies performed did not have prostate cancer.[1] Efforts to improve PSA testing have generally been directed towards enhancing the specificity: false-positive tests create an abundance of anxiety in patients and they are increasing the costs. New markers such as PSA density (PSAD), PSA density of the transition zone (PSA-TZ), PSA velocity (PSAV), age-related PSA parameters and f/tPSA have been introduced. However, initial enthusiasm was followed by drawbacks and concerns with regard to the real clinical value.

PSAD (the quotient of serum PSA level divided by prostate volume) has been reported to offer significant enhancement in cancer detection. The theory is that the majority of the prostate volume is associated with enlargement of the transition zone caused by benign prostatic hyperplasia.[2–4] Numerous investigators have demonstrated enhanced specificity with this adjustment.[5,6] Others have not been able to reproduce these findings,[7] and showed that using PSA density for testing was no better than using PSA alone.[8] This discrepancy may have a number of etiologies, including differences in ultrasound measurement error, variability in histologic make-up in the cohort undergoing biopsy, biopsy sampling error and PSA assay variability. An important factor is a decrease in cancer detection observed in glands larger than 55 cm^3.[9] Djavan et al.[10] as well as Maeda and coworkers[11] showed enhanced performance with this measurement. Lin and others[12] were unable to reproduce these results. Most clinicians have abandoned the use of PSA density in clinical practice.

The age-specific PSA reference range, a PSA derivative that has been used widely, is based on the concept that an arbitrary cut-off level (e.g. 4 ng/ml) may not be appropriate for all men. Oesterling and coworkers[13] suggested that, because PSA levels increase with advancing age, it would be better to use a lower cut-off level in younger men and a higher cut-off level in older men. He suggested that this would provide greater test sensitivity in younger men and fewer negative biopsy results in older men. Reissigl and associates[14] found that age-specific cut-offs resulted in enhancement of both sensitivity and specificity in men between the ages of 45 and 59; in older men, specificity was increased and cancer detection was slightly diminished. Unfortunately, others have reported dissimilar results. Borer and associates[15] observed that the majority of cancers that would be missed in men between the ages of 60 and 79 had life-threatening characteristics. The Northwest Prostate Institute (NWPI) group demonstrated in a screening study that, with a cut-off of 4 ng/ml for all men, there was only a marginal increase in the positive biopsy rate and a significant reduction in cancer detection.

Theoretically PSA velocity in prostate cancer patients will lead to a more rapid increase in total PSA level over time than in men without cancer. Carter et al.[16] evaluated a cut-off value of 0.75 ng/ml per year. Other authors have been unable to reproduce the results of Carter using relatively short PSA intervals.[17] Catalona et al.[18] confirmed the inability to use PSA velocity in a clinically meaningful way. Carter and associates[19] originally had a minimum of 7 years between determinations, which other investigators did not find useful.

Many studies have shown that the ratio of the free to total PSA (f/tPSA) was lower in men with carcinoma. Catalona and associates reported the definitive trial of the role of f/tPSA.[20] This seven-institution study evaluated men with benign glands and total PSA (tPSA) levels between 4 and 10 ng/ml. An f/tPSA ratio of less than 25% showed a sensitivity of 95% at a specificity improvement of 20% over tPSA. Concerns about f/tPSA developed, because of the limited *in vitro* stability of fPSA, particularly in serum. This necessitates very strict sample handling, including separation of serum/plasma from the blood cells within a few hours of sample collection and analysis of serum on the same day as sample collection; otherwise, the sample has to be kept frozen (ideally at –70 °C for long-term storage) to provide optimal analysis.[1] Lack of uniformity between different manufacturers in achieving similar results using the same specimen is one major problem. This bias is compounded when a quotient is obtained.

COMPLEXED PSA

It has been recognized that the majority of PSA found in men with prostate cancer is complexed to α-1-antichymotrypsin as opposed to PSA found in healthy men.[17] Once PSA gains access to the systemic circulation, the majority becomes complexed to protease inhibitors including α-1-antichymotrypsin (ACT) and α-2-macroglobulin. The former actually occurs to a greater proportion in men with malignancy. This latter observation has been recognized for a number of years. However, early work to measure PSA–ACT complexes encountered difficulties demonstrating non-specific binding or over-recovery due to technical problems in the accurate measurement of complexed PSA.[17,21] Indeed, the reason we used f/tPSA is because such assays were not available, despite the fact that it was recognized that complexed PSA was the most important moiety to measure. It is intriguing that men with cancer had less variation of serum PSA than men without cancer, perhaps because men with cancer have a higher proportion of PSA complexed to ACT (PSA–ACT) than men without cancer.[22,23] The previous technical problems in designing specific PSA–ACT complex assay procedures have largely been overcome.[24] The benefits of a specific assay are numerous. One would expect that such an assay would provide the same specificity enhancement realized with the f/tPSA but would require only the measurement of a single analysate. In addition to the economic advantage of only a single test over measuring both the free and the total form of PSA, several other benefits accure. One is avoiding the variability of different manufacturers' assays, which leads to the lack of uniformity in total PSA or free PSA measured by different manufactures. However, decomposition of the PSA–ACT complex during storage at unfavourable conditions may release enzymatically active PSA *in vitro*.[24] This may result in formation of either AMG–PSA complexes, or false elevatation of free PSA, if the AMG fraction in blood has become inactivated. This could explain some of the data reported on samples stored at –20 °C for more than 20 years.[25] Also the clearance rate by the kidneys of the PSA–ACT complex is much slower than the clearance rate of free, non-complexed PSA (fPSA).[26] Problems associated with the f/tPSA ratio, particularly assay variability and the increase magnitude of error when the quotient is derived, would be obviated if the complex PSA (cPSA) were available.

Recently, the Bayer Corporation has developed a specific PSA assay directed against complexed PSA. The design of this assay reflects the level of PSA complexed to different serpin-inhibitor ligands, such as ACT, API and PCI: the PSA–ACT complexed fraction is by far the predominant PSA-complexed form measured by this PSA-complex assay. A multicenter evaluation established the analytical performance and clinical effectiveness of the Bayer Immuno 1 cPSA assay in monitoring prostate cancer patients during the course disease and therapy. The cPSA assay demonstrated an increased trend in clinical sensitivity (71–86%) for prostate carcinoma with increasing stage of disease. Clinical specificity for patients with benign urogenital disease was 74.8%. For other non-prostatic diseases specificity ranged from 91.1% to 100%.

Retrospective serial monitoring of 155 patients with prostate carcinoma showed concordance of the complexed PSA measurements to clinical status for 97% of the patients analyzed. The results from the multicenter clinical evaluation using the Bayer Immuno 1 cPSA assay were comparable to the results obtained with the Bayer Immuno 1 PSA assay.

In an initial retrospective study of the NWPI series, 300 men who had undergone ultrasound-guided prostate needle biopsy because of a suspect digital rectal examination and/or a PSA >4 ng/ml were analyzed. They ranged in age from 43 to 93 years, with a mean age of 67.7 ± 7.72 years. There was no difference in the age of men with and without malignancy. After centrifugation, the sera were immediately decanted and stored at –80 °C. A total of 75 had biopsy-proved prostate cancer and 225 benign histology. They compared the results of the Bayer Immuno-1 cPSA assay with the Hybritech Tandem R and total assays. To access the capture of the complex form of PSA, they performed a regression analysis comparing the Hybritech total PSA and the sum of the complex PSA by Bayer with the Hybritech free level. While there was a close approximation to a slope of one, there was a slight positive bias, perhaps indicating measurement of some complex PSA with the free assay or, conversely, some free PSA with the complex assay. As expected, the mean and median, tPSA (10 and 8.02 vs. 6.08 and 5.05 ng/ml, respectively; $P < 0.0001$) and cPSA (8.85 and 7.56 vs. 4.95 and 3.94 ng/ml, respectively; $P < 0.0001$) levels were higher in men with cancer, and the f/tPSA ratio (0.15 and 0.17 vs. 0.2 and 0.18, respectively; $P < 0.0001$) was significantly lower. At the 95% sensitivity, the specificity of tPSA, cPSA and f/tPSA were 21.8%, 26.7% and 15.6%, respectively. A cPSA cut-off of 2.52 demonstrated an enhanced specificity of 26.7%. To further understand the utility of cPSA, Brawer and colleagues[27] expanded the Seattle series by incorporating data from the Johns Hopkins University, as well as a cohort derived from a multicenter screening trial that is under way. In this investigation, 272 men with carcinoma detected on ultrasound-guided biopsy were compared with 385 men with benign findings. Again, the Bayer cPSA and the Hybritech tPSA and fPSA assays were used. At the 95% sensitivity level, tPSA provided 18% specificity at a cut-off of 3.06 ng/ml. The cPSA afforded a significant increase: 24% specificity at a cut-off value of 2.75 ng/ml. The f/tPSA ratio at this sensitivity level and a cut-off of 23.9% provided 23% specificity.

Recently Maeda et al. reported that the determination of cPSA was essentially more valuable than tPSA for prostate cancer detection.[28] All of 137 patients enrolled in their study underwent at least a systematic sextant transrectal biopsy under guidance of transrectal ultrasound. Indications for biopsy were suspicious digital rectal examination and/or a serum tPSA greater than 4 ng/ml. In all of the patients, the transition zone volume was evaluated by transurethral resection of the prostate (TRUS). However, the incidence of prostate cancer (16.8%) was lower than in other studies. Based on earlier data reported by Djavan et al. for tPSA and PSA density parameters,[29] they compared cPSA with the volume of the total and transition zone volume, cPSAD and cPSA-TZ, retrospectively. The areas under the receiver operating characteristic (ROC) curves were 85.5%, 83%, 78.2%, 59.2%, 68.5% and 68.3% for cPSA-TZ, cPSAD, cPSA, f/tPSA, f/cPSA and tPSA, respectively. The differences between cPSA, cPSA-TZ and cPSAD were not significant. Based on their data, the authors concluded that cPSA was not dependent on prostate volume. Furthermore, they concluded that cPSA was even more specific than tPSA and f/tPSA in men with intermediate tPSA levels.

In an other study, Okihara et al. investigated the value of cPSA in 151 patients with PSA levels between 2.5 and 4 ng/ml. The cancer detection rate was 24.5%. Complexed PSA was more specific than the f/tPSA ratio. At 93% sensitivity, a cut-off value of 2.3 ng/ml for cPSA and 31% for f/tPSA ratio provided 42% and 11% specificity. In men with prostate glands larger than 30 cm^3, cPSA density (55%) and cPSA-TZ (55%) had a significant higher sensitivity than f/tPSA (8%) at 92% specificity.

The greater enhancement in test performance with cPSA is shown in the range of 4–10 ng/ml ('gray zone') for tPSA. In most studies, cPSA provides essentially the equivalent performance of the f/tPSA ratio, but there are two exceptions.

In the study by Jung and associates,[30] cPSA performed significantly worse than f/tPSA. The investigators did a significant subset analysis of 40 patients with cancer and 40 without who were selected to have the same average total PSA to obviate the effect of tPSA between those men with and without cancer. This data manipulation may have introduced significant bias into their report. The study by Stamey and Yemoto[25] is different in a number of ways from the studies of other investigators.

The NWPI group is currently prospectively evaluating the role of cPSA in early detection of screening. To date, 2361 men have been evaluated. Using a cut-off of 4 ng/ml for tPSA, 401 men (17%) would require biopsy. With cPSA and a cut-off of 3.75 ng/ml, 311 men (13.2%) would require biopsy (a reduction in biopsy rate of 22.4%). In 91 men who underwent prostate biopsy because of a tPSA higher than 4 ng/ml, cancer was detected in 39.6%. Using cPSA with a cut-off of 3.75 ng/ml, two cancers would be missed and 11% of benign biopsies would be avoided. These data from a prospective screening cohort demonstrate for the first time that cPSA provides nearly equivalent sensitivity and significantly enhanced specificity, compared with tPSA.

The cPSA test has been improved for the monitoring of men with established malignancy. Food and Drug Administration (FDA) approval is pending for its use in early detection. Because cPSA is equivalent to tPSA for monitoring and staging purposes, and seems to show enhancement in early detection, it is likely that this analyses will be the PSA test of choice in the future.

Currently, the European prostate cancer detection group (EPCDG) carried out a prospective study with the Bayer cPSA assay. Initial reports indicate a significant improvement of specificity and a remarkable increase in sensitivity. Assuming that similar results are found in other currently ongoing studies, it would appear that complexed PSA would offer a significant advance over total PSA for early detection and screening in men with low and intermediate serum PSA levels. Complex PSA has recently been approved for monitoring men with prostate cancer. cPSA for staging and other applications should provide the same information as total PSA.

COMPLEX PSA FOLLOWING RADICAL PROSTATECTOMY

PSA has become the most valuable tool for detecting, staging and monitoring prostate cancer in the past decade.[31] After suggesting that the half-life of PSA might be a possible factor for the prognosis after radical prostatectomy, the pharmacokinetics of PSA gained more attention.[32,33] PSA half-life is 2–3 days[34] and a biphasic model for t/fPSA elimination was established.[34] Lein et al. confirmed the biphasic elimination and suggested the first phase of free PSA loss might be the result of its rapid binding to α_2-macroglobulin and ACT.[35] Other investigators have assumed that the initial free PSA decrease was related to renal clearance or caused by the operation itself.[36] In another study, the authors determined whether the assumed binding of PSA to its protease inhibitors after release into the circulation could be a possible reason for the elimination of free PSA. During the first 6 hours after radical retropubic prostatectomy, they found nearly constant levels of ACT–PSA and cPSA, in contrast to the rapid elimination of free PSA and a significant decrease in total PSA. From days 1 to 10, a continuous and nearly identical decrease of ACT–PSA and cPSA occurred compared to total PSA; free PSA was eliminated more rapidly. The first PSA decrease might be an effect of the operation itself or caused by renal elimination alone. The findings indicate that the initial rapid decrease of free PSA immediately after surgery could be caused by new complex forming of PSA with ACT and other serum protease inhibitors.[37]

INSULIN-LIKE GROWTH FACTOR 1

Insulin-like growth factors (IGFs), including IGF-1 and IGF-2, are mitogenic peptides involved in regulation of cell proliferation, differentiation and apoptosis in the prostate. IGF-1 and IGF-2 share a structural homology with proinsulin.[38] Both growth factors, IGF-1 and IGF-2, are present in high concentration in serum and nearly every mammalian cell type can synthesize and export them. Main production sites are the liver and bone marrow. The IGF regulatory system in each organ is tissue specific, but all systems share certain components, including ligands (IGF-1 and IGF-2), insulin-like growth factor binding proteins (IGFBPs) 1–6, IGF receptors (1 and 2), and IGFBP specific proteases. Extracellular IGF-1 is bound to a family of IGFBPs. A total of 75% or more of circulating IGF-1 is carried in a trimeric complex composed of IGFBP-3 and a liver-derived glycoprotein known as the acid labile subunit (ALS).[39] One of the most important IGFBP-specific proteases is prostate specific antigen. PSA cleaves IGFBP-3, it is unregulated by testosterone and it is likely to be critical in skeletal and distant neoplastic metastasis.[40] Recently, two studies used animal models to explore the changes in the IGF-axis during initiation, progression and metastasis of prostate cancer.[41,42] Both studies indicate that IGF-1 may be important for the development of benign prostatic hyperplasia, that dysplastic cells may start to produce IGF on their own and that cancer cells are able to secrete IGF-1 into serum. The IGF-1R receptor is overexpressed in hyperplasia, dysplasia and prostate cancer.

The interaction of IGF-1 and IGF-2 with IGF-1R is regulated by the IGFBPs 2–6, which are expressed by epithelial cells and to lesser degree by stromal cells.[43,44] Trasher et al.[45] found that IGFBP-2 is increased in malign tissue and that IGFBP-3 is decreased. The effects of protease activity may explain the decreased IGFBP-3 in prostate cancer cells. PSA is one of several serine proteases cleaving IGFBP-3, thus regulating the availability of the binding protein.[46] IGFBP-3 decreases the mitogenic effects of IGF-1 in normal prostate epithelial cells in culture and cleavage of the binding protein by PSA decreases its affinity for the IGF-1 receptor, thus reversing this inhibitory effect.[47] Recently, it was shown that IGFBP-2, -3 and -5 expression changes with Gleason score.[48] There is differential expression of IGFBP-2, -3 and -5 mRNA between high and low Gleason score tumors. Figueroa et al.[48] found statistically significant higher levels of IGFBP-2 and -5 mRNA, but lower expression of IGFBP-3 protein in tumors with high Gleason score compared to low or intermediate score lesions and benign tissues.

Population studies provide evidence for a relationship between circulating levels of both IGF-1 and IGFBP-3 and the risk of prostate cancer.[49–51] A cohort study by Wolk et al.[50] demonstrated an association between serum levels of IGF-1 and risk of prostate cancer. In 1998, results from a prospective case-control study nested within the Physicians Health Study, of IGF-1 and IGFBP-3, were published. In 1982, 14 916 US male physicians aged 40–82 provided a blood sample, which was then frozen and archived.[49] By 1992, 520 cases of prostate cancer were diagnosed in this cohort. A total of 152 of them provided plasma for IGF analyses. Then 152 age-matched controls were selected and plasma IGF was measured using an enzyme-linked immunosorbent assay. In a univariate analysis, IGF-1 was significantly related to prostate cancer risk ($P=0.006$). Comparing top to bottom quartiles, IGF-1 was now associated with a relative risk of 4.32, and an inverse association between IGFBP-3 and prostate cancer risk emerged. No significant difference was noted among men with high grade/stage disease versus low grade/stage cancers. The association appeared to be somewhat stronger for older men, but significant positive trends were noted for both older and younger men, using a cut-point at the age of 60. IGF-2 was not related to risk of prostate cancer. Overall, these studies show a 2–3-fold increase in the risk of developing cancer when IGF-1 is in the higher quartiles. Subsequently, Djavan et al.[52] evaluated prospectively the value of IGF-1 and IGF-derivate in early detection of prostate cancer. To optimize the statistical analysis, rebiopsies were performed in patients with negative first biopsy results. The positive rebiopsy rate was 18%. IGF-1 and IGFD were able to discriminate between prostate cancer and benign prostatic hyperplasia (BPH) ($P=0.03$ and 0.045) but were inferior to PSA alone. The combination of IGF-1 and PSA improved the predictive accuracy of PSA for prostate cancer significantly. The IGF-1/PSA ratio was the most promising parameter (AUC0 71.0%). An IGF-1/PSA cut-point of 25 provided a 95% sensitivity for detecting prostate cancer, while avoiding unnecessary biopsies in 24.1% of cases. In this study, age was not an independent predictor of IGF-1 level. Patients diagnosed with acromegaly have an elevated serum level of IGF-1 and are at an increased risk of developing colon cancer. In contrast, the incidence of prostate cancer has not been reported to be different in these patients. One of the possible reasons is a rise in serum IGFBP-3 levels paralleling the rise in IGF-1. ‘Free’ IGF-1 levels are similar to those of normal healthy subjects. IGF-1 is part of new therapy strategies for treating uncontrollable diabetes mellitus and GH-deficiency. Whether such therapies will lead to further long-term malignancies remains to be elucidated. Risk must be weighed against short-term benefits and life expectancy.

Cancer risk seems to be positively correlated with IGF-1 levels and negatively correlated with IGFBP-3 levels. Particularly high and low risks may be seen in individuals who are outliers on population plots of IGF-1 versus IGFBP-3. Individuals at higher risk may be those with circulating IGFBP-3 inappropriately low for their IGF-1 level. The timing of the association between chronically high IGF bioactivity and cancer risk is still unclear. There is some discussion about whether high IGF-1 levels are associated with a higher risk of initiating and completing early steps of prostate carcinogenesis, or that they are associated with unchanged risk of completing early steps but a higher rate of neoplastic progression from asymptomatic disease to aggressive and symptomatic disease.

In patients with prostate cancer, an alteration in serum IGFBPs occurs.[53] IGFBP-2 increases. This increase is highly correlated with the rise in serum PSA ($r=0.619$; $P<0.002$). The high and consistent correlation between PSA and IGFBP-2 serum levels suggests a prostate source for the increase in IGFBP-2.

A decrease in serum IGFBP-3 levels is the second most prominent alteration in serum IGFBPs in patients with prostate cancer. There is a high inverse correlation between IGF-1 and IGFBP-3 serum level. Additionally, when sera from patients with high PSA and low IGFBP-3 are mixed with sera from control healthy individuals, a decrease in IGFBP-3 level can be observed.[54]

PSA has IGFBP-3 as a target, with a fairly specific proteolytic activity.[46] It spares IGFBP-2 and IGFBP-4 and only partially affects IGFBP-1. The proteolysis of IGFBP-3 by PSA also fully reverses the inhibitory effects of this binding protein on prostate cell culture growth. Cleavage fragments of IGFBP-3 can be found in medium containing PSA, and its cleavage presumably releases IGF for mitogenic activity on the epithelial cells. *In vivo* this growth-stimulating effect of PSA, if it can be confirmed, could have significant implications. Therapy that reduces production of PSA by prostate tumor cells, or reduces PSA enzymatic activity, might help to inhibit progression of prostate cancer in the patient. The PSA/IGF/IGFBP-3 axis could work as follows: IGF-1 is able to stimulate or activate the androgen receptor, resulting in PSA production, which was shown *in vitro* in DU-145 and LNCaP cell lines.[55] PSA specifically cleaves IGFBP-3. This in turn results in a greater level of free IGF-1 to induce a resistance to apoptosis in prostate cancer cells, an increase in mitosis and, ultimately, an increase in PSA production. Subsequently, the now increased PSA can liberate more bound IGF-1. The tissue level of zinc, a potent inhibitor of PSA proteolytic activity found in the prostate, has been reported to be reduced in prostate cancer patients.[56] The resulting higher local availability of enzymatically active PSA may lead to the liberation of more IGF-1. This hypothesis suggests that IGF-1 may play an active role in prostate cancer progression.

IMPLICATIONS FOR PROSTATE CANCER TREATMENT AND PREVENTION

Elevated levels of circulating IGF-1, which are strongly associated with the risk of developing prostate cancer, are measured long before presentation of this cancer. IGF may be a new tool for prostate cancer risk assessment and reduction. Recent research projects indicated that current androgen-targeting therapies alter IGF physiology within androgen-responsive cells in a manner that contributes to apoptosis associated with androgen withdrawal. Several IGF binding proteins undergo an impressive up-regulation following castration but preceding the onset of castration-induced apoptosis.[57] Similar results were observed for antiandrogens[58] and growth inhibitory vitamin D analogs.[58,59] Recent data support the view that, in androgen-dependent tissues, IGFBP expression is suppressed by androgens and that, at the time of androgen withdrawal, IGFBP expression rises, leading to a massive decline in IGF bioactivity, thereby promoting apoptotic cell death, given the known antiapoptotic IGF activity.

Currently, the hypothesis that the efficacy of androgen-targeting therapies can be improved by combining them with IGF targeting therapies is under investigation. Other possible IGF-related approaches include the use of suppressors of IGF-1 expression (including GHRH antagonists and somatostatin analogs), IGF receptor blocking or antisense strategies, growth hormone antagonists and the use of IGF binding protein protease inhibitors. Inducers of expression of IGF binding proteins, including vitamin D-related compounds and the enhancement of cytotoxic chemotherapy by coadministration of agents that reduce IGF bioactivity, given the known antiapoptotic properties of IGF, are also under active investigation. This line of research has been stimulated by the results of preclinical studies and subsequent clinical trials that demonstrate enhanced efficacy of cytotoxic drugs in the presence of blockers of the HER2 receptor.[58–60] Small molecule inducers of IGF binding protein expression and of inhibitors of IGFBP proteolysis might have activity even for tumors with IGF autocrine loops, as would approaches that have the IGF receptor itself as a target. GHRH antagonists show surprisingly strong activity in models of aggressive prostate cancer, raising the possibility of novel mechanisms of action mediated by GHRH receptors on neoplastic cells, including the inhibition of IGF-2 expression.[61] β,β-Dimethyl acryl shikonin, an extract from the root of plant *Arnebia nobilis* has been shown to have anticancer properties but was found to be toxic. Several analogs of β,β-dimethyl acryloyl shikonin were synthesized and one of them, shikonin analog 93/637 (SA), was significantly less toxic. SA exerts an inhibitory effect on cellular growth and IGFs (up-regulating IGFBP-3, down-regulating local IGF-1 and IGF-1R), specifically in PC-3 cells.[62] The inhibitory effects of GHRH antagonists on PC-3 androgen-independent prostate cancer can be potentiated by concomitant use of bombesin/gastrin-releasing peptide antagonists. The combination of both types of analog apparently interferes with the IGF (a decrease in IGF-1) and bombesin/EGF pathways.[63]

The recent epidemiological observations that link IGF-1 to prostate cancer risk may have implications for future prevention strategies.[64,65] IGF represents a individual host characteristic that appears to be related to the probability of future carcinogenesis. Relative to other hormones, IGF-1 levels fluctuate little over time.

Men with high IGF-1 levels might benefit from enhanced screening procedures, including PSA screening. It has been proposed that a single IGF-1 measurement might be used to define a subset of the population for whom annual PSA screening for prostate cancer detection might be advisable at an early age. It might be premature to suggest intervention studies that would evaluate potential benefits of pharmacological reduction of IGF-1 levels from high-normal to mid-normal values. Long-term safety data are available for somatostatin analogs, which are one of several candidate molecules that reduce IGF-1 levels. Octreotide, a somatostatin analog, decreases the hepatic production of IGF-1 and glucocorticoids, such as dexamethasone, have been demonstrated to decrease the tissue levels of IGF-1 in bone.[66,67] However, their potential to lower levels from the 'high-normal' to 'mid-normal' range is still under investigation. It is possible that dose–response relationships might be influenced by pretreatment IGF-1 level. There is no rationale for proposing to lower levels to below the normal range for the purpose of prevention. The goal would be to move IGF serum level in men from 'high-normal' to 'low-normal'. Right now current monthly depot formulations are not the optimum for long-term use. However, the development of orally active IGF-1 suppressing drugs should be possible in the near future.

Potential IGF-1 targeting prevention strategies would be suitable only for the proportion of the population (10–20%) at increased risk specifically associated with high IGF-1 level or low IGFBP-3 level. A strategy aimed at a rise in IGFBP-3 level could also lower risk. Practical lifestyle modifications that could be used to lower levels have not been described to date.

In the past few years, the axis of ligands, receptors and binding proteins that forms the IGF system has received substantial attention. Considerable evidence has accumulated that links this complex system to the pathophysiology of prostate cancer. The accumulated data suggest novel treatment and prevention strategies for this very common cancer. Future research may concentrate on the manipulation of IGF in the management of prostate cancer or in its chemoprevention either by therapeutic regulation of IGF-1 levels in the upper quartile or by dietary modification.

HUMAN GLANDULAR KALLIKREIN 2 (HK2)

Prostate specific antigen (PSA = human glandular kallikrein 3/hK3) and hK2 are proteases that, along with human kallikrein 1 (hK1), comprise the kallikrein family, a subgroup of a large family of serine proteases.[68–72] There are structural similarities between hK1, hK2 and PSA.[73–75] The first known protein in this family is a true kallikrein, termed tissue kallikrein or hK1, which was discovered over 60 years ago.[76,77] The kallikrein function of hK1 is to cleave polypeptide substrates from low-molecular-weight (l)-kininogen to release a vasoactive decapeptide Lys-bradykinin (kallidin).[78] Initial studies indicated that PSA was capable of generating small amounts of kinin-like substances when incubated with seminal vesicle fluid.[79] However, recent studies with PSA preparations, free of trypsin-like activity, demonstrated that PSA does not appear to act as a true kallikrein despite its structural similarity to hK1.[80] A small amount of kininogenase activity has been observed with hK2 on high-molecular-weight kininogen, although this activity is 1000-fold lower than hK1.[80]

The primary structure of the hK2 protein is approximately 79% identical to PSA[75] and is one of the newer serum markers for prostate cancer detection. HK2 is a prostate specific, PSA-like kallikrein produced by the epithelial cells of the prostate,[81,82] which can convert the inactive zymogen (proPSA) to enzymatic PSA, which is a prerequisite for the formation of the PSA–ACT complex and of other complexed forms of PSA.[83] A unique feature of hK2 is that it seems to be increasingly expressed in transformed epithelium as compared to PSA, which begins to be less translated as cancer becomes more aggressive.

Several newly developed immunoassays specific for hK2[84,85] have demonstrated that hK2 levels are: (1) in the μg/ml range (1–2% of the total PSA concentrations)[84,86]; and (2) not directly proportional to PSA and may hence provide additional clinical information. Low analytic concentration and immunological cross-reactivity relating to similarity[84,87] with PSA initially hampered progress in this field, but can now be overcome. Presently, hK2 assays are reported that detect 0.03 ng/ml of hK2, insignificant (<0.01%) cross-reaction with PSA, and equimolar detection of free and complexed forms of hK2.[88] The majority of hK2 in serum is free, only about 5–20% of hK2 is believed to be present in complexed form, mainly in complex with ACT.[84,88,89] The serum levels of hK2 are very low in females and in prostate cancer patients following radical prostatectomy, whereas the hK2 levels are significantly higher in men with benign prostatic disease than in young healthy men. Further, the levels are significantly elevated in the men with localized prostate cancer compared to those with BPH.[85,90] Another study compared the hK2 levels of patients with localized prostate cancer, treated by radical prostatectomy. Patients who had a organ-confined disease in the histologic specimen had significantly lower levels of hK2 than patients with extraprostatic tumor extension in the specimen, whereas tPSA showed no statistically significant difference in these two groups.[91] Recently, it has been proposed that hK2 measurements in combination with f/tPSA can improve the sensitivity and specificity of cancer detection, and avoid unnecessary biopsies, also in total PSA levels from 2.5 to 4.0 ng/ml and digital rectal examination (DRE) results are not suspicious for prostate cancer. In this study a model for cancer detection using %free PSA and the thK2/fPSA ratio when PSA is 2–4 ng/ml is proposed that would identify as many as 40% of the cancers, and would require biopsy in only 16.5% of the men in this PSA range.[92] They also presented evidence that the %free PSA and thK2/fPSA might be useful when PSA values are 4–10 ng/ml and DRE results are not suspicious for prostate cancer. In this group, men are routinely recommended for biopsy. However, the false-negative rate of biopsy has been shown to be approximately 20%.[93] A thK2/fPSA greater than 0.18 might be used to identify those men who are at higher risk of cancer and should be followed up more aggressively after an initial negative biopsy. In an evaluation of men from the Goteborg screening study, where hK2 was compared to fPSA and tPSA,[94] 144 of 604 men with PSA levels higher than 3 ng/ml had prostate cancer in the following prostate biopsy. Medium values for all analyzed combinations as well as t-test comparisons statistically showed significantly higher hK2 levels as well as tPSA levels in men with prostate cancer. As expected, the f/tPSA ratio was significantly lower in men with malignancy.

Obviously, larger studies are needed to confirm the utility of the hK2 measurements in the serum and its value in combination and/or versus f/tPSA. This may require appropriate algorithms, such as logistic regression analysis and artificial neural networks, which are still not available. hK2 seems to offer complementary information to PSA. Its potential for replacing the latter, however, is doubtful.

In summary, PSA has revolutionized our management of men with prostate cancer. It offers the best means of detecting early-stage disease. Significant advances, particularly in the realm of increased specificity, will undoubtedly continue to unfold. Novel markers, such as human kallikrein type 2, prostate specific membrane antigen and other tests, will surely aid clinicians and their patients.

CONCLUSION

New tumor markers for prostate cancer detection, staging and monitoring are urgently required. In the past few years, considerable attention has been drawn to various ligands, receptors and binding proteins linked to prostrate cancer. Complex PSA is certainly the most promising marker currently under investigation. The IGF system has received substantial attention and evidence has accumulated that links this complex system to the pathophysiology of prostate cancer. Future research may concentrate on the manipulation of IGF in the management of prostate cancer or in its chemoprevention either by therapeutic regulation of IGF-1 levels in the upper quartile or by dietary modification.

The human glandular kallikrein 2 has to show its potential in further prospective studies. Early data initiated much controversy, but multicenter prospective early detection and screening studies are still lacking and thus required.

The currently best established new marker is complexed PSA. Its potential of a higher stability (as compared to free PSA) and its performance in the low serum PSA range will certainly improve our detection and screening efforts. Significant improvements in sensitivity and specificity have caused enthusiasm but further studies, which are currently ongoing, will define its precise role in cancer detection, staging and monitoring.

REFERENCES

1. Piironen T, Pettersson K, Suonpaa M et al. In vitro stability of free prostate-specific antigen (PSA) and prostate-specific antigen (PSA) complexed to alpha 1-antichymotrypsin in blood samples. *Urology* 1996; 48(6A Suppl.):81–7.
2. Stamey TA, Yang N, Hay AR et al. Prostate-specific antigen as a serum marker for adenocarcinoma of the prostate. *N. Engl. J. Med.* 1987; 317:909–16.
3. Babaian RJ, Fritsche HA, Evans RB. Prostate-specific antigen and prostate gland volume: correlation and clinical application. *J. Clin. Lab. Anal.* 1990; 4:135–7.
4. Benson MC, Whang IS, Olsson CA et al. The use of prostate specific antigen density to enhance the predictive value of intermediate levels of serum prostate specific antigen. *J. Urol.* 1992; 147:817–21.
5. Littrup PJ, Kane RA, Mettlin CJ et al. Cost-effective prostate cancer detection: reduction of low-yield biopsies. *Cancer* 1994; 74:3146–58.
6. Bangma CH, Kranse R, Blijenberg BG et al. The value of screening tests in the detection of prostate cancer. Part I: results of a retrospective evaluation of 1726 men. *Urology* 1995; 46:773–8.
7. Ohori M, Dunn JK, Scardino PT. Is prostate-specific antigen density more useful than prostate-specific antigen levels in the diagnosis of prostate cancer? *Urology* 1995; 46:666–71.
8. Brawer MK, Aramburu EA, Chen GL et al. The inability of prostate specific antigen density to enhance the predictive value of prostate specific antigen in the diagnosis of prostatic carcinoma. *J. Urol.* 1993; 150:369–73.
9. Letran J, Meyer G, Loberiza F et al. The effect of prostate volume on the yield of needle biopsy. *J. Urol.* 1998; 160:1718–21.
10. Zlotta AR, Djavan B, Marberger M et al. Prostate specific antigen density of the transition zone: a new effective parameter for prostate cancer prediction. *J. Urol.* 1997; 157:1315–21.
11. Maeda H, Ishitoya S, Maekawa S et al. Prostate specific antigen density of the transition zone in the detection of prostate cancer. *J. Urol.* 1997; 158:58A.
12. Lin DW, Gold MH, Ransom S et al. Transition zone prostate specific antigen density: lack of use in prediction of prostatic carcinoma. *J. Urol.* 1998; 160:77–82.
13. Oesterling JE, Cooner WH, Jacobsen SJ et al. Influence of patient age on the serum PSA concentration: an important clinical observation. *Urol. Clin. North Am.* 1993; 20:671–80.
14. Reissigl A, Pointner J, Horninger W et al. Comparison of different prostate-specific antigen cut-points for early detection of prostate cancer: results of a large screening population. *Urology* 1995; 46:662–5.
15. Borer JG, Serman J, Solomon MC et al. Age-specific reference ranges for prostate-specific antigen and digital rectal examination may not safely eliminate further diagnostic procedures. *J. Urol.*
16. Carter HB, Morrell CH, Pearson JD et al. Estimation of prostatic growth using serial prostate-specific antigen measurements in men with and without prostate disease. *Cancer Res.* 1992; 52:3323–8.
17. Porter JR, Hayward R, Brawer MK. The significance of short-term PSA change in men undergoing ultrasound guided prostate biopsy. *J. Urol.* 1994; 264(Suppl.):293A.
18. Catalona WJ, Smith DS, Ratliff TL. Value or measurement of the rate of change of serum PSA levels in prostate cancer screening. *J. Urol.* 1993; 150(Suppl.):300A.
19. Carter HB, Pearson JD, Chan DW et al. Prostate-specific antigen variability in men without prostate cancer: effect of sampling interval on prostate-specific antigen velocity. *Urology* 1995; 45:591–6.
20. Catalona WJ, Partin AW, Slawin KM et al. Use of the percentage of free prostate-specific antigen to enhance differentiation of prostate cancer from benign prostatic disease: a prospective multicenter clinical trial. *JAMA* 1998; 279:1542–7.
21. Nixon RG, Gold MH, Blase AB et al. Comparison of three investigational assays for the free form of prostate specific antigen. *J. Urol.* 1998; 160:420–5.
22. Stenman Uh, Leinonen J, Alfthan H et al. A complex between prostate-specific antigen and alpha-1-antichymotrypsin is the major form of prostate-specific antigen in serum of patients with prostatic cancer: assay of the complex improves clinical sensitivity for cancer. *Cancer Res.* 1991; 51:222–6.
23. Christensson A, Bjork T, Nilsson O et al. Serum prostate specific antigen complexed to alpha 1-antichymotrypsin as an indicator of prostate cancer. *J. Urol.* 1993; 150:100–5.
24. Lilja H, Haese A, Bjork T et al. Significance and metabolism of complexed and noncomplexed prostate specific antigen forms and human glandular kallikren 2 in clinically localized prostate cancer before and after radical prostatectomy. *J. Urol.* 1999; 162:2029–35.
25. Stamey TA, Yemoto CE. Examination of the three molecular forms of serum prostate specific antigen for distinguishing

negative from positive biopsy: relationship to transition zone volume. *J. Urol.* 2000; 163:119–26.
26. Bjork T, Ljungberg B, Piironen T et al. Rapid exponential elimination of free prostate-specific antigen contrasts the slow, capacity-limited elimination of PSA complexed to alpha 1-antichymotrypsin from serum. *Urology* 1998; 51:57–62.
27. Brawer MK, Cheli CD, Neaman IE et al. Complexed prostate specific antigen provides significant enhancement of specificity compared with total prostate specific antigen for detecting prostate cancer. *J. Urol.* 2000; 163:1476–80.
28. Maeda H, Arai Y, Aoki Y et al. Complexed prostate-specific antigen and its volume indexes in the detection of prostate cancer. *Urology* 1999; 54:225–8.
29. Djavan B, Zlotta AR, Byttebier G et al. Prostate specific antigen density of the transition zone for early detection of prostate cancer. *J. Urol.* 1998; 160:411–18.
30. Jung K, Elgeti U, Lein M et al. Ratio of free or complexed prostate-specific antigen (PSA) to total PSA: which ratio improves differentiation between benign prostatic hyperplasia and prostate cancer? *Clin. Chem.* 2000; 46:55–62.
31. Polascik TJ, Oesterling JE, Partin AW. Prostate specific antigen: a decade of discovery – what we have learned and where we are going. *J. Urol.* 1999; 162:293–306.
32. Oesterling JE, Chan DW, Epstein JI et al. Prostate specific antigen in the preoperative and postoperative evaluation of localized prostatic cancer treated with radical prostatectomy. *J. Urol.* 1988; 139:766–72.
33. Semjonow A, Hamm M, Rathert P. Half-life of prostate-specific antigen after radical prostatectomy: the decisive predictor of curative treatment? *Eur. Urol.* 1992; 21:200–5.
34. Partin AW, Piantadosi S, Subong ENP et al. Clearance rate of serum-free and total PSA following radical retropubic prostatectomy. *Prostate* 1996; 7(Suppl.):35–9.
35. Lein M, Brux B, Jung K et al. Elimination of serum free and total prostate-specific antigen after radical retropubic prostatectomy. *Eur. J. Clin. Chem. Clin. Biochem.* 1997; 35:591–5.
36. Brändle E, Hautmann O, Bachem M et al. Serum half-life time determination of free and total prostate-specific antigen following radical prostatectomy – a critical assessment. *Urology* 1999; 53:722–30.
37. Stephan C, Jung K, Brux B et al. ACT-PSA and complexed PSA elimination kinetics in serum after radical retropubic prostatectomy: proof of new complex forming of PSA after release into circulation. *Urology* 2000; 55:560–3.
38. Zapf J, Schmid C, Froesch ER. Biological and immunological properties of IGF-1 and IGF-2. *Clin. Endocrinol. Metab.* 1984; 13:7–12.
39. Holman SR, Baxter RC. IGFBP-3: factors affecting binary and ternary complex formation. *Growth Regul.* 1996; 6:42–7.
40. Fowkles J, Enghild J, Suzuki K et al. Matrix metalloproteinases degrade IGFBP-3 in dermal fibroblast cultures. *J. Biol. Chem.* 1994; 269:25742–6.
41. Kaplan PJ, Subburaman M, Cohen P. The Insulin-like Growth factor axis and prostate cancer: lessons from the transgenic adenocarcinoma of mouse prostate (Tramp) model. *Cancer Res.* 1999; 59:2203–9.
42. Wang YZ, Wong YC. Sex hormone-induced prostatic carcinogenesis in the Noble Rat: the role of insulin-like growth factor 1 (IGF-1) and vascular endothelial growth factor (VEGF) in the development of prostate cancer. *Prostate* 1998; 35:165–77.
43. Tennant M, Trasher J, Twomey P et al. Insulin-like growth factor-binding proteins (IGFBP)-4, -5 , -6 in the benign and malignant human prostate: IGFBP-5 messenger ribonucleic acid localization differs from IGFBP-5 protein localization. *J. Clin. Endocrinol. Metab.* 1996; 81:3783–90.
44. Tennant M, Trasher J, Twomey P et al. Insulin-like growth factor-binding protein-2 and -3 expression in benign human epithelium, prostate intraepithelial neoplasia, and adenocarcinoma of the prostate. *J. Clin. Endocrinol. Metab.* 1996; 81:411–7.
45. Delhanty PJ, Han VKM. An RGD to RGE mutation in the putative membrane binding domain of insulin-like growth factor binding protein-2 inhibits its potentation of IGF-2 induced thymidene uptake by SPC cells. Read at the *Annual Meeting of the Endocrine* Society, 1993, p.56.
46. Cohen P, Graves HC, Peehl DM et al. Prostate-specific antigen (PSA) is an insulin-like growth factor binding protein-3 protease found in seminal plasma. *J. Clin. Endocrinol. Metab.* 1992; 75:1046–53.
47. Cohen P, Peehl DM, Graves HC et al. Biological effects of prostate specific antigen as an insulin-like growth factor binding protein-3 protease. *J. Endocrinol.* 1994; 142:407–15.
48. Figueroa JA, De Raad S, Tadlock L et al. Differential expression of insulin-like growth factor binding proteins in high versus low gleason score prostate cancer. *J. Urol.* 1998; 159:1379–83.
49. Chan JM, Stampfer MJ, Giovannuci E et al. Plasma insulin-like growth factor-1 and prostate cancer risk: a prospective study. *Science* 1998; 279:563–6.
50. Wolk A, Mantzoros CS, Andersson SO et al. Insulin-like growth factor-1 and prostate cancer risk: a population based, case-control study. *J. Natl Cancer Inst.* 1998; 90:911–15.
51. Mantzoros CS, Tzonou A, Signorello LB et al. Insulin-like growth factor 1 in relation to prostate cancer and benign prostatic hyperplasia. *Br. J. Cancer* 1997; 76:115–18.
52. Djavan B, Bursa B, Seitz C et al. Insulin-like growth factor 1 (IGF-1), IGF-1 density, and IGF-1/PSA ratio for prostate cancer detection. *Urology* 1999; 54:603–6.
53. Kanety H, Madjar Y, Dagan Y et al. Serum insulin-like growth factor-binding protein-2 (IGFBP- 2) is increased and IGFBP-3 is decreased in patients with prostate cancer: correlation with serum prostate-specific antigen. *J. Clin. Endocrinol. Metab.* 1993; 77:229–33
54. Cohen P, Peehl DM, Stamey TA et al. Elevated levels of insulin-like growth factor binding protein-2 (IGFBP-2) in the serum of prostate cancer patients. *J. Clin. Endocrinol. Metab.* 1993; 76:1031–5.
55. Davenport ML, Isley WL, Pucilowska JB et al. Insulin-like growth factor binding protein-3 proteolysis is induced after elective surgery. *J. Clin. Endocrinol. Metab.* 1992; 75:590–5.
56. Culig Z, Hobisch A, Cronauer MV et al. Androgen receptor activation in prostatic tumor cell lines by insulin-like growth factor-1, keratinocyte growth factor and epidermail growth factor. *Cancer Res.* 1994; 54:5474–8.
57. Sutkowski DM, Goode RL, Baniel J et al. Growth regulation of prostatic stromal cells by prostate-specific antigen. *J. Natl Cancer Inst.* 1999; 91:1663–9.
58. Nickerson T, Pollak M, Huynh H. Castration-induced apoptosis in the rat ventral prostate is associated with increased expression of genes encoding insulin-like growth factor binding proteins 2, 3, 4 and 5. *Endocrinology* 1998; 139:807–10.
59. Nickerson T, Pollak M. Bicalutamide (Casodex)-induced prostate regression involves increased expression of genes encoding insulin-like growth factor binding proteins. *Urology* 1999; 54:1120–5.
60. Nickerson T, Huynh H. Vitamin D analogue EB1089-induced prostate regression is associated with increased gene expression of insulin-like growth factor binding proteins. *J. Endocrinol.* 1999; 160:223–39.
61. Rozen F, Yang X, Huynh HT et al. Antiproliferative action of vitamin-d-related compound and insulin-like growth factor binding protein 5 accumulation. *J. Natl Cancer Inst.* 1997; 89:652–6.
62. Baselga J, Norton L, Albanell J et al. Recombinant humanized anti-HER2 antibody (herceptin) enhances the antitumor activity of doxyrubicin against Her2/neu overexpression in human breast cancer cancer cenografts. *Cancer Res.* 1998; 58:2825–31.
63. Pegram MD, Lipton A, Hayes DF et al. Phase 2 study of receptor-enhanced chemosensitivity using recombinant humanized

anti-p185HER2/neu monoclonal antibody plus cisplatin in patients with HER2/neu-overexpressing metastatic breast cancer refractory to chemotherapy treatment. *J. Clin. Oncol.* 1998; 16:2659–71.

64. Samson DJ, Leyland-Jones B, Shak S. Addition of Herceptin (humanized anti-HER2 antibody) to first-line chemotherapy for HER2 overexpressing metastatic breast cancer (HER2+/MBC) markedly increases anticancer activity: a randomised, multinational controlled phase 3 trial. *Proc. Am. Soc. Clin. Oncol.* 1998; 17 (abstract).
65. Lamharzi M, Schally AV, Koppan M et al. Growth hormone-releasing antagonist Mz 5-15 inhibits growth of DU-145 human androgen-independent prostate carcinoma in nude mice and suppresses the level and mRNA expression of insulin-like growth factor 2. *Proc. Natl Acad. Sci. USA* 1998; 95:8864–8.
66. Gaddipati JP, Mani H, Shefali Raj K et al. Inhibition of growth and regulation of IGFs and VEGF in human prostate cancer cell lines by shikonin analogue 93/637 (SA). *Anticancer Res.* 2000; 20:2547–52.
67. Plonowski A, Schally AV, Varga JL et al. Potentiation of the inhibitory effect of growth hormone-releasing hormone antagonists on PC-3 human prostate cancer by bombesin antagonists indicative of interference with both IGF and EGF pathways. *Prostate* 2000; 44:172–80.
68. Lilja H. A kallikrein-like prostate protease in prostatic fluid cleaves the predominant seminal vesical protein. *J. Clin. Invest.* 1985; 76:1899–03.
69. Watt KW, Lee PJA, Timkulu T et al. Human prostate-specific antigen: structural and functional similarity with serine proteases. *Proc. Natl Acad. Sci. USA* 1986; 83:3166–70.
70. Berg T, Bratshaw RA, Carretero OA et al. A common nomenclature for members of the tissue (glandular) kallikrein gene families. *Agents Actions* Suppl. 1996; 38:19–25.
71. Lilja H. Structure, function and regulation of the enzyme activity of prostate-specific antigen. *World J. Urol.* 1993; 11:188–91.
72. Ban Y, Wang MC, Watt KW et al. The proteolytic activity of human prostate-specific antigen. *Biochem. Biophys. Res. Commun.* 1984; 123:482–8.
73. Carbini LA, Scicli GA, Carretero OA. The molecular biology of the kallikrein-chinin system. III. The human kallikrein gene family and kallikrein substrate. *J. Hypertens.* 1993; 11:893–8.
74. Oesterling JE. Prostatic tumor markers. *J. Urol. Clin. North Am.* 1993; 20:575–7.
75. Schedlich LJ, Bennetts BH, Morris BJ. Primary structure of human glandular kallikrein gene. *DNA* 1987; 6:429–37.
76. Frey EK, Kraut H, Werle E. Über die blutzuckersenkende Wirkung des Kallikreins (Padutins). *Klin. Wochenschr.* 1932; 11:846–9.
77. McDonald RJ, Margolius HS, Erdös EG. Molecular biology of tissue kallikrein. *Biochem. J.* 1988; 253:313–21.
78. Bhoola KD, Figueroa CD, Worthy K. Bioregulation of kinins: kallikreins, kininogens, and kininases. *Pharmacol. Rev.* 1992; 44:1–80.
79. Fichtner J, Graves HCB, Thatcher K et al. Prostate-specific antigen releases a kinin-like substance on proteolysis of seminal vesical fluid that stimulates smooth muscle contraction. *J. Urol.* 1996; 155:738–42.
80. Deperthes D, Marceau S, Frenette G et al. Human kallikrein hK2 has low kininogenase activity while prostate-specific antigen (hK3) has none. *Biochem. Biophys. Acta* 1997; 1343:102–6.
81. Dube JY, Tremblay RR. Biochemistry and potential roles of prostatic kallikrein hK2. *Mol. Urol.* 1997; 1:279–85.
82. Rittenhouse HG, Finlay JA, Mikolajczyk SD et al. Human kallikrein 2 (hK2) and prostate-specific antigen (PSA): two closely related, but distinct, kallikreins in the prostate. *Crit. Rev. Clin. Lab. Sci.* 1998; 35:275–368.
83. Lovgren J, Rajakoski K, Karp M et al. Activation of the zymogen form of prostate-specific antigen by human glandular kallikrein 2. *Biochem. Biophys. Res. Commun.* 1997; 238:549–55.
84. Piironen T, Lovgren J, Karp M et al. Immunofluorometric assay for sensitive and specific measurement of human prostatic glandular kallikrein (hK2) in serum. *Clin. Chem.* 1996; 42:1034–41.
85. Charlesworth MC, Young Cyf, Klee GG et al. Detection of a prostate-specific protein, hK2, in sera of patients with elevated prostate-specific antigen levels. *Urology* 1997; 49:487–93.
86. Kwiatkowski MK, Recker F, Piironen T et al. In prostatism patients the ration of human glandular kallikrein to free PSA improves the discrimination between prostate cancer and benign hyperplasia within the diagnostic grey zone of total PSA 4–10 ng/mL. *Urology* 1998; 52:360–5.
87. Lovgren J, Piironen T, Overmo C et al. Production of recombinat PSA and hK2 and analysis of their immunologic cross-reactivity. *Biochem. Biophys. Res. Commun.* 1995; 213:888–95.
88. Becker C, Piironen T, Kiviniemi J et al. Sensitive and specific immunodetection of human glandular kallikrein 2 (hK2) in serum. *Clin. Chem.* 2000; 46:198–206.
89. Grauer LS, Finlay JA, Mikolajczyk SD et al. Detection of human glandular kallikrein hK2 as its precursor form, and in complex with protease inhibitors in prostate carcinoma serum. *J. Androl.* 1998; 19:407–11.
90. Becker C, Piirone T, Pettersson K et al. Discrimination of men with prostate cancer from those with benign disease by measurements of human glandular kallikrein (hK2) in serum. *J. Urol.* 2000; 163:311–16.
91. Haese A, Becker C, Noldus J et al. Human glandular kallikrein 2: a potential serum marker for predicting the organ confined versus non-organ confined growth of prostate cancer. *J. Urol.* 2000; 163:1491–7.
92. Partin AW, Catalona WJ, Finaly JA et al. Use of human glandular kallikrein 2 for the detection of prostate cancer: preliminary analysis. *Urology* 1999; 54:839–45.
93. Keetch DW, Catalona WJ, Smith DS. Serial pros-tatic biopsies in men with persistently elevated serum prostate specific antigen values. *J. Urol.* 1994; 151:1571–4.
94. Becker C, Piironen T, Pettersson K et al. Clinical value of human glandular kallikrein 2 and free and total prostate-specific antigen in serum from a population of men with prostate-specific antigen levels 3.0 ng/ml or greater. *Urology* 2000; 55:694–9.

Are Vaccinations for Prostate Cancer Realistic?

D. Robert Siemens
Department of Urology, Queen's University, Kingston, Ontario, Canada

Timothy L. Ratliff
University of Iowa Department of Urology, Iowa City, IA, USA

INTRODUCTION

The need for novel and effective adjuvant treatment options for prostate cancer seems obvious. Despite stage migration of incident cases with the widespread implementation of prostate specific antigen (PSA) testing, present-day biochemical failure rates after management with prostatectomy, external beam radiotherapy and brachytherapy remain substantially high. For these patients, as well as those who present with more advanced disease, our ability to extend survival with chemotherapy and hormonal manipulation is limited at best, and associated with potentially significant morbidity.

Our immune system facilitates our survival in a remarkably hostile environment by the appropriate recognition and destruction of that which is deemed to be foreign to the body (non-self), while also recognizing and tolerating self. The study of tumor immunology is based on two propositions: (1) tumor cells express distinct antigens that are found in only negligible amounts in normal cells; and (2) these antigens can be recognized by the immune system leading to destruction of the 'foreign' cell. The effector mechanisms of the immune system are distributed virtually throughout the entire body and, therefore, the appeal of a vaccine targeted to neoplastic disease is enormous. Yet, for reasons that are still unclear (and under intense scrutiny) most tumors 'escape' this immune surveillance and can progress and eventually metastasize in patients with intact immune systems.

For this reason, and others, the development of an effective cancer vaccine is significantly different from that for infectious agents. Viral genes are relatively simple, dictating a limited number of defined antigens that need to be manipulated for immunization. Tumor cells possess a virtually unlimited number of antigens, the majority of which have not been identified. Vaccination against infectious agents occur prior to exposure in order to prevent disease, whereas the immune system has theoretically already been exposed to the tumor antigens in the case of a cancer vaccine and, therefore, may already be 'tolerant' to that antigen. Finally, the majority of cancer vaccines focus on stimulating a different arm of the immune response, cell-mediated immunity, compared to the desired humoral response to infectious agents.

The interest in immunotherapy for the treatment of neoplastic disease has dramatically increased over the last decade. Much of this has been due to advances in our understanding of tumor immunology as well as technological advances in molecular biology and gene transfer technology. Those interested in urologic malignancies, including prostate cancer, have pioneered and expanded on many of these advances, making the prospect of a prostate cancer vaccine indeed realistic.

ADVANCES IN TUMOR IMMUNOLOGY

The goal of tumor immunology is to understand the immune response to malignant cells and to be able to use this knowledge to create novel therapeutic strategies. In humans, the concept of tumor surveillance by the immune system is somewhat vindicated by the increase of some rare tumors with chronic immunosuppression. As well, in rare cases of melanoma or renal cell carcinoma, spontaneous regression of tumors support the ability of the immune system to decrease the progression of tumors.[1] Furthermore, studies have demonstrated a decreased cell-mediated immunity in some cancer patients, and it has been observed that decreased level of natural killer (NK) cell cytotoxic activity may play an important role in prostate cancer development[2] and metastases.[3] Although this impaired cellular immunity can be demonstrated in cancer patients (especially in patients with advanced disease), it is likely that a generalized immunologic deficit is not causative but that it reflects a secondary phenomenon. In fact, the vast majority of cancers are not increased in the immunodeficient host, suggesting that the normal immune response is unable to control most forms of cancer.

In the early 20th century, Coley used heat-killed bacterial infections to initiate an antitumor response.[4] These observations led to the supposition that one could over-ride tumor escape mechanisms in order to induce an antitumor response. However, clinically effective treatment failed to materialize until the advent of bacille Calmette–Guérin (BCG) for the treatment of superficial bladder cancer.[5] As well, Rosenberg and associates revitalized interest in immunotherapy with their work on lymphokine-activated killer cells and tumor-infiltrating lymphocytes.[6,7] Although the response rates in renal cell carcinoma have not been fully realized, these studies demonstrate the ability of immune cells to eliminate tumors previously considered to be resistant to immune surveillance.

Although an exhaustive review of tumor immunology is beyond the scope of this chapter, current concepts, including antigen presentation and the effector arms of the immune system, will be discussed to help understand the immunization strategies undertaken in the many preclinical and clinical vaccine trials for prostate cancer.

Cell-mediated immunity

Since the discovery of tumor-specific transplantation antigens, cell-mediated immunity has been recognized as the predominant immune effector response in tumor immunology. The cells that are involved in the immune response include lymphocytes, granulocytes and specialized antigen-presenting cells (APC). The granulocytes include neutrophils, which, along with macrophages and monocytes, are important for phagocytosis of antigens targeted by antibodies. The specialized APCs include monocytes, macrophages, Langerhan cells, Kupffer cells and dendritic cells. These cells express major histocompatability complex (MHC) molecules, both class I and class II, in order to display antigens appropriately to T lymphocytes (T cells). T cells, B lymphocytes (B cells) and natural killer cells make up the three major populations of lymphocytes and can be defined by the presence and type of transmembrane antigen receptors. The NK cells are large granular lymphocytes that are capable of killing certain cancer cells with no prior immunization and without restriction by MHC glycoproteins.

Evidence continues to mount to suggest that T cells will play the most important role in developing an antitumor vaccine. Mature T cells are made up of two major populations defined by the cell surface expression of CD4 or CD8 molecules.[8] The CD4+ cells recognize antigen when presented in association with MHC class II molecules while CD8+ T-cell antigen recognition is restricted by MHC class I expression.

T cells can also be classified by their functional role in modulating the immune response. Classically, CD4+ cells (helper T cells) are thought to play a helper role in stimulating other cells, such as B cells, other T cells and macrophages; whereas the CD8+ T cells (cytotoxic T cell or CTL) are cytotoxic to cells that display the antigen that they recognize. Furthermore, when native T cells are activated, they produce interleukin-2 (IL-2) and then differentiate into two distinct subpopulations that can be categorized based upon the different cytokines that they produce.[9,10] The T-helper subset, T_H1 cells, secrete predominantly IL-2, interferon-γ (IFNγ) and tumor necrosis factor-β (TNFβ) and promote cell-mediated immune responses. During antigen presentation, APCs produce IL-12, which drives T cells to differentiate toward a T_H1 response. IFNγ secretion by T_H1 cells acts as an autocrine agonist and will inhibit differentiation of T_H2 cells, another T-helper subset. T_H2 cells are the principal T-helper cell for B-cell function and propagation of the humoral allergic immune response. The cytokines that they produce are predominantly IL-4, IL-5, IL-9, IL-10 and IL-13. IL-4 is known to enhance the differentiation to a T_H2 response and at the same time antagonize the T_H1 response.

Antigen presentation

In order for T cells to recognize antigens and discriminate self from non-self, the antigenic peptides must be presented in association with the MHC molecules. The MHC is a genetic locus on chromosome 6 that encodes the cell surface structures, human leukocyte antigens (HLA).[11] With the aid of the MHC molecules, T cells can recognize foreign antigens in the context of self as the TCR recognizes a composite of both antigen associated with the correct MHC molecule. There are two structural classes of HLA molecules, each of which restricts antigen recognition to different T cells.[12]

Class I HLA molecules are found on virtually all nucleated cells and restrict CTLs. HLA-A, -B and -C are the three major class I genes and there are numerous alleles defined for each of the genes.[13] Class II molecules are found only on the specialized APCs such as macrophages, dendritic cells and B cells, and restrict antigen recognition to T-helper cells. There are also only three major categories of class II molecules, HLA-DR, -DQ and -DP and each molecule is encoded within the MHC.[14]

The ability of the T cell to recognize antigen is a property of the T-cell receptor (TCR)[15] and is specific for a certain peptide antigen/MHC combination. The TCR is a disulfide-linked heterodimer, very similar in structure to the Fab fragment of an immunoglobulin molecule. However, the engagement of peptide antigen/MHC molecule to the TCR is insufficient to initiate the transduction of these intracellular signals. Important costimulatory molecules separate from the TCR complex are also required for appropriate T-cell activation. The most studied costimulatory signal is the ligation of a T-cell surface molecule, CD28, to the CD80 (B7-1) or CD86 (B7-2) of the APC as this is a requisite second signal for T-cell activation.[16] Other cell-surface molecules, such as ICAM-3, LFA-1 and CD-2, are also important in the T-cell interactions with APCs.

Tumor escape mechanisms

Given the number of genetic alterations in cancer cells, a vast number of potential tumor antigens may exist, including any amino acid sequence in any membrane-bound or intracellular protein. Therefore, it seems that, for any tumor to develop, progress and eventually metastasize, it must first evade the immune system (tumor escape). One possible mechanism of tumor escape is the selection of tumor cell clones that express fewer immunodominant antigens, selected by the pressure of the normal host immune surveillance. Several studies have documented the outgrowth of tumor cell lines with few tumor-specific antigens and this loss of antigen expression could be due to antibody-induced internalization or antigenic variation. Fortunately, even immunoselected tumor cell variants have been shown to express a number of unique antigens that could potentially serve as targets for immunotherapy protocols.[17]

Even if tumor antigens are present, the tumor-bearing host may subsequently become tolerant. This tolerance may be due to improper antigen presentation or neonatal exposure to the antigen, such as in the case of carcinoma embryonic antigen (CEA). Fortunately, studies have demonstrated that immunization with some of these antigens can overcome tolerance and, subsequently, they may be used for immunotherapy trials.[18]

Decreased expression of the MHC molecules may also lead to tumor escape, as the lack of MHC input would result in ineffective presentation of antigen to immune cells in the context of self. Wallich showed that transfecting the MHC class I genes can inhibit the metastatic ability of tumor cells.[19] Bander et al. has confirmed lost or diminished class I expression in a number of cell lines as well as in frozen tissue specimens of prostate cancer.[20] Levin has demonstrated that more poorly differentiated prostate tumors expressed significantly less MHC class I molecules and, when both class I expression and degree of differentiation was considered, those with higher expression had better survival.[21] It is possible that HLA class I status may be an important prognostic factor, as well as an important target for future immunotherapy strategies.

As discussed earlier, effective CTL activation is dependent on stimulatory signals from the T_H1 subpopulation of helper T cells. As most solid tumors do not express class II MHC molecules that are necessary for antigen presentation to the helper T cells, tumor development may take advantage of the lack of activating signals. Recent studies have revealed that, in a number of advanced malignancies, the zeta (ζ)-chain of the T-cell receptor in tumor-infiltrating lymphocytes show deceased expression[22] and that this loss is associated with poorer prognosis.[23] Similarly, Healy has similarly demonstrated impaired expression and function of ζ-chains in the peripheral blood lymphocytes in patients with advanced prostate cancer.[24] The best characterized of the costimulatory molecules is the B7–CD28 receptor ligand pair and, in animal studies, have been shown to augment the antitumor immune response when amplified. Other molecules necessary to the binding of lymphocytes to antigen presenting cells include LFA-1, CD2 and ICAM-3 on the T cell as well as ICAM-1, ICAM-2, LFA-1 and LFA-3 on the APCs.

The resistance of certain populations of neoplastic cells to cytotoxic cell killing mechanisms may be another mechanism of tumor escape. The inability of tumor cells to undergo Fas-mediated death may contribute to evasion of immune surveillance. Lehmann et al. have described a novel method of tumor escape to NK-mediated killing. They describe the impaired binding of perforin on the cell surface of tumor cells, with subsequent resistance of granule-mediated cell death from the cytotoxic effector cells.[25]

Tumor cells may also be able to affect their environment potentially to decrease the effectiveness of any host antitumor immune response. Examples include release of free antigens that can interfere with the effective response of NK cells and helper T cells. As well, tumor cell production of cytokines (such as IL-10 and IL-18) has been shown to limit the effectiveness of immune surveillance. Transforming growth factor-β (TGFβ) has been demonstrated to inhibit IL-2-dependent immune responses of macrophages and T cells.[26] Another mechanism of tumor escape may involve Fas/Fas ligand interaction. Several human tumors have been found to express Fas ligand that can induce Fas receptor-mediated apoptosis of activated immune cells.[27] Kim and associates have found high expression of Fas ligand a relatively low expression of Fas in renal cell carcinoma, which may be involved in the evasion of immune effector cells.[28]

Finally, it is possible that host cells can be recruited to the tumor site in order actively to down-regulate antitumor T-cell immune responses.[29] There is some experimental evidence for the existence of T-cell subsets that may act in an antigen-specific suppressor role in tumor immunology. These T cells have been shown to be CD4+ and may have T_H2 characteristics, although their role in human cancers has been difficult to determine.

TECHNOLOGICAL ADVANCES

The interest in vaccine therapy for cancer has increased tremendously over the last decade, mostly due to significant advances in molecular biologic and gene transfer techniques. These technological advances have inspired numerous preclinical and clinical gene therapy strategies, facilitated the detection of novel tumor-associated antigens and furthered our understanding of tumor immunology.

Gene therapies for cancer

The use of therapeutic gene transfer for cancer is often described as being either ‘cytoreductive’ or ‘corrective’, and its prospects in the management of prostate cancer has been well summarized recently by Harrington et al.[30] ‘Corrective’ gene therapy refers to the replacement or inactivation of defective genes in neoplastic cells. Numerous preclinical investigations and several clinical evaluations have begun utilizing the replacement of tumor suppressor genes such as p53, p21, p16, progressive multifocal leukoencephalopathy

(PML) gene and the breast cancer susceptibility gene (BRCA1).[31–34] Another strategy to disable the cancer cell involves targeting the overexpression of particular oncogenes by introducing antisense mRNA into a malignant cell. These nucleic acid sequences potentially bind to the oncogene transcription products in order to inhibit the translation into protein. Investigations employing antisense oligonucleotides for prostate cancer have shown several interesting targets including c-*myc* and the antiapoptotic *bcl*-2 oncogene.[35,36] A serious drawback of these strategies is that, despite the identification of numerous tumor suppresser and growth-promoting genes in human prostate cancer, no lone genetic locus has been consistently associated with the disease. Often multiple mutations in many redundant pathways make gene selection for manipulation difficult. However, based on the many successes described above, 'corrective' gene therapy techniques may play an important role as an adjuvant to other cytotoxic therapies.[37,38]

An alternative strategy to replacing defective genes is 'cytoreductive' gene therapy, eradicating tumor cells directly (i.e. 'suicide' genes, induction of apoptosis) or indirectly (i.e. genes that elicit antitumor immune responses). So-called suicide gene therapy is perhaps the most widely studied direct 'cytoreductive' gene therapy for neoplastic disease. The gene-directed enzyme prodrug therapy utilizing the HSV-tk/gangcyclovir system is well described for the treatment of prostate cancer both *in vitro* and *in vivo*.[39,40] The specific transfer of genetic material (HSV-thymidine kinase) to neoplastic cells allows for the localized conversion of a relativley non-toxic prodrug (i.e. gangcyclovir) to an active cytotoxic drug, thereby decreasing the systemic toxicity. Several clinical studies have been published describing the safety and responses of such enzyme prodrug therapy for prostate cancer.[41–43] Other directly 'cytoreductive' gene therapy approaches include the direct injection of diphtheria toxin A gene, transfer of antiangiogenic genes and the genetic induction of apoptosis.[30] Indirect 'cytoreductive' gene therapies refer to the transfer of genes either *in situ* or *ex vivo* in order to stimulate an immune response. These immunomodulatory gene therapy techniques are the subject of intense research and will be discussed further on in the chapter.

Gene transfer techniques

Much of the success of the abovementioned gene therapy protocols have been due to the significant advances made in the ability to transfer particular genetic material to cells of interest both *in vitro* and *in vivo*. In general, the transfer of DNA or RNA requires some vehicle to allow efficient gene transfer. The choice of gene transfer vehicle, either viral or non-viral, depends on a number of factors, including the goal of the therapy (i.e. length of time gene expression is required), cells to be targeted for gene transfer and the size of the gene to be delivered. Non-viral gene therapy vectors include iontophoresis, liposomal transfer and the transfer of naked DNA. Although larger segments of DNA can be transferred using these means, they are substantially less efficient than viral vectors. Ternary complexes, however, utilize a combination of DNA in a complex with inactivated viral particles to allow higher gene transfer efficiency.

Gene transfer by viral vectors takes advantage of the natural ability of viruses to enter cells and express transgenes by those infected cells. Retroviral vectors were used in the first clinical gene therapy trials[44] and result in stable long-term transgene expression in many cell types by integrating into the host genome. However, their use in malignant disease may be limited by the need for actively dividing cells for infection, relative difficulties in viral titer production and a theoretical concern for insertional mutagenesis after incorporation into the host genome.[45,46] The adeno-associated virus vectors are based on a dependent, replication-deficient virus that requires coinfection with a helper virus (such as adenovirus). These vectors impart some advantages, such as substantially less inflammatory response to infection (compared to adenoviral and poxvirus vectors) and stable integration into the genome, although their space for gene insertion is limiting (5 kb). Poxvirus vectors are double-stranded DNA viruses that can carry large amounts of genetic material and are unique in their ability to express therapeutic genes without requiring transport of the vector to the target nucleus. Although the poxvirus vectors can be highly efficient, gene transfer is transient and, similar to adenoviral vectors, they may produce a potentially competitive antivector immune response.[47]

One of the most common viral vectors used in both preclinical studies and human clinical trials is the DNA adenovirus, which has the advantages of infecting a wide range of non-dividing cells, high titer virion production and a large amount of space for therapeutic gene insertion (7–10 kb). Gene expression after adenoviral vector infection is transient, as it does not integrate into the host genome but may be highly efficient because of an active receptor-mediated process. Recently, several groups have reported that the coxsackie and adenovirus receptor (CAR) is a common receptor for adenovirus type 2 and 5[48,49] and that low expression levels of this receptor portends poor gene transfer efficiency of recombinant adenoviral vectors.[50] The expression of this receptor may well be paramount to any preclinical models or clinical trials utilizing an adenoviral vector. The adenovirus vectors used in many preclinical and clinical studies are generally rendered replication defective, although others have utilized direct viral oncolysis by replication-competent viruses. So-called 'gutless' vectors have had the adenoviral genes removed from their genome, allowing insertion of larger transgenes (~35 kb).

Although there have been tremendous technological advances in the area of gene transfer that have quickly translated to the use of these vectors in numerous clinical trials, there remain many difficulties that need to be overcome before the widespread application to human disease. The non-viral vectors are very useful *in vitro* to allow transfer of large DNA segments but their use *in vivo* is limited owing to poor gene transfer efficiency and targeting. The viral vectors, although more efficient, have significant disadvantages, including the propensity of most viruses to induce both a cellular and humoral immune response, constraining *in vivo* gene transfer by eliminating the vector. This immune response may lead to significant limitations in human applications.[51,52] Furthermore, the non-specific toxicity from utilizing these vectors can be significant. Viremia, hepatotoxicity, hematotoxicity, infection of unintended cells and host immune activation all have been described.[53]

Undoubtedly, the efficient and accurate delivery of therapeutic genes to specific cells of interest remains the greatest obstacle for the future of gene therapy, particularly in solid tumor oncology. Efforts to decrease the immunogenicity of the vectors used to transfer genetic material could allow higher viral titers to be used as well as facilitate the ability to use the same vector for repeat injections.[54–56] Tissue-specific promoters (such as PSA,[57] PSMA,[58] the rat probasin promoter[59] and human glandular kallikrein promoter) and modification of viral vectors to take advantage of tissue/cell specific receptors should allow for greater cell targeting and decrease toxicity. Finally, as numerous groups have reported the poor efficiency of gene transfer when employing direct injections into organs or tumors, methods have been investigated to deliver the vectors themselves better to both neoplastic and non-neoplastic cells.[60–62]

Detection of tumor-associated antigens

The concept behind active immunotherapy is based on the theory that tumors possess specific antigens that can be recognized by the immune system, identified as being foreign and then destroyed. Indeed, numerous gene therapy studies have confirmed the hypothesis that most theoretically 'non-immunogenic' tumors are indeed immunogenic.[63,64] Although few truly 'foreign' antigens have been identified, the term tumor-associated antigen (TAA) describes antigens that are found on human tumors but are also expressed in varying degrees by normal cells. Since the first convincing demonstration of their existence,[65] there has been a great deal of interest in identifying these TAAs for therapeutic use in human cancers. Although most advances in this field have come from work with melanoma cells (probably the most immunogenic human tumor), prostate cancer TAAs are beginning to become identified, further stimulating the investigations toward an effective prostate cancer vaccine.

There are various sources of tumor-associated antigens that have the potential to be recognized by the immune system.[66,67] Probably the largest group of known potential antigens for immunotherapy are those that are tissue- or organ-specific (i.e. PSA). In melanoma research, differentiation antigens (i.e. tyrosinase, MART-1, gp 100) can be found on both normal melanocytes and many melanoma cells, and are targets undergoing clinical testing in antigen-specific vaccine studies. Technical advances have facilitated the identification of many of these potential antigens. For example, serial analysis of gene expression (SAGE) can be used to quantify gene expression and compare levels in neoplastic and non-neoplastic tissue.

Other categories of potential tumor-associated antigens include viral gene products (Epstein–Barr virus, human papilloma virus), activated oncogene products (*ras*, p53), rearranged normal gene products (CDK4-R24C) and over-expressed (HER-2/neu) or reactivated gene products (CEA, α-fetoprotein). An important example of a reactivated embryonic gene product is *MAGE1*, found to be expressed in about 50% of melanomas but not on any adult normal tissue except for the testicle. *MAGE1* was the first gene to be identified as encoding an antigen recognized by a human melanoma-specific T-cell clone.[68] This development of novel techniques to define MHC class I-restricted antigens has facilitated an explosion in the hunt for new and novel antigens to be used for cancer vaccines. Since the mid-1990s, T cells have been employed to find tumor peptide antigens that are restricted by class I molecules of the MHC. Two strategies, one biochemical and another genetic, identify an antigen already recognized by CD8+ T cells obtained *in vivo*.[69,70] The obvious value of detecting antigens in this fashion is that they may allow identification of targets that have a higher chance of escaping tolerance and thus be used in generating successful immunotherapies.

Although most of the investigation described above has taken place with melanoma models, prostate cancer is emerging as an exciting immunotherapy model. Not only do animal models exist but a variety of tumor-associated antigens (both glycoprotein and carbohydrate) have been identified including PSA, hK2, PSMA, MUC-1, MUC-2, GLOBO-H, GM-2 and Lewis^y.[71–76] Many of these antigens may well serve as targets for immune recognition.[77–79]

Although the identified antigens in the prostate cancer models are not truly tumor specific, the rationale for their use in cancer vaccines remains. The potential for generating autoimmunity is less ominous given that the prostate is relatively dispensable. Also, the normal tissue counterparts of many tumors express low levels of MHC class I, allowing a possible window in the therapeutic index, if a tissue-specific response can be elicited.[80]

TUMOR IMMUNOTHERAPY

Numerous novel cancer therapies utilizing the immune system have been developed and reported. Early attempts have focused on non-specific therapies and still the best example of cancer immunotherapy is the use of bacille Calmette-Guérin (BCG) for carcinoma *in situ* of the bladder. Morales et al. are also utilizing non-specific immunotherapies for prostate cancer with the direct injection of mycobacterial cell wall-DNA complexes (personal communication). More specific strategies include both passive and active immunization, and a brief description, including those published on prostate cancer, will help illustrate the great strides made in the area of tumor immunotherapy over a relatively short period of time.

Passive immunotherapy

Passive immunization involves administrating antibodies to the patient that will bind specific tumor antigens to kill those cells selectively. Prostate-specific membrane antigen (PSMA) is a cell surface glycoprotein expressed by prostate epithelial cells and is overexpressed in prostate cancer, including advanced stages. PSMA-specific monoclonal antibodies have been developed and may have a novel therapeutic application.[81,82] Kahn and associates have recently reported on a phase II clinical radioimmunotherapy trial utilizing capromab pendetide (PSMA monoclonal antibody) labeled with yttrium.[83] Unfortunately, serum PSA was not lowered in the eight men with biochemical failure after prostatectomy and significant bone marrow toxicity was observed. Katzenwadel et al. have also described the construction of a bispecific monoclonal antibody to PSA and the T-cell associated CD3 antigen to increase the antigen-specific cytotoxicity of T cells.[84] They demonstrated specific lysis of PSA expressing cells *in vitro* as well as reduction of tumor growth *in vivo*. Sinha and associates have also used an anti-PSA immunoglobulin G as a carrier for chemotherapeutic drugs (5-fluoro-2′-deoxyuridine) and have shown antitumor effects on LNCaP xenografts in nude mice.[85]

Adoptive immunotherapy is another passive, but non-specific, strategy involving the transfer of immunologically activated lymphoid cells. Clinical experience with renal cell carcinoma has revealed that the activation of human lymphoid cells (i.e. IL-2 activated LAK cells) is feasible and that systemic administration is safe.[86–88] Lubaroff et al. have shown that a severe combined immunodeficiency (SCID) mouse model is a viable system for studying adoptive therapies for human prostate cancer[89] and that antitumor activity has been demonstrated utilizing autologous IL-2-activated tumor-infiltrating lymphocytes (TILs) (personal communication). A novel approach to adoptive therapy was reported by Cesano et al. whereby the human T-cell line (TALL-104), which is marked by an MHC non-restricted cytotoxic activity against a wide range of tumors across several species, demonstrated significant antitumor effect against DU-145 tumors in SCID mice.[90] Gong and associates recently reported on the transduction of artificial receptors (specific for PSMA) into the T lymphocytes of patients with prostate cancer.[91] They were able to demonstrate lysis of cancer cells *in vitro* and support the possibility of adoptive immunotherapy in prostate cancer patients.

An alternative immunotherapeutic strategy to simply administering cytokines or activated immune cell populations systemically is generating a vaccine that can elicit a specific antitumor response *in vivo*. There are generally two ways of categorizing cancer vaccines, depending on the source of the immunizing antigens: antigen-specific or tumor cell vaccines.[67]

Antigen-specific vaccines

Antigen-specific vaccines refer to the delivery of recombinant peptides or proteins to a host in order to elicit an antitumor immune response. As previously described, there has been a great deal of interest in the discovery of novel tumor-associated antigens for the use in cancer vaccines. In any immunotherapy protocol, there are several different means of delivering these antigens to the host: (1) DNA-based vaccines; (2) peptide- or protein-based vaccines; (3) recombinant viral or bacterial vaccines; and (4) antigen-pulsed dendritic cell vaccines.

DNA vaccines encoding tumor antigens can be delivered as naked DNA or encapsulated by liposomes.[92] Despite the possibility of rapid degradation of DNA when given systemically, there has been renewed interest in DNA-based vaccines for both infectious agents as well as neoplastic diseases.[93,94]

The advantages of administering a cancer vaccine as a peptide or protein would include improved safety compared to other vector-based vaccines. Tumor antigens could be delivered simply as defined peptides or as proteins specifically designed to access the class 1 pathway of APCs *in vivo*. Several phase I clinical trials have been completed utilizing peptide-based vaccines for malignant melanoma, demonstrating minimal toxicity and some clinical and *in vitro* evidence of response.[95–97] Segments of the PSA protein have been shown to be immunogenic and stimulate a specific CTL reaction when used as a vaccine. Several peptide antigens, two 10-mer peptides from PSA (PSA-1 and PSA-3) and a PSA oligoepitope peptide (PSA-OP), have also been shown to elicit CTL responses *in vitro*.[98] From this initial work, several phase I/II clinical studies are under way in which patients are vaccinated against the PSA protein, with the hope that systemic immunity against prostate cancer will occur. Meidenbauer et al. reported on a cancer vaccine consisting of recombinant PSA with lipid A formulated in liposomes.[99] After injection in patients with prostate cancer, PSA reactive T-cell frequencies were detectable, albeit at very low levels.

As shown previously, advances in recombinant technology have stimulated interest in the incorporation of genes encoding relevant antigens into vectors such as viruses to augment their immune response. Vectors such as adenovirus and the poxviruses (including vaccinia) impart a number advantages including high efficiency gene transfer and targeting the MHC antigen-processing pathways. Sanda et al. reported on a limited phase I trial on PROSTVAC, a vaccina-PSA vaccine, in patients with biochemical recurrence after prostatectomy.[100] Similarly, Eder et al. published their experience with a vaccina-PSA vaccine in a phase I trial revealing minimal toxicity with virion concentrations up to 2.65×10^8 plaque-forming units.[101] They also demonstrated some clinical effect with stabilization of serum PSA in 14 of 33 patients.

While viruses are effective in generating CTL and ultimately antitumor activity in naïve mice, the use of viruses in settings where the host has been previously exposed to the virus significantly reduces transgene expression.[102,103] Studies by Yang and associates formally demonstrated that T-cell responses, specifically CTL, to viral proteins were responsible for the destruction of cells expressing the transgene.[102] Furthermore, antibodies in the serum have been shown to diminish the ability of viruses to deliver the transgene[103] and the resulting decrease in gene expression reduces subsequent CTL activation.[104,105] The diminished ability to generate CTL when antibodies to the viral vector are present has been suggested to be an important reason for the lack of CTL after adenovirus delivery of melanoma antigens in clinical trials.[106]

In order to facilitate the presentation of antigenic peptides as a cancer vaccine, much interest has been focused on the use of dendritic cells. These bone marrow-derived cells have the ability to process antigens and present them in the context of MHC molecules and other costimulatory molecules such as B7. They are the most potent APCs identified and are capable of activating naïve T cells. A potentially powerful strategy in cancer gene therapy involves the genetic engineering of dendritic cells with defined tumor antigens and their use as vaccines.[107] Dendritic cell vaccines can be produced by loading these cells *in vitro* with peptides, proteins, whole tumor cells or by infecting them *ex vivo* by viral vectors.[108] The first few clinical trials utilizing a dendritic cell-based vaccine for metastatic melanoma and follicular B-cell lymphoma have demonstrated both immunologic and clinical responses.[109,110] Phase 1 clinical trials for prostate cancer have also been performed using autologous dendritic cells associated with the HLA-2.1-specific PSMA. Little adverse effect was observed and several hormone-refractory patients showed significant responses to the vaccine.[111] The phase II trial using this dendritic cell vaccine has demonstrated an overall response rate of 30% (19/62) with 11 durable responses.[112]

Tumor cell vaccines

Active immunization can also be achieved by the administration of autologous or allogenic tumor cells modified to secrete cytokines or other immunostimulatory molecules capable of recruiting antitumor effector cells. Examples include *in vitro* tumor cell transfection by MHC class II genes,[113] transfection of lost MHC class I alleles[19] and transfection of the costimulatory molecule B7 to stimulate T cells.[114] The strategy of using tumor cell vaccines as a source of tumor antigens eliminates the need to identify and develop immunodominant antigens. As well, engineered tumor cells expressing cytokines may function by providing necessary growth factors to activated CTLs in the absence of helper T cells. Numerous different cytokines have been introduced into tumor cells *in vitro* including IL-2, IL-4, IL-6, IL-12, TNFα and granulocyte–macrophage colony-stimulating factor (GM-CSF). Many of these preclinical models have shown great success in controlling local tumor outgrowth as well as some measure of systemic protection. Numerous phase I–II clinical trials utilizing *ex vivo* transduced whole-tumor cell vaccines (most often involving malignant melanoma) have been published.[115–117]

Early studies by Sanda and associates showed that GM-CSF transfected rat prostatic adenocarcinomas grew more slowly than parental tumors.[46] Subsequently, Vieweg showed that IL-2 transfected rat R3327-MatLyLu also induced antitumor activity.[118] Existing tumors had a decreased rate of tumor outgrowth and protection was gained against subsequent tumor challenge. We have previously reported one approach to cancer immunotherapy involving the transfer of genes encoding the cytokines IL-2 and TNFα utilizing the canarypox viral vector, ALVAC.[119] The ALVAC virus was shown efficiently to infect murine prostate cancer cells, RM-1, and produce high levels of extrinsic gene product. As well, antitumor immunity was induced when tumor cells were infected by ALVAC cytokine recombinants and injected subcutaneously in the flanks of male C57BL/6 mice. Based on the optimistic results of such preclinical studies, an autologous vaccine approach is being investigated in a phase I/II study at Johns Hopkins in patients with extracapsular disease following radical prostatectomy. An allogeneic vaccine, using MHC class I-matched allogeneic cells transduced to secrete IL-2 and IFNγ, is employed in an NCI phase I/II study at the Memorial Sloan Kettering Institute. Likewise, a phase I clinical protocol for the intratumoral (*in situ*) injection of IL-2 and GM-CSF genes are underway for prostate cancer.

Cancer gene therapy strategies involving the transfer of genetic material *in situ* by viral and non-viral vectors have also been successful in stimulating active antitumor immunization. These *in vivo* schemes offer a number of practical advantages over their *ex vivo* counterparts, including the cost reduction of bypassing the isolation of target cells from the patient. Putzer has demonstrated impressive regression of pre-established tumors through the intratumoral injection of adenovirus recombinant for IL-12 and the B7-1 molecule.[120] We have also observed significant antitumor activity of murine prostate cancer nodules after intratumoral injection of the canarypox virus, ALVAC, recombinant for murine IL-2, IL-12 and TNFα. Nasu et al. have shown potent antitumor effects after infecting orthotopic mouse prostate cancers with adenovirus-mediated gene transfer of IL-12.[121] Phase I clinical protocols for the intratumoral injection of IL-2 and GM-CSF genes are under way for prostate cancer. Preliminary results of the intraprostatic injection of IL-2 packaged within a liposome delivery vehicle have been encouraging with decreases in serum PSA levels in 80% of the patients treated.[122]

CONCLUSIONS

Dramatic advances in available molecular techniques and increasing understanding of antigen presentation and recognition has led to much renewed interest in the possibility of a cancer vaccine. Novel preclinical studies and early clinical trials have resulted in sufficient successes to suggest a 'proof-of-principle' that immunotherapy may indeed become a useful adjunct in the treatment of prostate cancer.

However, we must obviously take care to interpret available data cautiously as tumor immunotherapy emerges from its infancy. Tumor escape from the immune surveillance remains largely unexplained and despite the tremendous interest in identifying antigens for cancer vaccines, few have been found that are truly unique to tumors. As well, much work must still be done to improve the efficacy and safety of the clinical translation of our technological advances, including gene therapy techniques. Finally, the exact role that immunotherapy will have in the management of prostate cancer has yet to be defined, and will remain that way for the near future given the time required to complete phase III trials for this disease.

REFERENCES

1. Montie JE, Stewart BH, Straffon RA et al. The role of adjunctive nephrectomy in patients with metastatic renal cell carcinoma. *J. Urol.* 1977; 117:272.
2. Lahat N, Alexander N, Levin DR et al. The relationship between clinical stage, natural killer activity and related immunological parameters in adenocarcinoma of the prostate. *Cancer Immunol. Immunother.* 1989; 28:208–13.
3. Santori A, Palmeieri G, Procopio A. Mechanism of target cell killing by natural killer cells. *Ital. J. Med.* 1989; 4:59–66.
4. Goodfield J. Dr. Coley's toxins. *Science* 1984; 226:68–73.
5. Lamm DL, Blumenstein BA, Crawford ED et al. A randomized trial of intravesical doxorubicin and immunotherapy with bacille Calmette–Guéerin for transitional-cell carcinoma of the bladder, *N. Engl. J. Med.* 1991; 325:1205–9.
6. Alexander RB, Rosenberg SA. Long-term survival of adoptively transferred tumor-infiltrating lymphocytes in mice. *J. Immunol.* 1999; 145:1615–20.
7. Spiess PJ, Yang JC, Rosenberg SA. Tumor-specific cytolysis by lymphocytes infiltrating lymphocytes expanded in recombinant interleukin-2. *J. Natl Cancer Inst.* 1987; 79:1067–75.
8. Haynes B, Denning SM, Le PT et al. Human intrathymic T cell differentiation. *Semin. Immunol.* 1990; 12:67–77.
9. Abbas AK, Murphy KM, Sher A. Functional diversity of helper T lymphocytes. *Nature* 1996; 383:787–93.
10. Mosmann TR, Sad S. The expanding universe of T-cell subsets: Th1, Th2 and more. *Immunol. Today* 1996; 17:138–46.
11. Huston DP, The biology of the immune system. *JAMA* 1997; 278:1804–14.
12. Bjorkman PJ, Saper MA, Samraoui B et al. Structure of the human class I histocompatability antigen, HLA-A2. *Nature* 1987; 329:506–12.
13. Trowsdale J, Tagoussis J, Campbell RD. Map of the human MHC. *Immunol. Today* 1991; 12:443–6.
14. Brown JH, Jardetsky TS, Gorga JC et al. The three-dimensional structure of the human class II histocompatability antigen, HLA-DR-1. *Nature* 1993; 364:33–9.
15. Bjorkman PJ, Davis MM. Model for the interaction of T cell receptors with the peptide/MHC complexes. *Cold Spring Harb. Symp. Quant. Biol.* 1989; 54:365–73.
16. June CH, Bluestone JA, Nadler LM et al. The B7 and CD28 receptor families. *Immunol. Today* 1994; 11:191–212.
17. Dudley ME, Roopenian DC. Loss of a unique tumor antigen by cytotoxic T lymphocyte immunoselection from a 3-methylcholanthrene-induced mouse sarcoma reveals secondary unique and shared antigens. *J. Exp. Med.* 1996; 184:441–7.
18. McLaughlin JP, Schlom J, Kantor JA et al. Improved immunotherapy of a recombinant carinoembryonic antigen vaccinia vaccine when given in combination with interleukin-2. *Cancer Res.* 1996; 56:2361–7.
19. Wallich R, Bulbuc N, Hammerling GJ et al. Abrogation of metastatic properties of tumor cells by de novo expression of H-2K antigens following H-2 gene transfection. *Nature* 1985; 315:301–5.
20. Bander NH, Yao D, Liu H et al. MHC class I and II expression in prostate carcinoma and modulation by interferon-alpha and -gamma. *Prostate* 1997; 33:233–39.
21. Levin I, Klein T, Kuperman O et al. The expression of HLA class I antigen in prostate cancer in relation to tumor differentiation and patient survival. *Cancer Detect. Prevent.* 1994; 18:443–5.
22. Finke JH, Zea AH, Stanley J et al. Loss of T cell receptor ζ chain and $p56^{lck}$ in T cells infiltrating human renal cell carcinoma. *Cancer Res.* 1993; 53:5613–16.
23. Zea A, Curti B, Longo D et al. Alterations in T cell receptor and signal transduction molecules in melanoma patients. *Clin. Cancer Res.* 1995; 1:1327–35.
24. Healy CG, Simons JW, Carducci MA et al. Impaired expression and function of signal-transducing zeta chains in peripheral T cells and natural killer cells in patients with prostate cancer. *Cytometry* 1998; 32:109–19.
25. Lehmann C, Zeis M, Schmitz N et al. Impaired binding of perforin on the surface of tumor cells is a cause of target cell resistance against cytotoxic effector cells. *Blood* 2000; 96:594.
26. Fu YX, Watson GA, Kassahara M et al. The role of tumor-derived cytokines on the immune system of mice bearing a mammary adenocarcinoma, I: induction of regulatory macrophages in normal mice by the in vivo administration of rGM-CSF. *J. Immunol.* 1991; 146:783–9.
27. Bennett MW, O'Connell J, O'Sullivan GC et al. Expression of Fas ligand by human gastric adenocarcinomas: a potential mechanism of immune escape in stomach cancer. *Gut* 1999; 44:156.
28. Kim YS, Kim KH, Choi JA et al. Fas (APO-1/CD95) ligand and Fas expression in renal cell carcinomas: correlation with the prognostic factors. *Arch. Pathol. Lab. Med.* 2000; 124:687.
29. Radoja S, Frey AB. Cancer-induced defective cytotoxic T lymphocyte effector function: another mechanism how antigenic tumors escape immnue-mediated killing. *Mol. Med.* 2000; 6:465.
30. Harrington KJ, Spitzwig C, Bateman AR et al. Gene therapy for prostate cancer: current status and future prospects. *J. Urol.* 2001; 166:1220–33.
31. Eastham JA, Hall SJ, Sehgal I et al. In vivo gene therapy with p53 or p21 adenovirus for prostate cancer. *Cancer Res.* 1995; 55:5151.
32. He D, Mu ZM, Le X et al. Adenovirus-mediated expression of PML suppresses growth and tumorgenicity of prostate cancer cells. *Cancer Res.* 1997; 57:1868–72.
33. Steiner MS, Lerner J, Greenberg M et al. Clinical phase 1 gene therapy trial using BRCA1 retrovirus is safe (abstract). *J. Urol.* 1998; 159:132.
34. Steiner MS, Zhang Y, Farooq F et al. Adenoviral vector containing wild-type p16 suppresses prostate cancer cell growth and prolongs survival by inducing cell senescence. *Cancer Gene Ther.* 2000; 7:360.
35. Steiner MS, Anthony CT, Lu Y et al. Antisense c-myc retroviral vector suppresses established human prostate cancer. *Human Gene Ther.* 1998; 9:747.
36. Gleave M, Tolcher A, Miayake H et al. Progression to androgen independence is delayed by adjuvant treatment with antisense Bcl-2 oligodeoxynucleotides after castration in the LNCaP prostate tumor model. *Clin. Cancer Res.* 1999; 5:2891.
37. Miayake H, Tolcher A, Gleave ME. Chemosensitization and delayed androgen-independent recurrence of prostate cancer with the use of antisense Bcl-2 oligodeoxynucleotides. *J. Natl Cancer Inst.* 2000; 92:34.
38. Geiger T, Muller M, Monia BP et al. Antitumor activity of a C-raf antisense oligonucleotide in combination with standard chemotherapeutic agents against various human tumors transplanted subcutaneuosly into nude mice. *Clin. Cancer Res.* 1997; 3:1179.
39. Eastham JA, Chen S-H, Sehgal I et al. Prostate cancer gene therapy: herpes simplex virus thymidine kinase transduction followed by gangcyclovir in mouse and human prostate cancer models. *Human Gene Ther.* 7:515.
40. Hall SJ, Mutchnik SE, Chen SH et al. Adenovirus mediated herpes simplex virus thymidine kinase gene and gangcyclovir therapy leads to systemic activity against spontaneous and induced metastasis in an orthotopic mouse model of prostate cancer. *Int. J. Cancer* 70:183.
41. Ayala G, Wheeler TM, Shalev M et al. Cytopathic effects of in situ gene therapy in prostate cancer. *Human Pathol.* 31:866.
42. Herman JR, Adler HL, Aguilar-Cordova E et al. In situ gene therapy for adenocarcinoma of the prostate: a phase I clinical trial. *Human Gene Ther.* 10:1239.

43. Shalev M, Kadmon D, The BS et al. Suicide gene therapy toxicity after multiple and repeat injections in patients with localized prostate cancer. *J. Urol.* 163:1747.
44. Rosenberg SA, Aebersold P, Cornetta K et al. Gene transfer into human – immunotherapy of patients with advanced melanoma, using tumor-infiltrating lymphocytes modified by retroviral gene trasduction. *N. Engl. J. Med.* 323:570–8.
45. Kasahara N, Dozy AM, Kan YW. Tissue specific targeting of retroviral vectors through ligand-receptor interactions. *Science* 266:1373.
46. Sanda MG, Ayyagari SR, Jaffee EM et al. Demonstration of a rational strategy for human prostate cancer gene therapy. *J. Urol.* 151:622–8.
47. Wang M, Bronte V, Chen PW et al. Active immunotherapy of cancer with a non-replicating recombinant fowlpox virus encoding a model tumor-associated antigen. *J. Immunol.* 154:4685–92.
48. Tomko RP, Xu R, Phillipson L. HCAR and MCAR: the human and mouse cellula receptors for subgroup C adenoviruses and group B coxsackieviruses. *Proc. Natl Acad. Sci. USA* 94:3352–6.
49. Bergelson JM et al. The murine CAR homolog is a receptor for coxsackie B viruses and adenoviruses. *J. Virol.* 72:415–19.
50. Li Y et al. Loss of adenoviral receptor expression in human bladder cancer cells: a potential impact on the efficacy of gene therapy. *Cancer Res.* 59:325–30.
51. Yang Y, Nunes F, Berencsi K et al. Cellular immunity to viral antigens limits E1-deleted adenoviruses for gene therapy. *Proc. Natl Acad. Sci. USA* 1994; 91:4407–41.
52. Crystal RG. Transfer of genes to humans: early lessons and obstacles to success. *Science* 1995; 270:404–10.
53. Rodriquez R, Simons JW. Urologic applications of gene therapy. *Urology* 1999; 54:401–6.
54. Chen HH, Mack LM, Kelly R et al. Persistence in muscle of an adenoviral vector that lacks all viral genes. *Proc. Natl Acad. Sci. USA* 1997; 94:1645.
55. Haeker SE, Stedman HH, Balice-Gordon RJ et al. In vivo expression of full-length human dystrophin from adenoviral vectors deleted of all viral genes. *Hum. Gene Ther.* 1995; 6:1417.
56. Siemens DR, Elzey BD, Lubaroff DM et al. Restoration of the ability to generate CTL in mice immune to adenovirus by delivery of virus in a collagen-based matrix. *J. Immunol.* 2000; 166:731–5.
57. Gotoh A, Ko SC, Shirakawa T et al. Development of prostate-specific antigen promoter-based gene therapy for an androgen-independent human prostate cancer. *J. Urol.* 1998; 160:220–9.
58. O'Keefe DS, Su SL, Bacich DJ et al. Mapping, genomic organization and promoter anlaysis of the human prostate-specific membrane antigen gene. *Biochim. Biophys. Acta* 1998; 1443:113–27.
59. Greenberg NM, DeMayo FJ, Sheppard PC et al. The rat probasin gene promoter directs hormonally and developmentally regulated expression of a heterologous gene specifically to the prostate in transgenic mice. *Mol. Endocrinol.* 1994; 8:230–9.
60. Beer SJ, Hilfinger JM, Davidson BL. Extended release of adenovirus from polymer microspheres: potential use in gene therapy for brain tumor. *Adv. Drug Delivery Rev.* 1997; 27:59–66.
61. Feldman LJ, Pastore CJ, Aubailly N et al. Improved efficiency of arterial gene transfer by use of poloxamer 407 as a vehicle for adenoviral vectors. *Gene Ther.* 1997; 4:189–98.
62. Siemens DR, Austin JC, Hedican SP et al. Viral delivery in a gelatin sponge matrix enhances gene expression and antitumor activity in a murine prostate cancer model. *J. Natl Cancer Inst.* 2000; 92:403–12.
63. McCune CS, Schapira DV, Mastrangelo MJ. Specific immunotherapy of advanced renal carcinoma: evidence for the polyclonality of metastases. *Cancer* 1981; 27:1984.
64. Connor J, Bannerij R, Saito S et al. Regression of bladder tumors in mice treated with interleukin-2 gene-modified tumor cells. *J. Exp. Med.* 1993; 177:1127–34.
65. Prehn RT, Main JM. Immunity to methylcholanthrene induced sarcomas. *J. Natl Cancer Inst.* 1957; 18:769–78.
66. Shu S, Plautz GE, Krauss J et al. *Tumor Immunol.* 1997; 28:1972–81.
67. Greten TF, Jaffee EM. Cancer vaccines. *J. Clin. Oncol.* 1999; 17:1047–60.
68. Van Pel A, van der Bruggen P, Coulie PG et al. Genes coding for tumor antigens recognized by cytolytic lymphocytes. *Immunol. Rev.* 1995; 145:229–50.
69. Cox AL, Skipper J, Chen Y et al. Identification of a peptide recognized by five melanoma-specific human cytotoxic T cell lines. *Science* 1994; 264:716–19.
70. Vandeneyende BJ, Vanderbruggen P. T cell defined tumor antigens. *Curr. Opin. Immunol.* 1997; 9:684–93.
71. Isreaeli RS, Powell T, Corr JG et al. Expression of prostate-specific membrane antigen. *Cancer Res.* 1994; 54:1807–11.
72. Carrato C, Balague C, De Bolos C et al. Differential apomucin expression in normal and neoplastic tissues. *Gastroenterology* 1994; 107:160–72.
73. Gambus G, Bolos CD, Andreu D et al. Detection of the MUC2 apomucin tandem repeat with a mouse monoclonal antibody. *Gastroenterology* 1993; 104:93–102.
74. Kim J, Park TK, Hu S et al. Defining the molecular recognition of GLOBO H (human breast cancer) antigen through probe structures prepared by total synthesis. *J. Org. Chem.* 1995; 60:7716–30.
75. Kiamura K, Livingston PO, Fortunato SF. Serological response patterns of melanoma patients immunized with a GM2 ganglioside conjugate vaccine. *Proc. Natl Acad. Sci. USA* 1995; 92:2805–9.
76. Hellstrom I, Garrigues HJ, Garrigues U et al. Highly tumor-reactive, internalizing, mouse monoclonal antibodies to Le(y)-related cells surface antigens. *Cancer Res.* 1990; 50:2183–7.
77. Slovin SF, Livingston PO, Rosen N et al. Targeted therapy for prostate cancer: The Memorial Sloan–Kettering Cancer Center Approach *Semin. Oncol.* 1996; 23:41–8.
78. Zhu ZY, Zhong CP, Xu WF et al. PSMA mimotope isolated from phage displayed peptide library can induce PSMA specific immune reponse. *Cell Res.* 1999; 9:271–80.
79. Lodge PA, Childs RA, Monahan SJ et al. Expression and purification of prostate-specific membrane antigen in the baculavirus expression system and recognition by prostate-specific membrane antigen-specific T cells. *J. Immunother.* 1999; 22:346–55.
80. Pardoll D. Tumor antigens: a new look for the 1990s. *Nature* 1994; 369:357.
81. Gong MC, Chang SS, Sadelain M et al. Prostate-specific membrane antigen (PSMA)-specific monoclonal antibodies in the treatment of prostate and other cancers. *Cancer Metast. Rev.* 1999; 18:483–90.
82. Elgamal AA, Holmes EH, Su SL et al. Prostate-specific membrane antigen (PSMA): current benefits and future value. *Semin. Surg. Oncol.* 2000; 18:10–16.
83. Kahn D, Austin JC, Maguire RT et al. A phase II study of [90Y] yttrium-capromab pendetide in the treatment of men with prostate cancer recurrence following radical prostatectomy. *Cancer Biother. Radiopharm.* 1999; 14:99–111.
84. Katzenwadel A, Schleer H, Gierschner D et al. Construction and in vivo evaluation of an anti-PSA × anti-CD3 bispecific antibody for the immunotherapy of prostate cancer. *Anticancer Res.* 2000; 20:1551–5.
85. Sinha AA, Quast BJ, Reddy PK et al. Intravenous injection of an immunoconjugate (anti-PSA-IgG conjugated to 5-fluoro-2′-deoxyuridine) selectively inhibits cell proliferation and induces cell death in human prostate cancer cell tumors grown in nude mice. *Anticancer Res.* 1999; 19:893–902.

86. Rosenberg SA, Lotze MT, Yang JC. Prospective randomized trial of high dose interleukin-2 alone or in combination with lymphokine-activated killer cells for the treatment of patients with advanced cancer. *J. Natl Cancer Inst.* 1993; 85:622–32.
87. Belldegrun A, Pierce W, Kaboo R et al. Interferon-alpha primed tumor infiltrating lymphocytes combined with interleukin-2 and interferon-alpha as therapy for metastatic renal cell carcinoma. *J. Urol.* 1993; 150:1384–90.
88. Figlin RA, Pierce WC, Kaboo R et al. Treatment of metastatic renal cell carcinoma with nephrectomy, interleukin-2 and cytokine-primed or CD8(+) selected tumor infiltrating lymphocytes from primary tumor. *J. Urol.* 1997; 158:740–5.
89. Lubaroff DM, Cohen MB, Schultz LD et al. Survival of human prostate carcinoma, benign hyperplastic tissues, and IL-2-activated lymphocytes in scid mice. *Prostate* 1995; 27:32–41.
90. Cesano A, Visnneau S, Santoli D. TALL-104 cell therapy of human solid tumors implanted in immunodeficient (SCID) mice. *Anticancer Res.* 1998; 18:2289–05.
91. Gong MC, Latouche JB, Krause A et al. Cancer patient T cells genetically targeted to prostate-specific membrane antigen specifically lyse prostate cancer cells and release cytokines in response to prostate-specific membrane antigen. *Neoplasia* 1999; 1:123–7.
92. Norman A, Parker S, Lew D et al. Preclinical pharmacokinetics, manufacturing and safety studies supporting a multicenter cancer gene therapy trial. *Hum. Gene Ther.* 1995; 6:549.
93. Tang DC, DeVit M, Johnston SA. Genetic immunization is a simple method for eliciting an immune response. *Nature* 1992; 356:152.
94. Xiang R, Lode HN, Chao TH et al. An autologous oral DNA vaccine protects against murine melanoma. *Proc. Natl Acad. Sci. USA* 2000; 97:5492.
95. Marchand M, Weynants P, Rankin E et al. Tumor regression responses in melanoma patients treated with a peptide encoded by gene MAGE-3. *Int. J. Cancer* 1995; 63:883–5.
96. Parkhurst MR, Salgaller ML, Southwood S et al. Improved induction of melanoma-reactive CTL with peptides from the melanoma antigen gp100 modified at HLA-A*0201-binding residues. *J. Immunol.* 1996; 157:2539–48.
97. Rosenberg SA, Yang JC, Schwartzentruber DJ et al. Immunologic and therapeutic evaluation of a synthetic peptide vaccine for the treatment of patients with metastatic melanoma. *Nature Med.* 1998; 321–7.
98. Correale P, Walmsley K, Zaremba S et al. Generation of human cytolytic T lymphocyte lines directed against prostate-specific antigen (PSA) employing a PSA oligoepitope peptide. *J. Immunol.* 1998; 161:3186–94.
99. Meidenbauer N, Harris DT, Spitle LE et al. Generation of PSA-reactive effector cells after vaccination with a PSA-based vaccine in patients with prostate cancer. *Prostate* 2000; 43:88–100.
100. Sanda MG, Smith DC, Charles LG et al. Recombinant vaccinia-PSA (PROSTVAC) can induce a prostate-specific immune response in androgen-modulated human prostate cancer. *Urology* 1999; 53:260–6.
101. Eder JP, Kantoff PW, Roper K et al. A phase I trial of a recombinant vaccinia virus expressing prostate-specific antigen in advanced prostate cancer. *Clin. Cancer Res.* 2000; 6:1632–8.
102. Yang Y, Ertl HC, Wilson JM. MHC class I-restricted cytotoxic T lymphocytes to viral antigens destroy hepatocytes in mice infected with E1-deleted recombinant adenoviruses. *Immunity* 1994; 1:433.
103. Kuriyama S, Tominaga K, Kikukawa M et al. Inhibitory effects of human sera on adenovirus-mediated gene transfer into rat liver. *Anticancer Res.* 1998; 18:2345.
104. Mullbacher A, Bellett A, Hla R. The murine cellular immune response to adenovirus type 5. *Immunol. Cell Biol.* 1989; 67:31.
105. Juilliard V, Villefroy P, Godfrin D et al. Long-term humoral and cellular immunity induced by a single immunization with replication-defective adenovirus recombinant vector. *Eur. J. Immunol.* 1995; 25:3467.
106. Rosenberg SA, Zhai Y, Yang JC et al. Immunizing patients with metastatic melanoma using recombinant adenoviruses encoding MART-1 or gp100 melanoma antigens. *J. Natl Cancer Inst.* 1998; 90:1894.
107. Tjoa B, Boynton A, Kenny G et al. Presentation of prostate tumor antigens by dendritic cells stimulates T-cell proliferation and cytotoxicity. *Prostate.* 1995; 28:65–9.
108. Specht JM, Wang G, Do MT et al. Dendritic cells retrovirally transduced with a model tumor antigen gene are therapeutically effective against established pulmonary metastases. *J. Exp. Med.* 186:1213–21.
109. Hsu FJ, Benike C, Fagoni F et al. Vaccination of patients with B-cell lymphoma using autologous antigen-pulsed dendritic cells. *Nature Med.* 1996; 2:52–8.
110. Nestle FO, Alijagic S, Gilliet M et al. Vaccination of melanoma patients with peptide- or tumor lysate-pulsed dendritic cells. *Nature Med.* 1998; 4:328–32.
111. Salgaller ML, Tjoa BA, Lodge PA et al. Dendritic cell-based immunotherapy of prostate cancer. *Crit. Rev. Immunother.* 1998; 18:109–19.
112. Tjoa BA, Simmons SJ, Elgamal A et al. Follow-up evaluation of a phase II prostate cancer vaccine trial. *Prostate* 1999; 40:125–9.
113. Orstrand-Rosenberg S, Thakur A, Clements V. Rejection of mouse sarcoma cells after transfection of MHC class II genes. *J. Exp. Med.* 1983; 144:4068–71.
114. Baskar S, Ostrand-Rosenberg S, Nabavi N et al. Constitutive expression of B7 restores immunogenicity of tumor cells expressing truncated major histocompatability complex class II molecules. *Proc. Natl Acad. Sci. USA* 1993; 90:5687–90.
115. Gansbacher B, Houghton G, Livingston P. A pilot study of immunization with HLA-A2 matched allogeneic melanoma cells that secrete interleukin-2 in patients with metastatic melanoma. *Hum. Gene Ther.* 1992; 3:677.
116. Dranoff G, Soiffer R, Lynch T et al. A phase 1 study of vaccination with irradiated melanoma cells engineered to secrete human granulocyte-macrophage colony stimulating factor. *Hum. Gene Ther.* 1997; 8:111.
117. Seigler HF, Darrow TL, Abdel-Wahab Z et al. A phase 1 trial of human gamma interferon transduced autologous tumor cells in patients with disseminated malignant melanoma. *Hum. Gene Ther.* 1994; 5:761.
118. Vieweg J, Rosenthal FM, Bannerji R et al. Immunotherapy of prostate cancer in the dunning rat model: use of cytokine gene modified tumor vaccines. *Cancer Res.* 1994; 54:1760–5.
119. Kawakita M, Rao GS, Ritchey JK et al. Effect of canarypox virus (ALVAC)-mediated cytokine expression on murine prostate tumor growth. *J. Natl Cancer Inst.* 1997; 89:428–36.
120. Putzer BM, Hitt M, Mueller WJ et al. Interleukin-12 and B7-1 costimulatory molecule expressed by an adenoviral vector act synergistically to facilitate tumor regression. *Proc. Natl Acad. Sci. USA* 1997; 94:10889–94.
121. Nasu Y, Bangma CH, Hull GW et al. Adenovirus-mediated interleukin-12 gene therapy for prostate cancer: suppression of orthotopic tumor growth and pre-established lung metastases in an orthotopic model. *Gene Ther.* 1999; 6:338–49.
122. Naitoh J, Tso CL, Kaboo R et al. Intraprostatic interleukin-2 (IL-2) gene therapy: preliminary results of a phase I clinical trial for the treatment of locally advanced prostate cancer. *J. Urol.* 1998; 159(Suppl.):254A.

Index